Sixteenth Edition

PHYSICAL Geology

Charles C. Plummer
Emeritus of California State University at Sacramento

Diane H. Carlson
Emeritus of California State University at Sacramento

Lisa Hammersley
California State University at Sacramento

Mc Graw Hill Education

PHYSICAL GEOLOGY, SIXTEENTH EDITION

Published by McGraw-Hill Education, 2 Penn Plaza, New York, NY 10121.

This book is printed on acid-free paper.

1 2 3 4 5 6 7 8 9 LWI 21 20 19 18

ISBN 978-1-259-91682-3
MHID 1-259-91682-0

Portfolio Manager: *Michael Ivanov, Ph.D.*
Product Developers: *Jodi Rhomberg*
Marketing Manager: *Kelly Brown*
Content Project Managers: *Jeni McAtee, Rachael Hillebrand*
Buyer: *Susan Culbertson*
Design: *Tara McDermott*
Content Licensing Specialist: *Melissa Homer*
Cover Image: *©Frank Krahmer/Radius Images/Getty Images*
Compositor: *Lumina Datamatics*

All credits appearing on page or at the end of the book are considered to be an extension of the copyright page.

Design elements: header bar (Canyon outside of Cedar City, Utah) ©Dave Moyer; Preface and Index Header bar image (clouds over snow covered mountain) ©Glow Images RF; Meet the authors bar image (Clouds over a hill, Texas) ©Glow Images RF; Title page (Snake River at Grand Tetons at sunset) ©Chad Ehlers/Getty Images RF; Appendis A header (High view of landscape) ©Rodrigo Torres/Glow Images RF; Appendis B header (Machu Pucchu) ©The Power of Forever Photography/Getty Images RF; Glossary header (Rub al Khali sand dunes) ©Berno Blüchel/age fotostock RF; computer icon (tablet) ©McGraw Hill Education; computer icon (pickaxe) ©Kangholanna58/Shutterstock RF; computer icon (pyrite) ©Siede Preis/Getty Images RF

Library of Congress Cataloging-in-Publication Data

Names: Plummer, Charles C., 1937- author. | Carlson, Diane H., author. | Hammersley, Lisa, author.

Title: Physical geology.

Description: Sixteenth edition / Charles C. Plummer (emeritus of California State University at Sacramento), Diane H. Carlson (emeritus of California State University at Sacramento), Lisa Hammersley (California State University at Sacramento). | New York, NY : McGraw-Hill Education, [2019]

Identifiers: LCCN 2017049322 | ISBN 9781259916823 (alk. paper)

Subjects: LCSH: Physical geology–Textbooks. | Geology–Textbooks.

Classification: LCC QE28.2 .P58 2019 | DDC 551–dc23 LC record available at https://lccn.loc.gov/2017049322

About the Cover

The cover photo shows the Moeraki Boulders found on Koekohe Beach in New Zealand. These large, almost spherical boulders are concretions that formed in marine mud just below the sea floor approximately 60 million years ago. Over millions of years, the loose mud was differentially cemented into solid rock (see chapter 6). Where the cement was concentrated, hard, spherical concretions formed.

Later, the region was uplifted and wave action (see chapter 14) eroded away the softer sandstone beds, leaving the harder concretions exposed at the beach. The Moeraki Boulders are not unique; similar spherical concretions can be found in a number of locations around the world, including Bowling Ball Beach in Northern California.

BRIEF CONTENTS

CONTENTS

©Carlos Gutierrez/UPI/Landov

4 Volcanism and Extrusive Rocks 78

©Doug Sherman/Geofile RF

5 Weathering and Soil 106

©Doug Sherman/Geofile RF

6 Sediment and Sedimentary Rocks 128

©John A. Karachewski

7 Metamorphism and Metamorphic Rocks 158

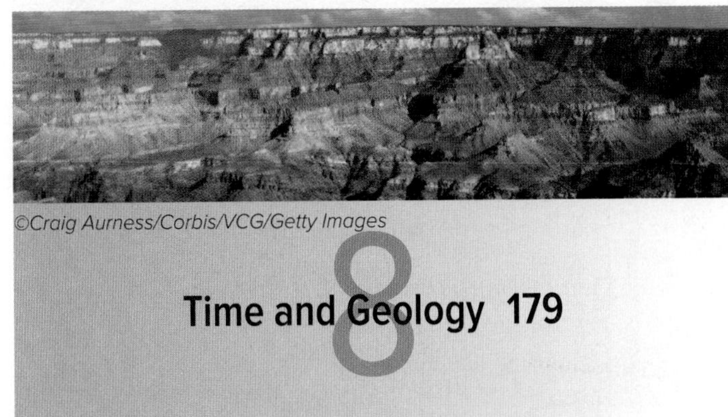
©Craig Aurness/Corbis/VCG/Getty Images

8 Time and Geology 179

©Alessandro della Valle/AP Images

9 Mass Wasting 205

©Bill Davis/Bossier City Sheriff's Office/AP Images

10 Streams and Floods 232

©Forcellini Danilo/Shutterstock

©Sarah Aciego

©Lee Frost/Robert Harding World Imagery/Getty Images

©Bogdan Bratosin/Moment/Getty Images

14 Waves, Beaches, and Coasts 343

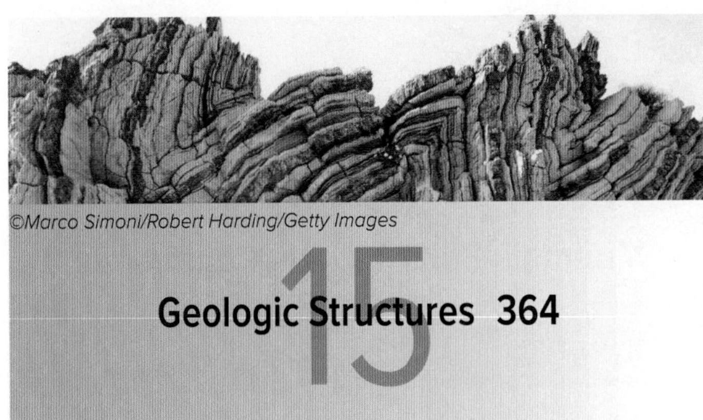
©Marco Simoni/Robert Harding/Getty Images

15 Geologic Structures 364

©The Asahi Shimbun/Corbis/Getty Images

16 Earthquakes 385

Graphic by Nathan Simmons/LLNL

17 Earth's Interior and Geophysical Properties 415

©Richard Carey/iStock 360/Getty Images

Source: NOAA/NGDC

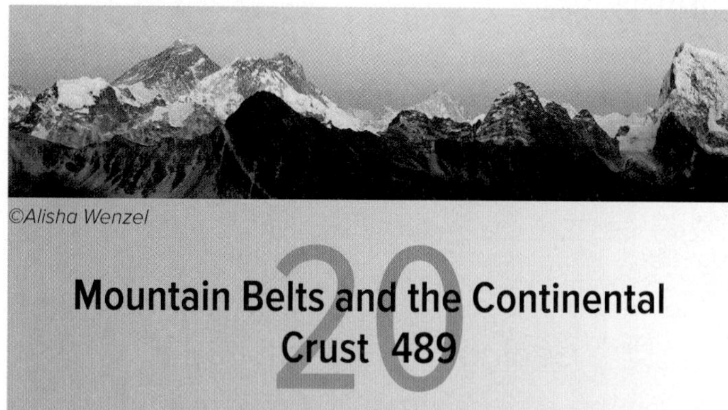

©Alisha Wenzel

Source: Image Science and Analysis Laboratory, NASA-Johnson Space Center

©Peter Hendrie/Photographer's Choice/Getty Images

Source: JPL/NASA/University of Arizona

WHY USE THIS BOOK?

One excellent reason is that it's tried and true. Since the book was first published in 1979, over 1,000,000 students have read this text as an introduction to physical geology. Proportionately, geology instructors have relied on this text for over 5,000 courses to explain, illustrate, and exemplify basic geologic concepts to both majors and non-majors. Today, the sixteenth edition continues to provide contemporary perspectives that reflect current research, recent natural disasters, unmatched illustrations, and unparalleled learning aids. We have worked closely with contributors, reviewers, and our editors to publish the most accurate and current text possible.

APPROACH

Our purpose is to clearly present the various aspects of physical geology so that students can understand the logic of what scientists have discovered as well as the elegant way the parts are interrelated to explain how Earth, as a whole, works.

This approach is epitomized by our treatment of plate tectonics. Plate tectonics is central to understanding how the Earth works. Rather than providing a full-fledged presentation of plate tectonics at the beginning of the textbook and overwhelming students, *Physical Geology* presents the essentials of plate tectonics in the first chapter. Subsequent chapters then detail interrelationships between plate tectonics and major geologic topics. For example, chapter 3, on igneous activity, includes a thorough explanation of how plate tectonics accounts for the generation of magma and resulting igneous rocks. Chapter 19, typically covered late in the course, presents a full synthesis of plate tectonics. By this time, students have learned the many aspects of physical geology and can appreciate the elegance of plate tectonics as a unifying paradigm.

CHANGES TO THE SIXTEENTH EDITION

New to the Sixteenth Edition

Each chapter has been revised and updated, and an overview of notable changes made to each chapter is given below:

Chapter 1 has been updated to reflect the current status of the U.S. petroleum industry. A new photo in box 1.3 shows a more modern image of geologists at work. We have added groundwater to the discussion of the hydrosphere. The section on the Earth's interior has been rewritten to include the concepts of mechanical layers and compositional layers.

Chapter 2 includes some new figures and updated text. The discussion of asbestos in Box 2.4 has been completely rewritten with a new emphasis on mineralogy and health hazards, and new photos have been added.

Chapter 3 has been updated and new web links added. Changes made to figures 3.3 and 3.14B will make them clearer for the reader.

Chapter 4 has been updated with new photos and web links.

Chapter 5 includes a new photo of scenic cliffs formed by differential weathering, and the figure illustrating spheroidal weathering has been revised. Questions at the end of the chapter have been revised and reorganized to better reflect the learning objectives and more clearly follow the flow of the chapter.

Chapter 6 includes a new photo of rounded sediment, and the discussion of how detrital rocks are classified and identified has been expanded. The end-of-chapter questions have been revised to more closely follow the learning objectives.

Chapter 7 has been updated with new photos and web links. A new section on metamorphic facies has been added to the section on plate tectonics and metamorphism. This section shows how the mineral assemblages in metamorphic rocks can provide information on the tectonic setting in which the rock formed.

Chapter 8 has been updated to improve readability, and new web links have been added.

Chapter 9 has minor rewrites to improve readability. Figure 9.3 is a new image showing an example of a landslide triggered by an earthquake. Figure 9.13 shows the effects of a recent mudslide.

Chapter 10 contains new photos of alluvial fans, stream terraces, high-discharge streams, and the recent flooding in Louisiana. Box 10.1 has been updated to include the latest controlled floods on the Colorado River, and the tables and graphs in box 10.3 have been revised to include the past 10 years of peak discharges along the Cosumnes River. The difficulty of estimating the size of a 100-year flood due to the lack of long-term records and the extreme weather events associated with climate change are discussed. Questions at the end of the chapter were revised to more closely follow the learning objectives.

Chapter 11 includes a new photo of groundwater contamination at a landfill, and a new photo of the Geysers Field in California. The fracking box has been updated and revised, and figure 1 of box 11.1 now more accurately reflects hydraulic gradient. We have also included new figures and a discussion of renewed subsidence in the Central Valley of California due to overpumping of deep aquifers during the recent historic drought. New web links have also been added.

Chapter 12 has been updated and new photos have been added. Box 12.1 has been rewritten to incorporate the potential impact of climate change on water availability. New web links have also been added to box 12.1.

Chapter 13 includes a revised discussion of flash floods and mudflows in deserts and a new photo of the catastrophic mudflow in southern California that buried more than 100 vehicles on Highway 58 in the Tehachapi Mountains. Box 13.2 includes minor revisions, and figure 13.12 has been replaced with a new photo of alluvial fans and playa lakes. Box 13.4 has been updated to include the first up-close study of sand dunes on a planet other than Earth. The Mars Science Laboratory rover, *Curiosity*, found miniature sand dunes that are attributed to the thin atmosphere on Mars. Similar large ripples preserved in 3.7-billion-year-old sandstone on Mars suggest the planet may have lost its atmosphere early in its history.

Chapter 14 has been updated and includes new photos of a barrier island along the Atlantic Coast and effects of rising sea level along the Gulf Coast. Web Resources at the end of the chapter have been updated.

Chapter 15 contains minor edits throughout the chapter to help clarify material for the student and improve readability. New photos of deformation along the San Andreas fault and a panoramic photo of Chief Mountain and the Lewis thrust fault in Montana have been added.

Chapter 16 has been updated to include the 2016 Kaikoura, New Zealand, and Amatrice, Italy, earthquakes as well as the human-induced earthquakes in Oklahoma caused by the deep injection of wastewater from oil and gas drilling operations. Spectacular new photos and drone footage of the ground rupture from the Kaikoura earthquake that ripped across the South Island of New Zealand have been added to the Earthquake-Related Hazards section. We have also added a new photo of a trench wall exposing offset layers of sediment along the San Andreas fault. The box "Waiting for the Big One in California" has been revised and updated to include new earthquake probabilities from the 2015 Uniform California Rupture Forecast (UCERF3). New web links detailing the earthquake forecast and simulations of ground motion during earthquakes in northern and southern California have also been added to box 16.3.

Chapter 17 opens with a new seismic tomography image of Earth that shows a large slab of subducted plate that sank through the entire mantle and is preserved below the Indian Ocean. The chapter has been updated to include the new attempt to drill through the oceanic crust to reach the mantle in the southeast Indian Ocean, and also the discovery of a possible new stiff layer in the upper part of the lower mantle based on high-pressure mineral experiments and on seismic tomography showing subducted plates pooling at 1500 km.

Chapter 18 has been updated and contains new photos and web links. Box 18.1 contains new research on tidal-triggered earthquakes on the East Pacific Rise, and the correlation of shallow earthquakes and tidal forces before the 2011 Tohoku earthquake in Japan and the Sumatra-Andaman earthquake in 2004. The "Turbidity Currents" section now contains web links to dramatic new video footage of turbidity currents in submarine canyons along the Baja and Mendocino coasts, and also turbidity currents modeled in laboratory settings. We have also expanded the discussion of submarine cable breaks caused by turbidity currents and the potential risk to the global economy caused by broken telecommunication cables that carry almost all of the digital and voice communications worldwide.

Chapter 19 has minor editing throughout the chapter to update content and improve clarity. The "Continent-Continent Convergence" section has been revised to reflect new mass balance calculations that suggest half of the Indian continent was subducted back into the mantle. The "What Causes Plate Motions?" section has been updated to include recent studies on mantle plumes.

Chapter 20 has undergone minor editing for improved readability.

Chapter 21 has been updated to reflect the rapid changes in the study of climate change. Figures 21.11, 21.12, and 21.18 have been updated to include the most recent data available. New web links have been added throughout.

Chapter 22 has been updated to reflect changes in the demand for, and price of, various resources. New photos and web links have been added.

Chapter 23 has been revised to reflect the current state of knowledge of the solar system. New images have been added where recent missions have produced improved imagery of the planets.

KEY FEATURES

Superior Photo and Art Programs

Geology is a visually oriented science, and one of the best ways to learn it is by studying illustrations and photographs. The outstanding photo and art programs in this text feature accuracy in scale, realism, and aesthetic appeal that provides students with the best visual learning tools available in the market. We strive to have the best photographs possible so that they are the next best thing to seeing geology on a field trip. We are again pleased to feature aerial photography from award-winning photographer/geologist Michael Collier, who gives students a birds-eye view of spectacular geology from western North America.

©Michael Collier

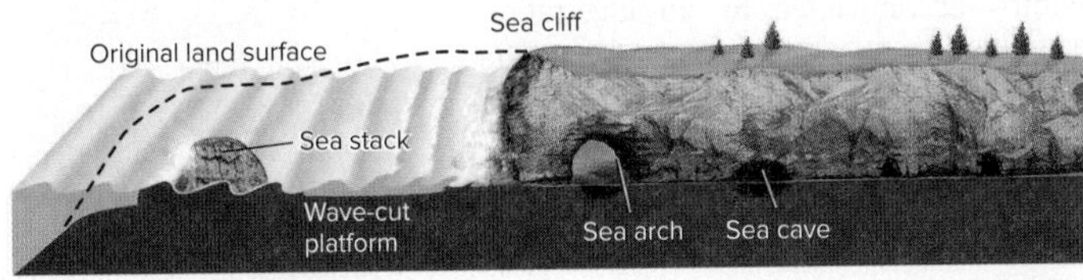

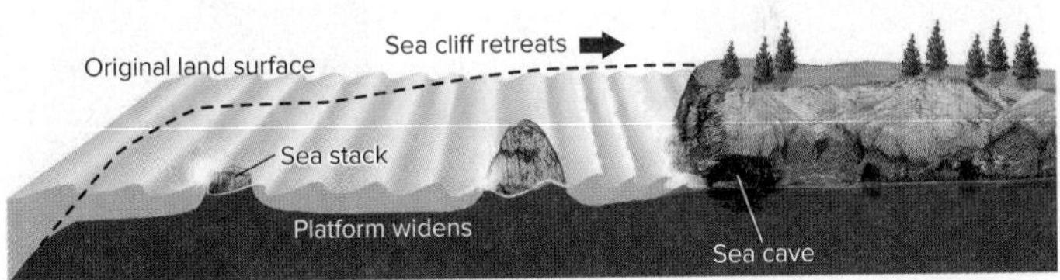

©David McGeary

Learning Objectives

Each chapter begins with a bulleted list of learning objectives to help students focus on what they should know and understand after reading the chapter.

> **LEARNING OBJECTIVES**
>
> - Differentiate between effusive and explosive eruptions, and describe the eruptive products associated with them.
> - Explain the relationship between magma composition, temperature, dissolved gas, and viscosity and relate them to eruptive violence.
> - Describe the five major types of volcanoes in terms of shape and eruptive style.

Environmental Geology Boxes

Discuss topics that relate the chapter material to environmental issues, including impact on humans (e.g., "Radon–A Radioactive Health Hazard").

In Greater Depth Boxes

Discuss phenomena that are not necessarily covered in a geology course (e.g., "Precious Gems") or present material in greater depth (e.g., "Calculating the Age of a Rock").

Earth Systems Boxes

Highlight the interrelationships between the geosphere, the atmosphere, and other Earth systems (e.g., "Oxygen Isotopes and Climate Change").

Planetary Geology Boxes

Compare features elsewhere in the solar system to their Earthly counterparts (e.g., "Stream Features on the Planet Mars").

A Geologist's View

Photos accompanied by an illustration depicting how a geologist would view the scene are featured in the text. Students gain experience understanding how the trained eye of a geologist views a landscape in order to comprehend the geologic events that have occurred.

©Diane Carlson

Animations

Figures representing key concepts such as plate tectonics, fault movement, earthquakes, isostasy, groundwater movement, sediment transport, glacial features, Earth movement, and other processes enhanced by animation are included online at McGraw-Hill Connect.

Study Aids are found at the end of each chapter and include:

- *Summaries* bring together and summarize the major concepts of the chapter.
- *Terms to Remember* include all the boldfaced terms covered in the chapter so that students can verify their understanding of the concepts behind each term.
- *Testing Your Knowledge Quizzes* allow students to gauge their understanding of the chapter and are aligned with the learning objectives presented at the beginning of each chapter. (The answers to the multiple choice portions are posted on Connect.)
- *Expanding Your Knowledge Questions* stimulate a student's critical thinking by asking questions with answers that are not found in the textbook.
- *Exploring Web Resources* describe some of the best sites on the Web that relate to the chapter.

McGraw-Hill Connect® is a highly reliable, easy-to-use homework and learning management solution that utilizes learning science and award-winning adaptive tools to improve student results.

Homework and Adaptive Learning

- Connect's assignments help students contextualize what they've learned through application, so they can better understand the material and think critically.
- Connect will create a personalized study path customized to individual student needs through SmartBook®.
- SmartBook helps students study more efficiently by delivering an interactive reading experience through adaptive highlighting and review.

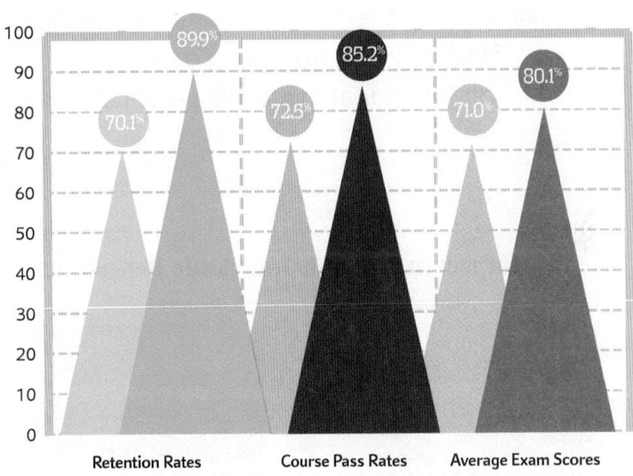

Using **Connect** improves retention rates by **19.8** percentage points, passing rates by **12.7** percentage points, and exam scores by **9.1** percentage points.

Over **7 billion questions** have been answered, making McGraw-Hill Education products more intelligent, reliable, and precise.

73% of instructors who use **Connect** require it; instructor satisfaction **increases** by 28% when **Connect** is required.

Quality Content and Learning Resources

- Connect content is authored by the world's best subject matter experts, and is available to your class through a simple and intuitive interface.
- The Connect eBook makes it easy for students to access their reading material on smartphones and tablets. They can study on the go and don't need internet access to use the eBook as a reference, with full functionality.
- Multimedia content such as videos, simulations, and games drive student engagement and critical thinking skills.

Robust Analytics and Reporting

©Hero Images/Getty Images

- Connect Insight® generates easy-to-read reports on individual students, the class as a whole, and on specific assignments.
- The Connect Insight dashboard delivers data on performance, study behavior, and effort. Instructors can quickly identify students who struggle and focus on material that the class has yet to master.
- Connect automatically grades assignments and quizzes, providing easy-to-read reports on individual and class performance.

Impact on Final Course Grade Distribution

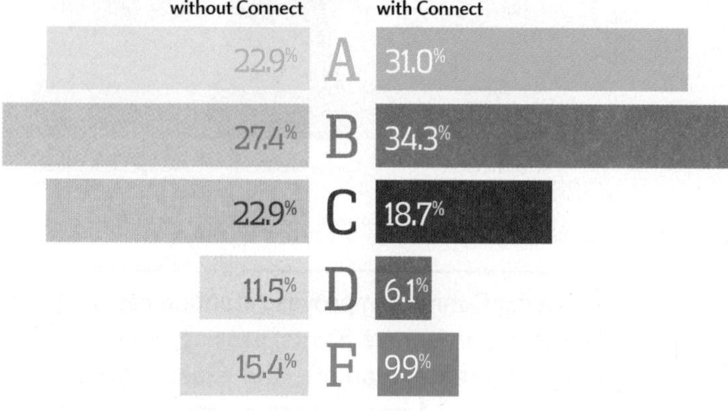

More students earn **As** and **Bs** when they use **Connect**.

Trusted Service and Support

- Connect integrates with your LMS to provide single sign-on and automatic syncing of grades. Integration with Blackboard®, D2L®, and Canvas also provides automatic syncing of the course calendar and assignment-level linking.
- Connect offers comprehensive service, support, and training throughout every phase of your implementation.
- If you're looking for some guidance on how to use Connect, or want to learn tips and tricks from super users, you can find tutorials as you work. Our Digital Faculty Consultants and Student Ambassadors offer insight into how to achieve the results you want with Connect.

www.mheducation.com/connect

Instructor Resources

The following resources can be found on Connect:

- ***Presentation Tools*** Everything you need for outstanding presentations.
  - Animations–Numerous full-color animations illustrating important processes are provided. Harness the visual impact of concepts in motion by importing these files into classroom presentations or online course materials.
  - Lecture PowerPoints–with animations fully embedded.
  - JPEG images–Full-color digital files of all illustrations that can be readily incorporated into presentations, exams, or custom-made classroom materials.
  - Tables–Tables from the text are available in electronic format.
- ***Google Earth and Virtual Vista Exercises***–Descriptions and questions to help students visualize and analyze geologic features.
- ***Instructor's Manual***–The instructor's manual contains chapter outlines, lecture enrichment ideas, and critical thinking questions.
- ***Computerized Test Bank***–A comprehensive bank of test questions is provided within a computerized test bank. Instructors can select questions from multiple McGraw-Hill test banks or author their own, and then either print the test for paper distribution or give it online.

ACKNOWLEDGMENTS

We have tried to write a book that will be useful to both students and instructors. We would be grateful for any comments by users, especially regarding mistakes within the text or sources of good geological photographs.

Although he is no longer listed as an author, this edition bears a lot of the writing style and geologic philosophy of the late David McGeary. He was coauthor of the original edition, published in 1979. His authorship continued through the seventh edition, after which he retired and turned over revision of his half of the book to Diane Carlson. We greatly appreciate his role in making this book successful way beyond what he or his original coauthor ever dreamed of.

Tom Arny wrote the planetary geology chapter for the tenth edition. This chapter was revised and updated by Steve Kadel for the eleventh and twelfth editions and by Mark Boryta for the fifteenth edition. Chris Cappa and Delphine Farmer wrote the chapter on climate change for the fourteenth edition, and Professor Cappa revised chapter 21 for the fifteenth edition. We greatly appreciate the publisher's "book team" whose names appear on the copyright page. Their guidance, support, and interest in the book were vital for the completion of this edition.

Thank you also to Cindy Shaw for her contribution to the superior art program of the eleventh and twelfth editions.

Diane Carlson would like to thank her husband Reid Buell for his tireless support, and for his technical assistance with engineering geology and hydrogeology material, in several chapters. Charles Plummer thanks his wife Beth Strasser for assistance with photography in the field, and for her perspective as a paleontologist and anthropologist. Lisa Hammersley would like to thank her husband Chris Cappa for his support, and for agreeing to find time to write the climate change chapter for this book.

The following individuals helped write and review learning goal-oriented content for LearnSmart for Geology:

Sylvester Allred, *Northern Arizona University*
Lisa Hammersley, *California State University, Sacramento*
Arthur C. Lee, *Roane State Community College*

Through each edition of *Physical Geology*, we have had outstanding feedback from reviewers who have provided careful evaluations and useful suggestions for improvement.

Dave Berner, *Normandale Community College*
Barbara L. Brande, *University of Montevallo*
Melissa Davis, *Ivy Tech Community College*
Don B. DeYoung, *Grace College*
Aley El-Shazly, *Marshall University*
Kevin P. Hefferan, *University of Wisconsin-Stevens Point*
Jennifer G. Kidder, *Tompkins Cortland Community College*
Jason Mayfield, *Diablo Valley College*
Ernie Morrison, *Cowley County College*
Ravindra Nandigam, *South Texas College*
Doreena Patrick, *The Richard Stockton College of New Jersey*
Charles G. Patterson, *Red Rocks Community College*
Christian Poppeliers, *East Carolina University*
Jeffery G. Richardson, *Columbus State Community College*
Adil M. Wadia, *The University of Akron Wayne College*
Stephanie Welch, *Southeastern Louisiana University*

MEET THE AUTHORS

Courtesy of C.C. Plummer

Charles Plummer at Tengboche, in the Himalayan Mountains of Nepal.

Courtesy of Reid Buell

Diane Carlson at Convict Lake in the Sierra Nevada Mountains of California.

Courtesy of Christopher Cappa

Lisa Hammersley on the coast of Northern California

CHARLES PLUMMER Professor Charles "Carlos" Plummer grew up in the shadows of volcanoes in Mexico City. There, he developed a love for mountains and mountaineering that eventually led him into geology. He received his B.A. degree from Dartmouth College. After graduation, he served in the U.S. Army as an artillery officer. He resumed his geological education at the University of Washington, where he received his M.S. and Ph.D. degrees. His geologic work has been in mountainous and polar regions, notably Antarctica (where a glacier is named in his honor). He taught at Olympic Community College in Washington and worked for the U.S. Geological Survey before joining the faculty at California State University, Sacramento.

At CSUS, he taught optical mineralogy, metamorphic petrology, and field courses as well as introductory courses. He retired from teaching in 2003. He skis, has a private pilot license, and is certified for open-water SCUBA diving. (plummercc@csus.edu)

DIANE CARLSON Professor Diane Carlson grew up on the glaciated Precambrian shield of northern Wisconsin and received an A.A. degree at Nicolet College in Rhinelander and a B.S. in geology at the University of Wisconsin at Eau Claire. She continued her studies at the University of Minnesota-Duluth, where she focused on the structural complexities of high-grade metamorphic rocks along the margin of the Idaho batholith for her master's thesis. The lure of the West and an opportunity to work with the U.S. Geological Survey to map the Colville batholith in northeastern Washington led her to Washington State University for her Ph.D. Dr. Carlson accepted a position at California State University, Sacramento, after receiving her doctorate and taught physical geology, structural geology, environmental geology, field techniques, and field geology. Professor Carlson is a recipient of the Outstanding Teacher Award from the CSUS School of Arts and Sciences. She is also engaged in researching the structural and tectonic evolution of part of the Foothill Fault System in the northern Sierra Nevada of California. (carlsondh@csus.edu)

LISA HAMMERSLEY Dr. Lisa Hammersley hails originally from England and received a B.Sc. in geology from the University of Birmingham. After graduating, she traveled the world for a couple of years before returning to her studies and received a Ph.D. in geology from the University of California at Berkeley. She joined the faculty at California State University, Sacramento in 2003, where she taught natural disasters, physical geology, geology of Mexico, mineralogy, and metallic ore deposits, receiving the Outstanding Teacher Award from the College of Natural Sciences and Mathematics in 2011. Dr. Hammersley specializes in igneous petrology with an emphasis on geochemistry. Her interests involve understanding magma chamber processes and how they affect the evolution of volcanic systems. She has worked on volcanic systems in Ecuador, Mexico, and the United States. Dr. Hammersley has also worked in the field of geoarcheology, using geologic techniques to identify the sources of rocks used to produce stone grinding tools found near the pyramids of Teotihuacan in Mexico. She is currently serving as the Associate Dean of the College of Natural Sciences and Mathematics. (hammersley@csus.edu)

CHAPTER 1

Introducing Geology, the Essentials of Plate Tectonics, and Other Important Concepts

Mount Robson, 3,954 meters (12,972 feet) above sea level, is the highest peak in the Canadian Rocky Mountains.
©J. A. Kraulis/Masterfile

LEARNING OBJECTIVES

- Know what physical geology is, and describe some of the things it is used for.
- Define a system, and describe the four Earth systems (spheres).
- Distinguish between the Earth's internal and external heat engines and list the processes driven by them.
- List the three major internal zones of the Earth.
- Describe the lithosphere and the asthenosphere.
- Sketch and label the different types of plate boundaries.
- Summarize the scientific method, and define the meaning of the word *theory*.
- Know the age of the Earth.

Have you ever looked out of the window of an airplane and wondered about the landforms that you see below you, or examined a pebble on a beach and wondered how it got there? Have you ever listened to a news report about a major natural disaster such as an earthquake, flood, or volcanic eruption, and asked yourself why it happened and what you would do if you found yourself in such a situation? What about the materials used to manufacture the electronics you use every day or the gasoline used to fuel your car–have you ever thought about where they come from, how they formed, and how we exploit them? These topics are all parts of **geology**–the scientific study of the Earth. Geologists use the scientific method to explain natural aspects of the Earth, such as what it is made of and the processes that affect it, and to interpret Earth's history. This chapter is an introduction to geology. We will first explore the uses of geology before introducing some of the important concepts such as the modern theory of plate tectonics and geologic time. These concepts form a framework for the rest of the book. Understanding the "big picture" presented here will aid you in comprehending the chapters that follow.

Strategy for Using This Textbook

- As authors, we try to be thorough in our coverage of topics so the textbook can serve you as a resource. Your instructor may choose, however, to concentrate only on certain topics for *your* course. Find out which topics and chapters you should focus on in your studying and concentrate your energies there.
- Your instructor may present additional material that is not in the textbook. Take good notes in class.
- Do not get overwhelmed by terms. (Every discipline has its own language.) Don't just memorize each term and its definition. If you associate a term with a concept or mental picture, remembering the term comes naturally when you understand the concept. (You remember names of people you know because you associate personality and physical characteristics with a name.) You may find it helpful to learn the meanings of frequently used prefixes and suffixes for geological terms. These can be found in appendix G.
- **Boldfaced** terms are ones you are likely to need to understand because they are important to the entire course.
- *Italicized* terms are not as important but may be necessary to understand the material in a particular chapter.
- Pay particular attention to illustrations. Geology is a visually oriented science, and the photos and artwork are at least as important as the text. You should be able to sketch important concepts from memory.
- Find out to what extent your instructor expects you to learn the material in the boxes. They offer an interesting perspective on geology and how it is used, but much of the material might well be considered optional for an introductory course and not vital to your understanding of major topics. Many of the In Greater Depth boxes are meant to be challenging–do not be discouraged if you need your instructor's help in understanding them.
- Read through the appropriate chapter before going to class. Reread it after class, concentrating on the topics covered in the lecture or discussion. Especially concentrate on concepts that you do not fully understand. Return to previously covered chapters to refresh your memory on necessary background material.
- Use the end-of-chapter material for review. The Summary is just that, a summary. Don't expect to get through an exam by only reading the summary and not the rest of the chapter. Use the Terms to Remember to see if you can visually or verbally associate the appropriate concept with each term. Answer the Testing Your Knowledge questions in writing. Be honest with yourself. If you are fuzzy on an answer, return to that portion of the chapter and reread it. Remember that these are just a sampling of the kinds of questions that might be on an exam.
- Geology, like most science, builds on previously acquired knowledge. You must retain what you learn from chapter to chapter. If you forget or did not learn significant concepts covered early in your course, you will find it frustrating later in the course. (To verify this, turn to chapter 20 and you will probably find it intimidating; but if you build on your knowledge as you progress through your course, the chapter material will fall nicely into place.)
- Explore the web links provided in this book. You will find they provide additional useful information.
- Be curious. Geologists are motivated by a sense of discovery. We hope you will be, too.

1.1 WHO NEEDS GEOLOGY?

Geology benefits you and everyone else on this planet. The clothes you wear, the food you eat, your smart phone, and your car exist because of what geologists have discovered about the Earth. The Earth can also be a killer. You might have survived an earthquake, flood, or other natural disaster thanks to action taken based on what scientists have learned about these hazards. Before getting into important scientific concepts, we will look at some of the ways geology has benefited you and will continue to do so.

Supplying Things We Need

We depend on the Earth for energy resources and the raw materials we need for survival, comfort, and pleasure. Every manufactured object relies on Earth's resources—even a pencil (figure 1.1). The Earth, at work for billions of years, has localized material into concentrations that humans can mine or extract. By learning how the Earth works and how different kinds of substances are distributed and why, we can intelligently search for metals, sources of energy, and gems. Even maintaining a supply of sand and gravel for construction purposes depends on geology.

The economic systems of Western civilization currently depend on abundant and cheap energy sources. Nearly all our vehicles and machinery are powered by petroleum, coal, or nuclear power and depend on energy sources concentrated unevenly in the Earth. The U.S. economy, in particular, is geared to petroleum and natural gas as cheap sources of energy. It is important to remember, however, that these resources took hundreds of millions of years to form, and they are being rapidly depleted. When fuel prices jump, people who are not aware that petroleum is a nonrenewable resource become upset and are quick to blame oil companies, politicians, and oil-producing countries. (The Gulf Wars of 1991 and 2003 were at least partially fought because of the industrialized nations' petroleum requirements.) In recent years, the United States has been able to reduce its reliance on imported oil by developing technology to access oil that was previously too difficult or too expensive to extract. Finding more of this diminishing resource will require more money and increasingly sophisticated knowledge of geology. Although many people are not aware of it, we face similar problems with diminishing resources of other materials, notably metals such as iron, aluminum, copper, and tin, each of which has been concentrated in a particular environment by the action of the Earth's geologic forces.

Just how much of our resources do we use? According to the Minerals Education Coalition, approximately 17,936 kilograms (39,543 pounds; for metric conversions, go to appendix E) of resources, including energy resources, must be mined annually to provide for every person in the United States. The amount of each commodity mined per person per year is 4,115 kilograms stone, 3,093 kilograms sand and gravel, 279 kilograms limestone for cement, 66 kilograms clays, 191 kilograms salt, 243 kilograms other nonmetals, 150 kilograms iron ore, 30 kilograms aluminum ore, 6 kilograms copper, 8 kilograms lead and zinc, 3 kilograms manganese, and 11 kilograms other metals. Americans' yearly per capita consumption of energy resources includes 3,463 liters (915 gallons) of petroleum, 2,609 kilograms of coal, 2,389 cubic meters (84,348 cubic feet) of natural gas, and 0.08 kilograms of uranium.

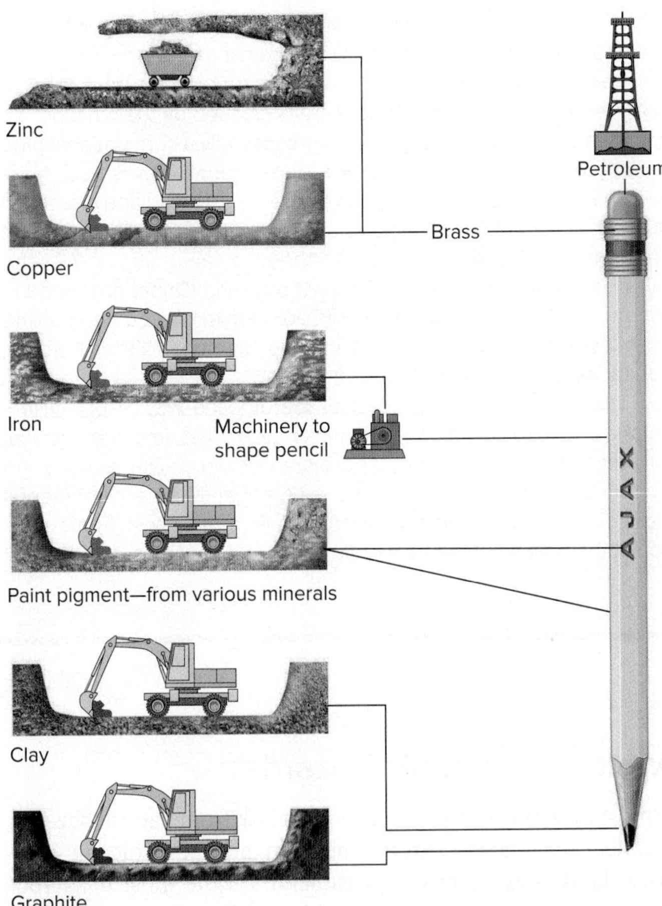

FIGURE 1.1
Earth's resources needed to make a wooden pencil.

Protecting the Environment

Our demands for more energy and metals have, in the past, led us to extract them with little regard for effects on the balance of nature within the Earth and therefore on us, Earth's residents. Mining of coal, if done carelessly, for example, can release acids into water supplies. Understanding geology can help us lessen or prevent damage to the environment—just as it can be used to find the resources in the first place.

The environment is further threatened because these are nonrenewable resources. Petroleum and metal deposits do not grow back after being harvested. As demands for these commodities increase, so does the pressure to disregard the ecological damage caused by the extraction of the remaining deposits. As the supply of resources decreases, we are forced to exploit them from harder-to-reach locations. The Deepwater Horizon oil spill in the Gulf of Mexico in 2010 was due in part to the very deep water in which drilling was taking place (see box 22.2).

Geology has a central role in these issues. Oil companies employ geologists to discover new oil fields, while the public and

ENVIRONMENTAL GEOLOGY 1.1

Delivering Oil—The Environment versus the Economy

In the 1960s, geologists discovered oil beneath the coast of the Arctic Ocean on Alaska's North Slope at Prudhoe Bay (box figure 1). It is now the United States' third largest oil field. Thanks to the Trans-Alaska pipeline, completed in 1977, Alaska has at times supplied as much as 25% of the United States' domestic oil, although it currently supplies only 7%.

In the late 1970s before Alaskan oil began to flow, the United States was importing almost half its petroleum, at a loss of billions of dollars per year to the national economy. At its peak, over 2 million barrels of oil a day flowed from the Arctic oil fields. Despite its important role in the American economy, some considered the Alaska pipeline and the use of oil tankers to be unacceptable threats to the area's ecology.

The 1,287-kilometer-long pipeline crosses regions of ice-saturated, frozen ground and major earthquake-prone mountain ranges that geologists regard as serious hazards to the structure.

Building anything on frozen ground creates problems. The pipeline presented enormous engineering problems. If the pipeline were placed on the ground, the hot oil flowing through it could melt the frozen ground. On a slope, mud could easily slide and rupture the pipeline. Careful (and costly) engineering minimized these hazards. Much of the pipeline is elevated above the ground (box figure 2).

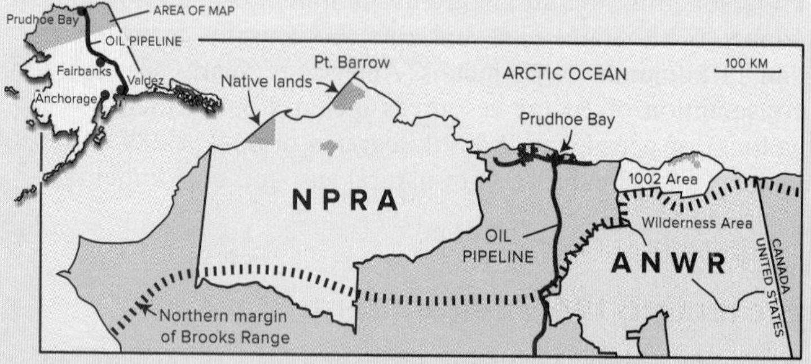

BOX 1.1 ■ FIGURE 1

Map of northern Alaska showing locations and relative sizes of the National Petroleum Reserve in Alaska (NPRA) and the Arctic National Wildlife Refuge (ANWR). "1002 Area" is the portion of ANWR being proposed for oil exploitation. Current oil production is taking place at Prudhoe Bay.

Source: U.S.G.S. Fact Sheet 045-02 and U.S.G.S. Fact Sheet 014-03

BOX 1.1 ■ FIGURE 2

The Alaska pipeline.

©David Applegate

Radiators conduct heat out of the structure. In some places, refrigeration equipment in the ground protects against melting.

Records indicate that a strong earthquake can be expected every few years in the earthquake belts crossed by the pipeline. An earthquake could rupture a pipeline—especially a conventional pipe as in the original design. When the Alaska pipeline was built, however, in several places sections were specially jointed and placed on slider beams to allow the pipe to shift as much as 6 meters without rupturing. In 2002, a major earthquake (magnitude 7.9—the same strength as the May 2008 earthquake in China, described in chapter 16, that killed more than 87,000 people) caused the pipeline to shift several meters, resulting in minor damage to the structure, but the pipe did not rupture (box figure 3).

The original estimated cost of the pipeline was $900 million, but the final cost was $7.7 billion, making it, at that time, the costliest privately financed construction project in history. The redesigning and construction that minimized the potential for an environmental disaster were among the reasons for the increased cost. Some spills from the pipeline have occurred. In January 1981, 5,000 barrels of oil

government depend on other geologists to assess the potential environmental impact of petroleum's removal from the ground, the transportation of petroleum (see box 1.1), and disposal of any toxic wastes from petroleum products.

The consumption of resources, in particular energy resources, is also affecting the Earth's climate. Chapter 21 covers the evidence for global climate change and its connection to greenhouse gases released by burning fossil fuels.

Avoiding Geologic Hazards

Almost everyone is, to some extent, at risk from natural hazards, such as earthquakes or hurricanes. Earthquakes, volcanic eruptions, landslides, floods, and tsunami are the most dangerous *geologic hazards.* Each is discussed in detail in appropriate chapters. Here, we will give some examples to illustrate the role that geology can play in mitigating geologic hazards.

BOX 1.1 ■ FIGURE 3

The Alaska pipeline where it was displaced along the Denali fault during the 2002 earthquake. The pipeline is fastened to teflon shoes, which are sitting on slider beams. Go to http://pubs.usgs.gov/fs/2003/fs014-03/pipeline.html for more information.
Source: Alyeska Pipeline Service Company/U.S. Geological Survey

were lost when a valve ruptured. In 2001, a man fired a rifle bullet into the pipeline, causing it to rupture and spill 7,000 barrels of oil into a forested area. In March 2006, a British Petroleum Company (BP) worker discovered a 201,000 gallon spill from that company's feeder pipes to the Trans-Alaska Pipeline. This was the largest oil spill on the North Slope to date. Subsequent inspection by BP of its feeder pipes revealed much more corrosion than expected. As a result, it made a very costly scaling back of its oil production to replace pipes and make major repairs.

Recently, two other large oil pipeline projects have caused much debate. The Keystone Pipeline delivers oil from Canada to refineries in the Midwest and the Gulf Coast of Texas. Although parts of the pipeline system are already operational, a proposed extension from Canada to Nebraska with a shorter route and larger-diameter pipe faced strong criticism from environmentalists, and in 2015, the plan was rejected by the U.S. government. In 2017, however, the U.S. government changed course and approved the pipeline.

In August 2016, Native American protests in North Dakota halted construction of a section of the Dakota Access pipeline, which is intended to span over 1,000 miles between North Dakota and Illinois. The protests were sparked by concerns about negative impacts on the environment and damage to sites of cultural importance.

The alternative to pipelines is transporting oil by rail, which can be hazardous. On December 30, 2013, a train carrying crude oil collided with another train in North Dakota. The collision caused a large explosion and fire, leading to a partial evacuation of the nearby town of Casselton. Earlier in 2013, a train carrying crude oil derailed in Quebec, Canada, killing more than 40 people in the town of Lac-Megantic.

Oil can also be transported by sea. When the tanker *Exxon Valdez* ran aground in 1989, more than 240,000 barrels of crude oil were spilled into the waters of Alaska's Prince William Sound. The spill, with its devastating effects on wildlife and the fishing industry, dramatically highlighted the conflicts between maintaining the energy demands of the American economy and conservation of the environment. Statistical studies of tanker accidents worldwide revealed the frequency with which large oil spills could be expected. The *Exxon Valdez* spill should not have been a surprise.

Additional Resources

The Alyeska pipeline company's site.

- www.alyeska-pipe.com/

U.S. Geological Survey fact sheet on the Arctic National Wildlife Refuge.

- http://pubs.usgs.gov/fs/2002/fs045-02/

Geotimes article on the 2006 oil spill. Links at the end of this and other articles lead to older articles published by the magazine.

- www.geotimes.org/aug06/WebExtra080706.html

On Tuesday, January 12, 2010, a magnitude 7 earthquake struck close to Port-au-Prince, the capital city of Haiti. The city and other parts of Haiti were left in ruins (figure 1.2*A*). Responses to the emergency were severely hampered because roads were blocked by debris, hospitals were heavily damaged, the seaport in Port-au-Prince was rendered unusable, and the control tower at the airport was damaged. This not only made it difficult for Haitian emergency workers to rescue those trapped or injured, but also made it difficult for international relief to reach the country quickly. The Haitian government estimates that over 300,000 people were killed and a million were left homeless. However, due to the immense damage and the difficulties involved in the response, the true impact in terms of casualties may never be known.

Just one month later, on February 10, a magnitude 8.8 earthquake hit off the coast of central Chile. The earthquake

A

B

FIGURE 1.2

Damage caused by earthquakes in (*A*) Haiti and (*B*) Chile in 2010. Notice how many of the buildings in Haiti were reduced to rubble. Although many buildings were destroyed in Chile, strict building codes meant that many, such as the high-rise apartment building in the background of (*B*), survived the massive magnitude 8.8 earthquake.
(A) Source: Tech. Sgt. James L. Harper, Jr., U.S. Air Force (B) Source: Walter D. Mooney, U.S. Geological Survey

was the sixth largest ever recorded, releasing 500 times as much energy as the Haitian earthquake, and was felt by 80% of the population. Movement of the sea floor due to the earthquake generated a tsunami that caused major damage to some coastal communities and prompted the issuance of a Pacific-wide tsunami warning. It is estimated that 525 people were killed and 1.5 million people were displaced.

Although the impact on Chile was significant (figure 1.2*B*), this enormous earthquake killed far fewer people than the earthquake that struck Haiti. Why is this, and could the deaths in Haiti have been avoided? As described later in this chapter, geologists understand that the outer part of the Earth is broken into large slabs known as *tectonic plates* that are moving relative to each other. Most of the Earth's geologic activity, such as earthquakes and volcanic eruptions, occurs along boundaries between tectonic plates. Both Chile and Haiti are located on plate boundaries, and both have experienced large earthquakes in the past. In fact, the largest earthquake ever recorded happened in Chile in 1960. The impact of earthquakes can be reduced, or *mitigated,* by engineering buildings to withstand shaking. Chile has strict building codes, which probably saved many lives. Haiti, however, is one of the poorest countries in the Western Hemisphere and does not have the stringent building codes of Chile and other wealthy nations. Because of this, thousands of buildings collapsed and hundreds of thousands lost their lives.

Japan is seen as a world leader in earthquake engineering, but nothing could prepare the country for the events of March 11, 2011. At 2:46 P.M., a devastating magnitude 9.0 earthquake hit the east coast of Japan. The earthquake was the largest known to have hit Japan. Soon after the earthquake struck, tsunami waves as high as 38.9 meters (128 feet) inundated the coast. Entire towns were destroyed by waves that in some cases traveled up to 10 kilometers (6 miles) inland. The death toll from this disaster was almost 16,000, and almost half a million people were left homeless. Things could have been much worse. Due to the high building standards in Japan, the damage from the earthquake itself was not severe. Japan has an earthquake early warning system, and after the earthquake struck, a warning went out to millions of people. In Tokyo, the warning arrived one minute before the earthquake was felt. This early warning is believed to have saved many lives. Japan also has a tsunami warning system, and coastal communities have clearly marked escape routes and regular drills for their citizens. Concrete seawalls were built to protect the coast. Unfortunately, the walls were not high enough to hold back a wave of such great height, and some areas designated as safe areas were not on high enough ground. Still, without the safety precautions in place, many more thousands of people could have lost their lives. In some communities, lives were saved by the actions of their ancestors. Ancient stone markers along the coastline, some more than 600 years old, warn people of the dangers of tsunami. In the hamlet of Aneyoshi, one of these stone markers reads, "Remember the calamity of the great tsunami. Do not build any homes below this point." The residents of Aneyoshi heeded the warning, locating their homes on higher ground, and the community escaped unscathed.

Volcanic eruptions, like earthquakes and tsunami, are products of Earth's sudden release of energy. Unlike earthquakes and tsunami, however, volcanic eruptions can last for extended periods of time. Volcanic hazards include lava flows, falling debris, and ash clouds (see box 1.2). The most deadly volcanic hazards are pyroclastic flows and volcanic mudflows. As described in chapter 4, a *pyroclastic flow* is a hot, turbulent mixture of expanding gases and volcanic ash that flows rapidly down the side of a volcano. Pyroclastic flows often reach

ENVIRONMENTAL GEOLOGY 1.2

A Volcanic Eruption in Iceland Shuts Down European Airspace for Over a Week

The hazards associated with volcanic eruptions are not necessarily localized. Volcanic ash spewed into the atmosphere presents a hazard to air traffic. Particles of ash can sandblast the windows and clog a plane's sensors. When fine particles of ash are sucked into jet engines, they melt and fuse onto the blades, causing the engines to fail. In 1985, a British Airways flight from London, England, to Auckland, New Zealand, flew into a cloud of ash flung up from Mount Galunggung in Indonesia. All four engines failed, and the plane dropped 14,000 feet before the engines could be restarted. This and other incidents have shown aviation authorities that extreme caution must be taken during a volcanic eruption.

In March 2010, Eyjafjallajökull (pronounced ay-uh-fyat-luh-yoe-kuutl-ul), a relatively small volcano in Iceland, began erupting lava from fissures on the side of the mountain. On the morning of April 14, the eruption shifted to new vents buried under the ice cap that covers the summit of the volcano and increased in intensity. The ice melted, adding cold water to the hot lava, causing it to cool rapidly and to fragment into ash particles. The ash was carried up into the atmosphere by an eruption plume where it encountered the jet stream, a band of high-speed winds that blow from west to east (box figure 1). The jet stream carried the ash cloud over much of northern Europe. Because of the hazard to air traffic, much of Europe's airspace was closed from April 15 to April 23, the largest disruption to air traffic since World War II. Flights into and out of Europe were canceled, leaving millions of passengers stranded around the world.

The cost to the airline industry is estimated to have been around $200 million a day. Total losses are estimated at $1.7 billion. The industry complained that the restrictions were too tight and that ash levels were low enough for safe flight.

BOX 1.2 ■ FIGURE 1

An ash plume from Iceland's Eyjafjallajökull volcano spreads south toward Europe. Notice that the southern end of the plume is being blown eastward by the polar jet stream.
Source: Jeff Schmaltz, MODIS Rapid Response Team, NASA/GSFC

Additional Resources

Amazing images of the eruption can be found at

- http://www.boston.com/bigpicture/2010/04/more_from_eyjafjallajokull.html

The Institute of Earth Sciences Nordic Volcanological Center, University of Iceland—lots of great information about the eruptions

- http://earthice.hi.is/eruption_eyjafjallajokull_2010

speeds of over 100 kilometers per hour and are extremely destructive. A *mudflow* is a slurry of water and rock debris that flows down a stream channel.

Mount Pinatubo's eruption in 1991 was the second largest volcanic eruption of the twentieth century. Geologists successfully predicted the climactic eruption (figure 1.3) in time for Philippine officials to evacuate people living near the mountain. Tens of thousands of lives were saved from pyroclastic flows and mudflows.

By contrast, one of the worst volcanic disasters of the twentieth century took place after a relatively small eruption of Nevado del Ruiz in Colombia in 1985. Hot volcanic debris blasted out of the volcano and caused part of the ice and snow capping the peak to melt. The water and loose debris turned into a mudflow. The mudflow overwhelmed the town of Armero at the base of the volcano, killing 23,000 people (figure 1.4). Colombian geologists had previously predicted such a mudflow could occur, and they published maps showing the location and

FIGURE 1.3

The major eruption of Mount Pinatubo on June 15, 1991, as seen from Clark Air Force Base, Philippines.
Source: Robert LaPointe, U.S. Air Force

FIGURE 1.4

Most of the town of Armero, Colombia, and its residents are buried beneath up to 8 meters of mud from the 1985 mudflow.
©Jacques Langevin/Corbis/Getty Images

extent of expected mudflows. The actual mudflow that wiped out the town matched that shown on the geologists' map almost exactly. Unfortunately, government officials ignored the map and the geologists' report; otherwise, the tragedy could have been averted.

Understanding Our Surroundings

It is a uniquely human trait to want to understand the world around us. Most of us get satisfaction from understanding our cultural and family histories, or learning how things such as car engines or computers work. Music and art help link our feelings to that which we have discovered through our life. The natural sciences involve understanding the physical and biological universe in which we live. Most scientists get great satisfaction from their work because, besides gaining greater knowledge from what has been discovered by scientists before them, they can find new truths about the world around them. Even after a basic geology course, you can use what you learn to explain and be able to appreciate what you see around you, especially when you travel. If, for instance, you were traveling through the Canadian Rockies, you might see the scene in this chapter's opening photo and wonder how the landscape came to be.

You might wonder: (1) why there are layers in the rock exposed in the cliffs; (2) why the peaks are so jagged; (3) why there is a glacier in a valley carved into the mountain; (4) why this is part of a mountain belt that extends northward and southward for thousands of kilometers; (5) why there are mountain ranges here and not in the central part of the continent. After completing a course in physical geology, you should be able to answer these questions as well as understand how other kinds of landscapes formed.

1.2 EARTH SYSTEMS

The awesome energy released by an earthquake or volcano is a product of forces within the Earth that move firm rock. Earthquakes and volcanoes are only two consequences of the ongoing changing of Earth. Ocean basins open and close. Mountain ranges rise and are then very slowly worn back down to plains. Studying how the Earth works can be as exciting as watching a great theatrical performance. The purpose of this book is to help you understand how and why those changes take place. More precisely, we concentrate on *physical geology,* which is the division of geology concerned with Earth materials, changes in the surface and interior of the Earth, and the dynamic forces that cause those changes. Put another way, physical geology is about how the Earth works.

But to understand geology, we must also understand how the solid Earth interacts with water, air, and living organisms. For this reason, it is useful to think of the Earth as being part of a system. A *system* is an arbitrarily isolated portion of the universe that can be analyzed to see how its components interrelate. For example, the *solar system* is a part of the much larger universe. The solar system includes the Sun, planets, the moons orbiting planets, and asteroids (see chapter 23).

The **Earth system** is a small part of the larger solar system, but it is, of course, very important to us. The Earth system has its components, which can be thought of as its subsystems. We refer to these as *Earth systems* (plural). These systems, or "*spheres,*" are the atmosphere, the hydrosphere, the biosphere, and the geosphere. You, of course, are familiar with the **atmosphere**, the gases that envelop the Earth. The **hydrosphere** is the water on or near Earth's surface. The hydrosphere includes the oceans, rivers, lakes, and glaciers of the world. It also includes **groundwater**, which is water that lies beneath the ground surface. Earth is unique among the planets in that two-thirds of its surface is covered by oceans. The **biosphere** is all of the living or once-living material on Earth. The **geosphere**, or **solid Earth system**, is the rock and other inorganic Earth material that make up the bulk of the planet. This book mostly concentrates on the geosphere; to understand geology, however, we must understand the interaction between the solid Earth and the other systems (spheres).

The Japanese tsunami involved the interaction of the geosphere and the hydrosphere. The earthquake took place in the geosphere. Energy was transferred into giant waves in the hydrosphere. The hydrosphere and geosphere again interacted when waves inundated the shores. Can you think of other ways in which the four spheres interacted, either during or as a result of the tsunami?

All four of the Earth systems interact with each other to produce soil, such as we find in farms, gardens, and forests. The solid "dirt" is a mixture of decomposed and disintegrated rock and organic matter. The organic matter is from decayed plants—from the biosphere. The geosphere contributes the rock that has broken down while exposed to air (the atmosphere) and water (the hydrosphere). Air and water also occupy pore space between the solid particles.

IN GREATER DEPTH 1.3

Geology as a Career

If someone says that she or he is a geologist, that information tells you almost nothing about what he or she does. This is because geology encompasses a broad spectrum of disciplines. Perhaps what most geologists have in common is that they were attracted to the outdoors. Most of us enjoyed hiking, skiing, climbing, or other outdoor activities before getting interested in geology. We like having one of our laboratories being Earth itself.

Geology is a collection of disciplines. When someone decides to become a geologist, she or he is selecting one of those disciplines. The choice is very large. Some are financially lucrative; others may be less so but might be more satisfying. Following are a few of the areas in which geologists work.

Petroleum geologists work at trying to determine where existing oil fields might be expanded or where new oil fields might exist (box figure 1). A petroleum geologist can make over $90,000 a year but may have to spend months at a time on an offshore drilling platform. Mining geologists might be concerned with trying to determine where to extend an existing mine to get more ore or trying to find new concentrations of ore that are potentially commercially viable. Environmental geologists might work at mitigating pollution or preventing degradation of the environment. Marine geologists are concerned with understanding the sea floor. Some go down thousands of meters in submersibles to study geologic features on the sea floor. Hydrogeologists study surface and underground water and assist in either increasing our supply of clean water or isolating or cleaning up polluted water. Glaciologists work in Antarctica studying the dynamics of glacier movement or collecting ice cores through drilling to determine climate changes that have taken place over the past 100,000 years or more. Other geologists who work in Antarctica might be deciphering the history of a mountain range, working on skis and living in tents. Volcanologists sometimes are killed or injured while trying to collect gases or samples of lava from a volcano. Some sedimentologists scuba dive in places like the Bahamas, skewering lobsters for lunch while they collect sediment samples. One geologist was the only scientist to work on the Moon. Geophysicists interpret earthquake waves or gravity measurements to determine the nature of Earth's interior. Seismologists are geophysicists who specialize in earthquakes.

BOX 1.3 ■ FIGURE 1

Petroleum geologists examine geological information.
©Monty Rakusen/Getty Images

Engineering geologists determine whether the rock or soil upon which structures (dams, bridges, buildings) are built can safely support those structures. Paleontologists study fossils and learn about when extinct creatures lived and the environment in which they existed.

Teaching is an important field in which geologists work. Some teach at the college level and are usually involved in research as well. Some teach Earth science (which includes meteorology, oceanography, and astronomy as well as geology) at the middle or high school level.

Many geologists enjoy the challenge and adventure of field work, but some work comfortably behind computer screens or in laboratories with complex analytical equipment. Usually, a geologist engages in a combination of field work, lab work, and computer analysis.

Geologists tend to be happy with their jobs. In surveys of job satisfaction in a number of professions, geology rates near or at the top. A geologist is likely to be a generalist who solves problems by bringing in information from beyond his or her specialty. Chemistry, physics, and life sciences are often used to solve problems. Problems geologists work on tend to be ones in which there are few clues. So the geologist works like a detective, piecing together the available data to form a plausible solution. In fact, some geologists work at solving crimes—forensic geology is a branch of geology dedicated to criminal investigations.

Not all people who major in geology become professional geologists. Physicians, lawyers, and businesspeople who have majored in geology have felt that the training in how geologists solve problems has benefited their careers.

Additional Resource

For more information, go to the American Geological Institute's career site at

- www.earthscienceworld.org/careers/brochure.html

1.3 AN OVERVIEW OF PHYSICAL GEOLOGY—IMPORTANT CONCEPTS

The remainder of this chapter is an overview of physical geology that should provide a framework for most of the material in this book. Although the concepts probably are new to you, it is important that you comprehend what follows. You may want to reread portions of this chapter while studying later chapters when you need to expand or reinforce your comprehension of this basic material. You will especially want to refresh your understanding of plate tectonics when you learn about the plate-tectonic setting for the origin of rocks in chapters 3 through 7.

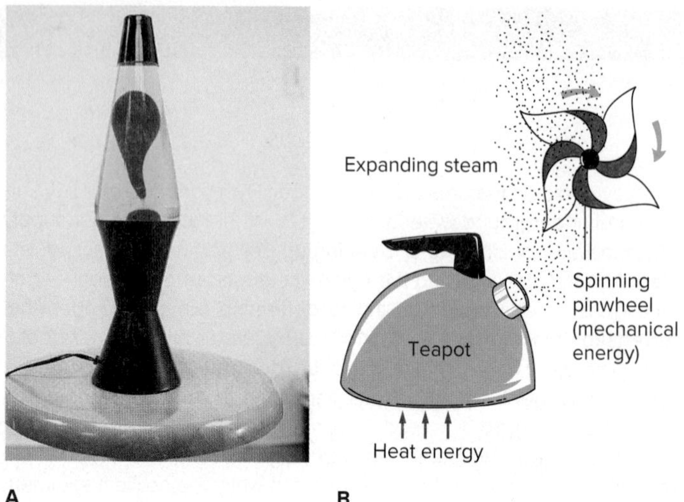

FIGURE 1.5

Two examples of simple heat engines. (*A*) A lava lamp. Blobs are heated from below and rise. Blobs cool off at the top of the lamp and sink. (*B*) A pinwheel held over steam. Heat energy is converted to mechanical energy.
(A) ©C.C. Plummer

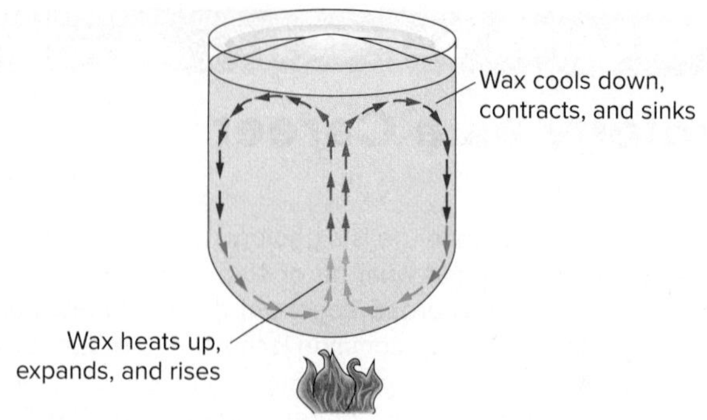

FIGURE 1.6

Convection of wax due to density differences caused by heating and cooling (shown schematically).

The Earth can be visualized as a giant machine driven by two engines, one internal and the other external. Both are *heat engines,* devices that convert heat energy into mechanical energy. Two simple heat engines are shown in figure 1.5. An automobile is powered by a heat engine. When gasoline is ignited in the cylinders, the resulting hot gases expand, driving pistons to the far ends of their cylinders. In this way, the heat energy of the expanding gas has been converted to the mechanical energy of the moving pistons, then transferred to the wheels, where the energy is put to work moving the car.

Earth's *internal* heat engine is driven by heat moving from the hot interior of the Earth toward the cooler exterior. Volcanic eruptions and earthquakes are products of this heat engine.

Earth's *external* heat engine is driven by solar energy. Heat from the Sun provides the energy for circulating the atmosphere and oceans. Water, especially from the oceans, evaporates because of solar heating. When moist air cools, we get rain or snow.

Over long periods of time, moisture at the Earth's surface helps rock disintegrate. Water washing down hillsides and flowing in streams loosens and carries away the rock particles. In this way, mountains originally raised by Earth's internal forces are worn away by processes driven by the external heat engine.

We will look at how the Earth's heat engines work and show how some of the major topics of physical geology are related to the *internal* and *surficial* (on the Earth's surface) processes powered by the heat engines.

Internal Processes: How the Earth's Internal Heat Engine Works

The Earth's internal heat engine works because hot, buoyant material deep within the Earth moves slowly upward toward the cool surface and cold, denser material moves downward–a process called **convection**. Visualize a vat of hot wax, heated from below (figure 1.6). As the wax immediately above the fire gets hotter, it expands, becomes less dense (that is, a given volume of the material will weigh less), and rises. Wax at the top of the vat loses heat to the air, cools, contracts, becomes denser, and sinks. A similar process takes place in the Earth's interior. Rock that is deep within the Earth and is very hot rises slowly toward the surface, while rock that has cooled near the surface is denser and sinks downward. Instinctively, we don't want to believe that rock can flow like hot wax. However, experiments have shown that under the right conditions, rocks are capable of being molded (like wax or putty). Deeply buried rock that is hot and under high pressure can deform, like taffy or putty. But the deformation takes place very slowly. If we were somehow able to strike a rapid blow to the deeply buried rock with a hammer, it would fracture, just as rock at Earth's surface would.

Earth's Interior

We divide the Earth's interior into *compositional layers* based upon changes in chemical composition and density, and into *mechanical layers* based upon changes in mechanical behavior, or strength. These layers are shown in figure 1.7. The three compositional layers of the earth are the **crust**, the **mantle**, and the **core**. The Earth's core and mantle and the lower parts of the crust are inaccessible to direct observation. No mine or oil well has penetrated through the crust, so our concept of the Earth's interior is based on indirect evidence. Chapter 17 explores the evidence used to understand the interior of the Earth.

The crust, the thin, outermost layer of the Earth, is analogous to the skin on an apple. The thickness of the crust is insignificant compared to the whole Earth. We have direct access to only the crust, and not much of the crust at that. We are like microbes crawling on an apple, without the ability to penetrate its skin. Because it is our home and we depend on it for resources, we are concerned more with the crust than with the inaccessible mantle and core.

The two major types of crust are *oceanic crust* and *continental crust.* Oceanic crust underlies the oceans and is relatively thin (on average, approximately 7 km thick). It is made of basalt, a volcanic rock that is somewhat denser than the rock that underlies the continents. Continental crust is much thicker than oceanic crust, averaging approximately 35 kilometers thick. Unlike oceanic crust, continental crust is made up of many

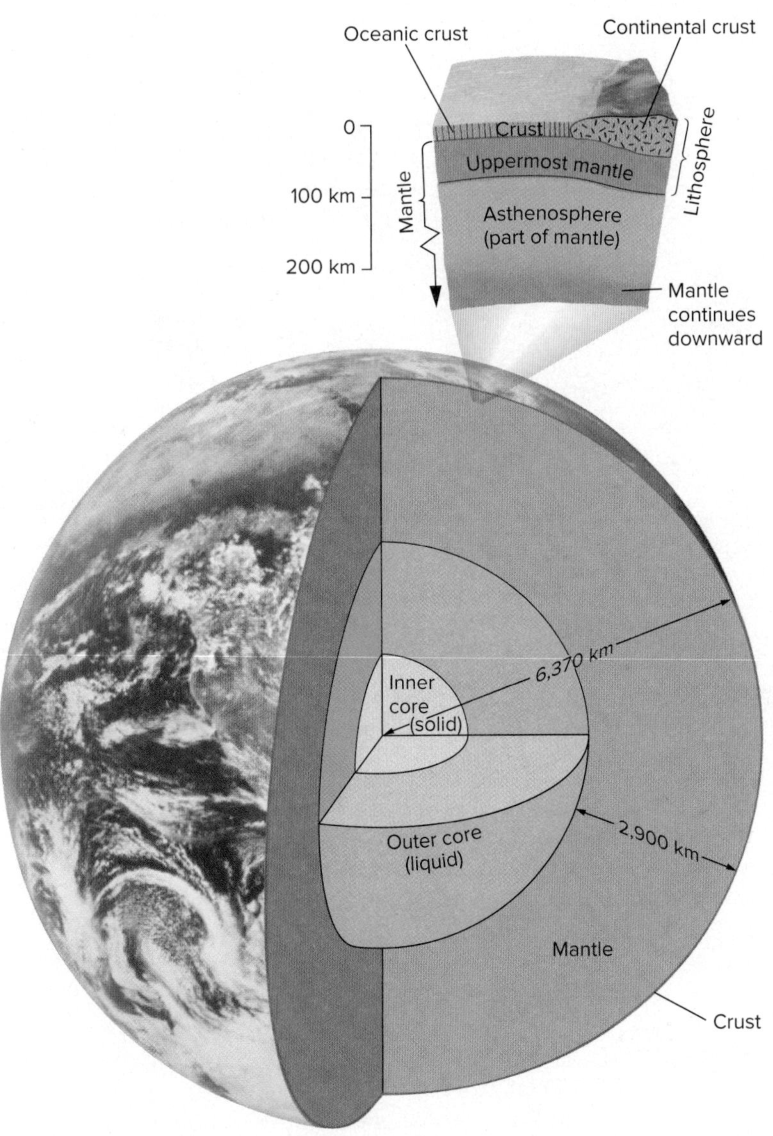

FIGURE 1.7

Cross section through the Earth. Expanded section shows the relationship between the two types of crust, the lithosphere and the asthenosphere, and the mantle. The crust ranges from 5 to 75 kilometers thick.
Source: NASA

different types of rock. Its average composition is equal to that of granite, a rock you may have seen in many kitchens because it is a popular material used to make countertops.

As described in more detail in chapter 17, the mantle is the thickest and most voluminous of these zones, making up more than 80% of the Earth's volume. The mantle is composed of rock that contains more iron and magnesium than crustal rocks and thus are more dense. Although the mantle is solid rock, parts of it flow slowly, generally upward or downward, depending on whether it is hotter or colder than adjacent mantle.

The core, the innermost and densest layer of the Earth, is believed to be made of metal—not rock like the mantle and crust—mostly iron and nickel. The core is divided into two mechanical layers, the solid inner core and the liquid outer core. It is the convection of liquid iron in the outer core that generates the Earth's magnetic field.

The expanded section of figure 1.7 shows two very important mechanical layers of the Earth. The crust and the uppermost part of the mantle are relatively rigid. Collectively, they make up the **lithosphere**. (To help you remember terms, the meanings of commonly used prefixes and suffixes are given in appendix G. For example, *lith* means "rock" in Greek. You will find *lith* to be part of many geologic terms.) The uppermost mantle underlying the lithosphere, called the **asthenosphere**, is soft and therefore flows more readily than the underlying mantle. It provides a "lubricating" layer over which the lithosphere moves (*asthenos* means "weak" in Greek). Where hot mantle material wells upward, it will uplift the lithosphere. Where the lithosphere is coldest and densest, it will sink down through the asthenosphere and into the deeper mantle, just as the wax does in figure 1.6. The effect of this internal heat engine on the crust is of great significance to geology. The forces generated inside the Earth, called **tectonic forces**, cause deformation of rock as well as vertical and horizontal movement of portions of the Earth's crust. The existence of mountain ranges indicates that tectonic forces are stronger than gravitational forces. (Mount Everest, the world's highest peak, is made of rock that formed beneath an ancient sea.) Mountain ranges are built over extended periods as portions of the Earth's crust are squeezed, stretched, and raised.

Most tectonic forces are mechanical forces. Some of the energy from these forces is put to work deforming rock, bending and breaking it, and raising mountain ranges. The mechanical energy may be stored (an earthquake is a sudden release of stored mechanical energy) or converted to heat energy (rock may melt, resulting in volcanic eruptions). The working of the machinery of the Earth is elegantly demonstrated by plate tectonics.

The Theory of Plate Tectonics

From time to time a theory emerges within a science that revolutionizes that field. (In common usage, the word *theory* is used for what scientists call a *hypothesis*—that is, a tentative answer to a question or solution to a problem. In science, however, as explained in box 1.4, a *theory* is a concept that has been highly tested and in all likelihood is true.) The theory of plate tectonics is as important to geology as the theory of relativity is to physics, the atomic theory to chemistry, or evolution to biology. It is a unifying theory that accounts for many seemingly unrelated geologic phenomena. Some of the disparate phenomena that plate tectonics explains are where and why we get earthquakes, volcanoes, mountain belts, deep ocean trenches, and mid-oceanic ridges.

Plate tectonics was seriously proposed as a hypothesis in the early 1960s, though the idea was based on earlier work—notably, the hypothesis of *continental drift*. In the chapters on igneous, sedimentary, and metamorphic rocks, as in the chapter on earthquakes, we will expand on what you learn about the theory here to explain the origin of some rocks and why volcanoes and earthquakes occur. Chapter 19 tells the story of how the theory of plate tectonics was developed. You will find that what you learned in many previous chapters helps you understand the evidence for plate tectonics and also that many geologic processes are interrelated and explained by plate tectonic theory.

Plate tectonics regards the lithosphere as broken into *plates* that are in motion (see figure 1.8). The plates, which are much

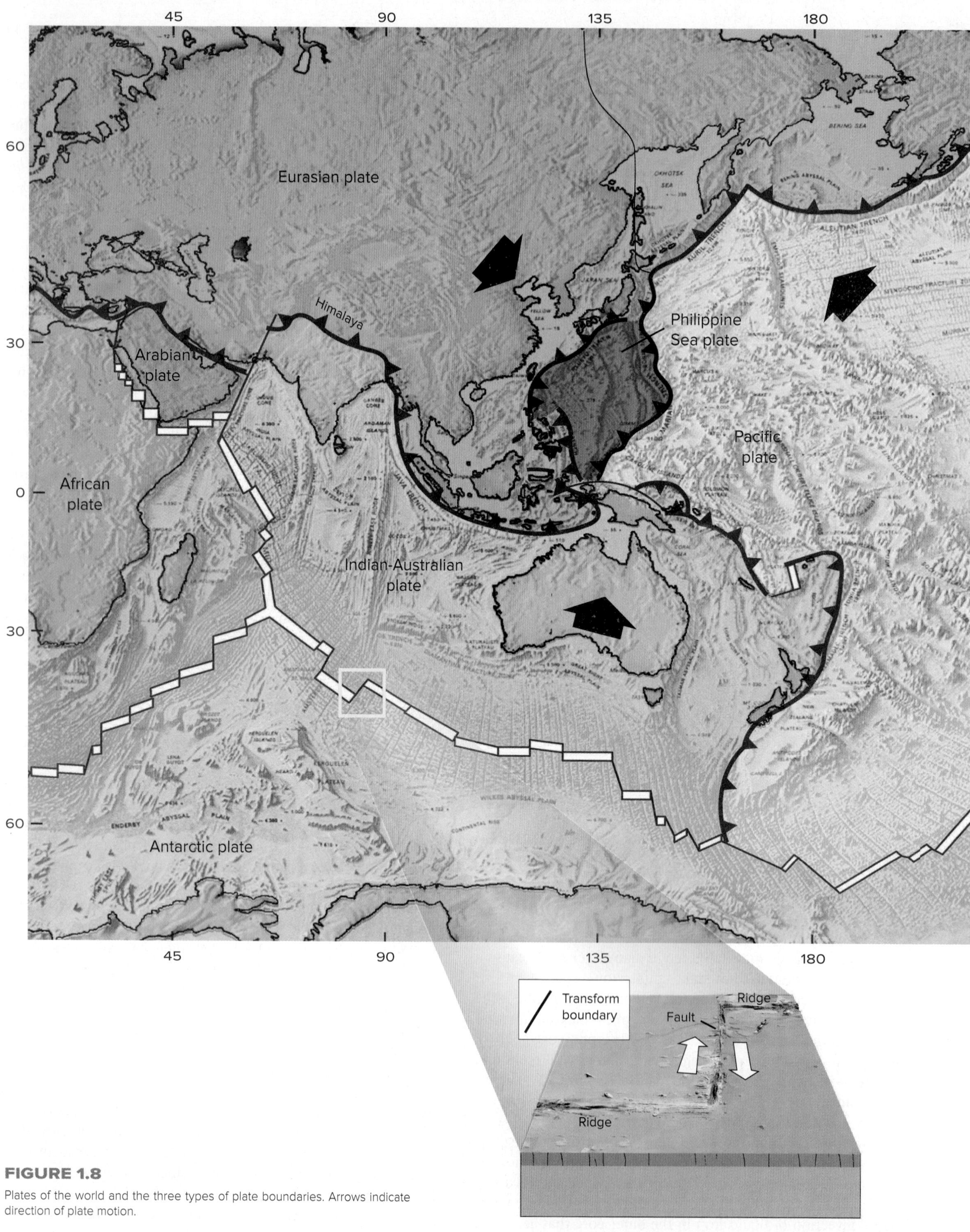

FIGURE 1.8

Plates of the world and the three types of plate boundaries. Arrows indicate direction of plate motion.

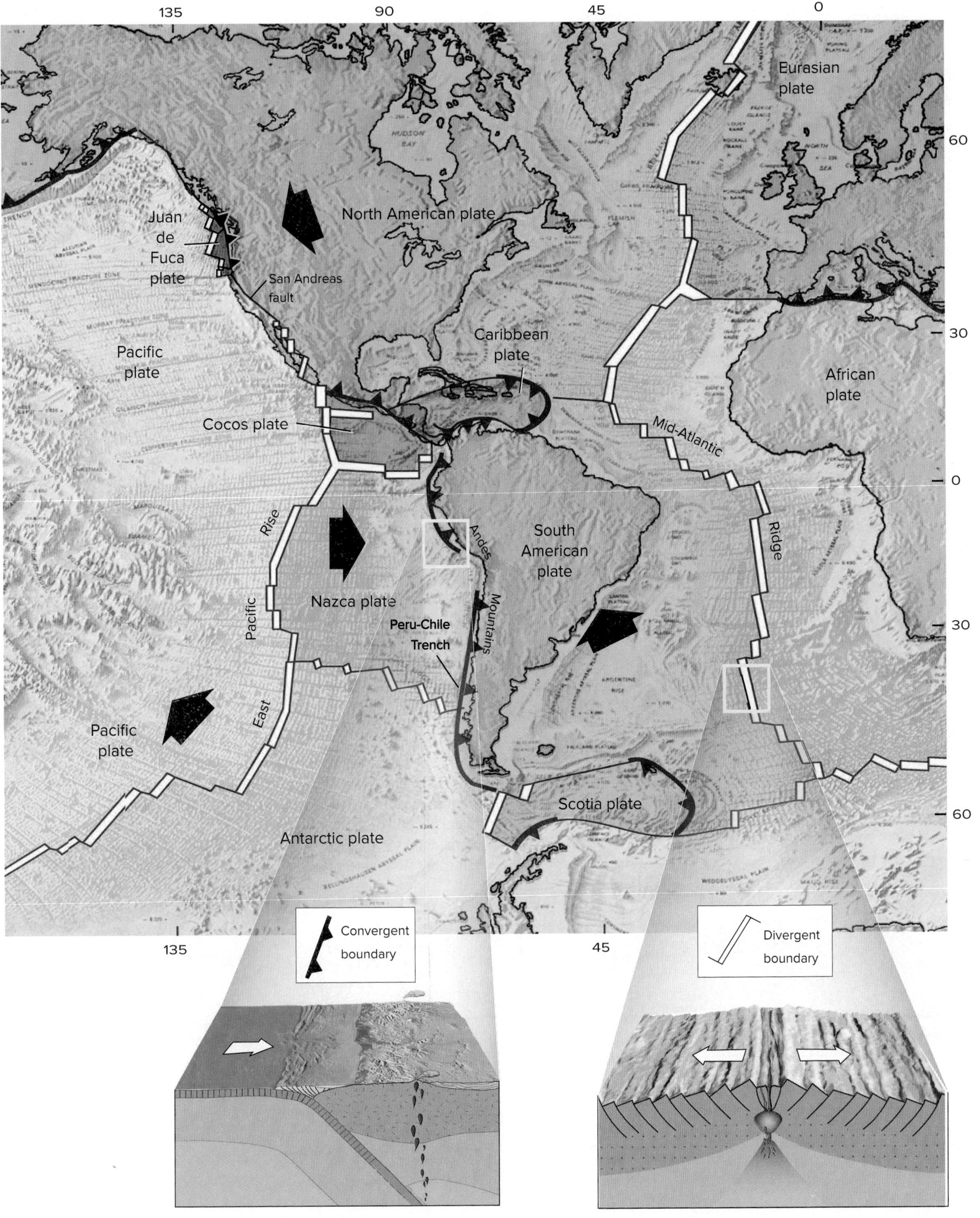

135
90
45
0
60
30
0
30
60
135
45
Eurasian plate
North American plate
Juan de Fuca plate
San Andreas fault
Pacific plate
Caribbean plate
African plate
Cocos plate
Mid-Atlantic Ridge
South American plate
Andes Mountains
Nazca plate
Peru-Chile Trench
East Pacific Rise
Pacific plate
Scotia plate
Antarctic plate
Convergent boundary
Divergent boundary

TABLE 1.1 Three Types of Plate Boundaries

Boundary	What Takes Place	Result
Divergent	Plates move apart	Creation of new ocean floor with submarine volcanoes; mid-oceanic ridge; small to moderate earthquakes
Convergent	Plates move toward each other	Destruction of ocean floor; creation and growth of mountain range with volcanoes; subduction zone; Earth's greatest earthquakes and tsunami
Transform	Plates move sideways past each other	No creation or destruction of crust; small to large earthquakes

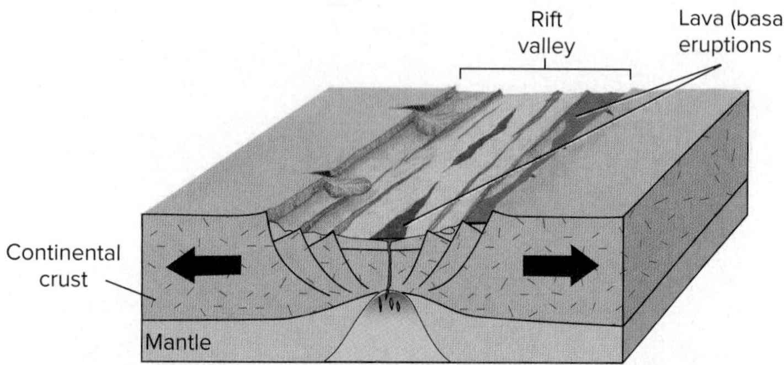

A–Continent undergoes extension. The crust is thinned and a rift valley forms.

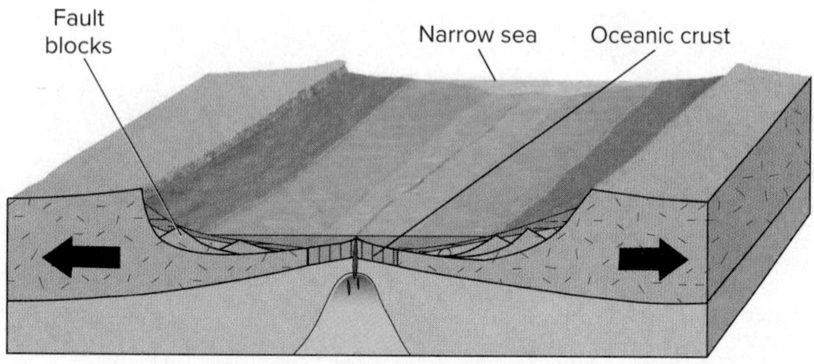

B–Continent tears in two. Continent edges are faulted and uplifted. Basalt eruptions form oceanic crust.

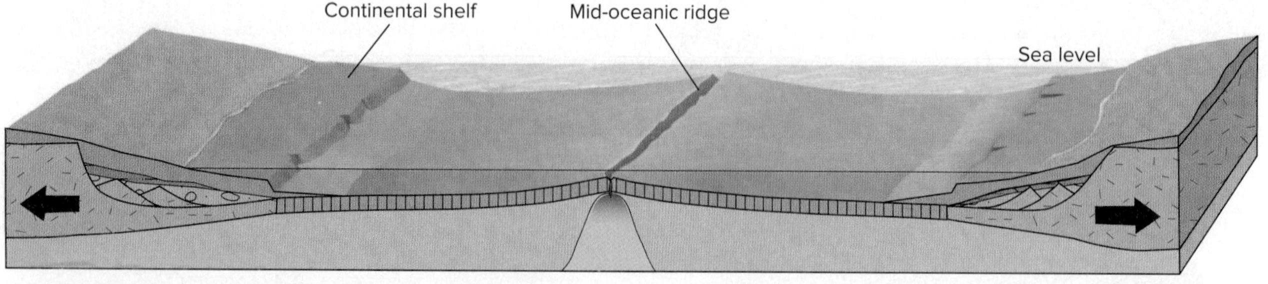

C–Continental sediments blanket the subsiding margins to form continental shelves. The ocean widens and a mid-oceanic ridge develops, as in the Atlantic Ocean.

FIGURE 1.9

A divergent boundary begins as a continent is pulled apart. As separation of continental crust proceeds, oceanic crust develops, and an initially narrow sea floor grows larger in time.

like segments of the cracked shell on a boiled egg, move relative to one another along *plate boundaries,* sliding upon the underlying asthenosphere. Much of what we observe in the rock record can be explained by the type of motion that takes place along plate boundaries. Plate boundaries are classified into three types based on the type of motion occurring between the adjacent plates. These are summarized in table 1.1.

Divergent Boundaries

The first type of plate boundary, a **divergent boundary**, involves two plates that are moving apart from each other. Most divergent boundaries coincide with the crests of submarine mountain ranges, called **mid-oceanic ridges** (figure 1.8). The mid-Atlantic ridge, which runs for approximately 16,000 kilometers (10,000 miles) from northeast of Greenland to the South Atlantic, is a classic, well-developed example. Motion along a mid-oceanic ridge causes small to moderate earthquakes.

Although most divergent boundaries present today are located within oceanic plates, a divergent boundary typically initiates within a continent. It begins when a split, or *rift,* in the continent is caused either by extensional (stretching) forces within the continent or by the upwelling of hot asthenosphere from the mantle below (figure 1.9*A*). Either way, the continental plate pulls apart and thins. Initially, a narrow valley is formed. Fissures extend into a magma chamber. **Magma** (molten rock) flows into the fissures and may erupt onto the floor of the rift. With continued separation, the valley deepens, the crust beneath the valley sinks, and a narrow sea floor is formed (figure 1.9*B*). Underlying the new sea floor is rock that has been newly created by underwater eruptions and solidification of magma in fissures. Rock that forms when magma solidifies is **igneous rock**. The igneous rock that solidifies on the sea floor and in the fissures becomes *oceanic crust.* As the two sides of the split continent continue to move apart, new fissures develop, magma fills them, and more oceanic crust is formed. As the ocean basin widens, the central zone where new crust is created remains relatively high. This is the mid-oceanic ridge that will remain as the divergent boundary as the continents continue to move apart and the ocean basin widens (figure 1.9*C*).

A mid-oceanic ridge is higher than the deep ocean floor (figure 1.9*C*) because the rocks, being hotter at the ridge, are less dense. A *rift valley*, bounded by tensional cracks, runs along the crest of the ridge. The magma in the chamber below the ridge that squeezes into fissures comes from partial melting of the underlying asthenosphere. Continued pulling apart of the ridge crest develops new cracks, and the process of filling and cracking continues indefinitely. Thus, new oceanic crust is continuously created at a divergent boundary. Not all of the mantle material melts—a solid residue remains under the newly created crust. New crust and underlying solid mantle make up the lithosphere that moves away from the ridge crest, traveling like the top of a conveyor belt. The rate of motion is generally 1 to 18 centimeters per year (0.4 to 7 inches per year—approximately the growth rate of a fingernail), slow in human terms but quite fast by geologic standards.

The top of a plate may be composed exclusively of oceanic crust or might include a continent or part of a continent. For example, if you live on the North American plate, you are riding westward relative to Europe because the plate's divergent boundary is along the mid-oceanic ridge in the North Atlantic Ocean (figure 1.8). The western half of the North Atlantic sea floor and North America are moving together in a westerly direction away from the mid-Atlantic ridge plate boundary.

Convergent Boundaries

The second type of boundary, one resulting in a wide range of geologic activities, is a **convergent boundary**, wherein plates move toward each other (figure 1.10). By accommodating the addition of new sea floor at divergent boundaries, the destruction of old sea floor at convergent boundaries ensures the Earth does not grow in size. Examples of convergent boundaries include the Andes mountain range, where the Nazca plate is descending or *subducting* beneath the South American plate, and

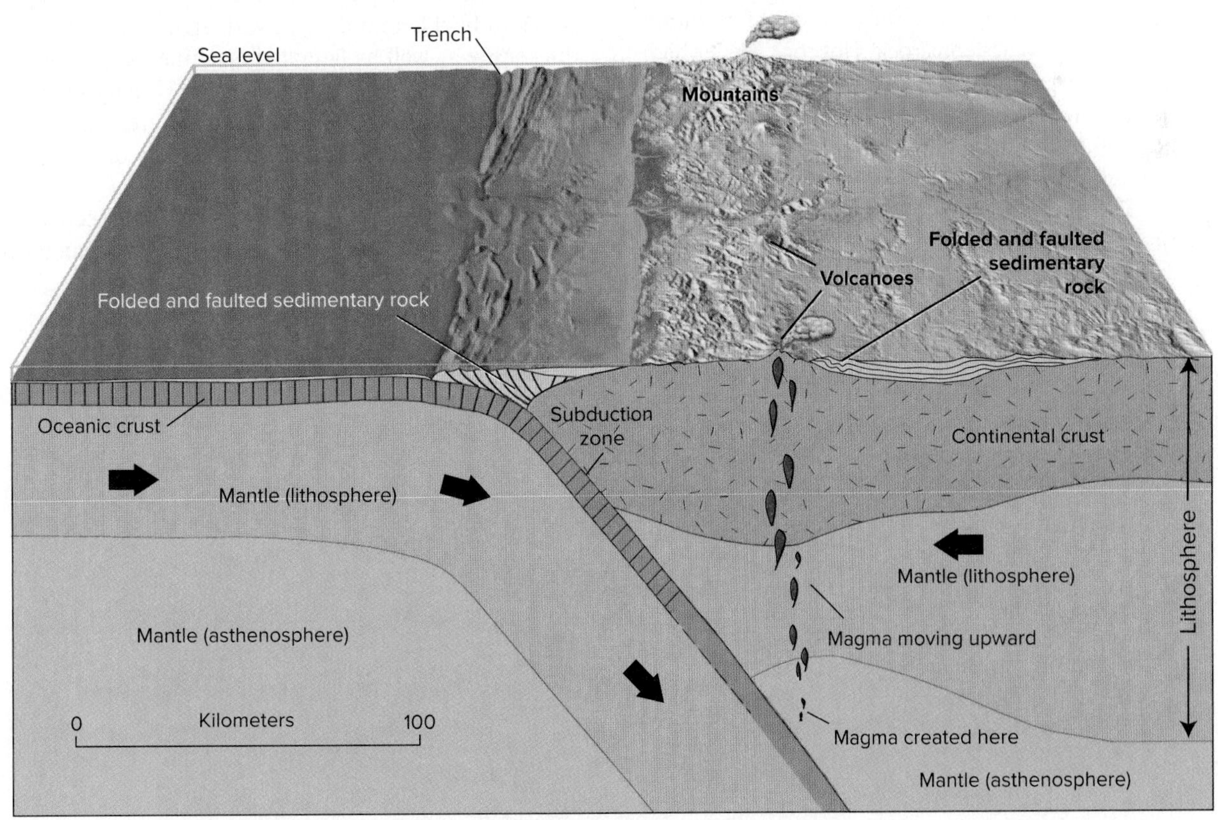

FIGURE 1.10

Block diagram of an ocean-continent convergent boundary. Oceanic lithosphere moves from left to right and is subducted beneath the overriding continental lithosphere. Magma is created by partial melting of the asthenosphere.

the Cascade Range of Washington, Oregon, and northern California, where the Juan de Fuca plate is subducting beneath the North American plate. Convergent boundaries, due to their geometry, are the sites of the largest earthquakes on Earth.

It is useful to describe convergent boundaries by the character of the plates that are involved: **ocean-continent**, **ocean-ocean**, and **continent-continent**. The difference in density of oceanic and continental lithosphere explains the contrasting geological activities caused by their convergence.

Ocean-Continent Convergence

If one plate is capped by oceanic crust and the other by continental crust, the less-dense, more-buoyant continental plate will override the denser, oceanic plate (figure 1.10). The oceanic plate bends beneath the continental plate and sinks along what is known as a **subduction zone**, a zone where an oceanic plate descends into the mantle beneath an overriding plate. Deep *oceanic trenches* are found where oceanic lithosphere bends and begins its descent. These narrow, linear troughs are the deepest parts of the world's oceans.

In the region where the top of the subducting plate slides beneath the asthenosphere, melting takes place and magma is created. Magma is less dense than the overlying solid rock. Therefore, the magma created along the subduction zone works its way upward and either erupts at volcanoes on the Earth's surface to solidify as *extrusive* igneous rock, or solidifies within the crust to become *intrusive* igneous rock. Hot rock, under high pressure, near the subduction zone that does not melt may change in the solid state to a new rock—**metamorphic rock**.

Near the edge of the continent, above the rising magma from the subduction zone, a major mountain belt, such as the Andes or Cascades, forms. The mountain belt grows due to the volcanic activity at the surface, the emplacement of bodies of intrusive igneous rock at depth, and intense compression caused by plate convergence. Layered sedimentary rock that may have formed on an ocean floor especially shows the effect of intense squeezing (for instance, the "folded and faulted sedimentary rocks" shown in figure 1.10). In this manner, rock that may have been below sea level might be squeezed upward to become part of a mountain range.

Ocean-Ocean Convergence

If both converging plates are oceanic, the denser plate will subduct beneath the less-dense plate (figure 1.11). A portion of a plate becomes colder and denser as it travels farther from the mid-oceanic ridge where it formed. After subduction begins, molten rock is produced just as it is in an ocean-continent subduction zone; however, in this case, the rising magma forms volcanoes that grow from an ocean floor rather than on a continent. The resulting mountain belt is called a *volcanic island arc.* Examples include the Aleutian Islands in Alaska and the islands that make up Japan, the site of the great earthquake and tsunami of 2011, described earlier.

Continent-Continent Convergence

If both converging plates are continental, a quite different geologic deformation process takes place at the plate boundary. Continental lithosphere is much less dense than the mantle below and, therefore, neither plate subducts. The buoyant nature of continental lithosphere causes the two colliding continental plates to buckle and deform with significant vertical uplift and thickening as well as lateral shortening. A spectacular example of continent-continent collision is the Himalayan mountain belt. The tallest peaks on Earth are located here, and they continue to grow in height due to continued collision of the Indian subcontinent with the continental Eurasian plate.

Continent-continent convergence is preceded by oceanic-continental convergence (figure 1.12). An ocean basin between

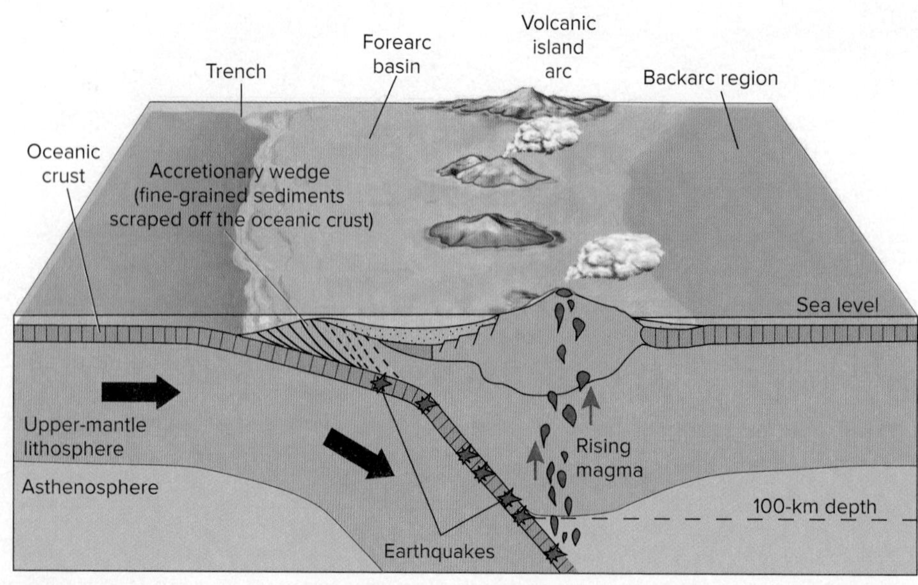

FIGURE 1.11
A volcanic island arc forms as a result of oceanic-oceanic plate convergence.

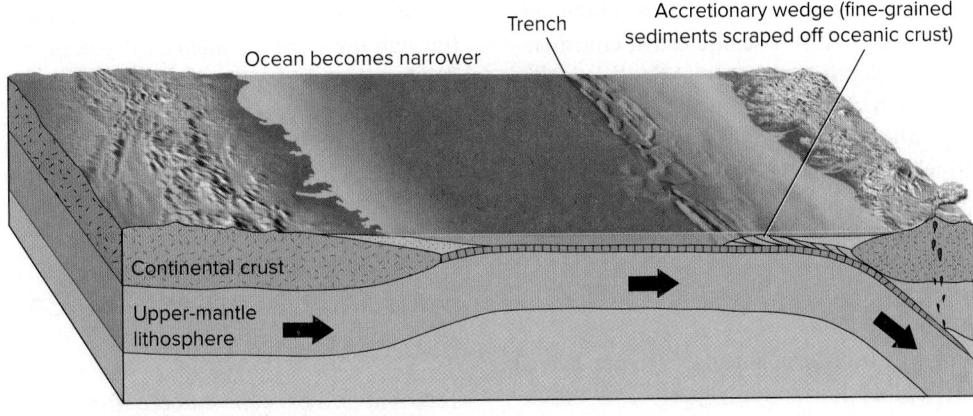

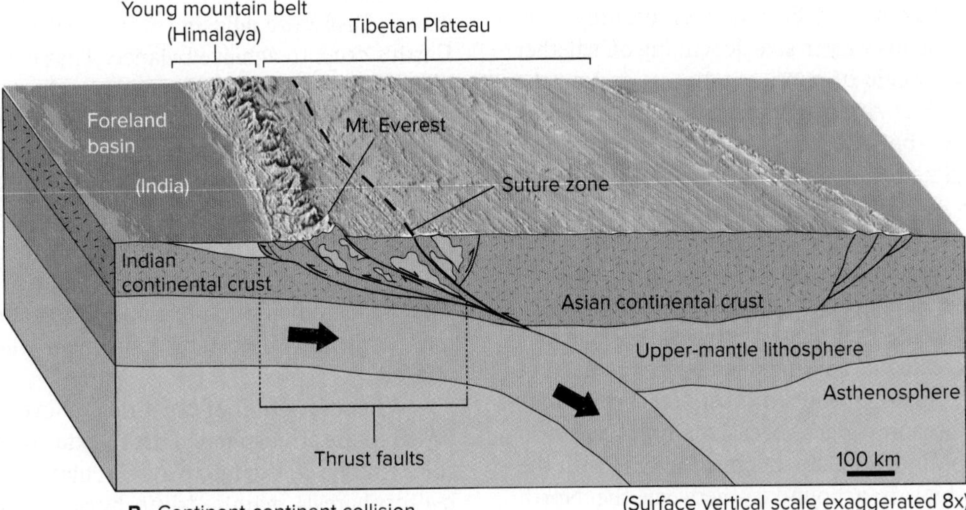

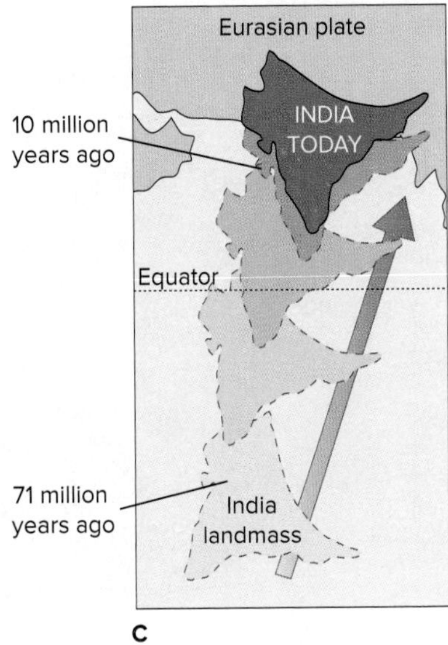

FIGURE 1.12

Continent-continent convergence is preceded by the closing of an ocean basin while ocean-continent convergence takes place. Figure 1. 12 (C) shows the position of India relative to the Eurasian plate in time. The convergence of the two plates created the Himalaya. Some of the features shown, such as accretionary wedge and foreland basin, are described in chapters 6 and 19.

two continents closes because oceanic lithosphere is subducted beneath one of the continents. Unlike oceanic crust, continental crust is too buoyant to be subducted into mantle. When the continents collide, subduction ceases and they become wedged together. India collided with Asia around 40 million years ago, yet the forces that propelled them together are still in effect. The rocks continue to be deformed and squeezed into higher mountains.

Transform Boundaries

The third type of boundary, a **transform boundary** (figure 1.13), occurs where two plates slide horizontally past each other, rather than toward or away from each other. The San Andreas fault in California and the Alpine fault of New Zealand are two examples of this type of boundary. Earthquakes resulting from motion along transform faults vary in size depending on whether the fault cuts through oceanic or continental crust and on the length of the fault. The San Andreas transform fault has generated large earthquakes, but the more numerous and much shorter transform faults within ocean basins generate much smaller earthquakes.

The significance of transform faults was first recognized in ocean basins. Here they occur as fractures perpendicular to mid-oceanic ridges, which are offset (figure 1.8). As shown in figure 1.13, the motion on either side of a transform fault is a result of rock that is created at and moving away from each of the displaced oceanic ridges. Although most transform faults are found along mid-oceanic ridges, occasionally a transform fault cuts through a continental plate. Such is the case with the San Andreas fault, which is a boundary between the North American and the Pacific plates.

Box 1.4 outlines how plate tectonic theory was developed through the *scientific method.* If you do not have a thorough understanding of how the scientific method works, be sure to study the box.

The U.S. Geological Survey's online publication, *This Dynamic Earth,* is an excellent supplement for learning about plate tectonics. Access it as described in "Exploring Web Resources" at the end of this chapter.

Surficial Processes: The Earth's External Heat Engine

Tectonic forces can squeeze formerly low-lying continental crustal rock along a convergent boundary and raise the upper part well above sea level. Portions of the crust also can rise because of **isostatic adjustment**, vertical movement of sections of Earth's crust to achieve balance. That is to say, lighter rock will "float" higher than denser rock on the underlying mantle. Isostatic adjustment is why an empty ship floats higher on the water than an identical one that is full of cargo. Continental crust, which is less dense than oceanic crust, will tend to float higher over the underlying mantle than oceanic crust (which is why the oceanic crust is below sea level and the continents are above sea level). After a portion of the continental crust is pulled downward by tectonic forces, it is out of isostatic balance. It will then rise slowly due to isostatic adjustment when tectonic forces are relaxed.

When a portion of crust rises above sea level, rocks are exposed to the atmosphere. Earth's external heat engine, driven by solar energy, comes into play. Circulation of the atmosphere and hydrosphere is mainly driven by solar energy. Our weather is largely a product of the solar heat engine. For instance, hot air

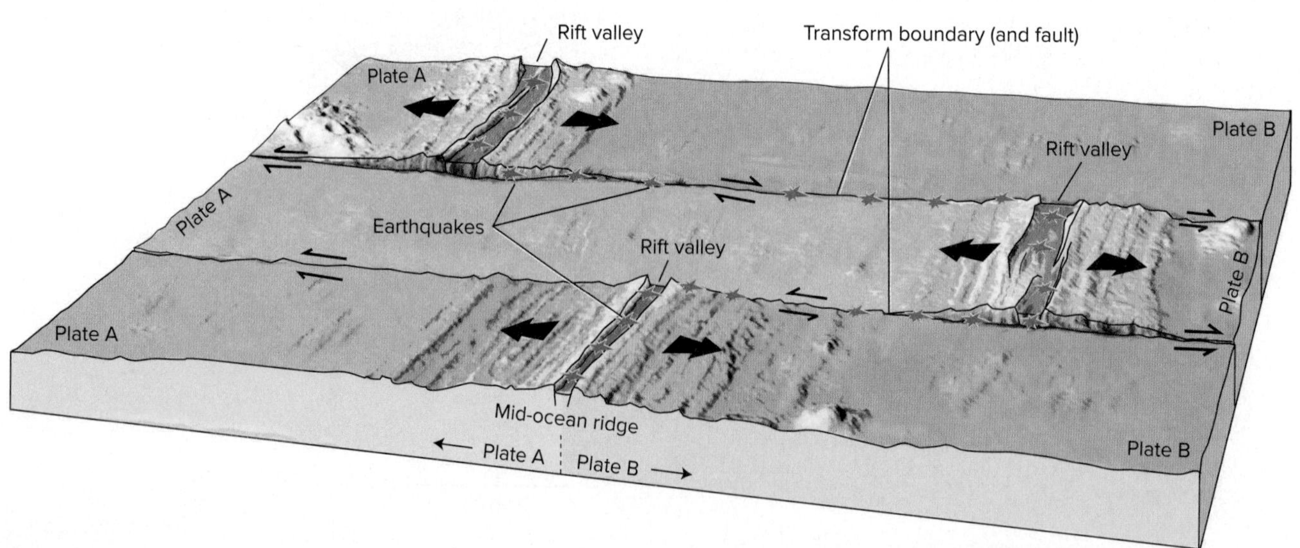

FIGURE 1.13

Transform faults (transform boundaries between plates) are the segments of the fractures between offset ridge crests. Oceanic crust is created at the ridge crests and moves away from the crest as indicated by the heavy arrows. The pairs of small arrows indicate motion on adjacent sides of fractures. Earthquakes take place along the transform fault because rocks are moving in opposite directions. The fractures extend beyond the ridges, but here the two segments of crust are moving in the same direction and rate and there are no earthquakes—these are not part of transform faults.

rises near the equator and sinks in cooler zones to the north and south. Solar heating of air creates wind; ocean waves are, in turn, produced by wind. When moist air cools, it rains or snows. Rainfall on hillsides flows down slopes and into streams. Streams flow to lakes or seas. Glaciers grow where there is abundant snowfall at colder, high elevations and flow downhill because of gravity.

Where moving water, ice, or wind loosens and removes material, **erosion** is taking place. Streams flowing toward oceans remove some of the land over which they run. Crashing waves carve back a coastline. Glaciers grind and carry away underlying rock as they move. In each case, rock originally brought up by the Earth's internal processes is worn down by surficial processes (figure 1.14). As material is removed through erosion, isostasy works to move the landmass upward, just as part of the submerged portion of an iceberg floats upward as ice melts. Or, going back to our ship analogy, as cargo is unloaded, the ship rises in the water.

Rocks formed at high temperature and under high pressure deep within the Earth and pushed upward by isostatic and tectonic forces are unstable in their new environment. Air and water tend to cause the once deep-seated rocks to break down and form new materials. The new materials, stable under conditions at the Earth's surface, are said to be in **equilibrium**–that is, adjusted to the physical and chemical conditions of their environment so that they do not change or alter with time. For example, much of an igneous rock (such as granite) that formed at a high temperature tends to break down chemically to clay. Clay is in equilibrium–that is to say it is stable–at the Earth's surface.

The product of the breakdown of rock is **sediment**, loose material. Sediment may be transported by an agent of erosion, such as running water in a stream. Sediment is deposited when the transporting agent loses its carrying power. For example, when a river slows down as it meets the sea, the sand being transported by the stream is deposited as a layer of sediment.

In time, a layer of sediment deposited on the sea floor becomes buried under another layer. This process may continue, burying our original layer progressively deeper. The pressure from overlying layers compresses the sediment, helping to consolidate the loose material. With the cementation of the loose particles, the sediment becomes *lithified* (cemented or otherwise consolidated) into a **sedimentary rock**. Sedimentary rock that becomes deeply buried in the Earth may later be transformed by heat and pressure into metamorphic rock.

1.4 GEOLOGIC TIME

We have mentioned the great amount of time required for geologic processes. As humans, we think in units of time related to personal experience–seconds, hours, years, a human lifetime. It stretches our imagination to contemplate ancient history that involves 1,000 or 2,000 years. Geology involves vastly greater amounts of time, often referred to as *deep time.*

Try to comprehend the vastness of deep time by going to the "Comprehending Geologic Time" section at the end of chapter 8. There we relate a very slow and very long movie to Earth's history. Figure 8.25 compares deep time to a trip across the United States at the speed of 1 kilometer per 1 million years.

To be sure, some geologic processes occur rapidly, such as a great landslide or an earthquake. These events occur when stored energy (like the energy stored in a stretched rubber band) is suddenly released. Most geologic processes, however, are slow but relentless, reflecting the pace at which the heat engines work. It is unlikely that a hill will visibly change in shape or height during your lifetime (unless through human activity). However, in a geologic time frame, the hill probably is eroding away quite rapidly. "Rapidly" to a geologist may mean that within a few million years, the hill will be reduced nearly to a plain. Similarly, in the geologically "recent" past of several million years ago, a sea may have existed where the hill is now. Some processes are regarded by geologists as "fast" if they are begun and completed within a million years.

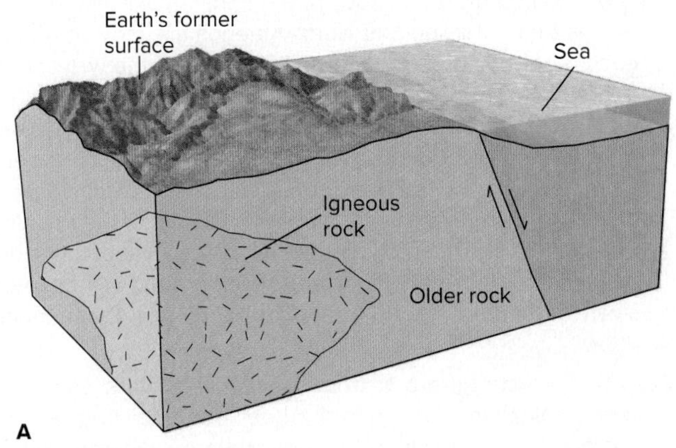

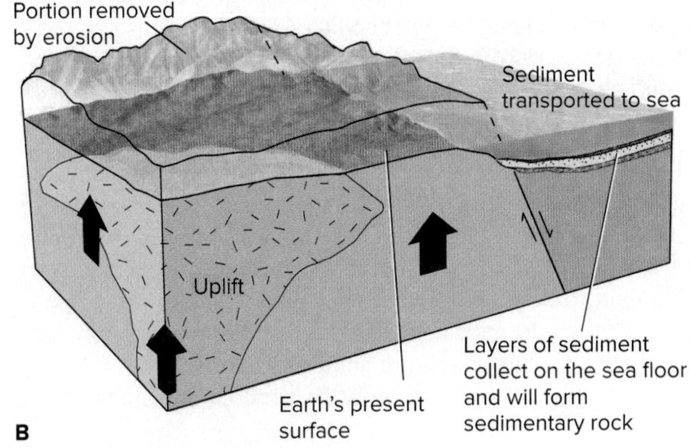

FIGURE 1.14

Erosion, deposition, and uplift. (*A*) Magma has solidified deep underground to become igneous rock. (*B*) As the surface erodes, sediment is transported to the sea to become sedimentary rock. Isostatic adjustment causes uplift of the continent. Erosion and uplift expose the igneous rock at the surface.

IN GREATER DEPTH 1.4

Plate Tectonics and the Scientific Method

Although the hypothesis was proposed only a few decades ago, plate tectonics has been so widely accepted and disseminated that most people have at least a rough idea of what it is about. Most nonscientists can understand the television and newspaper reports that include plate tectonics in reports on earthquakes and volcanoes. Our description of plate tectonics implies little doubt about the existence of the process. The theory of plate tectonics has been accepted as scientifically verified by geologists. Plate tectonic theory, like all knowledge gained by science, has evolved through the processes of the **scientific method**. We will illustrate the scientific method by showing how plate tectonics has evolved from a vague idea into a theory that is so likely to be true that it can be regarded as "fact."

The basis for the scientific method is the belief that the universe is orderly and that by *objectively* analyzing phenomena, we can discover their workings. Science is a deeply human endeavor that involves creativity. A scientist's mind searches for connections and thinks of solutions to problems that might not have been considered by others. At the same time, a scientist must be aware of what work has been done by others so that science can build on those works. Here, the scientific method is presented as a series of steps. A scientist is aware that his or her work must satisfy the requirements of the steps but does not ordinarily go through a formal checklist.

1. A question is raised or a problem is presented.
2. Available information pertinent to the question or problem is analyzed. Facts, which scientists call **data**, are gathered.
3. After the data have been analyzed, tentative explanations or solutions that are consistent with the observed data, called **hypotheses**, are proposed.
4. One predicts what would occur in given situations if a hypothesis were correct.
5. Predictions are tested. Incorrect hypotheses are discarded.
6. A hypothesis that passes the testing becomes a **theory**, which is regarded as having an excellent chance of being true. In science, however, nothing is considered proven absolutely. All scientifically derived knowledge is subject to being proven false. (Can you imagine what could prove that atoms and molecules don't exist?) A thoroughly and rigorously tested theory becomes, for all intents and purposes, a fact, even though scientists still call it a theory (e.g., atomic theory).

Like any human endeavor, the scientific method is not infallible. Objectivity is needed throughout. Someone can easily become attached to the hypothesis he or she has created and so tend subconsciously to find only supporting evidence. As in a court of law, every effort is made to have observers objectively examine the logic of both procedures and conclusions. Courts sometimes make wrong decisions; science, likewise, is not immune to error.

The following outline shows how the concept of plate tectonics evolved:

Step 1: A question asked or problem raised. Actually, a number of questions were being asked about seemingly unrelated geological phenomena.

What caused the submarine ridge that extends through most of the oceans of the world? Why are rocks in mountain belts intensely deformed? What sets off earthquakes? What causes rock to melt underground and erupt as volcanoes? Why are most of the active volcanoes of the world located in a ring around the Pacific Ocean?

Step 2: Gathering of data. Early in the twentieth century, the amount of data was limited. But through the decades, the information gathered increased enormously. New data, most notably information gained from exploration of the sea floor in the mid-1900s, forced scientists to discard old hypotheses and come up with new ideas.

Step 3: Hypotheses proposed. Most of the questions being asked were treated as separate problems wanting separate hypotheses. Some appeared interrelated. One hypothesis, **continental drift**, did address several questions. It was advocated by Alfred Wegener, a German scientist, in a book published in the early 1900s.

Wegener postulated that the continents were all once part of a single supercontinent called Pangaea. The hypothesis explained why the coastlines of Africa and South America look like separated parts of a jigsaw puzzle. Some 200 million years ago, this supercontinent broke up, and the various continents slowly drifted into their present positions. The hypothesis suggested that the rock within mountain belts becomes deformed as the leading edge of a continental crust moves against and over the stationary oceanic crust. Earthquakes were presumably caused by continuing movement of the continents.

Until the 1960s, continental drift was not widely accepted. It was scoffed at by many geologists who couldn't conceive of how a continent could be plowing over oceanic crust. During the 1960s, after new data on the nature of the sea floor became available, the idea of continental drift was incorporated into the concept of plate tectonics. What was added in the plate tectonic hypothesis was the idea that oceanic crust, as well as continental crust, was shifting.

Step 4: Prediction. An obvious prediction, if plate tectonics is correct, is that if Europe and North America are moving away from each other, the distance measured between the two continents is greater from one year to the next. But we cannot stretch a tape measure across oceans, and, until recently, we have not had the technology to accurately measure distances between continents. So, in the 1960s, other testable predictions had to be made. Some of these predictions and results of their testing are described in the chapter on plate tectonics. One of these predictions was that the rocks of the oceanic crust will be progressively older the farther they are from the crest of a mid-oceanic ridge.

Step 5: Predictions are tested. Experiments were conducted in which holes were drilled in the deep-sea floor from a specially designed ship. Rocks and sediment were collected from these holes, and the ages of these materials were determined. As the hypothesis predicted, the youngest sea floor (generally less than a million years old) is near the mid-oceanic ridges, whereas the

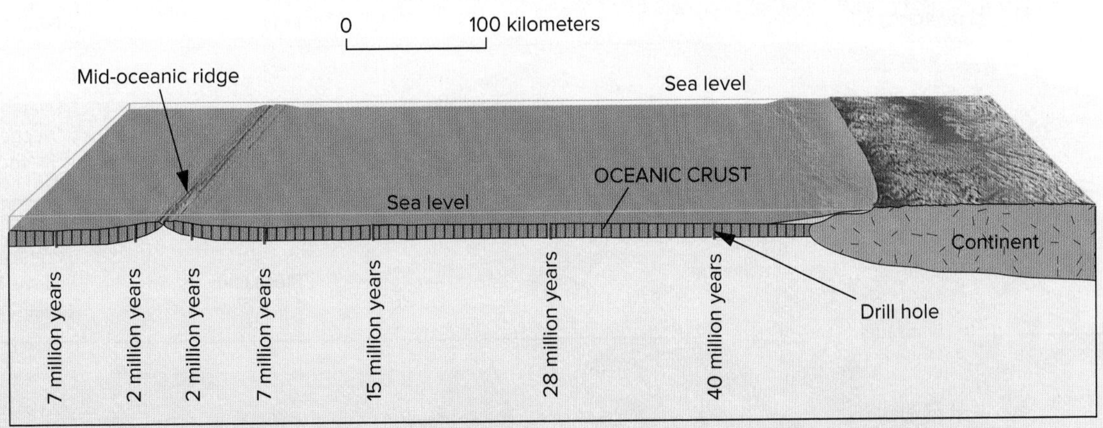

BOX 1.4 ■ FIGURE 1

Ages of rocks from holes drilled into the oceanic crust. (Vertical scale of diagram is exaggerated.)

oldest sea floor (up to about 200 million years old) is farthest from the ridges (box figure 1).

This test was only one of a series. Various other tests, described in some detail later in this book, tended to confirm the hypothesis of plate tectonics. Some tests did not work out exactly as predicted. Because of this, and more detailed study of data, the original concept was, and continues to be, modified. The basic premise, however, is generally regarded as valid.

Step 6: The hypothesis becomes a theory. Most geologists in the world considered the results of this and other tests to be positive, indicating that the concept is not reasonably disputable and is very probably true. It then became the plate tectonic theory.

During the last few years, plate tectonic theory has been further confirmed by the results of very accurate satellite surveys that determine where points on separate continents are relative to one another. The results indicate that the continents are indeed moving relative to one another. Europe and North America *are* moving farther apart.

Although it is unlikely that plate tectonic theory will be replaced by something we haven't thought of yet, aspects that fall under plate tectonics' umbrella (for instance, exactly how does magma form at a convergent plate boundary?) continue to be analyzed and revised as new data become available.

Important Note

Words used by scientists do not always have the same meaning when used by the general public. A case in point is the word *theory.* To most people, a "theory" is what scientists regard as a "hypothesis." For example, following the disappearance of Malaysia flight MH370 in 2014, several theories (including conspiracy theories) about the location of the plane and the cause of its disappearance were reported in the press. Clearly, each "theory" was a hypothesis in the scientific sense of the word. This has led to considerable confusion for nonscientists about science. You have probably heard the expression, "It's just a theory." Statements such as, "Evolution is just a theory," are used to imply that scientific support is weak. The reality is that theories such as evolution and plate tectonics have been so overwhelmingly verified that they come as close as possible to what scientists accept as being indisputable facts. They would, in laypersons' terms, be "proven."

The rate of plate motion is relatively fast. If new magma erupts and solidifies along a mid-oceanic ridge, we can easily calculate how long it will take that igneous rock to move 1,000 kilometers away from the spreading center. At the rate of 1 centimeter per year, it will take 100 million years for the currently forming part of the crust to travel the 1,000 kilometers.

Although we will discuss geologic time in detail in chapter 8, table 1.2 shows some reference points to keep in mind. The Earth is estimated to be about 4.55 (usually rounded to 4.5 or 4.6) billion years old (4,550,000,000 years). Fossils in rocks indicate that complex forms of animal life have existed in abundance on Earth for about the past 541 million years. Reptiles became abundant about 230 million years ago. Dinosaurs evolved from reptiles and became extinct about 66 million years ago. Humans have been here only about the last 5 million years. The eras and periods shown in table 1.2 comprise a kind of calendar for geologists into which geologic events are placed (as explained in the chapter on geologic time).

TABLE 1.2 Some Important Ages in the Development of Life on Earth

Millions of Years Before Present	Noteworthy Life	Eras	Periods
5	Earliest hominids	Cenozoic	Quaternary *Neogene *Paleogene
66	First important mammals		
	Extinction of dinosaurs		
		Mesozoic	Cretaceous Jurass ic Triassic
252	First dinosaurs		
300	First reptiles	Paleozoic	Permian Pennsylvanian Mississippian Devonian Silurian Ordovician Cambrian
400	Fishes become abundant		
541	First abundant fossils		
600	Some complex, soft-bodied life	Precambrian**	(The Precambrian accounts for the vast majority of geologic time.)
3,500	Earliest single-celled fossils		
4,550	Origin of the Earth		

*Not only are the immense spans of geologic time difficult to comprehend, but very slow processes are impossible to duplicate. A geologist who wants to study a certain process cannot repeat in a few hours a chemical reaction that takes a million years to occur in nature. As Mark Twain wrote in *Life on the Mississippi*, "Nothing hurries geology."

**Mark Twain (1835–1910) wrote in *Life on the Mississippi*.

Summary

Geology is the scientific study of Earth. We benefit from geology in several ways: (1) We need geology to find and maintain a supply of minable commodities and sources of energy; (2) geology helps protect the environment; (3) applying knowledge about geologic hazards (such as volcanoes, earthquakes, tsunami, landslides) saves lives and property; and (4) we have a greater appreciation of rocks and landforms through an understanding of how they form.

A *system* is an arbitrarily isolated portion of the universe that can be analyzed to see how its components interrelate. Earth systems are the atmosphere, the hydrosphere, the biosphere, and the geosphere (or solid Earth system). The Earth system is part of the solar system.

Geologic investigations indicate that Earth is changing because of internal and surficial processes. Internal processes are driven mostly by temperature differences within Earth's mantle. Surficial processes are driven by solar energy. Internal forces cause the crust of Earth to move.

Plate tectonic theory visualizes the lithosphere (the crust and uppermost mantle) as broken into plates that move relative to each other over the asthenosphere. The plates are moving *away* from divergent boundaries usually located at the crests of mid-oceanic ridges where new crust is being created. Divergent boundaries can develop in a continent and split the continent. Plates move *toward* convergent boundaries. In ocean-continent convergence, lithosphere with oceanic crust is subducted under lithosphere with continental crust. Ocean-ocean convergence involves subduction in which both plates have oceanic crust and the creation of a volcanic island arc. Continent-continent convergence takes place when two continents collide. Plates slide past one another at transform boundaries. Plate tectonics and isostatic adjustment cause parts of the crust to move up or down.

Erosion takes place at Earth's surface where rocks are exposed to air and water. Rocks that formed under high pressure and temperature inside Earth are out of equilibrium at the surface and tend to change to substances that are stable at the surface. Sediment is transported to a lower elevation, where it is deposited (commonly on a sea floor in layers). When sediment is cemented, it becomes sedimentary rock.

Although Earth is changing constantly, the rates of change are generally extremely slow by human standards.

Terms to Remember

asthenosphere 11
atmosphere 8
biosphere 8
continent-continent convergence 16
continental drift 20
convection 10
convergent boundary 15
core 10
crust 10
data 20
divergent boundary 14
Earth system 8
equilibrium 19
erosion 10
geology 2
geosphere (solid Earth system) 8
groundwater 8
hydrosphere 8
hypothesis 20
igneous rock 15
isostatic adjustment 18
lithosphere 11
magma 14
mantle 10
metamorphic rock 16
mid-oceanic ridges 14
ocean-continent convergence 16
ocean-ocean convergence 16
plate tectonics 11
scientific method 20
sediment 19
sedimentary rock 19
subduction zone 16
tectonic forces 11
theory 20
transform boundary 18

Testing Your Knowledge

Use the following questions to prepare for exams based on this chapter.

1. Examine the photo of the Canadian Rockies on the first page of this chapter. Which of the four Earth systems can you see? How are they interacting with each other?
2. What is the most likely geologic hazard in your part of your country?
3. Draw a cross section of the Earth and label each of the compositional layers and mechanical layers.
4. What are the relationships among the mantle, the crust, the asthenosphere, and the lithosphere?
5. What tectonic plate are you currently on? Where is the nearest plate boundary, and what kind of boundary is it?
6. Draw a sketch of each of the major types of plate boundaries. Show the direction of plate motion.
7. Subduction occurs at ocean-continent and ocean-ocean convergent boundaries but not at continent-continent convergent boundaries. Why is this?
8. What would the surface of Earth be like if there were no tectonic activity?
9. Explain why prehistoric cave dwellers never saw a dinosaur.
10. The division of geology concerned with Earth materials, changes in the surface and interior of the Earth, and the dynamic forces that cause those changes is
    a. physical geology. b. historical geology.
    c. geochemistry. d. paleontology.
11. Which is a geologic hazard?
    a. earthquake b. volcano
    c. mudflows d. floods
    e. wave erosion at coastlines f. landslides
    g. all of the preceding
12. Plate tectonics is a result of Earth's internal heat engine, powered by (choose all that apply)
    a. the magnetic field. b. the Sun.
    c. gravity. d. heat flowing from Earth's interior outward.
13. A typical rate of plate motion is
    a. 3–4 meters per year. b. 1 kilometer per year.
    c. 1–10 centimeters per year. d. 1,000 kilometers per year.
14. Volcanic island arcs are associated with
    a. transform boundaries.
    b. divergent boundaries.
    c. ocean-continent convergence.
    d. ocean-ocean convergence.
15. Oceanic and continental crust differ in
    a. composition. b. density.
    c. thickness. d. all of the preceding.
16. The forces generated inside Earth that cause deformation of rock as well as vertical and horizontal movement of portions of Earth's crust are called
    a. erosional forces. b. gravitational forces.
    c. tectonic forces. d. all of the preceding.
17. Plate tectonics is a(n)
    a. conjecture. b. opinion.
    c. hypothesis. d. theory.
18. The lithosphere is
    a. the same as the crust.
    b. the layer beneath the crust.
    c. the crust and uppermost mantle.
    d. only part of the mantle.
19. Erosion is a result of Earth's external heat engine, powered by (choose all that apply)
    a. the Sun.
    b. gravity.
    c. heat flowing from Earth's interior outward.
    d. the Earth's magnetic field.
20. The age of the Earth is approximately
    a. 4,550 years. b. 4,550,000 years.
    c. 4,550,000,000 years. d. 4,550,000,000,000 years.

Expanding Your Knowledge

1. Think about something you use every day—your cell phone. What materials is your cell phone made of? Where do these materials come from? How do we locate these materials?
2. Why are some parts of the lower mantle hotter than other parts?
3. According to plate tectonic theory, where are crustal rocks created? Why doesn't Earth keep getting larger if rock is continually created?
4. What percentage of geologic time is accounted for by the last century?

5. What would Earth be like without solar heating?
6. What are some of the technical difficulties you would expect to encounter if you tried to drill a hole to the center of the Earth?

Exploring Web Resources

http://pubs.usgs.gov/publications/text/dynamic.html

This Dynamic Earth by the U.S. Geological Survey is an online, illustrated publication explaining plate tectonics. You may want to go to the section "Understanding plate motion." This will help reinforce what you read about plate tectonics in this chapter. It goes into plate tectonics in greater depth, however, covering material that is in chapter 19 of this textbook.

www.uh.edu/~jbutler/anon/anontrips.html

Virtual Field Trips. The site provides access to geologic sites throughout the world. Many are field trips taken by geology classes. Check the alphabetical listing and see if there are any sites near you. Or watch a video clip in one of the QuickTime field trips.

www.usgs.gov

The *U.S. Geological Survey's* home page. Use this as a gateway to a wide range of geologic information.

http://geology.com/

Geology.com, a geoscience news and information website.

CHAPTER 2

Atoms, Elements, and Minerals

Crystals of tourmaline (variety: elbaite). Differences in color within each crystal are due to small changes in chemical composition incorporated into the minerals as they grew.
©Dr. Parvinder Sethi

LEARNING OBJECTIVES

- Know the definitions for the terms *mineral* and *rock* and understand how they differ.
- List the three subatomic particles.
- Compare and contrast the different types of bonding.
- Describe the silica tetrahedron and explain how silica tetrahedra combine to form the common silicate structures.
- Distinguish between the different chemical groups of minerals.
- Name the most common minerals in the Earth's crust.
- Recall the physical properties used to describe and identify minerals.
- Describe the different environments in which minerals form.

Have you ever wondered what all those different-colored spots in your granite countertop are? Or where we get the materials that we use to make everyday objects like cars, bicycles, and televisions? Or what exactly are all those pretty gemstones we use in jewelry? The answer to all of these questions and more is minerals! The importance of minerals to human life as we know it is immeasurable. Minerals are the source of many of the resources we use in everyday life such as lead, copper, iron, or gold. They are also the source of many of our dietary supplements such as magnesium, iron, or calcium (the calcium in soy milk comes from crushed limestone, a rock made of the mineral calcite). Some minerals are sought after because of their shape, color, or rarity. To geologists, minerals are important because they are the building blocks of the rocks that make up the Earth. The minerals in rocks tell a very important story about the origin of our world and, indeed, about all Earth-like planets. The considerable amount of information conveyed by minerals enriches our appreciation for nature.

The study of minerals is called *mineralogy.* As of May 2017, 5,256 minerals have been identified and approved by the International Mineralogical Association (http://www.ima-mineralogy.org/). Approximately 100 new minerals are added to the list every year. That's a lot of minerals, but only a couple hundred are really common and, of those, only about twenty form the majority of all rocks. Each type of mineral is distinguished by a combination of properties, some of which we can see with the unaided eye, others of which are discernable only at the microscopic and atomic levels. Examples of these properties include color, luster, hardness, chemical composition, and the transmission of light under a microscope. Minerals are so important and so easily distinguishable that geologists use them as the basis for classifying almost all rocks.

This chapter is the first of six on the material of which Earth is made. The following chapters are mostly about rocks. Nearly all rocks are made of minerals. Therefore, to be ready to learn about rocks, you must first understand what minerals are as well as the characteristics of some of the most common minerals.

In this chapter, after learning the difference between minerals and rocks, you are introduced to some basic principles of chemistry (this is for those of you who have not had a chemistry course). This will help you understand material covered in the chapters on rocks, weathering, and the composition of Earth's crust and its interior. You will discover that each mineral is composed of specific chemical elements, the atoms of which are in a remarkably orderly arrangement. A mineral's chemistry and the architecture of its internal structure determine the physical properties used to distinguish it from other minerals. You should learn how to readily determine physical properties and use them to identify common minerals. (Appendix A is a further guide to identifying minerals.)

2.1 MINERALS AND ROCKS

For most people, the term *mineral* brings to mind gemstones or dietary supplements. Often the term is used to identify something that is inorganic, as in "animal, vegetable, or mineral." When vitamin advertisers and nutritional specialists talk about "minerals," they are generally referring to single *elements*—such as magnesium, iron, or calcium—that have certain dietary benefits. Gemstones are minerals that are valued for their beauty and have been cut and polished. For geologists, the term **mineral** has a very specific definition:

> *A mineral is a naturally occurring, inorganic, crystalline solid that has a specific chemical composition.*

What does all of this mean? *Naturally occurring* tells us that a mineral must form through natural geologic processes. Manufactured diamonds, while possessing all of the other attributes of a mineral (inorganic, crystalline, specific chemical composition) cannot be considered true minerals because they are not formed naturally. Manufactured minerals are referred to as synthetic minerals. *Inorganic* means that minerals are not composed of the complex hydrocarbon molecules that are the basis of life-forms such as humans and plants. Minerals have a *specific chemical composition* that can be described by a chemical formula. Chemical formulas tell you which elements are in the mineral and in what proportion. For example, the common mineral halite (rock salt) has a chemical composition of NaCl. It is made of the two elements sodium (Na) and chlorine (Cl) with one sodium atom for every atom of chlorine. Many minerals contain more than just two elements. Potassium feldspar, a very common mineral in the Earth's crust, is made up of the elements potassium (K), aluminum (Al), silicon (Si), and oxygen (O). The formula for potassium feldspar is written $KAlSi_3O_8$. This means that for every atom of potassium in the mineral, there is one atom of aluminum, there are three of silicon, and there are eight of oxygen. All minerals have a *crystalline structure* where the atoms that make up the mineral are arranged in an orderly,

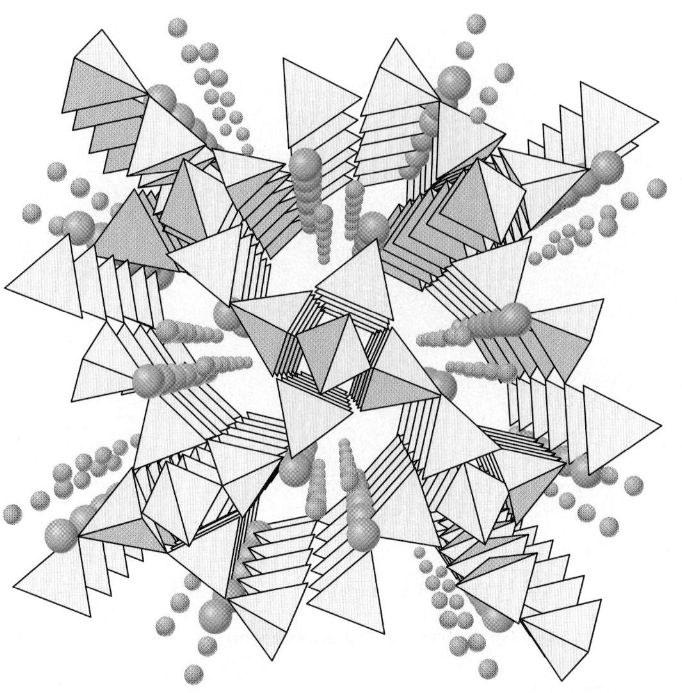

FIGURE 2.1

Model of the crystal structure of the mineral natrolite. The small (gray) spheres represent sodium; the large (blue) spheres are water molecules. The "pyramids" are silicon-oxygen tetrahedra (explained in the text).
Source: M. Ross, M. J. K. Flohr, and R. R. Ross, "Crystalline Solution Series and Order-Disorder Within the Natrolite Mineral Order," American Mineralogist, 77, 1992, 685–703.

repeating, three-dimensional pattern. The repeating nature of a crystalline structure can be seen clearly in figure 2.1, which shows a model of the crystal structure of the mineral natrolite as determined by the way X rays travel through the mineral (described later in the chapter).

Now that we have considered the definition of a mineral, it is important to consider the difference between minerals and rocks. Figure 2.2*A* is a picture of the common rock-type granite. Notice the different colors in the granite. When you look closer (figure 2.2*B*), you can see that the different colors are different minerals. The large pink crystals are the mineral potassium feldspar, which we have already talked about. The large white crystals are the mineral plagioclase feldspar, which is related to potassium feldspar but contains calcium and sodium instead of potassium. The smaller, glassy-looking crystals are the mineral quartz. The small, dark mineral grains are biotite mica. From this picture, it is clear that granite is composed of more than one type of mineral and, thus, the definition of a mineral would not fit this rock.

Rocks are defined as *naturally formed aggregates of minerals or mineral-like substances.* The granite in figure 2.2*A* and *B*, therefore, is a rock that is made up of the minerals quartz, plagioclase feldspar, potassium feldspar, and biotite. A rock can be composed of a single mineral. For example, limestone is composed of the mineral calcite. The reason that limestone is a rock and not defined simply as the mineral calcite is that limestone is made up of multiple crystals of calcite either grown in an

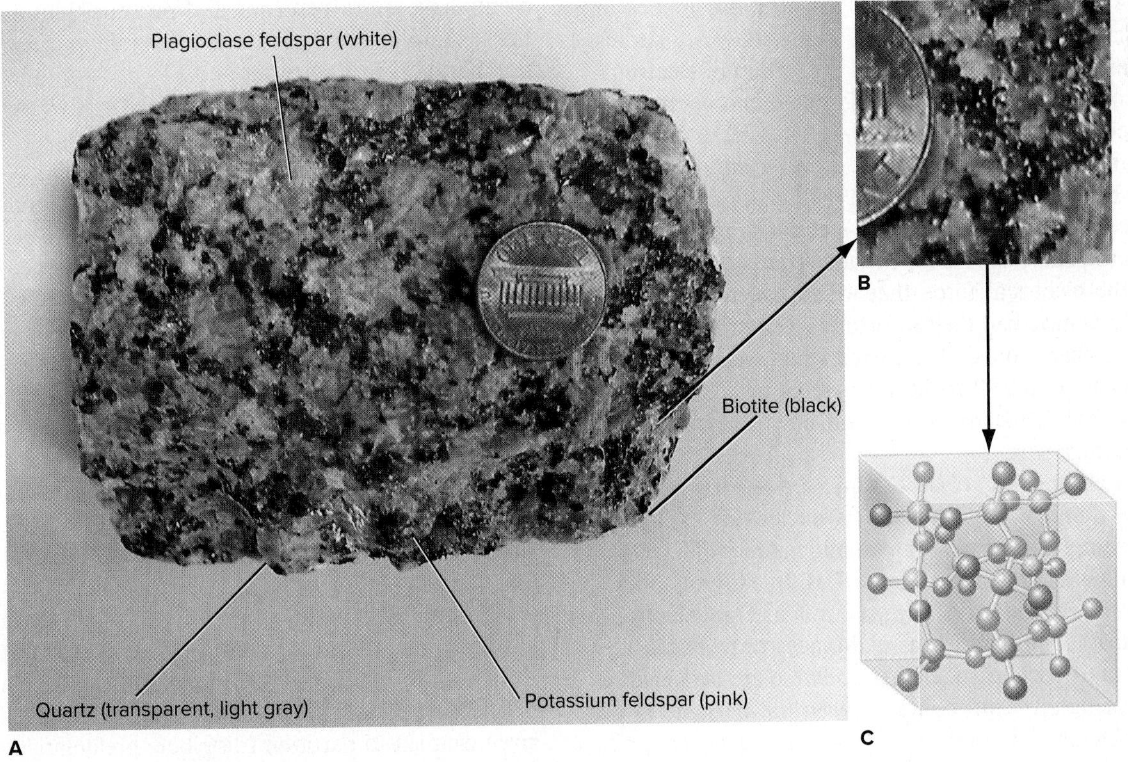

FIGURE 2.2

Rocks are aggregates of minerals. The *rock* granite (*A*), when examined closely (*B*), can be seen to be made up of crystals of different minerals. (*C*) The *mineral* quartz (SiO_2) is made up of atoms of the *elements* silicon (purple) and oxygen (red) bonded together.
(A) and (B) ©C. C. Plummer

interlocking pattern or cemented together. Although limestone is made up of a single mineral type, it is still an aggregate of many mineral grains. Some rocks can be composed of non-mineral substances. For example, coal is made of partially decomposed organic matter. Obsidian is made of silica glass, which is not crystalline and therefore not a mineral. Chapters 3, 6, and 7 cover the different types of rocks.

It is very important to keep the distinction between elements, minerals, and rocks clear when learning about geology. Rocks are composed of minerals, and minerals are composed of atoms of elements bonded together in an orderly crystalline structure. Look at figure 2.2*C* and notice how the close-up image of quartz shows the atoms bonded together in a repeating crystalline structure. How do the atoms in a mineral like quartz stick together? Why are minerals crystalline at all? In the next section, we will review the basic structure of atoms and consider why atoms bond together to form minerals. We will see how science reveals an underlying order to physical reality that is breathtaking and largely hidden from view when we look at the apparent randomness and chaos of the world.

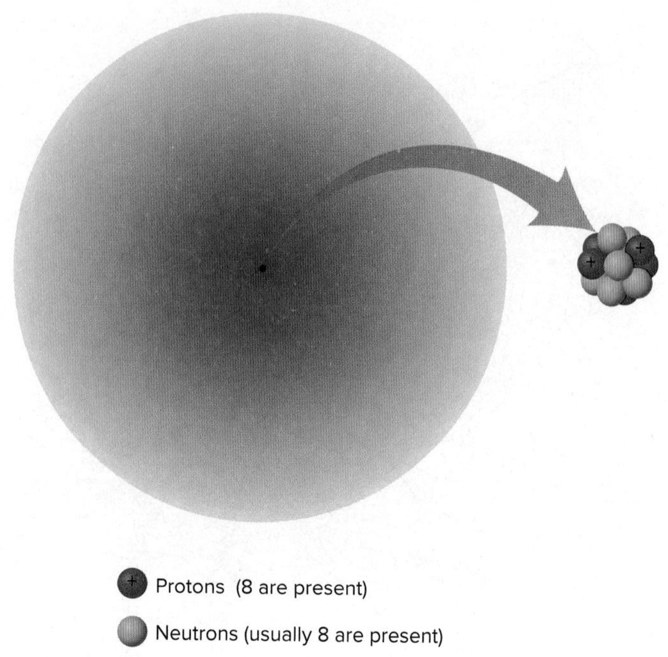

FIGURE 2.3

Model of an oxygen atom and its nucleus. The red area around the nucleus represents the electron cloud.

2.2 ATOMS AND ELEMENTS

To better understand the nature of minerals and answer the questions just posed, we need to look at what is happening at an extremely small, or atomic, scale.

Atoms are the smallest, electrically neutral assemblies of energy and matter that we know exist in the universe. Atoms consist of a central **nucleus** surrounded by a cloud of electrons. The nucleus contains positively charged **protons** and neutral particles called **neutrons**. Surrounding the nucleus is a cloud of negatively charged **electrons**. Electrons move in directions that allow them to balance out, or neutralize their charges. In atoms, electron charges are neutralized as the electrons crowd around the protons in the nucleus. It is the negative charges of electrons that provide the electrical force that we exploit to power the world. Some of us have had the misfortune of experiencing electrical force as a sharp jolt that occurred when we accidentally touched a live wire or a wall socket. This force results when the tiny, negatively charged electrons flow from one place to another, for example, along a wire.

We call each "species" of atom an *element.* An **element** is defined by the number of protons in its nucleus or its **atomic number**. For example, oxygen has an atomic number of 8, which tells you that it has eight protons (figure 2.3). In addition to having eight protons, each atom of oxygen contains eight electrons and, in its most abundant form, eight neutrons. There are ninety-two naturally occurring elements. These are arranged in order of increasing size and complexity on the periodic table (see appendix D).

The **atomic mass number** is the total number of neutrons and protons in an atom. Compared to protons and neutrons, electrons are so tiny that their mass does not contribute to the atomic mass of the atom. The atomic mass number of the oxygen atom shown in figure 2.3 is 16 (eight protons plus eight neutrons) and is indicated by the symbol ^{16}O. Heavier elements have more neutrons and protons than do lighter ones. For example, the heavy element gold has an atomic mass number of 197 (79 protons and 118 neutrons), whereas helium has an atomic mass number of only 4 (two protons and two neutrons).

Isotopes of an element are atoms containing different numbers of neutrons but the same number of protons. Isotopes are either stable or unstable. An unstable, or *radioactive, isotope* is one in which protons or neutrons are, over time, spontaneously lost or gained by the nucleus. The subatomic particles that unstable isotopes emit are what Geiger counters detect. This is *radioactivity,* which we know can be hazardous in high doses. Unstable isotopes of uranium and a few other elements are very important to geology because they are used to determine the ages of rocks. These isotopes decay at a known rate and, as described in chapter 8, are used as a kind of geologic stopwatch that starts running at the time some rocks form.

A *stable isotope* is an isotope that will retain all of its protons and neutrons through time. During recent years, stable isotopes have become increasingly important to geology and related sciences. Among the stable isotopes studied in geology are those of carbon, nitrogen, oxygen, sulfur, and hydrogen. Their usefulness in scientific investigations is due to the tendency of isotopes of a given element to partition (distribute preferentially between substances) in different proportions due to their minute weight difference. For instance, oxygen and hydrogen isotopes can be used as a proxy for the surface temperature of the Earth because when liquid water evaporates, the water vapor will have a slightly higher ratio of lighter to heavier isotopes compared to the isotopes that remain in the liquid water. Box 2.1 describes this in more detail.

An element's atomic weight is closely related to the mass number. **Atomic weight**, or **atomic mass**, is the weight of an *average* atom of an element, given in atomic mass units. Because sodium has only one naturally occurring isotope, its atomic mass number and its atomic weight are the same—23. On the other hand, chlorine has two common isotopes, with mass numbers of 35 and 37. The atomic weight of chlorine, which takes into account the abundance of each isotope, is 35.5 because the lighter isotope is more common than the heavier one.

The electrons in an atom are continuously on the move, like bees buzzing around a hive. Some are more energetic than others and move farther away from the nucleus as they move in the

EARTH SYSTEMS 2.1

Oxygen Isotopes and Climate Change

Oxygen has three stable isotopes. ^{16}O (the 16 tells us there are 16 protons and neutrons in the nucleus) is most abundant, making up 99.762% of Earth's oxygen. ^{17}O constitutes 0.038%, and ^{18}O, 0.200%. The ratio of ^{18}O to ^{16}O in a substance is determined using very accurate instruments called *mass spectrometers*. The ratio of ^{18}O to ^{16}O is 0.0020:1, or 0.002 atoms of ^{18}O for every one atom of ^{16}O. If partitioning did not take place, we would expect to find the same ratio of isotopes in any substance containing oxygen. However, there is considerable deviation because of the tendency of lighter and heavier atoms to partition or separate.

When water (H_2O) evaporates, molecules containing the lighter isotope (^{16}O) will evaporate more readily than those containing the heavier isotope (^{18}O). The water vapor will have a slightly higher abundance of ^{16}O relative to ^{18}O than the water left behind. Colder water will have a higher ratio of ^{18}O to ^{16}O than warmer water.

Oxygen isotope studies have allowed scientists to identify climate changes during relatively recent geologic time by determining the temperature changes of ocean water. Because we cannot sample past oceans, we use fossil shells to determine the oxygen isotope ratios at the time the organisms were alive. *Foraminifera* are microscopic and nearly microscopic organisms that live in considerable abundance just beneath an ocean surface. While they are alive, they grow their shells of calcite ($CaCO_3$), incorporating oxygen from the seawater. The oxygen in the shells has the $^{18}O/^{16}O$ ratio that is the same as that of the seawater. The particular isotopic ratio reflects the temperature of the seawater.

When foraminifera die, their shells settle onto the deep ocean floor, where they form a thin layer upon older layers of tiny shells. Deep-sea drilling retrieves cores of these layers of sediment. Foraminifera from each layer are analyzed and the $^{18}O/^{16}O$ ratios determined. The ages of the layers are also determined. From these data, the temperature of the ocean's surface water is inferred for the times the foraminifera were alive. Box figure 1 shows the fluctuation in temperature during the past 800,000 years.

These studies show how an Earth systems approach has been useful in determining knowledge about the atmosphere, the geosphere, the biosphere, and the hydrosphere. We can see that global warming and cooling are natural occurrences in the context of geologic time. What the data do not tell us is what effect humans are having on the climate. Is the present warming part of a natural cycle, or is the rapid increase in greenhouse gases (notably CO_2) reversing what would be a natural cooling cycle? Chapter 21 discusses climate change in detail.

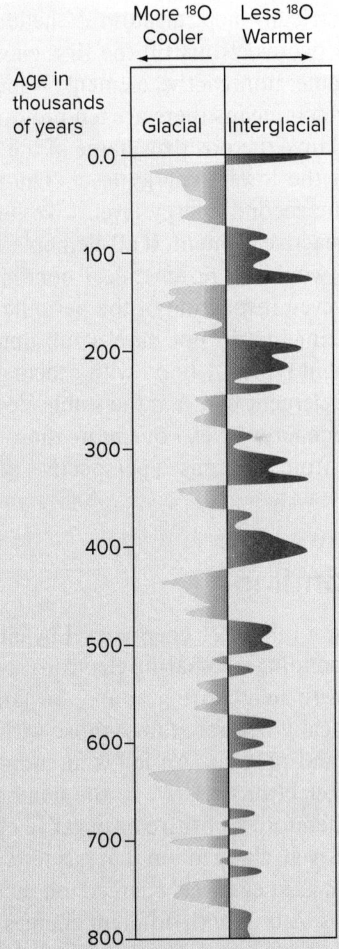

BOX 2.1 ■ FIGURE 1

Changes in climate during the past 800,000 years as determined by oxygen isotope ratios of ocean water as recorded in the shells of foraminifera found in deep-sea sediment cores. Blue—glacial times; red—interglacial times.

space around it. Although each electron moves throughout the space surrounding the nucleus, it will spend most of its time as part of an *energy level.* As a visual aid for comprehension, energy levels are often shown as concentric spherical *shells,* but it should be noted that chemists regard this as misleading. Each energy level can hold a specific number of electrons. The electrons will fill each level in order from the lowest to the highest energy. The most stable configuration for an atom is to have a full outer energy level or "shell." On a periodic table, this is represented by the elements in the right-hand column, known as the *noble gases.*

The first energy level is complete when it contains two electrons. The second and third energy levels are each complete with eight electrons. Consider the element helium. It has an atomic number of 2, which means there are two protons in its nucleus. An electrically neutral atom of helium contains two electrons. These two electrons fill the first energy level, so helium is a very stable, nonreactive element. Neon has an atomic number of 10, which means there are ten protons in its nucleus. Ten electrons balance the positive charge of the protons. Two of the electrons fill the lowest energy level. The remaining eight completely fill the second energy level. Like helium, neon is a very stable, nonreactive element. If all elements were like helium and neon, there would be no chemical bonding, no minerals, and no life! However, inspection of the periodic table will show you that most elements do not have a full outer energy level. These elements will typically bond with others or with other atoms of the same element to attain the stable electron configuration of a full outer energy level. For a more thorough explanation of atomic theory from a chemist's perspective, go to *Understanding Chemistry,* www.chemguide.co.uk/atommenu.html.

Ions and Bonding

Atoms can attain a full outer energy level by either exchanging electrons (ionic bonding) or sharing electrons (covalent and metallic bonding) with neighboring atoms. So far, we have been discussing electrically neutral atoms–those with an equal number of electrons and protons. An **ion** is an atom that has a surplus or deficit of electrons relative to the number of protons in its nucleus and therefore a positive or negative electrical charge. A **cation** is a positively charged ion that has fewer electrons than protons. An **anion** is a negatively charged ion that has more electrons than protons. Atoms with different charges are attracted to one another, and this forms the basis for **ionic bonding**.

Consider the elements chlorine and sodium that make up halite. Sodium has an atomic number of 11, which means there are eleven protons in its nucleus. A neutrally charged atom of sodium has eleven electrons to balance the positive charge of the eleven protons. Two of the electrons fill the lowest energy level, eight electrons fill the second energy level, and the final electron will exist in the third energy level. This energy configuration is not stable, so the sodium atom will give up its last electron if it can be taken up by *other* electron-deficient atoms. In each sodium ion, then, the eleven protons (11^+) and ten electrons (10^-) add up to a single positive charge ($^+1$). Chemists customarily abbreviate the sodium cation as Na^+. Chlorine has an atomic number of 17. An electrically neutral atom of chlorine will have seventeen protons and seventeen electrons. The seventeen electrons completely fill the first and second energy levels, but the third energy level will contain only seven of the eight electrons it needs to be filled. Chlorine will capture an electron and incorporate it in its outer energy level to attain a stable electron configuration. This produces an anion of chlorine with a single negative charge (Cl^-). Thus, when sodium and chlorine atoms are close together, sodium gives up an electron to chlorine (figure 2.4), and the resultant positive charge on sodium and negative charge on chlorine bonds them together in ionic bonding (figure 2.5).

Ionic bonding is the most common type of bonding in minerals. However, in most minerals the bonds between atoms are not purely ionic. Atoms are also commonly bonded together by **covalent bonding**, or bonding in which adjacent atoms *share* electrons. Diamond is composed entirely of covalently bonded carbon atoms (figure 2.6). Carbon has an atomic number of 6, which means that it has six protons and six electrons. Two electrons fill the lowest energy level, leaving four in the second energy level. Carbon atoms therefore need four more electrons to

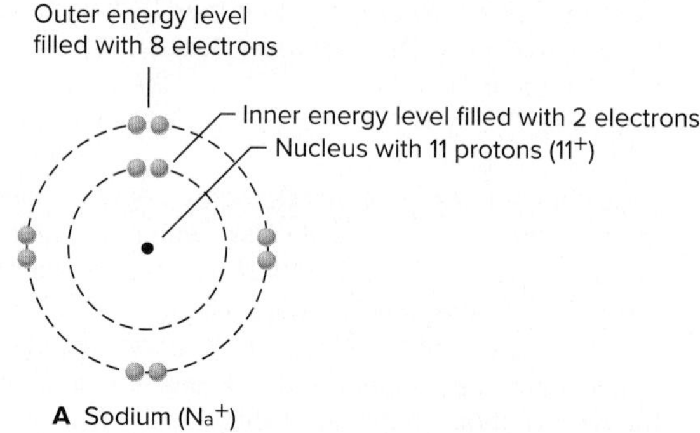

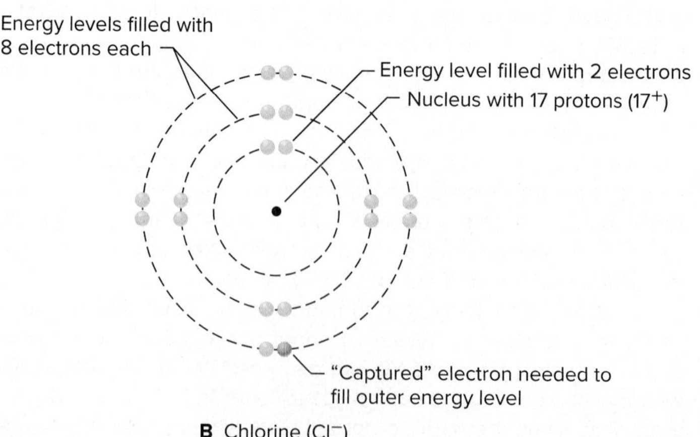

FIGURE 2.4

Diagrammatic representation of (*A*) sodium and (*B*) chlorine ions. The dots represent electrons in energy levels within an ion. Sodium has lost the electron that would have made it electrically neutral because a single electron in a higher energy level would be unstable. Chlorine has gained an electron to complete its outer energy level and make it stable.

fill their outer energy level. When carbon atoms are packed closely together, electrons can be shared with neighboring atoms. Each of the outer energy level electrons will spend half of its time in one atom and half in an adjacent atom. Electrical neutrality is maintained, and each atom, in a sense, has eight electrons in the outer energy level (even though they are not all there at the same time). Covalent bonds in diamond are very strong, and diamond is the hardest natural substance on Earth.

Graphite, like diamond, is pure carbon. (That is to say graphite and diamond are *polymorphs*—different crystal structures having the same composition.) Graphite is used in pencils and as a lubricant. Amazingly, the hardest mineral and one of the softest have the same composition. The distinction is in the bonding. In diamond, the covalent bonds form a three-dimensional structure. In graphite, the covalent bonds form sheets that are held together by much weaker electrostatic bonds (known as van der Waals bonds). It is these weak bonds that make graphite so soft. You can examine this in more detail by following the instructions in box 2.2, "Recommended Web Investigation."

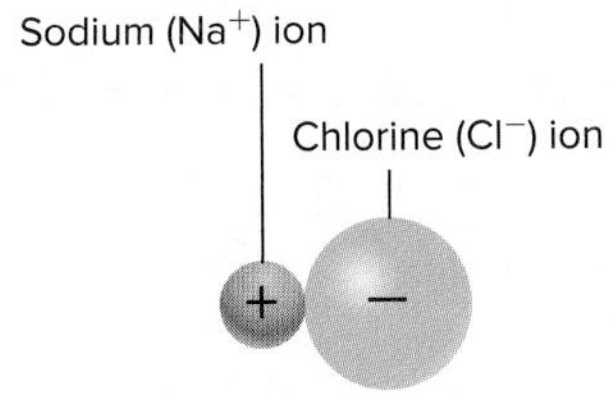

FIGURE 2.5

Ionic bonding between a sodium cation (Na^+) and a chlorine anion (Cl^-).

A third type of bonding, **metallic bonding**, is found in metals, such as copper or gold. The atoms are closely packed, and the electrons move freely throughout the crystal so as to hold the atoms together. The ease with which electrons move accounts for the high electrical conductivity of metals.

WEB BOX 2.2

Recommended Web Investigation

The minerals graphite and diamond are both made of covalently bonded carbon atoms, and yet diamond is the hardest mineral and graphite is one of the softest. The reason for this is the way in which the carbon atoms are ordered within the crystalline structure. To see in 3-D how graphite and diamond crystal structures differ, go to http://webmineral.com and, from the alphabetical listing of minerals, look up graphite. In the graphite crystallography section of the graphite page, you will find a model of the structure of graphite. Click "Large Pop-Up" and a larger image will open that you can drag and rotate to see from any perspective. The rods represent bonds. Each carbon atom is bonded to three others to form a hexagonal pattern in a sheet. There are no bonds shown between adjacent sheets. Thus, the sheets of bonded carbon will easily slide over one another. Now look up diamond and examine the crystal structure of diamond. Rotate the image and notice the three-dimensional bonding between the carbon atoms. These strong covalent bonds in all directions give diamond its hardness.

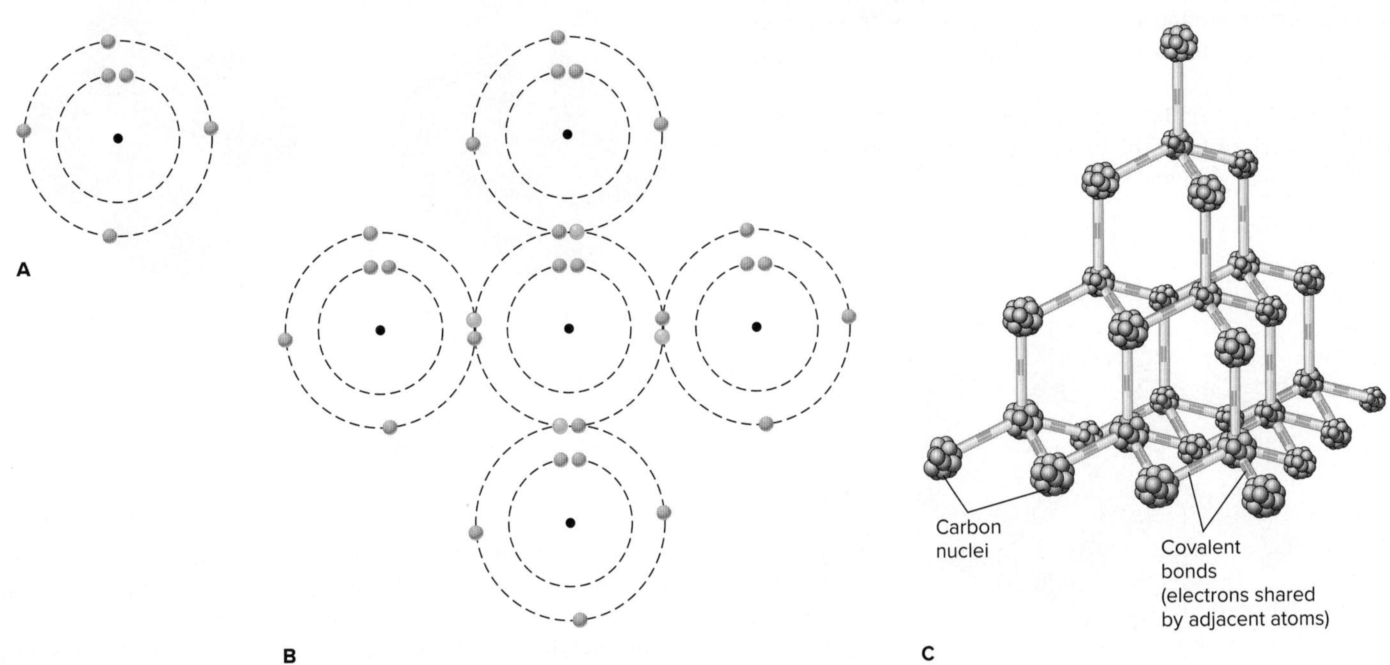

FIGURE 2.6

Covalent bonding in diamond. (*A*) A single atom of carbon has four electrons in its outer energy level and needs four more to fill that energy level. (*B*) Covalently bonded carbon atoms share electrons so that each has a full outer energy level. (*C*) In diamond, the carbon atoms are arranged in a crystalline structure that allows each carbon to bond covalently with four other carbon atoms.

Finally, after all atoms have bonded together, there may be weak, attractive forces remaining. This is the very weak force that holds adjacent sheets of mica or graphite together. It is also the force that holds water molecules together in ice (see In Greater Depth 2.8).

Crystalline Structures

A requirement for all three types of bonding that we have discussed is that the atoms are in close proximity to each other. Consider ionically bonded chlorine and sodium in halite. Under ordinary circumstances, like-charged ions repel one another and quickly move apart. They come close together only to form a stable mineral structure because they are "glued" into place by bonding with ions of the opposite charge. In other words, the need to neutralize electrical charges, while at the same time keeping like-charges apart, works to create a regular arrangement of atoms. Examine the halite in figure 2.7 and notice how the sodium and chlorine ions alternate so that each cation is in contact only with anions. Covalent and metallic bonding require that the atoms are packed together tightly enough that electrons can migrate between the atoms.

The field of swarming electrons extends farther out from the atomic nuclei of some elements than it does for others. The "size" of an atom (or an ion) is essentially the radius of its electron field; its *ionic radius*, in other words (figure 2.8). Ionic radii play an important role in the arrangement of atoms in a crystalline structure as well. When ions come together they tend to pack as efficiently as possible. No irregular holes may exist in the arrangement. A large number of *anions* (negatively charged ions) may crowd around a single, large *cation* (positively charged ion), while only a few anions may cluster about a small cation (as in figure 2.9).

Of particular importance in this respect are the crystal structures derived from the two most common elements in the Earth's crust–oxygen and silicon (box 2.3).

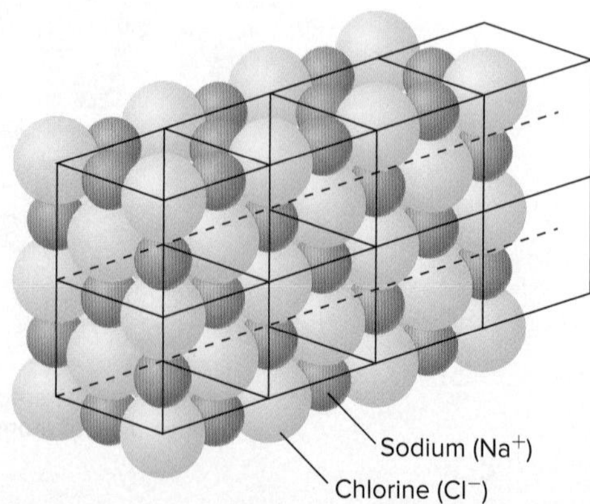

FIGURE 2.7

Model of the atomic structure of halite. The alternating three-dimensional stacking of atoms creates a box-like grid that is expressed in the cubic form of halite crystals.

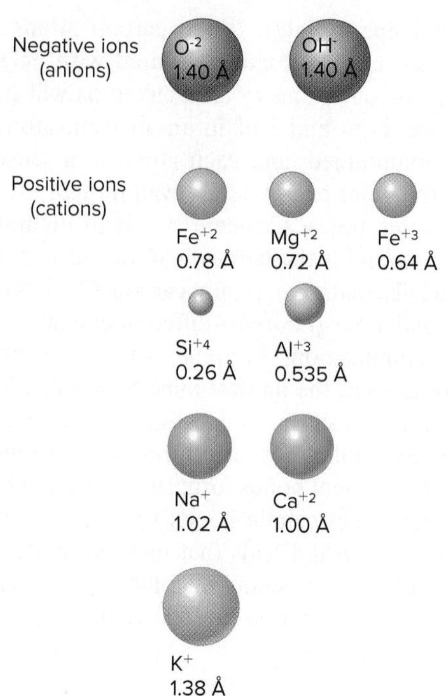

FIGURE 2.8

Sizes of most common ions in minerals, given in angstroms (Å). An angstrom is 10^{-8} cm. The ions that are close in size are in the same row, and these can replace one another in a crystal structure.

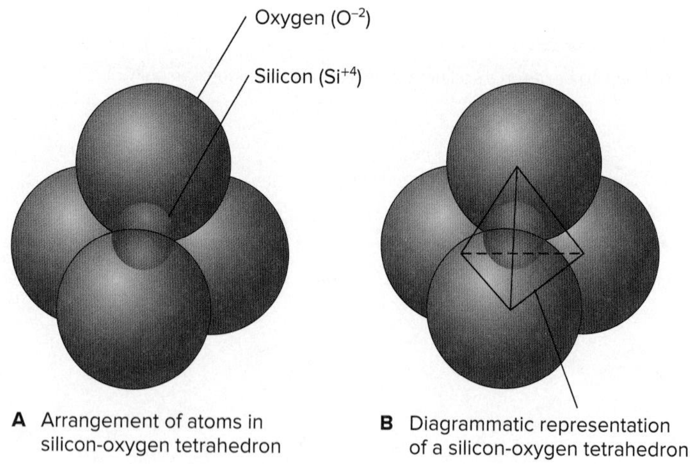

FIGURE 2.9

(*A*) The silicon-oxygen tetrahedron. (*B*) The silicon-oxygen tetrahedron showing the corners of the tetrahedron coinciding with the centers of oxygen ions.

Silicon is the *element used to make computer chips.* **Silica** is a term for *oxygen combined with silicon.* Because oxygen and silicon are the two most abundant elements in the crust, most minerals contain silica. The common mineral quartz (SiO_2) is pure silica that has crystallized. Quartz is one of many minerals that are **silicates**, substances that contain silica (as indicated by their chemical formulas). Most silicate minerals also contain one or more other elements.

IN GREATER DEPTH 2.3

Elements in the Earth

Estimates of the chemical composition of Earth's crust are based on many chemical analyses of the rocks exposed on Earth's surface. (Models for the composition of the interior of the Earth—the core and the mantle—are based on more indirect evidence.) Table 1 lists the generally accepted estimates of the abundance of elements in the Earth's crust. At first glance, the chemical composition of the crust (and, therefore, the average rock) seems quite surprising.

We think of oxygen as the O_2 molecules in the air we breathe. Yet most rocks are composed largely of oxygen because it is the most abundant element in the Earth's crust. Unlike the oxygen gas in air, oxygen in minerals is strongly bonded to other elements. By weight, oxygen accounts for almost half the crust, but it takes up 93% of the volume of an average rock. This is because the oxygen atom takes up a large amount of space relative to its weight. (Note how much bigger oxygen atoms are relative to other atoms in figure 2.9 and others.) It is not an exaggeration to regard the crust as a mass of oxygen with other elements occupying positions in crystalline structures between oxygen atoms.

Note that the third most abundant element is aluminum, which is more common in rocks than iron. Knowing this, one might assume that aluminum would be less expensive than iron, but of course this is not the case. Common rocks are not mined for aluminum because it is so strongly bonded to oxygen and other elements. The amount of energy required to break these bonds and separate the aluminum makes the process too costly for commercial production. Aluminum is mined from the uncommon deposits where aluminum-bearing rocks have been weathered, producing compounds in which the crystalline bonds are not so strong.

Collectively, the eight elements listed in table 1 account for more than 98% of the weight of the crust. All the other elements total only about 1.5%. Absent from the top eight elements are such vital elements as hydrogen (tenth in abundance by weight) and carbon (seventeenth in abundance by weight).

The element copper is only twenty-seventh in abundance, but our industrialized society is highly dependent on this metal. Most of the wiring in electronic equipment is copper, as are many of the telephone and power cables that crisscross the continent. However, the Earth's crust is not homogeneous, and geologic processes have created concentrations of elements such as copper in a few places. Exploration geologists are employed by mining companies to discover where (as well as why) ore deposits of copper and other metals occur (see chapter 22).

BOX 2.3 ■ TABLE 1

Crustal Abundance of Elements

Element	Symbol	Percentage by Weight	Percentage by Volume	Percentage of Atoms
Oxygen	O	46.6	93.8	60.5
Silicon	Si	27.7	0.9	20.5
Aluminum	Al	8.1	0.8	6.2
Iron	Fe	5.0	0.5	1.9
Calcium	Ca	3.6	1.0	1.9
Sodium	Na	2.8	1.2	2.5
Potassium	K	2.6	1.5	1.8
Magnesium	Mg	2.1	0.3	1.4
All other elements		1.5	—	3.3

The Silicon-Oxygen Tetrahedron

Silicon and oxygen combine to form the atomic framework for most common minerals on Earth. The basic structural unit consists of four oxygen anions packed together around a single, much smaller silicon cation, as shown in figure 2.9*A*. The four-sided, pyramidal, geometric shape called a *tetrahedron* is used to represent the four oxygen atoms surrounding a silicon atom. Each *corner* of the tetrahedron represents the *center* of an oxygen atom (figure 2.9*B*). This basic building block of a crystal is called a **silicon-oxygen tetrahedron** (also known as a *silica tetrahedron*). Take a look at figure 2.1 and see how geometric tetrahedra are used to represent oxygen and silicon. Imagine how impossible it would be to depict the crystalline structure if you had to draw in four oxygen atoms for each of the yellow tetrahedra.

The atoms of the tetrahedron are strongly bonded together. Within a silicon-oxygen tetrahedron, the negative charges exceed the positive charges (see figure 2.10*A*). A single silicon-oxygen tetrahedron is a complex ion with a formula of SiO_4^{-4} because the silicon cation has a charge of +4, and the four oxygen ions have eight negative charges (−2 for each oxygen anion).

A silicon-oxygen tetrahedron can bond either with positively charged ions, such as iron or aluminum, or with other silicon-oxygen tetrahedra. In other words, for the silicon-oxygen tetrahedron to be stable within a crystal structure, it must

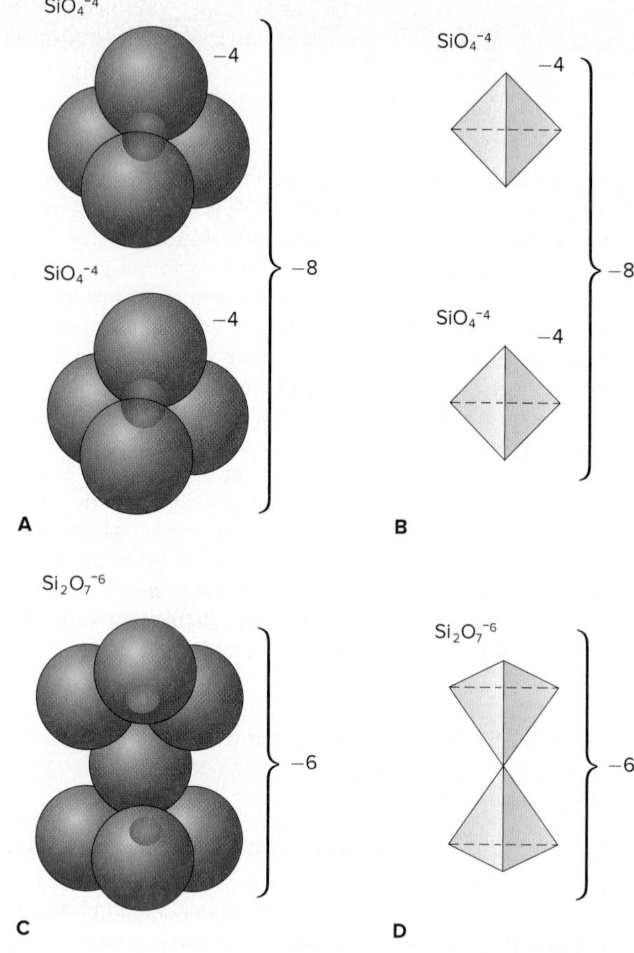

FIGURE 2.10

Two single tetrahedra (*A* and *B*) require more positively charged ions to maintain electrical neutrality than two tetrahedra sharing an oxygen atom (*C* and *D*). *B* and *D* are the schematic representations of *A* and *C*, respectively.

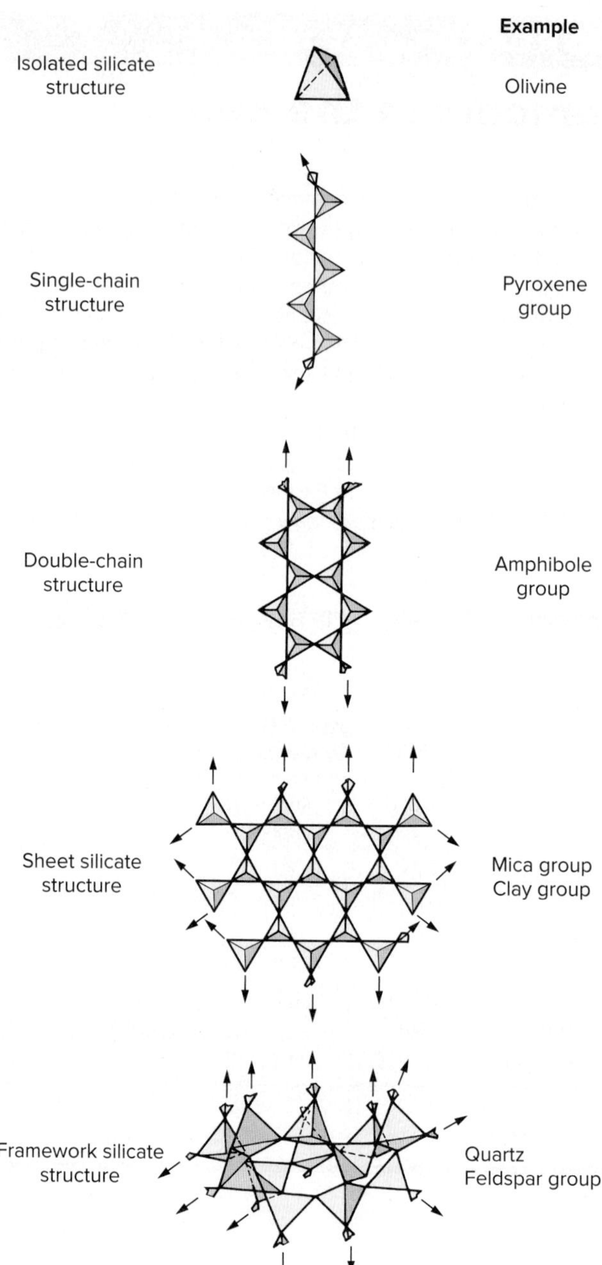

FIGURE 2.11

Common silicate structures. Arrows indicate directions in which structure repeats indefinitely. Common mineral examples for each structure are given. The physical properties of these minerals can be found in appendix A.

either (1) be balanced by enough positively charged ions or (2) share oxygen atoms with adjacent tetrahedra (as shown in figure 2.10*C* and *D*) and therefore reduce the need for extra, positively charged ions. The structures of silicate minerals range from an *isolated silicate structure*, which depends entirely on positively charged ions to hold the tetrahedra together, to *framework silicates* (quartz, for example), in which all oxygen atoms are shared by adjacent tetrahedra. The most common types of silicate structures are shown diagrammatically in figure 2.11 and are discussed next.

Isolated Silicate Structure

Silicate minerals that are structured so that none of the oxygen atoms are shared by tetrahedra have an **isolated silicate structure**. The individual silicon-oxygen tetrahedra are bonded together by positively charged ions (figure 2.12). The common mineral *olivine*, for example, contains two ions of either magnesium (Mg^{+2}) or iron (Fe^{+2}) for each silicon-oxygen tetrahedron. The formula for olivine is $(Mg,Fe)_2SiO_4$.

Chain Silicates

A **chain silicate structure** forms when two of a tetrahedron's oxygen atoms are shared with adjacent tetrahedra to form a chain (figures 2.11 and 2.13). Each chain, which extends indefinitely, has a net excess of negative charges. Minerals may have a *single-* or *double-chain structure.* For single-chain silicate structures, the ratio of silicon to oxygen (as figure 2.13 shows) is 1:3; therefore, each mineral in this group (the *pyroxene* group) incorporates

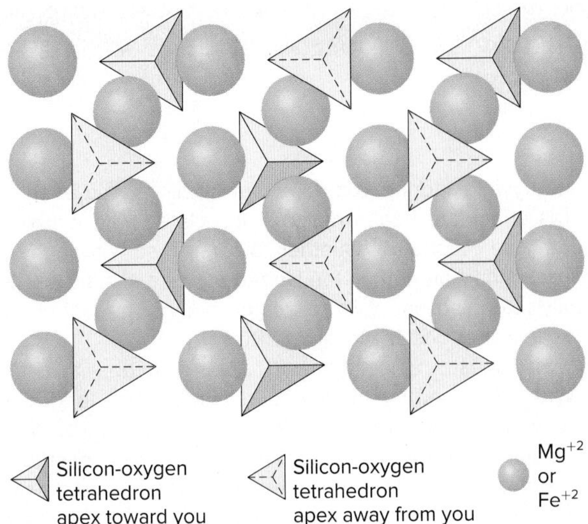

FIGURE 2.12
Diagram of the crystal structure of olivine, as seen from one side of the crystal.

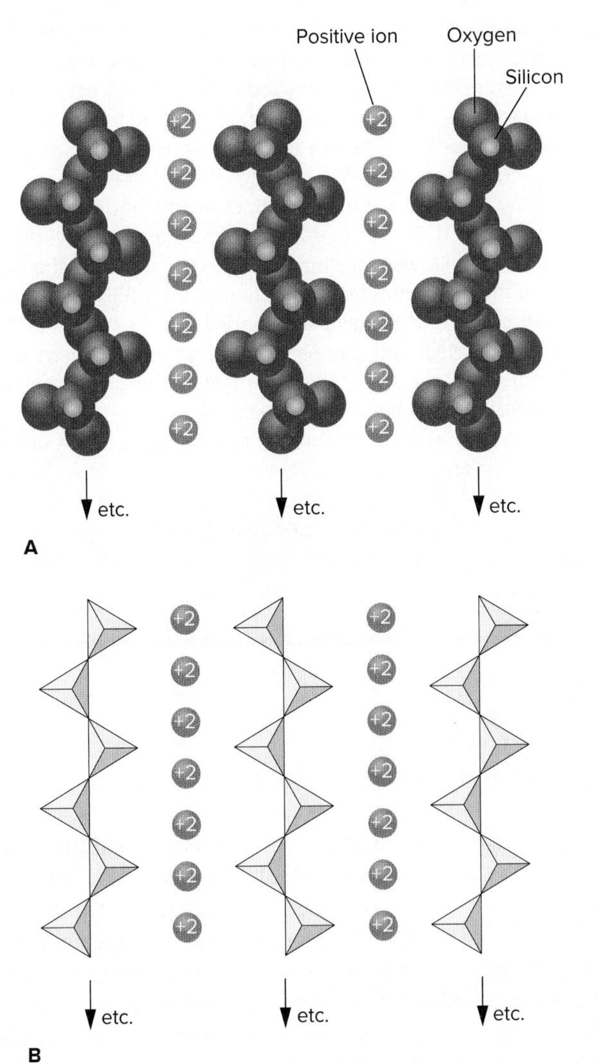

FIGURE 2.13
Single-chain silicate structure. (*A*) Model of a single-chain silicate mineral. (*B*) The same chain silicate shown diagrammatically as linked tetrahedra.

SiO_3^{-2} in its formula, and it must be electrically balanced by the positive ions (e.g., Mg^{+2}) that hold the parallel chains together. If a pyroxene has magnesium as the +2 ions bonding the chains shown in figure 2.13*A*, it has a formula of $MgSiO_3$ (commonly written as $Mg_2Si_2O_6$).

A double-chain silicate is essentially two adjacent single chains that are sharing oxygen atoms. The *amphibole* group is characterized by two parallel chains in which every other tetrahedron shares an oxygen atom with the adjacent chain's tetrahedron (figure 2.11). In even a small amphibole crystal, millions of parallel double chains are bonded together by positively charged ions.

Chain silicates tend to be shaped like columns, needles, or even fibers. The long structure of the external form corresponds to the linear dimension of the chain structure. Fibrous aggregates of certain chain silicates are called *asbestos* (see box 2.4).

Sheet Silicates

In a **sheet silicate structure**, each tetrahedron shares three oxygen atoms to form a sheet (figure 2.11). The *mica* group and the *clay* group of minerals are sheet silicates. The positive ions that hold the sheets together are sandwiched between the silicate sheets (box 2.5).

Framework Silicates

When all four oxygen ions are shared by adjacent tetrahedra, a **framework silicate structure** is formed. *Quartz* is a framework silicate mineral. The *feldspars* are a group of framework silicates. However, the feldspar structure is slightly more complex because aluminum substitutes for some of the silicon ions in some of the tetrahedra. Note from figure 2.8 that the ionic radius for aluminum is close to that of silicon, therefore Al^{+3} substitutes readily for Si^{+4}. This means that additional positive ions must be incorporated into the crystal structure to compensate for the aluminum's lower charge. For feldspars, these will be Na^{+}, K^{+}, or Ca^{+2}. Hence, feldspars, which collectively are the most abundant mineral group in Earth's crust, have formulas of $NaAlSi_3O_8$, $KAlSi_3O_8$, and $CaAl_2Si_2O_8$. The same kind of substitution also takes place in amphiboles and micas, which helps account for the wide variety of silicate minerals.

Nonsilicate Minerals

Although not as abundant in Earth, *nonsilicates,* minerals that do not contain silica, are nevertheless important. The *carbonates* have CO_3 in their formulas. Calcite, $CaCO_3$, is a member of this group and is one of the most abundant minerals at the Earth's surface where it occurs mainly in the sedimentary rock limestone. In dolomite, $CaMg(CO_3)_2$, also a carbonate, magnesium replaces some calcium in the calcite formula. Gypsum, $CaSO_4 \cdot 2H_2O$, is a *sulfate* (containing SO_4). *Sulfides* have S but not O in their formulas (pyrite, FeS_2, is an example). Hematite, Fe_2O_3, is an *oxide*–that is, it contains oxygen not bonded to Si, C, or S. Halite, NaCl, is a member of the *chloride* group. *Native elements* have only one element in their formulas. Some examples are gold (Au), copper (Cu), and the two minerals that are composed of pure carbon (C), diamond and graphite.

ENVIRONMENTAL GEOLOGY 2.4

Asbestos—How Hazardous Is It?

Asbestos is a generic name for the fibrous variety of six naturally occurring silicate minerals (box figure 1). A fibrous habit is one where the crystals are long and very thin. Five of the six varieties of asbestos are amphiboles (double-chain silicates), known commercially as "brown" and "blue" asbestos. The sixth variety is *chrysotile*, which is not a chain silicate. It belongs to the *serpentine* family of minerals (sheet silicates) and is more commonly known as "white asbestos." White asbestos is, by far, the most commonly used, making up approximately 90 percent of world consumption.

Asbestos has a number of desirable physical properties, including resistance to fire and high tensile strength, and it was widely used by manufacturers and builders. Asbestos has been used in insulation, drywall, flooring, roof tiles, brake linings, and even fabric. Concerns about the health hazards associated with exposure to asbestos have caused 55 countries to ban its use. Although asbestos is not banned outright in the United States and Canada, its use is heavily restricted and it is no longer mined. The last chrysotile mine in the United States closed in 2002.

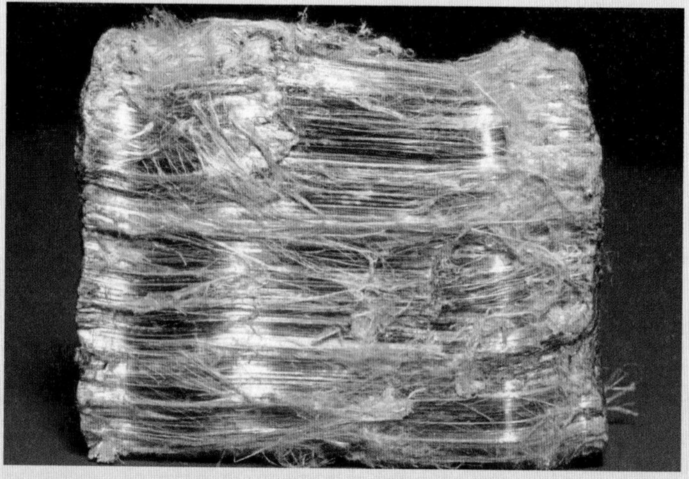

BOX 2.4 ■ FIGURE 1
Chrysotile asbestos.
©Doug Sherman/Geofile RF

Why is asbestos a health hazard? It has all to do with its fibrous habit. When breathed in, the tiny fibers can become lodged in the lungs, causing irritation and cellular damage. The most common diseases associated with asbestos exposure are asbestosis and mesothelioma. The connection between asbestos and these lung diseases was made in the early 1900s when it was noticed that people working in asbestos mines or living in towns close to asbestos mines showed a higher rate of death due to lung problems. Asbestosis is similar to silicosis contracted by miners; essentially, the lungs become clogged with asbestos dust after prolonged heavy exposure. Some of these workers, who were covered with fibers, were called snowmen. In Manville, New Jersey, children would catch the "snow" (asbestos particles released from a nearby asbestos factory) in their mouths. Although people experiencing heavy exposure for extended periods of time are at the most risk, any exposure to asbestos is considered unsafe.

Following the collapse of New York's World Trade Center towers on September 11, 2001, the dust in the air from destroyed buildings contained high levels of asbestos—much higher than the safety levels set by the Environmental Protection Agency (EPA). Faced with widespread panic and a mass exodus from the city, the EPA reversed itself and declared the air safe. Unfortunately, many emergency responders have since suffered from lung problems, including mesothelioma, which is linked to asbestos exposure.

What are the hazards of asbestos to an individual in a building where walls, floors, or ceilings contain asbestos? The risks are generally minimal because the fibers are locked up in the material and there is little risk of them becoming airborne. Special care must be taken however, when disturbing or removing materials

2.3 VARIATIONS IN MINERAL STRUCTURES AND COMPOSITIONS

It stands to reason that only a limited number of mineral compositions exist in nature because atoms cannot be combined randomly, and they can only come together to form a restricted number of crystalline structures. This does not mean, however, that each kind of mineral is compositionally different, or that individual mineral types can't show some internal compositional variation.

Ions of like size and charge may freely substitute for one another in the atomic structures of minerals. Iron (Fe^{2+}) and magnesium (Mg^{2+}), for example, interchangeably substitute to create a range of compositions in the common silicate mineral olivine. This is represented by the parentheses in the formula of olivine–$(Mg,Fe)_2SiO_4$. Olivine (see figure 2.12) is an example of a *solid solution series,* with pure magnesium olivine, Mg_2SiO_4, forming the bright green variety forsterite (or peridot, as a gem),

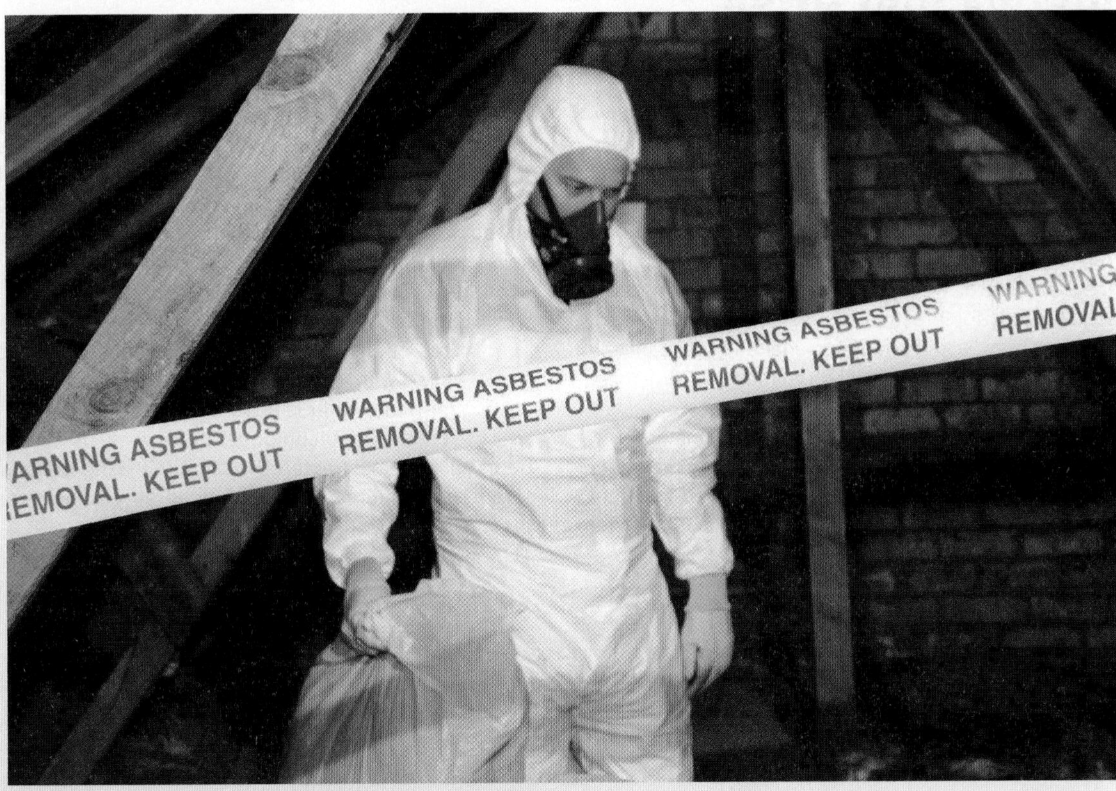

BOX 2.4 ■ FIGURE 2

A worker in protective clothing performs asbestos abatement.
©Alistair Forrester Shankie/Getty Images

that contain asbestos. Asbestos abatement (box figure 2) is done by professionals and involves sealing off the area being treated, removal of the materials taking care to minimize dust, careful packaging of removed materials in plastic, and disposal in an authorized landfill.

Additional Resources

Environmental Protection Agency

- https://www.epa.gov/asbestos

National Cancer Institute

- www.cancer.gov/cancertopics/factsheet/Risk/asbestos

and pure iron olivine forming the jet black variety fayalite, Fe_2SiO_4. The crystal structures of forsterite and fayalite are virtually identical.

Some minerals that show solid solution, like plagioclase feldspar and augite (a pyroxene), also show *compositional zoning,* with the centers of crystals dominated by one type of cation and the rims dominated by another. The grains of plagioclase in certain igneous rocks typically have calcium-rich centers and sodium-rich rims (figure 2.14). The change is due to the cooling of the molten rock from which the plagioclase crystallizes. Calcium-rich plagioclase is more stable at the high temperatures in which the crystals start growing. The crystals then develop sodium-rich rims as the remaining melt crystallizes.

Some minerals can have the same chemical composition but have different crystalline structures–described earlier as *polymorphism.* For example, calcite and aragonite both have the same formula $CaCO_3$. Their atomic crystal structures differ greatly, however. As you might expect, these two similar but distinctive mineral types result from separate conditions and

ENVIRONMENTAL GEOLOGY 2.5

Clay Minerals that Swell

Clay minerals are very common at Earth's surface; they are a major component of soil. There are a great number of clay minerals. What they all have in common is that they are sheet silicates. They differ by which ions hold sheets together and by the number of sheets "sandwiched" together.

Ceramic products and bricks are made from clay. Surprisingly, some clay minerals are edible; some are used in the manufacturing of pills. *Kaolinite*, a clay mineral, was, until recently, the main ingredient in Kaopectate, a remedy for intestinal distress. Popular fast-food chains use clay minerals as a thickener for shakes (you can tell which ones because the chains do not call them milk shakes—they do not use milk).

Montmorillonite is one of the more interesting clay minerals. It is better known as *expansive clay* or *swelling clay*. If water is added to the montmorillonite, the water molecules are adsorbed into the spaces between silicate layers (box figure 1). This results in a large increase in volume, sometimes up to several hundred percent. The pressure generated by the swelling can be up to 50,000 kilograms per square meter. This is sufficient to lift a good-sized building.

If a building is erected on expansive clay that subsequently gets wet, a portion of the building will be shoved upward. In all likelihood, the foundation will break. Some people think that expansive soils have caused more damage than earthquakes and landslides combined. Damage in the United States is estimated to cost $2 billion a year.

On the other hand, swelling clays can be put to use. Montmorillonite, mixed with water, can be pumped into fractured rock or concrete. When the water is adsorbed, swelling clay expands to fill and seal the crack. The technique is particularly useful where dams have been built against fractured bedrock. Sealing the cracks with expansive clays ensures that water will stay in the reservoir behind the dam.

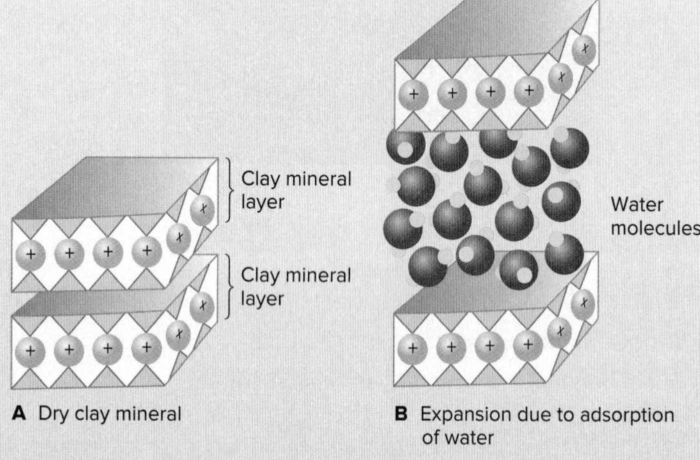

BOX 2.5 ■ FIGURE 1

(*A & B*) Expansive clays. (The orange ion represents aluminum in the clay layers and is not drawn to scale.) (*C*) The roller-coaster road is the result of uneven swelling and heaving of steeply dipping bedrock layers.
Courtesy of David C. Noe, Colorado Geological Survey

processes of formation, with aragonite usually being an indicator of high-pressure crystallization.

Graphite and diamond, as discussed earlier, are another particularly spectacular example of polymorphism. Both minerals are made up of elemental carbon. They are unusual in that there is no other element involved in their structures. Besides their extreme differences in hardness, graphite is dark and appears metallic while diamond is usually transparent and has a brilliant luster. Graphite's crystal structure is sheetlike, and it forms within the crust, while diamond originates much deeper, under the higher pressure conditions of the mantle.

It is important to note that the physical characteristics of minerals that we can observe without complex laboratory equipment, such as color, hardness, and luster, are linked closely with the crystalline structures and chemical compositions of the minerals.

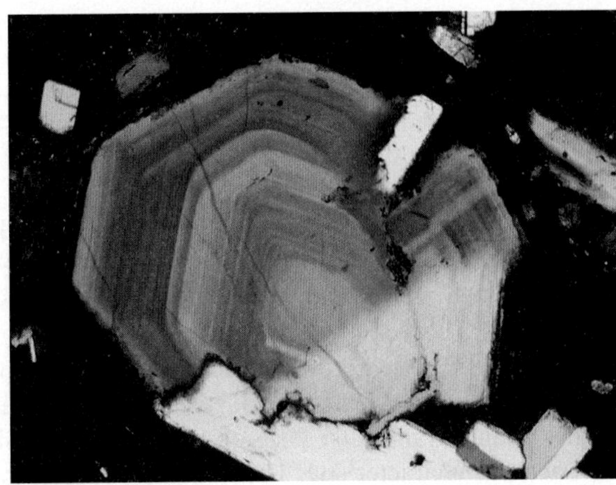

FIGURE 2.14

Zoning in plagioclase feldspar, as seen under a polarizing microscope. The concentric color bands each indicate different amounts of Ca and Na in the crystal structure.
©*Lisa Hammersley*

TABLE 2.1 Minerals of the Earth's Crust

Name	Chemical Composition	Type of Silicate Structure or Chemical Group
The most common rock-forming minerals (These make up more than 90% of the Earth's crust.)		
Feldspar Group		
Plagioclase	$CaAl_2Si_2O_8$ $NaAlSi_3O_8$	Framework silicate
Potassium feldspar (orthoclase, microcline)	$KAlSi_3O_8$	Framework silicate
Pyroxene Group (augite most common)	$(Ca,Na)(Mg,Fe)(Si,Al)_2O_6$	Single-chain silicate
Amphibole Group (hornblende most common)	Complex Fe, Mg, Al silicate hydroxide	Double-chain silicate
Quartz	SiO_2	Framework silicate
Mica Group		
Muscovite	$KAl_3Si_3O_{10}(OH)_2$	Sheet silicate
Biotite	$K(Mg,Fe)_3AlSi_3O_{10}(OH)_2$	Sheet silicate
Other common rock-forming minerals		
Silicates		
Olivine	$(Mg, Fe)_2SiO_4$	Isolated silicate
Garnet group	Complex silicates	Isolated silicate
Clay minerals group	Complex Al silicate hydroxides	Sheet silicate
Nonsilicates		
Calcite	$CaCO_3$	Carbonate
Dolomite	$CaMg(CO_3)_2$	Carbonate
Gypsum	$CaSO_4.2H_2O$	Sulfate
Much less common minerals of commercial value		
Halite	NaCl	Chloride
Diamond	C	Native element
Gold	Au	Native element
Hematite	Fe_2O_3	Oxide
Magnetite	Fe_3O_4	Oxide
Chalcopyrite	$CuFeS_2$	Sulfide
Sphalerite	ZnS	Sulfide
Galena	PbS	Sulfide

2.4 THE PHYSICAL PROPERTIES OF MINERALS

The best approach to understanding physical properties of minerals is to obtain a sample of each of the most common rock-forming minerals named in table 2.1. The properties used to describe and identify minerals can then be identified in your samples. The diagnostic properties of the minerals in table 2.1 can be found in appendix A.

To identify an unknown mineral, you should first determine its physical properties, then match the properties with the appropriate mineral, using a mineral identification key or chart such as the ones included in appendix A of this book. With a bit of experience, you may get to know the diagnostic properties for each common mineral and no longer need to refer to an identification table.

Color

The first thing most people notice about a mineral is its color. For some minerals, color is a useful property. *Muscovite* mica is silvery white or colorless, whereas *biotite mica* is black or dark brown. Most of the **ferromagnesian minerals** (iron/magnesium-bearing), such as *augite, hornblende,* olivine, and biotite, are either green or black.

Because color is so obvious, beginning students tend to rely too heavily on it as a key to mineral identification. Unfortunately, color is also apt to be the most ambiguous of physical properties (figure 2.15). If you look at a number of quartz crystals, for instance, you may find specimens that are white, pink, gray, brown, yellow, or purple. Color is extremely variable in quartz and many other minerals because even minute chemical impurities can strongly influence it. Obviously, it is poor procedure to attempt to identify quartz strictly on the basis of color.

Another way to consider how color is not always a good diagnostic property is to consider the minerals quartz, gypsum, calcite, and plagioclase feldspar. All of these minerals can have a white color. How then can you tell them apart? You have to determine other physical properties in addition to color.

Streak

Streak is the color of the powder formed when a mineral is crushed. A mineral's streak can be observed by scraping the edge of the sample across an unglazed porcelain plate known as a streak plate. This streak color is often very different from the

FIGURE 2.15

Why color may be a poor way of identifying minerals. These are all corundum gems, including ruby and sapphire.
©The Natural History Museum/Alamy

FIGURE 2.16

Hematite leaves a red-brown streak when scraped on a streak plate.
©Doug Sherman/Geofile RF

color of the mineral and is usually more reliable than color as a diagnostic property. For instance, hematite always leaves a reddish brown streak though the sample may be brown or red or silver in color (figure 2.16).

Many metallic minerals leave a dark-colored streak whereas most nonmetallic minerals leave a white or pale-colored streak. Many minerals, in particular many of the silicate minerals, are harder than the streak plate and, thus, it can be very difficult to obtain their streak.

Luster

The quality and intensity of *light* that is reflected from the surface of a mineral is termed **luster**. (A photograph cannot always show this quality.) The luster of a mineral is described by comparing it to familiar substances.

Luster is either *metallic* or *nonmetallic*. A **metallic luster** gives a substance the appearance of being made of metal. Metallic luster may be very shiny, like a chrome car part, or less shiny, like the surface of a broken piece of iron.

Nonmetallic luster is more common. The most important type is **glassy** (also called **vitreous) luster**, which gives a substance a glazed appearance, like glass or porcelain. Most silicate minerals have this characteristic. The feldspars, quartz, the micas, and the pyroxenes and amphiboles all have a glassy luster.

Less common is an **earthy luster**. This resembles the surface of unglazed pottery and is characteristic of the various clay minerals. Other nonmetallic lusters include *resinous* luster (appearance of resin), *silky* luster, and *pearly* luster.

Hardness

The property of scratchability, or **hardness**, can be tested fairly reliably. For a true test of hardness, the harder mineral or substance must be able to make a groove or scratch on a smooth, fresh surface of the softer mineral. For example, quartz can always scratch calcite or feldspar and is thus said to be harder than both of these minerals. Substances can be compared to **Mohs' hardness scale**, in which ten minerals are designated as standards of hardness (figure 2.17). The softest mineral, talc (used for talcum powder because it is softer than skin), is designated as 1. Diamond, the hardest natural substance on Earth, is 10 on the scale. (Its polymorph, graphite, has a hardness of 1.5.) Mohs' scale is a relative hardness scale. Figure 2.17 shows the absolute hardness for the ten minerals. The absolute hardness is obtained using an instrument that measures how much pressure is required to indent a mineral. Note that the difference in absolute hardness between corundum (9) and diamond (10) is around six times the difference between corundum and topaz (8).

Rather than carry samples of the ten standard minerals, a geologist doing field work usually relies on common objects to test for hardness (figure 2.17). A fingernail usually has a hardness of about 2.5. If you can scratch the smooth surface of a mineral with your fingernail, the hardness of the mineral must be less than 2.5. A copper coin or a penny has a hardness between 3 and 4; however, the brown, oxidized surface of most pennies is much softer, so check for a groove into the coin. A knife blade or a steel nail generally has a hardness slightly greater than 5, but it depends on the particular steel alloy used. A geologist uses a knife blade to distinguish between softer minerals, such as calcite, and similarly appearing harder minerals, such as quartz. Ordinary window glass, usually slightly harder than a knife blade (although some glass, such as that containing lead, is much softer), can be used in the same way as a knife blade for hardness tests. A file (one made of tempered steel for filing

IN GREATER DEPTH 2.6

Precious Gems

Diamond engagement rings are a tradition in our society. The diamond, often a significant financial commitment, symbolizes perpetual love. Jewelry with valuable gemstones is glamorous. We expect to see the rich and famous heavily bedecked with gemstones set in gold and other precious metals.

Gemstones are varieties of certain minerals that are valuable because of their beauty. Precious gemstones, or, simply, precious stones, are particularly valuable; semiprecious stones are much less valuable. Diamond, sapphire, ruby, emerald, and aquamarine are regarded as precious stones. What they all have in common is that they are transparent with even coloration and have a hardness greater than quartz (7 on the hardness scale). Their hardness ensures that they are durable.

Diamonds are usually clear, although some tinted varieties are particularly valuable. (The famous Hope diamond is blue.) Diamond's appeal is largely due to its unique, brilliant luster (called *adamantine* luster). This results from the way that light reflects from within the crystal and is dispersed into rainbowlike colors. The facets that you see on a diamond (see box figure 1) have been cut (or, more correctly, ground, using diamond dust) to enhance the gem's brilliance. The cut facets are not related to diamond's natural form, which is octahedral.

Sapphire and ruby are both varieties of the aluminum oxide corundum (9 on Mohs' scale). Sapphire can be various colors (except red), but blue sapphires are most valuable. Minute amounts of titanium and iron in the crystal structure give sapphire its blue coloration. Rubies are red due to trace amounts of chromium in corundum.

Emerald and aquamarine are varieties of beryl (hardness of 7.5). Emerald is the most expensive of these and owes its green color to chromium impurities. Aquamarine's blue color is due to iron impurities in the crystal structure.

BOX 2.6 ■ FIGURE 1

Cut diamond with facets that were ground into it using diamond dust on a lap. ©*Steve Hamblin/Alamy RF*

Recommended Web Investigation

The Image

- www.theimage.com/

Click on "Gemstone Gallery" and then "Beryl." You can read about the properties of beryl and details about emerald and aquamarine. Below the description, you can access images of these gems. You can go back and click "Sapphire" for information on sapphire and corundum. A photo of sapphires and a ruby is accessible at the bottom of the text. This site also contains information on how gems are faceted.

metal, not a fingernail file) can be used for a hardness of between 6 and 7. A porcelain streak plate also has a hardness of around 6.5.

External Crystal Form

The **crystal form** of a mineral is a set of faces that have a definite geometric relationship to one another (figure 2.18). A well-formed crystal of pyrite, for example (figure 2.18*C*), consists of six faces all square and joined at right angles. The crystal form of pyrite is a *cube,* in other words.

Crystals more commonly consist of several types of forms combined to generate the full body of each specimen. As a rule of thumb, if two or more faces on a crystal are identical in shape and size, they belong to the same crystal form.

Minerals displaying well-developed crystal faces have played an important role in the development of chemistry and physics. Nicolas Steno, a Danish naturalist of the seventeenth century, first noted that the angle between two adjacent faces of quartz is always exactly the same, no matter what part of the world the quartz sample comes from or the color or size of the quartz. As shown in figure 2.19, the angle between any two adjacent sides of the six-sided "pillar" (which is called a prism by mineralogists) is always exactly 120 degrees, while between a face of the "pillar" and one of the "pyramid" faces (forming the point of the crystal) the angle is always exactly 141 degrees 45 minutes.

The discovery of such regularity in nature usually has profound implications. When minerals other than quartz were studied, they too were found to have sets of angles for adjacent faces that never varied from crystal to crystal. This observation

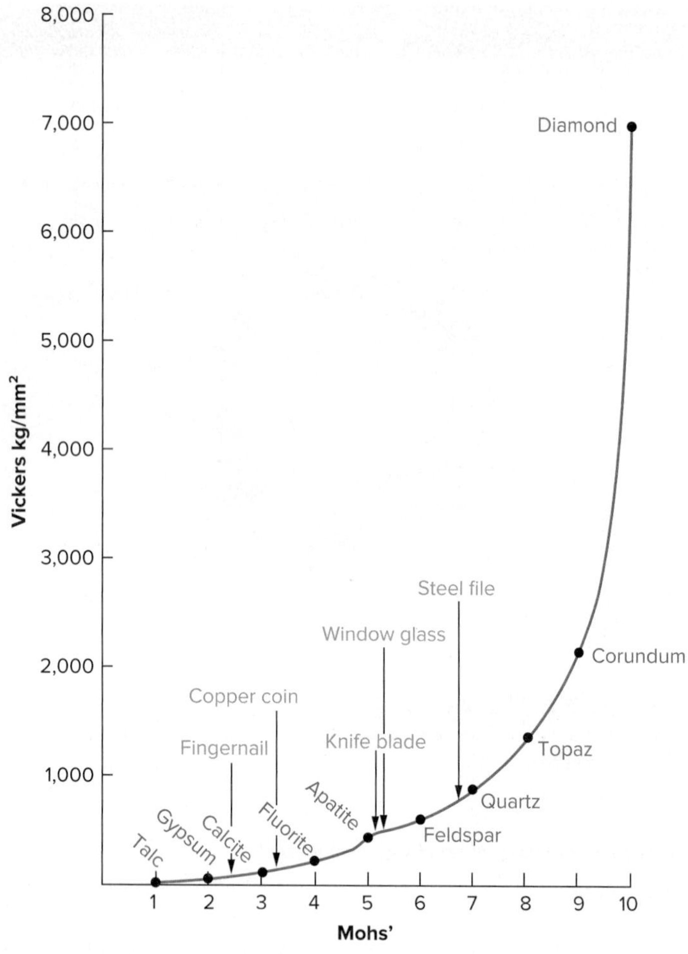

FIGURE 2.17

Mohs' hardness scale plotted against Vickers indentation values (kg/mm^2). Indentation values are obtained by an instrument that measures the force necessary to make a small indentation into a substance.

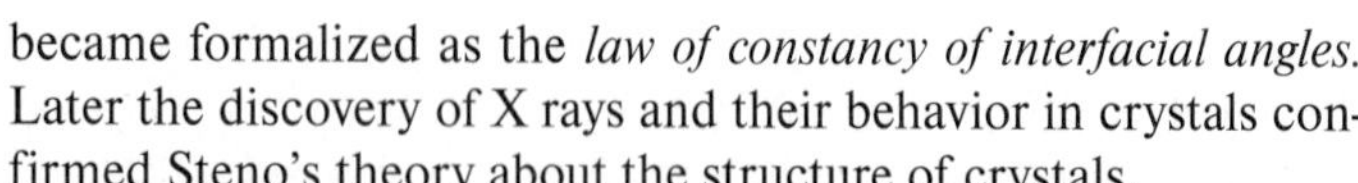

became formalized as the *law of constancy of interfacial angles.* Later the discovery of X rays and their behavior in crystals confirmed Steno's theory about the structure of crystals.

Steno suspected that each type of mineral was composed of many tiny, identical building blocks, with the geometric shape of the crystal being a function of how these building blocks are put together. If you are stacking cubes, you can build a structure having only a limited variety of planar forms. Likewise, stacking rhombohedrons in three dimensions limits you to other geometric forms (figure 2.20).

Steno's law was really a precursor of atomic theory, developed centuries later. Our present concept of crystallinity is that atoms are clustered into geometric forms–cubes, bricks, hexagons, and so on–and that a crystal is essentially an orderly, three-dimensional stacking of these tiny geometric forms. Halite, for example, may be regarded as a series of cubes stacked in three dimensions (see figure 2.7). Because of the cubic "building block," its usual crystal form is a cube with crystal faces at 90-degree angles to each other.

A

B

C

FIGURE 2.18

Characteristic crystal forms of three common minerals: (*A*) Cluster of quartz crystals. (*B*) Crystals of potassium feldspar. (*C*) Intergrown cubic crystals of pyrite.
(A) ©Doug Sherman/Geofile RF (B) ©Dr. Parvinder Sethi (C) ©Doug Sherman/Geofile RF

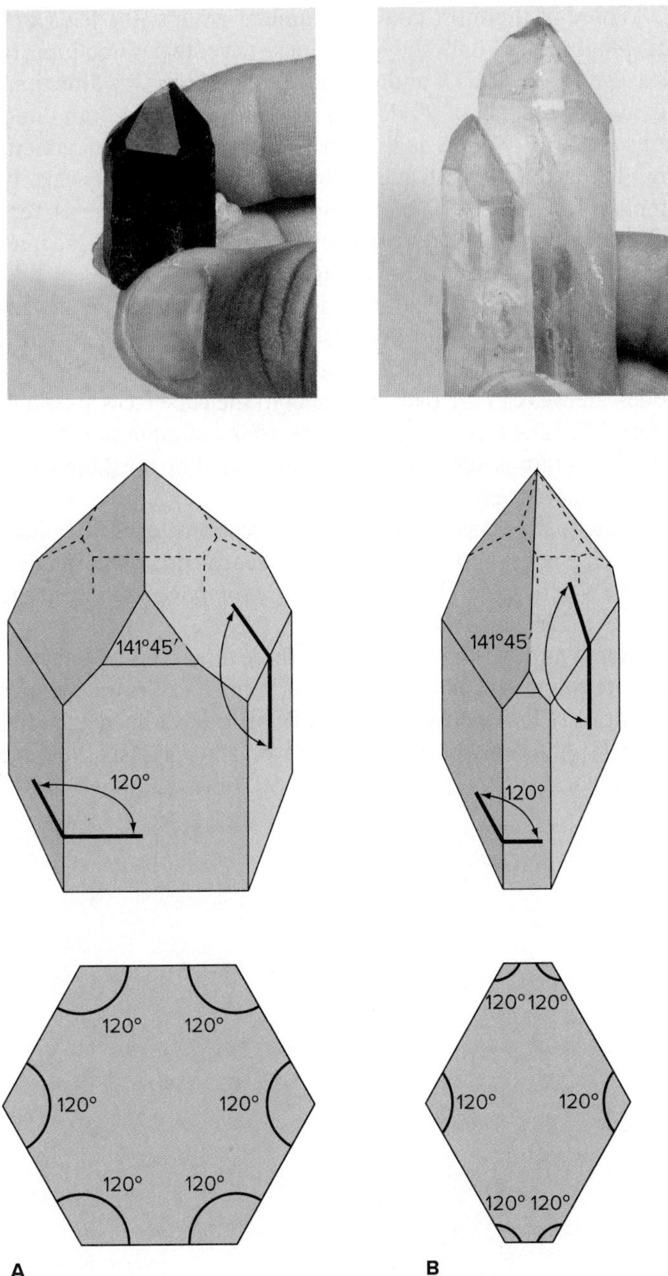

FIGURE 2.19

Quartz crystals showing how interfacial angles remain the same in perfectly proportioned (*A*) and misshapen (*B*) crystals. Cuts perpendicular to the prisms show that all angles are exactly 120°.
©*Dr. Parvinder Sethi*

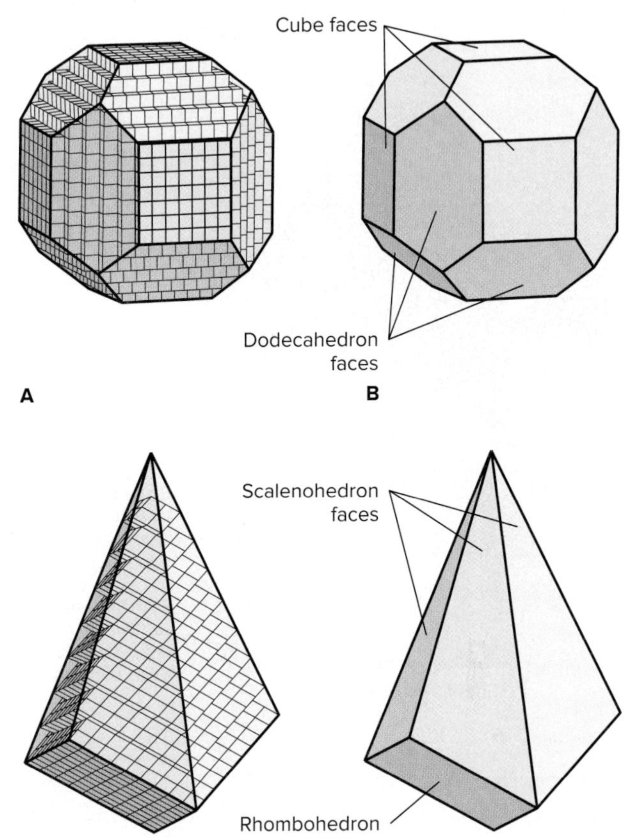

FIGURE 2.20

Geometric forms built by stacking cubes (*A*, *B*) and rhombohedrons (*C*, *D*). *A* and *B* are from a diagram published in 1801 by Haüy, a French mathematician. *A* and *B* show how cubes can be stacked for cubic and dodecahedral (12-sided) crystal forms. *C* and *D* show the relationship of stacked rhombohedrons to a "dog tooth" (scalenohedron) form and a rhombohedral face.

Cleavage

The internal order of a crystal may be expressed externally by crystal faces, or it may be indicated by the mineral's tendency to split apart along certain preferred directions. **Cleavage** is the ability of a mineral to break, when struck or split, along preferred planar directions.

A mineral tends to break along certain planes because the bonding between atoms is weaker there. In quartz, the bonds are equally strong in all directions; therefore, quartz has no cleavage. The micas, however, which are sheet silicates, are easily split apart into sheets (figure 2.21). If we could look at the arrangement of atoms in the crystalline structure of micas, we would see that the individual silicon-oxygen tetrahedra are strongly bonded to one another within each of the silicate sheets. The bonding *between* adjacent sheets, however, is very weak. Therefore, it is easy to split the mineral apart parallel to the plane of the sheets.

Cleavage is one of the most useful diagnostic tools because it is identical for a given mineral from one sample to another. Cleavage is especially useful for identifying minerals when they occur as small grains in rocks.

The wide variety of combinations of cleavage and *quality* of cleavage also increases the diagnostic value of this property. Mica has a single direction of cleavage, and its quality is perfect (figures 2.21*A* and 2.22*A*). Other minerals are characterized by one, two, or more cleavage directions; the quality can range from perfect to poor. (Poor cleavage is very hard for anyone but a well-trained mineralogist to detect.)

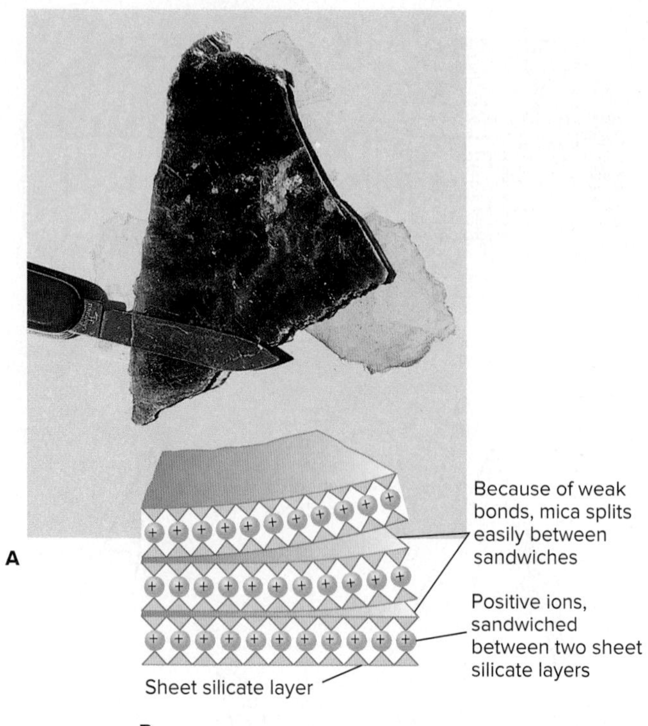

FIGURE 2.21

(*A*) Mica cleaves easily parallel to the knife blade. (*B*) Relationship of mica to cleavage. Mica crystal structure is simplified in this diagram.
(A) ©C. C. Plummer

Three of the most common mineral groups–the feldspars, the amphiboles, and the pyroxenes–have two directions of cleavage (figure 2.22*B* and *C*). In feldspars, the two directions are at angles of about 90 degrees to each other, and both directions are of very good quality. In pyroxenes, the two directions are also at about right angles, but the quality is only fair. In amphiboles (figure 2.23), the quality of the cleavage is very good and the two directions are at an angle of 56 degrees (or 124 degrees for the obtuse angle).

Halite is an example of a mineral with three excellent cleavage directions, all at 90 degrees to each other. This is called *cubic cleavage* (figure 2.22*D*). Halite's cleavage tells us that the bonds are weak in the planes parallel to the cube faces shown in figure 2.7. Take a close look at some grains of table salt. Notice that each grain is actually a tiny cube formed by breaking along halite's cleavage planes during crushing.

Calcite also has three cleavage directions, each excellent. But the angles between them are clearly not right angles. Calcite's cleavage is known as *rhombohedral* cleavage (figures 2.22*E* and 2.24).

Some minerals have more than three directions of cleavage. Fluorite, which grows into cubes, has very good cleavage in four directions. Cleavage fragments of fluorite have an octahedral shape. Sphalerite, the principal ore of zinc, has six cleavage directions.

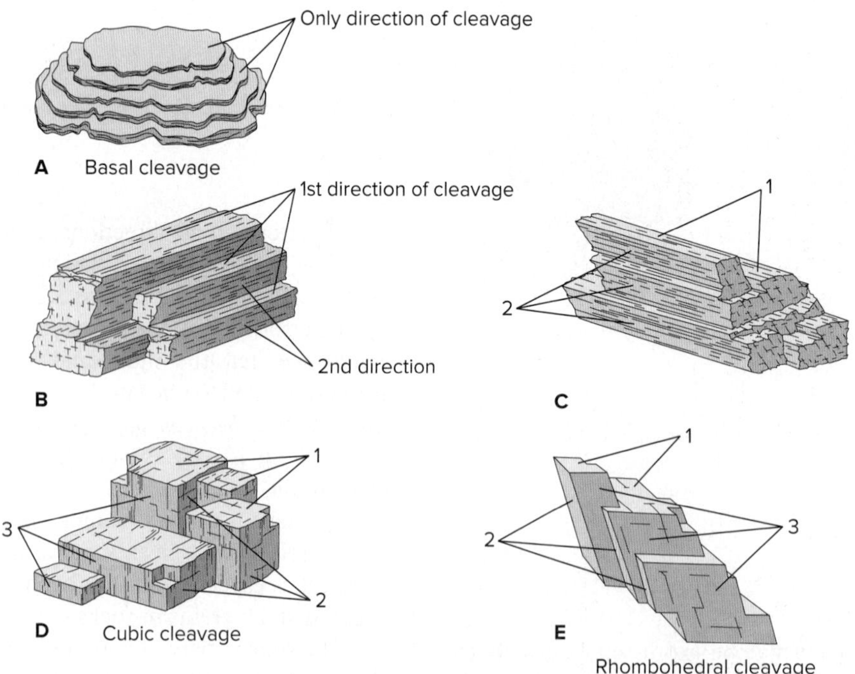

FIGURE 2.22

Most common types of mineral cleavage. Straight lines and flat planes represent cleavage. (*A*) One direction of cleavage (basal cleavage). Mica is an example. (*B*) Two directions of cleavage that intersect at 90° angles. Feldspar is an example. (*C*) Two directions of cleavage that do not intersect at 90° angles. Amphibole is an example. (*D*) Three directions of cleavage that intersect at 90° angles (cubic cleavage). Halite is an example. (*E*) Three directions of cleavage that do not intersect at 90° angles. Calcite is an example (calcite shows rhombohedral cleavage). Not shown are the two other possible types of cleavage—four directions (such as in fluorite) and six directions (as in sphalerite).

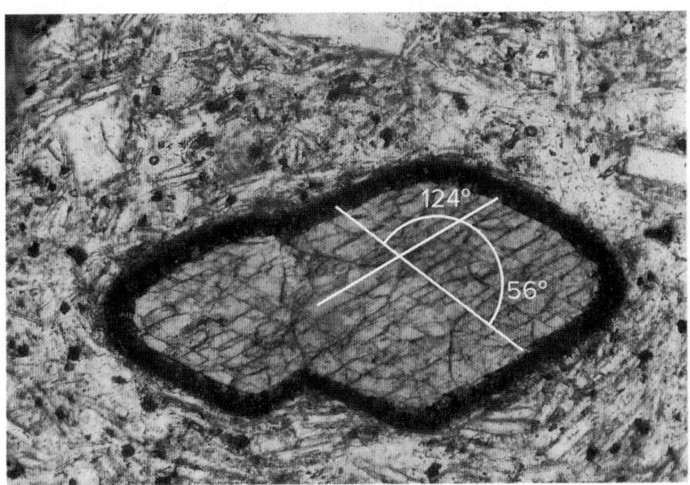

FIGURE 2.23

Amphibole cleavage as seen in a polarizing microscope.
©C. C. Plummer

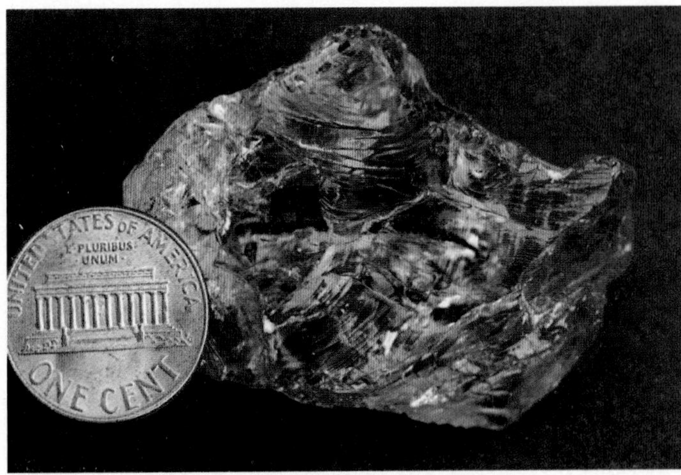

FIGURE 2.25

Conchoidal fracture in quartz.
©Marli Miller

FIGURE 2.24

Cleavage fragments of calcite.
©C. C. Plummer

Recognizing cleavage and determining angular relationships between cleavage directions take some practice. Students new to mineral identification tend to ignore cleavage because it is not as immediately apparent to the eye as color. But determining cleavage is frequently the key to identifying a mineral, so the small amount of practice needed to develop this skill is worthwhile.

Fracture

Fracture is the way a substance breaks where not controlled by cleavage. Minerals that have no cleavage commonly have an *irregular fracture.*

Some minerals break along curved fracture surfaces known as *conchoidal fractures* (figure 2.25). These look like the inside of a clam shell. This type of fracture is commonly observed in quartz and garnet (but these minerals also show irregular fractures). Conchoidal fracture is particularly common in glass, including obsidian (volcanic glass).

Minerals that have cleavage can fracture along directions other than that of the cleavage. The mica in figure 2.21*A* has irregular edges, which are fractures due to being torn perpendicular to the cleavage direction.

Specific Gravity

It is easy to tell that a brick is heavier than a loaf of bread just by hefting each of them. The brick has a higher **density**, weight per given volume, than the bread. Density is commonly expressed as **specific gravity**, the ratio of a mass of a substance to the mass of an equal volume of water.

Liquid water has a specific gravity of 1. (Ice, being lighter, has a specific gravity of about 0.9.) Most of the common silicate minerals are about two and a half to three times as dense as equal volumes of water: quartz has a specific gravity of 2.65; the feldspars range in specific gravity from 2.56 to 2.76. Special scales are needed to determine specific gravity precisely. However, a person can easily distinguish by hand very dense minerals such as galena (a lead sulfide with a specific gravity of 7.5) from the much less dense silicate minerals.

Gold, with a specific gravity of 19.3, is much denser than galena. Because of its high density, gold can be collected by panning. While the lighter clay and silt particles in the pan are sloshed out with the water, the gold lags behind in the bottom of the pan.

Special Properties

Some properties apply to only one mineral or to only a few minerals. Smell is one. Some clay minerals have a characteristic earthy smell when they are moistened. A few minerals have a distinctive taste. If you lick halite, it tastes salty because it is, of course, halite (more commonly known as salt).

Pyrite commonly exhibits **striations**–straight, parallel lines on the flat surfaces of crystal faces (figure 2.18*C*). The tourmaline crystal in the opening photograph for this chapter also displays striations.

The mineral *magnetite* (an iron oxide) owes its name to its characteristic physical property of being attracted to a magnet. Where large bodies of magnetite are found in the Earth's crust, compass needles point toward the magnetite body rather than to magnetic north. Airplanes navigating by compass have become lost because of the influence of large magnetite bodies. Some other minerals are weakly magnetic; their magnetism can be detected only by specialized magnetometers, similar to metal detectors in airports. Magnetism is important to modern civilization. We use magnets in computer hard drives, cell phones, and some speakers. In later chapters, you will see how magnetite in igneous rocks has preserved a record of Earth's magnetic field through geologic time; this has been an important part of the verification of plate tectonic theory. Some bacteria create magnetite, and this has been used to support the hypothesis that life has existed on Mars. Some researchers believe that migrating birds and animals have cells in their brains that contain small amounts of magnetite. They exploit the magnetic properties of magnetite to help them navigate during their migration.

Quartz has the property of generating electricity when squeezed in a certain crystallographic direction. This property, called piezoelectricity, relates to its use in quartz watches (see Web box 2.7).

Minerals have many other properties, including melting point, electrical and heat conductivity, and so on. Most are not relevant to introductory geology. Two categories of properties that are important are optical properties and the effects of X rays on minerals.

A clear crystal of calcite exhibits an unusual optical property. If you place transparent calcite over an image on paper, you will see two images (figure 2.26). This phenomenon is known as *double refraction* and is caused by light splitting into two components when it enters some crystalline materials. Each of the components is traveling through the mineral at different velocities. Most minerals possess double refraction, but it is usually slight and can only be observed using polarizing filters, notably in polarizing microscopes. Polarizing microscopes are very useful to professional geologists and advanced students for identifying minerals and interpreting how rocks formed. Photomicrographs elsewhere in this book were taken through polarizing microscopes (for example, figures 2.14 and 3.6*B*). Explaining optical phenomena such as this is beyond the scope of this book but, if interested, you can go to the Molecular Expressions Microscopy Primer site at http://micro.magnet.fsu.edu/primer/virtual/virtualpolarized.html.

Specialized equipment is needed to determine some properties. Perhaps most important are the characteristic effects of minerals on X rays, which we can explain only briefly here. X rays entering a crystalline substance are deflected by planes of atoms within the crystal. The X rays leave the crystal at precise and measurable angles controlled by the orientation of the planes of atoms that make up the internal crystalline structure (figure 2.27). The pattern of X rays exiting can be recorded on photographic film or by various recording instruments. Each mineral has its own pattern of reflected X rays, which serves as an identifying fingerprint.

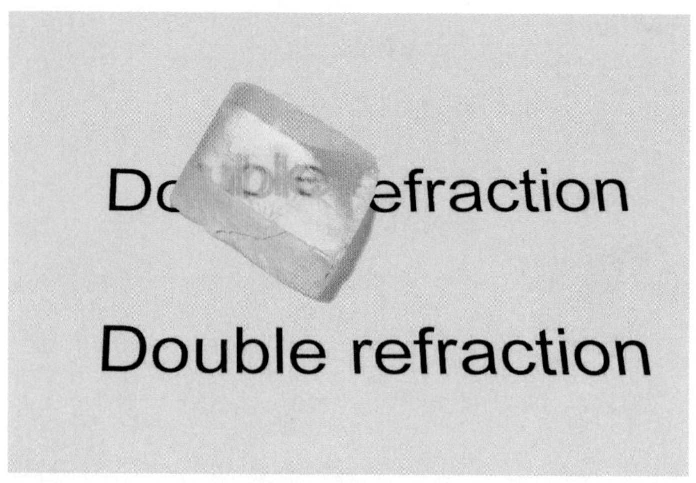

FIGURE 2.26

Double refraction in calcite. Two images of the letters are seen through the transparent calcite crystal.
©Dr. Parvinder Sethi

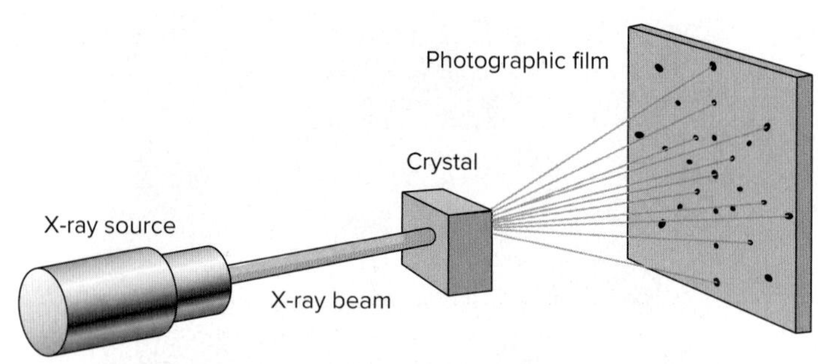

FIGURE 2.27

An X-ray beam passes through a crystal and is deflected by the rows of atoms into a pattern of beams. The dots exposed on the film are an orderly pattern used to identify the particular mineral.

WEB BOX 2.7

On Time with Quartz

Ever wonder why your watch has "quartz" printed on it? A small slice of quartz in the watch works to keep incredibly accurate time. This is because a small electric current applied to the quartz causes it to vibrate at a very precise rate (close to 100,000 vibrations per second).

To view an excellent video on how quartz is used to keep time, visit https://www.youtube.com/watch?v=1pM6uD8nePo.

IN GREATER DEPTH 2.8

Water and Ice—Molecules and Crystals

Earth is often called the *blue planet* because oceans cover 70% of its surface. Ice dominates our planet's polar regions. Perhaps "Aqua" rather than "Earth" would be a more appropriate name for our planet. It is fortunate that water is so abundant because life would be impossible without it. In fact, we humans are made up mostly of water. The nature and behavior of water molecules helps explain why water is vital to life on Earth.

In a water molecule, the two hydrogen atoms are covalently bonded tightly to the oxygen atom. However, the shape of the molecule is asymmetrical, with the two hydrogen atoms on the same side of the atom (box figure 1). This means the molecule is polarized, with a slight excess positive charge at the hydrogen side of the molecule and a slight excess negative charge at the opposite side. Because of the slight electrical attraction of water molecules, other substances are readily attracted to the molecules and dissolved or carried away by water. Water has been called the universal solvent. Dirt washes out of clothing; water, in blood, carries nutrients to our muscles and transports waste to our kidneys and out of our bodies.

When water is in its liquid state, the molecules are moving about. Because of the polarity, molecules are slightly attracted to one another. For this reason, water molecules are closer together than molecules in most other liquids. However, in ice the water molecules are not as tightly packed together as in liquid water.

When water freezes, positive ends of the water molecules are attracted to negative ends of adjacent water molecules (box figure 2). (This type of bonding is known as *hydrogen bonding*.) The result is an orderly, three-dimensional pattern that is hexagonal, as in a honeycomb (which explains the hexagonal shape of snowflakes). The openness of the honeycomblike, crystalline structure of ice contrasts with the more closely packed molecules in liquid water. This is the reason ice is less dense than liquid water. This is an unusual solid-liquid relationship. For most substances, the solid is denser than its liquid phase.

The fact that ice is less dense than liquid water has profound implications. Ice floats rather than sinks in liquid water. Icebergs float in the ocean. Lakes freeze from the top down. Ice on a lake surface acts as an insulating layer that retards the freezing of underlying water. If ice sank, lakes would freeze much more readily and thaw much more slowly. Our climate would be very different if ice sank. The Arctic Ocean surface freezes during the winter but only at its surface. If the ice were to sink, more ocean water would be exposed to the cold atmosphere and would freeze and sink. Eventually, the entire Arctic Ocean would freeze and would not thaw during the summer. If this were the case, life, as we know it, probably would not exist.

When water freezes, it expands. A bottled beverage placed in a freezer breaks its container upon freezing. When water trapped in cracks in rock freezes, it will expand and will help break up the rock (as explained in the chapter on weathering).

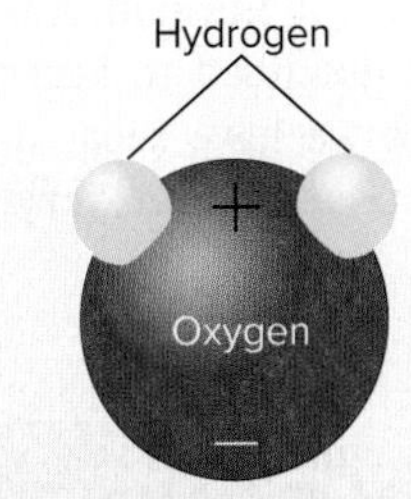

BOX 2.8 ■ FIGURE 1
Water molecule.

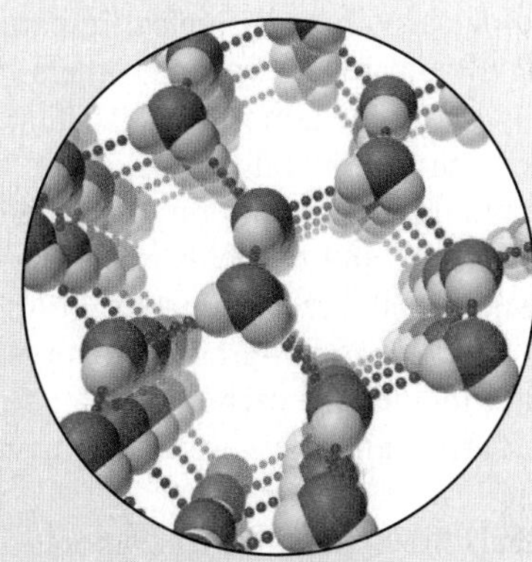

BOX 2.8 ■ FIGURE 2
Hexagonal structure of ice. Small, black dots represent the attraction between hydrogen atoms and oxygen atoms for adjacent water molecules.

Additional Resources

Ice research

The website for the Center for Ice and Climate in Denmark describes their ice-related research projects.

- http://www.iceandclimate.nbi.ku.dk/research/

Snow Crystals

Caltech's site. More about ice and nice pictures of snow crystals. Click Ice Properties under Snowflake Physics to see a model of the arrangement of oxygen and hydrogen atoms in the crystal structure of ice.

- www.its.caltech.edu/~atomic/snowcrystals/

Chemical Tests

One chemical reaction is routinely used for identifying minerals. The mineral calcite, as well as some other carbonate minerals (those containing CO_3 in their formulas), reacts with a weak acid to produce carbon dioxide gas. In this test, a drop of dilute hydrochloric acid applied to the sample of calcite bubbles vigorously, indicating that CO_2 gas is being formed.

Normally, this is the only chemical test that geologists do during field research.

Chemical analyses of minerals and rocks are done in labs using a wide range of techniques. A chemical analysis can accurately tell us the amount of each element present in a mineral. However, chemical analysis alone cannot be used to conclusively identify a mineral. We also need to know about the mineral's crystalline structure. As we have seen, diamond and graphite have an identical composition but very different crystalline structures.

2.5 THE MANY CONDITIONS OF MINERAL FORMATION

Minerals form under an enormously wide variety of conditions–most purely geological; others biological in nature. Some form tens of kilometers beneath the surface; others right at the surface and virtually out of the atmosphere itself.

The most common minerals are silicates, which incorporate the most abundant elements on Earth. Silicate minerals such as quartz, olivine, and the feldspars (plagioclase and potassium feldspar) crystallize primarily from molten rock (magma). They are *precipitates*–products of crystallizing liquid. Other precipitates include the carbonates calcite and aragonite, which grow in spring and cave waters and precipitate from ocean water.

Some minerals precipitate due to evaporation (e.g., halite). The very thick salt deposits underlying central Europe and the southern Great Plains exist because of the evaporation of seas millions of years ago.

Ice may be regarded as a very transient mineral at all but the coldest parts of Earth's surface. (Ice is a major crust-forming mineral on planets of the outer solar system, where it cannot melt; box 2.8).

Some minerals result from biological activity; for example, the building of coral reefs creates huge masses of calcite-rich limestone. Many organisms, including human beings, create magnetite within their skull cases. Bacteria also form huge amounts of sulfur by processing preexisting sulfate minerals. Most of our commercial supply of sulfur, in fact, comes from the mining of these *biogenic* deposits.

Some minerals crystallize directly from volcanic gases around volcanic vents–a process termed *sublimation.* Examples include ordinary sulfur, ralstonite, and thenardite (used as a natural rat poison). Sublimates are much less common than precipitates, though on planets and moons with intense volcanic activity, like Venus and Io, they cover wide swaths of planetary surface in thick beds.

We can understand the conditions of formation of most minerals with varying degrees of accuracy and precision using the tools of chemistry, especially with an understanding of thermodynamics and solutions. In fact, as implied at the beginning of this chapter, a good grasp of chemistry is a necessity for any advanced study of minerals.

Summary

Atoms are composed of *protons* (+), *neutrons,* and *electrons* (−). A given element always has the same number of protons. An atom in which the positive and negative electric charges do not balance is an *ion.*

Ions or atoms bond together in very orderly, three-dimensional structures that are *crystalline.*

A *mineral* is a crystalline substance that is naturally occurring, has a specific chemical composition, and forms through geologic processes. Minerals are the building blocks of rocks.

The two most abundant elements in the Earth's crust are oxygen and silicon. Most minerals are silicates, having the silicon-oxygen tetrahedron as their basic building block.

Minerals are usually identified by their physical properties. Cleavage is perhaps the most useful physical property for identification purposes. Other important physical properties are external crystal form, fracture, hardness, luster, color, streak, and specific gravity.

Terms to Remember

anion 30
atom 28
atomic mass 29
atomic mass number 28
atomic number 28
atomic weight 29
cation 30
chain silicate structure 34
cleavage 43
covalent bonding 30
crystal form 41
density 45
earthy luster 40
electron 28
element 28
ferromagnesian mineral 39
fracture 45
framework silicate structure 35
glassy (vitreous) luster 40
hardness 40
ion 30
ionic bonding 30
isolated silicate structure 34
isotope 28
luster 40
metallic bonding 31
metallic luster 40
mineral 26
Mohs' hardness scale 40
neutron 28
nonmetallic luster 40
nucleus 28
polymorph 31
proton 28
sheet silicate structure 35
silica 32
silicates 32
silicon-oxygen tetrahedron 33
specific gravity 45
streak 39
striations 46

Testing Your Knowledge

Use the following questions to prepare for exams based on this chapter.

1. Refer to the periodic table of the elements in appendix D and answer the following questions:
   a. Which element has an atomic number of 12?
   b. How many neutrons does it have?

c. How many electrons does a neutral atom of this element have?

d. Is it more likely to become an anion or a cation? Why?

2. The mineral fluorite (CaF_2) is composed of calcium (Ca^{2+}) and fluorine (F^-). What kind of bonding holds them together? Explain how you know this.

3. Using triangles to represent silica tetrahedra, start with a single triangle (to represent isolated silicate structure) and, by drawing more triangles, build on the triangle to show a single-chain silicate structure. By adding more triangles, convert that to a double-chain structure. Turn your double-chain structure into a sheet silicate structure.

4. Compare feldspar and quartz.

   a. How do they differ chemically?

   b. What type of silicate structure does each have?

   c. How would you distinguish between them on the basis of cleavage?

5. How do the crystal structures of pyroxenes and amphiboles differ from one another? Which physical property is used to distinguish between them?

6. Write out the definitions for the terms *element, mineral,* and *rock;* then for each of the following substances, determine whether it is an element, a mineral, or a rock.

   a. Iron b. Quartz
   c. Granite d. Limestone
   e. Calcium f. Calcite
   g. Obsidian h. Diamond
   i. Carbon

7. How would you distinguish between the following pairs of minerals on the basis of physical properties? (You might refer to appendix A.)

   olivine/pyroxene
   calcite/quartz
   mica/halite
   amphibole/hematite

8. Name the nonsilicate mineral groups and describe their defining chemical characteristics.

9. What are the eight most abundant elements in the Earth's crust? Table 2.1 gives the formulas for the most common minerals in the Earth's crust. How do these relate to the eight most abundant elements in the Earth's crust?

10. Describe the different conditions under which minerals form.

11. A substance that cannot be broken down into other substances by ordinary chemical methods is a(n)

    a. crystal. b. element.
    c. molecule. d. acid.

12. Which of these is *not* part of the definition of a mineral?

    a. Organic b. Crystalline
    c. Specific chemical composition d. Naturally occurring

13. The subatomic particle that contributes a single negative electrical charge is the

    a. proton. b. neutron.
    c. electron.

14. Atoms of an element containing different numbers of neutrons but the same number of protons are called

    a. ions. b. covalent.
    c. isotopes. d. neutral.

15. Atoms with either a positive or a negative charge are called

    a. compounds. b. ions.
    c. elements. d. isotopes.

16. The bonding between Cl and Na in halite is

    a. ionic. b. covalent.
    c. metallic.

17. Which is *not* true of a single silicon-oxygen tetrahedron?

    a. The atoms of the tetrahedron are strongly bonded together.
    b. It has a net negative charge.
    c. The formula is SiO_4.
    d. It has four silicon atoms.

18. In a single chain silicate, how many of the oxygen atoms in each silica tetrahedron are shared with neighboring silica tetrahedra?

    a. None b. One
    c. Two d. Three
    e. Four

19. Which of these common minerals is *not* a silicate?

    a. Quartz b. Feldspar
    c. Gypsum d. Mica

20. On Mohs' hardness scale, a fingernail has a hardness of about

    a. 2–3. b. 3–4.
    c. 5–6. d. 7–8.

21. The ability of a mineral to break along preferred directions is called

    a. fracture. b. crystal form.
    c. hardness. d. cleavage.

22. Crystalline substances are always

    a. ionically bonded. b. minerals.
    c. made of repeating patterns of atoms. d. made of glass.

23. Which of these minerals commonly forms as an evaporite (precipitate from evaporating water)?

    a. Quartz b. Hematite
    c. Halite d. Feldspar

Expanding Your Knowledge

1. Why are nonsilicate minerals more common on the surface of the Earth than within the crust?

2. How does oxygen in the atmosphere differ from oxygen in rocks and minerals?

3. What happens to the atoms in water when it freezes? Is ice a mineral? Is a glacier a rock?

4. How would you expect the appearance of a rock high in iron and magnesium to differ from a rock with very little iron and magnesium?

Exploring Web Resources

http://www.rockhounds.com/

Bob's Rock Shop. Contains a great amount of information for mineral collectors. Click on table of contents and then scroll down to find a mineral identification key and other useful pages, including "crystallography and mineral crystal systems," which is a more in-depth study of crystallography than that which is presented in this book.

http://www.webmineral.com/

Mineralogy Database. There are descriptions of close to 4,000 mineral species. The descriptions include mineral properties beyond the scope of an introductory geology course; however, there are links to other sites that include pictures of minerals. If you click on "mineral structures," then pick a mineral you have heard about from the list (e.g., calcite), you can click on various options. "Spin" will rotate the crystal structure. If you click on one or more of the elements listed, it will show those atoms in the structure.

http://www.mindat.org/

The *online mineralogy resource.* Another comprehensive source for mineral information. You can search for a mineral by name or by locality.

http://www.theimage.com/

The Image. Photos of minerals and gems. Click on Mineral Gallery and choose a mineral to view photos and properties of that mineral. The Gemstone Gallery has photos of gem minerals.

www.webelements.com/

Web elements periodic table. The periodic table of elements. You can click on an element to determine its properties.

www.uky.edu/Projects/Chemcomics/

The comic book periodic table of elements. An entertaining site in which you click on an element and see examples of comic book stories about that element.

CHAPTER 3

Igneous Rocks, the Origin and Evolution of Magma, and Intrusive Activity

The distinctive cliffs and domes of Yosemite Valley, including Half Dome shown on the right of this image, are all formed of the igneous rock granite.
©Dr. Parvinder Sethi

LEARNING OBJECTIVES

- Use the rock cycle to understand the relationships between igneous, sedimentary, and metamorphic rocks.
- Classify the most common igneous rock types using their texture and chemical composition.
- Understand the conditions under which rocks in the Earth's interior melt to form magma.
- Describe the magmatic processes that produce igneous rocks of varying composition.
- Differentiate between the various types of intrusive igneous structures.
- Relate magmatic activity to plate tectonic theory.

In chapter 2 you learned all about minerals. In the next few chapters you will learn about the three rock types (igneous, metamorphic, and sedimentary), how they form, and what they can tell us about processes in the interior of the Earth and on its surface. In this chapter we will explore **igneous rocks**–those formed when magma cools and solidifies. The study of igneous rocks provides us with an insight into the composition of the Earth's mantle and the processes that cause it to melt. The wide variety of igneous rocks found on the Earth's surface tells us about the processes that occur within magma chambers in the crust. Magmatic processes are the cause of spectacular volcanic eruptions (covered in chapter 4), and magma that has cooled and crystallized deep in the Earth's crust forms the backbones of many large mountain chains. For example, the awe-inspiring formations of Yosemite Valley shown in the opening image of this chapter are formed of granite, a rock that forms from magma cooling slowly beneath the Earth's surface.

We begin this chapter by introducing the rock cycle, a conceptual device that shows the interrelationship between igneous, sedimentary, and metamorphic rocks. We then discuss how igneous rocks are described in terms of their texture and composition, and how these properties are used to classify igneous rocks. Following this, we will focus on how igneous rocks are formed by examining the conditions under which the mantle melts to form magma, and the magmatic processes that produce the wide variety of igneous rocks seen on Earth's surface. We will then focus on intrusive igneous structures and their relationship to other rocks in the Earth's crust. Finally, we will conclude by relating igneous activity to plate tectonic theory.

3.1 THE ROCK CYCLE

A **rock** is naturally formed, consolidated material usually composed of grains of one or more minerals. You will see in chapter 5 how some minerals break down chemically and form new minerals when a rock finds itself in a new physical setting. For instance, feldspars that may have formed at high temperatures deep within the Earth can react with surface waters to become clay minerals at the Earth's surface.

As mentioned in chapter 1, the Earth changes because of its internal and external heat engines. If the Earth's internal engine had died (and tectonic forces had therefore stopped operating), the external engine plus gravity would long ago have leveled the continents virtually at sea level. The resulting sediment would have been deposited on the sea floor. Solid Earth would not be changing (except when struck by a meteorite or other extraterrestrial body). The rocks would be at rest. The minerals, water, and atmosphere would be in *equilibrium* (and geology would be a dull subject). But this is not the case. The internal and external forces continue to interact, forcing substances out of equilibrium. Therefore, the Earth has a highly varied and ever-changing surface. And minerals and rocks change as well.

A useful aid in visualizing these changing relationships is the **rock cycle** shown in figure 3.1. The three major rock types–igneous, metamorphic, and sedimentary–are shown. As you see, each may form at the expense of another if it is forced out of equilibrium with its physical or climatic environment by either internal or surficial forces. It is important to be aware that rock moves from deep to shallow, and from high to low temperature and pressure in response to tectonic forces and isostasy (covered in chapter 1).

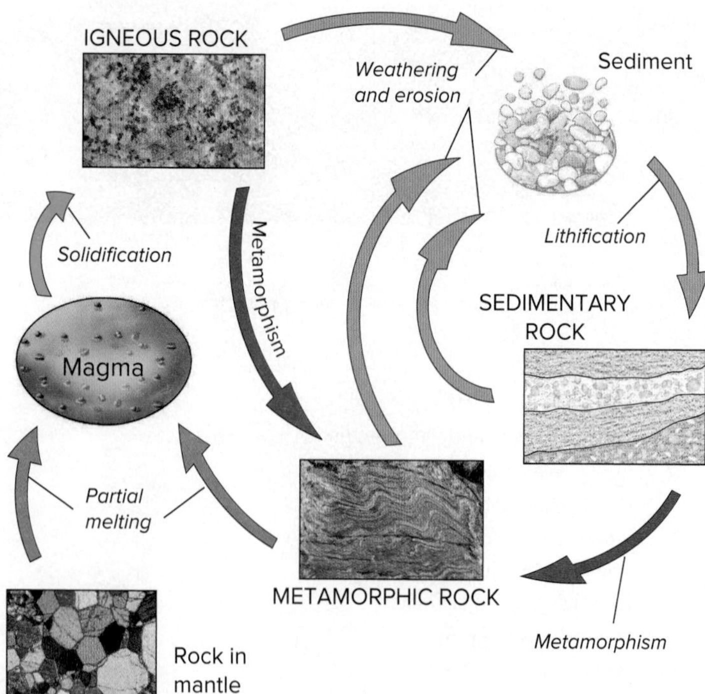

FIGURE 3.1

The rock cycle. The arrows indicate the processes whereby one kind of rock is changed to another. For clarity, arrows are not used to show that metamorphic rock can be re-metamorphosed to a different metamorphic rock or that igneous rock can be remelted to form new magma.

(A) ©Doug Sherman/Geofile RF (B) ©reneh9999/Shutterstock (C) ©Doug Sherman/Geofile RF

As described in chapter 1, *magma* is molten rock. (Magma may contain suspended solid crystals and gas.) *Igneous rocks* form when magma solidifies. If the magma is brought to the surface (where it is called lava) by a volcanic eruption, it will solidify into an *extrusive* igneous rock. Magma may also solidify very slowly beneath the surface. The resulting *intrusive* igneous rock may be exposed later after uplift and erosion remove the overlying rock (as shown in figure 1.14). The igneous rock, being out of equilibrium, may then undergo *weathering* and *erosion*, and the debris produced is transported and ultimately deposited (usually on a sea floor) as *sediment.* If the unconsolidated sediment becomes *lithified* (cemented or otherwise consolidated into a rock), it becomes a *sedimentary rock.* The rock is buried by additional layers of sediment and sedimentary rock. This process can only bury layered rock in the uppermost crust to a depth of several kilometers. Tectonic forces are required to transport sedimentary (and volcanic) rock to lower levels in the crust. Heat and pressure increase with increasing depth of burial. If the temperature and pressure become high enough, as occurs in the middle and lower levels of continental crust, the original sedimentary rock is no longer in equilibrium and recrystallizes. The new rock that forms is called a *metamorphic rock.* If the temperature gets very high, the rock partially melts, producing magma and completing the cycle.

The cycle can be repeated, as implied by the arrows in figure 3.1. However, there is no reason to expect all rocks to go through each step in the cycle. For instance, sedimentary rocks might be uplifted and exposed to weathering, creating new sediment.

We should emphasize that the rock cycle is a conceptual device to help students place the common rocks and how they form in perspective. As such, it is a simplification and does not encompass all geologic processes. For instance, most magma comes from partial melting of the mantle, rather than from recycled crustal rocks.

The Rock Cycle and Plate Tectonics

One way of relating the rock cycle to plate tectonics is illustrated by the example of what happens at a convergent plate boundary (figure 3.2). *Magma* is created in the zone of melting above the subduction zone. The magma, being less dense than adjacent rock, migrates upward toward the surface. A volcanic eruption takes place if magma reaches the surface. The magma solidifies into *igneous rock.* The igneous rock is exposed to the atmosphere

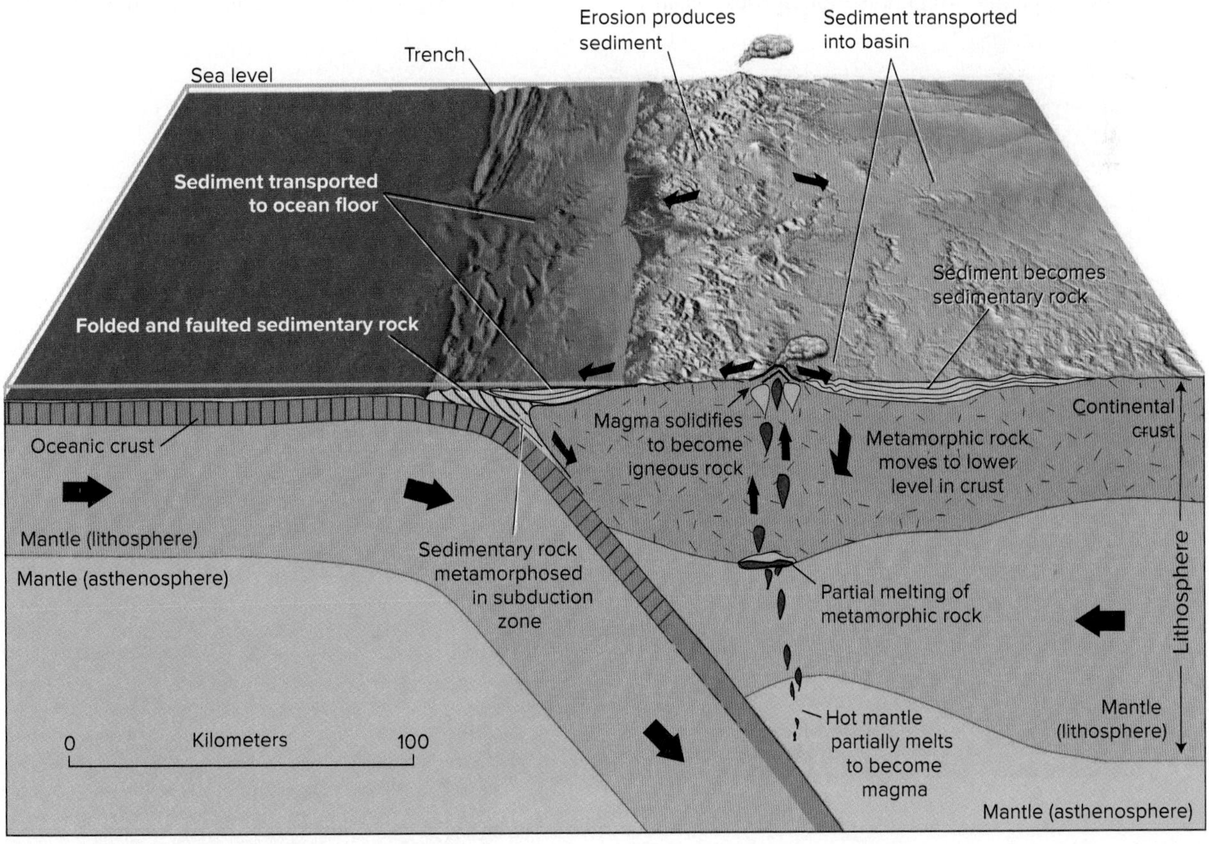

FIGURE 3.2

The rock cycle with respect to a convergent plate boundary. Magma formed within the mantle solidifies as igneous rock at the volcano. Sediment from the eroded volcano collects in the basin to the right of the diagram. Sediment converts to sedimentary rock as it is buried by more sediment. Deeply buried sedimentary rocks are metamorphosed. The most deeply buried metamorphic rocks partially melt, and the magma moves upward. An alternate way the rock cycle works is shown on the left of the diagram. Sediment from the continent (and volcano) becomes sedimentary rock, some of which is carried down the subduction zone. It is metamorphosed as it descends. It may contribute to the magma that forms in the mantle above the subduction zone.

and subjected to *weathering* and *erosion.* The resulting *sediment* is transported and then deposited in low-lying areas. In time, the buried layers of sediment lithify into *sedimentary rock.* The sedimentary rock becomes increasingly deeply buried as more sediment accumulates and tectonic forces push it deeper. After the sedimentary rock is buried to depths exceeding several kilometers, the heat and pressure become too great, and the rock recrystallizes into a *metamorphic rock.* As the depth of burial becomes even greater (several tens of kilometers), the metamorphic rock may find itself in a zone of melting. Temperatures are now high enough so that the metamorphic rock partially melts. Magma is created, thus completing the cycle.

The rock cycle diagram reappears on the opening pages of chapters 5 through 7. The highlighted portion of the diagram will indicate where the material covered in each chapter fits into the rock cycle.

3.2 IGNEOUS ROCKS

If you go to the island of Hawaii, you might observe red-hot lava flowing over the land and, as it cools, solidifying into the fine-grained (the grains are too small to be seen with the naked eye) black rock we call basalt. Basalt is an igneous rock, rock that has solidified from magma. **Magma** is molten rock, usually rich in silica and containing dissolved gases. (**Lava** is magma on the Earth's surface.) Igneous rocks may be either **extrusive** if they form at the Earth's surface (e.g., basalt) or **intrusive** if magma solidifies underground. **Granite**, a coarse-grained (the crystals are large enough to be seen with the naked eye) rock composed predominantly of feldspar and quartz, is an intrusive rock. In fact, granite is the most abundant intrusive rock found in the continents.

Unlike the volcanic rock in Hawaii, nobody has ever seen magma solidify into intrusive rock. So what evidence suggests that bodies of granite (and other intrusive rocks) solidified underground from magma?

- Mineralogically and chemically, intrusive rocks are essentially identical to volcanic rocks.
- Volcanic rocks are fine-grained or glassy due to their rapid solidification; intrusive rocks are generally coarse-grained, which indicates that the magma crystallized slowly underground. Experiments show that the slower cooling of liquids results in larger crystals.
- Experiments have confirmed that most of the minerals in these rocks can form only at high temperatures. Other experiments indicate that some of the minerals could have formed only under high pressures, implying they were deeply buried. More evidence comes from examining *intrusive*

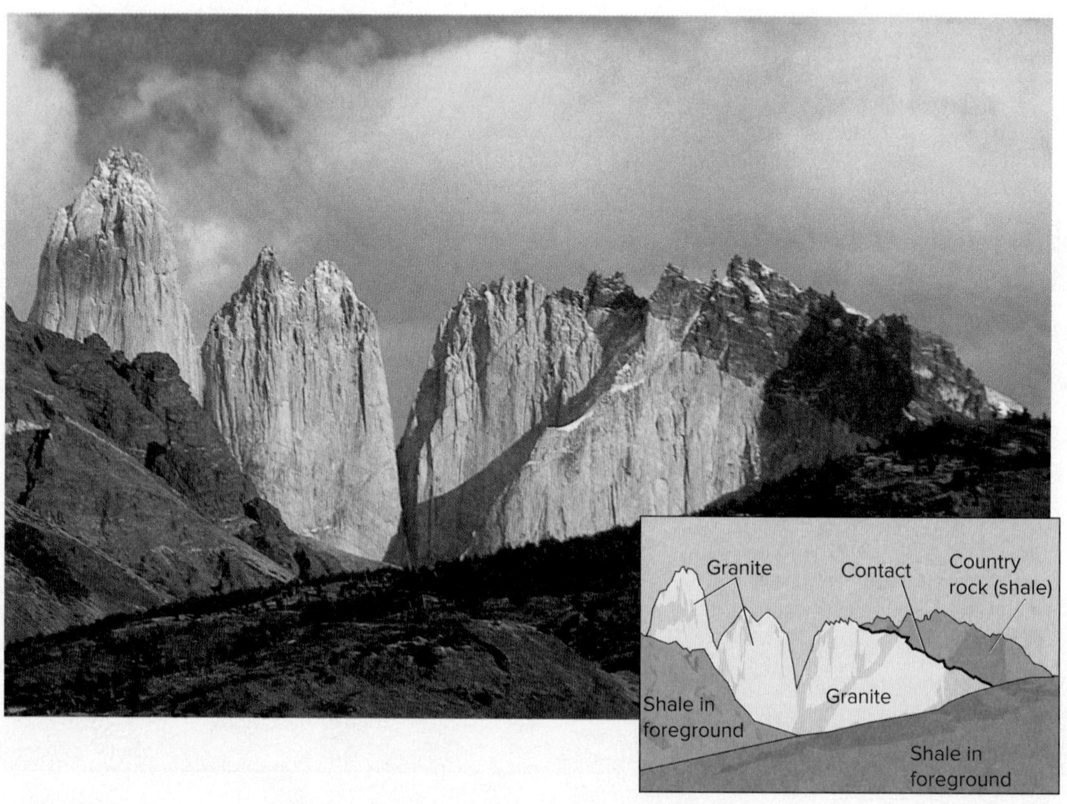

FIGURE 3.3

Granite (light-colored rock) solidified from magma that intruded dark-colored country rock in Torres del Paine, Chile. The dark-colored country rock is shale deposited in a marine environment. The spires are erosional remnants of rock that were once deep underground.
©Kay Kepler

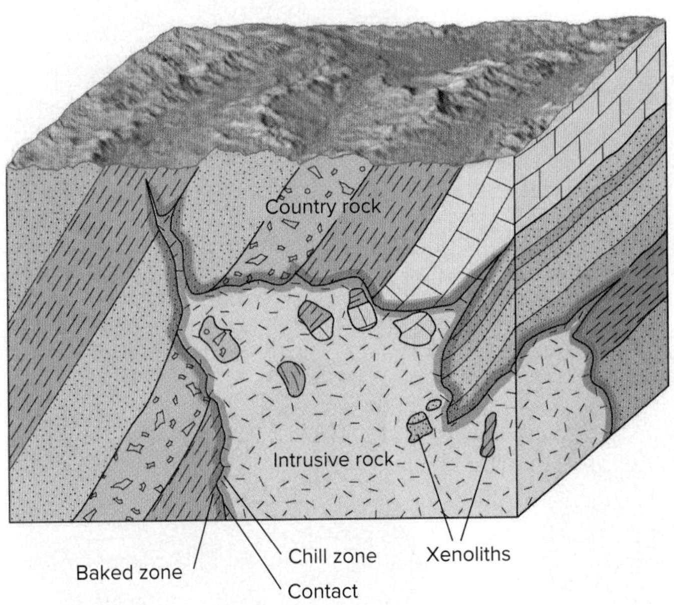

FIGURE 3.4

Igneous rock intruded preexisting rock (country rock) as a liquid. (Xenoliths are usually much smaller than indicated.)

contacts, such as those shown in figures 3.3 and 3.4. (A **contact** is a surface separating different rock types. Other types of contacts are described elsewhere in this book.)

- Preexisting solid rock, *country rock,* appears to have been forcibly broken by an intruding liquid, with the magma flowing into the fractures that developed. **Country rock**, incidentally, is an accepted term for any older rock into which an igneous body intruded.
- Close examination of the country rock immediately adjacent to the intrusive rock usually indicates that it appears "baked," or *metamorphosed,* close to the contact with the intrusive rock.
- Rock types of the country rock often match **xenoliths**, fragments of rock that are distinct from the body of igneous rocks in which they are enclosed.
- In the intrusive rock adjacent to contacts with country rock are **chill zones**, finer-grained rocks that indicate magma solidified more quickly here because of the rapid loss of heat to cooler rock.

Laboratory experiments have greatly increased our understanding of how igneous rocks form. However, geologists have not been able to make coarsely crystalline granite artificially. Only very fine-grained rocks containing the minerals of granite have been made from artificial magmas, or "melts." The temperature and pressure at which granite forms can be duplicated in the laboratory–but not the time element. According to calculations, a large body of magma requires over a million years to solidify completely. This very gradual cooling causes the coarse-grained texture of most intrusive rocks. Yet another problem in trying to apply experimental procedures to real rocks is determining the role of water and other gases in the crystallization of rocks such as granite. Only a small amount of gas is retained in rock crystallized underground from a magma, but large amounts of gas (especially water vapor) are released during volcanic eruptions. Laboratory experiments that involve melt solidification under gaseous conditions provide us with insight into the role that gases in underground magma might have played before they escaped. For example, laboratory studies have shown that the amount of water present in a granitic magma can control both the temperature range over which the granite crystallizes and the minerals that form from the magma as it cools. Later in this chapter, we will discuss how magma is formed and the processes that occur as it cools and crystallizes. But first, let's familiarize ourselves with igneous rocks and how they are described and named.

Classification of Igneous Rocks

Igneous rocks are named based on their texture and chemical composition. The texture of an igneous rock (its appearance in terms of grain size and presence of gas bubbles, for example) gives you information on where that rock formed, whether beneath the surface as an intrusive igneous rock or on the surface as an extrusive igneous rock. The chemistry of an igneous rock tells you about the origin of the magma and how it evolved before finally solidifying. Geologists infer the chemistry of an igneous rock by looking at the minerals it is composed of or, if the rock is too fine-grained to see individual minerals, by its color. Because magma of a single chemical composition can either cool slowly in a magma chamber, forming an intrusive rock, or cool rapidly on the surface, forming an extrusive rock, one name is given to the coarse-grained intrusive version and another to the fine-grained extrusive version. For example, earlier we described granite, a coarse-grained rock composed predominantly of feldspar and quartz. If a magma of the same composition as granite were to erupt onto the surface and cool rapidly, it would still form a rock composed predominantly of feldspar and quartz, but it would be fine-grained. This fine-grained rock would be called rhyolite. Some rocks have a very distinctive texture and are named for that texture. For example, some rhyolite cools so quickly it forms a black glass. This rock is known as obsidian.

Figure 3.5 shows the relationship between texture, chemistry, mineral content, and igneous rock names. Pictures of the most common igneous rocks are shown. You should examine this figure carefully while reading the following sections on igneous rock textures, chemistry of igneous rocks, and identifying igneous rocks.

Igneous Rock Textures

Texture refers to a rock's appearance with respect to the size, shape, and arrangement of its grains or other constituents. Igneous textures can be divided into three groups: **crystalline rocks** are made up of interlocking crystals (of, for instance, the minerals quartz and feldspar); **glassy rocks** are composed primarily of glass and contain few, if any, crystals; and **fragmental rocks** are

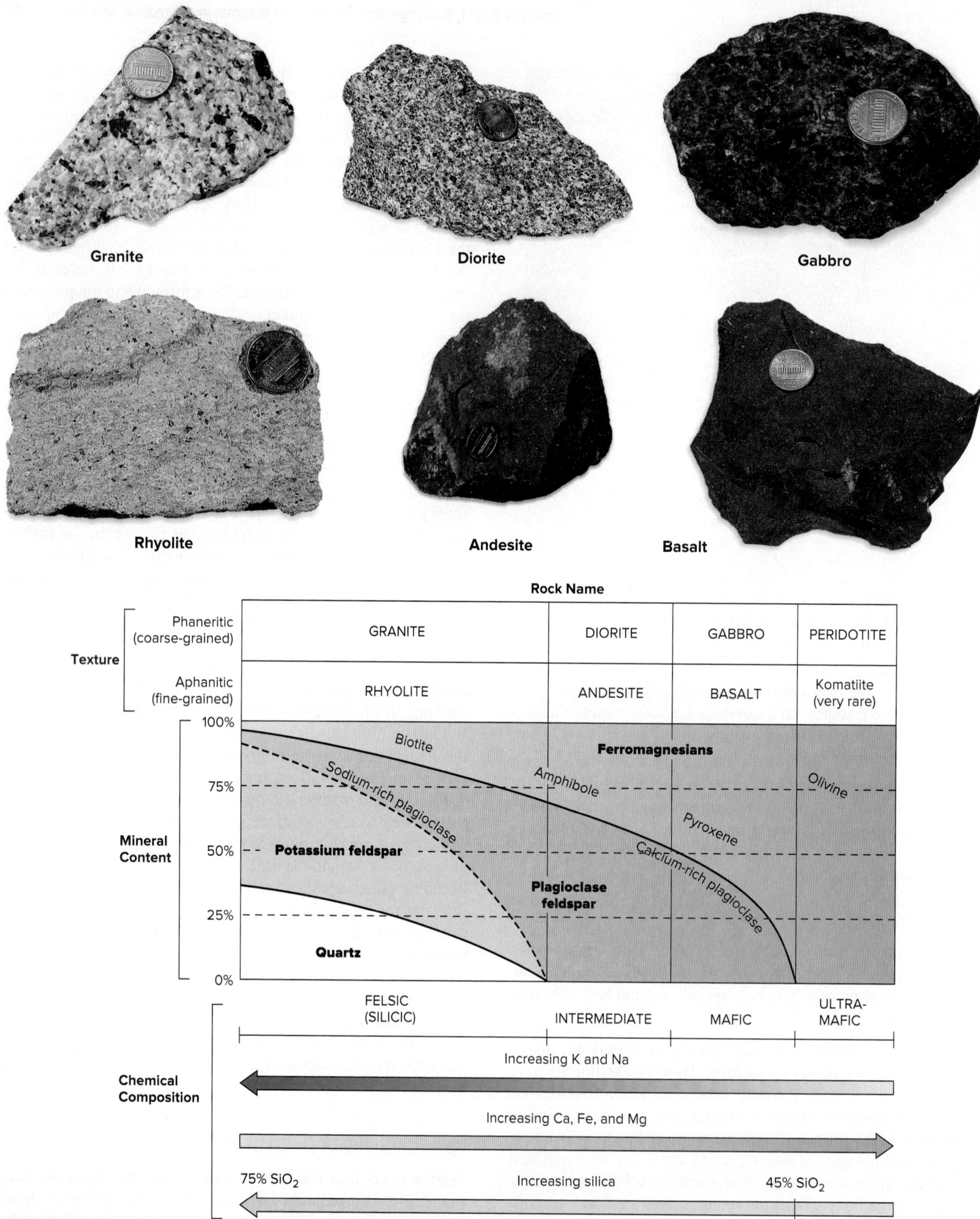

FIGURE 3.5

Classification chart for the most common igneous rocks. Rock names based on special textures are not shown. Sodium-rich plagioclase is associated with silicic rocks, whereas calcium-rich plagioclase is associated with mafic rocks. The names of the particular ferromagnesian minerals (biotite, etc.) are placed in the diagram at the approximate composition of the rocks in which they are most likely to be found. Samples of common igneous rocks are shown for reference. Do not try to identify real rocks by simply comparing them to photos—use the properties, such as identifying minerals and their amounts. ©C. C. Plummer

composed of fragments of igneous material. As you will see, the texture of an igneous rock can tell you a lot about where and how it formed.

Crystalline Textures

Most igneous rocks form from the cooling and solidification of magma. Crystals form and grow as the melt cools, and the result is a crystalline texture. The most significant aspect of texture in crystalline igneous rocks is grain (or crystal) size. Two critical factors determine grain size during the solidification of igneous rocks: rate of cooling and viscosity. If magma cools rapidly, the atoms have time to move only a short distance; they bond with nearby atoms, forming only small crystals. Grain size is controlled to a lesser extent by the viscosity–or resistance to flow–of the magma. Atoms in highly viscous lava cannot move as freely as those in a more fluid lava. Hence, a rock formed from viscous lava is more likely to be finer grained than one formed from more fluid lava.

Extrusive rocks are typically fine-grained or **aphanitic** (*a*, "not"; *phaner*, "visible"); that is, their crystals are too small to see easily with the naked eye. The grains, if they are crystals, are small because magma cools rapidly at the Earth's surface, so they have less time to form. Some intrusive rocks are also fine-grained; these occur as smaller bodies that apparently solidified near the surface upon intrusion into relatively cold country rock (probably within a couple kilometers of the Earth's surface). Igneous rocks that formed at considerable depth–usually more than several kilometers–are called **plutonic rocks** (after Pluto, the Roman god of the underworld). Characteristically, these rocks are coarse-grained, reflecting the slow cooling and solidification of magma.

For our purposes, coarse-grained or **phaneritic** (from *phaner*, meaning visible) rocks are defined as those in which the crystals are large enough to be seen easily with the naked eye. The crystals of plutonic rocks are commonly interlocked in a mosaic pattern (figure 3.6). An extremely coarse-grained (crystals ranging in size from a few centimeters to several meters in length) igneous rock is called a **pegmatite** (see box 3.1).

So, for practical purposes, if you can see the individual crystals in an igneous rock, you can regard it as phaneritic. If not, consider it aphanitic.

Some rocks contain **phenocrysts**, larger crystals that are enclosed in a *groundmass* of finer-grained crystals or glass. This is called a **porphyritic** texture. A milk chocolate bar containing whole almonds has the appearance of porphyritic texture where the chocolate represents the finer-grained groundmass and the almonds represent the phenocrysts. If the groundmass is aphanitic, the rock is considered to be extrusive. For instance, figure 3.7 shows an extrusive rock called *porphyritic andesite.* Porphyritic extrusive rocks are usually interpreted as having begun crystallizing slowly underground within a magma chamber (allowing time for some crystals to grow), followed by eruption and rapid solidification of the remaining magma at the Earth's surface. Some porphyritic rocks have a phaneritic groundmass. The larger phenocrysts enclosed in the groundmass are much bigger, usually 2 or more centimeters across. These rocks are considered to be intrusive. *Porphyritic granite* is an example.

A

B

FIGURE 3.6

(*A*) Phaneritic (coarse-grained) texture characteristic of plutonic rock. Feldspars are white and pink. Although this quartz is transparent, it appears light gray. Biotite mica is black. A U.S. penny is used for a scale because the roof of the monument is 1 millimeter thick and 1 centimeter wide. (*B*) A similar rock seen through a polarizing microscope. Note the interlocking crystal grains of individual minerals.
(A) ©C. C. Plummer (B) ©Lisa Hammersley

Glassy Textures

With extremely rapid or almost instantaneous cooling, individual atoms in the lava are "frozen" in place, forming glass rather than crystals. Remember, you learned in chapter 2 that, by definition, minerals have a crystalline structure where the atoms are arranged in an orderly, repeating, three-dimensional pattern. In a glass, the atoms are not ordered, so minerals are not present. **Obsidian** (figure 3.8), which is a dark volcanic glass, is one of the few rocks that is not composed of minerals. Because obsidian breaks with a conchoidal fracture (see chapter 2), it can be broken in such a way as to produce very sharp edges. The manufacture of obsidian tools dates back to the Stone Age. Today, some surgeons use obsidian for scalpel blades as the cutting edge can be many times sharper than steel surgical scalpels.

IN GREATER DEPTH 3.1

Pegmatite—A Rock Made of Giant Crystals

Pegmatites are extremely coarse-grained igneous rocks (see box figure 1). In some pegmatites, crystals are as large as 10 meters across. Strictly speaking, a pegmatite can be of diorite, gabbro, or granite. However, the vast majority of pegmatites are silica-rich, with very large crystals of potassium feldspar, sodium-rich plagioclase feldspars, and quartz. Hence, the term *pegmatite* generally refers to a rock of granitic composition (if otherwise, a term such as *gabbroic pegmatite* is used). Pegmatites are interesting as geological phenomena and important as minable resources.

The extremely coarse texture of pegmatites is attributed to both slow cooling and the low viscosity (resistance to flow) of the fluid from which they form. Lava solidifying to rhyolite is very viscous. Magma solidifying to granite, being chemically similar, should be equally viscous. Pegmatites, however, probably crystallize from a fluid composed largely of water. Geologists believe the following sequence of events accounts for most pegmatites.

As a granitic pluton cools, increasing amounts of the magma solidify into the minerals of a granite. By the time the pluton is well over 90% solid, the remaining (or *residual*) magma contains a very high amount of silica and ions of elements that will crystallize into potassium and sodium feldspars. Also present are elements that could not be accommodated into the crystal structures of the common minerals that formed during the normal solidification phase of the pluton. Fluids, notably water, that were in the original magma are left over as well. If no fracture above the pluton permits the fluids to escape, they are sealed in, as in a pressure cooker. The watery residual magma has a low viscosity, which allows atoms to migrate easily toward growing crystals. The crystals add more and more atoms and grow very large.

BOX 3.1 ■ FIGURE 1

Pegmatite in northern Victoria Land, Antarctica. The knife is 8 centimeters long. The black crystals are tourmaline. Quartz and feldspar are light colored.
©C. C. Plummer

Pegmatite bodies are generally quite small. Many are podlike structures, located either within the upper portion of a granite pluton or within the overlying country rock near the contact with granite, the fluid body evidently having squeezed into the country rock before solidifying. Pegmatite dikes are fairly common, especially within granite plutons, where they apparently filled cracks that developed in the already solid granite. Some pegmatites form small dikes along contacts between granite and country rock, filling cracks that developed as the cooling granite pluton contracted.

Most pegmatites contain only quartz, feldspar, and perhaps mica. Minerals of considerable commercial value are found in a few pegmatites. Large crystals of muscovite mica are mined from pegmatites. These crystals are called "books" because the cleavage flakes (tens of centimeters across) look like pages. Because muscovite is an excellent insulator, the cleavage sheets are used in electrical devices, such as toasters, to separate uninsulated electrical wires. Even the large feldspar crystals in pegmatites are mined for various industrial uses, notably the manufacture of ceramics.

Many rare elements are mined from pegmatites. These elements were not absorbed by the minerals of the main pluton and so were concentrated in the residual pegmatitic magma, where they crystallized as constituents of unusual minerals. Minerals containing the element lithium are mined from pegmatites. Lithium becomes part of a sheet silicate structure to form a pink or purple variety of mica called lepidolite. Uranium ores, similarly concentrated in the residual melt of magmas, are also extracted from pegmatites.

Some pegmatites are mined for gemstones. Emerald and aquamarine, varieties of the mineral beryl, occur in pegmatites that crystallized from a solution containing the element beryllium. A large number of the world's very rare minerals are found only in pegmatites, many of these in only one known pegmatite body. These rare minerals are mainly of interest to collectors and museums.

Hydrothermal veins (described in chapter 7) are closely related to pegmatites. Veins of quartz are common in country rock near granite. Many of these are believed to be formed from water that escapes from the magma. Silica dissolved in the very hot water crystallizes on the walls of cracks as the water cools while traveling surfaceward. Sometimes valuable metals such as gold, silver, lead, zinc, and copper are deposited with the quartz in veins. (See chapter 7 for more on veins.)

Textures Due to Trapped Gas

Magma stored deep underground is under high pressure, generally high enough to keep all its gases in a dissolved state. On eruption, the pressure is suddenly released and the gases come out of solution. This is analogous to what happens when a bottle of beer or soda is opened. Because the drink was bottled under pressure, the gas (carbon dioxide) is in solution. Uncapping the drink relieves the pressure, and the carbon dioxide separates from the liquid as gas bubbles. If you freeze the newly opened drink very quickly, you have a piece of ice with small,

A

B

FIGURE 3.7

Porphyritic andesite. A few large crystals (phenocrysts) are surrounded by a great number of fine grains. (*A*) Hand specimen. Grains in groundmass are too fine to see. (*B*) Photomicrograph (using polarized light) of a similar rock. The black-and-white striped phenocrysts are plagioclase, and the yellow ones are pyroxene.
(A) ©Dr. Parvinder Sethi (B) ©Lisa Hammersley

FIGURE 3.8

Obsidian. Notice the curved surfaces formed by conchoidal fracture.
©C. C. Plummer

FIGURE 3.9

Vesicular basalt.
©Dr. Parvinder Sethi

bubble-shaped holes. Similarly, when a lava solidifies while gas is bubbling through it, holes are trapped in the rock, creating a distinctive vesicular texture. **Vesicles** are cavities in extrusive rock resulting from gas bubbles that were in lava, and the texture is called *vesicular.* A vesicular rock has the appearance of Swiss cheese (whose texture is caused by trapped carbon dioxide gas). *Vesicular basalt* is quite common (figure 3.9). **Scoria**, a highly vesicular basalt, actually contains more gas space than rock.

In more viscous lavas, where the gas cannot escape as easily, the lava is churned into a froth (like the head on a glass of beer). When cooled quickly, it forms **pumice** (figure 3.10), a frothy glass with so much void space that it floats in water. Powdered pumice is used as an abrasive because it can scratch metal or glass. For example, powdered pumice is used in the production of pencil erasers. You may also have a piece of pumice in your shower that you use to buff the bottom of your feet.

Fragmental Texture

When trapped gases are released during an eruption, it can lead to lava being blasted out of the vent as fragments of rock known as **pyroclasts** (from the Greek *pyro,* "fire," and *clast,* "broken"). *Pyroclastic debris* is also known as *tephra.* When pyroclastic

A

B

mm

FIGURE 3.10

(*A*) A boulder of pumice can be easily carried because it is mostly air. (*B*) Seen close up, pumice is a froth of volcanic glass.
(*A*) ©*Diane Carlson* (*B*) ©*C. C. Plummer*

FIGURE 3.11

Volcanic tuff with a quarter for scale.
©*Dr. Parvinder Sethi*

material (ash, pumice, or crystalline rock fragments) accumulates and is cemented or otherwise consolidated, the new rock is called **tuff**, or **volcanic breccia**, depending on the size of the fragments. A tuff (figure 3.11) is a rock composed of fine-grained pyroclastic particles (dust and ash). A volcanic breccia is a rock that includes larger pieces of volcanic rock.

Chemistry of Igneous Rocks

The chemical composition of magma determines which minerals and how much of each will crystallize when an igneous rock forms. Because magma contains quite a lot of silicon and oxygen, igneous rocks are composed primarily of the silicate minerals quartz, plagioclase feldspar, potassium feldspar, amphibole, pyroxene, biotite, and olivine. The lower part of figure 3.5 shows the relationship of chemical composition to mineral composition. A magma that is rich in silica (SiO_2), aluminum, potassium, and sodium will crystallize minerals that contain those elements (feldspar and quartz). A magma that is rich in iron, magnesium and calcium will contain a lot of the dark-colored ferromagnesian minerals (pyroxene, amphibole, olivine, and biotite).

Chemical analyses of rocks are reported as weight percentages of oxides (e.g., SiO_2, MgO, Na_2O, etc.) rather than as separate elements (e.g., Si, O, Mg, Na). For virtually all igneous rocks, SiO_2 (silica) is the most abundant component. The amount of SiO_2 varies from about 45% to 75% of the total weight of common igneous rocks. Based on the amount of silica, igneous rocks are classified into four groups, which are, in order of decreasing silica content, felsic, intermediate, mafic, and ultramafic. Figure 3.12 shows the average chemical composition for felsic, intermediate, and mafic rocks. It is also useful to refer to figure 3.5 as you read the following sections because it shows the minerals associated with each of the four groups of igneous rocks as well as pictures of the major igneous rock types described next.

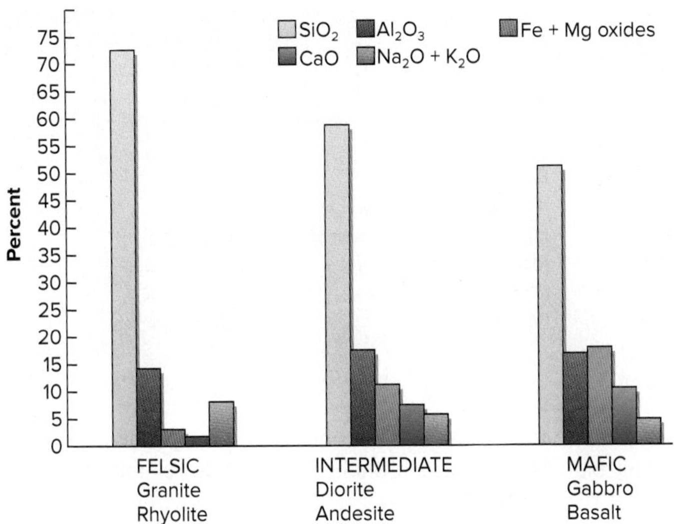

FIGURE 3.12

The average chemical composition of felsic, intermediate, and mafic rocks. Composition is given in weight percent of oxides. Note that as the amount of silica decreases, the oxides of Na and K decrease, and the oxides of Ca, Fe, and Mg increase. Al oxide does not vary significantly.

Felsic Rocks

Rocks with a silica content of 65% or more (by weight) are considered to be silica-rich (figure 3.12). The remaining 25% to 35% of these rocks is mostly aluminum oxide (Al_2O_3) and oxides of sodium (Na_2O) and potassium (K_2O), and they tend to contain only very small amounts of the oxides of calcium (CaO), magnesium (MgO), and iron (FeO and Fe_2O_3). These are called **felsic** rocks–silica-rich igneous rocks with a relatively high content of potassium and sodium (the *fel* part of the name comes from *feldspar,* which crystallizes from the potassium, sodium, aluminum, and silicon oxides; *si* in *felsic* is for silica). **Rhyolite** is a fine-grained, extrusive felsic rock. Granite is the coarse-grained intrusive equivalent of rhyolite.

Felsic rocks tend to be dominated by the light-colored minerals quartz, potassium feldspar, and plagioclase feldspar, with only small amounts of the dark ferromagnesian minerals. Because of this, felsic rocks are commonly light in color. A notable exception to this rule is the black glassy volcanic rock *obsidian.* So why is obsidian black–a color we usually associate with mafic rocks, such as basalt? If you look at a very thin edge of obsidian, it is transparent. Obsidian is, indeed, a form of stained glass. The black overall color is due to dispersion of extremely tiny magnetite crystals throughout the rock. Collectively, they act like pigment in ink or paint and give an otherwise clear substance color. For red obsidian, the magnetite (Fe_3O_4) has been exposed to air and has been oxidized to hematite (Fe_2O_3), which is red or red-brown.

Intermediate Rocks

Rocks with a silica content of between 55% and 65% are classified as **intermediate rocks**. Intermediate rocks contain significant amounts (30–50%) of dark ferromagnesian minerals like pyroxene and amphibole as well as light-colored plagioclase feldspar and small amounts of quartz. These can easily be discerned in **diorite**, the coarse-grained intermediate rock. Examine the picture of diorite in figure 3.5. You can see that about half the mineral grains are dark in color and the other half are white. **Andesite**, the fine-grained intermediate rock, is typically medium-gray or greenish-gray in color.

Mafic Rocks

Rocks with a silica content of between 45% and 55% (by weight) are considered *silica-poor,* even though SiO_2 is, by far, the most abundant constituent (figure 3.12). The remainder is composed mostly of the oxides of aluminum, calcium, magnesium, and iron. Rocks in this group are called **mafic**–silica-deficient igneous rocks with a relatively high content of magnesium, iron, and calcium. (The term *mafic* comes from *magnesium* and *ferrum,* the Latin word for iron.) Mafic rocks are made up predominantly of gray plagioclase feldspar and the ferromagnesian minerals pyroxene and olivine and tend to be dark in color. Mafic magma that cools slowly beneath the surface forms the coarse-grained, intrusive rock **gabbro**. If mafic magma erupts on the surface, it forms the dark, fine-grained, extrusive rock **basalt**.

Ultramafic Rocks

An **ultramafic rock** is one that contains less than 45% silica and is rich in iron, magnesium, and calcium. Ultramafic rocks are typically composed almost entirely of the ferromagnesian minerals olivine and pyroxene. No feldspars are present and, of course, no quartz. Note from the chart in figure 3.5 that **komatiite**, the volcanic ultramafic rock, is very rare. Ultramafic extrusive rocks are mostly restricted to the very early history of the Earth. For our purposes, they need not be discussed further. **Peridotite**, the coarse-grained intrusive rock, is composed of olivine and pyroxene and is the most abundant ultramafic rock.

Most ultramafic rocks come from the mantle rather than from the Earth's crust. Where we find large bodies of ultramafic rocks, the usual interpretation is that a part of the mantle has traveled upward as solid rock. In some cases, ultramafic rocks can form when ferromagnesian minerals crystallize in a mafic magma and settle down to the base of the magma chamber where they accumulate. This process, called crystal settling, is discussed later in this chapter.

Identifying Igneous Rocks

In order to identify an igneous rock, you must consider both its texture and the minerals it contains. Because of their larger mineral grains, plutonic rocks are the easiest to identify. The physical properties of each mineral in a plutonic rock can be determined more readily. And, of course, knowing what minerals are present makes rock identification a simpler task. For instance, gabbro is formed of coarse-grained ferromagnesian minerals and gray plagioclase feldspar. One can positively identify the feldspar on the basis of cleavage and, with practice, verify that no quartz is present. Gabbro's aphanitic counterpart is

TABLE 3.1 Identification of Most Common Igneous Rocks

	Felsic	Intermediate	Mafic	Ultramafic
Phaneritic (coarse-grained)	Granite	Diorite	Gabbro	Peridotite
Aphanitic (fine-grained)	Rhyolite	Andesite	Basalt	—
Mineral Content	Quartz, feldspars (white, light gray, or pink). Minor ferromagnesian minerals.	Feldspars (white or gray) and about 35–50% ferromagnesian minerals. No quartz.	Predominance of ferro-magnesian minerals. Rest of rock is plagioclase feldspar (medium to dark gray).	Entirely ferromagnesian minerals (olivine and pyroxene).
Color of Rock (most commonly)	Light-colored	Medium-gray or medium-green	Dark gray to black	Green to black

basalt, which is also composed of ferromagnesian minerals and plagioclase. The individual minerals cannot be identified by the naked eye, however, and one must use the less reliable attribute of color—basalt is usually dark gray to black.

As you can see from figure 3.5, *granite* and *rhyolite* are composed predominantly of feldspars (usually white or pink) and quartz. Granite, being phaneritic, can be positively identified by verifying that quartz is present. Rhyolite is usually cream-colored, tan, or pink. Its light color indicates that ferromagnesian minerals are not abundant. *Diorite* and *andesite* are composed of feldspars and significant amounts of ferromagnesian minerals (30–50%). The minerals can be identified and their percentages estimated to indicate diorite. Andesite, being aphanitic, can tentatively be identified by its medium-gray or medium-green color. Its appearance is intermediate between light-colored rhyolite and dark basalt.

Use the chart in figure 3.5 along with table 3.1 to identify the most common igneous rocks. You may also find it helpful to turn to appendix B, which includes a key for identifying common igneous rocks.

3.3 HOW MAGMA FORMS

Now that you've learned about the common igneous rocks and how they vary in terms of texture and composition, it is time to explore how magma forms and what happens to it as it moves through the Earth's crust. Let's begin with the question of how magma forms.

A common misconception is that the lava erupted from volcanoes comes from an "ocean of magma" beneath the crust. However, as you have already learned in chapter 1, the mantle is not molten but solid rock. In order to understand how magma forms, we must consider the source of heat for melting rock, the conditions at depth beneath the Earth's surface, and the conditions under which rocks in the mantle and lower crust will melt.

Heat for Melting Rock

Most of the heat that contributes to the generation of magma comes from the very hot Earth's core (where temperatures are estimated to be greater than 5,000°C). Heat is conducted toward the Earth's surface through the mantle and crust. This is comparable to the way heat is conducted through the metal of a frying pan. Heat is also brought from the lower mantle when part of the mantle flows upward, either through convection (described in chapters 1 and 19) or by hot mantle plumes.

The Geothermal Gradient and Partial Melting

A miner descending a mine shaft will notice a rise in temperature. This is due to the **geothermal gradient**, the rate at which temperature increases with increasing depth beneath the surface. Data show the geothermal gradient, on average, to be about 3°C for each 100 meters (30°C/km) of depth in the upper part of the crust, decreasing in the mantle. Figure 3.13 shows the geothermal gradient for the crust and upper mantle.

Unlike ice, which has a single melting point, rocks melt over a range of temperatures. This is because they are made up of more than one mineral, and each mineral has its own melting point. Figure 3.13 shows the range of temperatures over which rocks will

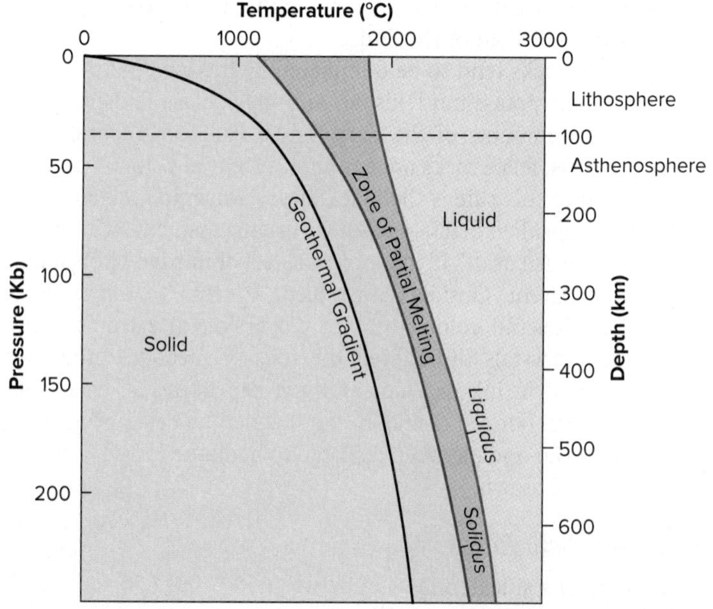

FIGURE 3.13

Geothermal gradient and zone of partial melting for mantle peridotite.

melt below the Earth's surface. The dark blue line labeled *solidus* shows the temperature at any given depth, below which the rocks are completely solid. The dark orange line labeled *liquidus* shows the temperature above which the rocks are completely liquid. Between the solidus and the liquidus lies the zone of partial melting, within which the rocks are partly solid and partly molten.

It is important to notice that in figure 3.13 the geothermal gradient does not intersect the zone of partial melting. At all depths, the temperature is not high enough to allow the rock to melt, and no magma is forming. This is typical of the mantle in most locations. In order for magma to form, conditions must change so that the geothermal gradient can intersect the zone of partial melting. The two most common mechanisms believed to create these conditions are decompression melting and the addition of water.

Decompression Melting

The melting point of a mineral generally increases with increasing pressure. Pressure increases with depth in the Earth's crust, just as temperature does. So a rock that melts at a given temperature at the surface of the Earth requires a higher temperature to melt deep underground. **Decompression melting** takes place when a body of hot mantle rock moves upward and the pressure is reduced. Figure 3.14*A* shows the effect of decompression melting. Consider point *a*. At this pressure and temperature, the rock is below the zone of partial melting and will not melt. If, however, the pressure decreases, the geothermal gradient will move up as the pressure at any given temperature decreases. The rock at point *a* will now be at point *a'*, which is within the zone of partial melting, and magma will begin to form. As you will learn at the end of this chapter, decompression melting is an important process at divergent plate boundaries, where mantle material is rising and melting beneath mid-oceanic ridges.

Addition of Water (Flux Melting)

If enough water is present and under high pressure, a dramatic change occurs in the melting process. Water sealed in under high pressure helps break the silicon-oxygen bonds in minerals, causing the crystals to liquify. A rock's melting temperature is significantly lowered by water under high pressure. This process is known as **flux melting**. Figure 3.14*B* shows the effect of water on the melting point of rocks in the mantle. The "dry" curve shows the temperature needed to melt rock that contains no water. The "wet" curve shows the temperature needed to melt rock that contains water. Consider "dry" mantle rock at point *b*. At this depth (D), the mantle needs to be above temperature T_1 in order to melt. Point *b* lies to the right of the geothermal gradient, so the temperature in the mantle is not high enough for melting to occur. Addition of water to the mantle moves the melting curve to the left. Point *b'* represents the new melting point of the mantle (T_2), which lies to the left of the geothermal gradient. "Wet" mantle at this depth will therefore undergo melting. As you will learn at the end of this chapter, flux melting is an important process at convergent plate boundaries, where subduction carries water down into the mantle.

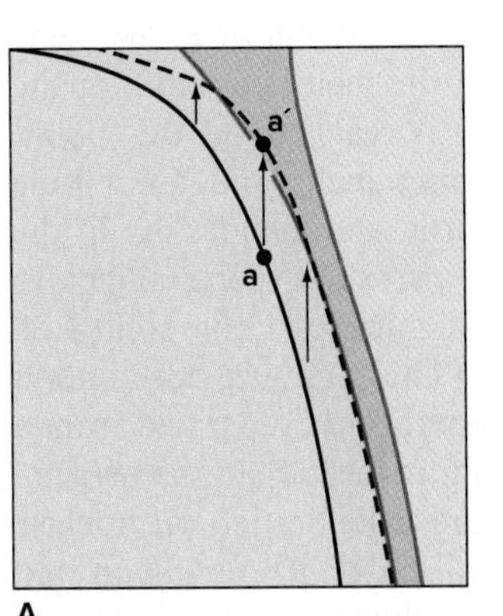

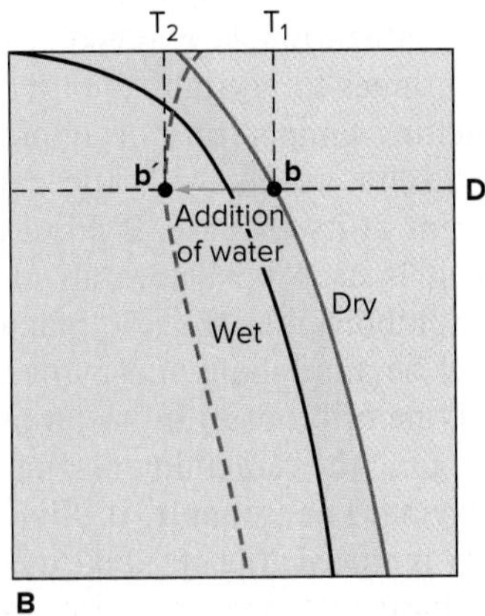

FIGURE 3.14

Mechanisms for melting rocks in the mantle. (*A*) Decompression melting. (*B*) Flux melting.

3.4 HOW MAGMAS OF DIFFERENT COMPOSITIONS EVOLVE

A major topic of investigation for geologists is why igneous rocks are so varied in composition. On a global scale, magma composition is clearly controlled by geologic setting. But why? Why are basaltic magmas associated with oceanic crust, whereas granitic magmas are common in the continental crust? On a local scale, igneous bodies often show considerable variation in rock type. For instance, individual plutons typically display a considerable range of compositions, mostly varieties of granite, but many also will contain minor amounts of gabbro or diorite. In this section, we describe processes that result in differences in the composition of magmas. The final section of this chapter relates these processes to plate tectonics for the larger view of igneous activity.

Sequence of Crystallization and Melting

Early in the twentieth century, N. L. Bowen conducted a series of experiments that determined the sequence in which minerals crystallize in a cooling magma. Bowen's experiments involved melting powdered rock, cooling the melt to a given temperature, and then observing the mineral present in the cooled rock. By repeating this process at different temperatures, he was able to observe that as you cool a basaltic melt, minerals tend to crystallize in a sequence determined by their melting temperatures. This sequence became known as **Bowen's reaction series** and is shown in figure 3.15. A simplified explanation of the series and its importance to igneous rocks is presented next.

Bowen's experiments showed that in a cooling magma, certain minerals are stable at higher temperatures and crystallize before those that are stable at lower temperatures. At higher temperatures, the sequence is broken into two branches. The *discontinuous branch* on the left-hand side of the diagram describes the formation of the ferromagnesian

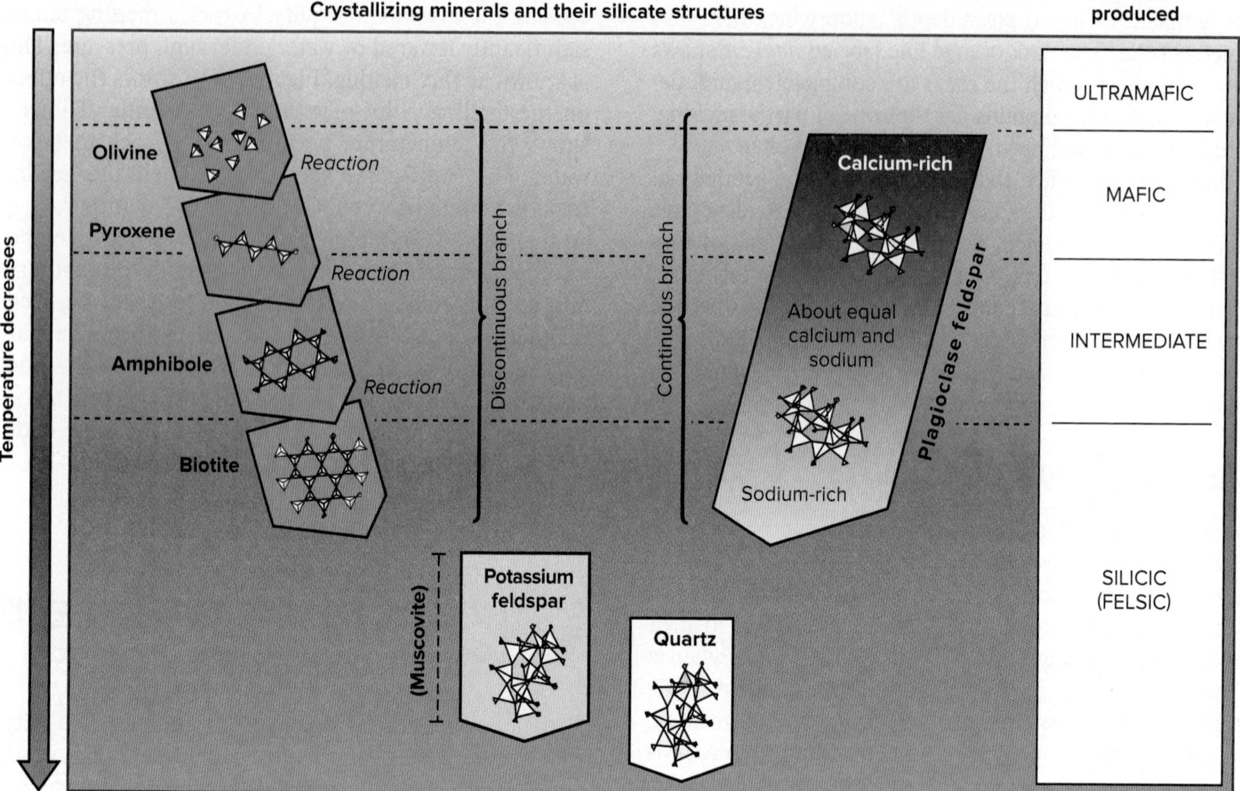

FIGURE 3.15

Bowen's reaction series. The reaction series as shown is very generalized. Moreover, it represents Bowen's experiments that involved melting a relatively silica-rich variety of basalt.

minerals olivine, pyroxene, amphibole, and biotite. This series is called the discontinuous series because these minerals have very different crystalline structures. The *continuous branch* on the right-hand side of the diagram contains only plagioclase feldspar. Plagioclase is a *solid solution* mineral (discussed in chapter 2 on minerals) in which either sodium or calcium atoms can be accommodated in its crystal structure, along with aluminum, silicon, and oxygen. The continuous branch describes the evolution of plagioclase from more calcium-rich compositions to more sodium-rich compositions with decreasing temperature.

Imagine a mafic magma as it cools and crystallizes. In figure 3.15, we can see that olivine crystallizes first, followed by calcium-rich plagioclase feldspar. Further cooling leads to the crystallization of pyroxene, then amphibole. The composition of plagioclase changes as magma is cooled and earlier-formed crystals react with the melt. The first plagioclase crystals to form as a hot melt cools contain calcium but little or no sodium. As cooling continues, the early-formed crystals grow and incorporate progressively more sodium into their crystal structures. As we progress through the reaction series, the formation of minerals rich in magnesium, iron, and calcium causes the remaining magma to become depleted in those elements and, at the same time, enriched in silicon, aluminum, and potassium. Any magma left at the point where the two branches meet is richer in silicon than the original magma; it also contains abundant potassium and aluminum. The potassium and aluminum combine with silicon to form *potassium feldspar.* (If the water pressure is high, *muscovite* may also form at this stage.) Excess SiO_2 crystallizes as *quartz.*

A complication is that early-formed crystals *react* with the remaining melt and recrystallize as cooling proceeds. For instance, early-formed olivine crystals react with the melt and recrystallize to pyroxene when pyroxene's temperature of crystallization is reached. Upon further cooling, pyroxene continues to crystallize until all of the melt is used up or the melting temperature of amphibole is reached. At this point, pyroxene reacts with the remaining melt, and amphibole forms at its expense. If all of the iron and magnesium in the melt is used up before all of the pyroxene recrystallizes to amphibole, then the ferromagnesian minerals in the solid rock will be amphibole and pyroxene. (The rock will not contain olivine or biotite.) If, however, minerals are separated from a magma, the remaining magma is more felsic than the original magma. For example, if olivine and calcium-rich plagioclase are removed, the residual melt will be richer in silicon and sodium and poorer in iron and magnesium. This process, known as *differentiation,* which can produce magmas of different composition from an initially mafic magma, is discussed further in the following section.

ENVIRONMENTAL GEOLOGY 3.2

Harnessing Magmatic Energy

Buried magma chambers indirectly contribute the heat for today's geothermal electric generating plants. As explained in chapter 11 (Groundwater), water becomes heated in hot rocks. The heat source is usually presumed to be an underlying magma chamber. The rocks containing the hot water are penetrated by drilling. Steam exiting the hole is used to generate electricity.

Why not drill into and tap magma itself for energy? The amount of energy stored in a body of magma is enormous. The U.S. Geological Survey estimates that magma chambers in the United States within 10 kilometers of Earth's surface contain about 5,000 times as much energy as the country consumes each year. Our energy problems could largely be solved if significant amounts of this energy were harnessed.

There are some formidable technical difficulties in drilling into a magma chamber and converting the heat into useful energy. Despite these difficulties, the United States has considered developing magmatic energy. Experimental drilling has been carried out in Hawaii through the basalt crust of a lava lake that formed in 1960.

As drill bits approach a magma chamber, they must penetrate increasingly hot rock. The drill bit must be made of special alloys to prevent it from becoming too soft to cut rock. The rock immediately adjacent to a basaltic magma chamber is around 1,000°C, even though that rock is solid. Drilling into the magma would require a special technique. One that was experimented with is a jet-augmented drill. As the drill enters the magma chamber, it simultaneously cools and solidifies the magma in front of the drill bit. Thus, the drill bit creates a column of rock that extends downward into the magma chamber and simultaneously bores a hole down the center of this column. Once the hollow column is deep enough within the magma chamber, a boiler is placed in the hole. The boiler is protected from the magma by the jacket of the column of rock. Water would be pumped down the hole and turned to water vapor in the boiler by heat from the magma. Steam emerging from the hole would be used to generate electricity.

In principle, the idea is fairly simple, but there are serious technical problems. For one thing, high pressures would have to be maintained on the drill bit during drilling and while the boiler system was being installed; otherwise, gases within the magma might blast the magma out of the drill hole and create a human-made volcano. (The closest thing to a human-made volcano occurred in Iceland when a small amount of magma broke into a geothermal steam well and erupted briefly at the well head, showering the area with a few tons of volcanic debris.)

Bowen's reaction series can also be used to consider the formation of magma by melting rock. If you heat a rock, the minerals will melt in reverse order. In other words, you would be going up the series as diagrammed in figure 3.15. Any quartz and potassium feldspar in the rock would melt first. If the temperature were raised further, biotite and sodium-rich plagioclase would contribute to the melt. Any minerals higher in the series would remain solid unless the temperature were raised further.

Differentiation

The process by which different ingredients separate from an originally homogeneous mixture is called **differentiation**. An example is the separation of whole milk into cream and nonfat milk. Differentiation in magmas takes place through a number of processes. As previously described, crystals that remain in contact with the melt as it cools will react to form minerals lower in Bowen's reaction series. If, however, crystals are no longer in contact with the melt, they will not react with it, and the elements they contain are effectively removed from the melt. For example, if olivine, pyroxene, and calcium-rich plagioclase crystallize from a mafic magma, they take up magnesium, iron, and calcium from the melt. If these early-formed crystals are removed from the melt, the magma becomes depleted in those elements and enriched in silicon, aluminum, sodium, and potassium. In other words, it becomes more felsic in composition. This differentiation process, often called *fractional crystallization*, can occur in a number of ways. One example is **crystal settling**, the downward movement of minerals that are denser (heavier) than the magma from which they crystallized.

If crystal settling takes place in a mafic magma chamber, olivine and, perhaps, pyroxene crystallize and settle to the bottom of the magma chamber (figure 3.16). This makes the remaining magma more felsic. Calcium-rich plagioclase also separates as it forms. The remaining magma is, therefore, depleted of calcium, iron, and magnesium. Because these minerals were economical in using the relatively abundant silica, the remaining magma becomes richer in silica as well as in sodium and potassium.

It is possible that by removing enough mafic components, the residual magma would be felsic enough to solidify into granite (or rhyolite). But it is more likely that only enough mafic components would be removed to allow an intermediate residual magma, which would solidify into diorite or andesite. The lowermost portions of some intrusions are composed predominantly of olivine and pyroxene, whereas upper levels are considerably less mafic. Even in large intrusions, however, differentiation has rarely progressed far enough to produce granite within the sill.

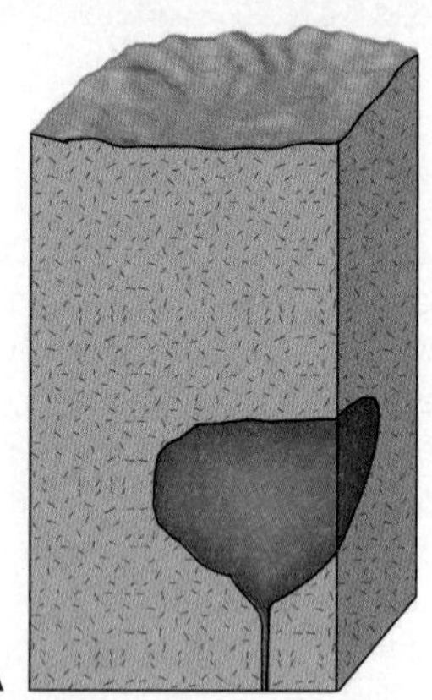

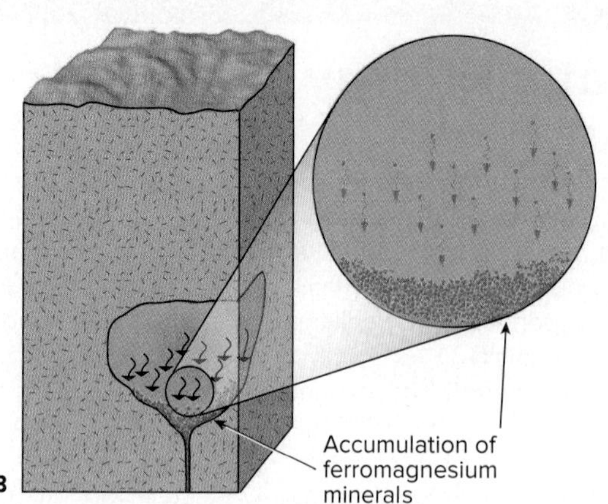

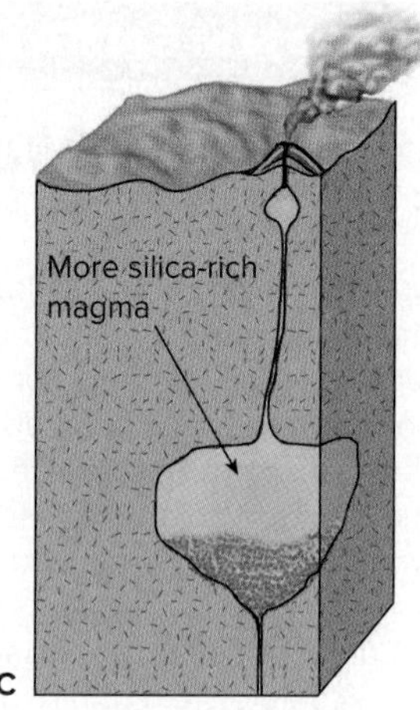

FIGURE 3.16

Differentiation of a magma body. (*A*) Recently intruded mafic magma is completely liquid. (*B*) Upon slow cooling, ferromagnesian minerals, such as olivine, crystallize and sink to the bottom of the magma chamber. The remaining liquid is now an intermediate magma. (*C*) Some of the intermediate magma moves upward to form a smaller magma chamber at a higher level that feeds a volcano.

Partial Melting

As mentioned earlier, progressing upward through Bowen's reaction series (going from cool to hot) gives us the sequence in which minerals in a rock melt. As might be expected, the first portion of a rock to melt as temperatures rise forms a liquid with the chemical composition of quartz and potassium feldspar. The oxides of silicon plus potassium and aluminum "sweated out" of the solid rock could accumulate into a pocket of felsic magma. If higher temperatures prevailed, more mafic magmas would be created. Small pockets of magma could merge and form a large enough mass to rise buoyantly through the lower crust and toward the surface. In nature, temperatures rarely rise high enough to melt a rock entirely.

Partial melting of the lower continental crust likely produces felsic magma. The magma rises and eventually solidifies at a higher level in the crust into granite, or rhyolite if it reaches Earth's surface.

Geologists generally regard basaltic magma (Hawaiian lava, for example) as the product of partial melting of ultramafic rock in the mantle, at temperatures hotter than those in the crust. The solid residue left behind in the mantle when the basaltic magma is removed is an even more silica-deficient ultramafic rock.

Assimilation

A very hot magma may melt some of the country rock and *assimilate* the newly molten material into the magma (figure 3.17). This is like putting a few ice cubes into a cup of hot coffee. The ice melts and the coffee cools as it becomes diluted. Similarly, if a hot basaltic magma, perhaps generated from the mantle, melts portions of the continental crust, the magma simultaneously becomes richer in silica and cooler. It is possible that some intermediate magmas, such as those associated with circum-Pacific andesite volcanoes, may derive from assimilation of crustal rocks by a basaltic magma.

Magma Mixing

Some of our igneous rocks may be "cocktails" of different magmas. The concept is quite simple. If two magmas meet and merge within the crust, the new, combined magma should have a composition intermediate to the two original magmas (figure 3.18). For example, if you had approximately equal amounts of a granitic magma mixing with a basaltic magma, the resulting magma would be intermediate in composition and crystallize underground as diorite or erupt on the surface to solidify as andesite. However, because of the profound differences in the properties of felsic and mafic magmas, most notably their respective temperature differences, magma mixing generally does not produce a homogeneously mixed magma. Consider the sequence shown in figure 3.18. The mafic magma likely has a temperature of over 1,100°C, whereas a felsic magma would likely be several hundred degrees cooler. The mafic magma would be quickly cooled when the two magmas meet, and most of it would solidify rapidly. Some of the mafic minerals would react with the felsic magma and be absorbed in it, but most of the mafic magma would become blobs of basalt or gabbro included in the more felsic magma. Overall, the pluton would have an average chemical composition that is intermediate, but the rock that forms would not be a homogeneous intermediate rock. Because of this, *magma mingling* might be a better term for the process.

3.5 INTRUSIVE BODIES

Intrusions, or **intrusive structures**, are bodies of intrusive rock whose names are based on their size and shape as well as their relationship to surrounding rocks. They are important aspects of the architecture, or *structure,* of the Earth's crust. The various intrusions are named and classified on the basis of the following considerations: (1) Is the body large or small? (2) Does it have a particular geometric shape? (3) Did the rock form at a

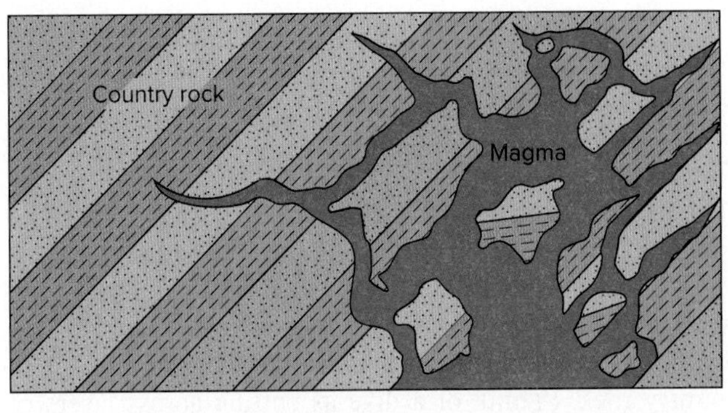

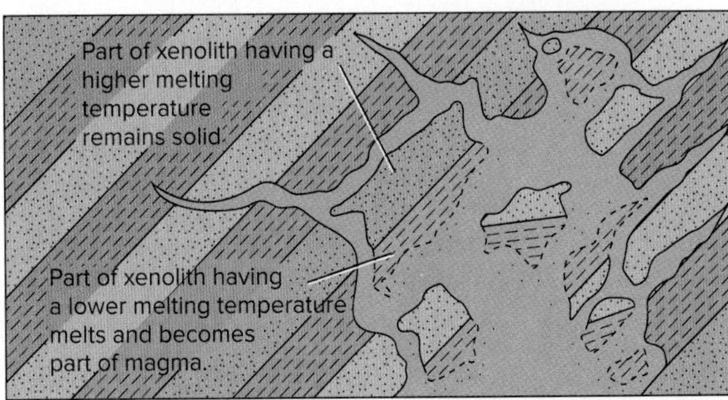

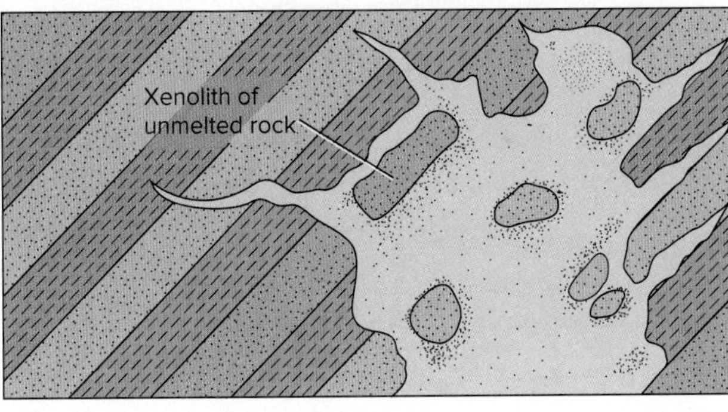

FIGURE 3.17

Assimilation. Magma formed is intermediate in composition between the original magma and the absorbed country rock. (*A*) Ascending magma breaks off blocks of country rock (the process called *stoping*). (*B*) Xenoliths of country rock with melting temperatures lower than the magma melt. (*C*) The molten country rock blends with the original magma, leaving unmelted portions as inclusions.

considerable depth, or was it a shallow intrusion? (4) Does it follow layering in the country rock?

Shallow Intrusive Structures

Some igneous bodies apparently solidified near the surface of the Earth (probably at depths of less than 2 kilometers). These bodies appear to have solidified in the subsurface "plumbing systems" of volcanoes or lava flows. Shallow intrusive structures tend to be relatively small compared with those that formed at considerable depth. Because the country rock near the Earth's surface generally is cool, intruded magma tends to chill and solidify relatively rapidly. Also, smaller magma bodies will cool

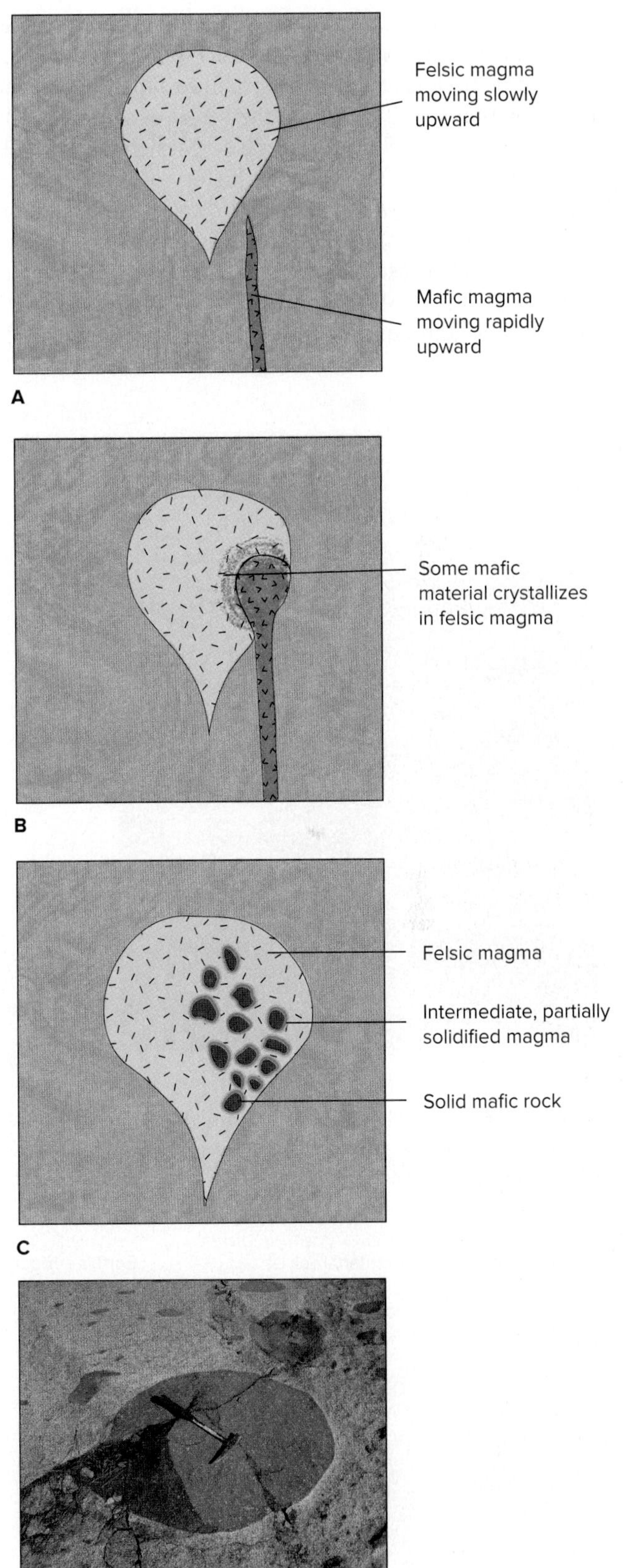

FIGURE 3.18

Mixing of magmas. (*A*) Two bodies of magma moving surfaceward. (*B*) The mafic magma catches up with the felsic magma. (*C*) An inhomogeneous mixture of felsic, intermediate, and mafic material. (*D*) "Blobs" of mafic magma in granite at Loch Leven Lakes, CA.
(D) ©Jonathan Meurer

faster than larger bodies, regardless of depth. For both of these reasons, shallow intrusive bodies are likely to be fine-grained.

A **volcanic neck** is an intrusive structure apparently formed from magma that solidified within the throat of a volcano. One of the best examples is Ship Rock in New Mexico (figure 3.19). Here is how geologists interpret the history of this feature. A volcano formed above what is now Ship Rock. The magma for the volcano moved upward through a more or less cylindrical conduit. Eruptions ceased and the magma underground solidified into what is now Ship Rock. In time, the volcano and its underlying rock—the country rock around Ship Rock—eroded away. The more resistant igneous body eroded more slowly into its present shape. Weathering and erosion are continuing. (Falling rock has been a serious hazard to rock climbers.)

Dikes and Sills

Another, and far more common, intrusive structure can also be seen at Ship Rock. The low, wall-like ridge extending outward from Ship Rock is an eroded dike. A **dike** is a tabular (shaped like a tabletop), discordant, intrusive structure (figure 3.20). *Discordant* means that the body is not parallel to any layering in the country rock. (Think of a dike as cutting across layers of

A

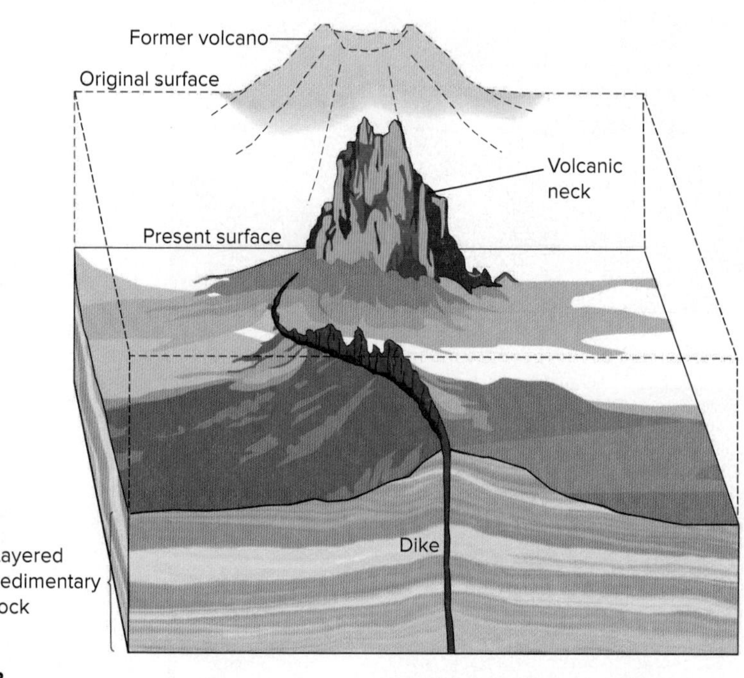

B

FIGURE 3.19

(A) Ship Rock in New Mexico, which rises 420 meters (1,400 feet) above the desert floor. *(B) Relationship to the former volcano.*
(A) ©Bill Hatcher/National Geographic/Getty Images

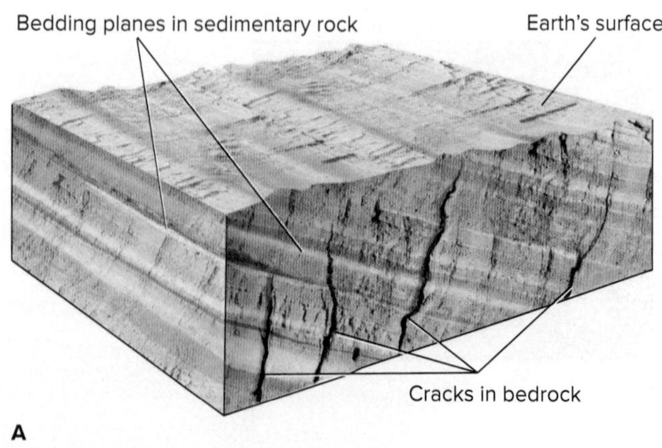

A

B

FIGURE 3.20

(A) Cracks and bedding planes are planes of weakness. (B) Concordant intrusions where magma has intruded between sedimentary layers are sills; discordant intrusions are dikes.

country rock.) Dikes may form at shallow depths and be fine-grained, such as those at Ship Rock, or form at greater depths and be coarser-grained. Dikes need not appear as walls protruding from the ground (figure 3.21). The ones at Ship Rock do so only because they are more resistant to weathering and erosion than the country rock.

A **sill** is also a tabular intrusive structure, but it is *concordant.* That is, sills, unlike dikes, are parallel to any planes or layering in the country rock (figures 3.20 and 3.22). Typically, the country rock bounding a sill is layered sedimentary rock. As magma squeezes into a crack between two layers, it solidifies into a sill.

If the country rock is not layered, a tabular intrusion is regarded as a dike.

Intrusives that Crystallize at Depth

A **pluton** is a body of magma or igneous rock that crystallized at considerable depth within the crust. Where plutons are exposed at the Earth's surface, they are arbitrarily distinguished by size. If the area of surface exposure of plutonic rock is indicated on a map to be greater than 100 square kilometers, the body is called a **batholith** (figure 3.23). Most batholiths extend over areas vastly greater than the minimum 100 square kilometers. A smaller pluton that has a surface exposure of less than 100 square kilometers is called a **stock**. As can be seen in figure 3.23, however, a stock may be the exposed portion of a larger batholith.

Although batholiths may contain mafic and intermediate rocks, they almost always are predominantly composed of granite. Detailed studies of batholiths indicate that they are formed of numerous, coalesced plutons. Apparently, large blobs of magma worked their way upward through the lower crust and collected 5 to 30 kilometers below the surface, where they solidified (figure 3.24). These blobs of magma, known as **diapirs**, are less dense than the surrounding rock that is pliable and shouldered aside as the magma rises. When viscous magma intrudes between two layers of sedimentary rock, it may generate enough pressure to lift the overlying layer into an arch. The magma then cools to form a concordant, mushroom-shaped or domed pluton known as a *laccolith.*

Batholiths occupy large portions of North America, particularly in the west. Over half of California's Sierra Nevada (figure 3.25) is a batholith whose individual plutons were emplaced during a period of over 100 million years. An even larger batholith extends almost the entire length of the mountain ranges of Canada's west coast and southeastern Alaska—a distance of 1,800 kilometers. Smaller batholiths are also found in eastern North America in the Piedmont east of the Appalachian Mountains and in New England and the coastal provinces of Canada. Laccoliths are common in the United States, often forming distinctive landforms. Pine Valley Mountain, near St. George, Utah, is one of the largest laccoliths in the United States.

Granite is considerably more common than rhyolite, its volcanic counterpart. Why is this? Silicic magma is much more *viscous* (that is, more resistant to flow) than mafic magma. Therefore, a silicic magma body will travel upward through the crust more slowly and with more difficulty than mafic or

FIGURE 3.21

Dikes (light-colored rocks) in northern Victoria Land, Antarctica. ©C.C. Plummer

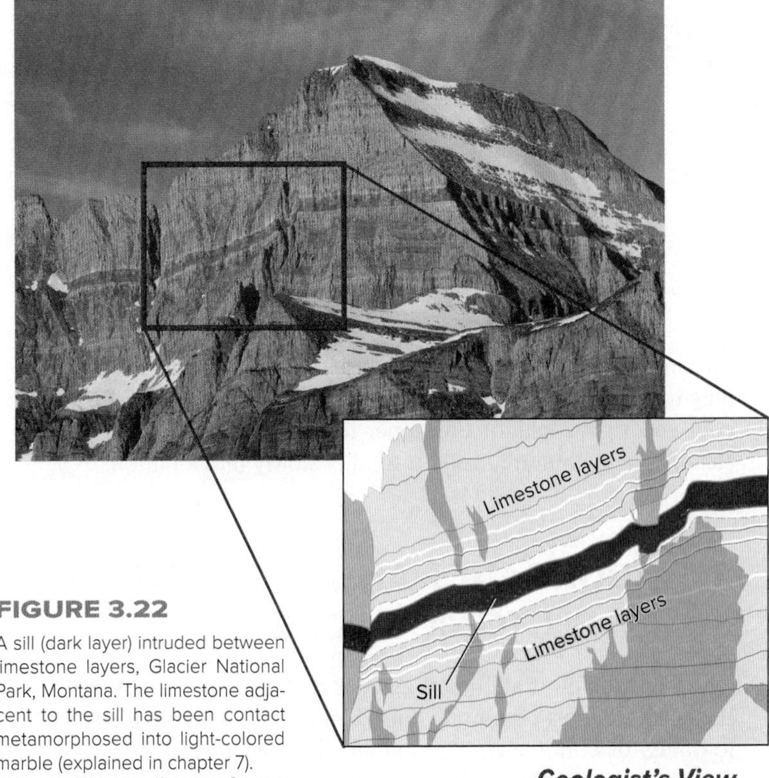

FIGURE 3.22

A sill (dark layer) intruded between limestone layers, Glacier National Park, Montana. The limestone adjacent to the sill has been contact metamorphosed into light-colored marble (explained in chapter 7). ©James Steinberg/Science Source

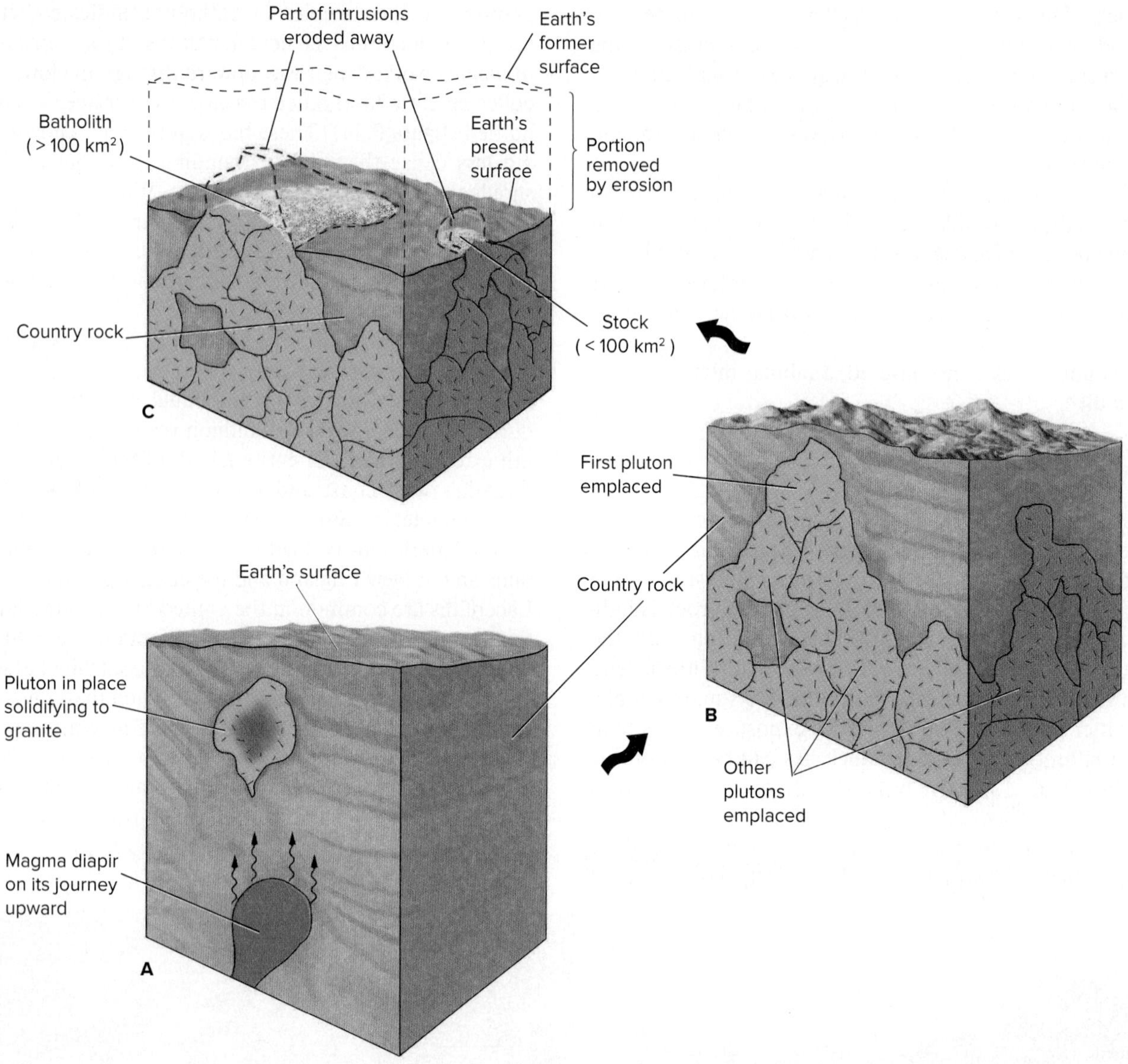

FIGURE 3.23

(*A*) The first of numerous magma diapirs has worked its way upward and is emplaced in the country rock. (*B*) Other magma diapirs have intruded, coalesced, and solidified into a solid mass of plutonic rock. (*C*) After uplift and erosion, surface exposures of plutonic rock are a batholith and a stock.

intermediate magma. Unless it is exceptionally hot, a silicic magma will not be able to work its way through the relatively cool and rigid rocks of the upper few kilometers of crust. Instead, it is much more likely to solidify slowly into a pluton.

3.6 ABUNDANCE AND DISTRIBUTION OF PLUTONIC ROCKS

Granite is the most abundant igneous rock in mountain ranges. It is also the most commonly found igneous rock in the interior lowlands of continents. Throughout the lowlands of much of Canada, very old plutons have intruded even older metamorphic rock. As explained in chapter 20 on mountains and the continental crust, very old mountain ranges have, over time, eroded and become the stable interior of a continent. Metamorphic and plutonic rocks similar in age and complexity to those in Canada are found in the Great Plains of the United States. Here, however, they are mostly covered by a veneer (a kilometer or so) of younger, sedimentary rock. These basement rocks are exposed to us in only a few places. In Grand Canyon, Arizona, the Colorado River has eroded through the layers of sedimentary rock to expose the ancient plutonic and metamorphic basement. In the Black Hills of South Dakota, local uplift and subsequent erosion have exposed similar rocks.

Granite, then, is the predominant igneous rock of the continents. As described in chapter 4, basalt and gabbro are the

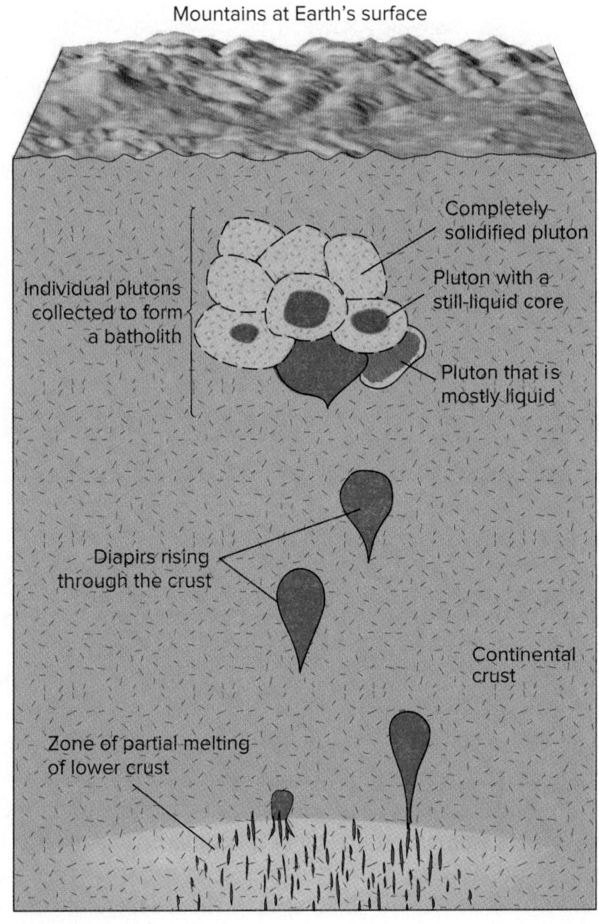

FIGURE 3.24

Diapirs of magma travel upward from the lower crust and solidify in the upper crust. (Not drawn to scale.)

predominant rocks underlying the oceans. Andesite (usually along continental margins) is the building material of most young volcanic mountains. Underneath the crust, ultramafic rocks make up the upper mantle.

3.7 EXPLAINING IGNEOUS ACTIVITY BY PLATE TECTONICS

One of the appealing aspects of the theory of plate tectonics is that it accounts reasonably well for the variety of igneous rocks and their distribution patterns. (Chapter 1 has an overview of plate tectonics.) Divergent boundaries are associated with the creation of basalt and gabbro of the oceanic crust. Andesite and granite are associated with convergent boundaries. Table 3.2 summarizes the relationships.

Igneous Processes at Divergent Boundaries

The crust beneath the world's oceans (over 70% of Earth's surface) is mafic volcanic and intrusive rock, covered to a varying extent by sediment and sedimentary rock. Most of this basalt and gabbro was created at mid-oceanic ridges, along divergent plate boundaries. Geologists agree that the mafic magma produced at divergent boundaries is formed by partial melting of the asthenosphere. The *asthenosphere,* as described in chapter 1, is the plastic zone of the mantle beneath the rigid *lithosphere* (the upper mantle and crust that make up a plate). Along divergent boundaries, the asthenosphere is relatively close (5 to 10 kilometers) to the surface (figure 3.26).

FIGURE 3.25

Part of the Sierra Nevada batholith. All light-colored rock shown here (including that under the distant snow-covered mountains) is granite. The extent of the Sierra Nevada batholith is shown in the inset. ©C.C. Plummer

TABLE 3.2 Relationships between Rock Types and Their Usual Plate-Tectonic Setting

Rock	Original Magma	Final Magma	Processes	Plate-Tectonic Setting
Basalt and gabbro	Mafic	Mafic	Partial melting of mantle (asthenosphere)	1. Divergent boundary—oceanic crust created 2. Intraplate • plateau basalt • volcanic island chains (e.g., Hawaii)
Andesite and diorite	Mafic (usually)	Intermediate	Partial melting of mantle (asthenosphere) followed by: • differentiation or • assimilation or • magma mixing	Convergent boundary
Granite and rhyolite	Felsic	Felsic	Partial melting of lower crust	1. Convergent boundary 2. Intraplate • over mantle plume

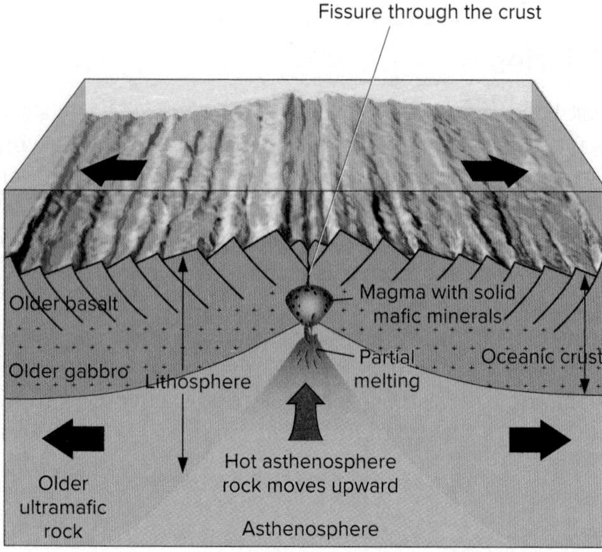

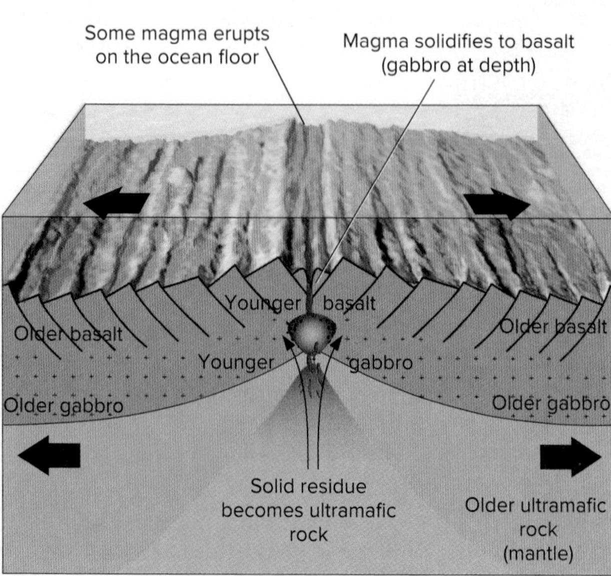

FIGURE 3.26

Schematic representation of how basaltic oceanic crust and the underlying ultramafic mantle rock form at a divergent boundary. The process is more continuous than the two-step diagram implies. (*A*) Partial melting of asthenosphere takes place beneath a mid-oceanic ridge, and magma rises into a magma chamber. (*B*) The magma squeezes into the fissure system. Solid mafic minerals are left behind as ultramafic rock.

The probable reason the asthenosphere is plastic or "soft" is that temperatures there are only slightly lower than the temperatures required for partial melting of mantle rock.

If extra heat is added, or pressure is reduced, partial melting should take place. The asthenosphere beneath divergent boundaries probably is mantle material that has welled upward from deeper levels of the mantle. As the hot asthenosphere gets close to the surface, decrease in pressure results in partial melting. In other words, *decompression melting* takes place. The magma that forms is mafic and will solidify as basalt or gabbro. The portion that did not melt remains behind as a silica-depleted, iron-and-magnesium-enriched ultramafic rock.

Some of the basaltic magma erupts along a submarine ridge to form pillow basalts (described in chapter 4), while some fills near-surface fissures to create dikes. Deeper down, magma solidifies more slowly into gabbro. The newly solidified rock is pulled apart by spreading plates; more magma fills the new fracture and some erupts on the sea floor. The process is repeated, resulting in a continuous production of mafic crust.

The basaltic magma that builds the oceanic crust is formed by partial melting of the underlying mantle, depleting the mantle beneath the ridge of much of its calcium, aluminum, and silicon oxides. The unmelted residue (olivine and pyroxene) becomes depleted mantle, but it is still a variety of ultramafic rock. The rigid ultramafic rock, the overlying gabbro and basalt, and any sediment that may have deposited on the basalt collectively are the lithosphere of an oceanic plate, which moves away from a spreading center over the asthenosphere. (The nature of the oceanic crust is described in more detail in chapter 18.)

Intraplate Igneous Activity

Igneous activity within a plate, a long distance from a plate boundary, is unusual. These hot spots have been hypothesized to be due to hot **mantle plumes**, which are narrow upwellings of hot material within the mantle. Examples include the long-lasting volcanic activity that built the Hawaiian Islands and the eruptions at Yellowstone National Park in Wyoming. The ongoing eruptions in Hawaii take place on oceanic crust, whereas eruptions at Yellowstone represent continental intraplate activity. The silicic eruptions at Yellowstone that took place some 600,000 years ago were much larger and more violent than any eruptions that have occurred in historical time.

The huge volume of mafic magma that erupted to form the Columbia plateau basalts of Washington and Oregon (described in chapter 4) is attributed to a past hot mantle plume, according to a recent hypothesis (figure 3.27). In this case, the large volume of basalt is due to the arrival beneath the lithosphere and decompression melting of a mantle plume with a large head on it.

Igneous Processes at Convergent Boundaries

Intermediate and felsic magmas are clearly related to the convergence of two plates and subduction. However, exactly what takes place is debated by geologists. Compared to divergent boundaries, there is less agreement about how magmas are generated at convergent boundaries. The scenarios that follow are currently regarded by geologists to be the best explanations of the data.

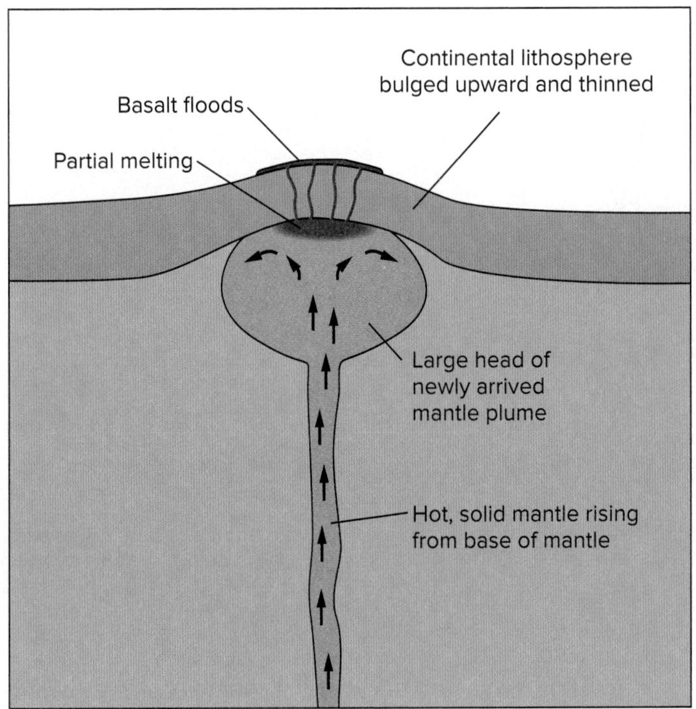

FIGURE 3.27

A hot mantle plume with a large head rises from the lower mantle. When it reaches the base of the lithosphere, it uplifts and stretches the overlying lithosphere. The reduced pressure results in decompression melting, producing basaltic magma. Large volumes of magma travel through fissures and flood the Earth's surface.

The Origin of Andesite

Magma for most of our andesitic composite volcanoes (such as those found along the west coast of the Americas) seems to originate from a depth of about 100 kilometers. This coincides with the depth at which the subducted oceanic plate is sliding under the asthenosphere (figure 3.28). Partial melting of the asthenosphere takes place, resulting in a mafic magma. In most cases, melting occurs because the subducted oceanic crust releases water into the asthenosphere. The water collected in the oceanic crust when it was beneath the ocean and is driven out as the descending plate is heated. The water lowers the melting temperature of the ultramafic rocks in this part of the mantle. In other words *flux melting* takes place. Partial melting produces a mafic magma.

But how can we keep producing magma from ultramafic rock after those rocks have been depleted of the constituents of the mafic magma? The answer is that hot asthenospheric rock continues to flow into the zone of partial melting. As shown in figure 3.28, asthenospheric ultramafic rock is dragged downward by the descending lithospheric slab. More ultramafic rock flows laterally to replace the descending material. A continuous flow of hot, fertile (containing the constituents of basalt) ultramafic rock is brought into the zone where water, moving upward from the descending slab, lowers the melting temperature. After being depleted of basaltic magma, the solid, residual, ultramafic rock continues to sink deeper into the mantle.

On its slow journey through the crust, the mafic magma evolves into an intermediate magma through differentiation and by assimilation of felsic crustal rocks.

Under special circumstances, basalt of the descending oceanic crust can partially melt to yield an intermediate magma. In most subduction zones, the basalt remains too cool to melt, even at a depth of over 100 kilometers. But geologists believe that partial melting of the subducted crust produces the magma for andesitic volcanoes in South America. Here, the oceanic crust is much younger and considerably hotter than normal. The spreading axis where it was created is not far from the trench. Because the lithosphere has not traveled far before being subducted, it is still relatively hot.

The reason that partial melting of subducted basalt is unusual is that this kind of subduction and magma generation is, geologically speaking, short-lived. Subduction will end when the overriding plate crashes into the mid-oceanic ridge. Most subduction zones are a long distance from the divergent boundaries of their plates, so steep subduction and magma production from the asthenosphere are the norm.

The Origin of Granite

To explain the great volumes of granitic plutonic rocks, many geologists think that partial melting of the lower continental crust must take place. The continental crust contains the high amount of silica needed for a felsic magma. As the felsic rocks

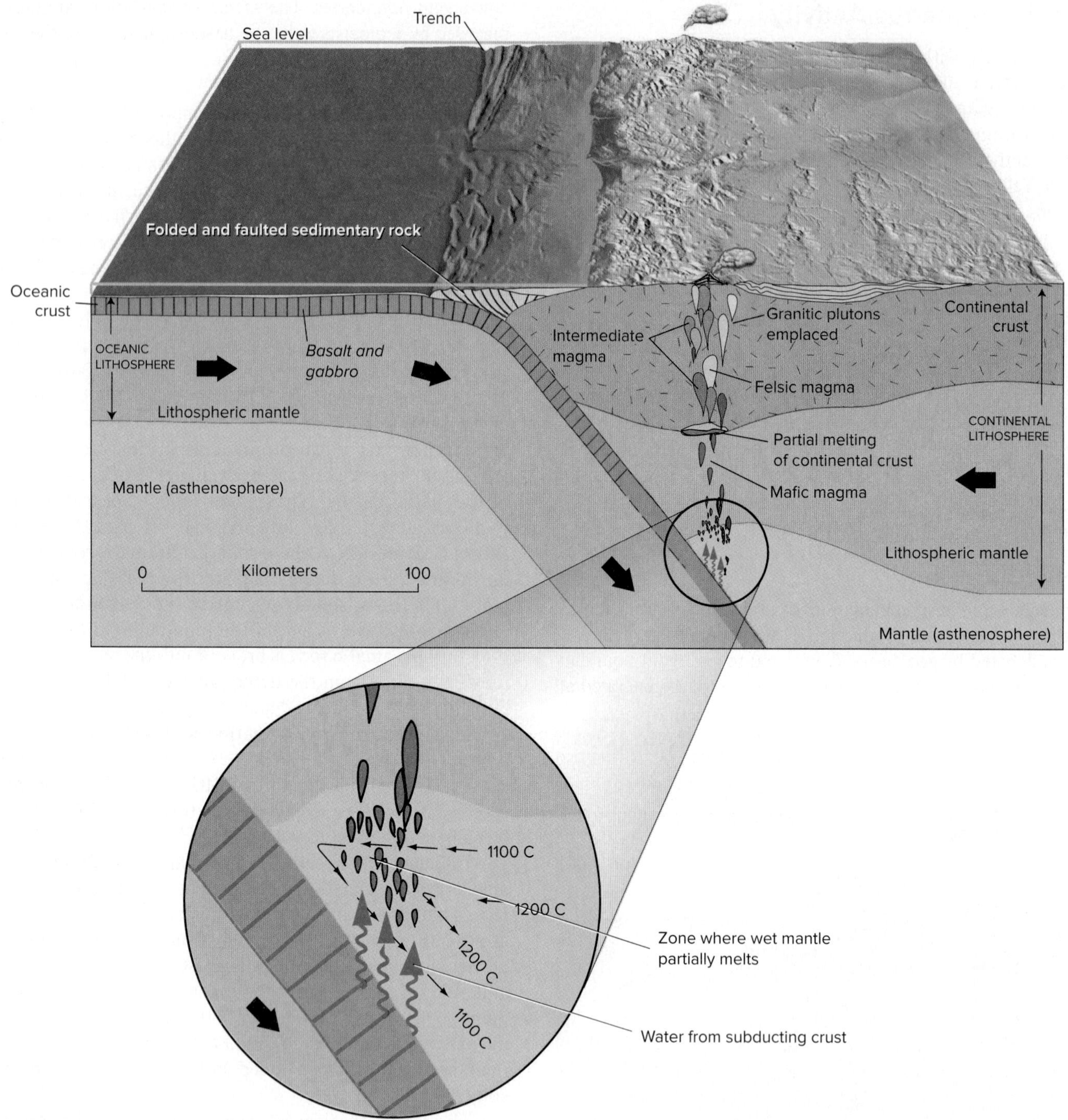

FIGURE 3.28

Generation of magma at a convergent boundary. Mafic magma is generated in the asthenosphere above the subducting oceanic lithosphere, and felsic magma is created in the lower crust. The inset shows the circulation of asthenosphere and lines of equal temperature (isotherms). Partial melting of wet ultramafic rock takes place in the zone where it is between 1,100°C and 1,200°C.

of the continental crust have relatively low melting temperatures (especially if water is present), partial melting of the lower continental crust is likely. However, calculations indicate that the temperatures we would expect from a normal geothermal gradient are too low for melting to take place. Therefore, we need an additional heat source.

Currently, geologists think that the additional heat is provided by mafic magma that was generated in the asthenosphere and moved upward. The process of *magmatic underplating* involves mafic magma pooling at the base of the continental crust, supplying the extra heat necessary to partially melt the overlying, silica-rich crustal rocks (figure 3.29). Mafic magma

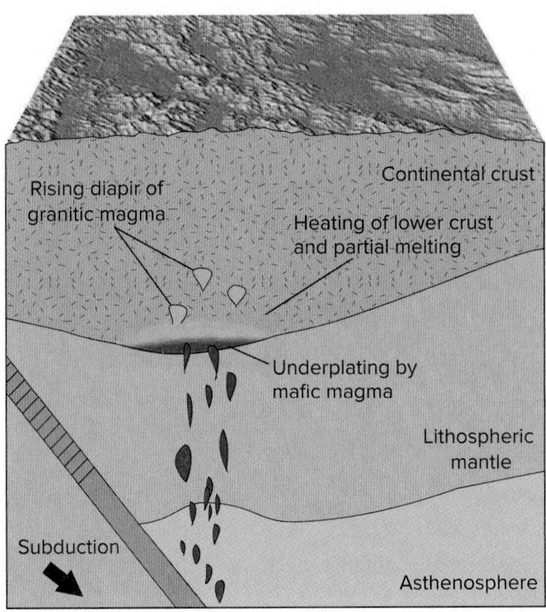

FIGURE 3.29

How mafic magma could add heat to the lower crust and result in partial melting to form a granitic magma. Mafic magma from the asthenosphere rises to underplate the continental crust.

generated in the asthenosphere rises to the base of the crust. The mafic magma is denser than the overlying silica-rich crust; therefore, it collects as a liquid mass that is much hotter than the crust. The continental crust becomes heated (as if by a giant hotplate). When the temperature of the lower crust rises sufficiently, partial melting takes place, creating felsic magma. The felsic magma collects and forms diapirs, which rise to a higher level in the crust and solidify as granitic plutons (or, on occasion, reach the surface and erupt violently).

Summary

The interaction between the internal and external forces of the Earth is illustrated by the rock cycle, a conceptual device relating igneous, sedimentary, and metamorphic rocks to each other, to surficial processes such as weathering and erosion, and to internal processes such as tectonic forces. Changes take place when one or more processes force Earth's material out of equilibrium.

Igneous rocks form from solidification of magma. If the rock forms at the Earth's surface, it is *extrusive. Intrusive rocks* are igneous rocks that formed underground. Igneous rocks are named based upon their texture and composition. Intrusive rocks have a *phaneritic* (coarse-grained) texture, and extrusive rocks have an *aphanitic* (fine-grained) texture. Felsic (or silicic) rocks are rich in silica, whereas mafic rocks are silica-poor. The mineral content of igneous rocks reflects the chemical composition of the magmas from which they formed. *Granite, diorite,* and *gabbro* are the coarse-grained equivalents of *rhyolite, andesite,* and *basalt,* respectively. *Peridotite* is an *ultramafic* rock made entirely of ferromagnesian minerals and is mostly associated with the mantle.

Basalt and gabbro are predominant in the oceanic crust. Granite predominates in the continental crust. Younger granite batholiths occur mostly within younger mountain belts. Andesite is largely restricted to narrow zones along convergent plate boundaries.

The *geothermal gradient* is the increase in temperature that occurs with increase in depth. Melting of the mantle occurs through *decompression melting* and *flux melting. Partial melting* of the mantle usually produces basaltic magma, whereas granitic magma is most likely produced by partial melting of the lower continental crust.

No single process can satisfactorily account for all igneous rocks. In the process of *differentiation,* based on *Bowen's reaction series,* a residual magma more silicic than the original mafic magma is created when the early-forming minerals separate out of the magma. In *assimilation,* a hot, original magma is contaminated by picking up and absorbing rock of a different composition. Rocks of intermediate composition can form when *magma mixing* occurs between a felsic magma and a mafic magma.

Some intrusive rocks have solidified near the surface as a direct result of volcanic activity. Volcanic *necks* solidified within volcanoes. Fine-grained *dikes* and *sills* may also have formed in cracks during local extrusive activity. A sill is *concordant*–parallel to the planes within the country rock. A dike is *discordant*–not parallel to planes in the country rock. Both are tabular bodies. Coarser grains in either a dike or a sill indicate that it probably formed at considerable depth. Most intrusive rock is *plutonic*–that is, coarse-grained rock that solidified slowly at considerable depth. Most plutonic rock exposed at the Earth's surface is in *batholiths*–large plutonic bodies. A smaller body is called a *stock.*

The theory of *plate tectonics* incorporates the preceding concepts. Basalt is generated where hot mantle rock partially melts, most notably along divergent boundaries. The fluid magma rises easily through fissures, if present. The ferromagnesian portion that stays solid remains in the mantle as ultramafic rock. Granite and andesite are associated with subduction. Differentiation, assimilation, and partial melting may each play a part in creating the observed variety of rocks.

Terms to Remember

felsic (silicic) 61
flux melting 63
fragmental rock 55
gabbro 61
geothermal gradient 62
glassy rock 55
granite 54
igneous rock 52
intermediate 61
intrusion (intrusive structure) 66
intrusive rock 54
komatiite 61
lava 54
mafic 61
magma 54
mantle plume 73
obsidian 57
pegmatite 57
peridotite 61
phaneritic (coarse-grained) 57
phenocryst 57
pluton 69
plutonic rock 57
porphyritic 57
pumice 59
pyroclasts (tephra) 59
rhyolite 61
rock 52
rock cycle 52
scoria 59
sill 69
stock 69
texture 55
tuff 60
ultramafic rock 61
vesicles 59
volcanic breccia 60
volcanic neck 68
xenolith 55

Testing Your Knowledge

Use the following questions to prepare for exams based on this chapter.

1. You find an igneous rock that has large crystals of plagioclase feldspar held within a fine-grained groundmass that is medium gray in color. What kind of rock is it likely to be? What is the name for this texture, and what does it tell you about how the rock formed?
2. Why do mafic magmas tend to reach the surface much more often than felsic magmas?
3. How would you distinguish, on the basis of minerals present, among granite, gabbro, and diorite?
4. Why is a higher temperature required to form magma at the oceanic ridges than in the continental crust?
5. Describe the differences between the continuous and the discontinuous branches of Bowen's reaction series.
6. What is the difference between feldspar found in gabbro and feldspar found in granite?
7. Describe the different processes that could lead to a mafic magma evolving into an intermediate magma.
8. How do batholiths form?
9. What process causes the asthenosphere to melt, generating magma at (*a*) a convergent boundary; (*b*) a divergent boundary?
10. What rock would probably form if magma that was feeding volcanoes above subduction zones solidified at considerable depth?
11. The major difference between intrusive igneous rocks and extrusive igneous rocks is
    a. where they solidify.
    b. chemical composition.
    c. type of minerals.
    d. all of the preceding.
12. Which is *not* an intrusive igneous rock?
    a. gabbro
    b. diorite
    c. rhyolite
    d. peridotite
13. Flux melting occurs when
    a. water is added to the asthenosphere.
    b. a mantle plume increases the temperature of the asthenosphere.
    c. mantle material undergoes depression.
    d. mantle material flows downward to greater depths and higher temperatures.
14. The continuous branch of Bowen's reaction series contains the mineral
    a. pyroxene. b. plagioclase.
    c. amphibole. d. quartz.
15. The discontinuous branch of Bowen's reaction series contains the mineral
    a. pyroxene. b. amphibole.
    c. olivine. d. all of the above.
16. A change in magma composition due to melting of surrounding country rock is called
    a. magma mixing. b. assimilation.
    c. differentiation. d. partial melting.
17. By definition, stocks differ from batholiths in
    a. shape. b. size.
    c. chemical composition. d. age.
18. A discordant shallow intrusive structure is called a
    a. stock. b. dike.
    c. sill. d. laccolith.
19. The most common igneous rock of the continents is
    a. basalt. b. granite.
    c. rhyolite. d. peridotite.
20. Mafic magma is generated at divergent boundaries because of
    a. water under pressure.
    b. decompression melting.
    c. magmatic underplating.
    d. melting of the lithosphere.

Expanding Your Knowledge

1. In parts of major mountain belts, there are sequences of rocks that geologists interpret as slices of ancient oceanic lithosphere. Assuming that such a sequence formed at a divergent boundary and was moved toward a convergent boundary by plate motion, what rock types would you expect to make up this sequence, going from the top downward?
2. What would happen, according to Bowen's reaction series, under the following circumstances: olivine crystals form and only the surface of each crystal reacts with the melt to form a coating of pyroxene that prevents the interior of olivine from reacting with the melt?

Exploring Web Resources

https://www.geolsoc.org.uk/ks3/gsl/education/resources/rockcycle/page3892.html

Rock Cycle Animation. This site shows each stage of the rock cycle in an easy-to-follow animation.

http://leggeo.unc.edu/Petunia/IgMetAtlas/mainmenu.html

Atlas of Rocks, Minerals, and Textures (from University of North Carolina). This site contains some photomicrographs of plutonic and volcanic rocks. The images are thin sections (slices of rock so thin that most minerals are transparent) seen in a polarizing microscope. Most images are taken from cross-polarized light, which causes many minerals to appear in distinctive, bright colors. For some of the rocks (gabbro, for instance), you can also see what they look like under plain polarized light by clicking the circle with the horizontal, gray lines.

https://web.csulb.edu/depts/geology/facultypages/bperry/basicgeo/igneous_tour.html

Igneous Rocks Tour. This site has some hand specimen images of common igneous rocks and should provide a useful review for rock identification.

CHAPTER

4

Volcanism and Extrusive Rocks

Volcanic lightning generated during the eruption of Chaitén volcano in Chile, May 2008.
©Carlos Gutierrez/UPI/Landov

LEARNING OBJECTIVES

- Differentiate between effusive and explosive eruptions, and describe the eruptive products associated with them.
- Explain the relationship between magma composition, temperature, dissolved gas, and viscosity and relate them to eruptive violence.
- Describe the five major types of volcanoes in terms of shape and eruptive style.
- Know the major hazards associated with volcanic eruptions.
- Describe the three components of volcanic hazard mitigation.
- Explain the role of plate tectonics in determining the location and eruptive style of volcanoes.

On May 2, 2008, Chaitén volcano, located in a remote and sparsely populated part of Chile, suddenly burst into violent eruption, sending a column of volcanic ash more than 20 kilometers into the air. By the next day, a plume of ash had spread across Chile and Argentina to the Atlantic Ocean, affecting water supplies, ground transport, and airline traffic. The eruption continued for almost three years, causing extensive damage to the nearby town of Chaitén and coating large areas of Argentina and Chile with ash. The opening photo for this chapter, which shows volcanic lightning generated in the ash cloud from Chaitén, demonstrates the combination of deadliness and beauty that makes volcanoes so compelling.

Volcanic eruptions are some of the most spectacular and deadly geologic phenomena. Not surprisingly, myths and religions relating gods to volcanoes flourish in cultures that live with volcanoes. In Iceland, Loki, of Norse mythology, is regarded as imprisoned underground, blowing steam and lava up through fissures. Pacific Northwest Indians regarded the Cascade volcanoes as warrior gods who would sometimes throw red-hot boulders at each other. In Hawaii, Madame Pele is regarded as a goddess who controls eruptions. According to legend, Pele and her sister tore up the ocean floor to produce the Hawaiian island chain. Today, many fervently believe that Pele dictates when and where an eruption will take place. In the 1970s, when Kilauea began erupting near a village, residents chartered an airplane and dropped flowers and a bottle of gin into the lava vent to appease Pele.

While awesome natural spectacles, volcanic eruptions also provide important information about the workings of Earth's interior. Eruptions vary in terms of eruptive style and degree of explosive violence. A strong correlation exists between the chemical composition of magma (or lava), its physical properties, and the violence of an eruption. The size and shape of volcanoes and lava flows and their pattern of distribution on Earth's surface also correlate with the composition of their lavas. Our observations of volcanic activity fit nicely into plate tectonic theory. Understanding volcanism also provides a background for theories relating to mountain building and the development and evolution of continental and oceanic crust (topics covered in later chapters). Landforms are created through volcanic activity, and portions of Earth's surface are built up by repeated lava flows. Less commonly, as at Mount St. Helens, landforms are destroyed by violent eruptions (box 4.1). As you learned in chapter 3, by studying the magma, gases, and rocks from eruptions, we can infer the chemical conditions as well as the temperatures and pressures within Earth's crust or underlying mantle.

In chapter 3 you learned about the origin of magma, igneous rock classification, and intrusive structures. In this chapter we will concentrate on extrusive (volcanic) igneous activity. We will begin by broadly defining eruptions into two categories: explosive and effusive. We will explore how the nature of volcanic eruptions is controlled by the physical properties of magma. We will then discuss the products of volcanic eruptions and describe different types of volcanoes. Following this we will examine volcanic hazards and the types of volcano monitoring that are used to mitigate these hazards. Finally, we will examine the link between the types of volcanoes (and their eruptive styles) and plate-tectonic settings.

4.1 WHAT ARE VOLCANOES AND WHY SHOULD WE STUDY THEM?

Volcanism occurs when magma makes its way to the Earth's surface. **Volcanoes** are landforms formed by the extrusion of lava or the ejection of rock fragments from a vent. Volcanoes come in many shapes and sizes, and eruptions can vary widely in their duration, violence, and the type of material erupted.

On October 25, 2010, after a short period of intense seismic unrest, Mount Merapi, Indonesia's most active volcano, erupted violently (figure 4.1*A*). The eruptions that occurred over the next few weeks sent eruption columns several kilometers up into the atmosphere, generating *pyroclastic flows* (fast-moving flows of hot ash and pyroclastic debris) that flowed down the slopes of the volcano. Over 350,000 people were evacuated from the affected area, and volcanic ash caused major disruption to air travel across Java. By early December, when the volcanic activity had subsided, 353 people had been reported killed. These were mostly people who had refused to evacuate or who had returned to their homes while the volcanic activity continued. In addition to the destruction of villages located on the slopes of Merapi, crops in the vicinity were ruined, and many livestock were killed. **Explosive eruptions**, also called *pyroclastic eruptions*, such as the eruption of Merapi, are dominated by the generation of solid volcanic fragments. Explosive eruptions vary enormously in size. Box 4.1 discusses the 1980 eruption of Mount St. Helens, a relatively large explosive eruption. A smaller explosive eruption of Eyjafjallajökull in Iceland in 2010 caused major disruption to European air traffic (see box 1.2).

ENVIRONMENTAL GEOLOGY 4.1

Mount St. Helens Blows Up

Before 1980, Mount St. Helens, in southern Washington, had not erupted since 1857. On March 27, 1980, ash and steam eruptions began and continued for the next six weeks. These were minor eruptions in which magma was not erupted. Rather, they were due to exploding gas blasting out the volcano's previously formed rock. However, the steam and the pattern of earthquakes indicated that magma was working its way upward beneath the volcano.

After several weeks, the peak began swelling—like a balloon being inflated—indicating that magma was now inside the volcano. The northern flank of the volcano bulged outward at a rate of 1.5 meters per day. Bulging continued until the surface of the northern slope was displaced outward over a hundred meters from its original position. The bulge was too steep to be stable, and the U.S. Geological Survey warned of another hazard—a mammoth landslide.

On May 18, a monumental blast destroyed the summit and north flank of Mount St. Helens. Seconds after the eruption began, an area extending northward 10 kilometers was stripped of all vegetation and soil.

Although the sequence of events was exceedingly rapid, it is now clear what happened (box figure 1). A fairly strong earthquake loosened the bulging north slope, triggering a landslide. The landslide, known as a *debris avalanche*, moved at speeds of over 160 kilometers per hour (100 mph). It was one of the largest landslides ever to occur, but it was eclipsed by the huge eruption that followed. The landslide stripped away the lid on the magma chamber, and because of the reduced pressure, the previously dissolved gases in the magma exploded (box figure 1C). The violent froth of gas and magma blasted away the mountain's north flank and roared outward at up to 1,000 kilometers per hour (600 mph). The huge lateral blast of hot gas and volcanic rock debris killed everything near the volcano and, beyond the 10-kilometer scorched zone, knocked down every tree in the forest.

For the next 30 hours, exploding gases propelled frothing magma and volcanic ash vertically into the high atmosphere. The mushroom-shaped cloud of ash was blown northeastward by winds. A rain of ash went on for days, causing damage as far away as

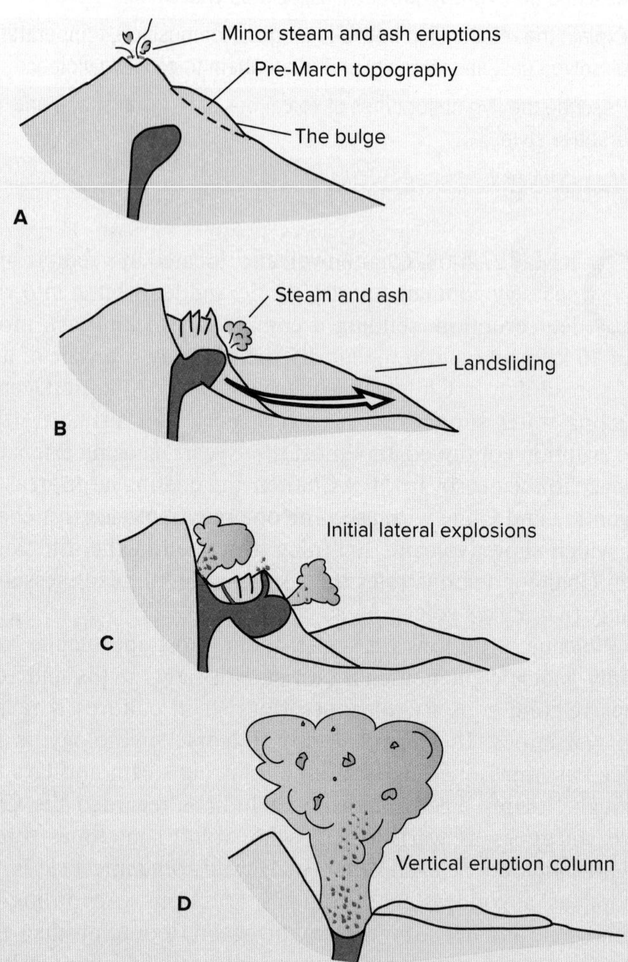

BOX 4.1 ■ FIGURE 1

Sequence of events at Mount St. Helens, May 18, 1980. (*A*) Just before the eruption. (*B*) The landslide relieves the pressure on the underlying magma. (*C*) Magma blasts outward. (*D*) Full vertical eruption.

Kilauea on the island of Hawaii has been erupting constantly since 1983. The style of eruption is very different from that seen at Mount Merapi or Mount St. Helens. In Hawaii, lava extrudes out of vents and fissures in the ground as lava flows. In March 2011, a new fissure, almost 500 meters long, opened on the eastern side of Kilauea (figure 4.1*B*). Lava spewed out of the fissure, in places fountaining 65 feet into the air. Eruptions dominated by lava flows are called **effusive eruptions** and are typically less dangerous than explosive eruptions. While the lava flows from Kilauea have destroyed some homes and roadways, very few people have been killed or injured.

While the dangers associated with volcanoes provide a compelling reason to study them, there are many other reasons. Volcanoes provide geologists with information on processes occurring within the Earth's mantle. Volcanic eruptions can affect the Earth's climate. Volcanoes can also be beneficial.

Creation of New Land

Although occasionally a highway or village is overrun by outpourings of lava, the overall effects of volcanism have been favorable to humans in Hawaii. Lava flowing into the sea and solidifying adds real estate to the island of Hawaii. Kilauea

Montana. Volcanic mudflows caused enormous damage during and after the eruption. The mudflows resulted when water from melted snow and glacier ice mixed with volcanic debris to form a slurry with the consistency of wet cement. Mudflows flowed down river valleys, carrying away steel bridges and other structures (see chapter 9, notably figure 9.14).

Damage was in the hundreds of millions of dollars, and 63 people were killed. The death toll might have been much worse had not scientists warned public officials about the potential hazards, causing them to evacuate the danger zone before the eruption. For comparison, 29,000 people were killed during an eruption of Mount Pelée (described later in this chapter), and 23,000 lives were lost in a 1985 volcanic mudflow in Colombia.

Mount St. Helens re-awoke in 2004. Lava oozed into the crater, forming a new dome (described later in this chapter). This eruption ended in 2008 and, although Mount St. Helens is considered the most likely Cascades volcano to erupt, a violent eruption is not considered likely. Other volcanoes in the Pacific Northwest, however, could erupt and be disastrous to nearby cities. Seattle and Tacoma are close to Mount Rainier. Mount Hood is practically in Portland, Oregon's suburbs. Vancouver, British Columbia, could be in danger if either Mount Garibaldi to the north or Mount Baker in Washington to the south erupts.

BOX 4.1 ■ FIGURE 2

Mount St. Helens, May 18, 1980. Looking north, we can see the last of the huge lateral explosion from the far side of the volcano. This was followed by vertical eruption of gases and pyroclasts from the top of the volcano.
Source: Robert Krimmel/U.S. Geological Survey

Additional Resources

USGS Cascade Volcano Observatory—Mount St. Helens

- http://volcanoes.usgs.gov/volcanoes/st_helens/

This website provides a wealth of information, maps, and photos of Mount St. Helens as well as current monitoring data.

USDA Forest Service—Mount St. Helens volcano cams

- http://www.fs.fed.us/gpnf/volcanocams/msh/

This website shows live images of Mount St. Helens and includes information on the current status of the volcano.

volcano has been erupting since 1983, spewing out an average of 325,000 cubic meters of lava a day. This is the equivalent of 40,000 dump truckloads of material. In twenty years, 2.5 billion cubic meters of lava were produced—enough to build a highway that circles the world over five times. Were it not for volcanic activity, Hawaii would not exist. The islands are the crests of a series of volcanoes that have been built up from the bottom of the Pacific Ocean over millions of years (the vertical distance from the summit of Mauna Loa volcano to the ocean floor greatly exceeds the height above sea level of Mount Everest). When lava flows into the sea and solidifies, more land is added to the islands. Hawaii is, quite literally, growing.

In addition to gaining more land, Hawaii benefits in other ways from its volcanoes. Weathered volcanic ash and lava produce excellent, fertile soils (think pineapples and papayas). Moreover, Hawaii's periodically erupting volcanoes (which are relatively safe to watch) are great spectacles that attract both tourists and scientists, benefiting the island's economy.

Geothermal Energy

In other areas of recent volcanic activity, underground heat generated by igneous activity is harnessed for human needs. Steam or superheated water trapped in layers of hot volcanic rock is

A

B

FIGURE 4.1

Contrasting styles of volcanic eruptions. (*A*) Explosive eruption from Mount Merapi, Indonesia, in November 2010. A pyroclastic flow can be seen descending the slopes in the foreground. (*B*) Effusive eruption of lave from a fissure that opened on the flanks of Kilauea volcano, Hawaii, in March 2011.
(A) ©Dwi Oblo/Reuters/Landov (B) Source: U.S. Geological Survey/Hawaiian Volcano Observatory

tapped by drilling and then piped out of the ground to power turbines that generate electricity. The United States is the biggest producer of geothermal power, followed by the Philippines, Indonesia, Mexico, and Italy. Naturally heated geothermal fluids can also be tapped for space or domestic water heating or industrial use, as in paper manufacturing. (For more information, go to http://geothermal.marin.org/, chapter 11 on groundwater, or chapter 22 on geologic resources.)

Effect on Climate

The atmosphere was created by degassing magma during the time following Earth's formation. Even now, gases and dust given off by major volcanic eruptions can profoundly alter worldwide climate. Occasionally, a volcano will spew large amounts of fine volcanic dust and gas into the high atmosphere. Winds can keep fine particles suspended over the Earth for years. The 1991 eruption of Mount Pinatubo in the Philippines produced noticeably more colorful sunsets worldwide. More significantly, it reduced solar radiation that penetrates the atmosphere. Measurements indicated that the worldwide average temperature dropped approximately one degree Celsius for a couple of years. While this may not seem like much, it was enough to temporarily offset the global warming trend of the past 100 years.

The 1815 eruption of Tambora in Indonesia was the largest single eruption in a millennium—40 cubic kilometers of material were blasted out of a volcanic island, leaving a 6-kilometer-wide depression. The following year, 1816, became known as "the year without summer." In New England, snow in June was widespread, and frosts throughout the summer ruined crops. Parts of Europe suffered famine because of the cold-weather effects on agriculture. (See chapter 21 for more information on the impact of volcanoes on the climate.)

4.2 ERUPTIVE VIOLENCE AND PHYSICAL CHARACTERISTICS OF LAVA

Box 4.2 describes the Volcanic Explosivity Index (VEI), which is used to indicate how powerful volcanic eruptions are. The scale goes from nonexplosive and gentle eruptions to mega-colossal eruptions. What determines the degree of violence associated with volcanic activity? Whether an eruption is effusive or explosive? Why can we state confidently that active volcanism in Hawaii poses only slight danger to humans, but we expect violent eruptions to occur around the margins of the Pacific Ocean? Whether eruptions are violently explosive or relatively "quiet" is largely determined by two factors: (1) the amount of gas in the lava or magma and (2) the ease or difficulty with which the gas can escape to the atmosphere. The **viscosity**, or resistance to flow, of a lava determines how easily the gas escapes. The more viscous the lava and the greater the volume of gas trying to escape, the more violent the eruption. Later we will show how these factors not only determine the degree of violence of an eruption but also influence the shape and height of a volcano.

Lava is a mixture of molten silicate rock, crystals, and gas. When we look at a volcanic rock, we commonly see these components. For example, examine the rock shown in figure 4.2. This basalt contains small *phenocrysts* of white plagioclase feldspar held in a finer-grained *groundmass.* The phenocrysts represent crystals held in molten lava. The *vesicles* in the rock represent gases expanding and escaping from the lava as it erupted. From active volcanoes we have learned that most of the gas released during eruptions is water vapor, which condenses as steam. Other gases, such as carbon dioxide, sulfur dioxide,

FIGURE 4.2

Vesicular basalt with phenocrysts of plagioclase feldspar.
©Dirk Wiersma/Science Source

hydrogen sulfide (which smells like rotten eggs), and hydrochloric acid, are given off in lesser amounts with the steam. If a lava is too viscous to allow the gas bubbles to form easily, it will fragment, causing large, explosive eruptions that can blast ash and rock debris kilometers into the atmosphere. The three factors that influence the viscosity of lava are (1) the silica (SiO_2) content of the lava; (2) the temperature of the lava; and (3) the amount of gas dissolved in magma. Although explosive eruptions are driven by the expansion of dissolved gases, the greater the dissolved gas content in a magma, the more fluid it is. If the lava being extruded is considerably hotter than its solidification temperature, the lava is less viscous (more fluid) than when its temperature is near its solidification point. Temperatures at which lavas solidify range from about 700°C for felsic rocks to 1,200°C for mafic rocks.

The silica content of magma is a major factor in determining viscosity. As described in chapter 3, volcanic rocks, and the magma from which they form, have silica contents that range from 45% to 75% by weight. **Felsic rocks** are silica-rich (65% or more SiO_2). *Rhyolite* is the most abundant felsic volcanic rock. **Mafic rocks** are *silica-deficient.* Their silica content is close to 50%. *Basalt* is the most common mafic rock. **Intermediate rocks** have a silica content between that of felsic and mafic rocks. The most common intermediate rock is *andesite.* Table 4.1 provides a review of the common extrusive rock types and their characteristics.

Mafic lavas, which are relatively low in SiO_2, tend to flow easily. Conversely, felsic lavas are much more viscous and flow sluggishly. You can picture the difference by imagining how maple syrup flows compared with toothpaste. Mafic lava is around 10,000 times as viscous as water, whereas felsic magma is around 100 million times the viscosity of water. Why are felsic lavas more viscous than mafic lavas? In chapter 2 you learned that the silicate minerals form when silica-oxygen tetrahedra bond together with other cations to form different crystal structures. Lavas rich in silica are more viscous because even before they

TABLE 4.1 Names for Extrusive Rocks

Names for Finely Crystalline Rocks Based on Chemical or Mineralogical Composition (See chapter 3 for pictures.)		
Rock Name	**Chemical Composition**	**Description**
Rhyolite	Silicic	Light colored. Usually cream-colored, tan, or pink. Mostly finely crystalline white or pink feldspar and quartz.
Andesite	Intermediate	Moderately gray or green color. A little over half of rock is light- to medium-gray plagioclase feldspar, while the rest is ferromagnesian minerals (usually *pyroxene* or *amphibole*).
Basalt	Mafic	Black or dark gray. The rock is made up mostly of ferromagnesian minerals (notably *olivine* and *pyroxene*) and calcium-rich *plagioclase feldspar.*
Adjectives Used to Modify Rock Names		
Porphyritic (e.g., porphyritic andesite)		Some crystals (phenocrysts) are larger than 1 millimeter (usually considerably larger). Most grains are smaller than 1 millimeter. Or phenocrysts are enclosed in glass.
Vesicular (e.g., vesicular basalt)		Holes (vesicles) in rock due to gas trapped in solidifying lava.
Names for Rocks Based on Texture		
Obsidian		Volcanic glass that is usually silicic. Black or reddish with a conchoidal fracture.
Pumice		Frothy volcanic glass.
Scoria		Vesicular basalt in which the volume of vesicles is greater than that of the solid rock.
Tuff		Consolidated, fine pyroclastic material.
Volcanic breccia		Consolidated, pyroclastic debris that includes coarse material (lapilli, blocks, or bombs).

IN GREATER DEPTH 4.2

Volcanic Explosivity Index

To indicate how powerful volcanic eruptions are, scientists use the **Volcanic Explosivity Index** or **VEI**. The index is on a scale of 0 to 8 (box table 1) and is based on a number of factors, including the volume of erupted pyroclastic material, the height of the eruption column, and how long the eruption lasts. Like the Richter magnitude scale for earthquakes (discussed in chapter 16), the VEI is logarithmic, meaning that each interval on the scale represents a tenfold increase in the size of the eruption. An eruption of VEI 3 is ten times bigger than a 2 and one hundred times smaller than a 5 (box figure 1).

BOX 4.2 ■ TABLE 1

The Volcanic Explosivity Index

VEI	Description	Plume Height	Volume	Classification	How Often	Example
0	nonexplosive	< 100 m	1,000s m^3	Hawaiian	daily	Kilauea
1	gentle	100–1,000 m	10,000s m^3	Haw/Strombolian	daily	Stromboli
2	explosive	1–5 km	1,000,000s m^3	Strom/Vulcanian	weekly	Galeras, 1992
3	severe	3–15 km	10,000,000s m^3	Vulcanian	yearly	Ruiz, 1985
4	cataclysmic	10–25 km	100,000,000s m^3	Vulcanian/Plinian	10s of years	Galunggung, 1982
5	paroxysmal	>25 km	1 km^3	Plinian	100s of years	St. Helens, 1980
6	colossal	>25 km	10s km^3	Plinian/Ultra-Plinian	100s of years	Krakatau, 1883
7	super-colossal	>25 km	100s km^3	Ultra-Plinian	1,000s of years	Tambora, 1815
8	mega-colossal	>25 km	1,000s km^3	Ultra-Plinian	10,000s of years	Yellowstone, 2 Ma

Source: Volcano World. http://volcano.oregonstate.edu/how-big-are-eruptions.

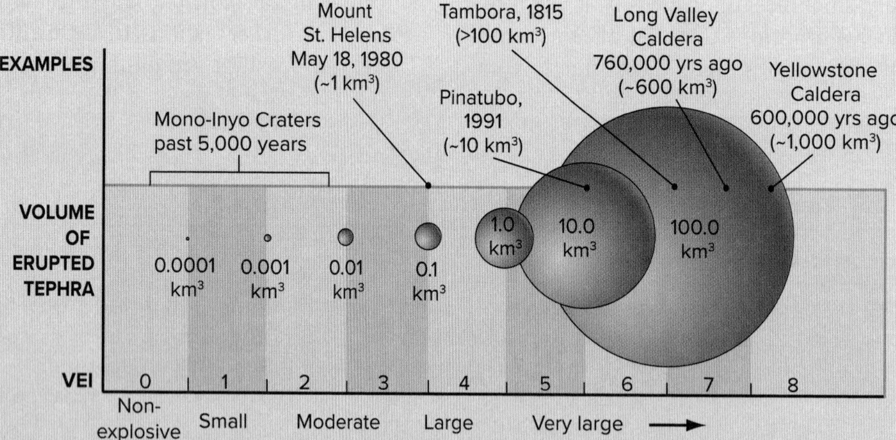

BOX 4.2 ■ FIGURE 1

VEI of past explosive eruptions. The volume for each eruption is given in parentheses. The relative increase in volume for each step on the scale is represented by the red circles. Note that the increase in volume is tenfold for each step.
Source: United States Geological Survey

have cooled enough to allow crystallization of minerals, silicon-oxygen tetrahedra have linked to form small, framework structures in the lava. Although too few atoms are involved for the structures to be considered crystals, the total effect of these silicate structures is to make the liquid lava more viscous, much the way that flour or cornstarch thickens gravy.

Because felsic magmas are the most viscous, they are associated with the most violent, explosive eruptions. Mafic magmas are the least viscous and commonly erupt as lava flows in effusive eruptions (such as in Hawaii). Eruptions associated with intermediate magma can be violent or can produce lava flows, depending on the amount of dissolved gas in the magma.

Surface water introduced into a volcanic system can greatly increase the explosivity of an eruption, as exemplified by the 2010 eruption of Eyjafjallajökull volcano in Iceland (described in box 1.2).

4.3 THE ERUPTIVE PRODUCTS OF VOLCANOES

A volcano is an opening in the earth's crust through which molten lava, ash, and gases are ejected. Volcanic material that is ejected from and deposited around a central vent produces the conical shape typical of volcanoes. The **vent** is the opening through which an eruption takes place. The **crater** of a volcano is a basinlike depression over a vent at the summit of the cone (figure 4.3). Material is not always ejected from the central vent. In a **flank eruption**, lava pours from a vent on the side of a volcano.

Effusive Eruptions

Effusive eruptions are characterized by lava flows. As described previously, the nature of an eruption is controlled by the characteristics of the lava, in particular the gas content and viscosity. Effusive eruptions are most commonly basaltic in composition because mafic basalts are less viscous, and gases can escape easily. Intermediate and felsic lava can erupt effusively if the gas content is quite low.

Mafic Lava Flows

Due to its low silica content, basaltic lava is typically low viscosity and flows easily. Volcanic activity in Hawaii is dominantly basaltic in composition, and Hawaiian names have been given to two distinctive surfaces of basalt lava flows. **Pahoehoe** (pronounced *pah-hoy-hoy*) is characterized by a ropy or billowy surface (figure 4.4). Pahoehoe is formed by the rapid cooling and solidification of the surface of the lava flow, rather like the skin that forms on the top of a cup of hot chocolate. As the lava below the solidified surface continues to flow, the "skin" is dragged along and becomes folded and rumpled, rather like what happens to the skin on the top of your hot chocolate when you tip the cup. **A'a** (pronounced *ah-ah*) is a flow that has a jagged, rubbly surface (figure 4.5). It forms when basalt is cool enough to have partially solidified and moves slowly as a pasty mass. Its largely solidified front is shoved forward as a pile of rubble. Pahoehoe flows often change into a'a flows farther from the vent as the lava cools and becomes more viscous.

FIGURE 4.3

Volcanic crater (200 meter diameter) on Karymsky volcano, Kamchatka, Russia. The lake in the background fills a large crater known as a caldera.
Source: C. Dan Miller, U.S. Geological Survey

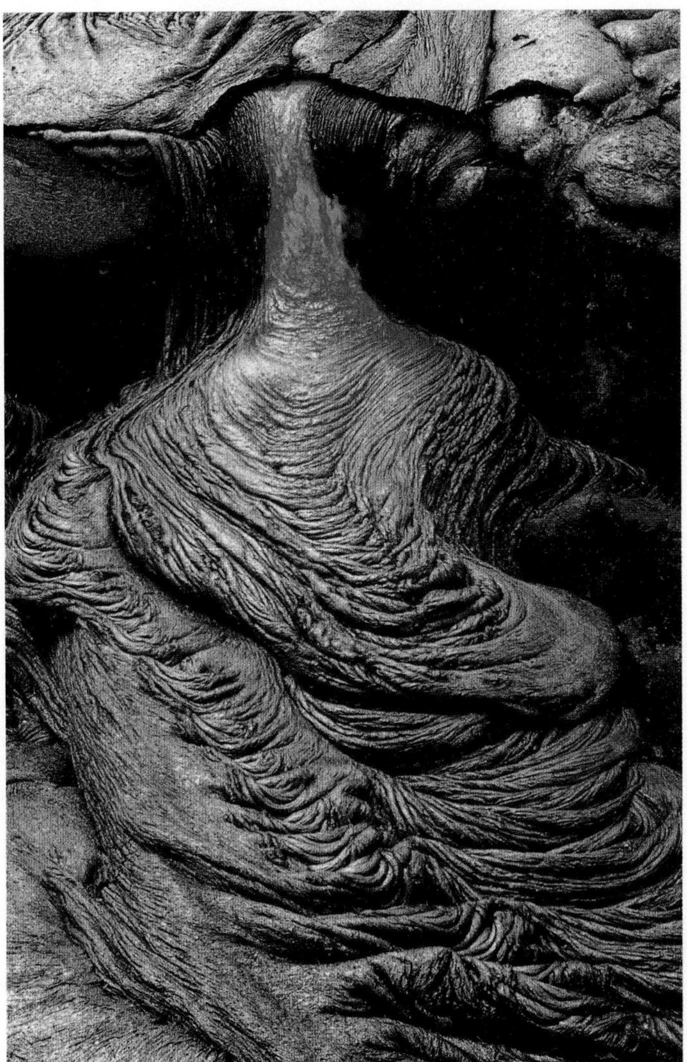

FIGURE 4.4

Flow of lava solidifying to pahoehoe in Hawaii.
Source: J. D. Griggs, U.S. Geological Survey

A (usually) minor feature called a *spatter cone,* a small, steep-sided cone built from lava sputtering out of a vent (figure 4.6), will occasionally develop on a solidifying lava flow. When a small concentration of gas is trapped in a cooling lava flow, lava is belched out of a vent through the solidified surface of the flow. Falling lava plasters itself onto the developing cone and solidifies. The sides of a spatter cone can be very steep, but they are rarely over 10 meters high. An exception to this is Pu'u O'o, the 250-meter-high, combined spatter and cinder cone on the

FIGURE 4.5

An *a'a* flow in Hawaii, 1983.
Source: J. D Griggs, U.S. Geological Survey

FIGURE 4.6

A spatter cone (approximately 1 meter high) erupting in Hawaii.
Source: J. B. Judd, U.S. Geological Survey

A

B

FIGURE 4.7

(*A*) Lava stream seen through a collapsed roof of a lava tube during a 1970 eruption of Kilauea volcano, Hawaii. Note the ledges within the tube, indicating different levels of flows. (*B*) Lava tube at Lava Beds National Monument, California. The narrow, dark shelf on either side of the tube marks the level of the lava stream, indicating where lava solidified against the walls of the tube.
(A) Source: J. B. Judd, U.S. Geological Survey (B) ©C. C. Plummer

eastern flank of Kilauea shield volcano. It is located at the vent for the ongoing (1983–onward) lava eruptions.

A **lava tube** is a tunnel-like conduit for lava that develops after most of a fluid, pahoehoe-type flow has solidified (figure 4.7*A*). The tube's roof and walls solidified along with the earlier, broader flow. The tube provides insulation so that the rapidly flowing lava loses little heat and remains fluid. Much of the lava in the ongoing Hawaiian eruptions flows underground in a lava tube, traveling about 7 kilometers from Pu'u O'o, the currently erupting vent, to the sea. Lava Beds National Monument in northeastern California (figure 4.7*B*) contains many lava tubes that formed within basaltic lava flows erupted 30,000 to 40,000 years ago from Medicine Lake volcano. When the eruption ended, lava drained from the tubes, leaving them hollow.

Flood Basalts

Lava that is very nonviscous and flows almost as easily as water does not build a cone around a vent. Rather, it flows out of long fissures that extend through Earth's crust. **Flood basalts** are vast outpourings of mafic lava from fissures that can cover wide areas with multiple lava flows, building thick *lava plateaus.*

The Columbia Plateau area of Washington, Idaho, and Oregon, for example, is constructed of layer upon layer of basalt (figure 4.8), in places as thick as 3,000 meters. The area covered is over 400,000 square kilometers. Each individual flood of lava added a layer, usually between 15 and 100 meters thick and sometimes thousands of square kilometers in extent. The outpourings of lava that built the Columbia Plateau took place from 17.5 to 6 million years ago, but 95% erupted between 17 and 15.5 million years ago. Similar huge, lava plateau-building events have not

EARTH SYSTEMS 4.3

The Largest Humanly Observed Fissure Eruption and Collateral Deadly Gases

Huge eruptions from fissures, such as those of the Columbia Plateau, have not occurred in historical time. The largest fissure eruption and basaltic flood documented by humans took place in Iceland in 1783. Eruptions began when 130 cinder cones built up along a 25-kilometer-long fissure when rising magma encountered groundwater. Eventually pyroclastic activity yielded to Hawaiian-type lava flows, creating the Laki flow. Fluid basalt flowed out of the fissure for several months. Over that time some 12 cubic kilometers of basalt lava covered 565 square kilometers of land.

Along with the lava, tremendous amounts of gases were released. These had a devastating effect on Iceland's biosphere. A blue haze of gas, called a "dry fog" or "vog," hung over Iceland and parts of northern Europe for months. Fluorine in the gas contaminated grass, and over 200,000 sheep, cattle, and horses died of fluoridosis. The resulting famine was made worse because fishermen couldn't get out to sea due to the "vog." Some 10,000 Icelanders died because of the famine. That represents one-fifth of Iceland's population at that time.

In Europe the "vog" was more of an irritant to people than a danger. The winter of 1783–84 was exceptionally severe. Ben Franklin, who was the American envoy to France at the time, became the first person to link volcanic eruptions to climate changes. He suggested that the gases and dust from the eruptions may have blocked enough sunshine to result in the severe cold.

Additional Resource

The Laki and Grimsvotn Eruptions of 1883–1885 (Volcano World site)

- http://volcano.oregonstate.edu/vwdocs/volc_images/europe_west_asia/laki.html

FIGURE 4.8

Flood basalt layers in the Columbia Plateau, Dry Falls State Park, Washington.
©Cynthia Shaw

occurred since then. (The hypothesis that these are due to the arrival of huge mantle plumes beneath the lithosphere is described in chapter 3.) Even relatively small basaltic floods not associated with shield volcanoes are a rarity (see box 4.3).

Even larger basalt plateaus are found in India and Siberia. Their times of eruption coincide with the two largest mass extinctions of life on Earth. The one in Siberia occurred 252 million years ago, around the time of the largest mass extinction, when up to 96% of marine species and 70% of terrestrial species were wiped out. The eruptions are a prime suspect because of the enormous amount of gases that must have been emitted. These would have changed the atmosphere and worldwide climate. The Indian eruptions occurred about 66 million years ago and coincided with the mass extinction in which the last of the dinosaurs died. Although this mass extinction is generally blamed on a large asteroid hitting Earth (see chapter 8), the intense volcanic activity may have been a contributing factor.

Basalt plateaus have their counterparts in the oceans. These were unknown until they were discovered through deep-ocean drilling a couple of decades ago. The largest of these *oceanic plateaus* is the Ontang Java Plateau in the western Pacific ocean. This plateau is approximately the size of Alaska. A thick sequence of sedimentary rocks covers the huge volume of basalt that formed the plateau around 120 million years ago.

Columnar Jointing

Some basaltic lava flows show a distinctive feature formed of parallel, mostly six-sided, vertical columns. This characteristic is called **columnar jointing** (figure 4.9). The columns can be explained by the way in which basalt contracts as it cools *after* solidifying. Basalt solidifies completely at temperatures below about 1,000°C. The hot layer of rock then continues to cool to temperatures normal for the Earth's surface. Like most solids, basalt contracts as it cools. The layer of basalt is easily able to accommodate the shrinkage in the narrow vertical dimension; but the cooling rock cannot "pull in" its edges, which may be many kilometers away. Instead, the rock contracts toward evenly spaced centers of contraction. Tension cracks develop halfway between neighboring centers. A hexagonal fracture pattern is the most efficient way in which a set of contraction centers can

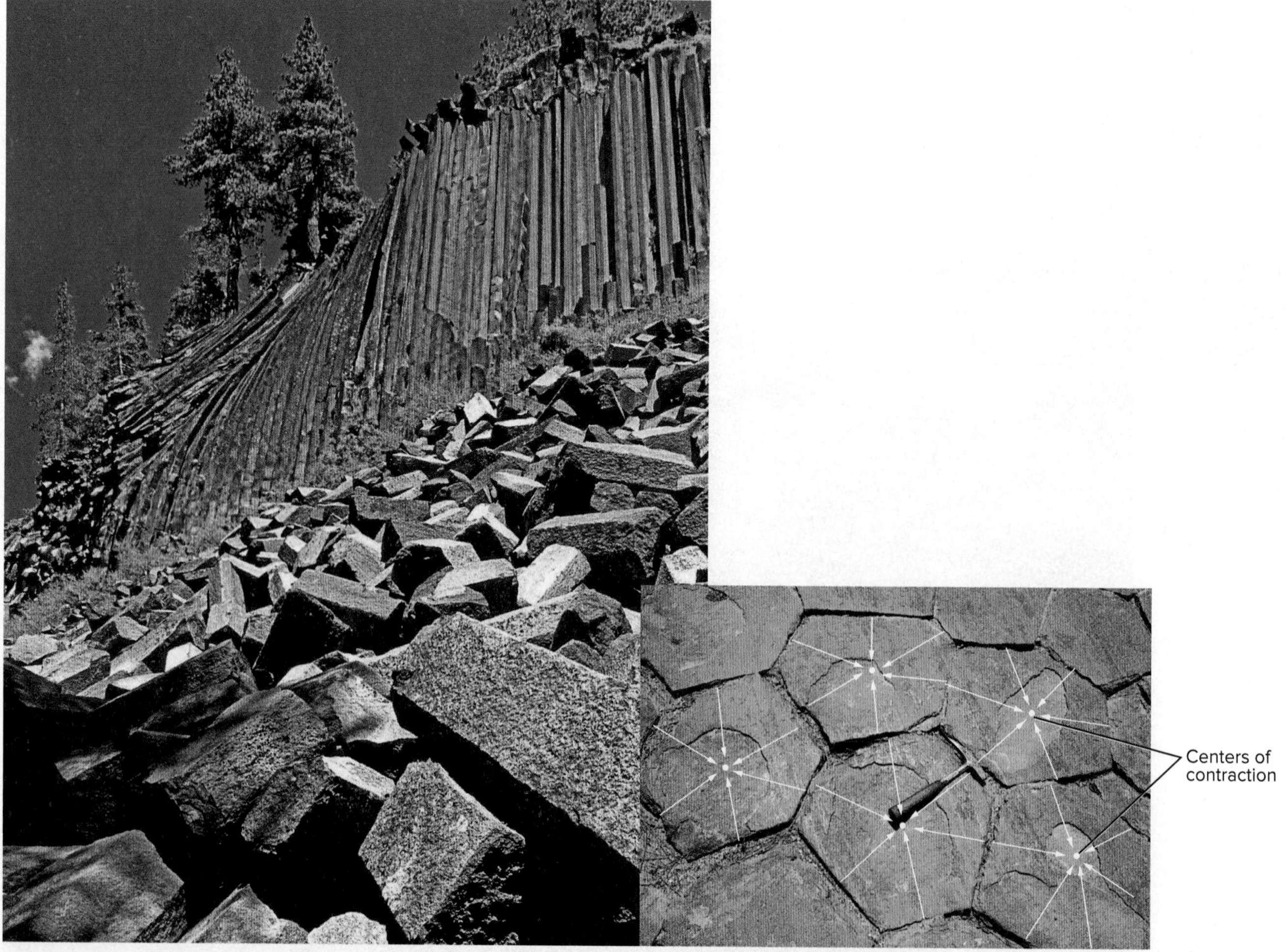

FIGURE 4.9

Columnar jointing at Devil's Postpile, California. Inset shows top view with centers of contraction drawn in. A rock hammer is used for scale. (Scratches were caused by glacial erosion as described in chapter 12.)

(A) ©Doug Sherman/Geofile RF (B) ©C. C. Plummer

FIGURE 4.10

Pillow basalt, Lillooet glacier, British Columbia, Canada.
©John J. Clague, Simon Fraser University

FIGURE 4.11

Pillow basalt erupted from the Galapagos Rift. Photo taken from a submersible vessel.
Source: Courtesy of the Galapagos Rift 2005 Expedition/NOAA Ocean Explorer

share fractures. Although most columns are six-sided, some are five- or seven-sided.

Submarine Lava Flows

When lava erupts into water it cools rapidly, forming a distinctive feature known as **pillow structure**–rocks, generally basalt, occurring as pillow-shaped, rounded masses closely fitted together (figure 4.10). From observations of submarine eruptions by divers, we know how these **pillow basalts** are produced: Fluid, pahoehoe-type lava flows into water. Elongate blobs of lava break out of a thin skin of solid basalt over the top of a flow that is submerged in water. Each blob is squeezed out like toothpaste, and its surface is chilled to rock within seconds. A new blob forms as more lava inside breaks out. Each new pillow settles down on the pile, with little space left in between. Some pillow basalt forms in lakes and rivers or where lava flows from land into the sea (as in Hawaii). However, most pillow basalt forms at mid-oceanic ridge crests.

According to plate tectonic theory, basalt magma flows up the fracture that develops at a divergent boundary (explained in chapter 3). The magma that reaches the sea floor solidifies as pillow basalt (figure 4.11). The rest solidifies in the fracture as a dike. Pillow basalt that is overlying a series of dikes is sometimes found in mountain ranges. These probably formed during sea-floor spreading in the distant past followed, much later, by uplift.

Intermediate and Felsic Lava Flows

Because intermediate and felsic lavas are much more viscous than mafic lava, flows tend to be thicker and flow over much shorter distances. Sometimes the lava is too viscous to flow and will build up into a lava dome (described later in this chapter). Intermediate lava, having intermediate viscosity, forms lava flows of intermediate thickness. Occasionally, more felsic lavas can flow over greater distances if they are very hot when they erupt.

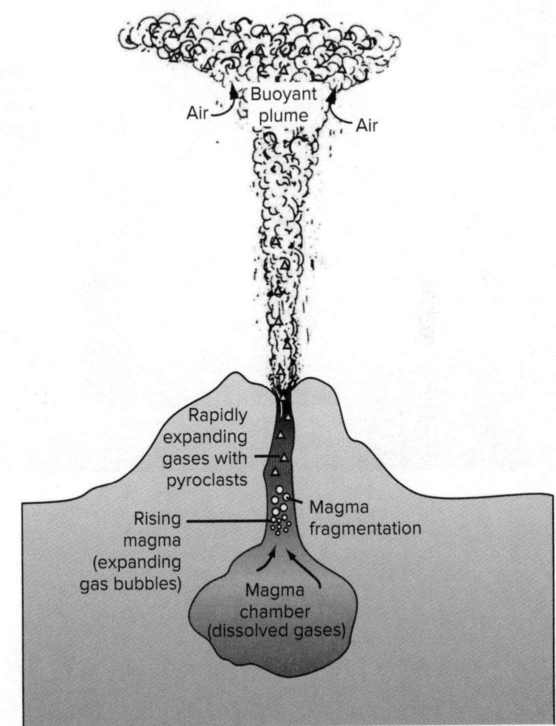

FIGURE 4.12

Generation of an explosive eruption. Gases dissolved in magma come out of solution and expand as magma rises from the magma chamber toward the surface. The expanding gases fragment the viscous magma, generating an upward blast of hot gas and pyroclasts. The jet draws in air as it rises, expanding to form a buoyant plume.

Explosive Eruptions

Explosive eruptions are driven by the expansion of gases in viscous magma (figure 4.12). When the magma is deep underground, the gases are kept dissolved by the pressure. When the magma rises toward the surface, the decrease in pressure causes the gas to come out of solution and expand. Imagine a bottle of soda. Before opening it, you see very few bubbles in the liquid because the gas (carbon dioxide) is under pressure. When you open the bottle, you release the pressure and gas comes out of solution and expands, forming bubbles. If you release the pressure too quickly, the bubbles expand so much that the mixture of soda and bubbles spills out of the bottle. During an explosive eruption, expanding, hot gases fragment the rapidly cooling magma into pieces and blast them into the air. These fragments are know as **pyroclasts** (from the Greek *pyro* "fire" and *klastos* "broken"). The hot gas and pyroclasts are blasted upward as a plume, which draws in air as it rises.

Pyroclastic Materials

Also known as *tephra,* pyroclasts can range in size from fine **dust** and **ash** (figures 4.13*A* and *B*) to blocks six meters or more in diameter (20 feet, the size of a recreational vehicle).

Pyroclasts are named according to their size:

Dust	<1/8 millimeter
Ash	1/8–2 millimeters
Cinder or lapilli	2–64 millimeters
Blocks and bombs	>64 millimeters

Cinder is often used as a less-restricted, general term for smaller pyroclasts. **Lapilli** is used for the 2- to 64-millimeter particles—a size range that extends from that of a grain of rice to a peach. When solid rock has been blasted apart by a volcanic explosion, the pyroclastic fragments are *angular,* with no rounded edges or corners, and are called **blocks.** If lava is ejected into the air, a molten blob becomes streamlined during flight, solidifies, and falls to the ground as a **bomb**, a spindle or lens-shaped pyroclast (figures 4.13*C* and 4.13*D*).

During an eruption, expanding hot gases can propel pyroclasts high into the atmosphere as a column rising from a

FIGURE 4.13

Pyroclastic material (*A*) Ash. (*B*) Planes covered with ash from the 1980 eruption of Mt. St. Helens. (*C*) Volcanic bombs. (*D*) Lava bomb trajectories are visible during a May 2008 nighttime eruption of Anak Kraktau volcano in Indonesia. Magma blobs that solidify in the air will land as bombs. If they are still molten upon landing, they will spatter.
(A) Source: David Wieprecht/U.S. Geological Survey (B) Source: P. W. Lipman/U.S. Geological Survey (C) ©C. C. Plummer; (D) ©Richard Roscoe/Stocktrek Images RF

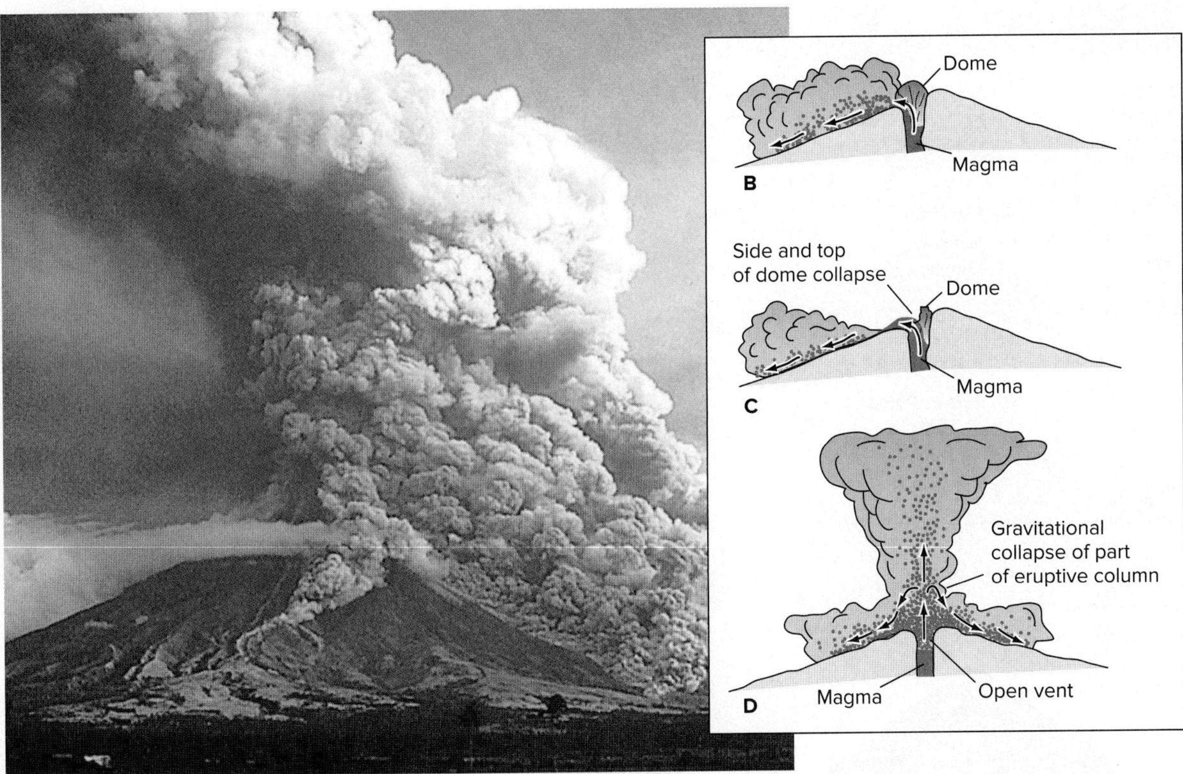

FIGURE 4.14

(*A*) Pyroclastic flow descending Mayon volcano, Philippines (elevation 2,460 meters), in 1984. Ways in which pyroclastic flows can form: (*B*) Blasting out from under a plug capping a volcano. (*C*) Collapse of part of a steep-sided dome. (*D*) Gravitational collapse of an eruptive column.
(A) Source: Chris Newhall, U.S. Geological Survey

volcano. At high altitudes, the pyroclasts often spread out into a dark mushroom cloud. The fine particles are transported by high atmospheric winds. Eventually, debris settles back to Earth under gravity's influence as *pyroclastic fall* (often called *ashfall* or *pumice fall*) deposits.

Pyroclastic Flows

A **pyroclastic flow** is a mixture of gas and pyroclastic debris that is so dense that it hugs the ground as it flows rapidly into low areas (figure 4.14). Pyroclastic flows develop in several ways. Some are associated with volcanic domes. An exploding froth of gas and magma can blast out of the side of a dome or viscous plug capping a volcano. A steep-sided dome might collapse, allowing violent release of magma and its gases. For some volcanoes, a pyroclastic flow results from gravitational collapse of a column of gas and pyroclastic debris that was initially blasted vertically into the air. These turbulent masses can travel up to 200 kilometers per hour and are extremely dangerous. In 1991, a pyroclastic flow at Japan's Mount Unzen killed 43 people, including three geologists and famous volcano photographers Maurice and Katia Krafft. Far worse was the destruction of St. Pierre on the Caribbean island of Martinique, where about 29,000 people were killed by a pyroclastic flow in 1902 (see box 4.4).

4.4 TYPES OF VOLCANOES

The major types of volcanoes (shield, cinder cone, composite, lava dome, and caldera) are markedly distinct from one another in size, shape, and, usually, composition. Table 4.2 provides a comparison of shield volcanoes, cinder cones, and composite volcanoes. Note from the scales that the shield volcano shown is vastly bigger than the other two, and the composite volcano is much bigger than the cinder cone.

Shield Volcanoes

Shield volcanoes are broad, gently sloping volcanoes constructed of solidified lava flows. Eruptive activity at shield volcanoes is dominated by effusive eruption of low viscosity, mostly basaltic lava. During eruptions, the lava spreads widely and thinly due to its low viscosity. Because the lava flows from a central vent, without building up much near the vent, the slopes are usually between 2 degrees and 10 degrees from the horizontal, producing a volcano in the shape of a flattened dome or "shield" (figure 4.15).

The islands of Hawaii are essentially a series of shield volcanoes, built upward from the ocean floor by intermittent eruptions over millions of years (figure 4.15*B*). Although spectacular

TABLE 4.2 Comparison of the Three Types of Volcanoes

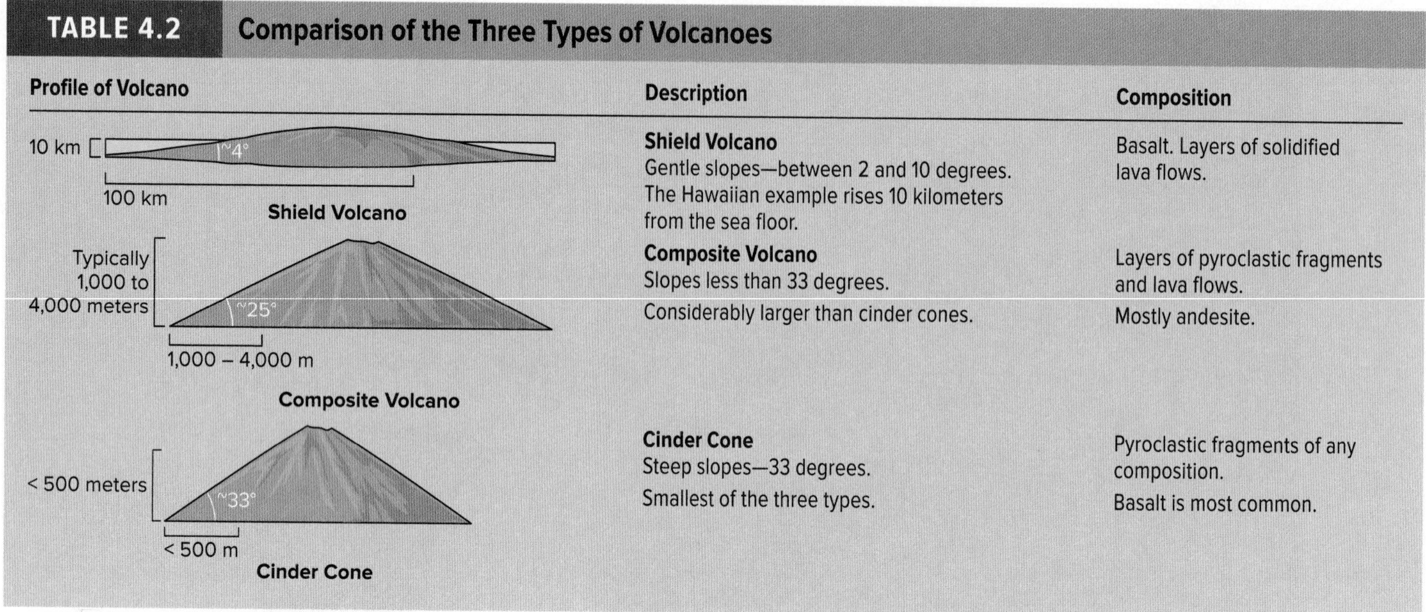

Profile of Volcano	Description	Composition
Shield Volcano	**Shield Volcano** Gentle slopes—between 2 and 10 degrees. The Hawaiian example rises 10 kilometers from the sea floor.	Basalt. Layers of solidified lava flows.
Composite Volcano	**Composite Volcano** Slopes less than 33 degrees. Considerably larger than cinder cones.	Layers of pyroclastic fragments and lava flows. Mostly andesite.
Cinder Cone	**Cinder Cone** Steep slopes—33 degrees. Smallest of the three types.	Pyroclastic fragments of any composition. Basalt is most common.

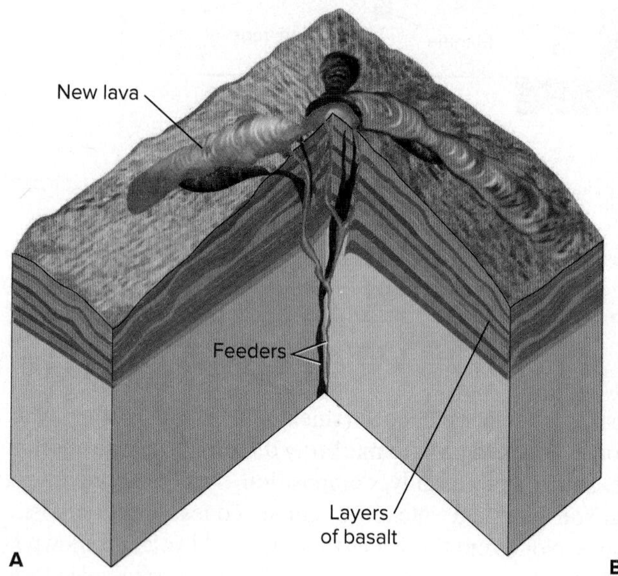

FIGURE 4.15

(*A*) Cutaway view of a shield volcano. (*B*) The top of Mauna Loa, a shield volcano in Hawaii, and its summit caldera, which is approximately 2 kilometers wide. The smaller depressions are pit craters.
(B) Source: D. W. Peterson/USGS

to observe, the effusive eruptions are relatively nonviolent because the lavas are fairly fluid (low viscosity). The features associated with effusive eruptions described earlier in the section on the eruptive products of volcanoes (pahoehoe, a'a, spatter cones, lava tubes, etc.) are all common on the islands of Hawaii. In fact, the names pahoehoe and a'a are Hawaiian names, and shield volcanoes got their name from their resemblance to Hawaiian warriors' shields.

Measuring nearly 25 kilometers (16 miles) in height and 624 kilometers (374 miles) across, the largest shield volcano in the solar system is Olympus Mons on Mars. For more information on extraterrestrial volcanic activity, see box 4.5.

Cinder Cones

A **cinder cone** (less commonly called a *pyroclastic cone*) is a volcano constructed of pyroclastic fragments ejected from a central vent (figure 4.16). Unlike a shield volcano, which is made up of lava flows, a cinder cone is formed exclusively of pyroclasts. In contrast to the gentle slopes of shield volcanoes, cinder cones commonly have slopes of about 30 degrees. Most of the ejected material lands near the vent during an eruption, building up the cone to a peak. The steepness of slopes of accumulating loose material is limited by gravity to about 33 degrees. Cinder cones tend to be very much smaller than shield volcanoes. In fact,

FIGURE 4.16

Cerro Negro, a 230-meter-high cinder cone in Nicaragua, erupting.
Source: Mark Hurd Aerial Surveys Corp. Courtesy California Geological Survey

cinder cones are commonly found on the flanks and in the calderas of Hawaii's shield volcanoes. Few cinder cones exceed a height of 500 meters.

Cinder cones form by pyroclastic material accumulating around a vent. They form because of a buildup of gases and are independent of composition. Most cinder cones are associated with mafic or intermediate lava. Felsic cinder cones, which are made of fragments of pumice, are also known as pumice cones.

The life span of an active cinder cone tends to be short. The local concentration of gas is depleted rather quickly during the eruptive periods. Moreover, as landforms, cinder cones are temporary features in terms of geologic time. The unconsolidated pyroclasts are eroded relatively easily.

Composite Volcanoes

A **composite volcano** (also called a **stratovolcano**) is one constructed of alternating layers of pyroclastic fragments and solidified lava flows (figure 4.17). The slopes are intermediate in steepness compared with cinder cones and shield volcanoes. Pyroclastic layers build steep slopes as debris collects near the vent, just as in cinder cones. However, subsequent lava flows partially flatten the profile of the cone as the downward flow builds up the height of the flanks more than the summit area. The solidified lava acts as a protective cover over the loose pyroclastic layers, making composite volcanoes less vulnerable to erosion than cinder cones.

Composite volcanoes are built over long spans of time. Eruption is intermittent, with hundreds or thousands of years of inactivity separating a few years of intense activity. During the quiet intervals between eruptions, composite volcanoes may be eroded by running water, landslides, or glaciers. These surficial processes tend to alter the surface, shape, and form of the cone. But because of their long lives and relative resistance to erosion, composite cones can become very large.

The extrusive material that builds composite cones is predominantly of intermediate composition, although there may be some felsic and mafic eruptions. Therefore, *andesite* is the rock most associated with composite volcanoes. If the lava is especially hot, the relatively low-viscosity fluid flows easily from the crater down the slopes. On the other hand, if enough gas pressure exists, an explosive eruption may litter the slopes with pyroclastic andesite, particularly if the lava has fully or partially solidified and clogged the volcano's vent.

The composition as well as the eruptive history of individual volcanoes can vary considerably, even within a single volcanic arc. Take for instance the Cascade volcanoes (figure 4.18) of Washington, Oregon, and northern California. Mount Rainier is composed of 90% lava flows and only 10% pyroclastic layers. Conversely, Mount St. Helens was built mostly from pyroclastic eruptions–reflecting a more violent history. As would be expected, the composition of the rocks formed during the 1980 eruptions of Mount St. Helens is somewhat higher in silica than average for Cascade volcanoes.

Distribution of Composite Volcanoes

Nearly all the larger and better-known volcanoes of the world are composite volcanoes. They tend to align along two major belts (figure 4.19). The **circum-Pacific belt**, or "Ring of Fire," is the larger. The Cascade Range volcanoes (figure 4.18), including Mount St. Helens, Crater Lake, and Lassen Peak, make up a small segment of the circum-Pacific belt.

Further south, the Mexican Volcanic Belt has several composite volcanoes, some that rise higher than 5,000 meters, including Orizaba (third-highest peak in North America) and Popocatépetl. Popocatépetl (affectionately called "Popo"), at 5,484 meters (17,991 feet) above sea level, is one of North America's highest mountains. It is 55 kilometers east of Mexico City, one of the world's most populous cities. Popo awakened from a long period of dormancy in 1994. In December 2000, Popo had its largest eruption in over 1,000 years, and 50,000 people near its flanks were evacuated to shelters. On January 31, 2001, a pyroclastic flow descended the volcano to within 8 kilometers of

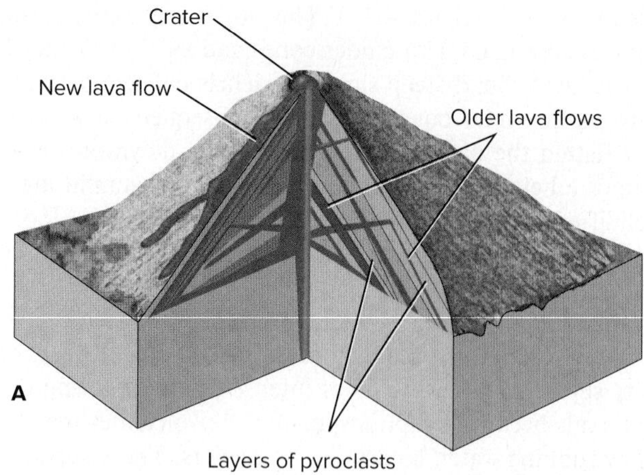

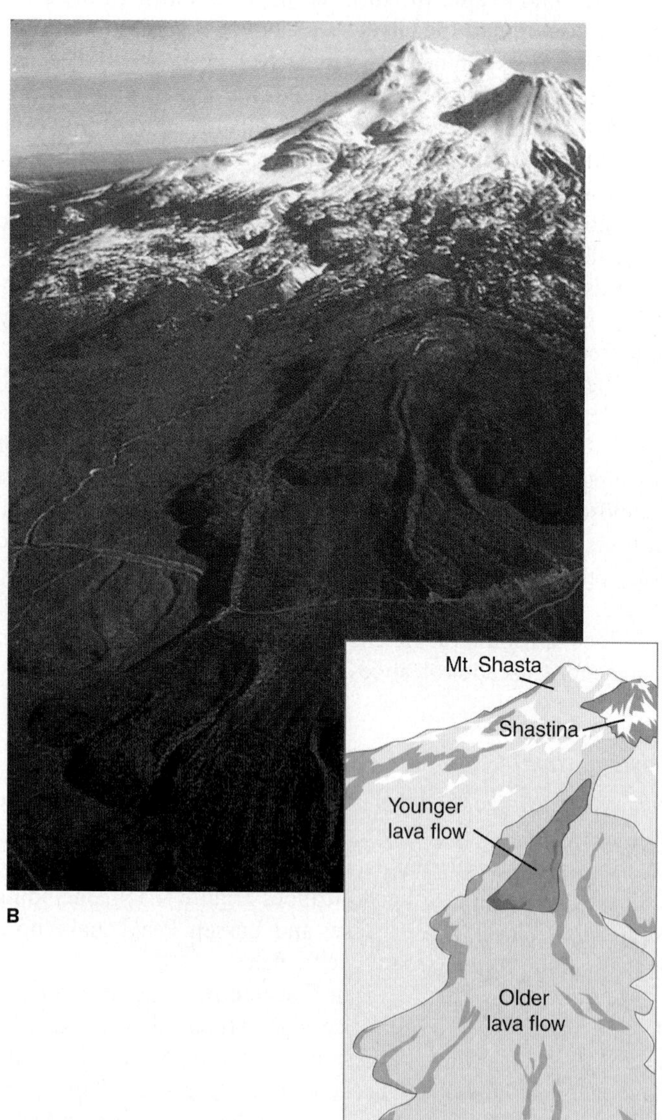

FIGURE 4.17

(*A*) Cutaway view of a composite volcano. Light-colored layers are pyroclasts. (*B*) Mount Shasta, a composite volcano in California. Shastina on Mount Shasta's flanks is a subsidiary cone, largely made of pyroclasts. Note the lava flow that originated on Shasta and extends beyond the volcano's base.
(B) ©B. Amundson

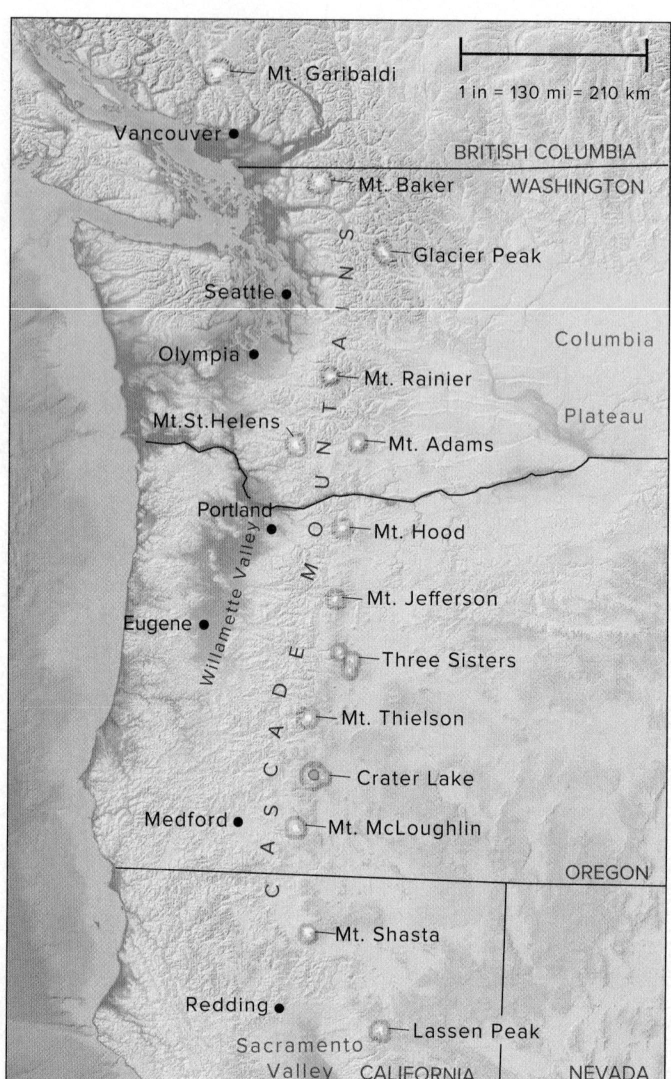

FIGURE 4.18

The Cascade volcanoes. The named volcanoes are ones that have erupted in geologically recent time.
Source: U.S. Geological Survey

a town. The volcano has continued to be sporadically active since then.

The circum-Pacific belt includes many volcanoes in Central America, western South America, and Antarctica. Mount Erebus, in Antarctica, is the southernmost active volcano in the world.

The western portion of the Pacific belt includes volcanoes in New Zealand, Indonesia (including Mount Merapi), the Philippines (with Pinatubo, whose 1991 caldera-forming eruption was the second-largest eruption of the twentieth century), and Japan. The beautifully symmetrical Fujiyama, in Japan, is probably the most frequently photographed and painted volcano in the world, as well as its most-climbed mountain. The northernmost part of the circum-Pacific belt includes active volcanoes in Russia (see figure 4.3) and on Alaska's Aleutian Islands. The 1912 eruption of Novarupta

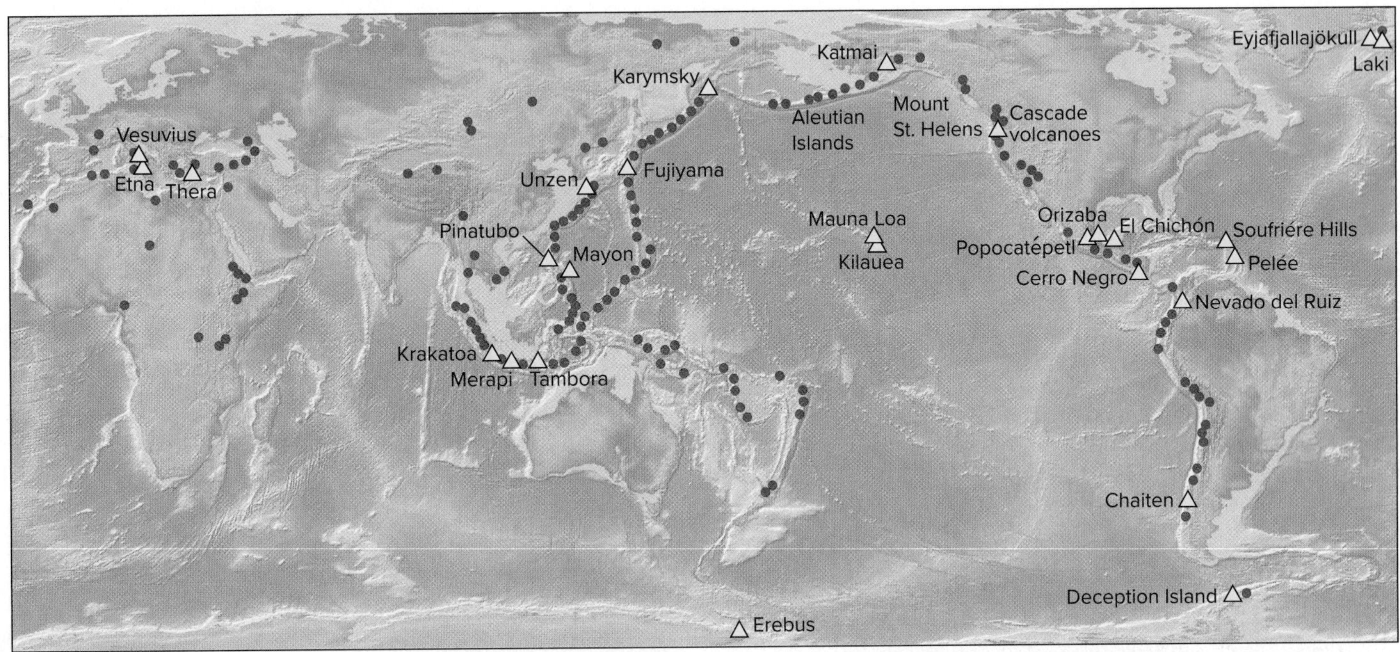

FIGURE 4.19

Map of the world showing recently active major volcanoes. Red dots represent individual volcanoes. Yellow triangles represent volcanoes mentioned in this chapter.

in Alaska's Katmai National Park was the world's largest in the twentieth century.

The second major volcanic belt is the **Mediterranean belt**, which includes Mount Vesuvius, Mount Etna, and Mount Thera. An exceptionally violent eruption of Mount Thera, an island in the Mediterranean, may have destroyed an important site of early Greek civilization. (Some archaeologists consider Thera the original "lost continent" of Atlantis.) Mount Etna, on the island of Sicily, is Europe's largest volcano and one of the world's most active volcanoes. Its largest eruption in 300 years began in 1991 and lasted for 473 days. Some 250 million cubic meters of lava covered 7 square kilometers of land. A town was saved from the lava by heroic efforts that included building a dam to retain the lava (the lava quickly overtopped it), plugging some natural channels, and diverting the lava into other, newly constructed channels. Mount Vesuvius, located near the Italian city of Naples, erupted in A.D. 79, destroying the Roman cities of Pompeii and Herculaneum (see section 4.5).

Lava Domes

Lava domes are steep-sided, dome- or spine-shaped masses of volcanic rock formed from viscous lava that solidifies in or immediately above a volcanic vent. Lassen Peak, the southernmost of the Cascade volcanoes, is a lava dome. Most of the viscous lavas that form volcanic domes are high in silica. They commonly solidify as obsidian, the glassy chemical equivalent of rhyolite (or, less commonly, andesite). If minerals do crystallize, the rock is rhyolite if it is from a felsic magma, or andesite if it is from an intermediate magma.

Volcanic domes often form within the craters of composite volcanoes. A volcanic dome grew within the crater of Mount St. Helens after the climactic eruption of May 1980. This was expected because of the high viscosity of the lava from the eruptions. In 1983 alone, the dome increased its elevation by 200 meters. After years of quiescence, dome growth resumed in October 2004. At that time, lava extrusion shifted, and a new dome began growing adjacent to the original dome. In 2005, 70 million cubic meters of lava were extruded to build seven domes in the crater. Lava extruded at a rate of one large pickup truck load per second. The volcano has been quiet again since January 2008. Go to http://volcanoes.usgs.gov/volcanoes/st_helens/st_helens_multimedia_11.html and scroll to the bottom of the page to find time-lapse movies of the dome growth activity.

Because the thick, pasty lava that squeezes from a vent is too viscous to flow, it builds up a steep-sided dome or spine (figure 4.20). Some volcanic domes act like champagne corks, keeping gases from escaping. If the plug is removed or broken, the gas and magma escape suddenly and violently, usually as a pyroclastic flow. Some of the most destructive volcanic explosions known have been associated with volcanic domes (see box 4.4).

Calderas

A **caldera** is a volcanic depression much larger than the original crater, having a diameter of at least 1 kilometer. A caldera can be created when a volcano's summit is blown off by exploding gases or when a volcano (or several volcanoes) collapses into a partially emptied magma chamber (see figure 4.21). Caldera-forming

A Viscous lava wells up into a crater.

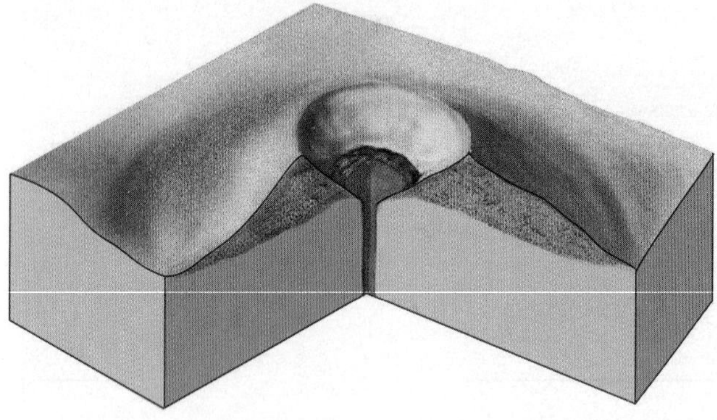

B A dome grows as more magma is extruded. The outer part is solid and breaks as the growing dome expands.

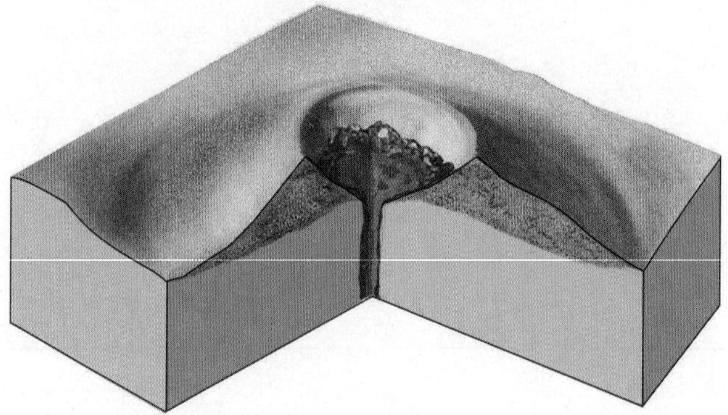

C If magma continues to be fed into the steep-sided dome, it may rise above the rim of the crater.

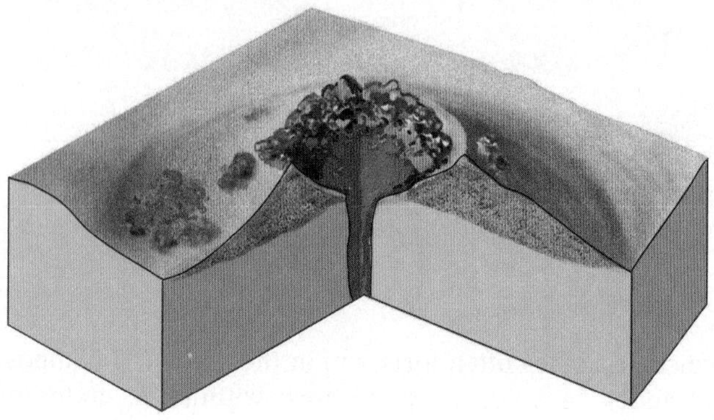

D

FIGURE 4.20

A volcanic dome forming in the crater of a pumice cone (*A, B, C*). (*D*) Mono Craters, eastern California, is a line of craters with lava domes. The dome in the crater in the foreground has not grown above the level of the crater's rim (like *B*). Some in the background have overtopped their rims. You can also see some short and steep lava flows, reflecting the very viscous silicic lava that erupted. The photo of pumice in figure 3.10*A* was taken on the flanks of the cinder cone in the foreground. A two-lane highway provides a scale for the photo. The Sierra Nevada range is on the skyline.
(D) Source: C. Dan Miller, U.S. Geological Survey

eruptions are extremely violent, blasting huge volumes of pyroclastic material into the atmosphere and generating pyroclastic flows that can blanket vast areas in sometimes hundreds of meters of hot volcanic debris.

The most famous caldera in the United States is misnamed "Crater Lake." A series of eruptions in prehistoric time (about 7,700 years ago) created the depression now occupied by Crater Lake in Oregon (figure 4.21). Approximately 50 cubic kilometers of volcanic debris was erupted and now covers more than a million square kilometers in Oregon and neighboring states. The original volcano, named Mount Mazama (now regarded as a cluster of overlapping volcanoes), is estimated to have been about 2,000 meters higher than the present rim of Crater Lake. For more on Crater Lake and Mount Mazama, go to http://pubs.usgs.gov/fs/2002/fs092-02/.

The eruptions that formed Crater Lake were relatively small when compared to two other calderas in the United States. About 640,000 years ago a massive eruption at Yellowstone formed a caldera 70 by 50 kilometers across. The eruption blasted 1,000 cubic kilometers of rhyolitic pyroclastic material into the atmosphere. For comparison, the 1980 eruption of Mount St. Helens erupted only about 1 cubic kilometer of lava. Enormous pyroclastic flows buried the surrounding area with thick deposits of hot pyroclastic material. Ash from the eruption covers much of central North America. The Long Valley caldera in California formed about 760,000 years ago with the eruption of approximately 600 cubic kilometers of material. Ash from this eruption is found as far away as Nebraska. Figure 4.22 shows the areas covered in at least 2 centimeters of ash from the Crater Lake, Yellowstone, and Long Valley eruptions. Ash from the 1980 eruption of Mount St. Helens is shown for comparison. Both Yellowstone and Long Valley are considered to be active, and while eruptions of the scale that formed the calderas are not believed to be imminent, volcanologists monitor them carefully.

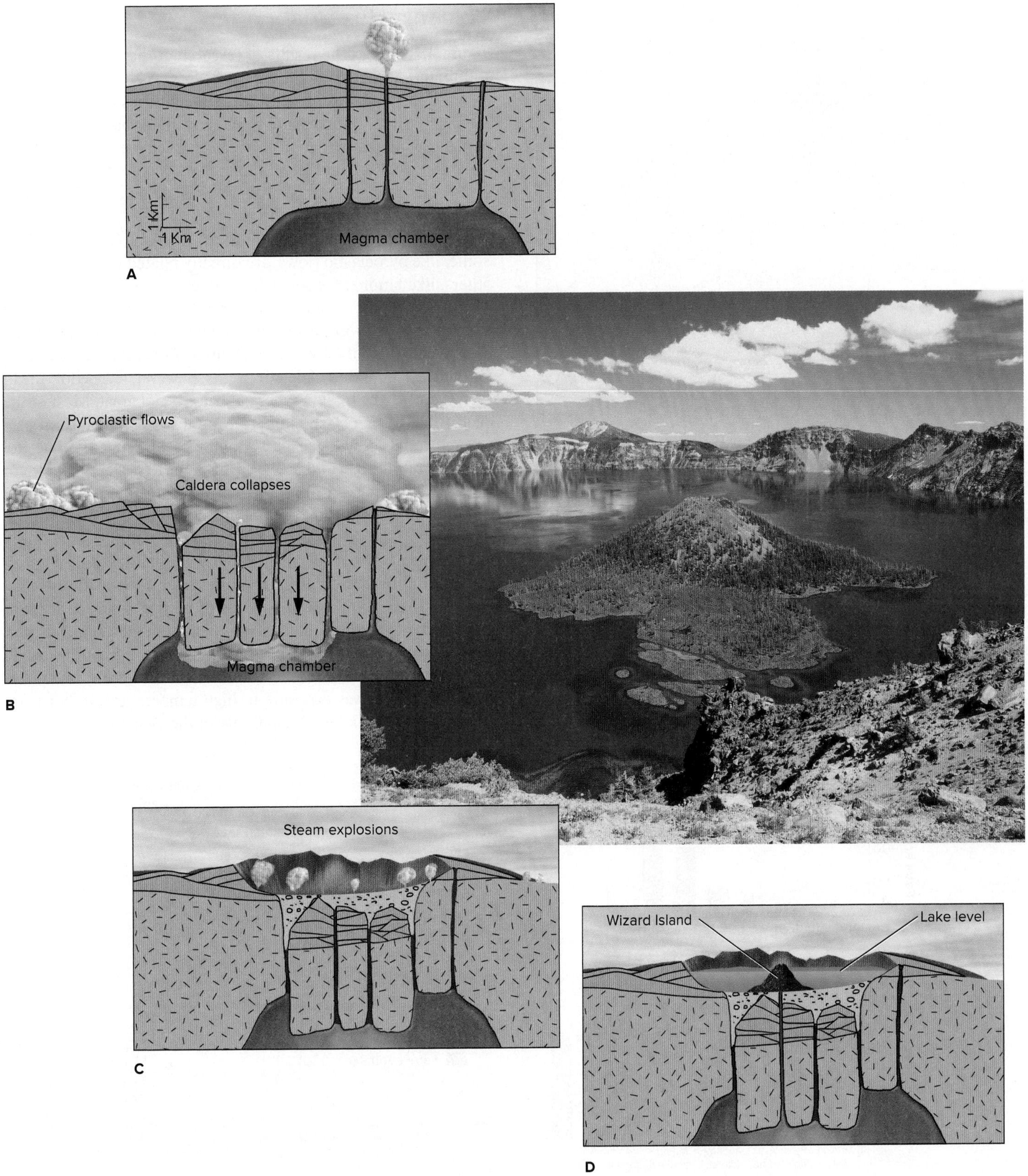

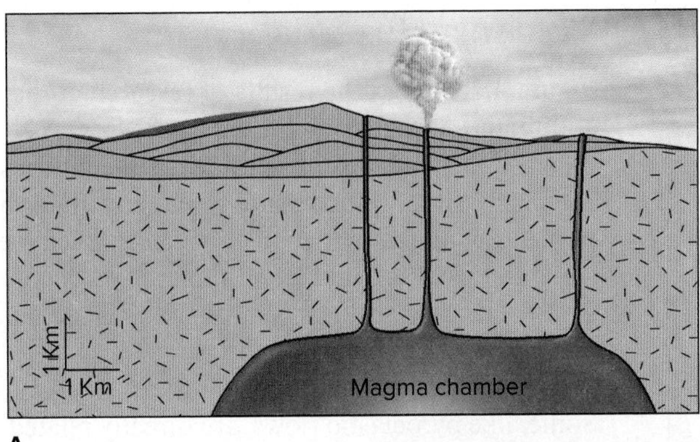

A

Pyroclastic flows

Caldera collapses

Magma chamber

B

Steam explosions

C

Wizard Island

Lake level

D

FIGURE 4.21

Crater Lake, Oregon. The lake is approximately 10 kilometers (6 miles) across. Its development and geologic history: (*A*) Cluster of overlapping volcanoes forms. (*B*) Collapse into the partially emptied magma chamber is accompanied by violent eruptions. (*C*) Volcanic activity ceases, but steam explosions take place in the caldera. (*D*) Water fills the caldera to become Crater Lake, and minor renewed volcanism builds a cinder cone (Wizard Island).
(A-D) Source: C. Bacon, U.S. Geological Survey; (C) ©Robert Glusic/Corbis RF

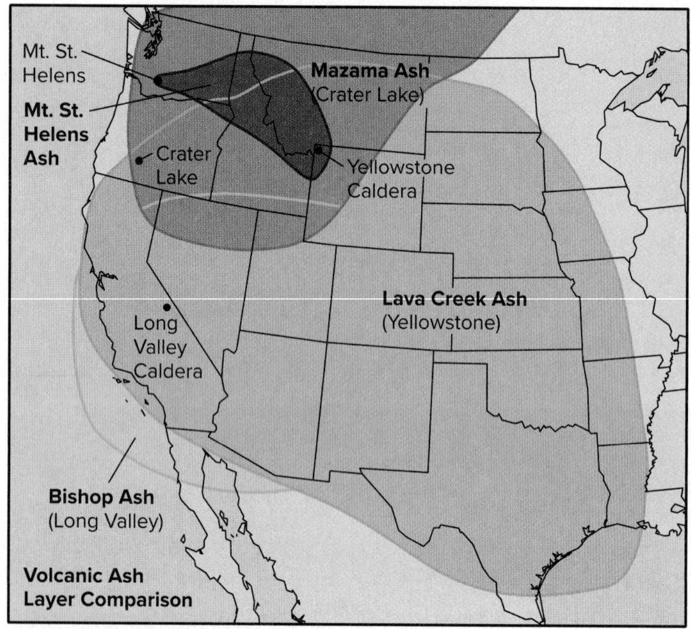

FIGURE 4.22

Volcanic ash deposits (>2 centimeters) from caldera eruptions at Crater Lake, Yellowstone, and Long Valley. Ash from the 1980 eruption of Mount St. Helens is shown for comparison.

4.5 LIVING WITH VOLCANOES

Volcanic eruptions present a number of challenges to human beings. Hazards include ash clouds, pyroclastic flows, and mudflows. Reducing the impact of volcanic eruptions (called *mitigation*) involves monitoring, hazard mapping, and alerting the population about a potential eruption.

Volcanic Hazards

Figure 4.23 shows the results of research at the Smithsonian Institute and Macquarie University, Australia. Note the dramatic increase in fatalities during the recent centuries (figure 4.23*A*). This is not due to increasing volcanic activity but to increasing population and more people living near volcanoes. Figure 4.23*B*, which shows the cumulative number of deaths during the last seven centuries, also shows that most of the fatalities have been caused by seven major eruptions.

Volcanoes can kill in a number of ways (figure 4.23*C*). Some, like pyroclastic flows, are directly related to the eruption; others, like famine, are secondary effects.

Pyroclastic flows, already described in detail earlier in this chapter, are the most deadly volcanic hazards, accounting for approximately 30% of all volcano fatalities. *Pyroclastic fall* accounts for the largest number of deadly events; however, few people die in each event, so the total number of deaths is not great. Most of the deaths due to pyroclastic fall are caused by the collapse of ash-covered roofs or by being hit by falling rock fragments. The Roman city of Pompeii and at least four other towns near Naples in Italy were destroyed in A.D. 79 when Mount Vesuvius erupted (figure 4.24). Before the eruption, vineyards on the flanks of the apparently "dead" volcano extended to the summit. After Vesuvius erupted without warning, Pompeii was buried under 5 to 8 meters of hot ash. Seventeen centuries later, the town was rediscovered. Excavation revealed molds of people, many with facial expressions of terror. Initially it was believed that the people had been suffocated by ashfall, but recent studies suggest that while 38% of the deaths were due to buildings collapsing under the weight of ash, it was exposure to high temperatures in hot pyroclastic surges that killed the majority of the victims.

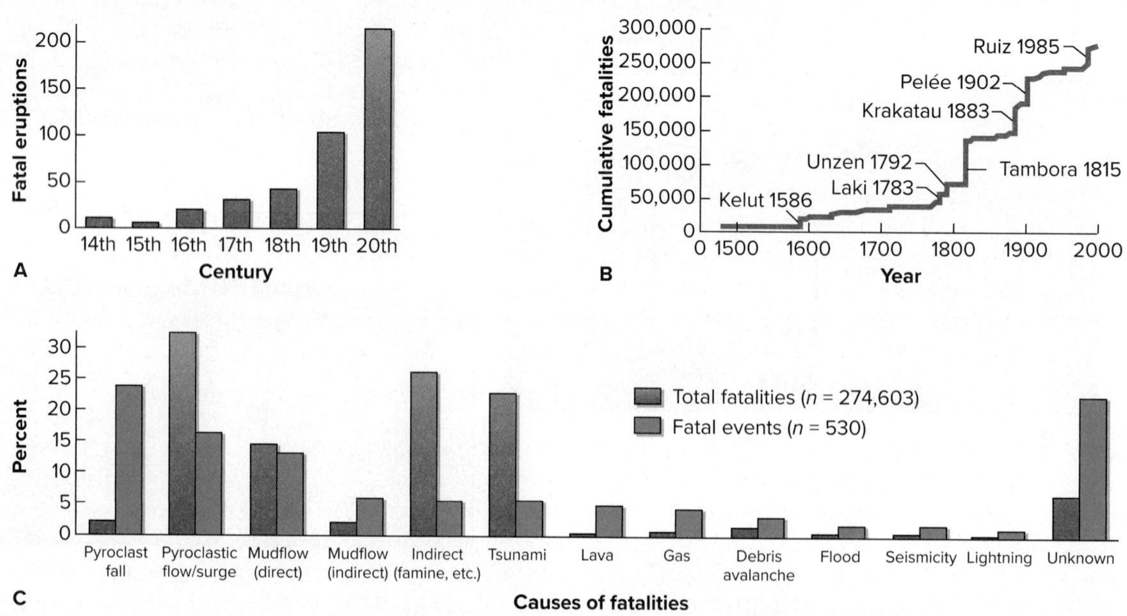

FIGURE 4.23

Volcano fatalities. (*A*) Fatal volcanic eruptions per century. (*B*) Cumulative volcano fatalities. Note the big jumps with the seven most deadly eruptions. These were eruptions that killed over 10,000 people and account for two-thirds of the total. (*C*) The causes of volcano fatalities.
Source: T. Simkin, L. Siebert, and R. Blong, "Volcano Fatalities," *Science, 291, 255. American Association for the Advancement of Science, 2001.*

FIGURE 4.24

(*A*) Pompeii with Mount Vesuvius in the background. (*B*) Casts of bodies of people who died in Pompeii, buried by ash from the eruption of Vesuvius, A.D. 79. The casts were made by pouring plaster into voids in the ash left by the dead.
(A) ©Olivier Goujon/Robert Harding World Imagery/ Getty Images (B) ©Bettmann/Corbis/Getty Images

Volcanic eruptions are often accompanied by the release of large amounts of gas such as water vapor, carbon dioxide, and sulfur dioxide. As you've learned, it is the expansion of gas, primarily water vapor, that drives explosive eruptions. Sometimes the gas itself can be a deadly hazard. Lake Nyos, Cameroon, in western Africa is an example of this type of event. Carbon dioxide seeped into the lake within the crater of an active volcano. The gas dissolved into the cold water at the bottom of the lake, slowly saturating the water with carbon dioxide. On August 21, 1986, the lake suddenly released a large cloud of carbon dioxide. Being heavier than air, the carbon dioxide flowed down the slopes of the volcano, engulfing nearby villages. About 1,700 people living within 25 kilometers (16 miles) of the lake were suffocated; at least 3,500 head of cattle also died.

Volcanic mudflows, called **lahars**, are responsible for about 15% of volcano-related fatalities. Explosive eruptions deposit large amounts of loose pyroclastic material over the slopes of a volcano and the surrounding area. When this material mixes with rainwater or snowmelt, it forms a dense slurry of water, ash, and even large boulders that flows rapidly down the steep slopes of the volcano. In 1985, 23,000 people in the town of Armero, Colombia, were killed when hot volcanic debris blasted from Nevado del Ruiz in Colombia melted ice and snow capping the peak. The meltwater mixed with the pyroclastic material, forming a lahar that buried the town.

Famine and other indirect causes account for 23% of fatalities. Widespread destruction of crops and farm animals can cause regional famine (as occurred with the eruption of Tambora in 1815). Note the large number of deaths attributable to famine were the result of relatively few events.

Lightning is a spectacular and sometimes deadly effect of volcanic eruptions. Volcanic lightning (see the image of Chaitén volcano in Chile at the beginning of this chapter) is generated by tiny particles of ash thrown out by the volcano. The ash is believed to cause friction that generates an electrical charge. During the eruption of Parícutin in Mexico that destroyed two villages, the only three fatalities were due to volcanic lightning.

Monitoring Volcanoes

A volcano is considered **active** if it is currently erupting or has erupted recently. Volcanoes that have not erupted in many thousands of years but that are expected to at some point in the future are considered **dormant**. **Extinct volcanoes** are those that have not erupted for a very long time and show no signs of ever erupting again. Volcanoes can lie dormant for very long periods of time, making the distinction between extinct and dormant difficult to make. For example, the Soufriére Hills volcano on Montserrat (see box 4.4) was thought to be extinct before waking up in 1995.

ENVIRONMENTAL GEOLOGY 4.4

A Tale of Two Volcanoes—Lives Lost and Lives Saved in the Caribbean

Montserrat and Martinique are two of the tropical islands that are part of a volcanic island arc (box figure 1). During the twentieth century, both islands had major eruptions that destroyed towns. Violent and deadly pyroclastic flows associated with the growth of volcanic domes caused most of the destruction. For one island, the death toll was huge, and for the other, it was minimal.

In 1902, the port city of St. Pierre on the island of Martinique was destroyed after a period of dome growth and pyroclastic flows on Mount Pelée (no relationship to Pele, Hawaii's goddess of volcanoes). A series of pyroclastic flows broke out of a volcanic dome and flowed down the sides of the volcano. Searingly hot pyroclastic flows can travel at up to 200 kilometers per hour and will destroy any living things in their paths. After the pyroclastic flows began, the residents of St. Pierre became fearful, and many wanted to leave the island. The authorities claimed there was no danger and prevented evacuation. There was an election coming up, and the governor felt that most of his supporters lived in the city. He did not want to lose their votes, but neither the governor nor any of the city's residents would ever vote. The climax came on the morning of May 8, when great fiery, exploding clouds descended like an avalanche down the mountainside, raced down a stream valley, through the port city and into the harbor. St. Pierre and the ships anchored in the harbor were incinerated. Temperatures within the pyroclastic flow were estimated at 700°C. Some of the dead had faces that appeared unaffected by the incinerating storm. However, the backs of their skulls were blasted open by their boiling brains. About 29,000 people were burned to death or suffocated (of the two survivors in St. Pierre, one was a condemned prisoner in a poorly ventilated dungeon).

Ninety-three years later, in July 1995, small steam-ash eruptions began at Soufriére Hills volcano on the neighboring island of Montserrat. As a major eruption looked increasingly likely, teams of volcanologists from France, the United Kingdom, the United States (including members of the U.S. Geological Survey's Volcano Disaster Assistance Team that had successfully predicted the eruption of Mount Pinatubo in the Philippines, as described in chapter 1), and elsewhere flew in to study the volcano and help assess the hazards. An unprecedented array of modern instruments (including seismographs, tiltmeters, and gas analyzers) were deployed around the volcano. In November 1995, viscous, andesitic lava built a dome over the vent. Pyroclastic flows began when the dome collapsed in March 1996. Pyroclastic flows continued with more dome building and collapsing. By 1997, nearly all of the people in the southern part of the island were evacuated, following advice from the scientific teams. In June 1997, large eruptions took place, and pyroclastic flows destroyed the evacuated capital city of Plymouth. In contrast to the tragedy of St. Pierre, only nineteen people were killed in the region.

In August 1997, major eruptions resumed. This time, the northern part of the island, previously considered safe, was faced with pyroclastic flows (box figure 1), and more people were evacuated from the island. Activity continued, at least into the mid-2000s, but with decreasing intensity. In May 2004, a volcanic mudflow went through the already uninhabitable town of Plymouth. Up to 6 meters of debris were deposited, partially burying buildings still left in the town.

Additional Resources

Mount Pelée, West Indies (Volcano World site)

This site contains some excellent photos from the 1902 eruptions. The second page has photos of the famous spine that grew in Mount Pelée after the tragic eruption.

- http://volcano.oregonstate.edu/pelee

Montserrat Volcano Observatory

Includes up-to-date reports on volcanic activity.

- http://www.mvo.ms/

There are approximately 500 active volcanoes in the world, 50 of which are in the United States. There are more than 1,500 potentially active volcanoes. With an estimated 500 million people living near active volcanoes, volcanic monitoring is essential for reducing loss of life. The active U.S. volcanoes are monitored by the USGS through a series of volcano observatories, including the Hawaiian Volcano Observatory and the Cascade Volcano Observatory.

Volcanic hazard mitigation has three important components: hazard mapping, monitoring, and alerts.

Hazard mapping involves the study of deposits from past eruptions of a volcano. This provides information on the frequency of eruptions of the volcano and the typical size and style of eruptive activity. Hazard maps allow populations living near a volcano to create evacuation plans in case of an eruption. Unlike earthquakes, which occur without warning, volcanoes often show signs of unrest weeks or even months prior to an eruption. The earliest sign that a volcano is entering an eruptive phase is often increased seismic activity beneath the volcano as magma makes its way toward the surface. Seismographs, described in detail in chapter 16, are used to monitor seismic activity beneath volcanoes that are thought likely to erupt in the future. As magma gets closer to the surface, it may cause the ground surface to bulge upward. Volcanologists use

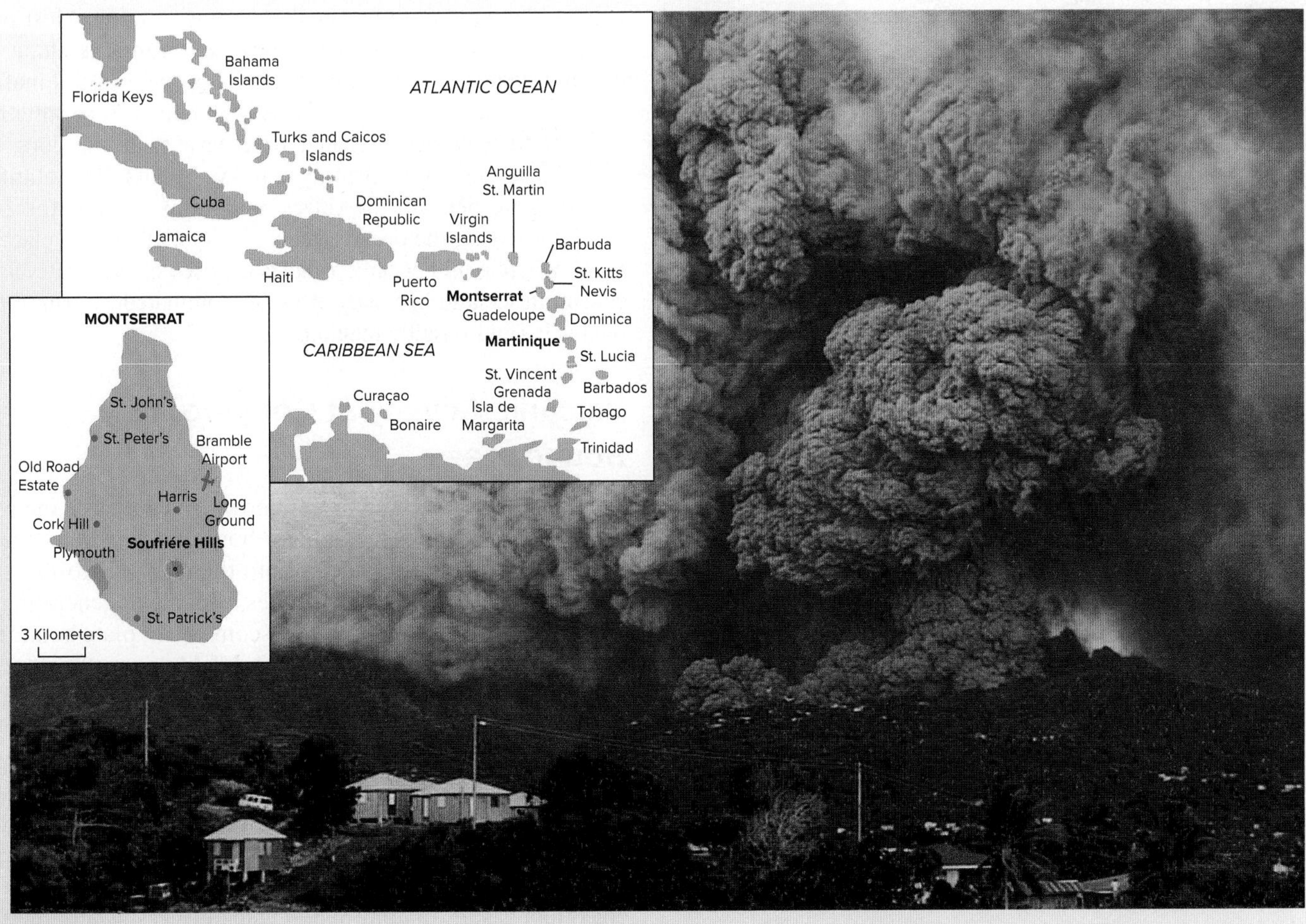

BOX 4.4 ■ FIGURE 1

Eruption of Soufriére Hills volcano on Montserrat, August 4, 1997. An ash cloud billows upward above a ground-hugging pyroclastic flow. Map of the West Indies showing location of Montserrat, Martinique, and Soufriére Hills volcano.
©Kevin West/AP Images

instruments on the ground such as tiltmeters as well as satellites to measure changes in the ground surface (figure 4.25). Volcanic gases may escape as magma moves closer to the surface. Volcanologists measure the amount of gas emitted from a volcano, looking for sudden rises or drops that may indicate an imminent eruption. Heat from magma can be measured at the surface and may produce visible signs such as increased hydrothermal springs activity or melting snow. Using all of this information, volcanologists can alert populations living close to a volcano that an eruption is about to occur. Alerts are also released to the aviation industry so that planes can avoid hazardous ash clouds (figure 4.26).

For more information on volcanic hazards and monitoring, visit the website of the USGS volcano hazards program at http://volcanoes.usgs.gov/.

4.6 PLATE TECTONICS AND VOLCANISM

Earlier in this chapter, we asked why it is that we can state confidently that active volcanism in Hawaii poses only slight danger to humans, but we expect violent eruptions to occur

FIGURE 4.25

U.S. Geological Survey scientist at a Global Positioning System (GPS) station on the east flank of Mount St. Helens. Mount Adams, another Cascade Range volcano, is visible in the distance. More than a dozen GPS stations were installed on or around Mount St. Helens to measure deformation of the ground surface when activity at the volcano resumed in 2004.
Source: Mike Poland, U.S. Geological Survey

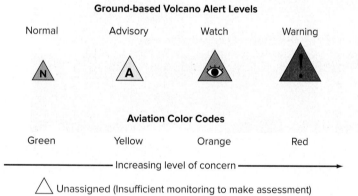

Standard Volcano Icons

Ground-based Volcano Alert Levels

Normal	Advisory	Watch	Warning
N	A	(eye)	!

Aviation Color Codes

Green	Yellow	Orange	Red

Increasing level of concern →

△ Unassigned (Insufficient monitoring to make assessment)

FIGURE 4.26

Icons for the U.S. Geological Survey Volcanic Activity Alert-Notification System.
USGS Volcano Hazards Program

around the margins of the Pacific Ocean. You learned that the composition of a magma and the amount of gas (primarily water vapor) it contains determine whether an eruption will be effusive or explosive. In chapter 3 you learned that most magma originates in the mantle at plate boundaries. Mantle rocks only melt under particular circumstances, primarily as a result of *decompression melting* and *flux melting.* The type of melting, the composition of the magma, the amount of gas it contains, and therefore the style of eruption, are all related to plate-tectonic setting.

Volcanic Activity at Divergent Boundaries

At divergent plate boundaries, decompression melting of the asthenosphere (see figures 3.14*A* and 3.26) generates basaltic magma that contains only small amounts of water. Most of the formation of the sea floor has involved eruptions along mid-oceanic ridges. The eruptions almost always consist of mafic lavas that create basalt. As described in chapter 3, basaltic rock, thought to have been formed from lava erupting along mid-oceanic ridges and forming pillow basalts or solidifying underground beneath the ridges, makes up virtually the entire crust underlying the oceans. Iceland is one of the few places on Earth where a mid-oceanic ridge is exposed above sea level. Volcanism in Iceland is mostly effusive, dominated by eruptions of relatively fluid basaltic magma.

Volcanic Activity at Convergent Boundaries

Nearly all the larger and better-known volcanoes of the world are located on convergent plate boundaries, where oceanic lithosphere is being subducted into the mantle. Most of these volcanoes are composite volcanoes, capable of generating violent explosive eruptions. Why are convergent plate boundaries associated with explosive volcanism? Melting at convergent boundaries occurs when the subducted oceanic crust releases water into the overlying asthenosphere, lowering its melting temperature (flux melting; see figures 3.14*B* and 3.28). When the hydrated asthenosphere partially melts, the magma contains significantly more water than magma generated at divergent margins. The majority of lavas erupted along convergent margins are andesitic in composition. Andesite is more viscous than basalt. In continental volcanic arcs, where the crust is thick, magma can evolve to rhyolite, which is even more viscous than andesite. This combination of viscous lava and large amounts of water vapor is what generates explosive eruptions.

Within-Plate Volcanic Activity

Volcanic activity that occurs away from plate boundaries, within tectonic plates, is related to mantle plumes (hot spots). Mantle plumes are narrow upwellings of hot mantle material, and partial melting occurs as a result of decompression (figure 3.27). Hot spot melting is associated with large volumes of basaltic magma. The eruption of basalt in Hawaii, generating large shield volcanoes, is an example of hot spot volcanism. The large flood basalts of the Columbia Plateau are also believed to be due to a mantle plume. Some continental hot spots generate very large, caldera-forming eruptions of rhyolite (e.g., Yellowstone). This is thought to be due to melting of large amounts of thick continental crust by mafic mantle melts.

PLANETARY GEOLOGY 4.5

Extraterrestrial Volcanic Activity

Volcanic activity has been a common geologic process operating on the Moon and several other bodies in the solar system. Approximately one-sixth of the Moon's surface consists of nearly circular, dark-colored, smooth, relatively flat lava plains. The lava plains, found mostly on the near side of the Moon, are called *maria* (singular, *mare;* literally, "seas"). They are believed to be huge meteorite impact craters that were flooded with basaltic lava during the Moon's early history. There are also a few extinct shield volcanoes on the Moon.

Elongate trenches or cracklike valleys called *rilles* are found mainly in the smoother portions of the lunar maria. They range in length from a few kilometers to hundreds of kilometers. Some are arc-shaped or crooked and are regarded as drained basaltic lava channels.

Mercury, the innermost planet, also has areas of smooth plains, suspected to be volcanic in origin.

Radar images of Venus show a surface that is young and probably still volcanically active. More than three-fourths of that surface is covered by continuous plains formed by enormous floods of lava. Close examination of these plains reveals extensive networks of lava channels and individual lava flows thousands of kilometers long.

Large shield volcanoes, some in chains along a great fault, have been identified on Venus, and molten lava lakes may exist. In other places, thick lavas have oozed out to form kilometer-high, pancake-shaped domes. Radar studies have shown that some of these domes are composed of a glassy substance mixed with bubbles of trapped gas. Fan-shaped deposits adjacent to some volcanoes may be pyroclastic debris.

Several of Venus's volcanoes emit large amounts of sulfur gases, causing the almost continuous lightning that has been observed by spacecraft.

Nearly half of the planet Mars may be covered with volcanic material. There are areas of extensive lava flows similar to the lunar maria and a number of volcanoes, some with associated lava flows.

Mars has at least nineteen large shield volcanoes, probably composed of basalt. The largest one, Olympus Mons (box figure 1), is almost three times the height of Mount Everest and wider than the state of Arizona. Its caldera is 80 kilometers across.

Hundreds of volcanoes have been discovered on Jupiter's moon Io (box figure 2), and some of those have erupted for periods of at least four months. Material rich in sulfur compounds is thrown at least 500 kilometers into space at speeds of up to 3,200 kilometers per hour. This material often forms umbrella-shaped clouds as it spreads out and falls back to the surface. Lakes of very hot silicate lava, perhaps mafic or ultramafic, are common. More than 100 calderas larger than 25 kilometers across have been observed, including one that vents sulfur gases. The energy source for Io's volcanoes may be the gravitational pulls of Jupiter and two of its other larger satellites, causing Io to heat up much as a piece of wire will do if it is flexed continuously.

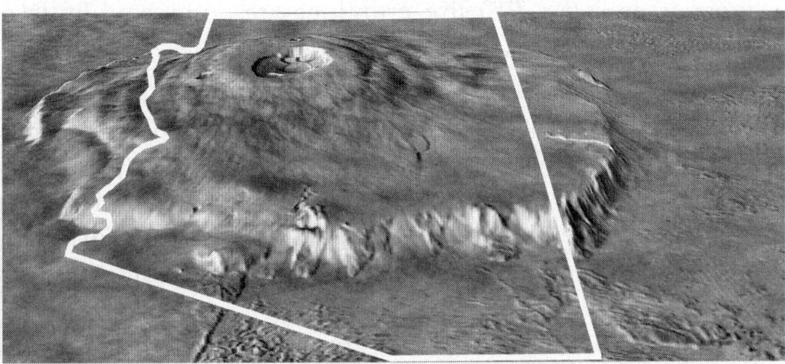

BOX 4.5 ■ FIGURE 1

Perspective view of Olympus Mons, the largest volcano and tallest mountain in the solar system. This Martian volcano is over 624 km wide and 25 km high. Note the outline of the state of Arizona for size comparison.
Source: NASA/MOLA Science Team

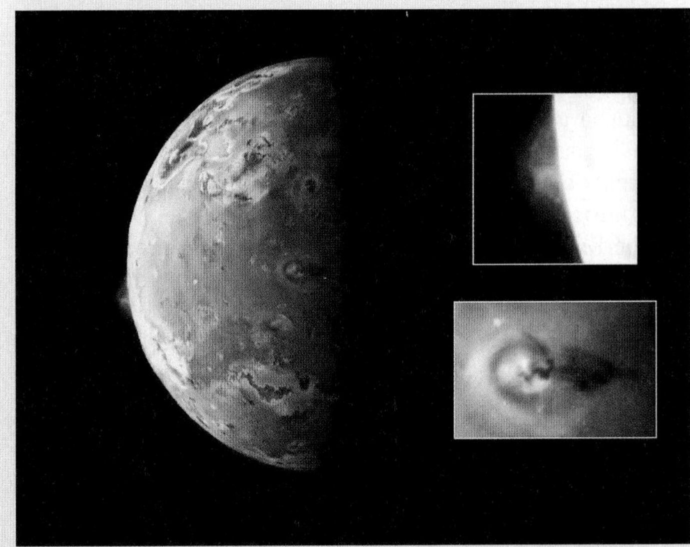

BOX 4.5 ■ FIGURE 2

Two volcanic plumes on Jupiter's moon Io. The plume on the left horizon (and upper inset) is 140 kilometers high; the one in the center (and lower inset) is 75 kilometers high. For details go to photojournal.jpl.nasa.gov/catalog/PIA00703.
Source: NASA/JPL

Neptune's moon Triton is the third object in the solar system that has active volcanoes. There, "ice volcanoes" erupt what is probably nitrogen frost.

Additional Resources

Active Volcanoes of Our Solar System

- http://geology.com/articles/active-volcanoes-solar-system.shtml

Io: Facts about Jupiter's Volcanic Moon

- http://www.space.com/16419-io-facts-about-jupiters-volcanic-moon.html

Summary

A volcano is an opening in the earth's crust through which molten lava, ash, and gases are ejected. Volcanic eruptions can be *effusive* or *explosive.* Effusive eruptions are dominated by lava flows. Explosive eruptions are dominated by pyroclastic material.

The style of eruption is controlled by the gas content and chemistry of the lava. Lava contains 45% to 75% *silica* (SiO_2). The more silica, the more viscous the lava is. Viscosity is also influenced by the temperature and gas content of the lava. Viscous lavas are associated with more violent eruptions than are fluid lavas.

The main products of effusive eruptions are lava flows. Basaltic lava flows are fluid. *Pahoehoe* lava flows have a ropy surface and *a'a* flows have a rubbly surface. *Flood basalts* are large eruptions of very fluid basalt that form thick sequences of lava flows. *Columnar jointing* develops in solidified basalt flows. Basalt that erupts underwater forms a *pillow structure.* Pillow basalts commonly form along the crests of mid-oceanic ridges. Rhyolite lava is very viscous and unless very hot, cannot flow for great distances. Felsic lava flows are very thick, often forming lava domes. Andesite lava has a viscosity intermediate between basaltic and rhyolitic lava and forms flows of intermediate thickness.

The main products of explosive eruptions are *pyroclasts,* fragments of lava that form as a result of volcanic explosions. Pyroclastic material is classified according to size. *Dust* and *ash* are the finest particles, while *lapilli* and *bombs* are the largest particles. Pyroclastic material can fall as *tephra* or can be deposited by *pyroclastic flows.*

There are five major types of volcanoes: shield, cinder cone, composite cone, lava dome, and caldera. A *shield volcano* is built up by successive eruptions of mafic lava. Its slopes are gentle, but its volume is generally large. A *cinder cone* is composed of loose pyroclastic mterial that forms steep slopes as it falls from the air back to near the crater. *Composite cones* are made of alternating layers of pyroclastic material and solidified lava flows. They are not as steep as cinder cones, but they are steeper than shield volcanoes. Lava domes are formed from very viscous lava that piles up over a volcanic vent. Calderas are large volcanic craters formed by the collapse of volcanoes into magma chambers.

Volcanic hazards include direct hazards such as *pyroclastic flow, pyroclastic fall,* and *volcanic mudflow* and indirect hazards such as famine. More people have been killed by pyroclastic flows and by famine than by other volcanic hazards. Volcanic hazard mitigation involves mapping of old volcanic deposits; monitoring of volcanic activity such as seismicity, gas emission, ground deformation, and heat flow; and issuing warnings to the public.

Plate tectonics helps us understand why certain types of volcanoes are common in particular regions. At divergent margins, melting of the asthenosphere driven by *decompression melting* produces basalt. Volcanic activity includes eruption of pillow basalts at mid-oceanic ridges and basaltic lava flows in Iceland. At convergent margins, explosive eruptions from composite volcanoes are common. *Flux melting* occurs when water is driven off the subducting plate into the overlying asthenosphere. Lavas erupted at convergent margins contain higher amounts of water and are commonly intermediate or felsic in composition. At hot spots, large volumes of basalt are generated in the asthenosphere. These can erupt at the surface, forming large shield volcanoes and flood basalts, or they can melt thick continental crust, forming large pyroclastic eruptions.

Terms to Remember

a'a 85
active volcano 99
ash 90
block 90
bomb 90
caldera 95
cinder 90
cinder cone 92
circum-Pacific belt 93
columnar jointing 88
composite volcano (stratovolcano) 93
crater 85
dormant volcano 99
dust 90
effusive eruption 80
explosive eruption (pyroclastic eruption) 79
extinct volcano 99
flank eruption 85
flood basalt (plateau basalt) 86
lahar 99
lapilli 90
lava dome 95
lava tube 86
Mediterranean belt 95
pahoehoe 85
pillow basalts (pillow structure) 89
pyroclast 90
pyroclastic flow 91
shield volcano 91
vent 85
viscosity 82
Volcanic Explosivity Index (VEI) 84
volcanism 79
volcano 79

Terms Covered in Chapter 3 that are Useful for Chapter 4

andesite 61
basalt 61
felsic rock 61
intermediate rock 61
lava 54
mafic rock 61
magma 54
obsidian 57
porphyritic 57
pumice 59
rhyolite 61
scoria 59
tuff 60
vesicles 59
volcanic breccia 60

Testing Your Knowledge

Use the following questions to prepare for exams based on this chapter.

1. Describe the difference between effusive and explosive eruptions in terms of their eruptive products and the hazards they pose to society.
2. What are the three main components of magma? What factors control the viscosity of magma?

3. On examining a basaltic lava flow in Hawaii, you discover that close to the vent, the lava flow has a pahoehoe texture, but further from the vent, it becomes an a'a flow. What can explain this?
4. What do pillow structures indicate about the environment of volcanism?
5. What roles do gases and viscosity play in the generation of explosive eruptions?
6. Consider the eruption of Mount Merapi, described at the beginning of this chapter and shown in figure 4.1. What kind of volcano is Merapi? What evidence did you use to determine this? Based upon your answer, what do you think the plate-tectonic setting is of Mount Merapi? What composition of lava do you think it mostly erupts?
7. Describe how a caldera forms.
8. Compare lava flows, pyroclastic flows, and lahars as volcanic hazards.
9. You are a volcanologist working on a volcano that has been active in the past but is currently quiet. What methods would you use to monitor this volcano, and what would those methods tell you about processes beneath the surface?
10. What are the three components of volcanic hazard mitigation? Provide examples of each.
11. Which of the following is an example of a shield volcano?
    a. Mount St. Helens, Washington State
    b. Mount Merapi, Indonesia
    c. Mauna Loa, Hawaii
    d. Cerro Negro, Nicaragua
12. Volcanic eruptions can affect the climate because
    a. they heat the atmosphere.
    b. volcanic dust and gas can reduce the amount of solar radiation that penetrates the atmosphere.
    c. they change the elevation of the land.
    d. all of the preceding.
13. The gas most commonly released during a volcanic eruption is
    a. water vapor. b. carbon dioxide.
    c. sulfur dioxide. d. hydrogen sulfide.
    e. oxygen.
14. _____ is a rock composed of frothy volcanic glass
    a. Obsidian b. Basalt
    c. Tuff d. Pumice
    e. Volcanic breccia
15. A lava flow with a ropy or billowy surface is called
    a. pahoehoe b. a'a
    c. pillow lava d. lahar
    e. lava tube
16. Which of these is *not* a type of pyroclastic material?
    a. ash b. dust
    c. lapilli e. bomb
    d. a'a
17. Which of these is *not* a major type of volcano?
    a. shield b. cinder cone
    c. composite d. stratovolcano
    e. spatter cone
18. An example of a composite volcano is
    a. Mount Rainier. b. Fujiyama.
    c. Mount Vesuvius. d. all of the preceding.
19. Which volcano is *not* usually made of basalt?
    a. shield b. composite cone
    c. spatter cone d. cinder cone
20. Which of the following is *not* a component of volcanic hazard mitigation?
    a. Mapping older volcanic deposits
    b. Preventing a volcanic eruption
    c. Monitoring earthquake activity around a volcano
    d. Alerting nearby residents of an imminent eruption

Expanding Your Knowledge

1. What might explain the remarkable alignment of the Cascade volcanoes?
2. What would the present-day environmental effects be for an eruption such as that which created Crater Lake?
3. Why are there no active volcanoes in the eastern parts of the United States and Canada?
4. Why are volcanic eruptions at convergent plate boundaries typically more explosive than those at divergent plate boundaries?

Exploring Web Resources

http://volcano.oregonstate.edu/

Volcano World. This is an excellent site to learn about volcanoes. Learn more about the volcanoes of the world by clicking on the Volcanoes tab. Explore volcanoes by clicking on "Fun Stuff" and selecting "Virtual Volcano Fieldtrips."

www.geo.mtu.edu/volcanoes/

Michigan Tech volcanoes page. The focus of this site is on scientific and educational information relative to volcanic hazard mitigation. Clicking on "volcanic humor" will show the lighter side of volcanology.

www.volcanolive.com/contents.html

Volcano Live. This well-organized site is maintained by an Australian volcanologist. You can link to live cameras at most of the volcanoes discussed in this chapter (Mount Fuji, Mount Erebus, Mount Etna, etc.). You can get up-to-date information on what is erupting in the world and much more.

http://hvo.wr.usgs.gov/kilauea/

Hawaii Volcano Observatory's Kilauea Website. You can find information about Kilauea and its past and present activity as well as photos taken today and during the past.

http://volcanoes.usgs.gov/publications/

Products and fact sheets of the U.S. Geological Survey's volcanic hazards program. Lists many of the USGS online fact sheets on volcanoes.

CHAPTER

5 Weathering and Soil

Differential weathering and erosion at Bryce Canyon National Park in Utah has produced spires in the sandstone beds.
©Doug Sherman/Geofile RF

LEARNING OBJECTIVES

- Why is weathering important to life on Earth?
- Explain the differences between and the processes involved in mechanical and chemical weathering.
- Name and describe the types of features formed by weathering.
- Discuss the role of acids in chemical weathering, and list the weathering products of the common rock-forming minerals.
- Describe the factors that affect the formation of soil.
- Name and sketch the soil horizons that form in a humid climate, and explain how the layers form.

In this chapter, you will study several visible signs of weathering in the world around you, ranging from the large rounded landforms in Yosemite National Park to the rounded edges of boulders. As you study these features, keep in mind that weathering processes make the planet suitable for human habitation. The weathering of rock affects the composition of Earth's atmosphere, helping to maintain a habitable climate. Weathering also produces soils, upon which grow the forests, grasslands, and agriculture of the world.

How does rock weather? You learned in chapters 3 and 4 that the minerals making up igneous rocks crystallize at relatively high temperatures and sometimes at high pressures as magma and lava cool. Although these minerals are stable when they form, most of them are not stable during prolonged exposure at the surface. In this chapter, you see how minerals and rocks change when they are subjected to the physical and chemical conditions existing at the surface. Rocks undergo mechanical weathering (physical disintegration) and chemical weathering (decomposition) as they are attacked by air, water, and microorganisms. Your knowledge of the chemical composition and atomic structure of minerals will help you understand the reactions that occur during chemical weathering.

Weathering processes create sediments (primarily mud and sand) and soil. Sedimentary rocks, which form from sediments, are discussed in chapter 6. In a general sense, weathering prepares rocks for erosion and is a fundamental part of the rock cycle (figure 5.1), transforming rocks into the raw material that eventually becomes sedimentary rocks. Through weathering, there are important links between the rock cycle and the atmosphere and biosphere.

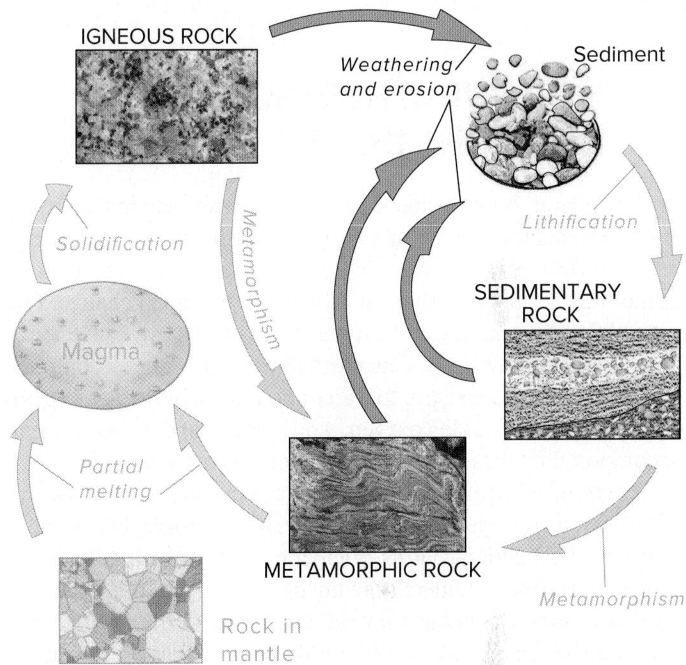

FIGURE 5.1

The rock cycle shows how rocks can be weathered and eroded and recycled into sediment to form new rocks.
(Igneous Rock) ©Doug Sherman/Geofile RF; (Metamorphic Rock) ©reneh9999/Shutterstock; (Sedimentary Rock) ©Doug Sherman/Geofile RF

5.1 WEATHERING, EROSION, AND TRANSPORTATION

Rocks exposed at Earth's surface are constantly being changed by water, air, varying temperature, and other environmental factors. Granite may seem indestructible, but given time and exposure to air and water, it can decompose and disintegrate into soil. The processes that affect rock are *weathering*, *erosion*, and *transportation.*

The term **weathering** refers to the processes that change the physical and chemical character of rock at or near the surface. For example, if you abandon a car, particularly in a wet climate, eventually the paint will flake off and the metal will rust. The car weathers. Similarly, the tightly bound crystals of any rock can be loosened and altered to new minerals when exposed to air and water during weathering. Weathering breaks down rocks that are either stationary or moving.

Erosion is the picking up or *physical removal* of rock particles by an agent such as ocean waves, running water, or glaciers. Weathering helps break down a solid rock into loose particles that are easily eroded. Rainwater flowing down a cliff or hillside removes the loose particles produced by weathering. Similarly, if you sandblast rust off a car, erosion takes place. Humans, of course, are awesome agents of erosion. A single pass by a bulldozer can do more to change a landscape than thousands of years of natural weathering and erosion.

After a rock fragment is picked up (eroded), it is transported. **Transportation** is the movement of eroded particles by agents such as rivers, waves, glaciers, or wind. Weathering processes continue during transportation. A boulder being transported by a stream can be physically worn down and chemically altered as it is carried along by the water. In the car analogy, transportation would take place when a stream of rust-bearing water flows away from a car in which rust is being hosed off.

Water is necessary for chemical weathering to take place. Oxygen dissolved in water oxidizes iron in rocks. Carbon dioxide mixed with water makes a weak acid that causes most minerals to decompose; this acid is the primary cause of chemical weathering. Running water contributes to weathering and erosion by loosening and removing particles and by abrading rocks during transportation in streams. Ice in glaciers is a very effective agent of erosion as rocks frozen in the base of a glacier grind down the underlying bedrock. Freezing and thawing of water in cracks in rock is also very effective at mechanically breaking them up.

5.2 HOW WEATHERING CHANGES ROCKS

Rocks undergo both mechanical weathering and chemical weathering. **Mechanical weathering** (physical disintegration) includes several processes that break rock into smaller pieces. The change in the rock is physical; there is little or no chemical change. For example, water freezing and expanding in cracks can cause rocks to disintegrate physically. **Chemical weathering** is the decomposition of rock from exposure to water and atmospheric gases (principally carbon dioxide, oxygen, and water vapor). As rock is decomposed by these agents, new chemical compounds form.

Mechanical weathering breaks up rock but does not change the composition. A large mass of granite may be broken into smaller pieces by frost action, but its original crystals of quartz, feldspar, and ferromagnesian minerals are unchanged. On the other hand, if the granite is being chemically weathered, some of the original minerals are chemically changed into different minerals. Feldspar, for example, will change into a clay mineral (with a crystal structure similar to mica). In nature, mechanical and chemical weathering usually occur together, and the effects are interrelated.

Weathering is a relatively long, slow process. Typically, cracks in rock are enlarged gradually by frost action or plant growth (as roots pry into rock crevices), and as a result, more surfaces are exposed to attack by chemical agents. Chemical weathering initially works along contacts between mineral grains. Tightly bound crystals are loosened as weathering products form at their contacts. Mechanical and chemical weathering then proceed together, until a once-tough rock slowly crumbles into individual grains.

Solid minerals are not the only products of chemical weathering. Some minerals—calcite, for example—dissolve when chemically weathered. We can expect limestone and marble, rocks consisting mainly of calcite, to weather chemically in quite a different way than granite.

A

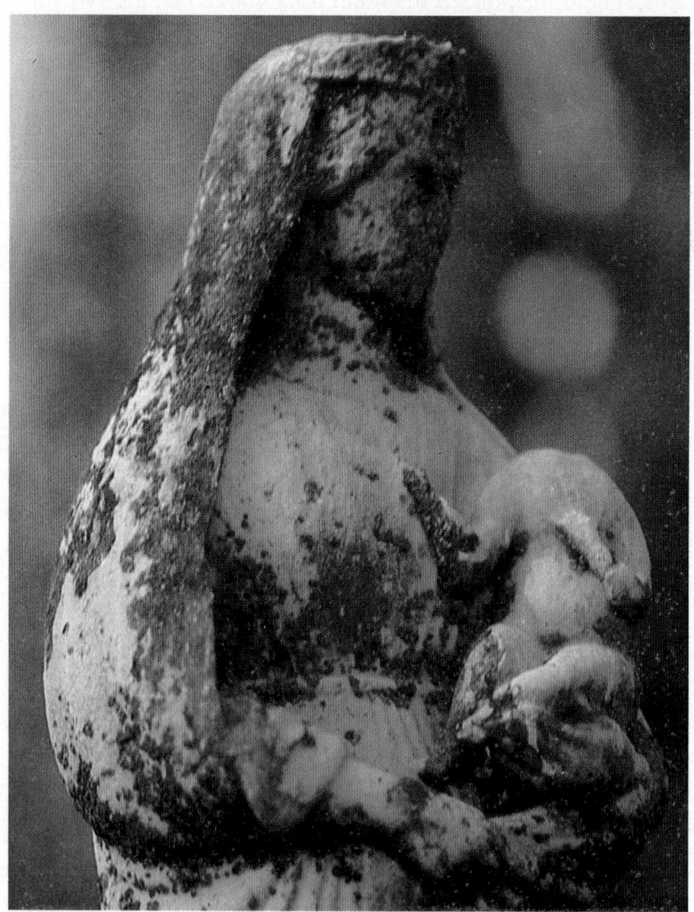

B

FIGURE 5.2

*(A)*The effects of chemical weathering are obvious in the marble gravestone on the right but not in the slate gravestone on the left, which still retains its detail. Both gravestones date to the 1780s. *(B)* This marble statue has lost most of the fine detail on the face and the baby's head has been dissolved by chemical weathering.
(A) ©C.C. Plummer (B) ©David McGeary

5.3 EFFECTS OF WEATHERING

The results of chemical weathering are easy to find. Look along the edges or corners of old stone structures for evidence. The inscriptions on statues and gravestones that have stood for several decades may no longer be sharp (figure 5.2). Building

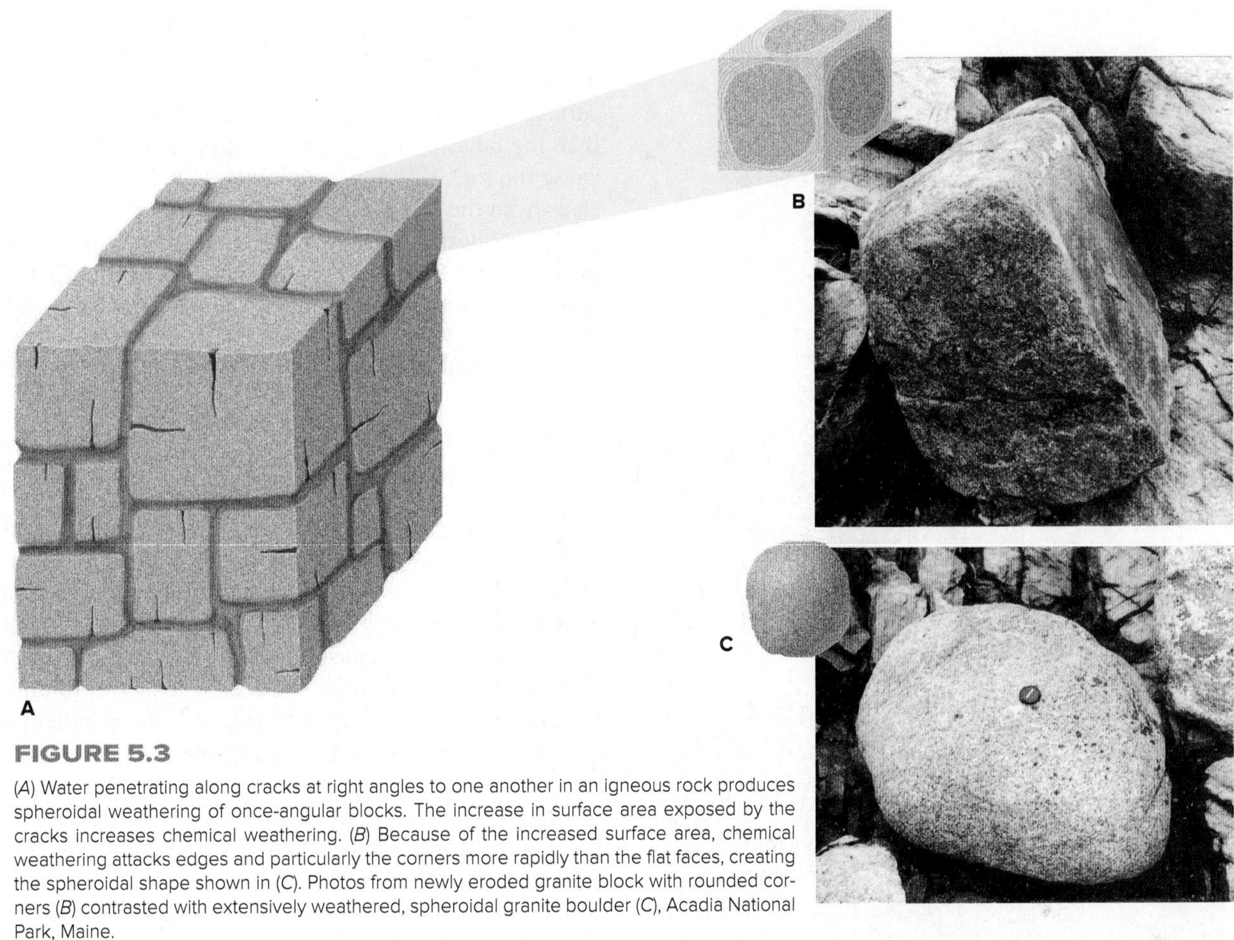

FIGURE 5.3

(*A*) Water penetrating along cracks at right angles to one another in an igneous rock produces spheroidal weathering of once-angular blocks. The increase in surface area exposed by the cracks increases chemical weathering. (*B*) Because of the increased surface area, chemical weathering attacks edges and particularly the corners more rapidly than the flat faces, creating the spheroidal shape shown in (*C*). Photos from newly eroded granite block with rounded corners (*B*) contrasted with extensively weathered, spheroidal granite boulder (*C*), Acadia National Park, Maine.
©Bret Bennington

blocks of limestone or marble exposed to rain and atmospheric gases may show solution effects of chemical weathering in a surprisingly short time. Granite and slate gravestones and building materials are much more resistant to weathering due to the strong silicon-oxygen bonds in the silicate minerals. However, after centuries, the mineral grains in granite may be loosened, cracks enlarged, and the surface discolored and dulled by the products of weathering. Surface discoloration is also common on rock *outcrops*, where rock is exposed to view, with no plant or soil cover. That is why field geologists carry rock hammers–to break rocks to examine unweathered surfaces.

We tend to think of weathering as destructive because it mars statues and building fronts. As rock is destroyed, however, valuable products can be created. Soil is produced by rock weathering, so most plants depend on weathering for the soil they need in order to grow. In a sense, then, all agriculture depends on weathering. Weathering products transported to the sea by rivers as dissolved solids make seawater salty and serve as nutrients for many marine organisms. Some metallic ores, such as those of copper and aluminum, are concentrated into economic deposits by chemical weathering.

Weathering has also had a dramatic impact on the composition of Earth's atmosphere. Chemical weathering removes carbon dioxide from the atmosphere, allowing it to be transformed into limestone and stored in the crust. Without chemical weathering, the elevated levels of carbon dioxide in the atmosphere would have long ago made Earth too hot to sustain life (see box 5.1).

Many weathered rocks display interesting shapes. **Spheroidal weathering** occurs where rock has been rounded by weathering from an initial blocky shape. It is rounded because chemical weathering acts more rapidly or intensely on the corners and edges of a rock than on the smooth rock faces (figure 5.3).

Differential weathering describes the tendency for different types of rock to weather at different rates. For example, shale (composed of soft clay minerals) tends to weather and erode much faster than sandstone (composed of hard quartz mineral). Figure 5.4 shows a striking example of differential weathering. Many scenic cliffs are the result of differential weathering (figure 5.5).

5.4 MECHANICAL WEATHERING

Of the many processes that cause rocks to disintegrate, the most effective are pressure release and frost action.

FIGURE 5.4
Pedestal rock near Lees Ferry, Arizona. Resistant sandstone cap protects weak shale pedestal from weathering and erosion. Hammer for scale is barely visible at base of pedestal.
©David McGeary

FIGURE 5.5
Differential weathering of more resistant red sandstone layers and softer shale layers form scenic cliffs at Red Rock Canyon State Park, California.
©Diane Carlson

Pressure Release

The reduction of pressure on a body of rock can cause it to crack as it expands; **pressure release** is a significant type of mechanical weathering. A large mass of rock, such as a batholith, originally forms under great pressure from the weight of several kilometers of rock above it. This batholith is gradually exposed by tectonic uplift of the region followed by erosion of the overlying rock (figure 5.6). The removal of the great weight of rock above the batholith, usually termed *unloading*, allows the granite to expand upward. Cracks called **sheet joints** develop parallel to the outer surface of the rock as the outer part of the rock expands more than the inner part (figure 5.6*B* and *C*). On slopes, gravity may cause the rock between such joints to break loose in concentric slabs from the underlying granite mass. This process of spalling off of rock layers is called **exfoliation**; it is somewhat similar to peeling layers from an onion. **Exfoliation domes** (figure 5.7) are large, rounded landforms developed in massive rock, such as granite, by exfoliation. Some famous examples of exfoliation domes include Stone Mountain in Georgia and Half Dome in Yosemite.

Frost Action

Did you ever leave a bottle of water in the freezer, coming back later to find the water frozen and the bottle burst open? When water freezes at 0°C (32°F), the individual water molecules jumbled together in the liquid align into an ordered crystal structure, forming ice. Because the crystal structure of ice takes up more space than the liquid, water expands 9% in volume when it freezes. This unique property makes water a potent agent of mechanical weathering in any climate where the temperature falls below freezing.

Frost action—the mechanical effect of freezing water on rocks—commonly occurs as frost wedging or frost heaving. In **frost wedging**, the expansion of freezing water pries rock apart. Most rock contains a system of cracks called *joints*, caused by the slow flexing of brittle rock by deep-seated Earth forces (see chapter 15). Water that has trickled into a joint in a rock can freeze and expand when the temperature drops below 0°C (32°F). The expanding ice wedges the rock apart, extending the joint or even breaking the rock into pieces (figure 5.8). Frost wedging is most effective in areas with many days of freezing and thawing (mountaintops and midlatitude regions with pronounced seasons). Partial thawing during the day adds new water to the ice in the crack; refreezing at night adds new ice to the old ice.

Frost heaving lifts rock and soil vertically. Solid rock conducts heat faster than soil, so on a cold winter day, the bottom of a partially buried rock will be much colder than soil at the same depth. As the ground freezes in winter, ice forms first under large rock fragments in the soil. The expanding ice layers push boulders out of the ground, a process well known to New England farmers and other residents of rocky soils. Frost heaving bulges the ground surface upward in winter, breaking up roads and leaving lawns spongy and misshapen after the spring thaw.

Other Processes

Several other processes mechanically weather rock but in most environments are less effective than frost action and pressure release. *Plant growth*, particularly roots growing in cracks (figure 5.9*A*), can break up rocks, as can *burrowing animals.*

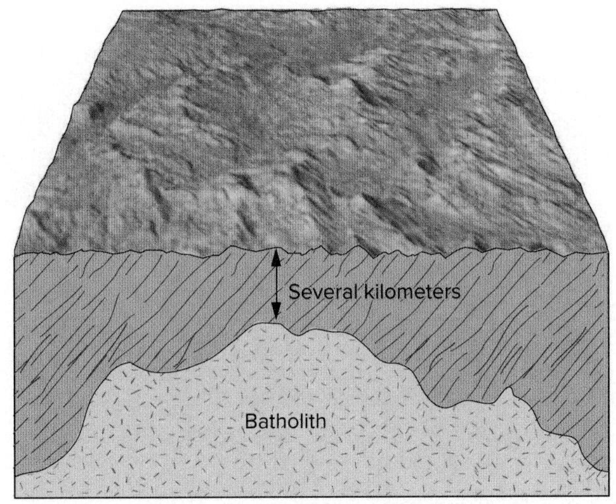

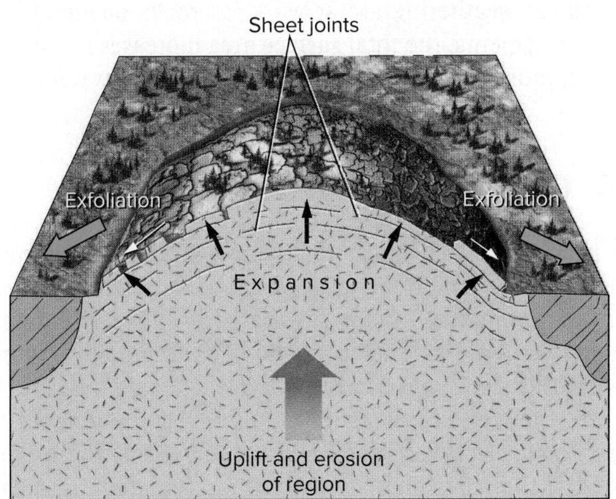

FIGURE 5.6

Sheet joints caused by pressure release. A granite batholith (*A*) is exposed by regional uplift followed by the erosion of the overlying rock (*B*). Unloading reduces pressure on the granite and causes outward expansion. Sheet joints are closely spaced at the surface where expansion is greatest. Exfoliation of rock layers produces rounded exfoliation domes. (*C*) Sheet joints in a granite outcrop near the top of the Sierra Nevada, California. The granite formed several kilometers below the surface and expanded outward when it was exposed by uplift and erosion.
(C) ©David McGeary

FIGURE 5.7

Half Dome in Yosemite National Park, California, is an example of an exfoliation dome. Note the onion-like layers of rock that are peeling off the dome, and the climbers on a cable ladder for scale.
©Dean Conger/Corbis/Getty Images

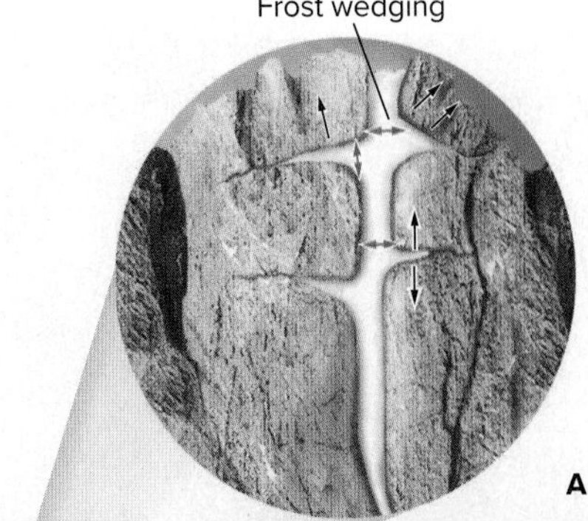

FIGURE 5.8

(*A*) Frost wedging occurs when water fills joints (cracks) in a rock and then freezes. The expanding ice wedges the rock apart. (*B*) Frost wedging has broken the rock and sculpted Crawford Mountain in Banff National Park, Alberta, Canada. The broken rock forms cone-shaped piles of debris (talus) at the base of the mountains.
(B) ©Marli Miller

A

B

FIGURE 5.9

(*A*) Tree roots will pry this rock apart as they grow within the rock joints, Sierra Nevada Mountains, California. (*B*) This rock is being broken by the extreme temperature variation in a desert, Mojave Desert, California. Note the tremendous increase in surface area that results from the rock being split into layers.
(A) ©Diane Carlson (B) ©Crystal Hootman and Diane Carlson

Such activities help to speed up chemical weathering by enlarging passageways for water and air. *Extreme changes in temperature*, as in a desert environment (figure 5.9*B*) or in a forest fire, can cause a rock to expand until it cracks. The *pressure of salt crystals* formed as water evaporates inside small spaces in rock

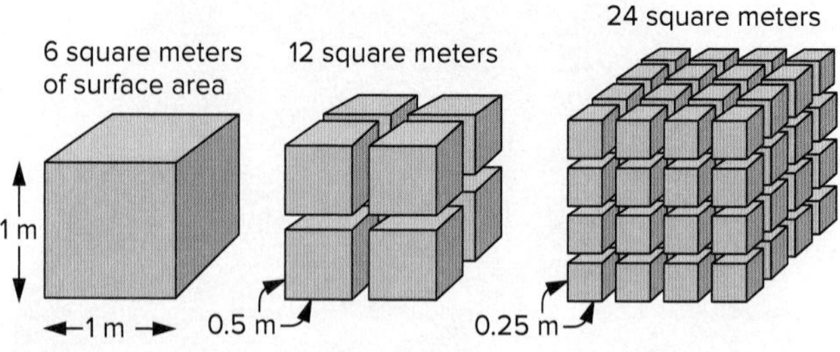

FIGURE 5.10

Mechanical weathering can increase the surface area of a rock, accelerating the rate of chemical weathering. As a cube breaks up into smaller pieces, its volume remains the same, but its surface area increases.

also helps to disintegrate desert rocks. Whatever processes of mechanical weathering are at work, as rocks disintegrate into smaller fragments, the total surface area increases (figure 5.10), allowing more extensive chemical weathering by water and air.

5.5 CHEMICAL WEATHERING

The processes of chemical weathering, or *rock decomposition*, transform rocks and minerals exposed to water and air into new chemical products. A mineral that crystallized deep underground from a water-deficient magma may eventually be exposed at the surface, where it can react with the abundant water there to form a new, different mineral. A mineral containing very little oxygen may react with oxygen in the air, extracting oxygen atoms from the atmosphere and incorporating them into its own crystal structure, thus forming a different mineral. These new minerals are weathering products. They have adjusted to physical and chemical conditions at (or near) Earth's surface. Minerals change gradually at the surface until they come into *equilibrium*, or balance, with the surrounding conditions.

Role of Oxygen

Oxygen is abundant in the atmosphere and quite active chemically, so it often combines with minerals or with elements within minerals that are exposed at Earth's surface.

The rusting of an iron nail exposed to air is a simple example of chemical weathering. Oxygen from the atmosphere combines with the iron to form iron oxide, the reaction being expressed as follows:

$$4Fe^{+3} + 3O_2 \rightarrow 2Fe_2O3$$
$$\text{iron} + \text{oxygen} \rightarrow \text{iron oxide}$$

Iron oxide formed in this way is a weathering product of numerous minerals containing iron, such as the ferromagnesian group (pyroxenes, amphiboles, biotite, and olivine). The iron in the ferromagnesian silicate minerals must first be separated from the silica in the crystal structure before it can oxidize. The iron oxide (Fe_2O_3) formed is the mineral **hematite**, which has

FIGURE 5.11

Sandstone has been colored red by hematite, released by the chemical weathering of ferromagnesian minerals, Thermopolis, Wyoming.
©Diane Carlson

FIGURE 5.12

A mudpot of boiling mud is created by intense chemical weathering of the surrounding rock by the acid gases dissolved in a hot spring, Yellowstone National Park, Wyoming.
©David McGeary

a brick-red color when powdered. If water is present, as it usually is at Earth's surface, the iron oxide combines with water to form **limonite**, which is the name for a group of mostly amorphous, hydrated iron oxides (often including the mineral *goethite*), which are yellowish-brown when powdered. The general formula for this group is $Fe_2O_3 \cdot nH_2O$ (the n represents a small, whole number such as 1, 2, or 3 to show a variable amount of water). The brown, yellow, or red color of soil and many kinds of sedimentary rock is commonly the result of small amounts of hematite and limonite released by the weathering of iron-containing minerals (figure 5.11).

Role of Acids

The most effective agent of chemical weathering is acid. Acids are chemical compounds that give off hydrogen ions (H^+) when they dissociate, or break down, in water. Strong acids produce a great number of hydrogen ions when they dissociate, and weak acids produce relatively few such ions.

The hydrogen ions given off by natural acids disrupt the orderly arrangement of atoms within most minerals. Because a hydrogen ion has a positive electrical charge and a very small size, it can substitute for other positive ions (such as Ca^{2+}, Na^+, or K^+) within minerals. This substitution changes the chemical composition of the mineral and disrupts its atomic structure. The mineral decomposes, often into a different mineral, when it is exposed to acid.

Some strong acids occur naturally on Earth's surface, but they are relatively rare. Sulfuric acid is a strong acid emitted during many volcanic eruptions. It can kill trees and cause intense chemical weathering of rocks near volcanic vents. The bubbling mud of Yellowstone National Park's mudpots (figure 5.12) is produced by rapid weathering caused by acidic sulfur gases that are given off by some hot springs. Strong acids also drain from some mines as sulfur-containing minerals such as pyrite oxidize and form acids at the surface (figure 5.13). Uncontrolled mine drainage can kill fish and plants downstream and accelerate rock weathering.

The most important natural source of acid for rock weathering at Earth's surface is dissolved carbon dioxide (CO_2) in water. Water and carbon dioxide form *carbonic acid* (H_2CO_3), a weak acid that dissociates into the hydrogen ion and the bicarbonate ion (see equation *A* in table 5.1). Even though carbonic acid is a weak acid, it is so abundant at Earth's surface that it is the single most effective agent of chemical weathering.

Earth's atmosphere (mostly nitrogen and oxygen) contains 0.03% carbon dioxide. Some of this carbon dioxide dissolves in rain as it falls, so most rain is slightly acidic when it hits the ground. Large amounts of carbon dioxide also dissolve in water that percolates through soil. The openings in soil are filled with a gas mixture that differs from air. Soil gas has a much higher content of carbon dioxide (up to 10%) than does air because carbon dioxide is produced by the decay of organic matter and the respiration of soil organisms in the biosphere, such as worms. Rainwater that has trickled through soil is therefore usually acidic and readily attacks minerals in the unweathered rock below the soil (figure 5.14).

Solution Weathering

Some minerals are completely dissolved by chemical weathering. *Calcite*, for instance, goes into solution when exposed to carbon dioxide and water, as shown in equation *B* in table 5.1 and in figure 5.14. The carbon dioxide and water combine to

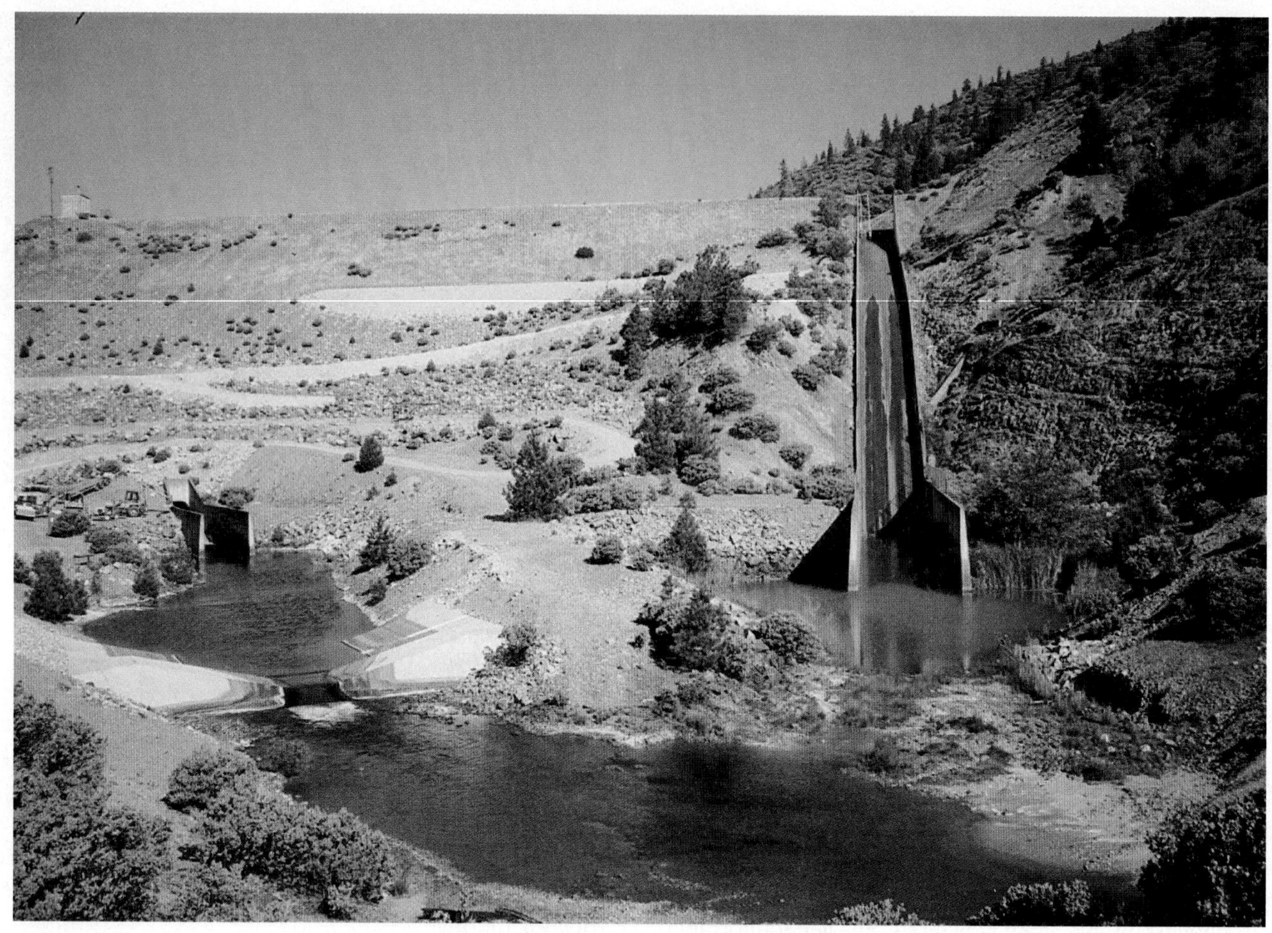

FIGURE 5.13

Spring Creek debris dam collects acid mine drainage from the Iron Mountain Mines Superfund site in northern California.
Source: Charles Alpers, U.S. Geological Survey

TABLE 5.1 Chemical Equations Important to Weathering

A. Solution of Carbon Dioxide in Water to Form Acid								
CO_2 carbon dioxide	+	H_2O water	$\rightleftharpoons$	H_2CO_3 carbonic acid	$\rightleftharpoons$	H^+ hydrogen ion	+	HCO_3^- bicarbonate ion
B. Solution of Calcite								
$CaCO_3$ calcite	+	CO_2 carbon dioxide	+	H_2O water	$\rightleftharpoons$	Ca^{2+} calcium ion	+	$2HCO_3^-$ bicarbonate ion
C. Solution of Calcite								
$CaCO_3$	+	H^+	+	HCO_3^-	$\rightleftharpoons$	Ca^{2+}	+	$2HCO_3^-$
D. Chemical Weathering of Feldspar to Form a Clay Mineral								
$2KAlSi_3O_8$ potassium feldspar	+	$\underbrace{2H^+ + 2HCO_3^-}$ (from CO_2 and H_2O) + H_2O	$\rightarrow$	$Al_2Si_2O_5(OH)_4$ clay mineral	+	$\underbrace{2K^+ + 2HCO_3^-}$ (soluble ions)	+	$4\ SiO_2$ silica in solution or as fine solid particles

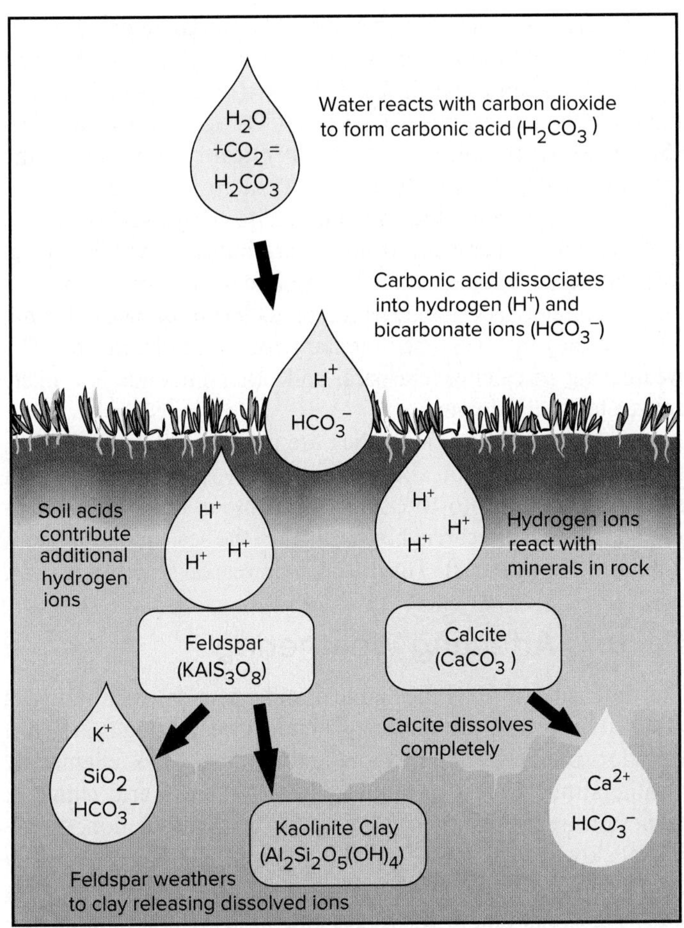

FIGURE 5.14

Chemical weathering of feldspar and calcite by carbonic and soil acids. Water percolating through soil weathers feldspar to clay and completely dissolves calcite. Soluble ions and soluble silica weathering products are washed away.

form carbonic acid, which dissociates into the hydrogen ion and the bicarbonate ion, as you have seen, so the equation for the solution of calcite can also be written as equation *C* in table 5.1.

There are no solid products in the last part of the equation, indicating that complete solution of the calcite has occurred. Caves can form underground when flowing groundwater dissolves the sedimentary rock limestone, which is mostly calcite. Rain can discolor and dissolve statues and tombstones carved from the metamorphic rock marble, which is also mostly calcite (see figure 5.2).

Chemical Weathering of Feldspar

The weathering of feldspar is an example of the alteration of an original mineral to an entirely different type of mineral as the weathered product. When feldspar is attacked by the hydrogen ion of carbonic acid (from carbon dioxide and water), it forms clay minerals. In general, a **clay mineral** is a hydrous aluminum silicate with a sheet-silicate structure like that of mica. Therefore, the entire silicate structure of the feldspar crystal is altered by weathering: feldspar is a framework silicate, but the clay mineral product is a sheet silicate, differing both chemically and physically from feldspar.

Let us look in more detail at the weathering of feldspar (equation *D* in table 5.1). Rainwater percolates down through soil, picking up carbon dioxide from the atmosphere and the upper part of the soil. The water, now slightly acidic, comes in contact with feldspar in the lower part of the soil (figure 5.14), as shown in the first part of the equation. The acidic water reacts with the feldspar and alters it to a clay mineral.

The hydrogen ion (H^+) attacks the feldspar structure, becoming incorporated into the clay mineral product. When the hydrogen moves into the crystal structure, it releases potassium (K) from the feldspar. The potassium is carried away in solution as a dissolved ion (K^+). The bicarbonate ion from the original carbonic acid does not enter into the reaction; it reappears on the right side of the equation. The soluble potassium and bicarbonate ions are carried away by water (groundwater or streams).

All the silicon from the feldspar cannot fit into the clay mineral, so some is left over and is carried away as silica (SiO_2) by the moving water. This excess silica may be carried in solution or as extremely small solid particles.

The weathering process is the same regardless of the type of feldspar: K-feldspar forms potassium ions; Na-feldspar and Ca-feldspar (plagioclase) form sodium ions and calcium ions, respectively. The ions that result from the weathering of Ca-feldspar are calcium ions (Ca^{2+}) and bicarbonate ions (HCO_3^-), both of which are very common in rivers and underground water, particularly in humid regions.

Chemical Weathering of Other Minerals

The weathering of ferromagnesian or dark minerals is much the same as that of feldspars. Two additional products are found on the right side of the equation—magnesium ions and iron oxides (hematite, limonite, and goethite).

The susceptibility of the rock-forming minerals to chemical weathering is dependent on the strength of the mineral's chemical bonding within the crystal framework. Because of the strength of the silicon-oxygen bond, quartz is quite resistant to chemical weathering. Thus, quartz (SiO_2) is the rock-forming mineral least susceptible to chemical attack at Earth's surface. Ferromagnesian minerals such as olivine, pyroxene, and amphibole include other positively charged ions such as Al, Fe, Mg, and Ca. The presence of these positively charged ions in the crystal framework makes these minerals vulnerable to chemical attack due to the weaker chemical bonding between these ions and oxygen, as compared to the much stronger silicon-oxygen bonds. For example, olivine—$(Fe, Mg)_2SiO_4$—weathers rapidly because its isolated silicon-oxygen tetrahedra are held together by relatively weak ionic bonds between oxygen and iron and magnesium. These ions are replaced by H^+ ions during chemical weathering similar to that described for the feldspars.

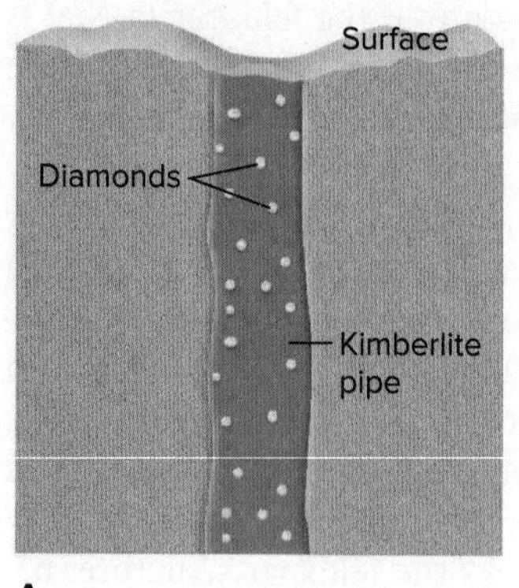

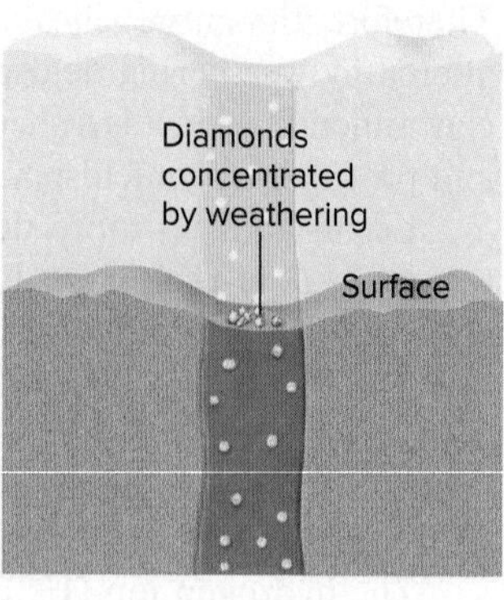

FIGURE 5.15

Residual concentration by weathering. (*A*) Cross-sectional view of diamonds widely scattered within kimberlite pipe. (*B*) Diamonds concentrated on surface by removal of rock by weathering and erosion.

Weathering and Diamond Concentration

Diamond is the hardest mineral known and is also extremely resistant to weathering. This is due to the very strong covalent bonding of carbon, as described in chapter 2. But diamonds are often concentrated by weathering, as illustrated in figure 5.15. Diamonds are brought to the surface of Earth in *kimberlite pipes,* columns of brecciated or broken ultramafic rock that have risen from the upper mantle. Diamonds are widely scattered in diamond pipes when they form. At the surface, the ultramafic rock in the pipe is preferentially weathered and eroded away. The diamonds, being more resistant to weathering, are left behind, concentrated in rich deposits on top of the pipes. Rivers may redistribute and reconcentrate the diamonds, as in South Africa and India. In Canada, kimberlite pipes have been eroded by glaciers, and diamonds may be found widely scattered in glacial deposits.

Weathering Products

Table 5.2 summarizes weathering products for the common minerals. Note that quartz and clay minerals commonly are left after complete chemical weathering of a rock. Sometimes other solid products, such as iron oxides, also are left after weathering.

The solution of calcite supplies substantial amounts of calcium ions (Ca^{2+}) and bicarbonate ions (HCO_3^-) to underground water. The weathering of Ca-feldspars (plagioclase) into clay minerals can also supply Ca^{2+} and HCO_3^- ions, as well as silica (SiO_2), to water. Under ordinary chemical circumstances, the dissolved Ca^{2+} and HCO_3^- can combine to form solid $CaCO_3$ (calcium carbonate), the mineral calcite. Dissolved silica can also precipitate as a solid from underground water. This is significant because calcite and silica are the most common materials precipitated as cement, which binds loose particles of sand, silt, and clay into solid sedimentary rock (see chapter 6). The weathering of calcite, feldspars, and other minerals is a likely source for such cement.

If the soluble ions and silica are not precipitated as solids, they remain in solution and may eventually find their way into a stream and then into the ocean. Enormous quantities of dissolved material are carried by rivers into the sea (one estimate is 4 billion tons per year). This is the main reason seawater is salty.

Factors Affecting Weathering

The intensity of both mechanical and chemical weathering is affected by a variety of factors. Chemical weathering is largely a function of the availability of liquid water. Rock chemically weathers much faster in humid climates than in arid climates. Limestone, which is extremely susceptible to dissolution, weathers quickly and tends to form valleys in wet regions such as the Appalachian Mountains. However, in the arid west, limestone is a resistant rock that forms ridges and cliffs. Temperature is also a factor in chemical weathering. The most intense chemical weathering occurs in the tropics, which are both wet and hot. Polar regions experience very little chemical weathering because of the frigid temperatures and the absence of liquid water. Mechanical weathering intensity is also related to climate (temperature and humidity), as well as to slope. Temperate climates, where abundant water repeatedly freezes and thaws, promote extensive frost weathering. Steep slopes cause rock to fall and break up under the influence of gravity. The most intense mechanical weathering probably occurs in high mountain peaks where the combination of steep slopes, precipitation, freezing and thawing, and flowing glacial ice rapidly pulverize the solid rock.

TABLE 5.2 Weathering Products of Common Rock-Forming Minerals

Original Mineral	Under Influence of CO_2 and H_2O	Main Solid Product		Other Products (Mostly Soluble)
Feldspar	→	Clay mineral	+	Ions (Na^+, Ca^{2+}, K^+), SiO_2
Ferromagnesian minerals (including biotite mica)	→	Clay mineral	+	Ions (Na^+, Ca^{2+}, K^+, Mg^{2+}), SiO_2, Fe oxides
Muscovite mica	→	Clay mineral	+	Ions (K^+), SiO_2
Quartz	→	Quartz grains (sand and silt)		
Calcite	→	—		Ions (Ca^{2+}, HCO_3^-)

EARTH SYSTEMS 5.1

Weathering, the Carbon Cycle, and Global Climate

Weathering has affected the long-term climate of Earth by changing the carbon dioxide content of the atmosphere through the inorganic carbon cycle (see box figure 1). Carbon dioxide is a "greenhouse gas" that traps solar heat near the surface, warming the Earth. The planet Venus has a dense atmosphere composed mostly of CO_2, which traps so much solar heat that the surface temperature averages a scorching 480°C (about 900°F—see chapter 22). Earth has comparatively very little CO_2 in its atmosphere (see box table 1)—enough to keep most of the surface above freezing but not too hot to support life. However, when Earth first formed, its atmosphere was probably very much like that of Venus, with much more CO_2. What happened to most of the original carbon dioxide in Earth's atmosphere? Geologists think that a quantity of CO_2 equal to approximately 65,000 times the mass of CO_2 in the present atmosphere lies buried in the crust and upper mantle of Earth. Some of this CO_2 was used to make organic molecules during photosynthesis and is now trapped as buried organic matter and fossil fuels in sedimentary rocks. However, the majority of the missing CO_2 was converted to bicarbonate ion (HCO_3^-) during chemical weathering and is locked away in carbonate minerals (primarily $CaCO_3$) that formed layers of limestone rock.

The inorganic carbon cycle helps to regulate the climate of Earth because CO_2 is a greenhouse gas, chemical weathering accelerates with warming, and the formation of limestone occurs mostly in warm, tropical oceans. When Earth's climate is warm, chemical weathering and the formation of limestone increase, drawing CO_2 from the atmosphere, which cools the climate. When the global climate cools, chemical weathering and limestone formation slow down, allowing CO_2 to accumulate in the atmosphere from volcanism, which warms the Earth. An increase in chemical weathering can also lead to global cooling by removing more CO_2 from the atmosphere. For example, the Cenozoic uplift and weathering of large regions of high mountains such as the Alps and the Himalaya may have triggered the global cooling that culminated in the glaciations of the Pleistocene epoch.

For more information on the carbon cycle, see box 21.2.

BOX 5.1 ■ TABLE 1

Carbon Dioxide in the Atmospheres of Earth, Mars, and Venus

	Earth	Mars	Venus
CO_2%	0.33	95.3	96.5
Total surface pressure, bars	1.0[a]	0.006	92

[a] Approximately 50 bars of CO_2 is buried in the crust of the Earth as limestone and organic carbon.

Additional Resource

- http://earthobservatory.nasa.gov/Features/CarbonCycle/

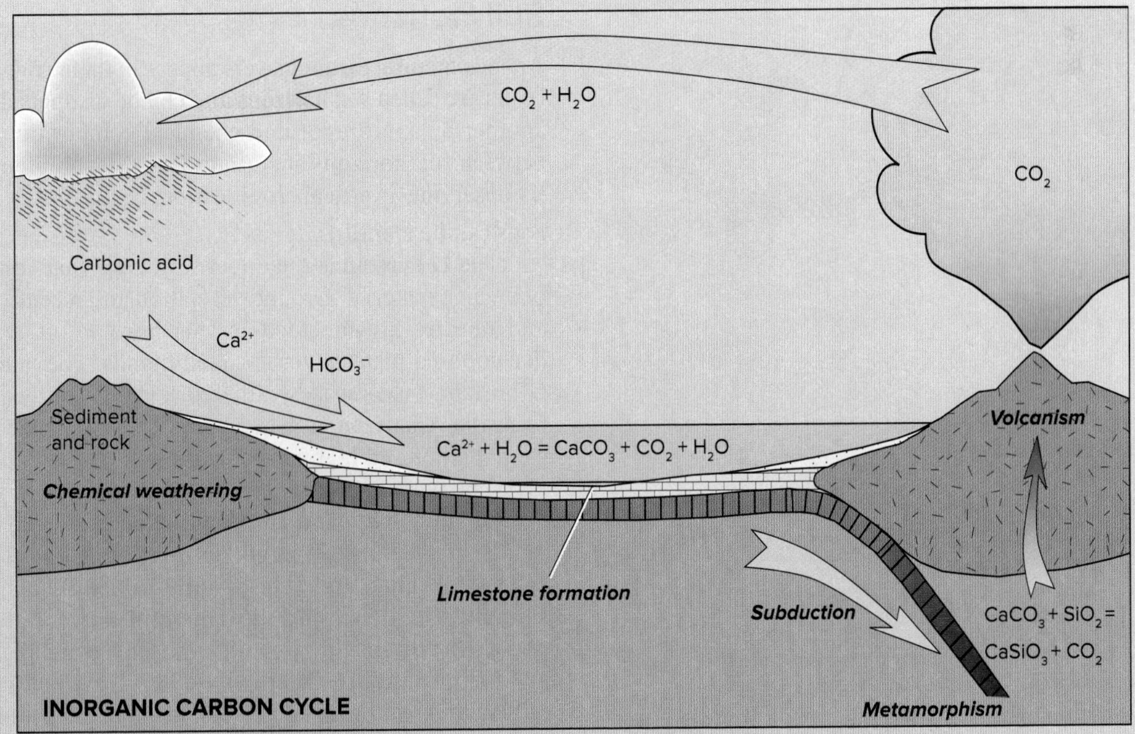

BOX 5.1 ■ FIGURE 1

Carbon dioxide dissolves in water to form carbonic acid in the atmosphere. Carbonic acid reacts with sediment and rocks during chemical weathering, releasing calcium ions and bicarbonate ions (HCO_3^-), which are carried by rivers into the sea. The precipitation of $CaCO_3$ mineral in the oceans (see chapter 6) forms layers of limestone rock. Deep burial of limestone leads to metamorphism, which causes silica and calcite to form calcium silicate minerals and carbon dioxide. The CO_2 remains trapped in Earth's interior until it is released during volcanic eruptions.

5.6 SOIL

In terms of Earth systems, soil forms an essential interface between the solid Earth (geosphere), biosphere, hydrosphere, and atmosphere. Soil is an incredibly valuable resource that supports life on Earth. In common usage, soil is the name for the loose, unconsolidated material that covers most of Earth's land surface. Geologists call this material *regolith*, however, and reserve the term **soil** for a layer of weathered, unconsolidated material that contains organic matter and is capable of supporting plant growth. A mature, fertile soil is the product of centuries of mechanical and chemical weathering of rock, combined with the addition and decay of plant and other organic matter.

An average soil is composed of 45% rock and mineral fragments (including clay), 5% decomposed organic matter, or humus, and 50% pore space. The rock and mineral fragments in a soil provide an anchoring place for the roots of plants. The clay minerals attract water molecules (figure 5.16) and plant-nutrient ions (figure 5.17), which are loosely held and available for uptake by plant roots. The humus releases weak acids that contribute to the chemical weathering of soil. Humus also produces plant nutrients and increases the water retention ability of the soil. The pore spaces are the final essential component of a fertile soil. Water and air circulate through the pore spaces, carrying dissolved nutrients and carbon dioxide, which is necessary for the growth of plants.

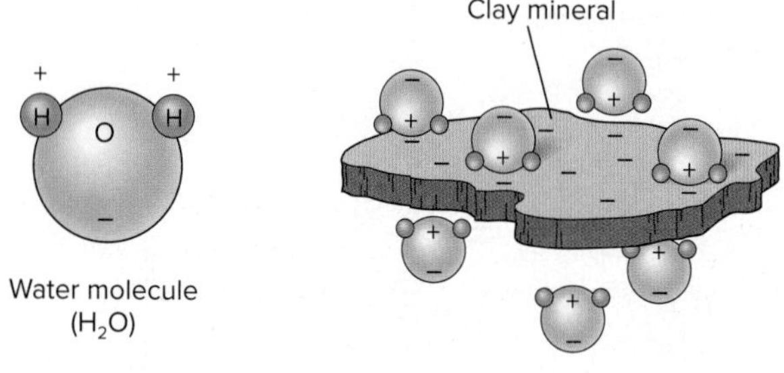

FIGURE 5.16

Negative charge on the outside of a platy clay mineral attracts the positive end of a water molecule.

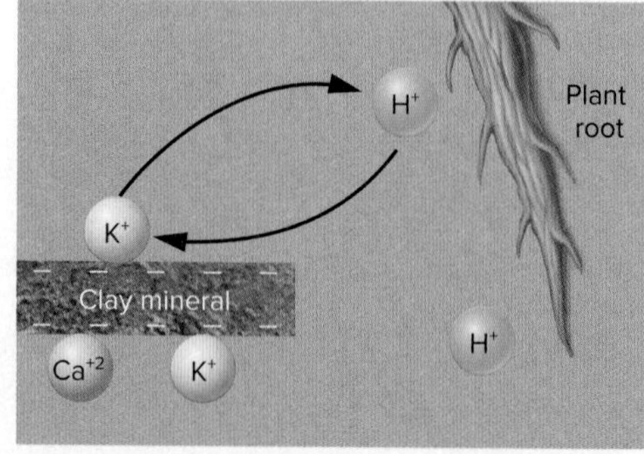

FIGURE 5.17

A plant root releases H^+ (hydrogen) ions from organic acids and exchanges them for plant-nutrient ions held by clay minerals.

The size and number of pore spaces, and therefore the ability of a soil to transmit air and water, are largely a function of the texture of a soil. *Soil texture* refers to the proportion of different-sized particles, generally referred to as sand, silt, and clay. Quartz generally weathers into sand grains that help keep soil loose and aerated, allowing good water drainage. Partially weathered crystals of feldspar and other minerals can also form sand-sized grains. Soils with too much sand, however, can drain too rapidly and deprive plants of necessary water.

Clay minerals occur as microscopic plates and help hold water and plant nutrients in a soil. Because of ion substitution within their sheet-silicate structure, most clay minerals have a negative electrical charge on the flat surfaces of the plates. This negative charge attracts water and nutrient ions to the clay mineral (figure 5.17). Plant nutrients, such as Ca^{2+} and K^+, commonly supplied by the weathering of minerals such as feldspar, are also held loosely on the surface of clay minerals. A plant root can release H^+ from organic acids and exchange it for the Ca^{2+} and K^+ that the plant needs for healthy growth (figure 5.17). Too much clay in a soil may pack together closely, though, causing pore spaces to be too small to allow water to drain properly. Too much water in the soil and not enough air may cause plant roots to rot and die.

Silt particles are between sand and clay in size. A soil with approximately equal parts sand, silt, and clay is referred to as **loam**. Loamy soils are well-drained, may contain organic matter, and are often very fertile and productive.

Soil Horizons

As soils mature, distinct layers appear in them (figure 5.18). Soil layers are called **soil horizons** and can be distinguished from one another by appearance and chemical composition. Boundaries between soil horizons are usually transitional rather than sharp. By observing a vertical cross section, or *soil profile*, various horizons can be identified.

The **O horizon** is the uppermost layer that consists entirely of organic material. Ground vegetation and recently fallen leaves and needles are included in this horizon, as well as highly decomposed plant material called humus. The humus from the O horizon mixes with weathered mineral matter just below to form the **A horizon**, a dark-colored soil layer that is rich in organic matter and high in biological activity, both plant and animal. The two upper horizons are often referred to as *topsoil.*

Organic acids and carbon dioxide produced by decaying plants in the topsoil percolate down into the **E horizon**, or **zone of leaching**, and help dissolve minerals such as iron and calcium. The downward movement of water in the E horizon carries the dissolved minerals, and fine-grained clay minerals as well, into the soil layer below. This leaching (or *eluviation*) of clay and soluble minerals can make the E horizon pale and sandy.

The material leached downward from the E horizon accumulates in the **B horizon**, or **zone of accumulation**. This layer is often quite clayey and stained red or brown by hematite and limonite. Calcite may also build up in B horizons.

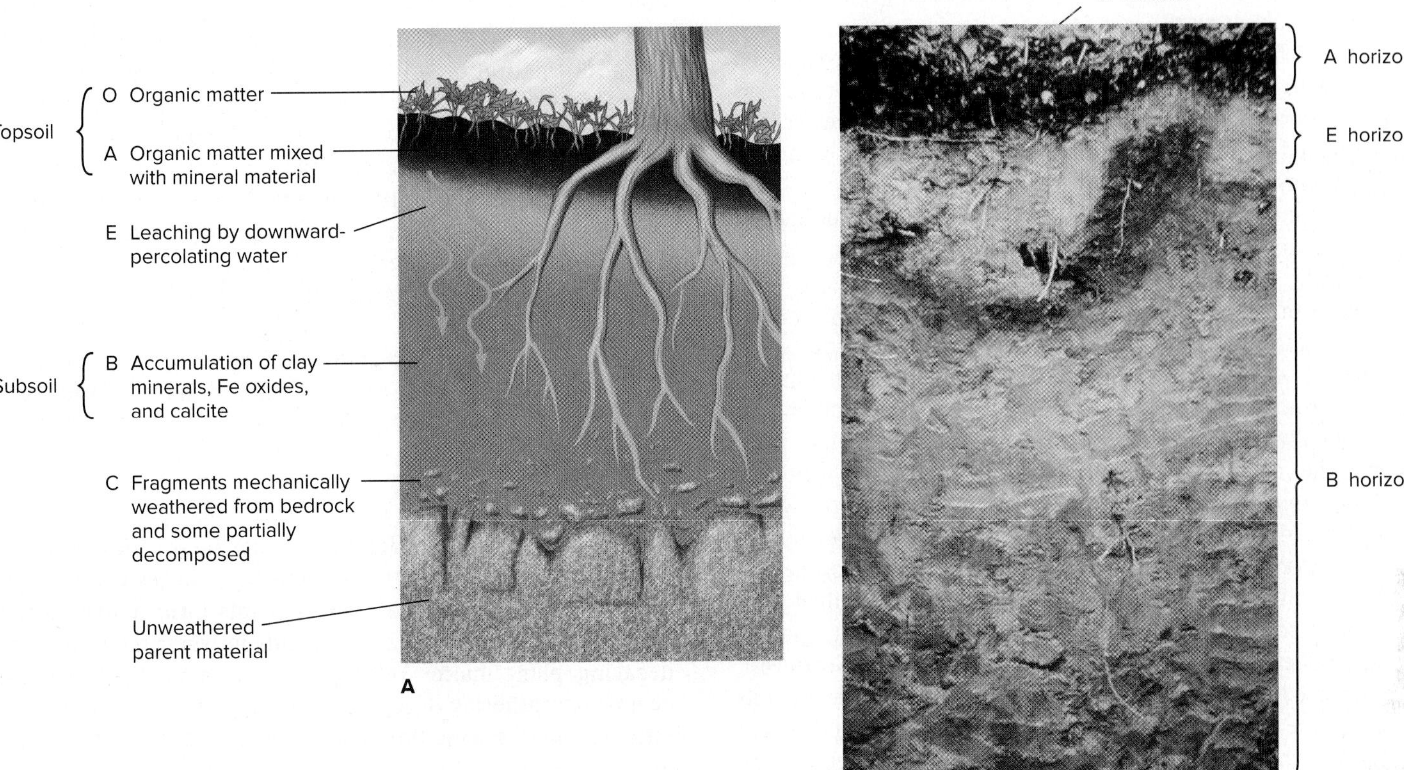

FIGURE 5.18

(*A*) Horizons (O, A, E, B, and C) in a soil profile that form in a humid climate. (*B*) Soil profile that shows the A horizon stained dark by humus. The E horizon is lighter in color, sandy, and crumbly. The clayey B horizon is stained red by hematite, leached downward from the E horizon.
(B) Source: USDA Natural Resources Conservation Service

This horizon is frequently called the *subsoil.* Within the B horizon, a hard layer of Earth material called *hardpan* may form in wet climates where clay minerals, silica, and iron compounds have accumulated in the B horizon from eluviation of the overlying E horizon. A hardpan layer is very difficult to dig or drill through and may even be too hard for backhoes to dig through; planting a tree in a lawn with a hardpan layer may require a jackhammer. Tree roots may grow laterally along rather than down through hardpan; such shallow-rooted trees are often uprooted by the wind.

The **C horizon** is incompletely weathered parent material that lies below the B horizon. The parent material is commonly subjected to mechanical and chemical weathering from frost action, roots, plant acids, and other agents. The C horizon is transitional between the unweathered rock or sediment below and the developing soil above.

Factors Affecting Soil Formation

Most soils take a long time to form. The rate of soil formation is controlled by rainfall, temperature, slope, and, to some extent, the type of rock that weathers to form soil. High temperature and abundant rainfall speed up soil formation, but in most places, a fully developed soil that can support plant growth takes hundreds or thousands of years to form.

It would seem that the properties of a soil should be determined by the rock (the *parent material*) from which it formed, and this is partly true. Several other factors, however, are important in the formation of a soil, sometimes playing a larger role than the rocks themselves in determining the types of soils formed. These additional factors, *slope*, *living organisms*, *climate*, and *time*, are discussed in the following paragraphs.

Parent Material

The character of a soil depends partly on the parent material from which it develops. The parent material is the source of the weathered mineral matter that makes up most of a soil. A soil developing on weathering granite will be sandy, as sand-sized particles of quartz and feldspar are released from the granite. As the feldspar grains weather completely, fine-grained clay minerals are formed. The resulting soil will contain a variety of grain sizes and will have drainage and water-retention properties conducive to plant growth.

A soil forming on basalt may never be sandy, even in its early stages of development. If chemical weathering processes are more prevalent than mechanical weathering processes, the fine-grained feldspars in the basalt will weather directly to fine-grained clay minerals. Since the parent rock had no coarse-grained minerals and no quartz to begin with, the resulting soil may lack sand. Such a soil may not drain well, although it can be quite fertile.

Both of these soils are called **residual soils**; they develop from weathering of the bedrock beneath them (figure 5.19). Figure 5.18*A* is a diagram of residual soil developing in a humid climate from a bedrock source. **Transported soils** do not develop from locally formed rock, but from regolith brought in from some other region (figure 5.19). (Keep in mind that it is not the soil itself that is transported, but the parent material from which it is formed.) For example, mud deposited by a river during times of flooding can form an excellent agricultural soil next to the river after floodwaters recede. Wind deposits called *loess* (see chapter 13) form the base for some of the most valuable food-producing soils in the Midwest and the Pacific Northwest. Transported soils are generally more fertile than residual soils because the parent material is transported from many different locations; there is more variety in the chemical makeup of the parent material, so a greater variety of minerals and nutrients are supplied to the resulting soil.

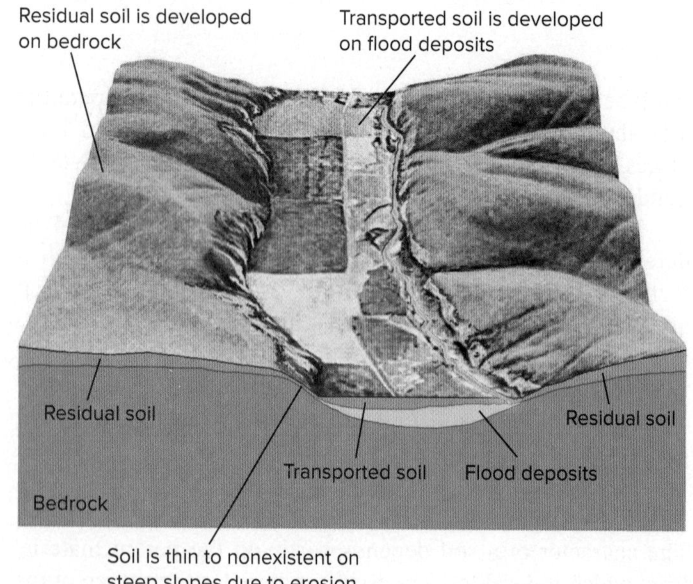
Residual soil is developed on bedrock
Transported soil is developed on flood deposits
Residual soil
Residual soil
Transported soil
Flood deposits
Bedrock
Soil is thin to nonexistent on steep slopes due to erosion

FIGURE 5.19
Residual soil develops from weathering of bedrock beneath the hills, whereas transported soil develops on top of flood plain deposits (regolith) in the stream valley. Soils are thin to nonexistent on the steeper slopes because of erosion.

Slope

The slope of the land surface provides an important control on the formation of soil (figure 5.19). Soils tend to be thin or nonexistent on steep slopes, where gravity keeps water and soil particles moving downhill. Vegetation is sparse on steep slopes, so there are not many roots to hold the weathering rock in place and little organic matter to provide nutrients. By contrast, soils in bottomlands may be very thick but poorly drained and waterlogged. Vegetation in the bottomlands does not decay completely, and thick, dark layers of peat may form.

The optimal topography for soil formation is flat or gently sloping uplands, allowing good drainage, minimal erosion, and healthy vegetation cover.

Living Organisms

The biosphere plays an important role in soil development. The chief function of living organisms is to provide organic material to the soil. Decomposing plants form humus, which supplies nutrients to the soil and aids in water retention. The decaying plant matter releases organic acids that increase chemical weathering of rocks. Growing plants send roots deep into the soil, breaking up the underlying bedrock and opening up pore spaces.

Burrowing organisms such as ants, worms, and rodents bring soil particles to the surface and mix the organic and mineral components of the soil. They create passageways that allow for the circulation of air and water, increasing chemical weathering and accelerating soil formation. Microorganisms such as bacteria, fungi, and protozoa promote the decomposition of organic matter to humus, and some bacteria fix nitrogen in the soil, making it available for uptake by plants. The interdependency of plants, animals, and soil is a mutually beneficial and delicately balanced system.

Climate

Climate is perhaps the most influential factor affecting soil thickness and character. The same parent materials in the same topography will form significantly different soil types under different climatic conditions. Temperature and precipitation determine whether chemical or mechanical weathering processes will dominate and strongly influence the rate and depth of weathering. The amount and types of vegetation and animal life that contribute to soil formation are also determined by climate.

Soils in temperate, moist climates, as in Europe and the eastern United States, tend to be thick and are generally characterized by downward movement of water through Earth materials (figure 5.18 shows such a soil). In general, these soils tend to be fertile, have a high content of aluminum and iron oxides and well-developed horizons, and are marked by effective downward leaching due to high rainfall and to the acids produced by decay of abundant humus.

FIGURE 5.20

Soil profile marked by upward-moving groundwater that evaporates underground in a drier climate, precipitating calcium carbonate within the soil, sometimes forming a light-colored layer.
Source: Donald Yost, USDA Natural Resources Conservation Service

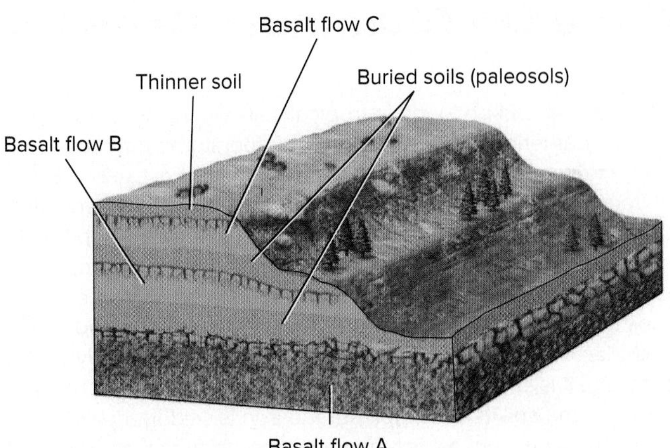

FIGURE 5.21

Because soils tend to become thicker with time, the thickness of soils developed on successive basalt flows can be used to estimate the length of time between eruptions. Based on the soil thickness, more time elapsed between the eruption of basalt flows A and B than between flows B and C. These soils have been covered by the youngest basalt flow (C) and are examples of buried soils or *paleosols*.

In arid climates, as in many parts of the western United States, soils tend to be thin and are characterized by little leaching, scant humus, and the *upward* movement of soil water beneath the land surface. The water is drawn up by subsurface evaporation and capillary action. As the water evaporates beneath the land surface, salts are precipitated within the soil (figure 5.20). An extreme example of salt buildup can be found in desert *alkali soils*, in which heavy concentrations of toxic sodium salts may prevent plant growth.

Another example of the control of climate on soil formation is found in the tropical rain forests of the world. The high temperatures and abundant rainfall combine to form extremely thick red soils called *oxisols*, or laterites, that are highly leached and generally infertile. See box 5.2 for a discussion of these soils.

Time

Note that the character of a soil changes with time. In a soil that has been weathering for a short period of time, the characteristics are largely determined by the parent material. Young soils can retain the structure of the parent rock, such as bedding layers. As time progresses, other factors become more important and climate eventually predominates. Soils forming from many different kinds of igneous, metamorphic, and sedimentary rocks can become quite similar, given the same climate and enough time. In the long term, the only characteristic of the parent rock to have significance is the presence or absence of coarse grains of quartz.

With time, soils tend to become thicker. In regions of ongoing volcanic activity, the length of time between eruptions can be estimated by the thickness of the soil that has formed on each flow (figure 5.21). A soil that has been buried by a lava flow, volcanic ash, windblown dust, glacial deposits, or other sediment is called a buried soil, or *paleosol* (*paleo* = ancient). Such soils may be distinctive and traceable over wide regions and may contain buried organic remains, making them useful for dating rocks and sediments and for interpreting past climates and topography.

Soil Erosion

Although soil accounts for an almost insignificant fraction of all Earth materials, it is one of the most significant resources in terms of its effect on life. Soil provides nourishment and physical support for plant life. It is the very base of the food chain that supports human existence. As such, soil is one of Earth's most vital resources, but it is also one of Earth's most abused.

The upper layers of a soil, the O and A horizons, are the most fertile and productive. These are the layers that are most vulnerable to erosion due to land mismanagement by poor farming and grazing practices. Scientists estimate that the Earth has lost about 10% of its productive value (the ability to provide crops, pasture, and forest products) over the last 50 years. If measures are not taken to curb the loss of fertile soils to erosion, an additional 10% of Earth's productivity could be lost in the next 25 years.

Where Do Aluminum Cans Come From?

The answer can be found in tropical soils where extreme chemical weathering occurs. The high temperatures and abundant rainfall in tropical regions produce some of the thickest soil on Earth. Lush vegetation grows over this soil, but the soil itself is very infertile. As water percolates down through the soil in this hot, humid climate, plant nutrients are dissolved and carried downward, out of reach of plant roots. Even silica is dissolved in this environment, and the soil that remains is composed almost entirely of iron and aluminum oxides. This highly leached soil is an oxisol, commonly called *laterite*, and it is characteristically red in color (box figure 1). The iron oxides give the soil its red color, but the iron is seldom rich enough to mine. The aluminum oxides, however, may form rich ore deposits near the surface.

As leaching proceeds, nearly pure layers of bauxite ($Al(OH)_3$), the principal ore of aluminum, are left near the surface in large deposits that are on average 4–6 meters thick and may cover many square kilometers (box figure 2). Eighty percent of the world's aluminum is mined from large blanket deposits in West Africa, Australia, South America, and India.

The bauxite ore is washed, crushed, and then dissolved under high temperature and pressure in a caustic solution of sodium hydroxide. The chemical equation for this process is:

$$Al(OH)_3 + NA^+ + OH^- \rightarrow Al(OH)_4^- + NA^+$$

BOX 5.2 ■ FIGURE 1

Laterite soil (oxisol) develops in very wet climates, where intense, downward leaching carries away all but iron and aluminum oxides. Many laterites are a rusty orange to deep red color from the oxidation of the iron oxides.
Source: USDA Natural Resources Conservation Service

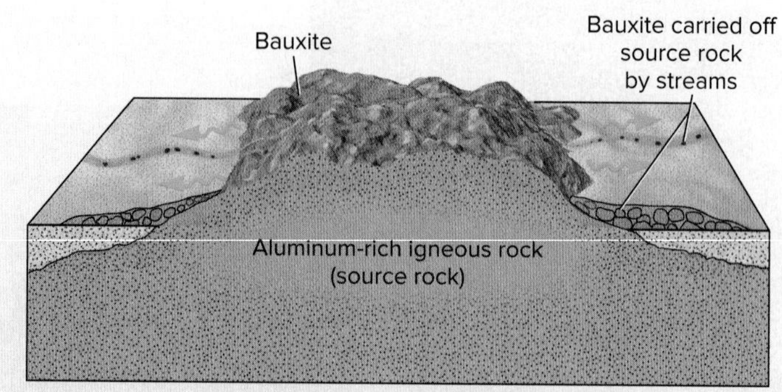

BOX 5.2 ■ FIGURE 2

Bauxite forms by intense tropical weathering of an aluminum-rich source rock such as a volcanic tuff.

The undissolved residue, mostly iron, silica, and titanium, settles to the bottom, and the sodium aluminate solution is pumped into precipitators, where the previous reaction is reversed.

$$Al(OH)_4^- + NA^+ \rightarrow Al(OH)_3 + NA^+ + OH^-$$

The sodium hydroxide is recovered and returned to the beginning of the process, and the pure bauxite crystals are passed to another process that drives off water to form a white alumina (aluminum oxide) powder.

$$2Al(OH)_3 \rightarrow Al_2O_3 + 3H_2O$$

The alumina is then smelted by passing electric currents through the powder to separate the metallic aluminum from the oxygen. Small amounts of other metals may be added to the molten aluminum to form alloys, and the aluminum is cast into blocks that are sent to factories for further processing. The aluminum alloy used for beverage cans contains manganese, which helps the metal become more ductile as it is rolled into the thin sheets from which the cans are formed.

Aluminum smelting is very energy-intensive; 15.7 kilowatt-hours of electricity are required to produce a single kilogram of aluminum (the average home in the United States uses 24 kilowatt-hours of electricity each day). To control costs, smelters are often located near hydroelectric power plants, which are built for the sole purpose of powering the smelters.

Aluminum can be recycled repeatedly. Recycling uses only 5% of the energy required to make "new" aluminum. Twenty recycled cans can be produced with the same amount of energy required to produce a single can from bauxite. Worldwide, over 40% of the aluminum demand is supplied by recycled material. The aluminum from the beverage can that you toss into the recycle bins is recovered, mixed with a small percent of new aluminum, and is back on the shelves as a new can of soda in 6 to 8 weeks.

Additional Resource

- www.world-aluminium.org/

Web page for the *International Aluminum Institute* contains additional information on the production and use of aluminum.

How Soil Erodes

Soil particles are small and are therefore easily eroded by water and wind. Raindrops strike unprotected soil like tiny bombs, dislodging soil particles in a process called *splash erosion* (figure 5.22*A*). As rain continues, a thin sheet of running water forms over the landscape, carrying the dislodged soil particles away (*sheet erosion*). Currents that form in the sheet of water cut tiny channels called *rills* in the exposed soil (figure 5.22*B*). The rills deepen into *gullies,* which merge into stream channels. Rivers that turn brown and muddy after rain storms are evidence of the significant amount of soil that can be transported by water.

A

B

FIGURE 5.22

Soil erosion. (*A*) Splash erosion dislodges soil particles, making them available for removal by sheet erosion and rill erosion. (*B*) Smaller channels (rills) merge into larger channels as water erodes the loose material on this slope, Petrified Forest, western United States.
(A) Source: US Navy/USDA-NRCS (B) ©moodboard/Glow Images RF

Wind erosion is generally less significant than erosion by water, but it is a particular problem in arid and semiarid regions. The wind picks up the lighter components of a soil, such as the clays, silts, and organic matter, and may transport them many kilometers. These components are the ones that contribute most to soil fertility. Agricultural soils that have been depleted by wind erosion require increased use of fertilizers to maintain their productivity.

Rates of Erosion

The rate of soil erosion is influenced by several factors: soil characteristics, climate, slope, and vegetation. Coarse-grained soils with organic content tend to have larger pore spaces and can absorb more water than soils dominated by clay-sized particles. Less runoff occurs on the coarser soils, and less of the soil is eroded away. The type of rainfall also influences the amount of erosion. A gentle rain over a long period of time produces less splash erosion than a short, heavy rainstorm. More water can infiltrate the soil during the gentle rainfall, and there is less likelihood of sheet erosion occurring. Slope also plays an important role in soil erosion. Water moves more slowly on gentle slopes and is more likely to percolate down into the soil. The faster-moving water on steeper slopes does not infiltrate and has a greater ability to dislodge and transport soil particles down from the slope.

A very significant control on soil erosion rate is the amount and type of vegetation present. Plant roots form networks in the O and A horizons that bind soil particles. The leaf canopy protects the soil from the impact of raindrops, lowering the risk of splash erosion. Thick vegetation can reduce the wind velocity near the ground surface, preventing the loss of soil due to wind erosion. Human activity in the last two centuries has done much to remove the natural vegetation cover on the world's land surface. Large-scale farming operations, grazing, logging, mining, and construction have disrupted prairies, forests, and other natural environments, such as rain forests, leaving the underlying soils vulnerable to the effects of wind and water (figure 5.23).

Consequences of Erosion

All of the soil particles that are eroded by wind and water have to go somewhere, and they end up being deposited as sediments in streams, flood plains, lakes, and reservoirs. Erosion and sedimentation are natural processes, but they have been proceeding at an unnatural rate since the advent of mechanized farming. Since colonial times, forest land in the Chesapeake Bay watershed has been cleared for farming and timber. Over the last 150 years, so much sediment from the cleared land has been carried into the Chesapeake Bay that 787 acres of new land have been added to Maryland. The average water depth in the bay has been reduced by almost a meter in some places, requiring increased dredging to keep shipping channels open. Fine-grained sediments remain suspended in the water column, reducing water clarity and

FIGURE 5.23

Soil erosion caused by clear cutting of rain forest near Lake Baringo, Kenya, Africa. *©Mark Boulton/Alamy*

preventing light from reaching the bottom of the bay. The aquatic vegetation that supports and protects the oysters and other shellfish for which the bay is famous cannot survive in the reduced light.

Perhaps one of the most devastating consequences of soil erosion occurred in the American Midwest during the 1930s. Agriculture had expanded nearly tenfold in the Great Plains region between the 1870s and the 1930s. Advances in farm equipment allowed farmers to practice "intensive row crop agriculture," in which more than 100 million acres of prairie were plowed under and planted in long rows of crops such as corn, soybeans, and wheat. After several years of drought in the 1930s, the row crops failed and the soil was left exposed to the high winds that came whipping across the plains. Huge dust clouds called "black rollers" billowed up, burying vehicles and drifting like snow against houses (figure 5.24). The clouds of sediment drifted east, darkening the sky and falling as muddy rain and snow on the East Coast states. Years of hardship and suffering followed for the inhabitants of the Dust Bowl states.

The fertile agricultural soils of the Canadian plains and the northern United States took more than 10,000 years to develop on glacial deposits after the thick continental ice sheet melted. These soils and many others around the world are eroding at an alarming rate, much faster than they are being replaced by newly formed soils. This essential resource, upon which all life depends, has become a nonrenewable resource. Conservation practices such as windbreaks, contour plowing, terracing, and crop rotation have been implemented in recent years to help reduce the amount of topsoil lost to wind and water. More must be done, especially in developing countries, to protect this fragile resource.

A

B

FIGURE 5.24

The Dust Bowl. (*A*) A "black roller" bears down on a truck on Highway No. 59, south of Lamar, Colorado, in 1937. (*B*) Drifts of dust buried vehicles and outbuildings in Gregory County, South Dakota, 1936.
(A) Source: U.S. Department of Agriculture (B) Source: U.S. Department of Agriculture

Soil Classification

Early soil classification efforts were based largely on the geology of the underlying rocks. It became apparent, however, that different types of soil could form on the same underlying rock, depending upon climate, topography, and the age of the soil. In many cases, the underlying rock was the least significant factor involved. Several different approaches were tried, and in 1975 a soil classification system was developed that grouped soils into twelve large *orders* based upon the characteristics of the horizons present in soil profiles. Brief descriptions of the orders are given in table 5.3, along with the factors most important in the formation of each soil. Figure 5.25 shows the worldwide distribution of the twelve major orders.

TABLE 5.3 World Soil Orders

Soil Orders	Description	Controlling Factors
Alfisols	Gray to brown surface horizon, subsurface horizon of clay accumulation; medium to high in plant nutrient ions, common in humid forests.	Climate Organisms
Andisols	Soils formed in volcanic ash.	Parent material
Aridisols	Soils formed in dry climates, low in organic matter, often having horizons of carbonate, gypsum, or salt.	Climate
Entisols	Soils that have no horizons due to young age of parent material or to constant erosion.	Time Topography
Gelisols	Weakly weathered soils with permafrost within 2 meters of the surface.	Climate
Histosols	Wet, organic soils with relatively little mineral material, such as peat in swamps and marshes.	Topography
Inceptisols	Very young soils that have weakly developed horizons and little or no subsoil clay accumulation.	Time Climate
Mollisols	Nearly black surface horizon rich in organic matter and plant nutrient ions; subhumid to semiarid midlatitude grasslands.	Climate Organisms
Oxisols	Heavily weathered soils low in plant nutrient ions, rich in aluminum and iron oxides; humid, tropical climates; also called laterites.	Climate Time
Spodosols	Acid soils low in plant nutrient ions with subsurface accumulation of humus that is complexed with aluminum and iron; cool, humid pine forests in sandy parent material.	Parent material Organisms Climate
Ultisols	Strongly weathered soils low in plant nutrient ions with clay accumulation in the subsurface; humid temperate and tropical acid forest environments.	Climate Time Organisms
Vertisols	Clayey soils that swell when wet and shrink when dry, forming wide, deep cracks.	Parent material

After E. Brevik, 2002, *Journal of Geoscience Education*, v. 50, n. 5.

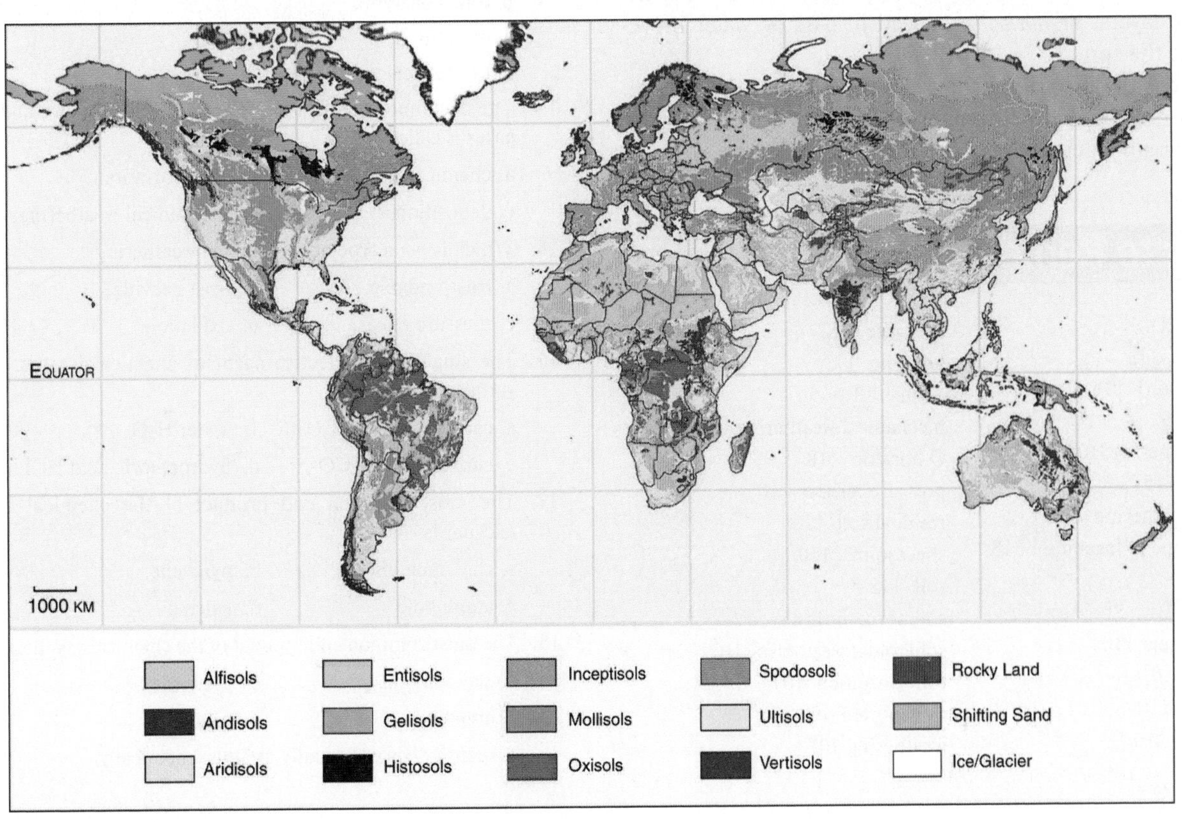

FIGURE 5.25

Worldwide distribution of soil orders.
Source: U.S. Department of Agriculture, Natural Resources Conservation Service, World Soil Resources Division, U.S. Department of Agriculture.

Summary

When rocks that formed deep in Earth become exposed at the Earth's surface, they are altered by *mechanical* and *chemical weathering.*

Weathering processes produce *spheroidal weathering, differentially weathered* landforms, *sheet joints,* and *exfoliation domes.*

Mechanical weathering, largely caused by *frost action* and *pressure release* after unloading, disintegrates (breaks) rocks into smaller pieces.

By increasing the exposed surface area of rocks, mechanical weathering helps speed chemical weathering.

Chemical weathering results when a mineral is unstable in the presence of water and atmospheric gases. As chemical weathering proceeds, the mineral's components recombine into new minerals that are more in equilibrium.

Weak acid, primarily from the solution of carbon dioxide in water, is an effective agent of chemical weathering.

Calcite dissolves when it is chemically weathered. Most of the silicate minerals form *clay minerals* when they chemically weather. Quartz is very resistant to chemical weathering.

Soil develops by chemical and mechanical weathering of a parent material. Some definitions of soil require that it contain organic matter and be able to support plant growth.

Soils, which can be *residual* or *transported,* usually have distinguishable layers, or *horizons,* caused in part by water movement within the soil.

Climate is the most important factor determining soil type. Other factors in soil development are parent material, time, slope, and organic activity.

Terms to Remember

A horizon 118
B horizon (zone of accumulation) 118
C horizon 119
chemical weathering 108
clay mineral 115
differential weathering 109
E horizon (zone of leaching) 118
erosion 107
exfoliation 110
exfoliation dome 110
frost action 110
frost heaving 110
frost wedging 110
hematite 112
limonite 113
loam 118
mechanical weathering 108
O horizon 118
pressure release 110
residual soil 120
sheet joints 110
soil 118
soil horizon 118
spheroidal weathering 109
transportation 107
transported soil 120
weathering 107

Testing Your Knowledge

Use the following questions to prepare for exams based on this chapter.

1. Why is weathering important to life on Earth?
2. Describe at least three processes that mechanically weather rock. How does mechanical weathering differ from chemical weathering?
3. How can mechanical weathering speed up chemical weathering?
4. Name and describe the types of features caused by weathering.
5. Name at least three natural sources of acid in solution. Which one is most important for chemical weathering?
6. Explain what happens chemically when calcite dissolves.
7. Describe what happens to each mineral within granite during the complete chemical weathering of granite in a humid climate. List the final products for each mineral.
8. Why do stone buildings tend to weather more rapidly in cities than in rural areas?
9. Describe the factors that affect the formation of soil.
10. Name and sketch the soil horizons that form in a humid climate. Explain how the layers (horizons) form.
11. What is the difference between a residual soil and a transported soil?
12. How do soils erode, and why is it important to minimize soil erosion?
13. Physical disintegration of rock into smaller pieces is called
    a. chemical weathering.
    b. transportation.
    c. deposition.
    d. mechanical weathering.
14. The decomposition of rock from exposure to water and atmospheric gases is called
    a. chemical weathering. b. transportation.
    c. deposition. d. mechanical weathering.
15. Which is *not* a type of mechanical weathering?
    a. frost wedging. b. frost heaving.
    c. pressure release. d. oxidation.
16. The single most effective agent of chemical weathering at Earth's surface is
    a. carbonic acid H_2CO_3. b. water H_2O.
    c. carbon dioxide CO_2. d. hydrochloric acid HCl.
17. The most common end product of the chemical weathering of feldspar is
    a. clay minerals. b. pyroxene.
    c. amphibole. d. calcite.
18. The most common end product of the chemical weathering of quartz is
    a. clay minerals. b. pyroxene.
    c. amphibole. d. calcite.
    e. quartz does not usually weather chemically.

19. Soil with approximately equal amounts of sand, silt, and clay along with a generous amount of organic matter is called
    a. loam. b. inorganic.
    c. humus. d. caliche.
20. The most important factor in determining the type of soil that forms is
    a. parent material. b. living organisms.
    c. slope. d. climate.
    e. time.
21. The soil horizon containing only organic material is the
    a. A horizon. b. B horizon.
    c. C horizon. d. O horizon.
    e. E horizon.
22. Hardpan forms in the
    a. A horizon. b. B horizon.
    c. C horizon. d. E horizon.

Expanding Your Knowledge

1. Which mineral weathers faster—hornblende or quartz? Why?
2. Compare and contrast the weathering rate and weathering products for Ca-rich plagioclase in the following localities:
   a. central Pennsylvania with 40 inches of rain per year;
   b. Death Valley with 2 inches of rain per year;
   c. an Alaskan mountaintop where water is frozen year-round.
3. The amount of carbon dioxide gas has been increasing in the atmosphere since the mid-twentieth century as a result of the burning of fossil fuels. What effect will the increase in CO_2 have on the rate of chemical weathering? The increase in CO_2 may cause substantial global warming in the future. What effect would a warmer climate have on the rate of chemical weathering? Give the reasons for your answers.
4. In a humid climate, is a soil formed from granite the same as one formed from gabbro? Discuss the similarities and possible differences with particular regard to mineral content and soil color.

Exploring Web Resources

http://soils.ag.uidaho.edu/soilorders/

University of Idaho Soil Science Division. Web page contains photos, descriptions, and surveys of the twelve major soil orders.

http://soils.usda.gov/

U.S. Department of Agriculture site provides information on soil surveys, photos, and comprehensive coverage of soils.

http://sis.agr.gc.ca/cansis/

Canadian Soil Information System provides links to detailed soil surveys and land inventories.

CHAPTER

6

Sediment and Sedimentary Rocks

Eroded sandstone formations formed from ancient sand dunes, North Coyote Buttes, Arizona.
©Doug Sherman/Geofile RF

LEARNING OBJECTIVES

- Know why sedimentary rocks are important to geologists and valuable in general.
- Using the rock cycle, explain how sediment is formed and may become a sedimentary rock.
- Describe how the three main types of sedimentary rocks are classified.
- Define *detrital sediment* and know how detrital sedimentary rocks are classified and identified.
- Explain how chemical sedimentary rocks are formed, and give several examples.
- Describe how coal forms and explain the origin of oil and gas.
- Illustrate how sedimentary structures identify how sediment was transported and used to interpret the environment of deposition.
- Describe what clues a geologist uses to interpret where and how a sedimentary rock formed and why this is important.

The rock cycle is a conceptual model of the constant recycling of rocks as they form, are destroyed, and then re-form (figure 6.1). We began our discussion of the rock cycle with igneous rock (chapters 3 and 4), and we now discuss sedimentary rocks. In this chapter, we first describe sediment and sedimentary rock and then discuss sedimentary structures and fossils. We also consider the importance of sedimentary rocks for interpreting the history of Earth and their tremendous economic importance. Metamorphic rocks, the third major rock type, are the subject of chapter 7.

In chapter 5, you saw how weathering produces sediment. In this chapter, we explain more about sediment origin, as well as the erosion, transportation, sorting, deposition, and eventual transformation of sediments to sedimentary rock. Because they have such diverse origins, sedimentary rocks are difficult to classify. We divide them into detrital, chemical, and organic sedimentary rocks, but this classification does not do justice to the great variety of sedimentary rock types. Furthermore, despite their great variety, only three sedimentary rocks are very common—shale, sandstone, and limestone.

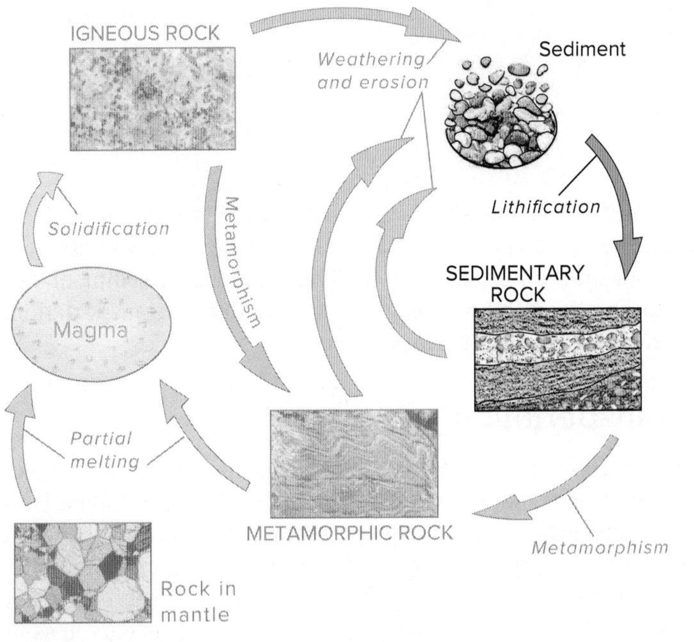

Sedimentary rocks contain sedimentary structures such as ripple marks, cross-beds, and mud cracks, as well as the fossilized remains of extinct organisms. These features, combined with knowledge of the sediment types within the rock and the sequence of rock layers, give geologists clues to interpret the environments in which the rocks were deposited. About three-fourths of the surface of the continents is blanketed by sedimentary rock, providing geologists with the information they need to reconstruct a detailed history of the surface of Earth and its biosphere.

Sedimentary rocks are also economically important. Most building materials such as stone, concrete, silica (glass), gypsum (plaster), and iron are quarried and mined from sedimentary rock (see box 6.1). Salt is also a sedimentary product and, in many places in the world, supplies of fresh water are pumped from sedimentary layers. Coal, crude oil, and natural gas, the fossil fuels that drove the Industrial Revolution, are all formed within and extracted from sedimentary rock.

FIGURE 6.1

The rock cycle shows how rocks can be weathered and eroded and recycled into sediment to form new rocks.
Igneous ©Doug Sherman/Geofile RF Metamorphic ©reneh9999/Shutterstock Mantle ©Doug Sherman/Geofile RF

6.1 SEDIMENT

Most sedimentary rocks form from loose grains of sediment. **Sediment** is the collective name for loose, solid particles of mineral that originate from:

1. Weathering and erosion of preexisting rocks (detrital sediments).
2. Precipitation from solution, including secretion by organisms in water (chemical sediments).

Sediment includes such particles as sand on beaches, mud on a lake bottom, boulders frozen into glaciers, pebbles in streams, and dust particles settling out of the air. An accumulation of clam shells on the sea bottom offshore is sediment, as are coral fragments broken from a reef by large storm waves.

These particles usually collect in layers on Earth's surface. An important part of the definition is that the particles are loose. Sediments are said to be *unconsolidated,* which means that the grains are separate, or unattached to one another.

TABLE 6.1 Sediment Particles and Detrital Sedimentary Rocks

Diameter (mm)	Sediment		Sedimentary Rock
256	Boulder		**Breccia** (angular particles) or
64	Cobble		**conglomerate** (rounded
	Pebble	Gravel	particles)
2	Sand		**Sandstone**
1/16	Silt	"Mud"	Siltstone (mostly silt)
1/256	Clay		**Shale** or mudstone (mostly clay)

Sandstone and shale are quite common; the others are relatively rare.

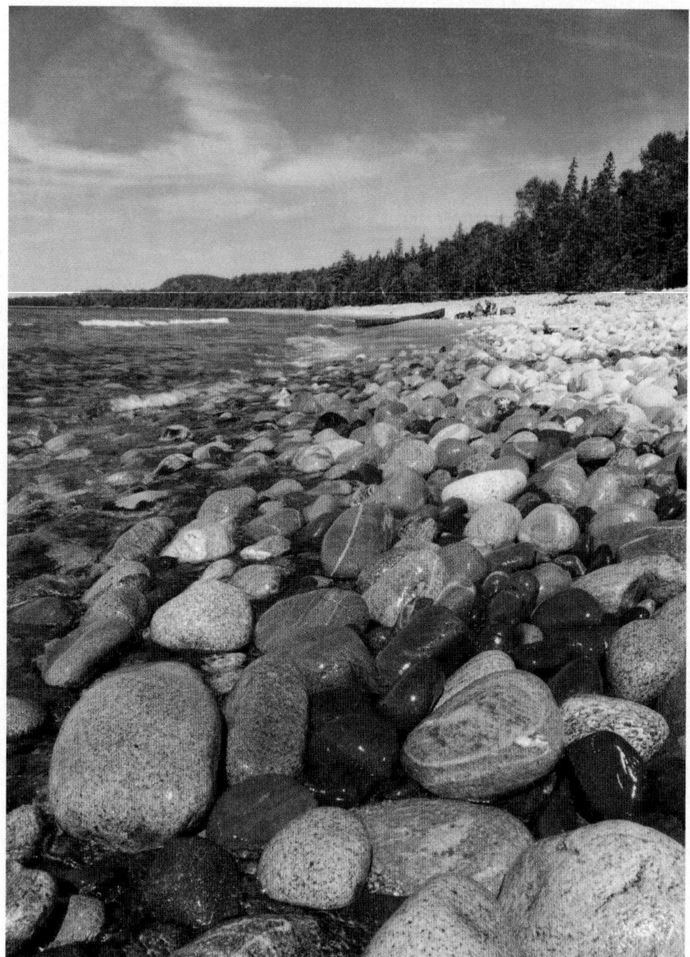

FIGURE 6.2
These boulders have been rounded by abrasion as wave action rolled them against one another on this beach.
©James Smedley/Alamy

Detrital sediment particles are classified and defined according to the size of individual fragments. Table 6.1 shows the precise definitions of particles by size.

Gravel includes all rounded particles coarser than 2 millimeters in diameter, the thickness of a U.S. nickel. (Angular fragments of this size are called *rubble.*) *Pebbles* range from 2 to 64 millimeters (about the size of a tennis ball). *Cobbles* range from 64 to 256 millimeters (about the size of a basketball), and *boulders* are coarser than 256 millimeters (figure 6.2).

Sand grains are from 1/16 millimeter (about the thickness of a human hair) to 2 millimeters in diameter. Grains of this size are visible and feel gritty between the fingers. **Silt** grains are from 1/256 to 1/16 millimeter. They are too small to see without a magnifying device, such as a geologist's hand lens. Silt does not feel gritty between the fingers, but it does feel gritty between the teeth (geologists commonly bite sediments to test their grain size). **Clay** is the finest sediment, at less than 1/256 millimeter, too fine to feel gritty to fingers or teeth. *Mud* is a term loosely used for a mixture of silt and clay.

Note that we have two different uses of the word *clay*—a *clay-sized particle* (table 6.1) and a *clay mineral.* A clay-sized particle can be composed of any mineral at all provided its diameter is less than 1/256 millimeter. A clay mineral, on the other hand, is one of a small group of silicate minerals with a sheet-silicate structure. Clay minerals usually form in the clay-size range.

Quite often the composition of sediment in the clay-size range turns out to be mostly clay minerals, but this is not always the case. Because of its resistance to chemical weathering, quartz may show up in this fine-size grade. (Most silt is quartz.) Intense mechanical weathering can break down a wide variety of minerals to clay size, and these extremely fine particles may retain their mineral identity for a long time if chemical weathering is slow. The great weight of glaciers is particularly effective at grinding minerals down to the silt- and clay-size range, producing "rock flour," which gives a milky appearance to glacial meltwater streams (see chapter 12).

Weathering, erosion, and transportation are some of the processes that affect the character of sediment. Both mechanically weathered and chemically weathered rock and sediment can be eroded, and weathering continues as erosion takes place. Sand being transported by a river also can be actively weathered, as can mud on a lake bottom. The character of sediment can also be altered by *rounding* and *sorting* during transportation, and even after eventual *deposition.*

Transportation

Most sediment is **transported** some distance by gravity, wind, water, or ice before coming to rest and settling into layers. During transportation, sediment continues to weather and change in character in proportion to the distance the sediment is moved.

Rounding is the grinding away of sharp edges and corners of rock fragments during transportation. Rounding occurs in sand and gravel as rivers, glaciers, or waves cause particles to hit and scrape against one another (figure 6.2) or against a rock surface, such as a rocky streambed. Boulders in a stream may show substantial rounding in less than 1 kilometer of travel.

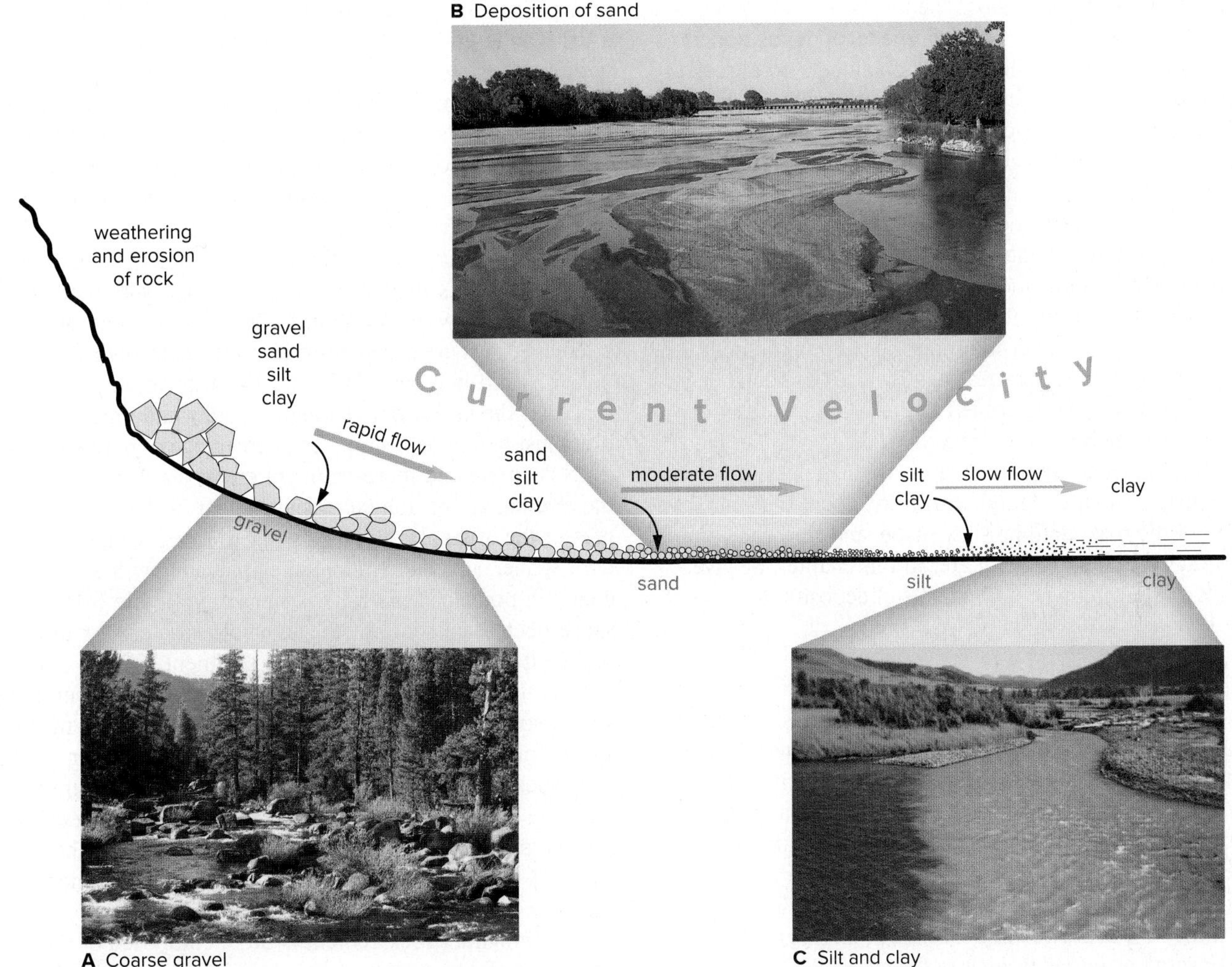

FIGURE 6.3

Cross-sectional (profile) view of sediment sorting by a river. (*A*) The coarsest material (gravel) is deposited first in the headwaters of the river where the flow of water is rapid. (*B*) Deposition of sand occurs as the river loses energy as it flows across a floodplain. (*C*) Silt and clay are carried and eventually deposited at the mouth of a river when the current velocity slows.

(A) ©Diane Carlson (B) ©David McGeary (C) ©C. W. Montgomery

Sorting is the process by which sediment grains are selected and separated according to grain size (or grain shape or specific gravity) by the agents of transportation, especially by running water. Because of their high viscosity and manner of flow, glaciers are poor sorting agents. Glaciers deposit all sediment sizes in the same place, so glacial sediment usually consists of a mixture of clay, silt, sand, and gravel. Such glacial sediment is considered *poorly sorted.* Sediment is considered *well-sorted* when the grains are nearly all the same size. A river, for example, is a good sorting agent, separating sand from gravel, and silt and clay from sand. Sorting takes place because of the greater weight of larger particles. Boulders weigh more than pebbles and are more difficult for the river to transport, so a river must flow more rapidly to move boulders than to move pebbles. Similarly, pebbles are harder to move than sand, and sand is harder to move than silt and clay.

Figure 6.3 shows the sorting of sediment by a river as it flows out of steep mountains onto a gentle flood plain, where the water loses energy and slows down. As the river loses energy, the heaviest particles of sediment are deposited. The boulders come to rest first (figure 6.3*A*). As the river continues to slow and becomes less turbulent, cobbles and then pebbles are deposited. Sand comes to rest as the river loses still more energy (figure 6.3*B*). Finally, the river is carrying only the finest sediment–silt and clay (figure 6.3*C*). The river has sorted the original sediment mix by grain size.

Deposition

When transported material settles or comes to rest, **deposition** occurs. Sediment is deposited when running water, glacial ice, waves, or wind loses energy and can no longer transport its load.

Deposition also refers to the accumulation of chemical or organic sediment, such as clam shells on the sea floor or plant material on the floor of a swamp. Such sediments may form as organisms die and their remains accumulate, perhaps with no transportation at all. Deposition of salt crystals can take place as seawater evaporates. A change in the temperature, pressure, or chemistry of a solution may also cause precipitation–hot springs may deposit calcite or silica as the warm water cools.

The **environment of deposition** is the location in which deposition occurs. A few examples of environments of deposition are the deep-sea floor, a desert valley, a river channel, a coral reef, a lake bottom, a beach, and a sand dune. Each environment is marked by characteristic physical, chemical, and biological conditions. You might expect mud on the sea floor to differ from mud on a lake bottom. Sand on a beach may differ from sand in a river channel. Some differences are due to varying sediment sources and transporting agents, but most are the result of conditions in the environments of deposition themselves.

One of the most important jobs of geologists studying sedimentary rocks is to try to determine the ancient environment of deposition. Geologists compare features found in modern environments of deposition with clues left in the rock record to interpret where the sedimentary rock may have formed. Such clues include the vertical sequence of rock layers in the field, the fossils and sedimentary structures found within the rock, the mineral composition of the rock, and the size, shape, and surface texture of the individual sediment grains. Later in this chapter, we give a few examples of interpreting environments of deposition.

Preservation

Not all sediments are preserved as sedimentary layers. Gravel in a river may be deposited when a river is low but then may be eroded and transported by the next flood on the river. Many sediments on land, particularly those well above sea level, are easily eroded and carried away, so they are not commonly preserved. Sediments on the sea floor are easier to preserve. In general, continental and marine sediments are most likely to be preserved if they are deposited in a *subsiding* (sinking) *basin* and if they are covered or *buried by later sediments.*

Lithification

Lithification is the general term for the processes that convert loose sediment into sedimentary rock. Most sedimentary rocks are lithified by a combination of *compaction,* which packs loose sediment grains tightly together, and *cementation,* in which the precipitation of cement around sediment grains binds them into a firm, coherent rock. *Crystallization* of minerals from solution, without passing through the loose-sediment stage, is another way that rocks may be lithified. Some layers of sediment persist for tens of millions of years without becoming fully lithified. Usually, layers of *partially lithified sediment* have been buried deep enough to become compacted but have not experienced the conditions required for cementation.

As sediment grains settle slowly in a quiet environment such as a lake bottom, they form an arrangement with a great deal of open space between the grains (figure 6.4*A*). The open spaces between grains are called *pores,* and in a quiet environment, a deposit of sand may have 40% to 50% of its volume as open **pore space**. (If the grains were traveling rapidly and impacting one another just before deposition, the percentage of pore space will be less.) As more and more sediment grains are deposited on top of the original grains, the increasing weight of this *overburden* compresses the original grains closer together, reducing the amount of pore space. This shift to a tighter packing, with a resulting decrease in pore space, is called **compaction** (figure 6.4*B*). As pore space decreases, some of the interstitial water that usually fills sediment pores is driven out of the sediment.

As underground water moves through the remaining pore space, solid material called **cement** can precipitate in the pore space and bind the loose sediment grains together to form a solid rock. The cement attaches very tightly to the grains, holding them in a rigid framework. As cement partially or completely fills the pores, the total amount of pore space is further reduced (figure 6.4*C*), and the loose sand forms a hard, coherent sandstone by **cementation.**

Sedimentary rock cement is often composed of the mineral calcite or of other carbonate minerals. Dissolved calcium and bicarbonate ions are common in surface and underground waters. If the chemical conditions are right, these ions may recombine to form solid calcite, as shown in the following reaction.

$$\underbrace{Ca^{2+} + 2HCO_3^-}_{\text{dissolved ions}} \rightarrow \underbrace{CaCO_3}_{\text{calcite}} + H_2O + CO_2$$

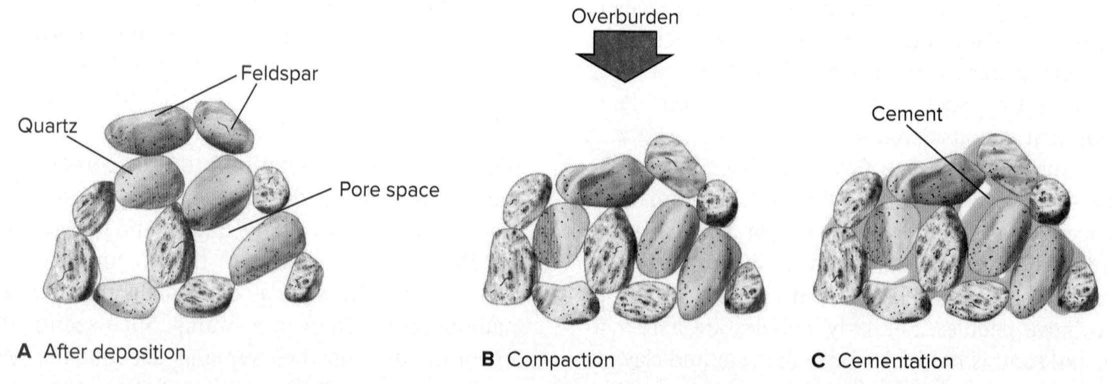

FIGURE 6.4

Lithification of sand grains to become sandstone. (*A*) Loose sand grains are deposited with open pore space between the grains. (*B*) The weight of overburden compacts the sand into a tighter arrangement, reducing pore space. (*C*) Precipitation of cement in the pores by groundwater binds the sand into the rock sandstone, which has a clastic texture.

FIGURE 6.5

Crystalline dolostone as seen through a polarizing microscope. Note the interlocking crystals of dolomite mineral that grew as they precipitated during recrystallization. Such crystalline sedimentary textures have no cement or pore spaces. ©Bret Bennington

Silica is another common cement. Iron oxides and clay minerals can also act as cement but are less common than calcite and silica. The dissolved ions that precipitate as cement originate from the chemical weathering of minerals such as feldspar and calcite. This weathering may occur within the sediments being cemented, or at a very distant site, with the ions being transported tens or even hundreds of kilometers by water before precipitating as solid cement.

A sedimentary rock that consists of sediment grains bound by cement into a rigid framework is said to have a **clastic texture**. Usually such a rock still has some pore space because cement rarely fills the pores completely (figure 6.4*C*).

Some sedimentary rocks form by **crystallization**, the development and growth of crystals by precipitation from solution at or near Earth's surface (the term is also used for igneous rocks that crystallize as magma cools). These rocks have a **crystalline texture**, an arrangement of interlocking crystals that develops as crystals grow and interfere with each other. Crystalline rocks lack cement. They are held together by the interlocking of crystals. Such rocks have minimal pore space because the crystals typically grow until they fill all available space. Some sedimentary rocks with a crystalline texture are the result of *recrystallization,* the growth of new crystals that form from and then destroy the original clastic grains of a rock that has been buried (figure 6.5).

6.2 TYPES OF SEDIMENTARY ROCKS

Sedimentary rocks are formed from (1) eroded mineral grains, (2) minerals precipitated from low-temperature solution, or (3) consolidation of the organic remains of plants. These different types of sedimentary rocks are called, respectively, *detrital, chemical,* and *organic* rocks.

Most sedimentary rocks are **detrital sedimentary rocks**, formed from cemented sediment grains that are fragments of preexisting rocks. The rock fragments can be either identifiable pieces of rock, such as pebbles of granite or shale, or individual mineral grains, such as sand-sized quartz and feldspar crystals loosened from rocks by weathering and erosion. Clay minerals formed by chemical weathering are also considered fragments of preexisting rocks. During transportation, the grains may have been rounded and sorted. Table 6.1 shows the detrital rocks, such as conglomerate, sandstone, and shale, and how these rocks vary in grain size.

Chemical sedimentary rocks are deposited by precipitation of minerals from solution. An example of inorganic precipitation is the formation of *rock salt* as seawater evaporates. Chemical precipitation can also be caused by organisms. The sedimentary rock *limestone* is often formed from the cementation of broken pieces of seashell and fragments of calcite mineral produced by corals and algae. Such a rock is called a *bioclastic* limestone.

Not all chemical sedimentary rocks accumulate as sediment. Some limestones are crystallized as solid rock by corals and coralline algae in reefs. Chert crystallizes in solid masses within some layers of limestone. Rock salt may crystallize directly as a solid mass or it may form from the crystallization of individual salt crystals that behave as sedimentary particles until they grow large enough to interlock into solid rock.

Organic sedimentary rocks are rocks that are composed of organic carbon compounds. *Coal* is an organic rock that forms from the compression of plant remains, such as moss, leaves, twigs, roots, and tree trunks.

Appendix B describes and helps you identify the common sedimentary rocks. The standard geologic symbols for these rocks (such as dots for sandstone, and a "brick-wall" symbol for limestone) are shown in appendix F and will be used in the remainder of the book.

6.3 DETRITAL ROCKS

Detrital sedimentary rocks are formed from the weathered and eroded remains (detritus) of bedrock. Detrital rocks are also often referred to as *terrigenous clastic rocks* because they are composed of *clasts* (broken pieces) of mineral derived from the erosion of the land. The classification of detrital sedimentary rocks is based mainly on clast or grain size, and also on the amount of rounding and sorting, and on the composition of the grains and cement.

Breccia and Conglomerate

Sedimentary breccia is a coarse-grained sedimentary rock formed by the cementation of coarse, angular fragments of rubble (figure 6.6). Because grains are rounded so rapidly during transport, it is unlikely that the angular fragments within breccia have moved very far from their source (see figure 6.3). Sedimentary breccia is commonly a talus slope deposit that forms at the

FIGURE 6.6

Breccia is characterized by coarse, angular fragments. The cement in this rock is colored by hematite. The wide black and white bars on the scale are 1 centimeter long; the small divisions are 1 millimeter. Note that most grains exceed 2 millimeters (table 6.1).
©David McGeary

FIGURE 6.7

An outcrop of a poorly sorted conglomerate. Note the rounding of cobbles, which vary in composition and size. The cement in this rock is also colored by hematite. Long scale bar is 10 centimeters; short bars are 1 centimeter.
©David McGeary

base of a steep rock cliff that is being mechanically weathered. Landslide deposits also might lithify into sedimentary breccia. This type of rock is not particularly common.

Conglomerate is a coarse-grained sedimentary rock formed by the cementation of rounded gravel. It can be distinguished from breccia by the definite roundness of its particles (figure 6.7). Because conglomerates are coarse-grained, the particles may not have traveled far; but some transport was necessary to round the particles. Angular fragments that fall from a cliff and then are carried a few kilometers by a river or pounded by waves crashing in the surf along a beach are quickly rounded. Gravel that is transported down steep submarine canyons or carried by glacial ice, however, can be transported tens or even hundreds of kilometers before deposition.

Sandstone

Sandstone is formed by the cementation of sand grains (figure 6.8). Any deposit of sand can lithify to sandstone. Rivers deposit sand in their channels, and wind piles up sand into dunes. Waves deposit sand on beaches and in shallow water. Deep-sea currents spread sand over the sea floor. As you might imagine, sandstones show a great deal of variation in mineral composition, degree of sorting, and degree of rounding.

Quartz sandstone is a sandstone in which more than 90% of the grains are quartz (figure 6.8*A*). Because quartz is resistant to chemical weathering, it tends to concentrate in sand deposits as the less-resistant minerals such as feldspar are weathered away. The quartz grains in a quartz sandstone are usually well-sorted and well-rounded because they have been transported for great distances (figure 6.9*A*). Most quartz sandstone was deposited as beach sand or dune sand.

A sandstone with more than 25% of the grains consisting of feldspar is called *arkose* (figure 6.8*B*). Because feldspar grains are preserved in the rock, the original sediment obviously did not undergo severe chemical weathering, or the feldspar would have been destroyed. Mountains of granite in a desert could be a source for such a sediment, for the rapid erosion associated with rugged terrain would allow feldspar to be mechanically weathered and eroded before it is chemically weathered (a dry climate slows chemical weathering). Most arkoses contain coarse, angular grains (figure 6.9*B*), so transportation distances were probably short. An arkose may have been deposited within an **alluvial fan**, a large, fan-shaped pile of sediment that usually forms where a stream emerges from a narrow canyon onto a flat plain at the foot of a mountain range (figure 6.10).

Sandstones may contain a substantial amount of **matrix** in the form of fine-grained silt and clay in the space between larger sand grains (figure 6.11). A matrix-rich sandstone is poorly sorted and often dark in color. It is sometimes called a "dirty sandstone."

Graywacke (pronounced "gray-wacky") is a type of sandstone in which more than 15% of the rock's volume consists of fine-grained matrix (figures 6.8*C* and 6.9*C*). Graywackes are commonly hard and dense, and they are generally dark gray or green. The sand grains may be so coated with matrix that they are difficult to see, but they typically consist of quartz, feldspar, and sand-sized fragments of other fine-grained sedimentary, volcanic, and metamorphic rocks. Most graywackes probably formed from sediment-laden turbidity currents that are deposited in deep water (see figure 6.29).

The Fine-Grained Rocks

Rocks consisting of fine-grained silt and clay are called *shale, siltstone, claystone,* and *mudstone.*

Shale is a fine-grained sedimentary rock notable for its ability to split into layers (called *fissility*). Splitting takes place along the surfaces of very thin layers (called *laminations*) within the shale (figure 6.12). Most shales contain both silt and clay (averaging about two-thirds clay-sized clay minerals and one-third

A

B

C

FIGURE 6.8

Types of sandstone. (*A*) Quartz sandstone; more than 90% of the grains are quartz. (*B*) Arkose; the grains are mostly feldspar and quartz. (*C*) Graywacke; the grains are surrounded by dark, fine-grained matrix. (Small scale divisions are 1 millimeter; most of the sand grains are about 1 millimeter in diameter.) *Photos by David McGeary*

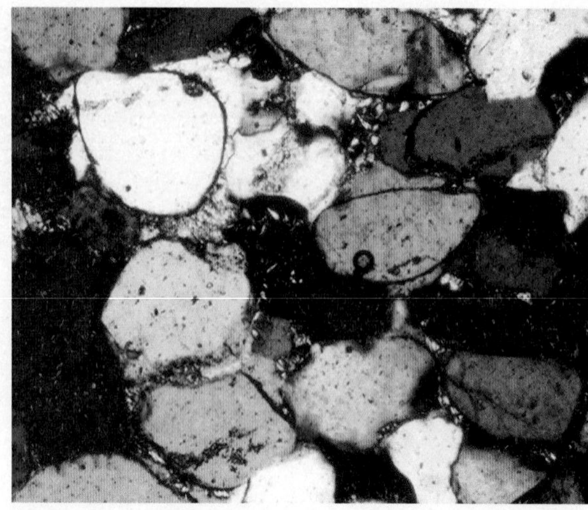
A

B

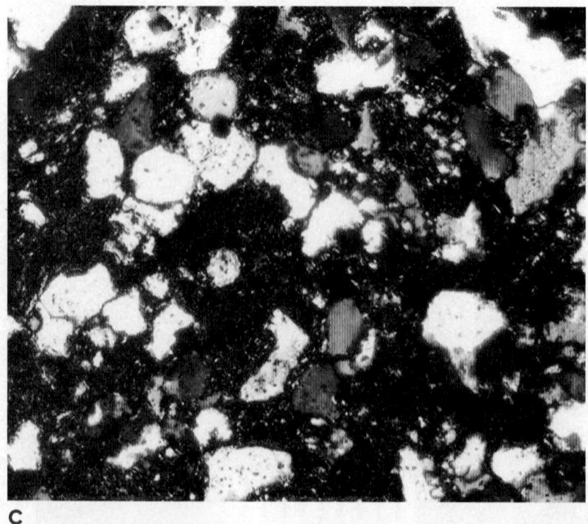
C

FIGURE 6.9

Detrital sedimentary rocks viewed through a polarizing microscope. (*A*) Quartz sandstone; note the well-rounded and well-sorted grains. (*B*) Arkose; large feldspar grain in center surrounded by angular quartz grains. (*C*) Graywacke; quartz grains surrounded by brownish matrix of mud.
©Bret Bennington

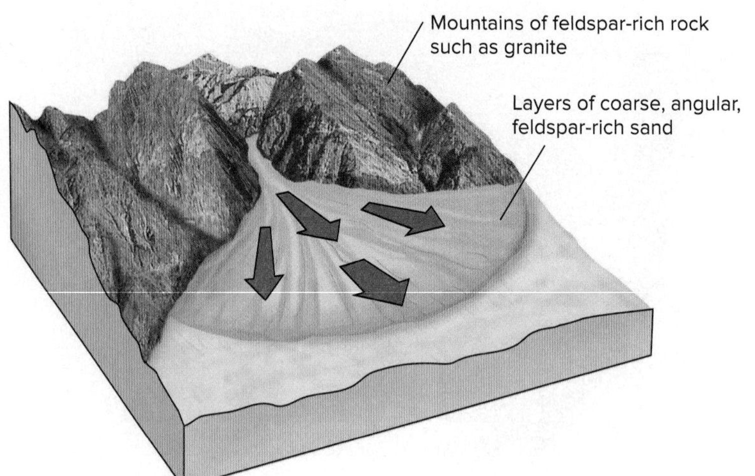

FIGURE 6.10

Feldspar-rich sand (arkose) may accumulate from the rapid erosion of feldspar-containing rock such as granite. Steep terrain accelerates erosion rates so that feldspar may be eroded before it is completely chemically weathered into clay minerals.

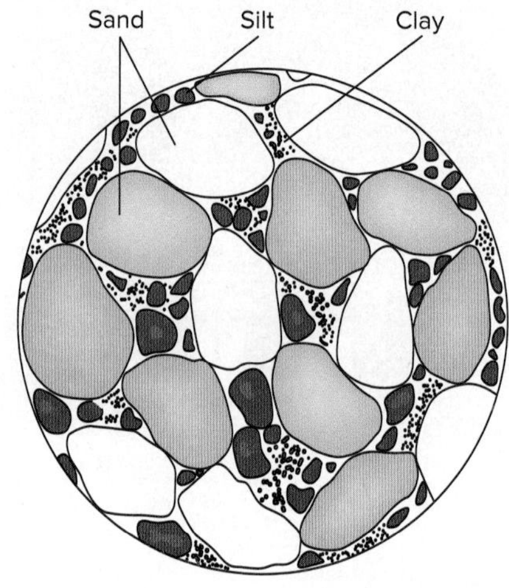

FIGURE 6.11

A poorly sorted sediment of sand grains surrounded by a matrix of silt and clay grains. Lithification of such a sediment would produce a "dirty sandstone."

silt-sized quartz) and are so fine-grained that the surface of the rock feels very smooth. The silt and clay deposits that lithify as shale accumulate on lake bottoms, at the ends of rivers in deltas, on river flood plains, and on quiet parts of the deep-ocean floor.

Fine-grained rocks such as shale typically undergo pronounced compaction as they lithify. Figure 6.13 shows the role of compaction in the lithification of shale from wet mud. Before compaction, as much as 80% of the volume of the wet mud may have been pore space filled with water. The flake-like clay minerals were randomly arranged within the mud. Pressure from overlying material packs the sediment grains together and reduces the overall volume by squeezing water out of the pores. The clay minerals are reoriented perpendicular to the pressure, becoming parallel to one another like a deck of cards. The fissility of shale is due to weaknesses between these parallel clay flakes.

A

B

FIGURE 6.12

(*A*) An outcrop of shale from Hudson Valley in New York. Note how this fine-grained rock tends to split into very thin layers. (*B*) Shale pieces; note the very fine grain (scale in centimeters), very thin layers (laminations) on the edge of the large piece, and tendency to break into small, flat pieces (fissility).
(A) ©John Buitenkant/Science Source (B) ©David McGeary

Compaction by itself does not generally lithify sediment into sedimentary rock. It does help consolidate clayey sediments by pressing the microscopic clay minerals so closely together that attractional forces at the atomic level tend to bind them together. Even in shale, however, the primary method of lithification is cementation.

A rock consisting mostly of silt grains is called *siltstone.* Somewhat coarser-grained than most shales, siltstones lack the fissility and laminations of shale. *Claystone* is a rock composed predominately of clay-sized particles but lacking the fissility of shale. *Mudstone* contains both silt and clay, having the same grain size and smooth feel of shale but lacking shale's laminations and fissility. Mudstone is massive and blocky, while shale is visibly layered and fissile.

6.4 CHEMICAL SEDIMENTARY ROCKS

Chemical sedimentary rocks are precipitated from a low-temperature aqueous environment. Chemical sedimentary rocks are precipitated either directly by inorganic processes or by the actions of organisms. Chemical rocks include *carbonates, chert,* and *evaporites.*

Carbonate Rocks

Carbonate rocks contain the CO_3^{2-} ion as part of their chemical composition. The two main types of carbonates are limestone and dolomite.

Limestone

Limestone is a sedimentary rock composed mostly of calcite ($CaCO_3$). Limestones are precipitated either by the actions of organisms or directly as the result of inorganic processes. Thus, the two major types of limestone can be classified as either *biochemical* or *inorganic limestone.*

Biochemical limestones are precipitated through the actions of organisms. Most biochemical limestones are formed on

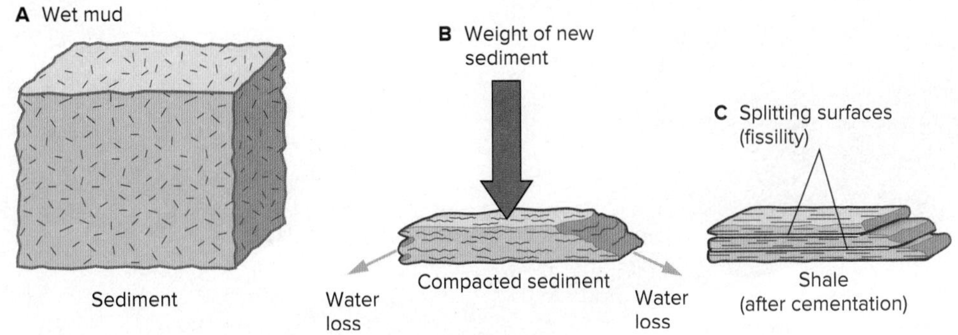

FIGURE 6.13

Lithification of shale from the compaction and cementation of wet mud. (*A*) Randomly oriented silt and clay particles in wet mud. (*B*) Particles reorient, water is lost, and pore space decreases during compaction caused by the weight of new sediment deposited on top of the wet mud. (*C*) Splitting surfaces in cemented shale form parallel to the oriented mineral grains.

continental shelves in warm, shallow seawater. Biochemical limestone may be precipitated directly in the core of a reef by corals, encrusting algae, or other shell-forming organisms (figure 6.14). Such a rock would contain the fossil remains of organisms preserved in growth position.

Biochemical limestones may also form from wave-broken fragments of algae, corals, and shells. The fragments may be of any size (gravel, sand, silt, and clay) and are commonly sorted and rounded as they are transported by waves and currents across the sea floor (figure 6.15). The action of these waves and currents and subsequent cementation of these fragments into rock give these limestones a clastic texture. These *bioclastic* (or *skeletal*) *limestones* have a variety of appearances. They may be relatively coarse-grained with recognizable fossils (figure 6.16) or uniformly fine-grained and dense from the accumulation of microscopic fragments of calcareous algae (figures 6.16 and 6.17). A variety of limestone called *coquina* forms from the cementation of shells and shell fragments that accumulated on the shallow sea floor near shore (figure 6.18). It has a clastic texture and is usually coarse-grained, with easily recognizable shells and shell fragments in it. *Chalk* is a light-colored, porous, very fine-grained variety of bioclastic limestone that forms from the sea-floor accumulation of microscopic marine organisms that drift near the sea surface (figure 6.19).

Inorganic limestones are precipitated directly as the result of inorganic processes. *Oölitic limestone* is a distinctive variety of inorganic limestone formed by the cementation of sand-sized *oöids,* small spheres of calcite inorganically precipitated in warm, shallow seawater (figure 6.20). Strong tidal currents roll the oölites back and forth, allowing them to maintain a nearly spherical shape as they grow. Wave action may also contribute to their shape. Oölitic limestone has a clastic texture.

Tufa and *travertine* are inorganic limestones that form from fresh water. Tufa is precipitated from solution in the water of a continental spring or lake, or from percolating groundwater. Travertine may form in caves when carbonate-rich water loses CO_2 to the cave atmosphere. Tufa and travertine both have a crystalline texture; however, tufa is generally more porous, cellular, or open than travertine, which tends to be more dense.

FIGURE 6.14

Corals precipitate calcium carbonate to form limestone in a reef. Water depth about 8 meters (25 feet), San Salvador Island, Bahamas.
©David McGeary

Limestones are particularly susceptible to **recrystallization**, the process by which new crystals, often of the same mineral composition as the original grains, develop in a rock. Calcite grains recrystallize easily, particularly in the presence of water and under the weight of overlying sediment. The new crystals that form are typically large and can be easily seen in a rock as slight reflections off their broad, flat cleavage faces. Recrystallization often destroys

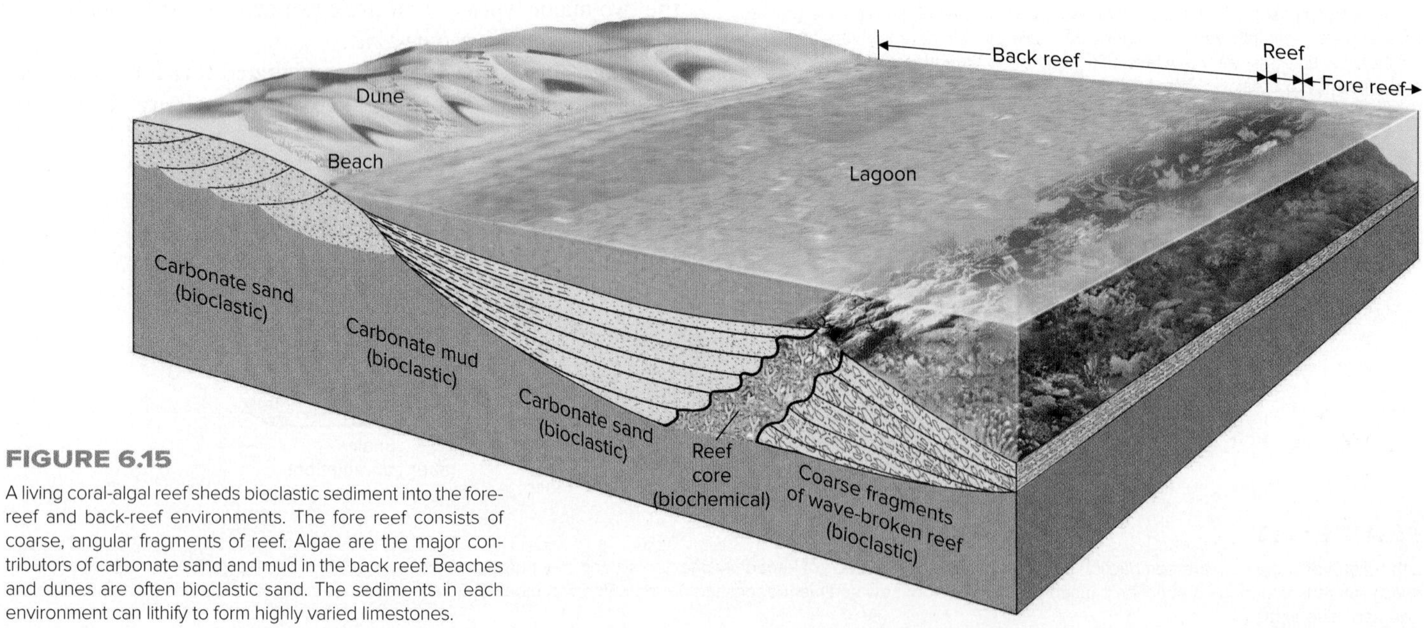

FIGURE 6.15

A living coral-algal reef sheds bioclastic sediment into the fore-reef and back-reef environments. The fore reef consists of coarse, angular fragments of reef. Algae are the major contributors of carbonate sand and mud in the back reef. Beaches and dunes are often bioclastic sand. The sediments in each environment can lithify to form highly varied limestones.

FIGURE 6.16

Bioclastic limestones. The two on the left are coarse-grained and contain visible fossils of corals and shells. The limestone on the right consists of fine-grained carbonate mud formed by calcareous algae.
©David McGeary

FIGURE 6.17

Green algae on the sea floor in 3 meters of water on the Bahama Banks. The "shaving brush" alga is *Penicillus,* which produces great quantities of fine-grained carbonate mud.
©David McGeary

FIGURE 6.18

Coquina, a variety of bioclastic limestone, is formed by the cementation of coarse shells.
©David McGeary

FIGURE 6.19

Chalk is a fine-grained variety of bioclastic limestone formed of the remains of microscopic marine organisms that live near the sea surface.
©David McGeary

the original clastic texture and fossils of a rock, replacing them with a new crystalline texture. Therefore, the geologic history of such a rock can be very difficult to determine.

Dolomite

The term **dolomite** (table 6.2) is used both as a rock name and as a mineral name. To avoid confusion, many geologists refer to a rock that is made mostly of the mineral dolomite as a *dolostone.* The term *dolomite* is reserved for the mineral with the chemical formula $CaMg(CO_3)_2$. Dolomite formation is important from an economic perspective because the porosity of carbonate rocks often increases as dolomite forms. Many important oil reservoirs are in dolostone layers that have enhanced porosity from dolomite formation.

A

B

FIGURE 6.20

(*A*) Aerial photo of underwater dunes of oöids chemically precipitated from seawater on the shallow Bahama Banks, south of Bimini. Tidal currents move the dunes. (*B*) An oölitic limestone formed by the cementation of oöids (small spheres). Small divisions on scale are 1 millimeter wide.
©David McGeary

TABLE 6.2 Chemical Sedimentary Rocks

Inorganic Sedimentary Rocks			
Rock	**Composition**	**Texture**	**Origin**
Limestone	$CaCO_3$	Crystalline Oölitic	May be precipitated directly from seawater. Cementation of oölites (oöids) precipitated chemically from warm, shallow seawater (*oölitic* limestone). Also forms in caves as *travertine* and in springs, lakes, or percolating groundwater as *tufa*
Dolomite	$CaMg(CO_3)_2$	Crystalline	Alteration of limestone by Mg-rich solutions (usually)
Evaporites			Evaporation of seawater or a saline lake
Rock salt	NaCl	Crystalline	
Rock gypsum	$CaSO_4 \cdot 2H_2O$	Crystalline	
Chert	SiO_2 (silica)	Crystalline	Precipitated as nodules or layers by silica-rich groundwater

Biochemical Sedimentary Rocks			
Rock	**Composition**	**Texture**	**Origin**
Limestone	$CaCO_3$ (calcite)	Clastic or crystalline	Cementation of fragments of shells, corals, and coralline algae (*bioclastic limestone* such as coquina and chalk). Also precipitated directly by organisms in reefs
Chert	SiO_2 (silica)	Crystalline (usually)	Cementation of microscopic marine organisms; rock usually recrystallized

Dolomite is relatively common in the ancient rock record, but geologists have not observed significant quantities of dolomite forming in modern environments. This discrepancy between ancient and modern dolomite abundance has led to what geologists refer to as the "dolomite problem." Geologists have proposed several models to explain how dolomite forms, and it is likely that dolomite forms by more than one mechanism. All models need to explain the observations that dolomite does not precipitate directly from seawater in modern environments, and that dolomite is more common in the geologic past.

Most dolomite that we see in the rock record probably forms as a replacement for existing carbonate material, and there are several models to explain the process. Studies show that although dolomite does not precipitate directly from seawater, it is common in the shallow subsurface in the Persian Gulf. In this arid region, dolomite forms in broad, flat tidal zones where the sediment is made of fine, highly reactive lime mud. The lime mud is composed of very small particles of calcite or aragonite ($CaCO_3$). Replacement reactions start when seawater evaporates from the tidal flat, leaving behind a dense, salty brine that percolates through the pore spaces of the lime mud. As the brine grows more concentrated, gypsum starts to precipitate, removing Ca ions from the pore water. The remaining pore water becomes relatively enriched in Mg, and small, poorly formed dolomite crystals begin to replace the original carbonate material. This is called "proto"-dolomite, and it forms as an early replacement for lime mud in shallow tidal flat environments. The reactions are summarized as follows:

1. Evaporation in hot, dry tidal flat environments produces salty brines that percolate through the pore spaces of the lime mud.
2. Gypsum and other evaporite minerals begin to precipitate from the dense, salty brine. This removes calcium ions from the brine.

$$Ca^{2+} + SO_4^{2-} + 2H_2O \rightarrow CaSO_4 \cdot 2H_2O$$

3. The brine becomes relatively enriched in Mg^{2+} ions as more gypsum precipitates. When the ratio of $Mg/Ca > 10/1$, dolomite becomes chemically stable, and Mg^{2+} starts to replace Ca^{2+} on an ion-by-ion basis.
4. Early, poorly crystalline "proto"-dolomite crystals form by replacement as Mg^{2+} ions substitute regularly for Ca^{2+} ions in the crystal structure.

$$Mg^{2+} + 2CaCO_3 \rightarrow (CaMgCO_3)_2 + Ca^{2+}$$

Much of the thinly interbedded dolostone and limestone that we see in the rock record is probably a result of the early replacement of reactive lime mud in shallow tidal flat environments.

Dolomite also forms when pressure and temperature increase as a result of basin subsidence and deep burial of the carbonate sediment. Geochemists have determined that higher temperatures and pressures increase the stability of dolomite over calcite or aragonite, so carbonate rocks that have been buried often change to dolostones. This process is enhanced by hot, salty pore fluids that migrate through the carbonate material. These deep-burial dolostones are also a replacement for the original carbonate material, and they may contain large, clear dolomite cement crystals. Deep-burial dolostones tend to be thick and widespread, and they often have increased porosity from the conversion of calcite or aragonite to dolomite. Deep-burial dolomite forms slowly in geologic terms, and this mechanism does not explain the shallow interbedded limestones and dolostones that are common in the rock record.

Chert

A hard, compact, fine-grained sedimentary rock formed almost entirely of silica, **chert** occurs in two principal forms—as irregular, lumpy nodules within other rocks and as layered deposits like other sedimentary rocks (figure 6.21). The nodules, often found in limestone, probably formed from inorganic precipitation

A

B

FIGURE 6.21

(A) Chert nodules in limestone near Bluefield, West Virginia. *(B)* Bedded chert from the coast ranges, California. Camera lens cap (5.5 centimeters) for scale.
(A) ©Parvinder Sethi (B) ©David McGeary

IN GREATER DEPTH 6.1

Valuable Sedimentary Rocks

Many sedimentary rocks have uses that make them valuable. *Limestone* is widely used as building stone and is also the main rock type quarried for crushed rock for road construction. Pulverized limestone is the main ingredient of cement for mortar and concrete and is also used to neutralize acid soils in the humid regions of the United States. *Coal* is a major fuel, used widely for generating electrical power and for heating. Plaster and plasterboard for home construction are manufactured from *gypsum,* which is also used to stabilize the shrink-swell characteristics of clay-rich soils in some areas. Huge quantities of *rock salt* are consumed by industry, primarily for the manufacture of hydrochloric acid. More familiar uses of rock salt are for table salt and melting ice on roads.

Some *chalk* is used in the manufacture of blackboard chalk, although most classroom chalk is now made from pulverized limestone. The filtering agent for beer brewing and for swimming pools is likely to be made of *diatomite,* an accumulation of the siliceous remains of microscopic diatoms.

Clay from *shale* and other deposits supplies the basic material for ceramics of all sorts, from hand-thrown pottery and fine porcelain to sewer pipe. *Sulfur* is used for matches, fungicides, and sulfuric acid; and *phosphates* and *nitrates* for fertilizers are extracted from natural occurrences of special sedimentary rocks (although other sources also are used). Potassium for soap manufacture comes largely from *evaporites,* as does boron for heat-resistant cookware and fiberglass and sodium for baking soda, washing soda, and soap. *Quartz sandstone* is used in glass manufacturing and for building stone.

Many *metallic ores,* such as the most common iron ores, have a sedimentary origin. The pore space of sedimentary rocks acts as a reservoir for groundwater (chapter 11), crude oil, and natural gas. In chapter 22, we take a closer look at these resources and other useful Earth materials.

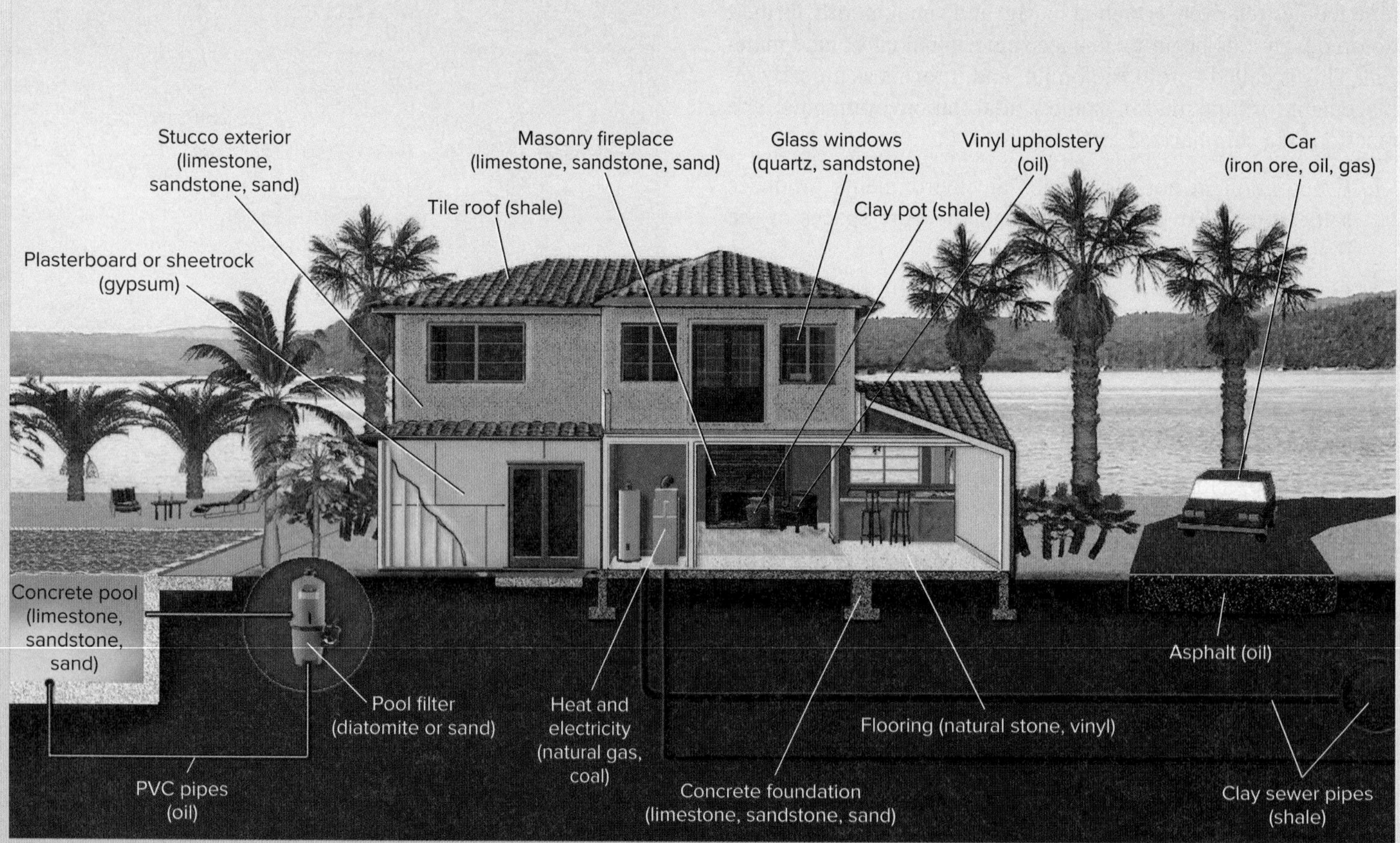

BOX 6.1 ■ FIGURE 1

Common uses of materials that are sedimentary in origin.

FIGURE 6.22

Salt deposited on the floor of a dried-up desert lake, Death Valley, California.
©Michael Collier

FIGURE 6.23

Coal bed in the Black Warrior Coal Basin, Alabama. Note the fossil tree stump preserved in place at the top of the coal.
©Bret Bennington

as underground water replaced part of the original rock with silica. The layered, or bedded cherts typically form from the accumulation of delicate, glasslike shells of microscopic marine organisms on the sea floor.

Microscopic fossils composed of silica are abundant in some cherts. But because chert is susceptible to recrystallization, the original fossils are easily destroyed, and the origin of many cherts remains unknown.

Evaporites

Rocks formed from crystals that precipitate during the evaporation of water are called **evaporites**. They form from the evaporation of seawater or a saline lake (figure 6.22), such as Great Salt Lake in Utah. *Rock gypsum,* formed from the mineral gypsum ($CaSO_4 \cdot 2H_2O$), is a common evaporite. *Rock salt,* composed of the mineral halite (NaCl), may also form if evaporation continues. Other less common evaporites include the borates, potassium salts, and magnesium salts. All evaporites have a crystalline texture. Extensive deposits of rock salt and rock gypsum have formed in the past where shallow, continental seas existed in hot, arid climates. Similarly, modern evaporite deposits are forming in the Persian Gulf and in the Red Sea.

6.5 ORGANIC SEDIMENTARY ROCKS

Coal is a sedimentary rock that forms from the compaction of plant material that has not completely decayed (figure 6.23). Rapid plant growth and deposition in water with a low oxygen content are needed, so shallow swamps or bogs in a temperate or tropical climate are likely environments of deposition. The plant fossils in coal beds include leaves, stems, tree trunks, and stumps with roots often extending into the underlying shales, so apparently most coal formed right at the place where the plants grew. Coal usually develops from *peat,* a brown, lightweight, unconsolidated or semiconsolidated deposit of plant remains that accumulate in wet bogs. Peat is transformed into coal largely by compaction after it has been buried by sediments.

Partial decay of the abundant plant material uses up any oxygen in the swamp water, so the decay stops and the remaining organic matter is preserved. Burial by sediment compresses the plant material, gradually driving out any water or other volatile compounds. The coal changes from brown to black as the amount of carbon in it increases. Several varieties of coal are recognized on the basis of the type of original plant material and the degree of compaction (see chapter 22).

6.6 THE ORIGIN OF OIL AND GAS

Oil and natural gas seem to originate from organic matter in marine sediment. Microscopic organisms, such as diatoms and other single-celled algae, settle to the sea floor and accumulate in marine mud. The most likely environments for this are restricted basins with poor water circulation, particularly on continental shelves. The organic matter may partially decompose, using up the dissolved oxygen in the sediment. As soon as the oxygen is gone, decay stops and the remaining organic matter is preserved.

Continued sedimentation buries the organic matter and subjects it to higher temperatures and pressures, which convert the organic matter to oil and gas. As muddy sediments compact, the gas and small droplets of oil may be squeezed out of the mud and may move into more porous and permeable sandy layers nearby. Over long periods of time, large accumulations of gas and oil can collect in the sandy layers. Both oil and gas are less dense than water, so they generally tend to rise upward through water-saturated rock and sediment. Natural gas represents the end point in petroleum maturation.

Details of the origin of coal, oil, and gas are discussed in chapter 22.

6.7 SEDIMENTARY STRUCTURES

Sedimentary structures are features found within sedimentary rock. They usually form during or shortly after deposition of the sediment but before lithification. Structures found in sedimentary rocks are important because they provide clues that help geologists determine the means by which sediment was transported and also its eventual resting place, or environment of deposition. Sedimentary structures may also reveal the orientation, or upward direction, of the deposit, which helps geologists unravel the geometry of rocks that have been folded and faulted in tectonically active regions.

One of the most prominent structures, seen in most large bodies of sedimentary rock, is **bedding**, a series of visible layers within rock (figure 6.24). Most bedding is horizontal because the sediments from which the sedimentary rocks formed were originally deposited as horizontal layers. The principle of **original horizontality** states that most water-laid sediment is deposited in horizontal or near-horizontal layers that are essentially parallel to Earth's surface. In many cases, this is also true for sediments deposited by ice or wind. If each new layer of sediment buries previous layers, a stack of horizontal layers will develop, with the oldest layer on the bottom and the layers becoming younger upward. Sedimentary rocks formed from such sediments preserve the horizontal layering in the form of beds (figure 6.24). A **bedding plane** is a nearly flat surface of deposition separating two layers of rock. A change in the grain size or composition of the particles being deposited, or a pause during deposition, can create bedding planes.

In sandstone, a thicker bed of rock will often consist of a series of thinner, inclined beds called **cross-beds** (figure 6.25). Cross-beds form because in flowing air and water, sand grains move as migrating ripples and dunes. Sand is pushed up the shallow side of the ripple to the crest, where it then avalanches down the steep side, forming a cross-bed. Cross-beds form one after the other as the ripple migrates downstream (figure 6.26). Ripples can also be preserved on the surface of a bed of sandstone, forming **ripple marks**, if they are buried by another layer of sediment (figure 6.27). Ripple marks produced by currents flowing in a single direction are asymmetrical (as discussed previously and in figures 6.27*B* and *D*). In waves, water moves

FIGURE 6.24

Bedding in sandstone and shale, Utah. The horizontal layers formed as one type of sediment buried another in the geologic past. The layers get younger upwards. ©*David McGeary*

PLANETARY GEOLOGY 6.2

Sedimentary Rocks: The Key To Mars' Past

Sedimentary rocks on Mars are currently being studied by planetary geologists to decipher its early history and determine if Mars was once a warmer, wetter planet. Currently, the atmosphere on Mars is too thin and its surface too cold to allow liquid water to exist (see chapter 23). But recent evidence shows that Mars was once wet enough to host lakes and seas. New observations from robotic spacecraft exploring Mars show evidence for extensive deposits of water-lain sedimentary rock. In orbit around Mars, the *Mars Global Surveyor, Mars Express,* and *Mars Reconnaissance Orbiter* spacecraft have taken thousands of high-resolution photographs, many of which reveal widespread, laterally continuous layers that appear to be sedimentary rock. For example, hundreds of layers of rock are exposed in parts of the walls of the Valles Marineris, a large chasm on Mars that resembles the Grand Canyon but is almost 4,000 kilometers (2,700 miles) long! In the Mawrth Vallis, the rock layers have been identified as thick beds of clay. Because clay minerals can form only in the presence of liquid water and because the clay beds are thick and cover such a wide geographic area, one interpretation is that large bodies of standing water may have existed on Mars. These extensive lakes would have formed very early in the planet's history and probably lasted for millions of years.

While the Mars Orbiters search from the sky, Mars Rovers and Landers have been exploring the surface of the planet. The Mars Exploration Rover named *Opportunity* landed inside a small crater with exposures of layered rock and later traversed the Martian surface to enter a larger crater with more layered rock (box figure 1). Detailed photographic and spectrographic analyses of these layered rocks have revealed sedimentary features such as cross-beds, hematite mineral concretions, and the presence of minerals such as jarosite that typically form in water. More recently, the *Phoenix Mars Lander* set down near the polar region and found frozen water in the soil under the landing site. During the *Phoenix* mission, the first wet chemical analyses done on any planet other than Earth found more evidence for the possible past occurrence of water on Mars in the form of magnesium, sodium, potassium, and chloride salts (evaporites). Subsequent analyses indicate the presence of calcium carbonate (limestone), an important discovery because carbon-containing compounds are necessary for life as we know it on Earth.

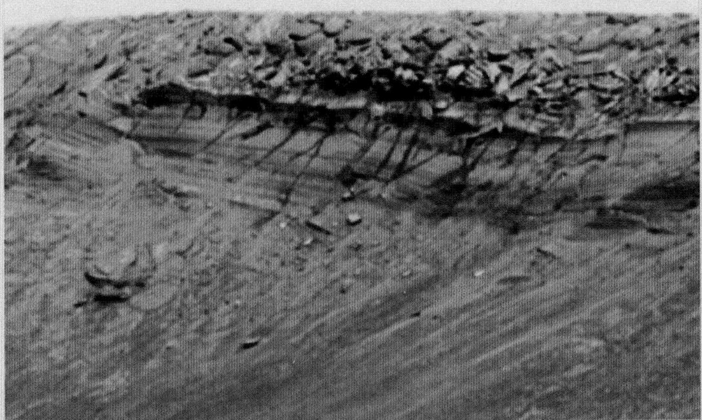

BOX 6.2 ■ FIGURE 1

Layers of sedimentary rock exposed inside the rim of Endurance Crater photographed by the Mars Exploration Rover *Opportunity.*
Source: NASA/JPL/Cornell

BOX 6.2 ■ FIGURE 2

Mosaic of images from the mast camera on NASA's *Curiosity* rover shows sedimentary deposits in the Glenelg area of Gale Crater.
©NASA/JPL-Caltech/MSSS

The most recent rover, the *Mars Science Laboratory* (also known as *Curiosity*), is exploring Gale Crater with a complex array of new capabilities. Gale was selected from high-resolution *Mars Reconnaissance Orbiter* images showing a long-lived and varied geologic history that included potential lake deposits. *Curiosity* has in fact found and characterized strong evidence for the long-lived existence of water at Gale, including a mudstone (lithified clay deposit) and other sedimentary materials deposited by water (box figure 2). *Curiosity* will continue to drill, sample, and analyze rocks in its trek through Mars' geologic history, and it will collaborate with *MAVEN (Mars Atmospheric* and *Volatile EvolutioN)* to investigate how the atmosphere has evolved from one dense enough to support a wetter climate to the thin atmosphere present today. More exciting discoveries are anticipated as Mars continues to be one of the most promising places to look for evidence of extraterrestrial life in our solar system.

Additional Resources

Information about the Mars exploration program at NASA, including the *Mars Science Laboratory (Curiosity)* and images and updates from ongoing missions, such as the *Mars Reconnaissance Orbiter* and the *Opportunity Lander,* is available from the Jet Propulsion Laboratory/NASA Mars Program website.
http://marsprogram.jpl.nasa.gov/

Spectacular images from the *Mars Science Laboratory (Curiosity)* can be found at the MSL website.
http://mars.jpl.nasa.gov/msl/

Visit the European Space Agency's *Mars Express* website for information about this ongoing mission.
http://www.esa.int/SPECIALS/Mars_Express/

Visit the Malin Space Science Systems website for an extensive collection of archived images from recent Mars missions.
http://www.msss.com/

FIGURE 6.25

Cross-bedded sandstone in Zion National Park, Utah. Note how the thin layers have formed at an angle to the more extensive bedding planes (also tilted) in the rock. This cross-bedding was formed in sand dunes deposited by the wind.
©David McGeary

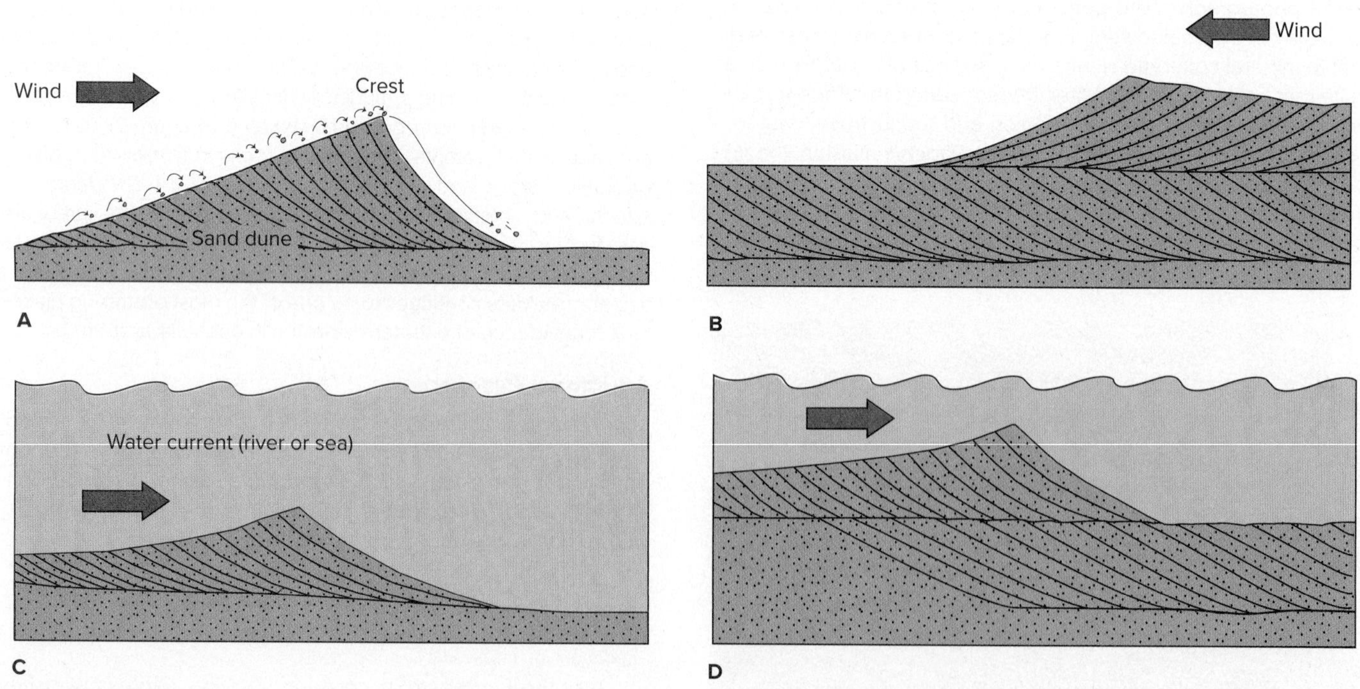

FIGURE 6.26

The development of cross-beds in wind-blown sand (*A* and *B*) and water-deposited sand (*C* and *D*). (*A*) Sand grains migrate up the shallow side of the dune and avalanche down the steep side, forming cross-beds. (*B*) Second layer of cross-beds forms as wind shifts and a dune migrates from the opposite direction. (*C*) Underwater current deposits cross-beds as ripple migrates downstream. (*D*) Continued deposition and migration of ripples produces multiple layers of cross-beds.

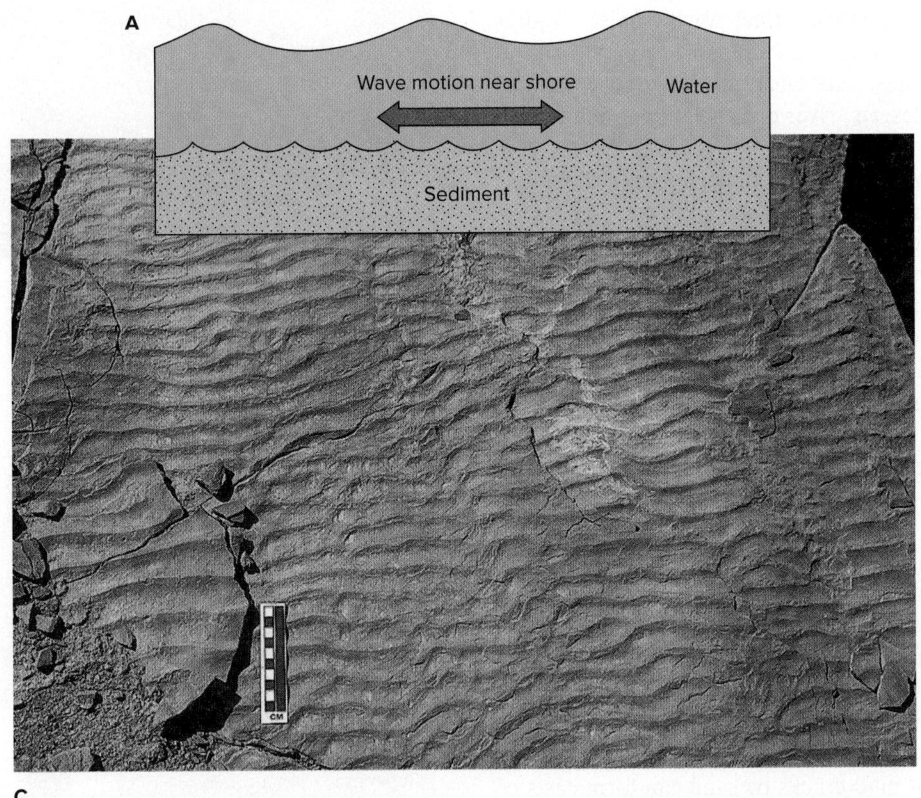

FIGURE 6.27

Development of ripple marks in loose sediment. (*A*) Symmetric ripple marks form beneath waves. (*B*) Asymmetric ripple marks, forming beneath a current, are steeper on their down-current sides. (*C*) Ripple marks on a bedding plane in sandstone, Capitol Reef National Park, Utah. Scale in centimeters. (*D*) Current ripples in wet sediment of a tidal flat, Baja California.

(C) ©David McGeary (D) ©Frank M. Hanna

back and forth, producing symmetrical wave ripples (figures 6.27*A* and *C*). Ripple marks and cross-beds can form in conglomerates, sandstones, siltstones, and limestones and in environments such as deserts, river channels, river deltas, and shorelines.

A **graded bed** is a layer with a vertical change in particle size, usually from coarse grains at the bottom of the bed to progressively finer grains toward the top (figure 6.28). A single bed may have gravel at its base and grade upward through sand and silt to fine clay at the top. A graded bed may be deposited by a turbidity current. A **turbidity current** is a turbulently flowing mass of sediment-laden water that is heavier than clear water and therefore flows downslope along the bottom of the sea or a lake. Turbidity currents are underwater avalanches and are typically triggered by earthquakes or submarine landslides. Figure 6.29 shows the development of a graded bed by turbidity-current deposition.

Mud cracks form as a polygonal pattern of cracks in very fine-grained sediment as it dries (figure 6.30). Because drying requires air, mud cracks form only in sediment exposed above water. Mud cracks may form in lake-bottom sediment as the lake dries up; in flood-deposited sediment as a river level drops; or in marine sediment exposed to the air, perhaps temporarily by a falling tide. Cracked mud can lithify to form shale, preserving the cracks. The filling of mud cracks by sand can form casts of the cracks in an overlying sandstone.

6.8 FOSSILS

Fossils are the remains of organisms preserved in sedimentary rock. Most sedimentary layers contain some type of fossil, and some limestones are composed entirely of fossils. Most fossils are preserved by the rapid burial in sediment of bones, shells, or teeth, which are the mineralized hard parts of animals most resistant to decay (figure 6.31). The original bone or shell is rarely preserved unaltered; the original mineral is often recrystallized or replaced by a different mineral such as pyrite or silica. Bone and wood may be *petrified* as organic material is replaced and pore spaces filled with mineral. Shells entombed within rock are commonly dissolved away by pore waters, leaving only impressions or *molds* of the original fossil. Leaves and undecayed organic tissue can also be preserved as thin films of carbon (figure 6.32*A*). *Trace fossils* are a type of sedimentary structure produced by the impact of an organism's activities on the sediment. Footprints, trackways, and burrows are the most common trace fossils (figure 6.32*B*).

Many *paleontologists* study fossils to learn about the evolution of life on Earth, but fossils are also very useful for interpreting depositional environments and for reconstructing the climates of the past. Fossils can be used to distinguish freshwater from marine environments and to infer the water depth at which a particular sedimentary layer was deposited. Tropical, temperate, and arid climates can be associated with distinctive types of fossil plants. Marine *microfossils,* the tiny shells produced by ocean-dwelling plankton, can be analyzed to determine the water temperature that surrounded the shell when it formed.

FIGURE 6.28

A graded bed has coarse grains at the bottom of the bed and progressively finer grains toward the top. Coin for scale.
©*David McGeary*

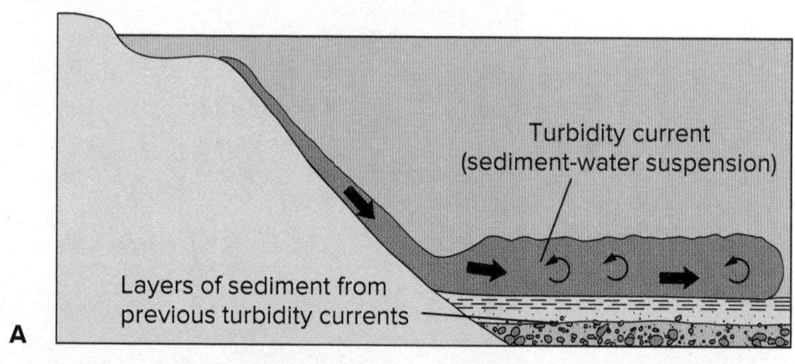

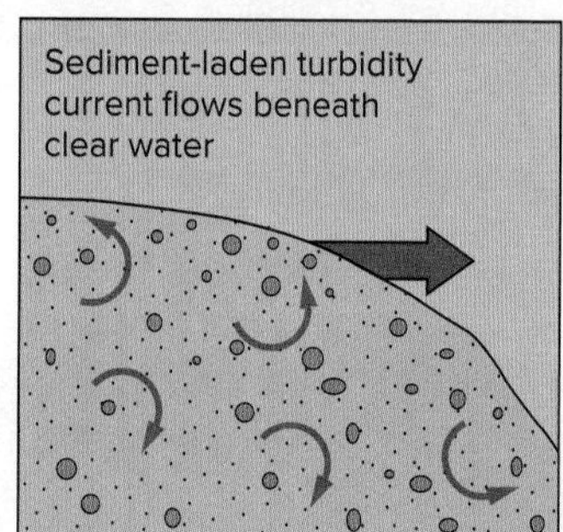

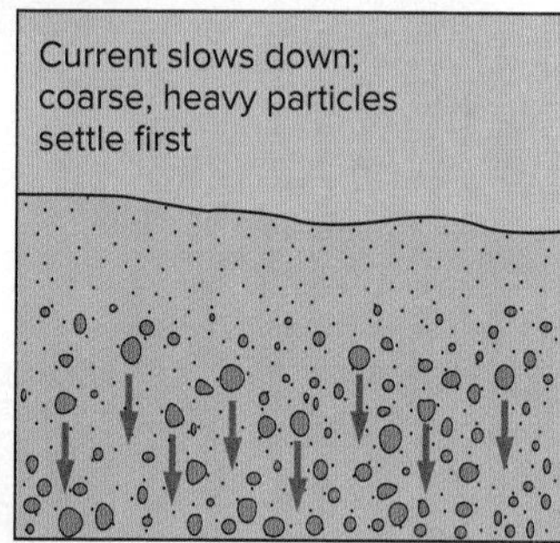

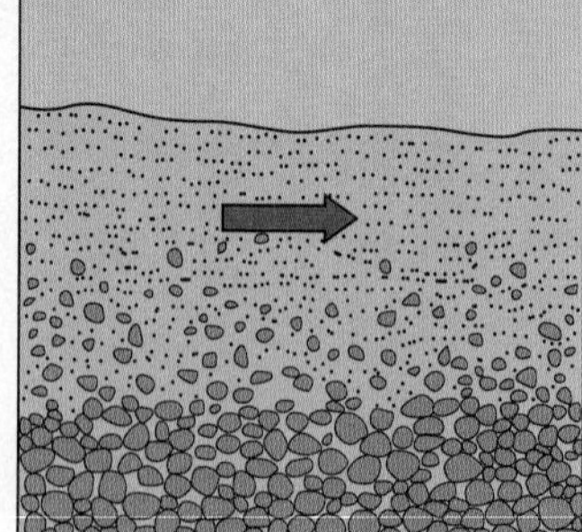

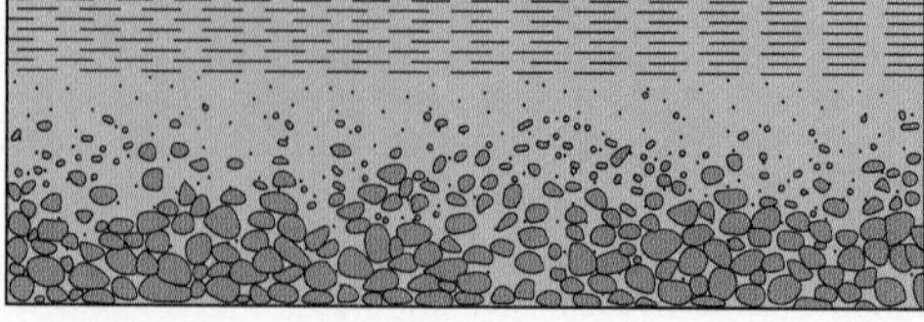

FIGURE 6.29

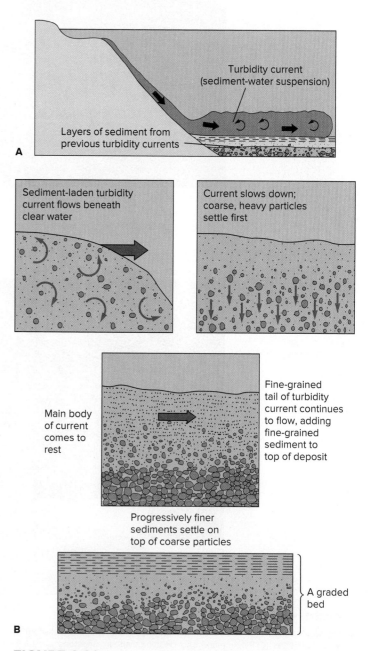

Formation of a graded bed by deposition from a turbidity current. (*A*) Slurry of sediment and water moves downslope along the sea floor. (*B*) As the turbidity flow slows down, larger grains are deposited first, followed by progressively finer grains, to produce a graded bed.

A

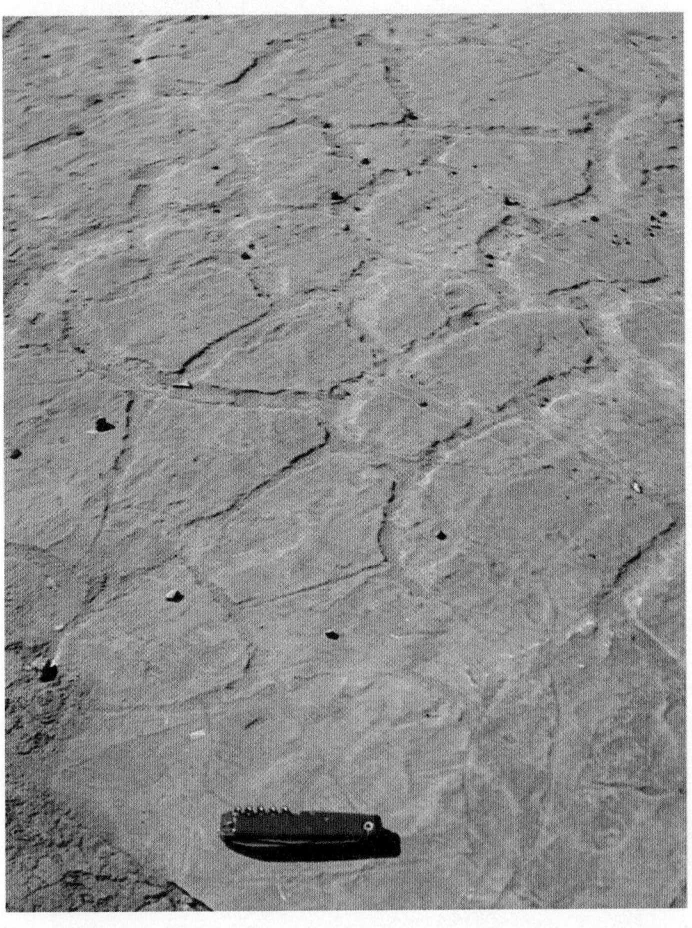

B

FIGURE 6.30

(*A*) Mud cracks in recently dried mud. (*B*) Mud cracks preserved in shale; they have been partially filled with sediment.
©David McGeary

FIGURE 6.31

Fossil clams, brachiopods, and trilobites in the Hamilton Shale of New York. Some of the fossils have their original shell material; other fossils are preserved as impressions.
©Bret Bennington

A

B

FIGURE 6.32

(*A*) Fossil fish in a rock from western Wyoming. (*B*) Dinosaur footprint in shale, Tuba City, Arizona. Scale in centimeters.
(A) ©Alan Morgan RF (B) ©David McGeary

Much of our detailed knowledge of Earth's climate changes over the last 150 million years has come from the study of microfossils extracted from layers of mud deposited on the deep-ocean floor.

6.9 FORMATIONS

A **formation** is a body of rock of considerable thickness that is large enough to be mappable, and with characteristics that distinguish it from adjacent rock units. Although a formation is usually composed of one or more beds of sedimentary rock, units of metamorphic and igneous rock are also called formations. It is a convenient unit for mapping, describing, or interpreting the geology of a region.

Formations are often based on rock type. A formation may be a single thick bed of rock such as sandstone. A sequence of several thin sandstone beds could also be called a formation, as could a sequence of alternating limestone and shale beds.

The main criterion for distinguishing and naming a formation is some visible characteristic that makes it a recognizable unit. This characteristic may be rock type or sedimentary structures or both. For example, a thick sequence of shale may be overlain by basalt flows and underlain by sandstone. The shale, the basalt, and the sandstone are each a different formation. Or a sequence of thin limestone beds, with a total thickness of many tens of meters, may have recognizable fossils in the lower half and distinctly different fossils in the upper half. The limestone sequence is divided into two formations on the basis of its fossil content.

Formations are given proper names: the first name is often a geographic location where the rock is well exposed, and the second is the name of a rock type, such as Navajo Sandstone, Austin Chalk, Baltimore Gneiss, Onondaga Limestone, or Chattanooga Shale. If the formation has a mixture of rock types, so that one rock name does not accurately describe it, it is called simply "formation," as in the Morrison Formation or the Martinsburg Formation.

A **contact** is the boundary surface between two different rock types or ages of rocks. In sedimentary rock formations, the contacts are usually bedding planes.

Figure 6.33 shows the three formations that make up the upper part of the canyon walls in Grand Canyon National Park in Arizona. The contacts between formations are also shown.

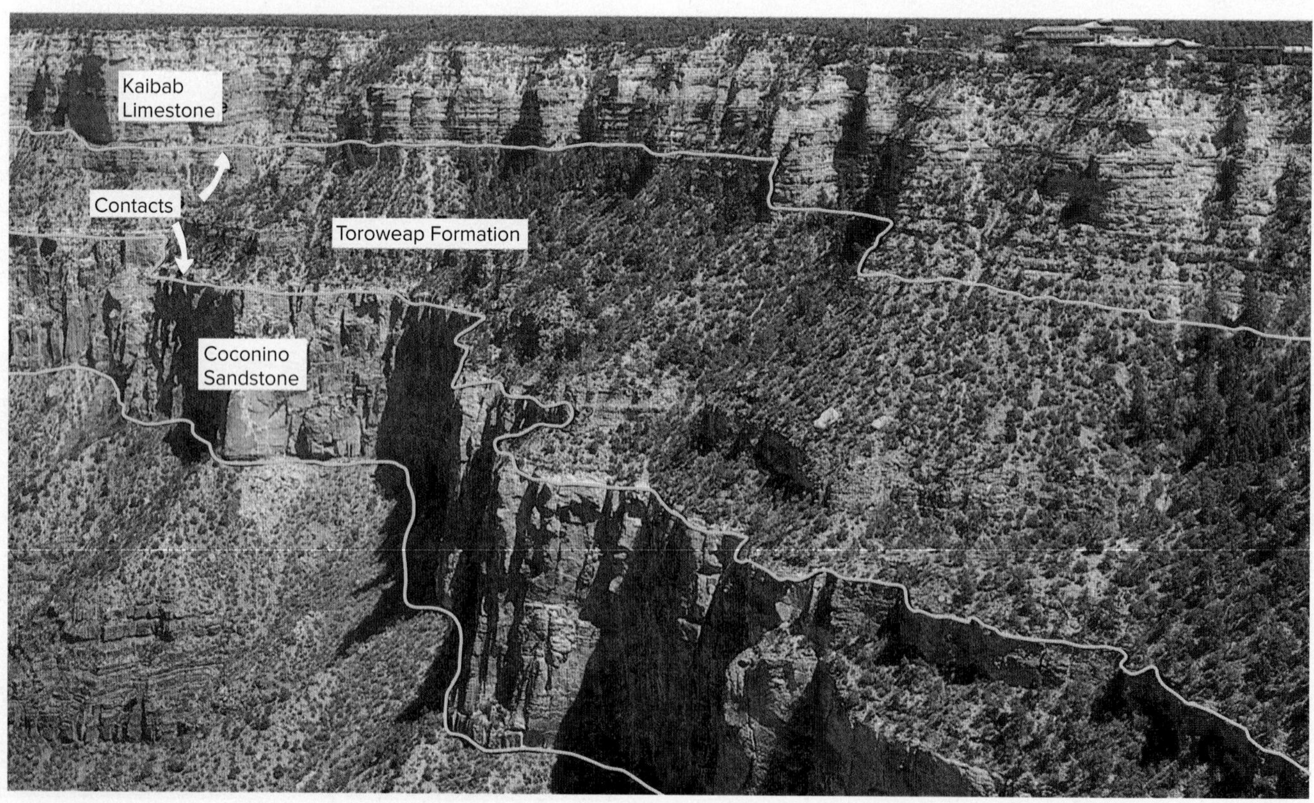

FIGURE 6.33

The upper three formations in the cliffs of the Grand Canyon, Arizona. The Kaibab Limestone and the Coconino Sandstone are resistant in the dry climate and form cliffs. The Toroweap Formation contains some shale and is less resistant, forming slopes. The gray lines are drawn to show the approximate contacts or boundaries between the formations.

©David McGeary

6.10 INTERPRETATION OF SEDIMENTARY ROCKS

Sedimentary rocks are important in interpreting the geologic history of an area. Geologists examine sedimentary formations to look for clues such as fossils; sedimentary structures; grain shape, size, and composition; and the overall shape and extent of the formation. These clues are useful in determining the source area of the sediment, the environment of deposition, and the possible plate-tectonic setting at the time of deposition.

Source Area

The **source area** of a sediment is the locality that eroded and provided the sediment. The most important things to determine about a source area are the types of rocks that were exposed in it and its location and distance from the site of eventual deposition.

The *rock type* exposed in the source area determines the character of the resulting sediment. The composition of a sediment can indicate the source area rock type, even if the source area has been completely eroded away. A conglomerate may contain cobbles of basalt, granite, and chert; these rock types were obviously in its source area. An arkose containing coarse feldspar, quartz, and biotite may have come from a granitic source area. Furthermore, the presence of feldspar indicates the source area was not subjected to extensive chemical weathering and that erosion probably took place in an arid environment with high relief. A quartz sandstone containing well-rounded quartz grains, on the other hand, probably represents the erosion and deposition of quartz grains from preexisting sandstone. Quartz is a hard, tough mineral very resistant to rounding by abrasion, so if quartz grains are well-rounded, they have undergone many cycles of erosion, transportation, and deposition, probably over tens of millions of years.

Sedimentary rocks are also studied to determine the *direction* and *distance* to the source area. Figure 6.34 shows how several characteristics of sediment may vary with distance from a source area. Many sediment deposits get thinner away from the source, and the sediment grains themselves usually become finer and more rounded.

Sedimentary structures often give clues about the directions of ancient currents (*paleocurrents*) that deposited sediments. Refer back to figure 6.25 and notice how cross-beds slope downward in the direction of current flow. Ancient current direction can also be determined from asymmetric ripple marks (figures 6.27*C* and *D*).

Environment of Deposition

Figure 6.35 shows the common environments in which sediments are deposited. Geologists study modern environments in great detail so that they can interpret ancient rocks. Clues to the ancient environment of deposition come from a rock's composition, the size and sorting and rounding of the grains, the sedimentary structures and fossils present, and the vertical sequence of the sedimentary layers.

Continental environments include alluvial fans, river channels, flood plains, lakes, and dunes. Sediments deposited on land are subject to erosion, so they often are destroyed. The great bulk of sedimentary rocks comes from the more easily preserved shallow marine environments, such as deltas, beaches, lagoons, shelves, and reefs. The characteristics of major environments are covered in detail in later chapters (10, 12 through 14, 18). In this section, we describe the main sediment types and sedimentary structures found in each environment.

Glacial Environments

Glacial ice often deposits narrow ridges and layers of sediment in valleys and widespread sheets of sediment on plains. Glacial sediment (*till*) is an unsorted mix of unweathered boulders, cobbles, pebbles, sand, silt, and clay. The boulders and cobbles may be scratched from grinding over one another under the great weight of the ice.

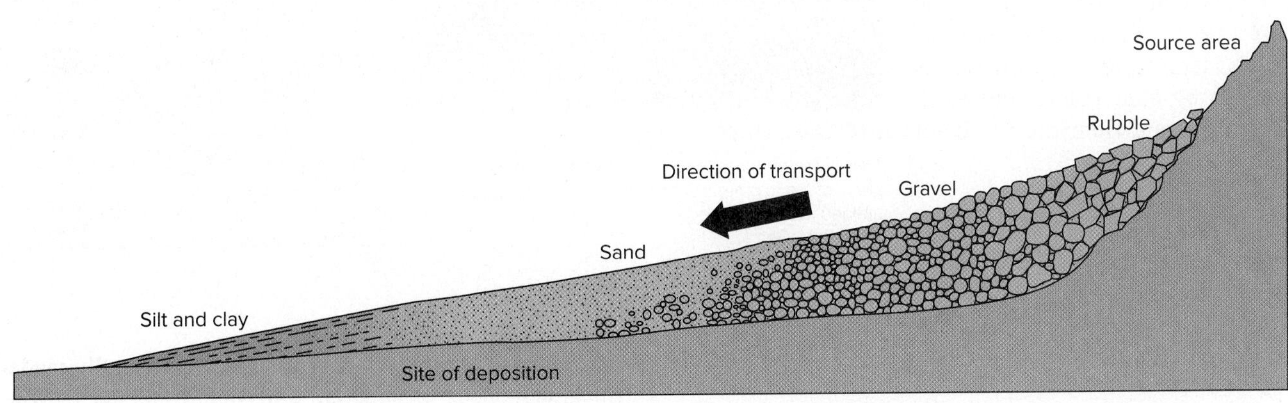

FIGURE 6.34

Sediment deposits often become thinner away from the source area, and sediment grains usually become finer and more rounded. The rocks that form from these sediments would change with distance from the source area from breccia to conglomerate to sandstone to shale. See appendix F for rock symbols.

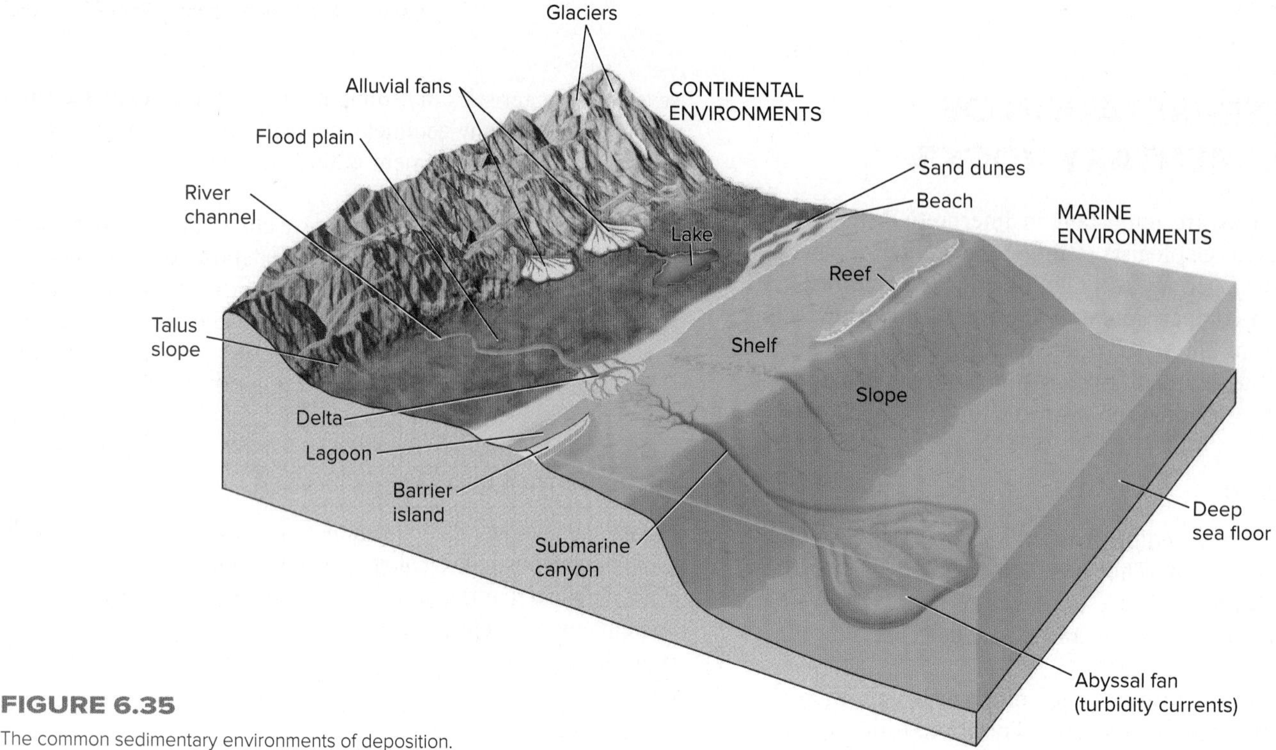

FIGURE 6.35

The common sedimentary environments of deposition.

Alluvial Fan

As streams emerge from mountains onto flatter plains, they deposit broad, fan-shaped piles of sediment. The sediment often consists of coarse, arkosic sandstones and conglomerates, marked by coarse cross-bedding and lenslike channel deposits (figure 6.36).

River Channel and Flood Plain

Rivers deposit elongate lenses of conglomerate or sandstone in their channels (figure 6.37). The sandstones may be arkoses or may consist of sand-sized fragments of fine-grained rocks. River channel deposits typically contain cross-beds and current ripple marks. Broad, flat flood plains are covered by periodic floodwaters, which deposit thin-bedded shales characterized by mud cracks and fossil footprints of animals. Hematite may color flood-plain deposits red.

FIGURE 6.36

Alluvial fan deposits, Baja, California. A channel deposit of conglomerate occurs within the coarse-grained sequence.
©David McGeary

Lake

Thin-bedded shale, perhaps containing fish fossils, is deposited on lake bottoms. If the lake periodically dries up, the shales will be mud-cracked and perhaps interbedded with evaporites such as gypsum or rock salt.

Delta

A delta is a body of sediment deposited when a river flows into standing water, such as the sea or a lake. Most deltas contain a great variety of subenvironments but are generally made up of thick sequences of siltstone and shale, marked by low-angle cross-bedding and cut by coarser channel deposits. Delta sequences may contain beds of peat or coal, as well as marine fossils such as clam shells.

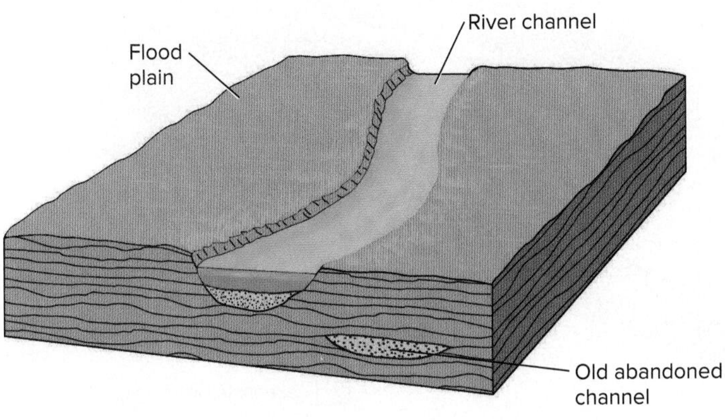

FIGURE 6.37

A river deposits an elongate lens of sand and gravel in its channel. Fine-grained silt and clay are deposited beside the channel on the river's flood plain.

Beach, Barrier Island, Dune

A barrier island is an elongate bar of sand built by wave action. Well-sorted quartz sandstone with well-rounded grains is deposited on beaches, barrier islands, and dunes. Beaches and barrier islands are characterized by cross-bedding (often low-angle) and marine fossils. Dunes have both high-angle and low-angle cross-bedding and occasionally contain fossil footprints of land animals such as lizards. All three environments can also contain carbonate sand in tropical regions, thus yielding cross-bedded clastic limestones.

Lagoon

A semienclosed, quiet body of water between a barrier island and the mainland is a lagoon. Fine-grained dark shale, cut by tidal channels of coarse sand and containing fossil oysters and other marine organisms, is formed in lagoons. Limestones may also form in lagoons adjacent to reefs (see figure 6.15).

Shallow Marine Shelves

On the broad, shallow shelves adjacent to most shorelines, sediment grain size decreases offshore. Widespread deposits of sandstone, siltstone, and shale can be deposited on such shelves. The sandstone and siltstone contain symmetrical ripple marks, low-angle cross-beds, and marine fossils such as clams and snails. If fine-grained *tidal flats* near shore are alternately covered and exposed by the rise and fall of tides, mud-cracked marine shale will result.

Reefs

Massive limestone forms in reef cores, with steep beds of limestone breccia forming seaward of the reef, and horizontal beds of sand-sized and finer-grained limestones forming landward (see figure 6.15). All these limestones are full of fossil fragments of corals, coralline and calcareous algae, and numerous other marine organisms.

Deep Marine Environments

On the deep-sea floor are deposited shale and graywacke sandstones. The graywackes are deposited by turbidity currents (figure 6.29) and typically contain graded bedding and current ripple marks.

Transgression and Regression

Sea level is not stable. Sea level has risen and fallen many times in the geologic past, flooding and exposing much of the land of the continents as it did so.

On a very broad, shallow marine shelf, several types of sediments may be deposited. On the beach and near shore, waves will deposit sand, which is usually derived from land. Farther from shore, in deeper, quieter water, land-derived silt and clay will be deposited. If the shelf is broad enough and covered with warm water, corals and algae may form carbonate sediments still farther seaward, beyond the reach of land-derived sediment. These sediments can lithify to form a seaward sequence of sandstone, shale, and limestone (figure 6.38*A*).

If sea level rises or the land sinks (subsides), large areas of land will be flooded, and these three environments of rock deposition will migrate across the land (figure 6.38*B*). This is a *transgression* of the sea as it moves across the land, and it can result in a bed of sandstone overlain by shale, which in turn is overlain by limestone (transgressive sequence). Note that different parts of a single rock bed are deposited at different times—the seaward edge of the sandstone bed, for example, is older than the landward edge.

In a *regression*, the sea moves off the land, and the three rock types are arranged in a new vertical sequence—limestone is overlain by shale and shale by sandstone (figure 6.38*C*). A drop in sea level alone will not preserve this regressive sequence of rocks. The land must usually subside rapidly to preserve these rocks so that they are not destroyed by continental erosion.

The angles shown in the figure are exaggerated—the rocks often appear perfectly horizontal. Geologists use these two contrasting vertical sequences of rock to identify ancient transgressions and regressions.

Tectonic Setting of Sedimentary Rocks

The dynamic forces that move plates on Earth are also responsible for the distribution of many sedimentary rocks. As such, the distribution of sedimentary rocks often provides information that helps geologists reconstruct past plate-tectonic settings.

In tectonically active areas, particularly along *convergent plate boundaries,* the thickening of the crust that forms a mountain belt also causes the adjacent crust to subside, forming basins (figure 6.39). Rapid erosion of the rising mountains produces enormous quantities of sediment that are transported by streams and turbidity currents to the adjacent basins. Continued subsidence of the basins results in the formation of great thicknesses of sedimentary rock that record the history of uplift and erosion in the mountain belt. For example, uplift of the ancestral Sierra Nevada and Klamath mountain ranges in California is recorded by the thick accumulation of turbidite deposits preserved in basins to the west of the mountains. There, graywacke sandstone deposited by turbidity currents

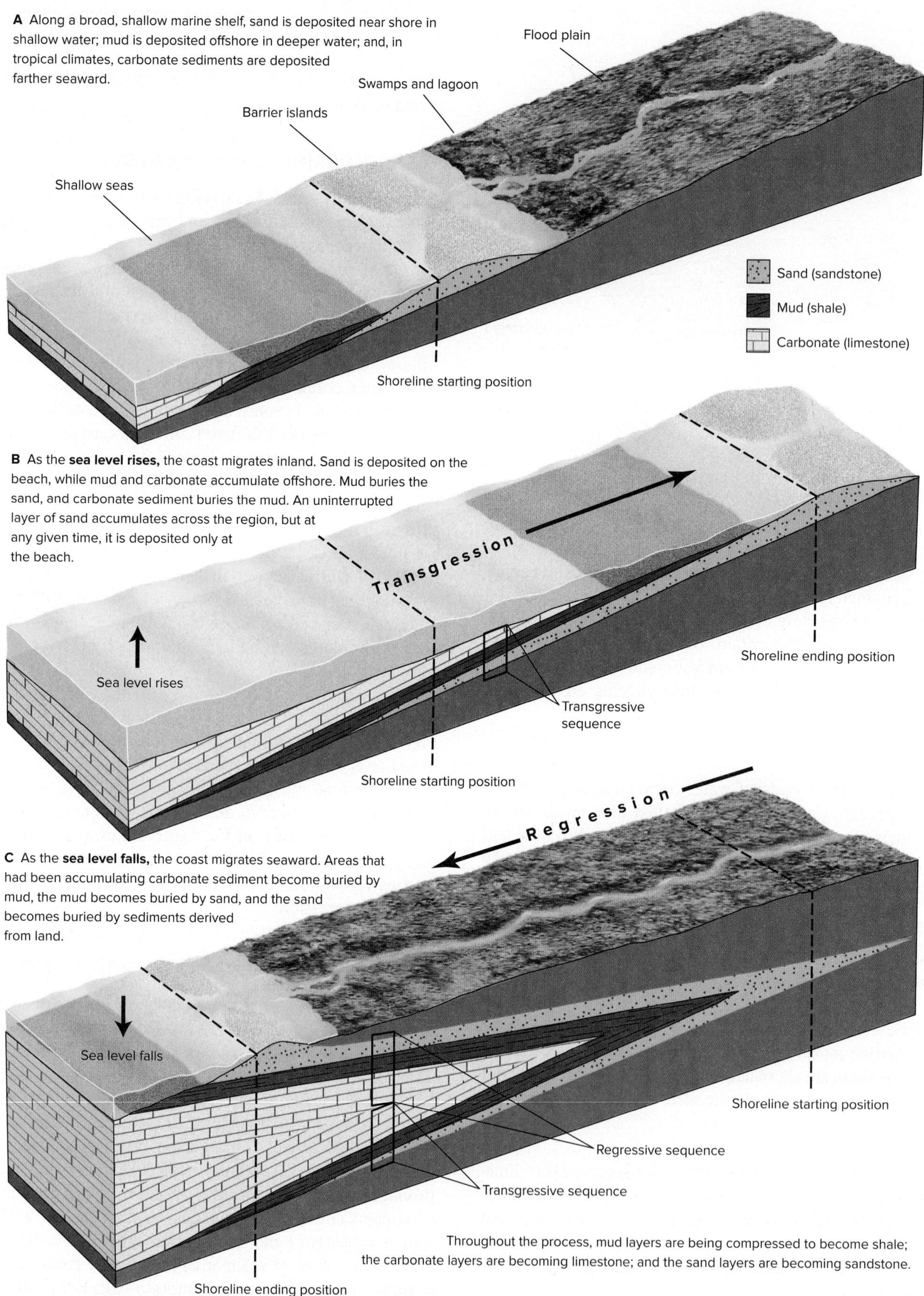
A Along a broad, shallow marine shelf, sand is deposited near shore in shallow water; mud is deposited offshore in deeper water; and, in tropical climates, carbonate sediments are deposited farther seaward.
Flood plain
Swamps and lagoon
Barrier islands
Shallow seas
Sand (sandstone)
Mud (shale)
Carbonate (limestone)
Shoreline starting position
B As the **sea level rises,** the coast migrates inland. Sand is deposited on the beach, while mud and carbonate accumulate offshore. Mud buries the sand, and carbonate sediment buries the mud. An uninterrupted layer of sand accumulates across the region, but at any given time, it is deposited only at the beach.
Transgression
Shoreline ending position
Sea level rises
Transgressive sequence
Shoreline starting position
Regression
C As the **sea level falls,** the coast migrates seaward. Areas that had been accumulating carbonate sediment become buried by mud, the mud becomes buried by sand, and the sand becomes buried by sediments derived from land.
Sea level falls
Shoreline starting position
Regressive sequence
Transgressive sequence
Throughout the process, mud layers are being compressed to become shale; the carbonate layers are becoming limestone; and the sand layers are becoming sandstone.
Shoreline ending position

FIGURE 6.38

Transgressions and regressions of the sea can form distinctive sequences of sedimentary rocks.

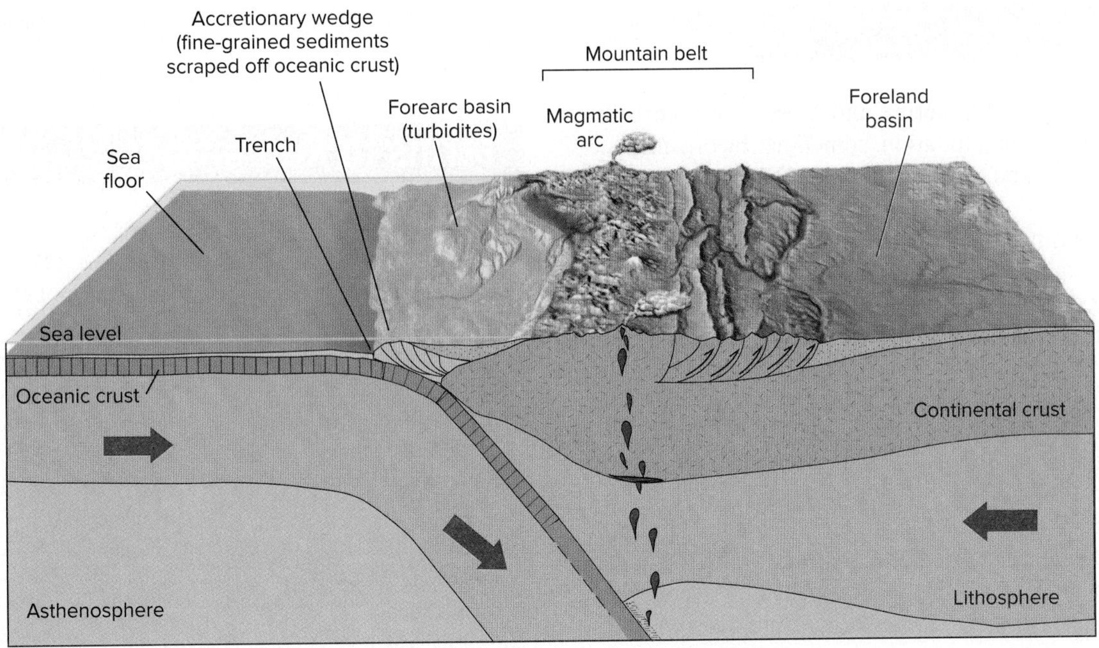

FIGURE 6.39

Sedimentary basins associated with convergent plate boundaries include a forearc basin on the oceanward side that contains mainly clastic sediments deposited by streams and turbidity currents from an eroding magmatic arc. Toward the craton (continent), a foreland basin also collects clastic sediment derived from the uplifted mountain belt and craton.

contains mainly volcanic clasts in the lower part of the sedimentary sequence and abundant feldspar clasts in the upper part of the sequence. This indicates that a cover of volcanic rocks was first eroded from the ancestral mountains, and then, as uplift and erosion continued, the underlying plutonic rocks were exposed and eroded. Other eroded mountains, such as the Appalachians, have left similar records of uplift and erosion in the sedimentary record.

It is not uncommon for rugged mountain ranges, such as the Canadian Rockies, European Alps, and Himalayas, that stand several thousand meters above sea level to contain sedimentary rocks of marine origin that were originally deposited below sea level. The presence of marine sedimentary rocks such as limestone, chert, and shale containing marine fossils at high elevations attests to the tremendous uplift associated with mountain building at convergent plate boundaries (see chapter 20).

Transform plate boundaries are also characterized by rapid rates of erosion and deposition of sediments as fault-bounded basins open and subside rapidly with continued plate motion. Because of the rapid rate of deposition and burial of organic material, fault-bounded basins are good places to explore for petroleum. Many of the petroleum occurrences in California are related to basins that formed as the San Andreas transform fault developed.

A *divergent plate boundary* may result in the splitting apart of a continent and formation of a new ocean basin. In the initial stages of continental divergence, a rift valley forms and fills with thick wedges of gravel and coarse sand along its fault-bounded margins; lake bed deposits and associated evaporite rocks may form in the bottom of the rift valley (figure 6.40). In the early stages, continental rifts will have extensive volcanics that contribute to the sediments in the rift. The Red Sea and adjacent East African Rift Zone have good examples of the features and sedimentary rocks formed during the initial stages of continental rifting.

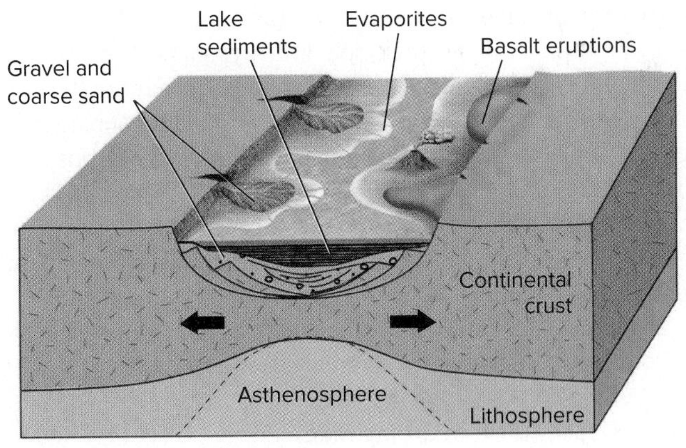

FIGURE 6.40

Divergent plate boundary showing thick wedges of gravel and coarse sand along fault-bounded margins of developing rift valley. Lake bed deposits and evaporite rocks are located on the floor of the rift valley. Refer to figures 19.20 and 19.22 for more detail of faulted margin and sediments deposited along a rifted continental margin.

Summary

Sediment forms by the weathering and erosion of preexisting rocks and by chemical precipitation, sometimes by organisms.

Gravel, sand, silt, and *clay* are sediment particles defined by grain size.

The composition of sediment is governed by the rates of chemical weathering, mechanical weathering, and erosion. During transportation, grains can become rounded and sorted.

Sedimentary rocks form by *lithification* of sediment, by *crystallization* from solution, or by consolidation of remains of organisms. Sedimentary rocks may be *detrital, chemical,* or *organic.*

Detrital sedimentary rocks form mostly by *compaction* and *cementation* of grains. *Matrix* can partially fill the *pore space* of detrital rocks.

Conglomerate forms from coarse, rounded sediment grains that often have been transported only a short distance by a river or waves. *Sandstone* forms from sand deposited by rivers, wind, waves, or turbidity currents. *Shale* forms from river, lake, or ocean mud.

Limestone consists of calcite, formed either as a chemical precipitate in a reef or, more commonly, by the cementation of shell and coral fragments or of oöids. *Dolomite* usually forms from the alteration of limestone by magnesium-rich solutions.

Chert consists of silica and usually forms from the accumulation of microscopic marine organisms. *Recrystallization* often destroys the original texture of chert (and some limestones).

Evaporites, such as rock salt and gypsum, form as water evaporates. *Coal,* a major fuel, is consolidated plant material.

Sedimentary rocks are usually found in *beds* separated by *bedding planes* because the original sediments are deposited in horizontal layers.

Cross-beds and *ripple marks* develop as moving sediment forms ripples and dunes during transport by wind, underwater currents, and waves.

A *graded bed* forms as coarse particles fall from suspension before fine particles due to decreasing water flow velocity in a *turbidity current.*

Mud cracks form in drying mud.

Fossils are the traces of an organism's hard parts or tracks preserved in rock.

A *formation* is a convenient rock unit for mapping and describing rock. Formations are lithologically distinguishable from adjacent rocks; their boundaries are *contacts.*

Geologists try to determine the *source area* of a sedimentary rock by studying its grain size, composition, and sedimentary structures. The source area's rock type and location are important to determine.

The *environment of deposition* of a sedimentary rock is determined by studying bed sequence, grain composition and rounding, and sedimentary structures. Typical environments include alluvial fans, river channels, flood plains, lakes, dunes, deltas, beaches, shallow marine shelves, reefs, and the deep-sea floor.

Plate tectonics plays an important role in the distribution of sedimentary rocks; the occurrence of certain types of sedimentary rocks is used by geologists to construct past plate-tectonic settings.

Terms to Remember

alluvial fan 134
bedding 144
bedding plane 144
cement 132
cementation 132
chemical sedimentary rocks 133
chert 141
clastic texture 133
clay 130
coal 143
compaction 132
conglomerate 134
contact 150
cross-beds 144
crystalline texture 133
crystallization 133
deposition 131
detrital sedimentary rocks 133
dolomite 140
environment of deposition 132
evaporite 143
formation 150
fossil 148
graded bed 148
gravel 130
limestone 137
lithification 132
matrix 134
mud crack 148
organic sedimentary rock 133
original horizontality 144
pore space 132
recrystallization 138
ripple marks 144
rounding 130
sand 130
sandstone 134
sediment 129
sedimentary breccia 133
sedimentary rocks 133
sedimentary structures 144
shale 134
silt 130
sorting 131
source area 151
transportation 130
turbidity current 148

Testing Your Knowledge

Use the following questions to prepare for exams based on this chapter.

1. Using the rock cycle, explain how sediment is formed and may become a sedimentary rock.
2. List the detrital sediment particles in order of decreasing grain size. How is grain size used to identify and classify detrital sedimentary rocks?
3. Sketch and describe the lithification of sand to sandstone.
4. How does a sedimentary breccia differ in appearance and origin from a conglomerate?
5. Describe with sketches how wet mud compacts before it becomes shale.
6. Explain how chemical sedimentary rocks are formed, and give several examples.
7. How does dolomite usually form?
8. How do evaporites form? Name two evaporites.
9. What is the origin of oil and gas? How does coal form?
10. Explain two ways that cross-bedding can form.
11. What do mud cracks tell us about the environment of deposition of a sedimentary rock?

12. How does a graded bed form?
13. Describe what clues a geologist uses to interpret where and how a sedimentary rock formed and why this is important.
14. Particles of sediment from 1/16 to 2 millimeters in diameter are of what size?
    a. gravel b. sand c. silt d. clay
15. Rounding is
    a. the rounding of a grain to a spherical shape.
    b. the grinding away of sharp edges and corners of rock fragments during transportation.
    c. a type of mineral.
    d. none of the preceding.
16. Compaction and cementation are two common processes of
    a. erosion. b. transportation.
    c. deposition. d. lithification.
17. Which is *not* a chemical or organic sedimentary rock?
    a. rock salt b. sandstone
    c. limestone d. gypsum
18. The major difference between breccia and conglomerate is
    a. size of grains.
    b. rounding of the grains.
    c. composition of grains.
    d. all of the preceding.
19. Which is *not* a type of sandstone?
    a. quartz sandstone b. arkose
    c. graywacke d. travertine
20. Shale differs from mudstone in that
    a. shale has larger grains.
    b. shale is visibly layered and fissile; mudstone is massive and blocky.
    c. shale has smaller grains.
    d. there is no difference between shale and mudstone.
21. The chemical element found in dolomite *not* found in limestone is
    a. Ca. b. Mg.
    c. C. d. O.
    e. Al.
22. In a graded bed, the particle size
    a. decreases upward.
    b. decreases downward.
    c. increases in the direction of the current.
    d. stays the same.
23. A body or rock of considerable thickness with characteristics that distinguish it from adjacent rock units is called a/an
    a. formation.
    b. contact.
    c. bedding plane.
    d. outcrop.
24. If sea level drops or the land rises, what is likely to occur?
    a. a flood
    b. a regression
    c. a transgression
    d. no geologic change will take place
25. Thick accumulations of graywacke and volcanic sediments can indicate an ancient
    a. divergent plate boundary.
    b. convergent boundary.
    c. transform boundary.
26. A sedimentary rock made of fragments of preexisting rocks is
    a. organic.
    b. chemical.
    c. detrital.
27. Clues to the nature of the source area of sediment can be found in
    a. the composition of the sediment.
    b. sedimentary structures.
    c. rounding of sediment.
    d. all of the preceding.

Expanding Your Knowledge

1. How might graded bedding be used to determine the tops and bottoms of sedimentary rock layers in an area where sedimentary rock is no longer horizontal? What other sedimentary structures can be used to determine the tops and bottoms of tilted beds?
2. Which would weather faster in a humid climate, a quartz sandstone or an arkose? Explain your answer.
3. A cross-bedded quartz sandstone may have been deposited as a beach sand or as a dune sand. What features could you look for within the rock to tell whether it had been deposited on a beach? On a dune?
4. Why is burial usually necessary to turn a sediment into a sedimentary rock?
5. Why are most beds of sedimentary rock formed horizontally?

Exploring Web Resources

http://pages.uoregon.edu/~rdorsey/SedResources.html

Web Resources for Sedimentary Geology site contains a comprehensive listing of resources available on the World Wide Web.

www.lib.utexas.edu/geo/folkready/folkprefrev.html

Online version of *Petrology of Sedimentary Rocks* by Professor Robert Folk at the University of Texas at Austin.

http://walrus.wr.usgs.gov/seds/

Visit the *U.S. Geological Survey Bedform and Sedimentology* site for computer and photographic images and movies of sedimentary structures.

CHAPTER

7 Metamorphism and Metamorphic Rocks

Contorted veins visible in a blueschist that formed under high-pressure conditions in a subduction zone. Sunol Regional Park, CA.
©John A. Karachewski

LEARNING OBJECTIVES

- Using the rock cycle, compare metamorphic rocks to sedimentary and igneous rocks in terms of the conditions under which they form.
- Describe the factors that control the characteristics of metamorphic rocks.
- Distinguish between nonfoliated and foliated metamorphic rocks in terms of their appearance and the conditions under which they form.
- Outline the characteristics used to classify metamorphic rocks and describe the most common metamorphic rocks.
- Compare and contrast contact metamorphism and regional metamorphism.
- Relate the pressure-temperature conditions of metamorphism to plate-tectonic setting.
- Describe the importance of hydrothermal processes during metamorphism.

This chapter on metamorphic rocks, the third major category of rocks in the rock cycle (figure 7.1), completes our description of Earth materials (rocks and minerals). The information on igneous and sedimentary processes in previous chapters should help you understand metamorphic rocks, which form from *preexisting* rocks.

For example, look at the photo on the opening page of this chapter. This rock started out as basalt, erupting at a mid-oceanic ridge and forming part of the sea floor. Later, the oceanic crust was carried into a subduction zone and transported to great depths beneath the surface. Under these high-pressure conditions, the minerals forming the basalt recrystallized into new minerals, giving the new metamorphic rock its blue color and distinctive foliated texture.

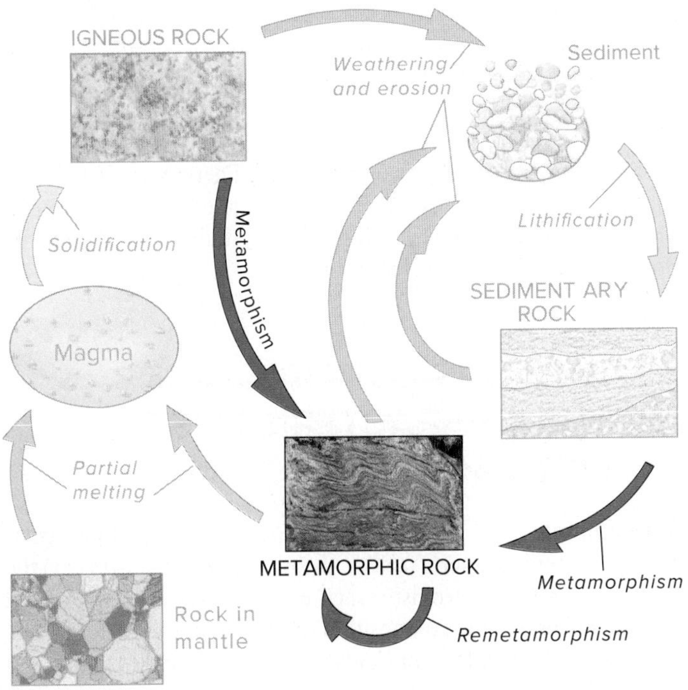

FIGURE 7.1

The rock cycle shows how igneous and sedimentary rocks undergo metamorphism to become metamorphic rocks. A metamorphic rock can also undergo remetamorphism to become a different metamorphic rock.

(Igneous): ©Doug Sherman/Geofile RF; (Metamorphic): ©reneh9999/Shutterstock; (Mantle): ©Doug Sherman/Geofile RF

Metamorphic rocks that form deep within Earth's crust provide geologists with many clues about conditions at depth. Therefore, understanding metamorphism will help you when we consider geologic processes involving Earth's internal forces. Metamorphic rocks are a feature of the oldest exposed rocks of the continents and of major mountain belts. They are especially important in providing evidence of what happens during subduction and plate convergence.

In this chapter, we will explore the process of metamorphism and the factors that control the characteristics of metamorphic rocks before learning to describe and name metamorphic rocks. Following this, we will explore some of the major types of metamorphism and the concept of metamorphic grade. We will conclude the chapter by exploring metamorphism and plate tectonics.

7.1 METAMORPHISM

From your study so far of Earth materials and the rock cycle, you know that rocks change, given enough time, when their physical environment changes radically. In chapter 6, you learned that as sediments are buried they undergo compaction and cementation, transforming them into sedimentary rocks. In chapter 3, you saw how deeply buried rocks will melt (or partially melt) to form magma when temperatures are high enough.

What happens to rocks that are deeply buried but are not hot enough to melt? They become metamorphosed.

Metamorphism (a word from Latin and Greek that means literally "changing of form") refers to changes to rocks that take place in Earth's interior. The changes may be new textures, new mineral assemblages, or both. Transformations occur in the *solid state* (meaning the rock does not melt). The new rock is a **metamorphic rock**.

The conversion of a slice of bread to toast is a solid-state process analogous to metamorphism of rock. When the bread (think "sedimentary rock") is heated, it converts to toast (think "metamorphic rock"). The toast is texturally and compositionally different from its parent material, bread. Although the rock remains solid during metamorphism, it is important to recognize that fluids, notably water, often play a significant role in the metamorphic process.

In nearly all cases, a metamorphic rock has a texture clearly different from that of the original **parent rock** (or **protolith**). When limestone is metamorphosed to marble, for example, the fine grains of calcite coalesce and recrystallize into larger calcite crystals. The calcite crystals are interlocked in a mosaic pattern that gives marble a texture distinctly different from that of the parent limestone. If the limestone is composed entirely of calcite, then metamorphism into marble involves no new minerals, only a change in texture.

More commonly, the various elements of a parent rock react chemically and crystallize into new minerals, thus making the metamorphic rock distinct both mineralogically and texturally from the parent rock. This is because the parent rock is unstable in its new environment. The old minerals recrystallize into new ones that are at *equilibrium* in the new environment. For example, clay minerals form at Earth's surface (see chapter 5). Therefore, they are stable at the low temperature and pressure conditions both at and just below Earth's surface. When subjected to the temperatures and pressures deep within Earth's crust, the clay minerals of a shale can recrystallize into coarse-grained mica.

As most metamorphism takes place at moderate to great depths in Earth's crust, metamorphic rocks provide us with a window to processes that take place deep underground, beyond our direct observation. Metamorphic rocks are exposed over large regions because of erosion of mountain belts and the accompanying uplift due to isostatic adjustment (the vertical movement of a portion of Earth's crust to achieve balance, described in chapter 1). In fact, the stable cores of continents, known as *cratons,* are largely metamorphic rocks and granitic plutons. As described in chapter 20 ("Mountain Belts and the Continental Crust"), the North American craton is the central lowlands between the Appalachians and the Rocky Mountains. Very ancient (Precambrian) complexes of metamorphic and intrusive igneous rocks are exposed over much of Canada (known as the *Canadian Shield*). The inside front cover shows the Canadian Shield as the region underlain by Precambrian rocks. In the Great Plains of the United States, also part of the craton, similar rocks form the *basement* underlying a veneer of younger sedimentary rocks (see appendix H for the geologic map of North America. The tan area on that legend indicates is "Platform deposits on Precambrian basement"). Ancient metamorphic and plutonic rocks form the cratons of the other continental landmasses (e.g., Africa, Antarctica, Australia) as well.

No one has observed metamorphism taking place, just as no one has ever seen a granite pluton form. What, then, leads us to believe that metamorphic rocks form in a solid state (i.e., without melting) at high pressure and temperature? Many metamorphic rocks found on Earth's surface exhibit contorted layering (figure 7.2). The layering can be demonstrated to have been either caused by metamorphism or inherited from original, flat-lying sedimentary bedding (even though the rock has since recrystallized). These rocks, now hard and brittle, would shatter if smashed with a hammer. But they must have been **ductile** (or **plastic**), capable of being bent and molded under stress, to have been folded into such contorted patterns. In a laboratory, we can reproduce high pressure and temperature conditions and demonstrate such

FIGURE 7.2
Metamorphic rock from Greenland. Metamorphism took place 3,700 million years ago—it is one of the oldest rocks on Earth.
©C.C. Plummer

ductile behavior of rocks on a small scale. Therefore, a reasonable conclusion is that these rocks formed at considerable depth, where such conditions exist. Moreover, crystallization of a magma would not produce contorted layering.

7.2 FACTORS CONTROLLING THE CHARACTERISTICS OF METAMORPHIC ROCKS

A metamorphic rock owes its characteristic texture and particular mineral content to several factors, the most important being (1) the conditions of metamorphism, such as temperature, pressure, and tectonic forces, (2) the composition of the parent rock before metamorphism, (3) the effects of fluids such as water, and (4) time.

Temperature

The heat needed for metamorphic reactions comes primarily from the outward flow of geothermal energy from Earth's deep interior. Usually, the deeper a rock is beneath the surface, the hotter it will be. (An exception to this is the temperature distribution along convergent plate boundaries due to subduction of cold crust, described later in this chapter.) The particular temperature for rock at a given depth depends on the local *geothermal gradient* (described in chapter 3). Additional heat could be derived from magma if magma bodies are locally present.

A mineral is said to be *stable* if, given enough time, it does not react with another substance or convert to a new mineral or substance. Any mineral is stable only within a given temperature range. The stability temperature range of a mineral varies with factors such as pressure and the presence or absence of other substances. Some minerals are stable over a wide temperature range. Quartz, if not mixed with other minerals, is stable at atmospheric pressure (i.e., at Earth's surface) up to about 800°C. At higher pressures, quartz remains stable to even higher temperatures. Other minerals are stable over a temperature range of only 100°C or 200°C.

By knowing (from results of laboratory experiments) the particular temperature range in which a mineral is stable, a geologist may be able to deduce the temperature of metamorphism for a rock that includes that mineral.

Minerals stable at higher temperatures tend to be less dense (or have a lower specific gravity) than chemically identical minerals (polymorphs) stable at lower temperatures. As temperature increases, the atoms vibrate more within their sites in the crystal structure. A more open (less tightly packed) crystal structure, such as high-temperature minerals tend to have, allows greater vibration of atoms. (If the heat and resulting vibrations become too great, the bonds between atoms in the crystal break and the substance becomes liquid.)

The upper limit on temperature in metamorphism overlaps the temperature of partial melting of a rock. If partial melting takes place, the component that melts becomes a magma; the solid residue (the part that doesn't melt) remains a metamorphic rock. Temperatures at which the igneous and metamorphic realms can coexist vary considerably. For an ultramafic rock (containing only ferromagnesian silicate minerals), the temperature will be over 1,200°C. For a metamorphosed shale under high water pressure, a granitic melt component can form in the metamorphic rock at temperatures as low as 650°C.

Pressure

Usually, when we talk about pressure, we mean **confining pressure**; that is, pressure applied equally on all surfaces of a substance as a result of burial or submergence. A diver senses confining pressure (known as *hydrostatic pressure*) proportional to the weight of the overlying water (figure 7.3). The pressure uniformly squeezes the diver's entire body surface. Likewise, an object buried deeply within Earth's crust is compressed by strong confining pressure, called *lithostatic pressure,* which forces grains closer together and eliminates pore space. For metamorphism, pressure is usually given in *kilobars.* A kilobar is 1,000 bars. A *bar* is very close (0.99 atmospheres) to standard atmospheric pressure, so that, for all practical purposes, a kilobar is the pressure equivalent of a thousand times the pressure of the atmosphere at sea level. The *pressure gradient,* the increase in lithostatic pressure with depth, is approximately 1 kilobar for every 3.3 kilometers of burial in the crust.

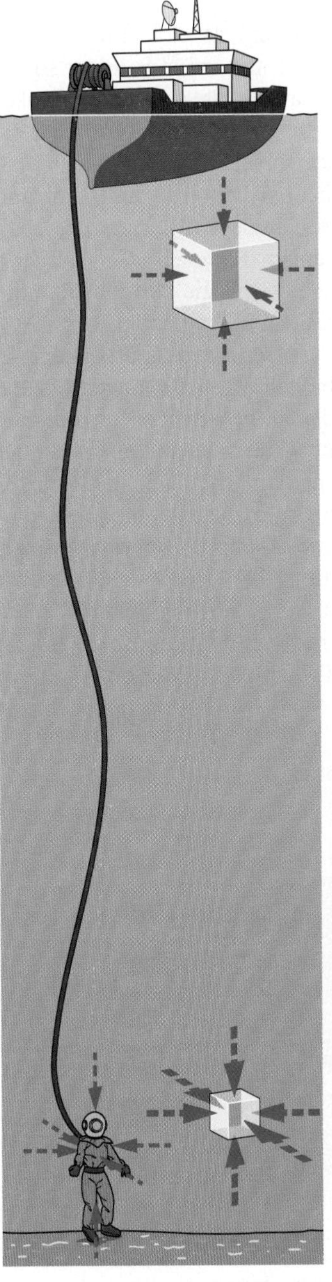

A

B

FIGURE 7.3

Confining pressure. (*A*) The diver's suit is pressurized to counteract hydrostatic pressure. Object (cube) has a greater volume at low pressure than at high pressure. (*B*) These Styrofoam cups were identical. The shrunken cup was carried to a depth of 2,250 meters by the submersible ALVIN in a biological sampling dive to the Juan de Fuca Ridge, off the coast of Washington state.

(B) Photo courtesy of the National Science Foundation-funded REVEL Project, University of Washington

Any new mineral that has crystallized under high-pressure conditions tends to occupy less space than did the mineral or minerals from which it formed. The new mineral is denser than its low-pressure counterparts because the pressure forces atoms closer together into a more closely packed crystal structure.

But what if pressure and temperature both increase, as is commonly the case with increasing depth into the Earth? If the effect of higher temperature is greater than the effect of higher pressure, the new mineral will likely be less dense. A denser new mineral is likely if increasing pressure effects are greater than increasing temperature effects.

Differential Stress

Most metamorphic rocks show the effects of tectonic forces. When forces are applied to an object, the object is under **stress**, force per unit area. If the forces on a body are stronger or weaker in different directions, a body is subjected to **differential stress.** Differential stress tends to deform objects into oblong or flattened forms. If you squeeze a rubber ball between your thumb and forefinger, the ball is under differential stress. If you squeeze a ball of dough (figure 7.4*A*), it will remain flattened after you stop squeezing, because dough is ductile (or plastic). To illustrate the difference between confining pressure and differential stress, visualize a drum filled with water. If you place a ball of putty underwater in the bottom of the drum, the ball will not change its shape (its volume will decrease slightly due to the weight of the overlying water). Now take the putty ball out of the water and place it under the drum. The putty will be flattened into the shape of a pancake due to the differential stress. In this case, the putty is subjected to *compressive differential stress* or, more simply, **compressive stress** (as is the dough ball shown in figure 7.4*A*).

Differential stress is also caused by **shearing**, which causes parts of a body to move or slide relative to one another across a plane. An example of shearing is when you spread out a deck of cards on a table with your hand moving parallel to the table. Shearing often takes place perpendicular to, or nearly perpendicular to, the direction of compressive stress. If you put a ball of putty between your hands and slide your hands while compressing the putty, as shown in figure 7.4*B*, the putty flattens parallel to the shearing (the moving hands) as well as perpendicular to the compressive stress.

Some rocks can be attributed exclusively to shearing during faulting (movement of bedrock along a fracture, described in chapter 15) in a process sometimes called *dynamic metamorphism.* Rocks in contact along the fault are broken and crushed when movement takes place. A *mylonite* is an unusual rock that is formed from pulverized rock in a fault zone. The rock is streaked out parallel to the fault in darker and lighter components due to shearing. Mylonites are believed to form at a depth of around a kilometer or so, where the rock is still cool and brittle (rather than ductile), but the pressure is sufficient to compress the pulverized rock into a compact, hard rock. Where found, they occupy zones that are only about a meter or so wide.

Foliation

Differential stress has a very important influence on the texture of a metamorphic rock because it forces the constituents of the rock to become parallel to one another. For instance, the pebbles in the metamorphosed conglomerate shown in figure 7.5 were originally more spherical but have been flattened by differential stress. When a rock has a planar texture, it is said to be *foliated.* **Foliation** is manifested in various ways. If a platy

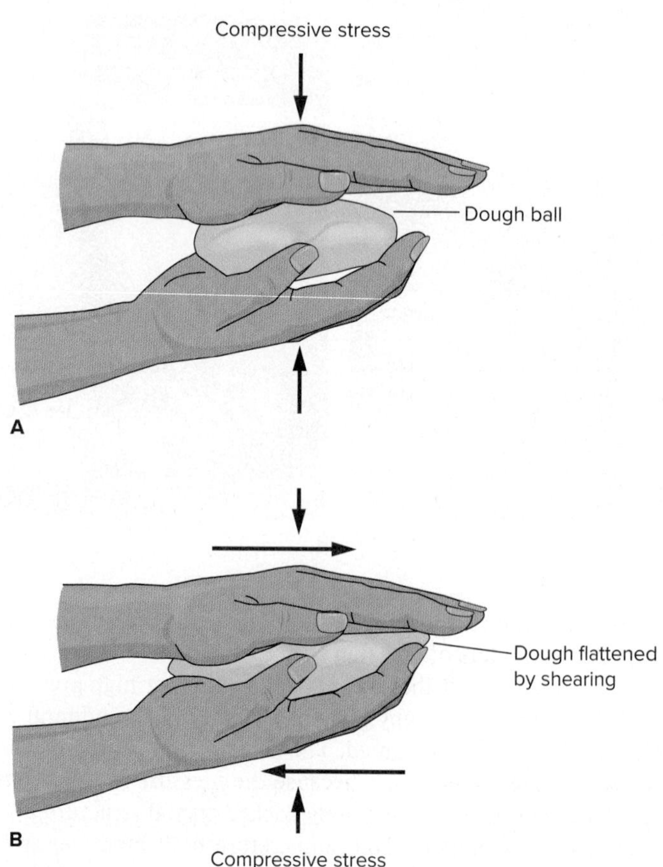

FIGURE 7.4

(*A*) Compressive stress exerted on a ball of dough by two hands. More force is exerted in the direction of arrows than elsewhere on the dough. (*B*) Shearing takes place as two hands move parallel to each other at the same time that some compressive force is exerted perpendicular to the flattening dough.

FIGURE 7.5

(A) Metamorphosed conglomerate in which the pebbles have been flattened (sometimes called a stretched pebble conglomerate). Compare to (B), an unmetamorphosed conglomerate (this is figure 6.7).
(A) ©C.C. Plummer (B) ©David McGeary

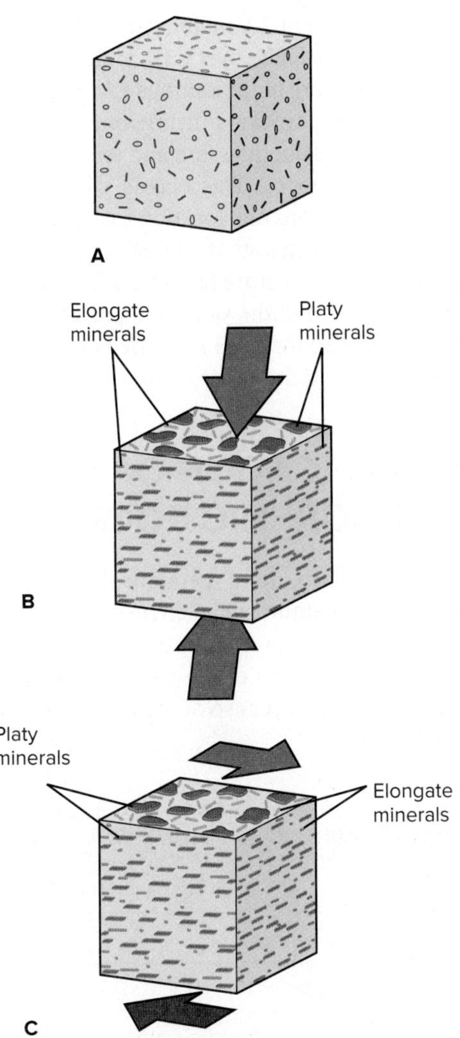

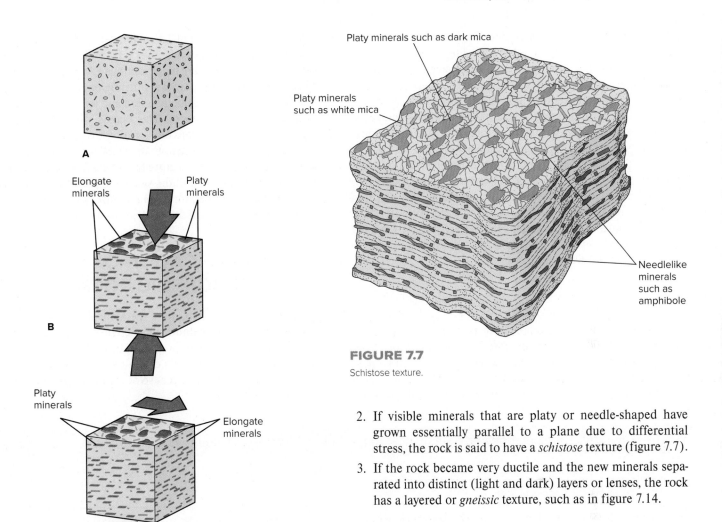

FIGURE 7.6

Orientation of platy and elongate minerals in metamorphic rock. (*A*) Platy minerals randomly oriented (e.g., clay minerals before metamorphism). No differential stress involved. (*B*) Platy minerals (e.g., mica) and elongate minerals (e.g., amphibole) have crystallized under the influence of compressive stress. (*C*) Platy and elongate minerals developed with shearing as the dominant stress.

mineral (such as mica) is crystallizing within a rock that is undergoing differential stress, the mineral grows in such a way that it remains parallel to the direction of shearing or perpendicular to the direction of compressive stress (figure 7.6). Any platy mineral attempting to grow against shearing is either ground up or forced into alignment. Minerals that crystallize in needlelike shapes (for example, hornblende) behave similarly, growing with their long axes parallel to the plane of foliation. The three very different textures described next (from lowest to highest degree of metamorphism) are all variations of foliation and are important in classifying metamorphic rocks:

1. If the rock is fine-grained and splits easily along nearly flat and parallel planes, indicating that preexisting, microscopic, platy minerals were realigned during metamorphism, we say it has a *slaty texture,* or that it possesses *slaty cleavage.*

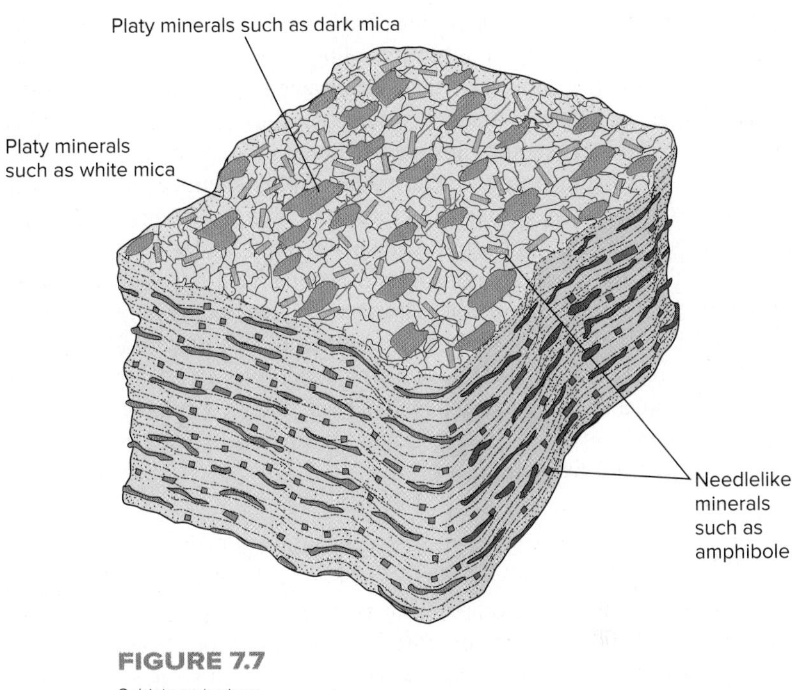

FIGURE 7.7

Schistose texture.

2. If visible minerals that are platy or needle-shaped have grown essentially parallel to a plane due to differential stress, the rock is said to have a *schistose* texture (figure 7.7).
3. If the rock became very ductile and the new minerals separated into distinct (light and dark) layers or lenses, the rock has a layered or *gneissic* texture, such as in figure 7.14.

Composition of the Parent Rock

If no new elements or chemical compounds are added to the rock during metamorphism, the mineral content of the metamorphic rock will be the same as the chemical composition of the parent rock. For example, a quartz sandstone with a silica cement is composed entirely of silica (SiO_2). During metamorphism, the silica will recrystallize to form interlocking grains of quartz, and the new metamorphic rock would be called a quartzite. New metamorphic minerals can form if the parent rock contains the chemical compounds needed to form those minerals. For example, the mineral wollastonite can form during metamorphism of a limestone that contains some silica. The reaction that creates wollastonite is as follows:

$CaCO_3$	+	SiO_2	→	$CaSiO_3$	+	CO_2
calcite		quartz		wollastonite (a mineral)		carbon dioxide

If fluids are involved in metamorphism, the composition of the parent rock, and thus the composition of the metamorphic rock, can be changed.

Fluids

Hot water (as vapor) is the most important fluid involved in metamorphic processes, although other gases, such as carbon dioxide, sometimes play a role. The water may have been trapped in a parent sedimentary rock or given off by a cooling pluton. Water may also be given off from minerals that have water in their crystal structure (e.g., clay, mica). As temperature rises during metamorphism and a mineral becomes unstable, its water is released.

Water is thought to help trigger metamorphic chemical reactions. Water, moving through fractures and along grain margins, is a sort of intrarock rapid transit for ions. Under high pressure, it moves between grains, dissolves ions from one mineral, and then carries these ions elsewhere in the rock where they can react with the ions of a second mineral. The new mineral that forms is stable under the existing conditions.

Fluids can also carry new ions into a rock, effectively changing its composition. This process, known as metasomatism, is discussed later in this section.

Time

The effect of time on metamorphism is hard to comprehend. Most metamorphic rocks are composed predominantly of silicate minerals, and silicate compounds are notorious for their sluggish chemical reaction rates. When garnet crystals taken from a metamorphic rock collected in Vermont were analyzed, scientists calculated a growth rate of 1.4 millimeters per million years. The garnets' growth was sustained over a 10.5-million-year period. Because the rate of metamorphic reactions can be so slow, it is possible to find metamorphic rocks with different mineral assemblages that formed under the same pressure and temperature conditions from identical parent rocks. This occurs when the duration of metamorphism varies from one location to another.

Many laboratory attempts to duplicate metamorphic reactions believed to occur in nature have been frustrated by the time element. The several million years during which a particular combination of temperature and pressure may have prevailed in nature are impossible to duplicate.

7.3 CLASSIFICATION OF METAMORPHIC ROCKS

As we noted before, the kind of metamorphic rock that forms is determined by the metamorphic environment (primarily the particular combination of pressure, stress, and temperature) and by the chemical constituents of the parent rock. Many kinds of metamorphic rocks exist because of the many possible combinations of these factors.

TABLE 7.1 Classification and Naming of Metamorphic Rocks (Based Primarily on Texture)

Nonfoliated

Name Based on Mineral Content of Rock

Usual Parent Rock	Rock Name	Predominant Minerals	Identifying Characteristics
Limestone	Marble	Calcite	Coarse interlocking grains of calcite (or, less commonly, dolomite).
Dolomite	Dolomite marble	Dolomite	Calcite (or dolomite) has rhombohedral cleavage; hardness intermediate between glass and fingernail. Calcite effervesces in weak acid.
Quartz sandstone	Quartzite	Quartz	Rock composed of interlocking small granules of quartz. Has a sugary appearance and vitreous luster; scratches glass.
Shale	Hornfels	Fine-grained micas	A fine-grained, dark rock that generally will scratch glass. May have a few coarser minerals present.
Basalt	Hornfels	Fine-grained ferromagnesian minerals, plagioclase	

Foliated

Name Based Principally on Kind of Foliation Regardless of Parent Rock. Adjectives Describe the Composition (e.g., biotite-garnet schist)

Texture	Rock Name	Typical Characteristic Minerals	Identifying Characteristics
Slaty	Slate	Clay and other sheet silicates	A very fine-grained rock with an earthy luster. Splits easily into thin, flat sheets.
Intermediate between slaty and schistose	Phyllite	Mica	Fine-grained rock with a silky luster. Generally splits along wavy surfaces.
Schistose	Schist	Biotite and muscovite amphibole	Composed of visible platy or elongated minerals that show planar alignment. A wide variety of minerals can be found in various types of schist (e.g., garnet-mica schist, hornblende schist, etc.).
Gneissic	Gneiss	Feldspar, quartz, amphibole, biotite	Light and dark minerals are found in separate, parallel layers or lenses. Commonly, the dark layers include biotite and hornblende; the light-colored layers are composed of feldspars and quartz. The layers may be folded or appear contorted.

Metamorphic rocks, just like igneous and sedimentary rocks, are classified according to their texture and composition. (Appendix B contains a systematic procedure for identifying common metamorphic rocks.) Let us first consider texture. The relationship of texture to rock name is summarized in table 7.1. The first question to ask when looking at the texture of a metamorphic rock is, *is it foliated or nonfoliated* (figure 7.8)?

A

B

FIGURE 7.8

Photomicrographs taken through a polarizing microscope of metamorphic rocks. (*A*) Nonfoliated rock and (*B*) Foliated rock. Multicolored grains are biotite mica; gray and white are mostly quartz.
©*Lisa Hammersley*

Nonfoliated Rocks

If the rock is nonfoliated, it is named on the basis of its composition. The two most common nonfoliated rocks are marble and quartzite, composed, respectively, of calcite and quartz.

Marble, a coarse-grained rock composed of interlocking calcite crystals (figure 7.9), forms when limestone recrystallizes during metamorphism. If the parent rock is dolomite, the recrystallized rock is a *dolomite marble.* Marble has long been valued as a building material and as a material for sculpture (figure 7.9*B*), partly because it is easily cut and polished and partly because it reflects light in a shimmering pattern, a result of the excellent cleavage of the individual calcite crystals. Marble is, however, highly susceptible to chemical weathering (see chapter 5).

Quartzite (figure 7.10) is produced when grains of quartz in sandstone are welded together while the rock is subjected to high temperature. This makes it as difficult to break along grain boundaries as through the grains. Therefore, quartzite, being as hard as a single quartz crystal, is difficult to crush or break. It is the most durable of common rocks used for construction, both because of its hardness and because quartz is not susceptible to chemical weathering.

Hornfels is a very fine-grained, nonfoliated, metamorphic rock whose parent rock is most commonly shale or basalt. If it forms from shale, characteristically only microscopically visible micas form from the shale's clay minerals. Sometimes a few minerals grow large enough to be seen with the naked eye; these are minerals that are especially capable of crystallizing under the particular temperature attained during metamorphism. If hornfels forms from basalt, amphibole, rather than mica, is the predominant fine-grained mineral produced.

Foliated Rocks

If the rock is foliated, you need to determine the type of foliation to name the rock. For example, a rock with a schistose texture is called a *schist.* But this name tells us nothing about what minerals are in this rock, so we add adjectives to describe the composition–for example, *garnet-mica schist.* The following are the most common foliated rocks progressing from lower grade (lower pressures and temperatures) to higher grade (higher pressures and temperatures):

Slate is a very fine-grained rock that splits easily along flat, parallel planes (figure 7.11). Although some slate forms from volcanic ash, the usual parent rock is shale. Slate develops under temperatures and pressures only slightly greater than those found in the sedimentary realm. The temperatures are not high enough for the rock to thoroughly recrystallize. The important controlling factor is differential stress. The original clay minerals partially recrystallize into equally fine-grained, platy minerals. Under differential stress, the old and new platy minerals are aligned, creating slaty cleavage in the rock. A slate indicates that a relatively cool and brittle rock has been subjected to intense tectonic activity.

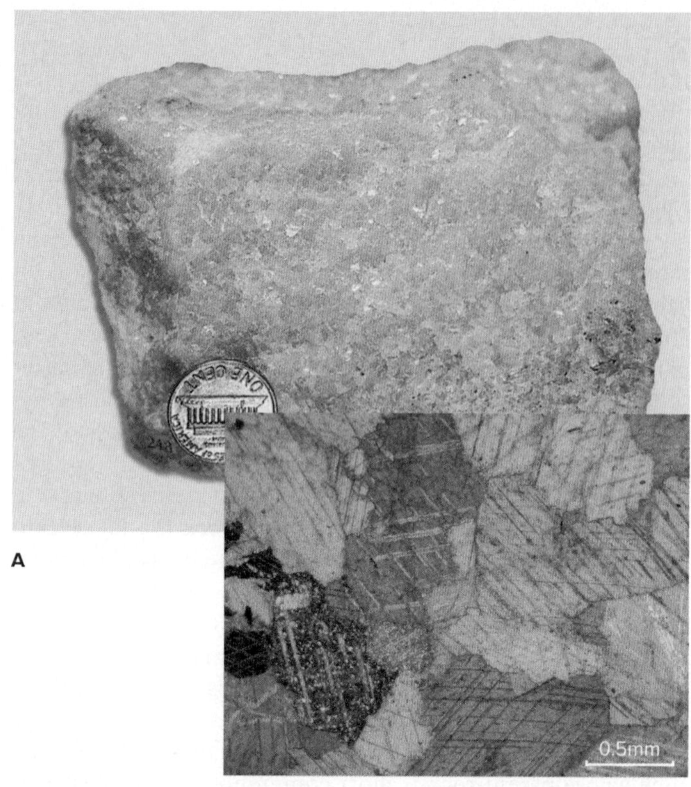

FIGURE 7.9

(*A*) Hand specimen of marble. Inset is a photomicrograph showing interlocking crystals of calcite. (*B*) Close-up of *David* by Michelangelo at the piazza della Signoria, Italy.
(A) ©C.C. Plummer (inset) ©Lisa Hammersley (B) ©Stefano Cellai/Shutterstock

FIGURE 7.10

Quartzite. Inset shows photomicrograph taken using a polarizing microscope.
©C.C. Plummer (inset) ©Lisa Hammersley

FIGURE 7.11

Slate outcrop in Antarctica. Inset is hand specimen of slate.
Source: P. D. Rowley, U.S. Geological Survey (inset) ©Tyler Boyes/Shutterstock

Because of the ease with which it can be split into thin, flat sheets, slate is used for making chalkboards, pool tables, and roof tiles.

Phyllite is a rock in which the newly formed micas are larger than the platy minerals in slate but still cannot be seen with the naked eye. This requires a further increase in temperature over that needed for slate to form. The very fine-grained mica imparts a satin sheen to the rock, which may otherwise closely resemble slate (figure 7.12). But the slaty cleavage may be crinkled in the process of conversion of slate to phyllite.

A **schist** is characterized by coarser-grained minerals that are approximately parallel-oriented. Platy or elongate minerals that crystallize from the parent rock are clearly visible to the naked eye. Which minerals form depends on the particular combination of temperature and pressure prevailing during recrystallization as well as the composition of the parent rock. Two, of several, schists that form from shale are *mica schist* and *garnet-mica schist* (figure 7.13). Although they both have the same parent rock, they form under different combinations of temperature and pressure. If the parent rock is basalt, the schists that form are quite different. If the predominant ferromagnesian mineral that forms during metamorphism of basalt is amphibole, the metamorphic rock is an *amphibole schist.* At a lower grade, the predominant mineral is chlorite, a green micaceous mineral, in a *chlorite schist.*

FIGURE 7.12

Phyllite, exhibiting a crinkled, silky-looking surface.
©C.C. Plummer

FIGURE 7.13

Garnet-mica schist. Small, subparallel flakes of muscovite mica reflect light. Garnet crystals give the rock a "fish-scale" appearance.
©C.C. Plummer

FIGURE 7.14

Gneiss.
©Deposit Photos/Glow Images RF

Gneiss is a rock consisting of light and dark mineral layers or lenses. It forms under temperature and pressure conditions that are high enough to cause the minerals to separate into layers. Platy or elongate minerals (such as mica or amphibole) in dark layers alternate with layers of light-colored minerals of no particular shape. Usually, coarse feldspar and quartz are predominant within the light-colored layers. In composition, a gneiss may resemble granite or diorite, but it is distinguishable from those plutonic rocks by its foliation (figure 7.14).

Temperature conditions under which a gneiss develops approach those at which granite solidifies. It is not surprising, then, that the same minerals are found in gneiss and in granite. In fact, a previously solidified granite can be converted to a gneiss under appropriate pressure and temperature conditions and if the rock is under differential stress.

7.4 TYPES OF METAMORPHISM

Metamorphism can occur in a wide range of environments, each of which results in different pressure and temperature conditions. Metamorphism can be broadly classified into types based upon similarities in setting and metamorphic conditions. There are many types of metamorphism, including shock metamorphism, which occurs during meteorite impacts (box 7.1). Here we discuss three important types of metamorphism: contact metamorphism, hydrothermal metamorphism, and regional metamorphism.

PLANETARY GEOLOGY 7.1

Impact Craters and Shock Metamorphism

The spectacular collision of the comet Shoemaker-Levy with Jupiter in 1994 served to remind us that asteroids and comets occasionally collide with a planet. Earth is not exempt from collisions. Large meteorites have produced impact craters when they have collided with Earth's surface. One well-known meteorite crater is Meteor Crater in Arizona, which is a little more than a kilometer in diameter (box figure 1). Many much larger craters are known in Canada, Germany, Australia, and other places.

Impact craters display an unusual type of metamorphism called *shock metamorphism*. The sudden impact of a large extraterrestrial body results in brief but extremely high pressures. Quartz may recrystallize into the rare SiO_2 minerals coesite and stishovite. Quartz that is not as intensely impacted suffers damage (detectable under a microscope) to its crystal lattice.

The impact of a meteorite also may generate enough heat to locally melt rock. Molten blobs of rock are thrown into the air and become streamlined in the Earth's atmosphere before solidifying into what are called *tektites*. Tektites may be found hundreds of kilometers from the point of meteorite impact.

BOX 7.1 ■ FIGURE 1
Meteor Crater in Arizona. Diameter of the crater is 1.2 kilometers.
©Getty RF

A large meteorite would blast large quantities of material high into the atmosphere. According to theory, the change in global climate due to a meteorite impact around 65 million years ago caused extinctions of many varieties of creatures (see box 8.2 on the extinction of dinosaurs). Evidence for this impact includes finding tiny fragments of shock metamorphosed quartz and tektites in sedimentary rock that is 65 million years old.

The intense shock caused by a meteorite creates large faults that can be filled with crushed and partially melted rocks. One of the largest such structures, at Sudbury, Ontario, is the host for very rich metallic ore deposits.

Shock metamorphosed rock fragments are much more common on the Moon than on Earth. There may be as many as 400,000 craters larger than a kilometer in diameter on the Moon. Mercury's surface is remarkably similar to that of the Moon. Our two neighboring planets, Venus and Mars, are not as extensively cratered as is the Moon. This is because these planets, like Earth, have been tectonically active since the time of greatest meteorite bombardment, about 4 billion years ago. If Earth had not been tectonically active and if we didn't have an atmosphere driving erosion, Earth would have around sixteen times the number of meteorite craters as the Moon and would appear just as pockmarked with craters.

Additional Resources

Meteor Crater
Website for Meteor Crater in Arizona
- www.meteorcrater.com/

Shock Metamorphism
Website including descriptions and photos of shock metamorphism features
- http://www.impact-structures.com/shock-metamorphism-page/

Contact Metamorphism

Contact metamorphism (also known as *thermal* metamorphism) is metamorphism in which high temperature is the dominant factor. Confining pressure may influence which new minerals crystallize; however, the confining pressure is usually relatively low. This is because contact metamorphism mostly takes place not too far beneath Earth's surface (less than 10 kilometers). Contact metamorphism occurs adjacent to a pluton when a body of magma intrudes relatively cool country rock (figure 7.15*A*). The process can be thought of as the "baking" of country rock adjacent to an intrusive contact; hence, the term *contact metamorphism.* The zone of contact metamorphism (also called an *aureole*) is usually quite narrow—generally from 1 to 100 meters wide. Differential stress is rarely significant. Therefore, the most common rocks found in an aureole are the nonfoliated rocks: marble when igneous rock intrudes limestone, quartzite when quartz sandstone is metamorphosed, hornfels when shale is baked.

Marble and quartzite also form under conditions of regional metamorphism. When grains of calcite or quartz recrystallize, they tend to be equidimensional, rather than elongate or platy. For this reason, marble and quartzite do not usually exhibit visible foliation, even though subjected to differential stress during metamorphism.

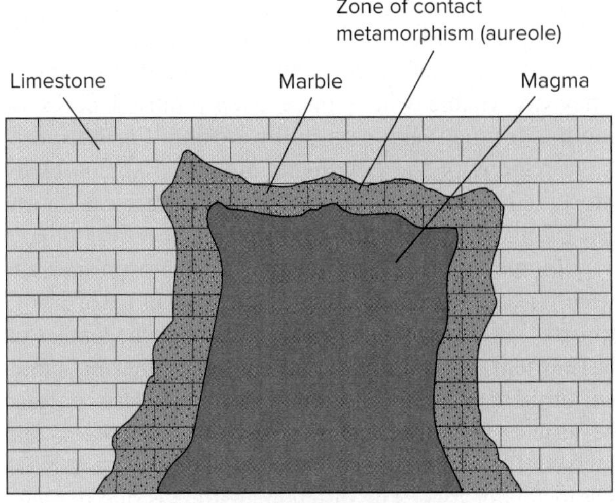

A

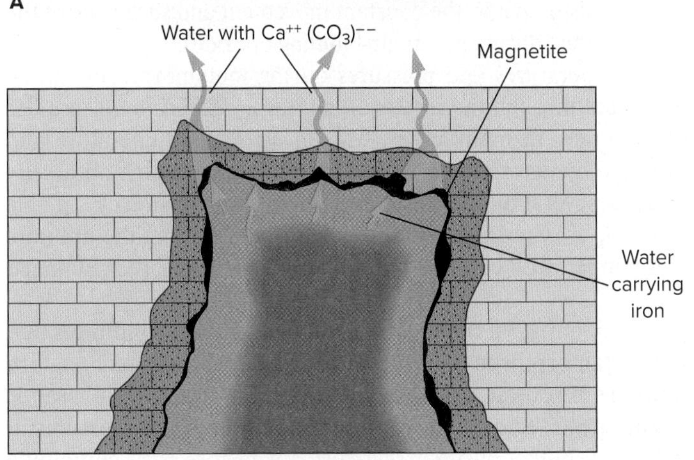

B

FIGURE 7.15

(*A*) Contact metamorphism. Magma intrudes country rock (limestone), and marble forms along contact. (*B*) Metasomatism. As magma solidifies, gases bearing ions of iron leave the magma, dissolve some of the marble, and deposit iron as magnetite.

Hydrothermal Metamorphism

As described earlier, hot water is involved to some extent in most metamorphic processes. As shown in table 7.2, hydrothermal fluids can metamorphose an existing rock or can deposit entirely new minerals in open spaces. Water is important during **hydrothermal metamorphism** because it can transport ions from one mineral to another and increase the rate of metamorphism.

Metasomatism is metamorphism coupled with the introduction of *ions* from an external source. The ions are brought in by water from outside the immediate environment and are incorporated into the newly crystallizing minerals. Often, metasomatism involves ion exchange. Newly crystallizing minerals replace preexisting ones as water simultaneously dissolves and replaces ions.

When metasomatism is associated with contact metamorphism, the ions are introduced from a cooling magma. Some important commercially mined deposits of metals such as iron, tungsten, copper, lead, zinc, and silver are attributed to metasomatism. Figure 7.15*B* shows how magnetite (iron oxide) ore bodies have formed through metasomatism. Ions of the metal are transported by water and react with minerals in the host rock. Elements within the host rock are simultaneously dissolved out of the host rock and replaced by the metal ions brought in by the fluid. Because of the solubility of calcite, marble commonly serves as a host for metasomatic ore deposits called skarns.

TABLE 7.2 Hydrothermal Processes

Role of Water	Name of Process or Product
Water transports ions between grains in a rock undergoing metamorphism. Some water may be incorporated into crystal structures.	Metamorphism
Water brings ions from outside the parent rock, and they are added to the rock during metamorphism. Other ions may be dissolved and removed.	Metasomatism
Water passes through cracks or pore spaces in rock and precipitates minerals on the walls of cracks and within pore spaces.	Hydrothermal veins and disseminated deposits

Hot water also plays an important role in creating new rocks and minerals. These form entirely by precipitation of ions derived from hydrothermal solutions. *Hydrothermal minerals* can form in void spaces or between the grains of a host rock. An aggregate of hydrothermal minerals may form within a preexisting fracture in a rock to form a **hydrothermal vein**.

Quartz veins are especially common where igneous activity has occurred (figure 7.16). Hot water can come from the cooling magma or from groundwater heated by a pluton and circulated by convection. Where the water is hottest, rock in contact with it is partially dissolved. As the hot water travels upward toward Earth's surface, usually along fractures, temperature and pressure decrease. Fewer ions can be carried in solution as the water cools, so minerals will precipitate onto the walls of the fractures. Most commonly, silica (SiO_2) dissolves in the very hot water, then will cake on the walls of cracks to form quartz veins.

Hydrothermal Processes and Ore Deposits

Hydrothermal veins are very important economically. In them, we find most of the world's great deposits of zinc, lead, silver, gold, tungsten, tin, mercury, and, to some extent, copper (see figure 7.17). Ore minerals containing these metals are usually found in quartz veins.

Some ore-bearing solutions percolate upward between the grains of the rock and deposit very fine grains of ore mineral throughout. These are called *disseminated ore deposits.* Usually, metallic sulfide ore minerals are distributed in very low concentration through large volumes of rock, both above and within a pluton. Most of the world's copper comes from disseminated deposits, also called *porphyry copper deposits,* because the associated pluton is usually porphyritic. Other metals, such as lead, zinc, molybdenum, silver, and gold may be deposited along with

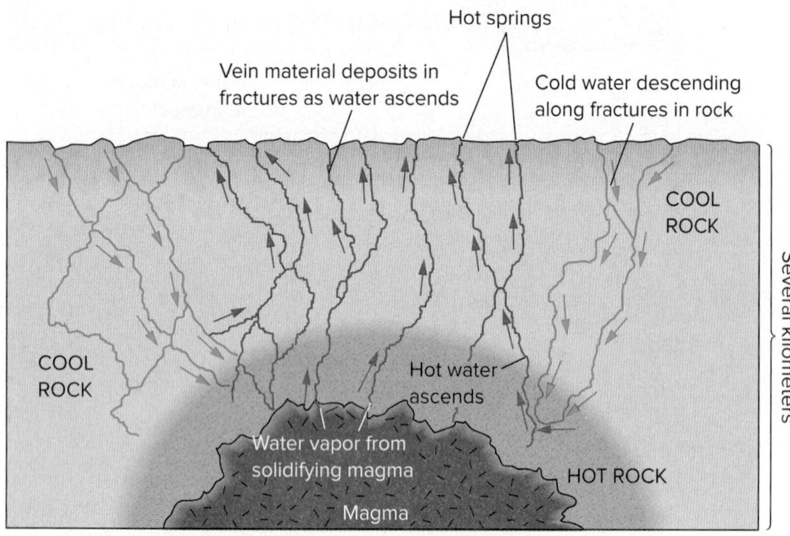

FIGURE 7.16

How veins form. Cold water descends, is heated, dissolves material, ascends, and deposits material as water cools and pressure drops upon ascending.

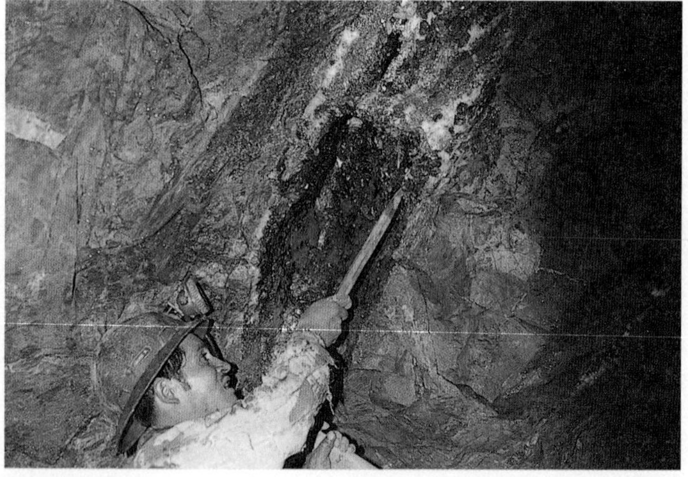

FIGURE 7.17

A wide vein that contains masses of sphalerite (dark), pyrite and chalcopyrite (both shiny yellow), as well as white quartz, in the Casapalca mine in Peru. It was mined for zinc and copper.
©Brian Skinner

copper. Some very large gold mines are also in disseminated ore deposits.

Regional Metamorphism

The great majority of the metamorphic rocks found on Earth's surface are products of **regional metamorphism**, which is metamorphism that takes place over wide areas and generally at considerable depth in the crust. Regional metamorphism is commonly associated with mountain building at convergent plate boundaries, and regional metamorphic rocks are almost always foliated, indicating differential stress during recrystallization. Metamorphic rocks are prevalent in the most intensely deformed portions of mountain ranges. They are visible where once deeply buried cores of mountain ranges are exposed by erosion. Furthermore, large regions of the continents are underlain by metamorphic rocks, thought to be the roots of ancient mountains long since eroded down to plains or rolling hills.

Temperatures during regional metamorphism vary widely. Usually, the temperatures are in the range of 300°C to 800°C. Temperature at a particular place depends to a large extent on depth of burial and the geothermal gradient of the region. Locally, temperature may also increase because of heat given off by nearby magma bodies. The high confining pressure is due to burial under five or more kilometers of rock. The differential stress is due to tectonism; that is, the constant movement and squeezing of the crust during mountain-building episodes.

Temperatures and pressures during metamorphism can be estimated through the results of laboratory experimental studies of minerals. In many cases, we can estimate temperature and pressure by determining the conditions under which an assemblage of several minerals can coexist. In some instances, a single mineral, or *index mineral,* suffices for determining the pressure and temperature combination under which a rock recrystallized (box 7.2).

Depending on the pressure and temperature conditions during metamorphism, a particular parent rock may recrystallize into one of several metamorphic rocks. For example, if basalt is metamorphosed at relatively low temperatures and pressures, it will recrystallize into a *greenschist,* a schistose rock containing chlorite (a green sheet-silicate), actinolite (a green amphibole), and sodium-rich plagioclase. Or it will recrystallize into a *greenstone,* a rock that has similar minerals but is not foliated. (A greenstone would indicate that the tectonic forces were not strong enough to induce foliation while the basalt was recrystallizing.) At higher temperatures and pressures, the same basalt would recrystallize into an *amphibolite,* a rock composed of hornblende, plagioclase feldspar, and, perhaps, garnet. Metamorphism of other parent rocks under conditions similar to those that produce amphibolite from basalt should produce the metamorphic rocks shown in table 7.3.

7.5 METAMORPHIC GRADE

The minerals present in a metamorphic rock indicate its *metamorphic grade.* Low-grade rocks formed under relatively cool temperatures and high-grade rocks at high temperatures, whereas medium-grade rocks recrystallized at around the middle of the range of metamorphic temperatures.

When a rock becomes buried to increasingly greater depths, it is subjected to increasingly greater temperatures and pressures and will undergo *prograde metamorphism*—that is, it progressively recrystallizes into a higher-grade rock. As an example, consider figure 7.18, which shows what happens to a shale as it

TABLE 7.3 Regional Metamorphic Rocks that Form Under Approximately Similar Pressure and Temperature Conditions

Parent Rock	Rock Name	Predominant Minerals
Basalt	Amphibolite	Hornblende, plagioclase, garnet
Shale	Mica schist	Biotite, muscovite, quartz, garnet
Quartz sandstone	Quartzite	Quartz
Limestone or dolomite	Marble	Calcite or dolomite

is buried deeper and deeper in the Earth's crust during regional metamorphism.

Slate, which looks quite similar to the shale from which it forms, is the first metamorphic rock to form, and the lowest grade. Its slaty cleavage develops as a result of differential stress during incipient recrystallization of clay minerals to other platy minerals. As described earlier, phyllite is a rock that is transitional between slate and schist and, as such, we expect it to have formed at a depth between where slate and schist form.

Schist forms at higher temperatures and usually higher pressures than does phyllite. However, schist with shale as a parent rock forms over a wide range of temperatures and pressures. Figure 7.18 indicates the metamorphic setting for two varieties of schist (there are a number of others) that form from the same shale. For this particular composition of shale, *mica schist* indicates a grade of metamorphism slightly higher than that of phyllite. Garnet requires higher temperatures to crystallize in a schist, so the *garnet-mica schist* probably formed at a deeper level than that of mica schist.

If schist is subjected to high enough temperatures, its constituents become more mobile and the rock recrystallizes into gneiss. The constituents of feldspar migrate (probably as ions) into planes of weakness caused by differential stress where feldspars, along with quartz, crystallize to form light-colored layers. The ferromagnesian minerals remain behind as the dark layers.

If the temperature is high enough, partial melting of rock may take place, and a magma collects in layers within the foliation planes of the solid rock. After the magma solidifies, the rock becomes a **migmatite**, a mixed igneous and metamorphic rock (figure 7.19). A migmatite can be thought of as a "twilight

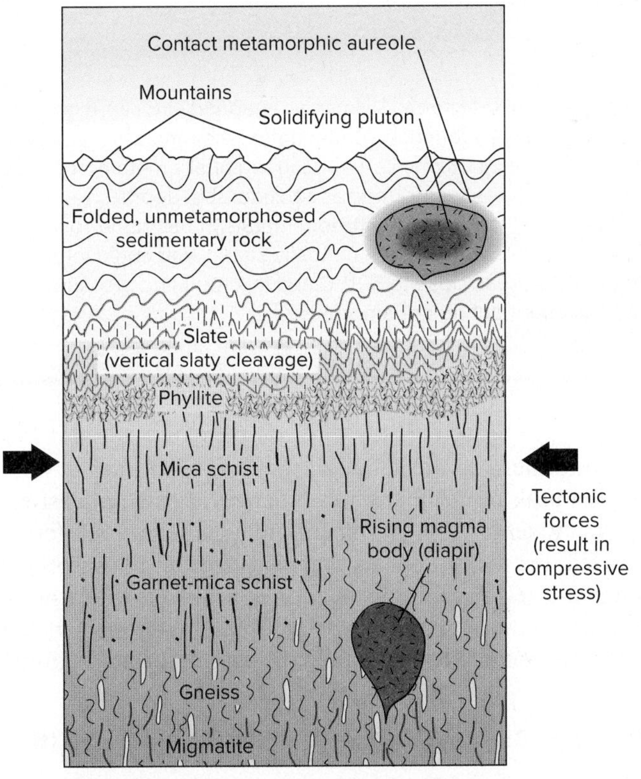

FIGURE 7.18

Schematic cross section representing an approximately 30-kilometer portion of Earth's crust during metamorphism. Rock names given are those produced from shale.

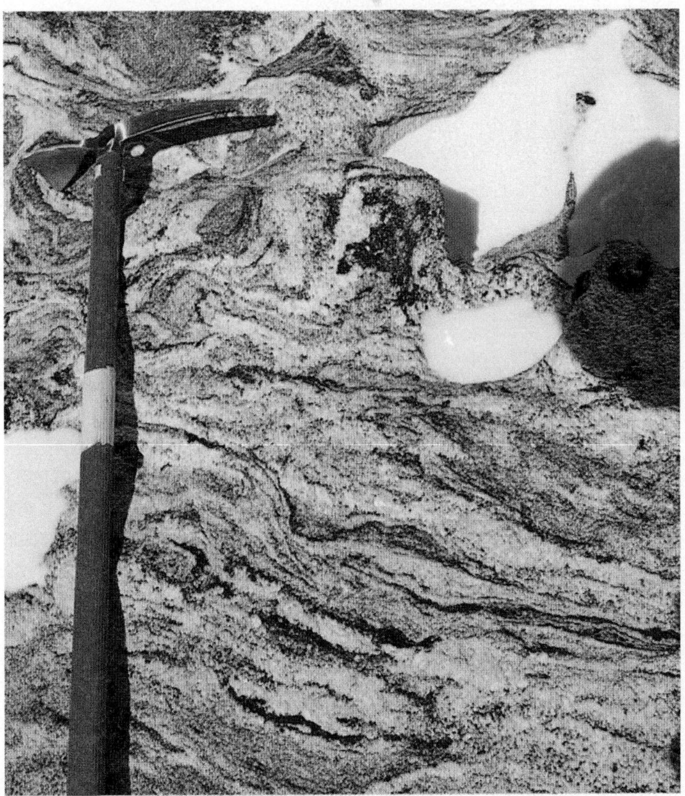

FIGURE 7.19

Migmatite in the Daniels Range, Antarctica.
©C.C. Plummer

IN GREATER DEPTH 7.2

Index Minerals

Certain minerals can only form under a restricted range of pressure and temperature. Stability ranges of these minerals have been determined in laboratories. When found in metamorphic rocks, these minerals can help us infer, within limits, what the pressure and temperature conditions were during metamorphism. For this reason, they are known as *index minerals.* Among the best known are *andalusite, kyanite,* and *sillimanite.* All three have an identical chemical composition (Al_2SiO_5) but different crystal structures (they are *polymorphs*). They are found in metamorphosed shales that have an abundance of aluminum. Box figure 1 is a phase diagram showing the pressure-temperature fields in which each is stable. Box figure 2 is a map showing metamorphic patterns across the Grenville Province of the Canadian Shield. These patterns were established using the minerals andalusite-sillimanite-kyanite.

If andalusite is found in a rock, this indicates that pressures and temperatures were relatively low. Andalusite is often found in contact metamorphosed shales (hornfels). Kyanite, when found in schists, is regarded as an indicator of high pressure; but note that the higher the temperature of the rock, the greater the pressure needed for kyanite to form. Sillimanite is an indicator of high temperature and can be found in some contact metamorphic rocks adjacent to very hot intrusions as well as in regionally metamorphosed schists and gneisses that formed at considerable depths.

Note that if you found all three minerals in the same rock and could determine that they were mutually stable, you could infer that the temperature was close to 500°C and the confining pressure was almost 4 kilobars during metamorphism.

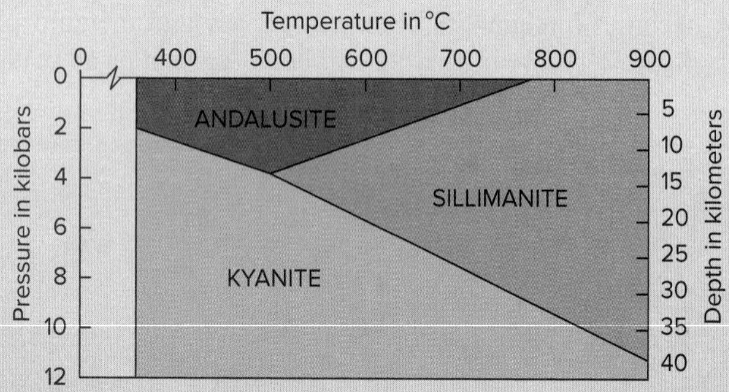

BOX 7.2 ■ FIGURE 1

Phase diagram showing the stability relationships for the Al_2SiO_5 minerals.
Source: M. J. Holdaway, American Journal of Science, 1971, Vol. 271.

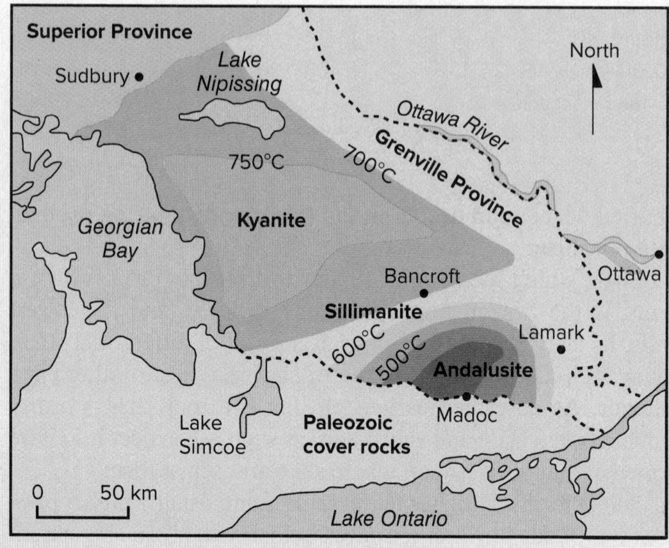

BOX 7.2 ■ FIGURE 2

Regional metamorphic patterns across the Grenville Province of the Canadian Shield. Colored bands represent reconstructed burial temperatures based on minerals present in the metamorphic rocks. Higher grades of metamorphism occur in the west of the Grenville Province and indicate deeper burial and higher temperatures in that area.
Nick Eyles

zone" rock that is neither fully igneous nor entirely metamorphic.

The metamorphic rocks that we see usually have minerals that formed at or near the highest temperature reached during metamorphism. But why doesn't a rock recrystallize to one that is stable at lower temperature and pressure conditions during its long journey to the surface, where we now find it? The answer is that water is usually available during prograde metamorphism, but it is driven off in high-temperature environments, and the rock may be relatively dry after reaching its peak temperatures. The absence of water means that chemical reactions will be very slow at the cooler temperatures, limiting the degree to which the rock can adjust to its new, lower-grade conditions. Substantial *retrograde metamorphism* only occurs if additional water is introduced to the rock after peak metamorphism. Tectonic forces at work during the peak of metamorphism fracture the rock extensively and permit water to get to the mineral grains. After tectonic forces are relaxed, the rocks move upward as a large block as isostatic adjustment takes place. Rocks that do indicate retrograde metamorphism are those that recrystallized under lower temperature and pressure conditions than during the peak of metamorphism.

Pressure and Temperature Paths in Time

Index minerals (box 7.2) and mineral assemblages in a rock can be used to determine the approximate temperature and pressure conditions that prevailed during metamorphism. Precise determination of the chemical composition of some minerals can determine the temperature or pressure present during the growth of a particular mineral. The usual basis for determining temperature (*geothermometry*) or pressure (*geobarometry*) during mineral growth is the ratio of pairs of elements (e.g., Fe and Al) within the crystal structure of the mineral.

Modern techniques allow us to determine chemical compositional changes across a grain of a mineral in a rock. An *electron microprobe* is a microscope that allows the user to focus on a tiny portion of a mineral in a rock, then shoot a very narrow beam of electrons into that point in the mineral. The extent and manner in which the beam is absorbed by the mineral are translated, using specialized software, into the precise chemical composition of the mineral at that point. If the mineral is *zoned* (that is, the chemical composition changes within the mineral, as described in chapter 2), the electron microprobe will indicate the differing composition within the mineral grain.

A mineral grows from the center outward, adding layers of atoms as it becomes larger. If pressure and temperature conditions change as the mineral grows, the chemical composition of the outer layers may be different from that of the inner layers, reflecting those changes. Figure 7.20 shows the results of one such study. The diagram shows the changes of temperature and pressure experienced by a rock during burial, metamorphism, and subsequent uplift. The line represents the temperature-pressure-time path. If chemical analysis of the zones in the mineral indicates that pressure and temperature are both increasing, this indicates the rock is being buried deeper while becoming hotter. If temperature and pressure are both decreasing, the rock is cooling down at the same time that pressure is being reduced because of erosion at Earth's surface.

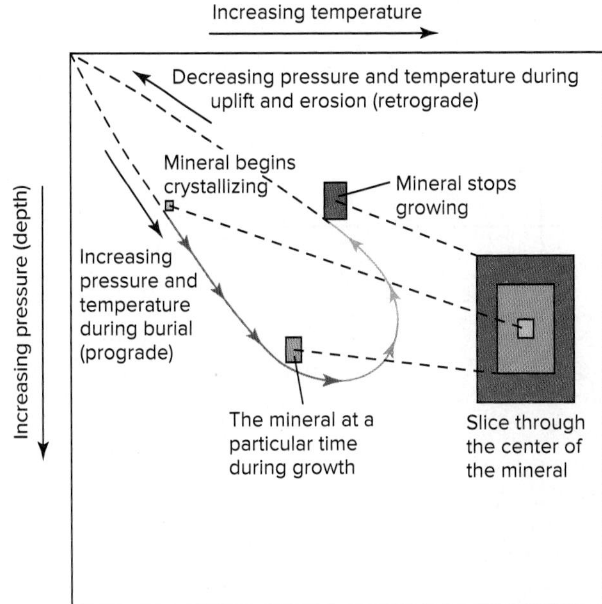

FIGURE 7.20

Pressure-temperature-time path for growth of a mineral during metamorphism. An electron microprobe is used to determine the precise chemical composition of the concentric zones of the mineral. The data are used to determine the pressure and temperature during the growth of the mineral. Three stages during the growth of the mineral are correlated to the graph—beginning of growth (center of crystal), an arbitrary point during its growth, and the end of crystallization (the outermost part of the crystal). The green segment of the path indicates increasing pressure and temperature during metamorphism. The orange segment indicates that pressure was decreasing while temperature continued to rise. The blue segment indicates temperature and pressure were both decreasing. The decrease in pressure is likely to be the result of uplift and erosion at the surface. The dashed lines are inferred pressure and temperature paths before and after metamorphism.

7.6 PLATE TECTONICS AND METAMORPHISM

Studies of metamorphic rocks have provided important information on conditions and processes within the lithosphere and have aided our understanding of plate tectonics. Conversely, plate tectonic theory has provided models that allow us to explain many of the observed characteristics of metamorphic rocks.

Foliation and Plate Tectonics

Figure 7.21 shows an oceanic-continental boundary (oceanic lithosphere is subducted beneath continental lithosphere). One of the things the diagram shows is where differential stress that is responsible for foliation is taking place. Shearing takes place in the subduction zone where the oceanic crust slides beneath continental lithosphere. For here, we infer that the sedimentary rocks and some of the basalt become foliated, during metamorphism, roughly parallel to the subduction zone (parallel to the lines in the diagram).

Within the thickest part of the continental crust shown in figure 7.21, flowage of rock is indicated by the purple arrows. The crust is thickest here beneath a growing mountain belt. The thickening is due to the compression caused by the two colliding plates. Within this part of the crust, rocks flow downward and then outward (as indicated by the arrows) in a process (described in chapter 20 on mountains) of *gravitational collapse and spreading*. Under this concept, the central part of a mountain belt becomes too high after plate convergence and is gravitationally unstable. This forces the rock downward and outward. Regional metamorphism takes place throughout, and we expect foliation in the recrystallizing rocks to be approximately parallel to the arrows.

Pressure-Temperature Regimes

Before the advent of plate tectonics, geologists were hard-pressed to explain how some rocks apparently were metamorphosed at relatively cool temperatures yet high pressures. We expect rocks to be hotter as they become more deeply buried. How could rocks stay cool, yet be deeply buried?

Figure 7.22 shows experimentally determined stability fields for a few metamorphic minerals. Line *x* indicates a common geothermal gradient during metamorphism. At the appropriate pressure and temperature, kyanite begins to crystallize in the rock. If it were buried deeper, its pressure and temperature would change along line *x*. Eventually, it would cross the stability boundary and sillimanite would crystallize rather than kyanite.

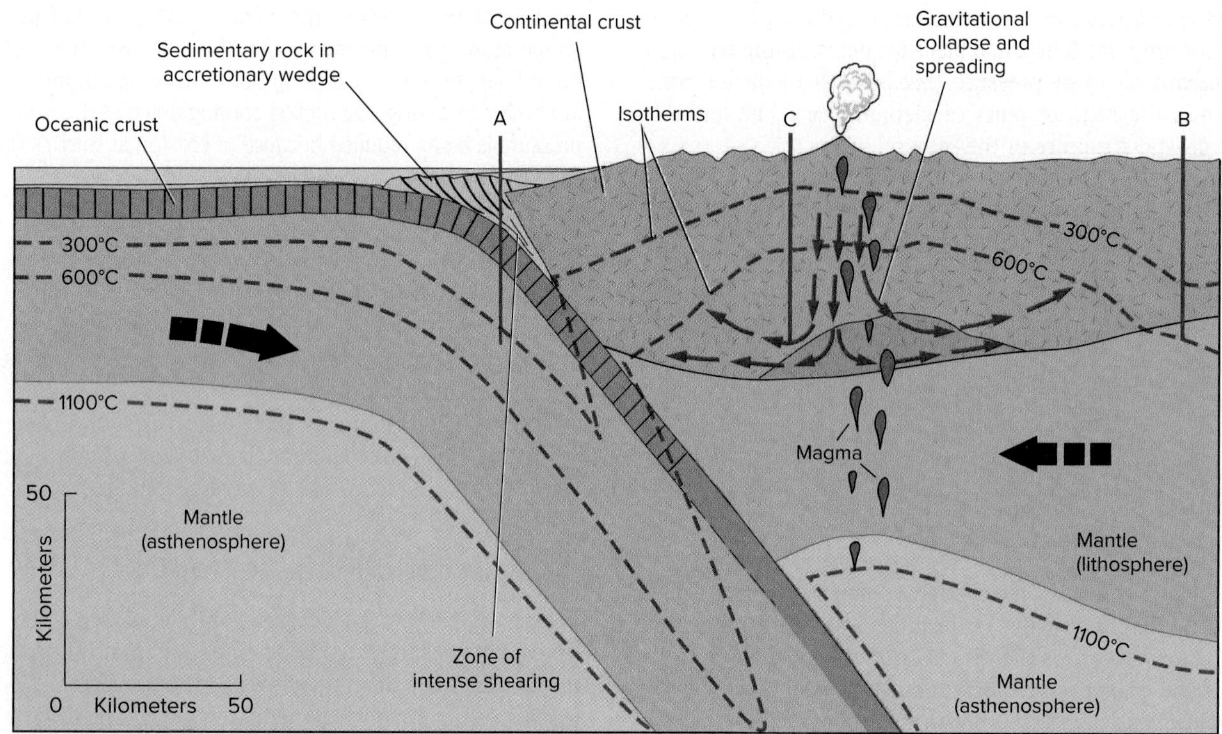

FIGURE 7.21

Metamorphism across a convergent plate boundary. All rock that is hotter than 300° or deeper than 5 kilometers is likely to be undergoing metamorphism.

Source: W. G. Ernst, Metamorphism and Plate Tectonics Regimes. Stradsburg, PA: Dowden, Hutchinson & Ross, 1975, p. 425.

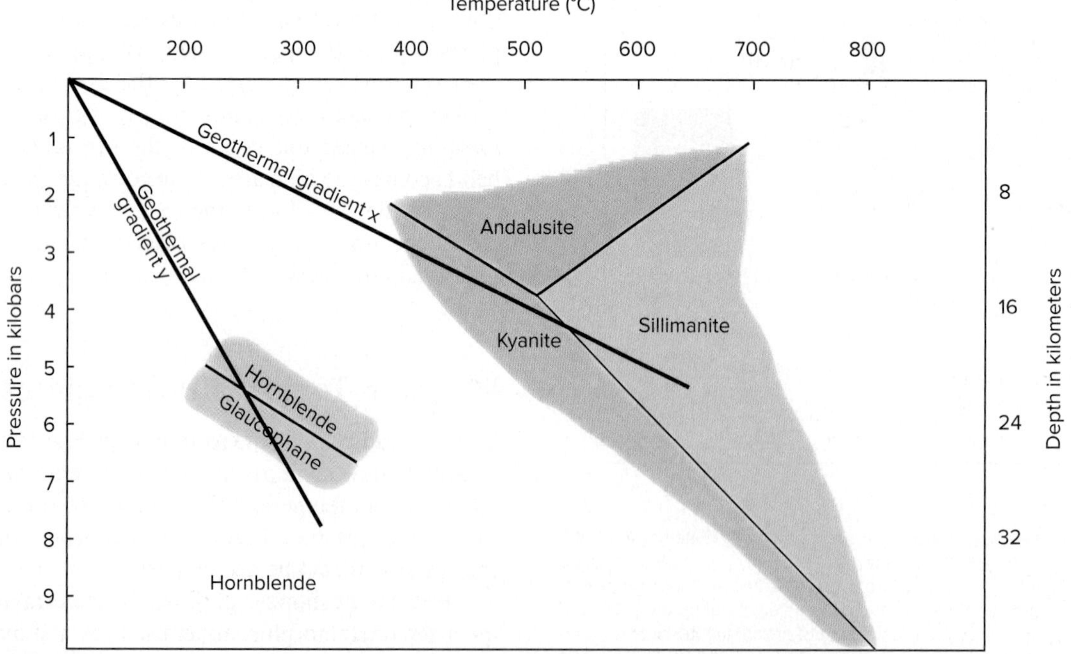

FIGURE 7.22

Stability fields for a few minerals. (Many more mineral stability fields can be used for increased accuracy.) The fields are based on laboratory research. Prograde metamorphism taking place with a geothermal gradient *x* involves a high temperature increase with increasing pressure. Prograde metamorphism under conditions of geothermal gradient *y* involves low temperature increase with increasing pressure. Hornblende is a calcium-bearing amphibole; glaucophane is a sodium-bearing amphibole.

By contrast, if a rock contains glaucophane (sodium-rich amphibole), rather than calcium-rich hornblende, the rock must have formed under high pressure but abnormally low temperature for its depth of burial. Line *y* represents a possible geothermal gradient that must have been very low, and the increase in temperature was small with respect to the increase in pressure.

If we return to figure 7.21, we can use it to see how plate tectonics explains these very different pressure-temperature regimes at a convergent boundary. Confining pressure is directly related to depth. For this reason, we expect the same pressure at any given depth. For example, the pressure corresponding to 20 kilometers is the same under a hot volcanic area as it is within the relatively cool rocks of a plate's interior. Temperature, however, is quite variable as indicated by the dashed red lines. Each of these lines is an **isotherm**, a line connecting points of equal temperature.

Each of the three places (*A, B,* and *C*) in figure 7.21 would have a different geothermal gradient. If you were somehow able to push a thermometer through the lithosphere, you would find the rock is hotter at shallower depths in areas with higher geothermal gradients than at places where the geothermal gradient is low. As indicated in figure 7.21, the geothermal gradient is higher, progressing downward through an active volcanic-plutonic complex (for instance, the Cascade Mountains of Washington and Oregon) than it is in the interior of a plate (e.g., beneath the Great Plains of North America). The isotherms are bowed upward in the region of the volcanic-plutonic complex because magma created at lower levels works its way upward and brings heat from the asthenosphere into the mantle and crust of the continental lithosphere. At point *C,* we would expect the metamorphism that takes place to result in minerals that reflect the high temperature relative to pressure conditions such as those along line *x* in figure 7.22.

If we focus our attention at the line at *A* in figure 7.21, we can understand how minerals can form under high pressure but relatively low-temperature conditions. You may observe that the bottom of line *A* is at a depth of about 50 kilometers, and if a hypothetical thermometer were here, it would read just over 300° because it would be just below the 300° isotherm. Compare this to vertical line *C* in the volcanic-plutonic complex. The confining pressure at the base of this line would be the same as at the base of line *A,* yet the temperature at the base of line *C* would be well over 600°. The minerals that could form at the base of line *A* would not be the same as those that could form at line *C.* Therefore, we would expect quite different metamorphic rocks in the two places, even if the parent rock had been the same.

So when we find high-pressure/low-temperature minerals (such as glaucophane) in a rock, we can infer that metamorphism took place while subduction carried basalt and overlying sedimentary rocks downward. Thus, plate tectonics accounts for the abnormally high-pressure/low-temperature geothermal gradients (such as line *y* in figure 7.22).

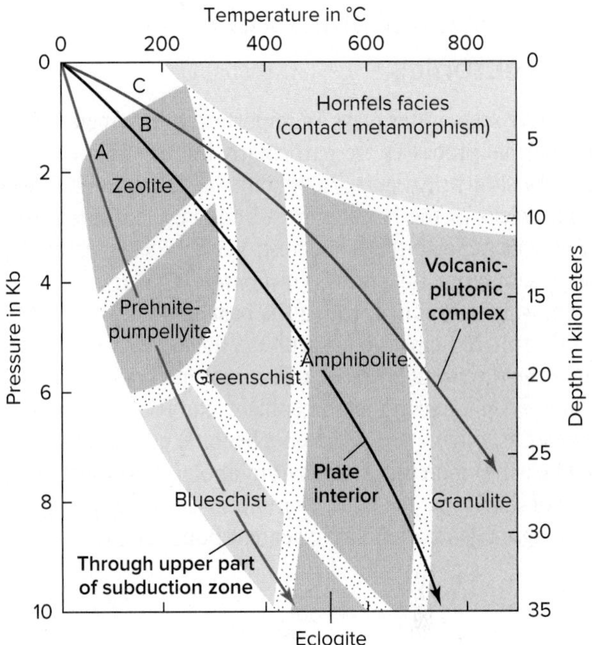

FIGURE 7.23

Metamorphic facies. Pressure and temperature stability fields for the metamorphic facies.

Metamorphic Facies

As you've seen above, the minerals present in a metamorphic rock can provide information about the tectonic setting in which metamorphism occurred. Rocks having the same mineral assemblage are regarded as belonging to the same **metamorphic facies**, implying that they formed under broadly similar pressure and temperature conditions (figure 7.23). The name for each facies is based on the assemblage of minerals or the name of a rock common to that facies. For instance, a rock composed mostly of the minerals chlorite, actinolite, and epidote (all of which are green minerals) belongs to the greenschist facies. On the other hand, rocks of the same chemical composition belonging to the amphibolite facies are largely made up of hornblende (a common type of amphibole) and garnet.

The concept of metamorphic facies is analogous to defining climatic zones by the combinations of plants found in each zone. A place where ferns, palm trees, and vines flourish corresponds to a climate with warm temperatures and abundant rainfall. On the other hand, a combination of palm trees, cactus, and sagebrush implies a hot, dry climate.

The concept of metamorphic facies preceded plate-tectonic theory by several decades. Although earlier geologists were able to relate the individual facies to pressure and temperature combinations, they had no satisfactory explanation for the environments that produced the various combinations. Figure 7.21, which relates the temperature of regional metamorphism to plate tectonics, may be used to infer the environment for each of the metamorphic facies shown in figure 7.23. The three paths marked *A, B,* and *C* in figure 7.23 correspond to the change in pressure and temperature conditions as you move from the top to the bottom of the three vertical lines marked *A, B,* and *C* in figure 7.21. As you move along each path, the grade of metamorphism increases.

Hydrothermal Metamorphism and Plate Tectonics

As you have seen, water plays an important role in metamorphism. Hydrothermal processes are particularly important at mid-oceanic ridges. As shown in figure 7.24, cold seawater moves downward through cracks in the basaltic crust and is cycled upward by heat from magma beneath the ridge crest. Very hot water returns to the ocean at submarine hot springs (*hydrothermal vents*).

Hot water traveling through the basalt and gabbro of the oceanic lithosphere helps metamorphose these rocks while they are close to the divergent boundary. This is sometimes called *seafloor metamorphism.* During metamorphism, the ferromagnesian igneous minerals, olivine and pyroxene, become converted to *hydrous* (water-bearing) minerals such as amphibole. An important consequence of this is that the hydrous minerals may eventually contribute to magma generation at convergent boundaries.

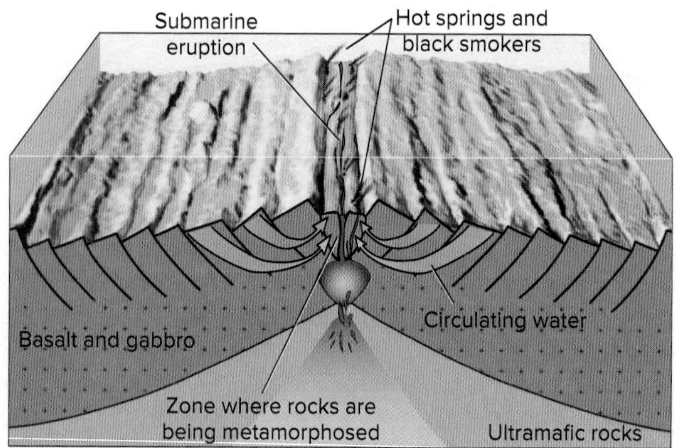

FIGURE 7.24

Cross section of a mid-oceanic ridge (divergent plate boundary). Water descends through fractures in the oceanic crust, is heated by magma and hot igneous rocks, and rises.

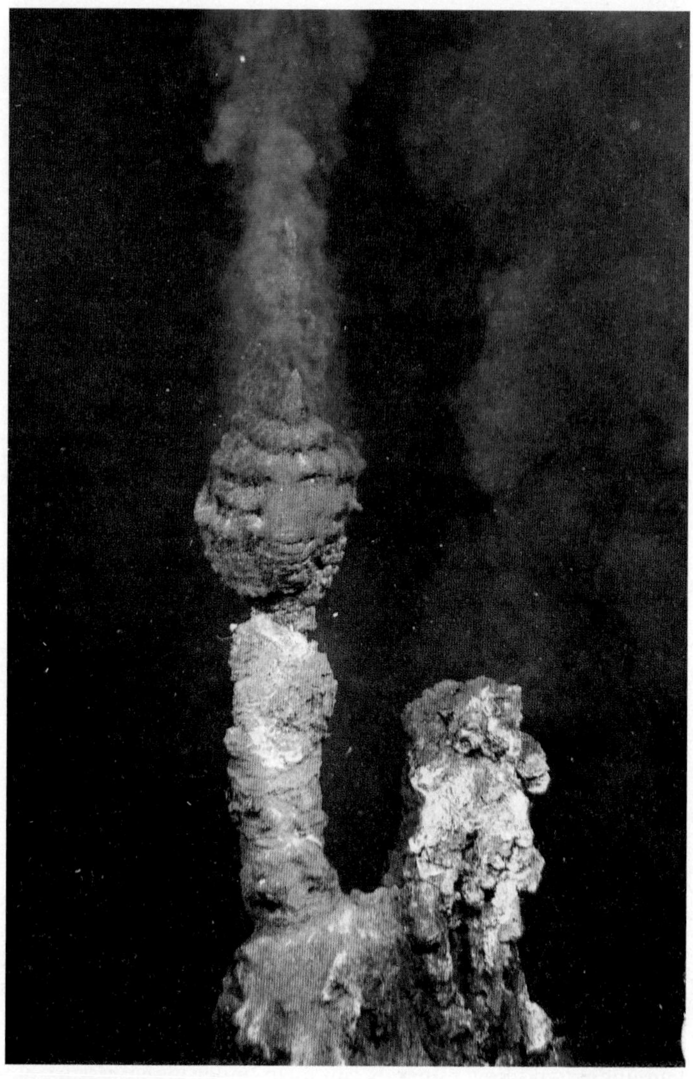

FIGURE 7.25

An example of a "black smoker" or hydrothermal vent. The "smoke" is a hot plume of metallic sulfide minerals being discharged into cold seawater from a chimney. *Source: Courtesy of New Zealand American Ring of Fire 2007 Exploration, NOAA Vents Program*

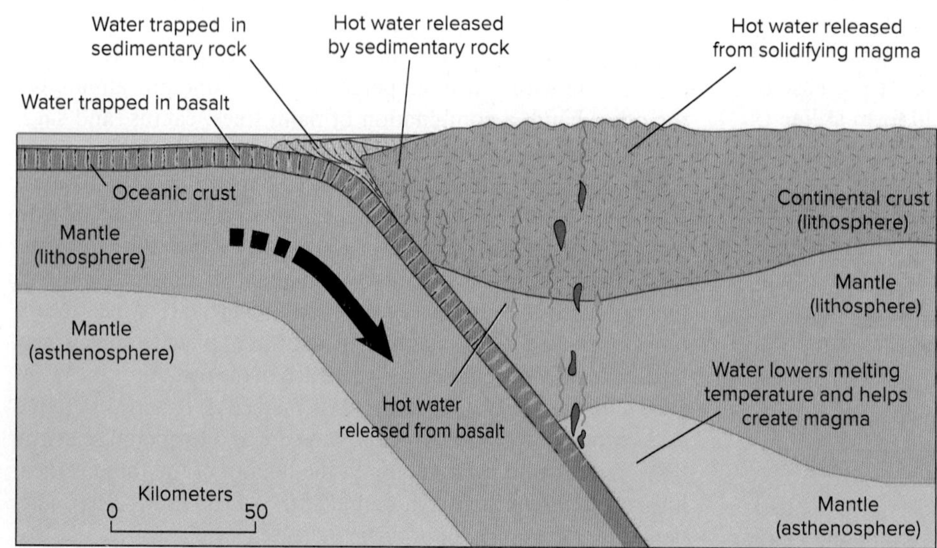

FIGURE 7.26

Water at a convergent boundary. Seawater trapped in the oceanic crust is carried downward and released upon heating at various depths within the subduction zone.

As the seawater moves through the crust, it also dissolves metals and sulfur from the crustal rocks and magma. When the hot, metal-rich solutions contact cold seawater, metal sulfides are precipitated in a mound around the hydrothermal vent. This process has been filmed in the Pacific, where some springs spew clouds of fine-grained ore minerals that look like black smoke (figure 7.25). To learn more about seafloor hydrothermal vents, go to http://www.whoi.edu/main/topic/hydrothermal-vents.

The metals in rift-valley hot springs are predominantly iron, copper, and zinc, with smaller amounts of manganese, gold, and silver. Although the mounds are nearly solid metal sulfide, they are usually small and widely scattered on the sea floor, so commercial mining of them may not be practical.

After oceanic crust is subducted, the minerals are dehydrated deep in a subduction zone (see figure 7.26). The water produced moves upward into the overlying asthenosphere and contributes to melting and magma generation, as described in chapter 3.

Summary

Metamorphic rocks form from other rocks that are subjected to high temperature, generally accompanied by high confining pressure. Although recrystallization takes place in the solid state, water, which is usually present, aids metamorphic reactions. Foliation in metamorphic rocks is due to *differential stress* (either *compressive stress* or *shearing*). Slate, phyllite, schist, and gneiss are foliated rocks that indicate increasing grade of regional metamorphism. They are distinguished from one another by the type of foliation.

Contact metamorphic rocks are produced during metamorphism usually without significant differential stress but with high temperature. Contact metamorphism occurs in rocks immediately adjacent to intruded magmas.

Regional metamorphism, which involves heat, confining pressure, and differential stress, has created most of the metamorphic rock of Earth's crust. Different parent rocks as well as widely varying combinations of pressure and temperature result in a large variety of metamorphic rocks. Combinations of minerals in a rock can indicate what the pressure and temperature conditions were during metamorphism. Extreme metamorphism, where the rock partially melts, can result in *migmatites.*

Hydrothermal veins form when hot water precipitates material that crystallizes into minerals. During *metasomatism,* hot water introduces ions into a rock being metamorphosed, changing the chemical composition of the metasomatized rock from that of the parent rock.

Plate tectonic theory accounts for the features observed in metamorphic rocks and relates their development to other activities in Earth. In particular, plate tectonics explains (1) the deep burial of rocks originally formed at or near Earth's surface; (2) the intense squeezing necessary for the differential stress, implied by foliated rocks; (3) the presence of water deep within the lithosphere; and (4) the wide variety of pressures and temperatures believed to be present during metamorphism.

Terms to Remember

compressive stress 162
confining pressure 161
contact metamorphism 168
differential stress 162
ductile (plastic) 160
foliation 162
gneiss 167
hornfels 165
hydrothermal metamorphism 169
isotherm 175
marble 165
metamorphic facies 175
metamorphic rock 159
metamorphism 159
metasomatism 169
migmatite 171
parent rock (protolith) 160
phyllite 167
quartzite 165
regional metamorphism 170
schist 167
shearing 162
slate 165
stress 162
vein 169

Testing Your Knowledge

Use the following questions to prepare for exams based on this chapter.

1. What are the effects on metamorphic minerals and textures of temperature, confining pressure, and differential stress?
2. Describe the role of fluids during metamorphism and metasomatism.
3. Compare the pressure and temperature conditions of regional and contact metamorphism. How do regional metamorphic rocks commonly differ in texture from contact metamorphic rocks?
4. What is metamorphic grade? Describe what happens to (*a*) a shale and (*b*) a basalt as they are subjected to increasingly high-grade metamorphism.
5. How would you distinguish between: (*a*) schist and gneiss? (*b*) slate and phyllite? (*c*) quartzite and marble? (*d*) granite and gneiss?
6. Why would a builder choose to construct a building from quartzite blocks? Why might the builder choose to use marble?
7. Compare the metamorphic conditions at a convergent plate boundary to those at a divergent plate boundary.
8. Another term for parent rock is
   a. starter. b. primer.
   c. protolith. d. sediment.
9. Which of these is regarded as a low-grade metamorphic rock?
   a. migmatite b. schist
   c. slate d. gneiss
10. Foliation forms as a result of
    a. confining pressure.
    b. lithostatic pressure.
    c. differential stress.
    d. contact metamorphism.

11. Metamorphic rocks are classified primarily on
    a. texture–the presence or absence of foliation.
    b. mineralogy–the presence or absence of quartz.
    c. environment of deposition.
    d. chemical composition.
12. Which is *not* a foliated metamorphic rock?
    a. gneiss b. schist
    c. phyllite d. hornfels
13. Limestone recrystallizes during metamorphism into
    a. hornfels. b. marble.
    c. quartzite. d. schist.
14. Quartz sandstone is changed during metamorphism into
    a. hornfels. b. marble.
    c. quartzite. d. schist.
15. A foliated metamorphic rock, with very small crystals of mica creating a crinkled, silky appearance, would be called
    a. hornfels. b. phyllite.
    c. quartzite. d. schist.
16. A foliated metamorphic rock, with clearly visible crystals of mica creating a fish-scale appearance, would be called
    a. hornfels. b. phyllite.
    c. quartzite. d. schist.
17. A schist that developed in a high-pressure, low-temperature environment likely formed
    a. in the lower part of the continental crust.
    b. in a subduction zone.
    c. in a mid-oceanic ridge.
    d. near a contact with a magma body.
18. The major difference between metamorphism and metasomatism is
    a. the temperature at which each takes place.
    b. the minerals involved.
    c. the area or region involved.
    d. metasomatism is metamorphism coupled with the introduction of ions from an external source.
19. Hydrothermal metamorphism is especially important at
    a. mid-oceanic ridges. b. continental collision zones.
    c. impact craters. d. fault zones.
20. A metamorphic rock that has undergone partial melting to produce a mixed igneous-metamorphic rock is a
    a. gneiss. b. hornfels.
    c. schist. d. migmatite.

Expanding Your Knowledge

1. What metamorphic rocks are used for commercial purposes?
2. Where were the metals before they were concentrated in hydrothermal vein ore deposits?
3. What happens to originally horizontal layers of sedimentary rock when they are subjected to the deformation associated with regional metamorphism?
4. Where in Earth's crust would you expect most migmatites to form?

Exploring Web Resources

www.geol.ucsb.edu/faculty/hacker/geo102C/lectures/part1.html

University of California Santa Barbara's *Metamorphic Petrology* website. This site is meant for a course on metamorphic rock. It is well illustrated and can be used for in-depth learning of particular topics.

http://geology.com/rocks/metamorphic-rocks.shtml

Information about common metamorphic rocks, including good photos.

CHAPTER

8

Time and Geology

Grand Canyon, Arizona. Horizontal Paleozoic beds (top of photo) overlie tilted Precambrian beds (Grand Canyon Series) and older, Precambrian metamorphic rock (Vishnu Schist).
©Craig Aurness/Corbis/VCG/Getty Images

LEARNING OBJECTIVES

- Differentiate between relative age and numerical age.
- Know the relative dating principles and use them to determine a sequence of geologic events.
- Discuss how fossils are used to determine relative age and how paleontology contributed to the development of the geologic time scale.
- Describe radioactive decay and how radiogenic isotopes can be used to determine numerical age.
- Know the age of the Earth and the major subdivisions of geologic time.

The immensity of geologic time is hard for humans to perceive. It is unusual for someone to live a hundred years, but a person would have to live 10,000 times that long to observe a geologic process that takes a million years. In this chapter, we try to help you develop a sense of the vast amounts of time over which geologic processes have been at work.

Geologists working in the field or with maps or illustrations in a laboratory are concerned with relative time—unraveling the sequence in which geologic events occurred. For instance, a geologist looking at the photo of Arizona's Grand Canyon in this introduction can determine that the tilted sedimentary rocks are older than the horizontal sedimentary rocks and that the lower layers of the horizontal sedimentary rocks are older than the layers above them. But this tells us nothing about how long ago any of the rocks formed. To determine how many years ago rocks formed, we need the specialized techniques of radioactive isotope dating. Through isotopic dating, we have been able to determine that the rocks in the lowermost part of the Grand Canyon are well over a billion years old.

This chapter explains how to apply several basic principles to decipher a sequence of events responsible for geologic features. These principles can be applied to many aspects of geology—as, for example, in understanding geologic structures (chapter 15). Understanding the complex history of mountain belts (chapter 20) also requires knowing the techniques for determining relative ages of rocks.

Determining age relationships between geographically separated rock units is necessary for understanding the geologic history of a region, a continent, or the whole Earth. Substantiation of the plate tectonics theory depends on intercontinental correlation of rock units and geologic events, piecing together evidence that the continents were once one great body.

Widespread use of fossils led to the development of the standard geologic time scale. Originally based on relative age relationships, the subdivisions of the standard geologic time scale have now been assigned numerical ages in thousands, millions, and billions of years through isotopic dating. Think of the geologic time scale as a sort of calendar to which events and rock units can be referred.

8.1 THE KEY TO THE PAST

Until the 1800s, people living in Western culture did not question the religious perception of Earth being only a few thousand years old. On the other hand, Chinese and Hindu cultures believed the age of Earth was vast beyond comprehension—more in line with what has now been determined scientifically. In the Christendom of the seventeenth and eighteenth centuries, formation of all rocks and other geologic events were placed into a biblical chronology. This required that features we observe in rocks and landscapes were created supernaturally and catastrophically. The sedimentary rocks with marine fossils (clams, fish, etc.) that we find in mountains thousands of meters above sea level were believed to have been deposited by a worldwide flood (Noah's flood) that inundated all of Earth, including its highest mountains, in a matter of days. Because no known physical laws could account for such events, they were attributed to divine intervention. In the eighteenth century, however, James Hutton, a Scotsman often regarded as the father of geology, realized that geologic features could be explained through present-day processes. He recognized that our mountains are not permanent but have been carved into their present shapes and will be worn down by the slow agents of erosion now working on them. He realized that the great thicknesses of sedimentary rock we find on the continents are products of sediment removed from land and deposited as mud and sand in ancient seas. The time required for these processes to take place had to be incredibly long. Hutton broke from conventional thinking that Earth is no more than a few thousand years old when he wrote in 1788, "We find no sign of a beginning—no prospect for an end." Hutton's writings were not widely read, but his concept of geological processes requiring vast amounts of time was popularized in the 1800s when Charles Lyell published his landmark book, *Principles of Geology,* which refined and expanded upon many of Hutton's ideas. Charles Darwin was among those influenced by Lyell's writing. His evolutionary theory involving survival of the fittest, published in the mid-1800s, required the great amount of time that the works of Hutton and Lyell proposed.

The concept that geologic processes operating at present are the same processes that operated in the past eventually became known as the principle of **uniformitarianism**. The principle is stated more succinctly as "The present is the key to the past." The term *uniformitarianism* is a bit unfortunate because it suggests that changes take place at a uniform *rate.* It does not allow for the fact that sudden, violent events, such as a major, short-lived volcanic eruption, can also influence Earth's history. Many geologists prefer **actualism** in place of uniformitarianism. The term *actualism* comes closer to conveying the principle that the same processes and natural laws that operated in the past are those we can actually observe or infer from observation as operating at present. It is based on the assumption, central to the sciences, that physical laws are independent of time and location. Under present usage, uniformitarianism has the same meaning as actualism for most geologists.

We now realize that geology involves time periods much greater than a few thousand years. But how long? For instance, were rocks near the bottom of the Grand Canyon (see chapter-opening photo) formed closer to 10,000 or 100,000 or 1,000,000 or 1,000,000,000 years ago? What geologists needed was some "clock" that began running when rocks formed. Such a "clock" was found shortly after radioactivity was discovered. Dating based on radioactivity (discussed later in this chapter) allows us to determine a rock's **numerical age** (also known as *absolute age*)–age given in years or some other unit of time. Geologists working in the field or in a laboratory with maps, cross sections, and photographs are more often concerned with **relative time**, the *sequence* in which events took place, rather than the number of years involved.

These statements show the difference between numerical age and relative time:

"The American Revolutionary War took place after the signing of the Magna Carta but before World War II." This statement gives the time of an event (the Revolutionary War) relative to other events.

But in terms of numerical age, we could say: "The Revolutionary War took place about two and a half centuries ago." Note that a numerical age does not have to be an *exact* age, merely age given in units of time. Because most geologic problems are concerned with the sequence of events, we discuss relative age first.

8.2 RELATIVE TIME

The geology of an area may seem, at first glance, to be hopelessly complex. A nongeologist might think it impossible to decipher the sequence of events that created such a geologic pattern; however, a geologist has learned to approach seemingly formidable problems by breaking them down to a number of simple problems. (In fact, a geologic education trains students in a broad spectrum of problem-solving techniques, useful for a wide variety of applications and career opportunities.) As an example, the geology of the Grand Canyon, shown in the chapter-opening photo, can be analyzed in four parts: (1) horizontal layers of rock; (2) inclined layers; (3) rock underlying the inclined layers (plutonic and metamorphic rock); and (4) the canyon itself, carved into these rocks.

After you have studied the following section, return to the photo of the Grand Canyon and see if you can determine the sequence of geologic events that took place.

Principles Used to Determine Relative Age

Most of the individual parts of the larger problem are solved by applying several simple principles while studying the exposed rock. In this way, the sequence of events or the relative time involved can be determined. Contacts are particularly useful for deciphering the geologic history of an area. (**Contacts**, as described in previous chapters, are the surfaces separating two different rock types or rocks of different ages.)

To explain various relative age principles, we will use a fictitious place that bears some resemblance to the Grand Canyon. We will call this place, represented by the block diagram of figure 8.1, Minor Canyon. The formation names are also fictitious. (**Formations**, as described in chapter 6, are bodies of rock of considerable thickness with recognizable characteristics that make each distinguishable from adjacent rock units. They are named after local geographic features, such as towns or landmarks. Grand Canyon's

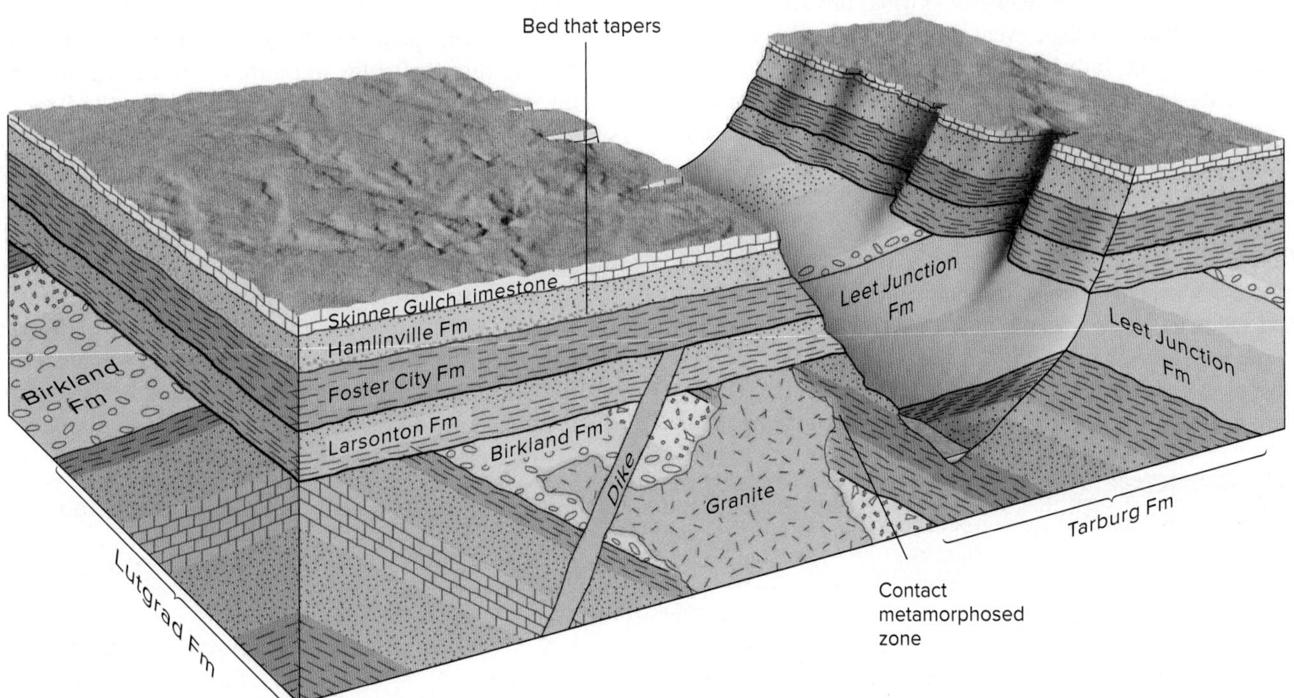

FIGURE 8.1

Block diagram representing the Minor Canyon area. (Tilted layers that are exposed in the canyon and are younger than the Leet Junction Formation are not named because they are not discussed or part of the figures that follow.) A key to the symbols representing rock types can be found in the back of this book.

formation names are shown on the chapter's opening photo.) The symbols in this diagram represent different rock types. For example, the dashes in the Foster City Fm represent shale. A key to these symbols is found in the back of this book. Note the contacts between the tilted formations, the horizontal formations, the granite, and the dike. What sequence of events might be responsible for the geology of Minor Canyon? (You might briefly study the block diagram and see how much of the geologic history of the area you can decipher before reading further.)

Our interpretations are based mainly on layered rock (sedimentary or volcanic). The subdiscipline of geology that uses interrelationships between layered rock (mostly) or sediment to interpret the history of an area or region is known as *stratigraphy* (from the Latin word *stratum,* meaning a thing spread out, or a cover). Four of stratigraphy's principles are used to determine the geologic history of a locality or a region. These are the principles of (1) original horizontality, (2) superposition, (3) lateral continuity, and (4) cross-cutting relationships. These principles can be used to interpret the geologic history of the Minor Canyon area shown in figure 8.1.

Original Horizontality

The principle of **original horizontality** (as described in chapter 6) states that beds of sediment deposited in water formed as horizontal or nearly horizontal layers. (The sedimentary rocks in figure 8.1 were originally deposited in a marine environment.)

Note in figure 8.1 that the Larsonton Formation and overlying rock units (Foster City Formation, Hamlinville Formation, and Skinner Gulch Limestone) are horizontal. Evidently, their original horizontal attitude has not changed since they were deposited. However, the Lutgrad, Birkland, Tarburg, and Leet Junction Formations must have been tilted after they were deposited as horizontal layers. By applying the principle of original horizontality, we have determined that a geologic event–tilting of bedrock–occurred after the Leet Junction, Tarburg, Birkland, and Lutgrad Formations were deposited on a sea floor. We can also see that the tilting event did not affect the Larsonton and overlying formations. (A reasonable conclusion is that tilting was accompanied by uplift and erosion, all before renewed deposition of younger sediment.)

Superposition

The principle of **superposition** states that within a sequence of undisturbed sedimentary or volcanic rocks, the oldest layer is at the bottom and layers are progressively younger upward in the stack.

Obviously, if sedimentary rock is formed by sediment settling onto the sea floor, then the first (or bottom) layer must be there before the next layer can be deposited on top of it. The principle of superposition also applies to layers formed by multiple lava flows, where one lava flow is superposed on a previously solidified flow.

Applying the principle of superposition, we can determine that the Skinner Gulch Limestone is the youngest layer of sedimentary rock in the Minor Canyon area. The Hamlinville Formation is the next oldest formation, and the Larsonton Formation is the oldest of the still-horizontal sedimentary rock units. Similarly, we assume that the inclined layers were originally horizontal (by the first principle). By mentally restoring them to their horizontal position (or "untilting" them), we can see that the youngest formation of the sequence is the Leet Junction Formation and that the Tarburg, Birkland, and Lutgrad Formations are progressively older.

Lateral Continuity

The principle of **lateral continuity** states that an original sedimentary layer extends laterally until it tapers or thins at its edges. This is what we expect at the edges of a depositional environment, or where one type of sediment interfingers laterally with another type of sediment as environments change. In figure 8.1, the bottom bed of the Hamlinville Formation (represented by orange dots) tapers as we would expect from this principle. We are not seeing any other layers taper, either because we are not seeing their full extent within the diagram or because they have been truncated (cut off abruptly) due to later events.

Cross-Cutting Relationships

The fourth principle can be applied to determine the remaining age relationships at Minor Canyon. The principle of **cross-cutting relationships** states that a disrupted pattern is older than the cause of disruption. A layer cake (the pattern) has to be baked (established) before it can be sliced (the disruption).

To apply this principle, look for disruptions in patterns of rock. Note that the valley in figure 8.1 is carved into the horizontal rocks as well as into the underlying tilted rocks. The sedimentary beds on either side of the valley appear to have been sliced off, or *truncated,* by the valley. (The principle of lateral continuity tells us that sedimentary beds normally become thinner toward the edges rather than stop abruptly.) So the event that caused the valley must have come after the sedimentation responsible for deposition of the Skinner Gulch Limestone and underlying formations. That is, the valley is younger than these layers. We can apply the principle of cross-cutting relationships to contacts elsewhere in figure 8.1, with the results shown in table 8.1.

TABLE 8.1 Relative Ages of Features in Figure 8.1 Determinable by Cross-Cutting Relationships

Feature	Is Younger Than	But Older Than
Valley (canyon)	Skinner Gulch Limestone	
Foster City Formation	Dike	Hamlinville Formation
Dike	Larsonton Formation	Foster City Formation
Larsonton Formation	Leet Junction Formation and granite	Dike
Granite	Tarburg Formation	Larsonton Formation

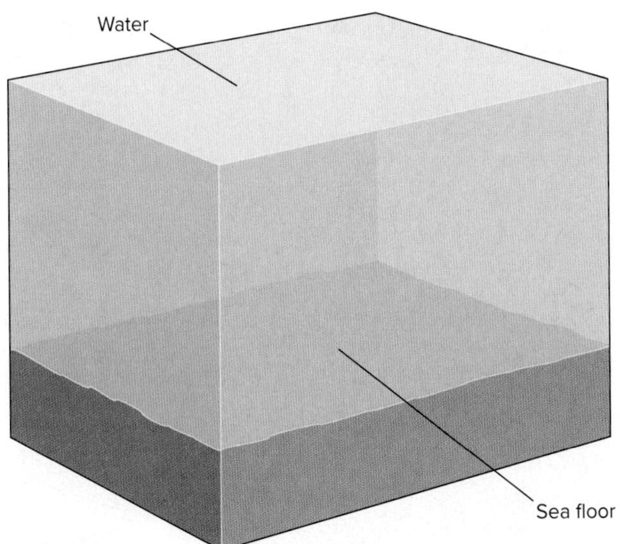

FIGURE 8.2
The area during deposition of the initial sedimentary layer of the Lutgrad Formation.

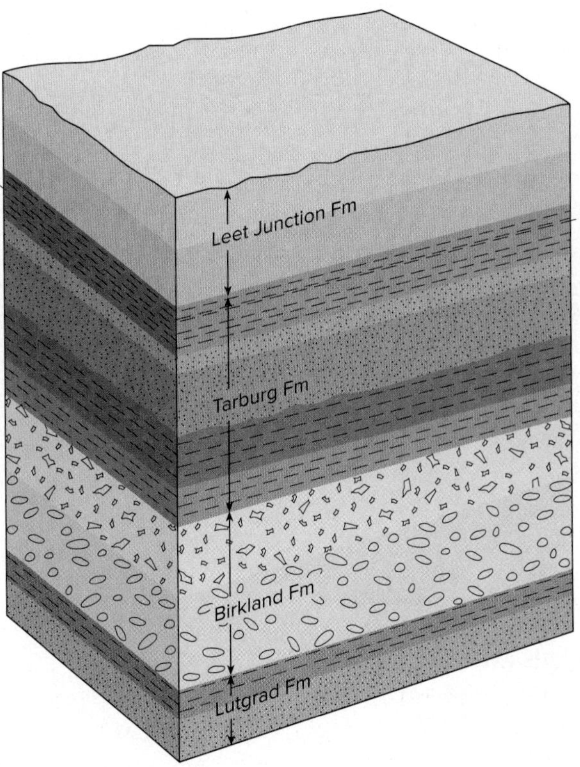

FIGURE 8.3
The area after deposition of the four formations shown but before intrusion of the granite.

We can now describe the geological history of the Minor Canyon area represented in figure 8.1 on the basis of what we have learned by applying the principles. Figures 8.2 through 8.11 show how the area changed over time, progressing from oldest to youngest events.

By *superposition,* we know that the Lutgrad Formation, the lowermost rock unit in the tilted sequence, must be the oldest of the sedimentary rocks as well as the oldest rock unit in the diagram. From the principle of *original horizontality,* we infer that these layers must have been tilted after they formed. Figure 8.2 shows initial sedimentation of the Lutgrad Formation taking place. If the entire depositional basin were shown, the layer would be tapered at its edges, according to the principle of *lateral continuity.*

Superposition indicates that the Birkland Formation was deposited on top of the Lutgrad Formation. Deposition of the Tarburg and Leet Junction Formations followed in turn (figure 8.3).

The truncation of bedding in the Lutgrad, Birkland, and Tarburg Formations by the granite tells us that the granite intruded sometime after the Tarburg Formation was formed (this is an *intrusive contact*). Although figure 8.4 shows that the granite was emplaced before tilting of the layered rock, we cannot determine from looking at figure 8.1 whether the granite intruded the sedimentary rocks before or after tilting. We can, however, determine through *cross-cutting relationships* that tilting and intrusion of the granite occurred before deposition of the Larsonton Formation. Figure 8.5 shows the rocks in the area have been tilted and erosion has taken place. Sometime later, sedimentation was renewed, and the lowermost layer of the Larsonton Formation was deposited on the erosion surface, as shown in figure 8.6. Contacts representing buried erosion surfaces such as these are called *unconformities* and are discussed in more detail in the "Unconformities" section of this chapter.

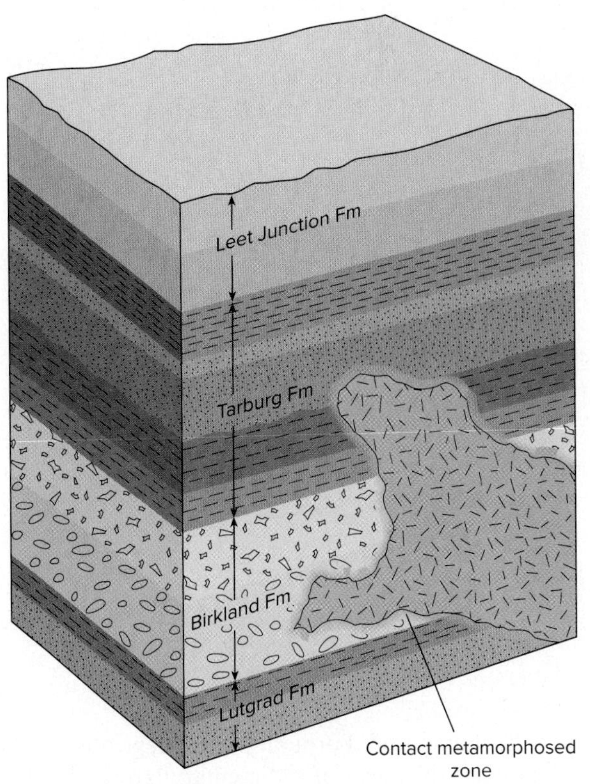

FIGURE 8.4
The area before layers were tilted and after intrusion of granite, if the intrusion took place before tilting.

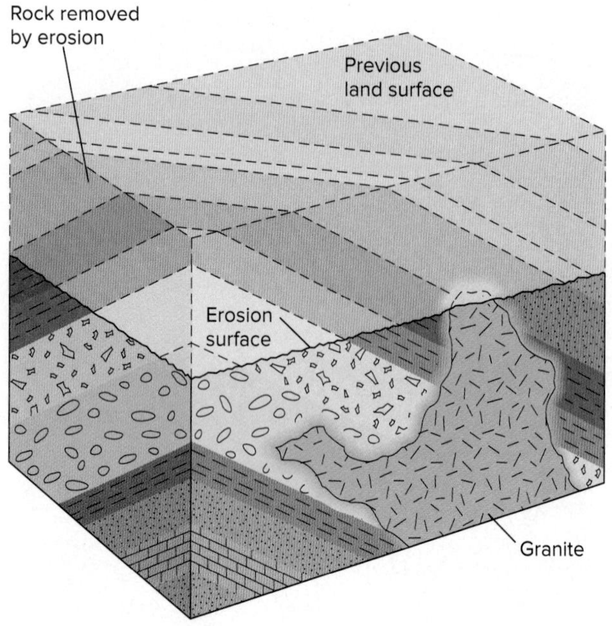

FIGURE 8.5

The area before deposition of the Larsonton Formation. Dashed lines show rock probably lost through erosion.

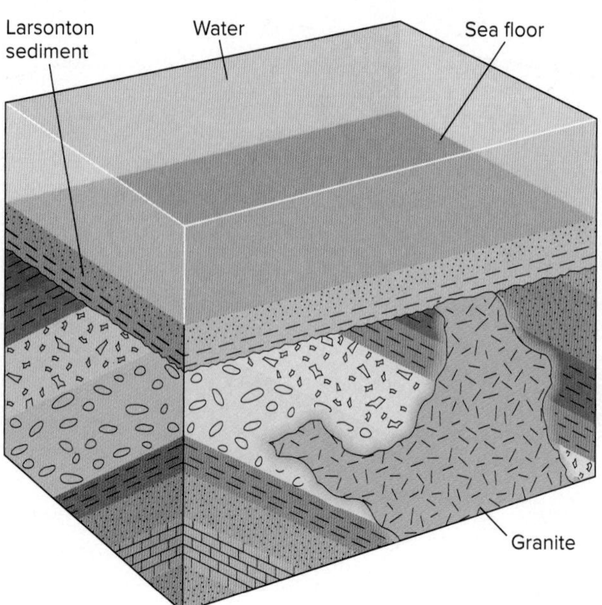

FIGURE 8.6

The area at the time the Larsonton Formation was being deposited.

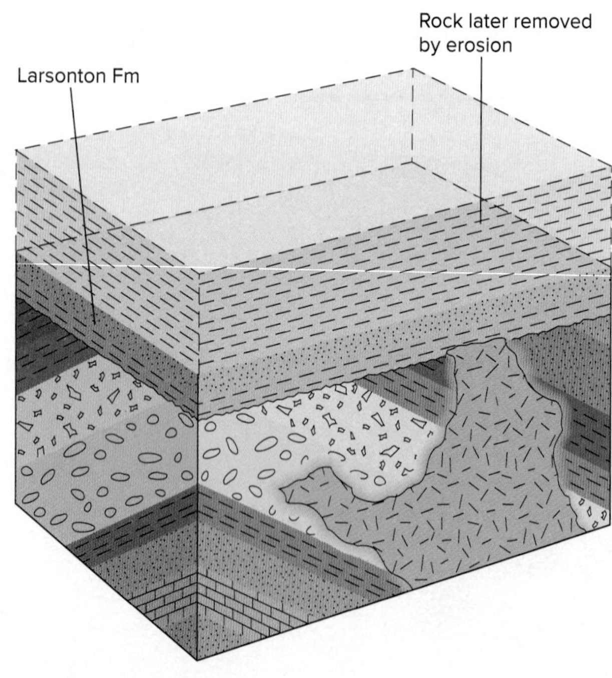

FIGURE 8.7

Area before intrusion of dike. Thickness of layers above the Larsonton Formation is indeterminate.

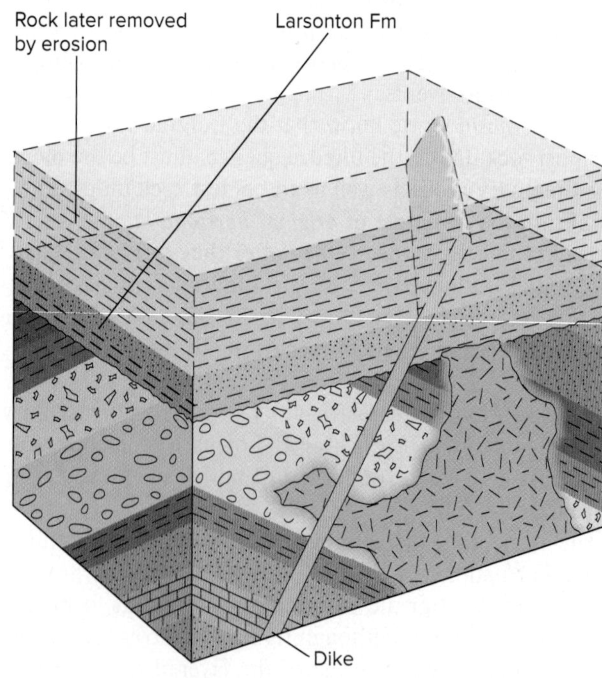

FIGURE 8.8

Dike intruded into the Larsonton Formation and preexisting, overlying layers of indeterminate thickness.

After the Larsonton Formation was deposited, an unknown additional thickness of sedimentary layers was deposited, as shown in figure 8.7. This can be determined through application of cross-cutting relationships. The dike is truncated by the Foster City Formation; therefore, it must have extended into some rocks that are no longer present, as shown in figure 8.8. Figure 8.9 shows the area after the erosion that truncated the dike took place.

Once again, sedimentation took place as the lowermost layer of the Foster City Formation blanketed the erosion surface

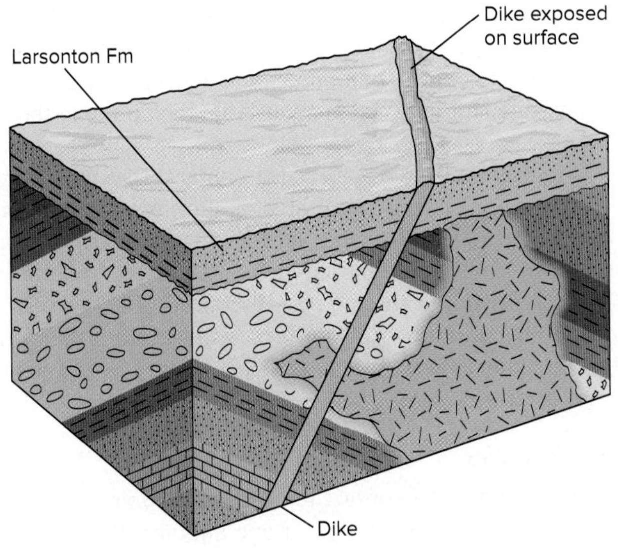

FIGURE 8.9

The area after rock overlying the Larsonton Formation, along with part of the dike, was removed by erosion.

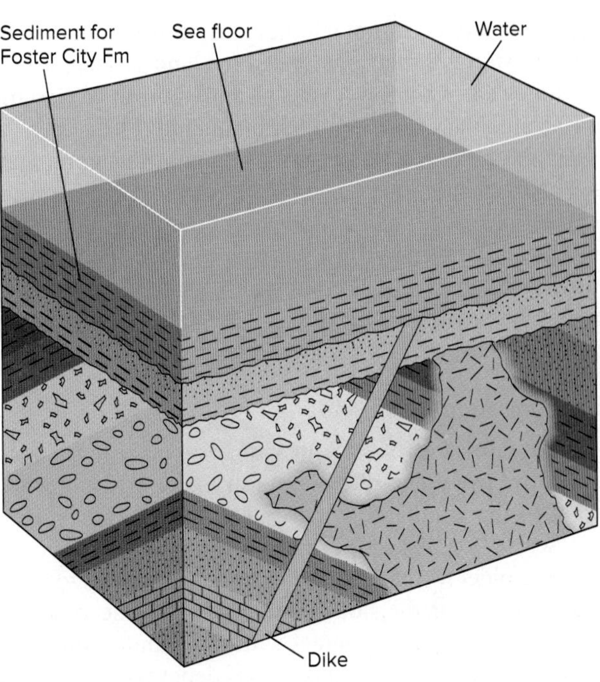

FIGURE 8.10

Sediment being deposited that will become part of the Foster City Formation.

(figure 8.10). Sedimentation continued until the uppermost layer (top of the Skinner Gulch Limestone) was deposited. At some later time, the area was raised above sea level, and the stream began to carve the canyon (figure 8.11). Because the valley sides truncate the youngest layers of rock, we can determine from figure 8.1 that the last event was the carving of the valley.

Note that there are limits on how precisely we can determine the relative age of the granite body. It definitely intruded *before* the Larsonton Formation was deposited and *after* the Tarburg Formation was deposited. As no contacts can be observed between the Leet Junction Formation and the granite, we cannot say whether the granite is younger or older than the Leet Junction Formation. Nor, as mentioned earlier, can we determine whether the granite formed before, during, or after the tilting of the lower sequence of sedimentary rocks.

Now, if you take another look at the chapter-opening photo of the Grand Canyon (and figure 8.16), you should be able to determine the sequence of events. The sequence (going from older to younger) is as follows: Regional metamorphism took place resulting in the Vishnu Schist of the lower part of the

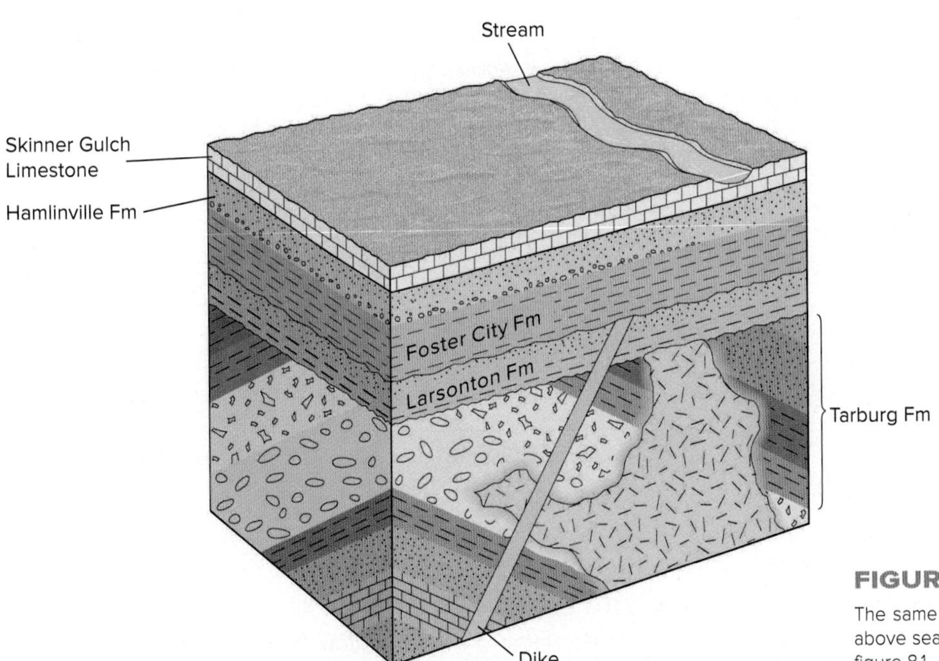

FIGURE 8.11

The same area after all of the rocks had formed and then had risen above sea level. The stream is beginning to form the valley visible in figure 8.1.

Grand Canyon (you cannot tell these are schists from the photograph). Erosion followed and leveled the land surface. Sedimentation followed, resulting in the Grand Canyon Series rocks. These sedimentary layers were subsequently tilted (they were also faulted, although this is not evident in the photograph). Once again, erosion took place. The lowermost of the presently horizontal layers of sedimentary rock was deposited (the Tapeats Sandstone followed by the Bright Angel Shale). Subsequently, each of the layers progressively higher up the sequence formed. Finally, the stream (the Colorado River) eroded its way through the rock, carving the Grand Canyon.

Other Time Relationships

Other characteristics of geology can be applied to help determine relative ages. The tilted layers in figure 8.12 immediately adjacent to the granite body have been *contact metamorphosed* (think "seared" or "baked"). This indicates that the Tarburg Formation and older formations shown in figure 8.1 had to be there before intrusion of the hot, granite magma. The base of the Larsonton Formation in contact with the granite would not be contact metamorphosed because it was deposited after the granite had cooled (and had been exposed by erosion).

The principle of **inclusion** states that fragments included in a host rock are older than the host rock. In figure 8.12, the granite contains inclusions of the tilted sedimentary rock. Therefore, the granite is younger than the tilted rock. The rock overlying the granite has granite pebbles in it. Therefore, the granite is older than the horizontal sedimentary rock.

Unconformities

In this and earlier chapters, we noted the importance of *contacts* for deciphering the geologic history of an area. In chapters 3 and 6 we described intrusive contacts and sedimentary contacts. Faults (described in chapter 15) are a third type of contact. The final important type of contact is an *unconformity.* Each type of contact has a very different implication about what took place in the geologic past.

An **unconformity** is a surface (or contact) that represents a *gap in the geologic record,* with the rock unit immediately above the contact being considerably younger than the rock beneath. Most unconformities are buried erosion surfaces. Unconformities are classified into three types–disconformities, angular unconformities, and nonconformities–with each type having important implications for the geologic history of the area in which it occurs.

Disconformities

In a **disconformity**, the contact representing missing rock strata separates beds that are parallel to one another. Probably what has happened is that older rocks were eroded away parallel to the horizontal bedding plane; renewed deposition later buried the erosion surface (figure 8.13).

Because it often appears to be just another sedimentary contact (or bedding plane) in a sequence of sedimentary rock, a disconformity is the hardest type of unconformity to detect in the field. Rarely, a telltale weathered zone is preserved immediately below a disconformity. Usually, the disconformity can be detected only by studying fossils from the beds in a sequence of sedimentary rocks. If certain fossil beds are absent, indicating that a portion of geologic time is missing from the sedimentary record, it can be inferred that a disconformity is present in the sequence. Although it is most likely that some rock layers are missing because erosion followed deposition, in some instances neither erosion nor deposition took place for a significant amount of geologic time.

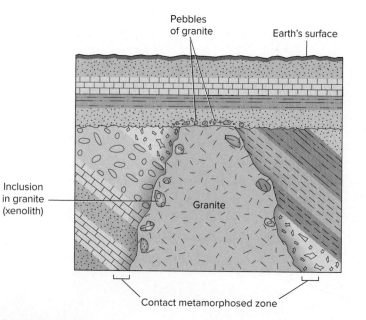

FIGURE 8.12

Age relationships indicated by contact metamorphism, inclusions (xenoliths) in granite, and pebbles of granite.

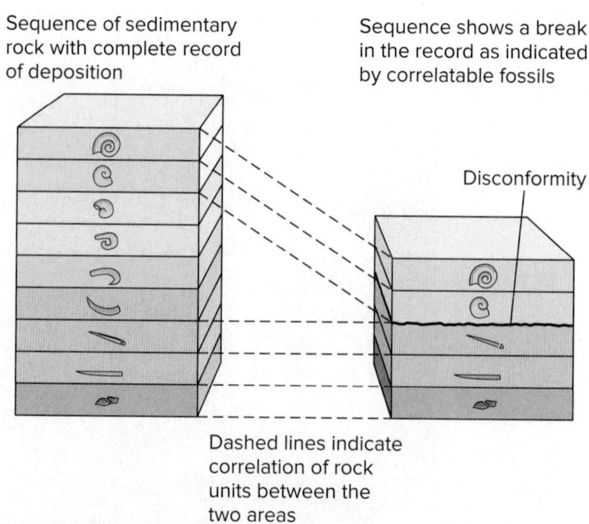

FIGURE 8.13

Schematic representation of a disconformity. The disconformity is in the block on the right.

Angular Unconformities

An **angular unconformity** is a contact in which younger strata overlie an erosion surface on tilted or folded layered rock. It implies the following sequence of events, from oldest to youngest: (1) deposition and lithification of sedimentary rock (or solidification of successive lava flows if the rock is volcanic); (2) uplift accompanied by folding or tilting of the layers; (3) erosion; and (4) renewed deposition (usually preceded by subsidence) on top of the erosion surface (figure 8.14). Figures 8.1 and 8.12 also show angular unconformities but with simple tilting rather than folding of the older beds.

Nonconformities

A **nonconformity** is a contact in which an erosion surface on plutonic or metamorphic rock has been covered by younger sedimentary or volcanic rock (figure 8.15). A nonconformity generally indicates deep or long-continued erosion before subsequent burial because metamorphic or plutonic rocks form at considerable depths in Earth's crust.

The geologic history implied by a nonconformity, shown in figure 8.15, is (1) crystallization of igneous or metamorphic rock at depth; (2) erosion of at least several kilometers of overlying rock (the great amount of erosion further implies considerable uplift of this portion of Earth's crust); and (3) deposition of new sediment, which eventually becomes sedimentary rock, on the ancient erosion surface. Figures 8.1 and 8.12 also show nonconformities; however, these represent erosion to a relatively shallow depth as the rocks intruded by the pluton have not been regionally metamorphosed, as was the case for those in figure 8.15.

Correlation

In geology, **correlation** usually means determining the time equivalency of rock units. Rock units may be correlated within a region, a continent, and even between continents. Various methods of correlation are described along with examples of how the principles we described earlier in this chapter are used to determine whether rocks in one area are older or younger than rocks in another area.

Physical Continuity

Finding **physical continuity**—that is, being able to trace physically the course of a rock unit—is one way to correlate rocks between two different places. The prominent white layer of cliff-forming rock in figure 8.16 is the Coconino Sandstone, exposed along the upper part of the Grand Canyon. It can be seen all the way across the photograph. You can physically follow this unit for several tens of kilometers, thus verifying that, wherever it is

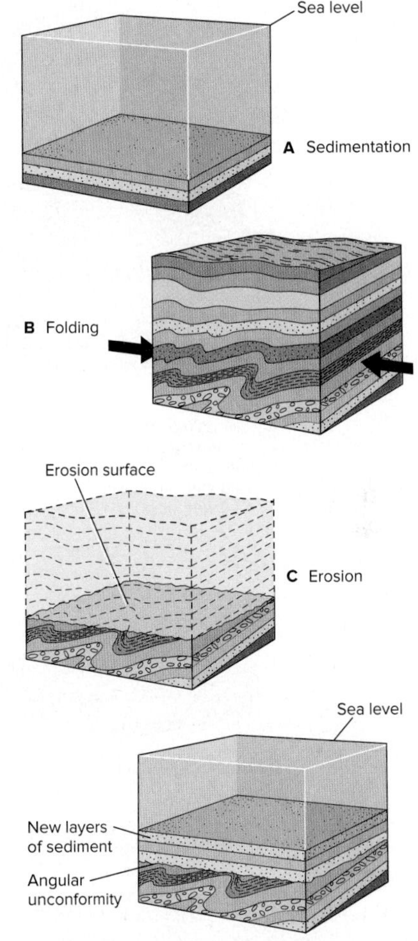

FIGURE 8.14

A particular sequence of events (*A–D*) producing an angular unconformity. Marine deposited sediments are uplifted and folded (probably during plate-tectonic convergence). Erosion removes the upper layers. The area drops below sea level (or sea level rises) and renewed sedimentation takes place. (An angular unconformity can also involve terrestrial sedimentation.) (*E*) is an angular unconformity at Cody, Wyoming.
(E) ©C.C. Plummer

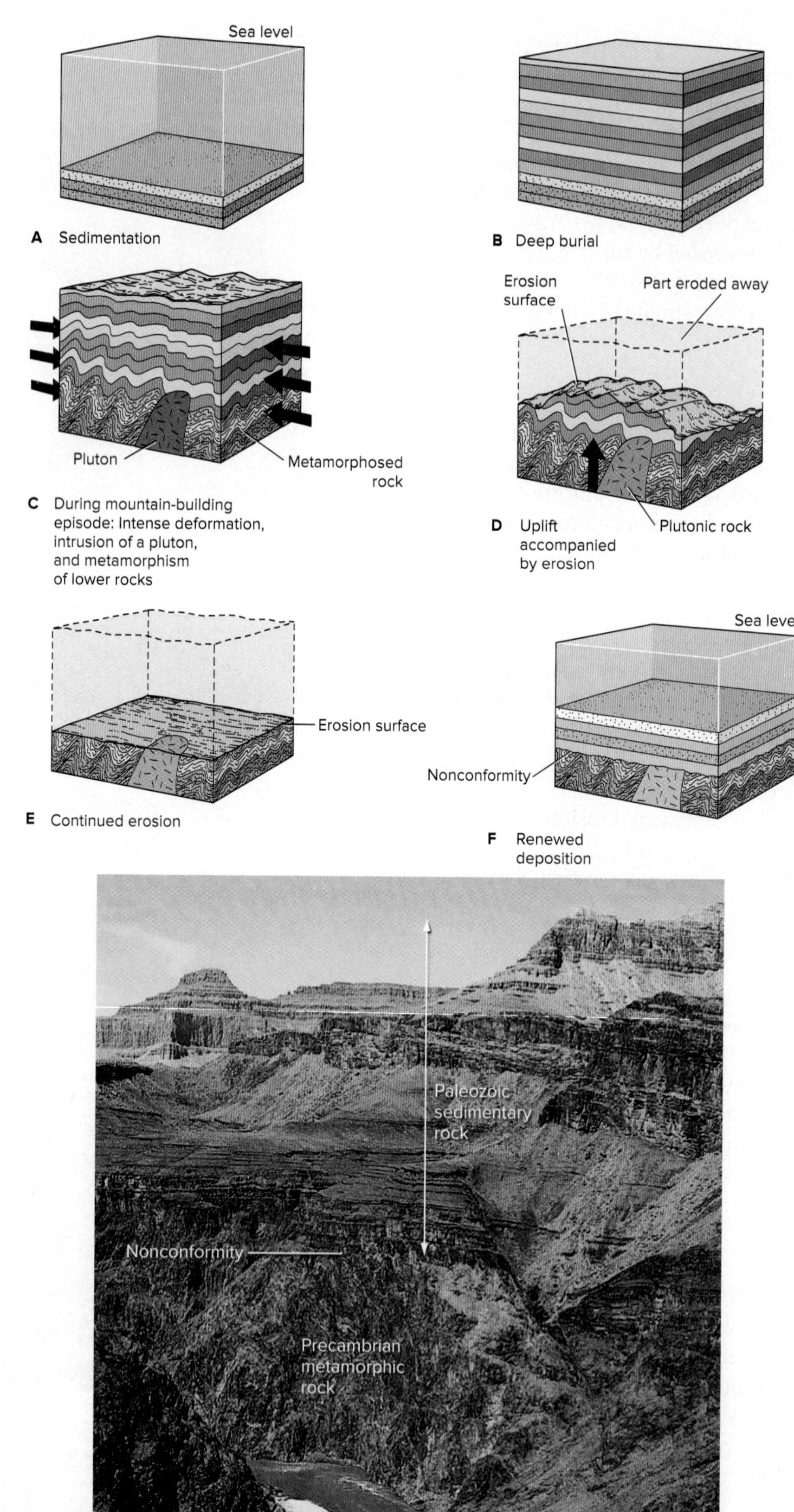

FIGURE 8.15

(*A–F*) Sequence of events implied by a nonconformity underlain by metamorphic and plutonic rock. (*G*) A nonconformity in Grand Canyon, Arizona. Paleozoic sedimentary rocks overlie vertically foliated Precambrian metamorphic rocks.
(G) ©C.C. Plummer

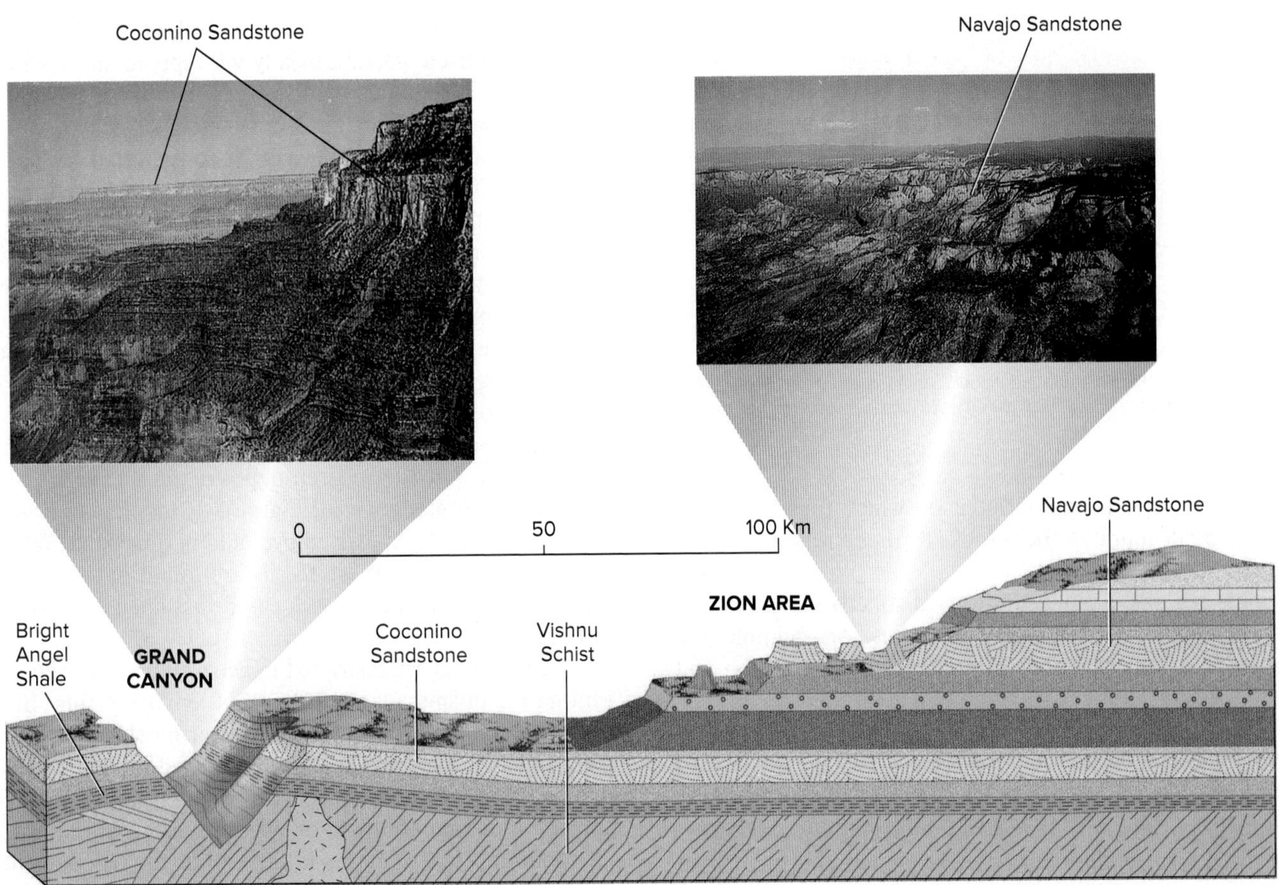

FIGURE 8.16

Schematic cross section through part of the Colorado Plateau showing the relationship of the Coconino Sandstone, the white cliff-forming unit in the left photo, in Grand Canyon to the Navajo Sandstone, white unit in the right photo, at Zion National Park.
©C.C. Plummer

exposed in the Grand Canyon, it is the same rock unit. The Grand Canyon is an ideal location for correlating rock units by physical continuity. However, it is not possible to follow this rock unit from the Grand Canyon into another region because it is not continuously exposed. We usually must use other methods to correlate rock units between regions.

Similarity of Rock Types

Under some circumstances, correlation between two regions can be made by assuming that similar rock types in two regions formed at the same time. This method must be used with extreme caution, especially if the rocks being correlated are common ones.

To show why correlation by similarity of rock type does not always work, we can try to correlate the white, cliff-forming Coconino Sandstone in the Grand Canyon with a rock unit of similar appearance in Zion National Park about 100 kilometers away (figure 8.16). Both units are white sandstone. Cross-bedding indicates that both were once a series of sand dunes. It is tempting to correlate them and conclude that both formed at the same time. But if you were to drive or walk from the rim of the Grand Canyon (where the Coconino Sandstone is *below* you), you would get to Zion by ascending a series of layers of sedimentary rock stacked on one another. In other words, you would be getting into progressively younger rock, as shown diagrammatically in figure 8.16. In short, you have shown through *superposition* that the sandstone in Zion (called the Navajo Sandstone) is younger than the Coconino Sandstone.

Correlation by similarity of rock types is more reliable if a very unusual sequence of rocks is involved. If you find in one area a layer of green shale on top of a red sandstone that, in turn, overlies basalt of a former lava flow and then find the same sequence in another area, you probably would be correct in concluding that the two sequences formed at essentially the same time.

When the hypothesis of continental drift was first proposed (see chapters 1 and 19), important evidence was provided by correlating a sequence of rocks (figure 8.17) consisting of glacially deposited sedimentary rock (tillites, described in chapter 12 on glaciation), overlain by continental sandstones, shales, and coal beds. These strata are in turn overlain by basalt flows. The sequence is found in parts of South America, Australia, Africa, Antarctica, and India. It is very unlikely that an identical sequence of rocks could have formed on each of the continents if they were widely separated, as they are at present. Therefore, the continents on which the sequence is found are likely to have

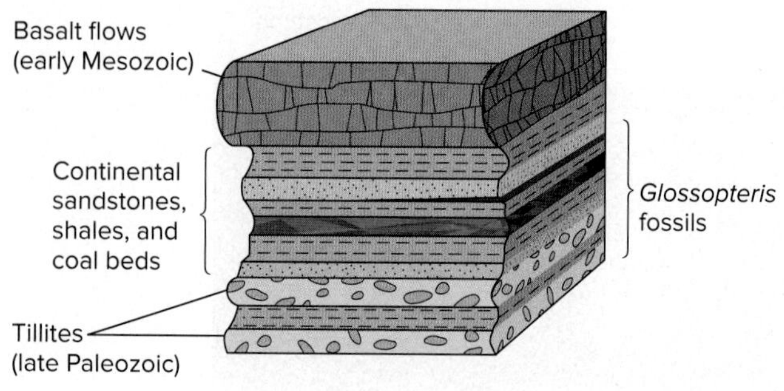

FIGURE 8.17

Rock sequences similar to this are found in India, Africa, South America, Australia, and Antarctica. The rocks in each of these localities contain the fossil plant *Glossopteris.*

been part of a single supercontinent on which the rocks were deposited. Fossils found in these rocks further strengthened the correlation.

In some regions, a *key bed,* a very distinctive layer, can be used to correlate rocks over great distances. An example is a layer of volcanic ash produced from a very large eruption and distributed over a significant portion of a continent.

Correlation by Fossils

Fossils are common in sedimentary rock, and their presence is important for correlation. Plants and animals that lived at the time the rock formed were buried by sediment, and their fossil remains are preserved in sedimentary rock. Most of the fossil species found in rock layers are now extinct–99.9% of all species that ever lived are extinct. (The concept of *species* for fossils is similar to that in biology.)

In a thick sequence of sedimentary rock layers, the fossils nearer the bottom (i.e., in the older rock) are more unlike today's plants and animals than are those near the top. As early as the end of the eighteenth century, naturalists realized that the fossil remains of creatures of a series of "former worlds" were preserved in Earth's sedimentary rock layers. In the early nineteenth century, a self-educated English surveyor named William Smith realized that different sedimentary layers are characterized by distinctive fossil species and that *fossil species succeed one another through the layers in a predictable order.* Smith's discovery of this principle of **faunal succession** allowed rock layers in different places to be correlated based on their fossils. We now understand that faunal succession works because there is an evolutionary history to life on Earth. Species evolve, exist for a time, and go extinct. Because the same species never evolves twice (extinction is forever), any period of time in Earth history can be identified by the species that lived at that time. *Paleontologists,* specialists in the study of fossils, have patiently and meticulously over the years identified many thousands of species of fossils and determined the time sequence in which they existed. Therefore, sedimentary rock layers anywhere in the world can be assigned to their correct place in geologic history by identifying the fossils they contain.

Ideally, a geologist hopes to find an **index fossil**, a fossil from a very short-lived, geographically widespread species known to exist during a specific period of geologic time. A single index fossil allows the geologist to correlate the rock in which it is found with all other rock layers that contain that fossil.

Many fossils are of little use in time determination because the species thrived during too large a portion of geologic time. Sharks, for instance, have been in the oceans for a long time, so discovering an ordinary shark's tooth in a rock is not very helpful in determining the rock's relative age.

A single fossil that is not an index fossil is not very useful for determining the age of the rock it is in. However, finding several species of fossils (a fossil assemblage) in a layer of rock is generally more useful for dating rocks than a single fossil is, because the sediment must have been deposited at a time when all the species represented existed. Figure 8.18 depicts five species of fossils, each of which existed over a long time span. Where various combinations of these fossils are found in three rocks, the time of formation of each rock can be assigned to a narrow span of time.

Some fossils are restricted in geographic occurrence, representing organisms adapted to special environments. But many former organisms apparently lived over most of the Earth, and fossil assemblages from these may be used for worldwide correlation. Fossils in the lowermost horizontal layers of the Grand Canyon are comparable to ones collected in Wales, Great Britain, and many other places in the world (the trilobites in figure 8.19 are an example). We can, therefore, correlate these rock units and say they formed during the same general span of geologic time.

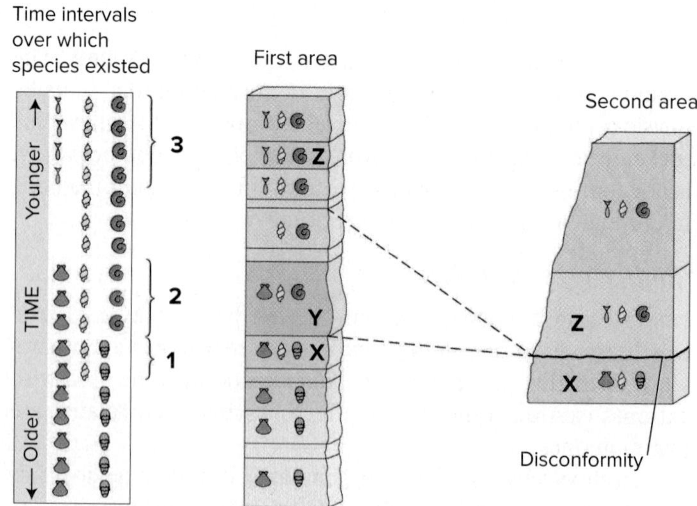

FIGURE 8.18

The use of fossil assemblages for determining relative ages.
Rock X contains [fossil symbols]. Therefore, it must have formed during time interval 1.
Rock Y contains [fossil symbols]. Therefore, it must have formed during time interval 2.
Rock Z contains [fossil symbols]. Therefore, it must have formed during time interval 3.
In the second area, fossils of time interval 2 are missing. Therefore, the surface between X and Z is a disconformity.

FIGURE 8.19

Elrathia kingii trilobites from the Middle Cambrian Wheeler Formation of Utah. The larger one is 10 mm in diameter.
©*Robert R. Gaines*

The Standard Geologic Time Scale

Geologists can use fossils in rock to refer the age of the rock to the **standard geologic time scale**, a worldwide relative time scale. Based on fossil assemblages, the geologic time scale subdivides geologic time. On the basis of fossils found, a geologist can say, for instance, that the rocks of the lower portion of horizontal layers in the Grand Canyon formed during the *Cambrian Period.* This implicitly correlates these rocks with certain rocks in Wales (in fact, the period takes its name from *Cambria,* the Latin name for Wales) and elsewhere in the world where similar fossils occur.

The geologic time scale, shown in a somewhat abbreviated form in table 8.2, has had tremendous significance as a unifying concept in the physical and biological sciences. The working out of the evolutionary chronology by successive generations of geologists and other scientists has been a remarkable human achievement. The geologic time scale consists of four **eons**. The **Phanerozoic Eon** (meaning "visible life") is divided into three **eras**, which are divided into **periods**, which are, in turn, subdivided into **epochs**. (Remember that this is a relative time scale. We will add dates later in this chapter after discussing isotopic dating.)

Precambrian denotes the vast amount of time that preceded the Paleozoic Era (which begins with the Cambrian Period). The **Paleozoic Era** (meaning "old life") began with the appearance of complex life (e.g., trilobites), as indicated by fossils. Rocks older than Paleozoic contain few fossils. This is because creatures with shells or other hard parts, which are easily preserved as fossils, did not evolve until the beginning of the Paleozoic.

The **Mesozoic Era** (meaning "middle life") followed the Paleozoic. On land, dinosaurs became the dominant animals of the Mesozoic. We live in the **Holocene** (or **Recent**) **Epoch** of the **Quaternary Period** of the **Cenozoic Era** (meaning "new life"). The Quaternary also includes the most recent ice ages, which were part of the **Pleistocene Epoch**.

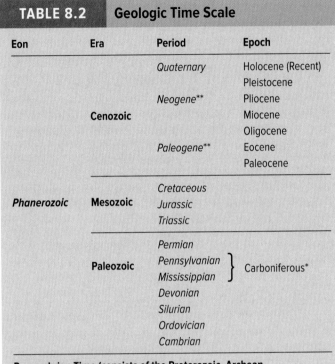

TABLE 8.2 Geologic Time Scale

Eon	Era	Period	Epoch
		Quaternary	Holocene (Recent)
			Pleistocene
		*Neogene***	Pliocene
	Cenozoic		Miocene
			Oligocene
		*Paleogene***	Eocene
			Paleocene
		Cretaceous	
Phanerozoic	Mesozoic	*Jurassic*	
		Triassic	
		Permian	
	Paleozoic	*Pennsylvanian*	Carboniferous*
		Mississippian	Carboniferous*
		Devonian	
		Silurian	
		Ordovician	
		Cambrian	

Precambrian Time (consists of the Proterozoic, Archean, and Hadean Eons)

*Outside of North America, Carboniferous Period is used rather than Pennsylvanian and Mississippian.

**In 2003, the International Commission on Stratigraphy recommended dropping Tertiary and Quaternary as periods and replacing them with Paleogene and Neogene. This proposal caused some controversy, and in 2009 the commission decided to replace the Tertiary with the Paleogene and Neogene while retaining the Quaternary as a separate era. Currently, the Geological Society of America annually updates the geologic time scale and posts it on www.geosociety.org/science/timescale/

It is noteworthy that the fossil record indicates that mass extinctions, in which a large number of species became extinct, occurred a number of times in the geologic past. The two greatest mass extinctions define the boundaries between the three eras (see boxes 8.1 and 8.2).

Fossils have been used to determine ages of the horizontal rocks in the Grand Canyon. All are Paleozoic. The lowermost horizontal formations (see chapter-opening photo) are Cambrian, above which are Devonian, Mississippian, Pennsylvanian, and Permian rock units. By referring to the geologic time scale (table 8.2), we can see that Ordovician and Silurian rocks are not represented. Thus, an unconformity (a disconformity) is present within the horizontally layered rocks of the Grand Canyon.

8.3 NUMERICAL AGE

Counting annual growth rings in a tree trunk will tell you how old a tree is. Similarly, layers of sediment deposited annually in glacial lakes can be counted to determine how long those lakes existed (*varves,* as these deposits are called, are explained in

EARTH SYSTEMS 8.1

Highlights of the Evolution of Life through Time

The following is a very condensed preview of what you are likely to learn about if you take a historical geology course.

The history of the biosphere is preserved in the fossil record. Through fossils, we can determine their place in the evolution of plants and animals as well as get clues as to how extinct creatures lived. The oldest readily identifiable fossils found are prokaryotes—microscopic, single-celled organisms that lack a nucleus. These date back to around 3.5 billion years (b.y.) ago, so life on Earth is at least that old. It is likely that even more primitive organisms date back further in time but are not preserved in the fossil record. Fossils of much more complex, single-celled organisms that contained a nucleus (eukaryotes) are found in rocks as old as 2.1 b.y. These are the earliest living creatures to have reproduced sexually. Colonies of unicellular organisms likely evolved into multicellular organisms. Multicellular algae fossils date back at least a billion years.

Imprints of larger multicellular creatures appear in rocks of late Precambrian age, about 700 to 550 million years ago (m.y.). These resemble jellyfish and worms.

Abundant fossil evidence of life does not appear until the beginning of the Paleozoic era, 541 million years ago. Large numbers of fossils appeared early in the Cambrian Period. Trilobites (see figure 8.19) evolved into many species and were particularly abundant during the Cambrian. Trilobites were arthropods that crawled on muddy sea floors and are the oldest fossils with eyes. They became less significant later in the Paleozoic, and finally, all trilobites became extinct by the end of the Paleozoic.

The most primitive fishes, the first vertebrates, date back to late in the Cambrian. Fishes similar to currently living species (including sharks) flourished during the Devonian (named after Devonshire, England). The Devonian is often called the "age of fishes." Amphibians evolved from air-breathing fishes late in the Devonian. These were the first land vertebrates. However, invertebrate land animals date back to the latest Cambrian, and land plants first appeared in the Ordovician. Reptiles and early ancestors of mammals evolved from amphibians in Pennsylvanian time or perhaps earlier.

The Paleozoic ended with the greatest mass extinction ever to occur on Earth. Around 80% of marine species died out as the Permian Period ended.

During the Mesozoic, new creatures evolved to occupy ecological domains left vacant by extinct creatures. Dinosaurs and mammals evolved from the animal species that survived the great extinction. Dinosaurs became the dominant group of land animals. Birds likely evolved from dinosaurs in the Mesozoic. Large, now extinct, marine reptiles lived in Mesozoic seas. Ichthyosaurs, for example, were up to 20 meters long, had dolphinlike bodies, and were probably fast swimmers. Flying reptiles, pterosaurs, some of which had wingspans of almost 10 meters, soared through the air.

The Cretaceous Period (and Mesozoic Era) ended with the second-largest mass extinction (around 75% of species, including the dinosaurs, were wiped out).

The Cenozoic is often called the age of mammals. Mammals, which were small, insignificant creatures during the Mesozoic, evolved into the many groups of mammals (whales, bats, canines, cats, elephants, primates, and so forth) that occupy Earth at present. Many species of mammals evolved and became extinct throughout the Cenozoic. Hominids (modern humans and our extinct ancestors) have a fossil record dating back 6 m.y. and likely evolved from a now extinct ancestor common to hominids, chimpanzees, and other apes.

We tend to think of mammals' evolution as being the great success story (because we are mammals); mammals, however, pale in comparison to insects. Insects have been around far longer than mammals and now account for an estimated 1 million species.

Additional Resources

University of California Museum of Paleontology
Find pictures of the fossils named in this box.

- www.ucmp.berkeley.edu/

The Paleontology Portal
Another site to find out about fossils. You can search by type of creature, by time, or by location.

- www.paleoportal.org/

chapter 12). But only within the few decades following the discovery of radioactivity in 1896 have scientists been able to determine numerical ages of rock units. We have subsequently been able to assign numerical values to the geologic time scale and determine how many years ago the various eras, periods, and epochs began and ended: The Cenozoic Era began some 66 million years ago, the Mesozoic Era started about 252 million years ago, and the Precambrian ended (or the Paleozoic began) about 541 million years ago. The Precambrian includes most of geologic time, because the age of Earth is commonly regarded as about 4.5 to 4.6 billion years.

In 2008, a rock from Hudson Bay in northern Canada (its location is indicated on the geologic map of North America in appendix H) was dated as being 4.28 billion years old. This rock is nearly 300 million years older than the previously dated oldest rock (age 4.03 billion years old). However, the age of 4.28 billion years has been challenged, with some suggesting that an age of 3.8 billion is more likely. In 2014, the oldest known mineral was dated at 4.4 billion years old, which is considerably older than the oldest rock dated so far. The mineral, a zircon crystal from Australia, was likely originally in a granite. Scientists who have studied this mineral think that its chemical makeup indicates that the granite formed from a magma that had a component of melted sedimentary rock. This would indicate that seas existed much earlier than geologists had previously thought possible.

Isotopic Dating

Radioactivity provides a "clock" that begins running when radioactive elements are sealed into newly crystallized minerals.

EARTH SYSTEMS 8.2

Demise of the Dinosaurs—Was It Extraterrestrial?

The story of the rise and fall of dinosaurs involves the biosphere (the dinosaurs and their ecosystem), the solar system (extraterrestrial objects), the atmosphere (which changed abruptly), and the hydrosphere (part of an ocean was vaporized). Dinosaurs dominated the continents during the Mesozoic Era. Now they prey on the imaginations of children of all ages and are featured in media ranging from movies to cartoons. It's hard to accept that beings as powerful and varied as dinosaurs existed and were wiped out. But the fossil record is clear—when the Mesozoic came to a close, dinosaurs became extinct. Not a single one of the numerous dinosaur species survived into the Cenozoic Era. Not only did the dinosaurs go, but about three-quarters of all plant and animal species, marine as well as terrestrial, were extinguished. This was one of Earth's "great dyings," or mass extinctions.

In 1980, geologist Walter Alvarez, his father, physicist Luis Alvarez, and two other scientists proposed a hypothesis that the dinosaur extinction was caused by the impact of an asteroid. This was based on the chemical analysis of a thin layer of clay marking the boundary between the Mesozoic and Cenozoic Eras (usually referred to as the K-T boundary—it separates the Cretaceous [K] and Tertiary [T] Periods). The Tertiary has since been replaced by the Paleogene and Neogene Periods, but the term K-T boundary persists. The K-T boundary clay was found to have about 30 times the amount of the rare element iridium as is normal for crustal rocks. Iridium is relatively abundant in meteorites and other extraterrestrial objects such as comets, and the scientists suggested that the iridium was brought in by an extraterrestrial body.

A doomsday scenario is visualized in which an asteroid 10 kilometers in diameter struck Earth. The asteroid would have blazed through the atmosphere at astonishing speed and, likely, impacted at sea. Part of the ocean would have been vaporized and a crater created on the ocean floor. There would have been an earthquake much larger than any ever felt by humans. Several-hundred-meter-high waves would crisscross the oceans, devastating life anywhere near shorelines. The lower atmosphere would have become intolerably hot, at least for a short period of time. The atmosphere worldwide would have been altered and the climate cooled because of the increased blockage of sunlight by dust particles suspended in the upper atmosphere.

For a while, the hypothesis was hotly debated. Other scientists hypothesized that the extinctions were caused by exceptionally large volcanic activity. Further evidence supporting the asteroid hypothesis accumulated. K-T layers throughout the world were found to have grains of quartz that had been subjected to shock metamorphism (see box 7.1). Microscopic spheres of glass that formed when rock melted from the impact and droplets were thrown high into the air were also found in the K-T layers. Sediment that appeared to have been deposited by giant sea waves was found in various locations.

The asteroid hypothesis advocates predicted that a large meteorite crater should be found someplace on Earth that could be dated as having formed around 66 million years ago, when the Mesozoic ended.

In 1990, the first evidence for the "smoking gun" crater was found. The now-confirmed crater is over 200 kilometers in diameter and centered along the coast of Mexico's Yucatan peninsula at a place called Chicxulub. The crater at Chicxulub, now buried beneath younger sedimentary rock, is the right size to have been formed by a 10-kilometer asteroid.

The existence of the crater was confirmed by geologists going over Mexican oil company records compiled during drilling for oil at Yucatan and finding breccias of the right age buried in the Chicxulub area. Breccias, due to meteorite impact, are common at known meteorite craters. The evidence for an asteroid impact is overwhelming.

Extinction was unfortunate, from the perspective of dinosaurs, but was very fortunate from a human's perspective. The only mammals in the Cretaceous were inconsequential, mostly rat-sized creatures that could stay out of the way of dinosaurs. They survived the extinction and, with dinosaurs no longer dominating the land, evolved into the many mammal species that populate Earth today, including humans.

Additional Resource

BBC—Dinosaurs Visit *Tyrannosaurus rex* and other famous dinosaurs and read more about the K-T extinction.

- www.bbc.co.uk/sn/prehistoric_life/dinosaurs/

The rates at which radioactive elements decay have been measured and duplicated in many different laboratories. Therefore, if we can determine the ratio of a particular radioactive element and its decay products in a mineral, we can calculate how long ago that mineral crystallized.

Determining the age of a rock through its radioactive elements is known as **isotopic dating** (previously, and somewhat inaccurately, called *radiometric dating*). Geologists who specialize in this important field are known as *geochronologists.*

Isotopes and Radioactive Decay

As discussed in chapter 2, every atom of a given element possesses the same number of protons in its nucleus. (Atomic numbers, which indicate the number of protons in the atom of an element, can be found in the periodic table of elements in appendix D.) The number of neutrons, however, need not be the same in all atoms of the same element. The **isotopes** of a given element have different numbers of neutrons but the same number of protons.

Uranium, for example, commonly occurs as two isotopes, uranium-238 (^{238}U) and uranium-235 (^{235}U). The former has a total of 238 protons and neutrons in its nucleus, whereas the latter has a total of 235. (^{238}U is, by far, the most abundant of naturally occurring uranium isotopes. Only 0.72% of uranium is ^{235}U; however, this is the isotope used for nuclear weapons and power generators.) Uranium has an atomic number of 92. Both isotopes therefore have 92 protons and the difference in their mass is due to ^{238}U containing 146 neutrons in its nucleus and ^{235}U containing only 143.

Radioactive decay is the spontaneous nuclear change of isotopes with unstable nuclei. Energy is produced with radioactive decay. Emissions from radioactive elements can be detected by a Geiger counter or similar device and, in high concentrations, can damage or kill humans (see box 8.3).

Nuclei of radioactive isotopes change primarily in three ways (figure 8.20). An *alpha* (α) *emission* is the ejection of 2 protons and 2 neutrons from a nucleus. When an alpha emission takes place, the *atomic number* of the atom is reduced by 2, because 2 protons are lost, and its *atomic mass number* is reduced by 4, because a total of 2 protons and 2 neutrons are lost. After an alpha emission, ^{238}U becomes ^{234}Th (thorium-234), which has an atomic number of 90. The original isotope (^{238}U) is referred to as the *parent isotope.* The new isotope (^{234}Th) is the *daughter product.*

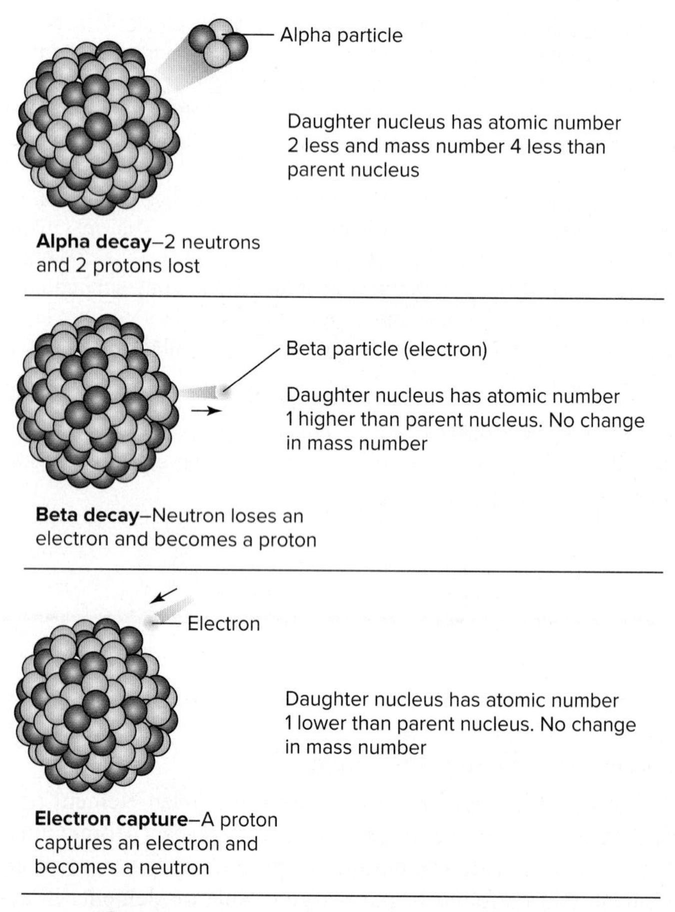
Alpha particle
Daughter nucleus has atomic number 2 less and mass number 4 less than parent nucleus
Alpha decay–2 neutrons and 2 protons lost
Beta particle (electron)
Daughter nucleus has atomic number 1 higher than parent nucleus. No change in mass number
Beta decay–Neutron loses an electron and becomes a proton
Electron
Daughter nucleus has atomic number 1 lower than parent nucleus. No change in mass number
Electron capture–A proton captures an electron and becomes a neutron
Proton
Neutron
Electron

FIGURE 8.20
Three modes of radioactive decay.

Beta (β) *decay* involves the release of an electron from a nucleus. To understand this, we need to explain that electrons, which have virtually no mass and are usually in orbit around the nucleus, are also in the nucleus as part of a neutron. A neutron is a proton with an electron inside of it; thus, it is electrically neutral. If an electron is emitted from a neutron during radioactive decay, the neutron becomes a proton and the atom's atomic number is increased by one. For example, when ^{234}Th (atomic number 90) undergoes a beta decay, it becomes ^{234}Pa (protactinium-234), an element with an atomic number of 91. Note that the atomic mass number has not changed. This is because atomic mass number is equal to the number of protons and neutrons in the nucleus. When a neutron converts to a proton by beta decay, the number of protons increases by one and the number of neutrons decreases by one, leaving the total the same. The weight of the ejected electron is negligible.

The third mode of change is *electron capture,* whereby a proton in the nucleus captures an orbiting electron. The proton becomes a neutron. The atom becomes a different element having an atomic number one less than its parent isotope. An example of this is the potassium-argon system in table 8.3, in which ^{40}K (potassium-40) becomes ^{40}Ar (argon-40). The parent isotope, potassium, has an atomic number of 19, and the atomic number of argon, the daughter product, is 18, because a proton was changed into a neutron.

Figure 8.21 shows how ^{238}U decays to ^{206}Pb (lead-206) in a series of alpha and beta emissions. The important point is not the intermediate steps but the starting and ending isotopes. In the process, ^{238}U loses 10 protons, so that the daughter product has an atomic number of 82 (which is lead), and loses a total of 32 protons and neutrons, so the new atomic mass number is 206. ^{206}Pb can only be produced by the decay of ^{238}U.

To understand how isotopic dating works, it is important to recognize that if a large number of atoms of a given radioactive isotope are present in a rock or mineral, the *proportion* (or percentage) of those atoms that will radioactively decay over a given time span is constant. For example, if you have 100,000 atoms of isotope X and over a period of a million years, a quarter of those atoms (25,000) radioactively decay, the proportion would be 1 in 4. You would have the same proportion of 1 in 4 if you started out with 300,000 atoms; after a million years, 75,000 of the atoms would have decayed. The proportional amount of atoms that decay in time is unaffected by chemical reactions or by the high pressures and high temperatures of Earth's interior.

The rate of proportional decay for isotopes is expressed as **half-life**, the time it takes for one-half of a given number of radioactive atoms to decay. The half-lives of some isotopes created in nuclear reactors are in fractions of a second. Naturally occurring isotopes used to date rocks have very long half-lives (table 8.3). ^{40}K has a half-life of 1.3 billion years. If you began with 1 milligram of ^{40}K, 1.3 billion years later one-half milligram of ^{40}K would remain. After another 1.3 billion years, half of the remaining one-half milligram of ^{40}K would decay, leaving one-fourth of

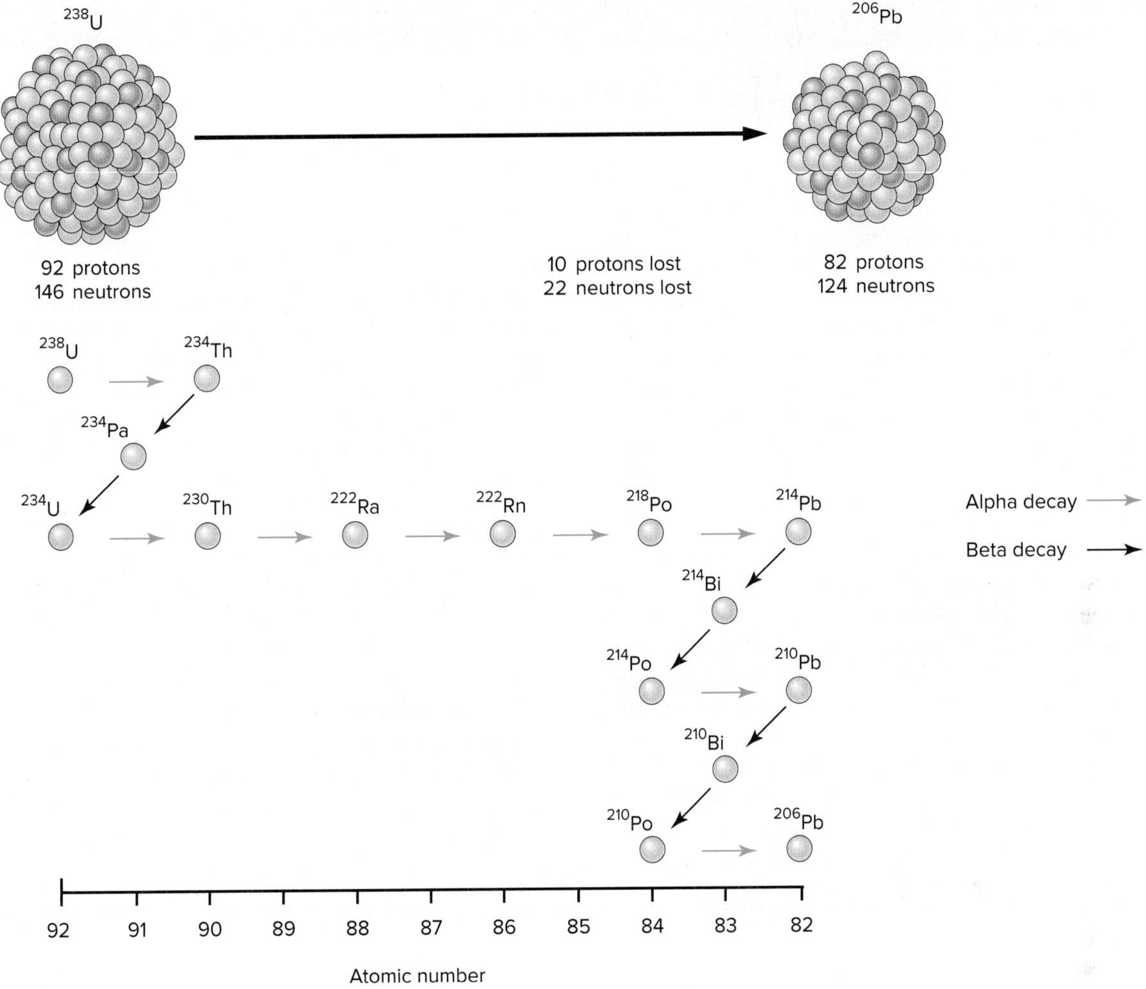

FIGURE 8.21

Uranium-238 decays to lead-206. The different intermediate steps in the process are shown below the models of the nuclei of ^{238}U and ^{206}Pb. Refer to appendix C or the periodic table of elements in appendix D for names of the elements shown.

a milligram, and after another half-life, only one-eighth of a milligram. Note that two half-lives do not equal a whole life. Compare this to a burning candle (figure 8.22*B*), which burns at a linear rate. The same amount of candle burns away for each period of time.

Different isotopes are used to date rocks of different ages. For very old rocks, isotopes with very long half-lives must be used (table 8.3). If the half-life is too short, all of the parent isotopes would have decayed away (and the clock stopped) long before the rock was collected. Conversely, for younger rocks, isotopes with shorter half-lives must be used. If the half-life is too long, not enough daughter isotopes will have been produced for an accurate analysis.

To determine the age of a rock by using ^{40}K, the amount of ^{40}K in that rock must first be determined by chemical analysis. The amount of ^{40}Ar (the daughter product) must also be determined. Adding the two values gives us how much ^{40}K was present when the rock formed. By knowing how much ^{40}K was

TABLE 8.3 Radioactive Isotopes Commonly Used for Determining Ages of Earth's Materials

Parent Isotope	Half-Life	Daughter Product	Effective Dating Range (years)
^{40}K	1.3 billion years	^{40}Ar	100,000–4.6 billion
^{238}U	4.5 billion years	^{206}Pb	10 million–4.6 billion
^{235}U	703 million years	^{207}Pb	10 million–4.6 billion
^{232}Th	14.1 billion years	^{208}Pb	10 million–4.6 billion
^{87}Rb	49 billion years	^{87}Sr	10 million–4.6 billion
^{14}C	5,730 years	^{14}N	100–40,000

ENVIRONMENTAL GEOLOGY 8.3

Radon, a Radioactive Health Hazard

Radon is an odorless, colorless gas. Every time you breathe outdoors, you inhale a harmless, minute amount of radon. If the concentration of radon that you breathe in a building is too high, however, you could, over time, develop lung cancer. It is one of the intermediate daughter products in the radioactive disintegration of ^{238}U to ^{206}Pb. It has a half-life of only 3.8 days.

Concentrations of radon are highest in areas where the bedrock is granite, gneiss, limestone, black shale, or phosphate-rich rock—rocks in which uranium is relatively abundant. Concentrations are also high where glacial deposits are made of fragments of these rocks. Even in these areas, radon levels are harmless in open, freely circulating air. Radon may dissolve in groundwater or build up to high concentrations in confined air spaces.

Radon was first recognized in the 1950s as a health hazard in uranium mines, where the gas would collect in poorly ventilated air spaces. Radon lodges in the respiratory system of an individual, and as it deteriorates into daughter products, the subatomic particles given off damage lung tissue. The U.S. Environmental Protection Agency (EPA) estimates that nearly 1 out of every 15 American homes have unacceptable radon levels in the air.

What should you do if you are living in a high radon area? First, have your house checked to see what the radon level is. (You may purchase a simple and inexpensive test kit at many home improvement centers.) Then, read up on what acceptable standards should be. In most buildings with a high radon level, the gas seeps in from the underlying soil through the building's foundation. If a building's windows are kept open and fresh air circulates freely, radon concentrations cannot build up. But houses are often kept sealed for air conditioning during the summer and heating during the winter. Air circulation patterns are such that a slight vacuum sucks the gases from the underlying soil into the house. Thus, radon concentrations might build up to dangerous levels.

The problem may be solved in several ways (aside from leaving windows open winter and summer). Basements can be made air tight so that gases cannot be sucked into the house from the soil. Air circulation patterns can be altered so that gases are not sucked in from underlying soil or are mixed with sufficient fresh, outside air.

If you are purchasing a new house, it would be a good idea to have it tested for radon before buying, particularly if the house is in an area of high-uranium bedrock or soil.

Additional Resource

US Environmental Protection Agency information site about radon. Includes a link to their radon zone map.

- http://www.epa.gov/radon

originally present in the rock and how much is still there, we can calculate the age of the rock on the basis of its half-life mathematically (see box 8.4). The graph in figure 8.22*A* applies the mathematical relationship between a radioactively decaying isotope and time and can be used to easily determine an isotopic age.

Radiocarbon Dating

Because of its short half-life of 5,730 years, radiocarbon dating is useful only in dating things and events accurately back to about 40,000 years–about seven half-lives. The technique is most useful in archaeological dating and for very young geologic events (i.e., Holocene, or Recent, volcanic and glacial features). It is also used to date historical artifacts. For instance, the Dead Sea Scrolls, the oldest of the surviving biblical manuscripts, were radiocarbon dated and their ages ranged from the third century B.C. to A.D. 68. These ages are consistent with estimates previously made by archaeologists and other scholars.

Radiocarbon dating is fundamentally different from the parent-daughter systems described previously in that ^{14}C is being created continuously in the atmosphere. Carbon (atomic number 6) in air is held in CO_2. It is mostly the stable isotope ^{12}C. However, ^{14}C is created in the atmosphere as follows:

- Neutrons as cosmic radiation bombard nitrogen (N), atomic number 7. A neutron is captured by an ^{14}N atom's nucleus.
- This causes a proton to be immediately expelled from the atom, and the atom becomes ^{14}C.
- The nucleus of the newly created carbon atom is unstable and will, sooner or later, through a beta emission (loss of an electron from a neutron), revert to ^{14}N.
- The rate of production of ^{14}C approximately balances the rate at which ^{14}C reverts to ^{14}N so that the level of ^{14}C remains essentially constant in the atmosphere.

Living matter incorporates ^{12}C and ^{14}C into its tissues; the ratios of ^{12}C and ^{14}C in the new tissues are the same as in the atmosphere. On dying, the plant or animal ceases to build new tissue. The ^{14}C disintegrates radioactively at the fixed rate of its half-life (5,730 years). The ratio of ^{12}C to radioactive ^{14}C in organic remains is determined in a laboratory. Using the ratio, the time elapsed since the death of the organism is calculated. We now know there has been some variation in the rate of

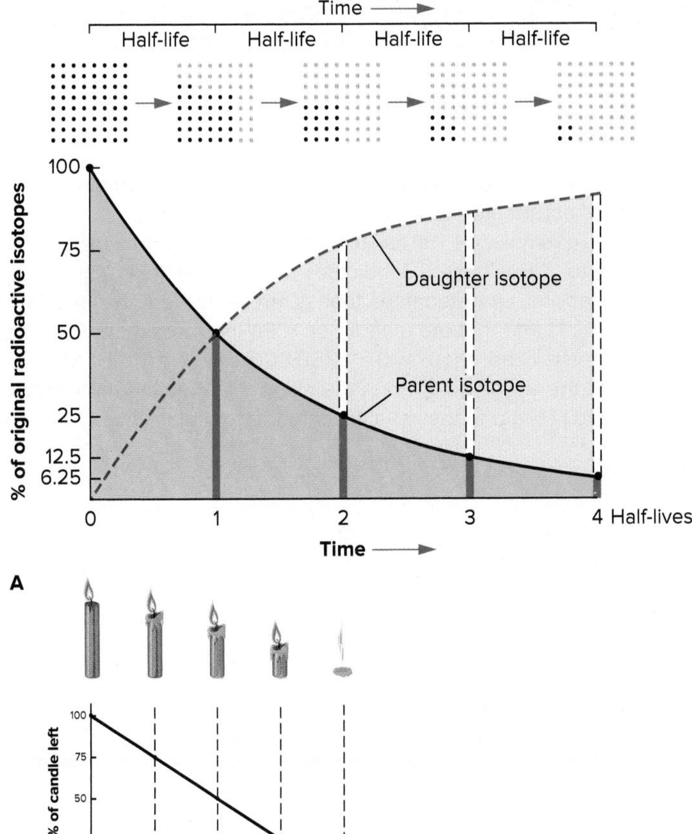

FIGURE 8.22

(*A*) The curve used to determine the age of a rock by comparing the percentage of radioactive isotope remaining in time to the original amount. Dark blue bars show the amount left after each half-life. Dashed red curve shows the amount disintegrated into daughter product and lost nuclear particles. The numbers of dots in the squares above the graph are proportional to the numbers of atoms. (*B*) For comparison, a candle burns at a linear rate. Note that for the candle, two "half lives" equal a whole life.

production of ^{14}C in the atmosphere in the past. Radiocarbon dates are now calibrated to account for those fluctuations.

Cosmogenic Isotope Dating

During the past couple decades, another dating technique has been added to geologists' numerical age determination arsenal. *Cosmogenic isotope dating,* or *surface exposure dating,* uses the effects of constant bombardment by neutron radiation coming from deep space (cosmogenic) of material at Earth's surface. The high-energy particles hit atoms in minerals and alter their nuclei. For instance, when the atoms in quartz are hit, oxygen is converted to beryllium-10 (^{10}Be) and silicon is changed to aluminum-26 (^{26}Al). The concentrations of these isotopes increase at a constant rate once a rock surface is exposed to the atmosphere because the influx of cosmogenic radiation is uniform over time. The length of time a rock surface has been exposed can be calculated by knowing the rate of increase of a cosmogenic isotope and determining the amount of that isotope in a mineral at a rock's surface.

One application of cosmogenic dating has been to determine how long ago boulders were deposited by advancing glaciers during the geologically recent ice ages (see chapter 12). However, dates obtained are minimum ages because snow that covered the boulders for part of the year reduced their exposure to cosmogenic radiation.

Uses of Isotopic Dating

When we are dating a rock, we are usually trying to determine how long ago that rock formed. But exactly what event in the rock's history is being dated depends on the type of rock and the isotopes analyzed. For a metamorphic rock, we are likely to be dating a time during the millions of years of the cooling of that rock rather than the peak of high temperature during metamorphism. For a sedimentary rock, we may be dating the older rock that weathered and eroded to form the sediment that eventually became the sedimentary rock. Some techniques determine isotopic ratios for a whole rock, while others use single minerals within a rock. Usually, an isotopic date determines how long ago the rock or mineral became a closed system; that is, how long ago it was sealed off so that neither parent nor daughter isotopes could enter or leave the mineral or rock. Each isotopic pair has a different *closure temperature*—the temperature below which the system is closed and the "clock" starts. For instance, the ^{40}K ^{40}Ar isotopic pair has closure temperatures ranging from 150°C to 550°C, depending on the mineral. (Ar is a gas and gets trapped in different crystal structures at different temperatures.)

Generally, the best dates are obtained from igneous rocks. For a lava flow, which cools and solidifies rapidly, the age determined is the precise time at which the rock formed. On the other hand, plutonic rocks, which may take over a million years to solidify, will not necessarily yield the time of intrusion but the time at which a mineral cooled below the closure temperature. Dating metamorphic rocks usually means determining when closure temperatures for particular minerals are reached during cooling. Sedimentary rocks are difficult to date reliably because they contain sediment, which is older than the sedimentary rock.

For an isotopic age determination to be accurate, several conditions must be met. To ensure that the isotopic system has remained closed, the rock collected must show no signs of weathering or hydrothermal alteration. Second, one should be able to infer there were no daughter isotopes in the system at the time of closure or make corrections for probable amounts of daughter isotopes present before the "clock" was set. Third, there must be enough parent and daughter atoms to be measurable by the instrument (a mass spectrometer) being used. And, of course, technicians and geochronologists must be highly skilled at working sophisticated equipment and collecting and processing rock specimens. (For more on mass spectrometry, go to www.youtube.com/watch?v=J-wao0O0_qM.)

Whenever possible, geochronologists will use more than one isotope pair for a rock. The two U-Pb systems (table 8.3) can usually be used together and provide an internal cross-check

IN GREATER DEPTH 8.4

Calculating the Age of a Rock

The relationship between time and radioactive decay of an isotope is expressed by the following equation (which is used to plot curves such as those shown in figure 8.22).

$$N = N_0 e^{-\lambda t}$$

N is the number of atoms of the isotope at time t, the time elapsed. N_0 is the number of atoms of that isotope present when the "clock" was set. The mathematical constant e has a value of 2.718. λ is a decay constant—a proportionality constant that relates the rate of decay of an isotope to the number of atoms of that isotope remaining.

The relationship between λ and the half-life (t_{hl}) is

$$\lambda = \frac{\ln 2}{t_{hl}} = \frac{0.693}{t_{hl}}$$

Replacing λ in the first equation and converting that equation to natural logarithmic (to the base e) form, we get

$$t = -\frac{t_{hl}}{.693} \ln\frac{N}{N_0}$$

N/N_0 is the ratio of parent atoms at present to the original number of parent atoms.

As an example, we will calculate the age of a mineral using ^{235}U decaying to ^{207}Pb. Table 8.3 indicates that the half-life is 713 million years. A laboratory determines that, at present, there are 440,000 atoms of ^{235}U and that the amount of ^{207}Pb indicates that when the mineral crystallized, there were 1,200,000 atoms of ^{235}U. (We assume that there was no ^{207}Pb in the mineral at the time the mineral crystallized.) Plugging these values into the formula, we get

$$t = -\frac{713{,}000{,}000}{.693} \ln\frac{440{,}000}{1{,}200{,}000}$$

Solving this gives us 1,032,038,250 years.

on the age determination. Because of their high closure temperatures, U-Pb systems are usually more realistic measures of the crystallization ages of rocks than K-Ar or Rb-Sr results.

Techniques for dating have been refined in recent years, reducing the uncertainties of dates. In 2008, scientists reported on calibration of the K-Ar system that gives dates closer to those obtained by U-Pb for a given rock or mineral. Because of this, the K-T boundary has been moved from 65.5 million years ago to 66.0 million years ago. The greatest mass extinction, at the close of the Paleozoic, has been moved from 251.0 million years ago to 252.2 million years ago. This new age places it at the time of huge flood basalt eruptions in Siberia.

How Reliable Is Isotopic Dating?

Half-lives of radioactive isotopes, whether short-lived, such as those used in medicine, or long-lived, such as those used in isotopic dating, have been found not to vary beyond statistical expectations. The half-life of each of the isotopes we use for dating rocks has not changed with physical conditions or chemical activity, nor could the rates have been different in the distant past. It would violate laws of physics for decay rates (half-lives) to have been different in the past. Moreover, when several isotopic dating systems are painstakingly done on a single ancient igneous rock, the same age is obtained within calculable margins of error. This confirms that the decay constants for each system are indeed constant.

Comparing isotopic ages with relative age relationships confirms the reliability of isotopic dating. For instance, a dike that cross-cuts rocks containing Cenozoic fossils gives us a relatively young isotopic age (less than 66 million years old), whereas a pluton truncated by overlying sedimentary rocks with earliest Paleozoic fossils yields a relatively old age (greater than 541 million years). Many thousands of similar determinations have confirmed the reliability of radiometric dating.

8.4 COMBINING RELATIVE AND NUMERICAL AGES

Radiometric dating can provide numerical time brackets for events whose relative ages are known. Figure 8.23 adds isotopic dates for each of the two igneous bodies in the fictitious Minor Canyon area of figure 8.1. The date obtained for the granite is 540 million years B.P. (before present), while the dike formed 78 million years ago. We can now state that the Tarburg Formation and older tilted layers formed before 540 million years ago (though we cannot say how much older they are). We still do not know whether the Leet Junction Formation is older or younger than the granite because of the lack of cross-cutting relationships. The Larsonton Formation's age is bracketed by the age of the granite and the age of the dike. That is, it is between 540 and 78 million years old. The Foster City and overlying formations are younger than 78 million years old; how much younger we cannot say.

Isotopic dates from volcanic ash layers or lava flows interlayered between fossiliferous sedimentary rocks have been used to assign numerical ages to the geologic time scale (figure 8.24). Isotopic dating has also allowed us to extend the time scale back into the Precambrian. There is, of course, a margin of

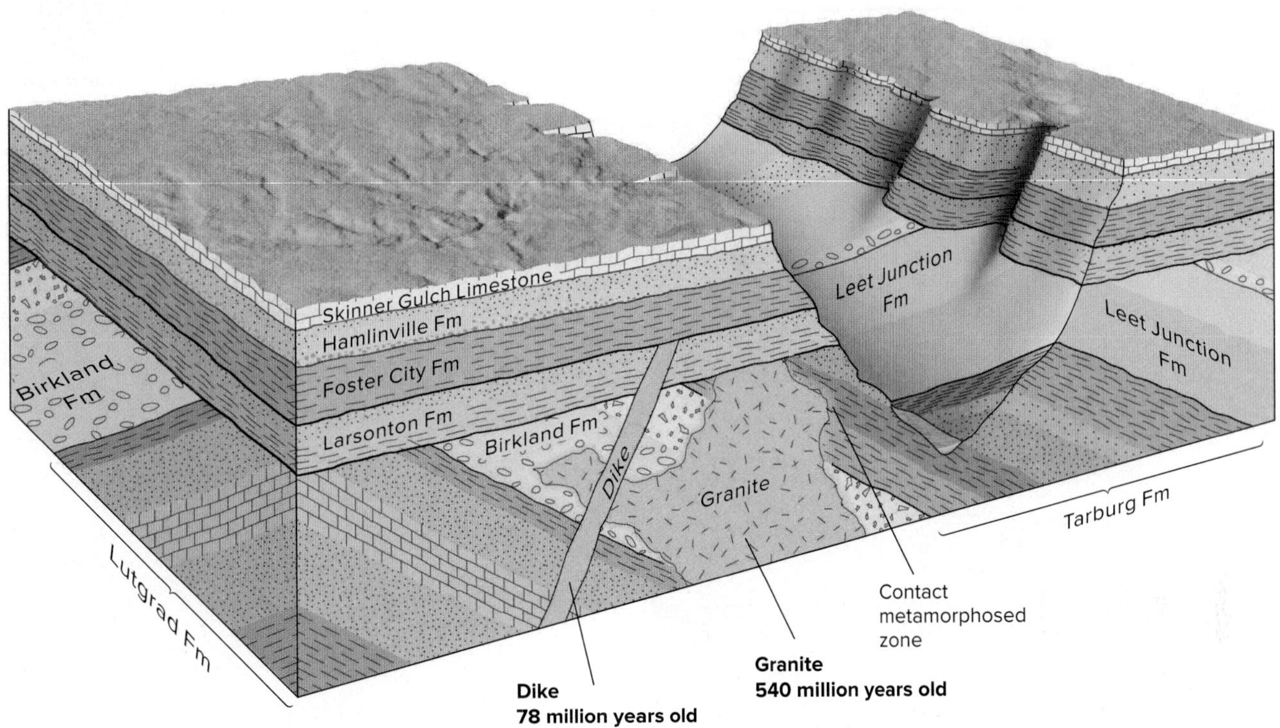

FIGURE 8.23

The Minor Canyon area as shown in figure 8.1 but with isotopic dates for igneous rocks indicated.

uncertainty in each of the given dates. The beginning of the Paleozoic, for instance, was regarded until recently to be 570 million years ago but with an uncertainty of ± 30 million years. Recent work has fixed the age as 542 ± 1 million years. There are inherent limitations on the dating techniques as well as problems in finding the ideal rock for dating. For instance, if you wanted to obtain the date for the end of the Paleozoic Era and the beginning of the Mesozoic Era, the ideal rock would be found where there is no break in deposition of sediments between the two eras, as indicated by fossils in the rocks. But the difficulties in dating sedimentary rock mean you would be unlikely to date such rocks. Therefore, you would need to date volcanic rocks interlayered with sedimentary rocks found as close as possible to the transitional sedimentary strata. Alternatively, isotopically dated intrusions, such as dikes, whose cross-cutting relationships indicate that the age of intrusion is close to that of the transitional sedimentary layers, could be used to approximate the numerical age of the transition.

Isotopic dating has shown that the Precambrian took up most (87%) of geologic time. Obviously, the Precambrian needed to be subdivided. The three major subdivisions of the Precambrian are the **Hadean** (*Hades* is, in Greek mythology, the underground place where the dead live; the name alludes to the hell-like nature of Earth's early surface), the **Archean**, and the **Proterozoic** (Greek for "beginning life"). Each is regarded as an eon, the largest unit of geologic time. A fourth, and youngest, eon is the **Phanerozoic** (Greek for "visible life"). The Phanerozoic Eon is all of geologic time with an abundant fossil record; in other words, it is made up of the three eras that followed the Precambrian.

8.5 AGE OF THE EARTH

In 1625, Archbishop James Ussher determined that Earth was created in the year 4004 B.C. His age determination was made by counting back generations in the Bible. This would make Earth 6,000 years old at present. That very young age of Earth was largely taken for granted by Western countries. By contrast, Hindus at the time regarded Earth as very old. According to an ancient Hindu calendar, the year A.D. 2000 would be year 1,972,949,101.

With the popularization of uniformitarianism in the early 1800s, Earth scientists began to realize that Earth must be very old—at least in the hundreds of millions of years. They were dealt a setback by the famous English physicist, Lord Kelvin. Kelvin, in 1866, calculated from the rate at which Earth loses heat that Earth must have been entirely molten between 20 and 100 million years ago. He later refined his estimate to between 20 and 40 million years. He was rather arrogant in scoffing at Earth scientists who believed that uniformitarianism indicated a much older age for Earth. The discovery of radioactivity in 1896 invalidated Kelvin's claim because it provided a heat source that he had not known about. When radioactive elements decay, heat is given off, and that heat is added to the heat already in Earth. The amount of radioactive heat given off at present approximates the heat Earth is losing. So, for all practical purposes, Earth is not getting cooler.

The discovery of radioactivity also provided the means to determine how old Earth is. In 1905, the first crude isotopic dates were done and indicated an age of 2 billion years. But since then, we have dated rocks on Earth that are twice that age.

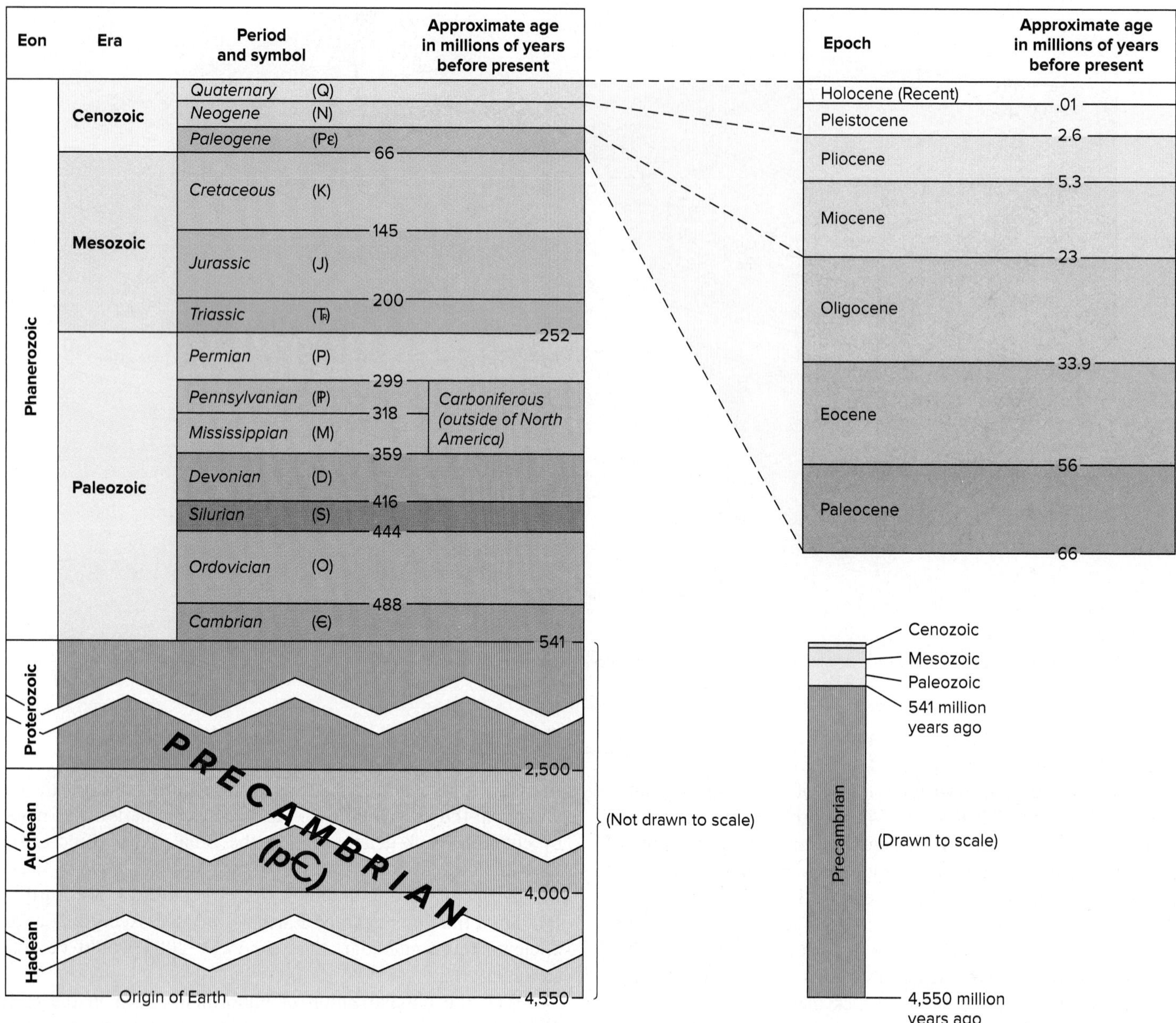

FIGURE 8.24

The geologic time scale. The small diagram to the right shows the Precambrian and the three eras at the same scale. Note that the Precambrian accounts for almost 90% of geologic time.

Source: After A. V. Okulitch, Geological Survey of Canada, Open File 3040 and International Commission on Stratigraphy, 2013. www.stratigraphy.org/index.php/ics-chart-timescale. Currently, the Geological Society of America annually updates the geologic time scale and posts it on www.geosociety.org/science/timescale/.

Earth is now regarded as being between 4.5 and 4.6 billion years old—much older than the oldest rock found. Because erosion and tectonic activity have recycled the original material at Earth's surface, we cannot determine Earth's age from its rocks. The age determination comes primarily from dates obtained from meteorites and lunar rocks. Most meteorites are regarded as fragments of material that did not coalesce into a planet. The oldest dates obtained from meteorites and lunar rocks are in the 4.5 to 4.6 billion-year range. It is highly likely that the planets and other bodies of the solar system, including Earth, formed at approximately the same time.

Comprehending Geologic Time

The vastness of geologic time (sometimes called deep time) is difficult for us to comprehend. One way of visualizing deep time is to imagine driving from Los Angeles to New York, a distance of approximately 4,500 kilometers, where each kilometer represents 1 million years—this is a very, very slow trip. The highlights of the trip corresponding to Earth's history are shown in figure 8.25. Note that if you live to be 100, your life is represented by less than the width of a curb at the edge of a sidewalk.

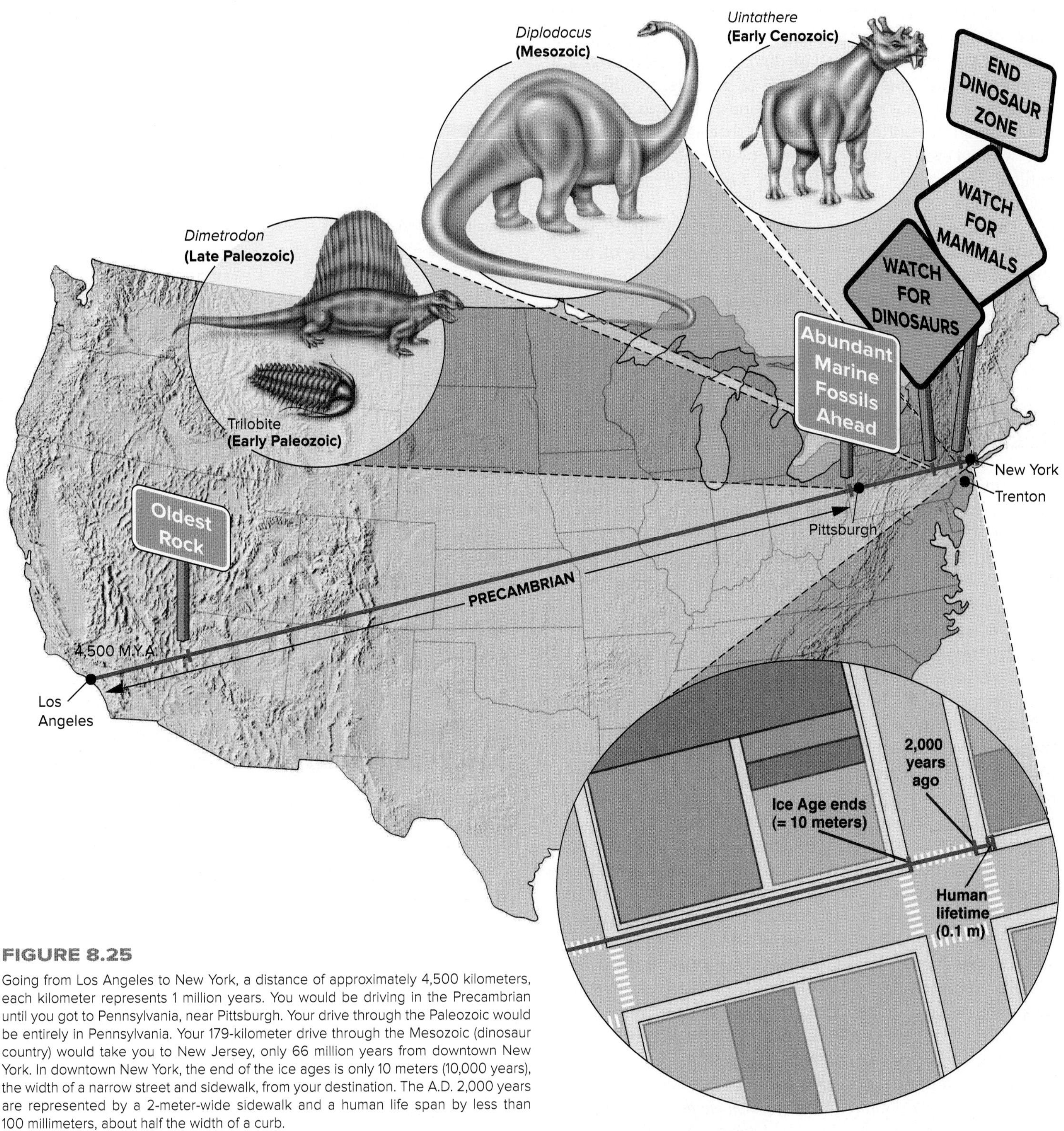

FIGURE 8.25

Going from Los Angeles to New York, a distance of approximately 4,500 kilometers, each kilometer represents 1 million years. You would be driving in the Precambrian until you got to Pennsylvania, near Pittsburgh. Your drive through the Paleozoic would be entirely in Pennsylvania. Your 179-kilometer drive through the Mesozoic (dinosaur country) would take you to New Jersey, only 66 million years from downtown New York. In downtown New York, the end of the ice ages is only 10 meters (10,000 years), the width of a narrow street and sidewalk, from your destination. The A.D. 2,000 years are represented by a 2-meter-wide sidewalk and a human life span by less than 100 millimeters, about half the width of a curb.

Another way to get a sense of geologic time is to compare it to a motion picture. A movie is projected at a rate of thirty-two frames per second; that is, each image is flashed on the screen for only 1/32 of a second, giving the illusion of continuous motion. But suppose that each frame represented 100 years. If you lived 100 years, one frame would represent your whole lifetime.

If we were able to show the movie on a standard projector, each 100 years would flash by in 1/32 of a second. It would take only 1/16 of a second to go back to the signing of the Declaration of Independence. The 2,000-year-old Christian era would be on screen for 3/4 of a second. A section showing all time back to the last major ice age would only be less than seven seconds long. However, you would have to sit through almost six hours of film to view a scene at the close of the Mesozoic Era (perhaps you would see the last dinosaur die). And to give a complete record from the beginning of the Paleozoic Era, this epic film would have to run continuously for two days. You would have to spend over two weeks (sixteen days) in the theater, without even a popcorn break between reels, to see a movie entitled *The Complete Story of Earth, from Its Birth to Modern Civilization.*

Thinking of our lives as taking less than a frame of such a movie can be very humbling. From the perspective of being stuck in that one last frame, geologists would like to know what the whole movie is like or, at least, get a synopsis of the most dramatic parts of the film.

Summary

The principle of *uniformitarianism* (or *actualism,*) a fundamental concept of geology, states that the present is the key to the past.

Relative time, or the sequence in which geologic events occur in an area, can be determined by applying the principles of *original horizontality, superposition, lateral continuity,* and *cross-cutting relationships.*

Unconformities are buried erosion surfaces that help geologists determine the relative sequence of events in the geologic past. Beds above and below a *disconformity* are parallel, generally indicating less intense activity in Earth's crust. An *angular unconformity* implies that folding or tilting of rocks took place before or around the time of erosion. A *nonconformity* implies deep erosion because metamorphic or plutonic rocks have been exposed and subsequently buried by younger rock.

Rocks can be correlated by determining the physical continuity of rocks between the two areas (generally, this works only for a short distance). A less useful means of correlation is similarity of rock types (which must be used cautiously).

The principle of *faunal succession* states that fossil species succeed one another in a definite and recognizable order. Fossils are used for worldwide correlation of rocks. Sedimentary rocks are assigned to the various subdivisions of the *geologic time scale* on the basis of fossils they contain, which are arranged according to the principle of faunal succession.

Numerical age—how many years ago a geologic event took place—is generally obtained by using *isotopic dating* techniques. Isotopic dating is accomplished by determining the ratio of the amount of a radioactive isotope currently in a rock or mineral being dated to the amount originally present. The time it takes for a given amount of an isotope to decay to half that amount is the *half-life* for that isotope. Rocks that are geologically old are usually dated by isotopes having half-lives of over a billion years. Radiocarbon dating of organic matter is used for dating events younger than forty thousand years. Cosmogenic isotopic dating is used to determine how long rock has been exposed at Earth's surface. Numerical ages have been determined for the subdivisions of the geologic time scale. The scientifically determined age of Earth is 4.5 to 4.6 billion years.

Terms to Remember

actualism 180
angular unconformity 187
Archean Eon 199
Cenozoic Era 191
contacts 181
correlation 187
cross-cutting relationships 182
disconformity 186
eon 191
epochs 191
eras 191
faunal succession 190
formations 181
Hadean Eon 199
half-life 194
Holocene (or Recent) Epoch 191
inclusion 186
index fossil 190
isotopes 193
isotopic dating 192
lateral continuity 182
Mesozoic Era 191
nonconformity 187
numerical age 181
original horizontality 182
Paleozoic Era 191
periods 191
Phanerozoic Eon 191
physical continuity 187
Pleistocene Epoch 191
Precambrian 191
Proterozoic Eon 199
Quaternary Period 191
radioactive decay 194
relative time 181
standard geologic time scale 191
superposition 182
unconformity 186
uniformitarianism 180

Testing Your Knowledge

Use the following questions to prepare for exams based on this chapter.

1. A dike cuts through a granite stock, which contains inclusions of the surrounding sandstone layer. In what order did these three units form?
2. By applying the various principles, draw a cross section of an area in which the following sequence of events occurred. The relative time relationship for all events should be clear from your single cross section that shows what the geology looks like at present.
   a. Metamorphism took place during the Archean. During later Precambrian time, uplift and erosion reduced the area to a plane.
   b. Three layers of marine sedimentary rock were deposited on the plain during Ordovician through Devonian time.
   c. Although sedimentation may have taken place during the Mississippian through Permian, there are currently no sedimentary rocks of that age in the area.
   d. A vertical dike intruded all rocks that existed here during the Permian.
   e. A layer of sandstone was deposited during the Triassic.
   f. All of the rocks were tilted 45 degrees during the early Cretaceous. This was followed by erosion to a planar surface.
   g. The area dropped below sea level, and two layers of sedimentary rock were deposited on the erosion surface during the Paleogene and Neogene.

h. Uplift and erosion during the Quaternary resulted in a slightly hilly surface.

i. Following erosion, a vertical dike fed a small volcano.

3. Why is it desirable to find an index fossil in a rock layer? In the absence of index fossils, why is it desirable to find several fossils in a rock unit to determine relative age?
4. Radioactive isotope X decays to daughter isotope Y with a half-life of 220,000 years. At present you have 1/4 gram of X in a rock. From the amount of daughter isotope Y currently in the rock, you determine that the rock contained 16 grams of isotope X when it formed. How many half-lives have gone by? How old is the rock?
5. Draw diagrams illustrating the three types of unconformity. How can you determine that:

   a. An unconformity you are examining is a nonconformity and not an angular unconformity?

   b. A disconformity is present in a sequence of sedimentary layers?
6. Using figure 8.23, suppose the base of the Hamlinville Formation has a layer of volcanic ash that is dated as being 49 million years old. How old is the Foster City Formation?
7. Explain why a geologist trying to date Devonian volcanic rocks would be more likely to use the potassium-argon system than radiocarbon dating.
8. "Geological processes operating at present are the same processes that have operated in the past" is the principle of

   a. correlation. b. catastrophism.
   c. uniformitarianism. d. absolutism.
9. "Within a sequence of undisturbed sedimentary rocks, the layers get younger going from bottom to top" is the principle of

   a. original horizontality. b. superposition.
   c. cross-cutting. d. unconformities.
10. If rock A cuts across rock B, then rock B is ____ rock A.

    a. younger than b. the same age as
    c. older than
11. A contact between parallel sedimentary rock layers that records missing geologic time is

    a. a disconformity. b. an angular unconformity.
    c. a nonconformity. d. a sedimentary contact.
12. Which is a method of correlation?

    a. physical continuity b. similarity of rock types
    c. fossils d. all of the preceding
13. Which is *not* a type of unconformity?

    a. disconformity b. angular unconformity
    c. nonconformity d. triconformity
14. A geologist could use the principle of inclusion to determine the relative age of

    a. fossils. b. metamorphism.
    c. shale layers. d. xenoliths.
15. Eras are subdivided into

    a. periods. b. ages.
    c. eons. d. epochs.
16. Periods are subdivided into

    a. eras. b. epochs.
    c. ages. d. time zones.
17. Which division of geologic time was the longest?

    a. Precambrian b. Paleozoic
    c. Mesozoic d. Cenozoic
18. The oldest abundant fossils of complex multicellular life with shells and other hard parts date from the

    a. Precambrian. b. Paleozoic.
    c. Mesozoic. d. Cenozoic.
19. The greatest mass extinction ever to occur on Earth took place at the end of the

    a. Cambrian. b. Mesozoic.
    c. Paleozoic. d. Cretaceous.

Expanding Your Knowledge

1. How much of the ^{238}U originally part of Earth is still present?
2. As indicated by fossil records, why have some ancient organisms survived through very long periods of time whereas others have been very short-lived?
3. To what extent would a composite volcano (see chapter 4) be subject to the three principles described in this chapter?
4. Suppose a sequence of sedimentary rock layers was tilted into a vertical position by tectonic forces. How might you determine (*a*) which end was originally up and (*b*) the relative ages of the layers?
5. Note that in table 8.2, the epochs are given only for the Cenozoic Era (as is commonly done in geology textbooks). Why are the epochs for the Mesozoic and Paleozoic considered less important and not given?
6. Why would you not be able to use the principle of superposition to determine the age of a sill (defined in chapter 3)?
7. Using information from box 8.4, calculate the age of a feldspar. At present, there are 1.2 million atoms of ^{40}K. The amount of ^{40}Ar in the mineral indicates that originally, there were 1.9 million ^{40}K atoms in the rock. Use a half-life of 1.3 billion years. (Hint: The answer is 862 million years.)

Exploring Web Resources

www.ucmp.berkeley.edu/exhibits/index.php

Online exhibits at UCMP. University of California Museum of Paleontology virtual exhibit. Click on "Tour of Geologic Time."

nemo.sciencecourseware.org/VirtualDating/

Virtual Dating. This site provides an excellent, interactive way of learning how isotopic dating works. You can change data presented and watch graphs and other illustrations change accordingly. Quizzes help you understand the material.

www.earth-time.org

Earthtime. A site for international collaborations among geochronologists and other geologists interested in refining numerical time. Check out the sections on education and mass spectrometry.

www.talkorigins.org/origins/faqs.html

Talk Origins. This is an excellent site for in-depth information on geologic time. Click on "Age of the Earth." Topics include isotopic dating, the geologic time scale, and changing views of the age of Earth. The site includes in-depth presentations of arguments for a young Earth and the scientific rebuttals to them.

www.asa3.org/ASA/resources/Wiens.html

Radiometric Dating: A Christian Perspective. At this website, you can get a very thorough knowledge of isotopic dating, how it works, and how it has been used to determine the age of Earth and other events. The author addresses concerns of people who feel that an old Earth is incompatible with their religious beliefs.

CHAPTER

9 Mass Wasting

A farm house and stable are damaged by a landslide outside the town of Entlebuch, central Switzerland, Tuesday, August 23, 2005. Note the displaced segment of the road. This landslide is an example of a slump-earthflow described in this chapter.

LEARNING OBJECTIVES

- Discuss the importance of mass wasting in shaping the Earth's surface.
- Assess the impact of mass wasting as a natural hazard.
- Describe the factors controlling slope stability.
- Identify triggers for mass wasting.
- Classify mass wasting according to type of movement, rate of movement, and type of material.
- Differentiate between the common types of mass wasting.
- Describe engineering techniques that can be used to prevent or minimize mass wasting.

You may recall from previous chapters that plate tectonics explains how rock is deformed and why we have mountains. If tectonism were not at work, the surfaces of the continents would long ago have been reduced to featureless plains by weathering and erosion. This chapter and chapters 10 through 14 are concerned with surficial processes, the interaction of rock, air, and water in response to gravity at or near the Earth's surface. Nearly all of the features we see as landforms–rounded or rugged mountains, river valleys, cliffs and beaches along seashores, caves, sand dunes, and so on–are products of surficial processes. Surficial processes involve weathering, erosion, transportation, and deposition. Subsequent chapters address the work of running water, groundwater (water that is beneath the surface), glaciers, wind, and ocean waves.

The mountains, rolling hills, coastlines, and river valleys that make up the landscape around us may appear to be permanent, but in fact they are constantly undergoing change. Material on mountain slopes or hillsides is considered to be out of equilibrium with respect to gravity. Under the force of gravity, the various agents of erosion (moving water, ice, and wind) are constantly working to move masses of rock or loose material downhill, making slopes gentler and therefore increasingly stable. This process, called mass wasting, includes the fast-moving, often catastrophic landslide events that appear in the news, but it also includes the almost imperceptibly slow movement of soil known as *creep*.

In this chapter, we will first define mass wasting and discuss how different mass wasting processes are classified. We will then explore the factors that control slope stability before describing some of the most common types of mass wasting in more detail. Finally, we will discuss how landslides can be prevented through engineering.

9.1 MASS WASTING

Mass wasting is movement in which bedrock, rock debris, or soil moves downslope in bulk, or as a mass, because of the pull of gravity. Mass wasting includes movement so slow that it is almost imperceptible (called *creep*) as well as **landslides**, a general term for the slow to very rapid descent of rock or soil (Table 9.1). The term *landslide* tells us nothing about the processes involved. As you will see, terms such as *earthflow* and *rockslide* are far more descriptive than *landslide*.

Mass wasting affects people in many ways. Its effects range from the devastation of a killer landslide (such as the debris avalanche described in box 9.1) to the nuisance of having a fence slowly pulled apart by soil creep. The cost in lives and property from landslides is surprisingly high. Damage and casualty reports for landslides are often overlooked because they are part of a larger disaster, such as an earthquake or heavy rain from a hurricane. According to a global data set compiled by the

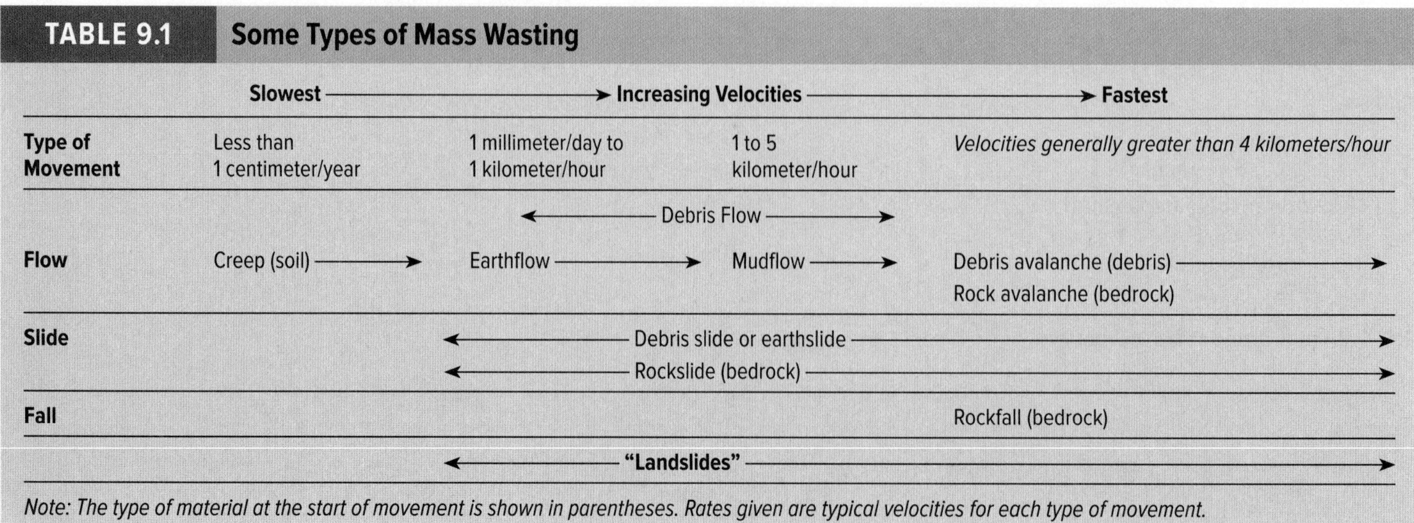

TABLE 9.1 Some Types of Mass Wasting

	Slowest ⟶	Increasing Velocities ⟶		Fastest
Type of Movement	Less than 1 centimeter/year	1 millimeter/day to 1 kilometer/hour	1 to 5 kilometer/hour	*Velocities generally greater than 4 kilometers/hour*
		⟵ Debris Flow ⟶		
Flow	Creep (soil) ⟶	Earthflow ⟶	Mudflow ⟶	Debris avalanche (debris) ⟶ Rock avalanche (bedrock)
Slide		⟵ Debris slide or earthslide ⟶ ⟵ Rockslide (bedrock) ⟶		
Fall				Rockfall (bedrock)
		⟵ **"Landslides"** ⟶		

Note: The type of material at the start of movement is shown in parentheses. Rates given are typical velocities for each type of movement.

International Landslide Center, 2,620 fatal landslides were recorded worldwide between 2004 and 2010, causing the death of 32,322 people. This data set does not include landslides triggered by earthquakes, and so the number of lives lost to landslides is probably much greater. Over time, landslides have cost Americans triple the combined costs of earthquakes, hurricanes, floods, and tornadoes. It is estimated that the annual cost of landslides in the United States has been over $1 billion and 25 to 50 lost lives. In many cases of mass wasting, a little knowledge of geology, along with appropriate preventive action, could have averted destruction.

9.2 CONTROLLING FACTORS IN MASS WASTING

Table 9.2 summarizes the factors that influence the likelihood and the rate of movement of mass wasting. The table makes apparent some of the reasons why the landslide (a debris avalanche) in Peru (box 9.1) occurred and why it moved so rapidly. (1) The slopes were exceptionally steep, and (2) the **relief** (the vertical distance between valley floor and mountain summit) was great, allowing the mass to pick up speed and momentum. (3) Water and ice not only added weight to the mass of debris, but made it more fluid. (4) Abundant loose rock and debris were available in the course of the moving mass. (5) Where the landslide began, there were no plants with roots to anchor loose material on the slope. Finally, (6) the region has earthquakes. Although the debris avalanche would have occurred eventually even without one, it was triggered by an earthquake. A *trigger* is the immediate cause of failure of already unstable ground.

Other factors influence susceptibility to mass wasting as well as its rate of movement. The orientation of planes of weakness in bedrock (bedding planes, foliation planes, etc.) is important if the movement involves bedrock rather than debris. Fractures or bedding planes oriented so that slabs of rock can slide easily along these surfaces greatly increase the likelihood of mass wasting.

Climatic controls inhibit some types of mass wasting and aid others (table 9.2). Climate influences how much and what kinds of vegetation grow in an area and what type of weathering occurs. A climate in which rain drizzles intermittently much of the year will have thick vegetation with roots that tend to inhibit mass wasting. But a prolonged period of rainfall or short-lived but heavy precipitation will make mass wasting more likely in any climate. In cold climates, freezing and thawing contribute to downslope movement.

Gravity

The driving force for mass wasting is gravity. Figures 9.1*A*–*C* show gravity acting on a block on a slope. The length of the red vertical arrow is proportional to the force—the heavier the material, the longer the arrow. The effect of gravity is resolvable into two component forces, indicated by the black arrows. One, the *normal force*, is perpendicular to the slope and is the component of gravity that tends to hold the block in place. The greater its length, the more force is needed to move the block. The other, called the **shear force**, is parallel to the slope and indicates the block's ability to move. The length of the arrows is proportional to the strength of each force. Compare figures 9.1*A* and 9.1*C*. When the slope is steeper, the shear force component of gravity is much greater than the normal force, and the block will be more likely to slide.

Friction counteracts the shear force. *Shear resistance* (represented by the brown arrow) is the force that would be needed to move the block. If that arrow is larger than the arrow representing shear force (as in figure 9.1*A*), the block will not move. The magnitude of the shear resistance (and the length of the brown arrow) is a function of friction and the size of the normal force. The brown arrow will be shorter (and the shear resistance lower)

TABLE 9.2 Summary of Controls of Mass Wasting

Driving Force: Gravity

Contributing Factors	Most Stable Situation	Most Unstable Situation
Slope angle	Gentle slopes or horizontal surface	Steep or vertical
Local relief	Low	High
Thickness of soil over bedrock	Slight thickness (usually)	Great thickness
Orientation of planes of weakness in bedrock	Planes at right angles to hillside slopes	Planes parallel to hillside slopes
Climatic factors:		
Ice in ground	Temperature stays above freezing	Freezing and thawing for much of the year
Water in soil or debris	Film of water around fine particles	Saturation of soil with water
Precipitation	Frequent but light rainfall	Episodes of heavy precipitation
Vegetation	Heavily vegetated	Sparsely vegetated

Triggers: (1) earthquakes; (2) weight added to upper part of a slope; (3) undercutting of bottom of slope; (4) heavy rainfall

ENVIRONMENTAL GEOLOGY 9.1

Disaster in the Andes

As a result of a tragic combination of geologic conditions and human ignorance of geologic hazards, one of the most devastating landslides (a debris avalanche) in history destroyed the town of Yungay in Peru in 1970. Yungay was one of the most picturesque towns in the Santa River Valley, which runs along the base of the highest peaks of the Peruvian Andes. Heavily glaciated Nevado Huascarán, 6,768 meters (22,204 feet) above sea level, rises steeply above the populated, narrow plains along the Santa River.

In May 1970, a sharp earthquake occurred. The earthquake was centered offshore from Peru about 100 kilometers from Yungay. Although the tremors in this part of the Andes were no stronger than those that have done only light damage to cities in the United States, many poorly constructed homes collapsed. Because of the steepness of the slopes, thousands of small rockfalls and rockslides were triggered.

The greatest tragedy began when a slab of glacier ice about 800 meters wide, perched near the top of Huascarán, was dislodged by the shaking. (A few years earlier, American climbers returning from the peak had warned that the ice looked highly unstable. The Peruvian press briefly noted the danger to the towns below, but the warning was soon forgotten.)

The mass of ice rapidly avalanched down the extremely steep slopes, breaking off large masses of rock debris and scooping out small lakes and loose rock that lay in its path. Eyewitnesses described the mass as a rapidly moving wall the size of a ten-story building. The sound was deafening. More than 50 million cubic meters of muddy debris traveled 3.7 kilometers (12,000 feet) vertically and 14.5 kilometers (9 miles) horizontally in less than four minutes, attaining speeds between 200 and 435 kilometers per hour (125 to 270 miles per hour). The main mass of material traveled down a steep valley until it came to rest, blocking the Santa River and burying about 1,800 people in the small village of Ranrahirca (box figure 1). A relatively small part of the mass of mud and debris that was moving especially rapidly shot up the valley sidewall at a curve and overtopped a ridge. The mass was momentarily airborne before it fell on the town of Yungay, completely burying it under several meters of mud and loose rock. Only the top of the church and tops of palm trees were visible, marking where the town center was buried (box figure 2). Ironically, the cemetery was not buried because it occupied the high ground. The few survivors were people who managed to run to the cemetery.

The estimated death toll at Yungay was 17,000. This was considerably more than the town's normal population because it was Sunday, a market day, and many families had come in from the country.

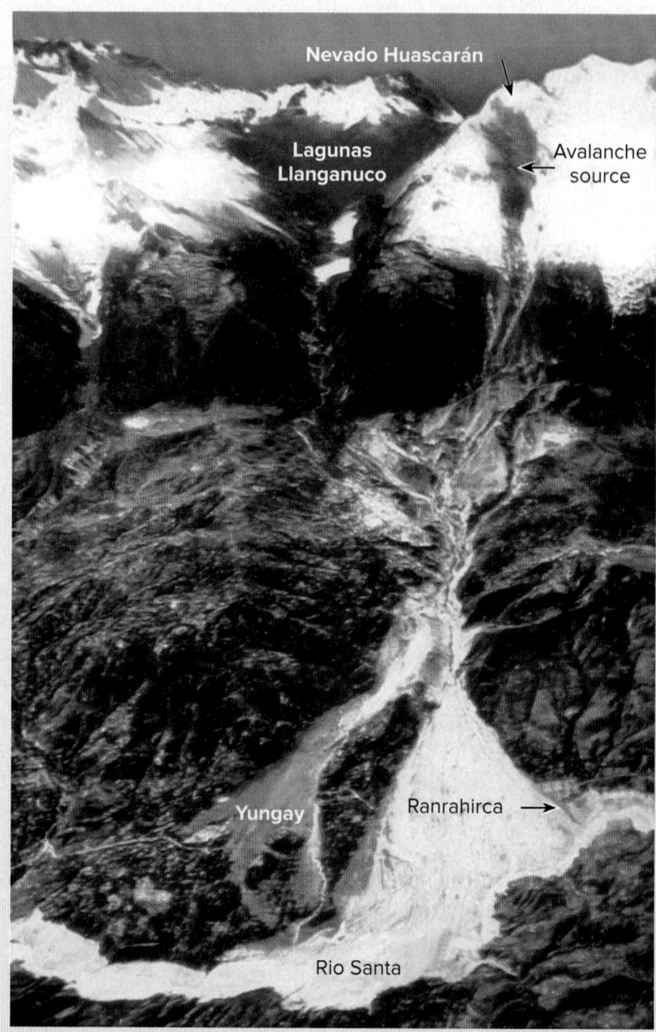

BOX 9.1 ■ FIGURE 1
Air photo showing the 1970 debris avalanche in Peru, which buried Yungay. The main mass of debris destroyed the small village of Ranrahirca.
Source: Servicio Aerfotografico de Peru, courtesy of the U.S. Geological Survey

For several days after the slide, the debris was too muddy for people to walk on, but within three years, grass had grown over the site. Except for the church steeple and the tops of palm trees that still protrude above the ground, and the crosses erected by families of those buried in the landslide, the former site of Yungay appears

A

B

C

BOX 9.1 ■ FIGURE 2

(*A*) Yungay is completely buried, except for the cemetery and a few houses on the small hill in the lower right of the photograph. (*B*) Behind the palm trees is the top of a church buried under 5 meters of debris at Yungay's central plaza. (*C*) Three years later.
(A) and (B) Source: George Plafker, U.S. Geological Survey (C) ©C. C. Plummer

to be a scenic meadow overlooking the Santa River. The U.S. Geological Survey and Peruvian geologists found evidence that Yungay itself had been built on top of debris left by an even bigger slide in the recent geologic past. More slides will almost surely occur here in the future.

Additional Resource

G. E. Ericksen, G. Plafker, and J. Fernandez Concha. 1970. *Preliminary Report On The Geologic Events Associated With The May 31, 1970, Peru Earthquake.* U.S. Geological Survey Circular 639.

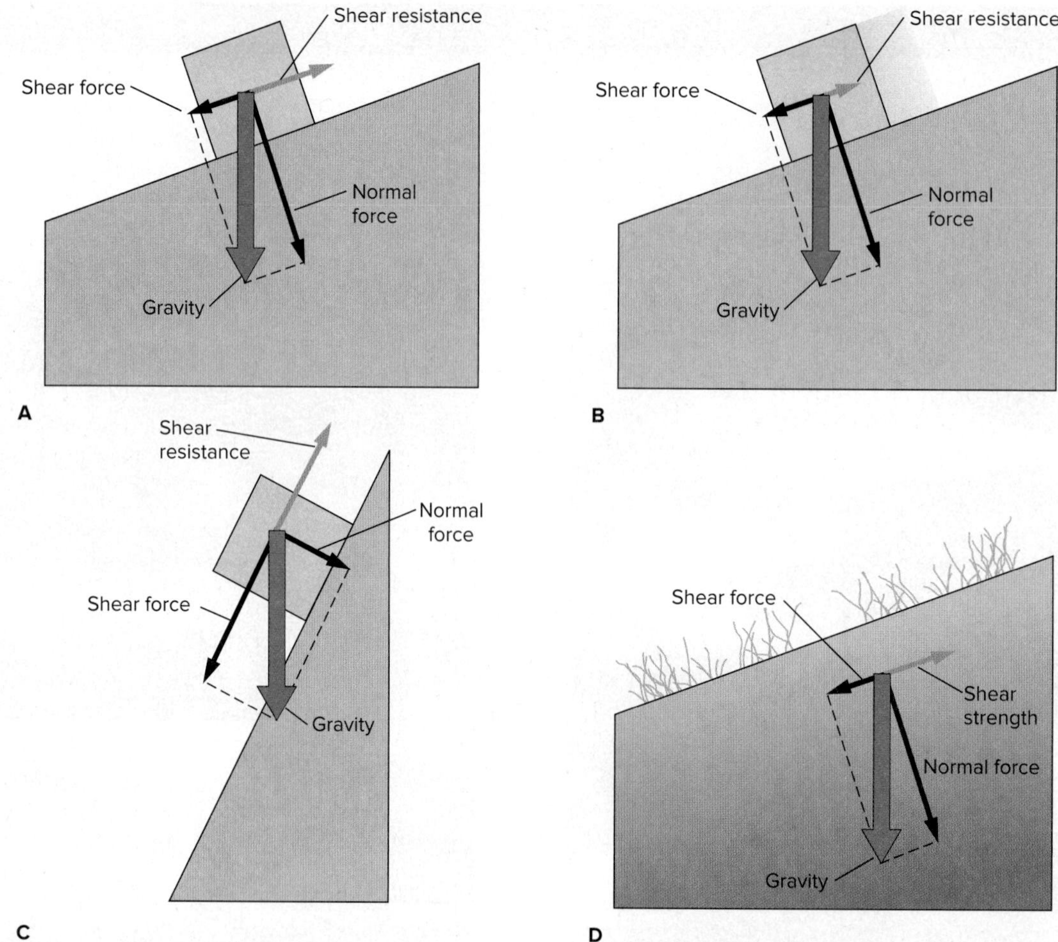

FIGURE 9.1

Relationship of shear force and normal force to gravity. (*A*) A block on a gently inclined slope in which the shear resistance (brown arrow) is greater than the shear force; therefore, the block will not move. (*B*) The same situation as in *A*, except that the shear resistance is less than the shear force; therefore, the block will be moving. (*C*) A block on a steep slope. Note how much greater the shear force is and how much larger the shear resistance has to be to prevent the block from moving. (*D*) Forces acting at a point in soil. Shear strength is represented by a brown arrow. If that arrow is longer than the one represented by shear force, soil at that point will not slide or be deformed.

if water or ice reduces the friction beneath the block. If the shear resistance becomes lower than the shear force, the block will slide (figure 9.1*B*). Similar forces act on soil on a hillside (figure 9.1*D*). The resistance to movement or deformation of that soil is its **shear strength**. Shear strength is controlled by factors such as the cohesiveness of the material, friction between particles, pore pressure of water, and the anchoring effect of plant roots. Shear strength is also related to the normal force. The larger the normal force, the greater the shear strength is. If the shear strength is greater than the shear force, the soil will not move or be deformed. On the other hand, if shear strength is less than shear force, the soil will flow or slide.

Building a heavy structure high on a slope demands special precautions. To prevent movement of both the slope and the building, pilings may have to be sunk through the soil, perhaps even into bedrock. Developers may have to settle for fewer buildings than planned if the weight of too many structures will make the slope unsafe.

Water

Water is a critical factor in mass wasting. When soil is saturated with water (as from heavy rain or melting snow), it becomes less viscous and is more likely to flow downslope. The added gravitational shear force from the increased weight is usually less important than the reduction in shear strength. This is due to increased *pore pressure* in which water forces soil grains apart.

Paradoxically, a small amount of water in soil can actually prevent downslope movement. When water does not completely fill the pore spaces between the grains of soil, it forms a thin film around each grain (as shown in figure 9.2). Loose grains adhere to one another because of the *surface tension* created by the film of water, and shear strength increases. Surface tension of water between sand grains is what allows you to build a sand castle. The sides of the castle can be steep or even vertical because surface tension holds the moist sand grains in place. Dry sand cannot be shaped into a sand castle because the sand grains

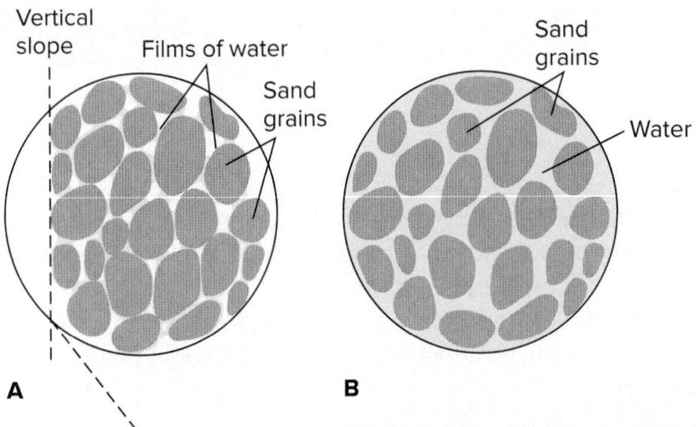

FIGURE 9.2

The effect of water in sand. (*A*) Unsaturated sand held together by surface tension of water. (*B*) Saturated sand grains forced apart by water; mixture flows easily. (*C*) A sand castle in Florida.
©*C. C. Plummer*

slide back into a pile that generally slopes at an angle of about 30 to 35 degrees from the horizontal. On the other hand, an experienced sand castle builder also knows that it is impossible to build anything with sand that is too wet. In this case, the water completely occupies the pore space between sand grains, forcing them apart and allowing them to slide easily past one another. When the tide comes in or someone pours a pail of water on your sand castle, all you have is a puddle of wet sand.

Similarly, as the amount of water in soil increases, the rate of movement tends to increase. Damp soil may not move at all, whereas moderately wet soil moves slowly downslope. Slow types of mass wasting, such as creep, are generally characterized by a relatively low ratio of water to earth. Mudflows always have high ratios of water to earth. A mudflow that continues to gain water eventually becomes a muddy stream.

Triggers

A sudden event may trigger mass wasting of a hillside that is unstable. Eventually, movement would occur without the triggering if conditions slowly became more unstable.

Earthquakes commonly trigger landslides (figure 9.3). The 1970 debris avalanche in Peru (see box 9.1) was one of thousands of landslides, mostly small ones, triggered by a quake. The May 2008 Sichuan earthquake in China (see chapter 16) killed almost 70,000 people. Much of the damage from the earthquake was from landslides. There were thousands of them, killing an estimated 20,000 people, including 158 earthquake relief workers who were trying to repair roads. Besides parts of towns being destroyed, landslide debris dammed parts of nine rivers, creating twenty-four new lakes. Because of potential catastrophic flooding from the failure of these unstable dams, bypass trenches were built to reduce the pressure from rising water.

Landslides often are triggered by heavy rainfall. The sudden influx of voluminous water quickly increases pore pressure in material. Rainstorms in normally dry southern California caused many landslides in January 2005 and parts of the two previous years. These are discussed in this chapter, as are other examples of water-triggered landslides, notably the Gros Ventre, Wyoming, and Vaiont, Italy, rockslides.

Construction sometimes triggers mass wasting. The extra weight of buildings on a hillside can cause a landslide, as can bulldozing a road cut at the base of a slope.

9.3 CLASSIFICATION OF MASS WASTING

A number of systems are used by geologists, engineers, and others for classifying mass wasting, but none has been universally accepted. Some are very complex and useful only to the specialist.

The classification system used here and summarized in table 9.1 is based on (1) rate of movement, (2) type of material, and (3) nature of the movement.

Rate of Movement

A landslide (debris avalanche) like the one in Peru (box 9.1) clearly involves rapid movement. Just as clearly, the movement of soil at a rate of less than a centimeter a year is slow movement. There is a wide range of velocities between these two extremes.

IN GREATER DEPTH 9.2

Landslides and Mudslides

As you have read in this chapter, the general term "landslide" tells us almost nothing about the nature of the mass wasting event that has taken place. It may or may not involve *sliding,* as defined in this book and used by geologists and engineers. During recent years, the news media have increasingly used the term "mudslide" rather than "landslide." In context with the terminology of this book, "mudslide" is an oxymoron (a combination of incongruous terms). Mud, which has a significant water content, will always flow, rather than slide.

When you read or hear about a mudslide, think about it as you would a landslide. It may or may not involve mud and it may or may not involve sliding.

FIGURE 9.3

A landslide triggered by the 2016 Kaikoura earthquake cuts off a highway and rail line near Ohau Point on the South Island of New Zealand. The magnitude 7.8 earthquake and hundreds of aftershocks triggered up to 100,000 landslides.
©Pool/Getty Images

Type of Material

Mass wasting processes are usually distinguished on the basis of whether the descending mass started as bedrock (as in a rockslide) or as unconsolidated material. For this discussion, we call any unconsolidated or weakly consolidated material at the Earth's surface, regardless of particle size or composition, **soil** (also called *engineering soil*–see chapter 5 for other definitions of soil). Soil can be debris, earth, or mud. **Debris** implies that coarse-grained fragments predominate in the soil. If the material is predominantly fine-grained (sand, silt, clay) it is called **earth** (not capitalized). **Mud**, as the name suggests, has a high content of water, clay, and silt.

The amount of water (or ice and snow) in a descending mass strongly influences the rate and type of movement.

Type of Movement

In general, the type of movement in mass wasting can be classified as mainly flow, slide, or fall (figure 9.4). A **flow** implies that the descending mass is moving downslope as a viscous fluid. **Slide** means the descending mass remains relatively intact, moving along one or more well-defined surfaces. A **fall** occurs when material free-falls or bounces down a cliff.

Two kinds of slides are shown in figure 9.4. In a **translational slide**, the descending mass moves along a plane approximately

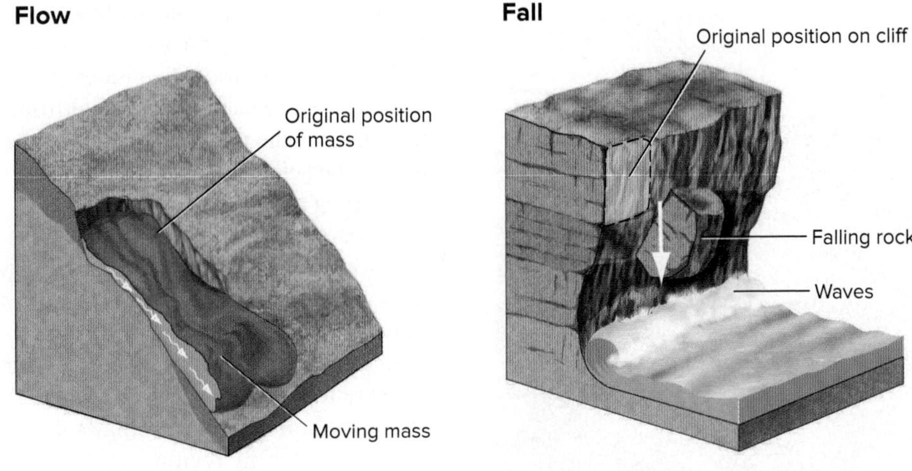

Slide

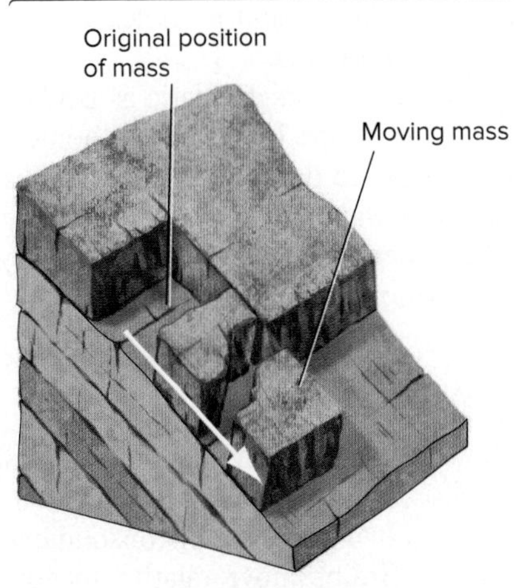

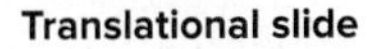

Translational slide

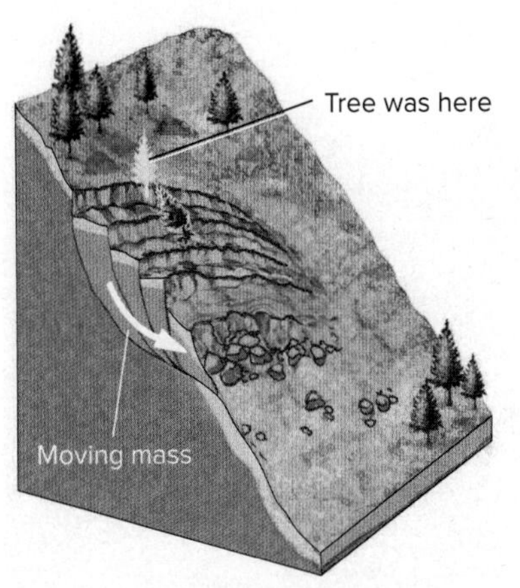

Rotational slide (slump)

FIGURE 9.4
Flow, slide, and fall.

parallel to the slope of the surface. A **rotational slide** (also called a **slump**) involves movement along a curved surface, the upper part moving downward while the lower part moves outward.

9.4 COMMON TYPES OF MASS WASTING

The types of mass wasting are shown in table 9.1. Here we will describe the most common ones in detail.

Creep

Creep (or **soil creep**) is very slow, downslope movement of soil. Shear forces, over time, are only slightly greater than shear strengths. The rate of movement is usually less than a centimeter per year and can be detected only by observations taken over months or years. When conditions are right, creep can take place along nearly horizontal slopes. Some indicators of creep are illustrated in figures 9.5 and 9.6.

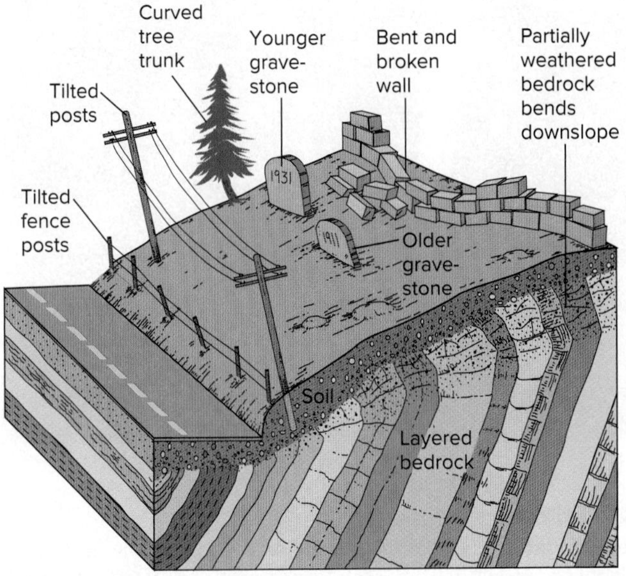

FIGURE 9.5
Indicators of creep.
Source: C. F. S. Sharpe

A

B

C

FIGURE 9.6

(*A*) Tilted gravestones in a churchyard at Lyme Regis, England (someone probably straightened the one upright gravestone). Grassy slope is inclined gently to the left. (*B*) Soil and partially weathered, nearly vertical sedimentary strata have crept downslope. (*C*) As a young tree grew, it grew vertically but was tilted by creeping soil. As it continued to grow, its new, upper part would grow vertically but in turn would be tilted.
(A) ©C. C. Plummer (B) ©Frank M. Hanna (C) ©Parvinder Sethi

Two factors that contribute significantly to creep are water in the soil and daily cycles of freezing and thawing. As we have said, water saturation facilitates movement of soil downhill. What keeps downslope movement from becoming more rapid in most areas is the presence of abundant grass or other plants that anchor the soil. (Understandably, overgrazing can severely damage sloping pastures.)

Several processes contribute to soil creep. Particles are displaced in cycles of wetting and drying. The soil tends to swell when wet and contract when dry so that movement takes place in a manner similar to that of a freeze-thaw cycle. Burrowing worms and other creatures "stir" the soil and facilitate movement under gravity's influence. Soil creep is more active where the soil freezes and thaws during part of the year. During the winter in regions such as the northeastern United States, the temperature may rise above and fall below freezing once a day. When there is moisture in the soil, each freeze-thaw cycle moves soil particles a minute amount downhill, as shown in figure 9.7.

Creep is not as dramatic as landsliding. However, it can be a costly nuisance. If you have a home on creeping ground, you likely will have doors that stick, cracks in walls, broken pipes, and a driveway that will need repaving often. You will find that you are spending more time and money on repairs than does a person who lives on stable ground.

Flows

A flow is a descending mass of material that moves downslope as a viscous fluid. Flow occurs when motion is taking place within a moving mass of unconsolidated or weakly consolidated material. Grains move relative to adjacent grains, or motion takes place along closely spaced, discrete fractures. The common varieties of flow–earthflow, debris flow, mudflow, and debris avalanche–are described in this section.

Earthflow

In an **earthflow**, earth moves downslope as a viscous fluid; the process can be slow or rapid. Earthflows usually occur on hillsides that have a thick cover of soil in which finer grains are predominant, often after heavy rains have saturated the soil. Typically, the flowing mass remains covered by a blanket of vegetation, with a *scarp* (steep cut) developing where the moving debris has pulled away from the stationary upper slope.

A landslide may be entirely an earthflow, as in figure 9.8, with soil particles moving past one another roughly parallel to the slope. Commonly, however, rotational sliding (slumping) takes place above the earthflow, as in figure 9.9 and the opening photo for this chapter. These figures each show a *rotational slide* (upper part) and an *earthflow* (lower part), and each can be called a *slump-earthflow.* In such cases, soil remains in a relatively coherent block or blocks that rotate downward and outward, forcing the soil below to flow.

A *hummocky* (characterized by mounds and depressions) lobe usually forms at the toe or front of the earthflow where soil has accumulated. An earthflow can be active over a period of

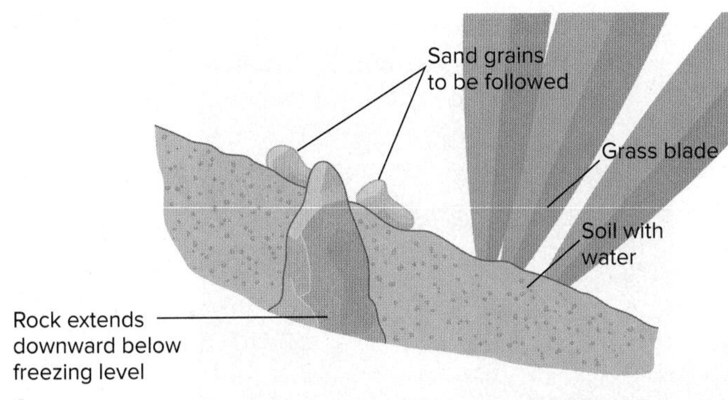

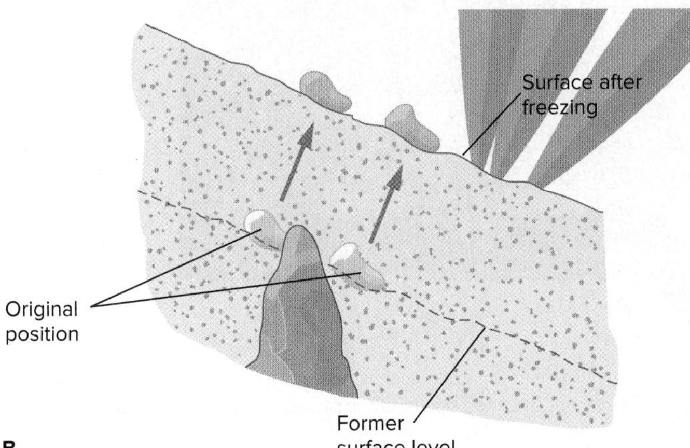

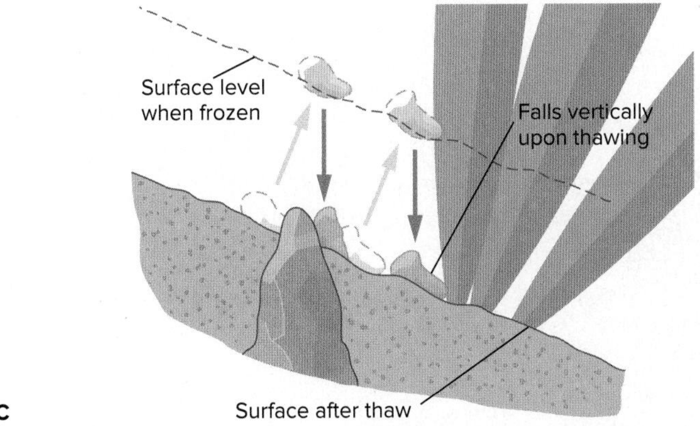

FIGURE 9.7

Downslope movement of soil, illustrated by following two sand grains (each less than a millimeter in size) during a freeze-thaw cycle. Movement downward might not be precisely vertical if adjacent grains interfere with each other.

hours, days, or months; in some earthflows, intermittent movement continues for years.

In March 1995, following an extraordinarily wet year, a slump-earthflow destroyed or severely damaged fourteen homes in the southern California coastal community of La Conchita (figure 9.9*B*). In January 2005, following fifteen days of record-breaking rainfall, around 15% of the 1995 landslide remobilized (see figure 9.12). A rapidly moving flow of soil killed ten people and severely damaged or destroyed thirty-six houses. Because future landslides are likely, the town of La Conchita was

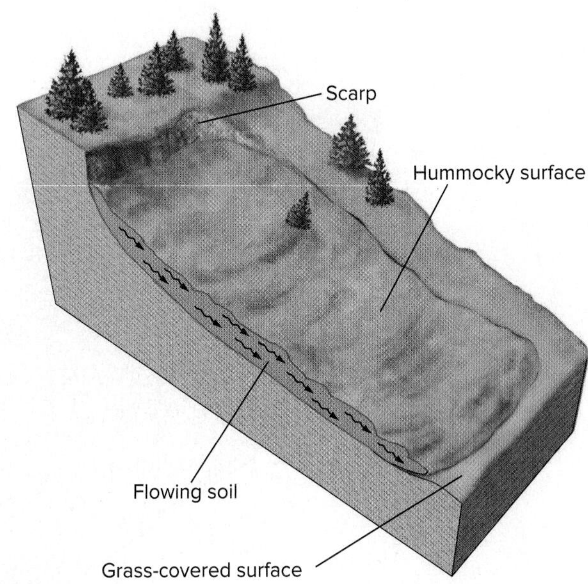

FIGURE 9.8

Earthflow. Soil flows beneath a blanket of vegetation.

abandoned. For details of the La Conchita landslides go to http://pubs.usgs.gov/of/2005/1067/ and click on either the PDF or HTML version.

People can trigger earthflows by adding too much water to soil from septic tank systems or by overwatering lawns. In one case, in Los Angeles, a man departing on a long trip forgot to turn off the sprinkler system for his hillside lawn. The soil became saturated, and both house and lawn were carried downward on an earthflow whose lobe spread out over the highway below.

Earthflows, like other kinds of landslides, can be triggered by undercutting at the base of a slope. The undercutting can be caused by waves breaking along shorelines or streams eroding and steepening the base of a slope. Along coastlines, mass wasting commonly destroys buildings. Entire housing developments and expensive homes built for a view of the ocean are lost. A home buyer who knows nothing of geology may not realize that the sea cliff is there because of the relentless erosion of waves along the shoreline. Nor is the person likely to be aware that a steepened slope creates the potential for landslides.

Bulldozers can undercut the base of a slope more rapidly than wave erosion, and such oversteepening of slopes by human activity has caused many landslides. Unless careful engineering measures are taken at the time a cut is made, road cuts or platforms carved into hillsides for houses (as illustrated in figure 9.21) may bring about disaster.

Solifluction and Permafrost

One variety of earthflow is usually associated with colder climates. **Solifluction** is the flow of water-saturated soil over impermeable material. Because the impermeable material beneath the soil prevents water from draining freely, the soil between the vegetation cover and the impermeable material becomes saturated (figure 9.10). Even a gentle slope is susceptible to movement under these conditions.

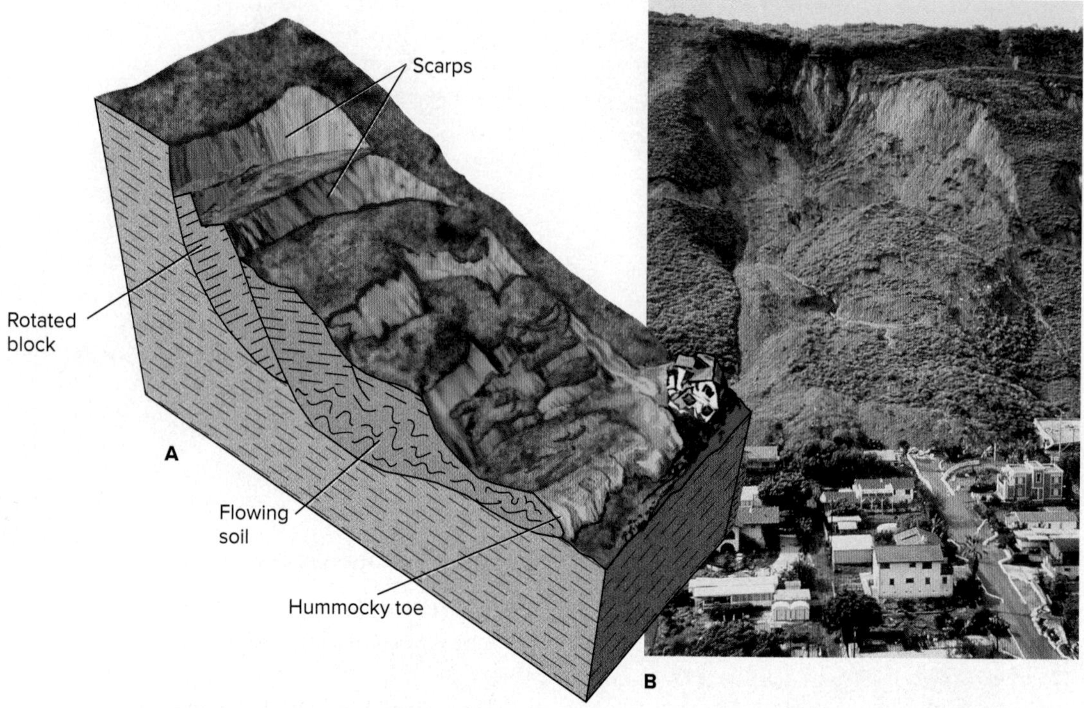

FIGURE 9.9

Earthflow with rotational sliding (slumping). (*A*) Soil in the upper part of the diagram remained mostly intact as it rotated downward in blocks. Soil in the lower portion flowed. (*B*) A slump-earthflow destroyed several houses in March 1995 at La Conchita, California.
Source: Robert L. Schuster, U.S. Geological Survey

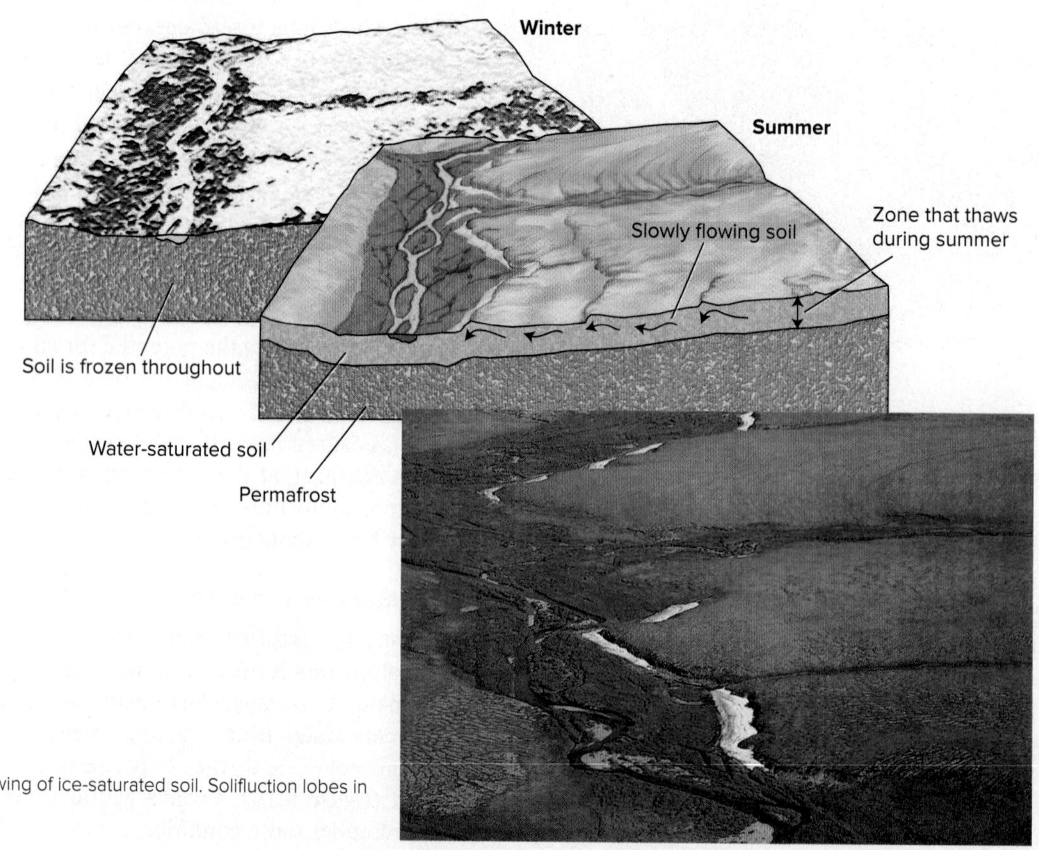

FIGURE 9.10

Solifluction due to thawing of ice-saturated soil. Solifluction lobes in northwestern Alaska.
©C. C. Plummer

The impermeable material beneath the saturated soil can be either impenetrable bedrock or, as is more common, **permafrost**, ground that remains frozen for many years. Most solifluction takes place in areas of permanently frozen ground, such as in Alaska and northern Canada. Permafrost occurs at depths ranging from a few centimeters to a few meters beneath the surface. The ice in permafrost is a cementing agent for the soil. Permafrost is as solid as concrete.

Permafrost in northern Canada was dated in 2008 as having formed approximately 740,000 years ago—much older than scientists believed possible. During that time, there have been two periods during which world climate was warmer than at present. This implies that the loss of permafrost during our ongoing global warming may not be as great as projected.

Above the permafrost is a zone that, if the soil is saturated, is frozen during the winter and indistinguishable from the underlying permafrost. When this zone thaws during the summer, the water, along with water from rain and runoff, cannot percolate downward through the permafrost, and so the slopes become susceptible to solifluction.

As solifluction movement is not rapid enough to break up the overlying blanket of vegetation into blocks, the water-saturated soil flows downslope, pulling vegetation along with it and forming a wrinkled surface. Gradually, the soil collects at the base of the slope, where the vegetated surface bulges into a hummocky lobe.

Solifluction is not the only hazard associated with permafrost. Great expanses of flat terrain in Arctic and subarctic climates become swampy during the summer because of permafrost, making overland travel very difficult. Building and maintaining roads is an engineering headache (figure 9.11). In the preliminary stages of planning the trans-Alaska pipeline, a road was bulldozed across permafrost terrain during the winter, removing the vegetation from the rock-hard ground. It was an excellent truck route during the winter, but when summer came, the road became a quagmire several hundred kilometers long. The strip can never be used by vehicles as planned, nor will the vegetation return for many decades. Building structures on permafrost terrain presents serious problems. For instance, heat from a building can melt underlying permafrost; the building then sinks into the mud. To learn more about permafrost, visit the Permafrost and Frozen Ground page of the National Snow & Ice Data Center's State of the Cryosphere website (http://nsidc.org/cryosphere/sotc/permafrost.html). This page provides an excellent description of permafrost as well as a discussion of the impact of climate change on permafrost and frozen ground.

FIGURE 9.11

A railroad built on permafrost terrain in Alaska.
Source: Lynn A. Yehle, U.S. Geological Survey

Debris Flow and Mudflow

A **debris flow** is flow involving soil in which coarse material (gravel, boulders) is predominant. A debris flow can be like an earthflow and travel relatively short distances to the base of a slope or, if there is a lot of water, a debris flow can behave like a mudflow and flow rapidly, traveling considerable distance in a channel. Rapidly moving debris flows can be extremely devastating.

The steep mountains that rise above Los Angeles and other southern California urban centers are sources of sometimes catastrophic debris flows (figure 9.12). In December 2003, dozens of debris flows took place in the San Bernadino Mountains. The debris flows followed a typical scenario that began during the hot, dry summer. Widespread forest fires scorched southern California, killing trees and ground cover. The anchoring effect of vegetation was gone. Geologists predicted that steep slopes underlain with thick debris were ripe for producing debris flows. Heavy rains would saturate the soil and trigger the debris flows. Heavy rains did indeed come in late December. On Christmas Day, one of many debris flows destroyed a church camp, killing fourteen people. You can see video footage of this debris flow by visiting http://www.youtube.com/watch?v=k3W-wDIR-Os.

The year 1978 was particularly bad for debris flows in southern California (box 9.3). One flow roared through a Los Angeles suburb carrying almost as many cars as large boulders. A sturdily built house withstood the onslaught but began filling with muddy debris. Two of its occupants were pinned to the wall of a bedroom and could do nothing as the room filled slowly with mud. The mud stopped rising just as it was reaching their heads. Hours later they were rescued. (John McPhee's *The Control of Nature* has a highly readable account of the 1978 debris flows in southern California.)

Debris flows (and mudflows) illustrate the interplay between Earth systems: Soil is produced through weathering—interaction of the atmosphere and the geosphere. Vegetation (the biosphere) grows in the soil and adds shear strength to stabilize the hillside. The atmosphere heats and dries the vegetation and produces thunderstorms, which ignite forest fires. Part of the biosphere is destroyed. Atmospheric conditions bring

ENVIRONMENTAL GEOLOGY—SATIRE 9.3

Los Angeles, A Mobile Society*

The following satirical newspaper column was written by the late humorist Art Buchwald in 1978, a year in which southern California had many landslides because of unusually wet weather.

Los Angeles—I came to Los Angeles last week for rest and recreation, only to discover that it had become a rain forest.

I didn't realize how bad it was until I went to dinner at a friend's house. I had the right address, but when I arrived there was nothing there. I went to a neighboring house where I found a man bailing out his swimming pool.

I beg your pardon, I said. Could you tell me where the Cables live?

"They used to live above us on the hill. Then, about two years ago, their house slid down in the mud, and they lived next door to us. I think it was last Monday, during the storm, that their house slid again, and now they live two streets below us, down there. We were sorry to see them go—they were really nice neighbors."

I thanked him and slid straight down the hill to the new location of the Cables' house. Cable was clearing out the mud from his car. He apologized for not giving me the new address and explained, "Frankly, I didn't know until this morning whether the house would stay here or continue sliding down a few more blocks."

Cable, I said, you and your wife are intelligent people, why do you build your house on the top of a canyon, when you know that during a rainstorm it has a good chance of sliding away?

"We did it for the view. It really was fantastic on a clear night up there. We could sit in our Jacuzzi and see all of Los Angeles, except of course when there were brush fires.

"Even when our house slid down two years ago, we still had a great sight of the airport. Now I'm not too sure what kind of view we'll have because of the house in front of us, which slid down with ours at the same time."

But why don't you move to safe ground so that you don't have to worry about rainstorms?

"We've thought about it. But once you live high in a canyon, it's hard to move to the plains. Besides, this house is built solid and has about three more good mudslides in it."

Still, it must be kind of hairy to sit in your home during a deluge and wonder where you'll wind up next. Don't you ever have the desire to just settle down in one place?

"It's hard for people who don't live in California to understand how we people out here think. Sure we have floods, and fire and drought, but that's the price you have to pay for living the good life. When Esther and I saw this house, we knew it was a dream come true. It was located right on the tippy top of the hill, way up there. We would wake up in the morning and listen to the birds, and eat breakfast out on the patio and look down on all the smog.

"Then, after the first mudslide, we found ourselves living next to people. It was an entirely different experience. But by that time we were ready for a change. Now we've slid again and we're in a whole new neighborhood. You can't do that if you live on solid ground. Once you move into a house below Sunset Boulevard, you're stuck there for the rest of your life.

"When you live on the side of a hill in Los Angeles, you at least know it's not going to last forever."

Then, in spite of what's happened, you don't plan to move out?

"Are you crazy? You couldn't replace a house like this in L.A. for $500,000."

What happens if it keeps raining and you slide down the hill again?

"It's no problem. Esther and I figure if we slide down too far, we'll just pick up and go back to the top of the hill, and start all over again; that is, if the hill is still there after the earthquake."

** Art Buchwald, "Los Angeles, A Mobile Society," 1978.*

heavy rain, and part of the hydrosphere mixes with the soil to produce the debris flows.

A **mudflow** is a flowing mixture of soil and water, usually moving down a channel. It differs from a debris flow in that fine-grained (sand, silt, clay) material is predominant. A mudflow can be visualized as a stream with the consistency of a thick milkshake. Most of the solid particles in the slurry are clay and silt (hence, the muddy appearance), but coarser sediment commonly is part of the mixture. A slurry of soil and water forms after a heavy rainfall or other influx of water and begins moving down a slope. Most mudflows quickly become channeled into valleys. They then move downvalley like a stream except that, because of the heavy load of sediment, they are more viscous. Mud moves more slowly than a stream but, because of its high viscosity, can transport boulders, automobiles, and even locomotives (figure 9.13). Houses in the path of a mudflow will be filled with mud, if not broken apart and carried away (box 9.4).

Mudflows are most likely to occur in places where soil is not protected by a vegetative cover. For this reason, mudflows are more likely to occur in arid regions than in wet climates. A hillside in a desert environment, where it may not have rained for many years, may be covered with a blanket of loose material. With sparse desert vegetation offering little protection, a sudden thunderstorm with drenching rain can rapidly saturate the soil and create a mudflow in minutes. Like debris flows, mudflows also occur after forest fires destroy slope vegetation that normally anchors soil in place.

Debris flows and mudflows frequently occur on young volcanoes that are littered with ash. Volcanic mudflows and debris

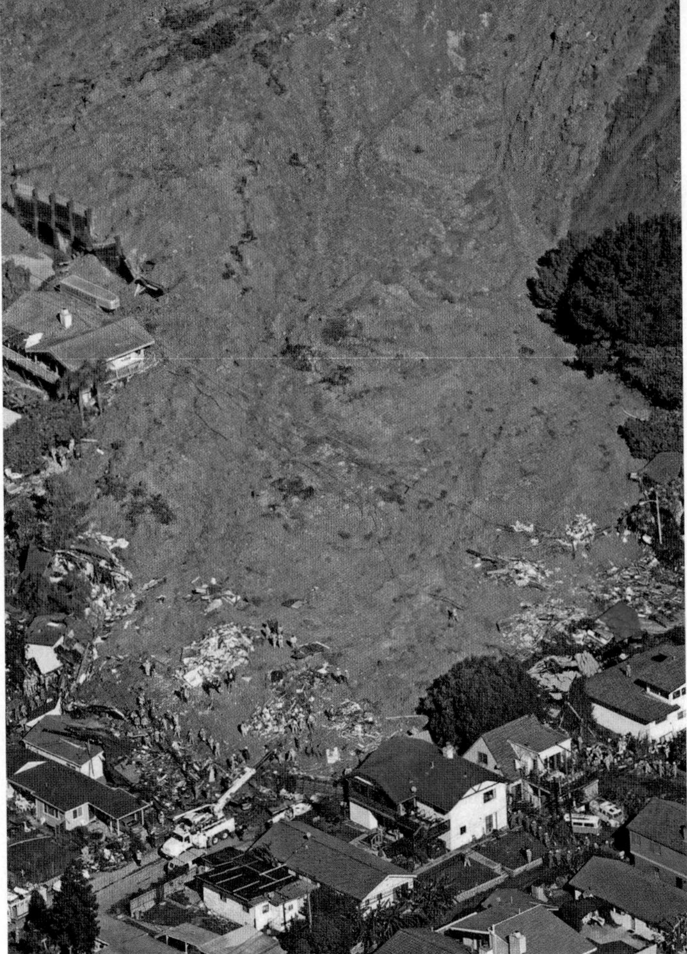

FIGURE 9.12

The portion of the La Conchita landslide that remobilized and killed ten people in January 2005. The U.S. Geological Survey classified this as a debris flow because of the abundance of coarse soil mixed with mud. Note that the flow overtopped the metal wall (upper left) put up to protect the town.
©Kevork Djansezian/AP Images

flows are called *lahars* (chapter 4). Water from heavy rains mixes with pyroclastic debris, as at the Philippines' Mount Pinatubo in 1991. For over a decade after the big eruption, lahars near Mount Pinatubo continued to cost lives and destroy property. Water also can come from glaciers that are melted by lava or hot pyroclastic debris, as occurred at Mount St. Helens in 1980 (figure 9.14) and at Colombia's Nevado del Ruiz in 1985, which cost 23,000 lives.

Debris Avalanche

The fastest variety of debris flow is a **debris avalanche**, a very rapidly moving, turbulent mass of debris, air, and water. The most deadly modern example is the one that buried Yungay (described in box 9.1).

Some geologists have suggested that in very rapidly moving rock avalanches, air trapped under the rock mass creates an air cushion that reduces friction. This could explain why some landslides reach speeds of several hundred kilometers per hour. But other geologists have contended that the rock mass is too turbulent to permit such an air cushion to form.

Falls

On May 3, 2003, New Hampshire lost its beloved symbol, the Old Man of the Mountain (figure 9.15), to rockfall. The Granite State's citizens associated resolute individualism with the rugged features outlined by the face high on a cliff. But the relentless work of water and frost-wedging enlarged the cracks in the granite until the overhanging rock broke apart.

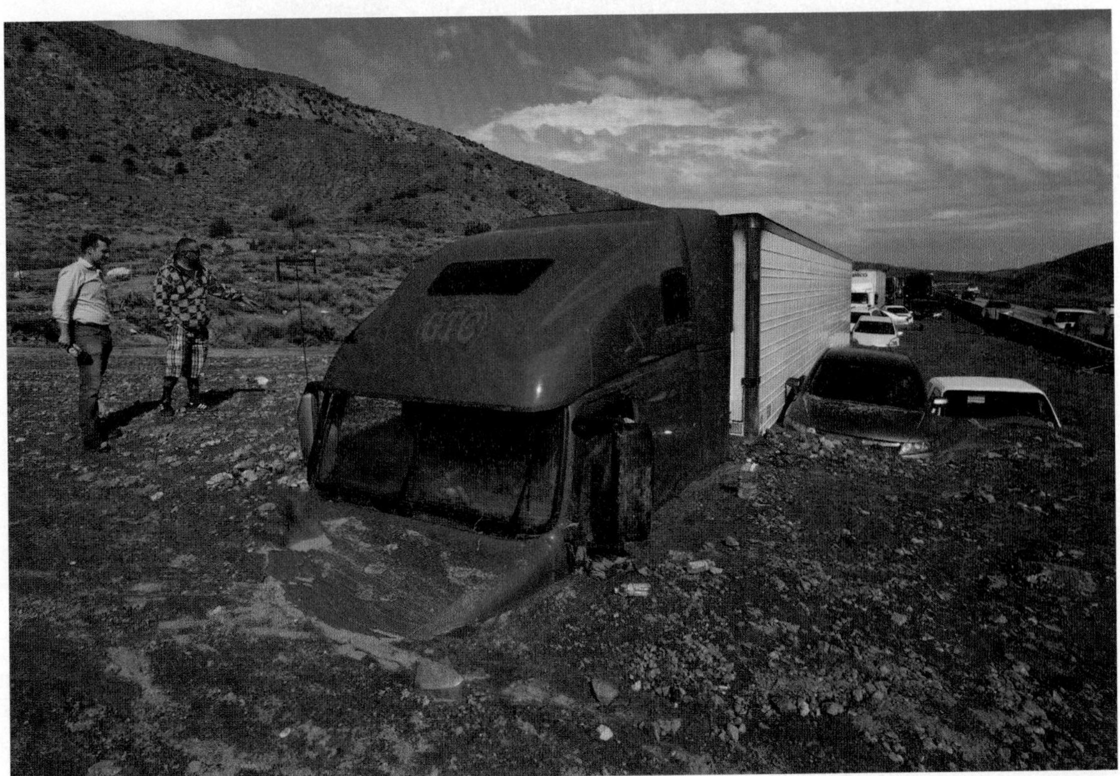

FIGURE 9.13

Vehicles trapped by a mudflow on California Highway 58 in Mojave, California, on October 16, 2015.
©Mark Ralston/AFP/Getty Images

ENVIRONMENTAL GEOLOGY 9.4

Surfing a Mudflow

On July 12, 2008, just north of the desert town of Independence, California, heavy rain from thunderstorms drenched the steep eastern flank of California's Sierra Nevada and adjoining Owens Valley. The water mixed with debris (including glacially deposited boulders), soil, and ash from the previous year's brush fire. Mudflows and debris flows were channeled into the north and south forks of Oak Creek before merging near a fish hatchery. Debris (including boulders) exceeded the volume of mud in parts of the flows, making them debris flows. Other portions are classified as mudflows because they had more mud than debris. To simplify the discussion we will lump them together and call them mudflows.

Don Rockwood, an accomplished ocean surfer, was preparing his supper in his motor home parked at Oak Creek campground. He was apprehensive over the very heavy rainfall. Suddenly, a tree carried by a 3- to 5-foot-high wave of mud crashed through the wall of his motor home. A few minutes later, an 8- to 12-foot-high wave tore apart the motor home (box figure 1) and swept Rockwood away. Instinctively, he body surfed the wave, riding on the mud's surface at an estimated 30 miles an hour. He rode the mud and debris wave approximately a mile downstream, when he saw his pickup truck, carried by the mudflow, bearing down on him. He managed to steer himself sideways and avoid a collision with the truck. He reached shallower and slower mud and managed to stand up. But a third wave, about 3 feet high, overpowered him and swept him away again. He was smashed against a rock and nearly suffocated. He was carried about half a mile further downstream before the mud slowed and he was again able to stand. He had lost his shoes and most of his clothing had been torn off. He looked for shelter, but there was none. He decided that he had to make it to the Mt. Whitney Fish Hatchery on the other side of the flowing mud. He picked up a loose branch to use as a pole and waded into the mudflow, heading toward the flashing red lights of a fire truck at the hatchery. (Mud and boulders had heavily damaged the fish hatchery as well as destroying twenty-five homes at and downstream from the hatchery.) Rockwood waded through mud up to his chest before emerging on the other side of the mudflow. Fire fighters were astonished to see a man caked with mud approach them. He was whisked to the hospital, where he was treated for major lacerations but was otherwise in good health.

Source: Preliminary report prepared for California Geologic Survey: D. Wagner, M. DeRose and J. Stroh, "The Oak Creek Mudflow of July 12, 2008, Independence, Inyo County, California," California Geological Survey.

BOX 9.4 ■ FIGURE 1
The destroyed remains of Don Rockwood's motor home after the 2008 mudflow. Note the boulders within the dried mud, deposited when the flow subsided.
D. Wagner, M. DeRose and J. Stroh, "The Oak Creek Mudflow of July 12, 2008, Independence, Inyo County, California," California Geological Survey. All rights reserved. Used with permission

When a block of bedrock breaks off and falls freely or bounces down a cliff, it is a **rockfall** (figure 9.16). Cliffs may form naturally by the undercutting action of a river, wave action, or glacial erosion. Highway or other construction projects may also oversteepen slopes. Bedrock commonly has cracks (joints) or other planes of weakness such as foliation (in metamorphic rocks) or sedimentary bedding planes. Blocks of rock will break off along these planes. In colder climates, rock is effectively broken apart by frost wedging (as explained in chapter 5).

FIGURE 9.14

Man examining a 75-meter-long bridge on Washington state highway 504, across the North Fork of the Toutle River. The bridge was washed out by mudflow during the May 18, 1980, eruption of Mount St. Helens. The steel structure was carried about 0.5 kilometer downstream and partially buried by the mudflow.
Source: Robert L. Schuster, U.S. Geological Survey

Commonly, an apron of fallen rock fragments, called **talus**, accumulates at the base of a cliff (figure 9.17).

A spectacular rockfall took place in Yosemite National Park in the summer of 1996, killing one man and injuring several other people. The rockfall originated from near Glacier Point (the place where the photo for figure 12.1 was taken). Two huge slabs (weighing approximately 80,000 tons) of an overhanging arch broke loose just seconds apart. (The arch was a product of exfoliation and broke loose along a sheet joint; see chapter 5.) The slabs slid a short distance over steep rock from which they were launched outward, as if from a ski jump, away from the vertical cliffs. The slabs fell free for around 500 meters (1,700 feet) and hit the valley floor 30 meters out from the base of the cliff (you would not have been hit if you had been standing at the base of the cliff). They shattered upon impact and created a dust cloud that obscured visibility for hours. A powerful air blast was created as air between the rapidly falling rock and the ground was compressed. The debris-laden blast of air felled a swath of trees between the newly deposited talus and a nature center building. In 1999, another rockfall in the same area killed one rock climber and injured three others. In October 2008, another rockfall landed at Yosemite's Curry Village. Three people were injured, and several tent cabins were destroyed at this lodging and dining complex (figure 9.18).

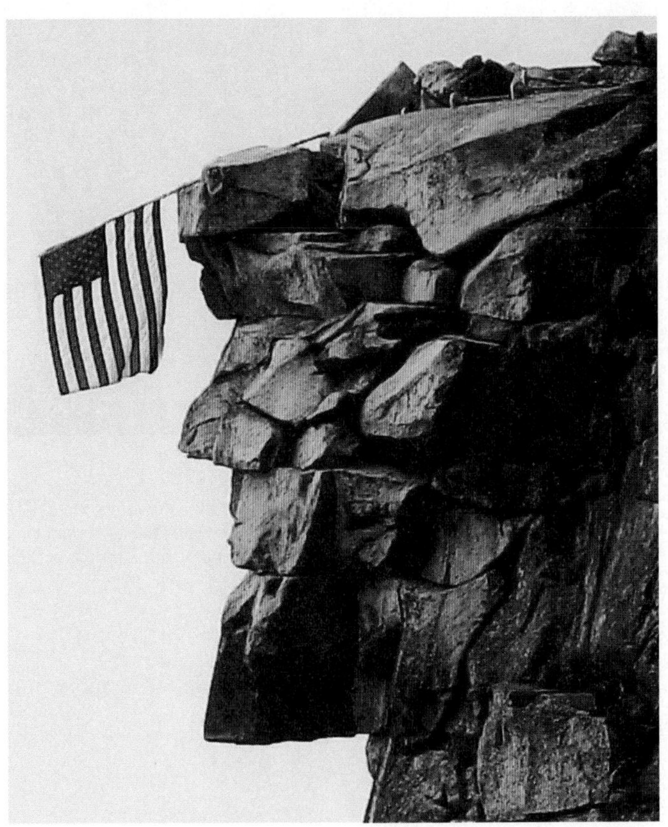

A

B

FIGURE 9.15

The Old Man of the Mountain in New Hampshire. (*A*) The profile of the face of a man was a product of weathering and erosion controlled largely by subhorizontal joints in granite. This is the profile that appeared on license plates and New Hampshire publications. (*B*) After succumbing to continuing erosion, features of the Old Man broke apart and became a rockfall on May 3, 2003.
©Jim Cole/AP Images

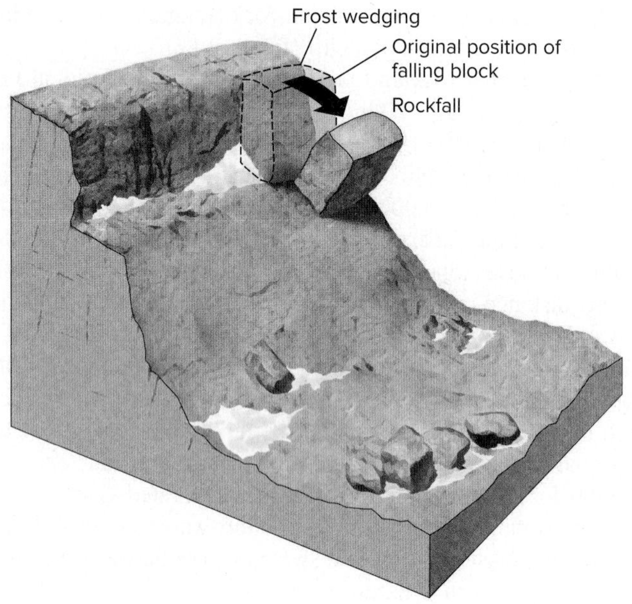

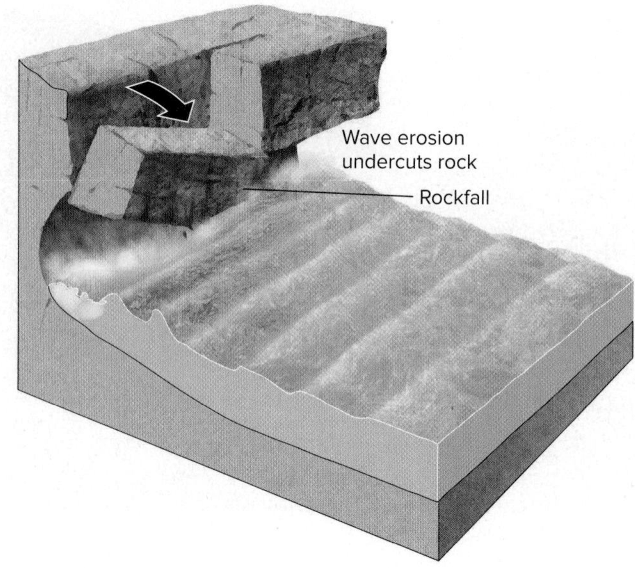

FIGURE 9.16

Two examples of rockfall.

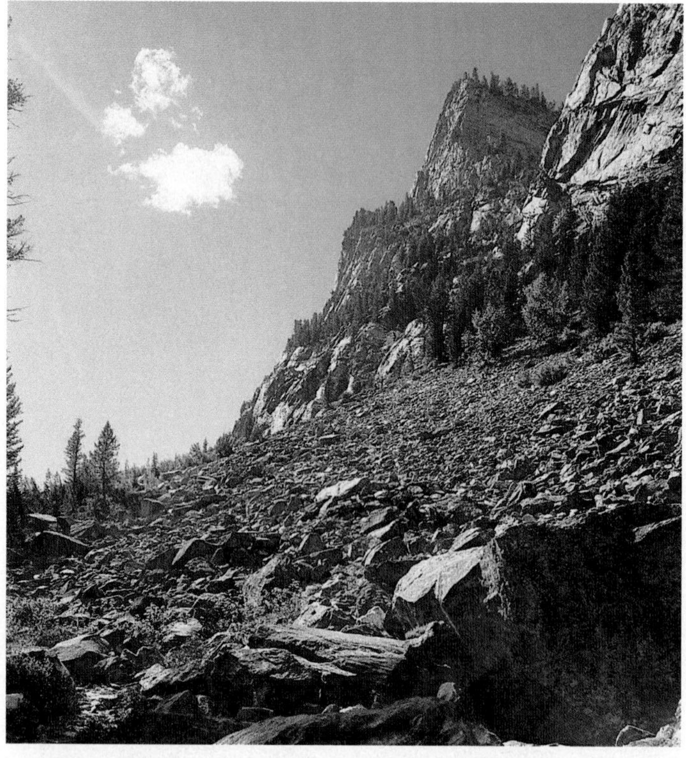

FIGURE 9.17

Talus.
©C. C. Plummer

FIGURE 9.18

A large boulder sits atop debris after it fell in Curry Village in Yosemite National Park in 2008, California. Falling boulders are a major hazard in Yosemite, a popular tourist destination. Parts of Curry Village were permanently closed after this rockfall.
©*Paul Sakuma/AP Images*

For an excellent and thorough report, go to the U.S. Geological Survey site *Rockfall in Yosemite*, http://pubs.usgs.gov/of/1999/ofr-99-0385/. To view an excellent video about rockfalls in Yosemite, visit the National Park Service's Yosemite Nature Notes website at http://www.nps.gov/yose/photosmultimedia/ynn.htm and look for Episode 10: Rockfall.

Slides

As described earlier, the term *slide* is used to describe a mass wasting event where the descending mass remains relatively intact, moving along one or more well-defined surfaces. When the surface over which sliding occurs is approximately parallel to the slope surface, the slide is called a translational slide. When the surface is curved, the slide is called a rotational slide,

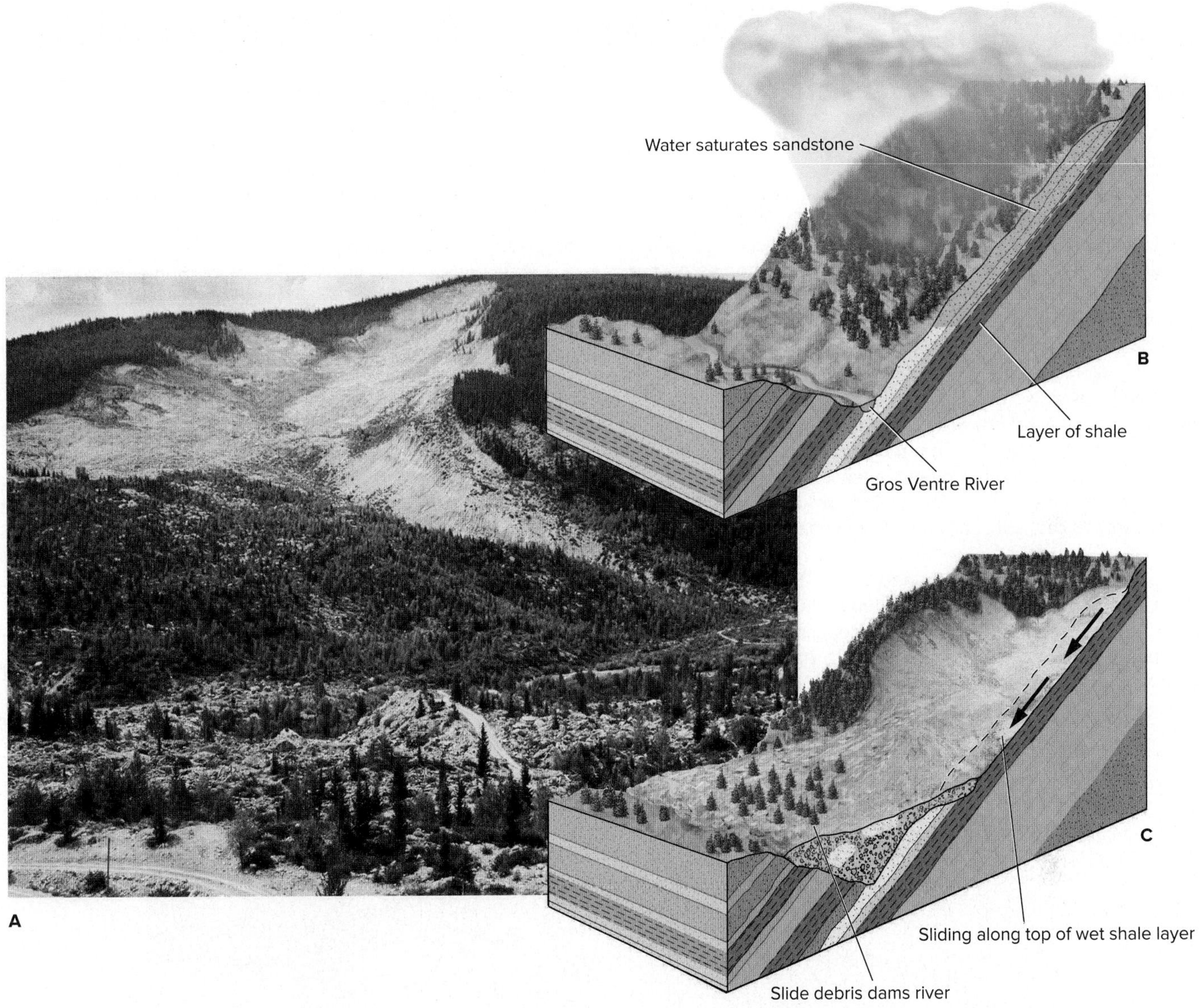

FIGURE 9.19

(*A*) Photo of Gros Ventre slide. (*B*) and (*C*) Diagram of the Gros Ventre, Wyoming, slide. *©D. A. Rahm, courtesy of Rahm Memorial Collection, Western Washington University*

or more commonly, a slump. The 1995 earthflow at La Conchita described in the section on flows started out as a slump (figure 9.9) but became a flow as the material broke apart. This is also true of the deadly *Oso slide*, which occurred in Washington state in 2014 (box 9.5). Next we discuss translational slides.

Rockslide and Rock Avalanche

A **rockslide** is, as the term suggests, the rapid sliding of a mass of bedrock along an inclined surface of weakness, such as a bedding plane, a major fracture in the rock, or a foliation plane. Once sliding begins, a rock slab usually breaks up into rubble. Like rockfalls, rockslides can be caused by undercutting at the base of the slope from erosion or construction. A classic example of a rockslide took place in 1925 in the Gros Ventre Mountains of Wyoming. Sliding occurred along a sedimentary bedding plane. Exceptionally heavy rains triggered the rockslide after water seeped into a layer of sandstone (figure 9.19). The high pore pressure of water in the sandstone had the effect of "lifting" the sandstone from the wet surface of the shale. Shear resistance of the sandstone was greatly reduced. The layers of sedimentary rock were inclined roughly parallel to the hillside. The rock layers overlying the shale and their soil cover slid into the valley, blocking the river. A rancher on horseback saw the ground beneath him begin to move and had to gallop to safe ground. The slide itself merely created a lake, but the natural dam broke two years later, and the resulting flood destroyed the small town of Kelly several kilometers downstream. Several residents who were standing on a bridge watching the floodwaters come down the valley were killed.

ENVIRONMENTAL GEOLOGY 9.5

The 2014 Oso, Washington, Slide—A Preventable Tragedy?

The residents of the Steelhead Haven were enjoying a quiet Saturday morning, when tragedy struck this rural neighborhood, near the town of Oso in Snohomish County, Washington. On March 22, 2014, a portion of an unstable hillside collapsed, sending debris across the Stillaguamish River and into the homes on the south side of the river. The slide started as a slump, but the soft material mobilized into a rapidly moving debris flow that deposited mud, debris, and trees over an area almost half a kilometer long and more than a kilometer wide (box figure 1). Approximately 30 homes in Steelhead Haven were buried under 10 to 20 meters (30 to 70 feet) of debris. After several weeks of searching in difficult conditions (box figure 2), 42 bodies were recovered. The body of the last missing person was finally discovered many months later.

What caused the Oso landslide, and could this tragedy have been prevented? Let us first consider the conditions that led to the slide.

Oso lies in a wide valley, formed by the Stillaguamish River, which is cutting down through the Whitman Bench, a large terrace formed from layers of gravel and sand deposited during the last glacial period. Lacking fine-grained clays to hold it together, the sand and gravel deposit is weak. Downcutting by the Stillaguamish River has created steep slopes in this weak material that are vulnerable to slumping. Infiltration of groundwater into the sediments further weakens the slopes. In the weeks prior to the landslide, the region had experienced more than twice the normal amount of rainfall. This heavy rain is thought to be the trigger for the failure that engulfed Steelhead Haven.

A recent survey by the U.S. Geological Survey shows that many landslides have occurred near Oso, many of them very large (box figure 3). The slope above Steelhead Haven had a history of instability dating back to 1937. Known then as the Hazel landslide, this large slump experienced repeated movements, and efforts to stabilize it date back to the early 1960s. In 2004, Snohomish County officials, concerned about the potential for a deadly slide in the Steelhead Haven area, considered buying the homes in the area and relocating the residents. This idea was rejected, however, due to resistance from the community. In 2010, a study that was commissioned by the county warned that the hillside above Steelhead Drive was one of the most dangerous in the county. And yet, people continued to live there and build new homes.

What Can We Learn from This Tragedy?

People all around the world live in the shadow of unstable slopes, and tragedies like the Oso landslide occur every year. How can they be prevented? Imagine you are a county official in an area like Oso.

BOX 9.5 ■ FIGURE 1

View of the Washington landslide from the east. Near the hillside, the scarps and blocks making up the slump are clearly visible. Out toward the toe, it can be seen that material from the slump mobilized, flowing outward as a debris flov. In the foreground, the Stillaguamish River is flooded due to its path being blocked by the deposit.
Source: Washington State Dept. of Transportation

BOX 9.5 ■ FIGURE 2

A search and rescue team searches through mud and debris from the Oso landslide.
©Justin Sullivan/Getty Images

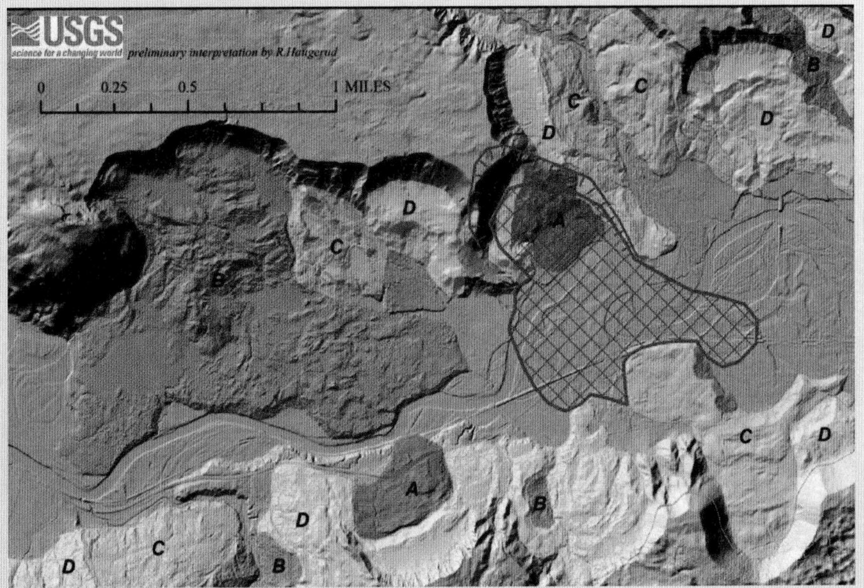

BOX 9.5 ■ FIGURE 3

A preliminary map of landslide deposits in the area around Oso, Washington. The March 22, 2014, Oso slide is marked by the red cross-hatched area. The colored areas show older landslide deposits, labeled with their relative ages, from A, youngest, to D, oldest. The Whitman Bench can be seen in the upper left part of the map. It is clear from this map that landslides are common in this valley, and can be very large.
Source: Ralph A. Haugerud, U.S. Geological Survey (Open-File Report 2014–1065)

When considering landslide hazards, you have to take into account the likelihood of a large event occurring and its probable impact and weigh them against the cost of expensive slope stabilization efforts, or the relocation of neighborhoods. The only information you have to work with is evidence of previous landslides. While officials were aware of the slide risk near Oso, no previous slides there had been anywhere near as large as the deadly slide of 2014.

Additional Resource

The U.S. Geological Survey has created a website containing information about the slide, monitoring efforts in the area, and links to photos and additional resources. Of particular interest is a computer simulation of the landslide.

- http://www.usgs.gov/blogs/features/usgs_top_story/landslide-in-washington-state/

Some rockslides travel only a few meters before halting at the base of a slope. In country with high relief, however, a rockslide may travel hundreds or thousands of meters before reaching a valley floor. If movement becomes very rapid, the rockslide may break up and become a rock avalanche. A **rock avalanche** is a very rapidly moving, turbulent mass of broken-up bedrock. Movement in a rock avalanche is flowage on a grand scale. The only difference between a rock avalanche and a debris avalanche is that a rock avalanche begins its journey as bedrock.

Ultimately, a rockslide or rock avalanche comes to rest as the terrain becomes less steep. Sometimes the mass of rock fills the bottom of a valley and creates a natural dam. If the rock mass suddenly enters a lake or bay, it can create a huge wave that destroys lives and property far beyond the area of the original landslide. Box 9.6 describes a disastrous rock avalanche that took place in Italy.

9.5 UNDERWATER LANDSLIDES

The steeper parts of the ocean floors sometimes have very large landslides. Prehistoric ones are indicated by large masses of jumbled debris on the deep-ocean floor. One, off the coast of the Hawaiian Islands, is much larger than any landslide mass on land. The debris from what is called the Nuuanu debris avalanche covers an area of 5,000 square kilometers (figure 9.20), larger than all of the present Hawaiian Islands combined, and includes volcanic rock blocks several kilometers across.

Another giant landslide off the south coast of the island of Hawaii took place around 100,000 years ago. This appears to have created an incredibly large tsunami that deposited coral fragments to elevations over 300 meters on some Hawaiian islands.

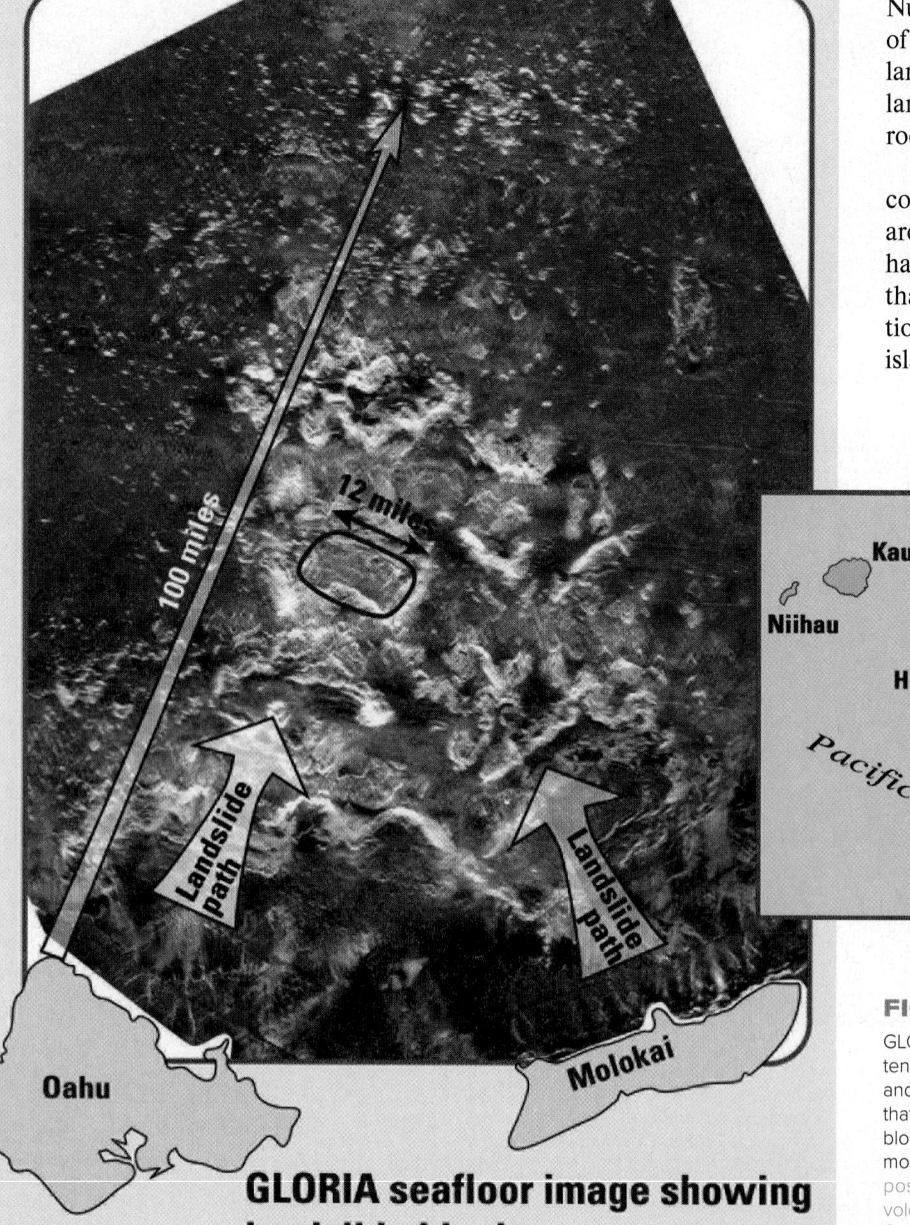

FIGURE 9.20

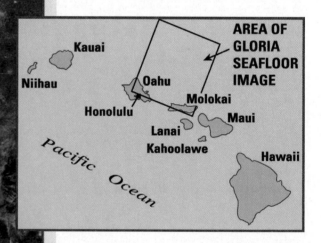

GLORIA seafloor image showing landslide blocks extending to over 100 miles (160 kilometers) from Oahu and Molokai islands. GLORIA is a long-range sonar that gives an oblique view of the sea floor. Individual blocks are up to 12 miles (20 kilometers) across. For more information go to: http://walrus.wr.usgs.gov/posters/underlandslides.html or www.mbari.org/volcanism/Hawaii/HR-Landslides.htm.

Source: Western Coastal & Marine Geology/U.S. Geological Survey

ENVIRONMENTAL GEOLOGY 9.6

A Rockslide Becomes a Rock Avalanche Which Creates a Giant Wave That Destroys Towns

A rock avalanche took place in the Italian Alps in 1963 that had tragic consequences.

A huge layer (1.8 kilometers long and 1 kilometer wide) of limestone broke loose parallel to its bedding planes (box figures 1 and 2). What began as a translational slide involved around 270 million cubic meters. Some of the original slab remained more or less intact as it traveled downward as a high-speed rockslide. However, most of the slabs broke into debris. The debris avalanche moved up to 100 kilometers per hour and, after plowing through the reservoir, deposited rock as high as 140 meters up the opposite side of the valley. Sudden displacement of water in the Vaiont Reservoir created a giant wave. The 245-meter (over two football fields) high wave overtopped the Vaiont Dam. It was the world's highest dam, rising 265 meters (870 feet) above the valley floor. Twenty-five hundred people were killed in the villages that the water flooded in the valleys below. The dam was not destroyed (box figures 1 and 2), a tribute to excellent engineering, but the men in charge of the building project were convicted of criminal negligence for ignoring the landslide hazards. The chief engineer committed suicide.

BOX 9.6 ■ FIGURE 1

Part of the 270 million cubic meters of rock from the rock avalanche that filled the former reservoir behind the Vaiont Dam. The dam, at the right of the photo, was the world's highest when built. Most of the dam face is buried under debris. Part of the bedding plane over which sliding took place can be seen on the mountainside. For a view from the upstream side of the landslide debris, go to http://web.csulb.edu/depts/geology/facultypages/bperry/Mass%20Wasting/VaiontDam.htm.
©Earle F. McBride

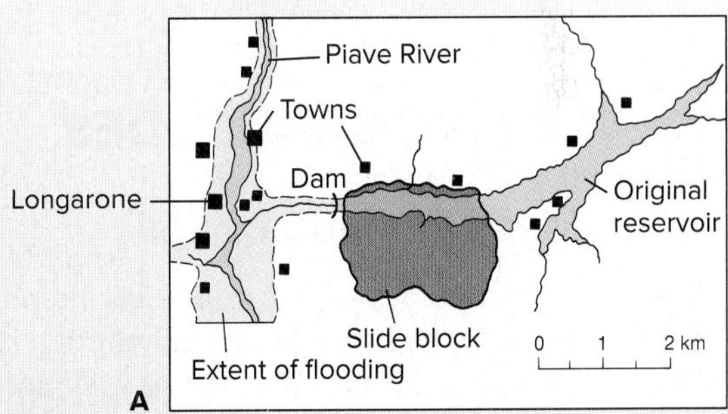

BOX 9.6 ■ FIGURE 2

(*A*) Map of the Vaiont area. (*B*) Photo showing the upper portion of the downstream face of the dam within a steep, narrow valley. It was taken in 2004 from Longarone, one of the largest towns along the Piave River destroyed by the flood. Visualize the 175-meter wall of water overtopping the dam.
©C. C. Plummer

One very large landslide evidently took place off the coast of northeastern Canada in 1929 following an earthquake. It systematically cut a series of trans-Atlantic telephone and telegraph lines. The existence and extent of the event were inferred decades later by analyzing the timing of the telephone conversation cutoffs and the distance of cables from the earthquake's epicenter. The underwater debris avalanche, described as a *turbidity current* in chapter 18 (see figure 18.11), traveled over 700 kilometers in thirteen hours at speeds from 15 to 60 kilometers per hour. The lengths of the sections of cable carried away indicate that the debris flow was up to 100 kilometers wide.

Scientists have recently found that a very large area of thick sediment off the central part of the East Coast of the United States is unstable and could become a giant submarine landslide. If it does go, it very likely will generate a giant tsunami that could be disastrous to coastal communities in Europe as well as North America.

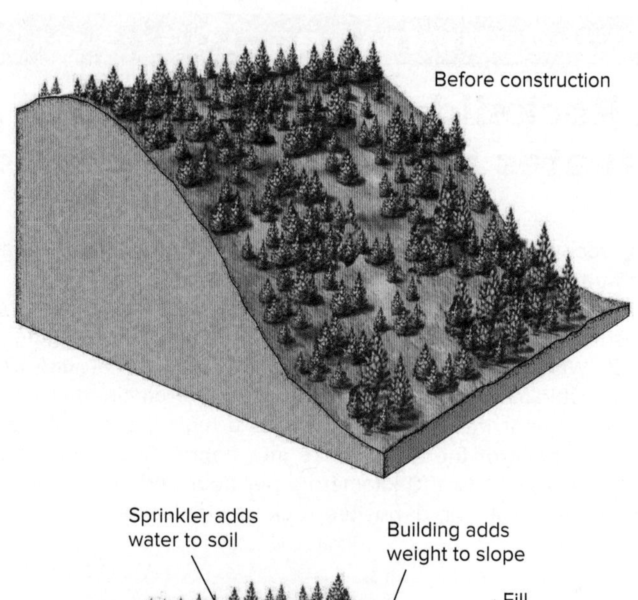

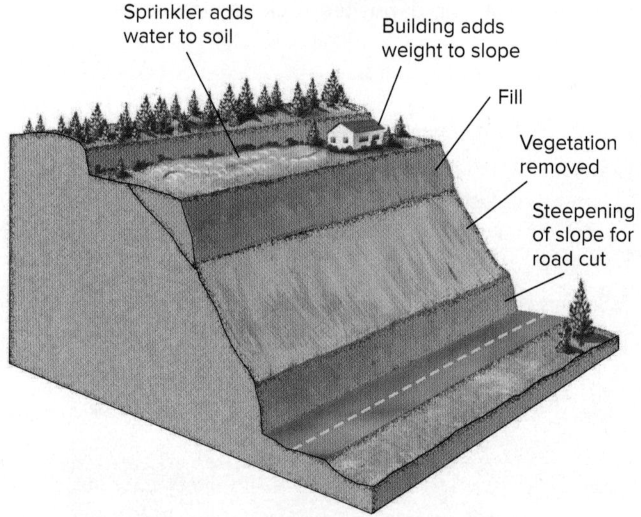

FIGURE 9.21

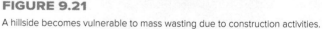

A hillside becomes vulnerable to mass wasting due to construction activities.

9.6 PREVENTING LANDSLIDES

Preventing Mass Wasting of Soil

Mass movements of soil usually can be prevented. Proper engineering is essential when the natural environment of a hillside is altered by construction. As shown in figure 9.21, construction generally makes a slope more susceptible to mass wasting of soil in several ways: (1) the base of the slope is undercut, removing the natural support for the upper part of the slope; (2) vegetation is removed during construction; (3) buildings constructed on the upper part of a slope add weight to the potential slide; and (4) extra water may be allowed to seep into the soil.

Some preventive measures can be taken during construction. A retaining wall is usually built where a cut has been made in the slope, but this alone is seldom as effective a deterrent to downslope movement as people hope. If, in addition, drain pipes are put through the retaining wall and into the hillside, water can percolate through and drain away rather than collecting in the soil behind the wall (figure 9.22). Without drains, excess water results in decreased shear strength, and the whole soggy mass can easily burst through the wall.

Another practical preventive measure is to avoid oversteepening the slope. The hillside can be cut back in a series of terraces rather than in a single steep cut. This reduces not only the

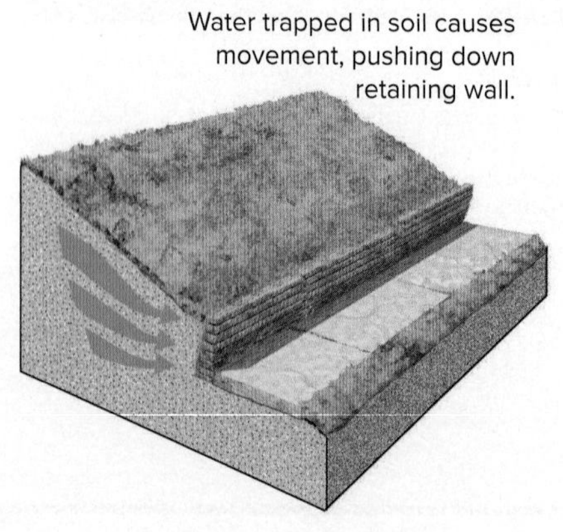

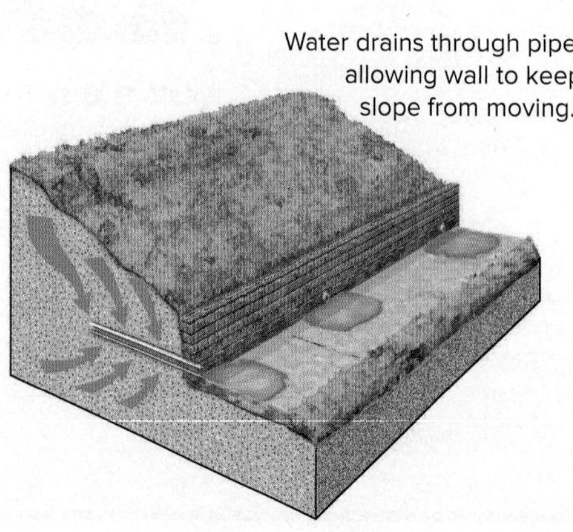

FIGURE 9.22

Use of drains to help prevent mass wasting.

slope angle, but also the shear force by removing much of the overlying material. It also prevents loose material (such as boulders dislodged from the top of the cut) from rolling to the base. Road cuts constructed in this way are usually reseeded with rapidly growing grass or plants whose roots help anchor the slope. A vegetation cover also minimizes erosion from running water.

Some roads and railroads in steep, mountainous areas are covered by sheds with sloping, reinforced concrete roofs (figure 9.23). Sliding debris and snow avalanches pass safely overhead rather than block a road and endanger lives.

FIGURE 9.23

A concrete shed with a sloping roof protects a highway from avalanches in the Canadian Rockies.
©C.W. Montgomery

Preventing Rockfalls and Rockslides on Highways

Rockslides and rockfalls are a major problem on highways built through mountainous country. Steep slopes and cliffs are created when road cuts are blasted and bulldozed into mountainsides. If the bedrock has planes of weakness (such as joints, bedding planes, or foliation planes), the orientation of these planes relative to the road cut determines whether there is a rockslide hazard (as in figure 9.24*A*). If the planes of weakness are inclined into the hill, there is no chance of a rockslide. On the other hand, where the planes of weakness are approximately parallel to the slope of the hillside, a rockslide may occur.

Various techniques are used to prevent rockslides. By doing a detailed geologic study of an area before a road is built, builders might avoid a hazard by choosing the least dangerous route for the road. If a road cut must be made through bedrock that appears prone to sliding, all of the rock that might slide could be removed (sometimes at great expense), as shown in figure 9.24*B*.

In some instances, slopes prone to rock sliding have been "stitched" in place by the technique shown in figure 9.25. Spraying a roadside exposure with concrete (also called *shotcrete*) may retard a landslide in some instances (figure 9.25*E*). Fences or railings on the side of a road can keep minor rockfalls from blocking the road.

Radio-transmitted, real-time monitoring of areas where mass wasting is active is valuable for predicting when a mass movement is about to speed up and be dangerous. Five sites along U.S. Highway 50 in California's Sierra Nevada are monitored by means of instruments placed at steep mass wasting sites by the U.S. Geological Survey, and the data is immediately available on the Internet. The instruments include buried pore pressure gauges as well as motion sensors that can tell when a slow-moving mass is starting to move faster.

The best way to avoid mass wasting damage to or destruction of your house is to get information on the susceptibility of the land to mass wasting before building or buying. A starting place, if you live in the United States, is to contact your state's geological survey (or equivalent organization) through www.stategeologists.org/.

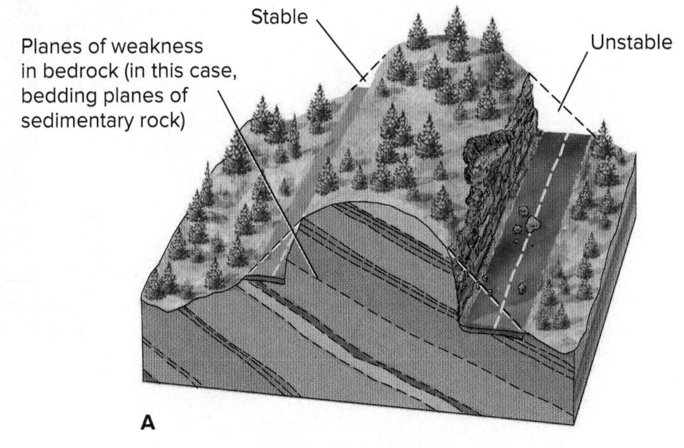

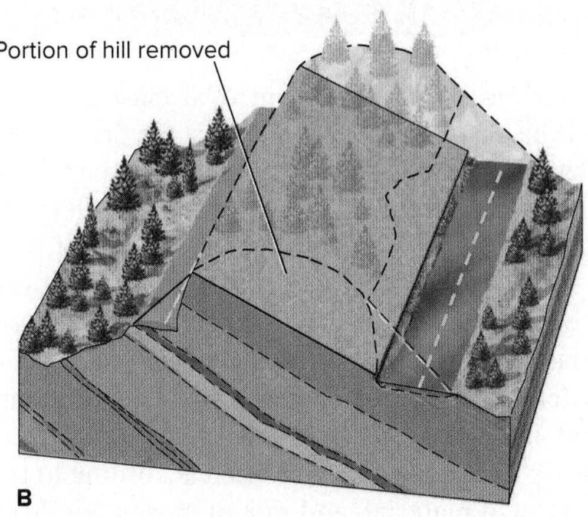

FIGURE 9.24

(*A*) Cross section of a hill showing a relatively safe road cut on the left and a hazardous road cut on the right. (*B*) The same hazardous road cut after removal of rock that might slide.

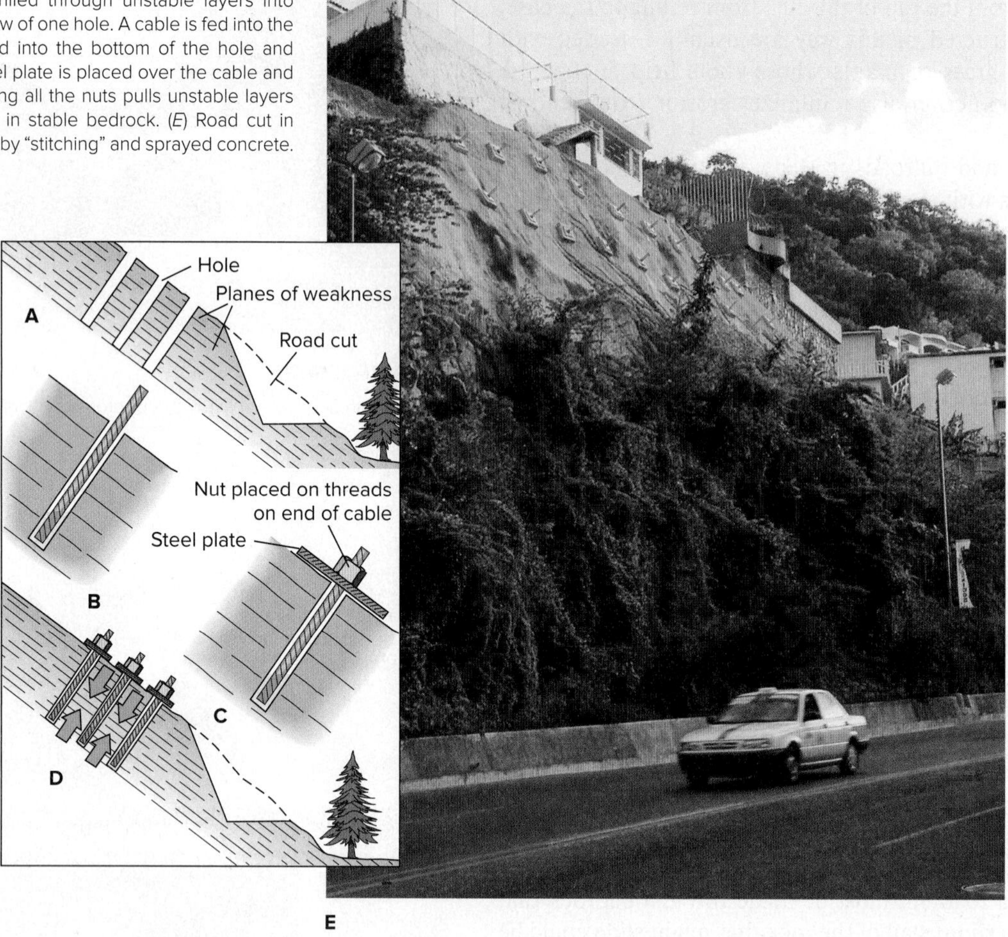

FIGURE 9.25

"Stitching" a slope to keep bedrock from sliding along planes of weakness. (*A*) Holes are drilled through unstable layers into stable rock. (*B*) Expanded view of one hole. A cable is fed into the hole and cement is pumped into the bottom of the hole and allowed to harden. (*C*) A steel plate is placed over the cable and a nut tightened. (*D*) Tightening all the nuts pulls unstable layers together and anchors them in stable bedrock. (*E*) Road cut in Acapulco, Mexico, stabilized by "stitching" and sprayed concrete. ©*C. C. Plummer*

Summary

Mass wasting is the movement of a mass of soil or bedrock toward the base of a slope. Gravity is the driving force for mass wasting. The component of gravitational force that propels mass wasting is the *shear force*, which occurs parallel to a slope. The resistance to that force is the *shear strength* of rock or soil. If shear force exceeds shear strength, mass wasting takes place. A number of other factors, including the presence of water, the materials and structures present in a hillside, vegetative cover, and external triggers, can determine whether movement will occur and, if it does, the rate of movement.

Mass wasting can be classified according to type of movement, type of material, and rate of movement. Movement can take place as a flow, slide, or fall. The material that moves can be bedrock or soil. *Soil*, as used in this chapter, is unconsolidated or weakly consolidated material, regardless of particle size. If soil is predominantly fine material, it is earth; if predominantly coarse, it is *debris*. *Mud* is a mixture of water, clay, and silt.

The slowest type of movement, *creep*, occurs mostly on relatively gentle slopes, usually aided by water in the soil. In colder climates, repeated freezing and thawing of water within the soil contribute to creep. *Landsliding* is a general term for more rapid mass wasting of rock, soil, or both. *Flows* include earthflows, debris flows, mudflows, and debris avalanches. *Earthflows*, in which finer-grained material is predominant, vary greatly in velocity, although they are not as rapid as *debris avalanches*, which are turbulent masses of debris, water, and air. *Solifluction*, a special variety of earthflow, usually takes place in arctic or subarctic climates, where ground is permanently frozen (*permafrost*). *Debris flows* involve coarser material than is present in earthflows. Typically, they travel farther than earthflows and, if a lot of water is present, travel long distances in channels and behave similarly to mudflows. A *mudflow* is a slurry of mostly clay, silt, and water. Most mudflows flow in channels much as streams do.

Rockfall is the fall of broken rock down a vertical or near-vertical slope. A *rockslide* is a slab of rock sliding down a less-than-vertical surface.

Landslides also take place underwater. The larger ones of these are vastly bigger than any that have occurred on land.

Terms to Remember

creep (soil creep) 230
debris 212
debris avalanche 219
debris flow 217
earth 212
earthflow 214
fall 212
flow 212
landslide 206
mass wasting 206
mud 212
mudflow 218
permafrost 215
relief 207
rock avalanche 223
rockfall 220
rockslide 223
rotational slide 213
shear force 207
shear strength 210
slide 212
slump 213
soil 212
solifluction 215
talus 221
translational slide 212

Testing Your Knowledge

Use the following questions to prepare for exams based on this chapter.

1. Draw diagrams similar to figure 9.1 to compare the shear force to the force of gravity for the following situations:
   a. a vertical cliff
   b. a flat horizontal plane
   c. a 45-degree slope
2. List and explain the key factors that control mass wasting.
3. Describe the three characteristics used to classify mass wasting.
4. Draw diagrams showing the four main types of mass wasting movement: flow, fall, translational slide, and rotational slide.
5. How does a rotational slide differ from a translational slide?
6. What role does water play in each of the types of mass wasting?
7. Describe at least three indicators of soil creep.
8. What is the difference between a mudflow and a debris flow?
9. Why is solifluction more common in colder climates than in temperate climates?
10. Using examples, demonstrate how slides can turn into flows.
11. What is the slowest type of mass wasting process?
    a. debris flow   b. rockslide
    c. creep   d. rockfall
    e. avalanche
12. The largest landslide has taken place
    a. on the sea floor.   b. in the Andes.
    c. on active volcanoes.   d. in the Himalaya.
13. A descending mass moving downslope as a viscous fluid is referred to as a
    a. fall.   b. landslide.
    c. flow.   d. slide.
14. The driving force behind all mass wasting processes is
    a. gravity.   b. slope angle.
    c. type of bedrock material.   d. presence of water.
    e. vegetation.
15. The resistance to movement or deformation of soil is its
    a. mass.   b. shear strength.
    c. shear force.   d. density.
16. Flow of water-saturated soil over impermeable material is called
    a. solifluction.   b. flow.
    c. slide.   d. creep.
17. A flowing mixture of soil and water, usually moving down a channel, is called a(n)
    a. mudflow.   b. slide.
    c. fall.   d. earthflow.
18. An apron of fallen rock fragments that accumulates at the base of a cliff is called
    a. bedrock.   b. regolith.
    c. soil.   d. talus.
19. How does construction destabilize a slope?
    a. Adds weight to the top of the slope
    b. Decreases water content of the slope
    c. Adds weight to the bottom of the slope
    d. Increases the shear strength of the slope
20. How can landslides be prevented during construction? (Choose all that apply.)
    a. Build retaining walls.   b. Cut steeper slopes.
    c. Install water drainage systems.   d. Add vegetation.

Expanding Your Knowledge

1. Why do people fear earthquakes, hurricanes, and tornadoes more than they fear landslides?
2. If you were building a house on a cliff, what would you look for to ensure that your house would not be destroyed through mass wasting?
3. Why isn't the land surface of Earth flat after millions of years of erosion by mass wasting as well as by other erosional agents?
4. Can any of the indicators of creep be explained by processes other than mass wasting?

Exploring Web Resources

https://landslides.usgs.gov/

Geologic hazards, landslides, U.S. Geological Survey. You can get to several useful sites from here. Reports on recent landslides can be accessed by clicking on the ones listed. Click on "National Landslide Information Center" for photos of landslides, including some described in this chapter. Watch animation of a landslide. You can access sources of information on landslides and other geologic features for any state, usually from a state's geologic survey.

http://www.nrcan.gc.ca/hazards/landslides

Landslides in Canada. Geological Survey of Canada's site has generalized descriptions of significant Canadian landslides.

http://www.landslideblog.org/

The Landslide Blog. A commentary on landslide events occurring worldwide written by Professor Dave Petley, Durham University in the United Kingdom.

CHAPTER

10 Streams and Floods

Aerial view of flooded houses along the Red River after record rainfall fell in Bossier City, Louisiana during August 2015.
©Bill Davis/Bossier City Sheriff's Office/AP Images

LEARNING OBJECTIVES

- Describe and sketch the movement of water in the hydrologic cycle.
- Know how the longitudinal profile and channel of a stream change from the headwaters to the mouth.
- Describe the nature of running water as it is channelized.
- Sketch and explain the different types of drainage patterns.
- Describe the factors that control stream erosion and deposition.
- Know how a stream transports its sediment load, and illustrate the types of landforms that develop from stream deposition.
- Discuss the development of stream valleys.
- Identify the types and causes of flooding and the measures used to reduce risk.

Running water, aided by mass wasting, is the most important geologic agent in eroding, transporting, and depositing sediment. Almost every landscape on Earth shows the results of stream erosion or deposition. Although other agents–groundwater, glaciers, wind, and waves–can be locally important in sculpting the land, stream action and mass wasting are the dominant processes of landscape development.

We begin by examining the relationship of running water to other water in the hydrologic cycle. The first part of this chapter also deals with the various ways that streams erode, transport, and deposit sediment. The second part describes landforms produced by stream action, such as valleys, flood plains, deltas, and alluvial fans, and shows how each of these is related to changes in stream characteristics. The chapter also includes a discussion of the causes and effects of flooding and various measures used to reduce flood risk.

10.1 THE HYDROLOGIC CYCLE

The interrelationship of the hydrosphere, geosphere, biosphere, and atmosphere is easy to visualize through the **hydrologic cycle**, the movement and interchange of water between the ocean, atmosphere, and land (figure 10.1). Solar radiation provides the

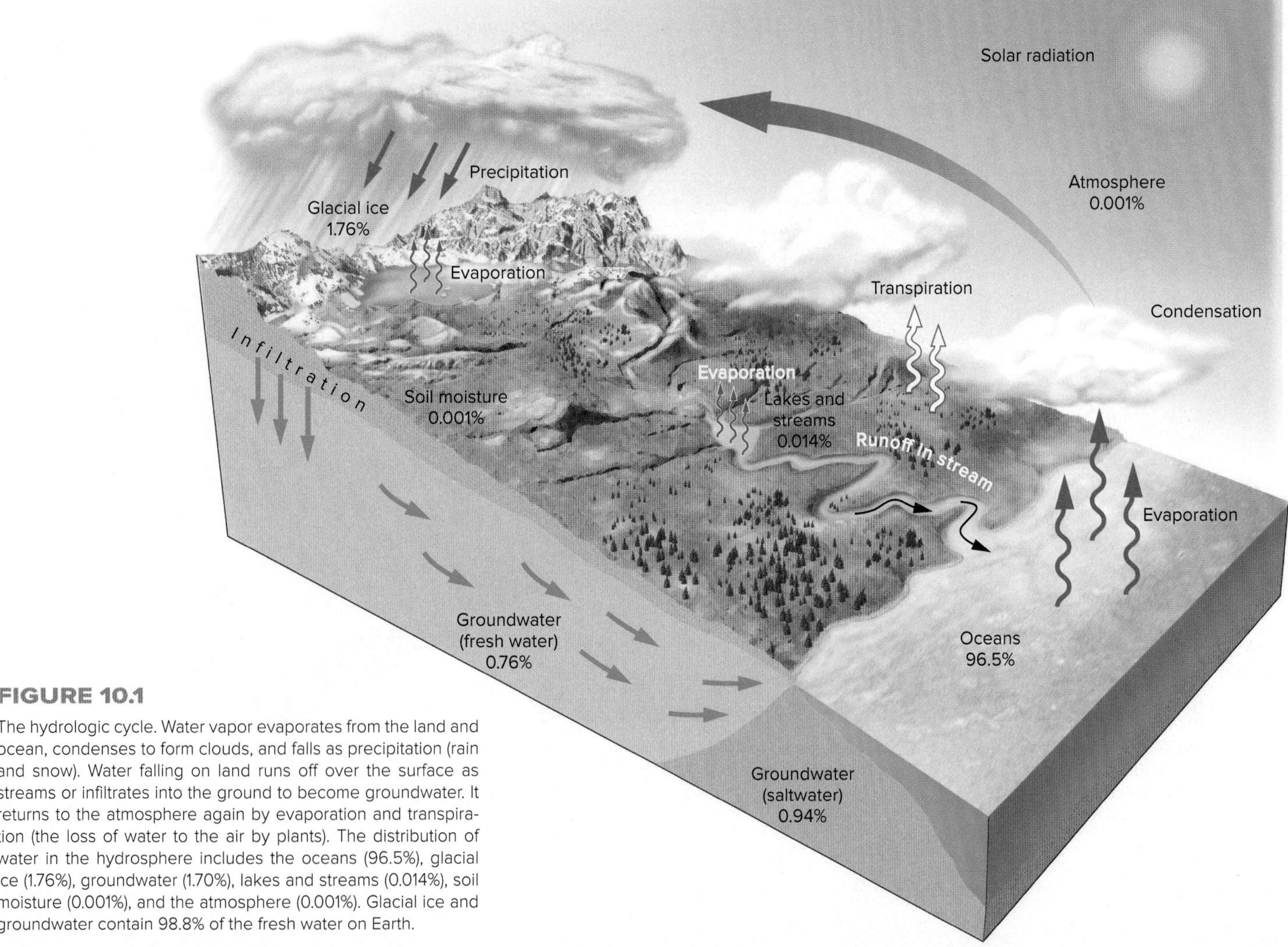

FIGURE 10.1

The hydrologic cycle. Water vapor evaporates from the land and ocean, condenses to form clouds, and falls as precipitation (rain and snow). Water falling on land runs off over the surface as streams or infiltrates into the ground to become groundwater. It returns to the atmosphere again by evaporation and transpiration (the loss of water to the air by plants). The distribution of water in the hydrosphere includes the oceans (96.5%), glacial ice (1.76%), groundwater (1.70%), lakes and streams (0.014%), soil moisture (0.001%), and the atmosphere (0.001%). Glacial ice and groundwater contain 98.8% of the fresh water on Earth.

necessary energy for *evaporation* of water vapor from the land and ocean. The oceans hold 97.5% of the water on Earth. When air becomes saturated with water (100% relative humidity), rises, and cools in the atmosphere, liquid droplets condense to form clouds. These droplets grow larger as more water leaves the gaseous state to form rain or snow, depending on the temperature. When rain (or snow) falls on the land surface as *precipitation*, more than half the water returns rather rapidly to the atmosphere by evaporation or *transpiration* from plants. A substantial amount of fresh water (1.76%) is held as ice in glaciers and snow pack. The remainder either flows over the land surface as *runoff* in streams, is held temporarily in lakes, or soaks into the ground by *infiltration* to form groundwater. Groundwater (the subject of chapter 11) moves, usually very slowly, underground and may flow back onto the surface a long distance from where it seeped into the ground. Groundwater is the second largest reserve of fresh water (0.76%) after glacial ice, and it represents an important source of drinking water along with surface water from streams and lakes (0.14%). Most water eventually reaches the ocean, where ongoing evaporation completes the cycle.

Only about 15% to 20% of rainfall normally ends up as surface runoff in rivers, although the amount of runoff can range from 2% to more than 25% with variations in climate, steepness of slope, soil and rock type, and vegetation. Steady, continuous rains can saturate the ground and the atmosphere, however, and lead to floods as runoff approaches 100% of rainfall.

10.2 RUNNING WATER

A **stream** is a body of running water that is confined in a channel and moves downhill under the influence of gravity. In some parts of the country, *stream* implies size: rivers are large, streams somewhat smaller, and brooks or creeks even smaller. Geologists, however, use *stream* for any body of running water, from a small trickle to a huge river.

Figure 10.2*A* shows a *longitudinal profile* of a typical stream viewed from the side. The stream begins in steep mountains and flows out across a gentle plain into the sea. The *headwaters* of a stream are the upper part of the stream near its source in the

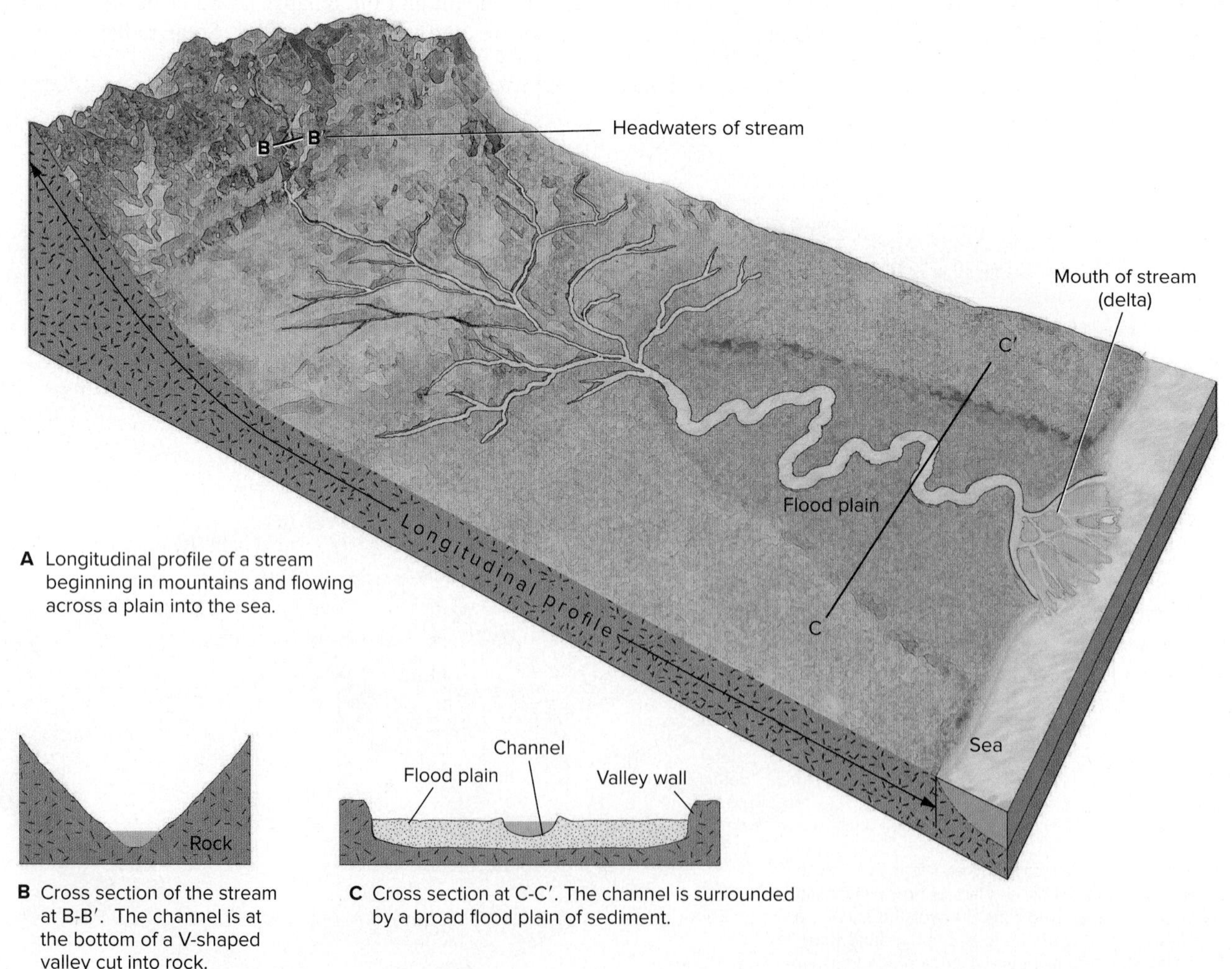

A Longitudinal profile of a stream beginning in mountains and flowing across a plain into the sea.

B Cross section of the stream at B-B′. The channel is at the bottom of a V-shaped valley cut into rock.

C Cross section at C-C′. The channel is surrounded by a broad flood plain of sediment.

FIGURE 10.2

Longitudinal profile and cross sections of a typical stream.

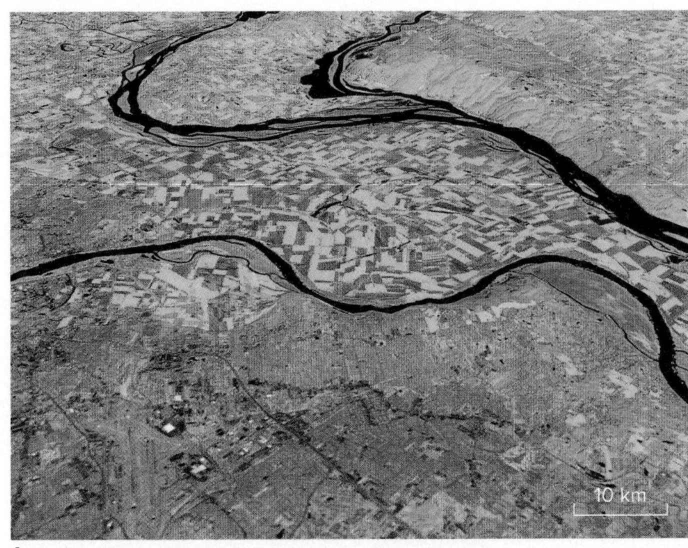

A

B

FIGURE 10.3

A stream normally stays in its channel, but during a flood it can spill over its banks onto the adjacent flatland (flood plain) as shown in these three-dimensional satellite images. (*A*) Before flooding image (August 14, 1991) of Missouri River (bottom), Mississippi River (upper left), and Illinois River (upper right). Vegetation is shown in green, and red indicates recently plowed fields (bare soil). (*B*) Image taken on November 7 after the huge floods of 1993 showing how the rivers spilled over their banks onto the flat flood plains. ©*NASA/GSFC/Science Source*

mountains. The *mouth* is the place where a stream enters the sea, a lake, or a larger stream. A *cross section* of a stream in steep mountains is usually a V-shaped valley cut into solid rock, with the stream channel occupying the narrow bottom of the valley; there is little or no flat land next to the stream on the valley bottom (figure 10.2*B*). Near its mouth, a stream usually flows within a broad, flat-floored valley. The stream channel is surrounded by a flat *flood plain* of sediment deposited by the stream (figure 10.2*C*).

A stream normally stays in its **stream channel**, a long, narrow depression eroded by the stream into rock or sediment. The stream *banks* are the sides of the channel; the *streambed* is the bottom of the channel. During a flood, the waters of a stream may rise and spill over the banks onto the flat flood plain of the valley floor (figure 10.3).

Not all water that moves over the land surface is confined to channels. Sometimes, particularly during heavy rains, water runs off as **sheetwash**, a thin layer of unchanneled water flowing downhill. Sheetwash is particularly common in deserts, where the lack of vegetation allows rainwater to spread quickly over the land surface. It also occurs in humid regions during heavy thunderstorms when water falls faster than it can soak into the ground. A series of closely spaced storms can also promote sheetwash; as the ground becomes saturated, more water runs over the surface.

Sheetwash, along with the violent impact of raindrops on the land surface, can produce considerable *sheet erosion*, in which a thin layer of surface material, usually topsoil, is removed by the flowing sheet of water. This gravity-driven movement of sediment is a process intermediate between mass wasting and stream erosion.

Overland sheetwash becomes concentrated in small channels, forming tiny streams called *rills*. Rills merge to form small streams, and small streams join to form larger streams. Most regions are drained by networks of coalescing streams.

10.3 DRAINAGE BASINS

Each stream, small or large, has a **drainage basin**, the total area drained by a stream and its tributaries (a **tributary** is a small stream flowing into a larger one) (figure 10.4*A*). A drainage basin can be outlined on a map by drawing a line around the region drained by all the tributaries to a river (figure 10.4*B*). The Mississippi River's drainage basin, for example, includes all the land area drained by the Mississippi River itself and by all its tributaries, including the Ohio and Missouri rivers. This great drainage system includes more than one-third the land area of the contiguous 48 states.

A ridge or strip of high ground dividing one drainage basin from another is termed a **divide** (figure 10.4*A*). The best known continental divide in the United States separates streams that flow to the Pacific Ocean from those that flow to the Atlantic and the Gulf of Mexico (figure 10.4*B*). It extends from the Yukon Territory down into Mexico and crosses Montana, Idaho, Wyoming, Colorado, and New Mexico in the United States. An eastern continental divide follows the Appalachian Mountains and separates those streams that flow into the Atlantic from those that flow into the Gulf of Mexico.

10.4 DRAINAGE PATTERNS

The arrangement, in map view, of a river and its tributaries is a **drainage pattern**. A drainage pattern can, in many cases, reveal the nature and structure of the rocks underneath it.

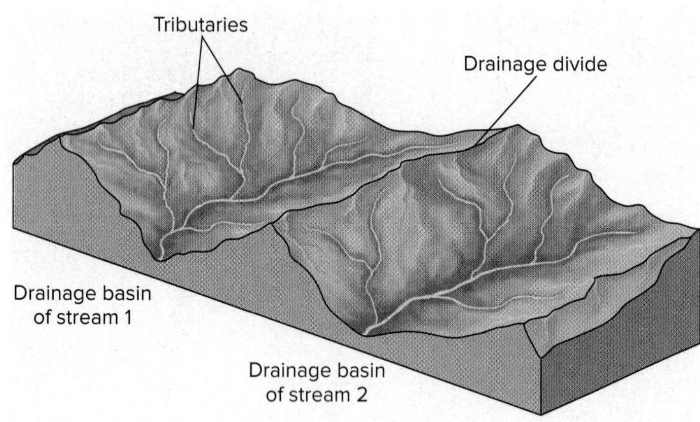

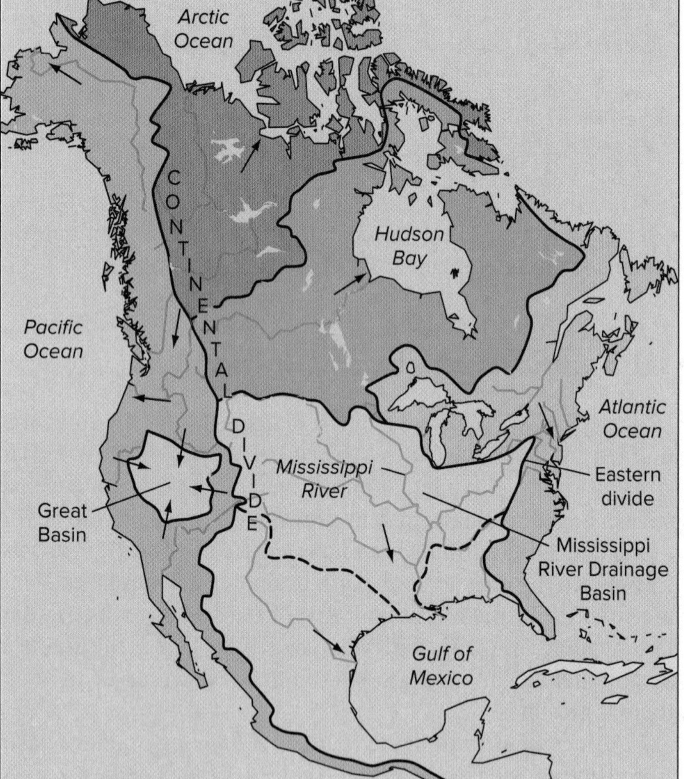

FIGURE 10.4

(*A*) The total area drained by each stream and its tributaries is called a drainage basin. A ridge, or drainage divide, separates two adjacent drainage basins. (*B*) Map of North America showing drainage divides (black lines) and major drainage basins. The continental divide separates rivers that flow into the Pacific from those that flow eastward into the Gulf of Mexico and Atlantic Ocean or Arctic Ocean. The Eastern divide separates rivers flowing into the Atlantic from those flowing into the Gulf of Mexico.

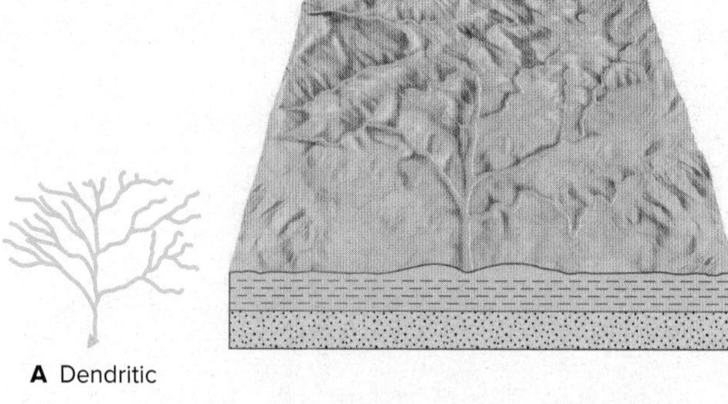

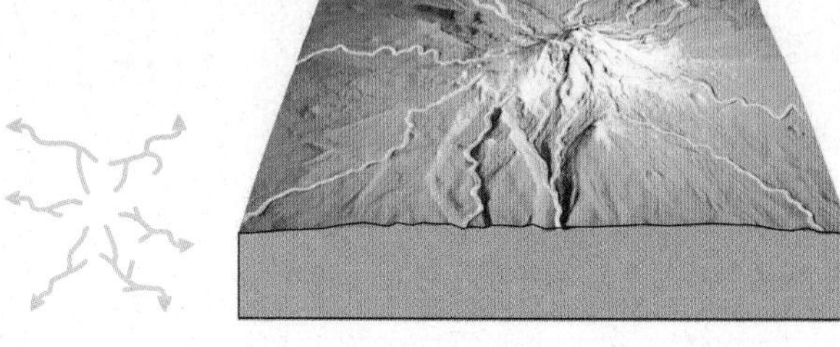

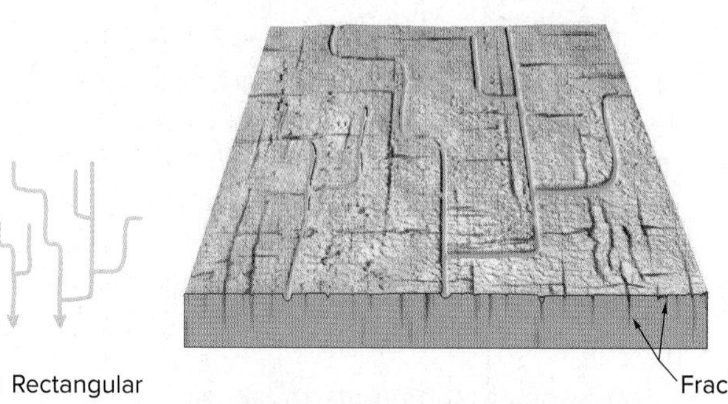

FIGURE 10.5

Drainage patterns can reveal something about the rocks underneath. (*A*) Dendritic pattern develops on uniformly erodible rock. (*B*) A radial pattern develops on a conical mountain or dome. (*C*) A rectangular pattern develops on regularly fractured rock. (*D*) A trellis pattern develops on alternating ridges and valleys caused by the erosion of resistant and nonresistant tilted rock layers.

Most tributaries join the main stream at an acute angle, forming a V (or Y) pointing downstream. If the pattern resembles branches of a tree or nerve dendrites, it is called **dendritic** (figures 10.4 and 10.5*A*). Dendritic drainage patterns develop on uniformly erodible rock or regolith and are the most common type of pattern. A **radial pattern**, in which streams diverge outward like spokes of a wheel, forms on high conical mountains, such as composite volcanoes and domes (figure 10.5*B*). A **rectangular pattern**, in which tributaries have frequent 90-degree bends and tend to join other streams at right angles, develops on regularly fractured rock (figure 10.5*C*). A network of fractures meeting at right angles forms pathways for streams because fractures are eroded more easily than unbroken rock. A **trellis pattern** consists of parallel main streams with short tributaries meeting them at right angles (figure 10.5*D*). A trellis pattern forms in a region where tilted layers of resistant rock such as sandstone alternate with nonresistant rock such as shale. Erosion of such a region results in a surface topography of parallel ridges and valleys.

10.5 FACTORS AFFECTING STREAM EROSION AND DEPOSITION

Stream erosion and deposition are controlled primarily by a river's *velocity* and, to a lesser extent, its *discharge*. Velocity is largely controlled by the stream *gradient*, channel shape, and channel roughness.

Velocity

The distance water travels in a stream per unit time is called the **stream velocity**. A moderately fast river flows at about 5 kilometers per hour (3 miles per hour). Rivers flow much faster during flood, sometimes exceeding 25 kilometers per hour (15 miles per hour).

The cross-sectional views of a stream in figure 10.6 show that a stream reaches its maximum velocity near the middle of the channel. When a stream goes around a curve, the region of maximum velocity is displaced by inertia toward the outside of the curve. Velocity is the key factor in a stream's ability to erode, transport, and deposit. High velocity (meaning greater energy) generally results in erosion and transportation; low velocity causes sediment deposition. Slight changes in velocity can cause great changes in the sediment load carried by the river.

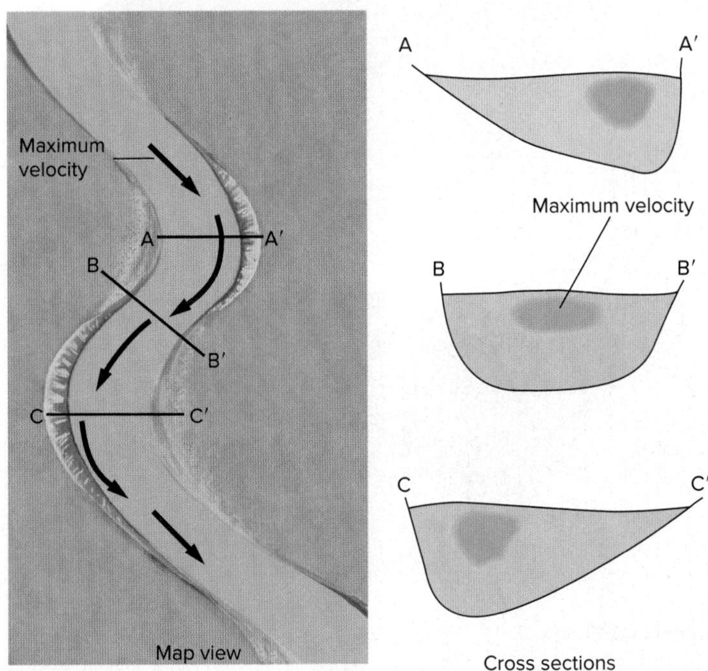

FIGURE 10.6

Regions of maximum velocity in a stream. Arrows on the map show how the maximum velocity shifts to the outside of curves. Sections show maximum velocity on outside of curves and in the center of the channel on a straight stretch of stream.

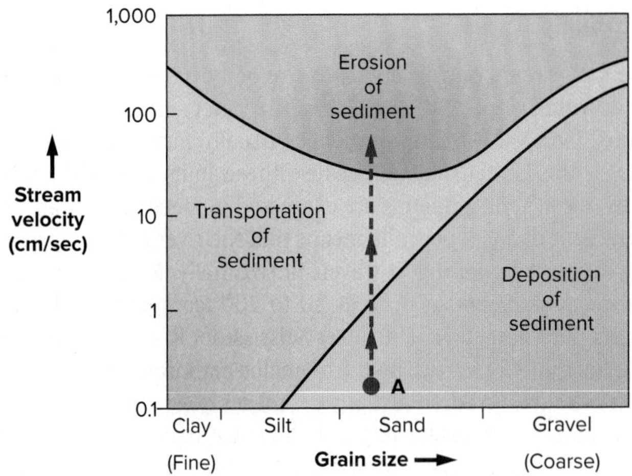

FIGURE 10.7

Logarithmic graph showing the stream velocities at which erosion and deposition of sediment occur. These velocities vary with the grain size of the sediment. See text for a discussion of point *A* and the dashed red line above it.

Figure 10.7 shows the stream velocities at which sediments are eroded, transported, and deposited. For each grain size, these velocities are different. The upper curve represents the minimum velocity needed to erode sediment grains. This curve shows the velocity at which previously stationary grains are first picked up by moving water. The lower curve represents the velocity below which deposition occurs, when moving grains come to rest. Between the two curves, the water is moving fast enough to transport grains that have already been eroded. Note that it takes a higher stream velocity to erode grains (set them in motion) than to transport grains (keep them in motion).

Point *A* on figure 10.7 represents fine sand on the bed of a stream that is barely moving. The vertical red arrows represent a flood with gradually increasing stream velocity. No sediment moves until the velocity is high enough to intersect the *upper* curve and move into the area marked "erosion." As the flood recedes, the velocity drops below the upper curve and into the transportation area. Under these conditions, the sand that was already eroded continues to be transported, but no new sand is eroded. As the velocity falls below the lower curve, all the sand is deposited again, coming to rest on the streambed.

The right half of the diagram shows that coarser particles require progressively higher velocities for erosion and transportation, as you might expect (boulders are harder to move than sand grains). The erosion curve also rises toward the left of the diagram, however. This shows that fine-grained silt and clay are actually harder to erode than sand. The reason is that molecular forces tend to bind silt and clay into a smooth, cohesive mass that resists erosion. Once silt and clay are eroded, however, they are easily transported. As you can see from the lower curve, the silt and clay in a river's suspended load are not deposited until the river virtually stops flowing.

Gradient

One factor that controls a stream's velocity is the **stream gradient**, the downhill slope of the bed (or of the water surface, if the stream is very large). A stream gradient is usually measured in feet per mile in the United States, because these units are used on U.S. maps (elsewhere, gradients are expressed in meters per kilometer). A gradient of 5 feet per mile means that the river drops 5 feet vertically for every mile that it travels horizontally. Mountain streams may have gradients as steep as 50 to 200 feet per mile (10 to 40 meters per kilometer). The lower Mississippi River has a very gentle gradient, 0.5 foot per mile (0.1 meter per kilometer) or less.

A stream's gradient usually decreases downstream. Typically, the gradient is greatest in the headwater region and decreases toward the mouth of the stream (see figure 10.2). Local increases in the gradient of a stream are usually marked by rapids.

Channel Shape and Roughness

The *shape of the channel* also controls stream velocity. Flowing water drags against the stream banks and bed, and the resulting friction slows the water down. In figure 10.8, the streams in *A* and *B* have the same cross-sectional area, but stream *B* flows more slowly than *A* because the wide, shallow channel in *B* has more surface for the moving water to drag against.

A stream may change its channel width as it flows across different rock types. Hard, resistant rock is difficult to erode, so a stream may have a relatively narrow channel in such rock. As a result, it flows rapidly (figure 10.9*A*). If the stream flows onto a softer rock that is easier to erode, the channel may widen, and the river will slow down because of the increased surface area dragging on the flowing water. Sediment may be deposited as the velocity decreases.

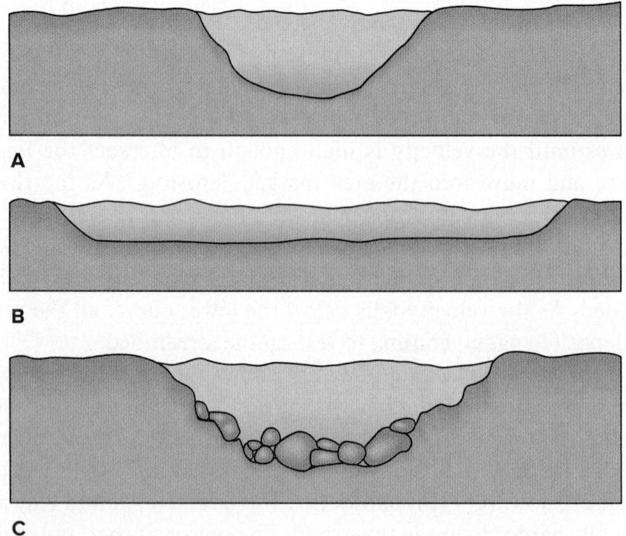

FIGURE 10.8

Channel shape and roughness influence stream velocity. (*A*) Semicircular channel allows stream to flow rapidly. (*B*) Wide, shallow channel increases friction, slowing river down. (*C*) Rough, boulder-strewn channel slows river.

The width of a stream may be controlled by factors external to the stream. A landslide may carry debris onto a valley floor, partially blocking a stream's channel (figure 10.9*B*). The constriction causes the stream to speed up as it flows past the slide, and the increased velocity may quickly erode the landslide debris, carrying it away downstream. Human interference with a river can promote erosion and deposition. Construction of a culvert or bridge can partially block a channel, increasing the stream's velocity (figure 10.9*C*). If the bridge was poorly de-

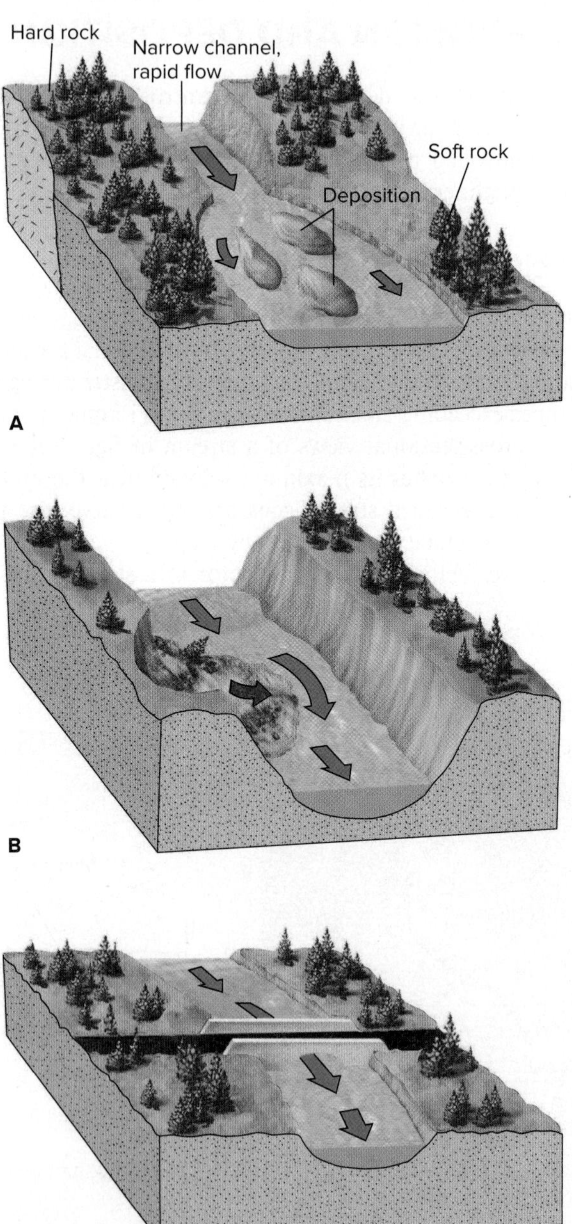

FIGURE 10.9

Channel width variations caused by rock type and obstructions. Length of arrow indicates velocity. (*A*) A channel may widen in soft rock. Deposition may result as stream velocity drops. (*B*) Landslide may narrow a channel, increasing stream velocity. Resulting erosion usually removes landslide debris. (*C*) Bridge piers (or other obstructions) will increase velocity and sometimes erosion next to the piers.

FIGURE 10.10

Increased discharge in mountain stream can move large boulders and provide a fast ride for rafters. Note the rounding of the boulders which marks the high-water mark.

©Javier Perini CM/Image Source

signed, it may increase velocity to the point where erosion may cause the bridge to collapse.

The *roughness of the channel* also controls velocity. A stream can flow rapidly over a smooth channel, but a rough, boulder-strewn channel floor creates more friction and slows the flow (see figure 10.8*C*). Coarse particles increase the roughness more than fine particles, and a rippled or wavy sand bottom is rougher than a smooth sand bottom.

Discharge

The **discharge** of a stream is the volume of water that flows past a given point in a unit of time. It is found by multiplying the cross-sectional area of a stream by its velocity (or width × depth × velocity). Discharge can be reported in cubic feet per second (cfs), which is standard in the United States, or in cubic meters per second (m^3/sec).

$$\begin{aligned}\text{Discharge (cfs)} &= \text{average stream width (ft)}\\ &\times \text{average depth (ft)}\\ &\times \text{average velocity (ft/sec)}\end{aligned}$$

A stream 100 feet wide and 15 feet deep flowing at 4 miles per hour (6 ft/sec) has a discharge of 9,000 cubic feet per second (cfs). In streams in humid climates, discharge increases downstream for two reasons: (1) water flows out of the ground into the river through the streambed and (2) small tributary streams flow into a larger stream along its length, adding water to the stream as it travels.

To handle the increased discharge, these streams increase in width and depth downstream. Some streams surprisingly increase slightly in velocity downstream, as a result of the increased discharge (the increase in discharge and channel size, and the typical downstream smoothness of the channel override the effect of a lessening gradient).

During floods, a stream's discharge and velocity increase, usually as a result of heavy rains over the stream's drainage basin. Flood discharge may be 50 to 100 times normal flow. Stream erosion and transportation generally increase enormously as a result of a flood's velocity and discharge. Swift mountain streams in flood can sometimes move large boulders, and they provide enough fast water for a thrilling ride for rafters (figure 10.10). Flooded areas may be intensely scoured, with river banks and adjacent lawns and fields washed away. As floodwaters recede, both velocity and discharge decrease, leading to the deposition of a blanket of sediment, usually mud, over the flooded area.

In a dry climate, a river's discharge can decrease in a downstream direction as river water evaporates into the air and soaks into the dry ground (or is used for irrigation). As the discharge decreases, the load of sediment is gradually deposited.

10.6 STREAM EROSION

A stream usually erodes the rock and sediment over which it flows. In fact, streams are among the most effective sculptors of the land. Streams cut their own valleys, deepening and widening them over long periods of time and carrying away the sediment that mass wasting delivers to valley floors. The particles of rock and sediment that a stream picks up are carried along to be deposited farther downstream. Streams erode rock and sediment in three ways—*hydraulic action*, *solution*, and *abrasion*.

Hydraulic action refers to the ability of flowing water to pick up and move rock and sediment (figure 10.11). The force of running water swirling into a crevice in a rock can crack the rock and break loose a fragment to be carried away by the stream. Hydraulic force can also erode loose material from a stream

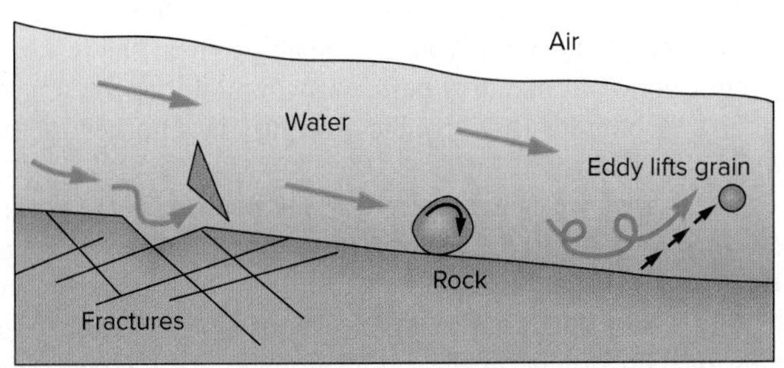

FIGURE 10.11

Hydraulic action can loosen, roll, and lift grains from the streambed.

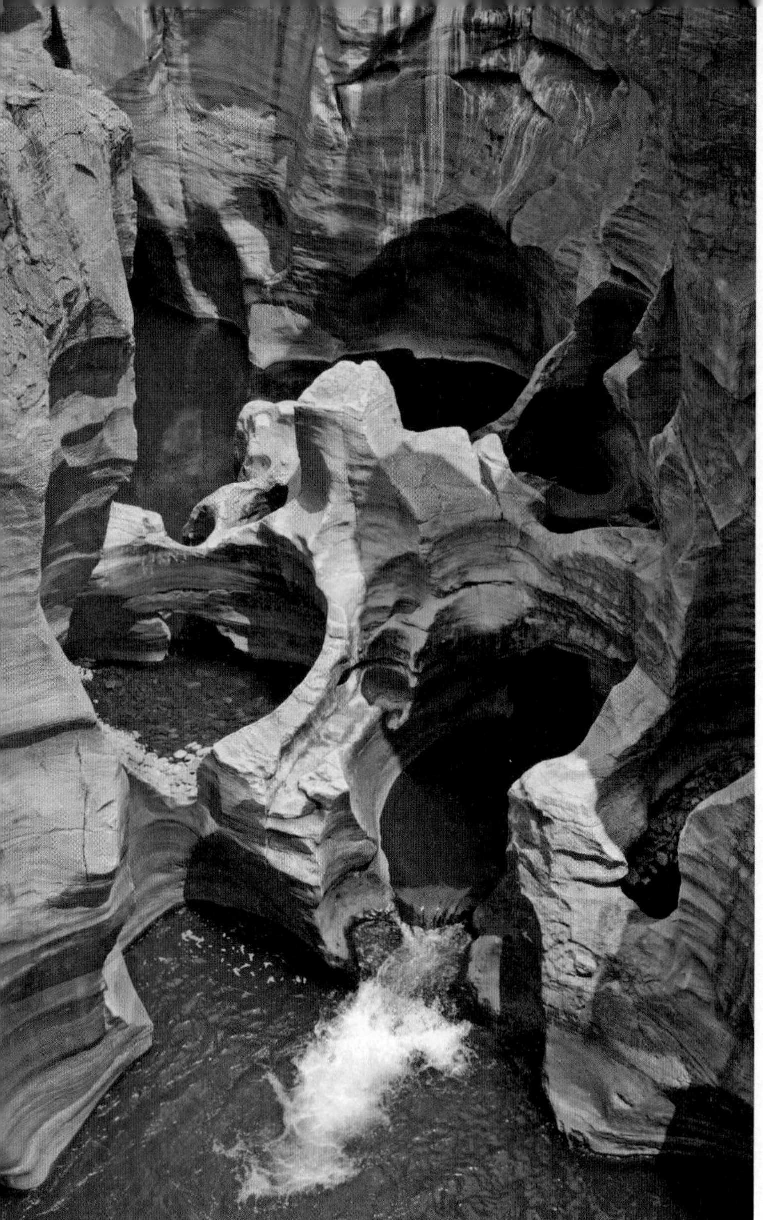

FIGURE 10.12

Deep potholes eroded by the Blyde River, Blyde River Canyon Nature Preserve, South Africa.
©Walter G Allgöwer/imageBROKER/age fotostock

bank on the outside of a curve. The pressure of flowing water can roll or slide grains over a streambed, and a swirling eddy of water may exert enough force to lift a rock fragment above a streambed. The great force of falling water makes hydraulic action particularly effective at the base of a waterfall, where it may erode a deep plunge pool. You may be able to hear the results of hydraulic action by standing beside a swift mountain stream and listening to boulders and cobbles hitting one another as they tumble along downstream.

From what you have learned about weathering, you know that some rocks can be dissolved by water. **Solution**, although ordinarily slow, can be an effective process of weathering and erosion (weathering because it is a response to surface chemical conditions; erosion because it removes material). A stream flowing over limestone, for example, gradually dissolves the rock, deepening the stream channel. A stream flowing over other sedimentary rocks, such as sandstone, can dissolve calcite cement, loosening grains that can then be picked up by hydraulic action.

The erosive process that is usually most effective on a rocky streambed is **abrasion**, the grinding away of the stream channel by the friction and impact of the sediment load. Sand and gravel tumbling along near the bottom of a stream wear away the streambed much as moving sandpaper wears away wood. The abrasion of sediment on the streambed is generally much more effective in wearing away the rock than hydraulic action alone. The more sediment a stream carries, the faster it is likely to wear away its bed.

The coarsest sediment is the most effective in stream erosion. Sand and gravel strike the streambed frequently and with great force, while the finer-grained silt and clay particles weigh so little that they are easily suspended throughout the stream and have little impact when they hit the channel.

Potholes are depressions that are eroded into the hard rock of a streambed by the abrasive action of the sediment load (figure 10.12). During high water when a stream is full, the swirling water can cause sand and pebbles to scour out smooth, bowl-shaped depressions in hard rock. Potholes tend to form in spots where the rock is a little weaker than the surrounding rock. Although potholes are fairly uncommon, you can see them on the beds of some streams at low water levels. Potholes may contain sand or an assortment of beautifully rounded pebbles.

10.7 STREAM TRANSPORTATION OF SEDIMENT

The sediment load transported by a stream can be subdivided into *bed load*, *suspended load*, and *dissolved load.* Most of a stream's load is carried in suspension and in solution.

The **bed load** is the large or heavy sediment particles that travel on the streambed (figure 10.13). Sand and gravel, which form the usual bed load of streams, move by either *traction* or *saltation.*

Large, heavy particles of sediment, such as cobbles and boulders, may never lose contact with the streambed as they move along in the flowing water. They roll or slide along the stream bottom, eroding the streambed and each other by abrasion. Movement by rolling, sliding, or dragging is called **traction**.

Sand grains move by traction, but they also move downstream by **saltation**, a series of short leaps or bounces off the bottom (see figure 10.13). Saltation begins when sand grains are momentarily lifted off the bottom by *turbulent* water (eddying, swirling flow). The force of the turbulence temporarily counteracts the downward force of gravity, suspending the grains in water above the streambed. The water soon slows down because the velocity of water in an eddy is not constant; then gravity overcomes the lift of the water, and the sand grain once again

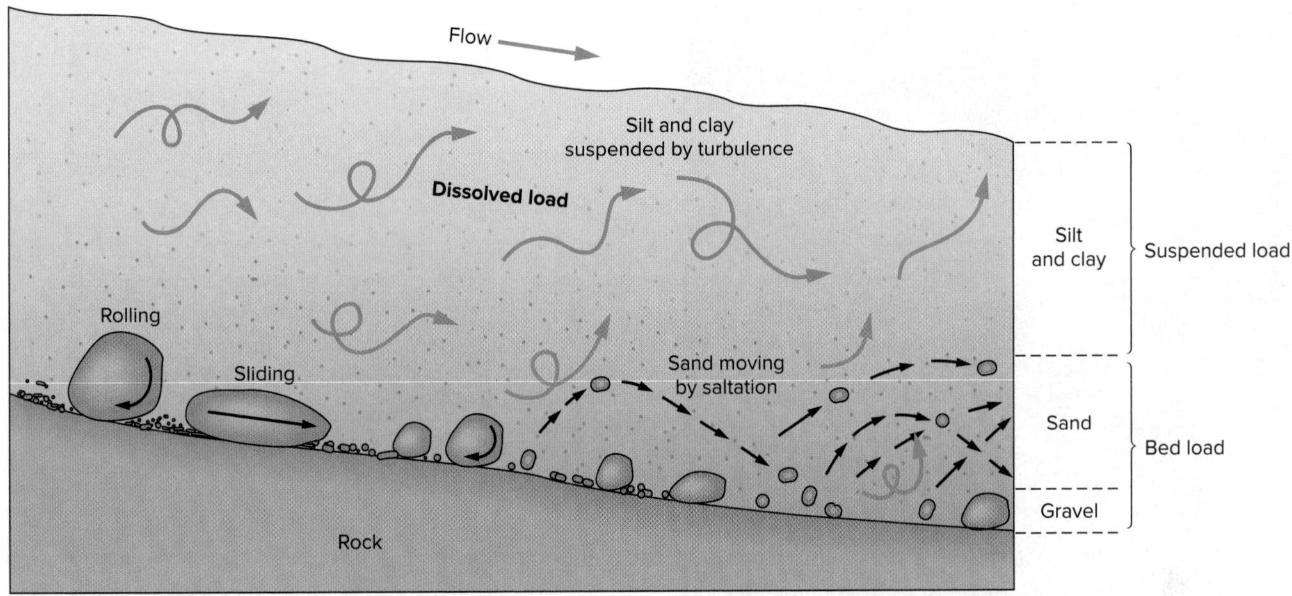

FIGURE 10.13

A stream's bed load consists of sand and gravel moving on or near the streambed by traction and saltation. Finer silt and clay form the suspended load of the stream. The dissolved load of soluble ions is invisible.

falls to the bed of the stream. While it is suspended, the grain moves downstream with the flowing water. After it lands on the bottom, it may be picked up again if turbulence increases, or it may be thrown up into the water by the impact of another falling sand grain. In this way, sand grains saltate downstream in leaps and jumps, partly in contact with the bottom and partly suspended in the water.

The **suspended load** is sediment that is light enough to remain lifted indefinitely above the bottom by water turbulence (see figure 10.13). The muddy appearance of a stream during a flood or after a heavy rain is due to a large suspended load (see figure 10.10). Silt and clay usually are suspended throughout the water, while the coarser bed load moves on the stream bottom. Suspended load has less effect on erosion than the less visible bed load, which causes most of the abrasion of the streambed. Vast quantities of sediment, however, are transported in suspension.

Soluble products of chemical weathering processes can make up a substantial **dissolved load** in a stream. Most streams contain numerous ions in solution, such as bicarbonate, calcium, potassium, sodium, chloride, and sulfate. The ions may precipitate out of water as evaporite minerals if the stream dries up, or they may eventually reach the ocean. Very clear water may in fact be carrying a large load of material in solution, for the dissolved load is invisible. Only if the water evaporates does the material become visible as crystals begin to form.

One estimate is that rivers in the United States carry about 250 million tons of solid load and 300 million tons of dissolved load each year. (It would take a freight train eight times as long as the distance from Boston to Los Angeles to carry 250 million tons.)

10.8 STREAM DEPOSITION

The sediments transported by a stream are often deposited temporarily along the stream's course (particularly the bed load sediments). Such sediments move sporadically downstream in repeated cycles of erosion and deposition, forming *bars* and *flood-plain deposits.* At or near the end of a stream, sediments may be deposited more permanently in a *delta* or an *alluvial fan.*

Bars

Stream deposits may take the form of a **bar**, a ridge of sediment, usually sand and gravel, deposited in the middle or along the banks of a stream (figure 10.14). Bars are formed by deposition when a stream's discharge or velocity decreases. During a flood, a river can move all sizes of sediment, from silt and clay up to huge boulders, because the greatly increased volume of water is moving very rapidly. As the flood begins to recede, the water level in the stream falls and the velocity drops. With the stream no longer able to carry all its sediment load, the larger boulders drop down on the streambed, slowing the water locally even more. Finer gravel and sand are deposited between

FIGURE 10.14

Sand and gravel bars deposited along the banks and middle of a stream. Green River at Horseshoe Bend, Utah.
©Michael Collier

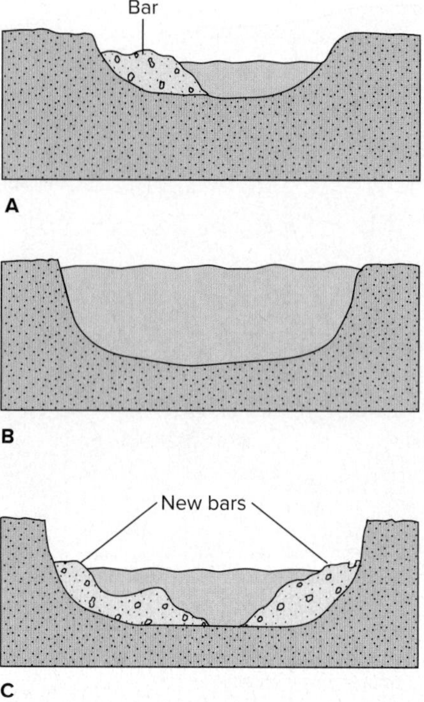

FIGURE 10.15

A flood can wash away bars in a stream, depositing new bars as the water recedes. (*A*) Normal water flow with sand and gravel bar. (*B*) Increased discharge and velocity during flood moves all sediment downstream. Channel deepens and widens if banks erode easily. (*C*) New bars are deposited as water level drops and stream slows down.

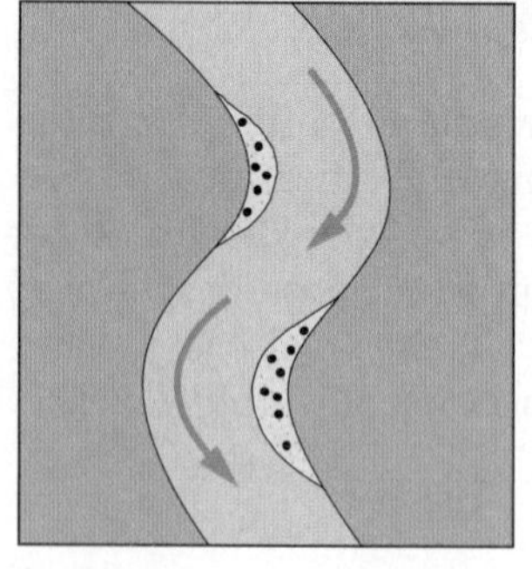

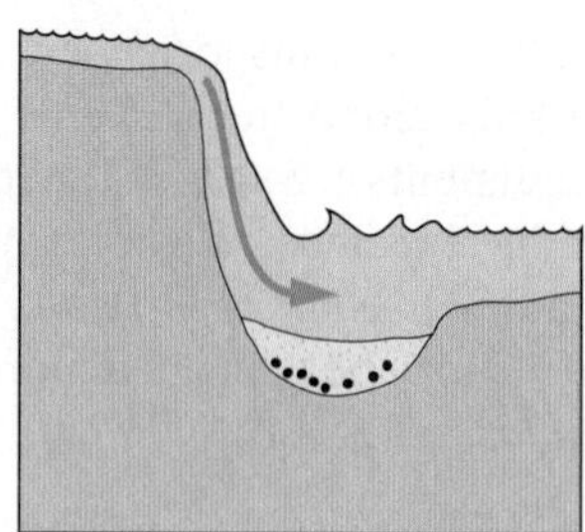

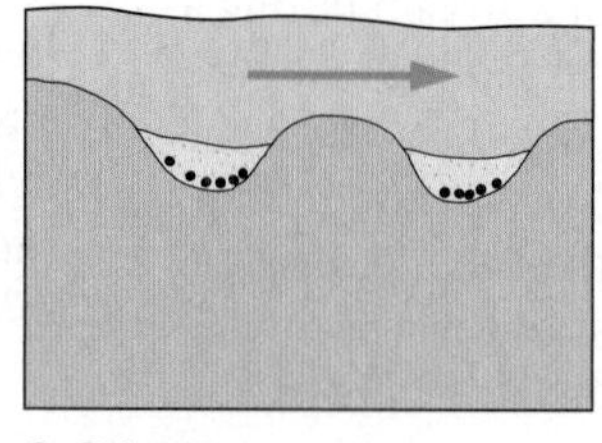

FIGURE 10.16

Types of placer deposits. (*A*) Stream bar. (*B*) Below waterfall. (*C*) Depressions on streambed. Valuable grains shown in black.

the boulders and downstream from them. In this way, deposition builds up a sand and gravel bar that may become exposed as the water level falls.

The next flood on the river may erode most of the sediment in this bar and move it farther downstream. But as the flood slows, it may deposit new gravel in approximately the same place, forming a new bar (figure 10.15). After each flood, river anglers and boat operators must relearn the size and position of the bars. Sometimes gold panners discover fresh gold in a mined-out river bar after a flood has shifted sediment downstream. A dramatic example of the shifting of sandbars occurred during the planned flood on the Colorado River downstream from the Glen Canyon Dam (box 10.1).

Placer Deposits

Placer deposits are found in streams where the running water has mechanically concentrated heavy sediment. The heavy sediment is concentrated in the stream where the velocity of the water is high enough to carry away lighter material but not the heavy sediment. Such places include river bars on the inside of meanders, plunge pools below waterfalls, and depressions on a streambed (figure 10.16). Grains concentrated in this manner include gold dust and nuggets, native platinum, diamonds and other gemstones, and worn pebble or sand grains composed of the heavy oxides of titanium and tin.

Braided Streams

Deposition of a bar in the center of a stream (a *midchannel bar*) diverts the water toward the sides, where it washes against the stream banks with greater force, eroding the banks and widening the stream (figure 10.17*A*). A stream heavily loaded with sediment may deposit many bars in its channel, causing the stream to widen continually as more bars are deposited. Such a stream typically goes through many stages of deposition, erosion, deposition, and erosion, especially if its discharge fluctuates. The stream may fill its main channel with sediment and become a **braided stream**, flowing in a network of interconnected rivulets around numerous bars (figure 10.17*B* and *C*). A braided stream characteristically has a wide, shallow channel.

A stream tends to become braided when it is heavily loaded with sediment (particularly bed load) and has banks that are easily eroded. The braided pattern develops in deserts as a sediment-laden stream loses water through evaporation and percolation into the ground. In meltwater streams flowing off glaciers, braided patterns tend to develop when the discharge from the melting glaciers is low relative to the great amount and ranges of size of sediment the stream has to carry.

Meandering Streams and Point Bars

Rivers that carry fine-grained silt and clay in suspension tend to be narrow and deep and to develop pronounced, sinuous curves called **meanders** (figure 10.18). In a long river, sediment tends to

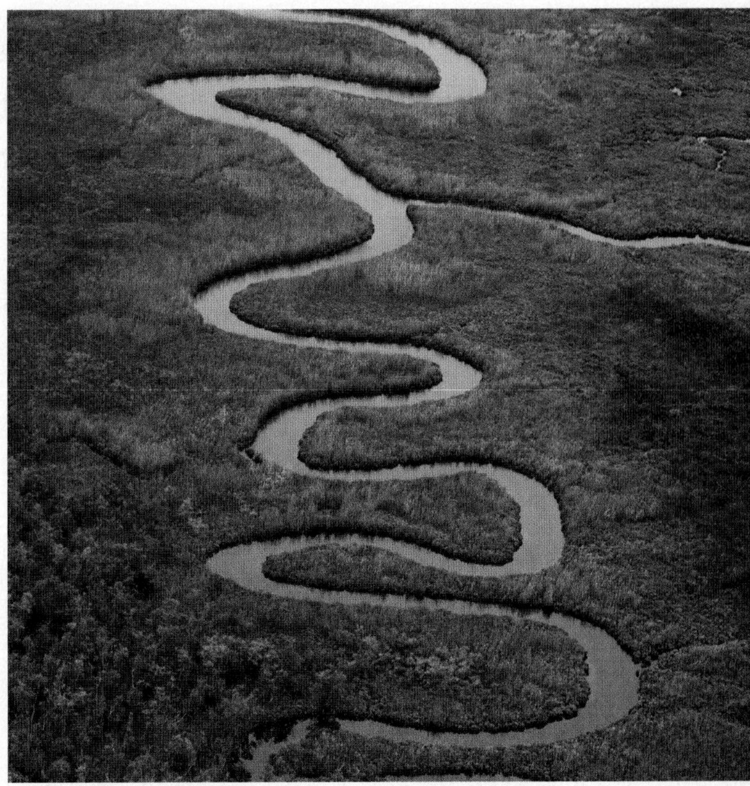

FIGURE 10.18
Meanders in a stream. These sinuous curves develop because a stream's velocity is highest on the outside of curves, promoting erosion there.
©Perry Thorsvik/America 24-7/Getty Images

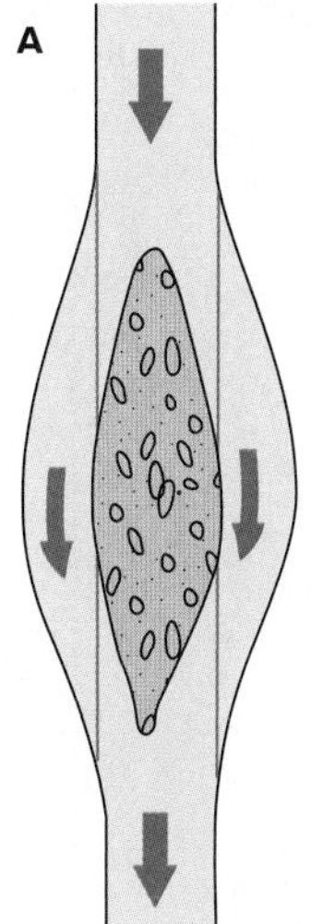

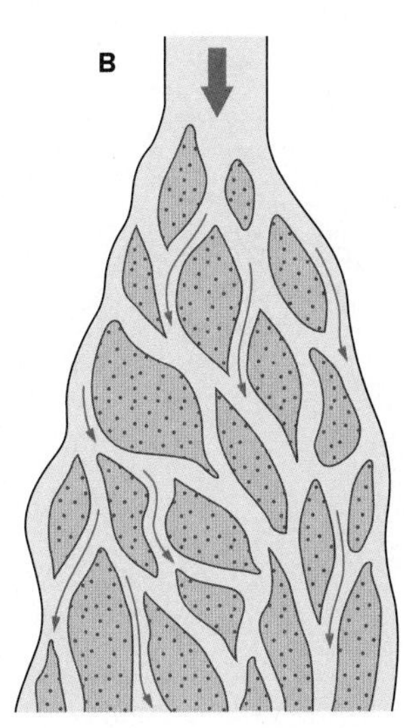

FIGURE 10.17
(*A*) A midchannel bar can divert a stream around it, widening the stream. (*B*) Braided stream occurs where there is an excess of sediment. Bars split main channel into many smaller channels, greatly widening the stream. (*C*) International Space Station view of a braided stream carrying a heavy suspended load of sand and gravel from melting glaciers, Brahmaputra, Tibet.
(c) Source: Earth Sciences and Image Analysis Laboratory at Johnson Space Center/NASA

ENVIRONMENTAL GEOLOGY 10.1

Controlled Floods in the Grand Canyon: Bold Experiments to Restore Sediment Movement in the Colorado River

The largest experiment ever conducted to restore sediment movement on a river is taking place along the Colorado River below the Glen Canyon Dam (box figure 1). The discharge from the Glen Canyon Dam was dramatically increased to simulate the effects of a flood on the Colorado River. One of the main goals of the controlled flooding experiment was to determine whether the higher flows would result in bed scour and redeposition of sandbars and beaches along the sides of the channel (box figure 2). Another goal was to measure and observe how rocks move along the bed of the river with increasing discharge and velocity of floodwaters.

The Colorado River had not experienced its usual summertime floods since the Glen Canyon Dam was completed in 1963. The construction of the dam decreased peak discharges or flows on the Colorado River, which resulted in sand being deposited mainly along the bed or bottom of the river and erosion of beaches along the banks of the river. The Glen Canyon Dam cuts off a significant percentage of the sand supply to the lower Colorado River such that most of the downstream sand is supplied by two tributary streams, the Paria and Little Colorado rivers. In August 1992, the Paria River flooded and deposited 330,000 tons of sand into the Colorado River, and in January 1993, a flood on the Little Colorado River deposited 10 million tons of sediment below its confluence with the Colorado River. The influx of sediment, coupled with the relatively low discharges from the dam, resulted in sand being concentrated along the bed of the Colorado River.

The first controlled flood experiment, conducted in 1996, resulted in sand initially being scoured from the bottom of the main channel and redeposited as bars and beaches. However, after three days of the higher flows, the bars and beaches began to erode, and sediment was once again deposited along the bottom of the river. Scientists had overestimated the amount of sand along the bottom of the Colorado River and the length of time necessary to redeposit it along the river banks.

After analyzing the data and results of the first flooding experiment, a second controlled flood was undertaken in 2004. In the second controlled flood, scientists waited for an adequate supply of sediment in the Colorado River from the tributary streams and shortened the length of time of higher flows. In the fall of 2004, tropical storms swept a million tons of sediment (*only a tenth of what the undammed Colorado once carried*) down the Paria River into the Colorado River. The conditions were now set for another attempt to simulate what happens to sediment during flood-level discharges on a river. This time, the bars and beaches were again temporarily restored, but they eroded quickly. The rapid erosion was due to an inadequate amount of sediment flushed downstream and to the daily variation of flows from the Glen Canyon Dam necessary to satisfy hydroelectricity needs. In 2008 and 2012, controlled flood experiments showed that the amount of sand, the grain size of the sand, and the velocity of the water are important variables in reestablishing sandbars. Less sand is required to rebuild the sandbars and beaches if the sand is finer-grained because it remains in suspension longer and is moved by back eddies to the river banks (box figure 2). However, the question remains whether controlled floods are effective in rebuilding sandbars with the amount and grain size of sand supplied by tributaries to the Colorado River below the Glen Canyon Dam (box figure 3).

Downstream at Lava Falls, another experiment was set up to determine how and whether large boulders deposited in the main channel from a debris flow would move with the increased discharge and velocity of the floodwater. Holes were drilled into 150 basalt boulders and radio tags were inserted so that their movement could be monitored and correlated with the increase in discharge and velocity of the river. Surface velocity measurements were taken by kayaking the river and charting the speed at which

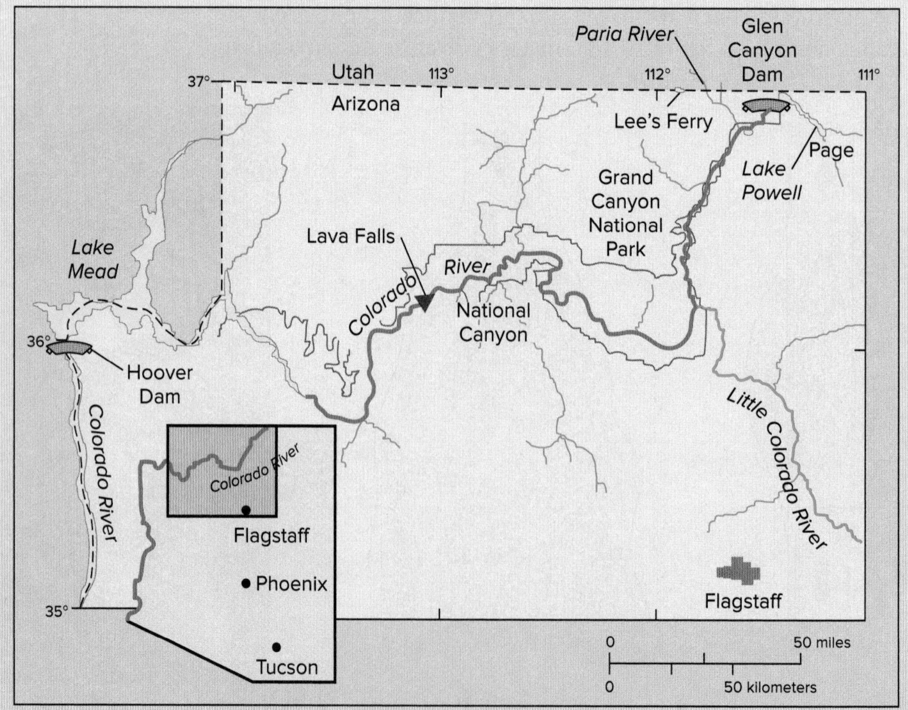

BOX 10.1 ■ FIGURE 1

Location map of the Grand Canyon controlled flood experiments.
Source: U.S. Geological Survey

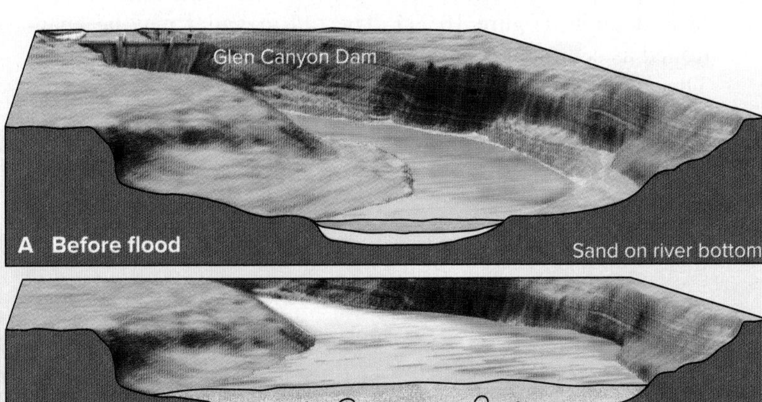

BOX 10.1 ■ FIGURE 2

Cross-sectional views of the distribution of sand before (*A*), during (*B*), and after (*C*) controlled floods on the Colorado River below Glen Canyon Dam. To view photos of sandbars before and after the 2012 controlled flooding and a map of the monitoring sites, visit www.gcmrc.gov.
Source: U.S. Geological Survey

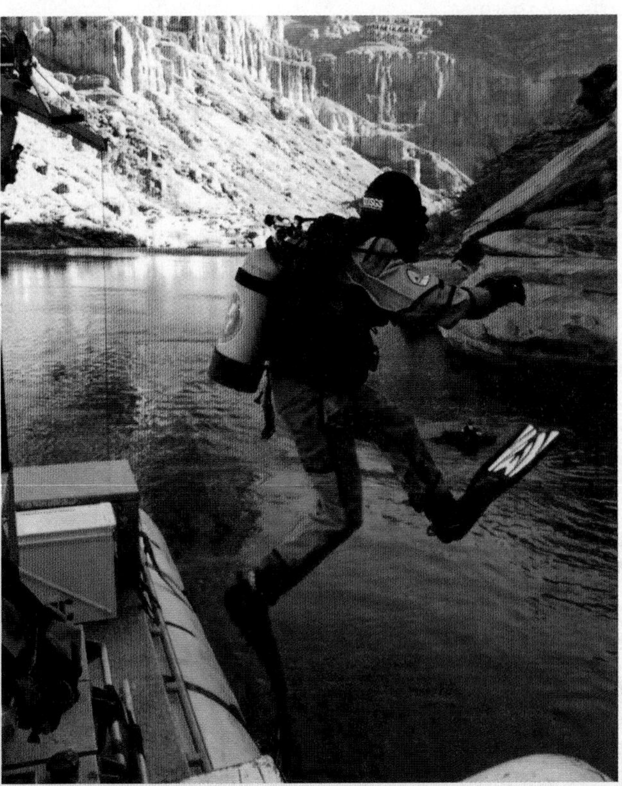

BOX 10.1 ■ FIGURE 3

A U.S. Geological Survey hydrologist jumps into the Colorado River to clean and maintain equipment in preparation for the 2012 controlled flood experiment.
Source: Bryan Smith, U.S. Geological Survey

floating balls moved. The surface velocities were used to calculate the velocity of the water close to the riverbed where the boulders were positioned. Dye was also injected into the river at peak flows to determine the average velocity of the water. The dye indicated that the velocity of the water increased downstream, particularly at the Lava Falls debris flow. This is because the floodwater accelerated as it flowed downstream, pushing the river water in front of it, which increased the downstream velocity. The first crest of water actually arrived behind Hoover Dam at Lake Mead a day ahead of the floodwater marked with a red dye.

The experimental floods, even though smaller than a naturally occurring flood, showed that beaches could be temporarily restored below a dam and that boulders could be moved out of rapids much like the process that occurs on an undammed river during a seasonal flood. However, in order to maintain the beaches, controlled releases from the Glen Canyon Dam need to occur more often. Based on these experiments on the Colorado River, the U.S. Department of Interior will now routinely release high flows through 2020 whenever adequate sand has accumulated in the river. In years when it is not possible to release a controlled flood because of water storage needs, flows will be kept low enough to prevent the sandbars from being eroded away. The results of this experiment suggest that other dammed rivers may benefit from periodic floods to help restore their natural conditions and thus minimize the adverse effects of damming a river.

Additional Resources

Terri Cook. 2013. "Releasing a Flood of Controversy on the Colorado River." *Earth.* Available online at http://www.earthmagazine.org/article/releasing-flood-controversy-colorado-river

P. E. Grams. 2013. *A Sand Budget for Marble Canyon, Arizona—Implications for Long-Term Monitoring of Sand Storage Change.* U.S. Geological Survey Fact Sheet 2013–3074, 4 pages. Available online at http://pubs.usgs.gov/fs/2013/3074/

Information, photos, podcasts, and videos of the November 18, 2012, controlled flood experiment. http://www.usbr.gov/uc/rm/gcdHFE/2012/

P. E. Grams and J. LaVista. 2015. Rebuilding Sandbars in the Grand Canyon. U.S. Geological Survey blog, http://www.usgs.gov/blogs/features/usgs_top_story/rebuilding-sandbars-in-the-grand-canyon/

become finer downstream, so meandering is common in the lower reaches of a river.

You have seen in figure 10.6 that a river's velocity is higher on the outside of a curve than on the inside. This high velocity can erode the river bank on the outside of a curve, often rapidly (figure 10.19).

The low velocity on the inside of a curve promotes sediment deposition. The sandbars in figure 10.20 have been deposited on the inside of curves because of the lower velocity there. Such a bar is called a **point bar** and usually consists of a series of arcuate ridges of sand or gravel.

The simultaneous erosion on the outside of a curve and deposition on the inside can deepen a gentle curve into a hairpin-like meander (see figure 10.20). Meanders are rarely fixed in position. Continued erosion and deposition cause them to migrate back and forth across a flat valley floor, as well as downstream, leaving scars and arcuate point bars to mark their former positions.

At times, particularly during floods, a river may form a **meander cutoff**, a new, shorter channel across the narrow neck of a meander (figure 10.21). The old meander may be abandoned as sediment separates it from the new, shorter channel. The cutoff meander becomes a crescent-shaped **oxbow lake** (figure 10.22). With time, an oxbow lake may fill with sediment and vegetation.

A

B

FIGURE 10.19

River erosion on the outside of a curve. Newaukum River, Washington. Pictures were taken in (*A*) January and (*B*) March 1965.
Source: P. A. Glancy, U.S. Geological Survey

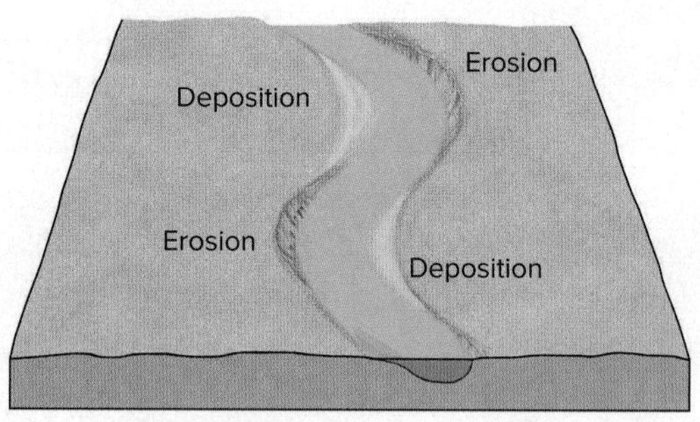

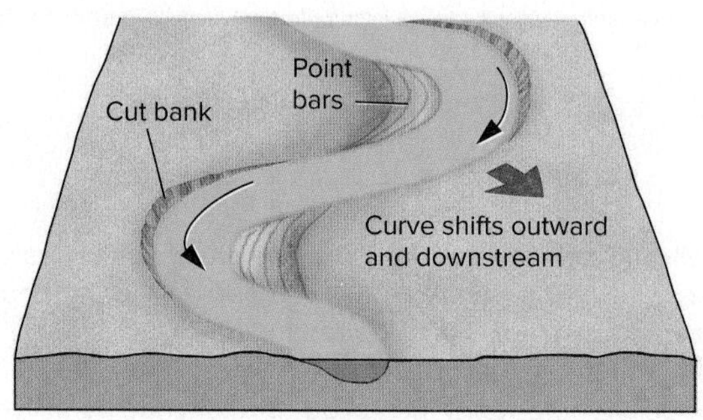

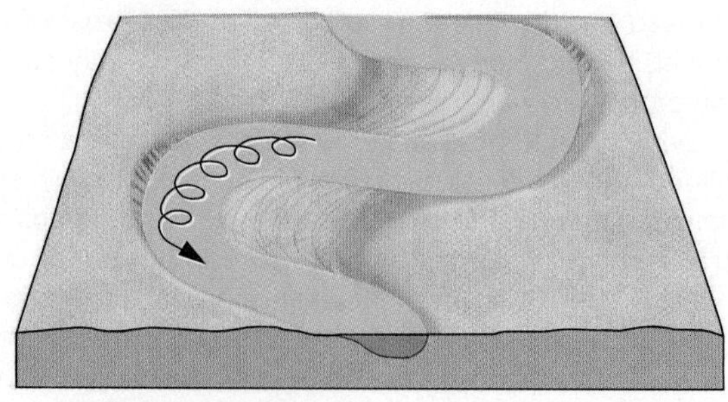

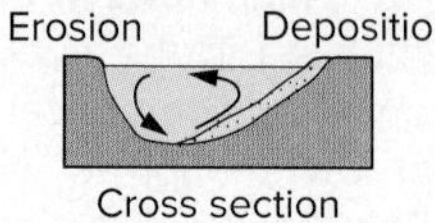

FIGURE 10.20

Development of river meanders and point bars by erosion and deposition on curves.

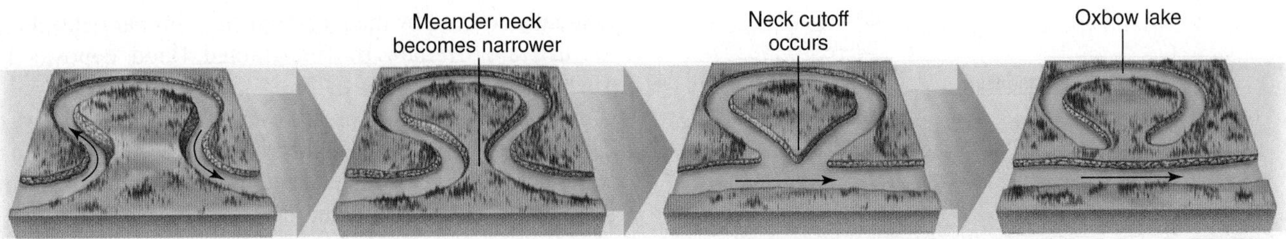

FIGURE 10.21

Creation of an oxbow lake by a meander neck cutoff. Old channel is separated from river by sediment deposition.

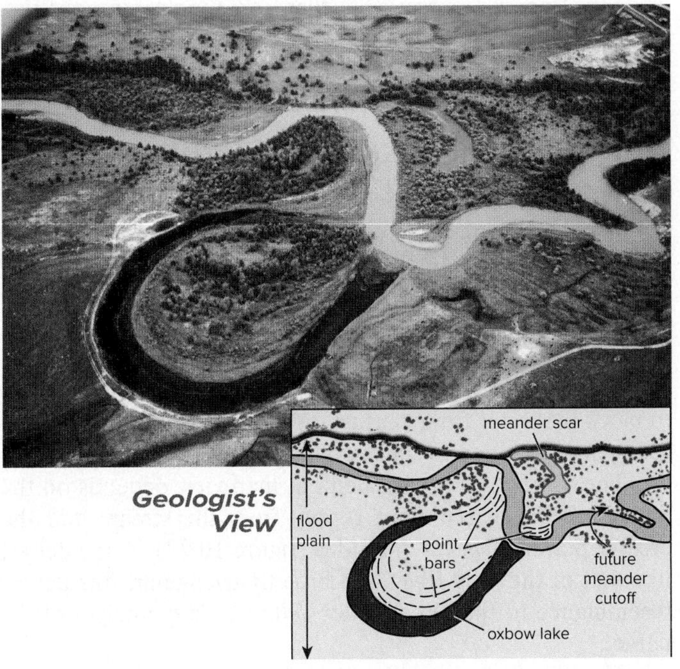

FIGURE 10.22

An oxbow lake marks the former position of a river meander, Blackfoot River near Vallet, Montana.
©James Steinberg/Science Source

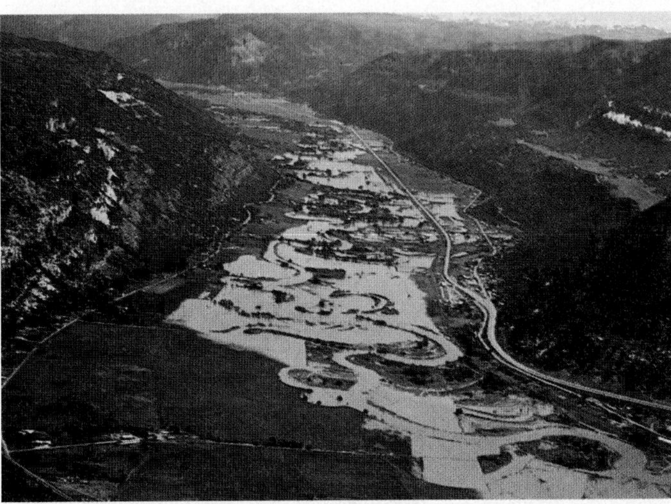

FIGURE 10.23

River flood plains. Flooded flood plain of the Animas River, Colorado.
©D. A. Rahm, courtesy Rahm Memorial Collection, Western Washington University

Flood Plains

A **flood plain** is a broad strip of land built up by sedimentation on either side of a stream channel. During floods, flood plains may be covered with water carrying suspended silt and clay (figure 10.23). When the floodwaters recede, these fine-grained sediments are left behind as a horizontal deposit on the flood plain.

Some flood plains are constructed almost entirely of horizontal layers of fine-grained sediment, interrupted here and there by coarse-grained channel deposits (figure 10.24*A*). Other flood plains are dominated by meanders shifting back and forth over the valley floor and leaving sandy point bar deposits on the inside of curves. Such a river will deposit a characteristic fining-upward sequence of sediments: coarse channel deposits are

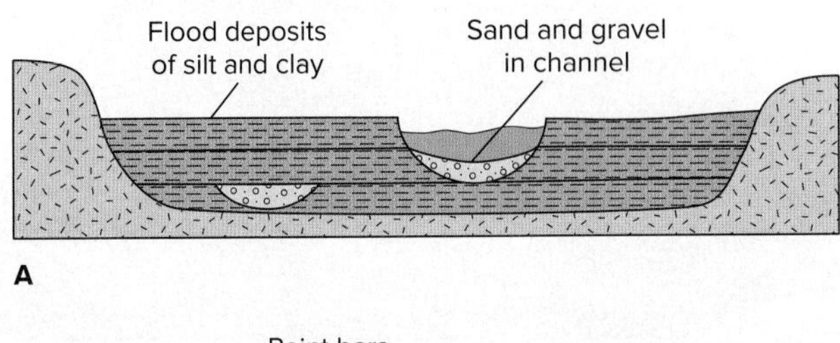

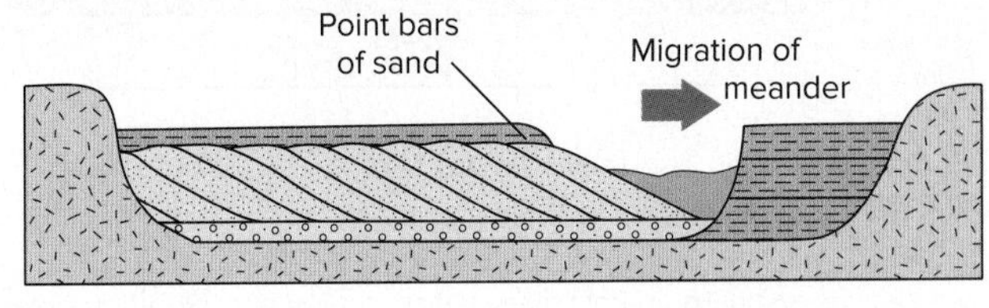

FIGURE 10.24

Flood plains. (*A*) Horizontal layers of fine-grained flood deposits with lenses of coarse-grained channel deposits. (*B*) A fining-upward sequence deposited by a migrating meander. Channel gravel is overlain by sandy point bars, which are overlain by fine-grained flood deposits.

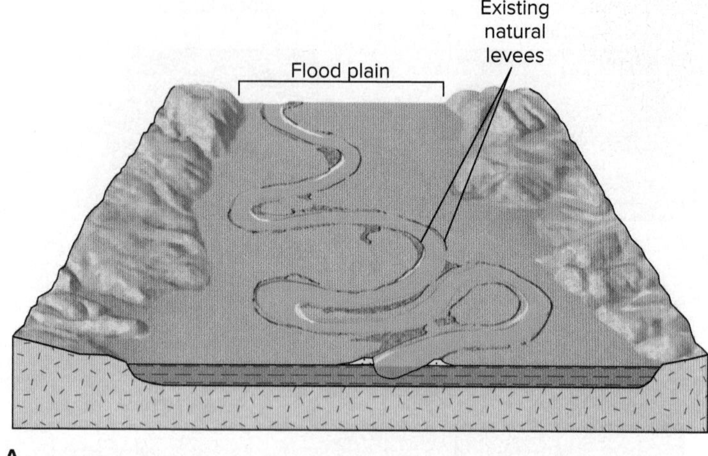

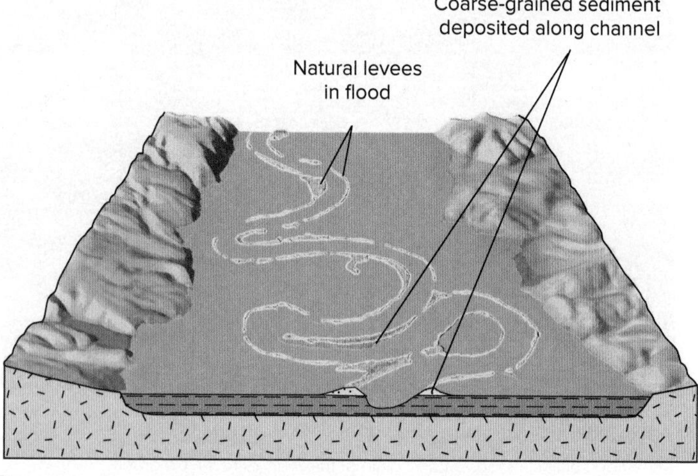

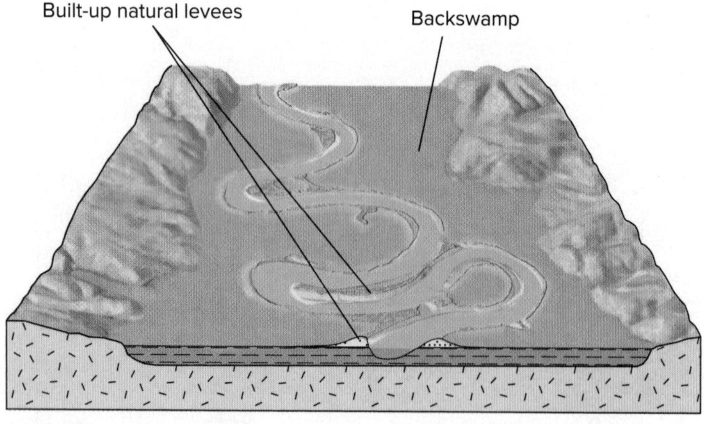

FIGURE 10.25

Natural levee deposition during a flood. Levees are thickest and coarsest next to the river channel and build up from many floods, not just one. (Relief of levees is exaggerated.) (*A*) Normal flow. (*B*) Flood. (*C*) After flood.

gradually covered by medium-grained point bar deposits, which in turn are overlain by fine-grained flood deposits (figure 10.24*B*).

As a flooding river spreads over a flood plain, it slows down. The velocity of the water is abruptly decreased by friction as the water leaves the deep channel and moves in a thin sheet over the flat valley floor. The sudden decrease in velocity of the water causes the river to deposit most of its sediment near the main channel, with progressively less sediment deposited away from the channel (figure 10.25). A series of floods may build up **natural levees**—low ridges of flood-deposited sediment that form on either side of a stream channel and thin away from the channel. The sediment near the river is coarsest, often sand and silt, while the finer clay is carried farther from the river into the flat, lowland area (the backswamp).

Deltas

Most streams ultimately flow into the sea or large lakes. A stream flowing into quiet water usually builds a **delta**, a body of sediment deposited at the mouth of a river when the river's velocity decreases (figure 10.26).

The surface of most deltas is marked by **distributaries**—small, shifting channels that carry water away from the main river channel and distribute it over the surface of the delta (figure 10.26). Sediment deposited at the end of a distributary tends to block the water flow, causing distributaries and their sites of sediment deposition to shift periodically.

The shape of a marine delta in map view depends on the balance between sediment supply from the stream and the erosive power of waves and tides (figure 10.27). Some deltas, like that of the Nile River, are broadly triangular; this delta's resemblance to the Greek letter *delta* (Δ) is the origin of the name.

The Nile Delta is a *wave-dominated delta* that contains barrier islands along its oceanward side (figure 10.27*A*); the barrier islands form by waves actively reworking the deltaic sediments. Some deltas form along a coast that is dominated by strong tides, and the sediment is reshaped into tidal bars that are aligned parallel to a tidal current (figure 10.27*B*). The Ganges-Brahmaputra Delta in Bangladesh is a good example of a *tide-dominated delta.* Other deltas, including that of the Mississippi River, are created when very large amounts of sediment are carried into relatively quiet water. Partly because dredging has kept the major distributary channels (locally called "passes") fixed in position for many decades, the Mississippi's distributaries have built long fingers of sediment out into the sea. The resulting shape has been termed a *birdfoot delta.* Because of the dominance of stream sedimentation that forms the fingerlike distributaries, birdfoot deltas like the Mississippi's are also referred to as *stream-dominated deltas* (figure 10.27*C*).

FIGURE 10.26

Internal construction of a small delta.

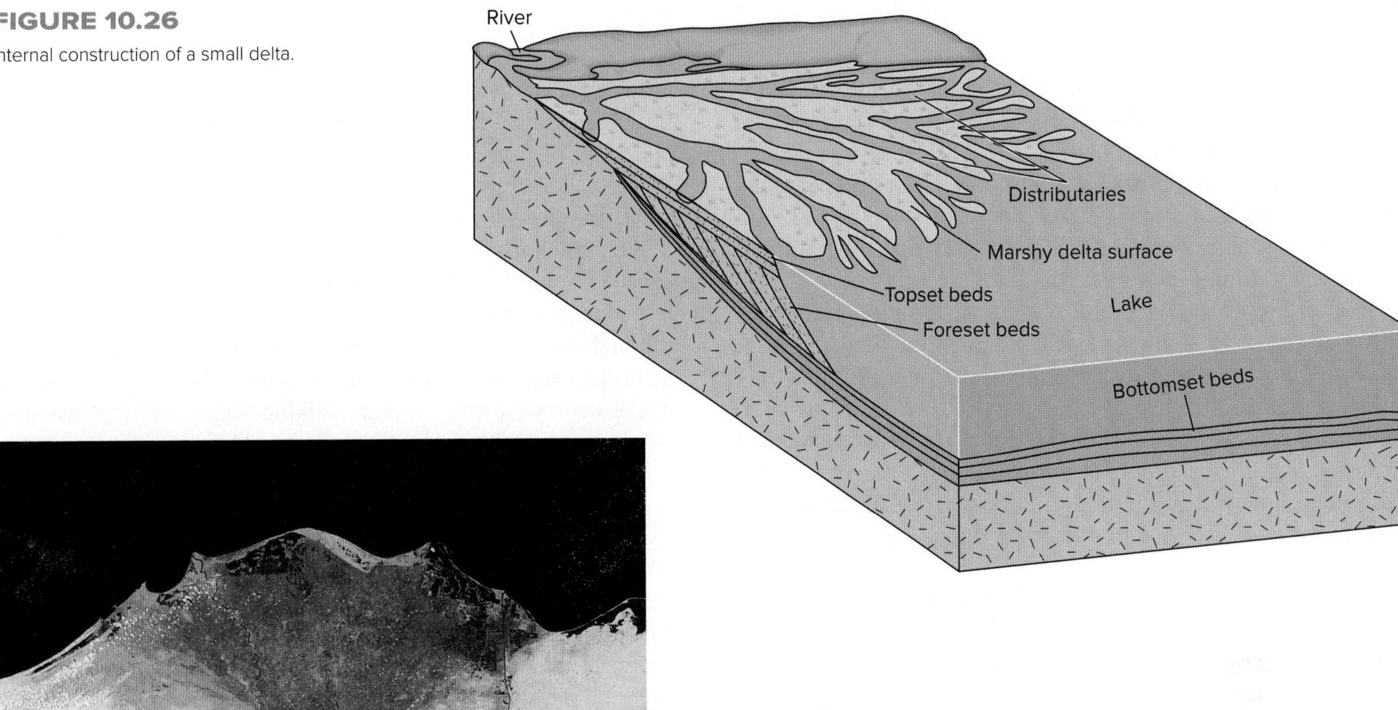

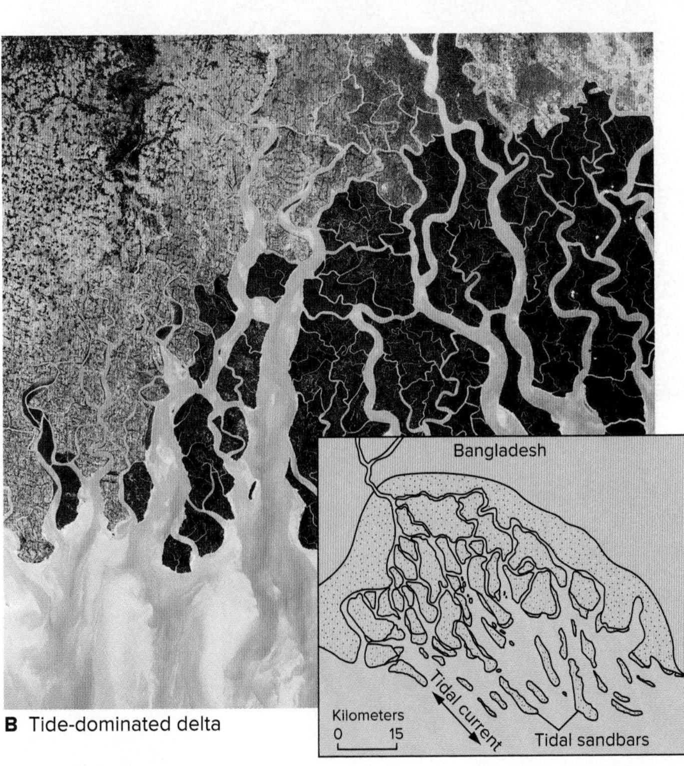

A Wave-dominated delta

B Tide-dominated delta

C Stream-dominated delta

FIGURE 10.27

The shape of a delta depends on the amount of sediment being carried by the river and on the vigor of waves and tides in the sea. (*A*) The Nile Delta is a wave-dominated delta with prominent barrier islands. (*B*) Because of the rich silt deposited from the Ganges River in the delta flood plain, the area is heavily cultivated and home to nearly 120 million people. The delta and its inhabitants are at particular risk from catastrophic floods during the heavy rains during the monsoon season. (*C*) Aster satellite photo of the Mississippi River delta taken in 2001. Note how sediment (shown in white) is carried by both river and ocean currents. *(A) Source: Jacques Descloitres, MODIS Land Science Team/NASA; (B) ©M-Sat Ltd/Science Source; (C) Source: Japan ASTER Science Team/GSFC/METI/ERSDAC/JAROS, and U.S./NASA*

Many deltas, particularly small ones in freshwater lakes, are built up from three types of deposits, shown in figure 10.26. *Foreset beds* form the main body of the delta. They are deposited at an angle to the horizontal. This angle can be as great as 20 to 25 degrees in a small delta where the foreset beds are sandy or less than 5 degrees in large deltas with fine-grained sediment. On top of the foreset beds are the *topset beds*, nearly horizontal beds of varying grain size formed by distributaries shifting across the delta surface. Out in front of the foreset beds are the *bottomset beds*, deposits of the finest silt and clay carried out into the lake by the river water flow or by sediments sliding downhill on the lake floor. Many of the world's great deltas in the ocean are far more complex than the simplified diagram shown in the figure. Shifting river mouths, wave energy, currents, and other factors produces many different internal structures.

The persistence of large deltas as relatively "dry" land depends on a balance between the rate of sedimentation and the rates of tectonic subsidence and compaction of water-saturated sediment. Many deltas are sinking, with seawater encroaching on once-dry land. The Mississippi Delta in Louisiana is sinking, as upstream dams catch sediment, reducing the delta's supply, and as extraction of oil and gas from beneath the delta accelerates subsidence. The flat surface of a delta is a risky place to live or farm, particularly in regions threatened by the high waves and raised sea level of hurricanes, such as the U.S. Gulf Coast and the countries of Bangladesh (the Ganges-Brahmaputra Delta) and Burma (Irrawaddy Delta) on the Indian Ocean. The Irrawaddy Delta was struck by the Asian equivalent of Hurricane Katrina on May 2, 2008, when Hurricane Nargis grew to a category 4 hurricane in the northern Indian Ocean and made landfall. Thousands of people lost their lives, and many were reported missing when the high waters from the storm surge struck the low-lying tidal delta. The damage and loss of life caused by flooding and erosion of the delta were catastrophic, making this one of the deadliest hurricanes to strike the northern Indian Ocean coast.

Alluvial Fans

Some streams, particularly in dry climates, do not reach the sea or any other body of water. They build alluvial fans instead of deltas. An **alluvial fan** is a large, fan- or cone-shaped pile of sediment that usually forms where a stream's velocity decreases as it emerges from a narrow mountain canyon onto a flat plain (figure 10.28). Alluvial fans are particularly well developed and exposed in the southwestern desert of the United States and in other desert regions, but they are by no means limited to arid regions.

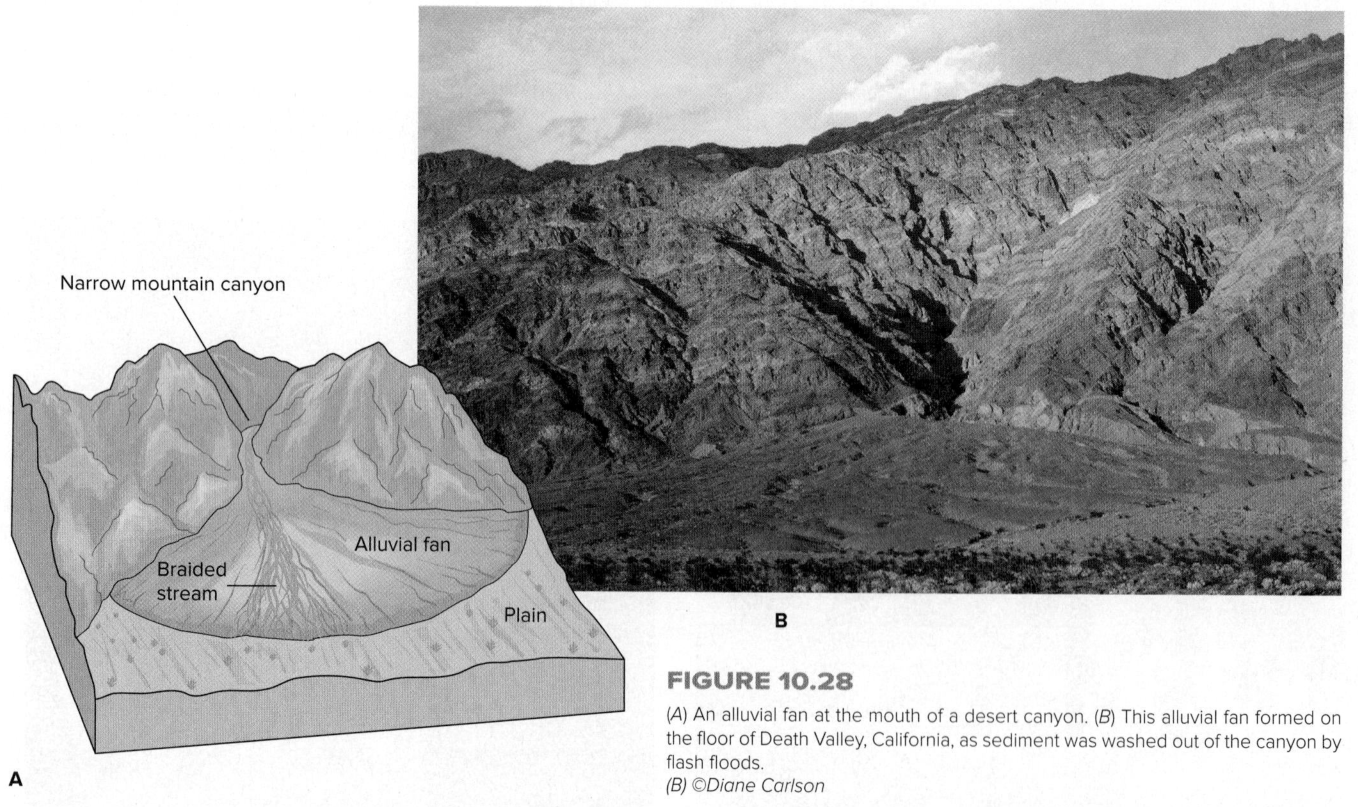

FIGURE 10.28

(*A*) An alluvial fan at the mouth of a desert canyon. (*B*) This alluvial fan formed on the floor of Death Valley, California, as sediment was washed out of the canyon by flash floods.
(B) ©Diane Carlson

An alluvial fan builds up its characteristic fan shape gradually as streams shift back and forth across the fan surface and deposit sediment, usually in a braided pattern. Deposition on an alluvial fan in the desert is discontinuous because streams typically flow for only a short time after the infrequent rainstorms. When rain does come, the amount of sediment to be moved is often greater than the available water, and material is moved as a debris flow before it comes to rest and is deposited.

The sudden loss of velocity when a stream flows from narrow mountain canyons onto a broad plain causes the sediment to deposit on an alluvial fan. The loss of velocity is due to the widening or branching of the channel as it leaves the narrow canyon. The gradual loss of water as it infiltrates into the fan also promotes sediment deposition. On large fans, deposits are graded in size within the fan, with the coarsest sediment dropped nearest the mountains and the finer material deposited progressively farther away. Small fans do not usually show such grading.

10.9 STREAM VALLEY DEVELOPMENT

Valleys, the most common landforms on the Earth's surface, are usually cut by streams. By removing rock and sediment from the stream channel, a stream deepens, widens, and lengthens its own valley.

Downcutting and Base Level

The process of deepening a valley by erosion of the streambed is called **downcutting**. If a stream removes rock from its bed, it can cut a narrow *slot canyon* down through rock (figure 10.29*A* and *B*). Such narrow canyons do not commonly form because mass wasting and sheet erosion usually remove rock from the valley walls. These processes widen the valley from a narrow, vertical-walled canyon to a broader, open, V-shaped canyon (figure 10.29*C* and *D*). Slot canyons persist, however, in very resistant rock with favorably oriented fractures or in regions where downcutting is rapid.

Downcutting cannot continue indefinitely because the headwaters of a stream cannot cut below the level of the streambed at the mouth. If a river flows into the ocean, sea level becomes the lower limit of downcutting. The river cannot cut below sea level, or it would have to flow uphill to get to the sea. For most streams, sea level controls the level to which the land can be eroded.

The limit of downcutting is known as **base level**; it is a theoretical limit for erosion of the Earth's surface (figure 10.30*A*). Downcutting will proceed until the streambed reaches base level. If the stream is well above base level, downcutting can be quite rapid; but as the stream approaches base level, the rate of downcutting slows down. For streams that reach the ocean, base level is close to sea level, but since streams need at least a gentle gradient to flow, base level slopes gently upward in an inland direction.

During the glacial fluctuations of the Pleistocene Epoch (see chapter 12), sea level rose and fell as water was removed from the sea to form the glaciers on the continents and returned to the sea when the glaciers melted. This means that base level rose and fell for streams flowing into the sea. As a result, the lower reaches of such rivers alternated between erosion (caused by low sea level) and deposition (caused by high sea level). Since the glaciers advanced and retreated several times, the cycle of erosion and deposition was repeated many times, resulting in a complex history of cutting and filling near the mouths of most old rivers.

Base levels for streams that do not flow into the ocean are not related to sea level. In Death Valley in California (figure 10.30*B*), base level for in-flowing streams corresponds to the lowest point in the valley, 86 meters (282 feet) *below* sea level (the valley has been dropped below sea level by tectonic movement along faults). On the other hand, base level for a stream above a high reservoir or a mountain lake can be hundreds or even thousands of meters above sea level. The surface of the lake or reservoir serves as temporary base level for all the water upstream (figure 10.30*C*). The base level of a tributary stream is governed by the level of its junction with the main stream. A ledge of resistant rock may act as a temporary base level if a stream has difficulty eroding through it.

The Concept of a Graded Stream

As a stream begins downcutting into the land, its longitudinal profile is usually irregular, with rapids and waterfalls along its course (figure 10.31 and see figure 10.29D). Such a stream, termed *ungraded*, is using most of its erosional energy in downcutting to smooth out these irregularities in gradient.

As the stream smooths out its longitudinal profile to a characteristic concave-upward shape, it becomes graded. A **graded stream** is one that exhibits a delicate balance between its transporting capacity and the sediment load available to it. This balance is maintained by cutting and filling any irregularities in the smooth longitudinal profile of the stream.

In this chapter's section on "Factors Affecting Stream Erosion and Deposition," you learned how changes in a stream's gradient can cause changes in its sediment load. An increase in gradient causes an increase in a stream's velocity, allowing the stream to erode and carry more sediment. A balance is maintained—the greater load is a result of the greater transporting capacity caused by the steeper gradient.

The relationship also works in reverse—a change in sediment load can cause a change in gradient. For example, a decrease in sediment load may bring about erosion of the stream's channel, thus lowering the gradient. Because dams trap sediment in the calm reservoirs behind them, most streams are almost completely sediment-free just downstream from dams. In some streams, this loss of sediment has caused severe channel erosion below a dam, as the stream adjusts to its new, reduced load.

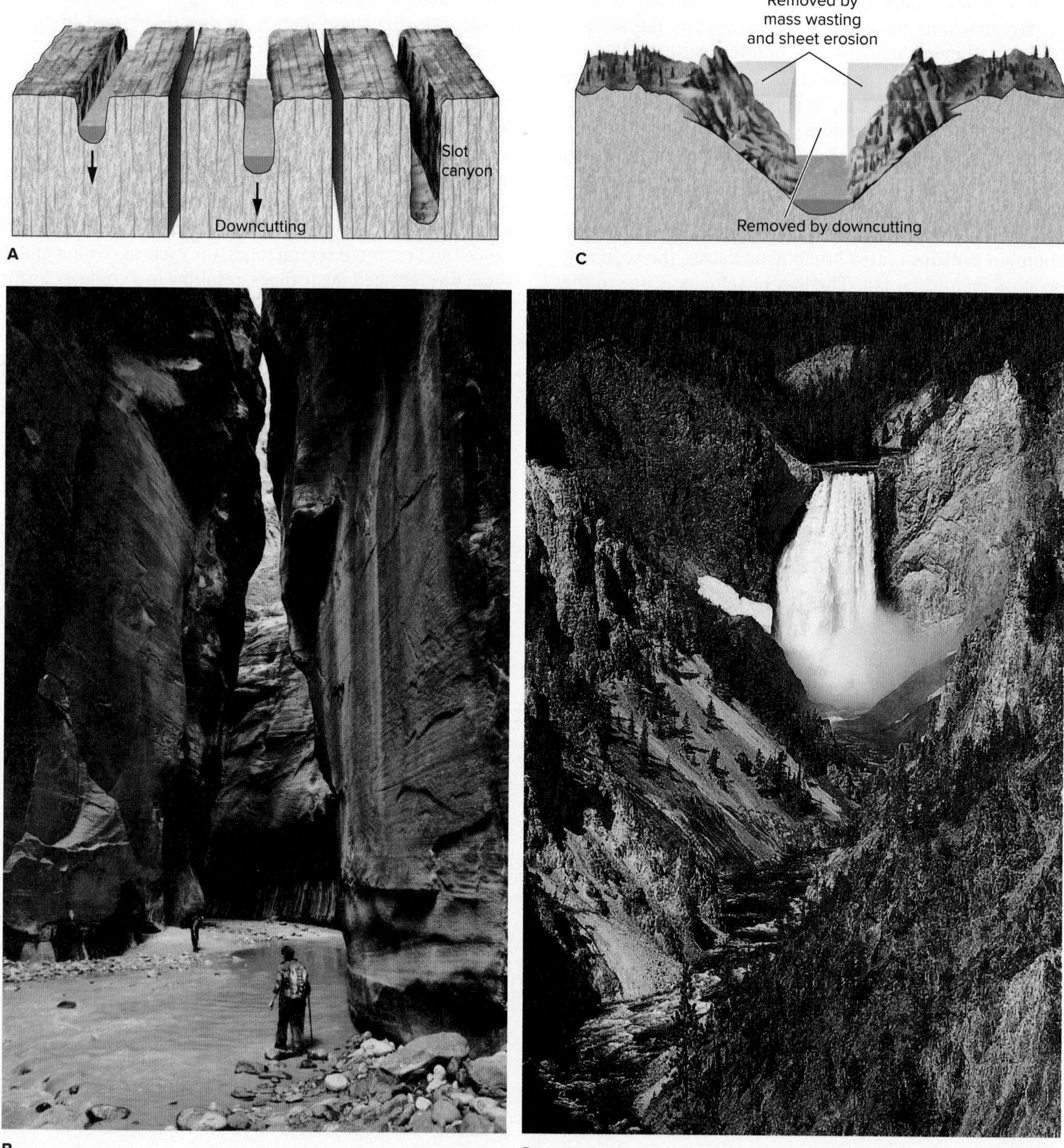

FIGURE 10.29

Downcutting, mass wasting, and sheet erosion shape canyons and valleys. (*A*) Downcutting can create slot canyons in resistant rock, particularly where downcutting is rapid during flash floods and fractures in the rock are favorably oriented. (*B*) Stream erosion has cut this unusual slot canyon through porous sandstone, Zion National Park, Utah. (*C*) Downslope movement of rock and soil on valley walls widens most canyons into V-shaped valleys. (*D*) The waterfall and rapids on the Yellowstone River in Wyoming indicate that the river is ungraded and actively downcutting. Note the V-shaped cross-profile and lack of flood plain due to the downslope movement of volcanic rock.
(B) ©Ethan Welty/Getty Images; (D) ©David McGeary

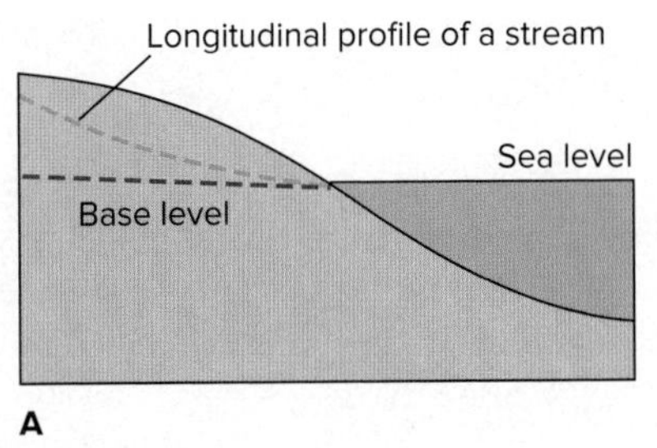

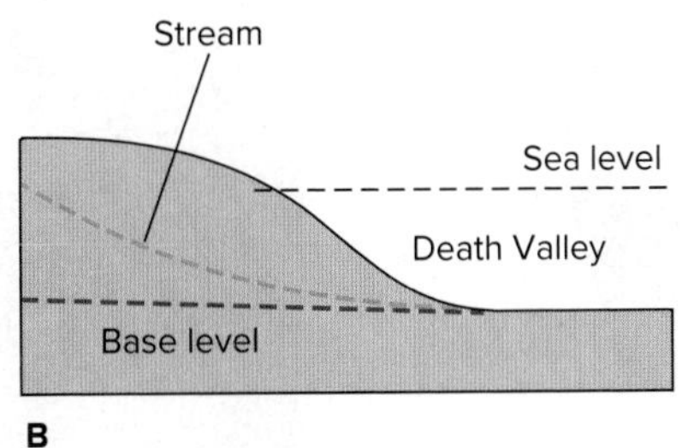

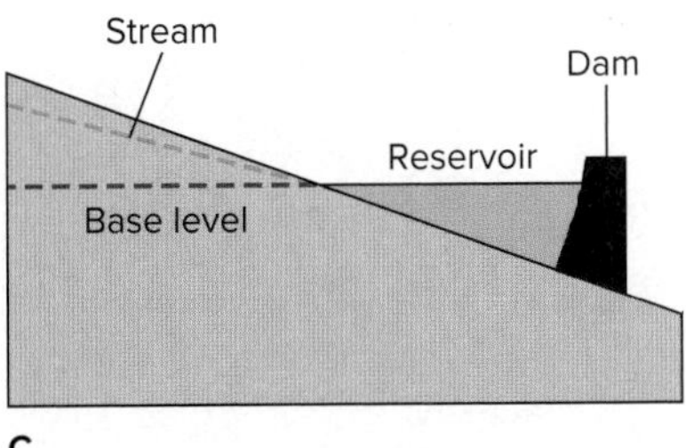

FIGURE 10.30

Base level is the lowest level of downcutting.

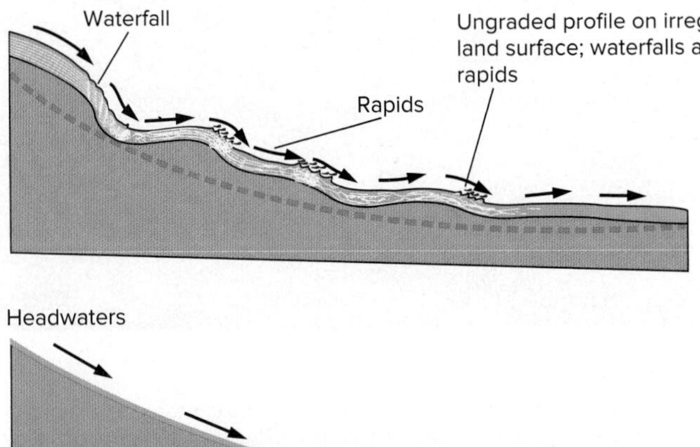

FIGURE 10.31

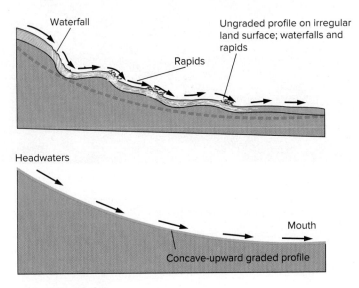

An ungraded stream has an irregular longitudinal profile with many waterfalls and rapids. A graded stream has smoothed out its longitudinal profile to a smooth, concave-upward curve.

A river's energy is used for two things—transporting sediment and overcoming resistance to flow. If the sediment load decreases, the river has more energy for other things. It may use this energy to erode more sediment, deepening its valley. Or it may change its channel shape or length, increasing resistance to flow, so that the excess energy is used to overcome friction. Or the river may increase the roughness of its channel, also increasing friction. The response of a river is not always predictable, and construction of a dam can sometimes have unexpected and perhaps harmful results.

Lateral Erosion

A graded stream can be deepening its channel by downcutting while part of its energy is also widening the valley by **lateral erosion**, the erosion and undercutting of a stream's banks and valley walls as the stream swings from side to side across its valley floor. The stream channel remains the same width as it moves across the flood plain, but the valley widens by erosion, particularly on the outside of curves and meanders where the stream impinges against the valley walls (figure 10.32). The valley widens as its walls are eroded by the stream and as its walls retreat by mass wasting triggered by stream undercutting. As a valley widens, the stream's flood plain increases in width also.

Headward Erosion

Building a delta or alluvial fan at its mouth is one way a river can extend its length. A stream can also lengthen its valley by **headward erosion**, the slow uphill growth of a valley above its original source through gullying, mass wasting, and sheet erosion (figure 10.33). This type of erosion is particularly difficult to stop. When farmland is being lost to gullies that are eroding headward into fields and pastures, farmers must divert sheet flow and fill the gully heads with brush and other debris to stop, or at least retard, the loss of topsoil.

Stream Terraces

Stream terraces are steplike landforms found above a stream and its flood plain (figure 10.34). Terraces may be benches cut in rock (sometimes sediment-covered), or they may be steps formed in sediment by deposition and subsequent erosion.

Figure 10.35 shows how a terrace forms as a river cuts downward into a thick sequence of its own flood-plain deposits. Originally, the river deposited a thick section of flood-plain sediments. Then the river changed from deposition to erosion and cut into its old flood plain, parts of which remain as terraces above the river.

Why might a river change from deposition to erosion? One reason might be regional uplift, raising a river that was once meandering near base level to an elevation well above base level. Uplift would steepen a river's gradient, causing the river to speed up and begin erosion. But there are several other reasons a river might change from deposition to erosion. A change from a dry to a wet climate may increase discharge and cause a river to begin eroding. A drop in base level (such as lowering of

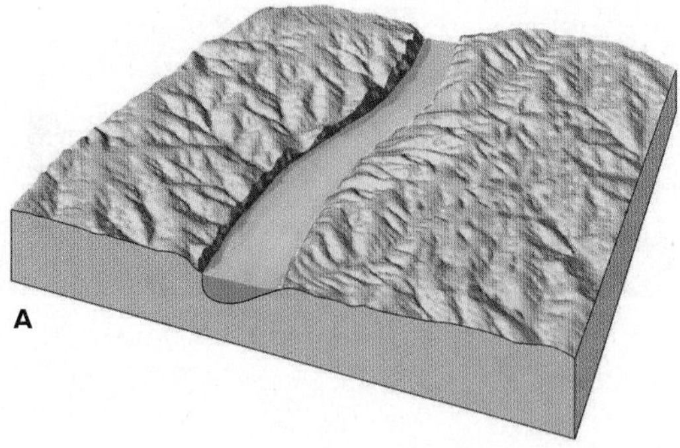

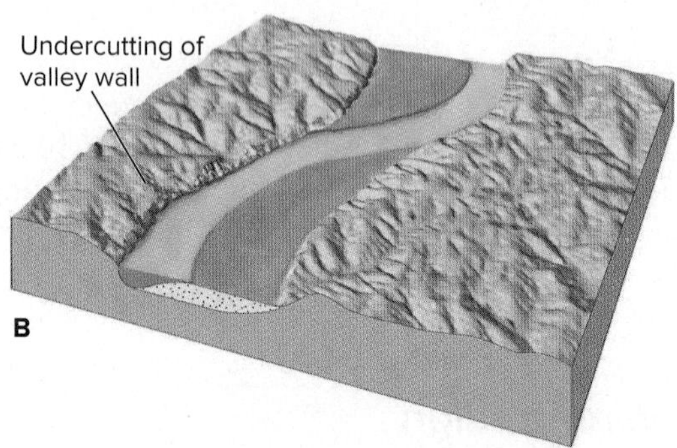

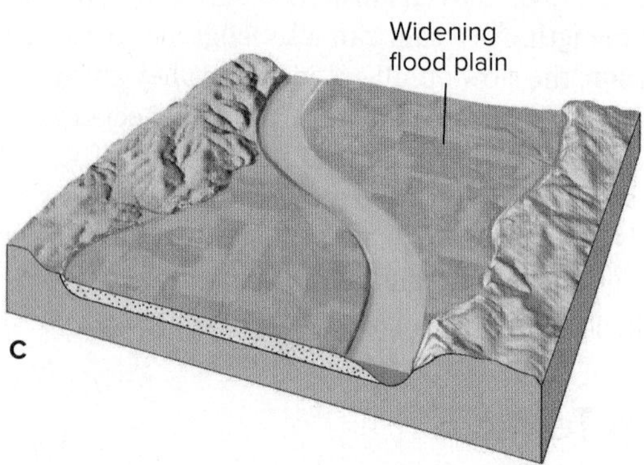

FIGURE 10.32
Lateral erosion can widen a valley by undercutting and eroding valley walls.

FIGURE 10.33
Headward erosion is lengthening this stream channel. Note the dendritic drainage pattern that is developing in the headwaters of the streams, New Plymouth, New Zealand.
©G. R. "Dick" Roberts/Natural Sciences Image Library

sea level) can have the same effect. A situation like that shown in figure 10.35 can develop in a recently glaciated region. Thick valley fill such as glacial outwash (see chapter 12) may be deposited in a stream valley and later, after the glacier stops producing large amounts of sediment, be dissected into terraces by the river.

Terraces can also develop from erosion of a bedrock valley floor. Bedrock benches are usually capped by a thin layer of flood-plain deposits.

Incised Meanders

Incised meanders are meanders that retain their sinuous pattern as they cut vertically downward below the level at which they originally formed. The result is a meandering *valley* with

FIGURE 10.34
Stream terraces near Jackson Hole, Wyoming. The stream has cut downward into its old flood plain.
©Diane Carlson

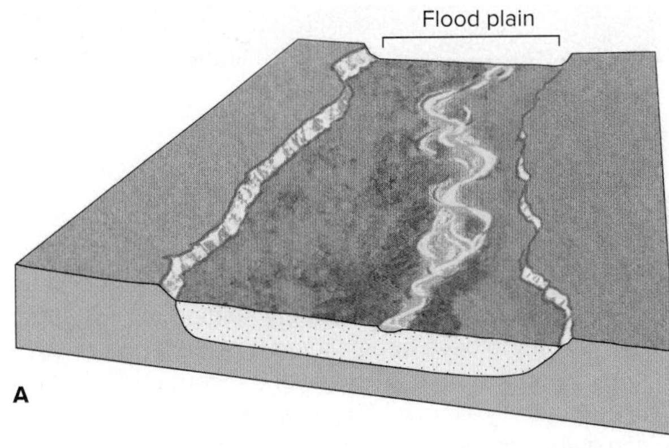

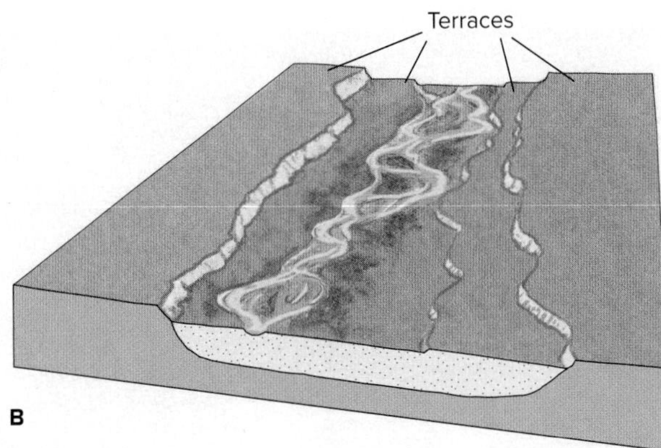

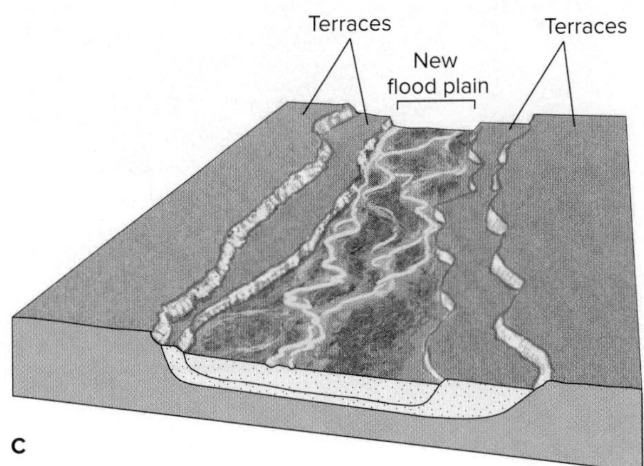

FIGURE 10.35

Terraces formed by a stream cutting downward into its own flood-plain deposits. (*A*) Stream deposits thick, coarse, flood-plain deposits. (*B*) Stream erodes its flood plain by downcutting. Old flood-plain surface forms terraces. (*C*) Lateral erosion forms new flood plain below terraces.

essentially no flood plain, cut into the land as a steep-sided canyon (figure 10.36*A*).

Some incised meanders may be due to the profound effects of a change in base level. They may originally have been formed as meanders in a laterally eroding river flowing over a flat flood plain, perhaps near base level. If regional uplift elevated the land high above base level, the river would begin downcutting and might be able to maintain its characteristic meander pattern

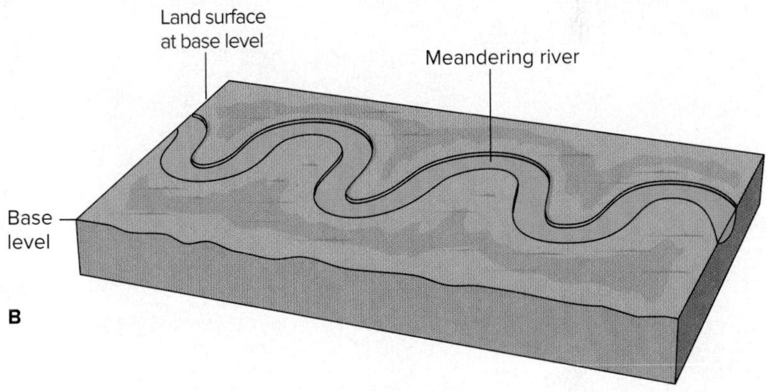

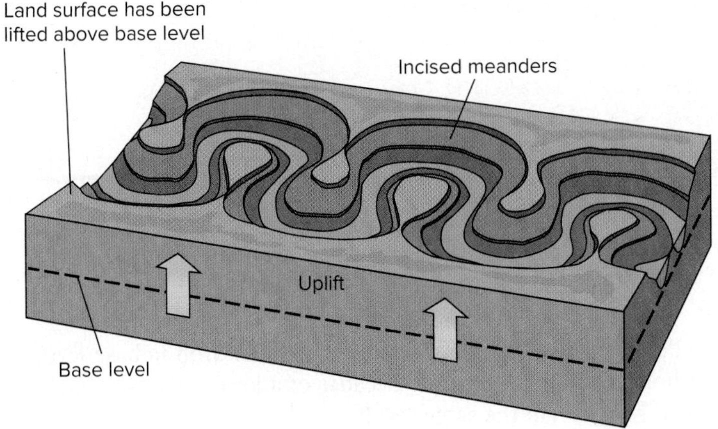

FIGURE 10.36

(*A*) Incised meanders of the Colorado River ("The Loop"), Canyonlands National Park, southwestern Utah. (*B*) Meandering river flowing over a flat plain cut to base level. (*C*) Regional uplift of land surface allows river to downcut and incise its meanders.

(A) ©Frank M. Hanna

PLANETARY GEOLOGY 10.2

Stream Features on the Planet Mars

Planetary geologists are fairly certain that liquid water does not exist for long on the surface of Mars today. With the present surface temperatures, atmospheric pressures, and water content in the Martian atmosphere, any liquid water would immediately evaporate. Recent evidence collected from the Mars orbiters and rovers indicate that conditions may have been different in the past and that liquid water once existed on Mars.

Certain features on Mars, called *channels,* closely resemble certain types of stream channels on Earth. Martian channels have tributary systems (box figure 1) and meanders, trend downslope, and tend to get wider toward their mouths. Many Martian channels that were originally thought to have been carved by running water have since been reinterpreted. Some could have been formed by surface collapse caused by the melting of frozen water underground. Others, particularly gullies with steep slopes, could be the result of gravity flow of dry surface material. Some channels, however, have

BOX 10.2 ■ FIGURE 1

Stream drainage system from the Southern Highlands of Mars, which resembles dendritic systems found on Earth.

Source: Courtesy NASA/From Mars Digital Image Map, image processing by Brian Fessler, Lunar and Planetary Institute

BOX 10.2 ■ FIGURE 2

A color-enhanced image from the *Mars Reconnaissance Orbiter* showing a delta in the Jezero Crater. Clay minerals are shown in green.

Source: Courtesy NASA/JPL/JHUAPL/MSSS/Brown University

while deepening its valley (figure 10.36*B*). A drop in base level without land uplift (possibly because of a lowering of sea level) could bring about the same result.

Although uplift may be a key factor in the formation of many incised meanders, it may not be *required* to produce them. Lateral erosion certainly seems to become more prominent as a river approaches base level, but some meandering can occur as soon as a river develops a graded profile. A river flowing on a flat surface high above base level may develop meanders early in its erosional history, and these meanders may become incised by subsequent downcutting. In such a case, uplift is not necessary.

10.10 FLOODING

Many cities, such as Pittsburgh, St. Louis, Sacramento, and Florence, Italy, are built beside rivers and therefore can be threatened by floods. Rivers are important transportation routes for ships and barges, and flat flood plains have excellent agricultural soil and offer attractive building sites for houses and industry.

Flooding does not occur every year on every river, but flooding is a natural process on all rivers; those who live in river cities and towns must be prepared. Heavy rains and the rapid melting of snow in the spring are the usual causes of floods. The rate and

slopes too shallow for material to move by gravity alone, and their braided stream channels indicate that they may have formed during periodic flooding events. Such flooding events could have been due to the sudden melting of ice in the Martian crust or polar ice caps during volcanic activity or impact events.

Images recently returned from the *Mars Reconnaissance Orbiter* and *Mars Science Laboratory* have provided exciting new evidence of flowing water on Mars early in the planet's history. The *Compact Reconnaissance Imaging Spectrometer for Mars* (CRISM), a highly sophisticated imaging system carried aboard the *Mars Reconnaissance Orbiter,* has the ability to analyze the mineral composition of rocks on the surface of the planet. The CRISM image in box figure 2 shows the distributary channels of a delta in the impact crater Jezero and also the presence of clay minerals. The clay minerals in the crater and the delta indicate that the crater was probably once occupied by a lake slightly larger than California's Lake Tahoe. The delta was fed by surface streams, which eroded upland rocks and transported clays into the lake. Box figure 3 shows alluvial conglomerate analyzed by the rover *Curiosity* (MSL), working in Gale Crater. The rounded shapes of the pebbles indicate that the particles traveled far, and Earth-based analogies suggest that the deposits were not formed during rare, catastrophic events, but rather that there was a fairly long history of flowing water forming the channels and depositing the gravels.

Planetary geologists continue to interpret the surface features on Mars and distinguish between those created by flowing water and those created by other processes. Continual advances in technology, such as the images provided by CRISM and MSL, and the analyses returned by the Mars Rovers, are adding exciting new information about wet environments on Mars that might have supported life.

BOX 10.2 ■ FIGURE 3

Cross-bedded conglomerate taken by the mast camera of NASA's *Curiosity* rover in Gale Crater, Mars. Rounded pebbles are too large for transport by wind, implying that water flowed continuously over a long time scale.
Source: NASA/JPL-CalTech/Malin Space Science Systems

Additional Resources

For more information on the possibility of life-supporting environments on Mars, visit the NASA blog site:
http://mars.jpl.nasa.gov/msl/blogs/index.cfm?FuseAction=ShowBlogs&BlogsID=283

Information about the *Mars Science Laboratory (Curiosity)* can be found at the Jet Propulsion Laboratory/NASA site:
http://mars.jpl.nasa.gov/msl/

Information about the *Mars Reconnaissance Orbiter,* including images from the Compact Reconnaissance Imaging Spectrometer for Mars (CRISM), may be found at:
http://marsprogram.jpl.nasa.gov/mro/

volume of rainfall and the geographic path of rainstorms often determine whether flooding will occur.

Floods are described by *recurrence interval*, the average time between floods of a given size. A "100-year flood" is one that can occur, on the *average,* every 100 years (box 10.3). A 100-year flood has a 1-in-100, or 1%, chance of occurring in any given year. It is perfectly possible to have two 100-year floods in successive years–or even in the same year. If a 100-year flood occurs this year on the river you live beside, you should not assume that there will be a 99-year period of safety before the next one.

Flood erosion is caused by the high velocity and large volume of water in a flood. Although relatively harmless on an uninhabited flood plain, flood erosion can be devastating to a city. As a river undercuts its banks, particularly on the outside of curves where water velocity is high, buildings, piers, and bridges may fall into the river. As sections of flood plain are washed away, highways and railroads are cut.

High water covers streets and agricultural fields and invades buildings, shorting out electrical lines and backing up sewers. Water-supply systems may fail or be contaminated. Water in your living room will be drawn upward in your walls by capillary action in wall plasterboard and insulation, creating a soggy mess that has to be torn out and replaced. High water on flat flood plains often drains away very slowly; street travel may be by boat for weeks. If floodwaters are deep enough, houses may float away.

IN GREATER DEPTH 10.3

Estimating the Size and Frequency of Floods

Because people have encroached on the flood plains of many rivers, flooding is one of the most universally experienced geologic hazards. To minimize flood damage and loss of life, it is useful to know the potential size of large floods and how often they might occur. This is often a difficult task because of the lack of long-term records for most rivers, and also the extreme weather events that have been occurring due to climate change. The U.S. Geological Survey monitors the stage (water elevation) and discharge of rivers and streams throughout the United States to collect data that can be used to attempt to predict the size and frequency of flooding and to make estimates of water supply.

Hydrologists designate floods based on their *recurrence interval,* or *return period.* For example, a 100-year flood is the largest flood expected to occur within a period of 100 years. This does not mean that a 100-year flood occurs once every century but that there is a 1-in-100 chance, or a 1% probability, each year that a flood of this size will occur. Usually, flood control systems are built to accommodate a 100-year flood because that is the minimum margin of safety required by the federal government if an individual wants to obtain flood insurance subsidized by the Federal Emergency Management Agency (FEMA).

To calculate the recurrence interval of flooding for a river, the annual peak discharges (largest discharge of the year) are collected and ranked according to size (box figure 1 and table 1). The largest annual peak discharge is assigned a rank (m) of 1, the second a rank of 2, and so on until all of the discharges are assigned a rank number. The *recurrence interval* **(R)** of each annual peak discharge is then calculated by adding 1 to the *number of years of record* **(n)** and dividing by its *rank* **(m).**

$$R = \frac{n + 1}{m}$$

BOX 10.3 ■ TABLE 1

Annual Peak Discharges and Recurrence Intervals in Rank Order for the Cosumnes River at Michigan Bar, California

Year	Peak Discharge (cfs)	Magnitude Rank (m)	Recurrence Interval
1997	93,000	1	110
1907	71,000	2	55
1986	45,100	3	33.66
1956	42,000	4	27.5
1963	39,400	5	22.0
1998	29,700	10	11.0
1928	22,900	20	5.5
1938	19,300	30	3.66
1952	12,500	40	27.5
1922	10,600	50	2.2
1915	8,200	60	1.83

Source: Peak Discharges from National Water Information System - U.S. Geological Survey 11335000 Cosumnes River at Michigan Bar.

For example, the Cosumnes River in California has 109 years of record **(n = 109)**, and in 1907, the second-largest peak discharge **(m = 2)** of 71,000 cfs occurred. The recurrence interval **(R)**, or expected frequency of occurrence, for a discharge this large is 55 years:

$$R = \frac{109 + 1}{2} = 55$$

That is, there is a 1-in-55, or slightly less than 2%, chance each year of a peak discharge of 71,000 cfs or greater occurring on the Cosumnes River.

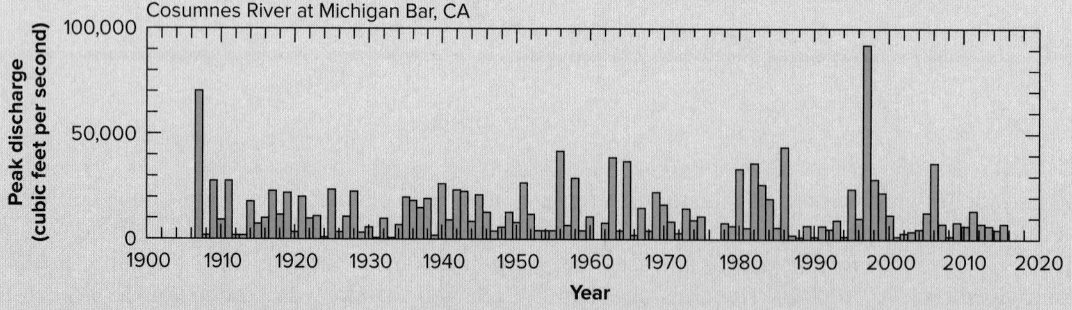

BOX 10.3 ■ FIGURE 1

Annual peak discharge for the Cosumnes River.
Source: U.S. Geological Survey Water-Data Report, CA-97-3 and National Water Information System - U.S. Geological Survey 11335000 Cosumnes River at Michigan Bar.

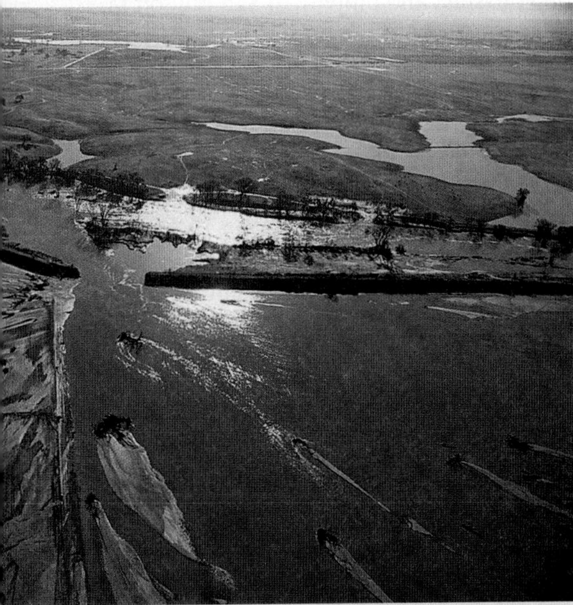

BOX 10.3 ■ FIGURE 2

Levee break along the Cosumnes River.
Source: Courtesy of Robert A. Eplett, Governor's Office of Emergency Services

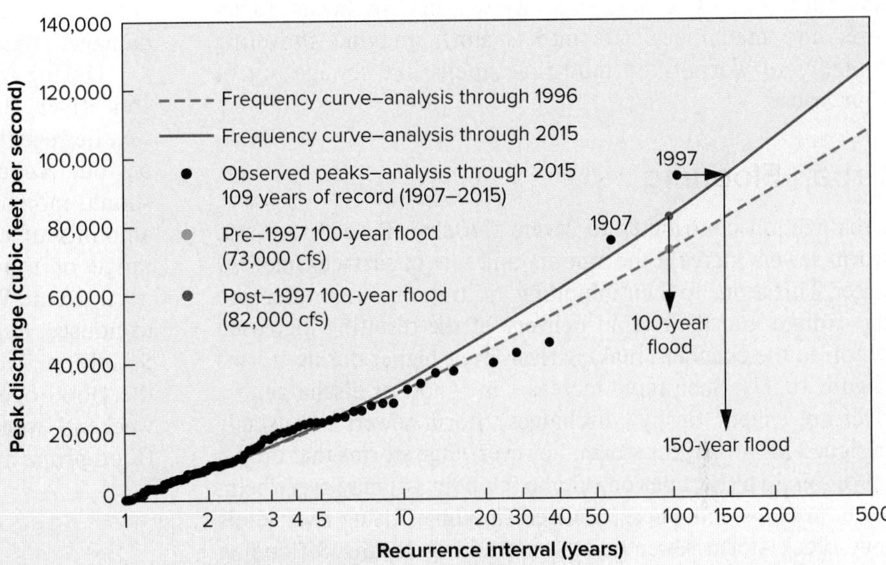

BOX 10.3 ■ FIGURE 3

Flood-frequency curves for the Cosumnes River.
Source: Richard Hunrichs, hydrologist, U.S. Geological Survey and U.S. Geological Survey Water-Data Report, CA-97-3 and National Water Information System.

The flood of record (largest recorded discharge) occurred on January 2, 1997, when heavy, unseasonably warm rains rapidly melted snow in the Sierra Nevada and caused flooding in much of northern California. A peak discharge of 93,000 cfs in the Cosumnes River resulted in levee breaks and widespread flooding of homes and agricultural areas (box figure 2). The recurrence interval for the 1997 flood (93,000 cfs) is 110 years:

$$R = \frac{109 + 1}{1} = 110 \text{ years}$$

A *flood-frequency curve* can be useful in providing an estimate of the discharge and the frequency of floods. The flood-frequency curve is generated by plotting the annual peak discharges against the calculated recurrence intervals (box figure 3). Because most of the data points defining the curve plot in the lower range of discharge and recurrence interval, there is some uncertainty in projecting larger flood events. Two flood-frequency curves are drawn in box figure 3; the red line represents the best-fit curve for all of the data, whereas the dashed blue line excludes the 1997 flood of record. Notice that the curve has a steeper slope when the 1997 data is included and that the size of the 100-year flood has increased based on the additional years of record. Because large floods do not occur as often as small floods, the rare large flood can have a dramatic effect on the shape of the flood-frequency curve and the estimate of a 100-year event. This is particularly true for a river like the Cosumnes that has had only two large events, one in 1907 and the other 90 years later in 1997.

The 100-year flood plain is based on the estimate of the discharge of the 100-year flood and on careful mapping of the flood plain. Changes in the estimated size of the 100-year flood could result in property that no longer has 100-year flood protection. In this case, property owners may be prevented from getting flood insurance or money to rebuild from FEMA unless new flood-control structures are built to provide additional protection or houses are raised or even relocated out of the flood plain.

Additional Resources

Water Resources Data for California, Water Year 1997. U.S. Geological Survey Water-Data Report CA-97-3.

H. C. Riggs. 1968. Frequency curves. *Techniques of Water-Resources Investigations of the U.S. Geological Survey. Book 4, Hydrologic Analysis and Interpretation.*

To find data sets to calculate the recurrence interval for rivers throughout the United States, access the U.S. Geological Survey Water Data Retrieval website:

- http://water.usgs.gov/usa/nwis/

Flood deposits are usually silt and clay. A new layer of wet mud on a flood plain in an agricultural region can be beneficial in that it renews the fields with topsoil from upstream, as used to be the case with the Nile River until the Aswan Dam was built. The same mud in a city will destroy lawns, furniture, and machinery. Cleanup is slow; imagine shoveling 4 inches of worm-filled mud that smells like sewage out of your house.

Urban Flooding

Urbanization contributes to severe flooding. Paved areas and storm sewers increase the amount and rate of surface runoff of water. This is due to their inhibiting infiltration of rainwater into the ground and their rapid delivery of the resulting increased runoff to the channels, making river levels higher during storms (figure 10.37). Such rapid increases in runoff or discharge to a river are called "flashy" discharges. Storm sewers are usually designed for a 100-year storm; however, large storms that drop a lot of rain in a short period of time (cloudburst) may overwhelm sewer systems and cause localized flooding. Rising river levels may block storm sewer outlets and add to localized flooding problems.

Bridges, docks, and buildings built on flood plains can also constrict the flow of floodwaters, increasing the water height and velocity and promoting erosion.

Flash Floods

Some floods occur rapidly and die out just as quickly. *Flash floods* are local, sudden floods of large volume and short duration, often triggered by heavy thunderstorms. A startling example occurred in 1976 in north-central Colorado along the Big Thompson River. Strong winds from the east pushed moist air up the front of the Colorado Rockies, causing thunderstorms in the steep mountains. The storms were unusually stationary, allowing as much as 30 centimeters (12 inches) of rain to fall in two days. Some areas received 19 centimeters in just over an hour. Little of this torrential rainfall could soak into the ground. The volume of water in the Big Thompson River swelled to four times the previously recorded maximum, and the river's velocity rose to an impressive 25 kilometers per hour for a few hours on the night of July 31. By the next morning, the flood was over, and the appalling toll became apparent—139 people dead, 5 missing, and more than $35 million in damages (figure 10.38*A*).

During the week of September 9, 2013, the Big Thompson River flooded again (figure 10.38*B*) as an extreme weather event dropped 43 centimeters of rain over a five-day period. An upper-level low-pressure system, fed by monsoonal moisture, stalled over Colorado and dropped record amounts of rain in all the drainage basins along the front range of the Colorado Rockies from Boulder, Colorado, north to the Wyoming border. Nine people died, and damage to houses, dams, bridges, and highways was estimated to be $1 billion. The loss of life would have been even greater had the flood occurred one week earlier during the Labor Day weekend when many people would have been camping in flood-prone areas.

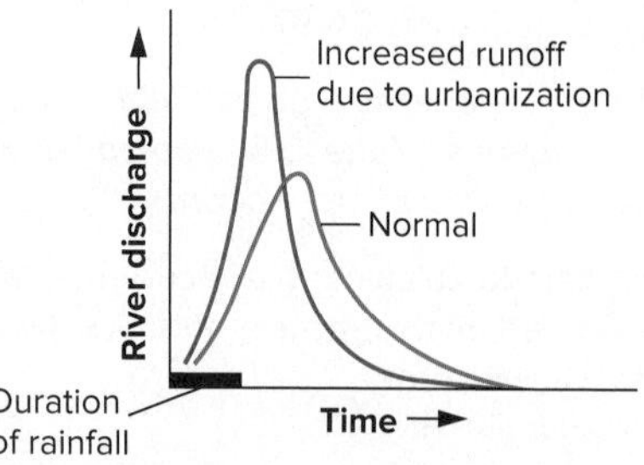

FIGURE 10.37

The presence of a city can increase the chance of floods. The blue curve shows the normal increase in a river's discharge following a rainstorm (black bar). The red curve shows the great increase in runoff rate and amount caused by pavement and storm sewers in a city.

A

B

FIGURE 10.38

(*A*) A cabin sits crushed against a bridge following the Big Thompson Canyon flash flood of 1976. (*B*) Birds-eye view of the devastation to U.S. Highway 34 and a bridge along the raging Big Thompson River during the catastrophic 2013 Colorado flash floods that dropped nearly 50 cm of rain in a five-day period.

(A) Source: W. R. Hansen, U.S. Geological Survey; (B) ©Andy Cross/The Denver Post/Getty Images

Reducing Flood Risk

The nature of all rivers is to periodically flood and overtop their levees. Wise land-use planning should go hand in hand with flood control. Wherever possible, buildings should be kept out of areas that might someday be flooded by 100-year floods.

The risk of flooding to river cities may be partially reduced by engineered flood-control structures (figure 10.39). Upstream dams can trap water and release it slowly after the storm. (A dam also catches sediment, which eventually fills its reservoir and ends its life as a flood-control structure.) Artificial levees are embankments built along the banks of a river channel to contain floodwaters within the channel. Protective walls of stone (riprap) or concrete are often constructed along river banks, particularly on the outside of curves, to slow erosion. Floodwalls, walls of concrete, may be used to protect cities from flooding; however, these flood-control structures may constrict the channel and cause water to flow faster with more erosive power downstream. Bypasses are also used along the Mississippi and other rivers to reduce the discharge in the main channel by diverting water through gates or weirs into designated basins in the flood plain. The bypasses serve to give part of the natural flood plain back to the river.

Dams and levees are designed to control certain specified floods. If the flood-control structures on your river were designed for 75-year floods, then a much larger 100-year flood will likely overtop these structures and may destroy your home. The disastrous floods along the Missouri and Mississippi rivers and their tributaries north of Cairo, Illinois, in 1993 resulted from many such failures in flood control (figure 10.40).

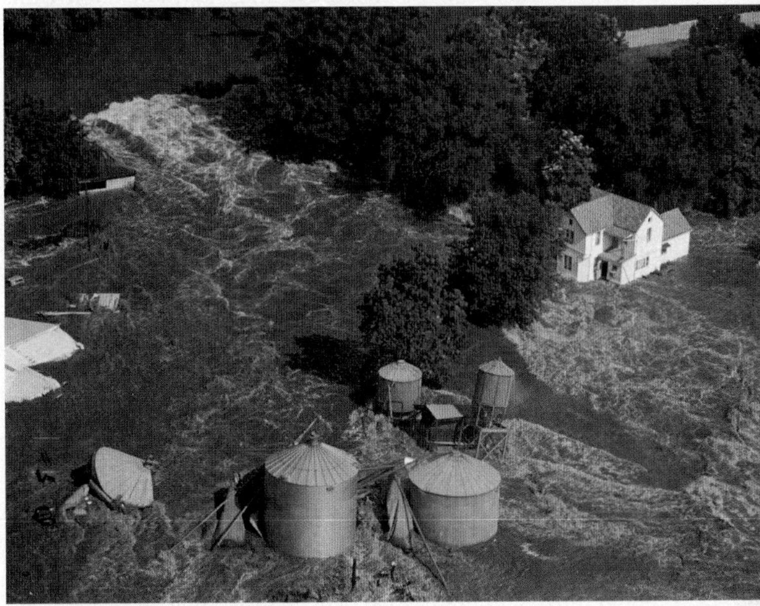

FIGURE 10.40

Mississippi River water pours through a broken levee onto a farm near Columbia, Illinois, during the Great Flood of 1993.
©James R. Finley/AP Images

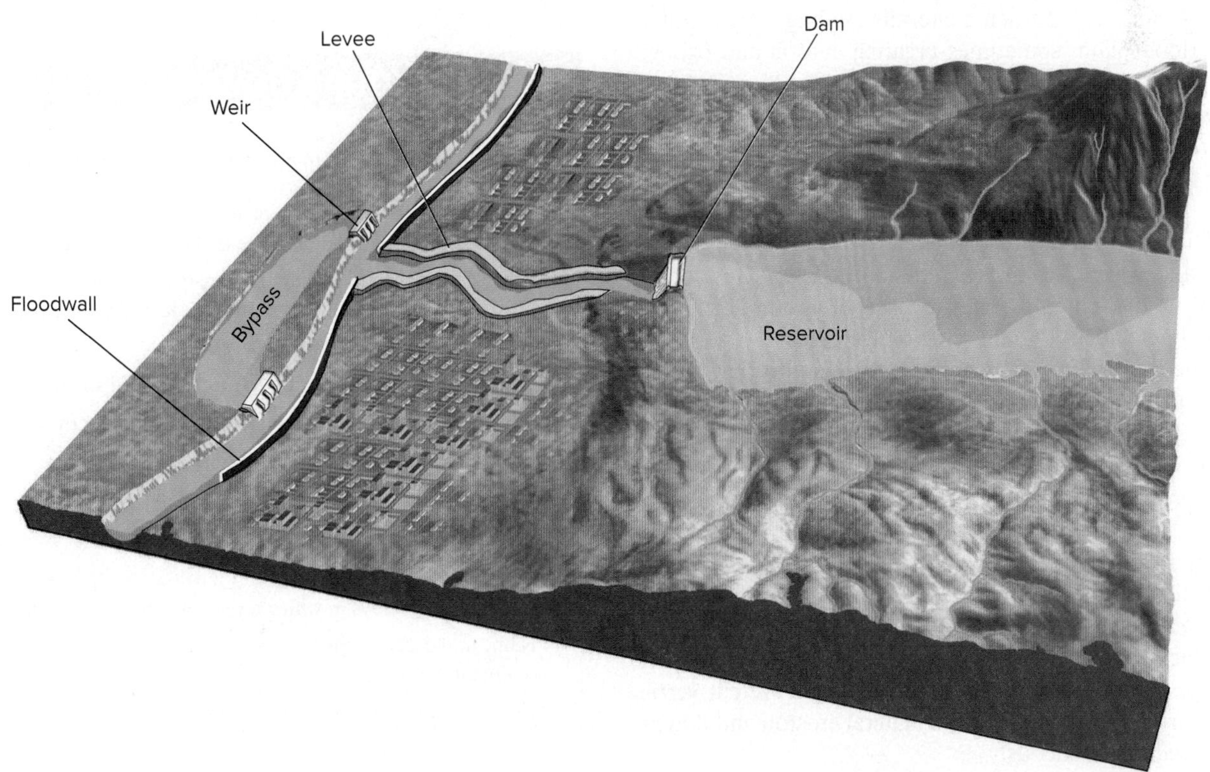

FIGURE 10.39

Examples of engineered flood-control structures.

Summary

Normally, stream *channels* are eroded and shaped by the streams that flow in them. Unconfined sheet flow can cause significant erosion.

Drainage basins are separated by *divides.*

A river and its tributaries form a *drainage pattern.* A *dendritic* drainage pattern develops on uniform rock, a *rectangular* pattern on regularly jointed rock. A *radial* pattern forms on conical mountains, while a *trellis* pattern usually indicates erosion of folded sedimentary rock.

Stream *velocity* is the key factor controlling sediment erosion, transportation, and deposition. Velocity is in turn controlled by several factors.

An increase in a stream's *gradient* increases the stream's velocity. *Channel shape* and *roughness* affect velocity by increasing or lessening friction. As tributaries join a stream, the stream's *discharge* increases downstream. Floods increase stream discharge and velocity.

Streams erode by *hydraulic action, abrasion,* and *solution.* They carry coarse sediment by *traction* and *saltation* as *bed load.* Finer-grained sediment is carried in *suspension.* A stream can also have a substantial *dissolved load.*

Streams create features by erosion and deposition. *Potholes* form by abrasion of hard rock on a streambed. *Bars* form in the middle of streams or on stream banks, particularly on the inside of curves where velocity is low (*point bars*). A *braided pattern* can develop in streams with a large amount of bed load.

Meanders are created when a laterally eroding stream shifts across the flood plain, sometimes creating cutoffs and oxbow lakes.

A *flood plain* develops by both lateral and vertical deposition. *Natural levees* are built up beside streams by flood deposition.

A *delta* forms when a stream flows into standing water. The shape and internal structure of deltas are governed by river deposition and wave and current erosion. *Alluvial fans* form, particularly in dry climates, at the base of mountains as a stream's channel widens and its velocity decreases.

Rivers deepen their valleys by *downcutting* until they reach *base level,* which is either sea level or a local base level.

A *graded stream* is one with a delicate balance between its transporting capacity and its available load.

Lateral erosion widens a valley after the stream has become graded.

A valley is lengthened by both *headward erosion* and sediment deposition at the mouth.

Stream terraces can form by erosion of rock benches or dissection of thick valley deposits during downcutting.

Incised meanders form as (1) river meanders are cut vertically downward following uplift or (2) lateral erosion and downcutting proceed simultaneously.

Flooding occurs when a river overtops its banks during periods of heavy rain and during rapid melting of snow in the spring.

The *recurrence interval* is an estimate of the average time between floods of a given size. Flood risk can be minimized by not allowing building in flood plains and by engineered structures such as dams, levees, floodwalls, and bypasses.

Terms to Remember

abrasion 240
alluvial fan 250
bar 241
base level 251
bed load 240
braided stream 243
delta 248
dendritic pattern 236
discharge 239
dissolved load 241
distributary 248
divide 235
downcutting 251
drainage basin 235
drainage pattern 235
flood plain 247
graded stream 253
headward erosion 253
hydraulic action 239
hydrologic cycle 233
incised meander 255
lateral erosion 253
meander 243
meander cutoff 246
natural levee 248
oxbow lake 246
point bar 246
pothole 240
radial pattern 236
rectangular pattern 236
saltation 240
sheetwash 235
solution 240
stream 234
stream channel 235
stream gradient 238
stream terrace 253
stream velocity 237
suspended load 241
traction 240
trellis pattern 237
tributary 235

Testing Your Knowledge

Use the following questions to prepare for exams based on this chapter.

1. Define and draw a sketch of a drainage basin. Include the main stream, tributaries, and drainage divides in your sketch.
2. What does a rectangular drainage pattern tell about the rocks underneath it? What is the most common type of drainage pattern? Why?
3. What factors control a stream's velocity?
4. Describe how bar deposition creates a braided stream.
5. In what part of a large alluvial fan is the sediment the coarsest? Why?
6. Describe one way that incised meanders form.
7. How does a meander neck cutoff form an oxbow lake?
8. How does a natural levee form?
9. Describe how stream terraces form.
10. Describe three ways in which a river erodes its channel.
11. Name and describe the three main ways in which a stream transports sediment.
12. How does a stream widen its valley?
13. What is base level?
14. What effect does a dam have on a river?
15. How does urbanization increase the risk of flooding?

16. The total area drained by a stream and its tributaries is called the
    a. hydrologic cycle.
    b. tributary area.
    c. divide.
    d. drainage basin.
17. Stream erosion and deposition are controlled primarily by a river's
    a. velocity.
    b. discharge.
    c. gradient.
    d. channel shape.
    e. channel roughness.
18. Hydraulic action, solution, and abrasion are all examples of stream
    a. erosion.
    b. transportation.
    c. deposition.
19. Cobbles are more likely to be transported in a stream's
    a. bed load.
    b. suspended load.
    c. dissolved load.
    d. all of the preceding.
20. A river's velocity is __________ on the outside of a meander curve compared to the inside.
    a. higher
    b. equal
    c. lower
21. Sandbars deposited on the inside of meander curves are called
    a. dunes.
    b. point bars.
    c. cutbanks.
    d. none of the preceding.
22. Which is *not* a drainage pattern?
    a. dendritic
    b. radial
    c. rectangular
    d. trellis
    e. none of the preceding
23. The broad strip of land built up by sedimentation on either side of a stream channel is
    a. a flood plain.
    b. a delta.
    c. an alluvial fan.
    d. a meander.
24. The average time between floods of a given size is
    a. the discharge.
    b. the gradient.
    c. the recurrence interval.
    d. the magnitude.
25. A platform of sediment formed where a stream flows into standing water is
    a. an alluvial fan.
    b. a delta.
    c. a meander.
    d. a flood plain.

Expanding Your Knowledge

1. Several rivers have been set aside as "wild rivers" on which dams cannot be built. Give at least four arguments against building dams on rivers. Give at least four arguments in favor of building dams.
2. Discuss the similarities between deltas and alluvial fans. Describe the differences between them.
3. How is the recurrence interval for a flood determined? How may new data affect the flood-frequency curve?
4. What effect would global warming have on the overall water budget in the hydrologic cycle? How might this influence the dynamics of a stream?

Exploring Web Resources

https://www.usgs.gov/science/science-explorer/Water

Contains extensive information on water issues throughout the United States and many links to U.S. Geological Survey data and online publications.

http://waterdata.usgs.gov/nwis/rt

Contains real-time streamflow data from U.S. Geological Survey gaging stations throughout the United States.

http://water.usgs.gov/usa/nwis/

Contains historical streamflow data from U.S. Geological Survey gaging stations throughout the United States.

http://floodobservatory.colorado.edu/index.html

The *Dartmouth Flood Observatory* website contains information on flood detection and satellite images of floods and flood damage from around the world.

http://www.sciencecourseware.org/VirtualRiver/

California State University, Los Angeles *Virtual River* exercise. Analyze streamflow data and observe animations of flowing streams.

CHAPTER 11

Groundwater

Hot groundwater comes to the surface in Morning Glory Pool at Yellowstone National Park. Blue and green bacteria color the interior of the pool where the water is hotter, and yellow and orange bacteria thrive in the cooler water along the edge.
©Forcellini Danilo/Shutterstock

LEARNING OBJECTIVES

- Define and sketch the saturated and unsaturated zones, the water table, and a perched water table.
- Discuss the difference between porosity and permeability.
- Define *aquifer*, and distinguish between a confined and an unconfined aquifer.
- Summarize the factors that control the storage and movement of groundwater.
- Sketch what happens to the water table near a pumped well; identify the conditions necessary for an artesian well.
- Illustrate the difference between a gaining stream and a losing stream.
- Describe several ways groundwater can be contaminated.
- Explain how caverns, sinkholes, and karst topography develop.
- Distinguish among springs, hot springs, and geysers. How do geysers erupt?

How much of the hydrosphere is groundwater? Compared to the oceans, not much. Approximately 1.7% of the world's water is groundwater, whereas 96.5% is ocean water. If we look at fresh water alone, we find that the amount of groundwater is approximately 100 times that of all rivers and freshwater lakes in the world. (However, the amount of fresh water stored in glaciers is more than twice that stored as fresh groundwater.)

Distribution of Water in the Hydrosphere (%)	
Oceans	96.5
Glaciers and other ice	1.76
Groundwater	1.70
Fresh	0.76
Saline	0.94
Lakes	0.013
Fresh	0.007
Saline	0.006
Soil moisture	0.001
Atmosphere	0.001
Rivers	0.0002

How water gets underground, where it is stored, how it moves while underground, how we look for it, and, perhaps most important of all, why we need to protect it are the main topics of this chapter.

Also important is how groundwater is related to surface rivers and springs. Groundwater can form distinctive geologic features, such as caves, sinkholes, and petrified wood. It also can appear as hot springs and geysers. Hot groundwater can be used to generate geothermal energy.

11.1 THE IMPORTANCE OF GROUNDWATER

Groundwater is a tremendously important resource. Managing our water resources becomes increasingly difficult as demands increase. Growing cities in arid climates are removing groundwater faster than it can be replenished. Pollution of groundwater by industrial wastes, agricultural pesticides, and other means can render the water unfit for human consumption. Growing population and improvements in lifestyles increasingly impact our water supply.

Many communities obtain the water they need from rivers, lakes, or reservoirs, sometimes using aqueducts or canals to bring water from distant surface sources. Another source of water lies directly beneath most towns. This resource is **groundwater**, the water that lies beneath the ground surface, filling the pore space between grains in bodies of sediment and clastic sedimentary rock, and filling cracks and crevices in all types of rock.

Groundwater is a major economic resource, particularly in the dry western areas of the United States and Canada, where surface water is scarce. Many towns and farms pump great quantities of groundwater from drilled wells. Even cities next to large rivers may pump their water from the ground because groundwater is commonly less contaminated and more economical to use than surface water.

11.2 THE STORAGE OF GROUNDWATER

As we saw from the hydrologic cycle described in chapter 10 (see figure 10.1), some of the water that precipitates from the atmosphere as rain and snow infiltrates the geosphere and becomes groundwater. How much precipitation soaks into the ground is influenced by climate, land slope, soil and rock type, and vegetation. In general, approximately 15% of the total precipitation ends up as groundwater, but that varies locally and regionally from 1% to 20%.

The Water Table

Responding to the pull of gravity, water percolates down into the ground through the soil and through cracks and pores in the rock. The rate of groundwater flow tends to decrease with depth

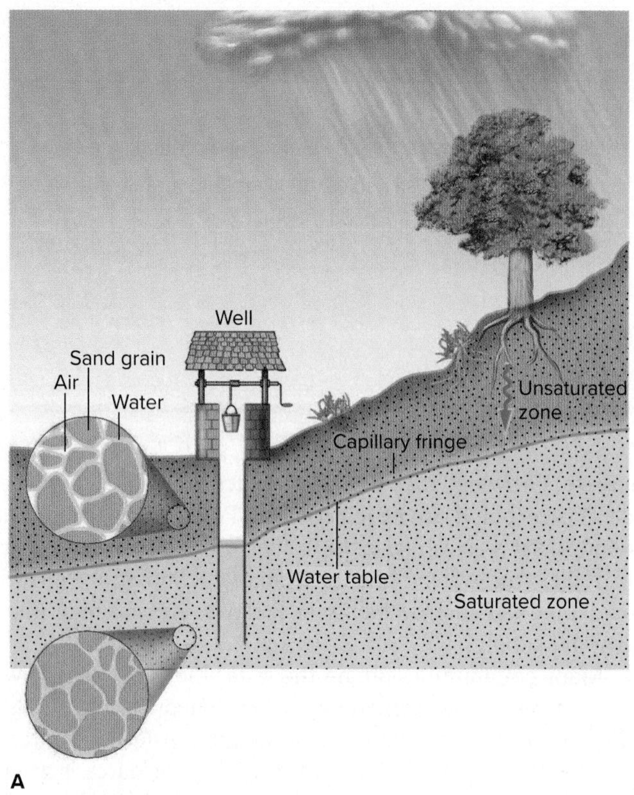

FIGURE 11.1

(*A*) The water table marks the top of the saturated zone in which water completely fills the rock pore space. Above the water table is the unsaturated zone in which rock openings typically contain both air and water. (*B*) Groundwater fills lakes that extend below the water table in northern Wisconsin. The surfaces of the lakes represent the top of the groundwater table.
(B) ©Reid Buell

because sedimentary rock pores tend to be closed by increasing amounts of cement and the weight of the overlying rock. Moreover, sedimentary rock overlies igneous and metamorphic crystalline basement rock, which usually contains even less groundwater.

The subsurface zone in which all rock openings are filled with water is called the **saturated zone** (figure 11.1*A*). If a well were drilled downward into this zone, groundwater would fill the lower part of the well. The water level inside the well marks the upper surface of the saturated zone; this surface is the **water table**.

Most rivers and lakes intersect the saturated zone. Rivers and lakes occupy low places on the land surface, and groundwater flows out of the saturated zone into these surface depressions. The water level at the surface of most lakes and rivers coincides with the water table (figure 11.1*B*). Groundwater also flows into mines and quarries cut below the water table.

Above the water table is a zone where not all the sediment or rock openings are filled with water. It is referred to as the **unsaturated zone** (figure 11.1*A*). Within the unsaturated zone, surface tension causes water to be held above the water table. The *capillary fringe* is a transition zone with higher moisture content at the base of the unsaturated zone just above the water table. Some of the water in the capillary fringe has been drawn or wicked upward from the water table (much like water rising up a paper towel if the corner is dipped in water). The capillary fringe is generally less than a meter thick but may be much thicker in fine-grained sediments and thinner in coarse-grained sediments such as sand and gravel.

Plant roots generally obtain their water from the belt of soil moisture near the top of the unsaturated zone, where fine-grained clay minerals hold water and make it available for plant growth. Most plants "drown" if their roots are covered by water in the saturated zone; plants need both water and air in soil pores to survive. (The water-loving plants of swamps and marshes are an exception.)

A **perched water table** is the top of a body of groundwater separated from the main water table beneath it by a zone that is not saturated (figure 11.2). It may form as groundwater collects above a lens of less permeable shale within a more permeable rock, such as sandstone. If the perched water table intersects the land surface, a line of springs can form along the upper contact of the shale lens. The water perched above a shale lens can provide a limited water supply to a well; it is an unreliable long-term supply.

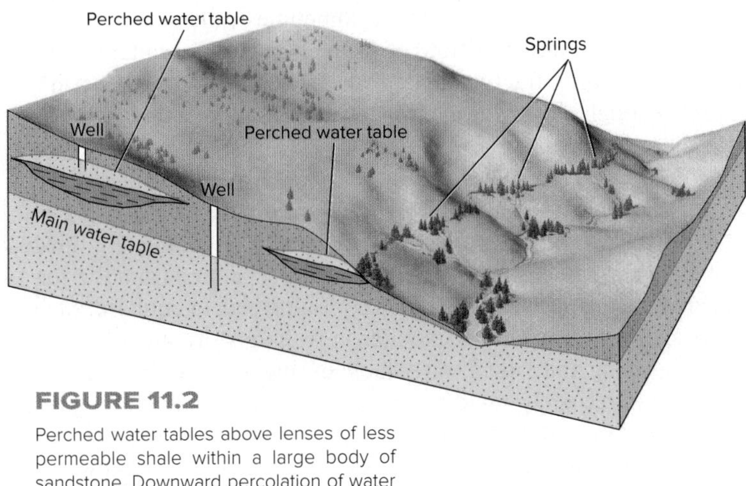

FIGURE 11.2

Perched water tables above lenses of less permeable shale within a large body of sandstone. Downward percolation of water is impeded by the less permeable shale.

Porosity and Permeability

Porosity, the percentage of rock or sediment that consists of voids or openings, is a measurement of a rock's ability to hold water. Most rocks can hold some water. Some sedimentary rocks, such as sandstone, conglomerate, and many limestones, tend to have a high porosity and therefore can hold a considerable amount of water. A deposit of loose sand may have a porosity of 30% to 50%, but this may be reduced to 10% to 20% by compaction and cementation as the sand lithifies (table 11.1). A sandstone in which pores are nearly filled with cement and fine-grained matrix material may have a porosity of 5% or less. Crystalline rocks, such as granite, schist, and some limestones, do not have pores but may hold some water in fractures and other openings.

Although most rocks can hold some water, they vary a great deal in their ability to allow water to pass through them. **Permeability** refers to the capacity of a rock to transmit a fluid such as water or petroleum through pores and fractures. In other words, permeability measures the relative ease of water flow and indicates the degree to which openings in a rock interconnect. The distinction between porosity and permeability is important. A rock that holds much water is called *porous*; a rock that allows water to flow easily through it is described as *permeable*. Most sandstones and conglomerates are both porous and permeable. An *impermeable* rock is one that does not allow water to flow through it easily. Unfractured granite and schist are impermeable. Shale can have substantial porosity, but it has low permeability because its pores are too small to permit easy passage of water.

Aquifers

An **aquifer** is a body of saturated rock or sediment through which water can move easily. Aquifers are both highly permeable and saturated with water. A well must be drilled into an aquifer to reach an adequate supply of water (figure 11.3). Good aquifers include sandstone, conglomerate, well-jointed limestone, bodies of sand and gravel, and some fragmental or fractured volcanic rocks such as columnar basalt (table 11.1).

TABLE 11.1 Porosity and Permeability of Sediments and Rocks

Sediment	Porosity (%)	Permeability
Gravel	25 to 40	Excellent
Sand (clean)	30 to 50	Good to excellent
Silt	35 to 50	Moderate
Clay	35 to 80	Poor
Glacial till	10 to 20	Poor to moderate
Rock		
Conglomerate	10 to 30	Moderate to excellent
Sandstone		
Well-sorted, little cement	20 to 30	Good to very good
Average	10 to 20	Moderate to good
Poorly sorted, well- cemented	0 to 10	Poor to moderate
Shale	0 to 30	Very poor to poor
Limestone, dolomite	0 to 20	Poor to good
Cavernous limestone	up to 50	Excellent
Crystalline rock		
Unfractured	0 to 5	Very poor
Fractured	5 to 10	Poor
Volcanic rocks	0 to 50	Poor to excellent

These favorable geologic materials are sought in "prospecting" for groundwater or looking for good sites to drill water wells.

Wells drilled in shale beds are not usually very successful because shale, although sometimes quite porous, is relatively impermeable (figure 11.3). Wet mud may have a porosity of 80% to 90% and, even when compacted to form shale, may still have a high porosity of 30%. Yet the extremely small size of the pores, together with the electrostatic attraction that clay minerals have for water molecules (see chapter 5), prevents water from moving readily through the shale into a well.

Because they are not very porous, crystalline rocks such as granite, gabbro, gneiss, schist, and some types of limestone are not good aquifers. The porosity of such rocks may be 1% or less. (Shale and crystalline rocks are sometimes called *aquitards* because they retard the flow of groundwater.) Crystalline rocks that are highly fractured, however, may be porous and permeable enough to provide a fairly dependable water supply to wells (figure 11.4).

Figure 11.5 shows the difference between an **unconfined aquifer**, which has a water table because it is only partly filled with water, and a **confined aquifer**, which is completely filled with water under pressure and is usually separated from the surface by a relatively impermeable confining bed, or aquitard, such as shale. An unconfined aquifer is recharged by precipitation, has a rising and falling water table during wet and dry seasons, and has relatively rapid movement of groundwater through it (figure 11.6). A confined aquifer is recharged slowly through confining shale beds. With very slow movement of groundwater, a confined aquifer may have no response at all to wet and dry seasons.

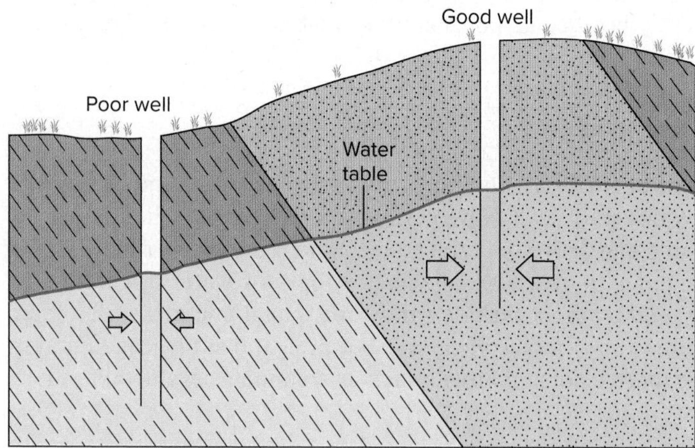

FIGURE 11.3

A well must be installed in an aquifer to obtain water. The saturated part of the highly permeable sandstone is an aquifer, but the less permeable shale is not. Although the shale is saturated, it will not readily transmit water.

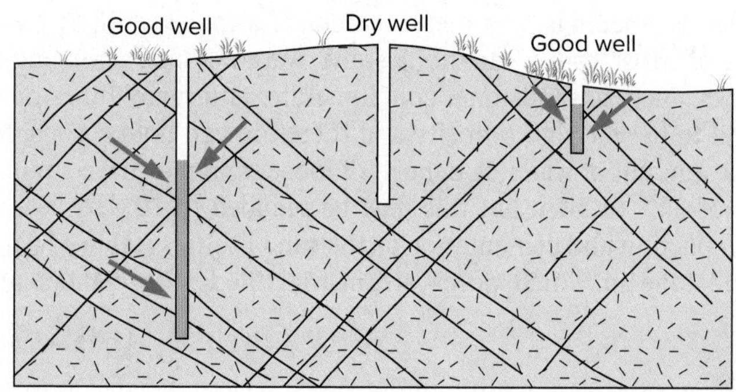

FIGURE 11.4

Wells can obtain some water from fractures in crystalline rock. Wells must intersect fractures to obtain water.

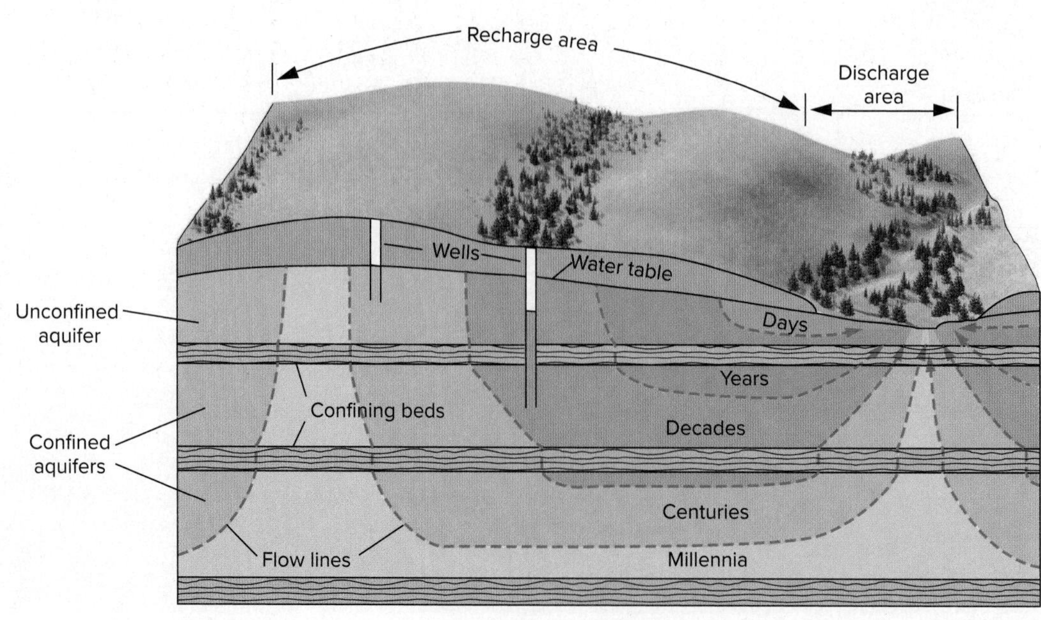

FIGURE 11.5

An unconfined aquifer is exposed to the surface and is only partly filled with water; water in a shallow well will rise to the level of the water table. A confined aquifer is separated from the surface by a confining bed and is completely filled with water under pressure; water in wells rises above the aquifer. Flow lines show direction of groundwater flow. Days, years, decades, centuries, and millennia refer to the time required for groundwater to flow from the recharge area to the discharge area. Water enters aquifers in recharge areas and flows out of aquifers in discharge areas.

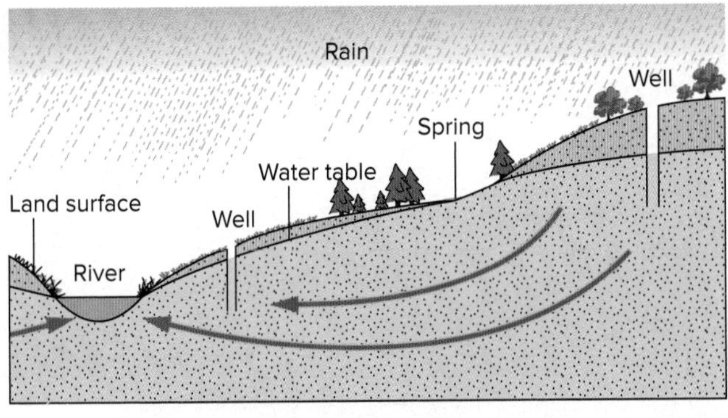

A

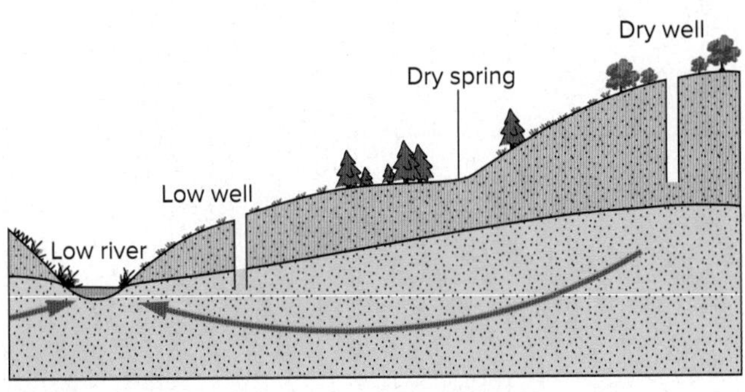

B

FIGURE 11.6

The water table in an unconfined aquifer rises in wet seasons and falls in dry seasons as water drains out of the saturated zone into rivers. (*A*) Wet season: water table and rivers are high; springs and wells flow readily. (*B*) Dry season: water table and rivers are low; some springs and wells dry up.

11.3 THE MOVEMENT OF GROUNDWATER

Compared to the rapid flow of water in surface streams, most groundwater moves relatively slowly through rock underground. Groundwater moves in response to differences in water pressure and elevation, causing water within the upper part of the saturated zone to move downward following the slope of the water table (figure 11.7). See box 11.1 for Darcy's law.

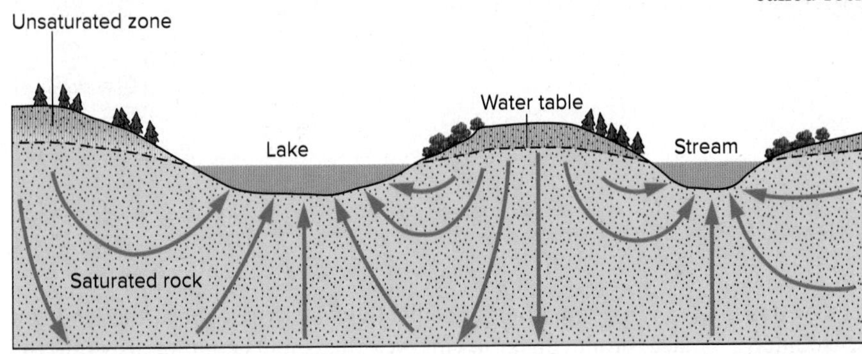

FIGURE 11.7

Movement of groundwater beneath a sloping water table in uniformly permeable rock. Near the surface, groundwater tends to flow parallel to the sloping water table.

The circulation of groundwater in the saturated zone is not confined to a shallow layer beneath the water table. Groundwater may move hundreds of meters vertically downward before rising again to discharge as a spring or seep into the beds of rivers and lakes at the surface (figure 11.7) due to the combined effects of gravity and the slope of the water table.

The *slope of the water table* strongly influences groundwater velocity. The steeper the slope of the water table, the faster groundwater moves. Water-table slope is controlled largely by topography—the water table roughly parallels the land surface (particularly in humid regions), as you can see in figure 11.7. Even in highly permeable rock, groundwater will not move if the water table is flat.

How fast groundwater flows also depends on the *permeability* of the rock or other materials through which it passes. If rock pores are small and poorly connected, water moves slowly. When openings are large and well connected, the flow of water is more rapid. One way of measuring groundwater velocity is to introduce a tracer, such as a dye, into the water and then watch for the color to appear in a well or spring some distance away. Such experiments have shown that the velocity of groundwater varies widely, averaging a few centimeters to many meters a day. Nearly impermeable rocks may allow water to move only a few centimeters per *year*, but highly permeable materials, such as unconsolidated gravel or cavernous limestone, may permit flow rates of hundreds or even thousands of meters per day.

11.4 WELLS

A **well** is a deep hole, generally cylindrical, that is dug or drilled into the ground to penetrate an aquifer within the saturated zone (see figure 11.3). Usually water that flows into the well from the saturated rock must be lifted or pumped to the surface. As figure 11.6 shows, a well dug in a valley usually has to go down a shorter distance to reach water than a well dug on a hilltop. During dry seasons, the water table falls as water flows out of the saturated zone into springs and rivers. Wells not deep enough to intersect the lowered water table go dry, but the rise of the water table during the next rainy season normally returns water to the dry wells. The addition of new water to the saturated zone is called **recharge**.

When water is pumped from a well, the water table is typically drawn down around the well into a depression shaped like an inverted cone known as a **cone of depression** (figure 11.8). This local lowering of the water table, called **drawdown**, tends to change the direction of groundwater flow by changing the slope of the water table. In lightly used wells that are not pumped, drawdown does not occur and a cone of depression does not form. In a simple rural well with a bucket lowered on the end of a rope, water cannot be extracted rapidly enough to significantly lower the water table. A well of this type is shown in figure 11.1*A*.

IN GREATER DEPTH 11.1

Darcy's Law and Fluid Potential

In 1856, Henry Darcy, a French engineer, found that the velocity at which water moves depends on the *hydraulic head* of the water and on the permeability of the material that the water is moving through.

The hydraulic head of a drop of water is equal to the elevation of the drop plus the water pressure on the drop:

$$\text{Hydraulic head} = \text{elevation} + \text{pressure}$$

In box figure 1*A*, the points A and B are both on the water table, so the pressure at both points is zero (there is no water above points A and B to create pressure). Point A is at a higher elevation than B, so A has a higher hydraulic head than B. The difference in elevation is equal to the difference in head, which is labeled **h**. Water will move from point A to point B (as shown by the dark blue arrow) because water moves from a region of high hydraulic head to a region of low head. The distance the water moves from A to B is labeled **L**. The *hydraulic* gradient is the difference in head between two points divided by the horizontal distance between the two points:

$$\textbf{Hydraulic gradient} = \frac{\textbf{difference in head}}{\textbf{distance}} = \frac{\Delta h}{L}$$

In box figure 1*B*, the two points have equal elevation, but the pressure on point C is higher than on point D (there is more water to create pressure above point C than point D). The head is higher at point C than at point D, so the water moves from C to D. In box figure 1*C*, point F has a lower elevation than point G, but F also has a higher pressure than G. The difference in pressure is greater than the difference in elevation, so F has a higher head than G, and water moves from F to G. Note that underground water may move downward, horizontally, or upward in response to

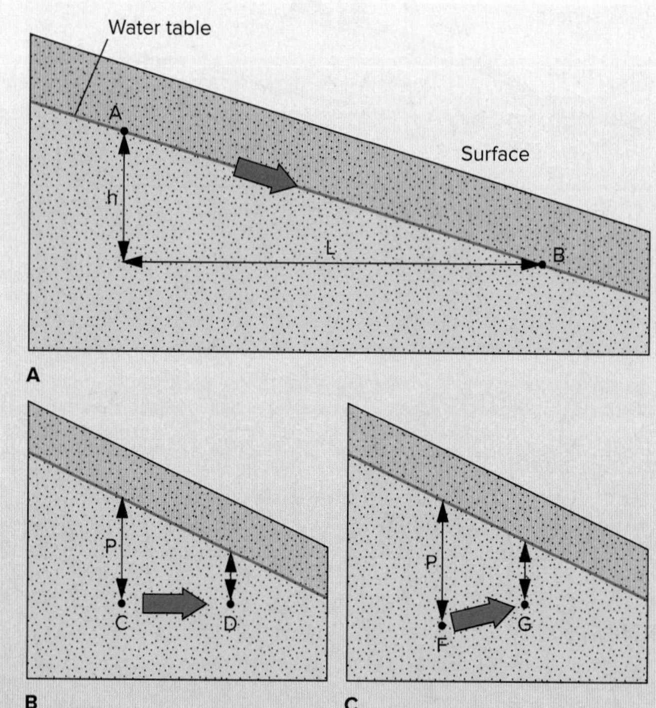

BOX 11.1 ■ FIGURE 1

Groundwater moves in response to hydraulic head (elevation plus pressure). Water movement shown by dark blue arrows. (*A*) Points A and B have the same pressure, but A has a higher elevation; therefore, water moves from A to B. (*B*) Point C has a higher pressure (arrow marked P) than D; therefore, water moves from C to D at the same elevation. (*C*) Pressure also moves water upward from F to G.

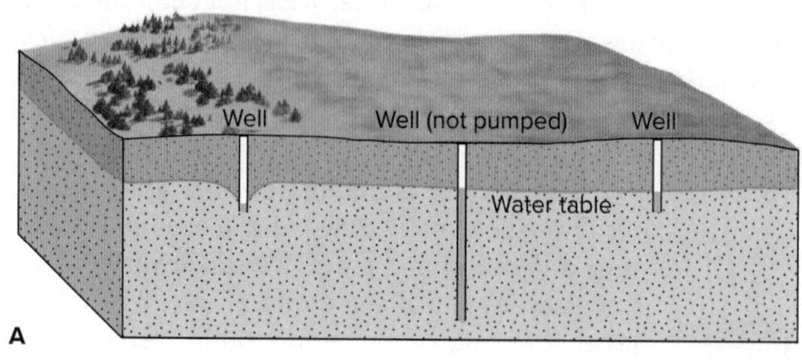

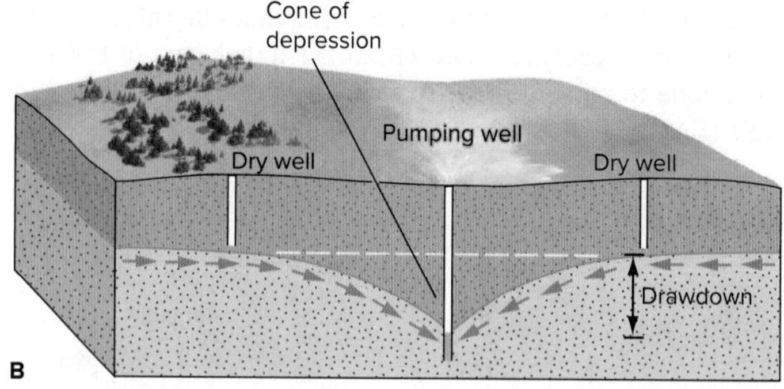

FIGURE 11.8

Pumping well lowers the water table into a cone of depression. If well is heavily pumped, surrounding shallow wells may go dry.

In unconfined aquifers, water rises in shallow wells to the level of the water table. In confined aquifers, the water is under pressure and rises in wells to a level above the top of the aquifer (see figure 11.5). Such a well is called an **artesian well**, and confined aquifers are also called *artesian aquifers.*

In some artesian wells, the water rises above the land surface, producing a flowing well that spouts continuously into the air unless it is capped (figure 11.9). Flowing wells used to occur in South Dakota, when the extensive Dakota Sandstone aquifer was first tapped (figure 11.10), but continued use has lowered

differences in head but that it always moves in the direction of the downward slope of the water table above it. One of the first goals of groundwater geologists, particularly in groundwater contamination investigations, is to find the slope of the local water table in order to determine the direction (and velocity) of groundwater movement.

The velocity of groundwater flow is controlled by both the permeability of the sediment or rock and the hydraulic gradient. *Darcy's law* states that the velocity equals the permeability multiplied by the hydraulic gradient. This gives the Darcian velocity (or the velocity of water flowing through an open pipe). To determine the actual velocity of groundwater, since groundwater only flows through the openings in sediment or rock, the Darcian velocity must be divided by the porosity.

Groundwater velocity = permeability/porosity × hydraulic gradient

$$\mathbf{V} = \frac{\mathbf{K}}{\mathbf{n}}\frac{\mathbf{\Delta h}}{\mathbf{L}}$$

(Darcy called **K** the hydraulic conductivity; it is a measure of permeability and is specific to a particular aquifer. The porosity is represented by **n** in the equation.)

Groundwater movement is shown in diagrams in relation to *equipotential lines* (lines of constant hydraulic head). Groundwater moves from regions of high head to regions of low head. Box figure 2 illustrates how *flow lines*, which show groundwater movement, cross equipotential lines at right angles as water moves from high to low head.

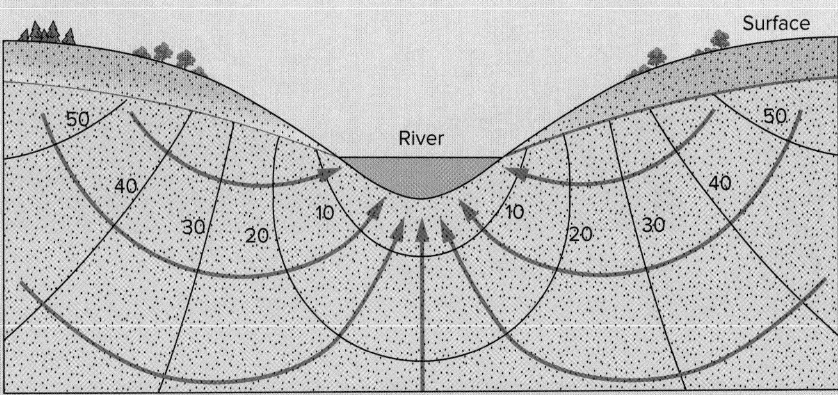

BOX 11.1 ■ FIGURE 2

Dark blue arrows are flow lines, which show direction of groundwater flow. Flow is perpendicular to equipotential lines (black lines with numbers), which show regions of equal hydraulic head. Groundwater generally flows from hilltops toward valleys, emerging from the ground as springs into streambeds and banks, lakes, and swamps.

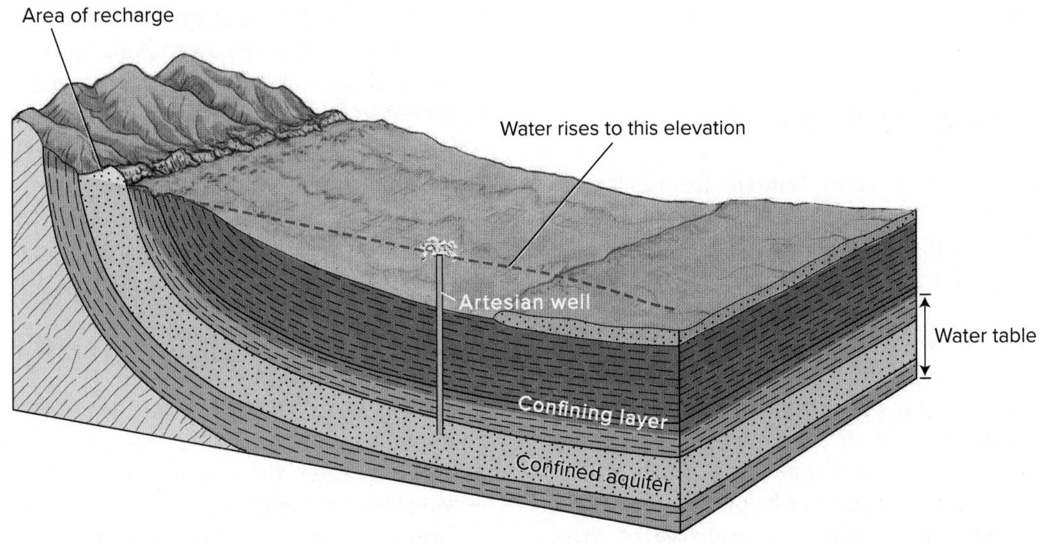

FIGURE 11.9

The Dakota Sandstone in South Dakota is a relatively unusual type of confined aquifer because it is tilted and exposed to the surface by erosion. Water in most wells rose above the land surface when the aquifer was first tapped in the 1800s.

ENVIRONMENTAL GEOLOGY 11.2

Hard Water and Soapsuds

"Hard water" is water that contains relatively large amounts of dissolved calcium (often from the chemical weathering of calcite or dolomite) and magnesium (from the ferromagnesian minerals or dolomite). Water taken from the groundwater supply or from a stream for home use may contain enough of these ions to prevent soap from lathering. Calcium and magnesium ions in hard water form gray curds with soap. The curd continues to form until all the calcium and magnesium ions are removed from the water and bound up in the curd. Only then will soap lather and clean laundry. Cleaning laundry in hard water, therefore, takes an excessively large amount of soap.

Hard water may also precipitate a scaly deposit inside teakettles and hot-water tanks and pipes (box figure 1). The entire hot-water piping system of a home in a hard-water area eventually can become so clogged that the pipes must be replaced.

"Soft water" may carry a substantial amount of ions in solution but not the ions that prevent soap from lathering. Water softeners in homes replace calcium and magnesium ions with sodium ions, which do not affect lathering or cause scale. But water containing a large amount of sodium ions, whether from a softener or from natural sources, may be harmful if used as drinking water by persons on a "salt-free" (low-sodium) diet for health reasons.

Additional Resources

For more information on the chemistry of hard water, and maps showing the location of groundwater and surface water in the United States where the water is hard, visit the U.S. Geological Survey website:

- http://water.usgs.gov/edu/hardness.html

BOX 11.2 ■ FIGURE 1
Scale caused by hard water coats the inside of this hot-water pipe.
©Sheila Terry/SPL/Science Source

the water pressure surface below the ground surface in most parts of the state. Water still rises above the aquifer but does not reach the land surface.

11.5 SPRINGS AND STREAMS

A **spring** is a place where water flows naturally from rock onto the land surface (figure 11.11). Some springs discharge where the water table intersects the land surface, but they also occur where water flows out from caverns or along fractures, faults, or rock contacts that come to the surface (figure 11.12).

Climate determines the relationship between streamflow and the water table. In rainy regions, most streams are **gaining streams**; that is, they receive water from the saturated zone (figure 11.13*A*). The surface of these streams coincides with the water table. Water from the saturated zone flows into the stream through the streambed and banks that lie below the water table. Because of the added groundwater, the discharge of these streams increases downstream. Where the water table intersects the land surface over a broad area, ponds, lakes, and swamps are found.

In drier climates, rivers tend to be **losing streams**; that is, they are losing water to the saturated zone (figure 11.13*B*). The water percolating into the ground beneath a losing stream causes the water table to slope away from the stream. In very dry climates, such as in a desert, a losing stream may be separated or *disconnected* from the underlying saturated zone, and a groundwater mound remains beneath the stream even if the streambed is dry (figure 11.13*C*).

11.6 CONTAMINATION OF GROUNDWATER

Groundwater in its natural state tends to be relatively free of contaminants in most areas. Because it is a widely used source of drinking water, the contamination of groundwater can be a very serious problem.

Pesticides and *herbicides* (such as diazinon, atrazine, DEA, and 2,4-D) applied to agricultural crops (figure 11.14*A*) can find their way into groundwater when rain or irrigation water leaches the contaminants downward into the soil. *Fertilizers* are

FIGURE 11.10

Historic photo showing artesian well spouting muddy water 30 meters above land surface in Woonsocket, South Dakota, in the early 1900s. Heavy use of this aquifer has reduced water pressure so much that spouts do not occur today.
Source: N. H. Darton, U.S. Geological Survey

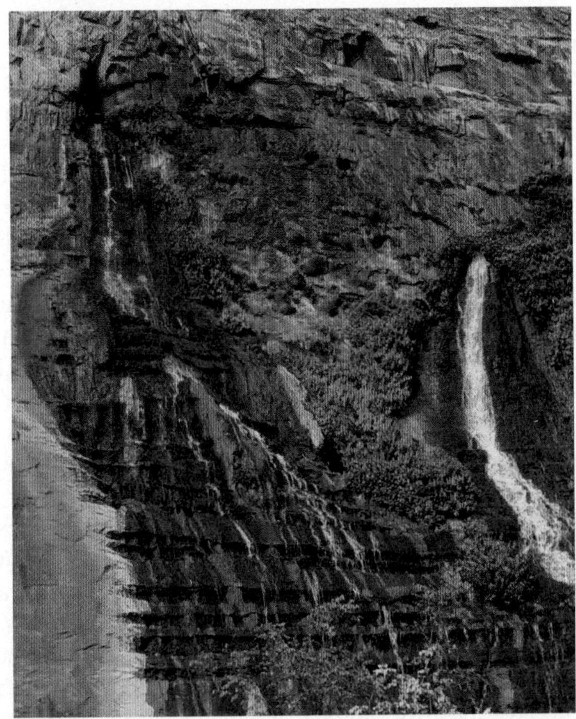

FIGURE 11.11

A large spring flowing from limestone in Vasey's Paradise, Marble Canyon, Arizona.
©Nature and Science/Alamy

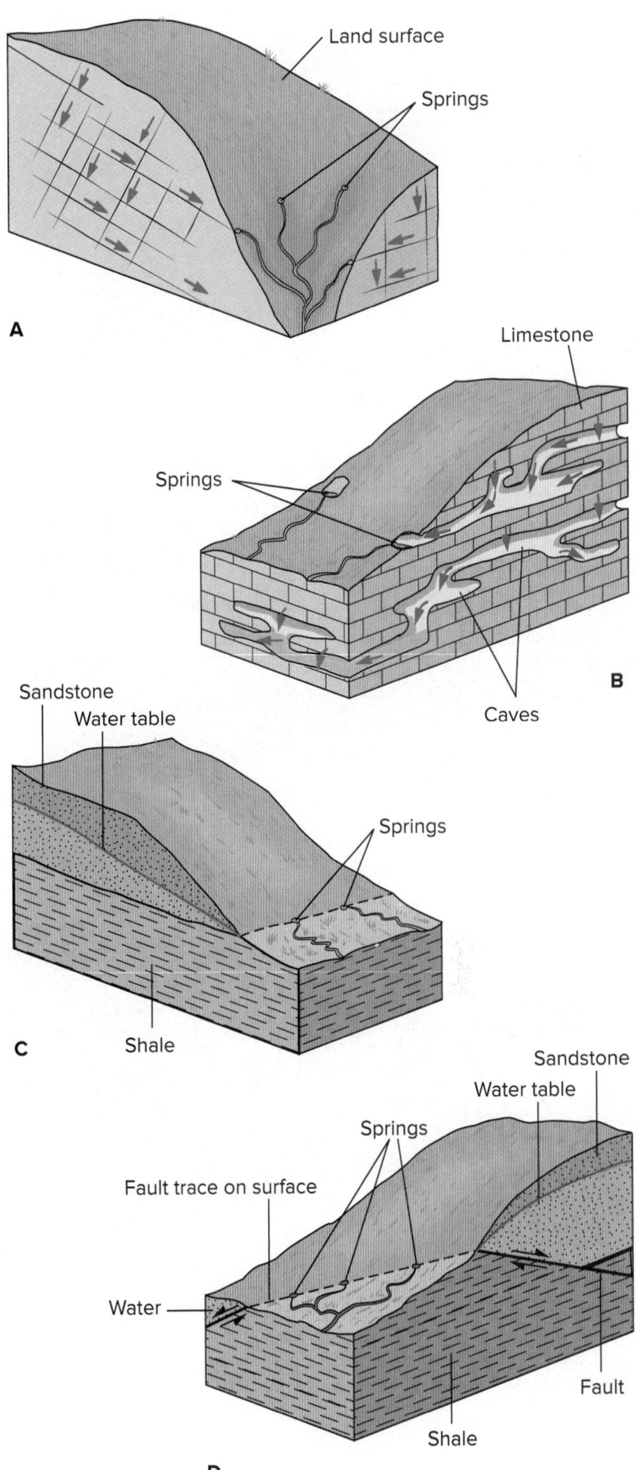

FIGURE 11.12

Springs can form in many ways. (*A*) Water moves along fractures in crystalline rock and forms springs where the fractures intersect the land surface. (*B*) Water enters caves along joints in limestone and exits as springs at the mouths of caves. (*C*) Springs form at the contact between a permeable rock such as sandstone and an underlying less permeable rock such as shale. (*D*) Springs can form along faults when permeable rock has been moved against less permeable rock. Arrows show relative motion along fault.

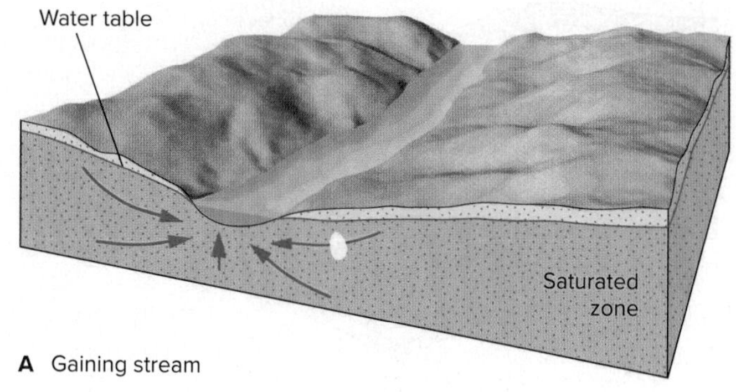

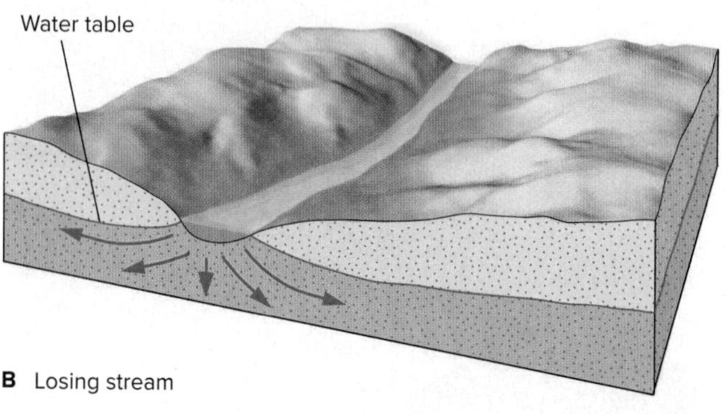

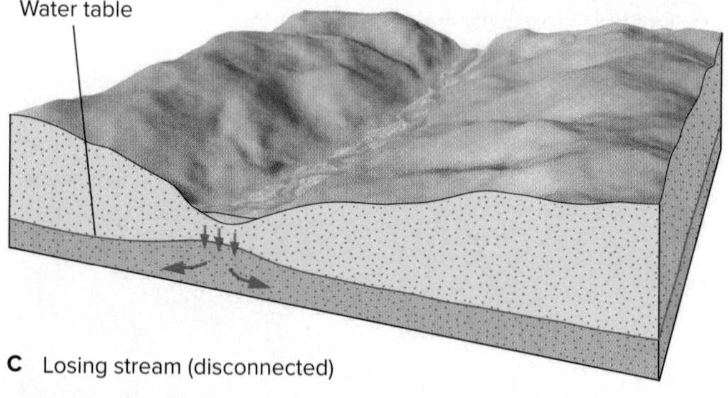

FIGURE 11.13

Gaining and losing streams. (*A*) Stream gaining water from saturated zone. (*B*) Stream losing water through streambed to saturated zone. (*C*) Water table can be close to the land surface, but disconnected from the surface water beneath a streambed that intermittently contains water.

also a concern. Nitrate, which forms from one of the most widely used fertilizers, is harmful in even small quantities in drinking water.

Rain can also leach pollutants from city landfills into groundwater supplies (figure 11.14B). Consider for a moment some of the things you threw away last year. A partially empty aerosol can of ant poison? The can will rust through in the landfill, releasing the poison into the ground and into the saturated zone below. A broken thermometer? The toxic mercury may eventually find its way to the groundwater supply. A half-used can of oven cleaner? The dried-out remains of a can of lead-based paint? *Heavy metals* such as mercury, lead, chromium, copper, and cadmium, together with household chemicals and poisons, can all be concentrated in groundwater supplies beneath landfills (figure 11.15). This is of greater concern in older, unlined landfills than in newer landfills that are engineered to contain contaminants with impermeable layers of clay and synthetic liners.

Liquid and solid wastes from septic tanks, sewage plants, and animal feedlots and slaughterhouses may contain *bacteria*, *viruses*, and *parasites* that can contaminate groundwater (figure 11.14*C*). Liquid wastes from industries (figure 11.14*D*) and military bases can be highly toxic, containing high concentrations of heavy metals and compounds such as cyanide and *PCBs* (polychlorinated biphenyls), which are widely used in industry. A degreaser called *TCE* (trichloroethylene) has been increasingly found to pollute both surface and underground water in many places. Toxic liquid wastes are often held in surface ponds or pumped down deep disposal wells. If the ponds leak, groundwater can become polluted. Deep wells may be safe for liquid waste disposal if they are deep enough, but contamination of drinking water supplies and even surface water has resulted in some localities from improper design of the disposal wells.

Acid mine drainage from coal and metal mines can contaminate both surface and groundwater (see figure 5.13). It is usually caused by sulfuric acid formed by the oxidation of sulfur in pyrite and other sulfide minerals when they are exposed to air by mining activity. Fish and plants are often killed by the acid waters draining from long-abandoned mines.

Radioactive waste is both an existing and a very serious potential source of groundwater contamination. The shallow burial of *low-level* solid and liquid radioactive wastes from the nuclear power industry has caused contamination of groundwater, particularly as liquid waste containers leak into the saturated zone and as the seasonal rise and fall of the water table at some sites periodically covers the waste with groundwater. The search for a permanent disposal site for solid, *high-level* radioactive waste (now stored temporarily on the surface) is a major national concern for the United States. The permanent site will be deep underground and must be isolated from groundwater circulation for thousands of years. Salt beds, shale, glassy tuffs, and crystalline rock deep beneath the surface have all been studied, particularly in arid regions where the water table is hundreds of meters below the land surface. The site selected for disposal of high-level waste, primarily spent fuel from nuclear reactors, is Yucca Mountain, Nevada, 180 kilometers northwest of Las Vegas. The site will be deep underground in volcanic tuff well above the current (or predicted future) water table and in a region of very low rainfall. The U.S. Congress, under intense political pressure from other candidate states that did not want the site, essentially chose Nevada by eliminating the funding for the study of all alternative sites. After much controversy over the ultimate safety of the site and objections from Nevada, Congress approved the Yucca Mountain site in 2002. Due to continued political pressure and new scientific studies, funding has been cut to prepare Yucca Mountain as the nation's permanent disposal site for

FIGURE 11.14

Some sources of groundwater pollution. (*A*) Pesticides. (*B*) Household garbage. (*C*) Animal waste. (*D*) Industrial toxic waste.
(A) ©Corbis/RF; (B) ©Diane Carlson; (C) Source: U.S. Department of Agriculture Soil Conservation Service; (D) Source: U.S. Department of Agriculture Soil Conservation Service

high-level nuclear waste. Until a permanent disposal site can be agreed upon and built, the United States continues to store 64,000 metric tons of spent fuel rods above ground at nuclear facilities across the country.

Not all groundwater contaminants form plumes within the saturated zone, as shown in figure 11.15. *Gasoline*, which leaks from gas station storage tanks at tens of thousands of U.S. locations, is less dense than water and floats upon the water table (figure 11.16). Some liquids such as TCE are heavier than water and sink to the bottom of the saturated zone, perhaps traveling in unpredicted directions on the surface of an impermeable layer (figure 11.16). Determining the extent and flow direction of groundwater pollution is a lengthy process requiring the drilling of tens, or even hundreds, of costly wells for each contaminated site.

Not all sources of groundwater contamination are manmade. Naturally occurring *minerals within rock and soil* may contain elements such as arsenic, selenium, mercury, and other toxic metals. Circulating groundwater can leach these elements out of the minerals and raise their concentrations to harmful levels within the water. Millions of people in Bangladesh are exposed to toxic levels of arsenic from well water, and arsenic contamination of wells is a serious problem in eastern New England.

Soil and rock can filter some contaminants out of groundwater. This filtering ability depends on the permeability and mineral composition of rock and soil. Under ideal conditions, human sewage can be purified by only 30 to 45 meters of travel through a sandy loam soil (a mixture of clay minerals, sand, and organic humus). The sewage is purified by filtration, ion absorption by clay minerals and humus, and decomposition by soil organisms in the biosphere (figure 11.17). On the other hand, extremely permeable rock, such as highly fractured granite or cavernous limestone, has little purifying effect on sewage. Groundwater flows so rapidly through such rocks that it is not purified even after hundreds

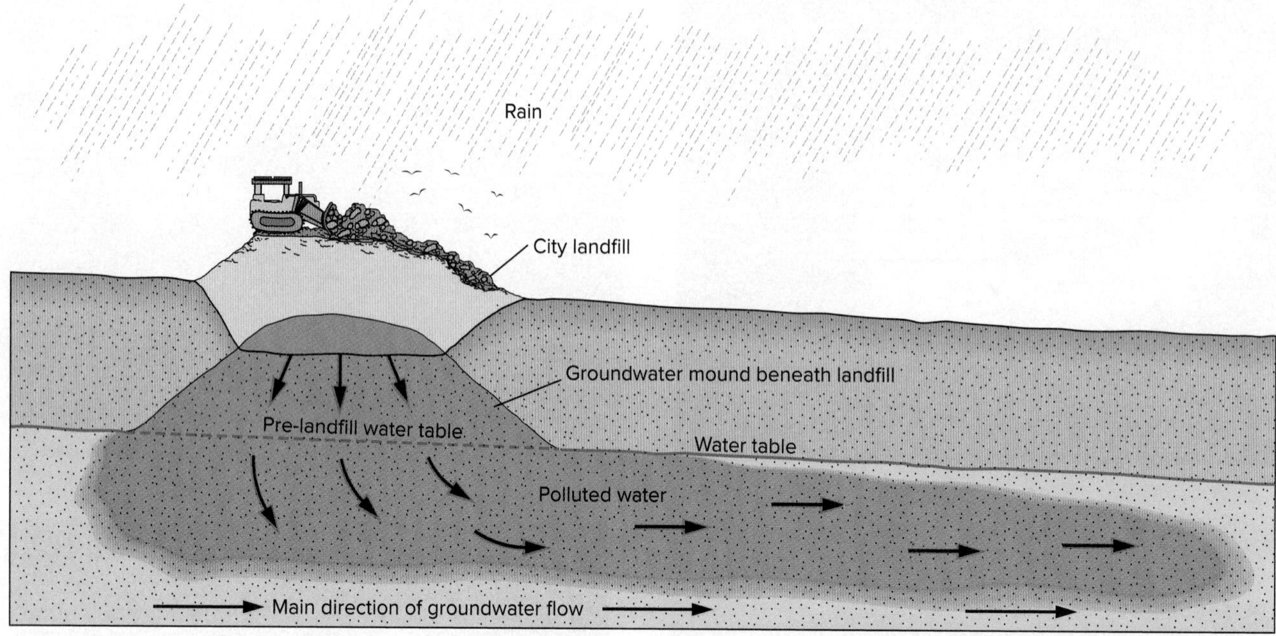

A Cross section

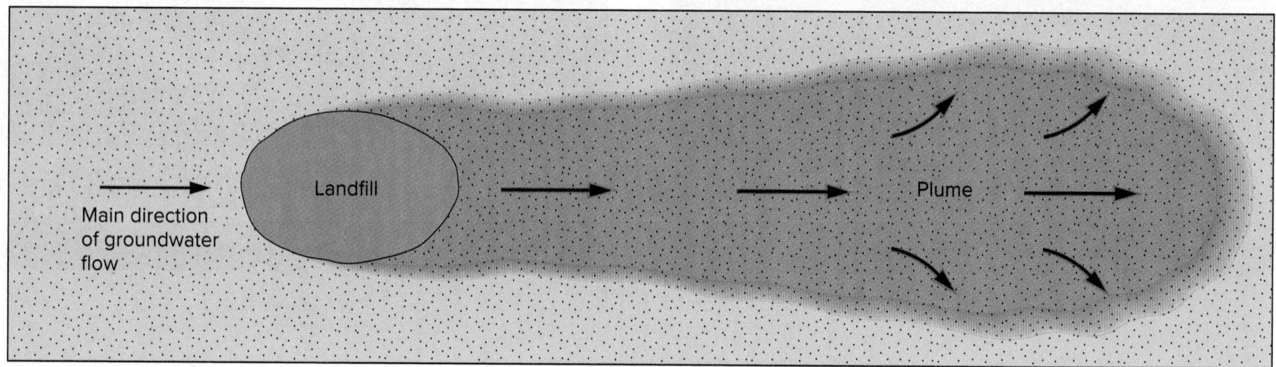

B Map view of contaminant plume. Note how it grows in size with distance from the pollution source.

FIGURE 11.15

(A) Waste piled on the land surface creates a groundwater mound beneath it because the landfill forms a hill and because the waste material is more porous and permeable than the surrounding soil and rock. Rain leaches pollutants into the saturated zone. *(B)* A plume of contaminated water will spread out in the direction of groundwater flow.

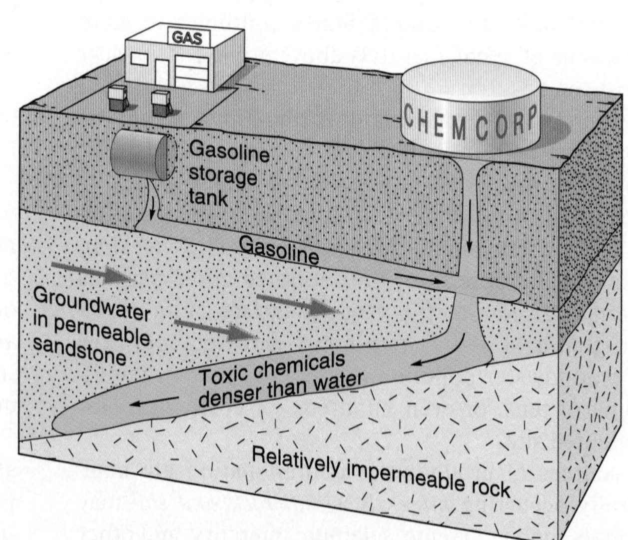

FIGURE 11.16

Not all contaminants move within the saturated zone, as shown in figure 11.15. Gasoline floats on water; many dense chemicals move along impermeable rock surfaces below the saturated zone.

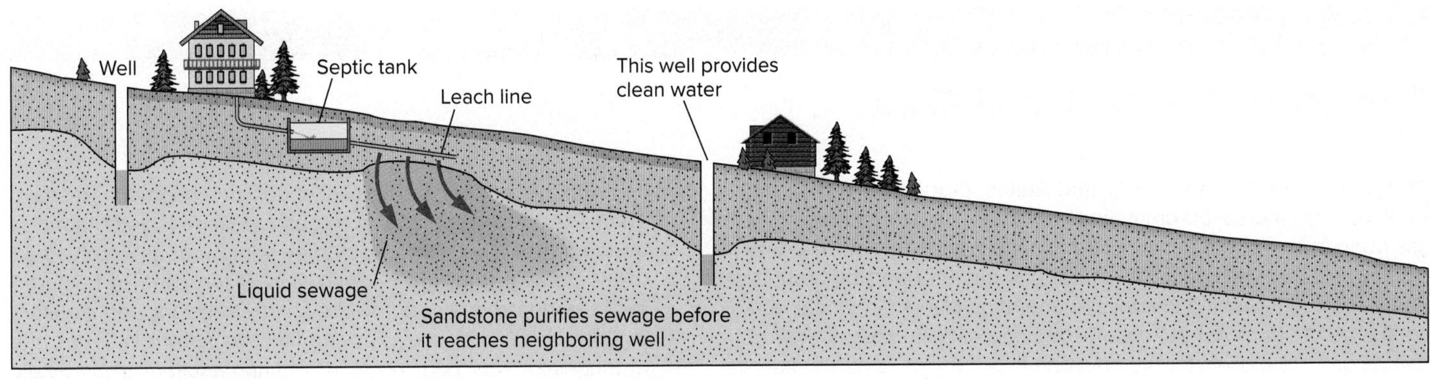

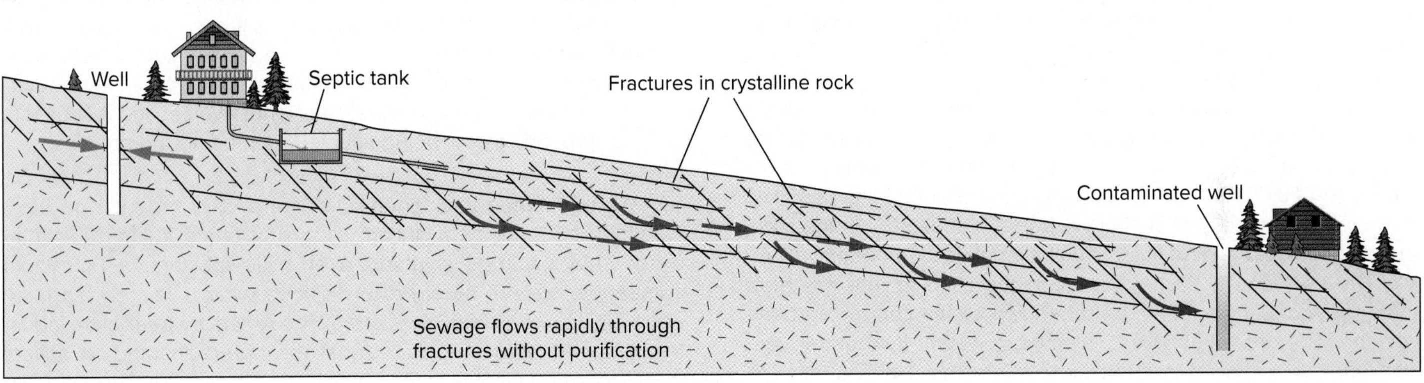

FIGURE 11.17

Rock type and distance control possible sewage contamination of neighboring wells. (*A*) As little as 30 meters of movement can effectively filter human sewage in sandstone and some other rocks and sediments. (*B*) If the rock has large open fractures, contamination can occur many hundreds of meters away.

of meters of travel. Some pesticides and toxic chemicals are not purified at all by passage through rock and soil, even soil rich in humus and clay minerals.

Contaminated groundwater is extremely difficult to clean up. Networks of expensive wells may be needed to pump contaminated water out of the ground and replace it with clean water. Because of the slow movement and large volume of groundwater, the cleanup process for a large region can take decades and tens of millions of dollars to complete.

Groundwater contamination can be largely prevented with careful thought and considerable expense. A city landfill can be sited high above the water table and possible flood levels or located in a region of groundwater discharge rather than recharge. A site can be sealed below by impermeable (and expensive) clay barriers and plastic liners, and sealed off from rainfall by an impermeable cover. Dikes can prevent surface runoff through or from the site. Although sanitary landfills are expensive, they are much cheaper than groundwater cleanup.

Pumping wells can cause or aggravate groundwater contamination (figure 11.18). Well drawdown can increase the slope of the water table locally, thus increasing the rate of groundwater flow and giving the water less time to be purified underground before it is used (figure 11.18*A*). Drawdown can even reverse the original slope of the water table, perhaps contaminating wells that were pure before pumping began (figure 11.18*B*). Heavily pumped wells near a coast can be contaminated by *saltwater intrusion* (figure 11.18*C* and *D*). Saltwater intrusion is becoming a serious problem as the demand for drinking water increases in rapidly growing coastal communities.

More recently, " fracking," the technique used to free natural gas by hydraulic fracturing, has been blamed for polluting nearby aquifers with hydrocarbons and with proprietary chemicals used in the process (see box 11.3).

11.7 BALANCING WITHDRAWAL AND RECHARGE

A local supply of groundwater will last indefinitely if it is withdrawn for use at a rate equal to or less than the rate of recharge to the aquifer. If groundwater is withdrawn faster than it is being recharged, however, the supply is being reduced and will one day be gone.

Heavy use of groundwater causes a regional water table to drop. In parts of western Texas and eastern New Mexico, the pumping of groundwater from the Ogallala aquifer has caused the water table to drop 30 meters over the past few decades. The lowering of the water table means that wells must be deepened and more electricity must be used to pump the water to

ENVIRONMENTAL GEOLOGY 11.3

Fracking for Natural Gas

During recent years, the United States' production of natural gas has increased dramatically. Within a decade, the country has gone from being a net importer of natural gas to a net exporter. During this time, the price of natural gas has dropped by around 30%. Many coal-burning electricity-generating plants have been reconfigured to burn natural gas. Replacing coal with clean-burning natural gas is a big win for the environment.

The newly recovered gas mostly comes from shale. Almost half of the natural gas produced in the United States is derived from shale, and the percentage is predicted to rise to 70% by 2040. You may recall from chapter 6 that shale is a very fine-grained sedimentary rock. It is regarded as an *aquitard* because fluid (gas and crude oil) will not readily flow through it. This is because individual grains of clay, and adjacent pore space, can only be seen with a microscope, and fluid in the pores tends to adhere to the clay particles. However, the porosity of shale may be as high as 30% of the rock, so shale can hold a lot of fluid. The challenge is to get it out.

The enormously successful technique used to free gas from shale is *hydraulic fracturing,* better known by the contracted term, *fracking.* The process involves injecting a watery fluid under very high pressure, applied in short pulses, into holes drilled into the shale (box figure 1). The shale is shattered and, after the pressure is reduced, gas (or oil) flows through the fracture system. Sand, added to the watery fluid, remains in the fractures and keeps them from closing. Besides water and sand, the fracking fluid contains "chemicals" (discussed below).

So, what is not to like about the new influx of natural gas? The two major environmental problems associated with fracking are:

Great volumes of water are used. Most wells being fracked use an average of 15 million liters of water per well. This could strain or decimate local water supplies, aquifers, and streams, which are used for drinking water, agriculture, and industry.

Contamination of aquifers. The Marcellus Shale is a large body of shale that underlies a significant portion of Pennsylvania, as well as parts of adjacent states. At one Pennsylvania town, fracking has been blamed for local contamination of an overlying aquifer. Some residents were able to ignite their tap water, implying that natural gas had escaped from the fractured shale. There were also reports of previously clear well water that had turned brown. The brown water may or may not have contained toxins derived from the fracking fluid. If the zone of fractured rock extends above or below the shale strata, adjacent aquifers may be contaminated by fracking fluid. As shown in figure 11.5, groundwater flow, particularly in confined aquifers, is extremely slow, so contaminated groundwater may not be detected for hundreds or thousands of years unless it is pumped out of a well drilled into the contaminated zone. A recent study has found a change in water chemistry in drinking water wells in lowland areas surrounding fracked wells in Pennsylvania, and it suggests that groundwater is being affected by the fracking process.

We mentioned that "chemicals" are part of the fracking fluid. "Chemicals" is the term drillers use in reporting fluids other than water. They could be anything, even toxic substances. Petroleum companies are not required to disclose what is in fracking fluid, even if it is toxic. In 2005, backed by the American petroleum industry, a federal law was enacted that exempted gas and oil drillers from disclosing what is in the fluids used for fracking. The rationale was that a company's recipe for fracking fluid is proprietary (a trade secret), and information shouldn't be available to competitors. Critics suggest that the companies don't want the public to know about toxic substances being injected into the ground and potentially polluting adjacent aquifers.

Known carcinogenic substances, such as formaldehyde and benzene, have been found at some recently fracked sites. This provides evidence that toxic substances are used in at least some fracking.

It would seem prudent to have laws that regulate fracking. The fracturing should take place entirely within the shale and not provide pathways for fluids to seep into adjacent aquifers or permeable bedrock. Regulating the composition of fracking fluid should take precedence over the petroleum industry wanting to protect a company's competitive advantage. California, in 2013, became the first state to require at least partial reporting of the composition of fracking fluids.

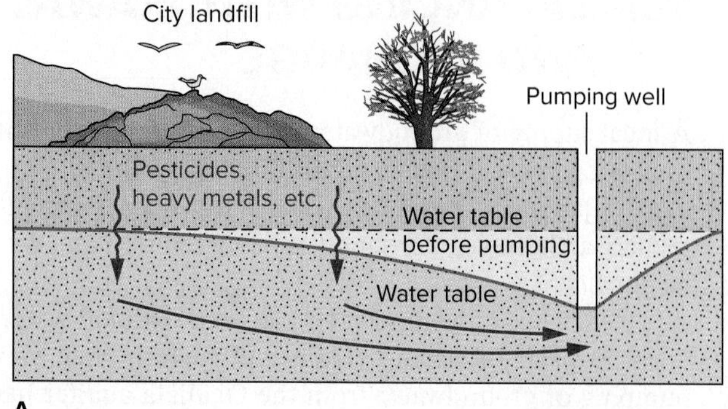

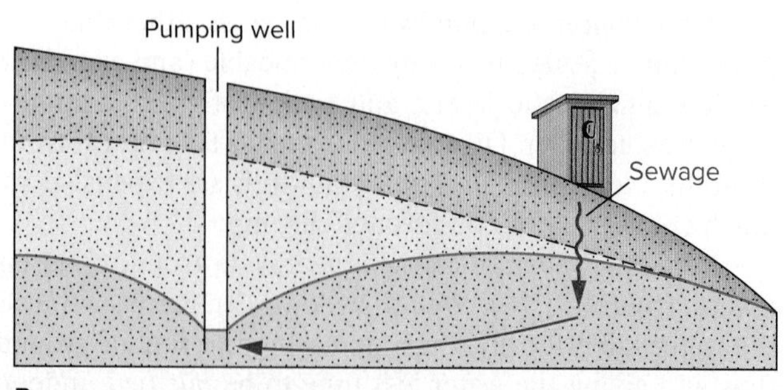

FIGURE 11.18

Groundwater pollution problems caused or aggravated by pumping wells. (*A*) Water table steepens near a landfill, increasing the velocity of groundwater flow and drawing contaminants into a well. (*B*) Water table slope is reversed by pumping, changing direction of the groundwater flow and contaminating the well.
(*C*) Well near a coast (before pumping). Fresh water floats on saltwater. (*D*) Well in *C* begins pumping, thinning the freshwater lens and drawing saltwater into the well.

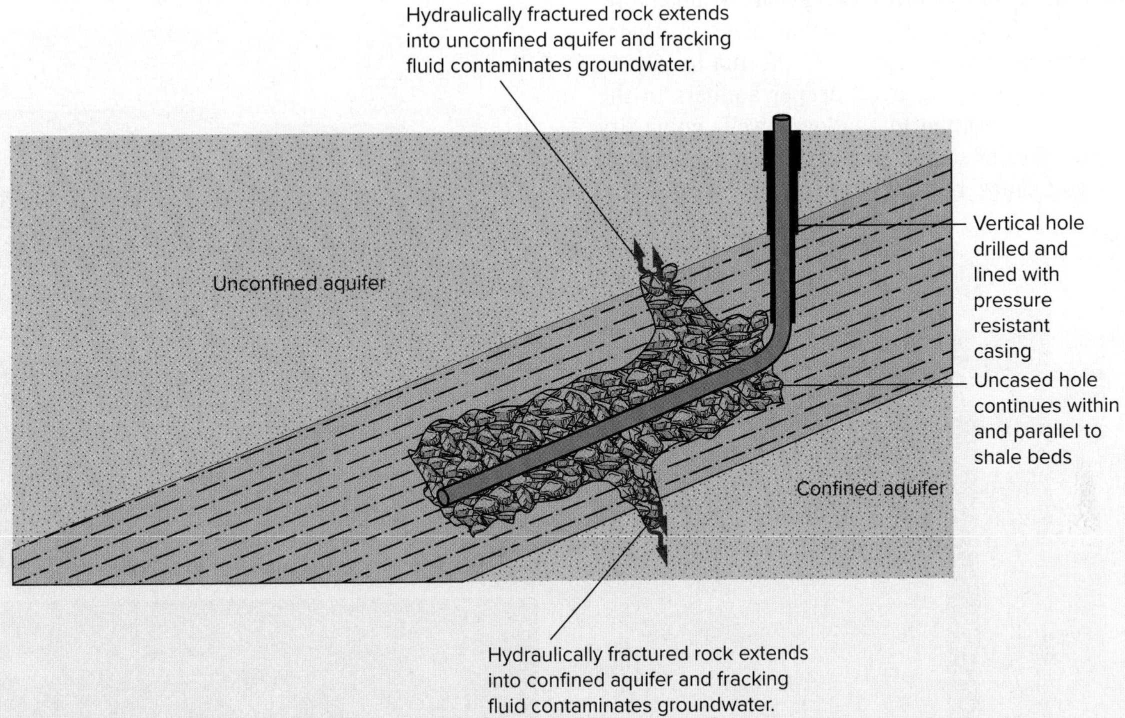

BOX 11.3 ■ FIGURE 1

An inclined bed of shale containing natural gas or petroleum is hydraulically fractured. Sandstone aquifers are polluted by fracking fluid where fracking extends into the sandstone or where the casing in the drill hole leaks. Once in the sandstone, the fracking fluids flow in unfracked rock because the sandstone is permeable. If the fracking had been confined to the shale, pollution of groundwater would not have occurred.

Additional Resources

- http://www.sciencemagazinedigital.org/sciencemagazine/27_june_2014?pg=44#pg44

The June 27, 2014, issue of *Science* contains a special section about the surge in production of natural gas, particularly in the United States. The surge is largely due to natural gas extracted from fracked shale.

- http://water.usgs.gov/owq/topics/hydraulic-fracturing/

General information about hydraulic fracturing and information where fracking is occurring in the United States.

- http://www.sciencenewsline.com/news/2016111602070054.html

New study links groundwater changes to fracking.

B. Yan and others. 2017. Association of groundwater constituents with topography in NE Pennsylvania, *Science of the Total Environment*, 577: 195–201.

- http://www.sciencedirect.com/science/article/pii/S0048969716323488

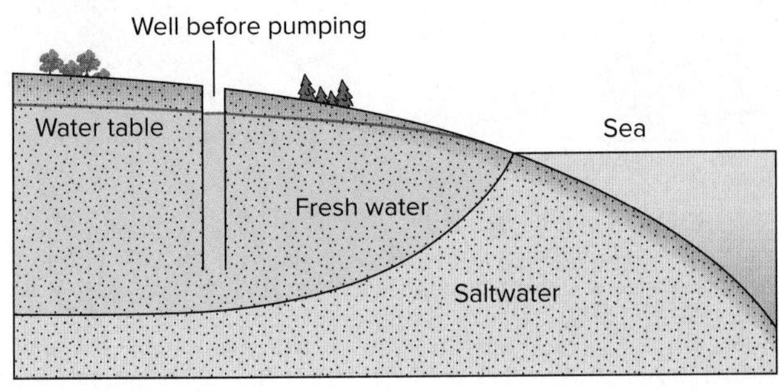

C

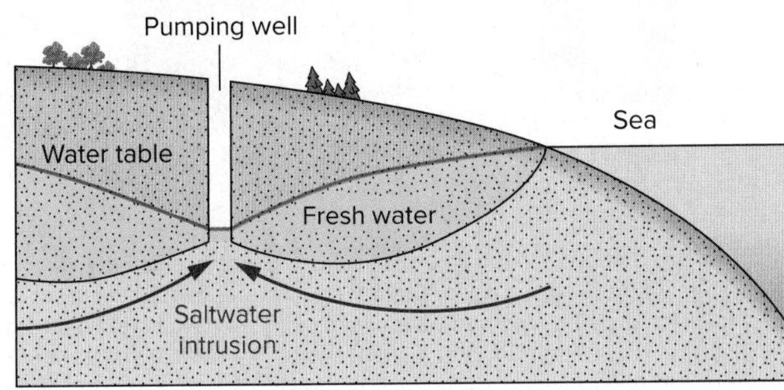

D

FIGURE 11.18 *(continued)*

the surface. Moreover, as water is withdrawn, the ground surface may settle because the water no longer supports the rock and sediment. Mexico City has subsided more than 13 meters, and portions of California's Central Valley sank 9 meters between 1925 and 1977 because of extraction of groundwater (figure 11.19*A*). The extreme drought in California has led to increased groundwater pumping from deeper aquifers in the Central Valley that has resulted in shallower wells going dry and unprecedented rates of subsidence (figure 11.19*B* and *C*). Such *subsidence* can crack building and bridge foundations,

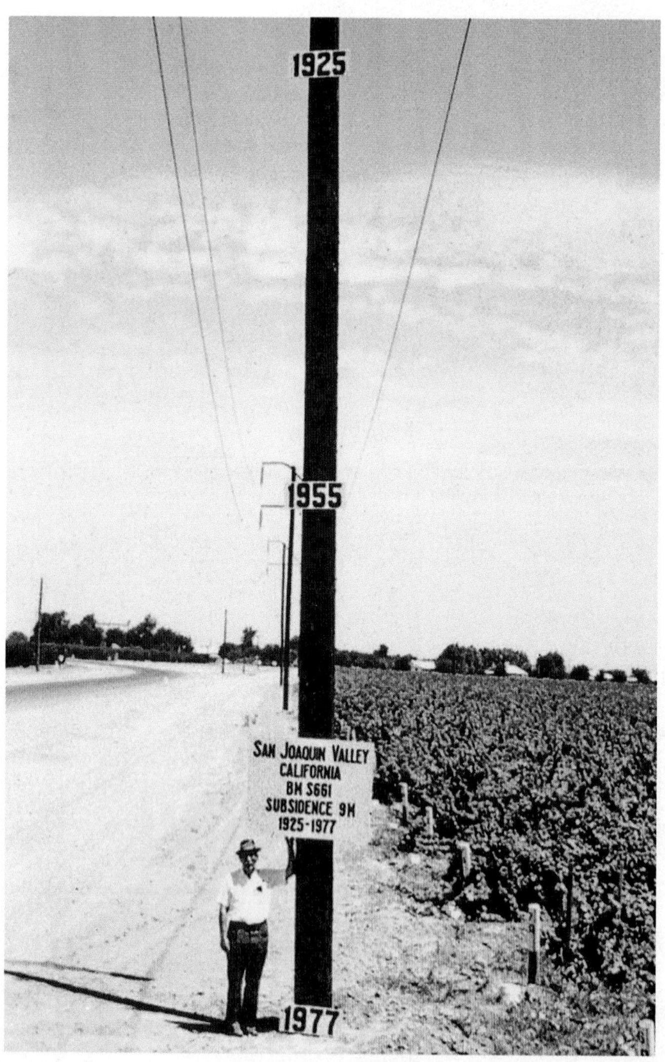

A

B

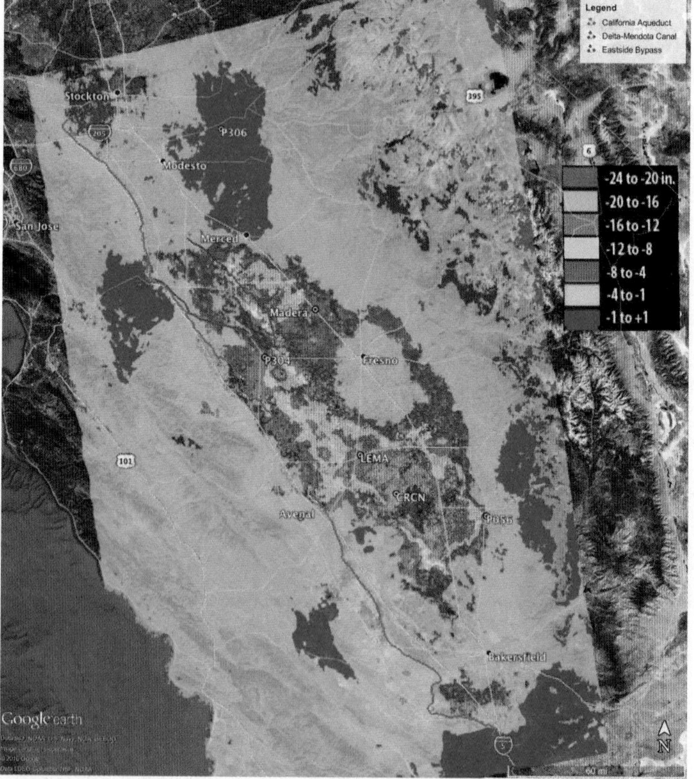

C

FIGURE 11.19

(*A*) Photo of legendary hydrogeologist Joseph Poland of the U.S. Geological Survey, who discovered that the land surface near Mendota, California, in the San Joaquin Valley sank 9 meters in 52 years due to the rapid withdrawal of groundwater to irrigate crops. To illustrate the tremendous amount of subsidence that had occurred, Poland used signs on a utility pole to indicate where a person would have been standing on the land surface in 1925, 1955, and 1977. In the late 1970s, subsidence decreased to less than a meter due to reduced groundwater pumping and increased use of surface water for irrigation. (*B*) The most recent drought in California led to an increased dependence on groundwater for irrigation and renewed subsidence of the land surface. To bring attention to the subsidence problem, U.S. Geological Survey hydrogeologist Michelle Rummler Sneed reenacted Poland's photo and posted signs on a surveying rod to show where the base of the bridge would have been located in 1965 and in 2016. (*C*) Satellite image shows unprecedented rates of subsidence from May 2014 to January 2015 in the Central Valley due to the overpumping of aquifers for irrigation. The ground surface sank over a foot in places in less than a year, which is the fastest rate of subsidence ever measured in the world.
(A) Source: Richard O. Ireland U.S. Geological Survey; (B) Source: U.S. Geological Survey; (C) Source: Produced from ESA remote sensing data. Tom Farr/ESA/NASA

ENVIRONMENTAL GEOLOGY 11.4

Sinkholes as a Geologic Hazard

In Winter Park, Florida, a suburb of Orlando (home of Disney World), on the evening of May 8, 1981, a sinkhole began forming. It continued to expand throughout the following day. A three-bedroom house was an early casualty of the growing pit. By the end of the second day, expansion stopped and the sinkhole was as wide as a football field and 25 meters deep. It displaced approximately 190,000 cubic meters of dirt and rock, and it is the largest sinkhole ever to be documented throughout its development and reported on nightly by national television.

Within a twenty-four-hour period, the sinkhole had swallowed the house, part of the municipal swimming pool, a portion of a street, and a pickup truck with a camper shell (box figure 1). The Winter Park sinkhole also swallowed five Porsches from an automobile agency that was also damaged by the sinkhole. Three of the Porsches were eventually recovered by a crane, but the other two were deeply buried and never located. The estimated cost of the damage from the now famous sinkhole was around $4 million. To mitigate the geologic hazard, the city filled the bottom of the sinkhole with dirt and concrete, and it is now a lake named Lake Rose for the woman who lost her house to the sinkhole.

Thirty-two years later, in June 2013, another new, smaller sinkhole developed in Winter Park. A woman lost her house and swimming pool to that sinkhole. Nearby in Clermont, Florida, a three-story villa was damaged in the early hours of August 12, 2013, when it partially collapsed into a sinkhole (box figure 2). Earlier in 2013, a sinkhole opening near Tampa, Florida, resulted in a rarely occurring human casualty. This sinkhole developed beneath a man's bedroom in the middle of the night. He was engulfed by debris, and rescue efforts were futile. His body was not recovered. Record drought conditions likely played a role in the 2013 sinkhole collapses.

Nearly a decade before the big sinkhole of Winter Park developed, a Florida agriculture agent wrote a letter to Orange County mayors warning them of the increased dangers of sinkhole activity due to overdevelopment and excessive use of groundwater. He pointed out that Winter Park was an especially high-risk area. In 1981, when the big sinkhole developed, the aquifers in central Florida had record low water table levels. When the groundwater is lower, it can no longer support the weight of the overlying rock, and collapse occurs.

Records of insurance claims filed for sinkhole damage in Florida show that in 2006, there were 2,300 claims filed. The numbers rose during the following years, and in 2010, there were 6,700. Insurance companies paid out $1.4 billion in claims during this five-year period. At first glance, it would seem that the rate of sinkhole activity is increasing with time in Florida. More likely, it is attributable to Florida's high population growth rate. As more and more rural land is taken over by urban development, the risk of sinkholes damaging buildings will increase.

BOX 11.4 ■ FIGURE 1

The sinkhole in Winter Park, Florida, that formed on May 8 and 9, 1981. The destroyed municipal swimming pool is at top left of photo; cut-off street is upper right. The pickup truck with camper is within the lower right sloping part of the hole. The surface of the water in the central part of the hole is at the water table level of the adjoining rocks.
Source: Rick Duerling, courtesy of Florida Geological Survey

BOX 11.4 ■ FIGURE 2

Partial collapse of a three-story villa in Clermont, Florida, when a 20-meter sinkhole opened up in the early morning hours of August 12, 2013.

roads, canals, and pipelines, and cause millions of dollars in damage. Overpumping of groundwater also causes compaction and porosity loss in rock and soil, and can permanently ruin good aquifers.

To avoid the problems of falling water tables, subsidence, and compaction, many towns use *artificial recharge* to increase the natural rate of recharge. Natural floodwaters or treated industrial or domestic wastewaters are stored in infiltration ponds on the surface to increase the rate of water percolation into the ground. Reclaimed, clean water from sewage treatment plants is commonly used for this purpose. In some cases, especially in areas where groundwater is under confined conditions, water is actively pumped down into the ground to replenish the groundwater supply. This is more expensive than filling surface ponds, but it reduces the amount of water lost through evaporation.

11.8 GEOLOGIC EFFECTS OF GROUNDWATER

In areas underlain by bedrock that is soluble, groundwater can easily dissolve rock. Geologic features formed by the dissolving ability of groundwater include caves and regions of karst topography. Groundwater can preserve fossils such as petrified wood, and it also carries the silica or calcite cement that lithifies sedimentary rock and precipitates concretions and geodes.

Caves

Caves (or **caverns**) are naturally formed, underground chambers. Most caves develop when slightly acidic groundwater dissolves limestone along joints and bedding planes, opening up cavern systems as calcite is carried away in solution (figure 11.20). Natural groundwater is commonly slightly acidic because of dissolved carbon dioxide (CO_2) from the atmosphere or from soil gases (see chapter 5).

Geologists disagree whether limestone caves form above, below, or at the water table. Most caves probably are formed by groundwater circulating below the water table, as shown in figure 11.20. If the water table drops or the land is elevated above the water table, the cave may begin to fill in again by calcite precipitation. The following equation can be read from left to right for calcite solution and from right to left for the calcite precipitation reaction (see also table 5.1).

Groundwater with a high concentration of calcium (Ca^{2+}) and bicarbonate (HCO_3^-) ions may drip slowly from the ceiling of an air-filled cave. As a water drop hangs on the ceiling of the cave, some of the dissolved carbon dioxide (CO_2) may be lost into the cave's atmosphere. The CO_2 loss causes a small amount of calcite to precipitate out of the water onto the cave ceiling. When the water drop falls to the cave floor, the impact may

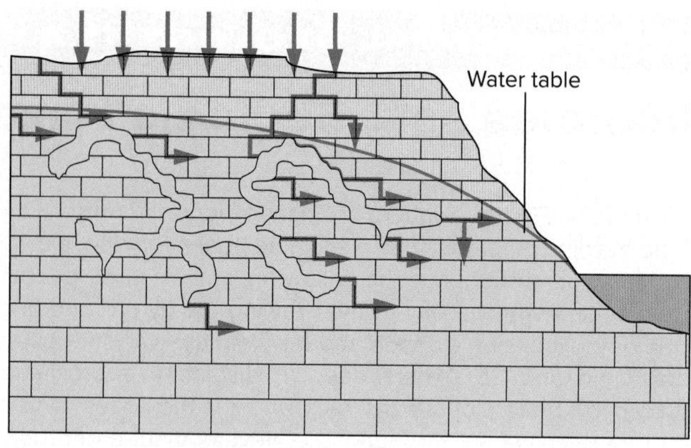

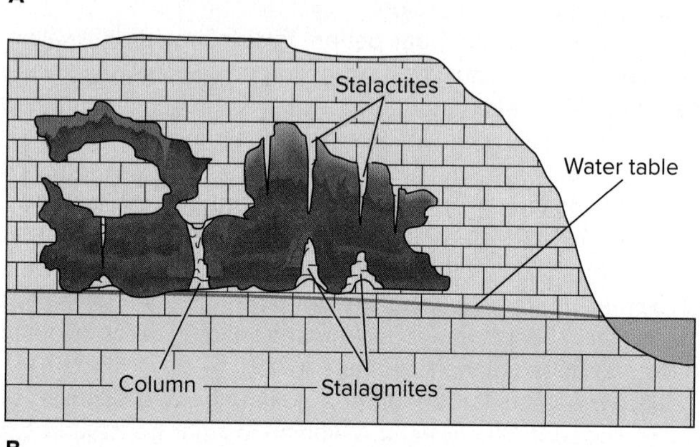

FIGURE 11.20

Solution of limestone to form caves. (*A*) Water moves along fractures and bedding planes in limestone, dissolving it to form caves below the water table. (*B*) Falling water table allows cave system, now greatly enlarged, to fill with air. Calcite precipitation forms stalactites, stalagmites, and columns above the water table.

cause more CO_2 loss, and another small amount of calcite may precipitate on the cave floor. A falling water drop, therefore, can precipitate small amounts of calcite on both the cave ceiling and the cave floor, and each subsequent drop adds more calcite to the first deposits.

$$\underset{\text{water}}{H_2O} + \underset{\text{carbon dioxide}}{CO_2} + \underset{\text{calcite in limestone}}{CaCO_3} \rightleftarrows \underset{\text{calcium ion}}{Ca^{2+}} + \underset{\text{bicarbonate ion}}{2HCO_3^-}$$

→ development of caves (solution)

← development of flowstone and dripstone (precipitation)

Deposits of calcite (and, rarely, other minerals) built up in caves by dripping water are called *dripstone* or **speleothems**. **Stalactites** are iciclelike pendants of dripstone hanging from cave ceilings (figure 11.20*B*). They are generally slender and are commonly aligned along cracks in the ceiling, which act as conduits for groundwater. **Stalagmites** are cone-shaped masses of dripstone formed on cave floors, generally directly below stalactites. Splashing water precipitates calcite over a large area on the cave floor, so stalagmites are usually thicker than the stalactites above them. As a stalactite grows downward and a stalagmite grows upward, they may eventually join to form a *column* (figure 11.20*B*). Figure 11.21 shows some of the intriguing features formed in caves.

In parts of some caves, water flows in a thin film over the cave surfaces rather than dripping from the ceiling. Sheetlike or ribbonlike *flowstone* deposits develop from calcite that is precipitated by flowing water on cave walls and floors.

The floors of most caves are covered with sediment, some of which is *residual clay,* the fine-grained particles left behind as insoluble residue when a limestone containing clay dissolves. (Some limestone contains only about 50% calcite.) Other sediment, including most of the coarse-grained material found on cave floors, may be carried into the cave by streams, particularly when surface water drains into a cave system from openings on the land surface.

Karst Topography

In areas where groundwater dissolves carbonate rock, a distinctive landscape may develop that is called **karst topography** (figure 11.22). Named for the Kras Plateau of western Slovenia and eastern Italy, areas of karst have a characteristic rocky surface marked by caverns and unique landforms. In the United States, areas of karst topography occur in humid areas that are underlain by limestone and include much of Florida, Missouri, Indiana, and Kentucky.

One of the most common landforms associated with karst topography is caused by the solution of soluble rock underground. Extensive cavern systems can undermine a region so that roofs collapse and form depressions. **Sinkholes** are closed depressions found on land surfaces typically underlain by limestone (figure 11.23). They form either by the *collapse* of a cave roof or by *solution* as descending water enlarges a crack in limestone. Sinkholes that collapse catastrophically, swallowing buildings and people, present a serious geologic hazard in areas of karst topography (see box 11.4). Sinkholes can also form in regions underlain by gypsum or rock salt, which is also soluble in water. The dramatic draining of Bayou Corne that sucked 30-meter-tall trees underwater and vented methane gas to the surface occurred in Louisiana on August 3, 2013, when the mining of a salt dome collapsed the sinkhole (view the video at: http://www.youtube.com/watch?v=a7cOSzEKvrQ).

Karst areas typically have a lack of surface streams, although one major river may flow at a level lower than the karst area. Streams sometimes disappear down sinkholes to flow through caves beneath the surface. In this specialized instance, a true *underground stream* exists. Such streams are quite rare, however, as most groundwater flows very slowly through pores and cracks in sediment or rock. You may hear people with wells describe the "underground stream" that their well penetrates, but this is almost never the case. Wells tap groundwater in the rock pores and crevices, not underground streams. If a well did tap a true underground river in a karst region, the water would probably be too polluted to drink, especially if it had washed down from the surface into a cavern without being filtered through soil or rock.

Some areas of karst topography look very different from the sinkhole-covered surfaces shown in figures 11.22 and 11.23. *Tower karst* is a spectacular type of karst topography that forms in warm, tropical climates where accelerated weathering produces steep or vertical-sided limestone towers. The best-known area of tower karst is in southern China, where picturesque 200-meter towers are exposed along the Li River valley (figure 11.24). Other areas of tower karst include Thailand, Viet Nam, and Malaysia.

Other Effects of Groundwater

Groundwater is important in the preservation of *fossils* such as **petrified wood**, which develops when porous buried wood is either filled in or replaced by inorganic silica carried in by groundwater (figure 11.25). The result is a hard, permanent rock, commonly preserving the growth rings and other details of the wood. Calcite or silica carried by groundwater can also replace the original material in marine shells and animal bones.

Sedimentary rock *cement*, usually silica or calcite, is carried into place by groundwater. When a considerable amount of cementing material precipitates locally in a rock, a hard, rounded mass called a **concretion** develops, typically around an organic nucleus such as a leaf, tooth, or other fossil (figure 11.26).

Geodes are partly hollow, globe-shaped bodies found in some limestones and locally in other rocks. The outer shell is amorphous silica, and well-formed crystals of quartz, calcite, or other minerals project inward toward a central cavity (figure 11.27). The origin of geodes is complex but clearly related to groundwater. Crystals in geodes may have filled original cavities or may have replaced fossils or other crystals.

In arid and semiarid climates, *alkali soil* may develop because of the precipitation of great quantities of sodium salts by evaporating groundwater. Such soil is generally unfit for plant growth. Alkali soil generally forms at the ground surface in low-lying areas (see chapter 5).

11.9 HOT WATER UNDERGROUND

Hot springs are springs in which the water is warmer than human body temperature. Water can gain heat in two ways while it is underground. First, and more commonly, groundwater may

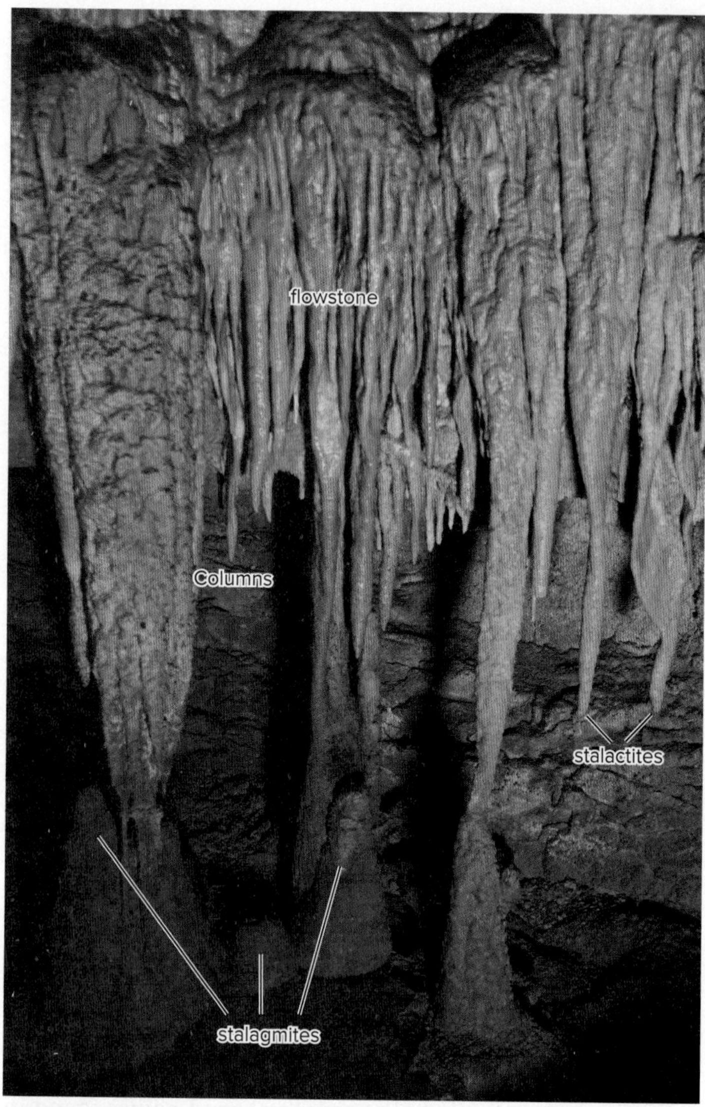

FIGURE 11.21

Stalactites cling to the ceiling and stalagmites grow from the ground in Mammoth Cave National Park, Kentucky. Columns form where stalactites and stalagmites join, and flowstone forms on the walls of the cavern.
©Nancy Nehring/Photodisc/Getty Images

FIGURE 11.22

Sinkholes formed in limestone near Timaru, New Zealand.
©G. R. Dick Roberts/Natural Sciences Image Library

circulate near a magma chamber or a body of cooling igneous rock. In the United States, most hot springs are found in the western states, where they are associated with relatively recent volcanism. The hot springs and pools of Yellowstone National Park in Wyoming are of this type.

Groundwater can also gain heat if it circulates unusually deeply in the Earth, perhaps along joints or faults. As discussed in chapter 3, the normal geothermal gradient (the increase in temperature with depth) is 25°C/kilometer (about 75°F/mile). Water circulating to a depth of 2 or 3 kilometers is warmed substantially above normal surface water temperature. The famous springs at Warm Springs, Georgia; Hot Springs, Arkansas; and Sulfur Springs, West Virginia have all been warmed by deep circulation. Warm water, regardless of its origin, is lighter than cold water and readily rises to the surface.

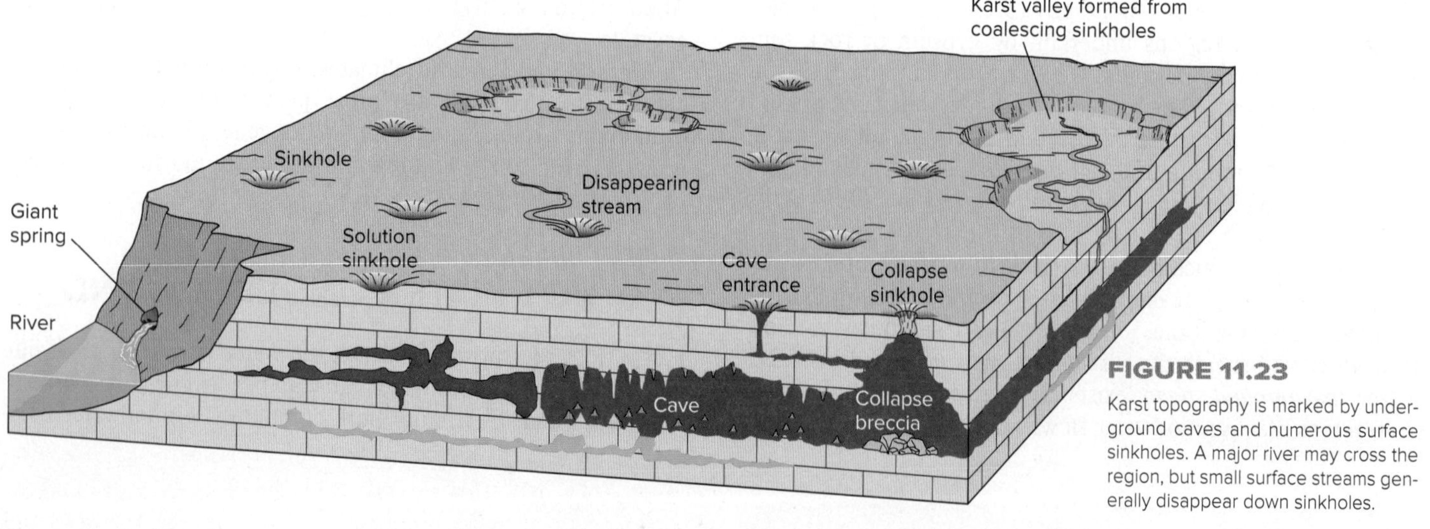

FIGURE 11.23

Karst topography is marked by underground caves and numerous surface sinkholes. A major river may cross the region, but small surface streams generally disappear down sinkholes.

FIGURE 11.24

Karst towers in China's Li River Valley. As indicated by the haze, this is a warm, humid climate. The towers are erosional remnants of pillars that separated caverns. The roofs of caverns collapsed and heavy, monsoon rainfall washed away the debris, leaving the present valleys between the pillars.
©C. C. Plummer

FIGURE 11.25

Petrified log in the Painted Desert, Arizona. The log was replaced by silica carried in solution by groundwater. Small amounts of iron and other elements color the silica in the log.
©Eric & David Hosking/Corbis/Getty Images

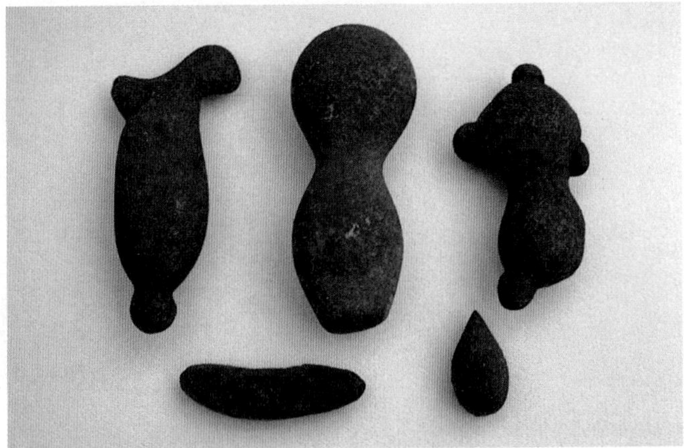

FIGURE 11.26

Concretions that have weathered out of shale. Concretions contain more cement than the surrounding rock and therefore are very resistant to weathering.
©David McGeary

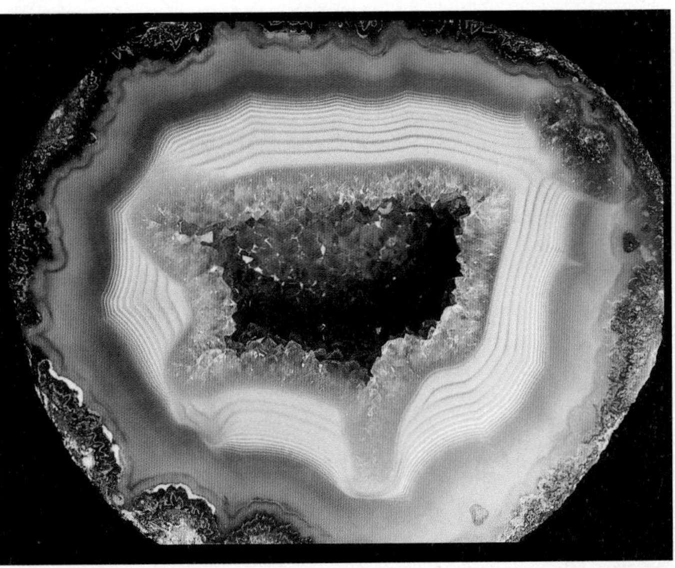

FIGURE 11.27

Concentric layers of amorphous silica are lined with well-formed amethyst (quartz) crystals growing inward toward a central cavity in a geode.
©Martin Land/Bruce Coleman, Inc./Avalon

A **geyser** is a type of hot spring that periodically erupts hot water and steam. The water is generally near boiling temperature (100°C). Eruptions may be caused by a constriction in the underground "plumbing" of a geyser, which prevents the water from rising and cooling. The events thought to lead to a geyser eruption are illustrated in figure 11.28. Water gradually seeps into a partially emptied geyser chamber, and heat supplied from below slowly warms the water. Bubbles of water vapor and other gases then begin to form as the temperature of the water rises. The bubbles may clog the constricted part of the chamber until the upward pressure of the bubbles pushes out some of the water above in a gentle surge, thus lowering the pressure on the water in the lower part of the chamber. This drop in pressure causes the chamber water, now very hot, to flash into vapor. The expanding vapor blasts upward out of the chamber, driving hot water with it and condensing into visible steam. The chamber, now nearly empty, begins to fill again, and the cycle is repeated. The entire cycle may be quite regular, as it is in Yellowstone's Old Faithful geyser, which averages about 79 minutes between eruptions (though it varies from about 45 to 105 minutes, depending on the amount of water left in the chamber after an eruption). Many geysers, however, erupt irregularly, some with weeks or months between eruptions.

As hot groundwater comes to the surface and cools, it may precipitate some of its dissolved ions as minerals. *Travertine* is a deposit of *calcite* that often forms around hot springs (figure 11.29), while dissolved *silica* precipitates as *sinter* (called *geyserite* when deposited by a geyser, as shown in figure 11.30). The composition of the subsurface rocks generally determines which type of deposit forms, although sinter can indicate higher subsurface temperatures than travertine because silica is harder to dissolve than calcite. Both deposits can be stained by the pigments of bacteria that thrive in the

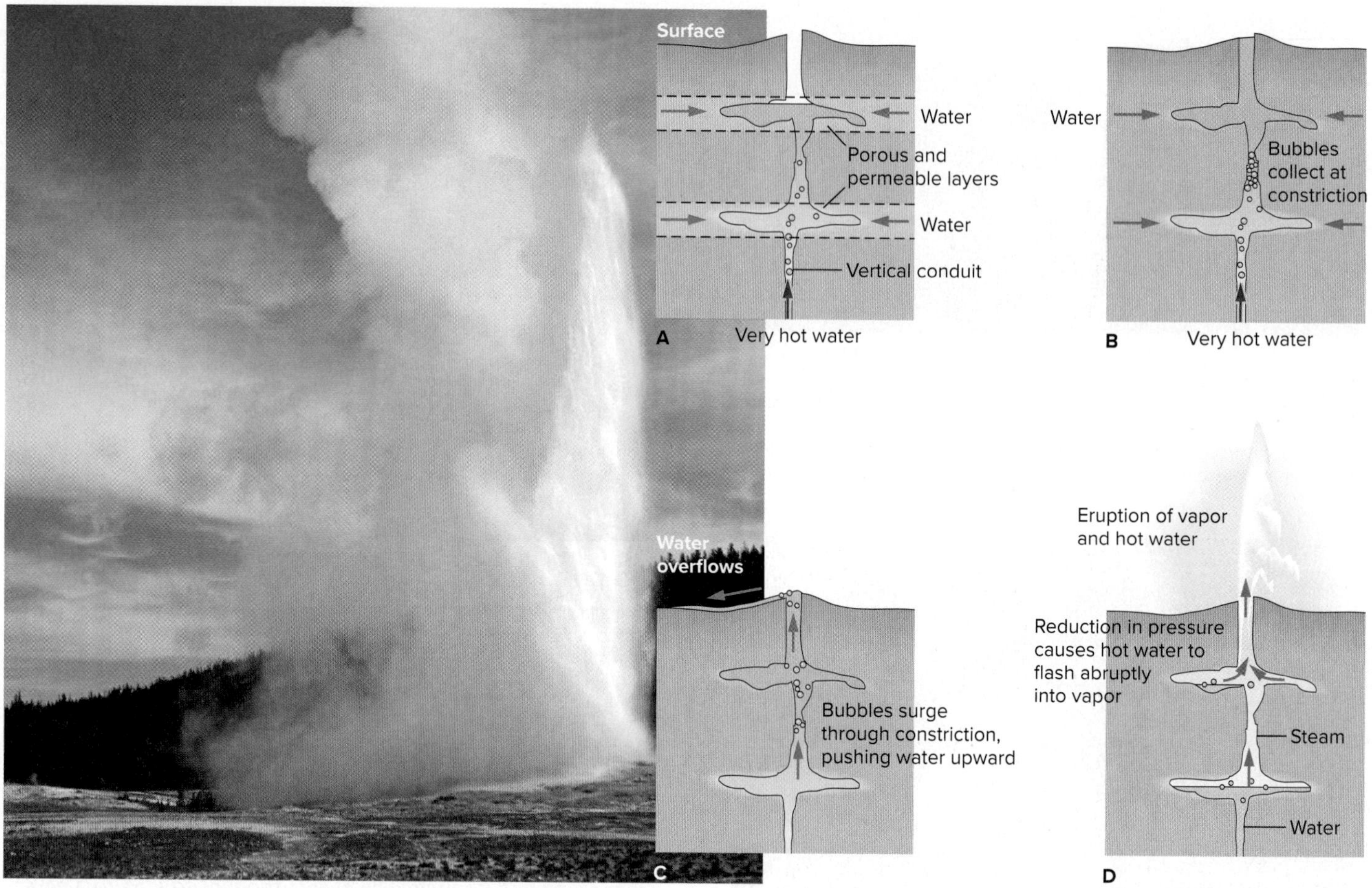

FIGURE 11.28

Eruptive history of a typical geyser in (*A*) through (*D*). Photo shows the eruption of Old Faithful geyser in Yellowstone National Park, Wyoming. See text for explanation. ©*Jeff Vanuga/Corbis RF*

hot water. These thermophilic bacteria are some of the most primitive of living bacteria in the biosphere and suggest that life may have arisen near hot springs.

A *mudpot* is a special type of hot spring that contains thick, boiling mud. Mudpots are usually marked by a small amount of water and strongly sulfurous gases, which combine to form strongly acidic solutions. The mud probably results from intense chemical weathering of the surrounding rocks by these strong acids (see figure 5.14).

Geothermal Energy

Electricity can be generated by harnessing naturally occurring steam and hot water in areas that are exceptionally hot underground. In such a *geothermal area*, wells can tap steam (or superheated water that can be turned into steam) that is then piped to a powerhouse, where it turns a turbine that spins a generator, creating electricity.

Geothermal energy production requires no burning of fuel, so the carbon dioxide emissions of power plants that burn coal, oil, or natural gas are not produced. Although geothermal

FIGURE 11.29

Precipitation of calcite in the form of travertine around a hot spring (Mammoth Hot Springs, Yellowstone National Park). Thermophilic bacteria living in the hot water provide the color. ©*Diane Carlson*

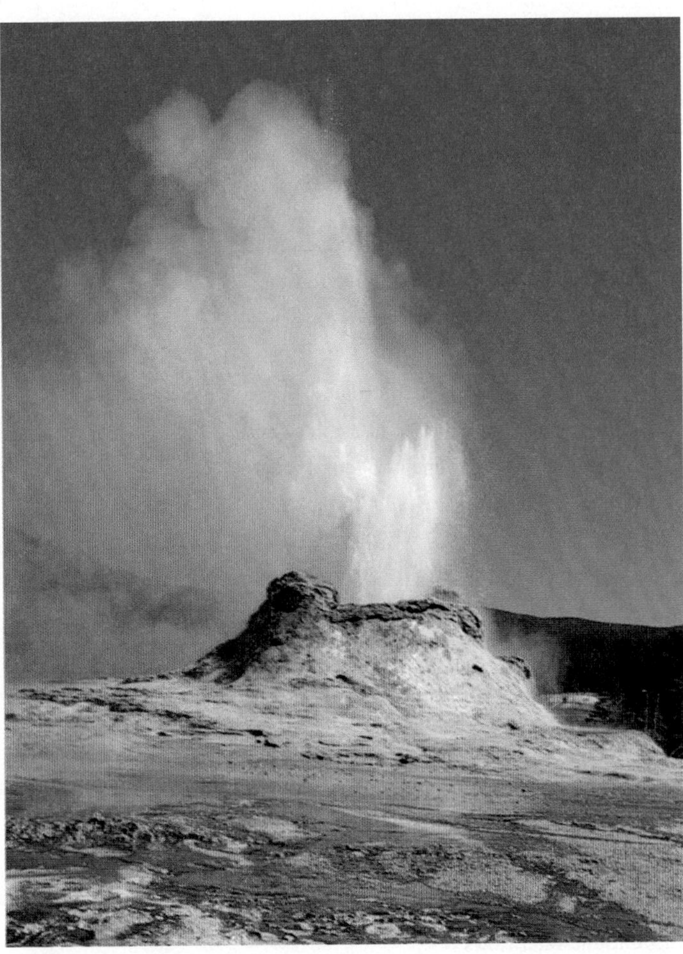

FIGURE 11.30

Geyserite deposits around the vent of Castle geyser, Yellowstone National Park. ©David McGeary

energy is relatively clean, it has some environmental problems. Workers need protection from toxic hydrogen sulfide gas in the steam, and the hot water commonly contains dissolved ions and metals, such as lead and mercury, that can kill fish and plants if discharged on the surface. Geothermal fluids are often highly corrosive to equipment, and their extraction can cause land subsidence. Pumping the cooled wastewater underground can help reduce subsidence problems.

Geothermal fields can be depleted. The largest field in the world is at The Geysers in California (figure 11.31), 120 kilometers north of San Francisco. The Geysers field decreased its capacity in recent years to just under 750 megawatts of electricity (enough for 750,000 people) because the field began running out of steam. As production declined, innovative solutions such as injecting wastewater from nearby communities were used to increase the steam capacity.

Nonelectric uses of geothermal energy include space heating (in Boise, Idaho; Klamath Falls, Oregon; and Reykjavik, the capital of Iceland), as well as paper manufacturing, ore processing, and food preparation.

FIGURE 11.31

Geothermal power plant at The Geysers, in northern California. Underground steam is piped to the power plant and is discharged from the cooling towers. *Source: Julie Donnelly-Nolan, U.S. Geological Survey*

Summary

About 15% of the water that falls on land percolates underground to become groundwater. Groundwater fills pores and joints in rock, creating a large reservoir of usable water in most regions.

Porous rocks can hold water. *Permeable* rocks permit water to move through them.

The *water table* is the top surface of the *saturated zone* and is overlain by the *unsaturated zone.*

Local variations in rock permeability may develop a *perched water table* above the main water table.

Groundwater velocity depends on rock permeability and the slope of the water table.

An *aquifer* is porous and permeable, and can supply water to wells. A *confined aquifer* holds water under pressure, which can create *artesian wells.*

Gaining streams, springs, and lakes form where the water table intersects the land surface. *Losing streams* contribute to the groundwater in dry regions.

Groundwater can be contaminated by city landfills, agriculture, industry, or sewage disposal. Some pollutants can be filtered out by passage of the water through moderately permeable geologic materials.

A pumped well causes a *cone of depression* that in turn can cause or aggravate groundwater pollution. Near a coast, it can cause *saltwater intrusion.*

Artificial *recharge* can help create a balance between withdrawal and recharge of groundwater supplies and help prevent subsidence.

Solution of limestone by groundwater forms *caves, sinkholes,* and *karst topography.* Calcite precipitating out of groundwater forms *speleothems* such as *stalactites* and *stalagmites* in caves.

Precipitation of material out of solution by groundwater helps form petrified wood, other fossils, sedimentary rock cement, concretions, geodes, and alkali soils.

Geysers and *hot springs* occur in regions of hot groundwater. *Geothermal energy* can be tapped to generate electricity.

Terms to Remember

Testing Your Knowledge

Use the following questions to prepare for exams based on this chapter.

1. What is the water table? Is it fixed in position?
2. What causes a perched water table?
3. Discuss the difference between porosity and permeability.
4. Name several geologic materials that make good aquifers. Define *aquifer.*
5. How does a confined aquifer differ from an unconfined aquifer?
6. What controls the velocity of groundwater flow?
7. What happens to the water table near a pumped well?
8. What conditions are necessary for an artesian well?
9. Describe several ways in which groundwater can become contaminated.
10. What chemical conditions are necessary for caves to develop in limestone? For stalactites to develop in a cave?
11. What is karst topography? How does it form?
12. Sketch four different origins for springs.
13. How does petrified wood form?
14. What distinguishes a geyser from a hot spring? Why does a geyser erupt?
15. The subsurface zone in which all rock openings are filled with water is called the
    a. saturated zone. b. water table.
    c. unsaturated zone. d. aquiclude.
16. Porosity is
    a. the percentage of a rock's volume that is openings.
    b. the capacity of a rock to transmit a fluid.
    c. the ability of a sediment to retard water.
    d. none of the preceding.
17. Permeability is
    a. the percentage of a rock's volume that is openings.
    b. the capacity of a rock to transmit a fluid.
    c. the ability of a sediment to retard water.
    d. none of the preceding.
18. An aquifer is
    a. a body of saturated rock or sediment through which water can move easily.
    b. a body of rock that retards the flow of groundwater.
    c. a body of rock that is impermeable.
19. Which rock type would make the best aquifer?
    a. shale b. mudstone
    c. sandstone d. all of the preceding
20. Which of the following determines how quickly groundwater flows?
    a. elevation b. water pressure
    c. permeability d. all of the preceding
21. Groundwater flows
    a. always downhill.
    b. from areas of high hydraulic head to low hydraulic head.
    c. from high elevation to low elevation.
    d. from high permeability to low permeability.
22. The drop in the water table around a pumped well is the
    a. drawdown. b. hydraulic head.
    c. porosity. d. fluid potential.

Expanding Your Knowledge

1. Describe any difference between the amounts of water that would percolate downward to the saturated zone beneath a flat meadow in northern New York and a rocky hillside in southern Nevada. Discuss the factors that control the amount of percolation in each case.
2. Where should high-level nuclear waste from power plants be stored? If your state or community uses nuclear power, where is your local waste stored?
3. Should all contaminated groundwater be cleaned up? How much money has been set aside by the federal government for cleaning

polluted groundwater? Who should pay for groundwater cleanup if the company that polluted the water no longer exists? Should some aquifers be deliberately left contaminated if there is no current use of the water or if future use could be banned?

4. Why are most of North America's hot springs and geysers in the western states and provinces?

Exploring Web Resources

http://toxics.usgs.gov/toxics/

Various sites and information about cleanup of toxics in surface and groundwater.

http://pubs.usgs.gov/ha/ha730/gwa.html

Groundwater Atlas for the United States. Good general information about aquifers.

http://water.usgs.gov/

Good general website that has a lot of links to water topics in the United States from the U.S. Geological Survey.

http://cavern.com/

Web page of the *National Caves Association* contains photos and information about caves and also a directory of caves by state and zip code.

www.caves.org/

Home page of the *National Speleological Society* contains links to web pages of local interest and to the society's bookstore.

CHAPTER 12

Glaciers and Glaciation

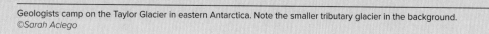

Geologists camp on the Taylor Glacier in eastern Antarctica. Note the smaller tributary glacier in the background.
©Sarah Aciego

LEARNING OBJECTIVES

- Distinguish between alpine glaciation and continental glaciation in terms of types of glaciers, movement of ice, and global distribution.
- Describe the erosional processes associated with glaciation.
- Identify glacial erosional features and explain how they form.
- Identify glacial depositional features and explain how they form.
- Describe the recent glacial history of North America and how it affected the landscape.

Most people are surprised to learn that the majority of the world's fresh water (approximately 70%) is held in glaciers. In chapters 9, 10, and 11, you have seen how the surface of the land is shaped by mass wasting, running water, and, to some extent, groundwater. Running water is regarded as the erosional agent most responsible for shaping the Earth's land surface. Where glaciers exist, however, they are far more effective agents of erosion, transportation, and deposition. Geologic features characteristic of glaciation are distinctly different from the features formed by running water. Once recognized, they lead one to appreciate the great extent of glaciation during the recent geologic past (that age popularly known as the Ice Age).

The spectacularly scenic areas in many North American national parks owe much of their beauty to glacial action. Yosemite Valley in California might have been another nondescript valley if glaciers had not carved it into its present shape (figure 12.1). Unlike stream-carved valleys, Yosemite is straight for long stretches. Its sides are steep, and the valley floor is flat (it is U-shaped rather than the characteristic V-shape of a stream-carved valley). The sediment beneath the vegetation in the valley floor is poorly sorted debris, unlike the sorted sediment deposited by a stream. All of these things are evidence that the Yosemite landscape has been carved by a glacier. But there is no glacier in Yosemite Valley today. Yosemite indicates, as does overwhelming evidence elsewhere in the world, that glaciation was more extensive in the geologically recent past.

Immense and extensive glaciers, covering as much as a third of Earth's land surface, had a profound effect on the landscape and our present civilization. Moreover, worldwide climatic changes during the Ice Age distinctively altered landscapes in areas far from the glacial boundaries. These episodes of glaciation took place within only the last two and a half million years, ending about 10,000 years ago. Preserved in the rock record, however, is evidence of extensive older glaciations approximately 300 million years ago. The record of this late Paleozoic glacial age is used as evidence for continental drift, as described in chapter 19 on plate tectonics.

Our lives and environment today have been profoundly influenced by the effects of past glaciation. For example, much of the fertile soil of the northern Great Plains of the United States developed on the loose debris transported and deposited by glaciers that moved southward from northern Canada during the last Ice Age. The thick blankets of sediment left in the Midwest store vast amounts of groundwater. The Great Lakes and the thousands of lakes in Minnesota and neighboring states and provinces are the products of past glaciation.

Before we can understand how a continental glacier was responsible for much of the soil in the Midwest or how a glacier confined to a valley could carve a Yosemite, we must learn something about present-day glaciers.

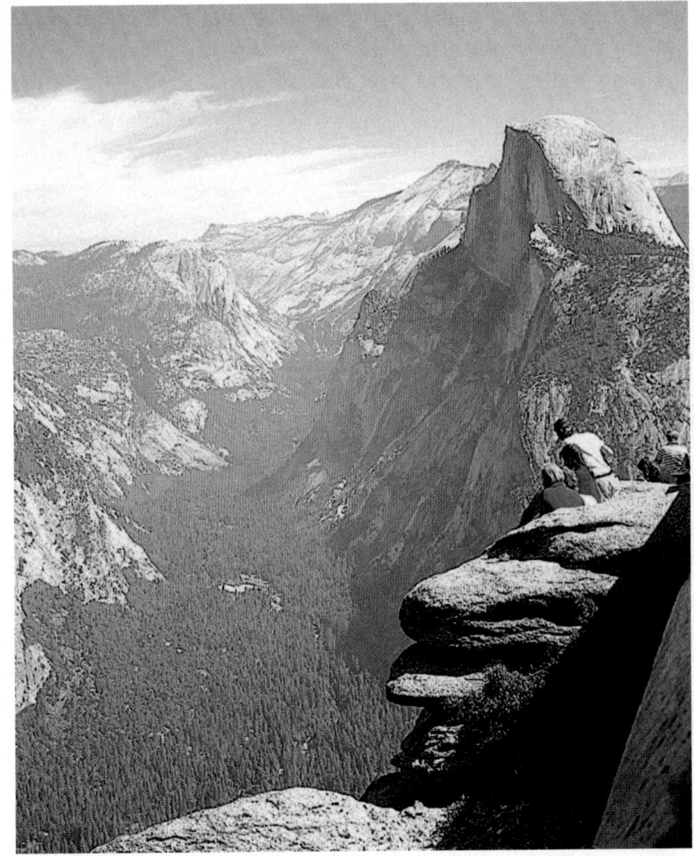

FIGURE 12.1

Yosemite Valley, as seen from Glacier Point, Yosemite National Park, California. Its U-shaped cross profile is typical of glacially carved valleys.
©C. C. Plummer

12.1 WHAT IS A GLACIER?

A **glacier** is a large, long-lasting mass of ice, formed on land, that moves under its own weight. It develops as snow is compacted and recrystallized. Glaciers can develop any place where, over a period of years, more snow accumulates than melts away or is otherwise lost.

There are two types of *glaciated* terrain on the Earth's surface. **Alpine glaciation** is found in mountainous regions, while **continental glaciation** exists where a large part of a continent (thousands of square kilometers) is covered by glacial ice. In both cases, the moving masses of ice profoundly and distinctively change the landscape.

Types of Glaciers

A simple criterion—whether or not a glacier is restricted to a valley—is the basis for classifying glaciers by form. A **valley glacier** or *alpine glacier* is a glacier that is confined to a valley and flows from a higher to a lower elevation. Like streams, small valley glaciers may be tributaries to a larger trunk system. Valley glaciers are prevalent in areas of alpine glaciation. As might be expected, most glaciers in the United States and Canada, being in mountains, are of the valley type (figure 12.2).

In contrast, an **ice sheet** is a mass of ice that is not restricted to a valley but covers a large area of land (over 50,000 square kilometers). Ice sheets are associated with continental glaciation. Only two places on Earth now have ice sheets: Greenland and Antarctica. A similar but smaller body is called an **ice cap**. Ice caps (as well as valley glaciers) are found in a few mountain highlands in Iceland and on islands in the Arctic Ocean, off Canada, Russia, and Scandinavia. An ice cap or ice sheet flows downward and outward from a central high point, as figure 12.3 shows.

FIGURE 12.2
Valley glacier in Alaska.
Source: Bruce Molina, U.S. Geological Survey

12.2 GLACIERS—WHERE THEY ARE, HOW THEY FORM AND MOVE

Distribution of Glaciers

Glaciers occur in temperate as well as polar climates. They are found where more snow falls during the cold time of year than can be melted during warm months.

Washington has more glaciers than any other state except Alaska because of the extensively glaciated mountains of western Washington. Washington's mountains have warmer winters but much more precipitation in the higher elevations than do the Rocky Mountains. There is more snow left after summer melting in Washington than in states to the east of it. Glaciers are common even near the equator in the very high mountains of South America and Africa because of the low temperatures at high altitudes.

Glaciation is most extensive in polar regions, where little melting takes place at any time of year. At present, about

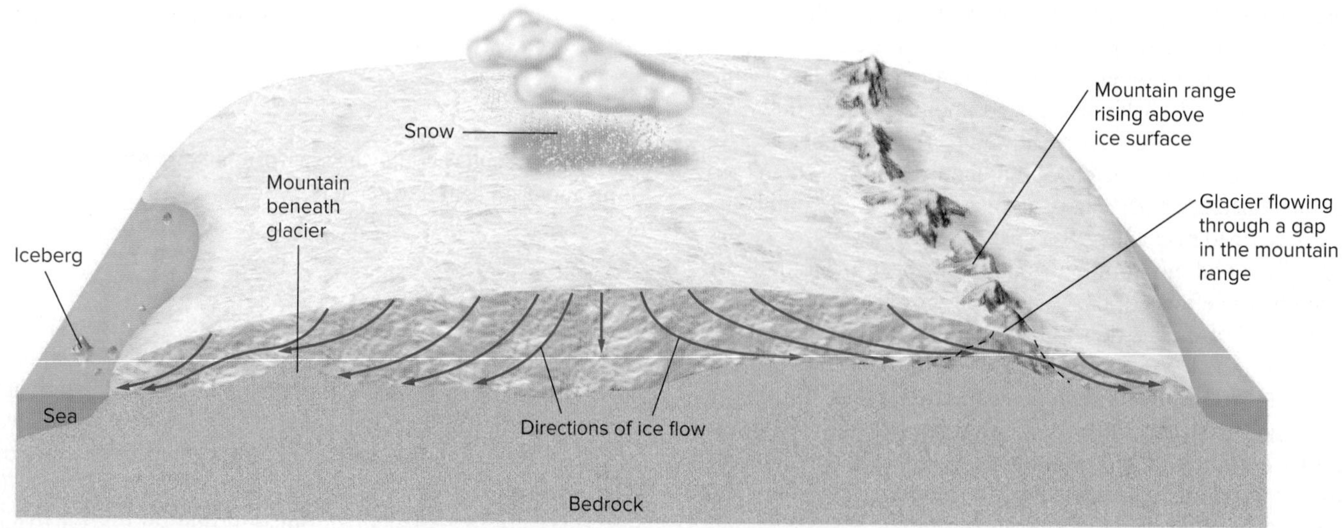

FIGURE 12.3
Diagrammatic cross section of an ice sheet. Vertical scale is highly exaggerated.

one-tenth of the land surface on Earth is covered by glaciers (compared with about one-third during the peak of the glacial ages). Approximately 85% of the present-day glacier ice is on the Antarctic continent, covering an area larger than the combined areas of western Europe and the United States; 10% is in Greenland. All the remaining glaciers of the world amount to only about 5% of the world's freshwater ice. This means that Antarctica is in fact storing most of Earth's fresh water in the form of ice. Some have suggested that ice from the Antarctic, towed as icebergs, could be brought to areas of dry climate to alleviate water shortages. It is worth noting that if all of Antarctica's ice were to melt, sea level around the world would rise about 65 meters (213 feet)—up to the Statue of Liberty's waist. This would flood the world's coastal cities and significantly decrease the land surface available for human habitation.

Formation and Growth of Glaciers

Snow converts to glacier ice in somewhat the same way that sediment turns into a sedimentary rock and then into metamorphic rock; figure 12.4 shows the process. Snowfall can be compared to sediment settling out of water. A new snowfall may be in the form of light "powder snow," which consists mostly of air trapped between many six-pointed snowflakes. In a short time, the snowflakes settle by compaction under their own weight, and much of the air between them is driven out. Meanwhile, the sharp points of the snowflakes are destroyed as flakes reconsolidate into granules. In warmer climates, partial thawing and refreezing result in coarse granules—the "corn snow" of spring skiing. In colder climates where little or no melting takes place, the snowflakes will recrystallize into fine granules. After the granular snow is buried by a new snowpack, usually during the following winter, the granules are compacted and weakly "cemented" together by ice. The compacted mass of granular snow, transitional between snow and glacier ice, is called *firn*. Firn is analogous to a sedimentary rock such as a weakly cemented sandstone.

Through the years, the firn becomes more deeply buried as more snow accumulates. More air is expelled, the remaining pore space is greatly reduced, and granules forced together recrystallize into the tight, interlocking mosaic of *glacier ice* (figure 12.4*B*). The recrystallization process involves little or no melting and is comparable to metamorphism. Glacier ice is texturally similar to the metamorphic rock, quartzite.

Under the influence of gravity, glacier ice moves downward and is eventually **ablated**, or lost. For glaciers in all but the coldest parts of the world, ablation is mostly due to melting, although some ice evaporates directly into the atmosphere (a process known as *sublimation*). If a moving glacier reaches a body of water, blocks of ice break off (or *calve*) and float free as **icebergs** (figure 12.5). In most of the Antarctic, ablation takes place largely through calving of icebergs and direct evaporation. Only along the coast does melting take place, and there for only a few weeks of the year.

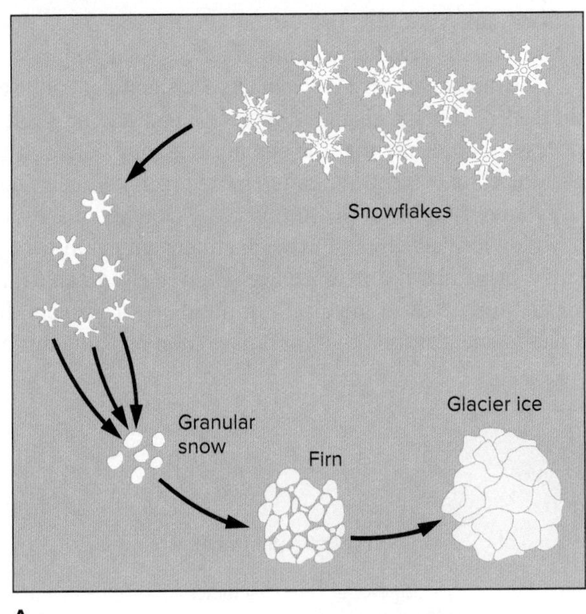

FIGURE 12.4

(*A*) Conversion of snow to glacier ice. (*B*) Thin slice of an ice core from a glacier (a core is shown in box 12.3, figure 1). The ice is between sheets of polarizing filters. In polarized light, the colors of individual ice grains vary depending on their crystallographic orientation. Without the polarizing filters, the ice would be transparent and clear. The ruler shows that many of the ice grains are over a centimeter in length.

(B) ©C. C. Plummer

FIGURE 12.5

Icebergs calved from Alaska's Bear Glacier drift in a glacial lagoon.
Source: Bruce Molina, U.S. Geological Survey

Glacial Budgets

If, over a period of time, the amount of snow a glacier gains is greater than the amount of ice and water it loses, the glacier's budget is *positive* and it expands. If the opposite occurs, the glacier decreases in volume and is said to have a *negative budget.* Glaciers with positive budgets push outward and downward at their edges; they are called **advancing glaciers**. Those with negative budgets grow smaller and their edges melt back; they are **receding glaciers**. Bear in mind that the glacial ice moves downvalley, as shown in figure 12.6, whether the glacier is advancing or receding. In a receding glacier, however, the rate of flow of ice is insufficient to replace all of the ice lost in the lower part of the glacier. If the amount of snow retained by the glacier equals the amount of ice and water lost, the glacier has a *balanced budget* and is neither advancing nor receding.

The upper part of a glacier, called the **zone of accumulation**, is the part of the glacier with a perennial snow cover (figure 12.6). The lower part is the **zone of ablation** (or *zone of wastage*), for there ice is lost, or ablated, by melting, evaporation, and calving.

The boundary between these two zones of a glacier is an irregular line called the **equilibrium line** (sometimes called the *snow line* or the *firn line*), which marks the highest point at which the glacier's winter snow cover is lost during a melt season (figure 12.7).

The equilibrium line may shift up or down from year to year, depending on whether there has been more accumulation or more ablation. Its location therefore indicates whether a glacier has a positive or negative budget. An equilibrium line migrating upglacier over a period of years is a sign of a negative budget, whereas an equilibrium line migrating downglacier indicates that the glacier has a positive budget. If an equilibrium line remains essentially in the same place year after year, the glacier has a balanced budget.

The **terminus** is the lower edge of a glacier. Its position reflects the glacier's budget. For a valley glacier, a positive budget results in the terminus moving downvalley. In a receding glacier, the terminus melts back upvalley. Because glacial ice moves slowly, migration of the terminus tends to lag several years behind a change in the budget.

If the terminus of an ice sheet is on land, it will advance or retreat in response to a positive or negative budget, just as for a valley glacier. If the terminus is at the continent's shoreline (as it is for our Antarctic and Greenland ice sheets), a positive budget results in a greater volume of icebergs calving into the sea.

Advancing or receding glaciers are significant and sensitive indicators of climatic change. However, an advancing glacier does not necessarily indicate that the climate is getting colder. It may mean that the climate is getting wetter, more precipitation is falling during the winter months, or the summers are cloudier. It is estimated that a worldwide decrease in the mean annual temperature of about 5°C could bring about a new ice age. Conversely, global warming can significantly reduce the size and numbers of glaciers. In general, valley glaciers around the world have been receding during the past

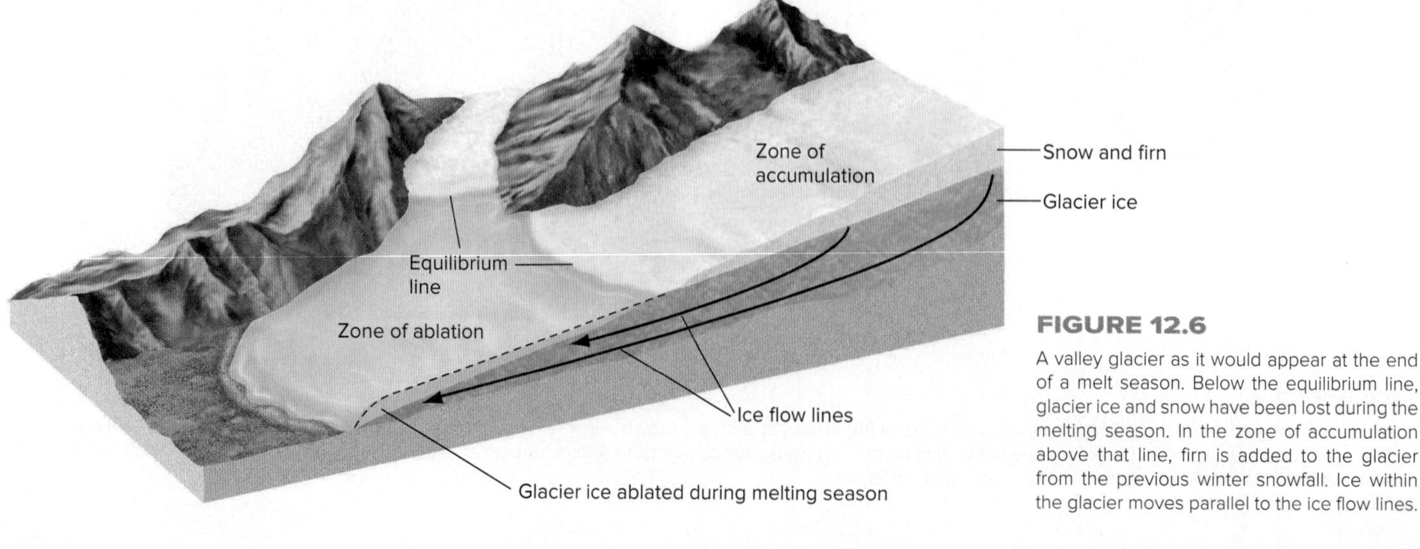

FIGURE 12.6

A valley glacier as it would appear at the end of a melt season. Below the equilibrium line, glacier ice and snow have been lost during the melting season. In the zone of accumulation above that line, firn is added to the glacier from the previous winter snowfall. Ice within the glacier moves parallel to the ice flow lines.

A

B

C

century. At present, glaciers at Glacier National Park in Montana and Mount Kilamanjaro in Africa are receding at a rate that, if sustained, will lead them to disappear in a few years. (Will Montana's park then be renamed Glacier-Free National Park?)

Glaciers can be an important source of water (box 12.1). With the ongoing global warming, the water supply from glaciers is diminishing as glaciers continue to recede.

Movement of Valley Glaciers

Valley glaciers move downslope under the influence of gravity at a variable rate, generally ranging from less than a few millimeters to 15 meters a day. Sometimes a glacier will move much faster for a brief period of time (see box 12.2). A glacier will flow faster where it is steeper. Also, the thicker parts of a glacier will flow faster than thinner parts. The upper part of a glacier tends to be steeper than at lower levels. If a glacier has an even gradient, the glacier will be thickest near the equilibrium line. So, except for locally steeper stretches, we expect the fastest moving ice to be near the equilibrium line. Below the equilibrium line, the glacier usually becomes progressively thinner and slower.

Glaciers in temperate climates–where the temperature of the glacier is at or near the melting point for ice–tend to move faster than those in colder regions–where the ice temperature stays well below freezing.

Velocity also varies within the glacier itself (figure 12.8). The central portion of a valley glacier moves faster than the sides (as water does in a stream), and the surface moves faster than the base. This is because of frictional drag at the base and along the sides of the glacier. How ice moves within a valley glacier has been demonstrated by studies in which holes are drilled through the glacier ice and flexible pipes inserted. Changes in the shape and position of the pipes are measured periodically. The results of these studies are shown diagrammatically in figure 12.8.

Note in the diagram that the base of the pipe has moved downglacier. This indicates **basal sliding**, which is the sliding of the glacier as a single body over the underlying rock. A thin film of meltwater that develops along the base from the pressure of the overlying glacier facilitates basal sliding. Think of a large bar of wet soap sliding down an inclined board.

Note that the lower portion of the pipe is bent in a downglacier direction. The bent pipe indicates **plastic flow** of ice, movement that occurs within the glacier due to the plastic or "deformable" nature of the ice itself. Visualize two neighboring grains of ice within the glacier, one over the other. Both are moving, carried along by the ice below them; however, the higher of the two ice grains slides over its underlying neighbor a bit farther.

FIGURE 12.7

South Cascade Glacier, Washington. If the photos were taken at the end of the melt season, the equilibrium line would be the boundary between white snow and darker glacier ice. Photo (*A*) was taken in 1965; note that the glacier extended into the lake. Photo (*B*) was taken in 2000; notice that the glacier has shrunk away from the lake. Photo (*C*), taken in 2010, shows that the glacier continued to recede.
(A & C) Source: U.S. Geological Survey; (B) Source: Robert Krimmel/U.S. Geological Survey

ENVIRONMENTAL GEOLOGY 12.1

Glaciers as a Water Resource

Few people think of glaciers as frozen reservoirs supplying water for irrigation, hydroelectric power, recreation, and industrial and domestic use. Yet, glacially derived water is an important resource in places such as Iceland, Norway, British Columbia, and the state of Washington. In Washington, streamflow from the approximately 800 glaciers there amounts to about 470 billion gallons of water during a summer, according to the U.S. Geological Survey. More water is stored in glacier ice in Washington than in all of the state's lakes, reservoirs, and rivers.

One important aspect of glacier-derived water is that it is available when needed most. In the Pacific Northwest, snow accumulates on glaciers during the wet winter months. During the winter, streams at lower elevations, where rain rather than snow falls, are full and provide plenty of water. During the summer, the climate in the Pacific Northwest is hotter and drier. Demand for water increases, especially for irrigation of crops. Streams that were fed by rainwater may have dried up. Yet, in the heat of summer, the period of peak demand, snow and ice on glaciers are melting, providing much-needed water to streams.

Climate change due to global warming may pose a threat to this important source of water. Glaciers in many places, including the Pacific Northwest, are retreating, and some are at risk of disappearing altogether within decades. What will the impact of this be? In Bolivia, the residents of La Paz rely on glacial meltwater to provide 27% of their water during the dry season. Studies have shown that the glaciated area around La Paz has shrunk in size by 50% since 1963. If the glaciers disappear entirely, the effect on dry season water supply will be significant.

Additional Resources

- https://www2.usgs.gov/climate_landuse/clu_rd/glacierstudies/benchmarkGlaciers.asp

Since 1957, the USGS has conducted a long-term investigation of three "benchmark" glaciers. Two are in Alaska; the third is South Cascade Glacier (figure 12.7) in Washington's North Cascade Range. A fourth benchmark glacier in Montana was added to the study in 2005. This website describes the project and provides links to the results of the study for each of the three glaciers.

- http://www2.ucar.edu/atmosnews/in-brief/7649/who-needs-glaciers

An article published in 2012 by the National Center for Atmospheric Research (NCAR) on research into the effects of glacial melt around the world.

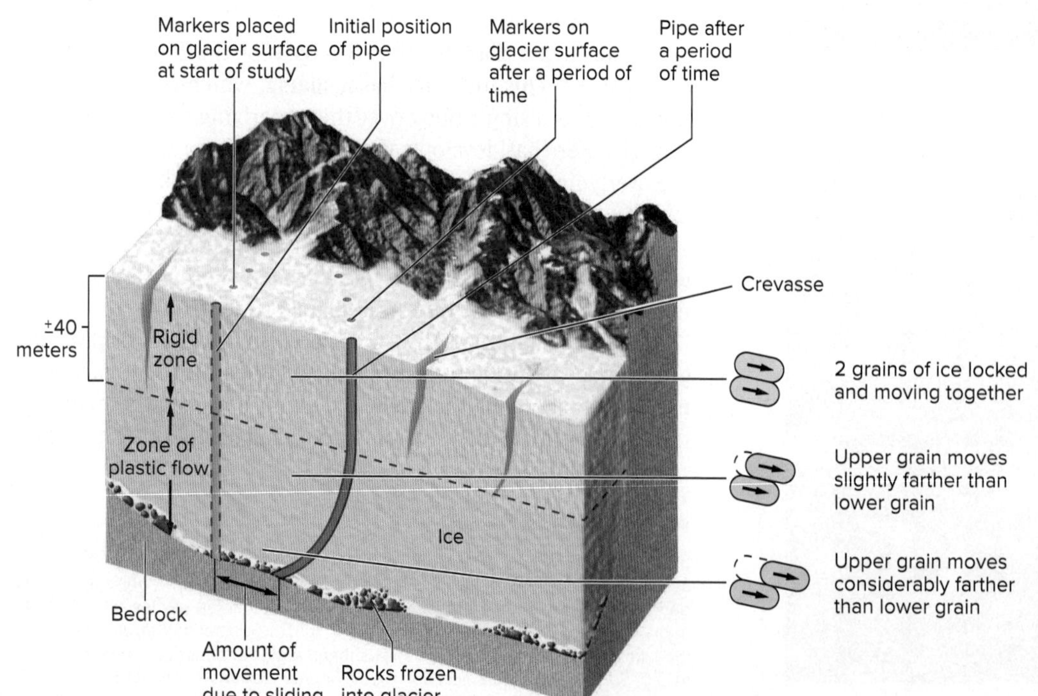

FIGURE 12.8

Movement of a glacier. Markers on the glacier indicate the center of the glacier moves faster than its side. Cross-sectional view shows movement within the glacier.

ENVIRONMENTAL GEOLOGY 12.2

Water Beneath Glaciers: Floods, Giant Lakes, and Galloping Glaciers

A Galloping Glacier

Glacial motion is often used as a metaphor for slowness ("The trial proceeded at a glacial pace"). But, some glaciers will *surge*—that is, move very rapidly for short periods following years of barely moving at all. The most extensively documented surge (or "galloping glacier") was that of Alaska's Bering Glacier in 1993–94. The Bering Glacier is the largest glacier in continental North America, and it surges on a 20–30 year cycle. After its previous surge in 1967, its terminus retreated 10 kilometers. In August 1993, the latest surge began. Ice traveled at velocities up to 100 meters per day for short periods of time and sustained velocities of 35 meters per day over a period of several months. The terminus advanced 9 kilometers by the time the surge ended in November 1994. When glaciers surge, the previously slow moving, lower part of a glacier breaks into a chaotic mass of blocks (box figure 1). Surges are usually attributed to a buildup of water beneath part of a glacier, floating it above its bed. In July 1994, a large flood of water burst from Bering Glacier's terminus, carrying with it blocks of ice up to 25 meters across.

A Flood

Glacial outburst floods are not always associated with surges. In October 1996, a volcano erupted beneath a glacier in Iceland. The glacier, which is up to 500 meters thick, covers one-tenth of Iceland. Emergency teams prepared for the flood that geologists predicted would follow the eruption. The expected flood took place early in November with a peak flow of 45,000 cubic meters per second (over 1.5 million cubic feet per second)! The flood lasted only a few hours; however, it caused between $10 and $15 million worth of damage. Three major bridges were destroyed or damaged, and 10 kilometers of roads were washed away. Because people had been kept away from the expected flood path, there were no casualties.

A Giant Lake

One of the world's largest lakes was discovered not long ago. But don't expect to take a dip in it or go windsurfing on it. It lies below the thickest part of the East Antarctic Ice Sheet and is named after the Russian research station, Vostok, which is 4,000 meters above the lake at the coldest and most remote part of Antarctica. Lake Vostok was discovered in the 1970s through ice-penetrating radar; however, its extent was unknown until 1996, when satellite-borne radar revealed how large it is. Studies indicate that the lake is 250 kilometers long and 50 kilometers wide—about the size of Lake Ontario. Its maximum depth is approximately 800 meters, placing it among the ten deepest lakes in the world. Lake Vostok is the largest of the almost 400 lakes known to lie beneath the Antarctic Ice Sheet.

By coincidence, the world's deepest ice hole was being drilled from Vostok Station above the lake when the size of Lake Vostok was being determined. The hole was completed in 1998, producing one of the world's longest ice cores (over 3,500 km). At that point, drilling was halted roughly 100 m from the top of the lake due to fear of contaminating it and harming whatever living organisms might be in the very old water. Drilling resumed in 2006 but did not reach the top of the lake surface until early 2012. A second borehole was successfully completed in 2015.

The lake has been sealed off from the rest of the world for approximately 15 million years. Remarkably, recent studies of ice from Lake Vostok ice cores have revealed the presence of over 3,500 species (mostly in the form of bacteria) in this high-pressure, freezing environment.

For more information on the Lake Vostok Drilling Project, visit http://www.southpolestation.com/trivia/10s/lakevostok.html.

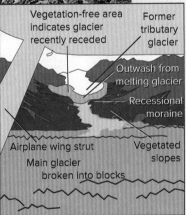

BOX 12.2 ■ FIGURE 1
Part of a glacier after a surge (lower part of photo). The debris-covered ice has been broken up into a chaotic mass of blocks. In the background is a small glacier that has retreated up its valley. Photo taken near the Canadian-Alaskan border.
©C. C. Plummer

The reason the pipe is bent more sharply near the base of the glacier is that pressure from overlying ice results in greater flowage with increasing depth. Deep in the glacier, ice grains are sliding past their underlying neighbors farther than similar ice grains higher up, where the pipe is less bent. We should point out that a glacier flows not only because ice grains slide past one another but also because ice grains deform and recrystallize.

In the **rigid zone**, or upper part of the glacier, the pipe has been moved downglacier; however, it has remained unbent. The ice nearer the top apparently rides along passively on the plastically moving ice closer to the base. In the rigid zone, grains of ice do not move relative to their neighbors because the pressure is not great enough to allow for plastic flow.

Crevasses

Along its length, a valley glacier moves at different rates in response to changes in the steepness of the underlying rock. Typically, a valley glacier rides over a series of rock steps. Where the glacier passes over a steep part of the valley floor, it moves faster. The upper rigid zone of ice, however, cannot stretch to move as rapidly as the underlying plastic-flowing ice. Being brittle, the ice of the rigid zone is broken by the tensional forces. Open fissures, or **crevasses**, develop (figure 12.9). Crevasses also form along the margins of glaciers in places where the path is curved, as shown in part of figure 12.9. This is because ice (like water) flows faster toward the outside of the curve. For glaciers in temperate climates, a crevasse should be no deeper than about 40 meters, the usual thickness of the rigid zone. If you are falling down a crevasse, it may be of some consolation that, as you are hurtling to death or injury, you realize on the way down that you will not fall more than 40 meters.

After the ice has passed over a steep portion of its course, it slows down, and compressive forces close the crevasses.

Movement of Ice Sheets

An ice sheet or ice cap moves like a valley glacier except that it moves downward and outward from a central high area toward the edges of the glacier (as shown in figure 12.3).

Glaciological research in Antarctica has determined how ice sheets grow and move. Antarctica has two ice sheets: the West Antarctic Ice Sheet is separated by the Transantarctic Mountains from the much larger East Antarctic Ice Sheet (figure 12.10). The two ice sheets join in the low areas between mountain ranges. Both are almost completely within a zone of accumulation because so little melting takes place (ablation is largely by calving of icebergs) and because occasional snowfalls nourish their high central parts. The ice sheets mostly overlie interior lowlands, but they also completely bury some mountain ranges. Much of the base of the West Antarctic Ice Sheet is on bedrock that is below sea level. At least one active volcano underlies the West Antarctic Ice Sheet (resulting in a depression in the ice sheet). Where mountain ranges are higher than the ice sheet, the ice flows between mountains as valley glaciers, known as *outlet glaciers*.

Although flowage in ice sheets is away from central highs where more snow accumulates, movement is not uniform. Most of the flowage in the East Antarctic Ice Sheet takes place in *ice streams*, zones where the movement is considerably faster than in adjoining ice, which is frozen to its bed. An ice stream is often heavily crevassed, and its boundaries are determinable by the transition from crevassed to crevasse-free ice.

At the South Pole (figures 12.10 and 12.11)—neither the thickest part nor the center of the East Antarctic Ice Sheet—the ice is 2,700 meters thick. The thickest part of the East Antarctic Ice Sheet is 4,776 meters. Research at ice sheets has

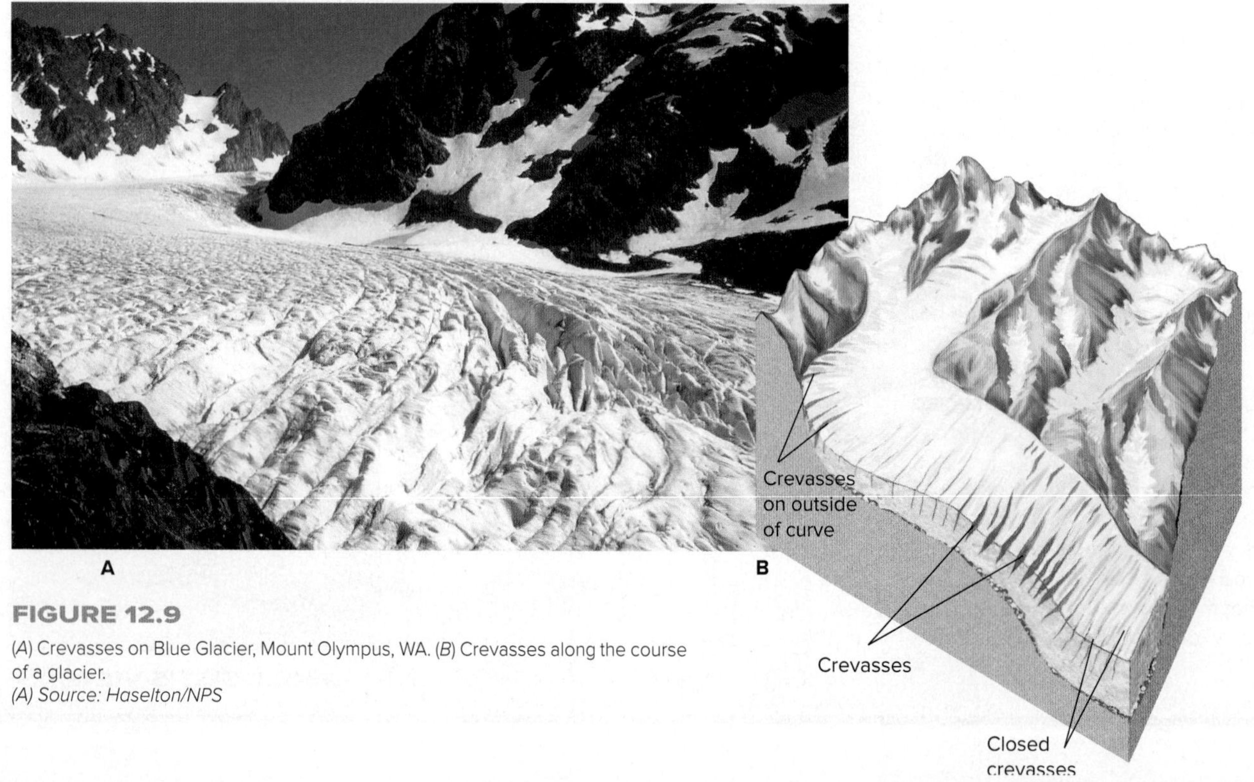

FIGURE 12.9

(*A*) Crevasses on Blue Glacier, Mount Olympus, WA. (*B*) Crevasses along the course of a glacier.
(A) Source: Haselton/NPS

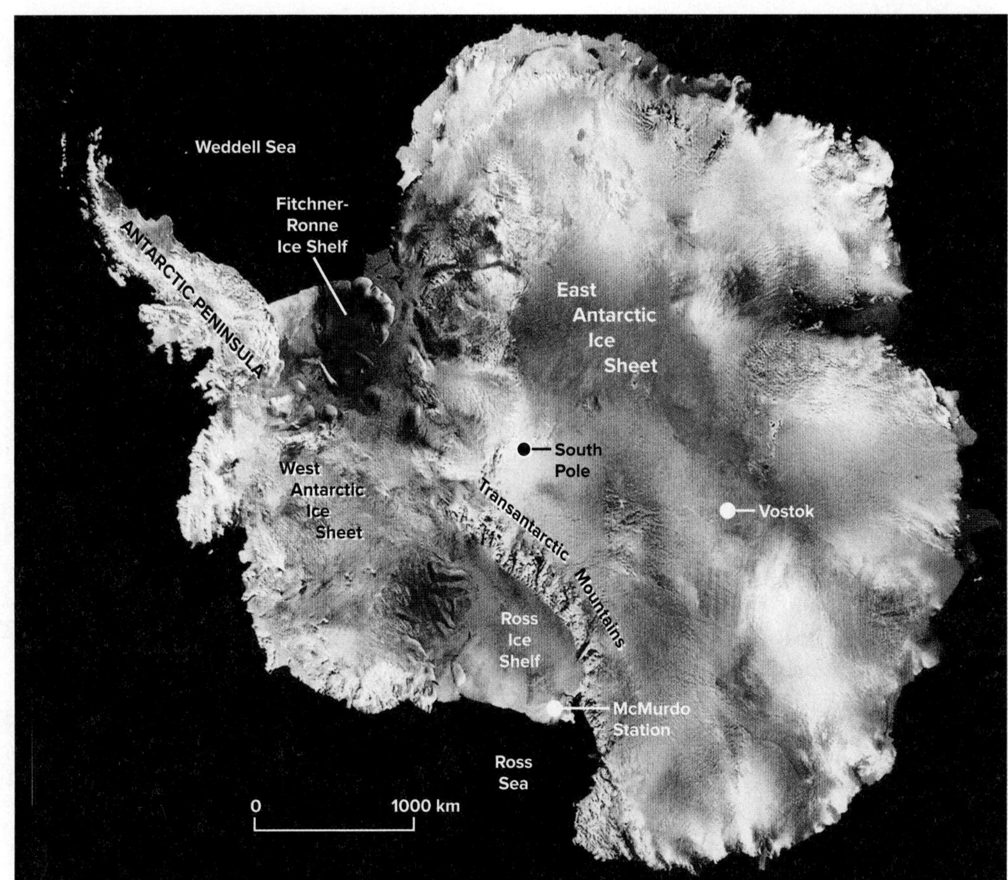

FIGURE 12.10

The Antarctic continent and its ice sheets. Vostok is at the highest part of the East Antarctic Ice Sheet. (False coloring is used to show variations among snow, ice, blue ice, and exposed rock.) To view an interactive map of Antarctica, go to http://lima.usgs.gov/. LIMA is short for Landsat Image Mosaic of Antarctica. You can zoom in on any part of the continent and see parts of mountain ranges or coastlines.
Source: U.S. Geological Survey/NASA

FIGURE 12.11

The South Pole. Actually, the true South Pole is several kilometers from here. The moving ice sheet has carried the striped pole away from the site of the true South Pole, where the pole was erected in 1956.
©C. C. Plummer

yielded important information regarding past climates (see box 12.3).

Most of the movement of the East Antarctic Ice Sheet is by means of plastic flow. It has been thought that most of the ice sheet is frozen to the underlying rocks and basal sliding takes place only locally. But the discovery of a giant lake and other smaller lakes beneath the thickest part of the East Antarctic Ice Sheet (see box 12.2) indicates that liquid water at its base is more widespread and basal sliding is more important than previously regarded. Also, scientists have detected subglacial streams with water flowing from one subglacial lake to another. (To see an animation of this phenomenon, go to http://svs.gsfc.nasa.gov/goto?20100.)

Recently, scientists have shown that an ice sheet grows not only by snowfall accumulation at its surface but also by accumulation of ice at its base. Liquid water collects in subglacial lakes and streams and then refreezes to become part of the overlying glacier. Box 12.4 describes a way in which this takes place.

12.3 GLACIAL EROSION

Wherever basal sliding takes place, the rock beneath the glacier is abraded and modified. As meltwater works into cracks in bedrock and refreezes, pieces of the rock are broken loose and frozen into the base of the moving glacier, a process known as *plucking.* While being dragged along by the moving ice, the rock within the glacier grinds away at the underlying rock (figure 12.12). The thicker the glacier, the more pressure on the rocks and the more effective the grinding and crushing.

EARTH SYSTEMS 12.3

Global Warming and Glaciers

Most of Earth's glaciers have been receding for a century (see figure 12.7). This is generally regarded as a consequence of global warming. That Earth's climate is warming is now clearly established. But some questions arise with regard to global warming: How does it compare to past episodes of warming? Is it part of a natural cycle? How much of it is anthropogenic (caused by humans)? What are the consequences of continued global warming on Earth systems? Can we do anything to reduce the global warming?

Glaciers, particularly the Antarctic and Greenland ice sheets, provide us with a means to answer these questions. Glaciers preserve records of precipitation, air temperatures, atmospheric dust, volcanic ash, carbon dioxide, and other atmospheric gases.

When snow becomes converted to glacier ice, some of the air that was mixed with the snowflakes becomes bubbles trapped in the glacier ice. By analyzing the air in these bubbles, we are analyzing the air that prevailed when an ancient ice layer formed. Drilling into glaciers and retrieving ice cores allows scientists to sample the environment at the time of ancient snowfalls. A cylindrical core of ice is extracted from a hollow drill after it has penetrated a glacier. The layers in an ice core represent the different layers of snow that converted to glacier ice (box figure 1). Each layer, when analyzed, can reveal information about conditions of the atmosphere at the time the snow accumulated and turned into ice.

An ambitious drilling project in Antarctica was completed in December 2004 by the European Project for Ice Coring in Antarctica (EPICA). Drilling retrieved ice core to a depth of 3,270 meters (10,728 feet), stopping 5 meters above the base of the ice sheet. The EPICA ice core yielded a record of the climate extending back 740,000 years.

The longest ice core was drilled at Vostok (described in box 12.2) in Antarctica in the 1990s. The Vostok core reached a depth of 3,701 meters and yielded a climate and atmospheric history of the past 420,000 years. Graphs derived from the research project are shown in box figure 2. The temperature variation is relative to the ice sheet's temperature during the past millennium. The team determined the temperature by studying hydrogen isotope variation (see chapter 2) within the ice layers. Methane and carbon dioxide are greenhouse gases. Note how the greenhouse gases correlate closely with the temperature variations. Also note the five periods when the temperature was warmest. These are the interglacial periods during which the North American and European ice sheets disappeared. Two of the five warm periods are emphasized for comparison—the Holocene interglacial epoch (which began about 12,000 years ago and is ongoing) and the previous interglacial, the Sangamonian (known as the Eemian in northern Europe).

Compare the Holocene temperature pattern to that of the Sangamonian. From this, you can understand why scientists infer that we should be in a period of declining temperatures leading

A B

BOX 12.3 ■ FIGURE 1

(*A*) An ice core being removed from an ice corer. (*B*) A one-meter length of ice core from the Greenland Ice Sheet showing layering of glacial ice. The bottom of the core was at a depth of 1,836 meters below the surface and formed in the year 14,248 B.C.
Source: National Ice Core Laboratory, National Science Foundation & U.S. Geological Survey

Pebbles and boulders that are dragged along are *faceted*, that is, given a flat surface by abrasion. Bedrock underlying a glacier is *polished* by fine particles and *striated* (scratched) by sharp-edged, larger particles. Striations and grooves on bedrock indicate the direction of ice movement (figure 12.13).

The grinding of rock across rock produces a powder called **rock flour**. Rock flour is composed largely of very fine (silt- and clay-sized) particles of unaltered minerals (pulverized from chemically unweathered bedrock). When *meltwater* washes rock flour from a glacier, the streams draining the glacier appear milky, and lakes into which glacial meltwater flows often appear a milky green color.

Not all glacier-associated erosion is caused directly by glaciers. Mass wasting takes place on steep slopes created by downcutting glaciers. Frost wedging breaks up bedrock ridges and cliffs above a glacier, causing frequent rockfalls. Snow avalanches bring down loose rocks onto the glacier. If rocks collect in the zone of accumulation, they may be incorporated into the

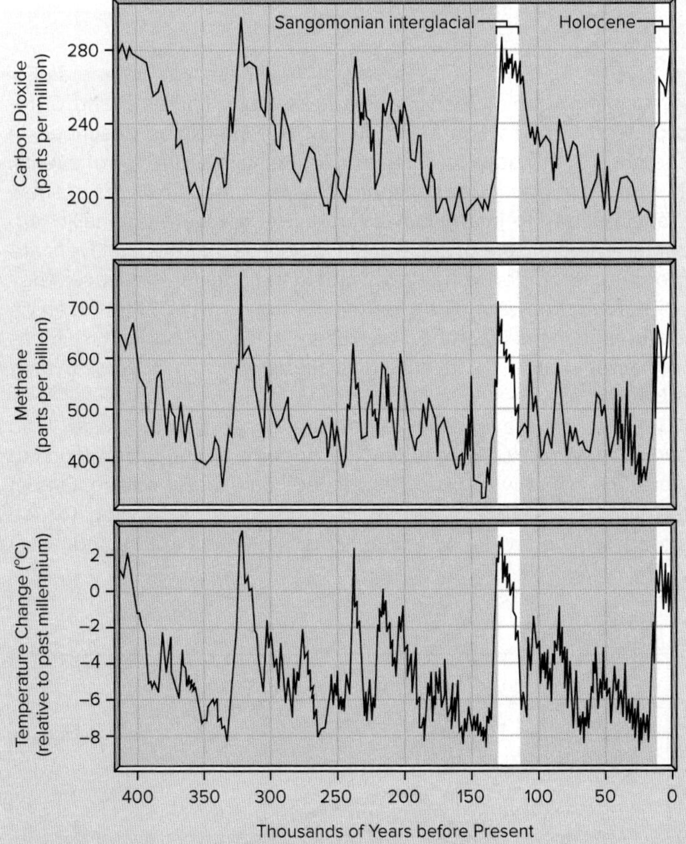

BOX 12.3 ■ FIGURE 2

Temperature, carbon dioxide, and methane content of air at Vostok on the East Antarctic Ice Sheet for the last 420,000 years. See text for explanation. Also note the rapid rising of temperature at the beginning of interglacial periods.

into the next glacial age. Instead, we have ongoing global warming. Further, there is now strong evidence that the global warming is due to anthropogenic contribution of greenhouse gases to our atmosphere. The two biggest culprits are methane and carbon dioxide, whose levels now far exceed the naturally derived levels reached during the warmer Sangamonian interglacial. Sea level worldwide during that time was several meters higher than at present. (This was caused by the expansion of the ocean waters due to heating, as well as melting of polar ice at a higher rate than at present.)

If computer models projecting higher rates of global warming are correct, we can expect the hydrosphere to be affected by gases in the atmosphere, and sea level will rise significantly. One group of researchers calculated that sea level could rise several tens of centimeters during this century. But this prediction is based mainly on expansion of the oceans due to being heated. If the higher temperatures also trigger disintegration of large parts of the ice sheets, sea levels could rise even higher. A higher sea level affects the biosphere, notably humans. A major proportion of the world's population lives at or close to a coastline. Houses would be destroyed by coastal erosion. Major cities, such as New York, would have to erect dikes to keep water out of buildings that are below sea level.

Additional Resources

Read chapter 21 of this book to learn more about climate change and how the paleoclimate record is constructed.

In February 2011, scientists and others completed the sixth field season at the West Antarctic Ice Sheet (WAIS) Divide Ice Core drilling site. The team reached its goal of drilling 3,330 meters vertically into the glacier—the second-deepest ice core hole ever. This site documents all aspects of the project.

- http://waisdivide.unh.edu/

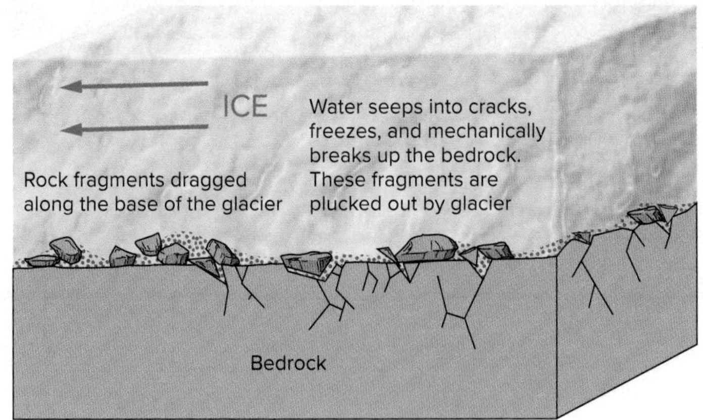

FIGURE 12.12

Plucking and abrasion beneath a glacier.

body of the glacier. If rocks fall onto the zone of ablation, they will ride on the glacial ice surface. Debris may also fall into crevasses to be transported within or at the base of a glacier, as shown in figure 12.19.

Erosional Landscapes Associated with Alpine Glaciation

We are in debt to glaciers for the rugged and spectacular scenery of high mountain ranges. Figure 12.14 shows how glaciation has radically changed a previously unglaciated mountainous region. The striking and unique features associated with mountain glaciation, described next, are due to the erosional effects of glaciers as well as frost wedging on exposed rock.

IN GREATER DEPTH 12.4

Growth of Ice Sheets from the Bottom Upward

During the International Polar Year (2007), scientists from six nations conducted coordinated studies of the Gamburtsev Mountains in the most remote part of the East Antarctic Ice Sheet. Although field parties endured temperatures of −30°C and high winds during the Antarctic summer, they never set foot on the mountains. This is because the Gamburtsevs were up to 3,000 meters below them, at the base of the ice sheet. Through geophysical techniques, notably ice-penetrating radar, scientists were able to produce a three-dimensional image of the mountains. The subglacial range is similar in size and topography to Europe's Alps. The topography indicates that Alpine glaciation took place before being overwhelmed by continental glaciation.

A very significant result of the expedition is the discovery that the ice sheet grew from the bottom up as well as from snowfall on its surface. Subglacial liquid water is detected at the base of the range as well as in its alpine mountain valleys. Water has moved upward, becoming a supercooled liquid as pressure from the overlying ice decreases, then flashes into ice when pressure is reduced sufficiently.

Go to www.LDEO.columbia.edu/news-events/some-antarctic-ice-forming-bottom. Read the text, then study figures *B* and *C* (reproduced here as figure 1). You can use principles described in chapter 8 of this book to determine the relative sequence of events. The *principle of original horizontality* tells us that the layers of snow that converted to internal layers of glacier ice were originally horizontal. The *principle of superposition* tells us that the ice layers are progressively older going downward in the original sequence. They were domed upward by a later event—the growth of the frozen-on ice (orange in the diagram). The frozen-on ice also broke through as well as deformed the layered glacial ice (*principle of cross-cutting relationships*).

Go to the previous figure on the website with the AGAP logo over the layered ice image. Play the video or, better yet, click on the word "vimeo" to watch two short videos on a larger screen. One of the videos is called "supercooling," the other is "analog for ice added from below" (or go to: http://vimeo.com/20690318 and http://vimeo.com/20602505).

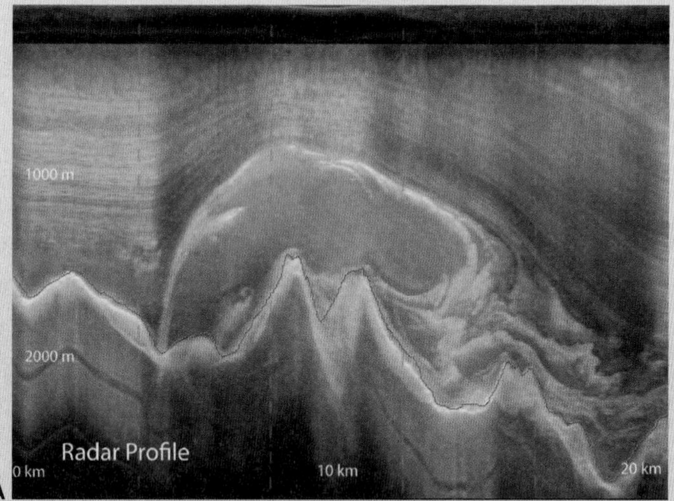

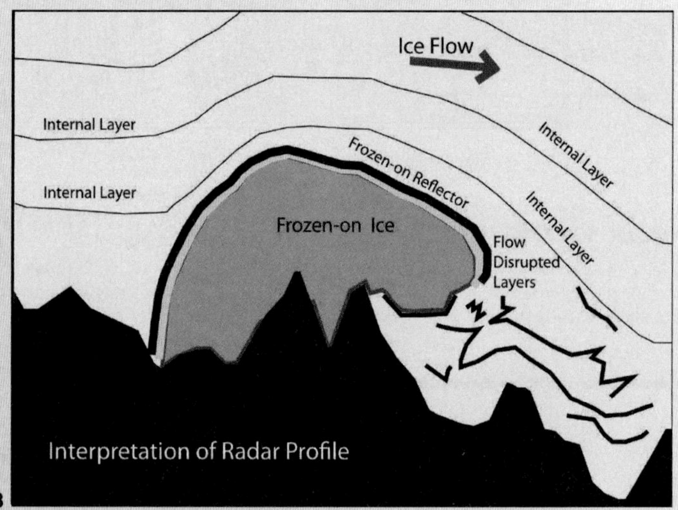

BOX 12.4 ■ FIGURE 1

Radar profile (*A*) and geologists' interpretation (*B*). A profile of the subglacial Gamburtsev Mountains shown by a red line in (*A*) and black area in (*B*). The internal layers formed from snow accumulating at the glacier's surface. The accretion plume of frozen-on ice is 1,100 meters thick. Intrusion of the frozen-on ice has folded and disrupted the glacier's internal layering to the right of the frozen-on ice.

Source: A and B reprinted with permission from **Science** *25 March 2011, Vol. 331 no. 6024 pp. 1592–1595. Robin E. Bell et al., "Widespread Persistent Thickening of the East Antarctic Ice Sheet by Freezing from the Base." © 2011 American Association for the Advancement of Science. Radar image A courtesy of Robin Bell, Lamont Doherty Earth Observatory, Columbia University.*

FIGURE 12.13

Striated and polished bedrock surface in south Australia. Unlike glacial striations commonly found in North America, these were caused by late Paleozoic glaciation. ©C. C. Plummer

Glacial Valleys

Glacially carved valleys are easy to recognize. A **U-shaped valley** (in cross profile) is characteristic of glacial erosion (figure 12.15), just as a V-shaped valley is characteristic of stream erosion.

The thicker a glacier is, the more erosive force it exerts on the underlying valley floor, and the more bedrock is ground away. For this reason, a large trunk glacier erodes downward more rapidly and carves a deeper valley than do the smaller tributary glaciers that join it. After the glaciers disappear, these tributaries remain as **hanging valleys** high above the main valley (figure 12.16).

Valley glaciers, which usually occupy valleys formerly carved by streams, tend to straighten the curves formed by running water. This is because the mass of ice of a glacier is too sluggish and inflexible to move easily around the curves. In the process of carving the sides of its valley, a glacier erodes or "truncates" the lower ends of ridges that extended to the valley. **Truncated spurs** are ridges that have *triangular facets* produced by glacial erosion at their lower ends (figure 12.14*B*).

Although a glacier tends to straighten and smooth the side walls of its valley, ice action often leaves the surface of the underlying bedrock carved into a series of steps. This is due to the variable resistance of bedrock to glacial erosion. Figure 12.17 shows what happens when a glacier abrades a relatively weak rock with closely spaced fractures. Water seeps into cracks in the bedrock, freezes there, and enlarges fractures or makes new ones. Rock frozen into the base of the glacier grinds and loosens more pieces. After the ice has melted back, a chain of **rock-basin lakes** (also known as **tarns**) may occupy the depressions carved out of the weaker rock. A series of such lakes, reminiscent of a string of prayer beads, is sometimes called *paternoster lakes.*

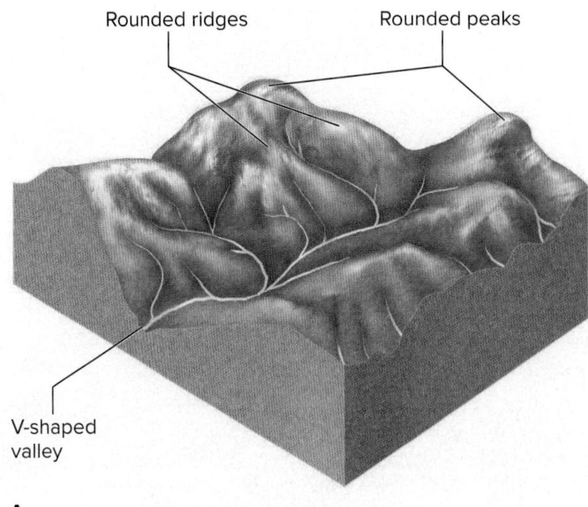

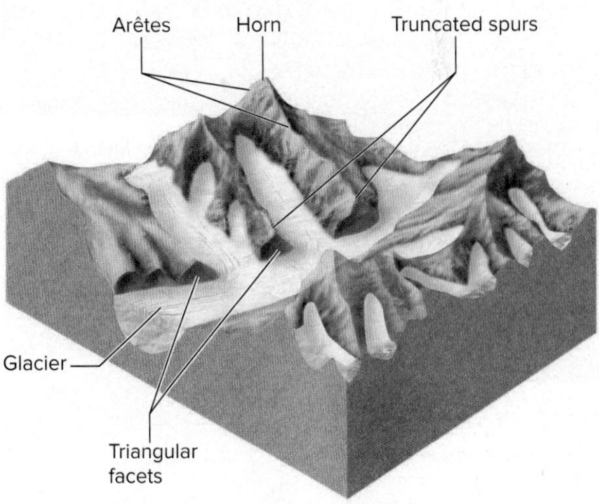

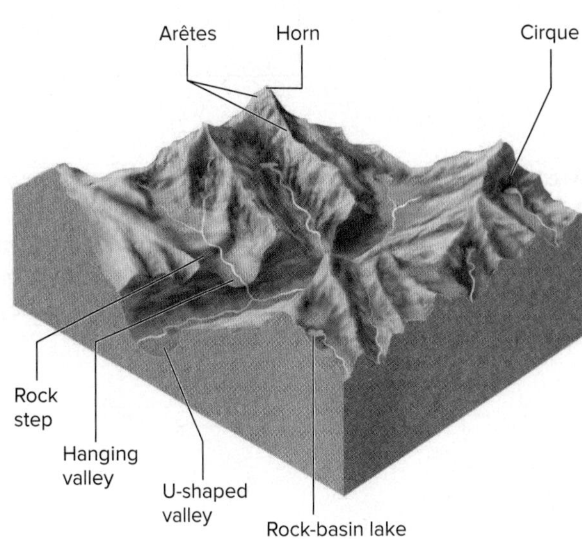

FIGURE 12.14

(*A*) A stream-carved mountain landscape before glaciation. (*B*) The same area during glaciation. Ridges and peaks become sharper due to frost wedging. (*C*) The same area after glaciation.

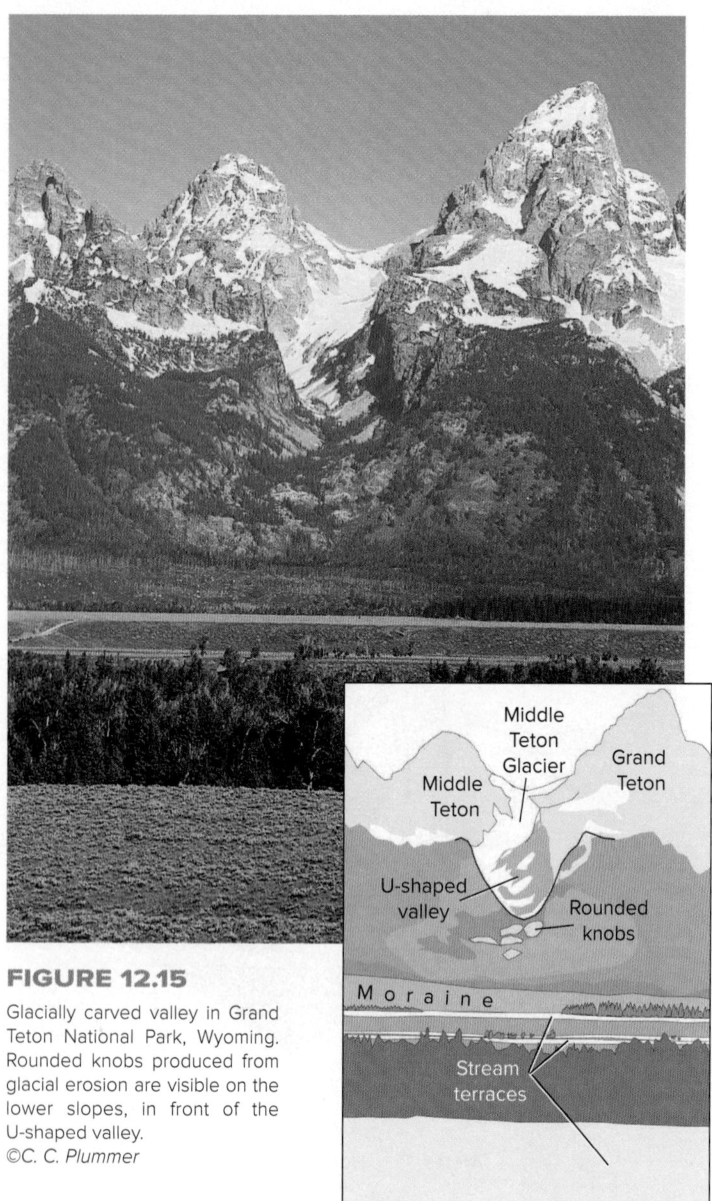

FIGURE 12.15

Glacially carved valley in Grand Teton National Park, Wyoming. Rounded knobs produced from glacial erosion are visible on the lower slopes, in front of the U-shaped valley.
©C. C. Plummer

FIGURE 12.16

A hanging valley in Yosemite National Park, California.
©C. C. Plummer

Areas where the bedrock is more resistant to erosion stand out after glaciation as *rounded knobs* (see figures 12.15 and 12.22), usually elongated parallel to the direction of glacier flow. These are also known as *roches moutonnées.* (In French, *roche* is rock and *moutonnée* means fleecy or curled.*)

Cirques, Horns, and Arêtes

A **cirque** is a steep-sided, half-bowl-shaped recess carved into a mountain at the head of a valley carved by a glacier (figure 12.18). In this unique, often spectacular, topographic feature, a large percentage of the snow accumulates that eventually converts to glacier ice and spills over the threshold as the valley glacier starts its downward course.

A cirque is not entirely carved by the glacier itself but is also shaped by the weathering and erosion of the rock walls above the surface of the ice. Frost wedging and avalanches break up the rock and steepen the slopes above the glacier. Broken rock tumbles onto the valley glacier and becomes part of its load, and some rock may fall into a crevasse that develops where the glacier is pulling away from the cirque wall (figure 12.19).

The headward erosional processes that enlarge a cirque also help create the sharp peaks and ridges characteristic of glaciated mountain ranges. A **horn** is the sharp peak that remains after cirques have cut back into a mountain on several sides (figure 12.20).

*The term was first used in the 1780s to describe an assemblage of rounded knobs in the Swiss Alps. It alluded to a fleecy wig called a *moutonné*, popular at that time, that was slicked down with sheep tallow. Later, the term came to refer to individual rounded knobs that resembled sheep—*mouton* is French for sheep.

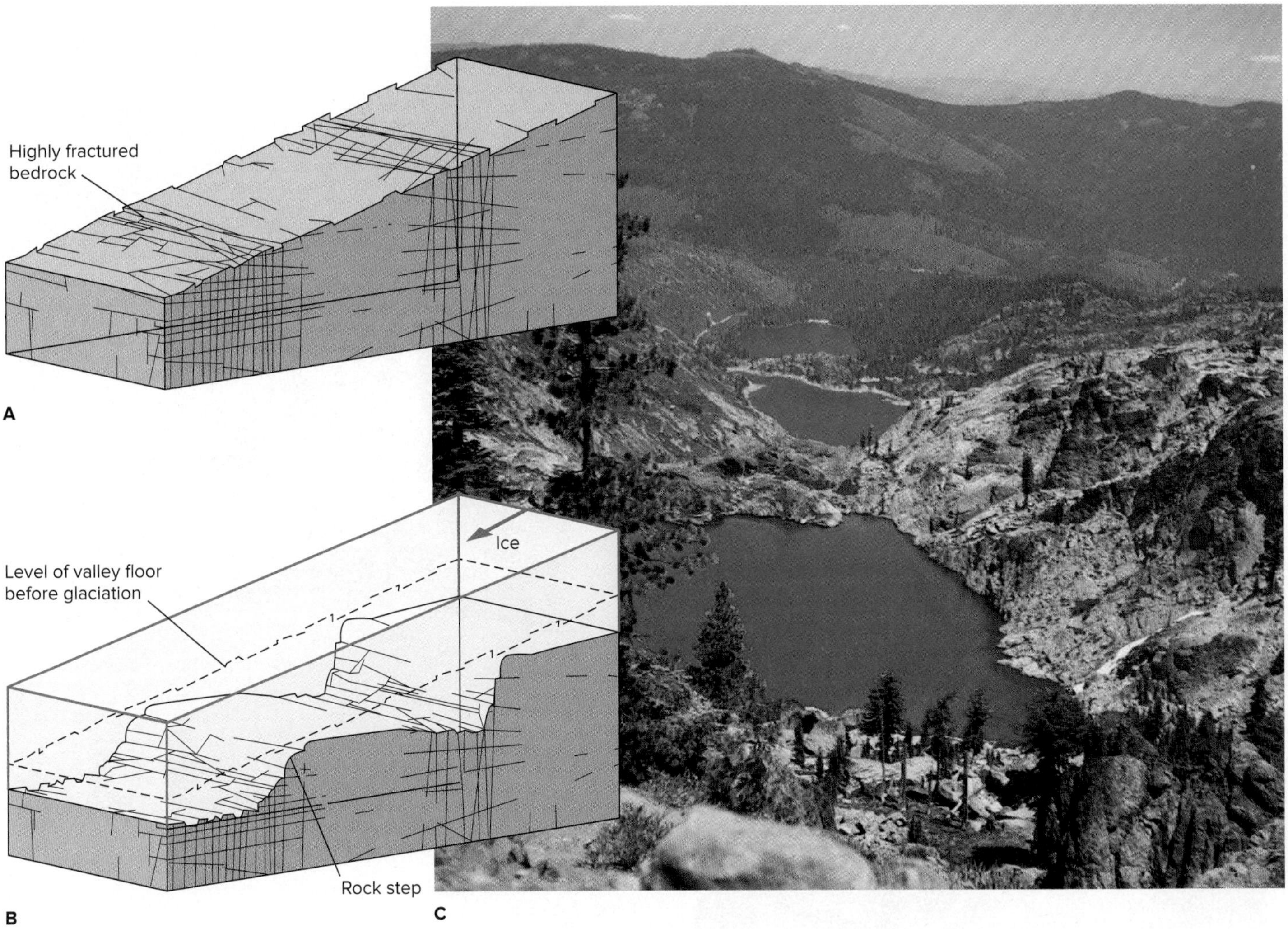

Highly fractured bedrock

A

Ice

Level of valley floor before glaciation

Rock step

B

C

FIGURE 12.17

Development of rock steps. (*A*) Valley floor before glaciation. (*B*) During glaciation. (*C*) Rock steps and rock-basin lakes. Sierra Nevada, California.
A and B after F. E. Matthes, 1930, U.S. Geological Survey; (C) ©C. C. Plummer

FIGURE 12.18

A cirque occupied by a small glacier in the Canadian Rocky Mountains. The glacier was much larger during the ice ages.
©C. C. Plummer

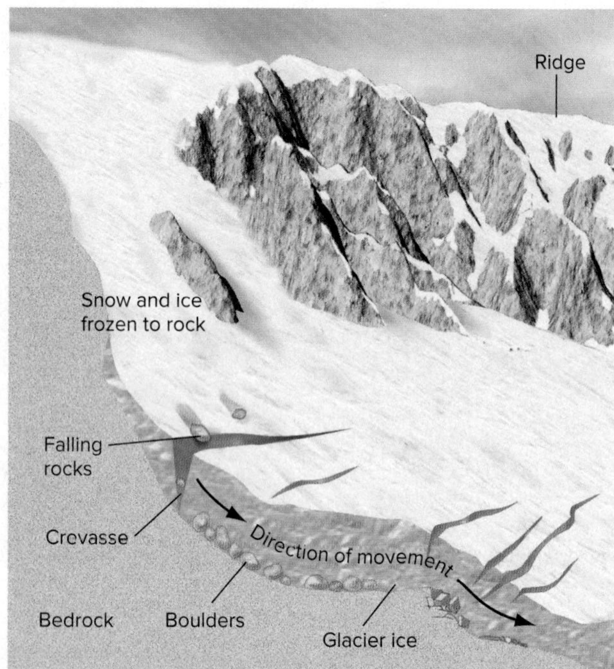

FIGURE 12.19
Cutaway view of a cirque.

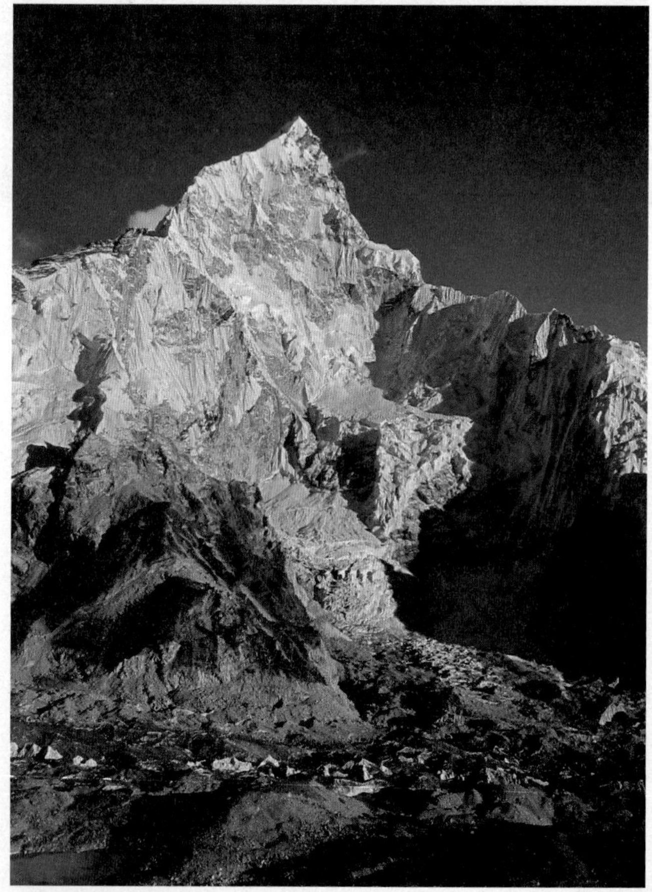

FIGURE 12.20
Ama Dablan, a horn in the Mount Everest region of the Himalaya in Nepal. Note the cirque below the peak.
©*Alisha Wenzel*

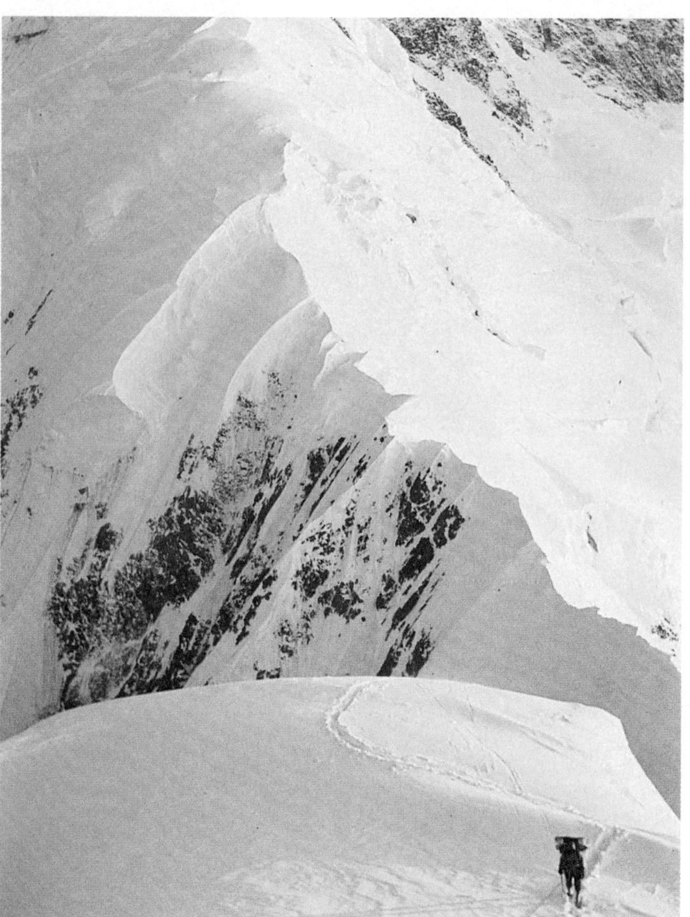

FIGURE 12.21
An arête on Mount Logan, Yukon, Canada.
©*C. C. Plummer*

Frost wedging works on the rock exposed above the glacier, steepening and cutting back the side walls of the valley. Sharp ridges called **arêtes** separate adjacent glacially carved valleys (figure 12.21).

Erosional Landscapes Associated with Continental Glaciation

In contrast to the rugged and angular nature of glaciated mountains, an ice sheet tends to produce rounded topography. The rock underneath an ice sheet is eroded in much the same way as the rock beneath a valley glacier; however, the weight and thickness of the ice sheet may produce more pronounced effects. Rounded knobs are common (figure 12.22), as are grooved and striated bedrock. Some grooves are actually channels several meters deep and many kilometers long. The orientation of grooves and striations indicates the direction of movement of a former ice sheet.

An ice sheet may be thick enough to bury mountain ranges, rounding off the ridges and summits and perhaps streamlining them in the direction of ice movement. Much of northeastern Canada, with its rounded mountains and grooved and striated

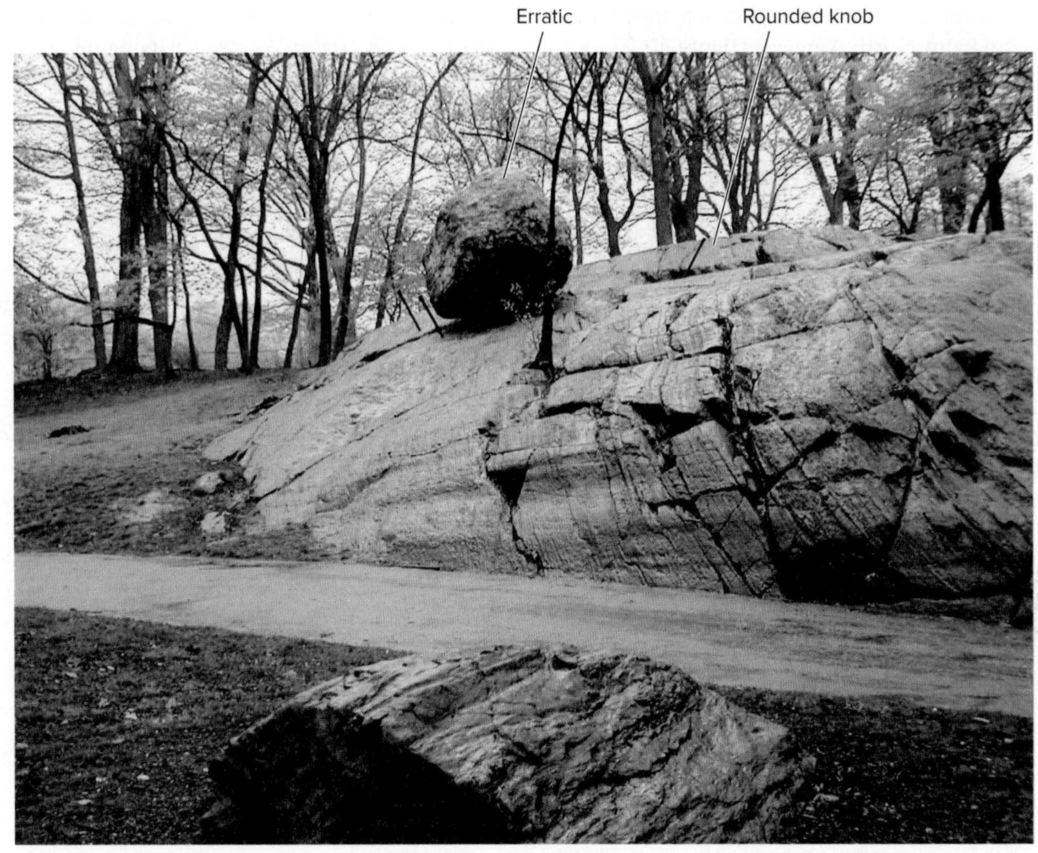

FIGURE 12.22

Rounded knob (*roche moutonnée*) with an erratic (the boulder) on it in Central Park, New York City.
©Charles Merguerian

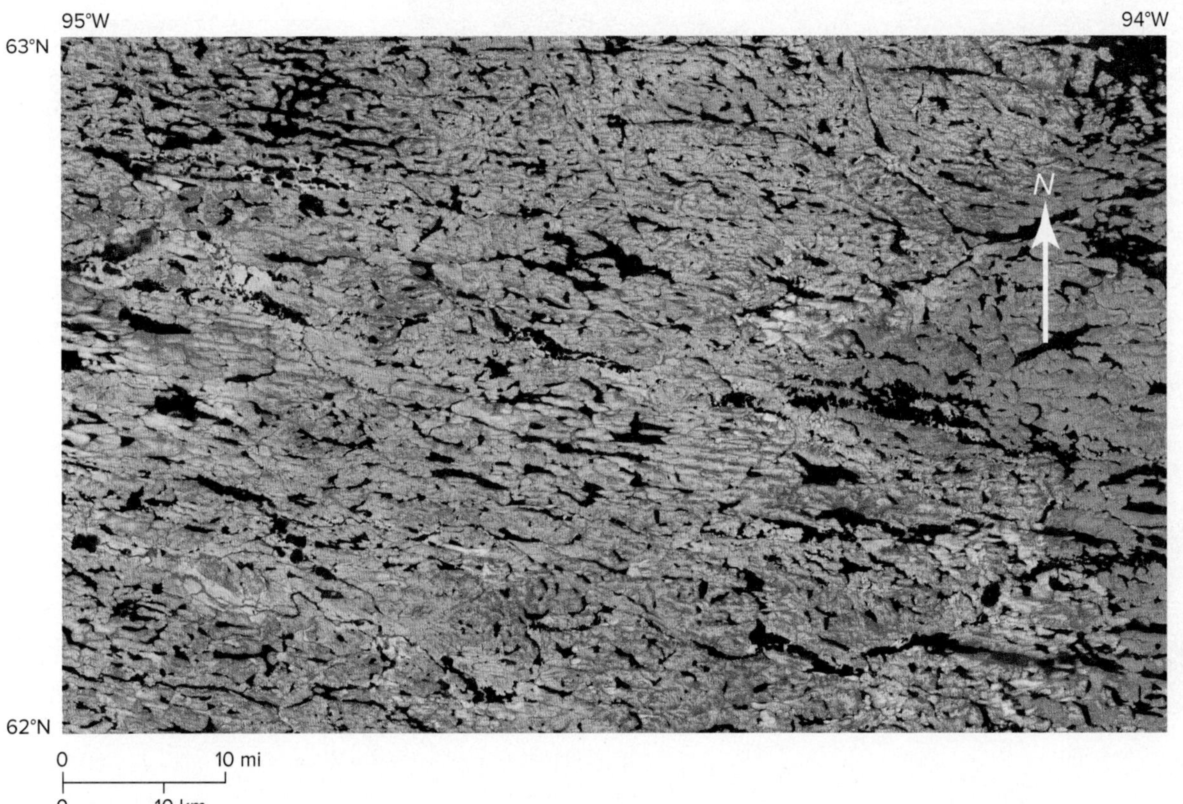

FIGURE 12.23

Glacially scoured terrain in the Canadian Shield area south of Clearwater Lakes, Quebec, Canada. In this region, glacial erosion was so complete that the underlying ancient bedrock was exposed. The generally northwest-southeast movement of the ice sheet is apparent in the alignment of the many scoured lakes and ponds.
Source: Jesse Allen and Robert Simmon, using Landsat 8 data from the USGS Earth Explorer

bedrock surface, shows the erosional effects of ice sheets that formerly covered that part of North America (figure 12.23).

12.4 GLACIAL DEPOSITION

The rock fragments scraped and plucked from the underlying bedrock and carried along at the base of the ice make up most of the load carried by an ice sheet but only part of a valley glacier's load. Much of a valley glacier's load comes from rocks broken from the valley walls.

Most of the rock fragments carried by glaciers are angular, as the pieces have not been tumbled around enough for the edges and corners to be rounded. The debris is unsorted, and clay-sized to boulder-sized particles are mixed together (figure 12.24). The unsorted and unlayered rock debris carried or deposited by a glacier is called **till**.

Glaciers are capable of carrying virtually any size of rock fragment, even boulders as large as a house. An **erratic** is an ice-transported boulder that has not been derived from underlying bedrock (figure 12.22). If its bedrock source can be found, the erratic indicates the direction of movement of the glacier that carried it.

Moraines

When till occurs as a body of unsorted and unlayered debris either on a glacier or left behind by a glacier, the body is regarded as one of several types of **moraines**. **Lateral moraines** are elongate, low mounds of till that form along the sides of a valley glacier (figures 12.24, 12.25, 12.26, and 12.27). Rockfall debris from the steep cliffs that border valley glaciers accumulates along the edges of the ice to form lateral moraines.

Where tributary glaciers come together, the adjacent lateral moraines join and are carried downglacier as a single long ridge of till known as a **medial moraine**. In a large trunk glacier that has formed from many tributaries, the numerous medial moraines give the glacier the appearance from the air of a multilane highway (figures 12.25 and 12.26).

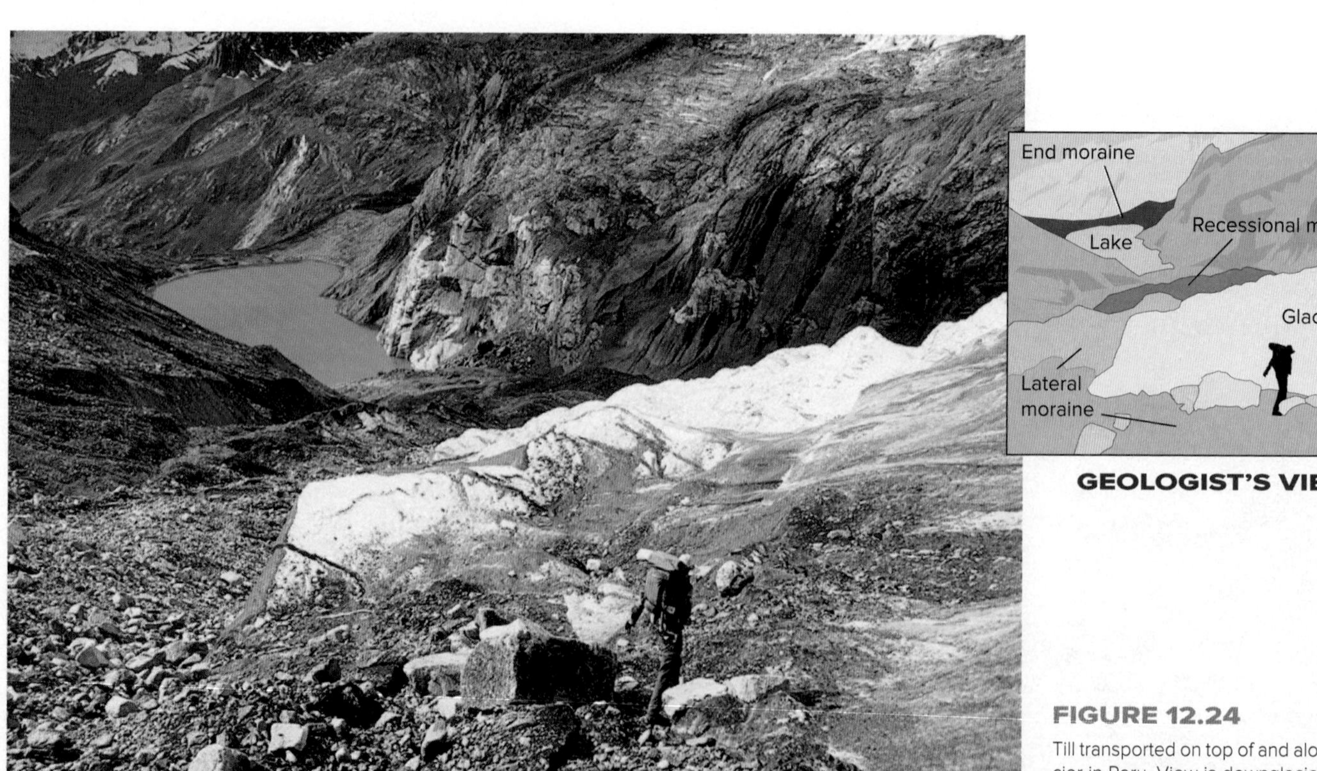

FIGURE 12.24

Till transported on top of and alongside a glacier in Peru. View is downglacier. The lake is dammed by an end moraine at its far end. ©C. C. Plummer

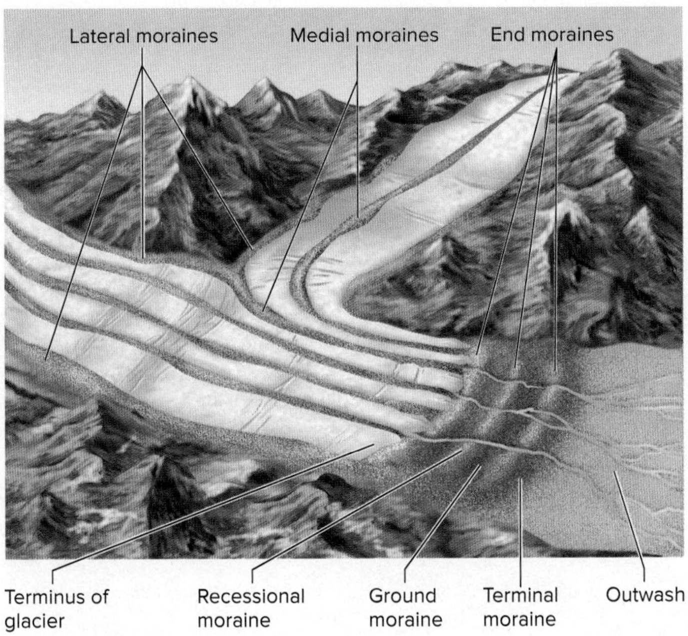

FIGURE 12.25

Moraines associated with valley glaciers.

An actively flowing glacier brings debris to its terminus. If the terminus remains stationary for a few years or advances, a distinct **end moraine**, a ridge of till, piles up along the front edge of the ice. Valley glaciers build end moraines that are crescent-shaped or sometimes horseshoe-shaped (figures 12.25 and 12.27). The end moraine of an ice sheet takes a similar lobate form but is much longer and more irregular than that of a valley glacier (figure 12.28).

Geologists distinguish two special kinds of end moraines. A *terminal moraine* is the end moraine marking the farthest advance of a glacier. A *recessional moraine* is an end moraine built while the terminus of a receding glacier remains temporarily stationary. A single receding glacier can build several recessional moraines (as in figures 12.24, 12.25, 12.27, and 12.28).

As ice melts, rock debris that has been carried by a glacier is deposited to form a **ground moraine**, a fairly thin, extensive layer or blanket of till (figures 12.25 and 12.28). Very large areas that were once covered by an ice sheet now have the gently rolling surface characteristic of ground moraine deposits.

In some areas of past continental glaciation, there are bodies of till shaped into streamlined hills called **drumlins** (figures 12.28 and 12.29). A drumlin is shaped like an inverted spoon aligned parallel to the direction of ice movement of the former

FIGURE 12.26

Medial and lateral moraines on valley glaciers, Yukon, Canada. Ice is flowing toward viewer and to lower right.
©C. C. Plummer

FIGURE 12.27

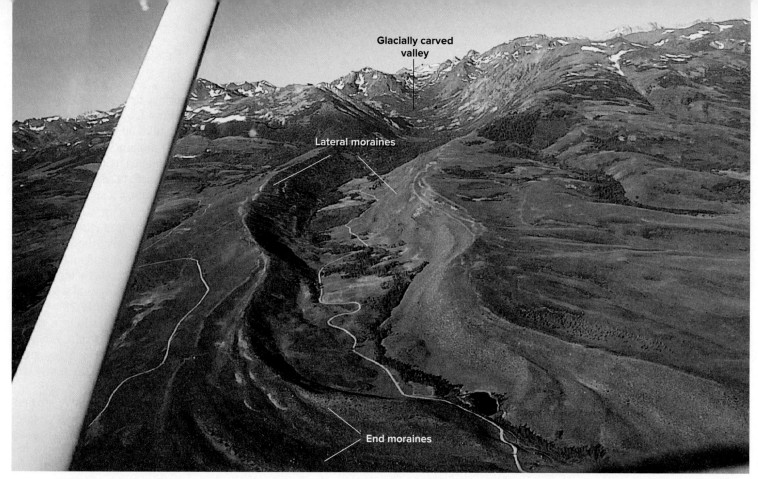

End moraines (recessional moraines) in the foreground curve into two long, lateral moraines. The two lateral moraines extend back to a glacially carved valley in the Sierra Nevada, California.
©C. C. Plummer

FIGURE 12.28

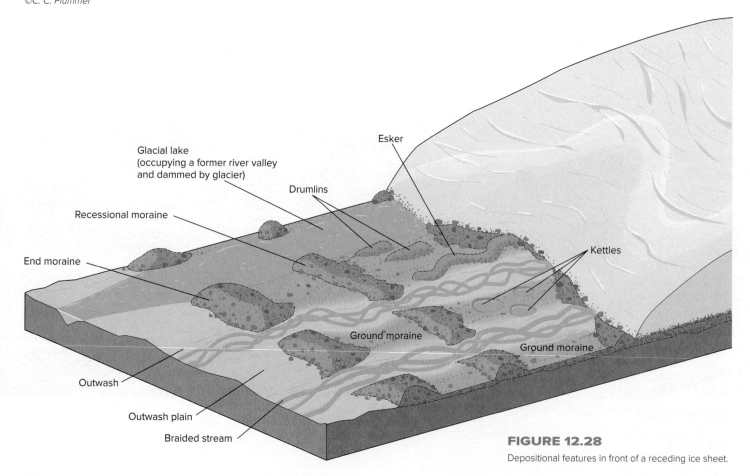

Depositional features in front of a receding ice sheet.

FIGURE 12.29

Drumlins near Fort Atkinson, Wisconsin. The glacier that formed them was moving from right to left of the photograph.
©Doug Sherman/Geofile

glacier. Its gentler end points in the downglacier direction. Because we cannot observe drumlins forming beneath present ice sheets, we are not certain how till becomes shaped into these streamlined hills.

Outwash

In the zone of ablation, large quantities of meltwater usually run over, beneath, and away from the ice. The material deposited by the debris-laden meltwater is called **outwash**. Because it has the characteristic layering and sorting of stream-deposited sediment, outwash can be distinguished easily from the unlayered and unsorted deposits of till. Because outwash is fairly well sorted and the particles generally are not chemically weathered, it is an excellent source of aggregate for building roadways and for mixing with cement to make concrete.

An outwash feature of unusual shape associated with former ice sheets and some very large valley glaciers is an **esker**, a long, sinuous ridge of water-deposited sediment (figures 12.28 and 12.30). Eskers can be up to 10 meters high and are formed of cross-bedded and well-sorted sediment. Evidently eskers are deposited in tunnels within or under glaciers, where meltwater loaded with sediment flows under and out of the ice.

As meltwater builds thick deposits of outwash alongside and in front of a retreating glacier, blocks of stagnant ice may be surrounded and buried by sediment. When the ice block finally melts (sometimes years later), a depression called a **kettle** forms (figures 12.28 and 12.31). Many of the small, scenic lakes in the upper middle west of the United States are kettle lakes. A *kame* is a low mound or irregular ridge formed of outwash deposits on a stagnating glacier. Sediment accumulates in depressions or troughs on a glacier's surface. When the ice melts, the sediment remains as irregular, moundlike hills. The irregular, bumpy landscape of hills and depressions associated with many moraines is known as *kame and kettle topography.*

FIGURE 12.30

An esker in northeastern Washington.
©D. A. Rahm, courtesy of Rahm Memorial Collection, Western Washington University

FIGURE 12.31

A kettle, a kame, and outwash (*background* and *left*) from a glacier, Yukon, Canada. Stagnant ice underlies much of the till.
©C. C. Plummer

The streams that drain glaciers tend to be very heavily loaded with sediment, particularly during the melt season. As they come off the glacial ice and spread out over the outwash deposits, the streams form a braided pattern (see chapter 10 on streams).

The large amount of rock flour that these streams carry in suspension settles out in quieter waters. In dry seasons or drought, the water may dry up, and the rock flour deposits may be picked up by the wind and carried long distances. Some of the best agricultural soil in the United States has been formed by rock flour that has been redeposited by wind. Such fine-grained, wind-blown deposits of dust are called *loess* (see chapter 13).

Glacial Lakes and Varves

Lakes often occupy depressions carved by glacial erosion, but they can also form behind dams built by glacial deposition. Commonly, a lake forms between a retreating glacier and an earlier end moraine (see figure 12.24).

In the still water of the lake, clay and silt settle on the bottom in two thin layers—one light-colored, one dark—that are characteristic of glacial lakes. Two layers of sediment representing one year's deposition in a lake are called a **varve** (figure 12.32). The light-colored layer consists of slightly coarser sediment (silt) deposited during the warmer part of the year when the nearby glacier is melting and sediment is transported to the lake. The silt settles within a few weeks or so after reaching the lake. The dark layer is finer sediment (clay)—material that sinks down more slowly during the winter after the lake surface freezes and the supply of fresh, coarser sediment stops due to lack of meltwater. The dark color is attributed to fine organic matter mixed with the clay.

FIGURE 12.32

Varves from a former glacial lake. Each pair of light and dark layers represents a year's deposition.
Source: U.S. Geological Survey

Because each varve represents a year's deposit, varves are like tree rings and indicate how long a glacial lake existed.

12.5 PAST GLACIATION

In the early 1800s, the hypothesis of past extensive continental glaciation of Europe was proposed. Among the many people who regarded the hypothesis as outrageous was the Swiss naturalist Louis Agassiz. But, after studying the evidence in Switzerland, he changed his mind. In 1837, he published a discourse arguing that Switzerland was, in the past, entirely glaciated. Subsequently, he, along with the eminent English geologist William Buckland, found the same evidence in the British Isles for past glaciation that was found in Switzerland. At present, there are no glaciers in Britain or Ireland. Agassiz, after further studies in northern Europe, concluded that a great glacier had covered most or all of Europe. Agassiz had to overcome

PLANETARY GEOLOGY 12.5

Mars on a Glacier

Meteorites are extraterrestrial rocks—fragments of material from space that have managed to penetrate Earth's atmosphere and land on Earth's surface. They are of interest not only to astronomers but to geologists, for they help us date Earth (chapter 8) and give us clues to what Earth's interior is like (see chapter 17) because many of the meteorites are thought to represent fragments of destroyed minor planets. Meteorites are rarely found; they usually do not look very different from Earth's rocks with which they are mixed.

The United States meteorite collecting program, known as ANSMET (Antarctic Search for Meteorites) has collected over 10,000 meteorites since it began in 1976. Collections from other countries' programs bring the total to over 20,000. This far exceeds the total collected elsewhere in the past two centuries. Over a thousand meteorites have been collected from one small area where the ice sheet abuts against the Transantarctic Mountains. The reason for this heavy concentration is that meteorites landing on the surface of the ice over a vast area have been incorporated into the glacier and transported to where ablation takes place. The process is illustrated in box figure 1.

A few of the meteorites are especially intriguing. Some almost certainly are rocks from the moon, while several others apparently came from Mars. Their chemistry and physical properties match what we would expect of a Martian rock. But how could a rock escape Mars and travel to Earth? Scientists suggest that a meteorite hit Mars with such force that fragments of that planet were launched into space. Eventually, some of the fragments reached Earth.

In 1996, researchers announced that they found what could be signs of former life on Mars in one of the meteorites collected twelve years earlier in Antarctica. The evidence included carbon-containing molecules that might have been produced by living organisms as well as microscopic blobs that could be fossil alien bacteria. But there are alternate explanations for each line of evidence, and a hot debate has ensued between scientists with opposing viewpoints.

Additional Resource

ANSMET The Antarctic Search for Meteorites

- http://caslabs.case.edu/ansmet/

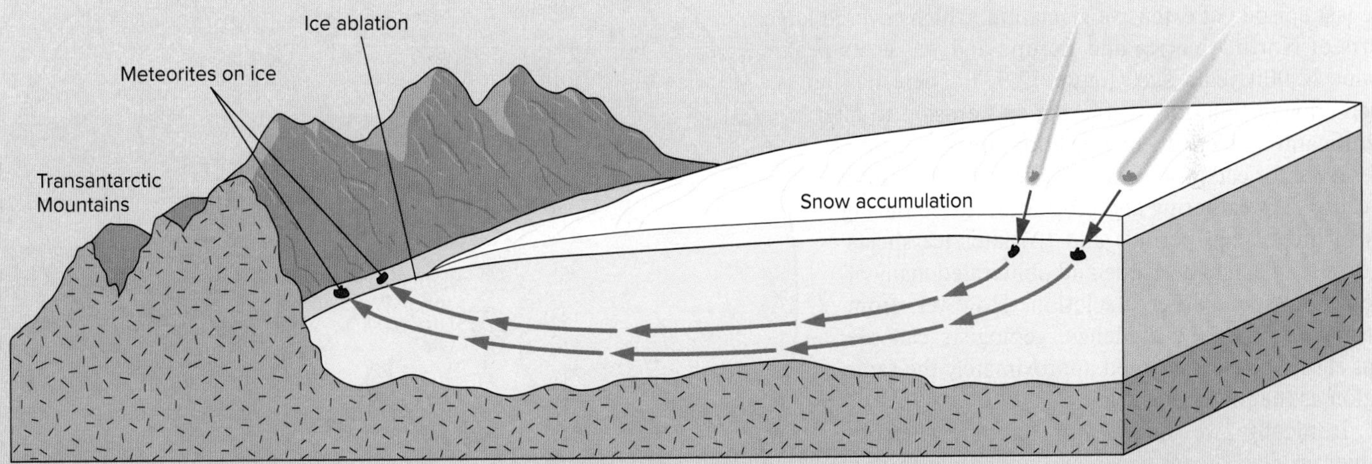

BOX 12.5 ■ FIGURE 1

Diagram showing the way in which meteorites are concentrated in a narrow zone of wastage along the Transantarctic Mountains. Two meteorites are shown as well as the paths they would have taken from the time they hit the ice sheet until they reached the zone of ablation. The vertical scale is greatly exaggerated.
Source: Antarctic Journal of the United States

skepticism over past climates being quite different from those of today. At the time, the hypothesis seemed to many geologists to be a violation of the principle of uniformitarianism. Agassiz later came to North America and worked with American geologists who had found similar indications of large-scale past glaciation on this continent.

As more evidence accumulated, the hypothesis became accepted as a theory that today is seldom questioned. The **theory of glacial ages** states that at times in the past, colder climates prevailed during which much more of the land surface of Earth was glaciated than at present.

As the glacial theory gained general acceptance during the latter part of the nineteenth century, it became clear that much of northern Europe and the northern United States as well as most of Canada had been covered by great ice sheets during the so-called Ice Age. It also became evident that even areas not covered by ice had been affected because of the changes in climate and the redistribution of large amounts of water.

We now know that the last of the great North American ice sheets melted away from Canada less than 10,000 years ago. In many places, however, till from that ice sheet overlies older tills, deposited by earlier glaciations. The older till is distinguishable from the newer till because the older till was deeply weathered during times of warmer climate between glacial episodes.

Geologists can reconstruct with considerable accuracy the last episode of extensive glaciation, which covered large parts of North America and Europe and was at its peak about 18,000 years ago (figure 12.33). There has not been enough time for weathering and erosion to alter significantly the effects of glaciation. Less evidence is preserved for each successively older glacial episode, because (1) weathering and erosion occurred during warm interglacial periods and (2) later ice sheets and valley glaciers overrode and obliterated many of the features of earlier glaciation. However, from piecing together the evidence, geologists can see that earlier glaciers covered approximately the same region as the more recent ones.

Ironically, we know more about when the numerous glacial ages began and ended not from glacial deposits but from deep-ocean sediment. As described in chapter 2, oxygen isotope studies ($^{18}O/^{16}O$ ratios in microfossil shells) have delineated changes in the temperature of near-surface ocean water. The changes of temperature of the seas have been correlated with periods of extensive worldwide glaciation and intervening, interglacial warm climates. Studies of ice cores from Antarctica and Greenland have also provided important information on climate change (see box 12.3).

Although the glacial ages are generally associated with the Pleistocene Epoch (see chapter 8), cooling actually began earlier. Recent work indicates that worldwide climate changes necessary for northern continental glaciation probably began at least 3 million years ago, late in the Tertiary Period, at least a million years before the Pleistocene. Moreover, Antarctica has been glaciated for 14 million years.

Earth has undergone episodic changes in climate during the last 2 to 3 million years. Actually, the climate changes necessary for a glacial age to occur are not so great as one might imagine. During the height of a glacial age, the worldwide average of annual temperatures was probably only about 5°C cooler than at present. Some of the intervening interglacial periods were probably a bit warmer worldwide than present-day average temperatures.

The likely causes of ice ages and their periodicity are included in chapter 21, on climate change. The pertinent discussion begins in the section A Brief History of Earth's Climate.

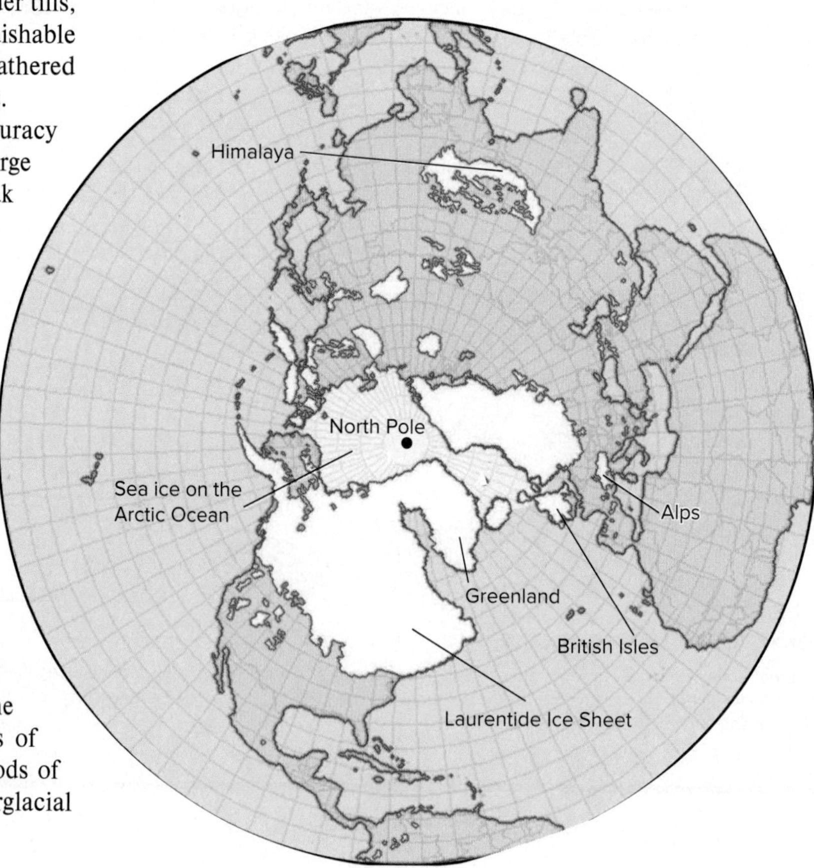

FIGURE 12.33

Maximum glaciation during the Pleistocene in the Northern Hemisphere. Small glaciated areas are in mountains. Note that ice sheets extended beyond present continental shorelines. This is because sea level was lower than at present. Note that the North Pole (center of map), which is in the Arctic Ocean, was not glaciated.

Source: J. Ehlers and P. L. Gibbard, Quaternary glaciations—extent and chronology, Parts I–III, Elsevier, 2004.

Direct Effects of Past Glaciation in North America

Moving ice abraded vast areas of northern and eastern Canada during the growth of the North American ice sheets. Most of the soil and sedimentary rock was scraped off, and underlying crystalline bedrock was scoured. Many thousands of future lake basins were gouged out of the bedrock.

The directions of ice flow can be determined from the orientation of striations and grooves in the bedrock and from elongate, rounded knobs. The largest ice sheet (the Laurentide Ice Sheet) moved outward from the general area now occupied by Hudson Bay, which is where the ice sheet was thickest. The present generally barren surface of the Hudson Bay area contrasts markedly with the Great Plains surface of southern Canada and northern United States, where vast amounts of till were deposited.

Most of the till was deposited as ground moraine, which, along with outwash deposits, has partially weathered to yield excellent soil for agriculture. Rock flour that originally washed out of ice sheets has been redistributed by wind, as *loess*, over large parts of the Midwest and eastern Washington to contribute to especially good agricultural land (see chapter 13). In many areas along the southern boundaries of land covered by ground moraines, broad and complex end moraines extend for many kilometers, indicating that the ice margin must have been close to stationary for a long time (figure 12.34). Numerous drumlins are preserved in areas such as Ontario, New England, and upstate New York. New York's Long Island is made of terminal and recessional moraines and outwash deposits. Erratics there come from metamorphic rock in New England. Cape Cod in Massachusetts was also formed from moraines.

Glaciers have a tremendous ability to form lakes through both erosion and deposition. Most states and provinces that were glaciated have thousands of lakes (figure 12.23). By contrast, Virginia, which was not glaciated, has only two natural lakes. Minnesota bills itself as "the land of 10,000 lakes." Most of those lakes are kettle lakes. The Finger Lakes in New York (figure 12.35) are in long, north-south glacially modified valleys that are dammed by recessional moraines at their southern ends. The Great Lakes are, at least in part, a legacy of continental

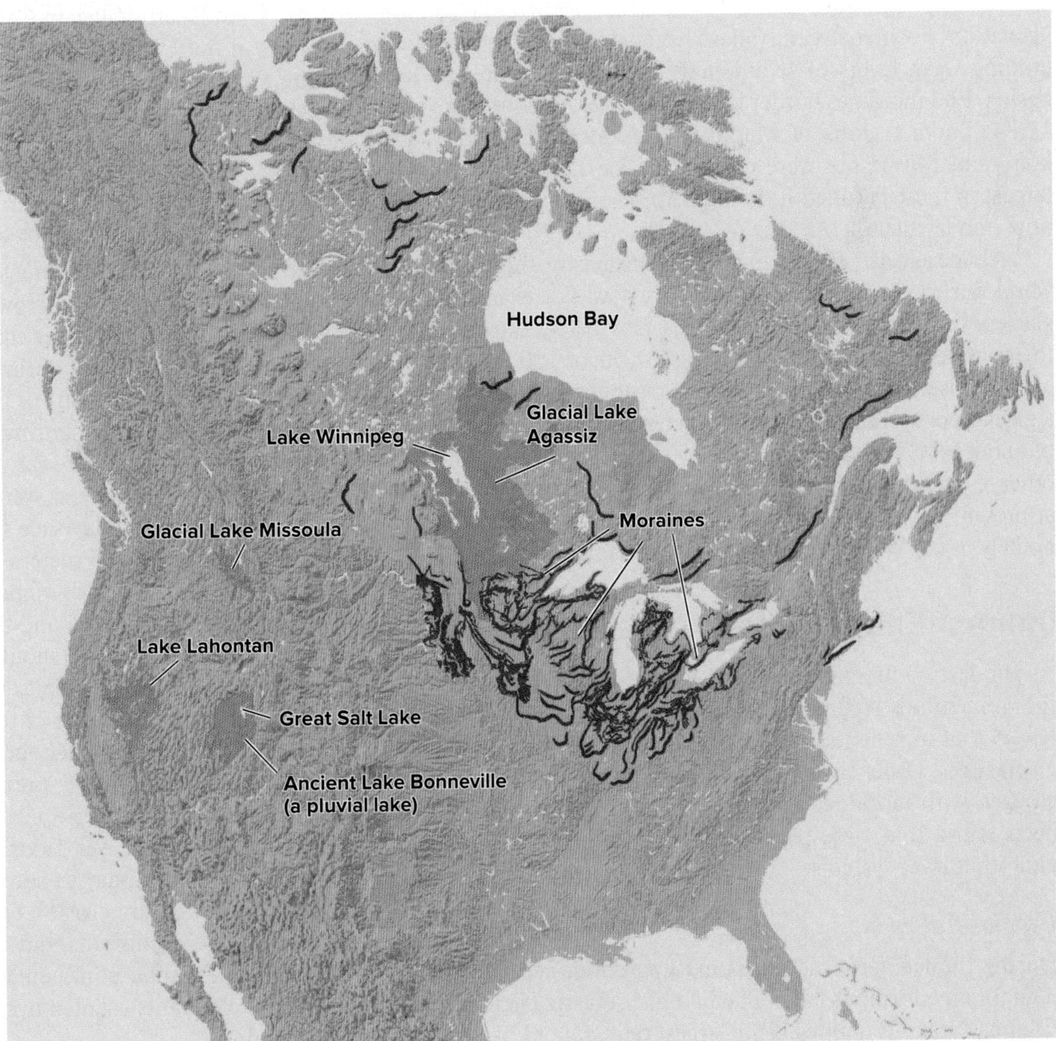

FIGURE 12.34

End moraines in the contiguous United States and Canada (shown by brown lines), and glacial lakes Agassiz and Missoula, as well as pluvial lakes in the western United States (magenta).

Source: C. S. Denny, U.S. Geological Survey, and the Geological Map of North America, Geological Society of America, and the Geological Survey of Canada

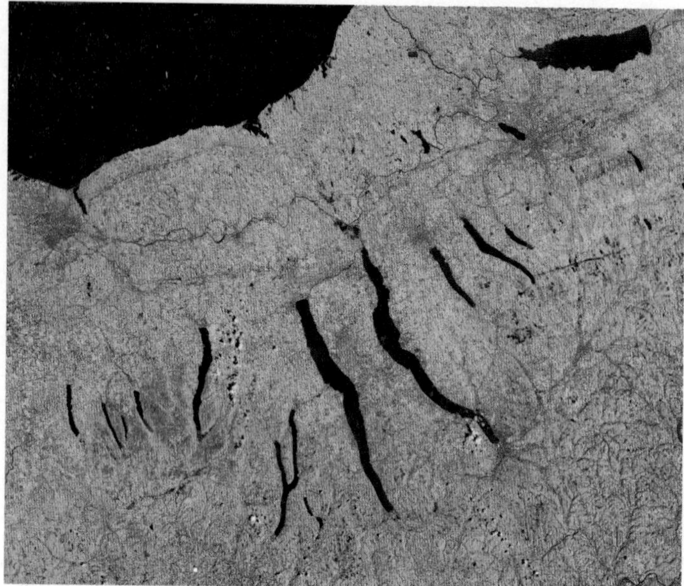

FIGURE 12.35

Satellite image of Finger Lakes in New York. Part of Lake Ontario is at the top. ©Marshall Ikonography/Alamy

glaciation. Former stream valleys were widened by the ice sheet eroding weak layers of sedimentary rock into the present lake basins. End moraines border the Great Lakes, as shown in figure 12.34. Large regions of Manitoba, Saskatchewan, North Dakota, and Minnesota were covered by ice-dammed lakes. The largest of these is called Lake Agassiz. The former lake beds are now rich farmland.

Alpine glaciation was much more extensive throughout the world during the glacial ages than it is now. For example, small glaciers in the Rocky Mountains that now barely extend beyond their cirques were then valley glaciers 10, 50, or 100 kilometers in length. Yosemite Valley, which is no longer glaciated, was filled by a glacier about a kilometer thick. Its terminus was at an elevation of about 1,300 meters above sea level. Furthermore, cirques and other features typical of valley glaciers can be found in regions that at present have no glaciers, such as the northern Appalachians—notably in the White Mountains of New Hampshire.

Indirect Effects of Past Glaciation

As the last continental ice sheet wasted away, what effects did the tremendous volume of meltwater have on American rivers? Rivers that now contain only a trickle of water were huge in the glacial ages. Other river courses were blocked by the ice sheet or clogged with morainal debris. Large dry stream channels have been found that were preglacial tributaries to the Mississippi and other river systems.

Glacial Lakes

During the ice ages, the retreating Laurentide Ice Sheet was a dam for an enormous lake—Glacial Lake Agassiz (figure 12.34). Canada's Lake Winnipeg is the largest remnant of Lake Agassiz.

Another noteworthy glacial lake was Glacial Lake Missoula. This lake formed when the ice sheet advanced into Montana. Ice dammed up a river system, creating a large lake. Eventually, the lake overtopped and destroyed the ice dam, resulting in a giant flood (box 12.6).

Pluvial Lakes

During the glacial ages, the climate in North America, beyond the glaciated parts, was more humid than it is now. Most of the now-arid regions of the western United States had moderate rainfall, as traces or remnants of numerous lakes indicate. These **pluvial lakes** (formed in a period of abundant rainfall) once existed in Utah, Nevada, and eastern California (figure 12.34). Some may have been fed by meltwater from mountain glaciers, but most were simply the result of a wetter climate.

Great Salt Lake in Utah is but a small remnant of a much larger body of fresh water called Lake Bonneville, which, at its maximum size, was nearly as large as Lake Michigan is today. Ancient beaches and wave-cut terraces on hillsides indicate the depth and extent of ancient Lake Bonneville. As the climate became more arid, lake levels lowered, outlets were cut off, and the water became salty, eventually leaving behind the Bonneville salt flats and the present very saline Great Salt Lake.

Even Death Valley in California—now the driest and hottest place in the United States—was occupied by a deep lake during the Pleistocene. The salt flats (see figure 6.22) that were left when this lake dried include rare boron salts that were mined during the pioneer days of the American West.

Lowering and Rising of Sea Level

All of the water for the great glaciers had to come from somewhere. The water was "borrowed" from the oceans, such that sea level worldwide was lower than it is today—at least 130 meters lower, according to scientific estimates. Figure 12.36 shows the present shoreline for part of North America and the position of the inferred coastline during the height of Pleistocene glaciation. It also shows where the shoreline would be if the Antarctic and Greenland ice sheets were to disappear.

Recall that if today's ice sheets were to melt, sea level worldwide would rise by 65 meters, and shorelines would be considerably farther inland. It's important to realize that our present shorelines are not fixed and that they are very much controlled by climate changes. We should also realize that we are still in a cooler-than-usual (relative to most of Earth's history) time, perhaps the lingering effects of the last ice age.

What is the evidence for lower sea level? Stream channels have been charted in the present continental shelves, the gently inclined, now submerged edges of the continents (described in chapter 18 on the sea floor). These submerged channels are continuations of today's major rivers and had to have been above sea level for stream erosion to take place. Bones and teeth from now-extinct mammoths and mastodons have been dredged up from the Atlantic continental shelf, indicating that these relatives of elephants roamed over what must have been dry land at the time.

IN GREATER DEPTH 12.6

The Channeled Scablands

In chapter 4, we described how the Columbia plateau in the Pacific Northwest (see figure 4.8) was built by a series of successive lava floods. The northeastern part of the plateau features a unique landscape, known as the channeled scablands, where the basalt bedrock has been carved into a series of large, interweaving valleys. From the air, the pattern looks like that of a giant braided stream. The channels, however, which range up to 30 kilometers wide and are usually 15 to 30 meters deep, are mostly dry.

The scablands are believed to have been carved by gigantic floods of water. Huge ripples in gravel bars (box figure 1) support this idea. To create these ripples, a flood would have to be about 10 times the combined discharge of all the world's rivers. This is much larger than any flood in recorded history.

What seems to have occurred is that, during the ice ages, a lobe of the ice sheet extended southward into northern Washington, Montana, and Idaho, blocking the head of the valley occupied by the Clark Fork River. The ice provided a natural dam for what is now known as Glacial Lake Missoula. Lake Missoula drowned a system of valleys in western Montana that extended hundreds of kilometers into the Rocky Mountains.

Ice is not ideal for building dams. Upon failure of the glacial dam, the contents of Lake Missoula became the torrential flood that scoured the Columbia plateau. There were dozens of giant floods. Advancing ice from Canada would reestablish the dam, only to be destroyed after the reservoir refilled.

Mars has what appear to be giant outflow channels that are similar to those of the channeled scablands. At present, Mars has no liquid water, but the channels suggest that there must have been a huge amount of water in the distant past.

Additional Resource

Ice Age Flood Home Site

- http://iafi.org/

BOX 12.6 ■ FIGURE 1

Giant ripples of gravel from the draining of Lake Missoula, Montana. For scale, see farm buildings in lower middle of photo. *Source: P. Weis, U.S. Geological Survey*

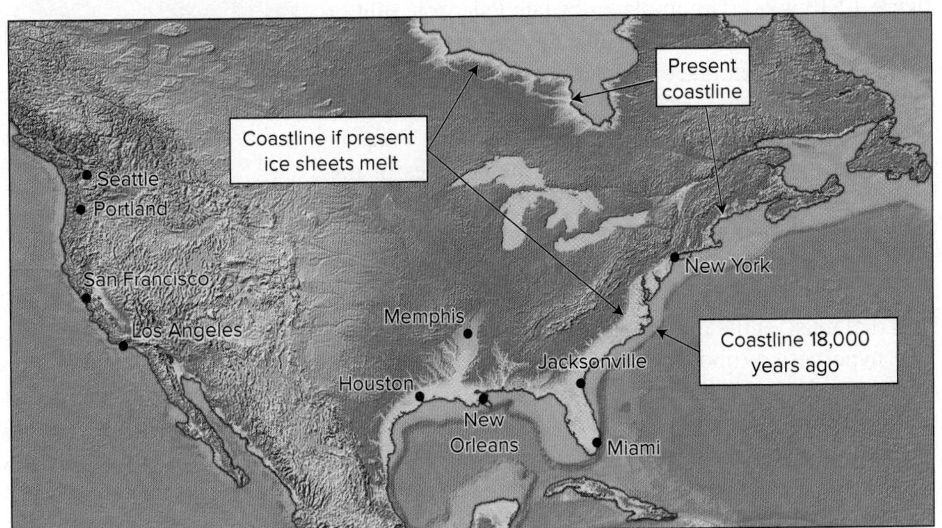

FIGURE 12.36

Map of North America showing the present coastline as well as the coastline at the height of the last glacial age, around 18,000 years ago. The map also shows the coastline we could expect if the Greenland and Antarctic ice sheets melt.

A **fiord** (also spelled fjord) is a coastal inlet that is a drowned glacially carved valley (figure 12.37). Fiords are common along the mountainous coastlines of Alaska, British Columbia, Chile, New Zealand, and Norway. Surprisingly, the lower reach of the Hudson River, just north of New York City, is a fiord. Fiords are evidence that valleys eroded by past glaciers were later partly submerged by rising sea level.

FIGURE 12.37

Tracy Arm Fiord, southeast Alaska. Underwater it is a deep, U-shaped valley. ©*Sunset Avenue Productions/DigitalVision/Getty Images RF*

Crustal Rebound

The weight of an ice sheet several thousand meters thick depresses the crust of Earth much as the weight of a person depresses a mattress. A land surface bearing the weight of a continental ice sheet may be depressed several hundred meters.

Once the glacier is gone, the land begins to rebound slowly to its previous height (see chapter 17 and figures 17.13 and 17.14). Uplifted and tilted shorelines along lakes are an indication of this process. The Great Lakes region is still rebounding as the crust slowly adjusts to the removal of the last ice sheet.

Evidence for Older Glaciation

Throughout most of geologic time, the climate has been warmer and more uniform than it is today. We think that the late Cenozoic Era is unusual because of the periodic fluctuations of climate and the widespread glaciations. However, glacial ages are not restricted to the late Cenozoic.

Evidence of older glaciation comes from rocks called tillites. A **tillite** is lithified till (figure 12.38). Unsorted rock particles, including angular, striated, and faceted boulders, have been consolidated into a sedimentary rock. In some places, tillite layers overlie surfaces of older rock that have been polished and striated. Tillites of the late Paleozoic and tillites representing a minor part of the late Precambrian crop out in parts of the southern continents. (The striated surface in Australia shown in figure 12.13 is overlain in places by late Paleozoic tillite.)

FIGURE 12.38

The Dwyka tillite in South Africa. It is of Permian (late Paleozoic) age. Similar Permian tillites found in South America, Australia, and India are used as evidence for the existence of a supercontinent and continental drift (see figure 19.4).
©*Robert J. Stull*

The oldest glaciation for which we have evidence appears to have taken place in what is now Ontario around 2.3 billion years ago.

Support is growing for the idea that a late Precambrian Ice Age was so extensive that the surfaces of the world's oceans were frozen. Although the concept was first proposed in the early twentieth century, scientists in the 1990s began taking it seriously and called it the *Snowball Earth hypothesis.* Evidence for the hypothesis includes tillites that must, at the time, have been deposited near the equator. The hypothesis proposes that the extreme cold was due to the Sun being weaker at the time and the absence of carbon dioxide and other greenhouse gases in the atmosphere. For more, go to www.snowballearth.org/.

Paleozoic glaciation provides strong support for plate tectonics. The late Paleozoic tillites in the southern continents (South Africa, Australia, Antarctica, South America) indicate that these landmasses were once joined (see chapter 19 on plate tectonics). Directions of striations show that an ice sheet flowed onto South America from what is now the South Atlantic Ocean. Because an ice sheet can build up only on land, it is reasonable to conclude that the former ice sheet was centered on the ancient supercontinent.

Summary

A *glacier* is a large, long-lasting mass of ice that forms on land and moves under its own weight. A glacier can form wherever more snow accumulates than is lost. *Ice sheets* and *valley glaciers* are the two most common types of glaciers. Glaciers move downward from where the most snow accumulates toward where the most ice is ablated.

A glacier moves by both basal sliding and internal flow. The upper portion of a glacier tends to remain rigid and is carried along by the ice flowing beneath it.

Glaciers advance and recede in response to changes in climate. A receding glacier has a *negative budget*, and an advancing one has a *positive budget*. A glacier's budget for the year can be determined by noting the relative position of the *equilibrium line*, which separates the *zone of accumulation* from the *zone of ablation*.

Snow recrystallizes into firn, which eventually becomes converted to glacier ice. Glacier ice is lost (or ablated) by melting, breaking off as icebergs, and direct evaporation of the ice into the air.

A glacier erodes by plucking and the grinding action of the rock it carries. The grinding produces rock flour and faceted and polished rock fragments. Bedrock over which a glacier moves is generally polished, striated, and grooved.

A mountain area showing the erosional effects of alpine glaciation has relatively straight valleys with U-shaped cross profiles. The floor of a glacial valley usually has a *cirque* at its head and descends as a series of rock steps. Small *rock-basin lakes* are commonly found along the steps and in cirques.

A *hanging valley* indicates that a smaller tributary joined the main glacier. A *horn* is a peak between several cirques. *Arêtes* usually separate adjacent glacial valleys.

A glacier deposits unsorted rock debris or *till*, which contrasts sharply with the sorted and layered deposits of glacial *outwash*. Till forms various types of *moraines*.

Fine silt and clay may settle as *varves* in a lake in front of a glacier, each pair of layers representing a year's accumulation.

Multiple till deposits and other glacial features indicate several major episodes of glaciation during the late Cenozoic Era. During each of these episodes, large ice sheets covered most of northern Europe and northern North America, and glaciation in mountain areas of the world was much more extensive than at present. At the peak of glaciation, about a third of Earth's land surface was glaciated (in contrast to the 10% of the land surface currently under glaciers). Warmer climates prevailed during interglacial episodes.

The glacial ages also affected regions never covered by ice. Because of wetter climate in the past, large lakes formed in now-arid regions of the United States. Sea level was considerably lower.

Glacial ages also occurred in the more distant geologic past, as indicated by late Paleozoic and Precambrian tillites.

Terms to Remember

ablation 293
advancing glacier 294
alpine glaciation 291
arête 306
basal sliding 295
cirque 304
continental glaciation 291
crevasse 298
drumlin 309
end moraine 309
equilibrium line 294
erratic 307
esker 311
fiord 318
glacier 291
ground moraine 309
hanging valley 303
horn 304
iceberg 293
ice cap 292
ice sheet 292
kettle 311
lateral moraine 307
medial moraine 307
moraine 307
outwash 311
plastic flow 295
pluvial lake 316
receding glacier 294
rigid zone 298
rock-basin lake 303
rock flour 300
tarns 303
terminus 294
theory of glacial ages 314
till 307
tillite 318
truncated spur 303
U-shaped valley 303
valley glacier 292
varve 312
zone of ablation 294
zone of accumulation 294

Testing Your Knowledge

Use the following questions to prepare for exams based on this chapter.

1. How do features caused by stream erosion differ from features caused by glacial erosion?
2. How does material deposited by glaciers differ from material deposited by streams? What is the name of a rock formed from this material?
3. How would you distinguish an alpine glaciated terrain from a continental glaciated terrain?
4. Describe the glacial erosion processes of plucking and abrasion. What features indicate these processes have occurred in the past?
5. Describe how arêtes, cirques, and horns form.
6. How does the glacial budget control the migration of the equilibrium line?
7. How do recessional moraines differ from terminal moraines?
8. How do kettles, eskers, and drumlins form?
9. Alpine glaciation
   a. is a type of glacier.
   b. exists where a large part of a continent is covered by glacial ice.
   c. is found in mountainous regions.
   d. is found only at high latitudes.
10. Continental glaciation
    a. is found in mountainous regions.
    b. only occurred during the last ice age.
    c. is a glacier found in the subtropics of continents.
    d. exists where a large part of a continent is covered by glacial ice.

11. At present, about ________% of the land surface of the Earth is covered by glaciers.
    a. 1/2 b. 1
    c. 2 d. 10
    e. 33 f. 50
12. Which is *not* a type of glacier?
    a. valley glacier b. ice sheet
    c. ice cap d. peak glacier
13. The boundary between the zone of accumulation and the zone of ablation of a glacier is called the
    a. firn. b. equilibrium line.
    c. ablation zone. d. moraine.
14. In a receding glacier,
    a. ice flows from lower elevations to higher elevations.
    b. the terminus moves upvalley.
    c. the equilibrium line moves to a lower elevation.
    d. all of the preceding.
15. In the interior of a glacier, beneath the rigid zone, movement occurs by
    a. basal sliding.
    b. plastic flow.
    c. passive sliding.
    d. melting.
16. How fast does the central part of a valley glacier move compared to the sides of the glacier?
    a. faster b. slower
    c. at the same rate
17. Glacially carved valleys are usually ________ shaped.
    a. V b. U
    c. Y d. all of the preceding
18. Which is *not* a type of moraine?
    a. medial b. terminal
    c. recessional d. ground
    e. esker
19. The last episode of extensive glaciation in North America was at its peak about ________ years ago.
    a. 2,000 b. 5,000
    c. 10,000 d. 18,000
20. During the ice ages, much of Nevada, Utah, and eastern California was covered by
    a. ice. b. huge lakes.
    c. deserts. d. the sea.

Expanding Your Knowledge

1. How might a warming trend cause increased glaciation?
2. How do, or do not, the Pleistocene glacial ages fit in with the principle of uniformitarianism?
3. Is ice within a glacier a mineral? Is a glacier a rock?
4. Could a rock that looks like a tillite have been formed by any agent other than glaciation?
5. What is the likelihood of a future glacial age? What effect might human activity have on causing or preventing a glacial age?

Exploring Web Resources

www.crevassezone.org/

Glacier movement studies on the Juneau Icefield, Alaska. Go to "Photo Gallery" to view photos of glacial features and other aspects of the project.

http://www.museum.state.il.us/exhibits/ice_ages/

Ice Ages. Illinois State Museum's virtual ice ages exhibit. The site features a tape clip showing the retreat of glaciers during the last ice age.

http://nsidc.org/cryosphere

The Cryosphere. General information on snow and ice. You can link to pages on glaciers, avalanches, and icebergs.

www.swisseduc.ch/glaciers/

Swiss Educ: Glaciers Online. A very nice website with stunning photographs illustrating a wide range of glacial features and processes. By glaciologists Jürg Alean (Swiss) and Michael Hambrey (British), authors of the book *Glaciers.*

CHAPTER

13

Deserts and Wind Action

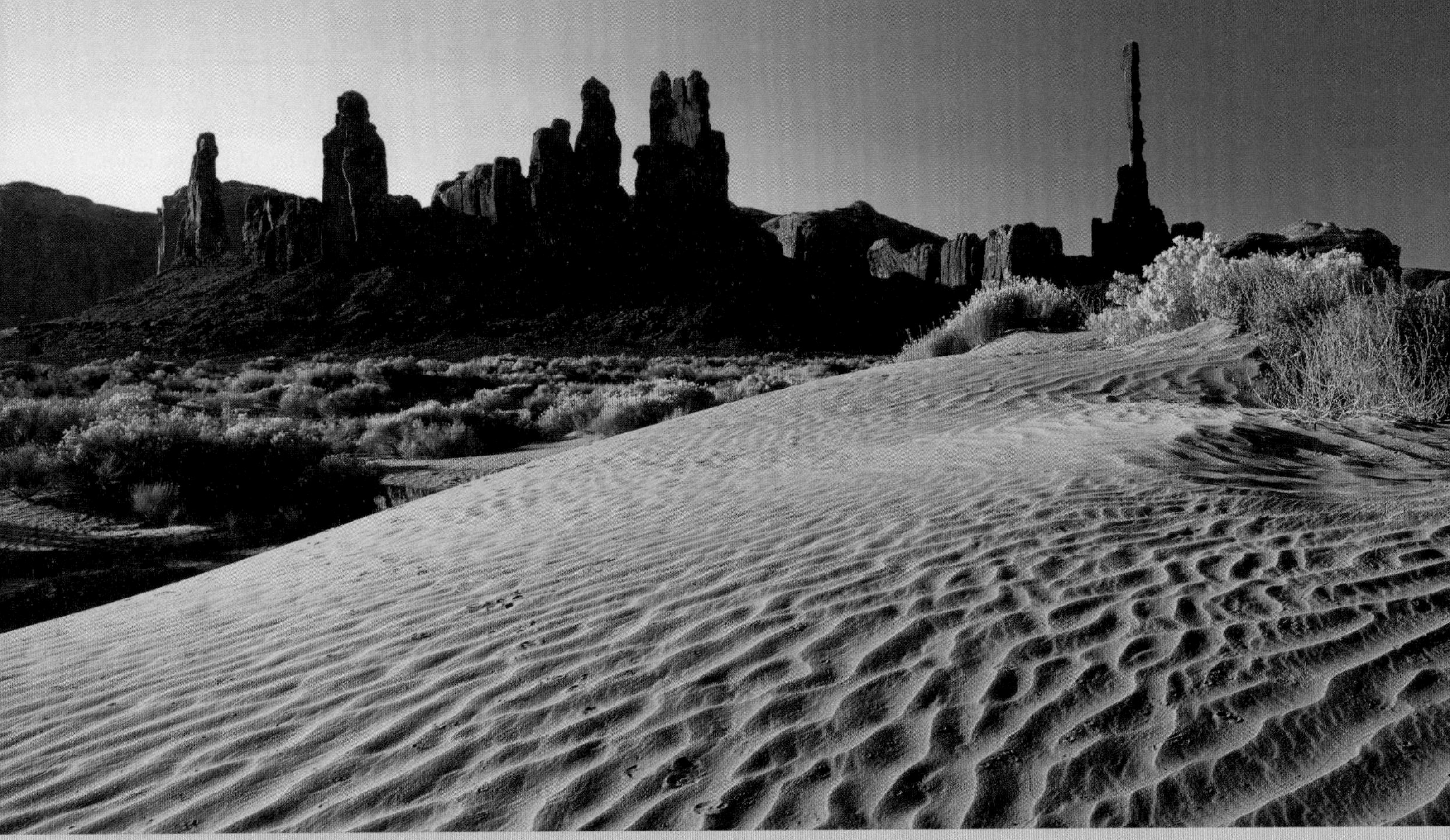

Totem Pole and Sand Springs, Monument Valley Tribal Park, Arizona.
©Lee Frost/Robert Harding World Imagery/Getty Images

LEARNING OBJECTIVES

- Explain where and how deserts form, and summarize the characteristics of deserts.
- Discuss the development of landforms and the features found in the desert southwest of the United States.
- What is the most effective agent of erosion in a desert? Why?
- Describe some features caused by wind erosion.
- Explain how loess is deposited and how it differs from sand.
- Describe how sand dunes form, and distinguish among the different types of dunes.

In chapters 9 through 12, you have seen how the land is sculpted by mass wasting, streams, groundwater, and glaciers. Here, we discuss the fifth agent of erosion and deposition: wind. Deserts and wind action are discussed together because of the wind's particular effectiveness in dry regions. But wind erosion and deposition can be very significant in other climates as well.

Deserts have a distinctive appearance because a dry climate controls erosional and depositional processes and the rates at which they operate. Although it seldom rains in the desert, running water is actually the dominant agent of land sculpture. Flash floods cause most desert erosion and deposition, even though they are rare events.

The word *desert* may suggest shifting sand dunes. Although moviemakers usually film sand dunes to represent deserts, only small portions of most deserts are covered with dunes. Actually, a **desert** is any region with low rainfall. A region is usually classified as a desert if it has a dry or *arid climate* with less than 25 centimeters of rain per year. The biosphere of deserts reflects the dryness of the air and the infrequency of precipitation. Few plants can tolerate low rainfall, so most deserts look barren.

Some specialized types of plants, however, grow well in desert climates despite the dryness. These plants are generally salt-tolerant, and they have extensive root systems to conserve water, so they often are widely spaced (figure 13.1). The leaves are usually very small, minimizing water loss by transpiration; they may even drop off the plants between rainstorms. During much of the year, many desert plants look like dead, dry sticks. When rain does fall on the desert, the plants become green, and many will bloom.

The lack of vegetation affects the geosphere because debris between plants is not anchored by roots, making loose material susceptible to mechanical weathering and erosion.

FIGURE 13.1

A scene from the Mojave Desert in southern California showing widely spaced plants that have adapted to less than 25 centimeters of rain per year.
©Diane Carlson

13.1 WHERE AND HOW DESERTS FORM

Obviously, deserts are related to the atmosphere, because the climate is caused by circulating air that usually is dry. Deserts, the atmosphere, and the hydrosphere are interrelated in various ways. Water from the hydrosphere must evaporate to become part of the atmosphere. Air over oceans tends to be moist. If conditions cause that air to become dry before it circulates to an arid region, precipitation will not take place.

The location of most deserts is related to descending air. The global pattern of air circulation is shown in simplified form in figure 13.2. The equator receives the Sun's heat more directly than the rest of the Earth. Air warms and rises at the equator, then moves both northward and southward to sink near 30° North latitude and 30° South latitude. The world's best-known deserts lie in a belt 10°–15° wide centered on 30° North and South latitude (figure 13.3).

Air sinking down through the atmosphere is compressed by the weight of the air above it. As air compresses, it warms up, and as it warms, it can hold more water vapor. Evaporation of water from the land surface into the warm, dry air is so great under belts of sinking air that moisture seldom falls back to Earth in the form of rain. The two belts at 30° North and South latitude characteristically have clear skies, much sunshine, little rain, and high evaporation.

In contrast to the belts centered on 30°, the equator is marked by rising air masses that expand and cool as they rise. In cooling, the air loses its moisture, causing cloudy skies and heavy precipitation. Thus, a belt of high rainfall at the equator separates the two major belts of deserts.

Not all deserts lie near the 30° latitude belts. Interaction with the geosphere also helps control the location of deserts.

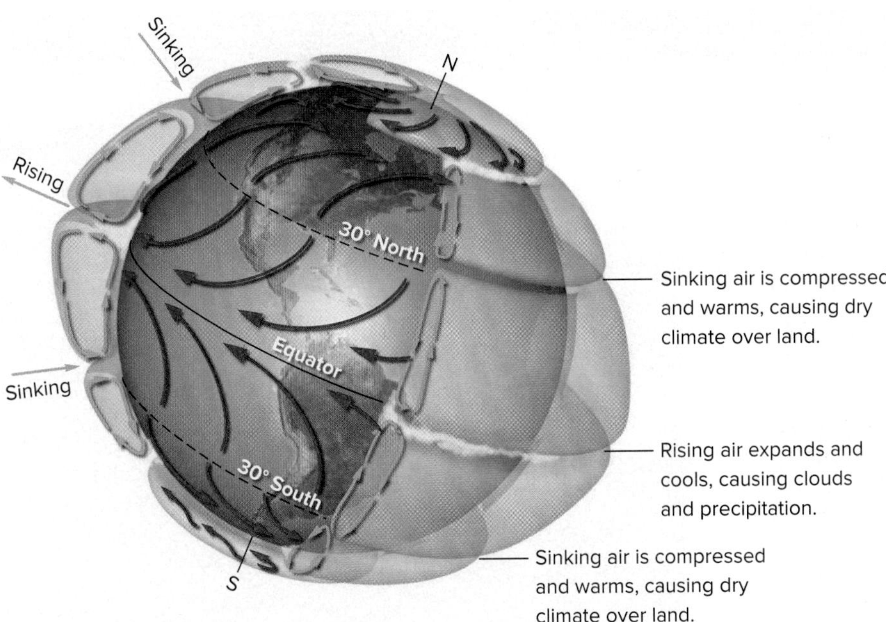

FIGURE 13.2

Global air circulation. Red arrows show surface winds. Blue arrows show vertical circulation of air. Air sinks at 30° North and 30° South latitude (and at the poles).

Some of the world's deserts are the result of the **rain shadow** effect of mountain ranges (figure 13.4). As moist air is forced up to pass over a mountain range, it expands and cools, losing moisture by condensation as it rises. The dry air coming down the other side of the mountain compresses and warms, bringing high evaporation with little or no rainfall to the downwind side of the range. This dry region downwind of mountains is the *rain shadow zone.* Parts of the southwestern United States desert in Nevada and northern Arizona are largely the result of the rain shadow effect of the Sierra Nevada range in eastern California.

Great distance from the ocean is another factor that can create deserts, since most rainfall comes from water evaporated from the sea. The dry climate of the large arid regions in China, well north of 30° North latitude, is due to their location in a continental interior and to the rain shadow effect of mountains such as the Himalayas.

Deserts also tend to develop on tropical coasts next to *cold ocean currents.* Cold currents run along the western edges of continents, cooling the air above them. The cold marine air warms up as it moves over land, causing high evaporation and little rain on the coasts. This effect is particularly pronounced on the Pacific coast of South America and the Atlantic coast of Africa and to a lesser extent western Australia.

Not all deserts are hot. The cold, descending air near the North and South Poles (figure 13.2) creates *polar deserts* that have an arid climate along with a snow or ice cover. The entire continent of Antarctica is a desert, as are most of Greenland and the northernmost parts of Alaska, Canada, and Siberia.

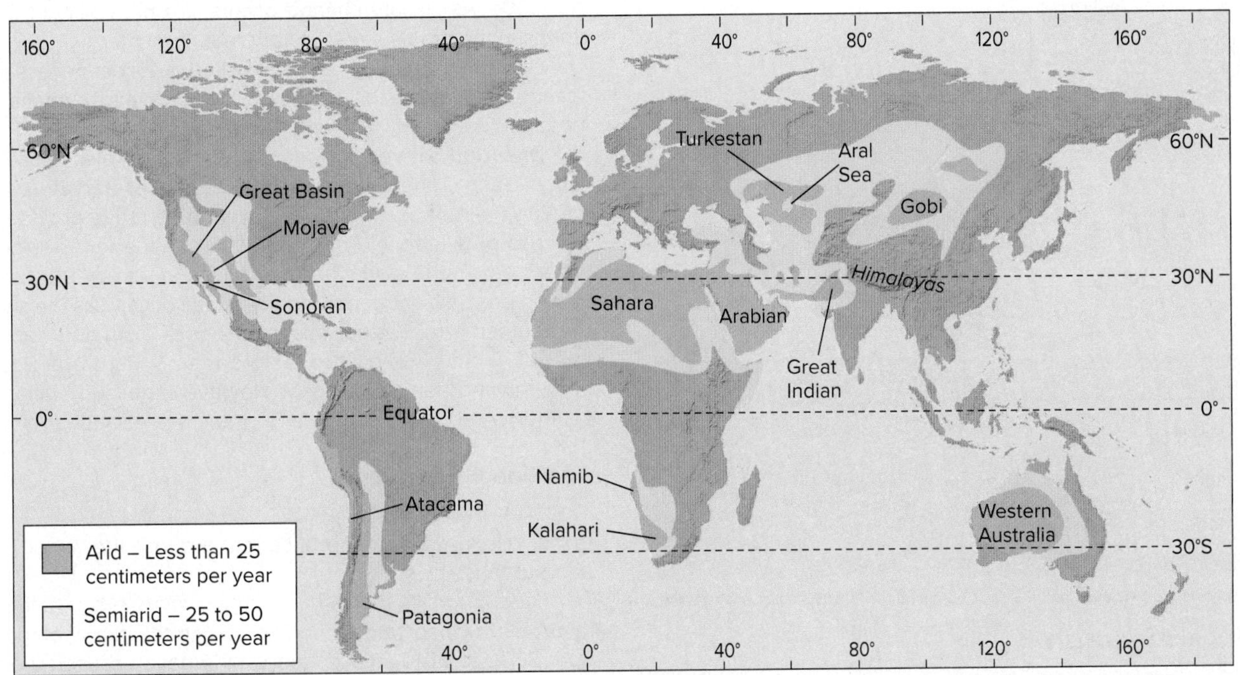

FIGURE 13.3

World distribution of nonpolar deserts. Most deserts lie in two bands near 30° North and 30° South.
Source: U.S. Department of Agriculture

ENVIRONMENTAL GEOLOGY 13.1

Expanding Deserts

Many geologists and geographers use a two-part definition of *desert.* A desert must have less than 25 centimeters of rain per year or must be so devoid of vegetation that few people can live there. Many dry regions have supported marginally successful agriculture and moderate human populations in the past but are being degraded into barren deserts today by overgrazing, overpopulation, and water diversions. The expansion of barren deserts into once-populated regions is called *desertification.*

Limited numbers of people can exist in dry regions through careful agricultural practices that protect water sources and limit grazing of sparse vegetation. Overuse of the land by livestock and humans, however, can strip it bare and make it uninhabitable. The large desert in northern Africa, shown in figure 13.3 and box figure 1, is the Sahara Desert, and the semiarid region (25 to 50 centimeters of rain per year) to the south of it is the Sahel. In the early 1960s, a series of abnormally wet years encouraged farmers in the Sahel to expand their herds and grazing lands.

A severe drought throughout the 1970s and 1980s caused devastation of the plant life of the region as starving livestock searched desperately for food, and humans gathered the last remaining sticks for firewood (box figure 1). Vast areas that were once covered with trees and sparse grass became totally barren, and an acute famine began, killing more than 100,000 people. The desert expanded southward, advancing in some places as much as 50 kilometers per year. The denuded soil in many regions became susceptible to wind erosion, leading to choking dust storms and new, advancing dune fields (some even migrating into cities).

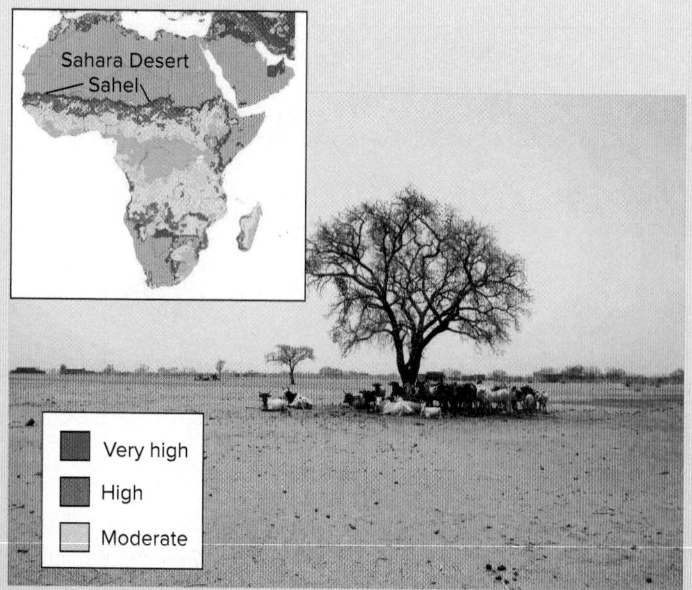

Source: U.S. Department of Agriculture. Inset: U.S. Dept. of Agriculture/NRCS/SSD/World Soil Resources

BOX 13.1 ■ FIGURE 1

Desertification in the Sahel area of Africa after a period of drought has left inhabitants and cattle desperately searching for water.
©Poncho/Photonica/Getty Images

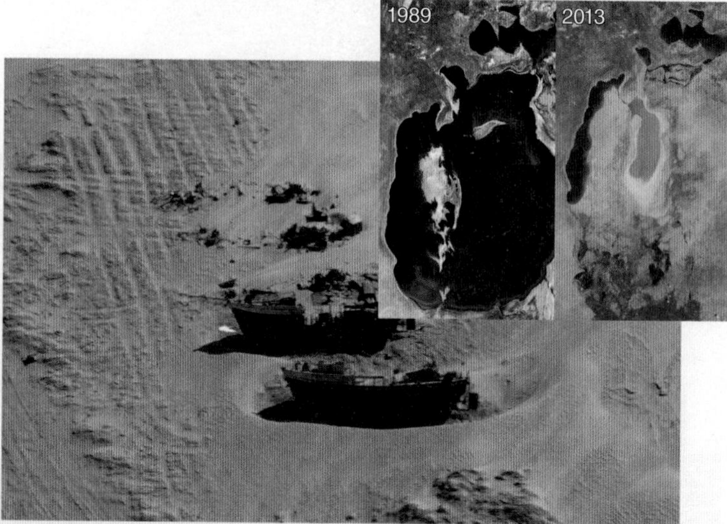

BOX 13.1 ■ FIGURE 2

Fishing boats marooned in a sea of sand as the southern Aral Sea decreases in size (inset) due to diversion of water for irrigation.
©David Turnley/Corbis/Getty Images. Inset (left & right): Source: University of Maryland Global Land Cover Facility and NASA Goddard Level 1 and Atmospheric Archive and Distribution System

Some of the same problems afflicted the midwestern United States in the 1930s, as intense land cultivation coupled with a prolonged drought produced the barren Dust Bowl during the time of the Great Depression. Improved soil-conservation practices have helped to reverse the trend in the United States, but the area was struck by heat waves and a severe drought in the 2010s that rivaled the 1930s. As climate change occurs, the possibility of future dust storms in the prairie states is very real.

Drought accelerates desertification but is not necessary for it to occur. Overloading the land with livestock and humans can strip marginal regions of vegetation even in wet years.

Diversion of rivers for agricultural use can also cause desertification. Such is the case in the Turkestan Desert, where two rivers feeding the Aral Sea (figure 13.3) were diverted to provide irrigation water to grow cotton. The Aral Sea, once the fourth-largest inland water body in the world, has decreased in size by 90% since 1960. Fishing boats are now marooned in a sea of sand as the shoreline of the southern Aral Sea has migrated tens of kilometers (box figure 2). In 2001, a dam was built to restore the northern Aral Sea and the once-thriving fishing industry. However, the dam will cause the southern Aral Sea to continue to shrink and disappear by 2020.

Additional Resources

- http://pubs.usgs.gov/gip/deserts/desertification/

Good overview of desertification around the world. Includes photographs.

- http://earthobservatory.nasa.gov/Features/WorldOfChange/aral_sea.php

NASA Earth Observatory photos showing the extent of the Aral Sea from 2000 to 2010 after decades of irrigation caused the southern Aral Sea to shrink by 90% of its original size. Site also includes history of the environmental degradation.

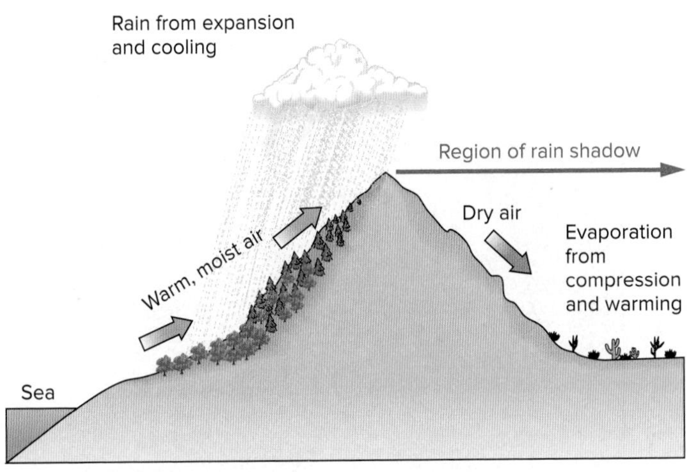

FIGURE 13.4

Rain shadow causes deserts on the downwind side of mountain ranges. Prevailing winds are from left to right.

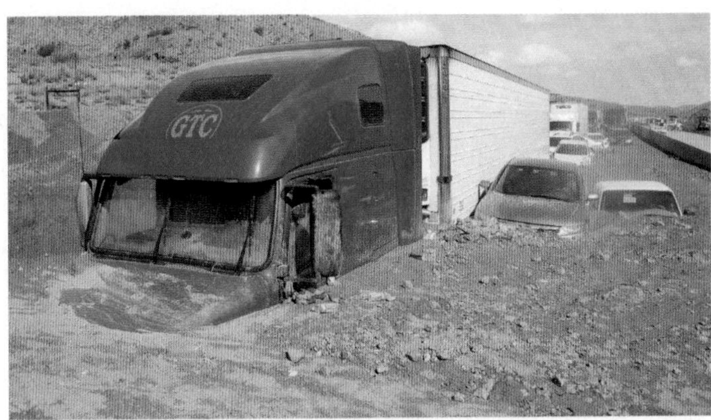

FIGURE 13.5

Vehicles swept away in a mudflow that occurred along Highway 58 in southern California on October 15, 2015, after a cloudburst dropped nearly two inches of rain in only 30 minutes.
©Gene Blevins/Polaris/Newscom

13.2 SOME CHARACTERISTICS OF DESERTS

Because of their low rainfall, deserts have characteristic drainage and topography that differ from those of humid regions. Desert streams usually flow intermittently. Water runs over the surface after storms, but during most of the year, streambeds are dry. As a result, most deserts *lack through-flowing streams.* The Colorado River in the southwestern United States and the Nile River in Egypt are notable exceptions. Both are fed by heavy rainfall in distant mountains. The runoff is great enough to sustain streamflow across dry regions with high evaporation.

Many desert regions have *internal drainage*; the streams drain toward landlocked basins instead of toward the sea. The surface of an enclosed basin acts as a local base level. Because each basin is generally filled to a different level than the neighboring basins, desert erosion may be controlled by many different *local base levels.* As a basin fills with sediment, its surface rises, leading to a *rising base level*, which is a rare situation in humid (wet) regions.

The limited rainfall that does occur in deserts often comes from violent thunderstorms, with a high volume of rain falling in a very short time. Desert thunderstorms may dump more than 13 centimeters of rain in one hour. Such a large amount of rain cannot soak readily into the sun-baked hardpan soil, so the water runs rapidly over the land surface, particularly where vegetation is sparse. This high runoff can create sudden local floods of high discharge and short duration called **flash floods**. Flash floods are more common in arid regions than in humid regions. They can turn normally dry streambeds into raging torrents for a short time after a thunderstorm. Because soil particles are not held in place by plant roots, these occasional floods can effectively erode the land surface in a desert region. As a result, desert streams normally are very heavily laden with sediment. Flash floods can easily erode enough sediment to become *mudflows* (see chapter 9). Such was the case on October 15, 2015, when almost two inches of rain fell in only 30 minutes and caused a catastrophic mudflow that closed Highway 58 in the high desert of southern California. More than 100 cars and semi-trailer trucks were swept away in a torrent of mud, and drivers narrowly escaped being buried alive in their vehicles (figure 13.5).

Desert stream channels are distinctive in appearance because of the great erosive power of flash floods and the intermittent nature of streamflow. Most stream channels are normally dry and covered with sand and gravel that is moved only during occasional flash floods. Rapid downcutting by sediment-laden floodwaters tends to produce narrow canyons with vertical walls and flat, gravel-strewn floors (see figure 10.29*A* and *B*). Such channels are often called *arroyos* or *dry washes.*

Newcomers to deserts sometimes get into serious trouble in desert canyons in rainy weather. Imagine for a moment that you have camped on the canyon floor to get out of the strong desert winds. Later that night, a towering thunderhead cloud forms, and heavy rain falls on the mountains several miles upstream from you. Although no rain has fallen on you, you are awakened several minutes later by a distant roar. The roar grows louder until a 3-meter "wall" of water rounds a bend in the canyon, heading straight for you at the speed of a galloping horse. Boulders, brush, and tree trunks are being swept along in this raging flash flood. The walls of the canyon are too steep to climb. Hikers died in Grand Canyon National Park in Arizona when a wall of water roared down a side canyon, and people have been killed in Death Valley National Park in California when their vehicles were carried away by flash floods that washed out most of the roads in the park. Stay out of desert canyons if there is any sign of rain; sleeping in such canyons is particularly dangerous.

FIGURE 13.6

Badlands topography (sharp ridges and V-shaped channels) eroded on shale in a dry region where plants are scarce. Badlands National Park, South Dakota.
©Diane Carlson

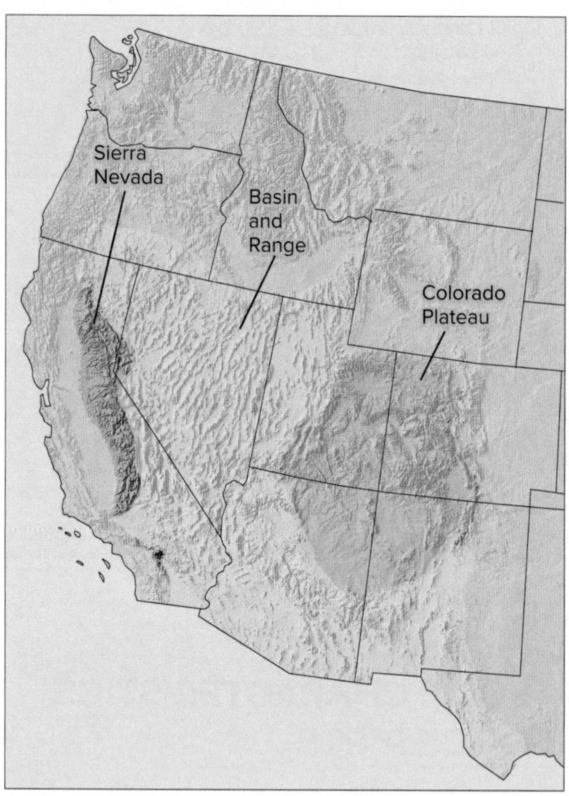

FIGURE 13.7

The Colorado Plateau and the Basin and Range province in the southwestern United States.
Source: Thelin and Pike, U.S. Geological Survey

The resistance of some rocks to weathering and erosion is partly controlled by climate. In a humid (wet) climate, limestone dissolves easily, forming low places on Earth's surface. In a desert climate, the lack of water makes limestone resistant, so it stands up as ridges and cliffs in the desert just as sandstone and conglomerate do. Lava flows and most igneous and metamorphic rock are also resistant. Shale is the least resistant rock in a desert, so it usually erodes more deeply than other rock types and forms gentler slopes or badlands topography (figure 13.6).

Although intersecting joints form angular blocks of rock in all climates, desert topography characteristically looks more angular than the gently rounded hills and valleys of a humid region. This may be due indirectly to the low rainfall in deserts. Shortage of water slows chemical weathering processes to the point where few minerals break down to form fine-grained clay minerals. Soils are coarse and rocky with few chemically weathered products. Plants, which help bind soil into a cohesive layer in humid climates, are rare in deserts, and so desert soils are easily eroded by wind and rainstorms. Downhill creep of thick, fine-grained soil is partly responsible for softening the appearance of jointed topography in humid climates. With thin, rocky soil and slow rates of creep, desert topography remains steep and angular.

Climate is only one of many things that determine the shape and appearance of the land. Rock structure is another. As an example, in the next section we will look closely at two different structural regions within the desert of the southwestern United States.

13.3 DESERT FEATURES IN THE SOUTHWESTERN UNITED STATES

Much of the southwestern United States has an arid (or semiarid) climate, partly because it is close to 30° North latitude and partly because of the rain shadow effect of the Sierra Nevada and other mountain ranges. Within this region of low rainfall are two areas of markedly different geologic structure. One area is the Colorado Plateau and the other is the Basin and Range province, a mountainous region centered on the state of Nevada. The boundaries of these two areas are shown in figure 13.7.

The *Colorado Plateau* centers roughly on the spot known as the Four Corners, where the states of Utah, Colorado, Arizona, and New Mexico meet at a common point. The rocks near the surface of the Colorado Plateau are mostly flat-lying beds of sedimentary rock over 1,500 meters above sea level. These rocks are well exposed at the Grand Canyon in Arizona.

Because the rock layers are well above sea level, they are vulnerable to erosion by the little rain that does fall in the region. Flat-lying layers of resistant rock, such as sandstone, limestone, and lava flows, form **plateaus**–broad, flat-topped areas elevated above the surrounding land and bounded, at least in part, by cliffs. As erosion removes the rock at its base, the cliff is gradually eroded back into the plateau (figure 13.8). Remnants of the resistant rock layer may be left behind, forming flat-topped mesas or narrow buttes (figure 13.8*A*). A **mesa** is a broad, flat-topped hill bounded by cliffs and capped with a resistant rock layer. A **butte** is a narrow hill of resistant rock with a flat top and very steep sides. Most buttes form by continued erosion of mesas. (The term *butte* is also used in other parts of the country for any isolated hill.)

A

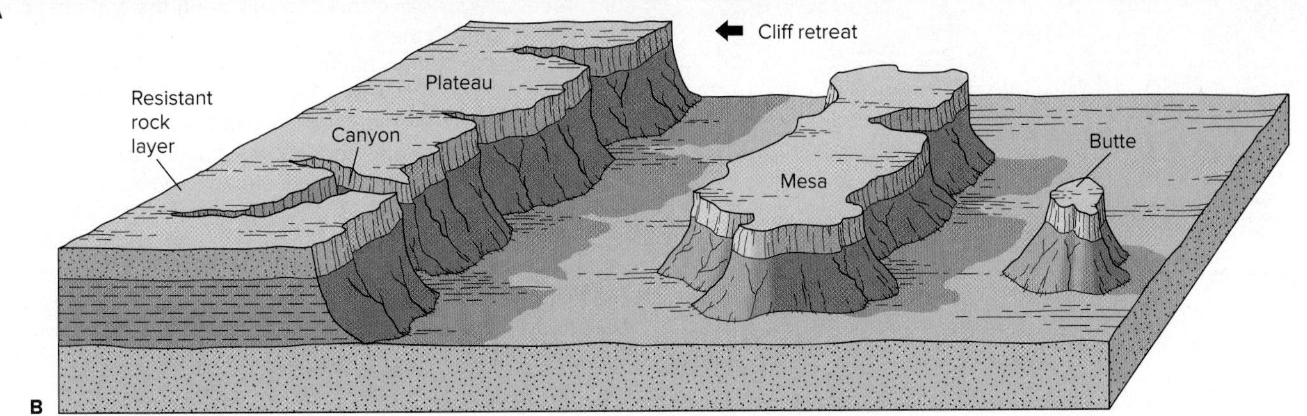

B

FIGURE 13.8

Characteristic landforms of the Colorado Plateau. (*A*) Mesas and buttes in Monument Valley, Arizona, an area of eroded, horizontal, sedimentary rocks. (*B*) Erosional retreat of a cliff at the edge of a plateau can leave behind mesas and buttes as erosional remnants of the plateau.
(A) ©David McGeary

The Colorado Plateau is also marked by peculiar, steplike folds (bends in rock layers) called *monoclines*. Erosion of monoclines (and other folds) leaves resistant rock layers protruding above the surface as ridges (figure 13.9). A steeply tilted resistant layer erodes to form a *hogback*, a sharp ridge that has steep slopes. A gently tilted resistant layer forms a *cuesta*, with one steep side and one gently sloping side.

Note that plateaus, monoclines, hogbacks, and cuestas are not unique to deserts. They are surface features found in all climates but are particularly well exposed in deserts because of thin soil and sparse vegetation.

The *Basin and Range province* is characterized by rugged mountain ranges separated by flat valley floors (figure 13.10). The blocks of rock that form the mountain ranges and the valley floors are bounded by **faults**, fractures in the ground along which some movement has taken place. (Chapter 15 discusses faults in more detail.) In the Basin and Range province, movement on the faults has dropped the valleys down relative to the adjacent mountain ranges to accommodate crustal thinning and extension (figure 13.11). Fault-controlled topography is found throughout the Basin and Range province, which covers almost all of Nevada and portions of bordering states as well as New Mexico and a small portion of Texas (figure 13.7). The numerous ranges in this province create multiple rain shadow zones and therefore a very dry climate.

EARTH SYSTEMS 13.2

Mysterious Sailboats of the Desert

It's an amazing sight—a mystery waiting to be "cracked." Go to Racetrack Playa in Death Valley National Park in California and you will see a variety of rocks that look out of place on the cracked clays of the dried lakebed or *playa*. The really odd features are the furrowed trails or tracks visible behind each misplaced rock (box figure 1). The mystery is, how did desert rocks make those tracks?

The tracks may be as narrow as a few inches or as wide as 3 feet and may be 300 meters (1,000 feet) long. The trails can be ruler straight, gently curved, angular zig-zags, or closed loops. The rocks at the ends of the furrows vary in shape and size and range from 500 grams (1 pound) to more than 270 kilograms (600 pounds) in weight. Closer inspection of the composition of the playa rocks indicates that the rocks are dolomite boulders that eroded from the mountains surrounding the edges of the dried lakebed.

BOX 13.2 ■ FIGURE 1

Sliding rocks and tracks on the mud-cracked surface of Racetrack Playa in Death Valley, California.
©*Paula Messina*

The puzzle of the moving playa rocks has perplexed people for more than half a century. Various agents of transportation of the playa rocks have been proposed, including magnetism, ground vibrations, gravity sliding, ice, floodwaters, and wind. One common idea was that wind propels the rocks across the slippery clay during wet, stormy weather. To most it seems inconceivable that winds alone could move a 600-pound rock.

Since 1948, scientists have designed experiments and studies to test the idea of wind as the prime-moving agent of the sliding rocks. Some tried to move rocks across the artificially muddied playa using wind created by airplane propellers, but were unable to budge large rocks in the "propeller" wind. Other scientists thought that cobble- and boulder-size rocks were skating across the playa on a thin sheet of ice.

Oddly enough, no one had witnessed the rocks in the process of skimming along the playa clay. The mystery as to the exact conditions necessary for playa rocks to sail along the surface attracted many researchers to the area.

Dr. Paula Messina brought the advances of the technological age—Global Positioning and geographic information systems (GPS, GIS)—to produce the first detailed map of the entire Racetrack Playa trails network. Dr. Messina thought the parallel trails she mapped were probably caused by the wind, and the non-parallel trails could have been produced by more erratic winds.

In 2010, Dr. Cynthia Cheung led a group of graduate and undergraduate interns from the Lunar and Planetary Sciences Academy at NASA's Goddard Space Flight Center to study the mysterious playa rocks (box figure 2). The interns spent a summer at Racetrack Playa and nearby Bonnie Claire Playas collecting data from buried

Heavy rainfall from occasional thunderstorms in the mountain ranges causes rapid erosion of the steep mountain fronts and resulting deposition on the valley floors (figure 13.11). Rock debris from the mountains, picked up by flash floods and mudflows, is deposited at the base of the mountain ranges in the form of **alluvial fans** (figure 13.12). Alluvial fans (described in chapter 10) build up where stream channels abruptly widen as they flow out of narrow canyons onto the open valley floors, causing a decrease in velocity and rapid deposition of sediment.

Although most of the sediment carried by runoff is deposited in alluvial fans, some fine clay may be carried in suspension onto the flat valley floor. If no outlet drains the valley, runoff water may collect and form a **playa lake** on the valley floor (figure 13.12). Playa lakes are usually very shallow and temporary, lasting for only a few days after a rainstorm. After the lake evaporates, a thin layer of fine mud may be left on the valley floor. The mud dries in the sun, forming a **playa**, a very flat surface underlain by hard, mud-cracked clay (see box 13.2). If the runoff contained a large amount of dissolved salt or if seeping groundwater brings salt to the surface, the flat playa surface may be underlain by a bright white layer of dried salt instead of mud.

Continued deposition near the base of the mountains may create a **bajada**, a broad, gently sloping depositional surface formed by the coalescing of individual alluvial fans (figures

hygrochrons or tiny sensors that measure temperature and humidity. They also collected photos of and GPS data from each rock and detailed measurements of each rock track.

The interns thought that ice may be the key. They hypothesized that ice "collars" form around the base of the rocks during freezing conditions. These collars of ice may allow the rocks to "float" and hence aid the movement of the rocks as the wind blows. When the ice collar melts, the rocks stop moving and settle into the wet playa clay.

BOX 13.2 ■ FIGURE 2

Several interns making notes about this moving rock that weighs about 25 pounds. *Source: Leva McIntire/LPSA intern/GSFC/NASA*

Researchers from Scripps Institution of Oceanography in La Jolla, California, think they may have finally cracked the mystery. They were the first to directly observe rock movements using GPS-instrumented rocks and time-lapse photography. They observed the rocks moving when a thin layer of ice formed on the playa, about the thickness of a glass windowpane, and then began melting in the late morning sun. The ice sheets broke into large pieces and pushed the rocks at low speeds in the direction of the wind and in the direction of water flowing under the ice. After the water was blown away in the playa, the characteristic tracks were visible in the mud. Does this same process also move the really big rocks? Or, does part of the mystery remain?

Additional Resources

R.D. Norris and others, 2014. Sliding rocks on Racetrack Playa, Death Valley National Park: First Observation of Rocks in Motion. PLoS ONE 9(8).

P. Messina and P. Stoffer. 2001. Using new technology to solve an old mystery. *California Geology*. Jan/Feb: 4–15.

- http://www.consrv.ca.gov/cgs/information/publications/cg_magazine/2001/Documents/JANFEB_2001.pdf

For a description of the research, time-lapse photography of the moving rocks, and videos of the mechanism, visit:

- https://scripps.ucsd.edu/news/mystery-solved-sailing-stones-death-valley-seen-action-first-time

Visit the NASA website to view interesting photos and a video of the mysterious roving rocks of Racetrack Playa.

- http://www.nasa.gov/topics/earth/features/roving-rocks.html

13.10 and 13.11). A bajada is much more extensive than a single alluvial fan and may have a gently rolling surface resulting from the merging of the cone-shaped fans.

Erosion of the mountain can eventually form a **pediment**, which is a gently sloping surface, commonly covered with a veneer of gravel, cut into the solid rock of the mountain (figure 13.11). A pediment develops uphill from a bajada as the mountain front retreats. It can be difficult to distinguish a pediment from the surface of the bajada downhill, because both have the same slope and gravel cover. The pediment, however, is an erosional surface, usually underlain by solid rock, while the bajada surface is depositional and may be underlain by hundreds of meters of sediment.

An abrupt change in slope marks the upper limit of the pediment, where it meets the steep mountain front. Many geologists who have studied desert erosion believe that as this steep mountain front erodes, it retreats uphill, maintaining a relatively constant angle of slope.

Notice that rock structure, not climate, largely controls the fact that plateaus and cliffs are found in the Colorado Plateau, while mountain ranges, broad valleys, alluvial fans, and pediments are found in the Basin and Range province. Features such as plateaus, mesas, and alluvial fans can also be found in humid climates wherever the rock structure is favorable to their development; they are not controlled by climate. Features such as steep canyons, playa lakes, thin soil, and sparse vegetation, however, *are* typically controlled by climate.

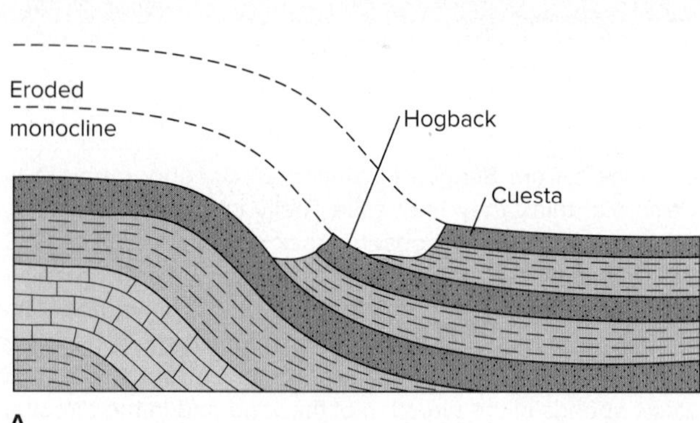

A

B

FIGURE 13.9

(*A*) Steplike monocline folds often erode so that resistant rock layers form hogbacks and cuestas (these features are not unique to deserts). (*B*) Raplee Ridge monocline in southeastern Utah.
(B) ©Stephen Reynolds

FIGURE 13.10

Basin and Range topography in Death Valley, California. In the distance, the fault-bounded Panamint Mountains rise more than 3 kilometers (11,000 feet) above Death Valley. Giant alluvial fans at the base of the mountains coalesce to form a bajada, and fine-grained sediments and salt deposits underlie the Badwater basin (playa lake) in the valley. The valley is 282 feet below sea level and is the lowest point in the Western Hemisphere.
©William Perry/Age Fotostock

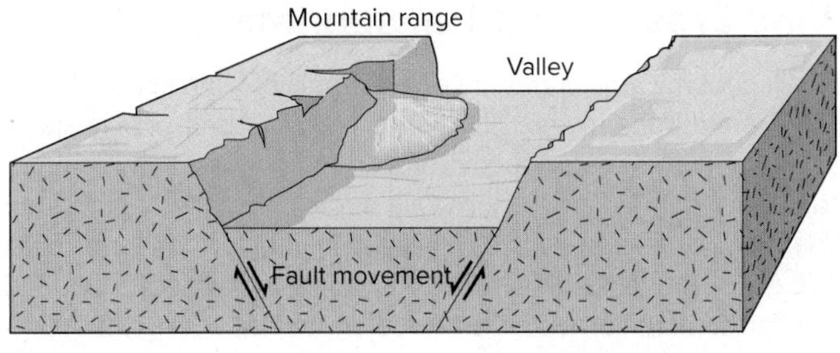

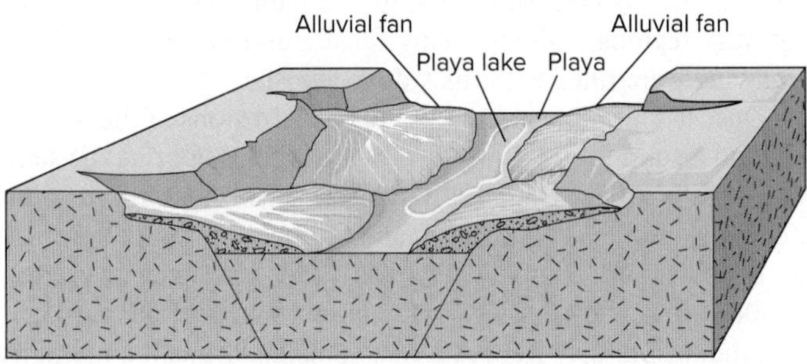

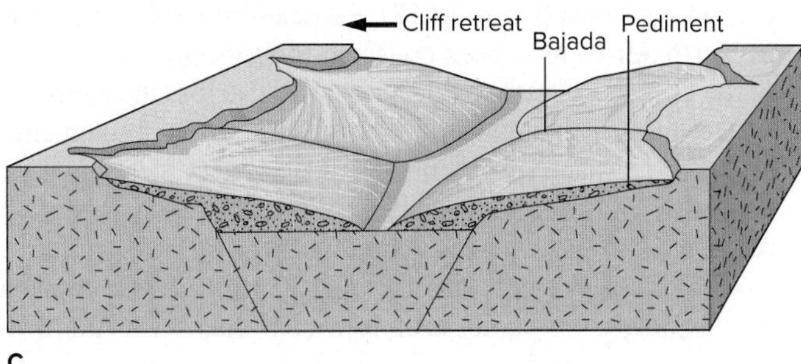

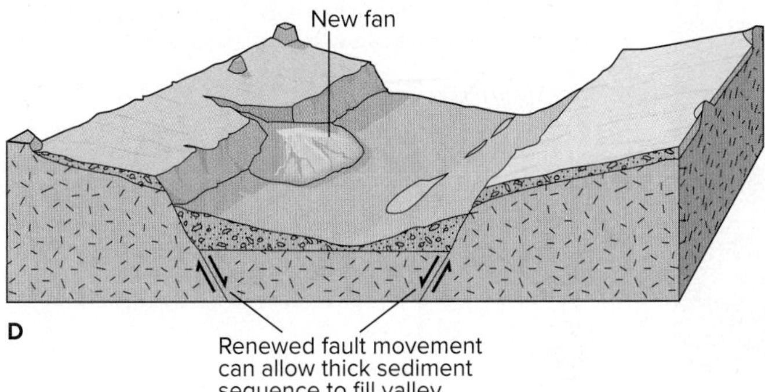

FIGURE 13.11

Development of landforms associated with Basin and Range topography.

FIGURE 13.12

Alluvial fans form at the base of canyons cut through the Black Mountains in Death Valley, California. White salt flats are visible at the edges of the alluvial fans in the playa valley, and playa lakes are visible in the center of the flat playa surface.
©Doug Sherman/Geofile

FIGURE 13.13

A wall of dust approaches a town in Kansas in October 1935. Because of the intensity and duration of the storms in the 1930s, parts of the Great Plains become known as the Dust Bowl.
Source: National Oceanic and Atmospheric Administration

13.4 WIND ACTION

Wind can be an important agent of erosion and deposition in any climate, as long as sediment particles are loose and dry. Wind differs from running water in two important ways. Because air is less dense than water, wind can remove only fine sediment–sand, silt, and clay. But wind is not confined to channels as running water is, so wind can have a widespread effect over vast areas.

In general, the faster the wind blows, the more sediment it can move. Wind velocity is determined by differences in air pressure caused by differences in air temperature. As air warms and cools, it changes density, and these density changes create air pressure differences that cause wind. Wet climates and cloud cover help buffer changes in air temperature, but in dry climates, daily temperature changes can be extreme. In a desert, the temperature may range from 10°C (50°F) at night to more than 40°C (100°F) in the daytime. Because of these temperature fluctuations, wind is generally stronger in deserts than in humid regions, commonly exceeding 100 kilometers per hour (60 miles per hour). The scarcity of vegetation in deserts to slow wind velocity by friction increases the effectiveness of desert winds.

Although strong winds are also associated with rainstorms and hurricanes, these winds seldom erode sediment because rain wets the surface sediment. Wet sediment is heavy and cohesive and will not be blown away. Strong winds in the desert, however, blow over loose, dry sediment, so wind is an effective erosional agent in dry climates. (As we said earlier in the chapter, running water in the form of flash floods is a far more important erosive agent than wind, even though the wind can be very strong.)

Wind Erosion and Transportation

Thick, choking *dust storms* are one example of wind action. "Dust Bowl" conditions in the 1930s in the agricultural prairie states lasted for several years due to droughts and poor soil-conservation practices (figure 13.13). Loose silt and clay are easily picked up from barren, dry soil, such as in a cultivated field. Wind erosion is even greater if the soil is disturbed by animals or vehicles. Silt and clay can remain suspended in turbulent air for a long time, so a strong wind may carry a dust cloud hundreds of meters upward and hundreds of kilometers horizontally. Dust storms of the 1930s frequently blacked out the midday sun; fertile soil was lost over vast regions, ruining many farms; and streets and rivers downwind were filled with thick dust deposits.

Dust storms are common in the desert regions of the Middle East and also in the southwestern United States. During the summer and fall of the early 2010s, large dust storms blew through the Phoenix, Arizona (figure 13.14), and Lubbock, Texas, areas, causing considerable property damage, fatal car accidents, and stranded motorists. These were the largest dust storms on record to hit the region; they were caused by ongoing drought conditions and high winds from monsoonal weather systems.

Wind-blown sediment is sometimes picked up on land and carried out to sea (figure 13.15). Particles from the Sahara Desert in North Africa have been collected from the air over the islands of the Caribbean after having been carried across the entire Atlantic Ocean. A substantial amount of the fine-grained sediment that settles to the sea floor is land-derived sediment that the wind has deposited on the sea surface. Ships 800 kilometers offshore have reported dustfalls a few millimeters thick covering their decks.

Volcanic ash can be carried by wind for very great distances. An explosive volcanic eruption can blast ash more than

FIGURE 13.14

A mile-high dust storm (haboob) swept across Phoenix, Arizona, on July 5, 2011, and was reminiscent of the dust storms from the 1930s Dust Bowl.

FIGURE 13.15

Satellite image of a dust storm from the Sahara Desert blowing off the coast of Africa northwestward out into the Atlantic Ocean on February 26, 2000. The thick plume of dust is about the size of Spain, and dust particles from this storm were blown all the way to the west side of the Atlantic. Such storms are fairly common in the Sahara Desert and are the world's greatest supplier of dust.
Source: Courtesy of Norman Kuring, SeaWiFS Project/NASA

15 kilometers upward into the air. Such ash may be caught in the high-altitude *jet streams*, narrow belts of strong winds with velocities sometimes greater than 300 kilometers per hour. Following the 1980 eruption of Mount St. Helens in western Washington, a visible ash layer blanketed parts of Washington, Idaho, and Montana to the east. At high altitudes, St. Helens ash could be detected blowing over New York and out over the Atlantic Ocean, 5,000 kilometers from the volcano. But the St. Helens eruption was a relatively small one. Ash from the 1883 Krakatoa eruption in Indonesia circled the globe for two years, causing spectacular sunsets and a slight, but measurable, drop in temperature as the ash reflected sunlight back into space. The 1815 eruption of Tambora, also in Indonesia, put so much ash into the air that there were summer blizzards, crop failures, and famine in New England and northern Europe in 1816, "the year without a summer." Lower temperatures (1°F) and brilliant sunsets also marked the 1991 eruption of Pinatubo in the Philippines.

Because sand grains are heavier than silt and clay, sand moves close to the ground in the leaping pattern called *saltation* (as does some sediment in streams). High-speed winds can cause *sandstorms*, clouds of sand moving rapidly near the land surface. The high-speed sand in such a storm can sandblast smooth surfaces on hard rock and scour the windshields and paint of automobiles. Because of the weight of the sand grains, however, sand rarely rises more than 1 meter above a flat land surface, even under extremely strong winds. Therefore, most of the sandblasting action of wind occurs close to the ground (figure 13.16). Telephone poles in regions of wind-driven sand often are severely abraded

A

B

FIGURE 13.16

(*A*) Wind erosion near the ground has sandblasted the lower 1 meter of this chemically weathered basalt outcrop, Death Valley, California. Hammer for scale. (*B*) Power pole with its base wrapped in an abrasion-resistant material to minimize wind erosion.
(A) ©David McGeary; (B) Photo courtesy of Paul Bauer

FIGURE 13.17

Ventifact eroded by sandblasting action of high winds in Death Valley, California. Predominant winds from the south and north (left and right) have sculpted the grooved (fluted) faces.
©Diane Carlson

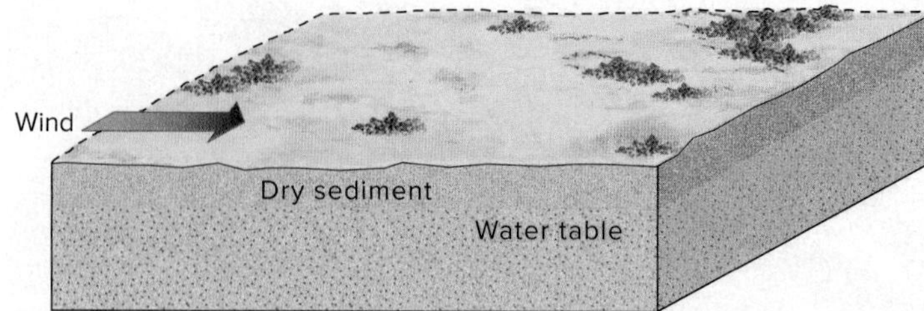

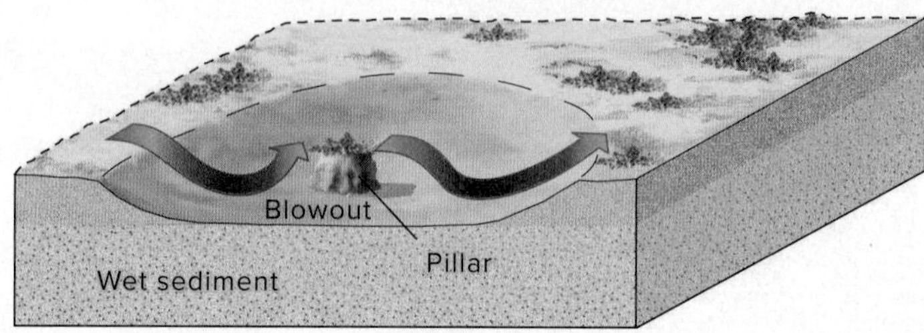

A

B

FIGURE 13.18

(A) Deflation by wind erosion can form a blowout in loose, dry sediment. Deflation stops at the water table. A pillar, or erosional remnant, may be found in the center of a blowout. *(B)* Large blowout near Harrison, Nebraska. Pillar top is the original level of land before wind erosion lowered the land surface by more than 3 meters. The pillar is the erosional remnant at the center of the blowout.
(B) Source: N. H. Darton, U.S. Geological Survey

near the ground. To prevent such abrasion, desert residents pile stones or wrap sheet metal around the bases of the poles.

Wind seldom moves particles larger than sand grains, but wind-blown sand may sculpt isolated pebbles, cobbles, or boulders into **ventifacts**–rocks with flat, wind-abraded surfaces (figure 13.17). If the wind direction shifts or the stone is turned, more than one flat face may develop on the ventifact.

Deflation

The removal of clay, silt, and sand particles from the land surface by wind is called **deflation**. If the sediment at the land surface is made up only of fine particles, the erosion of these particles by the wind can lower the land surface substantially. A **blowout** is a depression on the land surface caused by wind erosion (figure 13.18*A*). A *pillar*, or erosional remnant of the former land, may be left at the center of a blowout.

Blowouts are common in the Great Plains states (figure 13.18*B*). One in Wyoming measures 5 by 15 kilometers and is 45 meters deep. The enormous Qattara Depression in northwestern Egypt, more than 250 kilometers long and more than 100 meters *below* sea level, has been attributed to wind deflation. Deflation can continue to deepen a blowout in fine-grained sediment until it reaches wet, cohesive sediment at the water table.

Wind Deposition

Loess

Loess is a deposit of wind-blown silt and clay composed of unweathered, angular grains of quartz, feldspar, and other minerals weakly cemented by calcite. Loess has a high porosity, typically near 60%. Deposits of loess may blanket hills and valleys downwind of a source of fine sediment, such as a desert or a region of glacial outwash.

China has extensive loess deposits, more than 100 meters thick in places. Wind from the Gobi Desert carried the silt and clay that formed these deposits. Loess is easy to dig into and has the peculiar ability to stand as a vertical cliff without slumping (figure 13.19), perhaps because of its cement or perhaps because the fine, angular, sediment grains interlock with one another. For centuries, the Chinese have dug cavelike homes in loess cliffs. When a large earthquake shook China in 1920, however, many of these cliffs collapsed, burying alive about 100,000 people.

EARTH SYSTEMS 13.3

Desert Pavement and Desert Varnish

The interaction of the atmosphere and biosphere may result in two intriguing features that can be seen in many deserts, particularly on the surface of old alluvial fans no longer receiving new sediment.

Desert pavement is a thin, surface layer of closely packed pebbles (box figure 1). The pebbles were once thought to be lag deposits, left behind as strong winds blew away all the fine grains of a rocky soil. The pebbles are now thought to be brought to the surface by cycles of wetting and drying, which cause the soil to swell and shrink as water is absorbed and lost by soil particles. Swelling soil lifts pebbles slightly; drying soil cracks, and fine grains fall into the cracks. In this way pebbles move up, while fine grains move down. The surface layer of pebbles protects the land from wind erosion and deflation. When the desert pavement is disturbed, dust storms and new sand dunes may result.

Many rocks on the surface of deserts are darkened by a chemical coating known as *desert varnish.* Although the interior of the rocks may be light colored, a hard, often shiny, coating of dark iron and manganese oxides and clay minerals can build up on the rock surface over long periods of time (box figure 2). These paper-thin coatings can be used to obtain a numerical age of the exposed desert surface by measuring cosmogenic helium-3 isotopes preserved in the desert varnish. Numerical ages obtained from desert varnish are important to the study of desert landforms and regional climate change, and to the study of petroglyphs and early humans.

Although no one is quite certain how this coating develops, it seems to be added to the rocks from the outside, for even white quartzite pebbles with no internal source of iron, manganese, or clay minerals can develop desert varnish. One hypothesis is that the clay is wind-blown, perhaps sticking to rocks dampened by dew. A film of clay on a rock may draw iron- and manganese-containing solutions upward from the soil by capillary action, and the presence of the clay minerals may help deposit the dark manganese oxide that cements the clay to the rock, or silica from the dissolution of the rock surfaces may form the coating. Another hypothesis is that the oxide is deposited biologically by manganese-oxidizing bacteria. Regardless of how the varnish forms, the longer a rock is exposed on a desert land surface, the darker it becomes.

BOX 13.3 ■ FIGURE 1

Desert pavement on an old alluvial fan surface in Death Valley, California. The surface pebbles are closely packed; fine sand underlies the pebbles.
©Diane Carlson

BOX 13.3 ■ FIGURE 2

Petroglyphs carved on this rock cut through the dark desert varnish to show the lighter color of the interior of the rock, Valley of Fire, Nevada.
©Peter Ryan/National Geographic/Getty Images

PLANETARY GEOLOGY 13.4

Wind Action on Mars

Mars has an atmosphere only 1/200th as dense as Earth's but with very strong winds that have been recorded at more than 200 kilometers (120 miles) per hour. The sides of Olympus Mons (see **box 4.5,** figure 1) have been obscured by dust to a height of 15 kilometers (10 miles), a height made possible by the low gravity on Mars. Although dust storms occur throughout the year on Mars, the greatest number and the largest global dust events (those that cover the entire planet) occur during the southern spring and summer, when the southern polar cap of frozen carbon dioxide begins to sublimate. The difference in air temperature between the polar cap and the warmer surrounding landscape creates large pressure differences that produce high winds and trigger isolated dust storms. In June 2001, the *Mars Global Surveyor* spacecraft and Hubble Space Telescope recorded a sequence of dust storms that began along the retreating margin of the southern polar cap and near the Hellas impact crater (box figure 1). The individual storms intensified and moved north of the equator in only five days. This was the beginning of one of the largest global dust events observed on Mars in decades, and, for the first time, scientists were able to see how the development and progression of regional dust storms resulted in the entire planet being obscured by dust.

BOX 13.4 ■ FIGURE 1

Storm watch images of Mars from the Hubble Space Telescope show isolated dust storms on June 26, 2001, in the Hellas basin (lower right edge of Mars) and on the northern polar cap. By the end of July, the entire planet was clouded by dust that obscured its surface for several months.
Source: Courtesy NASA, James Bell (Cornell Univ.), Michael Wolff (Space Science Inst.), and the Hubble Heritage Team (STScI/AURA)

During the glacial ages of the Pleistocene Epoch, the rivers that drained what is now the Midwestern United States transported and deposited vast amounts of glacial outwash. Later, winds eroded silt and clay (originally glacial rock flour) from the flood plains of these rivers and blanketed large areas of the Midwest with a cover of loess (figure 13.20). Soils that have developed from the loess are usually fertile and productive. The grain fields of much of the Midwest and in the Palouse area in eastern Washington are planted on these rich soils. Wind erosion of cultivated, loess-covered hills in the Palouse region is a serious problem that has locally removed fertile soil from the hilltops.

Sand Dunes

Sand dunes are mounds of loose sand grains piled up by the wind. Dunes are most likely to develop in areas with strong winds that generally blow in the same direction. Patches of dunes are scattered throughout the southwestern United States desert. More extensive dune fields occur on some of the other deserts of the world, such as the Sahara Desert of Africa, which contains vast *sand seas.* Dunes are also commonly found just landward of beaches, where sand is blown inland. Beach dunes are common along the shores of the Great Lakes and along both coasts of the United States. Braided rivers (see chapter 10) can also be sources of sand for dune fields.

Images from the recent Mars Exploration Rover missions and Reconnaissance Orbiter show that the windswept surface of the planet contains features similar to those found on Earth. Barchan, star, transverse, and longitudinal sand dunes are prevalent, particularly on the floors of impact craters. Box figure 2 shows the four-meter high slip face of the Namib dune taken by the *Curiosity* rover in the Gale crater on December 17, 2015. The Namib dune is part of the Bagnold dune field, and orbiter images show that this active dune field migrates one meter per Earth year. *Curiosity* is the first rover to study sand dunes up close on the surface of Mars. On the windward top of the Bagnold dune field, *Curiosity* discovered unique, mid-size sand ripples containing smaller ripples (box figure 3). The mid-size ripples have sinuous crests spaced three meters apart and are more similar in size and form to current ripples (see figure 6.27*B* and *D*) formed by running water on Earth.

BOX 13.4 ■ FIGURE 2

Slip face of the Namib Dune taken by the Mars Science Laboratory Rover Curiosity visible in the foreground. Dune is 4 to 5 meters high and is part of the active Bagnold dune field in the Gale crater.
Source: JPL-Caltech/NASA

BOX 13.4 ■ FIGURE 3

Close-up of mid-size sand ripples that are unique to dune surfaces on Mars. Image taken by the *Curiosity* rover on December 13, 2015, on top of the Bagnold dune field.
Source: JPL-Caltech/MSSS/NASA

Planetary geologists think the mid-size sand ripples can form on Mars because of the low-density atmosphere. Mid-size ripples have also been preserved in 3.7-billion-year-old sandstone beds on Mars, and they suggest that the planet may have lost its atmosphere early in its history.

Additional Resources

M. Laporte and others. 2016. Large wind ripples on Mars: A record of atmospheric evolution. *Science* 353(6294): 55–58.

Information about the Mars exploration program at NASA, including the Mars Science Laboratory *(Curiosity)* and images and updates from ongoing missions, such as the Mars Reconnaissance Orbiter, is available from the Jet Propulsion Laboratory/NASA Mars Program websites.

- http://marsprogram.jpl.nasa.gov/
- http://mars.jpl.nasa.gov/msl

The mineral composition of the sand grains in sand dunes depends on both the character of the original sand source and the intensity of chemical weathering in the region. Many dunes, particularly those near beaches in humid regions, are composed largely of quartz grains because quartz is so resistant to chemical weathering. Inland dunes, such as the Great Sand Dunes National Park in Colorado, often contain unstable feldspar and rock fragments in addition to quartz. Some dunes are formed mostly of carbonate grains, particularly those near tropical beaches. At White Sands, New Mexico, dunes are made of gypsum grains, eroded by wind from playa lake beds.

Sand grains found in dunes are commonly well-sorted and well-rounded because wind is very selective as it moves sediment. Fine-grained silt and clay are carried much farther than sand, and grains coarser than sand are left behind when sand moves. The result is a dune made solely of sand grains, commonly all very nearly the same size. The prevalence of well-rounded grains in many dunes also may be due to selective sorting by the wind. Rounded grains roll more easily than angular grains, and so the wind may remove only the rounded grains from a source to form dunes. Wind will often selectively roll oolitic grains from a carbonate beach of mixed oolitic and skeletal grains.

FIGURE 13.19

Cavelike homes built into the loess hillside in Shanxi Province, China. ©*Liu Xiaoyang/China Images/Alamy*

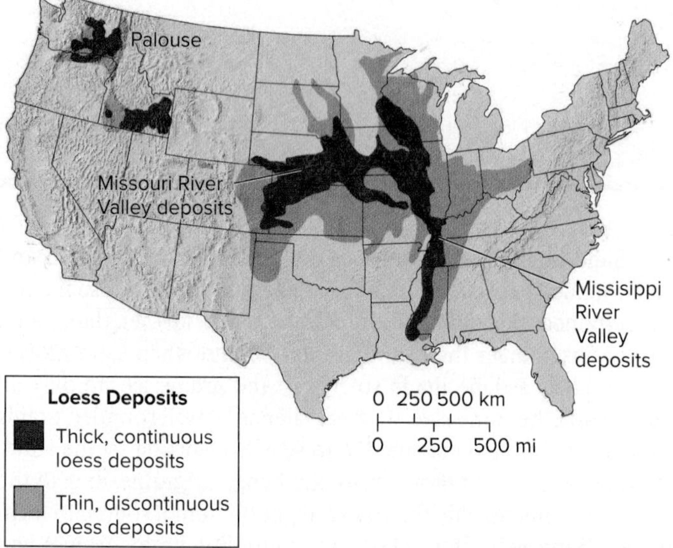

FIGURE 13.20

Major loess-covered areas in the United States. *Source: U.S. Department of Agriculture*

Most sand dunes are asymmetric in cross section, with a gentle slope facing the wind and a steeper slope on the downwind side. The steep downwind slope of a dune is called the **slip face** (figure 13.21). It forms from loose, cascading sand that generally keeps the slope at the *angle of repose*, which is about 34° for loose, dry sand. Sand is blown up the gentle slope and over the top of the dune. Sand grains fall like snow onto the slip face when they encounter the calm air on the downwind side of the dune. Loose sand settling on the top of the slip face may become oversteepened and slide as a small avalanche down the slip face. These processes form high-angle cross-bedding within the dune. When found in sandstone, such cross-bedding strongly suggests deposition as a dune (see figures 6.25 and 6.26).

In passing over a dune, the wind erodes sand from the gentle upwind slope and deposits it downwind on the slip face. As a result, the entire dune moves slowly in a downwind direction. The rate of dune motion is much slower than the speed of the wind, of course, because only a thin layer of sand on the surface of the dune moves at any one time. The dune may move only 10 to 15 meters per year. Over many years, however, the movement of dunes can be significant, a fact not always appreciated by people who build homes close to moving sand dunes.

If a dune becomes overgrown with grass or other vegetation, movement stops. The Sand Hills of north-central Nebraska are large dunes, formed during the Pleistocene or Holocene Epochs, that have become stabilized by vegetation. The migration of many beach dunes toward beach homes and roads has been stopped by planting a cover of beach grass over the dunes. Dune-buggy tires can uproot and kill the grass, however, and start the dunes moving again.

Sand moving over a dune surface typically forms *wind ripples*—small, low ridges of sand produced by saltation of the grains (figure 13.22). The ripples are similar to those formed in sediment by a water current (see chapter 6). Because sand moves perpendicularly to the long dimension of the ripples, a rippled sand surface indicates the direction of sand movement.

Types of Dunes

As figure 13.23 shows, dunes tend to develop certain characteristic shapes, depending on (1) the wind's velocity and direction (that is, whether constant or shifting); (2) the sand supply available; and (3) how the vegetation cover, if any, is distributed.

Where the sand supply is limited, a type of dune called a **barchan** generally develops. The barchan is crescent-shaped with a steep slip face on the inward or concave side. The horns on a barchan dune point in the downwind direction (figure 13.23*A*). Barchan dunes are usually separated from one another and move across a barren surface (figure 13.24). If more sand is available, the wind may develop a **transverse dune**, a relatively straight, elongate dune oriented perpendicular to the wind direction (figures 13.23*B* and 13.25).

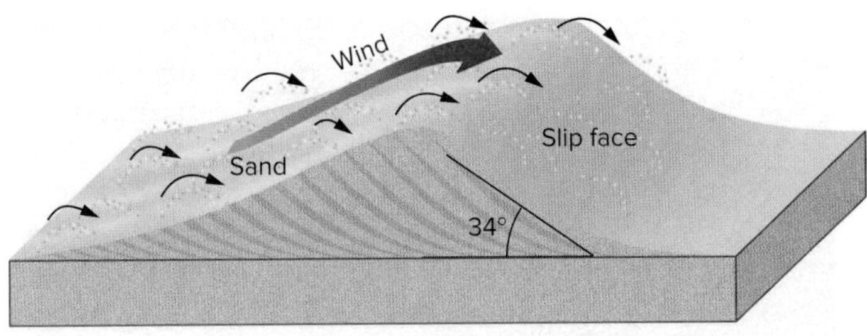

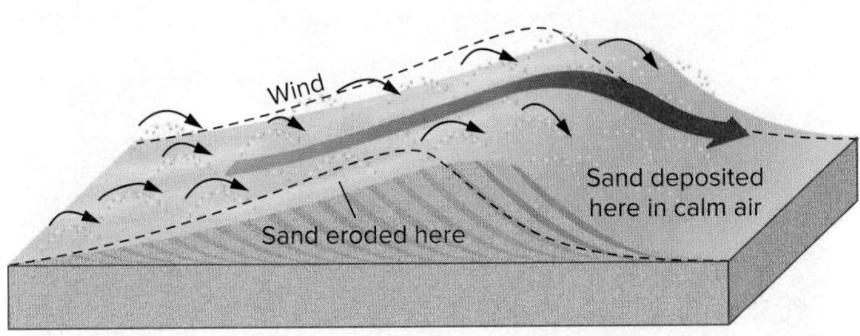

A

B

FIGURE 13.21

(*A*) A sand dune forms with a gentle upwind slope and a steeper slip face on the downward side. Sand eroded from the upwind side of the dune is deposited on the slip face, forming cross-beds. Movement of sand causes the dune to move slowly downwind. (*B*) Strong desert winds (60 miles per hour) blowing to the right remove sand from the gentle sloping upwind side of this dune. The sand settles onto the steep slip face on the right.
(B) ©David McGeary

FIGURE 13.22

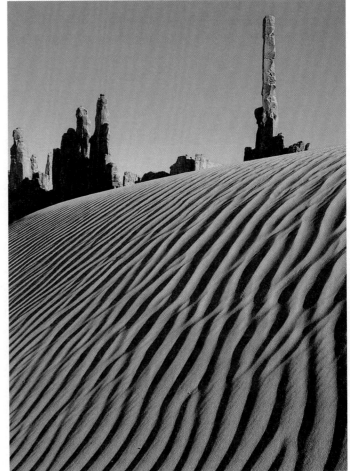

Wind ripples on sand surface, Monument Valley, Utah.
©Doug Sherman/Geofile

A **parabolic dune** is somewhat similar in shape to a barchan dune, except that it is deeply curved and is convex in the downwind direction. The horns point upwind and are commonly anchored by vegetation (figure 13.23*C*). The parabolic dune requires abundant sand and commonly forms around a blowout. Because they require abundant sand and strong winds, parabolic dunes are typically found inland from an ocean beach (figure 13.26). All three of these dune shapes develop in areas having steady wind direction, and all three have steep slip faces on the downwind side.

One of the largest types of dunes is the **longitudinal dune** or *seif* (figure 13.23*D*), which is a symmetrical ridge of sand that forms parallel to the prevailing wind direction. Longitudinal dunes occur in long, parallel ridges that are exceptionally straight and regularly spaced. They are typically separated by barren ground or desert pavement. Longitudinal dunes in the Sahara Desert (figure 13.27) are as high as 200 meters and more than 120 kilometers in length. Numerous hypotheses have been proposed to explain the development of longitudinal dunes, but none can adequately explain their spectacular size and regular spacing. It appears that crosswinds are important in piling up sand, which adds to the height of longitudinal dunes, whereas the more constant prevailing wind direction redistributes the sand down the length of the dunes. Smoke bomb experiments to analyze airflow have shown that the wind spirals down the intervening troughs between longitudinal dunes and may control the regularity of their spacing.

Not all dunes can be classified by an easily recognizable shape. Many of them are quite irregular.

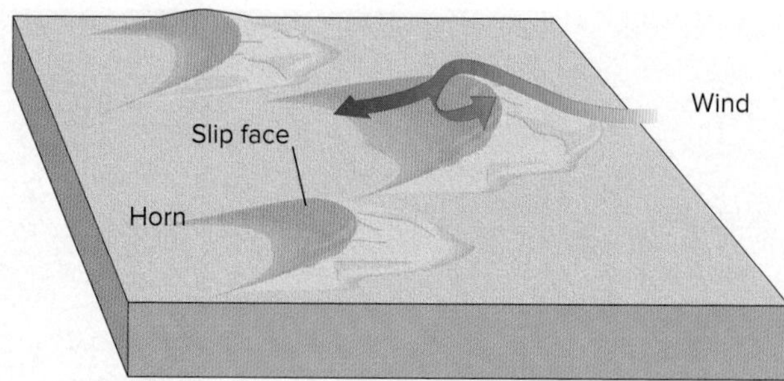

A Barchans

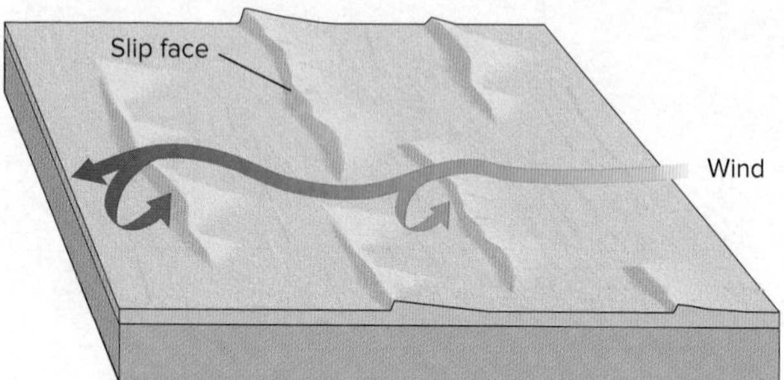

B Transverse dunes

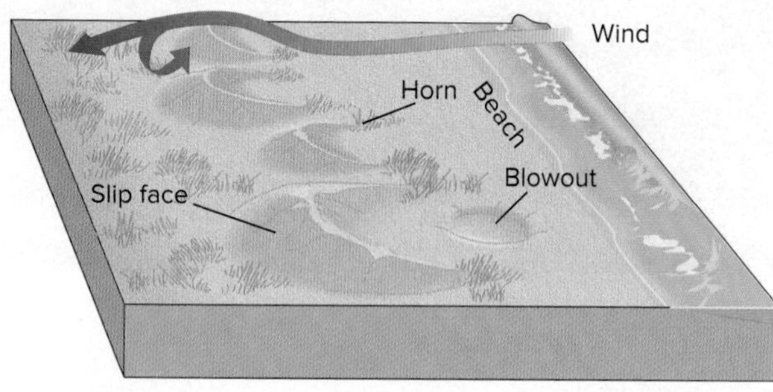

C Parabolic dunes

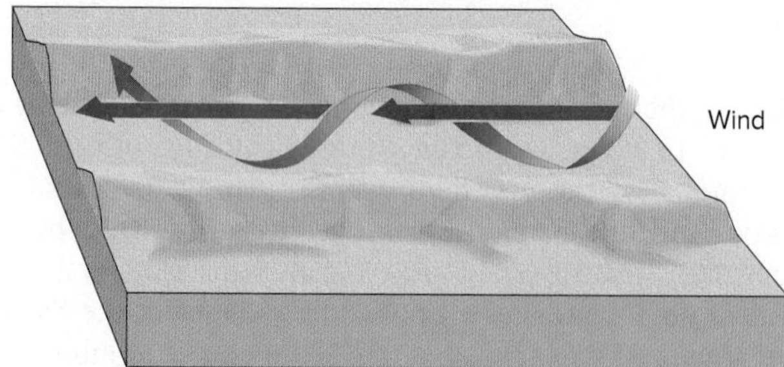

D Longitudinal dunes (seifs)

FIGURE 13.23

Types of sand dunes.

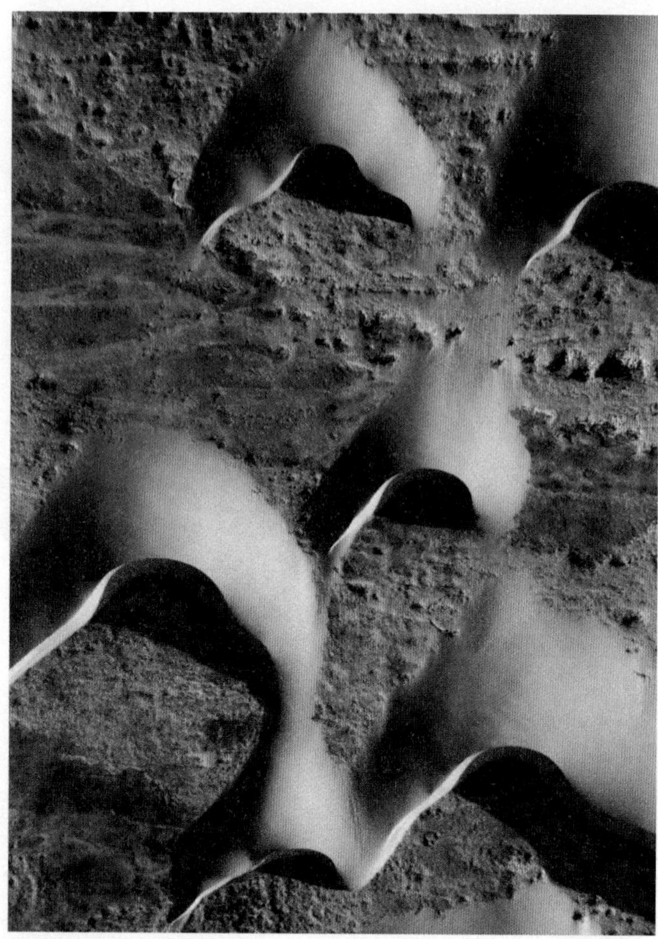

FIGURE 13.24

Barchan dunes formed in the Saharan desert from a limited supply of sand. Prevailing wind direction is toward the bottom of the photo.
©David Evans/National Geographic Creative

FIGURE 13.25

Transverse dunes with the slip face on the left side of the dunes indicates the wind blows from right to left.
©Corbis/Getty Images RF

FIGURE 13.26

Parabolic dunes near Pismo Beach, central California. Wind blows from left to right. The ocean and a sand beach are just to the left of the photo.
©Frank M. Hanna

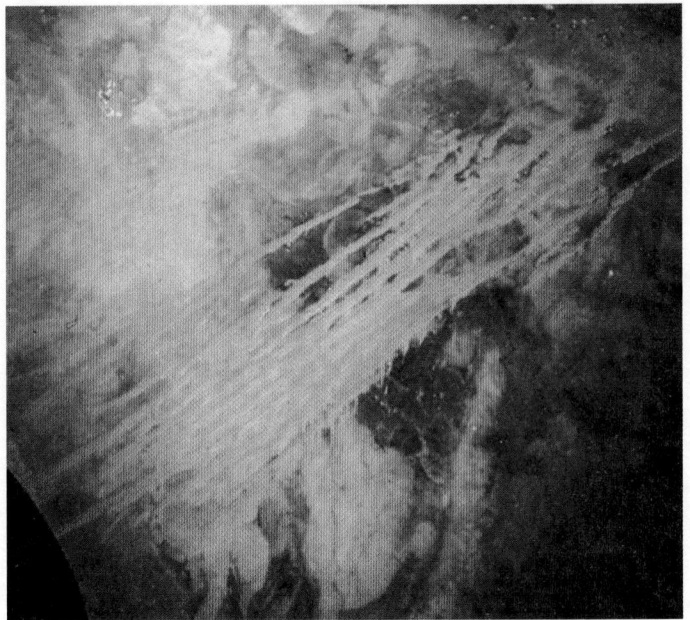

FIGURE 13.27

Longitudinal dunes in the Sahara Desert, Algeria. Photo from Gemini spacecraft at an altitude of about 100 kilometers.
Source: NASA

Summary

Deserts are located in regions where less than 25 centimeters of rain falls in a year. Such regions are found primarily in belts of descending air at 30° North and South latitude. Arid regions also may be due to the *rain shadow* of a mountain range, great distance from the sea, and proximity to a cold ocean current. Descending air forms cold deserts at the poles.

Desert landscapes differ from those of humid regions in lacking through-flowing streams and having internal drainage and many local, rising base levels. *Flash floods* caused by desert thunderstorms are effective agents of erosion despite the low rainfall. Limestone is resistant in deserts. Thin soil and slow rates of creep may give desert topography an angular look.

Parts of the southwestern United States are desert, the topography determined primarily by rock structure. Flat-lying sedimentary rocks of the Colorado Plateau are sculptured into cliffs, *plateaus*, *mesas*, and *buttes*. The fault-controlled topography of the Basin and Range province is marked by *alluvial fans*, *bajadas*, *playas*, and *pediments*.

Although wind erosion can be intense in regions of low moisture, streams are usually more effective than wind in sculpturing landscapes, regardless of climate.

Fine-grained sediment can be carried long distances by wind, even across entire continents and oceans.

Sand moves by *saltation* close to the ground, occasionally carving *ventifacts*.

Wind can *deflate* a region, creating a *blowout* in fine sediment.

Sand dunes move slowly downwind as sand is removed from the gentle upwind slope and deposited on the steeper *slip face* downwind.

Dunes are classified as *barchans*, *transverse dunes*, *parabolic dunes*, and *longitudinal dunes*, but many dunes do not resemble these types. Dune type depends on wind strength and direction, sand supply, and vegetation.

Terms to Remember

alluvial fan 328
bajada 328
barchan 338
blowout 334
butte 326
deflation 334
desert 322
fault 327
flash flood 325
loess 334
longitudinal dune 339
mesa 326
parabolic dune 339
pediment 329
plateau 326
playa 328
playa lake 328
rain shadow 323
sand dune 336
slip face 338
transverse dune 338
ventifact 334

Testing Your Knowledge

Use the following questions to prepare for exams based on this chapter.

1. What are two reasons why parts of the southwestern United States have an arid climate?
2. Describe the geologic structure and sketch the major landforms of
   a. the Colorado Plateau
   b. the Basin and Range province
3. How does a flash flood in a dry region differ from most floods in a humid region?
4. Give two reasons why wind is a more effective agent of erosion in a desert than in a humid region.
5. Sketch a cross section of an idealized dune, labeling the slip face and indicating the wind direction. Why does the dune move?
6. Name four types of sand dunes and describe the conditions under which each forms.
7. The defining characteristic of a desert is
   a. shifting sand dunes. b. high temperatures.
   c. low rainfall. d. all of the preceding.
   e. none of the preceding.
8. Which is characteristic of deserts?
   a. internal drainage b. limited rainfall
   c. flash floods d. slow chemical weathering
   e. all of the preceding
9. The major difference between a mesa and a butte is one of
   a. shape. b. elevation.
   c. rock type. d. size.
10. The Basin and Range province covers almost all of
    a. Utah. b. Nevada.
    c. Texas. d. Colorado.
11. Much of the southwestern United States is desert because (choose as many as apply)
    a. it is near 30° North.
    b. the western mountains create a rain shadow.
    c. cold ocean currents in the Pacific cause high evaporation rates in the land.
    d. it is a great distance from the ocean.
12. A very flat surface underlain by a dry lake bed of hard, mud-cracked clay is called a
    a. ventifact. b. plateau.
    c. playa. d. none of the preceding.
13. Rocks with flat, wind-abraded surfaces are called
    a. ventifacts. b. pediments.
    c. bajadas. d. none of the preceding.
14. The removal of clay, silt, and sand particles from the land surface by wind is called
    a. deflation. b. depletion.
    c. deposition. d. abrasion.
15. Which is *not* a type of dune?
    a. barchan b. transverse
    c. parabolic d. longitudinal
    e. all of the preceding are dunes
16. A broad ramp of sediment formed at the base of mountains when alluvial fans merge is
    a. a playa. b. a bajada.
    c. a pediment. d. an arroyo.
17. A surface layer of closely packed pebbles is called
    a. desert varnish. b. deflation.
    c. a blowout. d. desert pavement.

Expanding Your Knowledge

1. Study the photos of sand dunes in this chapter. Which way does the prevailing wind blow in each case?
2. Can deserts be converted into productive agricultural regions? How? Are there any environmental effects from such conversion?
3. At what relative depth is groundwater likely to be found in a desert? Why? Is the water likely to be drinkable? Why?

Exploring Web Resources

http://pubs.usgs.gov/gip/deserts/contents/

Online version of *Deserts: Geology and Resources* by A. S. Walker provides a good overview of deserts, processes, and mineral resources.

CHAPTER 14

Waves, Beaches, and Coasts

Sandy beach and sea stacks (erosional remnants of the headlands) form as the coast is being straightened by shoreline processes at Cannon Beach, Oregon.
©*Bogdan Bratosin/Moment/Getty Images*

LEARNING OBJECTIVES

- Explain how waves form, and discuss the factors that control wave height and wavelength. Compare and illustrate the movement of waves with the motion of water in a wave.
- Describe what happens when waves reach the shore.
- Describe how beaches develop, and explain the seasonal cycle of a beach. What are the sources of sand on beaches?
- Describe and sketch the features formed by the longshore drift of sand. Discuss the effects of human interference on the movement of sand.
- Identify and explain why there are different types of coasts, and illustrate the types of features that may develop in each.

Chapters 9 through 13 have dealt with the sculpturing of the land by mass wasting, streams, groundwater, glaciers, and wind. Water waves are another agent of erosion, transportation, and deposition of sediment. Earth's shorelines are continuously changing due to interaction between the hydrosphere and the geosphere. Along the shores of oceans and lakes, waves break against the land, building it up in some places and tearing it down in others.

Ordinary ocean waves (as opposed to tsunami) are created by wind–interaction between the atmosphere and the hydrosphere. Waves moving across an ocean transfer the energy derived from the wind to shorelines. This energy is used to a large extent in eroding and transporting sediment along the shoreline. Understanding how waves travel and move sediment can help you see how easily the balance of supply, transportation, and deposition of beach sediment can be disturbed. Such disturbances can be natural or human-made, and the changes that result often destroy beachfront homes and block harbors with sand.

Beaches have been called "rivers of sand" because breaking waves, as they sort and transport sediment, tend to move sand parallel to the shoreline. In this chapter, we look at how beaches are formed and examine the influence of wave action on such coastal features as sea cliffs, barrier islands, and terraces.

FIGURE 14.1
House along the coastline of the central Jersey shore is collapsed and buried by sand from the storm waves associated with the landfall of Superstorm Sandy that struck the Atlantic coast on October 31, 2012.
©Mike Groll/AP Images

14.1 SHORELINE DYNAMICS

If you spend a week at the shore during the summer, you may not notice any great change in the appearance of the beach while you are there. Even if you spend the whole summer at the seaside, nothing much seems to happen to the beach during those months. Tides rise and fall every day and waves strike the shore, but the sand that you walk on one day looks very much like the sand that you walked on the previous day. The shape of the beach does not appear to change, nor does the sand seem to move very much.

On most beaches, however, the sand *is* moving, in some cases quite rapidly. The shoreline is a very dynamic place. The beach looks the same from day to day only because new sand is being supplied at about the same rate that old sand is being removed.

Where is the sand going? Some sand is carried out to deep water. Some is piled up and stored high on the beach. But on most shores, much more of the sand moves along parallel to the beach in relatively shallow water. In this way, loose sand grains travel a hundred meters, or more, per day along some coasts, especially those subject to strong waves.

On some beaches, sand is being removed faster than it is being replenished. When this happens, beaches become narrower and less attractive for sunbathing and swimming. Where erosion is severe, buildings close to the beach can be undermined and destroyed by storm waves as the beach disappears or is shifted inland (figure 14.1). Such was the case when Superstorm Sandy struck the shore of New Jersey and New York in October 2012 and caused catastrophic damage. The sand moved from the beach may be redeposited in inconvenient places, such as roadways or across the mouth of a harbor, where it must be dredged out periodically. Because moving sand can create many problems for residents of coastal towns and cities, it is important to understand something of how and why the sand moves.

14.2 HOW DO WAVES FORM?

The energy that moves sand along a beach comes from the wind-driven water waves that break upon the shore. As wind blows over the surface of an ocean or a lake, some of the wind's energy is transferred to the water surface by friction, forming the waves that move through the water. The height of waves (and their length and speed) are controlled by the wind speed, the length of time that the wind blows, and the distance that the wind blows over the water (*fetch*). The largest waves form where high winds blow over a long expanse of open water for an extended period of time.

Wave shapes can vary. Short, choppy *seas* in and near a storm create a chaotic sea surface, often with considerable white foam as strong winds blow the tops off waves. Long, rolling *swells* form a regular series of similar-sized waves on shores that may be thousands of kilometers from the storms that generated the waves. (Summer surfing waves in southern California can be generated by large storms north of Australia in the Southern Hemisphere winter.) When waves break against the shore as *surf*, a large portion of their energy is spent moving sand along the beach.

The height of waves is the key factor in determining wave energy. **Wave height** is the vertical distance between the **crest**, which is the high point of a wave, and the **trough**, which is the low point (figure 14.2). In the open ocean, normal waves have heights of about 0.3 to 5 meters, although during violent storms, including hurricanes, waves can be more than 15 meters high. The highest wind wave ever measured was 34 meters by the anxious crew of a ship in the north Pacific in 1933. (The highest tsunami ever measured, caused by a submarine earthquake rather than wind, was 85 meters, in the Ryukyu island chain south of mainland Japan in 1971; see chapter 16.)

Wavelength is the horizontal distance between two wave crests (or two troughs). Most ocean wind waves are between 40 and 400 meters in length and move at speeds of 25 to 90 kilometers per hour (15 to 55 miles per hour) in deep water. Waves are also characterized by having a period, the time it takes for two crests (or two troughs) to pass by. Wind waves have periods of several seconds. Tsunamis have periods of 15 to 20 minutes.

The movement of water in a wave is like the movement of wheat in a field when wind blows across it. You can see the ripple caused by wind blowing across a wheat field, but the wheat does not pile up at the end of the field. Each stalk of wheat bends over when the wind strikes it and then returns to its original position. A particle of water moves in an *orbit*, a nearly circular path, as the wave passes (figure 14.3); the particle returns to its original position after the wave has passed. In deep water, when a wave moves across the water's surface, energy moves with the wave, but the water, like the wheat, does not advance with the wave; instead it moves in circular orbits.

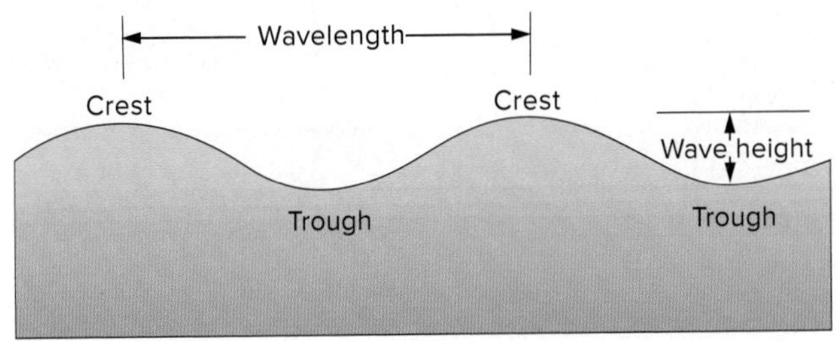

FIGURE 14.2

Wave height is the vertical distance between the wave crest and the wave trough. Wavelength is the horizontal distance between two crests.

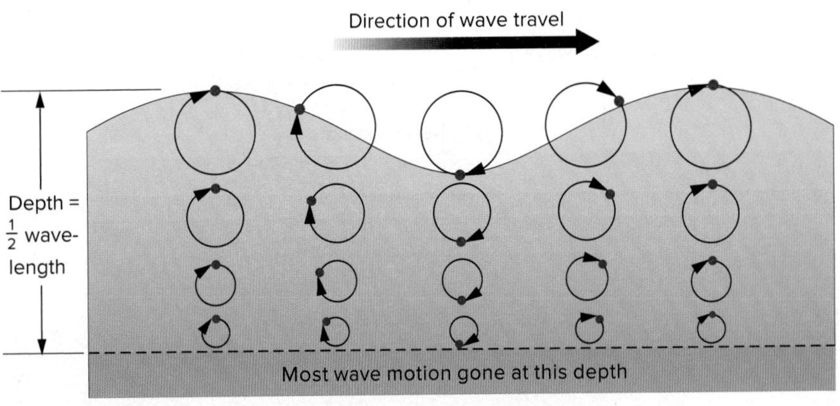

FIGURE 14.3

Orbital motion of water in waves dies out with depth. At the surface, the diameter of the orbits is equal to the wave height.

At the surface, the diameter of the orbital path of a water particle is equal to the height of the wave (figure 14.3). Below the surface, the orbits decrease in size until the motion is essentially gone at a depth equal to half the wavelength. This is why a submarine can cruise in deep, calm water beneath surface ships that are being tossed by the orbital motion of large waves. A person swimming just below the surface can feel the orbital motion of the water as the wave passes overhead.

Surf

As waves move from deep water to shallow water near shore, they begin to be affected by the ocean bottom. A wave first begins to "feel bottom" at the level of lowest orbital motion—that is, when the depth to the bottom equals half the wavelength. For example, a wave 150 meters long will begin to be influenced by the bottom at a water depth of 75 meters.

In shallow water, the presence of the bottom interferes with the circular orbits, which flatten into ovals (figure 14.4). The waves slow down and their wavelength decreases, though the period remains the same. Meanwhile, the sloping bottom wedges the moving water upward, increasing the wave height. Because the height is increasing while the length is decreasing, the waves become steeper and steeper until they break. A **breaker** is a wave that has become so steep that the crest of the wave topples forward, moving faster than the main body of the wave. The breaker then advances as a turbulent, often foamy, mass. Energy from the wind is transmitted by the wave and finally spent by breakers on the beach. Breakers collectively are called **surf**. Water in the surf zone has lost its orbital motion and moves back and forth, alternating between onshore and offshore flow.

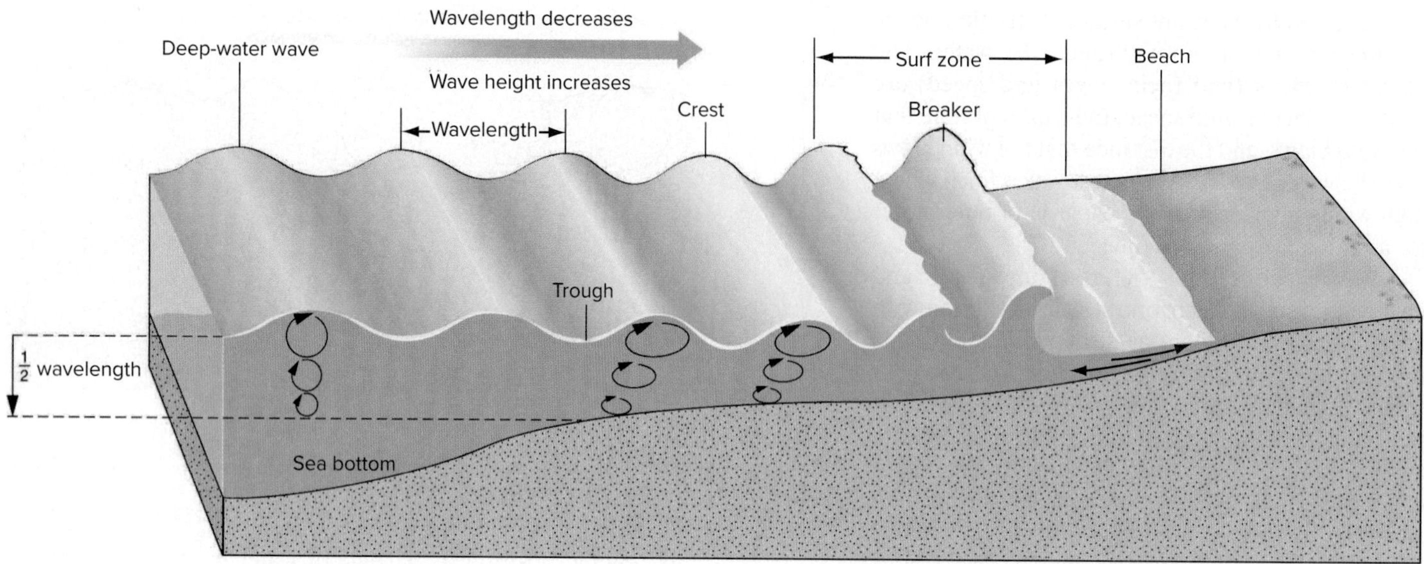

FIGURE 14.4

As a deep-water wave approaches shore, it begins to "feel" the sea bottom and slow down. Circular water orbits flatten, and the wave peaks and breaks. In the foamy surf zone, water moves back and forth rather than in orbits.

14.3 WHAT HAPPENS WHEN WAVES REACH THE SHORE?

Wave Refraction

Most waves do not come straight into shore. A wave crest usually arrives at an angle to the shoreline (figure 14.5*A*). One end of the wave breaks first, and then the rest breaks progressively along the shore.

The angled approach of a wave toward shore can change the direction of wave travel. One end of the wave reaches shallow water first, "feels bottom," and slows down while the rest of the wave continues at its deep-water speed (figure 14.5*B*). As more and more of the wave comes into contact with the bottom, more of the wave slows down. The wave slows progressively along its length, causing the wave crest to change direction and become more nearly parallel to the shoreline. This change in direction, or bending of waves as they enter shallow water, is called **wave refraction** (figure 14.5*B*).

A

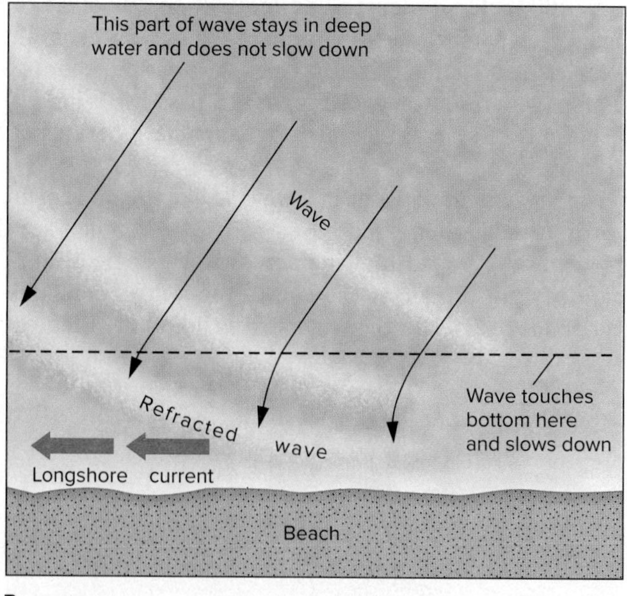

B

FIGURE 14.5

(*A*) These waves are arriving at an angle to the shoreline. They break progressively along the shore, from the upper right of the photo to the lower left. (*B*) Wave refraction changes the wave direction, bending the wave so it becomes more parallel to shore. The angled approach of waves to shore sets up a longshore current parallel to the shoreline. A longshore current, though probably small, would be expected to be moving from right to left in photo *A*.

(A) ©David McGeary

Longshore Currents

Although most wave crests become nearly parallel to shore as they are refracted, waves do not generally strike *exactly* parallel to shore. Even after refraction, a small angle remains between the wave crest and the shoreline. As a result, the water in the wave is pushed both *up* the beach toward land and *along* the beach parallel to shore.

Each wave that arrives at an angle to the shore pushes more water parallel to the shoreline. Eventually, a moving mass of water called a **longshore current** develops parallel to the shoreline (figure 14.5). The width of the longshore current is about equal to the width of the surf zone. The seaward edge of the current is the outer edge of the surf zone, where waves are just beginning to break; the landward edge is the shoreline. A longshore current can be very strong, particularly when the waves are large. Such a current can carry swimmers hundreds of meters parallel to shore before they are aware that they are being swept along. It is these longshore currents that transport most of the beach sand parallel to shore. The backwash of the waves and the flattened orbital motion of the waves scours sediment off the bottom that can be carried as suspended sediment by the longshore current.

Rip Currents

Rip currents are narrow currents that flow straight out to sea in the surf zone, returning water seaward that breaking waves have pushed ashore (figure 14.6). Rip currents travel at the water surface and die out with depth. They pulsate in strength, flowing most rapidly just after a set of large waves has carried a large amount of water onto shore. Rip currents can be important transporters of sediment, as they carry fine-grained sediment out of the surf zone into deep water.

As a single wave comes toward shore, its height varies from place to place. Rip currents tend to develop locally where wave height is low. Rip currents that are fixed in position are apt to be found over channels on the sea floor because depressions on the bottom reduce wave height. Complex wave interactions can also lower wave height, and rip currents that form because of wave interactions tend to shift position along the shore. Such shifting rip currents are usually spaced at regular intervals along the beach.

Rip currents are fed by water within the surf zone. They flow rapidly out through the surf zone and then die out quickly. Where waves are nearly parallel to a shoreline, longshore feeder currents of equal strength develop in the surf zone on either side

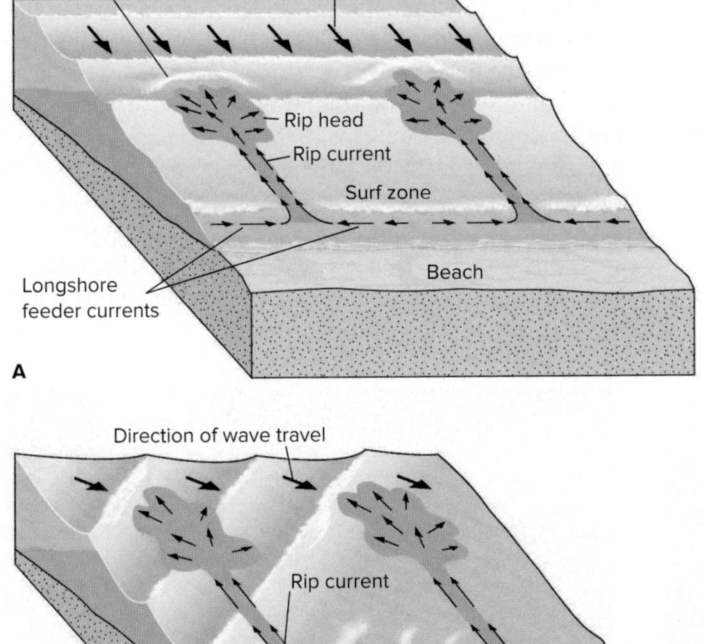

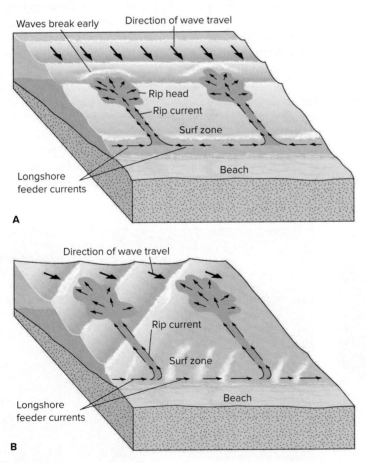

FIGURE 14.6

Rip currents and their feeder currents can develop regardless of the angle of approach of waves. (*A*) Waves approach parallel to shore; feeder currents on both sides of rip currents. (*B*) Waves approach at an angle to the shore; feeder current on only one side of rip current. (*C*) Rip currents along the coast of Baja, Mexico, carry dirty water and foam seaward; they can cause incoming waves to break early. Baja, California Sur, Mexico.

(C) ©Airphoto-Jim Wark

of a rip current (figure 14.6*A*). Where waves strike the shore at an angle and set up a strong, unidirectional, longshore current, a rip current is fed from one side by the longshore current, which increases in strength as it nears the rip current (figure 14.6*B*). Rip currents are also found alongside points of land and engineered structures such as jetties and piers, which can deflect longshore currents seaward.

You can easily learn to spot rip currents at a beach. Look for discoloration in the water where fine sediment is being picked up in the surf zone and moved seaward (figure 14.6*C*).

Getting caught in a rip current and being carried out to sea can panic an inexperienced swimmer—even though the trip will stop some distance beyond the surf zone as the rip dies out. A swimmer frightened by being carried away from land and into breaking waves can grow exhausted fighting the current to get back to shore. The thing to remember is that rip currents are narrow. Therefore, you can get out of a rip easily by swimming *parallel* to the beach instead of struggling against the current.

Surfers, on the other hand, often look for rip currents and paddle intentionally into them to get a quick ride out into the high breakers.

14.4 HOW DO BEACHES DEVELOP?

A **beach** is a strip of sediment (usually sand or gravel) that extends from the low-water line inland to a cliff or a zone of permanent vegetation. Waves break on beaches, and rising and falling tides may regularly change the amount of beach sediment that is exposed above water (figure 14.7*A*).

The steepest part of a beach is the **beach face**, which is the section exposed to wave action, particularly at high tide. Offshore from the beach face there is usually a **marine terrace**, a broad, gently sloping platform that may be exposed at low tide if the shore has significant tidal action. Marine terraces may be *wave-built* terraces constructed of sediment carried away from the shore by waves, or they may be *wave-cut* rock benches or platforms, perhaps thinly covered with a layer of sediment.

The upper part of the beach, landward of the usual high-water line, is the **berm**, a wave-deposited sediment platform that is flat or slopes slightly landward (figure 14.7*B*). It is usually dry, being covered by waves only during severe storms. This is where sunbathers set up their umbrellas and chairs.

Beach sediment is usually sand, typically quartz-rich because of quartz's resistance to chemical weathering. Heavy metallic minerals ("black sands") can also be concentrated on some beaches as less dense minerals such as quartz and feldspar are carried away by waves or wind (titanium-bearing sands are mined on some beaches in Florida and Australia). Tropical beaches may be made of bioclastic carbonate grains from offshore corals, algae, and shells. Some Hawaiian beaches are made of sand-sized fragments of basalt. Gravel beaches found on coasts, such as Maine, are attacked by the high energy of large waves (*shingle* is a regional name for disk-shaped gravel). Gravel beaches have a steeper face slope than sand beaches.

In seasonal climates, beaches often go through a summer-winter cycle (figure 14.8). This is due to the greater frequency of storms with strong winds during the winter months, which tend to produce high waves with short wavelengths. These high-energy waves tend to crash onshore and erode sand from the beach face and narrow the berm. Offshore, in less turbulent water, the sand settles to the bottom and builds an underwater sandbar (parallel to the beach) that serves as a "storage facility" for the next summer's sand supply. The following summer, or during calmer weather, low-energy waves with long wavelengths break over the sandbar and gradually push the sand back onto the beach face to widen the berm. Each season the beach changes in shape until it comes into equilibrium with the prevailing wave type.

Many winter beaches can be dangerous because of high waves and narrowed beaches. Several beaches along the Pacific

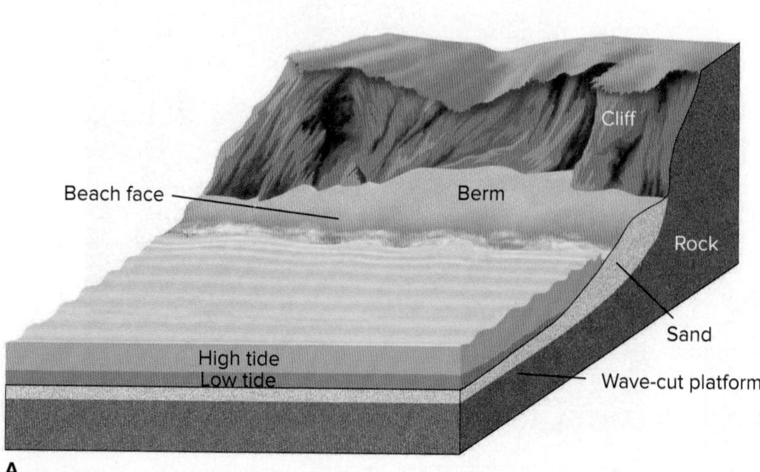

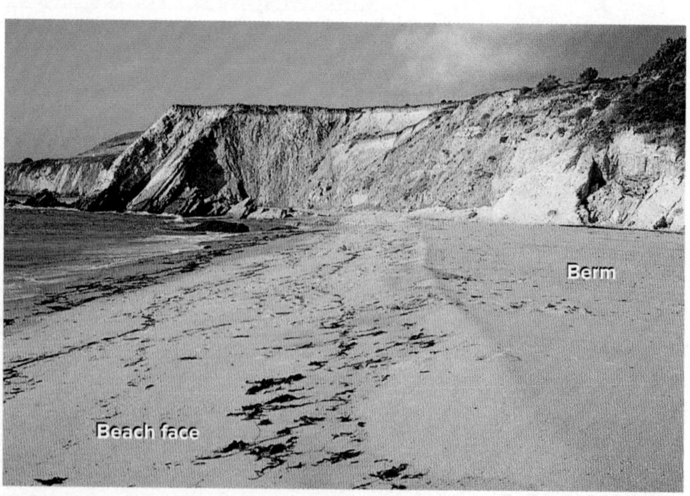

FIGURE 14.7

(*A*) Parts of a beach. (*B*) The beach face (on the left) and berm (on the right) on a northern California beach.
(*B*) ©*Diane Carlson*

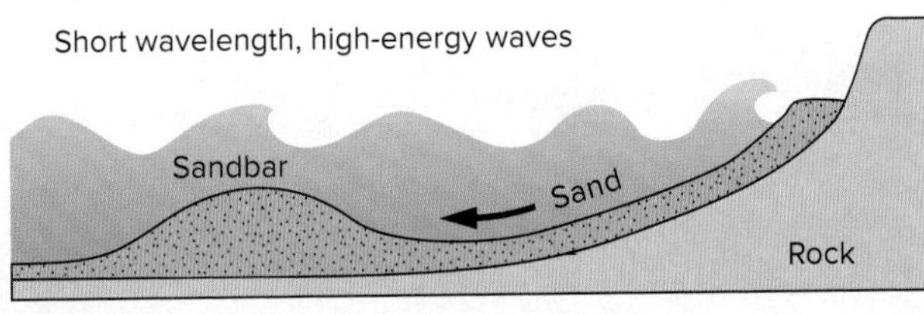

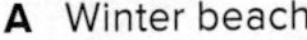

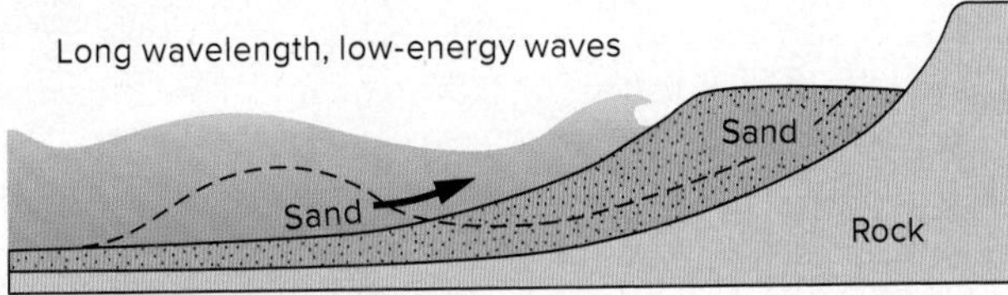

B Summer beach

FIGURE 14.8

Seasonal cycle of a beach caused by differing wave types. (*A*) Narrow winter beach. Waves may break once on the winter sandbar, then re-form and break again on the beach face. (*B*) Wide summer beach.

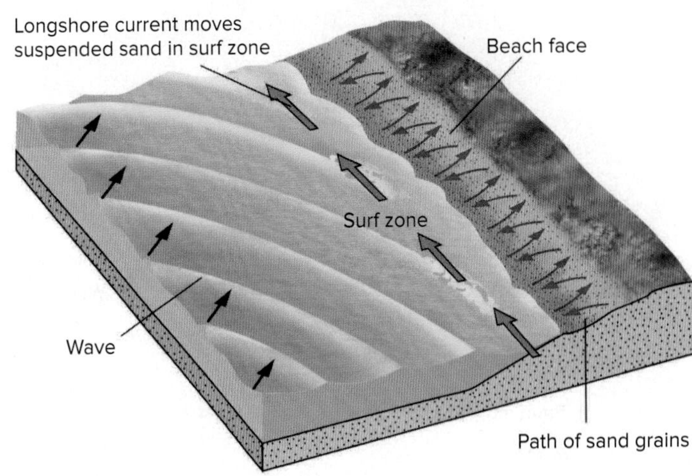

FIGURE 14.9

Longshore drift of sand on the beach face and by a longshore current within the surf zone.

coast of the United States and Canada are nearly free of accidents in the summer, when they are heavily used, but are regularly marked by drownings in the winter as beach walkers are swept off narrower beaches and out to sea by large storm waves.

14.5 WHY DOES SAND MOVE ALONG A SHORELINE?

Longshore drift is the movement of sediment parallel to shore when waves strike the shoreline at an angle. Figure 14.9 shows the two ways in which this movement of sediment (usually sand) occurs. Some longshore drift takes place directly on the beach face when waves wash up on land. A wave washing up on the beach at an angle tends to wash sand along at the same angle. After the wave has washed up as far as it can go, the water returns to the sea by running down the beach face by the shortest possible route, that is, straight downhill to the shoreline, not back along the oblique route it came up. (Wave run-up is known as *swash,* the return as *backwash.*) The effect of this zigzag motion is to move the sand in a series of arcs along the beach face.

Much more sand is moved by longshore transport in the surf zone, where waves are breaking into foam. The turbulence of the breakers erodes sand from the sea bottom and keeps it suspended. Even a weak longshore current can move the suspended sand parallel to the shoreline. The sand in the longshore current moves in the same direction as the sand drift on the beach face (figure 14.9).

Vast amounts of sand can be moved by longshore transport. The U.S. Army Corps of Engineers estimates that 436,000 cubic meters of sand per year are moved northward by waves at Sandy Hook, New Jersey, and 1 million cubic meters of sand per year are moved southward at Santa Monica, California.

Eventually, the sand that has moved along the shore by these processes is deposited. Sediment may build up off a point of land to form a **spit**, a fingerlike ridge of sediment that extends out into open water (figure 14.10*A* and *B*). A **baymouth bar**, a ridge of sediment that cuts a bay off from the ocean, is formed by sediment migrating across what was earlier an open bay (figure 14.10*A* and *C*). Off the western coast of the United States, a considerable amount of drifting sand is carried into the heads of underwater canyons, where the sediments slide down into deep, quiet water.

A striking but rare feature formed by longshore drift is a **tombolo**, a bar of sediment connecting a former island to the mainland. As shown in figure 14.11, waves are refracted around an island in such a way that they tend to converge behind the island. The waves sweep sand along the mainland (and from the island) and deposit it at this zone of convergence, forming a bar that grows outward from the mainland and eventually connects to the island.

Human Interference with Sand Drift

Several engineered features can interrupt the flow of sand along a beach (figure 14.12). *Jetties*, for example, are rock walls designed to protect the entrance of a harbor from sediment deposition and storm waves. Usually built in pairs, they protrude above the surface of the water. Figure 14.12*A* shows how sand piles up against one jetty while the beach next to the other, deprived of a sand supply, erodes back into the shore.

Groins are sometimes built in an attempt to protect beaches that are losing sand from longshore drifting. These short walls are built from a variety of materials perpendicular to shore to trap moving sand and widen a beach (figure 14.12*B*). However, once a groin is built to capture the sand, beaches down current will erode as the longshore current attempts to replenish its sediment load. The disappearance of neighboring beaches often results in lawsuits and in successive groins being constructed in an attempt to trap the remaining sand.

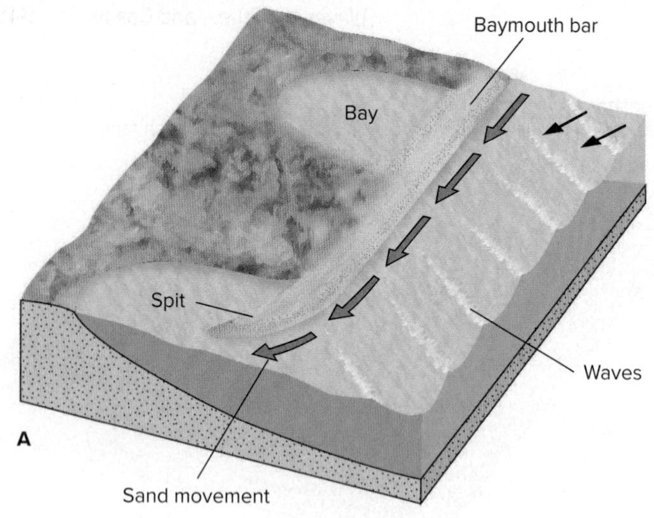

FIGURE 14.10

(*A*) Longshore drift of sand can form spits and baymouth bars. (*B*) Curved spit near Victoria, British Columbia. (*C*) A baymouth bar has sealed off this bay from the ocean as sand migrated across the mouth of the Russian River in northern California.
(B) ©D. A. Rahm, courtesy Rahm Memorial Collection, Western Washington University; (C) ©Diane Carlson

FIGURE 14.11

(*A*) A tombolo has connected this rock, once an island, to the shore. Note the waves bending around the two sides of the rock, near Santa Cruz, California. (*B*) Formation of a tombolo. Wave refraction around an island interrupts the longshore current and creates a sandbar that connects the island with the mainland.
(A) ©David McGeary

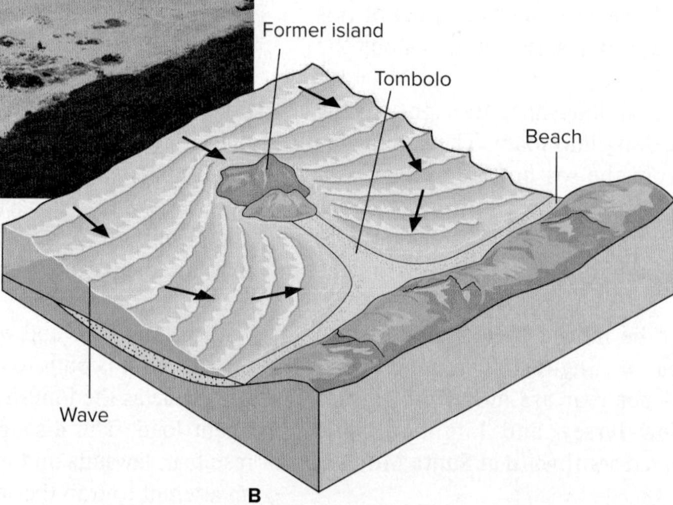

FIGURE 14.12

Sand piles up against obstructions and in areas deprived of wave energy. (*A*) Jetties at Manasquan Inlet, New Jersey. Sand drift to the right has piled sand against the left jetty and removed sand near the right jetty. (*B*) Groins at Ocean City, New Jersey. Sand drift is to the right. (*C*) Breakwater at Venice, California, has caused deposition of sand in the wave-protected zone.
(A & B) ©S. Jeffress Williams; (C) ©Bruce Perry, Department of Geological Sciences, CSU Long Beach

Sand deposition also occurs when a stretch of shore is protected from wave action by a *breakwater*, an offshore structure built to absorb the force of large, breaking waves and provide quiet water near shore. When the city of Santa Monica in California built a rock breakwater parallel to the shore to create a protected small-boat anchorage, the lessening of wave action on the shore behind the breakwater slowed the longshore current and allowed sand to build up there (figure 14.12*C*), threatening eventually to fill in the anchorage. The city had to buy a dredge to remove the sand from the protected area and redeposit it farther along the shore where the waves could resume moving sediment.

A beach attempts to come into equilibrium with the waves that strike it. The type and amount of sediment, the position of the sediment, and especially the movement of the sediment, adjust to the incoming wave energy. Whenever human activity interferes with sand drift or wave action, the beach responds by changing its configuration, usually through erosion or deposition in a nearby part of the beach.

Sources of Sand on Beaches

Some beach sand comes from the erosion of local rock, such as points of land or cliffs nearby. On a few beaches, replenishment comes from sand stored outside the surf zone in the deeper water offshore. Bioclastic carbonate beaches are formed from the remains of marine organisms offshore. But the greater part of the sand on most beaches comes from river sediment brought down to the ocean. Waves pick up this sediment and move it along the beach by longshore drift.

What happens to a beach if all the rivers contributing sand to it are dammed? Although damming a river may be desirable for many reasons (flood control, power generation, water supply, recreation), when a river is dammed, its sediment load no longer reaches the sea (see box 10.1). The sand that supplied the beach in the past now comes to rest in the quiet waters of the reservoir behind the dam. Longshore drift, however, continues to remove sand from beaches, even though little new sand is being supplied, and the result is a net loss of sand from beaches. Beaches without a sand supply eventually disappear. To prevent this, some coastal communities have set up expensive programs of building pipelines or draining reservoirs and trucking the trapped sand down to the beaches.

14.6 WHY ARE THERE DIFFERENT TYPES OF COASTS?

A beach is just a small part of the **coast**, which is all the land near the sea, including the beach and a strip of land inland from it. Coasts can be rocky, mountainous, and cliffed, as in northern New England and on the Pacific shore of North America, or they can be broad, gently sloping plains, as along much of the southeastern United States. Wave erosion and deposition can greatly modify coasts from their original shapes. Many coasts have been drowned during the past 15,000 years by the rise in sea level caused by the melting of the Pleistocene glaciers (see chapter 12). Other coasts have been lifted up by tectonic forces or by postglacial crustal rebound at a rate greater than the rise in sea level so that sea-floor features are now exposed on dry land.

Erosional Coasts

A great many steep, rocky coasts have been visibly changed by wave erosion. Soluble rocks such as limestone dissolve as waves wash against them, and more durable rocks such as granite are fractured or broken along fractures by the enormous pressures caused by waves slamming into rock (wave impact pressures have been measured as high as 60 metric tons per square meter).

An irregular coast with bays separated by rocky **headlands** (promontories), such as the southern California coastline, can be gradually straightened by wave action. Because wave refraction bends waves approaching such a coast until they are nearly parallel to shore, most of the waves' energy is concentrated on the headlands, while the bays receive smaller, diverging waves (figure 14.13). Rocky cliffs form from wave erosion on the headlands. The eroded material is deposited in the quieter water of nearby bays, forming broad beaches. **Coastal straightening** of an irregular shore gradually takes place through wave erosion of headlands and wave deposition in bays (figure 14.14).

Wave erosion of headlands produces **sea cliffs**, steep slopes that retreat inland by mass wasting as wave erosion undercuts them (figure 14.15). At the base of sea cliffs are sometimes found *sea caves*, cavities eroded by wave action along zones of weakness in the cliff rock. As headlands on irregular coasts are eroded landward, sea cliffs enlarge until the entire coast is marked by a retreating cliff (figure 14.14). On some exposed coasts, the rate of cliff retreat can be quite rapid, particularly if the rock is weakly consolidated. Some sea cliffs north of San Diego, California, and at Cape Cod National Seashore in Massachusetts are retreating at an average rate of 1 meter per year. Because sea-cliff erosion in weak rock is often in the form of large, infrequent slumps (see chapter 9), some portions of these coasts may retreat 10 to 30 meters in a single storm. Some of these cliffs have particularly vulnerable "ocean-view" homes and hotels at their very edges. Sea cliffs in hard, durable rock such as granite and schist retreat much more slowly.

Seawalls may be constructed along the bases of retreating cliffs to prevent wave erosion (figure 14.15). Seawalls of giant pieces of broken stone (riprap) or concrete tetrahedra are designed to absorb wave energy rather than allow it to erode cliff

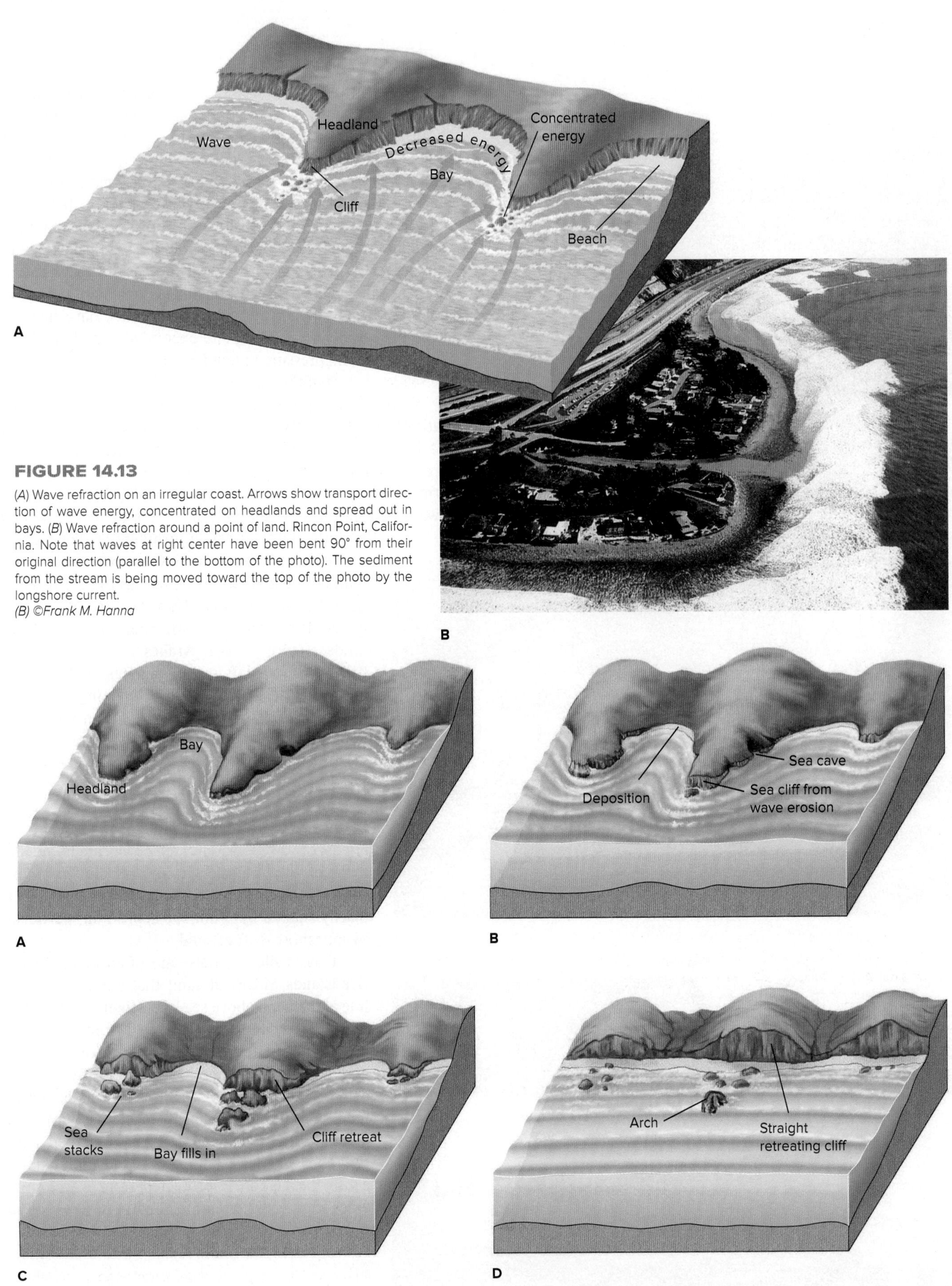

FIGURE 14.13

(*A*) Wave refraction on an irregular coast. Arrows show transport direction of wave energy, concentrated on headlands and spread out in bays. (*B*) Wave refraction around a point of land. Rincon Point, California. Note that waves at right center have been bent 90° from their original direction (parallel to the bottom of the photo). The sediment from the stream is being moved toward the top of the photo by the longshore current.
(B) ©Frank M. Hanna

FIGURE 14.14

Coastal straightening of an irregular coastline by wave erosion of headlands and wave deposition of sediment in bays. Continued erosion produces a straight, retreating cliff.

FIGURE 14.15

Retreating wave-cut cliff, north of Bodega Bay, Sonoma County, California. A concrete seawall has been built at the base of the cliff to slow wave erosion and help protect the cliff-edge homes. Note fragments of wave-destroyed structures near the seawall. Seawalls usually increase the erosion of sand beaches.
©David McGeary

rock. Vertical or concave seawalls of concrete are designed to reflect wave energy seaward rather than allow it to impact the shore. Some of the reflected energy, however, is focused at the base of the seawall, which eventually undermines it and causes the seawall to collapse. Reflection of waves from a seawall also increases the amount of wave energy just offshore, often increasing the amount of sand erosion offshore. Thus, a seawall designed to protect a sea cliff (and the buildings at its edge) may in some cases destroy a sand beach at the base of the cliff. However, seawalls engineered with openings, which are designed to minimize beach erosion in front of the walls, have had some success. Seawalls are difficult and expensive to build and maintain, but political pressure to build more of them will increase as the sea level rises in the future.

Wave erosion produces other distinctive features in association with sea cliffs. A **wave-cut platform** (or *terrace*) is a nearly horizontal bench of rock formed beneath the surf zone as a coast retreats by wave erosion (figure 14.16). The platform widens as the sea cliffs retreat. The depth of water above a wave-cut platform is generally 6 meters or less, coinciding with the depth at which turbulent breakers actively erode the sea bottom. **Stacks** are erosional remnants of headlands left behind as the coast retreats inland (figure 14.17). They form small, rocky islands off retreating coasts, often directly off headlands (figure 14.14). **Arches** (or *sea arches*) are bridges of rock left above openings eroded in headlands or stacks by waves. The openings are eroded in spots where the rock is weaker than normal, perhaps because of closely spaced fractures.

A

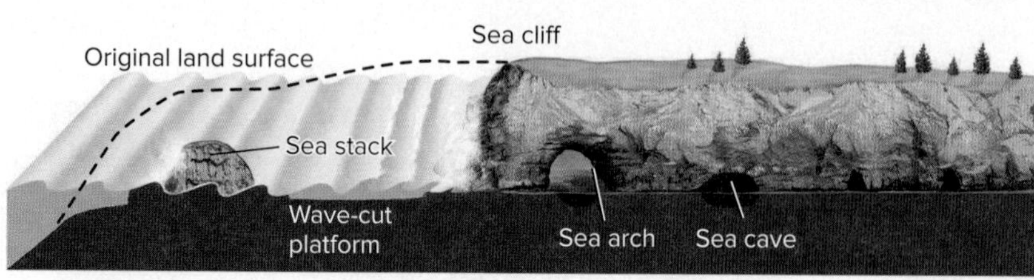

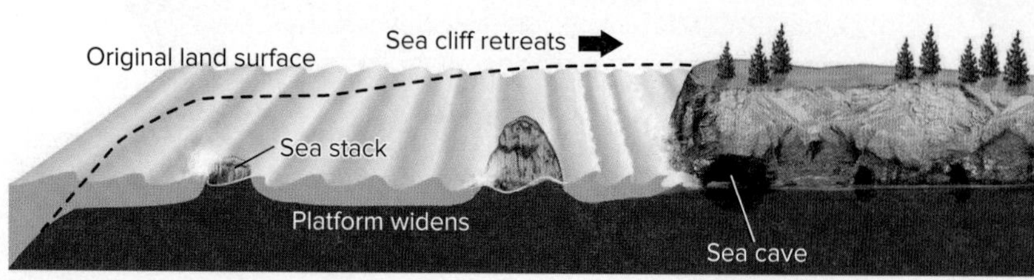

B

Depositional Coasts

Many coasts are gently sloping plains and show few effects of wave erosion. Such coasts are found along most of the Atlantic Ocean and Gulf of Mexico shores of the United States. These coasts are primarily shaped by sediment deposition, particularly by longshore drift of sand.

Coasts such as these are often marked by **barrier islands**–ridges of sand that parallel the shoreline and extend above sea level (figure 14.18). These barrier islands may have formed from sand eroded by waves from deeper water offshore, or they may be greatly elongated sand spits formed by longshore drift. The slowly rising sea level associated with the melting of the Pleistocene glaciers may have been a factor in their development. A protected lagoon separates barrier islands from the mainland. Because the lagoon is protected from waves, it provides a quiet waterway for boats. A series of such

FIGURE 14.16

(*A*) A wave-cut platform (the wide, horizontal bench of dark rock at the base of the cliffs) is exposed at low tide, La Jolla, California. (*B*) A wave-cut platform widens as a sea cliff retreats.
(A) ©David McGeary

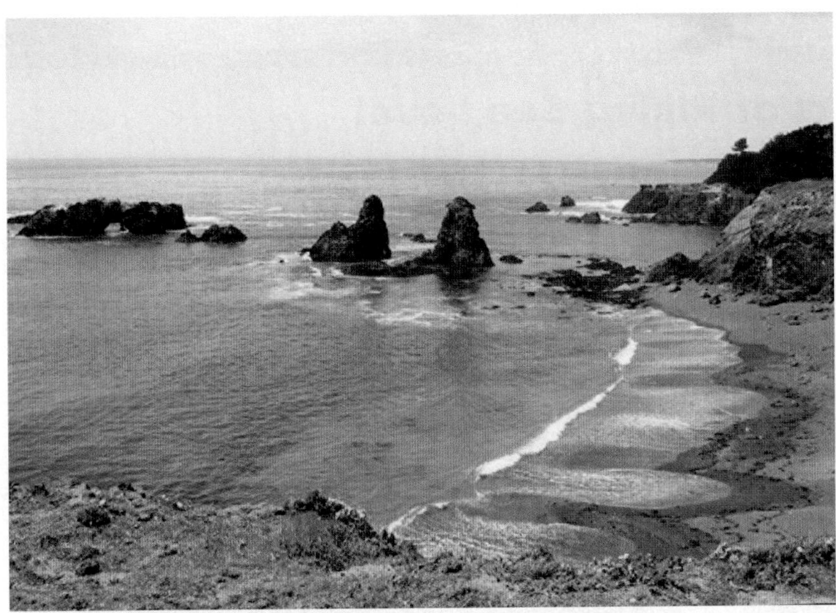

FIGURE 14.17

Sea stacks and an arch mark old headland positions on this retreating, wave-eroded coast in northern California.
©Diane Carlson

Some barrier islands along the Atlantic and Gulf coasts are densely populated. Atlantic City (New Jersey), Ocean City (Maryland), Miami Beach (Florida), and Galveston (Texas) are examples of cities built largely on barrier islands. In some of these cities, houses, luxury hotels, boardwalks, and condominiums are clustered near the edge of the sea; many are built upon the loose sand of the island (figure 14.19). These developed areas are vulnerable to late-summer hurricanes that sooner or later bring huge storm waves and storm surges onto these coasts, eroding the sand and undermining the building foundations at the water's edge.

Nonmarine deposition may also shape a coast. Rapid sedimentation in *deltas* by rivers can build a coast seaward (see figure 10.27). *Glacial deposition* can form shoreline features. Several islands, such as Martha's Vineyard, off the New England coast were glacially deposited; Long Island, New York, formed from the deposition of a recessional and end moraine.

FIGURE 14.18

(*A*) A barrier island on a gently sloping coast. A lagoon separates the barrier island from the mainland, and tidal currents flowing in and out of gaps in the barrier island deposit sediment as submerged tidal deltas. (*B*) A barrier island near Pensacola, Florida: open ocean to right, lagoon to left, and mainland Florida on far left. Light-colored lobes of sand within lagoon were eroded from the barrier island by hurricane waves, which washed entirely across the island and into the lagoon.
(B) ©Frank M. Hanna

lagoons stretches almost continuously from New York to Florida, and many also exist along the Gulf Coast, forming an important route for barge traffic. As tides rise and fall, strong tidal currents may wash in and out of gaps between barrier islands, distributing sand in submerged *tidal deltas* both landward and seaward of the gaps.

Drowned Coasts

Drowned (or *submergent*) coasts are common because sea level has been rising worldwide for the past 15,000 years (box 14.1). During the glacial ages of the Pleistocene, average sea level was 130 meters below its present level. The shallow sea floor near the continents was then dry land, and rivers flowed across it,

ENVIRONMENTAL GEOLOGY 14.1

Coasts in Peril—The Effects of Rising Sea Level

The thermal expansion of warming ocean water and the melting of glaciers and ice sheets have resulted in an increase in sea level. Sea level has risen, on average, 0.2 meters (8 inches) worldwide since 1880 and is predicted to rise at least 0.2 meters and no more than 2 meters (6.6 feet) by 2100. In the United States, along the coast of the Gulf of Mexico and the eastern Atlantic (box figure 1), the rise in sea level has been twice the global average due to subsidence along the coast.

The present rate of sea-level rise has caused erosion in more than 70% of the world's beaches in the past century. Rising sea level causes deeper offshore waters that generate higher, more powerful storm waves that erode sandy beaches. Rivers that empty into the ocean supply sediments that can maintain beaches if sea-level rise is not too rapid; however, because most of the world's rivers have been dammed, the amount of sediment they deliver to the coast has been greatly reduced. The faster rate of sea-level rise predicted for the next century, combined with the decreased sediment supply from the rivers, puts the world's remaining beaches at risk.

Also at great risk are gently sloping coastal areas because the higher waters are able to move farther inland. Coastal wetlands, tropical mangrove swamps, and cypress swamps are inundated by saltwater, which kills the vegetation and often converts the area into open ocean (box figure 2). These marshes and swamps act as sponges and buffer inland regions from storm waves, and their loss puts many populated areas at increased risk from storm damage.

BOX 14.1 ■ FIGURE 2

Rising sea level has turned coastal swamps in Grand Isle, Louisiana, into an island in the Gulf of Mexico. Bulkheads protect what is left of the dry land and houses from the open ocean.
©Matthew D White/Photolibrary/Getty Images

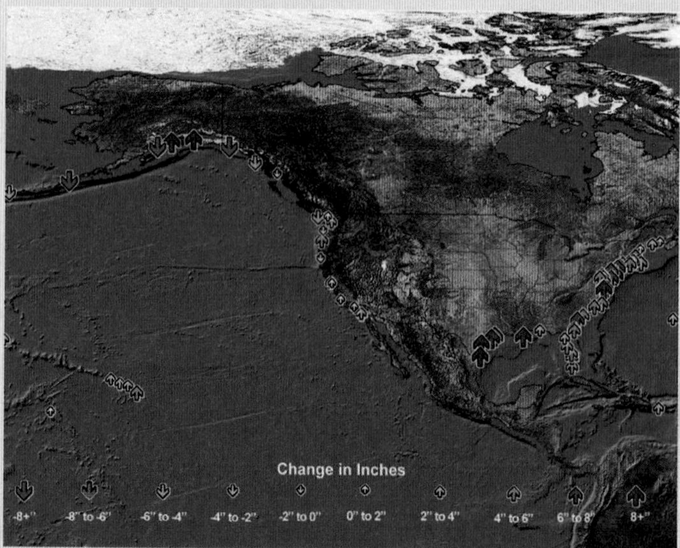

BOX 14.1 ■ FIGURE 1

Observed changes in relative sea level from 1958 to 2008 along the coast of the United States. Sea level has increased more than 8 inches in the past 50 years (shown by red arrows) along parts of the Atlantic and Gulf coasts.
Source: U.S. Global Change Research Program's 2009 National Climate Assessment (NCA)

Rising seas have also drowned barrier islands and mainland marshes in the Mississippi Delta that protect New Orleans from the full force of hurricanes. Since the 1930s, 5,000 square kilometers of Louisiana's coastal wetlands (an area nearly the size of Delaware) have been lost, and the shoreline is currently eroding at a rate of 9 meters per year. The erosion of the barrier islands that shelter New Orleans from hurricanes, coupled with elevation of the city below sea level, caused massive damage and loss of life when Hurricane Katrina struck in 2005.

Barrier islands are particularly vulnerable to sea-level rise. They are usually less than a meter above sea level on the landward side and, at most, a few meters above sea level on the ocean side. During storms, strong waves erode the ocean beaches, while high water inundates the landward beaches, possibly drowning the barrier island completely (box figure 3). Barrier islands can migrate landward, a process called "barrier rollover," as sand eroded from the seaward side is washed completely over the island and deposited on the landward side (box figure 4*A*). However, a great many barrier islands on the Atlantic Coast of the United States are developed as resort areas, with a dense concentration of hotels and houses covering an entire island. This development prevents the natural process of island migration. If rapidly rising seas drown the barrier islands, the mainland coasts will be subject to the full force of ocean waves. Coastal wetlands, river estuaries, and cities along the Atlantic Coast of the United States, from New Jersey to Florida, and along the Gulf of Mexico will be subject to increased flooding and storm damage. In addition, a higher sea level

BOX 14.1 ■ FIGURE 3

Before and after pictures of a barrier island at Mantoloking, New Jersey, that was washed over and eroded by storm waves and surge associated with Superstorm Sandy (2012). Notice how the sand beach was eroded on the open ocean side (bottom) and redeposited on the landward side (top), causing the barrier island to migrate landward. Construction crews with heavy machinery are visible in the bottom photo, clearing roads and moving the sand back to widen the beach to build a protective berm.
Source: U.S. Geological Survey

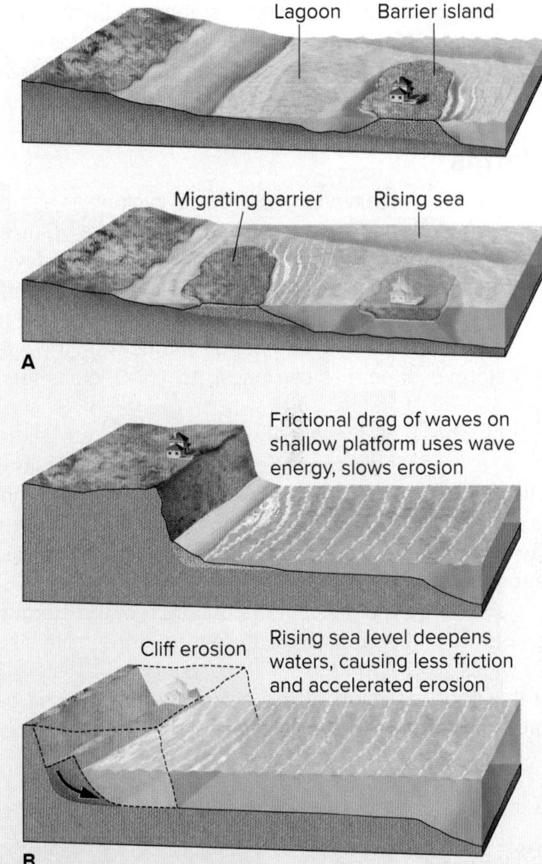

BOX 14.1 ■ FIGURE 4

Rising sea level can cause erosion of both gentle (*A*) and steep (*B*) coasts, leading to the destruction of buildings.

increases saltwater intrusion in coastal aquifers and also pushes saltwater farther up coastal rivers, endangering the water supply of cities.

Steep coasts, such as those along the Pacific coast of the United States, also face danger from the rising sea level. Narrow beaches in front of sea cliffs may be completely drowned, and the increased wave action that results can accelerate the erosion of sea cliffs (box figure 4*B*). Despite efforts to shore up the cliffs, more oceanfront property must be abandoned every year.

Careful planning for the use of coastal land is necessary to lessen the destruction caused by the effects of sea-level rise. As sea level rises, protective coastal wetlands and barrier islands will disappear, and coastal cliffs will be eroded away. Coastal communities will come under increasing threat from flooding, saltwater intrusion, and tropical storms. In the long term, decisions will need to be made about the future of coastal communities. In some cases, raising houses, wetland restoration, and "armoring" with seawalls may preserve oceanside development, but some communities may have to be abandoned, and those grim choices are not far off.

Additional Resources

Global Sea Level Rise Scenarios for the United States National Climate Assessment. 2012. NOAA Technical Report OAR CPO-1.

- http://cpo.noaa.gov/sites/cpo/Reports/2012/NOAA_SLR_r3.pdf
- https://www3.epa.gov/climatechange/impacts/coasts.html

U.S. Environmental Protection Agency *Climate Impacts and Coastal Areas* website discusses effects of climate change and sea-level rise on coastal zones.

EARTH SYSTEMS 14.2

Hurricanes—Devastation on the Coast

Hurricanes are the most powerful storms on Earth. With wind speeds up to 315 kilometers per hour, storm surges as high as 10 meters, and torrential rainfall, a hurricane can leave a coastal area completely devastated in a matter of hours.

The Storms

A hurricane, also called a typhoon or tropical cyclone, is a rotational storm with sustained wind speeds over 120 kilometers (75 miles) per hour. At the center of the storm is the eye, a calm area of warm air and clear skies about 40 kilometers in diameter. It is surrounded by the eye wall, where the highest winds, heaviest rains, and lowest pressures are found. Spiraling rain bands extend from the eye wall, creating a storm system that can be up to 1,600 kilometers wide (box figure 1).

Although the high winds and heavy rains are extremely destructive, a hurricane's most devastating weapon is its storm surge. The low pressure in the eye of the hurricane allows the sea surface to rise slightly. As the water depth decreases closer to the shore, the high winds in the eye wall amplify this temporary rise and push it shoreward.

The storm surge is highest in the quadrant of the storm where the winds blow onshore (box figure 2). In the Northern Hemisphere, hurricane winds spiral counterclockwise, so the highest storm surge is along the northeast side of the hurricane. A storm surge that occurs at high tide is especially devastating. Such was the case when Superstorm Sandy struck the coast of New Jersey and New York, sending a record high storm surge (9.9 meters) onshore. Sandy was the second largest Atlantic storm on record, and its sheer size spread winds over a large area that pushed a tremendous amount of water into bays and inlets along the northeast side of the storm. The numerous bays and inlets acted as funnels, channeling the water into narrower and narrower regions.

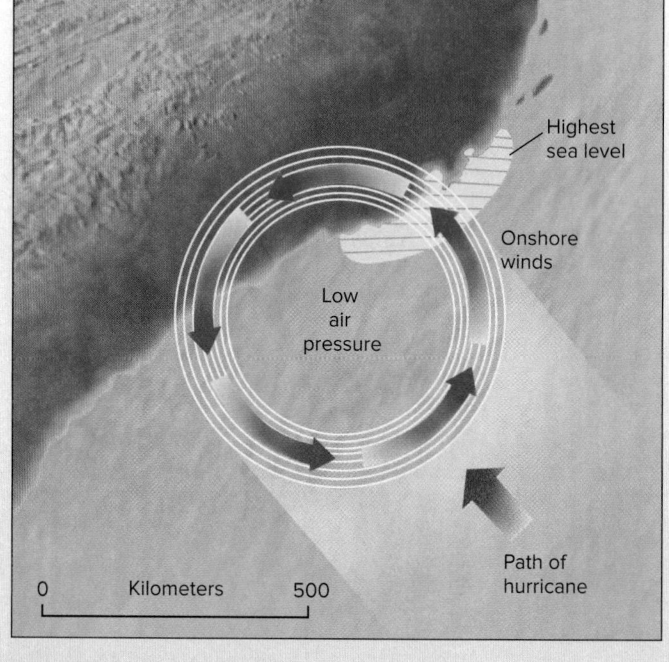

BOX 14.2 ■ FIGURE 2

Strong onshore winds in a hurricane pile water against the shore, forming a storm surge (high sea level) that may cause severe flooding on a low-lying coast. This is particularly a problem where irregular-shaped bays and estuaries (e.g., New York Harbor) funnel the water onshore. Damage is worse at high tide.

BOX 14.2 ■ FIGURE 1

Satellite image of Category 5 Super Typhoon Haiyan that struck the Philippines on November 7, 2013, with sustained maximum winds of 315 kilometers per hour (195 mph), making it the second strongest tropical cyclone ever recorded on Earth based on wind speed. Haiyan was the deadliest typhoon ever to hit the Philippines, killing at least 6,190 people and displacing nearly 4 million people from their homes.
Source: NOAA

An 8-meter-high storm surge struck Galveston, Texas, in 1900, completely covering the barrier island on which the city is built. Countless buildings were destroyed and 6,000 people died, many of whom were drowned or carried out to sea by the waves. Galveston was the target again in 2008 when Hurricane Ike stormed out of the Gulf of Mexico with winds of 177 kilometers per hour and a storm surge over 4 meters high. Galveston experienced extensive flooding, but the worst damage was sustained on the Bolivar Peninsula, just northeast of Galveston. Entire towns were carried off by the storm surge, with the debris washed up across Galveston Bay. Gilchrist and Crystal Beach, home to

BOX 14.2 ■ FIGURE 3

One house remains standing in Gilchrist, Texas, on Bolivar Peninsula. The house was rebuilt after Hurricane Rita swept through the area in 2005.
©Pool/AP Images

shrimpers and vacationers, were almost completely demolished (box figure 3).

New Orleans is another city that has been in the path of recent hurricanes. On August 29, 2005, the eye of Hurricane Katrina passed to the east of New Orleans, generating 9-meter-high storm surges in the Gulf of Mexico and Lake Pontchartrain. The surges were funneled through the canals of the city, increasing their height by up to 20% and their speed by 300%. Levees were overtopped and breached, and 80% of New Orleans was plunged under water. Three years later, New Orleans residents were evacuated again as Hurricane Gustav approached. The city was lucky this time; Gustav had weakened to a category 2 hurricane before it made landfall on September 1, 2008. Several of the newly rebuilt and reinforced levees were overtopped, but none were breached.

The Future

Hurricanes originate over tropical waters, requiring sea-surface temperatures greater than 26°C. Scientists studying climate change are concerned that the higher ocean temperatures predicted for the next century will cause an increase in the frequency and intensity of hurricanes and extreme weather events. Indeed, such an increase has been documented over the last decade, and the exceptionally powerful category 5 Super Typhoon Haiyan that struck the Philippines in 2013 was one of the strongest hurricanes ever recorded on Earth, resulting in catastrophic damage and record loss of life (box figure 1). The damage caused by hurricanes has been increasing at an alarming rate.

September 2017 was the most active month on record, in terms of hurricane strength and duration, in the North Atlantic. Hurricanes such as Harvey caused historic flooding in Houston, Texas, Irma affected the entire state of Florida, and Maria caused catastrophic damage to Dominica, and devastated much of Puerto Rico. With over 50% of the world's population and an increasing amount of its wealth concentrated on the coasts, every hurricane has the potential to cause widespread loss of life and property.

Additional Resources

- http://earthobservatory.nasa.gov/Features/Hurricanes/

Hurricanes: The Greatest Storms on Earth website gives an excellent overview of hurricanes and how they are formed.

- www.nasa.gov/mission_pages/hurricanes/main/index.html

NASA website gives the latest storm images and information about recent hurricanes.

- www.nhc.noaa.gov/

National Hurricane Center website shows information on current and past storms, as well as marine forecasts and hurricane alerts.

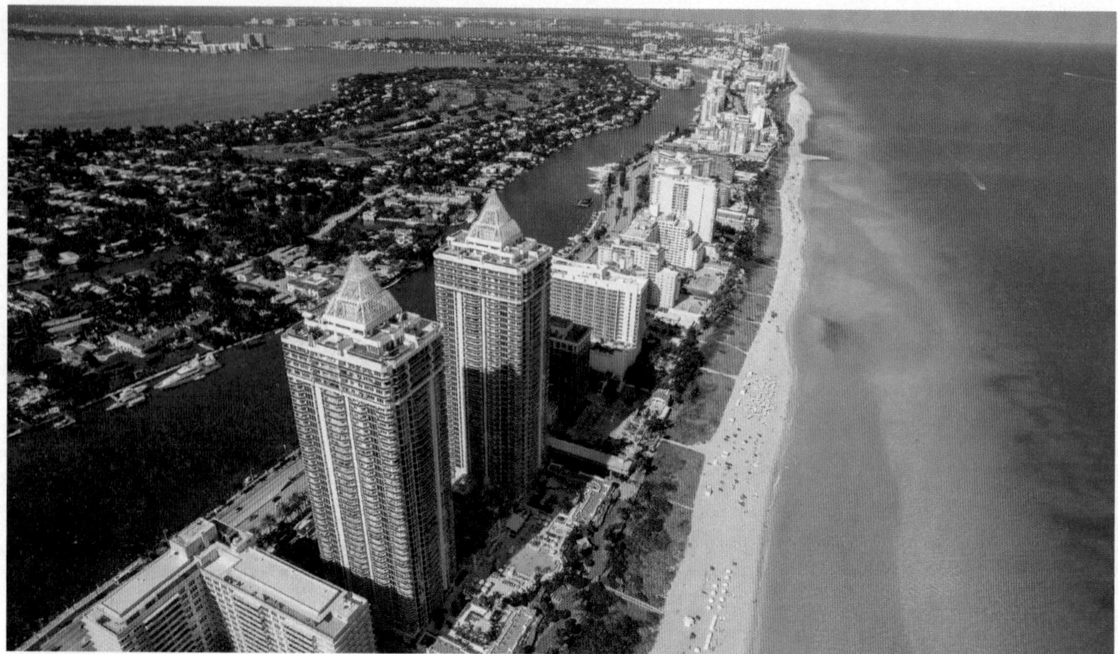

FIGURE 14.19

Hotels built upon the loose sand of a barrier island, Miami Beach, Florida.
©Miami2you/Shutterstock

cutting valleys. As the great ice sheets melted, sea level began to rise, drowning the river valleys. These drowned river mouths, called **estuaries**, mark many coasts today (figure 14.20), and they form very irregularly shaped coastlines. They extend inland as long arms of the sea. Fresh water from rivers mixes with the seawater to make most estuaries brackish. The quiet, protected environment of estuaries makes them very rich in marine life, particularly the larval forms of numerous species. Unfortunately, cities and factories built on many estuaries to take advantage of quiet harbors are severely polluting the water and the sediment of the estuaries. The poor circulation that characterizes most estuaries hinders the flushing away of this pollution, and estuary shellfish are sometimes not safe to eat as a result. For example, in the past 40 years the shellfish population in Chesapeake Bay has been drastically reduced.

Drowned coasts may be marked by **fiords**, glacially cut valleys flooded by rising sea level (see figure 12.37). They form in the same way as estuaries, except they were cut by glacial ice rather than rivers during low sea-level stands.

Uplifted Coasts

Uplifted (or *emergent*) coasts have been elevated by deep-seated tectonic forces or by crustal rebound caused by the removal of weight associated with the melting of continental ice sheets. The land has risen faster than sea level, so parts of the old sea floor are now dry land.

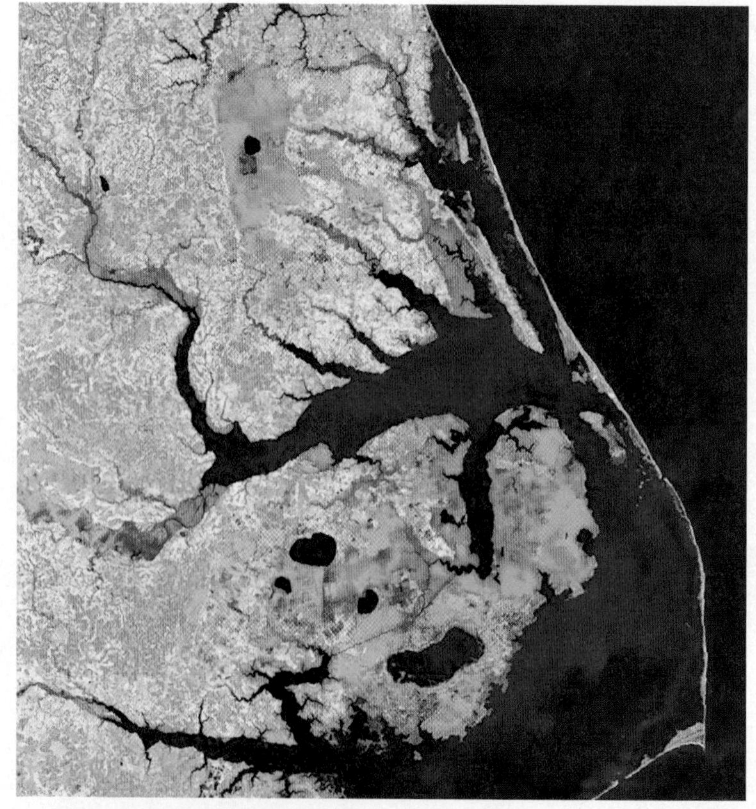

FIGURE 14.20

Landsat satellite photo of estuaries, Albemarle and Pamlico Sounds, North Carolina. Barrier islands are visible along the coastline.
Source: EROS Center, U.S. Geological Survey

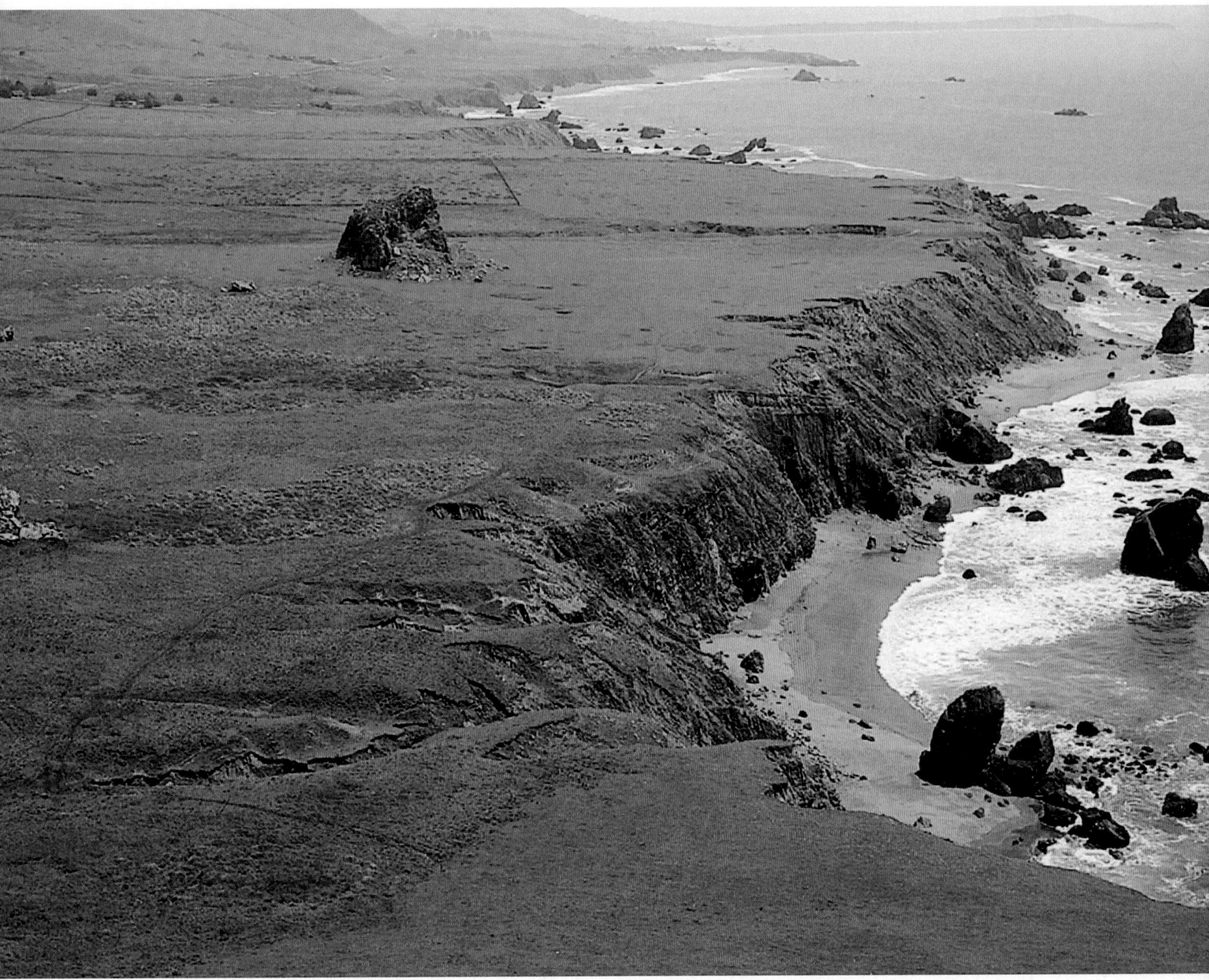

FIGURE 14.21

Uplifted marine terrace, northern California. The flat land surface at the top of the sea cliff was eroded by wave action, then raised above sea level by tectonic uplift. The rock knob on the terrace was once a stack.
©David McGeary

Marine terraces form just offshore from the beach face, as described in the "How Do Beaches Develop" section of this chapter. These terraces can be wave-cut platforms caused by erosion of rock associated with cliff retreat, or they can be wave-built terraces caused by the deposition of sediment. If the shore is elevated by tectonic uplift, these flat surfaces will become visible as *uplifted marine terraces* (figure 14.21). They formed below the ocean surface but are visible now because of uplift. The tectonically active Pacific coast of the United States has many areas marked by uplifted terraces, along with the erosional coast features described earlier.

The coasts of Canada, northern Britain, and Scandinavia have been uplifted by post-glacial crustal rebound. Along these coasts, the land had been depressed by the weight of Pleistocene ice sheets, several thousand meters thick, and it is now rising due to the removal of the weight of the ice. The rebound of the crust is much like the indentation made from the weight of your body when you sit on a soft bed or couch that gradually disappears or rebounds after you stand up.

The Biosphere and Coasts

The growth of coral and algal *reefs* offshore can shape the character of a coast. The reefs act as a barrier to strong waves, protecting the shoreline from most wave erosion (see figure 18.21). Carbonate sediments blanket the sea floor on both sides of a reef and usually form a carbonate sand beach on land (see figure 6.15). Southernmost Florida and the Bahamas have coasts of this type.

Branching *mangrove roots* dominate many parts of the southeastern United States coast. The roots dampen wave and current action, creating a quiet environment that provides a haven for the larval forms of many marine organisms and may trap fine-grained sediment. Mangroves also deposit layers of organic peat on low-lying coasts, such as the bayous of Louisiana.

Summary

Wind blowing over the sea surface forms waves, which transfer some of the wind's energy to shorelines. Orbital water motion extends to a depth equal to half the wavelength.

As a wave moves into shallow water, the ocean bottom flattens the orbital motion and causes the wave to slow and peak up, eventually forming a *breaker* whose crest topples forward. The turbulence of *surf* is an important agent of sediment erosion and transportation.

Wave refraction bends wave crests and makes them more parallel to shore. Few waves actually become parallel to the shore, and so *longshore currents* develop in the surf zone. *Rip currents* carry water seaward from the surf zone.

A beach consists of a *berm*, *beach face*, and *marine terrace.* Summer beaches have a wide berm and a smooth offshore profile. Winter beaches are narrow, with offshore bars.

Longshore drift of sand is caused by the waves hitting the beach face at an angle and by longshore currents.

Deposition of sand that is drifting along the shore can form *spits* and *baymouth bars.* Drifting sand may also be deposited against jetties or groins or inside breakwaters.

Rivers supply most sand to beaches, although local erosion may also contribute sediment. If the rivers' supply of sand is cut off by dams, the beaches gradually disappear.

Coasts may be erosional or depositional, drowned or uplifted, or shaped by organisms such as corals and mangroves.

Coastal straightening by wave refraction is caused by headland erosion and by deposition within bays.

A coast retreating under wave erosion can be marked by *sea cliffs*, a *wave-cut platform*, *stacks*, and *arches.*

Waves can form *barrier islands* off gently sloping coasts. River and glacial deposition can also shape coasts.

Drowned coasts are marked by *estuaries* and *fiords. Uplifted marine terraces* characterize coasts that have risen faster than the recent rise in sea level.

Terms to Remember

arch (sea arch) 354
barrier island 354
baymouth bar 349
beach 348
beach face 348
berm 348
breaker 345
coast 352
coastal straightening 352
crest (of wave) 345
estuary 360
fiord 360
headland 352
longshore current 347
longshore drift 349
marine terrace 348
rip current 347
sea cliff 352
spit 349
stack 354
surf 345
tombolo 349
trough (of wave) 345
wave-cut platform 354
wave height 345
wavelength 345
wave refraction 346

Testing Your Knowledge

Use the following questions to prepare for exams based on this chapter.

1. Describe the transition of deep-water waves into surf.
2. Show in a sketch the refraction of waves approaching a straight coast at an angle. Explain why refraction occurs.
3. What is a longshore current? Why does it occur? Examine the photo of waves arriving at an angle to the shoreline in figure 14.5*A* and predict which direction the longshore current would be moving.
4. What is a rip current? Why does it occur? How do you get out of a rip current?
5. How do summer beaches differ from winter beaches? Discuss the reasons for these differences.
6. What would happen to the beaches of most coasts if all the rivers flowing to the sea were dammed? Why?
7. In a sketch, show how and why sand moves along a beach face when waves approach a beach at an angle.
8. Show in a sketch how longshore drift of sand can form a baymouth bar. In which direction is the longshore drift of sand moving in figure 14.10*B?* Why?
9. Describe how waves can straighten an irregular coastline.
10. What does the presence of an estuary imply about the recent geologic history of a region?
11. The path a water particle makes as a wave passes in deep water is best described as
    a. elliptical.
    b. orbital.
    c. spherical.
    d. linear.
12. The surf zone is
    a. the area in which waves break.
    b. water less than one-half wavelength in depth.
    c. where the longshore current flows.
    d. all of the preceding.
13. The easiest method of escaping a rip current is to
    a. swim toward shore.
    b. swim parallel to the shore.
    c. swim away from the shore.

14. Why is beach sediment typically quartz-rich sand?
    a. Other minerals are not deposited on beaches.
    b. Quartz is the only mineral that can be sand-sized.
    c. Quartz is resistant to chemical weathering.
    d. None of the preceding
15. Longshore drift is
    a. the movement of sediment parallel to shore when waves strike the shoreline at an angle.
    b. a type of rip current.
    c. a type of tide.
    d. the movement of waves.
16. Which structures would interfere with longshore drift?
    a. jetties b. groins
    c. breakwaters d. all of the preceding
17. What is the most common source of sand on beaches?
    a. sand from river sediment brought down to the ocean
    b. land next to the beach
    c. offshore sediments
18. Which would characterize an erosional coast?
    a. headlands b. sea cliffs
    c. stacks d. arches
    e. all of the preceding
19. Which would characterize a depositional coast?
    a. headlands b. sea cliffs
    c. stacks d. arches
    e. barrier islands
20. A glacial valley drowned by rising sea level is
    a. a fiord. b. an estuary.
    c. a tombolo. d. a headland.
21. The storm surge of a hurricane is
    a. the highest winds.
    b. the tallest waves.
    c. the dome of high water in the center of the hurricane.
    d. the area of high pressure within the storm.

Expanding Your Knowledge

1. Sea level would rise by about 60 meters if *all* the glacial ice on Earth melted. How many U.S. cities would this affect?
2. What happens to a coast if its offshore reef dies?
3. Is a beach a good place to mine sand for construction? Explain your answer carefully.
4. The seaward tip of a headland may be the most rapidly eroding locality on a coast yet also the most expensive building site on a coast. Why is this so?

Exploring Web Resources

http://marine.usgs.gov/

Web page for the *Coastal and Marine Geology Program* of the U.S. Geological Survey contains information about numerous geologic studies of U.S. coastal areas.

http://woodshole.er.usgs.gov/

Web page for the U.S. Geological Survey *Woods Hole Coastal and Marine Science Center* for coastal and marine research contains information and data from ongoing scientific projects.

www.noaa.gov/ocean.html

Oceans web page from *National Oceanic and Atmospheric Administration* provides numerous links to oceanography research projects and data.

CHAPTER

15 Geologic Structures

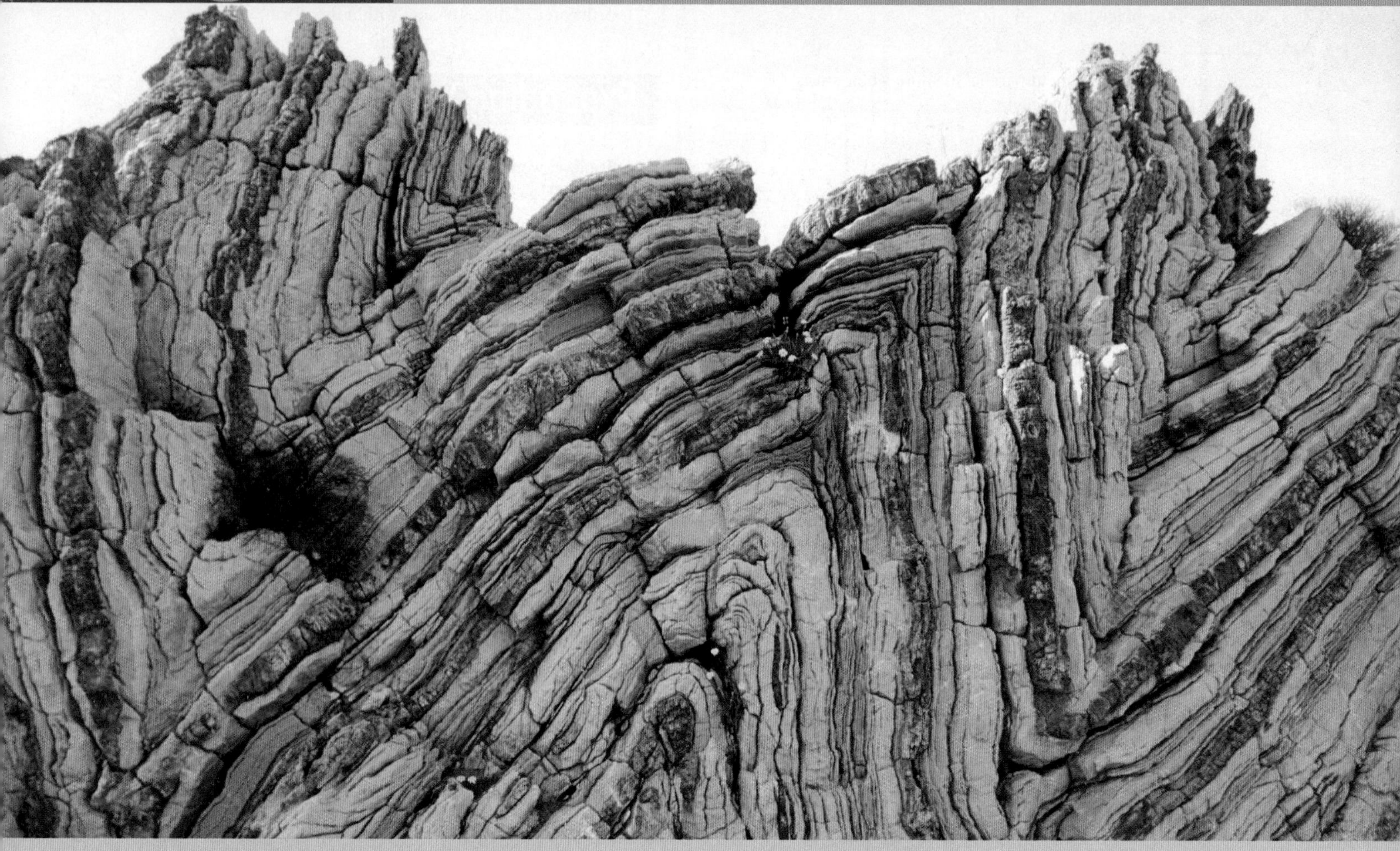

Rocks that were once horizontal have been contorted into folds during mountain building, Agio Pavlos in Southern Crete, Greece.
©Marco Simoni/Robert Harding/Getty Images

LEARNING OBJECTIVES

- Why are geologic structures important?
- Sketch and describe the three types of stress and the resulting strain.
- Explain how rocks behave when stressed.
- Define *strike*, *angle of dip*, and *direction of dip.*
- Sketch and describe the different types of folds, and explain how the shape and orientation of folds is used to interpret strain.
- Sketch and describe the relative displacement and type of stress for normal, reverse, and thrust faults as well as the two types of strike-slip faults.

In previous chapters, we have discussed how rock at the surface of Earth is affected by the atmosphere, hydrosphere, and biosphere. We now shift our focus to processes in the solid Earth system, or geosphere. In this chapter, we explain how rocks respond to tectonic forces caused by the movement of lithospheric plates and how geologists study the resulting geologic structures. Studying structural geology is very much like looking at the architecture of the crust and trying to relate how rocks that were once deposited under water in horizontal layers are now found bent (folded) and broken (faulted) many kilometers above sea level.

Subsequent chapters will require an understanding and knowledge of structural geology as presented in this chapter. To understand earthquakes, for instance, one must know about faults. Appreciating how major mountain belts and the continents have evolved (chapter 20) calls for a comprehension of faulting and folding. Understanding plate-tectonic theory as a whole (chapter 19) also requires a knowledge of structural geology. (Plate-tectonic theory developed primarily to explain certain structural features.) In areas of active tectonics, the location of geologic structures is important in the selection of safe sites for schools, hospitals, dams, bridges, and nuclear power facilities.

Also, understanding structural geology can help us more fully appreciate the problem of finding more of Earth's dwindling natural resources. Chapter 21 discusses the association of certain geologic structures with petroleum deposits and other valuable resources.

15.1 TECTONIC FORCES AT WORK

Stress and Strain in the Earth's Lithosphere

Tectonic forces deform parts of the lithosphere, particularly along plate margins. Deformation may cause a change in orientation, location, and shape of a rock body. In figure 15.1, originally horizontal rock layers have been deformed into wavelike folds that are broken by faults. The layers have been deformed, probably by tectonic forces that pushed or compressed the layers together until they were shortened by buckling and breaking.

When studying deformed rocks, structural geologists typically refer to **stress**, a force per unit area. Where stress can be measured, it is expressed as the force per unit area at a particular point; however, it is difficult to measure stress in rocks that are buried. We can observe the effects of past stress (caused by tectonic forces and confining pressure from burial) when rock bodies are exposed after uplift and erosion. From our observations, we may be able to infer the principal directions of stress that prevailed. We also can observe in exposed rocks the effect of forces on a rock that was stressed. **Strain** is the change in shape or size (volume), or both, in response to stress.

The relationship between stress and strain can be illustrated by deforming a piece of Silly Putty® (figure 15.2) or any other soft material such as pizza dough. If the Silly Putty is pushed together or squeezed from opposite directions, we say the stress is **compressive**. Compressive stress results in rocks being *shortened* or *flattened.* In figure 15.2*A*, an elongate piece of Silly Putty may shorten by bending, or folding, whereas a ball of Silly Putty will flatten by shortening in the direction parallel to the compressive stress and elongating or stretching in the direction perpendicular to it. Rocks that have been shortened or flattened are typically found along convergent plate boundaries where rocks have been pushed or shoved together.

A **tensional stress** is caused by forces pulling away from one another in opposite directions (figure 15.2*B*). Tensional stress results in a *stretching or extension* of material. If we apply a tensional stress on a ball of Silly Putty, it will elongate or stretch parallel to the applied stress. If the tensional stress is applied rapidly, the Silly Putty will first stretch and then break apart (figure 15.2*B*). At divergent plate boundaries, the lithosphere is undergoing extension as the plates move away from one another. Because rocks are very weak when pulled apart, fractures and faults are common structures.

When stresses act parallel to a plane, **shear stress** is produced. It is much like putting a deck of cards in your hands and shearing the deck by moving your hands in opposite directions (figure 15.3). A shear stress results in a *shear strain* parallel to the direction of the stresses. Shear stresses occur along actively moving faults.

How Do Rocks Behave When Stressed?

Rocks behave as elastic, ductile, or brittle materials, depending on the amount and rate of stress applied, the type of rock, and the temperature and pressure under which the rock is strained.

If a deformed material recovers its original shape after the stress is reduced or removed, the behavior is **elastic**. For example,

FIGURE 15.1

Deformed sedimentary beds exposed in a road cut near Palmdale, California. Squeezing due to movement along the San Andreas fault caused the originally horizontal sedimentary layers to be contorted into folds and broken by smaller faults.
©Doug Sherman/Geofile

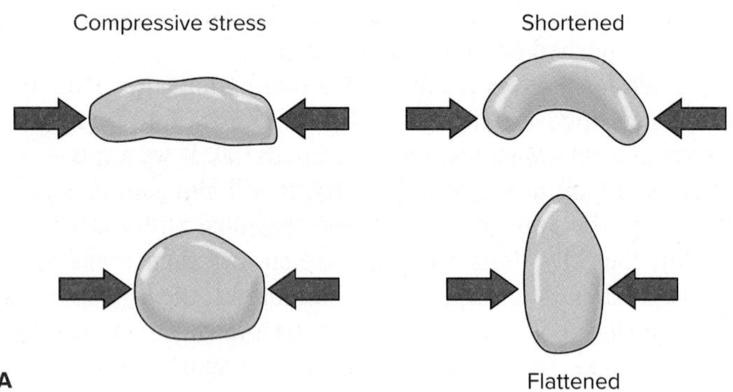

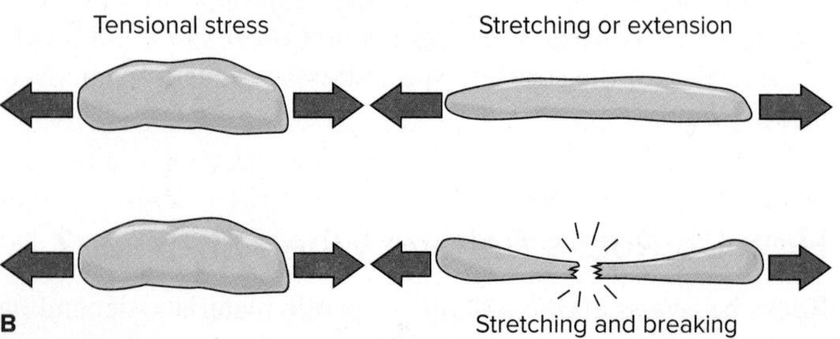

FIGURE 15.2

The effects of compressional and tensional stresses on Silly Putty®. (*A*) Compressing Silly Putty results in shortening either by folding or by flattening. (*B*) Pulling (tensional stress) Silly Putty causes stretching or extension; if pulled (strained) too fast, or chilled, the Silly Putty will break after first stretching.

FIGURE 15.3

Shear strain can be modeled by shearing a deck of cards.

if a tensional stress is applied to a rubber band, it will stretch as long as the stress is applied, but once the stress is removed, the rubber band returns (or recovers) to its original shape and its behavior is elastic. Silly Putty will behave elastically if molded into a ball and bounced. Most rocks can behave in an elastic way at very low stresses (a few kilobars). However, once the stress applied exceeds the **elastic limit** (figure 15.4), the rock will deform in a permanent way, just as the rubber band will break if stretched too far.

A rock that behaves in a **ductile** or plastic manner will bend while under stress and does not return to its original shape after the stress is removed. Silly Putty behaves as a ductile material unless the rate of strain is rapid. As discussed in chapter 7, rocks exposed to elevated pressure and temperature during regional metamorphism also behave in a ductile manner and develop a

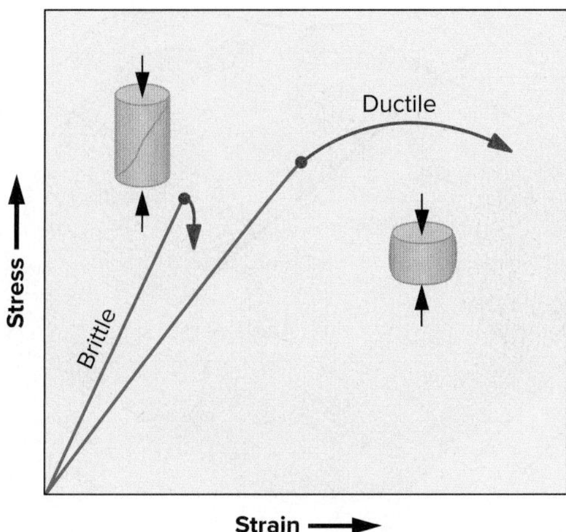

FIGURE 15.4

Graph shows the behavior of rocks with increasing stress and strain. Elastic behavior occurs along the straight-line portions (shown in blue) of the graph. At stresses greater than the elastic limit (red points), the rock will either deform as a ductile material or break, as shown in the deformed rock cylinders.

FIGURE 15.5

Geology students mapping tilted beds of rocks from a ridge-top vantage point, Mojave Desert, California.
©*Diane Carlson*

planar texture, or *foliation*, due to the alignment of minerals. As shown in figure 15.4, material behaving in a ductile manner does not require much of an increase in stress to continue to strain (relatively flat curve). Ductile behavior results in rocks that are permanently deformed mainly by folding or bending of rock layers (figure 15.1).

A rock exhibiting **brittle** behavior will fracture at stresses higher than its elastic limit, or once the stresses are greater than the strength of the rock. Rocks typically exhibit brittle behavior at or near Earth's surface, where temperatures and pressure are low. Under these conditions, rocks favor breaking rather than bending. Faults and joints are examples of structures that form by brittle behavior of the crust.

A sedimentary rock exposed at Earth's surface is brittle; it will fracture if you hit it with a hammer. How then do sedimentary rocks, such as those shown in figure 15.1, become bent (or deformed in a ductile way)? The answer is that either stress increased very slowly or the rock was deformed under considerable confining pressure (buried under more rock) and higher temperatures.

Note, however, that there are some fractures (faults) disrupting the bent layers in figure 15.1. This tells us that although the rock was ductile initially, the amount of stress increased or the rate of strain increased and the rock fractured.

15.2 HOW DO WE RECORD AND MEASURE GEOLOGIC STRUCTURES?

The study of geologic structures is of more than academic interest. The petroleum and mining industries, for example, employ geologists to look for and map geologic structures associated with oil and metallic ore deposits. Understanding and mapping geologic structures is also important for evaluating problems related to engineering decisions and seismic risk, such as determining the most appropriate sites for building dams, large bridges, or nuclear reactors, and even houses, schools, and hospitals.

Geologic Maps and Field Methods

In an ideal situation, a geologist studying structures would be able to fly over an area and see the local and regional patterns of bedrock from above. Sometimes this is possible, but very often soil and vegetation conceal the bedrock. Therefore, geologists ordinarily use observations from a number of individual *outcrops* (exposures of bedrock at the surface) in determining the patterns of geologic structures (figure 15.5). The characteristics of rock at each outcrop in an area are plotted on a map by means of appropriate symbols. With the data that can be collected, a geologist can make inferences about those parts of the area he or she cannot observe. A **geologic map**, which uses standardized symbols and patterns to represent rock types and geologic structures, is typically produced from the field map for a given area (for example, see the geologic map of North America at the end of the book). On such a map are plotted the type and distribution of rock units, the occurrence of structural features (folds, faults, joints, etc.), ore deposits, and so forth. Sometimes surficial features, such as deposits by former glaciers, are included, but these may be shown separately on a different type of geologic map.

Anyone trained in the use of geologic maps can gain considerable information about local geologic structures because standard symbols and terms are used on the maps and the accompanying reports. For example, the symbol ⊕ on a geologic

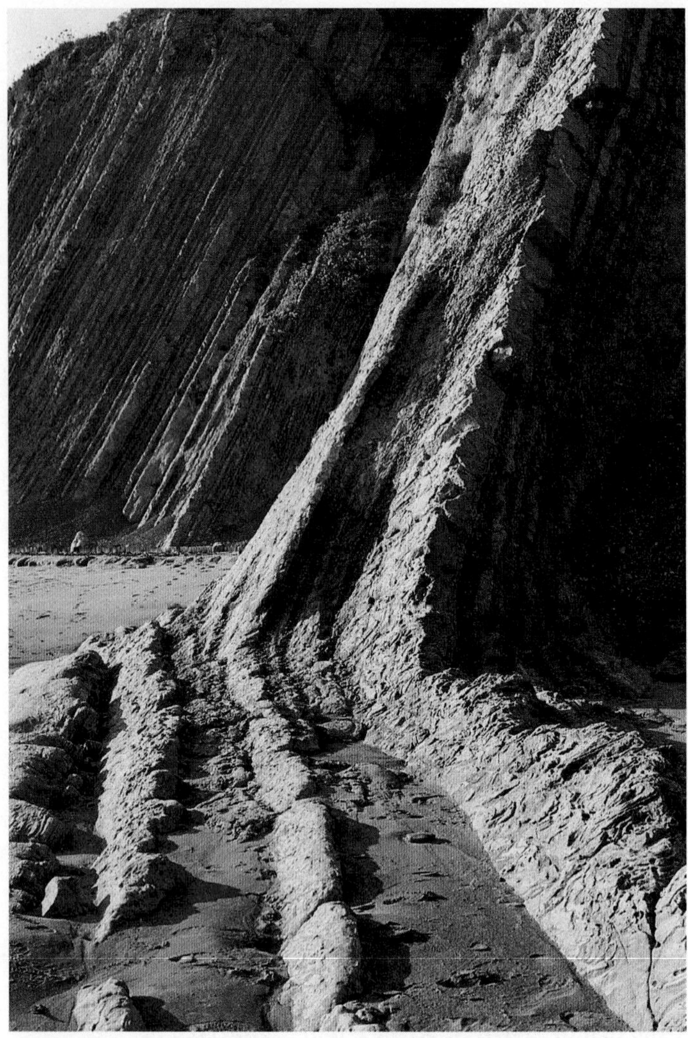

FIGURE 15.6

Tilted sedimentary beds along the coast of northern California near Point Arena. Here, the strike is the line formed by the intersection of the tilted sedimentary beds and the horizontal layer of sand in the foreground. The direction of dip is toward the left.
©Diane Carlson

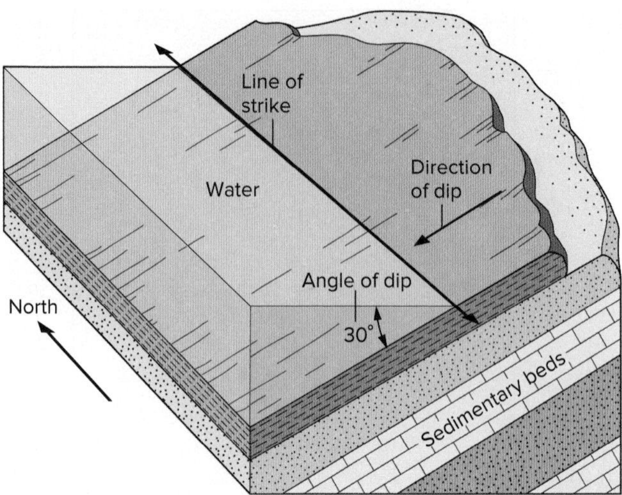

FIGURE 15.7

Strike, direction of dip, and angle of dip. The line of strike is found where an inclined bed intersects a horizontal plane (as shown here by the water). The dip direction is always perpendicular to the strike and in the direction that the bed slopes (or in the direction that a ball would roll down). The dip angle is the vertical angle of the inclined bed as measured from the horizontal plane.

map denotes horizontal bedding in an outcrop. Different colors and patterns on a geologic map represent distinct rock units.

Strike and Dip

According to the principle of *original horizontality*, sedimentary rocks and some lava flows and ashfalls are deposited as horizontal beds or strata. Where these originally horizontal rocks are found tilted, it indicates that tilting must have occurred after deposition and lithification (figure 15.6). Someone studying a geologic map of the area would want to know the extent and direction of tilting. By convention, this is determined by plotting the relationship between a surface of an inclined bed and an imaginary horizontal plane. You can understand the relationship by looking carefully at figure 15.7, which represents sedimentary beds exposed alongside a lake (the lake surface provides a convenient horizontal plane for this discussion).

Strike is the compass direction of a line formed by the intersection of an inclined plane with a horizontal plane. In this example, the inclined plane is a bedding plane. You can see from figure 15.7 that the beds are striking from north to south. Customarily, only the northerly direction (of the strike line) is given, so we simply say that beds strike north a certain number of degrees east or west (such as N50°E).

Observe that the **angle of dip** is measured downward from the horizontal plane to the bedding plane (an inclined plane). Note that the angle of dip (30° in the figure) is measured within a vertical plane that is perpendicular to both the bedding and the horizontal planes.

The **direction of dip** is the compass direction in which the angle of dip is measured. If you could roll a ball down a bedding surface, the compass direction in which the ball rolled would be the direction of dip.

The dip angle is always measured at a right angle to the strike—that is, perpendicular to the strike line as shown in figure 15.7. Because the beds could dip away from the strike line in either of two possible directions, the general direction of dip is also specified—in this example, west.

A specially designed instrument called a Brunton pocket transit (after the inventor) is used by geologists for measuring the strike and dip (figure 15.8). The pocket transit contains a compass, a level, and a device for measuring angles of inclination. Besides recording strike and dip measurements in a field notebook, a geologist who is mapping an area draws strike and dip symbols on the field map, such as ⅄ or ⅄ for each outcrop with dipping or tilted beds. On the map, the intersection of the two lines at the center of each strike and dip symbol represents

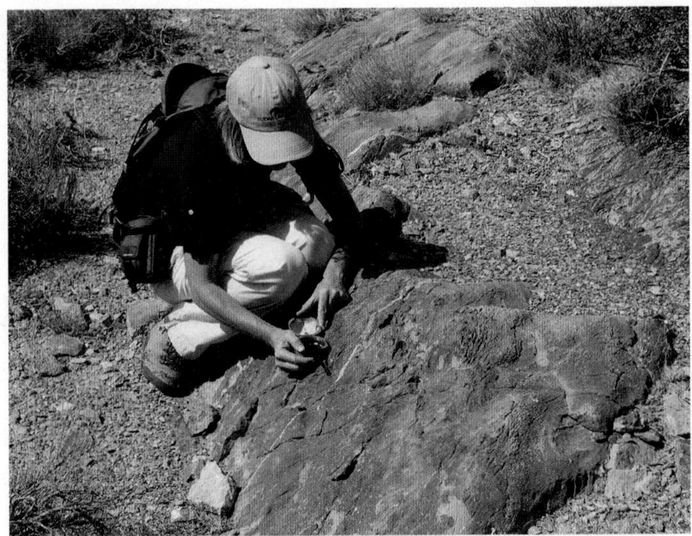

FIGURE 15.8

Geology student measuring the strike and dip of an inclined limestone bed in the White Mountains of southern California.
©*Diane Carlson*

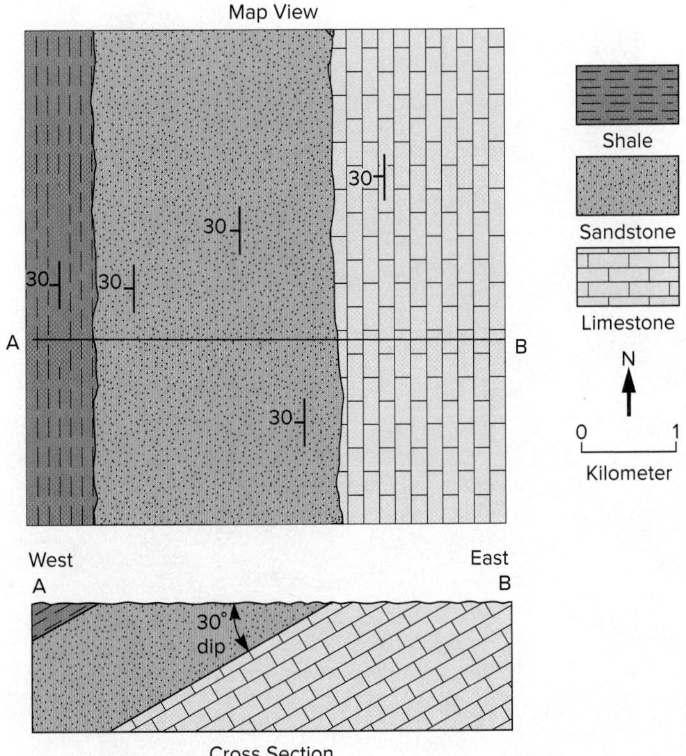

FIGURE 15.9

A geologic map and cross section of an area with three sedimentary formations. (Each formation may contain many individual sedimentary layers, as explained in chapter 6.) Beds strike north and dip 30° to the west. The geologic cross section (vertical cut) is constructed between points A and B on the map.

the location of the outcrop where the strike and dip of the bedrock were measured. The long line of the symbol is aligned with the compass direction of the strike. The small tick, which is always drawn perpendicular to the strike line, is put on one side or the other, depending on which of the two directions the beds actually dip. The angle of dip is given as a number next to the appropriate symbol on the map. Thus, 25 ⟋ indicates that the bed is dipping 25° from the horizontal toward the northwest and the strike is northeast (assuming that the top of the page is north). The orientation of the bed would be written N45°E, 25°NW. Figure 15.9 is a geologic map with cross section that shows all the sedimentary layers striking north and dipping 30° to the west (N0°, 30°W).

Beds with vertical dip require a unique symbol because they dip neither to the left nor to the right of the direction of strike. The symbol used is ✕, which indicates that the beds are striking northeast and that they are vertical (N30°E, 90°).

Geologic Cross Sections

A **geologic cross section** represents a vertical slice through a portion of Earth. It is much like a road cut (see figure 15.1) or the wall of a quarry in that it shows the orientation of rock units and structures in the vertical dimension. Geologic cross sections are constructed from geologic maps by projecting the dip of rock units into the subsurface (figure 15.9) and are quite useful in helping visualize geology in three dimensions. They are used extensively throughout this book as well as in professional publications.

15.3 FOLDS

Folds are bends or wavelike features in layered rock. Folded rock can be compared to several layers of rugs or blankets that have been pushed into a series of arches and troughs. Folds in rock often can be seen in road cuts or other exposures (figure 15.10). When the arches and troughs of folds are concealed (or when they exist on a grand scale), geologists can still determine the presence of folds by noticing repeated reversals in the direction of dip taken on outcrops in the field or shown on a geologic map.

The fact that the rock is folded or bent shows that it behaved as a ductile material. Yet the rock exposed in outcrops is generally brittle and shatters when struck with a hammer. The rock is not metamorphosed (most metamorphic rocks are intensely folded because they are ductile under the high pressure and temperature environment of deep burial and tectonic stresses). Perhaps folding took place when the rock was buried at a moderate depth where higher temperature and confining pressure favor ductile behavior. Alternatively, folding could have taken place close to the surface under a very low rate of strain.

Geometry of Folds

Determining the geometry or shape of folds may have important economic implications because many oil and gas deposits (see box 15.1) and some metallic mineral deposits are localized in folded rocks. The geometry of folds is also important in

IN GREATER DEPTH 15.1

Is There Oil Beneath My Property? First Check the Geologic Structure

An "oil pool" can exist only under certain conditions. Crude oil does not fill caves underground, as the term *pool* may suggest; rather, it simply occupies the pore spaces of certain sedimentary rocks, such as poorly cemented sandstone, in which void space exists between grains. Natural gas (less dense) often occupies the pore spaces above the crude oil, while water (more dense) is generally found saturating the rock below the oil pool (box figure 1). The oil and gas must be trapped by an overlying rock that is impermeable and will not let the petroleum seep to the surface.

A **source rock**, which is always a sedimentary rock, must be present for oil to form. The sediment of the source rock has to include remains of organisms buried during sedimentation. This organic matter partially decomposes into petroleum and natural gas. Once formed, the droplets of petroleum tend to migrate, following fractures and interconnecting pore spaces. Being less dense than the rock, the petroleum usually migrates upward, although horizontal migration does occur.

If it is not blocked by impermeable rock, the oil may migrate all the way to the surface, where it is dissipated and permanently lost. Natural oil seeps, where leakage of petroleum is taking place, exist both on land and offshore. Where impermeable rock blocks the oil droplets' path of migration, an oil pool may accumulate below the rock, much like helium-filled balloons might collect under a domed ceiling. For any significant amount of oil to collect, the rock below the impermeable rock must be porous as well as permeable. Such a rock, when it contains oil, is called a **reservoir rock**.

Another necessary condition is that the geologic structure must be one that favors the accumulation and retention of petroleum. An "anticlinal trap" is one of the best structures for holding oil. As oil became a major energy source and the demand for it increased, most of the newly discovered wells penetrated anticlinal traps. Geologists discovered these by looking for indication of anticlines exposed at the surface. As time went by, other types of structures were also found to be oil traps. Many of these were difficult to find because of the lack of telltale surface patterns indicating favorable underground structures. Box figure 2 illustrates some structures other than anticlinal traps that might have a potential for oil production.

Oil companies employ large numbers of geoscientists to complete detailed and sophisticated geologic studies of an area they hope may have the potential for an "oil strike." The geoscientists working in the petroleum industry also depend heavily on geophysical techniques (see chapter 17) for determining, by indirect means, the subsurface structural geology.

Even when everything indicates that conditions are excellent for oil to be present underground, there is no guarantee that oil will be found. Eventually, an oil company must commit a million dollars or more to drill a deep test well, or "wildcat" well. Statistics

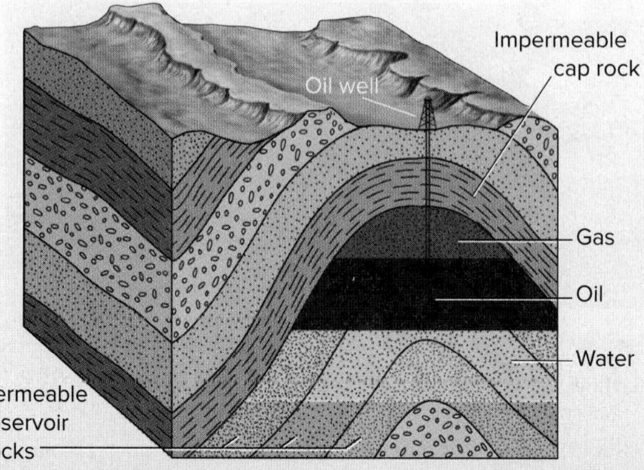

BOX 15.1 ■ FIGURE 1

(*A*) Oil and gas are concentrated or trapped in hinge of anticline. Gas and oil float on water in porous and permeable reservoir rock (sandstone) and are trapped, or held, by the overlying impermeable shale layer. (*B*) Eroded anticline forms trap in Lander oil field, Wyoming.
B (Top & Bottom) ©Diane Carlson

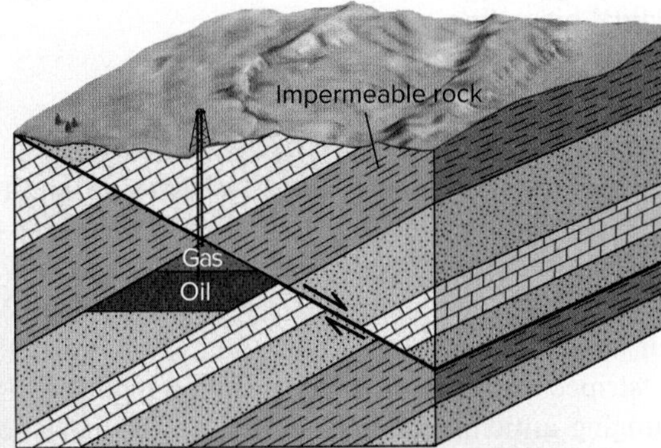

A Fault

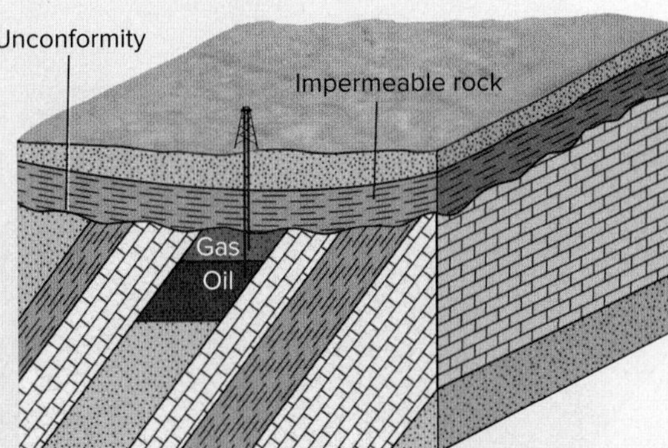

B Unconformity

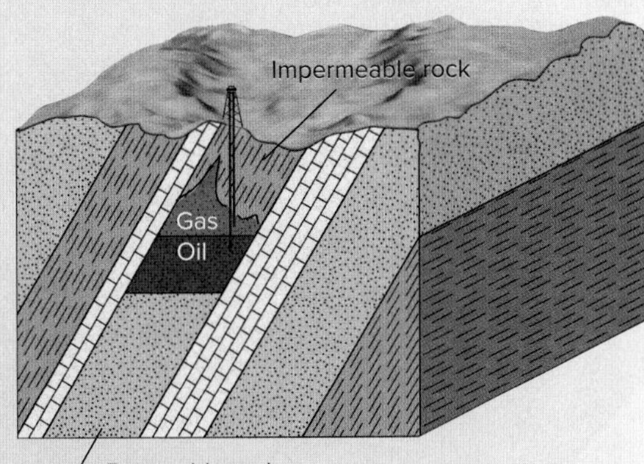

C Sedimentary facies change

BOX 15.1 ■ FIGURE 2

Structures other than anticlines that trap oil.

indicate that the chance of a test well yielding commercial quantities of oil is much less than 1 in 10. As more and more of the world's supply of petroleum is used up, what is left becomes more difficult and costly to find.

FIGURE 15.10

Folded sedimentary rock layers exposed at Lulworth Cove, Dorset, England. *©Martin Bond/Science Source*

unraveling how a rock was strained and how it might be related to the movement of tectonic plates. Folds are usually associated with shortening of rock layers along convergent plate boundaries, but they are also commonly formed where rock has been sheared along a fault.

Because folds are wavelike forms, two basic fold geometries are common—anticlines and synclines (figure 15.11).

An **anticline** is a fold shaped like an arch with the oldest rocks in the center of the fold. Usually the rock layers dip away from the **hinge line** (or *axis*) of the fold. The counterpart of an anticline is a **syncline**, a fold shaped like a trough with the youngest rocks in the center of the fold. The layered rock usually dips toward the syncline's hinge line. In the series of folds shown in figure 15.11, two anticlines are separated by a syncline. Each anticline and adjacent syncline share a **limb**. Note the hinge lines on the crests of the two anticlines and bottom of the syncline. Similar hinge lines could be located in the hinge areas at the contacts between any two adjacent folded layers. For each anticline and the syncline, the hinge lines are contained within the shaded vertical planes. Each of these planes is an **axial plane**, an

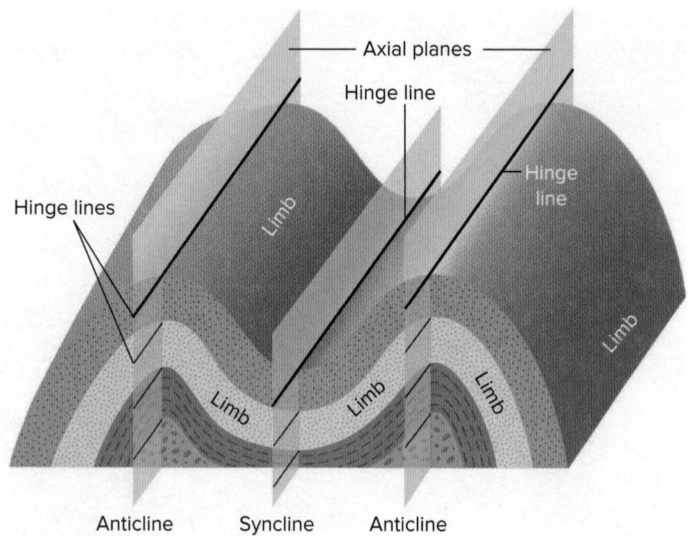

FIGURE 15.11

Diagrammatic sketch of two anticlines and a syncline illustrating the axial planes, hinge lines, and fold limbs.

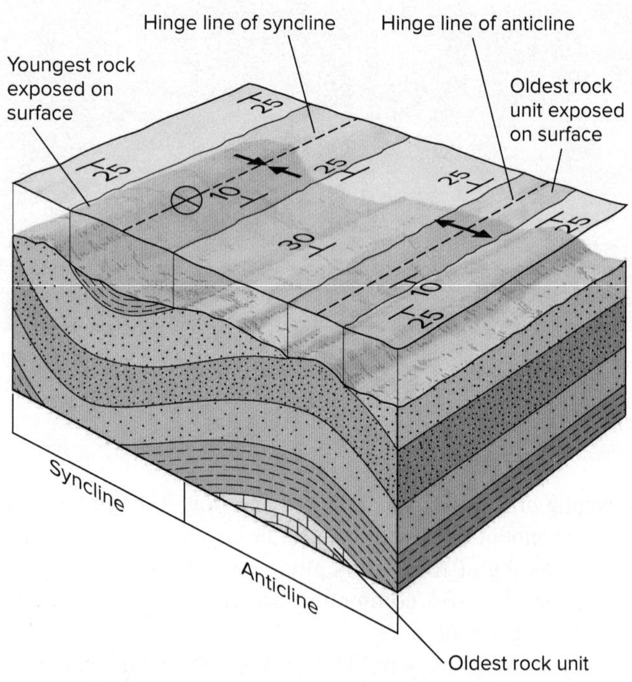

FIGURE 15.12

By measuring the strike and dip of exposed sedimentary beds in the field and plotting them on a geologic map (top surface), geologists can interpret the geometry of the geologic structure below the ground surface.

imaginary plane containing all of the hinge lines of a fold. The axial plane divides the fold into its two limbs.

It is important to remember that anticlines are not necessarily related to ridges nor synclines to valleys, because valleys and ridges are nearly always erosional features. In an area that has been eroded to a plain, the presence of underlying anticlines and synclines is determined by the direction of dipping beds in exposed bedrock, as shown in figure 15.12. (In the field, of course, the cross sections are not exposed to view as they are in the diagram.)

Figure 15.12 also illustrates how determining the relative ages of the rock layers, or beds, can tell us whether a structure is an anticline or a syncline. Observe that the oldest exposed rocks are along the hinge line of the anticline. This is because lower layers in the originally flat-lying sedimentary or volcanic rock were moved upward and are now in the core of the anticline. The youngest rocks, on the other hand, which were originally in the upper layers, were folded downward and are now exposed along the synclinal hinge line.

Plunging Fold

The examples shown so far have been of folds with horizontal hinge lines. These are the easiest to visualize. In nature, however, anticlines and synclines are apt to be **plunging folds**—that is, folds in which the hinge lines are not horizontal. On a surface leveled by erosion, the patterns of exposed strata (beds) resemble Vs or horseshoes (figures 15.13 and 15.14) rather than the parallel, striped patterns of layers in nonplunging folds. However, plunging anticlines and synclines are distinguished from one another in the same way as are nonplunging folds—by directions of dip or by the relative ages of beds.

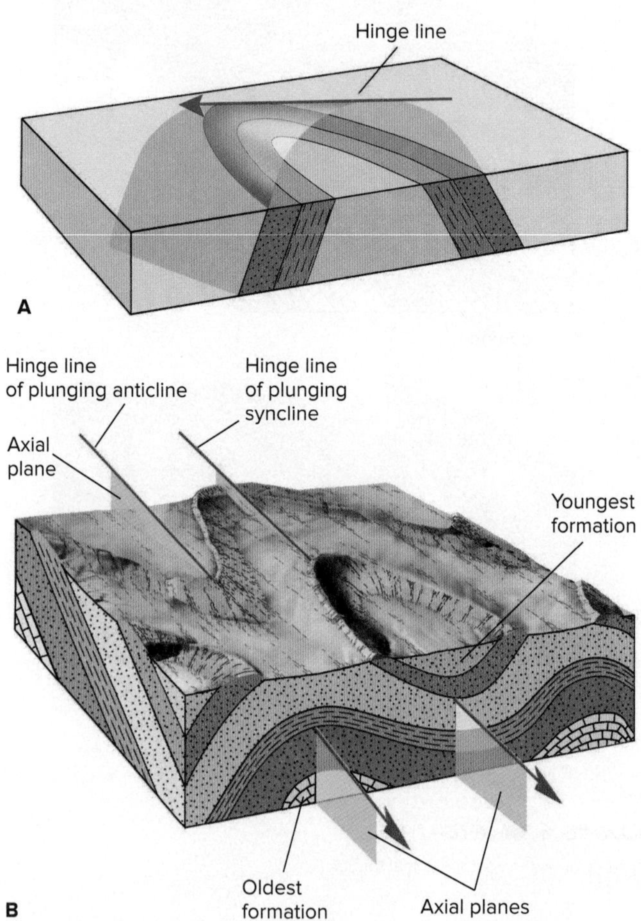

FIGURE 15.13

(*A*) Plunging fold that is cut by a horizontal plane has a V-shaped pattern. (*B*) Plunging anticline on left and right and plunging syncline in center. The hinge lines plunge toward the front of the block diagram and lie within the axial planes of the folds.

Geologist's view

FIGURE 15.14

Rock layers dip away from the center of a plunging anticline exposed at Sheep Mountain in Wyoming. Anticline plunges toward the bottom of the photo. *©Michael Collier*

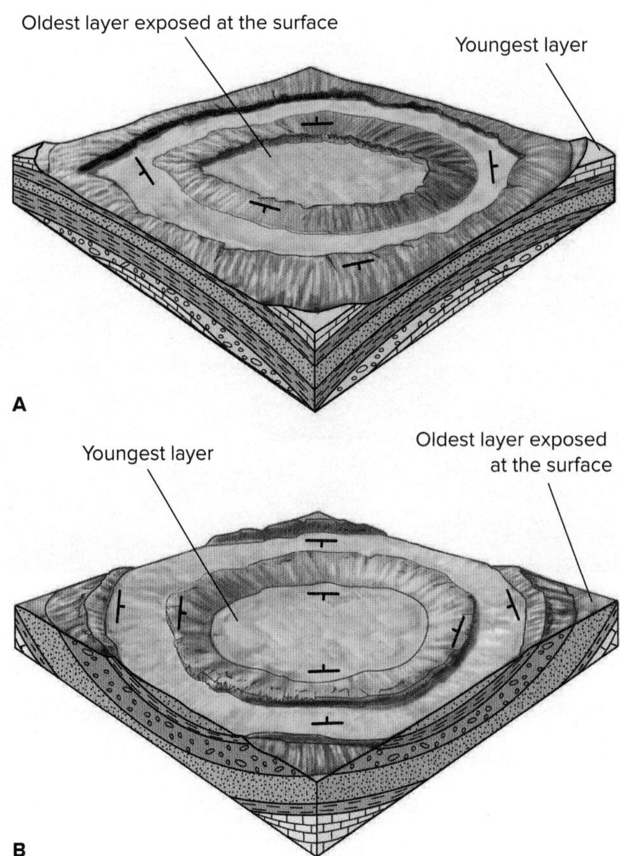

FIGURE 15.15

(*A*) Structural dome. (*B*) Structural basin.

A plunging syncline contains the youngest rocks in its center or core, and the V or horseshoe points in the direction opposite from the plunge. Conversely, a plunging anticline contains the oldest rocks in its core, and the V points in the same direction as the plunge of the fold.

Structural Domes and Structural Basins

A **structural dome** is a structure in which the beds dip away from a central point, and the oldest rocks are found in the center or core of the structure. In cross section, a dome resembles an anticline and is sometimes called a doubly plunging anticline. In a **structural basin** the beds dip toward a central point, and the youngest rocks are exposed in the center of the structure (figure 15.15); in cross section, it is comparable to a syncline (doubly plunging syncline). A structural basin is like a set of nested bowls. If the set of bowls is turned upside down, it is analogous to a structural dome.

Domes and basins tend to be features on a grand scale (some are more than a hundred kilometers across), formed by uplift somewhat greater (for domes) or less (for basins) than that of the rest of a region. Michigan's lower peninsula and parts of adjoining states and Ontario are on a large structural basin (see map at the end of the book). Domes of similar size are found in other parts of the Middle West. Smaller domes are found in the Rocky Mountains (figure 15.16).

FIGURE 15.16

Dome near Casper, Wyoming. The ridges are sedimentary layers that are resistant to erosion. Beds dip away from the center of the dome where the oldest rock layers are exposed.
©D. A. Rahm, courtesy of Rahm Memorial Collection, Western Washington University

Domes and anticlines (as well as some other structures) are important to the world's petroleum resources, as described in box 15.1 and in chapter 21.

Interpreting Folds

Folds occur in many varieties and sizes. Some are studied under the microscope, while others can have adjacent hinge lines tens of kilometers apart. Some folds are a kilometer or more in height. Figure 15.17 shows several of the more common types of folds. *Open folds* (figure 15.17*B*) have limbs that dip gently, and the angle between the limbs is large. All other factors being equal, the more open the fold, the less it has been strained by shortening. By contrast, if the angle between the limbs of the fold is small, then the fold is a *tight fold.* An *isoclinal fold*, one in which limbs are nearly parallel to one another, implies even larger shortening strain or shear strain (figure 15.17*C*).

A fold that has a vertical axial plane is referred to as an *upright fold.* However, where the axial plane of a fold is not vertical but is inclined or tipped over, the fold may be classified as *asymmetric.* If the axial plane is inclined to such a degree that the fold limbs dip in the same direction, the fold is classified as an *overturned fold* (figure 15.17*D*). Looking at an outcrop where only the overturned limb of a fold is exposed, you would probably conclude that the youngest bed is at the top. The principles of *superposition* (see chapter 8), however, cannot be applied to determine top and bottom for overturned beds. You must either see the rest of the fold or find primary sedimentary structures within the beds such as mud cracks that indicate the original top or upward direction.

Recumbent folds (figure 15.17*E*) are overturned to such an extent that the limbs are essentially horizontal. Recumbent folds are found in the cores of mountain ranges such as the Canadian Rockies, Alps, and Himalayas and record extreme shortening and shearing of the crust typically associated with plate convergence.

15.4 FRACTURES IN ROCK

If a rock is brittle, it will fracture. Commonly, there is some movement or displacement. If essentially no shear displacement occurs, a fracture or crack in bedrock is called a **joint**. If the rock on either side of a fracture moves parallel to the fracture plane, the fracture is a *fault.* Most rock at or near the surface is brittle, so nearly all exposed bedrock is jointed to some extent.

Joints

In discussing volcanoes, we described *columnar jointing*, in which hexagonal columns form as the result of contraction of a cooling, solidified lava flow (see figure 4.9). *Sheet jointing*, a type of jointing due to expansion (discussed along with weathering in chapter 5), is caused by the pressure release that accompanies the removal of overlying rock, and it has the effect of creating tensional stress perpendicular to the land surface.

Columnar and sheet joints are examples of fractures that form from nontectonic stresses and are therefore referred to as primary joints. In this chapter, we are concerned with joints that form not from cooling or unloading but from tectonic stresses.

Joints are among the most commonly observed structures in rocks (figure 15.18). A joint is a fracture or crack in a rock body along which essentially no displacement has occurred. Joints form at shallow depths in the crust where rock breaks in a brittle way and is pulled apart slightly by tensional stresses caused by bending or regional uplift. Where joints are oriented approximately parallel to one another, a **joint set** can be defined (figure 15.19).

Geologists sometimes find valuable ore deposits by studying the orientation of joints. For example, hydrothermal solutions may migrate upward through a set of joints and deposit quartz and economically important minerals such as gold, silver, copper, and zinc in the cracks. Accurate information about joints also is important in the planning and construction of large engineering projects, particularly dams and reservoirs. If the bedrock at a proposed location is intensely jointed, the possibility of dam failure or reservoir leakage may make that site too hazardous. The movement of contaminated groundwater from unlined landfills and abandoned mines may also be controlled by joints, which results in difficult and costly cleanups.

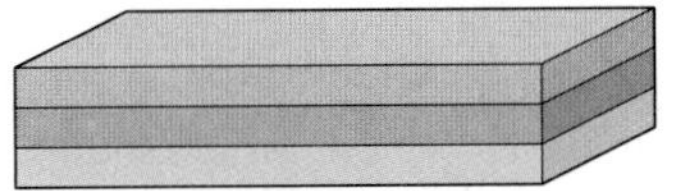

A Layers before folding

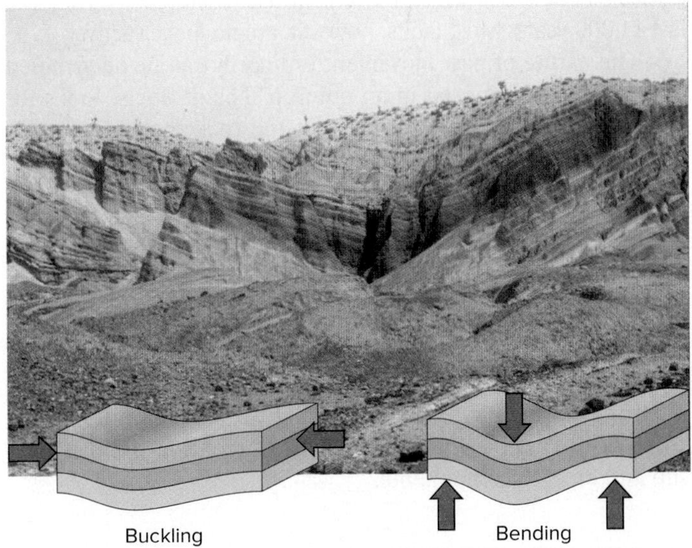

B Open folds—the two diagrams show alternate ways that stresses may have been distributed to have caused the folding

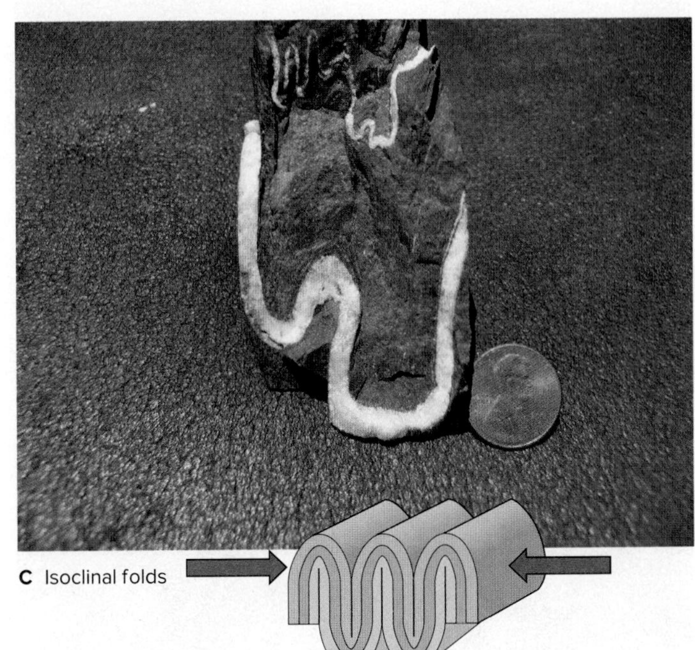

C Isoclinal folds

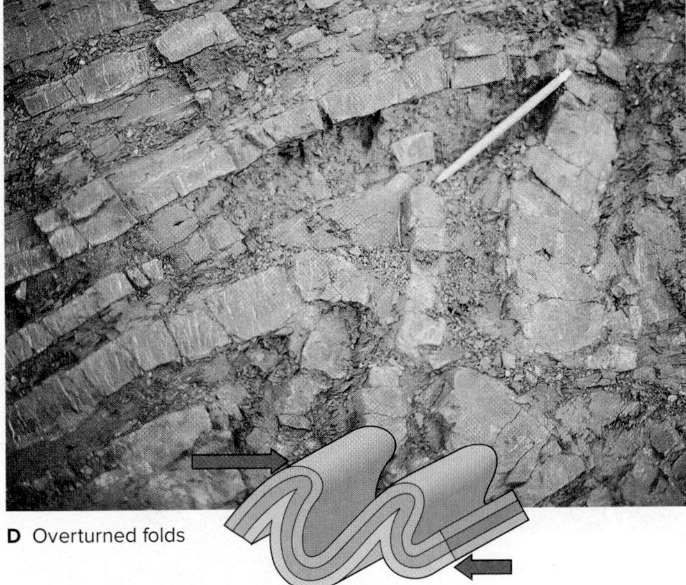

D Overturned folds

E Recumbent folds

FIGURE 15.17

Various types of folds. (*A*) Layers before folding. The length of the arrows in *B* through *E* is proportional to the amount and direction of shortening and shearing that caused folding. (*B*) Open fold from north of Barstow, California. (*C*) Isoclinal folds from the northern Sierra Nevada, California. (*D*) Overturned anticline from northern California. (*E*) Recumbent folds in the Alps.

(B) ©Diane Carlson; (C) ©Diane Carlson; specimen courtesy of Nathan Manley; (D) ©Diane Carlson; (E) Courtesy of Professor John Ramsay, from The Techniques of Modern Structural Geology, Vol. 2, J. G. Ramsay & M. J. Huber, ©1987 Academic Press

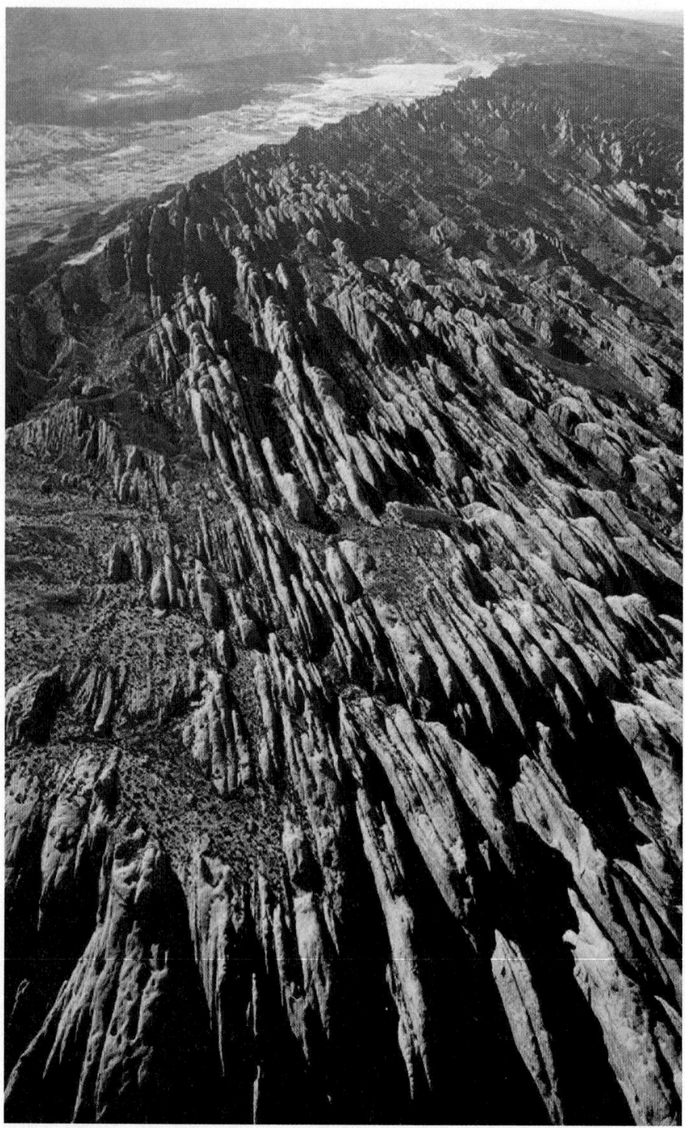

FIGURE 15.18

Vertical joints in sedimentary rock at Moab, Utah, formed in response to tectonic uplift of the region.
©Michael Collier

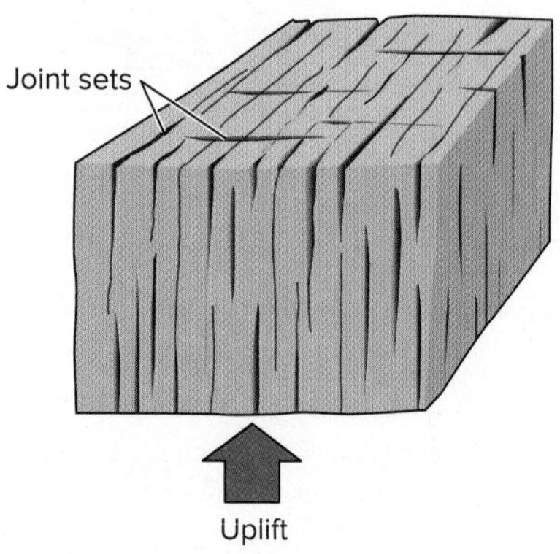

FIGURE 15.19

Two joint sets associated with uplift and a vertical compressional stress.

Faults

Faults are fractures in bedrock along which sliding has taken place. The displacement may be only several centimeters or may involve hundreds of kilometers. For many geologists, an active fault is regarded as one along which movement has taken place during the last 11,000 years. Most faults, however, are no longer active.

The nature of past movement ordinarily can be determined where a fault is exposed in an outcrop. The geologist looks for dislocated beds or other features of the rock that might show how much displacement has occurred and the relative direction of movement. In some faults, the contact between the two displaced sides is very narrow. In others, the rock has been broken or ground to a fractured or pulverized mass sandwiched between the displaced sides (figure 15.20).

Geologists describe fault movement in terms of direction of slippage: dip-slip, strike-slip, or oblique-slip (figure 15.21). In a **dip-slip fault**, movement is parallel to the dip of the fault surface. A **strike-slip fault** indicates *horizontal* motion parallel to the strike of the fault surface. An **oblique-slip fault** has both strike-slip and dip-slip components.

FIGURE 15.20

Fault in Big Horn Mountains, Wyoming, is marked by a 2-meter-wide zone of broken, red-stained rocks that offset rock layers. Rock layers have been folded into a syncline on the left side (footwall) of the fault.
©Diane Carlson

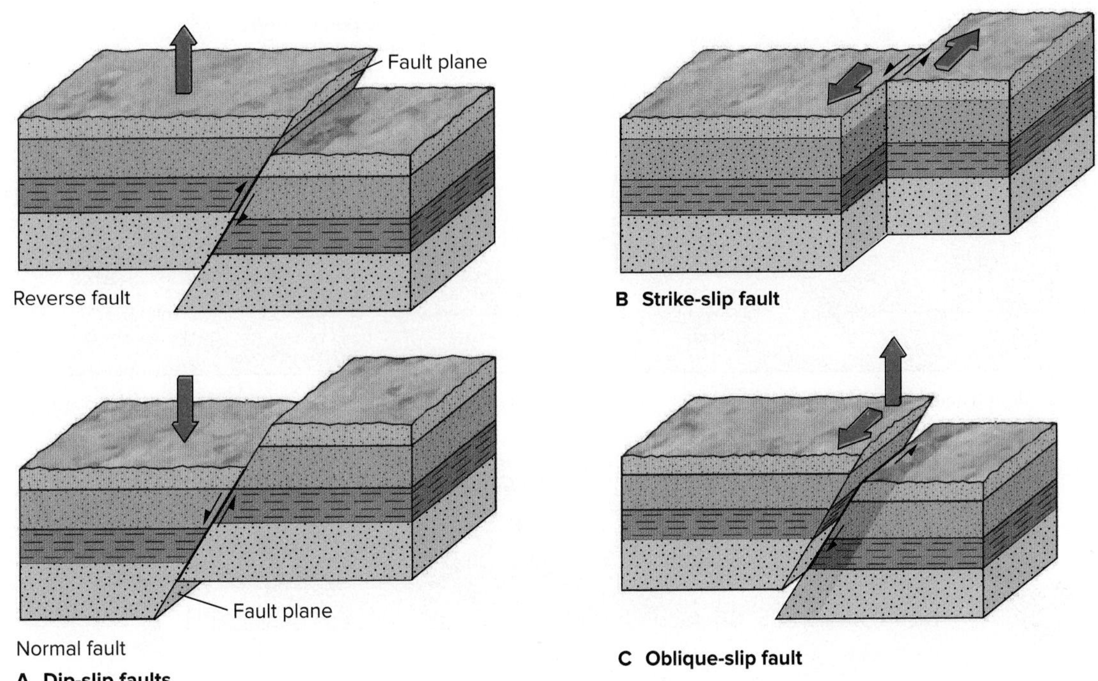

FIGURE 15.21

Three types of faults illustrated by displaced blocks. Although both blocks probably move when the fault slips, the heavier arrows show only the direction of movement on the left. (*A*) Dip-slip movement. (*B*) Strike-slip movement. (*C*) Oblique-slip movement. Black arrows show dip-slip and strike-slip components of movement.

Dip-Slip Faults

In a dip-slip fault, the movement is up or down parallel to the dip of the inclined fault surface. The side of the fault above the inclined fault surface is called the **hanging wall**, whereas the side below the fault is called the **footwall** (figure 15.22). These terms came from miners who tunneled along the fault looking for veins of mineralized rock (ore). As they tunneled, their feet were on the lower *footwall block* and they could hang their lanterns on the upper surface, or *hanging-wall block*.

Normal and reverse faults, the most common types of dip-slip faults, are distinguished from each other on the basis of the relative movement of the footwall block and the hanging-wall block. In a **normal fault** (figures 15.23 and 15.24), the hanging-wall block has moved down relative to the footwall block. The relative movement is represented on a geologic cross section by a pair of arrows, because geodetic measurement of normal faults indicates that both blocks move during slip. As shown in figure 15.23, a normal fault results in extension or lengthening of the crust. When there is extension of the crust, the hanging-wall block moves downward along the fault to compensate for the pulling apart of the rocks. Sometimes a block bounded by normal faults will drop down, creating a *graben*, as shown in figure 15.23*C*. (*Graben* is the German word for "ditch.") *Rifts* are grabens associated with divergent plate boundaries, either along mid-oceanic ridges or on continents (see chapters 18 and 19). The Rhine Valley in Germany and the Red Sea are examples of grabens.

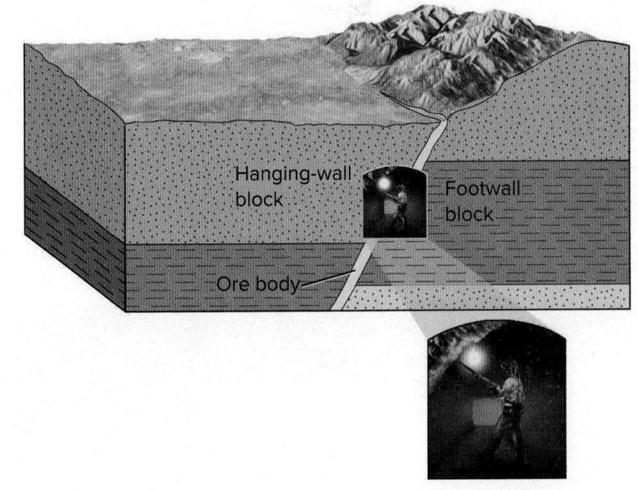

FIGURE 15.22

Relationship between the hanging-wall block and footwall block of a fault. The upper surface where a miner can hang a lantern is the hanging wall. The miner's feet are on the lower surface below the fault, which is the footwall.

If a block bounded by normal faults is uplifted sufficiently, it becomes a fault-block mountain range. (This is also called a *horst*, the opposite of a graben.) The Teton mountains and Sierra Nevada mountains are spectacular examples of fault-block mountain ranges. The Basin and Range province of Nevada and portions of adjoining states are also characterized by numerous mountain ranges (horsts) separated from adjoining valleys by

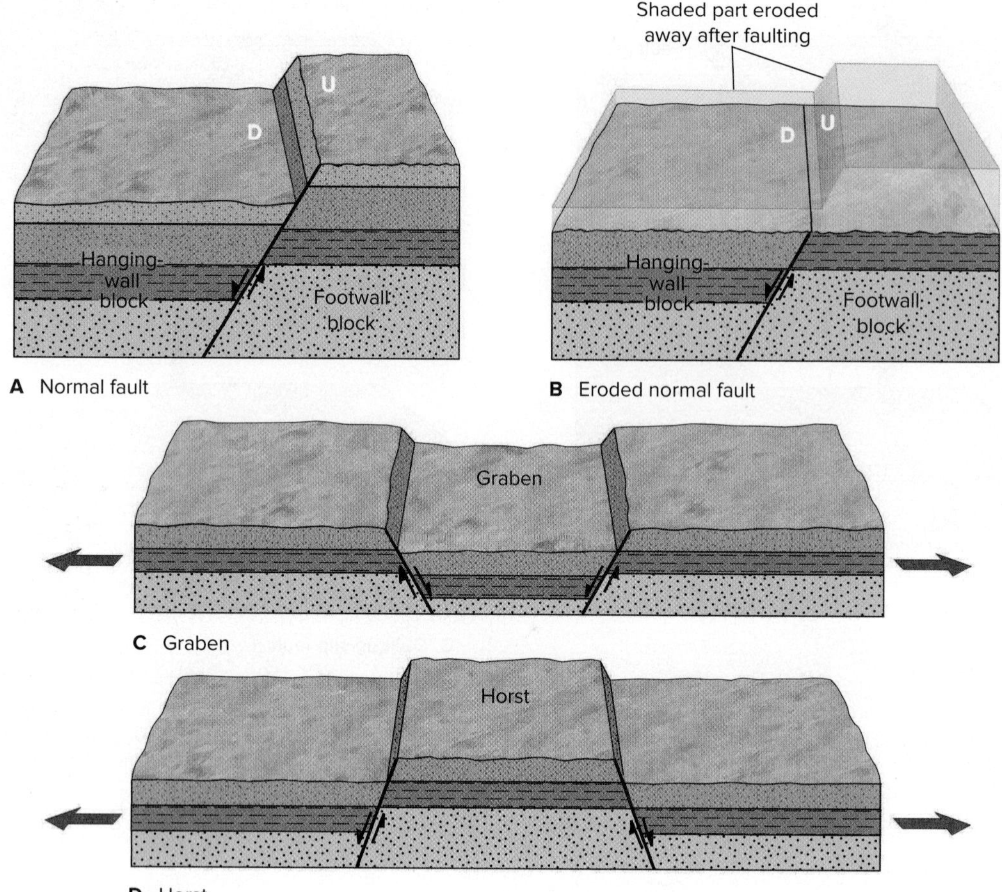

FIGURE 15.23

Normal faults. (*A*) Diagram shows the fault before erosion and the geometric relationships of the fault. (*B*) The same area after erosion. (*C*) A graben. (*D*) A horst. Arrows in *C* and *D* indicate horizontal extension of the crust.

FIGURE 15.24

Normal faults with prominent horst block offset volcanic ash layers in southern Oregon.
©*Diane Carlson*

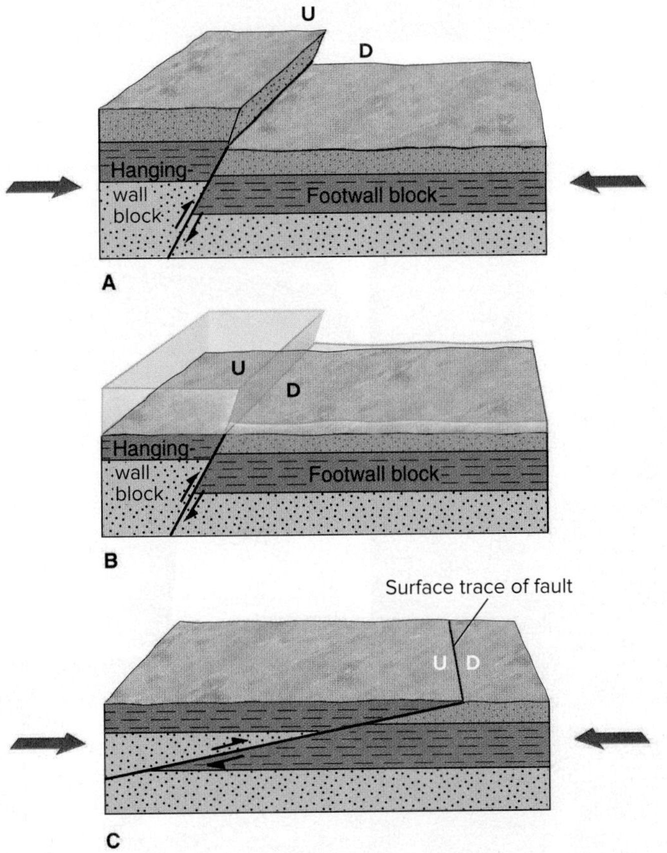

FIGURE 15.25

(*A*) A reverse fault. The fault is unaffected by erosion. Arrows indicate shortening direction. (*B*) Diagram shows area after erosion. (*C*) Thrust fault has a lower angle of dip and accommodates more shortening by stacking rock layers on top of one another.

normal faults (see chapter 20). Normal fault planes typically dip at steep angles (60°) at shallow depths but may become curved or even horizontal at depth (see figure 20.9).

In a **reverse fault**, the hanging-wall block has moved up relative to the footwall block (figures 15.25 and 15.26). As shown in figure 15.25, horizontal compressive stresses cause reverse faults. Reverse faults tend to shorten the crust.

A **thrust fault** is a reverse fault in which the dip of the fault plane is at a low angle (<30°) or even horizontal (figures 15.25*C* and 15.27). In some mountain regions, it is not uncommon for the upper plate (or hanging-wall block) of a thrust fault to have overridden the lower plate (footwall block) for several tens of kilometers. Thrust faults typically move or thrust older rocks on top of younger rocks (figure 15.27) and result in an extreme shortening of the crust. Thrust faults commonly form at convergent plate boundaries to accommodate the pushing together and shortening during convergence.

Strike-Slip Faults

A fault where the movement (or *slip*) is predominantly horizontal and therefore parallel to the strike of the fault is called a **strike-slip fault**. The displacement along a strike-slip fault is either left-lateral or right-lateral and can be determined by looking across the fault. For instance, if a recent fault displaced a stream (figure 15.28*A*), a person walking along the stream would stop where it is truncated by the fault. If the person looks across the fault and sees the stream displaced to the right, it is a **right-lateral fault**. In a **left-lateral fault**, a stream or other displaced feature would appear to the left across the fault. Again, we cannot tell which side actually moved, so pairs of arrows are used to indicate relative movement.

Large strike-slip faults, such as the San Andreas fault in California, typically define a zone of faulting that may be several kilometers wide and hundreds of kilometers long (see box 15.2). The surface trace of an active strike-slip fault is usually defined by a prominent linear valley that has been more easily eroded where the rock has been ground up along the fault during movement. The linear valley may contain lakes or sag ponds where the impermeable fault rock causes groundwater to pond at the surface. The trace of the fault may also be marked by offset surface features such as streams, fences, and roads or by distinctive rock units.

Strike-slip faults that have experienced a large amount of offset typically do not remain straight for long distances. They may either bend or step over to another fault that is parallel. Depending on the direction of the bend or stepover, the lithosphere is either pulled apart (*releasing bend*) or pushed together (*restraining bend*)

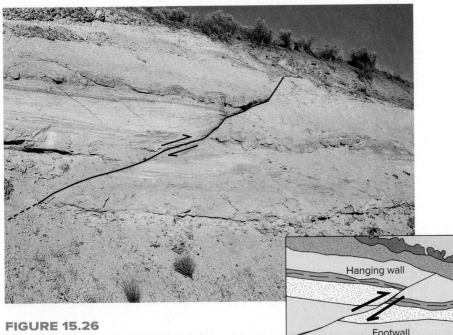

FIGURE 15.26

Reverse fault offsets volcanic ash beds, southern Oregon. Hanging wall has moved up relative to the footwall. Fault has been eroded and covered by younger sediments.
©Diane Carlson

IN GREATER DEPTH 15.2

California's Greatest Fault—The San Andreas

The San Andreas fault in California is the best-known geologic structure in the United States. Actually, the San Andreas is the longest of several, subparallel faults that transect western California (box figure 1). Collectively, these right-lateral faults are known as the San Andreas fault system. The system is in a zone approximately 100 kilometers wide and 1,300 kilometers long that extends into Mexico, ending at the Gulf of California. The San Andreas fault forms the transform plate boundary between the North American and Pacific plates (see figures 1.8 and 19.23).

Los Angeles, located on the Pacific plate, is slowly moving toward the San Francisco Bay area because of San Andreas fault motion. At an average rate of movement of 48 millimeters per year, Los Angeles could be a western suburb of San Francisco (or San Francisco an eastern suburb of Los Angeles) in some 25 million years. Earthquakes are produced by sudden movement within the fault system, as explained in chapter 16. Bedrock along the San Andreas fault shifted as much as 4.5 meters in association with the 1906 quake that destroyed much of San Francisco. Most geologists think strike-slip movement began about 30 million years ago, and that the total offset on the fault has been at least 560 kilometers.

The San Andreas fault is not a simple crack but a zone of broken and ground-up rock, usually a hundred meters or more wide. Its presence is easy to determine throughout most of its length. Along the fault trace are long, straight valleys (box figures 1 and 2) that show quite different rocks on either side. Stream channels follow much of the fault zone because the weak, ground-up rock along the fault is easily eroded. Locally, elongate lakes (called sag ponds) are found where the ground-up fault rock has dropped down or sagged more than the surface of adjacent parts of the fault zone, and groundwater has ponded in the impermeable fault rock. The fault was named after one of these sag ponds, San Andreas Lake, just south of San Francisco.

One can visually follow the trace of the San Andreas fault through the Carizzo Plain in central California (box figure 2). Long linear valleys and fault scarps are common surface expressions of the fault. Some of the most obvious features produced by the fault motion are displaced ridges, valleys, and streams. Box figure 3 shows how the channel of Wallace Creek has been offset and diverted by the right-lateral slip along the San Andreas fault.

The San Andreas fault continues to be the most extensively instrumented and carefully studied fault in the world. The northern and southern portions of the fault are locked (shown in red in box figure 1), and the central portion (shown in blue) creeps or slips at regular intervals. Just north of the locked portion at Parkfield, California, the San Andreas fault generates moderate-size earthquakes about every 22 years. Because of the occurrence of earthquakes at fairly regular intervals here, geologists have been closely monitoring the fault for the past two decades to learn about its behavior before, during, and after an earthquake—as part of the Parkfield Earthquake Experiment. More recently, the San Andreas Fault Observatory at Depth (SAFOD) was funded to drill a 4-kilometer-long borehole from the Pacific plate, through the fault zone, to the North American plate at Parkfield (box figure 1). This is the first time scientists have drilled into an active fault to reveal the processes that control the generation of earthquakes.

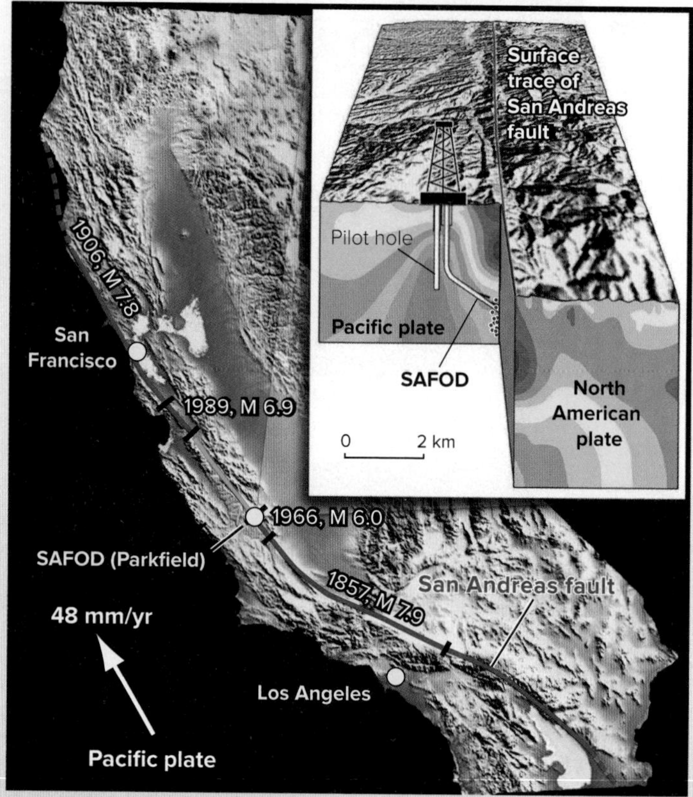

BOX 15.2 ■ FIGURE 1

Map showing the trace of the San Andreas fault (red lines show locked segments and blue line represents creeping portion) and the San Andreas Fault Observatory at Depth (SAFOD) located at Parkfield, California. The San Andreas fault separates the Pacific plate on the west from the North American plate on the east. The SAFOD inset diagram shows the locations of the pilot hole drilled in 2002 and the main borehole that directly samples and monitors the active San Andreas fault. The small purple dots represent recurring small earthquakes that signal slip of the fault. The colors below the ground surface represent electrical resistivity of the rocks. Red represents low resistivity and is interpreted to represent fluid along the fault.
Diagram courtesy of Earthscope/USGS

Rock cores retrieved from within the fault contain pieces of serpentine and talc, and the fault surfaces are coated with a thin layer of clay. The presence of these soft, hydrated minerals suggests that soft, "slippery" rocks facilitate the gradual slip, or *creep,* along the central segment of the San Andreas fault.

Sensitive seismic instruments have also been installed in the SAFOD borehole and will enable geologists to test hypotheses

BOX 15.2 ■ FIGURE 2

Trace of San Andreas fault in the Carrizo Plain, central California.
Source: Robert E. Wallace/U.S. Geological Survey

BOX 15.2 ■ FIGURE 3

Stream channel (Wallace Creek) offset by the San Andreas fault in the Carrizo Plain. The arrows on either side of the fault trace indicate the right-lateral motion.
©C. C. Plummer

about how earthquakes are generated, and to evaluate the roles of fluid pressure, rock friction, and state of stress in controlling fault strength. These instruments are designed to remain in operation for the next 10 to 20 years, providing valuable insights into the workings of California's greatest fault.

Additional Resources

U.S. Geological Survey Web page *The San Andreas Fault* by Sandra S. Schulz and Robert E. Wallace gives details about the fault.

- http://pubs.usgs.gov/gip/earthq3

An interactive map of the San Andreas fault.

- http://thulescientific.com/san-andreas-fault-map.html

Video describing the southern San Andreas fault system produced by the *Earthquake Country Alliance.*

- www.earthquakecountry.info/video/sanandreas.html

Web pages of the International Continental Scientific Drilling Program, Earthscope, and the U.S. Geological Survey that give details about the *San Andreas Fault Observatory at Depth* project.

- http://www-icdp.icdp-online.org/front_content.php?idcat=889
- http://earthquake.usgs.gov/research/parkfield/safod_pbo.php

A

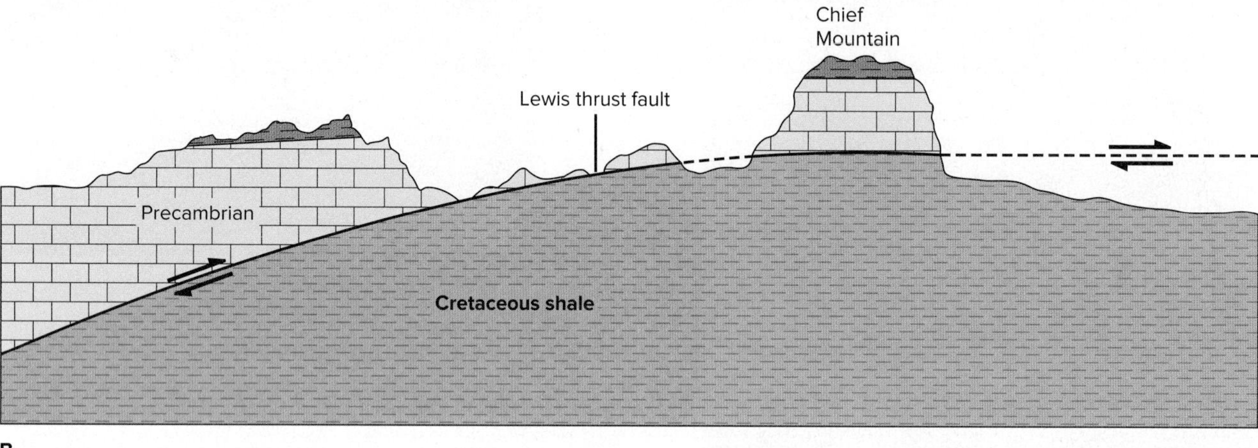

B

FIGURE 15.27

(*A*) Chief Mountain in Glacier National Park, Montana, is an erosional remnant of a major thrust fault. (*B*) Cross section of the area. Older (Precambrian) rocks have been thrust over younger (Cretaceous) rocks. Dashed lines show where the Lewis thrust fault has been eroded away.
(A) ©Diane Carlson

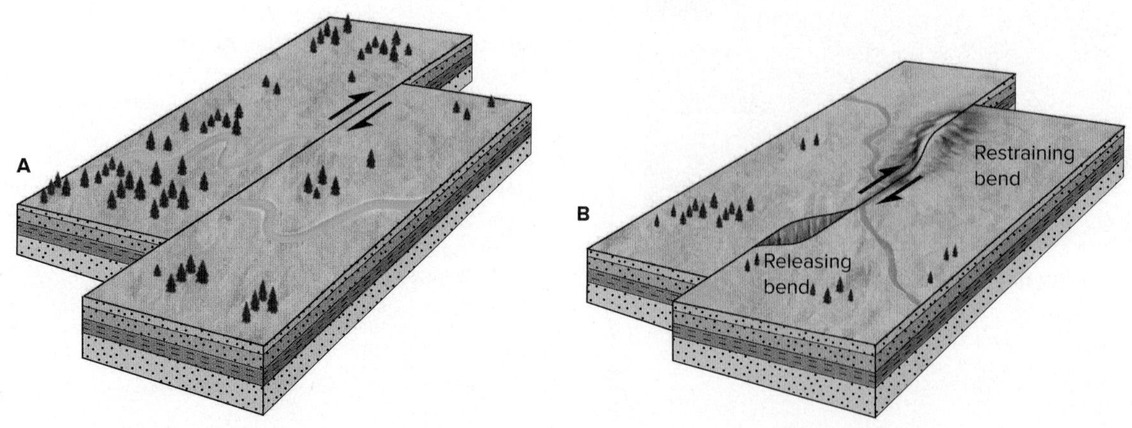

FIGURE 15.28

(*A*) Right-lateral strike-slip fault offsets a stream channel. Looking across the fault, you would need to walk to the right to find the continuation of the stream. (*B*) Strike-slip movement along curved faults produces gaps or basins at releasing bends where the lithosphere is pulled apart or shortening and hills where it is pushed together at restraining bends.

(figure 15.28*B*). Normal faults and grabens form in response to the pulling apart at the releasing bends and folds, and thrust faults form at the restraining bends to accommodate the pushing or pinching together of the lithosphere. Death Valley (see figure 13.10) is a good example of a deep graben formed along a releasing bend or stepover in the newly forming plate boundary along the eastern side of the Sierra Nevada Mountains in California. Where the San Andreas fault bends to the left, north of Los Angeles, the Transverse Mountains have been pushed up by folding and thrust faults.

Strike-slip faults accommodate shearing strain in the brittle, uppermost lithosphere, and they may also represent transform plate boundaries where plates slide past one another. One of the most famous examples of a transform fault is the San Andreas fault. The San Andreas fault is a right-lateral strike-slip fault that forms part of the boundary between the North American and Pacific plates.

Summary

Tectonic forces result in deformation of the Earth's crust. *Stress* (force per unit area) is a measure of the tectonic force and confining pressure acting on bedrock. Stress can be *compressive*, *tensional*, or *shearing*. *Strained* (changed in shape or size) rock records past stresses, usually as joints, faults, or folds.

A geologic map shows the structural characteristics of a region. *Strike* and *dip* symbols on geologic maps indicate the orientations of inclined surfaces such as bedding planes. The strike and dip of a bedding surface indicate the relationship between the inclined plane and a horizontal plane.

If rock layers bend (ductile behavior) rather than break, they become folded. Rock layers are folded into *anticlines* and *synclines* and recumbent folds. If the hinge line of a fold is not horizontal, the fold is *plunging*. Older beds exposed in the core of a fold indicate an anticline, whereas younger beds in the center of the structure indicate a syncline. In places where folded rock has been eroded to a plain, an anticline can usually be distinguished from a syncline by whether the beds dip toward the center (syncline) or away from the center (anticline). Also, the oldest rocks are found in the center of an eroded anticline, whereas the youngest rocks are found in the center or core of a syncline.

Fractures in rock are either *joints* or *faults*. A joint indicates that movement has not occurred on either side of the fracture; displaced rock along a fracture indicates a fault. *Dip-slip* faults are either *normal* or *reverse*, depending on the motion of the hanging-wall block relative to the footwall block. The relative motion of the hanging wall is upward in a reverse fault and downward in a normal fault. A reverse fault with a low angle of dip for the fault plane is a *thrust fault*. Reverse faults accommodate horizontal shortening of the crust, whereas normal faults accommodate horizontal stretching or extension.

In a *strike-slip* fault, which can be either left-lateral or right-lateral, horizontal movement parallel to the strike has occurred.

Terms to Remember

angle of dip 368
anticline 371
axial plane 371
brittle 367
compressive stress 365
dip-slip fault 376
direction of dip 368
ductile 366
elastic 366
elastic limit 366
fault 376
fold 369
footwall 377
geologic cross section 369
geologic map 367
hanging wall 377
hinge line 371
joint 374
joint set 374
left-lateral fault 379
limb 371
normal fault 377
oblique-slip fault 376
plunging fold 372
reservoir rock 370
reverse fault 379
right-lateral fault 379
shear stress 365
source rock 370
strain 365
stress 365
strike 368
strike-slip fault 376
structural basin 373
structural dome 373
syncline 371
tensional stress 365
thrust fault 379

Testing Your Knowledge

Use the following questions to prepare for exams based on this chapter.

1. Sketch and describe the three types of stress and the resulting strain.
2. What factors control whether a rock behaves as a brittle material or a ductile material?
3. What is the difference between strike, direction of dip, and angle of dip?
4. Draw a sketch of an anticline and label the limbs, axial plane, and hinge line.
5. Sketch and describe the different types of folds, and explain how the shape and orientation of folds is used to interpret strain.
6. On a geologic map, if no cross sections were available, how could you distinguish an anticline from a syncline?
7. Draw a simple geologic map using strike and dip symbols for a syncline plunging to the west.
8. How does a structural dome differ from a plunging anticline?
9. What is the difference between a joint and a fault?
10. Which of the following statements is true?
    a. When forces are applied to an object, the object is under stress.
    b. Strain is the change in shape or size (volume), or both, while an object is undergoing stress.
    c. Stresses can be compressive, tensional, or shear.
    d. All of the preceding.

11. Folds in a rock show that the rock behaved in a _____ way.
    a. ductile b. elastic
    c. brittle d. all of the preceding
12. The compass direction of a line formed by the intersection of an inclined plane with a horizontal plane is called
    a. strike. b. direction of dip.
    c. angle of dip. d. axis.
13. An anticline is
    a. a fold shaped like an arch with the youngest rocks exposed in the center of the fold.
    b. a trough-shaped fold with the oldest rocks exposed in the center of the fold.
    c. a fold shaped like an arch with the oldest rocks exposed in the center of the fold.
    d. a trough-shaped fold with the youngest rocks exposed in the center of the fold.
14. A syncline is
    a. a fold shaped like an arch with the youngest rocks exposed in the center of the fold.
    b. a trough-shaped fold with the oldest rocks exposed in the center of the fold.
    c. a fold shaped like an arch with the oldest rocks exposed in the center of the fold.
    d. a trough-shaped fold with the youngest rocks exposed in the center of the fold.
15. A structure in which the beds dip away from a central point and the oldest rocks are exposed in the center is called a(n)
    a. basin. b. anticline.
    c. structural dome. d. syncline.
16. Which is *not* a type of fold?
    a. open b. isoclinal
    c. overturned d. recumbent
    e. thrust
17. Fractures in bedrock along which movement has taken place are called
    a. joints. b. faults.
    c. cracks. d. folds.
18. If you locate a dip-slip fault while doing field work, what kind of evidence would you look for to determine whether the fault is normal or reverse?
19. In a normal fault, the hanging-wall block has moved _____ relative to the footwall block.
    a. upward b. downward
    c. sideways
20. Normal faults occur where
    a. there is horizontal shortening.
    b. there is horizontal extension.
    c. the hanging wall moves up.
    d. the footwall moves down.
21. Faults that typically move older rock on top of younger rock are
    a. normal faults. b. thrust faults.
    c. strike-slip faults.

Expanding Your Knowledge

1. Why do some rocks fold while others are faulted?
2. In what parts of North America would you expect to find the most intensely folded rock?
3. A subduction zone can be regarded as a very large example of what type of fault?
4. Looking at the San Andreas fault, shown in box 15.2, figure 1, where might restraining bends form? What kinds of structures might form there?
5. What features in sedimentary or volcanic rock layers would you look for to tell you that the rock was part of the overturned limb of a fold?

Exploring Web Resources

www.see.leeds.ac.uk/structure/learnstructure/

School of Earth Sciences, University of Leeds website is a dynamic collection of in-depth information and photos and diagrams of geologic structures and virtual field trips throughout the world.

www.geology.sdsu.edu/visualgeology/geology101/geo100/structure.htm

Structural Geology website by Gary Girty at San Diego State University contains animations of folds, faults, and interactive exercises dealing with the interpretation and description of geologic structures.

CHAPTER 16

Earthquakes

Homes and buildings in Kesennuma, Miyagi Prefecture, were destroyed by the tsunami caused by the magnitude 9.0 Tohoku earthquake that struck the northeast coast of Japan on March 11, 2011. Soon after the earthquake struck, tsunami waves as high as 38.9 meters (128 feet) flooded the coastal towns, leaving little time for people to escape to higher ground. The coastal communities were prepared for tsunami with a warning system in place and concrete seawalls. Unfortunately, the seawalls were not high enough, giving some people a false sense of security when the record-height tsunami waves struck and overtopped the seawalls and in places traveled up to 10 kilometers (6 miles) inland. The death toll from this disaster was over 15,000, with more than 400,000 homeless. The 2011 Tohoku earthquake was the strongest to hit Japan since modern instrumental recordings began and is the fourth-largest earthquake in the world since 1900.
©The Asahi Shimbun/Corbis/Getty Images

LEARNING OBJECTIVES

- Explain what causes earthquakes.
- Name and describe the various types of seismic waves.
- Describe how earthquakes are located and how the size of an earthquake is measured.
- List and describe the hazards associated with earthquakes.
- Discuss the distribution of earthquakes in the world with regard to location, depth of focus, size, and plate-tectonic setting.
- List the precursors that may occur before an earthquake. Describe how long-range forecasts of earthquake probability are determined.

This chapter will help you understand the nature and origin of earthquakes. We discuss the seismic waves created by earthquakes and how the quakes are measured and located by studying these waves. We also describe some effects of earthquakes, such as ground motion and displacement, damage to buildings, and quake-caused fires, landslides, and seismic sea waves (tsunami).

Earthquakes commonly affect other parts of Earth systems. Intense shaking associated with an earthquake not only can cause tremendous damage and loss of life but also can trigger landslides that may disperse pathogenic microbes into the atmosphere and cause additional human health concerns. Such was the case after the 1994 Northridge, California, earthquake. Another effect on the biosphere may be the unusual behavior of animals just before an earthquake, as reported by Chinese scientists. Ground breakage associated with earthquakes may affect the hydrosphere by creating new lakes (sag ponds), increasing groundwater flow from springs, and displacing stream channels. Tsunami generated by submarine earthquakes may cause tremendous damage to the coastal environment.

Earthquakes are largely confined to a few narrow belts on Earth. This distribution was once puzzling to geologists, but here we show how the concept of plate tectonics neatly explains it.

As geologists learn more about earthquake behavior, the possibility exists that we will be able to forecast earthquakes. We conclude the chapter with a look at this developing branch of study.

16.1 CAUSES OF EARTHQUAKES

What causes earthquakes? An **earthquake** is a trembling or shaking of the ground caused by the sudden release of energy stored in the rocks beneath Earth's surface. As described in chapter 15, tectonic forces acting deep in the Earth may put a *stress* on the rock, which may bend or change in shape (*strain*). If you bend a stick of wood, your hands put a stress (the force per unit area) on the stick; its bending (a change in shape) is the strain.

Like a bending stick, rock can deform only so far before it breaks. When a rock breaks, waves of energy are released and sent out through the Earth. These are **seismic waves**, the waves of energy produced by an earthquake. It is the seismic waves that cause the ground to tremble and shake during an earthquake.

The sudden release of energy when rock breaks may cause one huge mass of rock to slide past another mass of rock into a different relative position. As you know from chapter 15, the break between the two rock masses is a *fault*. The classic explanation of why earthquakes take place is called the **elastic rebound theory** (figure 16.1). It involves the sudden release of progressively stored strain in rocks, causing movement along a fault. Deep-seated internal forces (*tectonic forces*) act on a mass of rock over many decades. Initially, the rock bends but does not break. More and more energy is stored in the rock as the bending becomes more severe. Eventually, the energy stored in the rock exceeds the breaking strength of the rock, and the rock breaks suddenly, causing an earthquake. Two masses of rock move past one another along a fault. The movement may be vertical, horizontal, or both (figure 16.2). The strain on the rock is released; the energy is expended by moving the rock into new positions and by creating seismic waves.

The brittle behavior of breaking rock usually occurs near Earth's surface. Rocks at depth are subject to increased temperature and pressure, which tend to reduce brittleness. Deep rocks behave as ductile materials instead of breaking (*brittle* behavior); hence, there is a limit to the depth at which faults can occur. However, some of the deepest and largest earthquakes occur at convergent plate boundaries, where cold oceanic lithosphere is subducted and breaks in a brittle way, generating earthquakes.

Most earthquakes are associated with movement on faults, but in some quakes, the connection with faulting may be difficult to establish. Some quakes, including the 1994 Northridge, California, and 2010 Haiti earthquakes, occurred on buried thrust faults that did not cause surface displacement. Most earthquakes in the eastern United States are also not associated with surface displacement. Earthquakes also occur during explosive volcanic eruptions and as magma forcibly fills underground magma chambers prior to many eruptions; these quakes may not be associated with fault movement at all.

16.2 SEISMIC WAVES

The point within the Earth where seismic waves first originate is called the **focus** (or *hypocenter*) of the earthquake (figure 16.3). This is the center of the earthquake, the point of initial breakage and movement on a fault. Rupture begins at the focus and then spreads rapidly along the fault plane. The point on the Earth's surface directly above the focus is the **epicenter**.

Two types of seismic waves are generated during earthquakes. **Body waves** are seismic waves that travel through the Earth's interior, spreading outward from the focus in all directions.

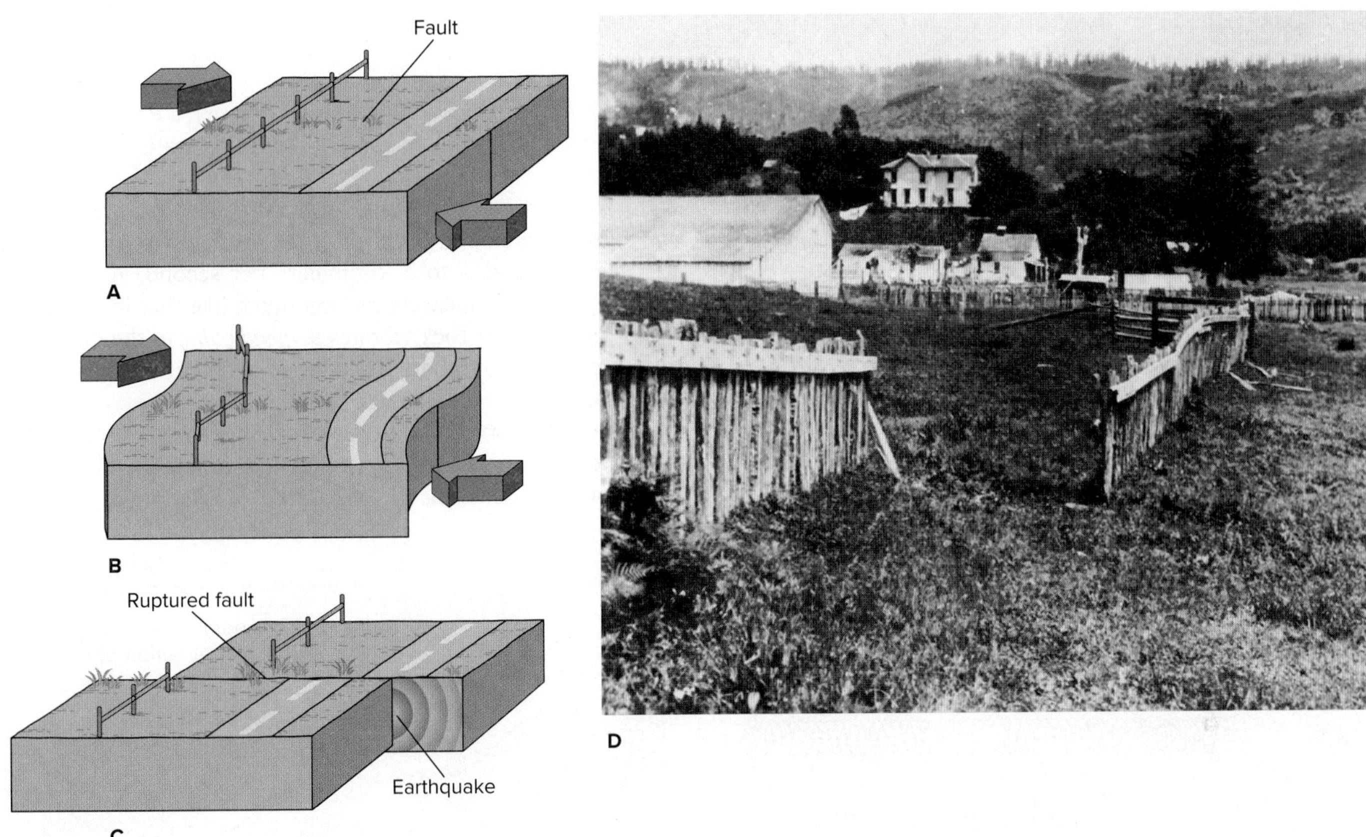

FIGURE 16.1

The elastic rebound theory of the cause of earthquakes. (*A*) Rock with stress acting on it. (*B*) Stress has caused strain in the rock. Strain builds up over a long period of time. (*C*) Rock breaks suddenly, releasing energy, with rock movement along a fault. Horizontal motion is shown; rocks can also move vertically or diagonally. (*D*) Fence offset nearly 3 meters after 1906 San Francisco earthquake.
(D) Source: G. K. Gilbert, U.S. Geological Survey

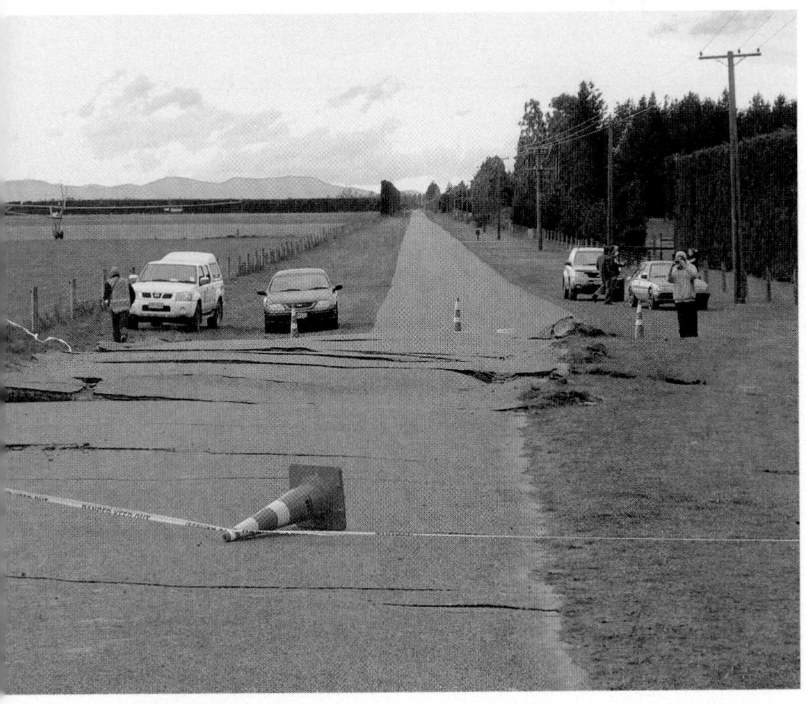

FIGURE 16.2

Horizontal offset of the once straight road caused by the magnitude 7.1 Canterbury earthquake that struck New Zealand on September 4, 2010. The displacement of the road indicates a *right-lateral* sense of offset along the fault that broke during the earthquake.

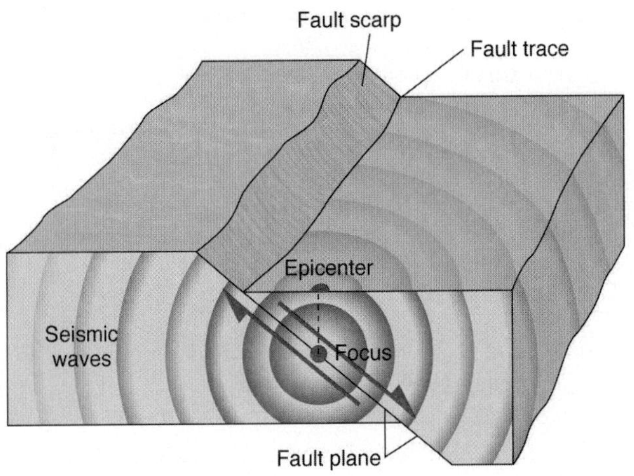

FIGURE 16.3

The focus of an earthquake is the point where rocks first break along a fault; seismic waves radiate from the focus. The epicenter is the point on the Earth's surface directly above the focus.

Surface waves are seismic waves that travel on Earth's surface away from the epicenter, like water waves spreading out from a pebble thrown into a pond. Rock movement associated with seismic surface waves dies out with depth into the Earth, just as water movement in ocean waves dies out with depth.

Body Waves

There are two types of body waves, both shown in figure 16.4. A **P wave** is a compressional (or longitudinal) wave in which rock vibrates back and forth *parallel* to the direction of wave propagation. Because it is a very fast wave, traveling through near-surface rocks at speeds of 4 to 7 kilometers per second (9,000 to more than 15,000 miles per hour), a P wave is the first (or *primary*) wave to arrive at a recording station following an earthquake.

The second type of body wave is called an **S wave** (*secondary*) and is a slower, transverse wave that travels through near-surface rocks at 2 to 5 kilometers per second. An S wave is propagated by a shearing motion much like that in a stretched, shaken rope. The rock vibrates *perpendicular* to the direction of wave propagation, that is, crosswise to the direction the waves are moving.

Both P waves and S waves pass easily through solid rock. A P wave can also pass through a fluid (gas or liquid), but an S wave cannot. We discuss the importance of this fact in chapter 17.

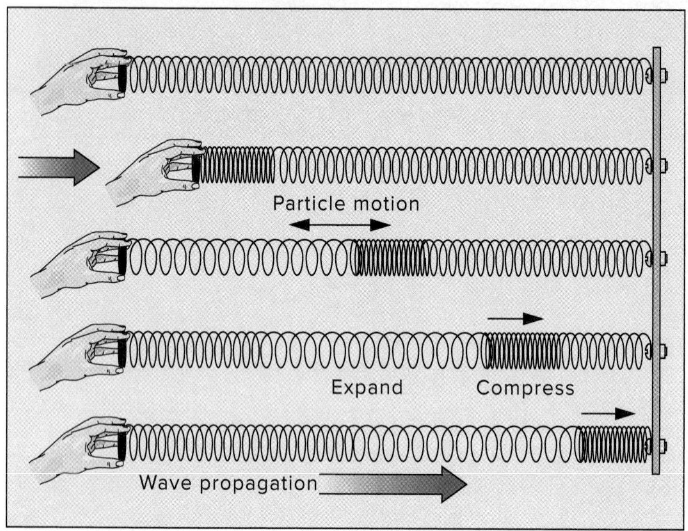

A Primary wave

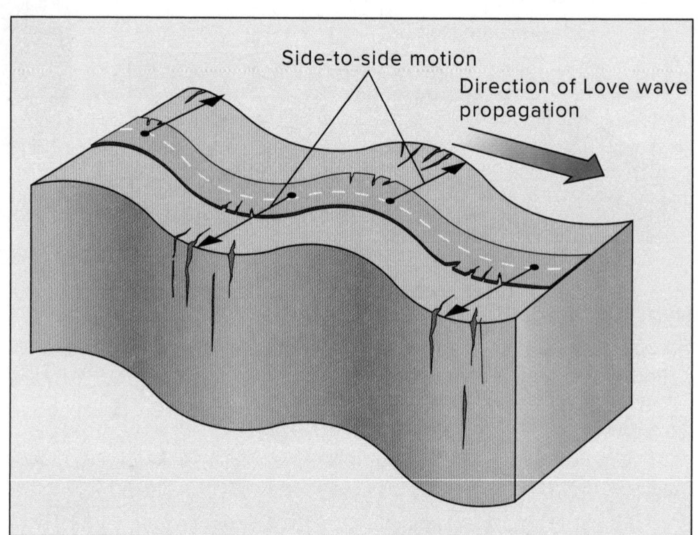

C Love wave

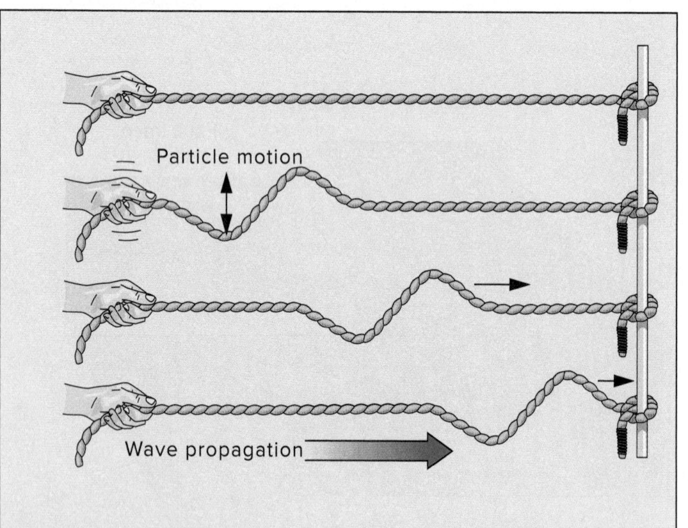
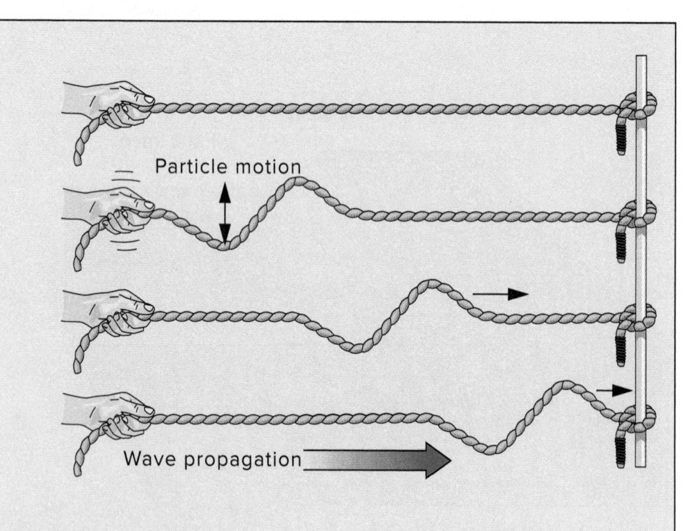

B Secondary wave

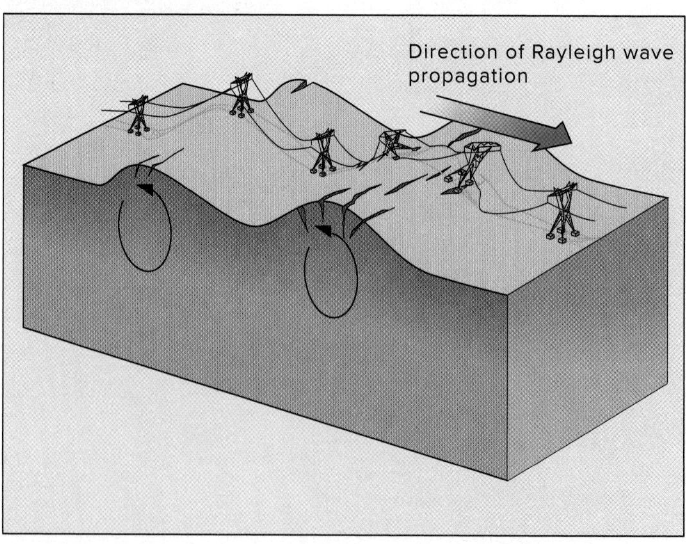

D Rayleigh wave

FIGURE 16.4

Particle motion in seismic waves. (*A*) A P wave is illustrated by a sudden push on the end of a stretched spring or Slinky®. The particles vibrate *parallel* to the direction of wave propagation. (*B*) An S wave is illustrated by shaking a loop along a stretched rope. The particles vibrate *perpendicular* to the direction of wave propagation. (*C*) Love waves behave like S waves in that the particle motion is perpendicular to the direction of wave travel along Earth's surface. (*D*) Rayleigh waves are like ocean waves and cause a rolling motion on Earth's surface. The particle motion is elliptical and opposite (counterclockwise) to the direction of wave propagation.

Surface Waves

Surface waves are the slowest waves set off by earthquakes. In general, surface waves cause more property damage than body waves because surface waves produce more ground movement and travel more slowly, so they take longer to pass. The two most important types of surface waves are Love waves and Rayleigh waves, named after the geophysicists who discovered them.

Love waves are most like S waves that have no vertical displacement. The ground moves side to side in a horizontal plane that is perpendicular to the direction the wave is traveling or propagating (figure 16.4*C*). Like S waves, Love waves do not travel through liquids and would not be felt on a body of water. Because of the horizontal movement, Love waves tend to knock buildings off their foundations and destroy highway bridge supports.

Rayleigh waves behave like rolling ocean waves. Unlike ocean waves, Rayleigh waves cause the ground to move in an elliptical path opposite to the direction the wave passes (figure 16.4*D*). Rayleigh waves tend to be incredibly destructive to buildings because they produce more ground movement and take longer to pass.

16.3 LOCATING AND MEASURING EARTHQUAKES

The invention of instruments that could accurately record seismic waves was an important scientific advance. These instruments measure the amount of ground motion and can be used to find the location, depth, and size of an earthquake.

The instrument used to measure seismic waves is a *seismometer.* The principle of the seismometer is to keep a heavy suspended mass as motionless as possible—suspending it by springs or hanging it as a pendulum from the frame of the instrument (figure 16.5). When the ground moves, the frame of the instrument moves with it; however, the inertia of the heavy mass suspended inside keeps the mass motionless to act as a point of reference in determining the amount of ground motion. Seismometers are usually placed in clusters of three to record the motion along the *x, y,* and *z* axes of three-dimensional space.

A seismometer by itself cannot record the motion that it measures. A **seismograph** is a recording device that produces a permanent record of Earth motion detected by a seismometer, historically recorded as wiggly lines drawn on a moving strip of paper, but now digitally recorded (figure 16.6). The record of Earth vibration is called a **seismogram**. The seismogram can be used to measure the strength of the earthquake.

A network of seismograph stations is maintained all over the world to record and study earthquakes (and nuclear bomb explosions). Within minutes after an earthquake occurs, distant seismographs begin to pick up seismic waves. A large earthquake can be detected by seismographs all over the world.

Because the different types of seismic waves travel at different speeds, they arrive at seismograph stations in a definite order: first the P waves, then the S waves, and finally the surface waves. These three different waves can be distinguished on the seismograms. By analyzing these seismograms, geologists can learn a great deal about an earthquake, including its location and size.

Determining the Location of an Earthquake

P and S waves start out from the focus of an earthquake at essentially the same time. As they travel away from the quake, the two kinds of body waves gradually separate because they are traveling at different speeds. On a seismogram from a station close to the earthquake, the first arrival of the P wave is separated from the first arrival of the S wave by a short distance on the paper record (figure 16.7). At a recording station far from

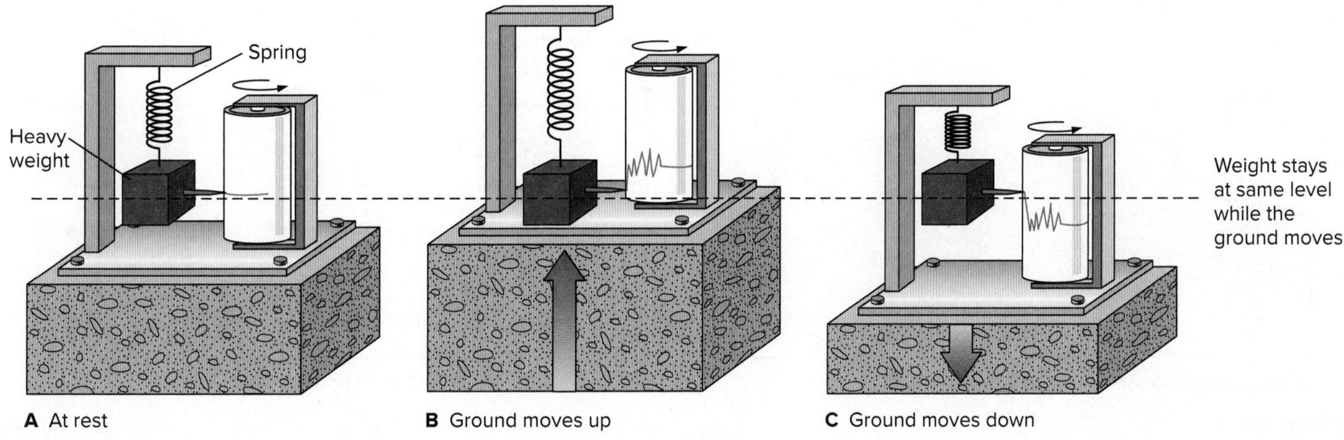

FIGURE 16.5

A simple seismograph for detecting vertical rock motion. The pen records the ground motion on the seismogram as the spring stretches and compresses with its up-and-down movement. Frame and recording drum move with the ground. Inertia of the weight keeps it and the needle relatively motionless.

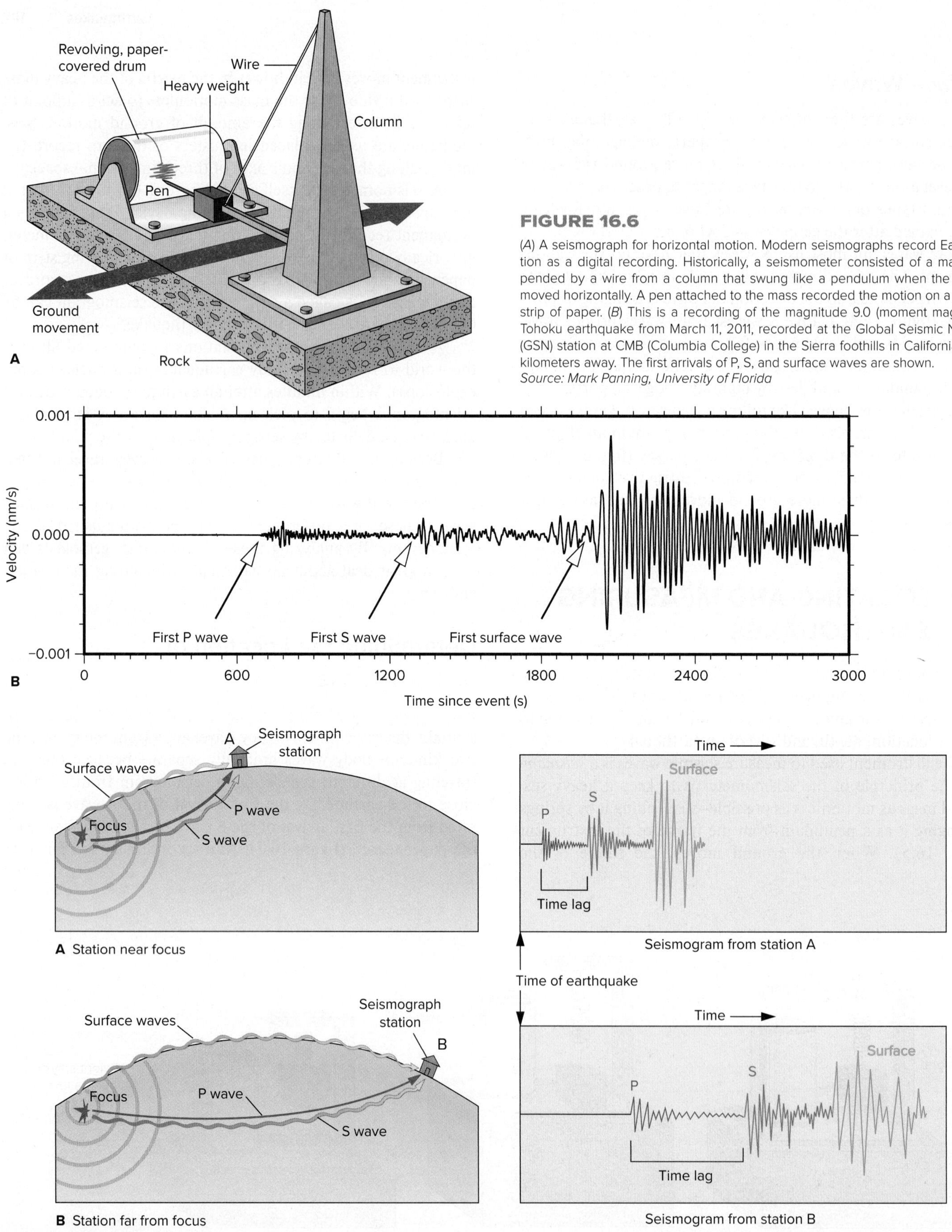

FIGURE 16.6

(*A*) A seismograph for horizontal motion. Modern seismographs record Earth motion as a digital recording. Historically, a seismometer consisted of a mass suspended by a wire from a column that swung like a pendulum when the ground moved horizontally. A pen attached to the mass recorded the motion on a moving strip of paper. (*B*) This is a recording of the magnitude 9.0 (moment magnitude) Tohoku earthquake from March 11, 2011, recorded at the Global Seismic Network (GSN) station at CMB (Columbia College) in the Sierra foothills in California 8,050 kilometers away. The first arrivals of P, S, and surface waves are shown.
Source: Mark Panning, University of Florida

FIGURE 16.7

Because of the difference in travel times, intervals between P waves, S waves, and surface waves increase with distance from the focus.

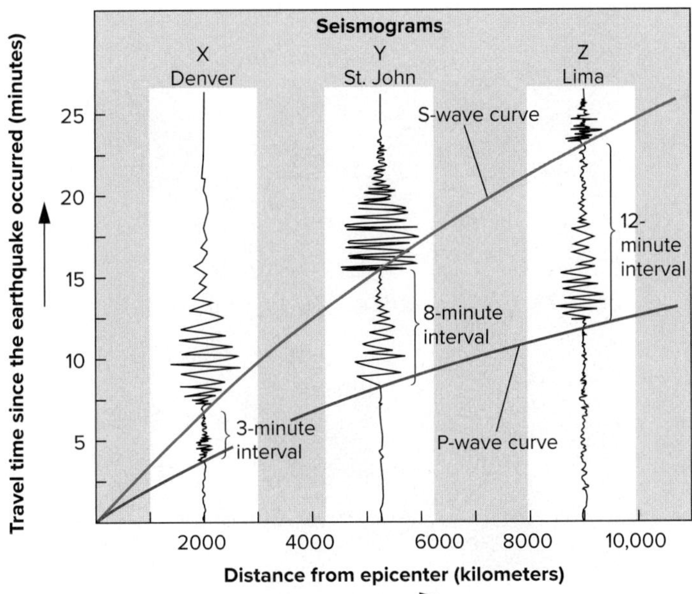

FIGURE 16.8

A travel-time curve is used to determine the distance to an earthquake. Note that the time interval between the first arrival of P and S waves increases with distance from the epicenter. Seismogram X has a 3-minute interval between P and S waves corresponding to a distance of 2,000 km from the epicenter, Y has an interval of 8 minutes, so the earthquake occurred 5,300 km away, and Z has an interval of 12 minutes, so it is 9,000 km from the epicenter.

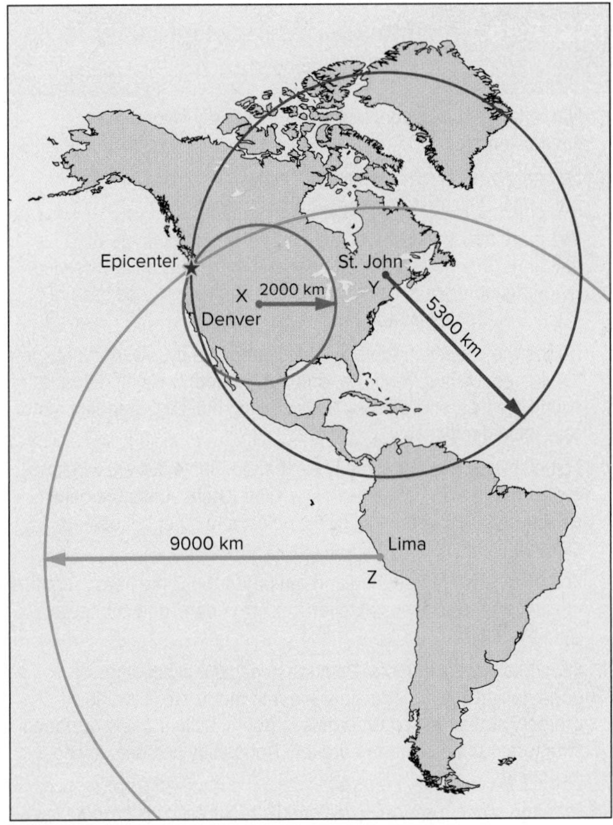

FIGURE 16.9

Locating an earthquake. The distance from each of three stations (Denver, St. John, and Lima) is determined from seismograms and the travel-time curves shown in figure 16.8. Each distance is used for the radius of a circle about the station. The location of the earthquake is just offshore of Vancouver, British Columbia, where the three circles intersect.

the earthquake, however, the first arrivals of these waves will be recorded much farther apart on the seismogram. The farther the seismic waves travel, the longer the time intervals between the arrivals of P and S waves and the more they are separated on the seismograms.

Because the time interval between the first arrivals of P and S waves increases with distance from the focus of an earthquake, this interval can be used to determine the distance from the seismograph station to a quake. The increase in the P-S interval is regular with increasing distance for several thousand kilometers and so can be graphed in a **travel-time curve**, which plots seismic-wave arrival time against distance (figure 16.8).

In practice, a station records the P and S waves from a quake, then a seismologist matches the interval between the waves to a standard travel-time curve. By reading directly from the graph, one can determine, for example, that an earthquake has occurred 5,300 kilometers (3,300 miles) away. This determination can often be made very rapidly, even while the ground is still trembling from the quake.

A single station can determine only the distance to a quake, not the direction. A circle is drawn on a globe with the center of the circle being the station and its radius the distance to the quake (figure 16.9). The scientists at the station know that the quake occurred somewhere on that circle, but from the information recorded, they are not able to tell where. With information from other stations, however, they can pinpoint the location of the quake. If three or more stations have determined the distance to a single quake, a circle is drawn for each station. If this is done on a map, the intersection of the circles locates the epicenter.

Analyses of seismograms can also indicate at what depth beneath the surface the quake occurred. Most earthquakes occur relatively close to Earth's surface, although a few occur much deeper. The maximum **depth of focus**—the distance between focus and epicenter—for earthquakes is about 670 kilometers (416 miles). Earthquakes are classified into three groups according to their depth of focus:

Shallow focus	0–70 kilometers deep
Intermediate focus	70–350 kilometers deep
Deep focus	350–670 kilometers deep

Shallow-focus earthquakes are most common; they account for 85% of total quake energy released. Intermediate- (12%) and deep- (3%) focus quakes are rarer because most deep rocks flow in a ductile manner when stressed or deformed; they are unable to store and suddenly release energy as brittle surface rocks do.

Measuring the Size of an Earthquake

The size of earthquakes is measured in two ways. One method is to find out how much and what kind of damage the quake has caused. This determines the **intensity**, which is a measure of an earthquake's effect on people and buildings. Intensities are expressed as

TABLE 16.1 Modified Mercalli Intensity Scale of 1931 (Abridged)

I. Not felt except by a very few under especially favorable circumstances.

II. Felt only by a few persons at rest, especially on upper floors of buildings. Delicately suspended objects may swing.

III. Felt quite noticeably indoors, especially on upper floors of buildings, but many people do not recognize it as an earthquake. Standing motor cars may rock slightly. Vibration like passing of truck. Duration estimated.

IV. During the day felt indoors by many, outdoors by few. At night some awakened. Dishes, windows, doors disturbed; walls made cracking sound. Sensation like heavy truck striking building. Standing motor cars rocked noticeably.

V. Felt by nearly everyone; many awakened. Some dishes, windows, etc., broken; a few instances of cracked plaster; unstable objects overturned. Disturbance of trees, poles, and other tall objects sometimes noticed. Pendulum clocks may stop.

VI. Felt by all; many frightened and run outdoors. Some heavy furniture moved; a few instances of fallen plaster or damaged chimneys. Damage slight.

VII. Everybody runs outdoors. Damage *negligible* in buildings of good design and construction; *slight* to moderate in well-built ordinary structures; *considerable* in poorly built or badly designed structures; some chimneys broken. Noticed by persons driving motor cars.

VIII. Damage *slight* in specially designed structures; *considerable* in ordinary substantial buildings with partial collapse; *great* in poorly built structures. Panel walls thrown out of frame structures. Fall of chimneys, factory stacks, columns, monuments, walls. Heavy furniture overturned. Sand and mud ejected in small amounts. Changes in well water. Persons driving motor cars disturbed.

IX. Damage *considerable* in specially designed structures; well-designed frame structures thrown out of plumb; *great* in substantial buildings, with partial collapse. Buildings shifted off foundations. Ground cracked conspicuously. Underground pipes broken.

X. Some well-built wooden structures destroyed; most masonry and frame structures destroyed with foundations; ground badly cracked. Rails bent. Considerable landslides from river banks and steep slopes. Shifted sand and mud. Water splashed (slopped) over banks.

XI. Few, if any (masonry), structures remain standing. Bridges destroyed. Broad fissures in ground. Underground pipelines completely out of service. Earth slumps and land slips in soft ground. Rails bent greatly.

XII. Damage total. Waves seen on ground surface. Lines of sight and level distorted. Objects thrown upward into the air.

Source: Wood and Neumann, Bulletin of the Seismological Society of America, 1931

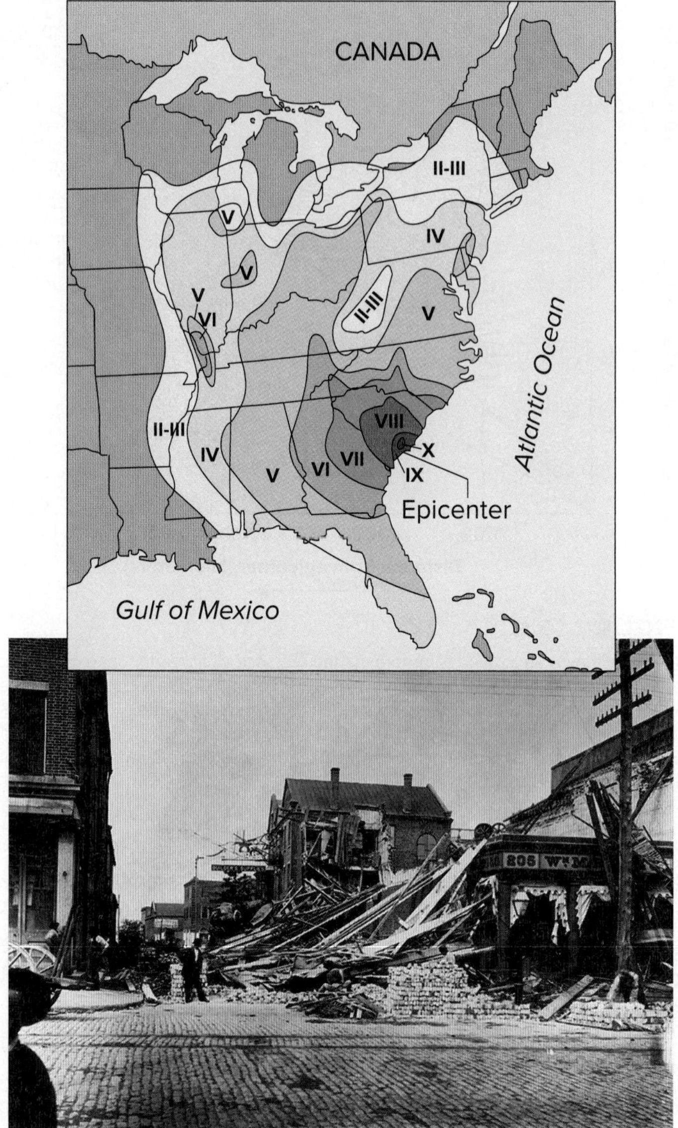

FIGURE 16.10

Zones of different intensity from the 1886 Charleston, South Carolina, earthquake. The map illustrates the general decrease in intensity with increasing distance from the epicenter, as well as the effect of different types of Earth materials. The photo shows damage in Charleston, South Carolina, from the 1886 earthquake.
Source: J. K. Hillers, U.S. Geological Survey

Roman numerals ranging from I to XII on the **modified Mercalli scale** (table 16.1); higher numbers indicate greater damage.

Although intensities are widely reported at earthquake locations throughout the world, using intensity as a measure of earthquake strength has a number of drawbacks. Because damage generally lessens with distance from a quake's epicenter, different locations report different intensities for the same earthquake (figure 16.10). Moreover, damage to buildings and other structures depends greatly on the type of geologic material on which a structure was built as well as the type of construction. Houses built on solid rock normally are damaged far less than houses built upon loose sediment, such as delta mud or bay fill. Brick and stone houses usually suffer much greater damage than wooden houses, which are somewhat flexible. Damage estimates are also subjective; people may exaggerate damage reports consciously or unconsciously. Intensity maps can be drawn for a single earthquake to show the approximate damage over a wide region (figure 16.10). Intensity maps are useful for assessing how different areas respond to seismic waves and provide valuable information for earthquake planning. But such maps cannot be drawn for uninhabited areas (the open ocean, for instance), so not all quakes can be assigned intensities. The one big advantage of intensity ratings is that no instruments are

TABLE 16.2 Comparison of Earthquake Magnitude, Description, Intensity, and Expected Annual World Occurrence

Richter Magnitude	Description	Maximum Expected Mercalli Intensity at Epicenter	Annual Expected Number
2.0	Very minor	I Usually detected only by instruments	600,000
2.0–2.9	Very minor	I–II Felt by some indoors, especially on upper floors	300,000
3.0–3.9	Minor	III Felt indoors	49,000
4.0–4.9	Light	IV–V Felt by most; slight damage	6,200
5.0–5.9	Moderate	VI–VII Felt by all; damage minor to moderate	800
6.0–6.9	Strong	VII–VIII Everyone runs outdoors; moderate to major damage	266
7.0–7.9	Major	IX–X Major damage	18
8.0 or higher	Great	X–XII Major and total damage	1 or 2

Source: U.S. Geological Survey

required, which allows seismologists to estimate the size of earthquakes that occurred before seismographs were available.

The second method of measuring the size of a quake is to calculate the amount of energy released by the quake. This method is usually done by measuring the height (amplitude) of one of the wiggles on a seismogram. The larger the quake, the more the ground vibrates and the larger the wiggle. After measuring a specific wave on a seismogram and correcting for the type of seismograph and for the distance from the quake, scientists can assign a number called the **magnitude**. It is a measure of the energy released during the earthquake.

For the past several decades, magnitude has been reported on the **Richter scale**, a numerical scale of magnitudes. The Richter scale is open-ended, meaning there are no earthquakes too large or too small to fit on the scale. The higher numbers indicate larger earthquakes. Very small earthquakes can have negative magnitudes, but these are seldom reported. The largest Richter magnitude measured so far is 8.6. Smaller earthquakes are much more common than large ones (table 16.2).

There are several methods of measuring magnitude, however. The original Richter scale applied only to shallow earthquakes in southern California. Different seismic waves (body or surface) can be measured to make the scale more useful over larger areas, so several different magnitudes are sometimes reported for a single quake. A further complication is that magnitudes calculated from seismograms tend to be inaccurate (usually too low) above magnitude 7.

A better method of calculating magnitude involves the use of the *seismic moment* of a quake, which is determined from the strength of the rock, surface area of the rupture, and the amount of rock displacement along the fault. The **moment magnitude** is the most objective way of measuring the energy released by a large earthquake. The 2011 Japan earthquake was a moment magnitude 9.0, the 1964 Alaska quake is estimated to have had a moment magnitude of 9.2, the 2004 Sumatra, Indonesia, quake a moment magnitude of 9.3, and the 1960 Chile quake 9.5. Unfortunately, the media rarely indicate which type of magnitude they are reporting, and scientists typically revise magnitudes for several weeks after a quake as they receive more information, so trying to find out the "real" magnitude of a recent quake can be confusing.

Because the Richter scale is logarithmic, the difference between two consecutive whole numbers on the scale means an increase of ten times in the amplitude of Earth's vibrations, particularly below magnitude 5. This means that if the measured amplitude of vibration for certain rocks is 1 centimeter during a magnitude-4 quake, these rocks will move 10 centimeters during a magnitude-5 quake occurring at the same location.

It has been estimated that a tenfold increase in the size of Earth vibrations is caused by an increase of roughly 32 times in terms of energy released. A quake of magnitude 5, for example, releases approximately 32 times more energy than one of magnitude 4. A magnitude-6 quake is about 1,000 times (32×32) more powerful in terms of energy released than a magnitude-4 quake. The actual energy released in earthquakes of varying magnitudes is shown in figure 16.11.

Although a seismograph is usually required to measure magnitude, this measure has many advantages over intensity as an indicator of earthquake strength. A worldwide network of standard seismograph stations now makes determining magnitude a routine matter. Eventually, a single magnitude number can be assigned to a single earthquake, whereas intensity varies for a single earthquake, depending on the amount and kind of local damage. Magnitudes can be reported for all quakes, even those in distant uninhabited areas where there is no property to affect.

Location and Size of Earthquakes in the United States

Figure 16.12 shows the locations of earthquakes that have occurred in the United States. Note that only a few localities are relatively free of earthquakes. According to the U.S. Geological Survey, earthquakes pose a significant risk to 75 million people in 39 states.

Most of the large earthquakes occur in the western states. Quakes in California, Nevada, Utah, Idaho, Montana, Washington, and other western states are related to known faults and usually

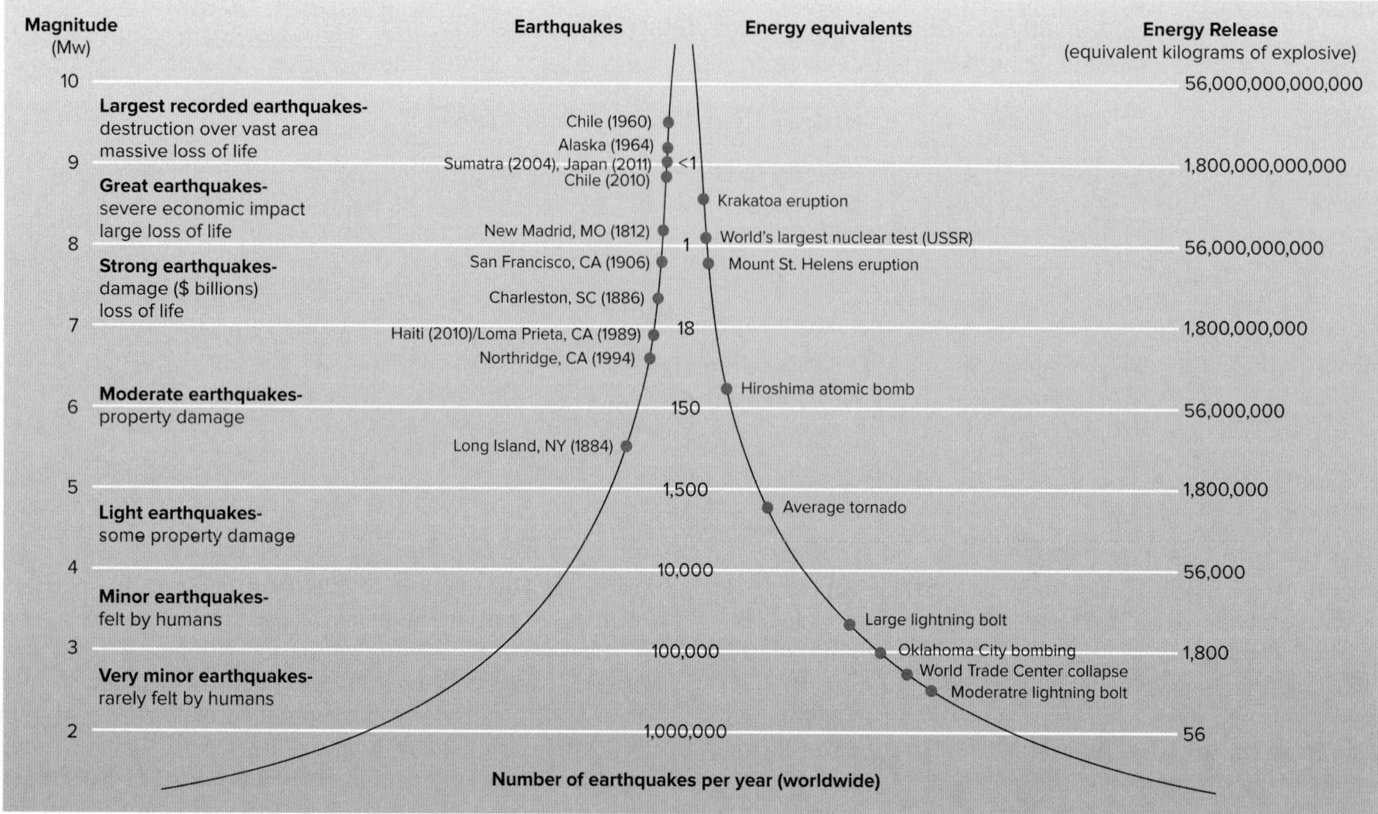

FIGURE 16.11

Diagram shows the relationship between the moment magnitude of an earthquake, the number of earthquakes per year throughout the world, and the energy released during an earthquake.
Source: IRIS Consortium (www.iris.edu/hq/)

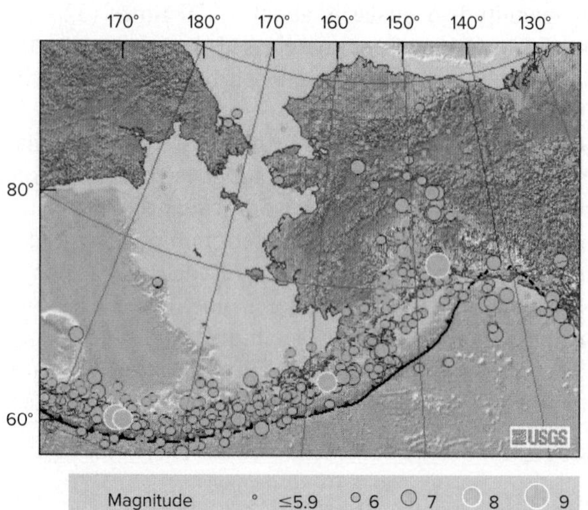

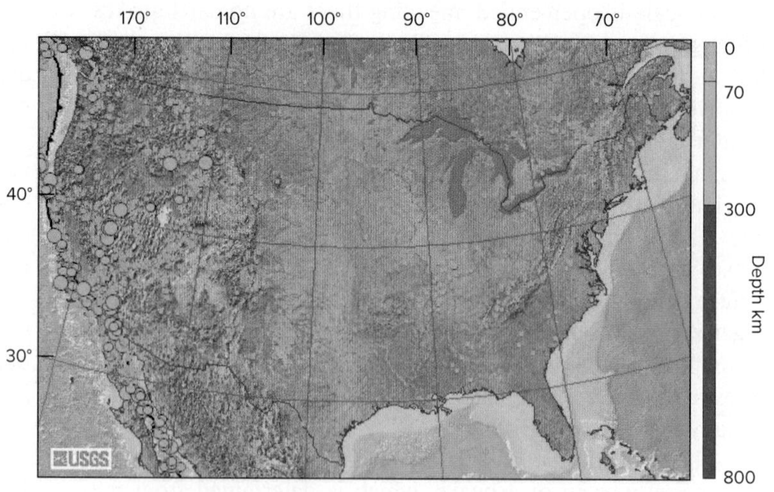

FIGURE 16.12

Earthquakes that have occurred in the United States and southern Canada.
Source: U.S. Geological Survey

(but not always) involve surface rupture of the ground. Earthquakes in Alaska occur mainly below the Aleutian Islands, where the Pacific plate is converging with and being subducted beneath the North American plate.

Earthquakes east of the Rocky Mountains are rarer and generally smaller and deeper than earthquakes in the western United States. They usually are not associated with surface rupture. The quakes may be occurring on the deeply buried, relatively inactive faults of old *divergent plate boundaries* and *failed rifts* (*aulacogens*), both of which are described in chapter 19.

Although large quakes are extremely rare in the central and eastern United States, when they do occur, they can be very destructive and widely felt, because Earth's crust is older, cooler, and more brittle in the east than in the west, and seismic waves travel more efficiently. The Saint Lawrence River Valley along the Canadian border has had several intensity IX and X earthquakes, most recently in 1944. Plymouth, Massachusetts, had an intensity IX quake in 1638, and a quake of intensity VIII occurred in 1775 near Cambridge, Massachusetts. In 1929, in Attica, New York, an earthquake of intensity IX knocked over 250 chimneys. A series of quakes (intensity XI) that occurred near New Madrid, Missouri, in the winter of 1811-1812 were the most widely felt earthquakes to occur in North America in recorded history. The quakes knocked over chimneys as far away as Richmond, Virginia, and rang church bells in Charleston, South Carolina.

The 1886 quake in Charleston, South Carolina, (intensity X) was felt throughout almost half the United States (figure 16.10) and killed sixty people; it was sharply felt in New York City. Moderate quakes hit southern Illinois in 2005 and again in 2008 and were felt throughout the Midwest, damaging buildings in Louisville, Kentucky. The East Coast of the United States was jolted by a magnitude 5.8 earthquake on August 23, 2011, that was located in central Virginia, only 84 miles southwest of Washington, D.C. The White House, Capitol, and monuments across National Mall were evacuated as were office buildings in New York City and as far away as Columbus, Ohio. Approximately $300 million in damage was caused by the earthquake, including cracks in the Washington Monument and landmark National Cathedral and collapse of buildings near the epicenter in Virginia.

Geologists have mapped regions of seismic risk in the United States (figure 16.13) and elsewhere throughout the world, primarily on the assumption that large earthquakes will occur in the future in places where they have occurred in the past. However, the seismic risk map does not include the rising number of earthquakes in Oklahoma, and neighboring states, caused by deep injection of wastewater from oil and gas well drilling operations. The volume of wastewater injected peaked in 2015, causing more than a thousand earthquakes in that year alone as the increased water pressure reactivated old basement faults. In 2016, the largest human-induced earthquake in Oklahoma (Mw5.8) struck near Pawnee and caused moderate to severe damage in town and damaged a courthouse 300 miles away in Kansas; the quake was felt as far away as Fargo, North Dakota, and San Antonio, Texas.

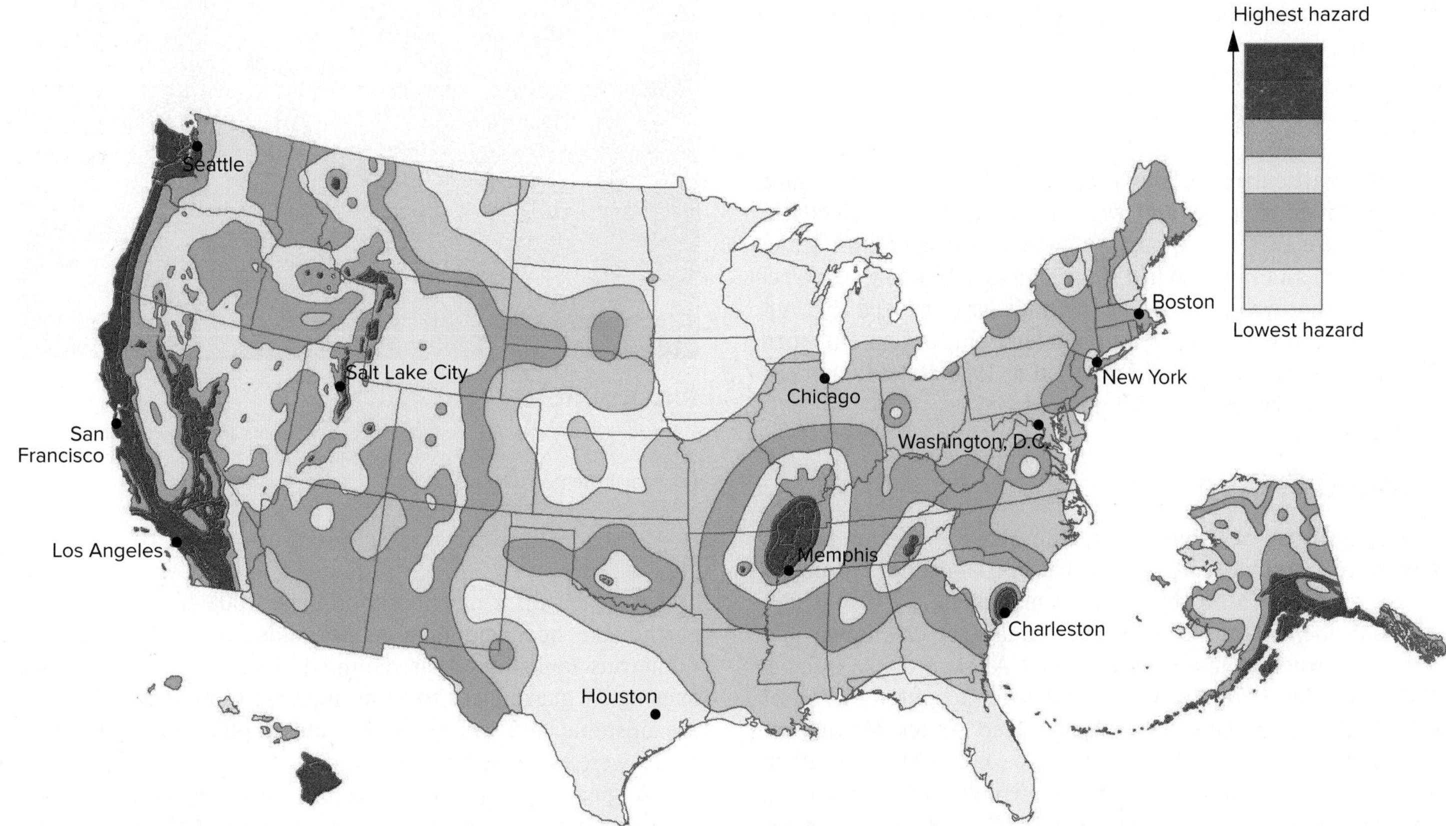

FIGURE 16.13

Map of seismic risk in the United States based on the expected amount of ground shaking and damage.
Source: USGS-National Seismic Hazard Mapping Project

16.4 EARTHQUAKE-RELATED HAZARDS

Hazards related to earthquakes include ground motion associated with shaking; fire; ground failure such as landslides, liquefaction, and fault rupture and scarps; aftershocks to already weakened buildings; and tsunami. The amount of damage is related to the size of the earthquake, the length of time shaking occurs, the type of ground—hard rock or soft sediments—and the type of building construction.

Ground Motion

Ground motion is the trembling and shaking of the land that can cause buildings to vibrate. During small quakes, windows and walls may crack from such vibration. In a very large quake, the ground motion may be visible. It can be strong enough to topple large structures such as office and apartment buildings and bridges (figure 16.14). Most people injured or killed in an earthquake are hit by falling debris from buildings. Because proper building construction can greatly reduce the dangers, building codes need to be both strict and strictly enforced in earthquake-prone areas. Much of the damage and loss of life in the 2010 Haiti earthquake was due to poorly constructed buildings that did not meet building codes. As we have seen, the location of buildings also needs to be controlled; buildings built on soft sediment are damaged more than buildings on hard rock.

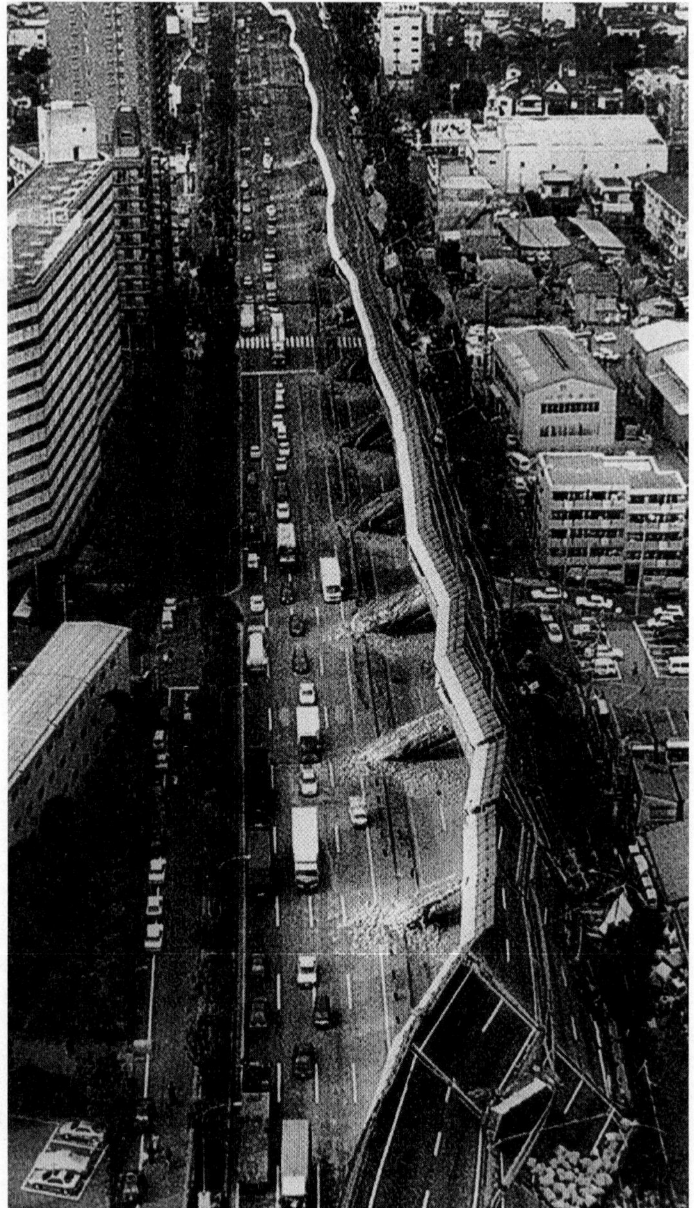

FIGURE 16.14

Elevated highway knocked over by a strong horizontal jolt during the 1995 Kobe, Japan, earthquake. Damage exceeded $400 billion and destroyed or severely damaged more than 88,000 buildings.
©Asahi Shimbun/AFP/Getty Images

Fire

Fire is a particularly serious problem just after an earthquake because of broken gas and water mains and fallen electrical wires (figure 16.15). Although fire was the cause of most of the damage to San Francisco in 1906, changes in building construction and improved fire-fighting methods have reduced (but not eliminated) the fire danger to modern cities. The stubborn Marina district fires in San Francisco in 1989 attest to modern dangers of broken gas and water mains.

Ground Failure

Landslides can be triggered by the shaking of the ground (figure 16.16). The 1959 Madison Canyon landslide in Montana was triggered by a nearby quake of magnitude 7.7. Landslides and subsidence caused extensive damage in downtown and suburban Anchorage during the 1964 Alaskan quake (magnitude 8.6). The 1970 Peruvian earthquake (magnitude 7.75) set off thousands of landslides in the steep Andes Mountains, burying more than 17,000 people (see box 9.1). In 1920 in China, over 100,000 people living in hollowed-out caves in cliffs of loess (described in chapter 13) were killed when a quake collapsed the cliffs. The 2001 El Salvador quake resulted in nearly 500 landslides, the largest of which occurred in Santa Tecla where 1,200 people were missing after tons of soil and rock fell on a neighborhood. The 2008 Sichuan, China, earthquake occurred in a mountainous area and triggered numerous landslides that dammed rivers in steep canyons, causing 34 quake lakes to form. Because these landslide dams are unstable and can result in catastrophic flooding, entire towns were evacuated downstream.

A special type of ground failure caused by earthquakes is *liquefaction*. This occurs when a water-saturated soil or sediment turns from a solid to a liquid as a result of earthquake shaking. Liquefaction may occur several minutes after an earthquake, causing buildings to sink (figure 16.17) and underground tanks

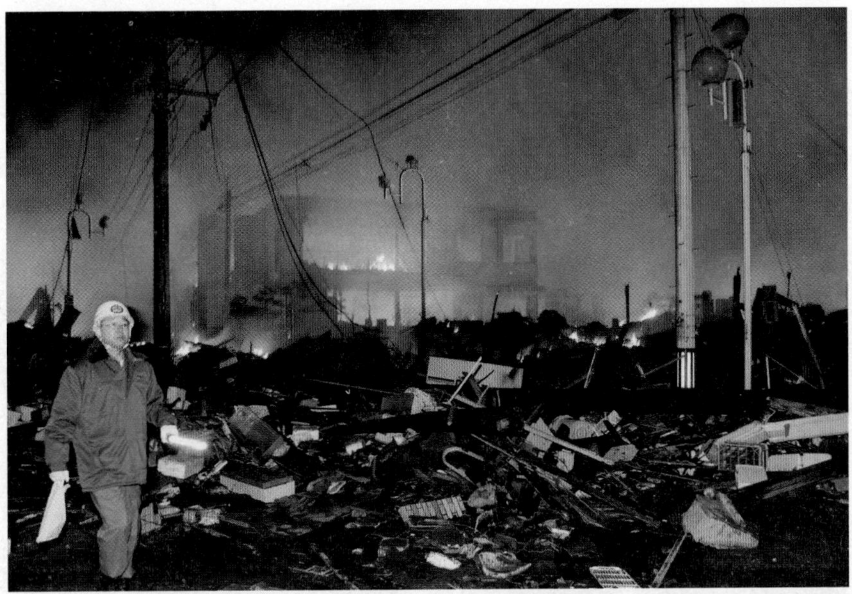

A

B

FIGURE 16.15

Fires caused by earthquakes. (*A*) Fires burn among the rubble in Iwaki city, Fukushima Prefecture, Japan, after the largest earthquake in Japan's recorded history struck the eastern coasts on March 11, 2011. (*B*) People on Sacramento Street watch the smoke rise from fires caused by the 1906 San Francisco earthquake; most of the damage from the earthquake was caused by fires that burned for days.
(A) ©Kyodo News/AP Images; (B) ©Arnold Genthe/AP Images

to float as once-solid sediment flows like water. Liquefaction was responsible for much of the damage in the 1989 Loma Prieta quake and contributed to the damage in the 1906 San Francisco, 1964 and 2002 Alaska, 1985 Mexico City, and 1995 Kobe (Japan) quakes.

Permanent displacement of the land surface may be the result of movement along a fault. Rocks can move vertically, those on one side of a fault rising while those on the other side drop. Rocks can also move horizontally–those on one side of a fault sliding past those on the other side. Diagonal movement with both vertical and horizontal components can also occur during a single quake (figure 16.2). Such movement can affect huge areas, although the displacement in a single earthquake seldom exceeds 8 meters. The trace of a fault on Earth's surface may appear as a low cliff, called a *scarp*, or as a closed tear in the ground (figure 16.18). In rare instances, small cracks open during a quake (but not to the extent that Hollywood films often portray). Ground displacement during quakes can tear apart buildings, roads, and pipelines that cross faults. Sudden subsidence of land near the sea can cause flooding and drownings. This was tragically seen during the 2011 Tohoku, Japan, earthquake and tsunami where the coast dropped 1 meter, allowing the tsunami waves to overtop seawalls built to protect low-lying communities.

FIGURE 16.16

Landslide in Redcliffs triggered by a magnitude 6.3 earthquake that struck Christchurch, New Zealand, on February 27, 2011.
©Torsten Blackwood/AFP/Getty Images

A

B

FIGURE 16.18

Ground displacement caused by the magnitude 7.8 Kaikora earthquake that struck the south island of New Zealand on November 13–14, 2016. (*A*) Geologists examining the scarp formed along the Papatea Fault that here uplifted the sea floor nearly three meters. (*B*) Tearing and offset of the ground along the Kekerengu Fault rupture. The right-lateral oblique slip fault offset the driveway and pulled the foundation out from under the house. To view spectacular drone footage of the incredible ground displacement caused by the New Zealand earthquake, go to https://www.youtube.com/watch?v=U3H8wlzXGYE, https://www.youtube.com/watch?v=ASqmadst42, and https://www.youtube.com/watch?v=Lno8Rpbe57c.
(A) ©Chris Madugo; (B) ©Alex Perrottet/www.rnz.co.nz

FIGURE 16.17

Liquefaction of soil by a 1964 quake in Niigata, Japan, caused earthquake-resistant apartment buildings to topple over intact.
Source: NOAA National Geophysical Data Center

IN GREATER DEPTH 16.1

Earthquake Engineering

Damage and loss of life can be substantially reduced by siting structures on solid bedrock or dense soils and by building structures that adhere to strict seismic building codes. In the catastrophic 7.0 magnitude earthquake that struck Haiti on January 10, 2010, 222,570 people lost their lives and 300,572 were injured when poorly constructed buildings collapsed (box figure 1*A*). In contrast, earthquake-resistant structures and enforcement of seismic building codes in Chile resulted in far less damage and loss of life when a magnitude 8.8 quake, releasing 500 times more energy than the Haiti quake, struck one month later on February 10, 2010.

Buildings that are constructed of strong, flexible, and light materials such as steel, wood, and reinforced concrete (strengthened by steel rebar) are the most resistant to damage by seismic shaking. Structures built with unreinforced concrete block or brick, which are only as strong as the mortar holding the blocks and bricks together, tend to lack flexibility and crumble in large earthquakes. This was the case when half the town of Amatrice, Italy, was leveled by a shallow, magnitude 6.2 earthquake that struck in the early morning hours on August 24, 2016, while people were sleeping. Unreinforced stone homes and buildings dating back to medieval times crumbled, killing 234 people, many of whom were children on holiday visiting their grandparents. This is in contrast to a town closer to the epicenter, Norcia, that saw no loss of life because it had seismically retrofitted buildings after being struck by a moderate earthquake twenty years earlier. During moderate-sized earthquakes, many houses lose their chimneys or brick facades. Buildings with heavy roofs made of tile or slate also tend to collapse. Engineers who examined the damaged buildings after the devastating Haiti earthquake noted that many of the buildings that collapsed were built of unreinforced or poorly reinforced concrete and with concrete with low cement content. In addition, many of the walls and slab floors and roofs were not connected or tied together. The shaking from the earthquake caused many of these buildings to crumble or topple, resulting in the tremendous number of deaths and injuries.

The 1985 Mexico City earthquake, which killed five thousand people and caused over $5 billion in damage, is a classic example of the effect soft soils have on the amplification of earthquake waves. The ground shaking was relatively mild in most parts of Mexico, but it was amplified in some buildings and by the lake-bed sediments beneath Mexico City. Most buildings are not rigid but slightly flexible; they sway gently like large pendulums when struck by wind or seismic waves. The time necessary for a single back-and-forth oscillation is called a period, and it varies with a building's height and mass. Small buildings have a shorter period of vibration than tall buildings. Rock and sediment vibrate the same way, with periods that vary with the size and density of the rock and sediment.

When earthquake waves struck Mexico City, many were vibrating with a two-second period. The body of lake sediment beneath the city has a natural period of two seconds also, so the wave motion was amplified by the sediments. This type of amplification, or *resonance*, occurs when you push a child on a swing—if you push gently in time with the swing's natural period, the swing goes higher and higher. The water-saturated sediment began to move like a sloshing waterbed. The ground moved back and forth by 40 centimeters (15.7 inches) every two seconds, and it did this fifteen to twenty times.

A

B

BOX 16.1 ■ FIGURE 1

(*A*) A school built of unreinforced concrete collapsed during the 2010 Haiti earthquake. (*B*) This fifteen-story building collapsed completely during the 1985 Mexico City earthquake, crushing all its occupants as its reinforced-concrete floors "pancaked" together. *(A) ©Andrew M. Freed/Purdue University; (B) Source: M. Celebi, U.S. Geological Survey*

This shaking was devastating to buildings with a natural two-second period (generally those five to twenty stories high), and several hundred buildings in this size range collapsed as they further resonated with the shaking (box figure 1*B*). Many structures weakened by the main shock collapsed during the large aftershock. On September 19, 2017, exactly 32 years from the date of the 1985 earthquake, Mexico City was shaken by a magnitude 7.1 earthquake that caused the collapse of more than 40 buildings and killed 370 people. Because the epicenter of this quake was closer, it released shorter period waves that caused poorly constructed buildings less than five stories to collapse.

Aftershocks

Aftershocks are small earthquakes that follow the main shock. Although aftershocks are smaller than the main quake, they can cause considerable damage, particularly to structures weakened by the powerful main shock. A long period of aftershocks can be extremely unsettling to people who have lived through the main shock. *Foreshocks* are small quakes that precede a main shock. They are usually less common and less damaging than aftershocks but can sometimes be used to help predict large quakes (although not all large quakes have foreshocks). The magnitude 9.0 earthquake that struck Japan on March 11, 2011, was preceded by a number of large foreshocks and hundreds of aftershocks. Unfortunately, the earlier earthquakes were not recognized as foreshocks until the main quake occurred two days later.

Tsunami

The sudden movement of the sea floor upward or downward during a submarine earthquake can generate very large sea waves, popularly called "tidal waves." Because the ocean tides have nothing to do with generating these huge waves, the Japanese term *tsunami* is preferred by geologists. **Tsunami** are also called **seismic sea waves**. They usually are caused by great earthquakes (magnitude 8+) that disturb the sea floor, but they also result from submarine landslides or volcanic explosions. When a large section of sea floor suddenly rises or falls during a quake, all the water over the moving area is lifted or dropped for an instant. As the water returns to sea level, it sets up long, low waves that spread very rapidly over the ocean (figure 16.19). Because vertical motion of the sea floor is most conducive to the formation

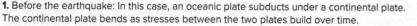

1. Before the earthquake: In this case, an oceanic plate subducts under a continental plate. The continental plate bends as stresses between the two plates build over time.

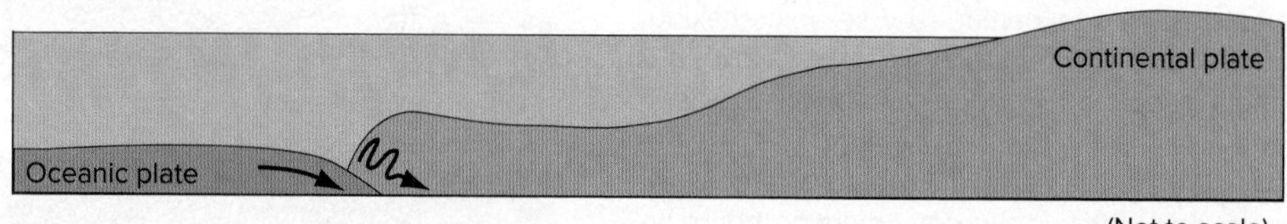

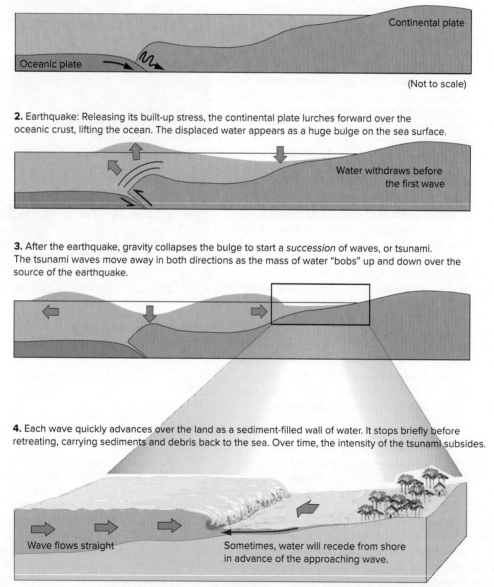

2. Earthquake: Releasing its built-up stress, the continental plate lurches forward over the oceanic crust, lifting the ocean. The displaced water appears as a huge bulge on the sea surface.

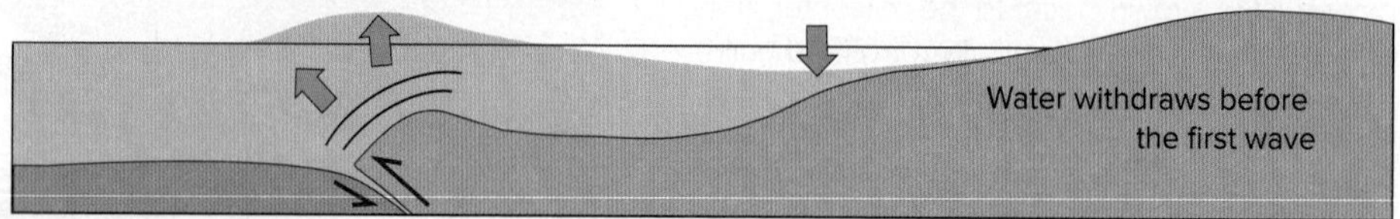

3. After the earthquake, gravity collapses the bulge to start a *succession* of waves, or tsunami. The tsunami waves move away in both directions as the mass of water "bobs" up and down over the source of the earthquake.

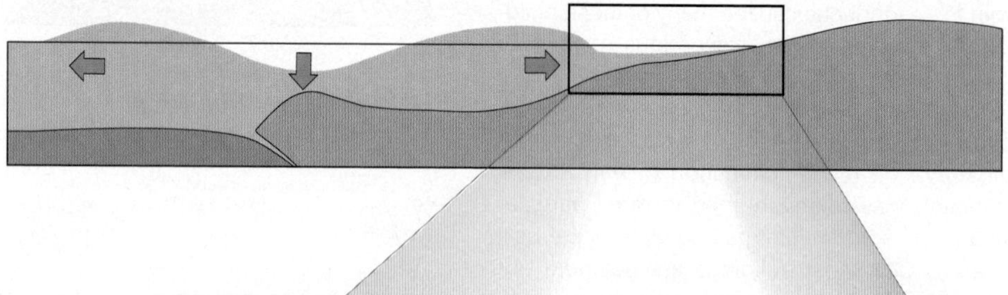

4. Each wave quickly advances over the land as a sediment-filled wall of water. It stops briefly before retreating, carrying sediments and debris back to the sea. Over time, the intensity of the tsunami subsides.

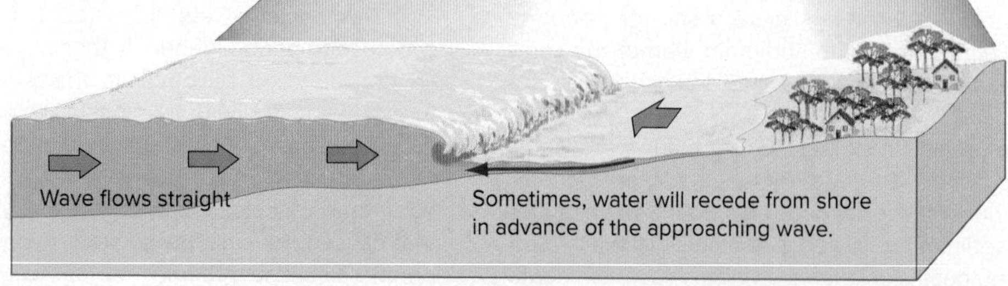

FIGURE 16.19

Tsunami waves are generated by a submarine earthquake that displaces the sea floor and water column above. Long, low waves are formed above the displaced sea floor to compensate for the momentary rise in sea level and spread very rapidly (at the speed of a jetliner) in the deep ocean. In shallower water, the tsunami slows to highway speeds and builds in height until it breaks and crashes onto the shore with incredible force, causing destructive flooding along low-lying coastal areas.

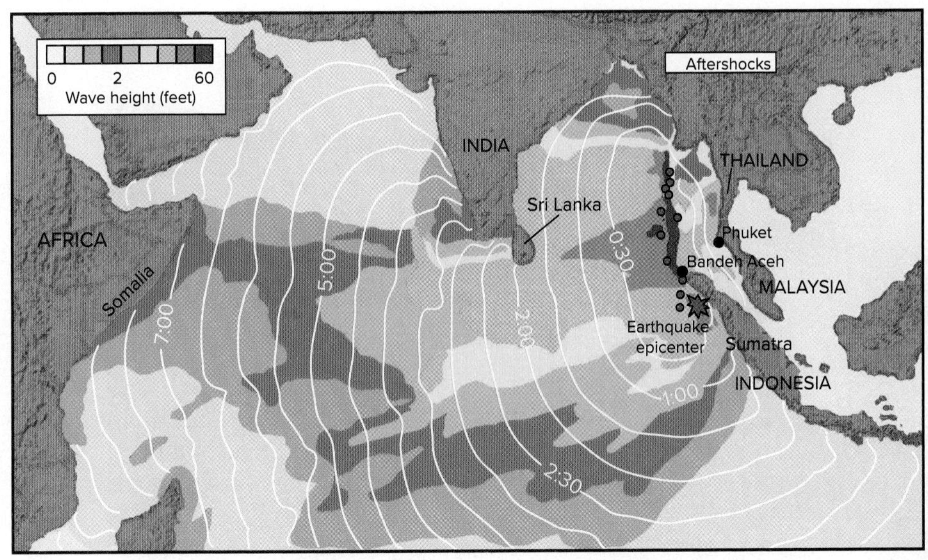

A

B

FIGURE 16.20

(*A*) Tsunami travel time and wave height (feet) map from December 26, 2004, Sumatra earthquake. The arrival times (hours) of the first wave of the tsunami are shown by the white contour lines. Bandeh Aceh was hit by the tsunami fifteen minutes after the earthquake, and it took seven hours to travel across the Indian Ocean to reach the coast of Africa. (*B*) Many people are surprised while others run for safety as the tsunami caused by the December 26, 2004, 9.3 magnitude Sumatra earthquake comes crashing onshore in Koh Raya, Thailand.
(B) ©John Russell/AFP/Getty Images

of tsunami, most are associated with subduction zone earthquakes, which tend to be some of the strongest quakes.

A tsunami is unlike an ordinary water wave on the sea surface. A large wind-generated wave may have a wavelength of 400 meters and be moving in deep water at a speed of 90 kilometers per hour (55 miles per hour). The wave height when it breaks on shore may be only 0.6 to 3 meters, although in the middle of hurricanes, the waves can be more than 15 meters (49 feet) high. A tsunami, however, may have a wavelength of 160 kilometers and may be moving more than 800 kilometers per hour (500 miles per hour). In deep water, the wave height may be only 0.6 to 2 meters, but near shore, the tsunami may peak at heights of 15 to 30 meters. This great increase in wave height near shore is caused by bottom topography; only a few localities have the combination of gently sloping offshore shelf and funnel-shaped bay that force tsunami to awesome heights (the record height was 85 meters in 1971 in the Ryukyu islands south of Japan). Along most coastlines tsunami height is very small.

Although the speed of the wave slows drastically as it moves through shallow water, a tsunami can still hit some shores as a very large, very fast wave. Because of its extremely long wavelength, a tsunami does not withdraw quickly as normal waves do. The water keeps on rising for five to ten minutes, causing great flooding before the wave withdraws. The long duration and great height of a tsunami can bring widespread destruction to the entire shore zone, as was witnessed in the devastating 2004 Indian Ocean and 2011 Tohoku, Japan, tsunami.

The most devastating tsunami in recorded history occurred on December 26, 2004, in the Indian Ocean. The tsunami was triggered by a 9.3-magnitude earthquake that struck off the northern coast of Sumatra, Indonesia, as the Indian plate shifted under the Eurasian plate, causing a subduction zone (*megathrust*) earthquake. After the 1960 9.5-magnitude quake in Chile, this is the second-largest earthquake ever recorded in the world. The shift in the sea floor caused a tsunami that hit the towns of Bandeh Aceh and Meulaboh, Sumatra, minutes later. The tsunami traveled across the Indian Ocean at the speed of a jetliner and caused massive deaths in low-lying coastal communities (figure 16.20*A*) that were filled with tourists during the Christmas holiday. The Indian Ocean tsunami caused more casualties than any other tsunami in recorded history. In total, more than 220,272 people were killed and over 20,000 were listed as missing. Horrific accounts of the tsunami coming on shore were told by countless people caught in the low-lying coastal areas (figure 16.20*B*). As in many large tsunami, the sea first retreated, luring many for a closer look at exposed coral reefs and stranded fish on the beach. Minutes later, the first wave struck with tremendous force, causing massive loss of life and damage. The first wave retreated with the same force it had when it came onshore, and many desperately hung onto trees so they were not swept out to sea. Then, successive waves struck with the same incredible force and caused additional damage and loss of life. At the time, no tsunami-warning system, or other means of warning communities of the possibility of a tsunami, was in place in the Indian Ocean as it is in the Pacific Ocean.

On March 11, 2011, a devastating tsunami caused by the magnitude 9.0 Tohoku, Japan, earthquake struck the northeast coast of Japan (figure 16.21), resulting in over $300 billion in losses and flooding at the Fukishima nuclear power plant, which resulted in three meltdowns and the release of radioactivity. The death toll was nearly 23,000—much less than that from the 2004 Indian

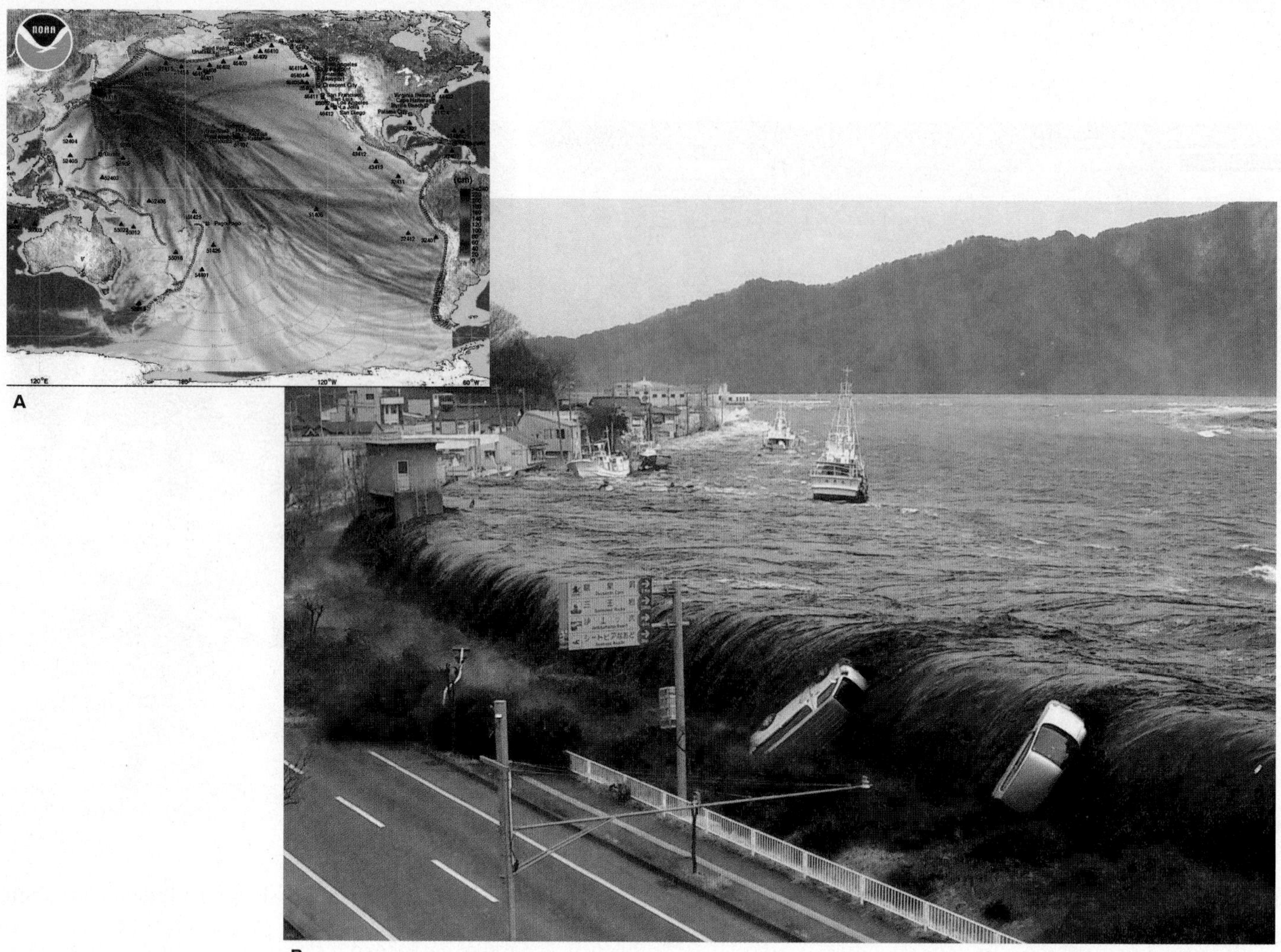

FIGURE 16.21

(*A*) Tsunami travel time and wave height map from the magnitude 9.0 Tohoku, Japan, earthquake that struck northeastern Japan on March 11, 2011. The tsunami arrived along the northeastern coast of Japan within thirty minutes of the megathrust earthquake, and it reached Hawaii seven hours later, causing tens of millions of dollars in damage. The tsunami eventually reached the west coast of the United States nine hours later, causing $48 million in damage and taking the life of one person in northern California who ventured too close to the ocean to photograph the wave and was swept away. (*B*) Photo of tsunami overtopping seawall and crashing onshore at Miyako, Japan.
(A) Source: NOAA; (B) ©JIJI PRESS/AFP/Getty Images

Ocean tsunami because of the earthquake early-warning system in Japan and the evacuation drills practiced by those who live in low-lying coastal communities. Many of the low-lying coastal towns built seawalls to protect against the threat of tsunami waves, but the seawalls gave a false sense of security to many who did not heed the warning sirens and run to higher ground. Many tried to drive to higher ground and were caught in traffic jams and could not escape the tremendous power of the unexpectedly high tsunami wave. A magnitude 7.1 earthquake that occurred two days before the main quake and caused a smaller tsunami, which the seawalls held back, led residents to think that another larger quake would not follow so soon or generate a tsunami of historic heights that would overtop the seawalls (figure 16.21*B*). What was not realized at the time was that the magnitude 7.1 quake was a foreshock to the magnitude 9.0 megathrust earthquake that would occur just 72 kilometers (45 miles) offshore two days later. Because of the size of the fault rupture and thrusting upward of the sea floor in deep water, a gigantic tsunami wave, up to 38.9 meters in height, crashed ashore only thirty minutes after the earthquake, leaving towns in piles of rubble and drowning those who could not escape fast enough to higher ground.

Other historic tsunami include the tsunami formed by the 1960 Chilean quake that crossed the Pacific Ocean and did extensive damage in Japan. Another destructive tsunami was generated April 1, 1946, by a 7.3-magnitude earthquake offshore from Alaska. It devastated the city of Hilo, Hawaii, causing 159 deaths. The tsunami following the 1964 Alaskan quake drowned twelve people in Crescent City, California, a small coastal town near the Oregon border. Wave damage near an epicenter can be awesome. The 1946 Alaska tsunami destroyed the Scotch Cap lighthouse on nearby Unimak Island, sweeping it off its concrete base, which was 10 meters above sea level, and killing its five occupants. The wave also swept away a radio tower, whose base was 31 meters above sea level.

16.5 WORLD DISTRIBUTION OF EARTHQUAKES

Most earthquakes are concentrated in narrow geographic belts (figure 16.22*A*), although *some* earthquakes have occurred in most regions on Earth. The boundaries of plates in the plate-tectonic theory are defined by these earthquake belts (figure 16.22*B*). The

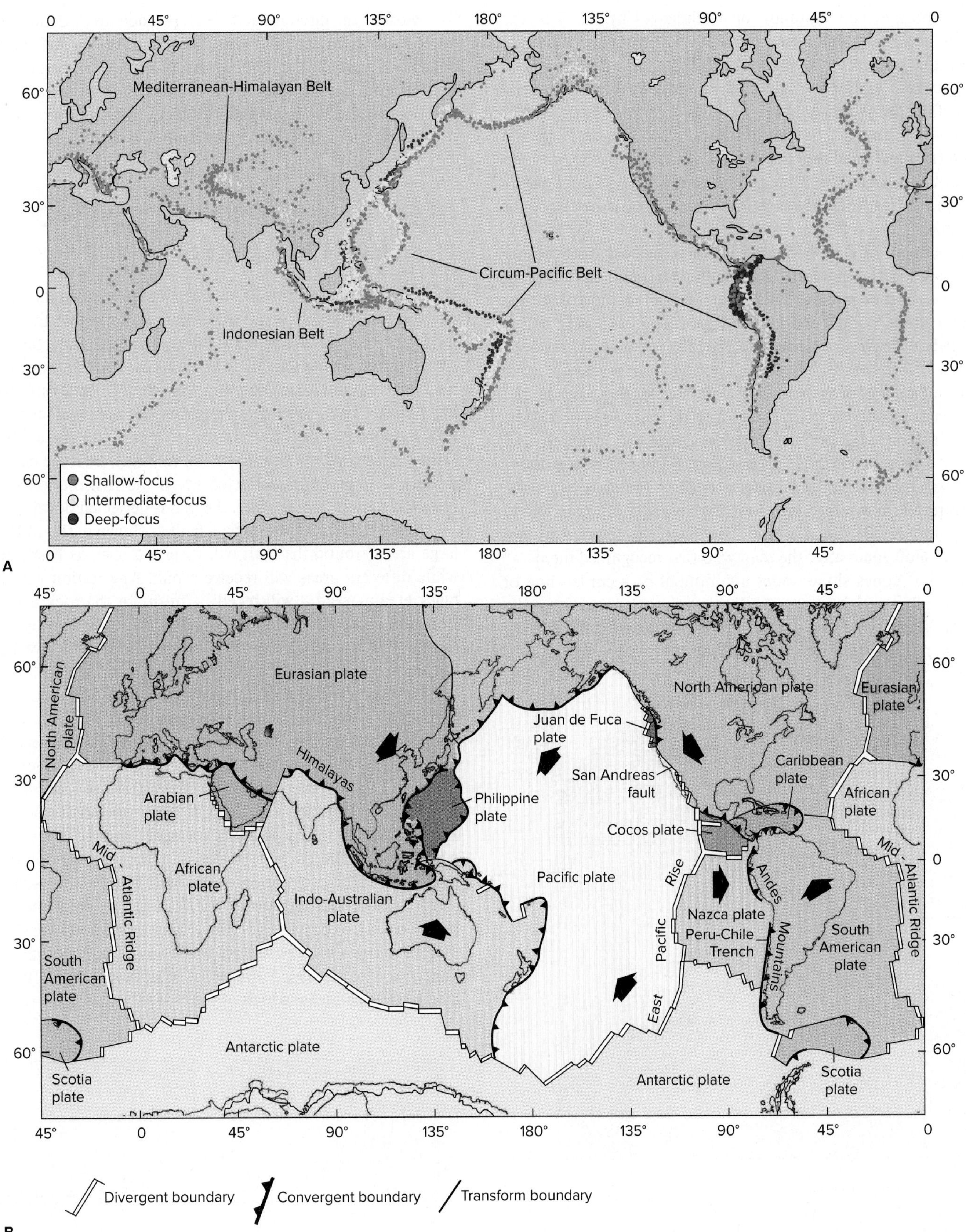

FIGURE 16.22

(*A*) Distribution of shallow-, intermediate-, and deep-focus earthquakes. (*B*) The major plates of the world in the theory of plate tectonics. Compare the locations of plate boundaries with earthquake locations shown in figure 16.22*A*. Double lines show diverging plate boundaries; single lines show transform boundaries. Heavy lines with triangles show converging boundaries; triangles point down subduction zone.
Source: U.S. Geological Survey, W. Hamilton

most important concentration of earthquakes by far is in the **circum-Pacific belt**, which encircles the rim of the Pacific Ocean. Within this belt occur approximately 80% of the world's shallow-focus quakes, 90% of the intermediate-focus quakes, and nearly 100% of the deep-focus quakes.

Another major concentration of earthquakes is in the **Mediterranean-Himalayan belt**, which runs through the Mediterranean Sea, crosses the Mideast and the Himalayas, and passes through the East Indies to meet the circum-Pacific belt north of Australia.

A number of shallow-focus earthquakes occur in two other significant locations on Earth. One is along the summit or crest of the *mid-oceanic ridge*, a huge underwater mountain range that runs through all the world's oceans (see figure 1.8 and chapter 18). A few earthquakes have also been recorded in isolated spots usually associated with basaltic volcanoes, such as those of Hawaii.

In most parts of the circum-Pacific belt, earthquakes, andesitic volcanoes, and *oceanic trenches* (see chapter 18) appear to be closely associated. Careful determination of the locations and depths of focus of earthquakes has revealed the existence of distinct *earthquake zones* that begin at oceanic trenches and slope landward and downward into Earth at an angle of about 30° to 60° (figure 16.23). Such zones of inclined seismic activity are called **Benioff zones** after the man who first recognized them.

Benioff zones slope under a continent or a curved line of islands called an **island arc**. Andesitic volcanoes may form the islands of the island arc, or they may be found near the edge of a continent that overlies a Benioff zone.

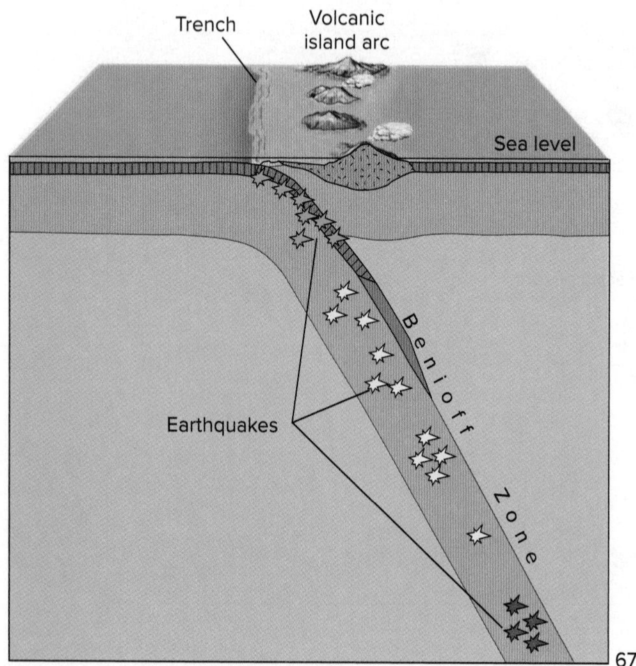

FIGURE 16.23

A Benioff zone of earthquakes begins at an oceanic trench and dips under a continent (such as South America) or a volcanic island arc. Upper part of Benioff zone may extend to a depth of 670 kilometers.

Most of the circum-Pacific belt is made up of Benioff zones associated in this manner with oceanic trenches and andesitic volcanoes. Parts of the Mediterranean-Himalayan belt represent Benioff zones, too, notably in the eastern Mediterranean Sea and the East Indies. Essentially all the world's intermediate- and deep-focus earthquakes occur in Benioff zones.

16.6 FIRST-MOTION STUDIES OF EARTHQUAKES

By studying seismograms of an earthquake on a distant fault, geologists can tell which way rocks moved along that fault. First-motion studies played an important role in determining the overall sense of movement along plate boundaries. Rock motion is determined by examining seismograms from many locations surrounding a quake. Each seismograph station can tell whether the first rock motion recorded there was a push or a pull (figure 16.24). If the rock moved toward the station (a push), then the pen drawing the seismogram is deflected up. If the first motion is away from the station (a pull), then the pen is deflected down.

If an earthquake occurs on a fault as shown in figure 16.25*A*, large areas around the fault will receive a push as first motion, while different areas will receive a pull. Any station within the black area marked *A* will receive a push, for the rock is moving from the epicenter toward those stations, as shown by the arrows in the figure. All stations in area *C* will also receive a push, but areas *B* and *D* will record a pull as the first motion.

In figure 16.25*B*, you can see the same pattern of pushes and pulls, but in this case, the pattern is caused by a fault with a different orientation. Either fault can cause the same pattern, if the rock moves in the direction shown by the arrows. In other words, there are two possible solutions to any pattern of first motions.

If the orientation of the fault trace on Earth's surface is known, as it is for most faults on land (and for a few on the ocean floor), the correct choice of the two solutions can be made. But if the orientation of the fault is not known—as is the usual case for earthquakes at sea or at great depth—the choice between the two possible solutions can be difficult. One solution may be more *likely*, based on the study of topography, other faults, or aftershocks. Patterns of aftershocks after the main quake often delineate which of the two solutions is correct.

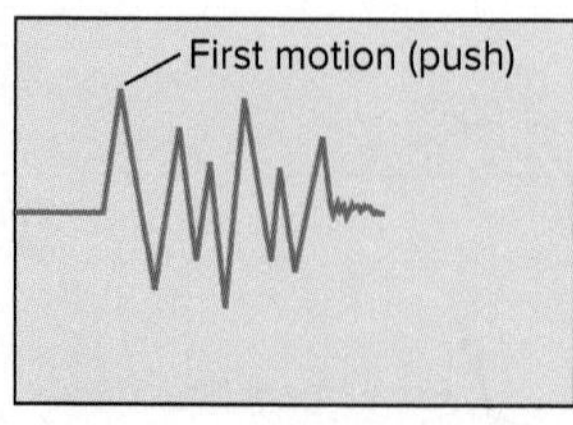

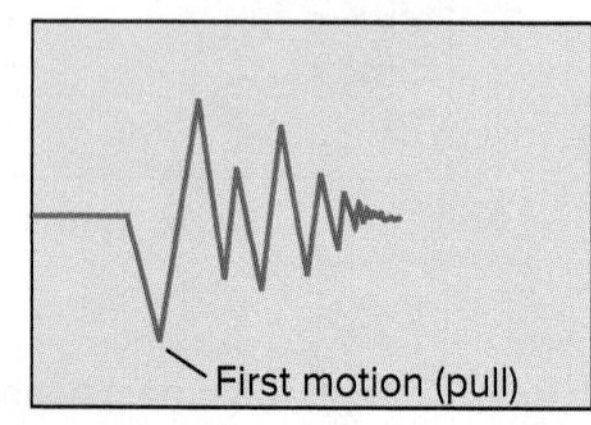

FIGURE 16.24

Seismograms showing how first horizontal motions of rocks along a fault are determined. (*A*) If the first motion is a push (from the epicenter to the seismograph station), the seismogram trace is deflected upward. (*B*) If it is a pull (away from the station), the deflection is downward.

EARTH SYSTEMS 16.2

Darwin and Chilean Earthquakes

Coastal Chile has a long history of very large earthquakes. As shown in figure 16.11, the largest earthquake on Earth since seismometers have been used to determine magnitude was the 1960 Chilean quake. This was a 9.5 moment magnitude earthquake. Chile's February 27, 2010, Maule earthquake of moment magnitude 8.8 was Earth's fifth-highest magnitude earthquake. Box figure 1 shows a portion of the Chilean coast that rose from below sea level during the 2010 quake. It is pale colored because of decaying algae, seaweed, and other intertidal life. It's likely that the photographer had to endure the stench of rotting organisms.

Flash back over a century and a half ago: In 1835, during the fourth year of the voyage of the HMS *Beagle*, the ship and its crew made an extended stay in Chile for resupply and refurbishment. Charles Darwin was the expedition's not yet famous naturalist. His observations and collection of organisms were part of the basis for his theory of biological evolution. *On the Origin of Species* would be published in 1859 and revolutionize biology.

Darwin spent his time in Chile wandering around the countryside making observations and collecting plant and animal specimens. On February 20, 1835 (almost exactly 175 years before the 2010 quake), Darwin was resting on the ground in a wooded area when violent shaking began. Ground motion was so strong that he could not stand up. The earthquake lasted for two minutes. Darwin noted in his journal that it felt a lot longer. Nearby towns were destroyed.

BOX 16.2 ■ FIGURE 1

The uplifted sea floor caused by the magnitude 8.8 Maule earthquake that struck Chile on February 27, 2010. The white coating on the rocks is from algae that have died after being uplifted more than a meter; the original coastline is visible along the base of the cliff.
Photo ©Marcelo Farias, Universidad de Chile

A day or so later, Darwin became aware of a stench along the shoreline. This was due to the putrefaction of dead mussels and other intertidal organisms that were now above sea level (as in box figure 1). Darwin reasoned that either sea level had dropped or land had risen. It was unlikely that the entire Pacific Ocean had suddenly been lowered, so he inferred that land below sea level must have risen. Clearly, the sudden rise of land was associated with the great earthquake that he had just experienced. Darwin appears to be the first person to link earthquakes with major, sudden displacement of land.

Darwin then hypothesized that the Andes, whose peaks he could see to the east, must have risen from beneath the sea accompanied by similar past earthquakes. This would explain reports of marine fossils found in rocks high in the Andes. In the spring of 1835, Darwin would trek in the Andes, attaining heights up to 4,000 meters above sea level. (The highest mountain in the Andes is 6,920 meters.) He found and collected marine fossils in sedimentary rocks at these elevations, adding support to his hypothesis.

Uplift of Andean rock from below sea level to over 4,000 meters, at the rate of around a meter per century, implied an inconceivably huge amount of time—at least for most people living in the early 1800s. Darwin was the first scientist to hypothesize that major mountain ranges rose from below sea level. Darwin had taken Charles Lyell's groundbreaking three volumes of *Principles of Geology* onboard HMS *Beagle* and was greatly influenced by Lyell's writings on uniformitarianism and deep time (as described in the beginning of chapter 8 of this book).

Although Darwin is an icon of biology, in his university education he concentrated on what was known as geology. Even late in his life he regarded himself primarily as a geologist.

For more on Chile's earthquakes and Darwin, hear (or read a transcript of) an interview on National Public Radio with USGS geophysicist Ross Stein. Go to: www.npr.org/templates/story/story.php?storyId=124361777

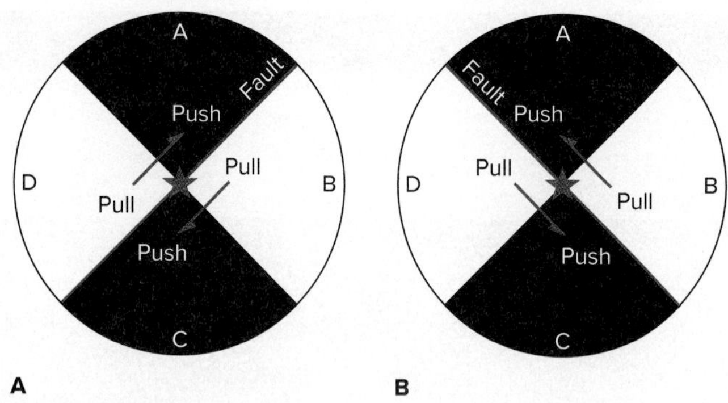

FIGURE 16.25

Map view of two possible solutions for the same pattern of first motion. Each solution has a different fault orientation. If the fault orientation is known, the correct solution can be chosen. The star marks the epicenter, and rock motion is shown by the arrows.

16.7 EARTHQUAKES AND PLATE TECTONICS

One of the great attractions of the theory of plate tectonics is its ability to explain the distribution of earthquakes and the rock motion associated with them.

As described briefly in chapter 1, the theory of plate tectonics states that Earth's lithosphere is divided into a few giant *plates.* Plates are rigid slabs of rock, thousands of kilometers wide and 70 to 125 (or more) kilometers thick, that move across Earth's surface. Because the plates include continents and sea floors on their upper surfaces, the plate tectonic theory explains why the continents and sea floors are moving. The plates change not only position but also size and shape (as we discuss in more detail in chapter 19).

Earthquakes occur commonly at the edges of plates but only occasionally in the middle of a plate. The close correspondence between plate edges and earthquake belts can be seen by comparing the map of earthquake distribution in figure 16.22*A* with the plate map in figure 16.22*B*.

This correspondence is hardly surprising—plate boundaries are identified and *defined* by earthquakes. According to plate tectonics, earthquakes are caused by the interactions of plates along plate boundaries. Therefore, narrow bands of earthquakes are used to outline plates on plate maps. This can be clearly seen in the east Pacific Ocean off South America, where the Nazca plate (figure 16.22*B*) is almost completely outlined by earthquake epicenters (figure 16.22*A*).

The earthquakes on the western border of the Nazca plate are shallow-focus quakes, and they occur in a narrow belt along the crest of the mid-oceanic ridge here, locally called the East Pacific Rise. The quakes along the eastern boundary occupy a broader belt that lies mostly within South America. This belt includes shallow-, intermediate-, and deep-focus earthquakes in a Benioff zone that begins at the Peru-Chile Trench just offshore and slopes steeply down under South America to the east. The Nazca plate moves eastward, away from the crest of the mid-oceanic ridge and toward the subduction zone at the trench, where the plate plunges down into the mantle. The plate's western boundary is located at the crest of the East Pacific Rise, and its eastern boundary is at the bottom of the Peru-Chile Trench.

Earthquakes at Plate Boundaries

As you have learned, there are three types of plate boundaries, *divergent boundaries* where plates move away from each other, *transform boundaries* where plates move horizontally past each other, and *convergent boundaries* where plates move toward each other. Each type of boundary has a characteristic pattern of earthquake distribution and rock motion.

Divergent Boundaries

At a divergent boundary, where plates move away from each other, earthquakes are shallow, restricted to a narrow band, and much lower magnitude than those that occur at convergent or transform boundaries. A divergent boundary on the sea floor is marked by the crest of the mid-oceanic ridge and the *rift valley* that is often (but not always) found on the ridge crest (figure 16.26). The earthquakes are located along the sides of the rift valley and beneath its floor. The rock motion that is deduced from first-motion studies shows that the faults here are normal faults, parallel to the rift valley. The ridge crest is under tension, which is tearing the sea floor apart, creating the rift valley, and causing the earthquakes.

A divergent boundary within a *continent* is usually also marked by a rift valley, shallow-focus quakes, and normal faults. The African Rift Valleys in eastern Africa (figure 16.22*B*) are such a boundary. Tensional forces are tearing eastern Africa slowly apart, creating the rift valleys, some of which contain lakes (see figure 19.21). Other areas where the continental crust is being pulled apart, such as the Basin and Range province in the western United States, are also marked by normal faults and shallow earthquakes.

Transform Boundaries

Where two plates move past each other along a transform boundary, the earthquakes are shallow. First-motion studies indicate strike-slip motion on faults parallel to the boundary. The earthquakes are aligned in a narrow band along the transform fault. Although most transform faults occur on the ocean floor and offset ridge segments, some are found in the continental crust. The San Andreas fault in California is the most famous example of a right-lateral transform fault (see box 16.3). The Alpine fault in New Zealand is another example of a right-lateral transform fault.

Convergent Boundaries

Convergent boundaries are of two general types, one marked by the *collision* of two continents, the other marked by *subduction* of the ocean floor under a continent (figure 16.26) or another piece of sea floor. Each type has a characteristic pattern of earthquakes.

Collision boundaries are characterized by broad zones of shallow earthquakes on a complex system of faults (figure 16.26). Some of the faults are parallel to the dip of the suture zone that

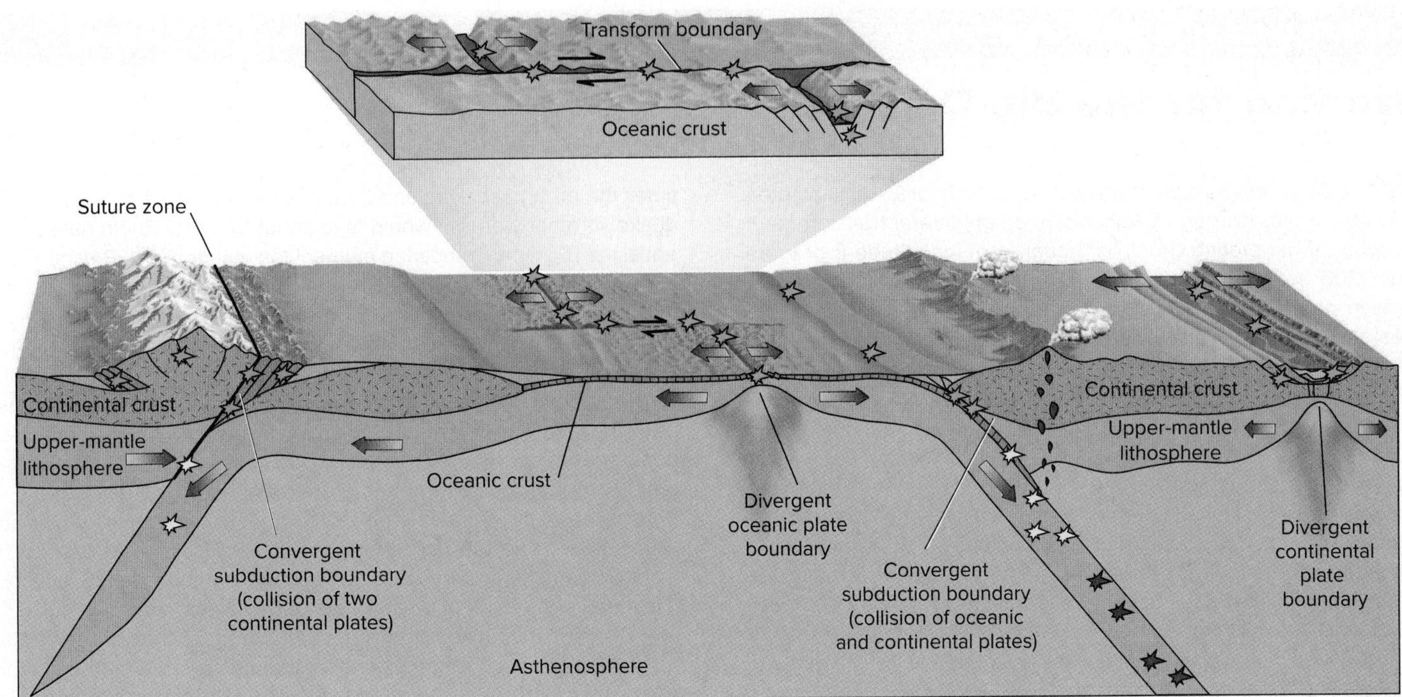

FIGURE 16.26

Distribution of earthquakes at plate boundaries. Shallow-focus earthquakes occur at divergent boundaries where the lithosphere is being pulled apart and also along transform boundaries where slip in the lithosphere accommodates the spreading between oceanic ridges. Shallow- to deep-focus earthquakes occur where a lithosphere subducts during collision of two plates.

marks the line of collision; some are not. One continent usually overrides the other slightly (continents are not easily subducted), creating thick crust and a mountain range. The Himalayas represent such a boundary (figure 16.22*B*). The seismic zone is so broad and complex at such boundaries that other criteria, such as detailed geologic maps, must be used to identify the position of the suture zone at the plate boundary.

During *subduction*, earthquakes occur for several different reasons. As a dense oceanic plate bends to go down at a trench, it stretches slightly at the top of the bend, and normal faults occur as the rocks are subjected to *tension*. This gives a block-faulted character to the outer (seaward) wall of a trench. For some distance below the trench, the subducting plate is in contact with the overlying plate. First-motion studies of earthquakes at these depths show that the quakes are caused by slip as one plate slides or is thrust beneath the other. These thrust faults, or *megathrusts*, may generate very large magnitude subduction zone earthquakes. The devastating 2011 Tohoku, Japan, and 2004 Sumatra earthquakes and associated tsunami were megathrust earthquakes caused by subduction. The largest of the megathrust earthquakes are thought to occur where convergence rates are fast, where young, buoyant oceanic plates are subducted, or where the downgoing plates are relatively flat and allow larger areas of the slab to break at the same time.

At greater depths, where the descending plate is not in direct contact with the overlying plate, earthquakes are common, but the reasons for them are not obvious. The quakes are confined to a thin zone, only 20 to 30 kilometers thick, within the lithosphere of the descending plate, which is about 100 kilometers thick. This zone is thought to be near the top of the lithosphere, where the rock is colder and more brittle.

Subduction Angle

The horizontal and vertical distribution of earthquakes can be used to determine the angle of subduction of a down-going plate. Subduction angles vary considerably from trench to trench. Many plates start subducting at a gentle angle, which becomes much steeper with depth. At a few trenches in the open Pacific Ocean, subduction begins (and continues) at almost a vertical angle.

Subduction angle is probably controlled by plate density and the rate of plate convergence. Older oceanic lithosphere, such as that in the southeast Pacific, tends to be colder and more dense and therefore subducts at a steeper angle; younger oceanic plates in close proximity to the oceanic ridge are warmer and more buoyant and subduct at a shallower angle. A faster rate of convergence may also result in a shallower angle of subduction.

ENVIRONMENTAL GEOLOGY 16.3

Waiting for the Big One in California

The San Andreas fault, running north–south for 1,300 kilometers (807 miles) through California, is a right-lateral transform fault capable of generating great earthquakes of magnitude 8 or more. The 1906 earthquake near San Francisco caused a 450-kilometer scar in northern California (box figure 1). The portion of the fault nearest Los Angeles last broke in 1857 in a quake that was probably of comparable size. The ground has not broken in either of these regions since these quakes. Each old break is now a seismic gap, where rock strain is being stored prior to the next giant quake.

Recent California quakes have been considerably smaller than the "Big One" long predicted by geologists to be in the magnitude-8 range. The 1906 quake in the north had an estimated Richter magnitude of 8.25, and the southern break in 1857 near Fort Tejon was estimated to have a moment magnitude (M) of 7.8. In contrast, the 1989 Loma Prieta quake on the San Andreas fault near San Francisco was a M7.2, and the 1994 Northridge quake (not on the San Andreas fault) was M6.7. So, recent California quakes have been about magnitude 7 or less, and the Big One should be 8. A magnitude-8 quake has 10 times the ground shaking and 32 times the energy of a magnitude-7 quake. In other words, it would take about 32 Loma Prieta quakes to equal the Big One. Comparing moment magnitudes for 1994 and 1857 in southern California, it would take nearly 64 Northridge quakes to equal the Fort Tejon quake.

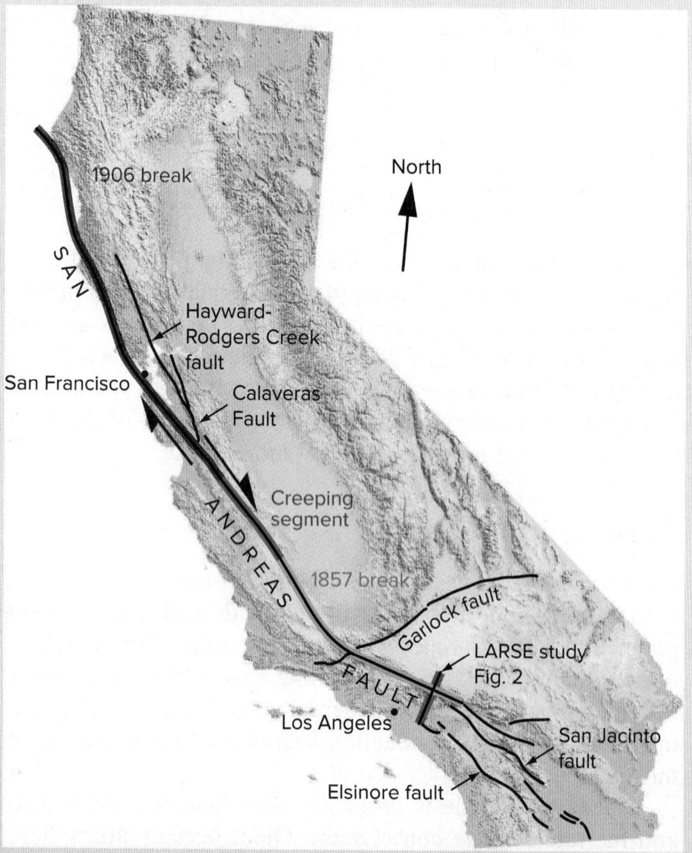

BOX 16.3 ■ FIGURE 1

The two major breaks on the San Andreas fault in California. Each break occurred during a giant earthquake (break from the 1857 earthquake is shown in green and the 1906 earthquake is shown in red). Each old break is now a seismic gap where the fault is locked and may be the future site for another major earthquake. A creeping segment (blue) separates the two locked portions.
Source: U.S. Geological Survey

A great earthquake of magnitude 8 could strike either the northern section or the southern section of the San Andreas fault. Which section will break first? Because the southern section has been inactive longer, it may be the likelier candidate. A magnitude-8 quake here could cause hundreds of billions of dollars in damage and kill thousands of people if it struck during weekday business hours when Los Angeles-area buildings and streets are crowded with people. The M6.7 Northridge quake caused more than $20 billion in damage and was the most costly earthquake in U.S. history. It is daunting to think of an earthquake 64 times more powerful.

Detailed paleoseismology studies suggest that great earthquakes have a recurrence interval of about 105 years on the southern portion of the San Andreas fault near San Bernardino. Historic records in California do not go very far back in time, and much of the evidence involves isotopic dating of broken beds of carbon-rich sediments. Because the time elapsed since the most recent 1857 earthquake is much longer than the 105-year average between quakes, geologists are concerned that the southern part of the fault near the Mojave Desert may rupture again in a M7.6–7.8 earthquake, putting the urban San Bernardino–Riverside area at great risk.

Because of the destructiveness of the Northridge earthquake and the earlier 1987 Whittier Narrows quake (M5.9), geologists are also concerned that a blind thrust fault (fault that cannot be seen at the surface) might rupture closer to downtown Los Angeles. To determine the underground configuration of the blind thrust faults and to investigate how deep sedimentary basins are that will amplify shaking in the region, the Los Angeles Regional Seismic Experiment (LARSE) was undertaken to predict where the strongest shaking will occur during future earthquakes. The LARSE project involved setting off underground explosive charges to generate sound waves that could be analyzed by powerful computers to produce images of the subsurface. The experiment revealed a main blind thrust fault 20 kilometers (12 miles) beneath the surface that extends from near the San Andreas fault and transfers stress and strain upward and southward under the San Gabriel Valley and the Los Angeles Basin (box figure 2). The images also show that the sedimentary basin under the San Gabriel Valley is nearly 5 kilometers (3 miles) deep—much deeper than originally thought—which will increase the potential for strong shaking during the next earthquake in this highly populated area.

According to the 2015 Uniform California Earthquake Rupture Forecast (UCERF3), the probability of a repeat of the 1906 quake (8+) on the locked northern portion of the San Andreas fault may be very low, less than 5% for the next thirty years (box figure 3). However, the new statewide forecast (UCERF3) estimates the chance of a magnitude 6.7 earthquake in northern California to be 95% over the next thirty years. A likely candidate for the quake is not the San Andreas, but the Hayward-Rodger Creek fault and the Calaveras fault across the bay from San Francisco. Such a quake near or under eastern Bay-area cities such as Oakland and Berkeley would

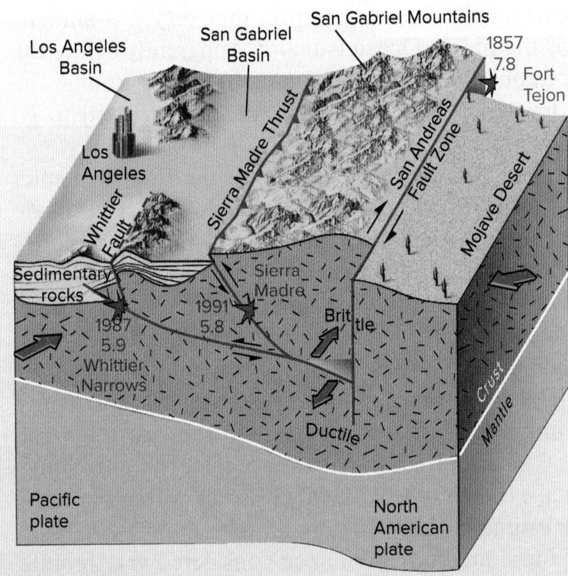

BOX 16.3 ■ FIGURE 2

Los Angeles Regional Seismic Experiment (LARSE). Diagram shows an interpretation of the subsurface structures imaged under the San Andreas fault zone westward under the San Gabriel and Los Angeles Basins.
Source: U.S. Geological Survey

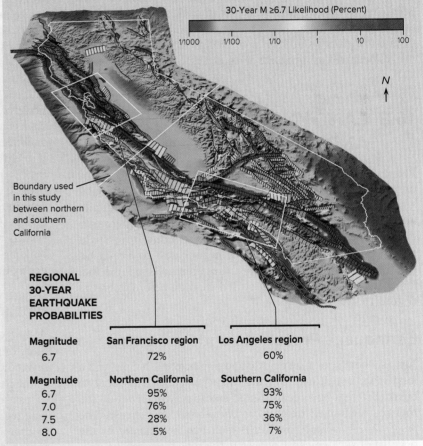

CALIFORNIA AREA EARTHQUAKE PROBABILITY

More than 99%

probability in the next 30 years for one or more magnitude 6.7 or greater quakes capable of causing extensive damage and loss of life. The map shows the distribution throughout the state of the likelihood of having a nearby earthquake rupture (within 3 or 4 miles).

30-Year M ≥6.7 Likelihood (Percent)

1/1000 1/100 1/10 1 10 100

N

Boundary used in this study between northern and southern California

REGIONAL 30-YEAR EARTHQUAKE PROBABILITIES

Magnitude	San Francisco region	Los Angeles region
6.7	72%	60%

Magnitude	Northern California	Southern California
6.7	95%	93%
7.0	76%	75%
7.5	28%	36%
8.0	5%	7%

BOX 16.3 ■ FIGURE 3

Map of California showing the probability of a magnitude 6.7 or greater earthquake occurring in the next thirty years (the typical duration of a homeowner mortgage), as determined by the Uniform California Earthquake Rupture Forecast, Version 3.
Source: U.S. Geological Survey, California Geological Survey, Southern California Earthquake Center

cause far greater death and destruction than the 1989 Loma Prieta earthquake.

The UCERF3 report estimates that the southern part of the San Andreas fault in the Los Angeles region has a 7% probability of a M8 earthquake and a 60% chance of a magnitude 6.7 earthquake within the next thirty years. However, the overall probability of a magnitude 6.7 earthquake in southern California is 93% for the next thirty years (box figure 3). The faults with the highest probabilities of generating a M6.7 or greater earthquake are the southern San Andreas, and the San Jacinto and Elsinore faults, which parallel the San Andreas (see box figure 1). The UCERF3 report estimates that there is a more than 99% chance of an M6.7 earthquake occurring somewhere in the California region within the next thirty years.

It is clear from the new studies that, even though the probability of a magnitude 8+ Big One along the San Andreas fault is low, California needs to be prepared for the near certainty of a magnitude 6.7 earthquake in the next thirty years.

Additional Resources

For more information about the likelihood of the San Andreas fault creating a large earthquake and video simulations of the shaking, visit U.S. Geological Survey websites:

- https://pubs.usgs.gov/fs/2015/3009/pdf/fs2015-3009.pdf
- https://earthquake.usgs.gov/regional/nca/simulations/

For more details on the Los Angeles Regional Seismic Experiment, visit:

- http://geopubs.wr.usgs.gov/fact-sheet/fs110-99/

Website for the Southern California Earthquake Center:

- www.scec.org/

In summary, earthquakes are very closely related to plate tectonics. Most plate boundaries are defined by the distribution of earthquakes, and plate motion can be deduced by the first motions of the quakes. Analysis of first motions can also help determine the type and orientation of stresses that act on plates, such as tension and compression. Quake distribution with depth indicates the angle of subduction and has shown that some plates change subduction angle and even break up as they descend.

A few quakes, such as those that occur in the center of plates, cannot easily be related to plate motion. These *intraplate* earthquakes probably occurred along older faults that are no longer plate boundaries but remain zones of crustal weakness. Some of the most destructive earthquakes in the United States, such as the 1811–1812 New Madrid, Missouri; 1886 Charleston, South Carolina; and 1755 Boston, Massachusetts, quakes occurred as intraplate earthquakes.

16.8 EARTHQUAKE PREDICTION AND FORECASTING

People who live in earthquake-prone regions are plagued by unscientific predictions of impending earthquakes by popular writers and self-proclaimed prophets. Several techniques are being explored for *scientifically* forecasting a coming earthquake. One group of methods involves monitoring slight changes, or *precursors*, that occur in rock next to a fault before the rock breaks and moves; these methods assume that large amounts of strain are stored in rock before it breaks (figure 16.1).

Earthquake Precursors and Prediction

Just as a bent stick may crackle and pop before it breaks with a loud snap, a rock may give warning signals that it is about to break. Before a large quake, small cracks may open within the rock, causing small tremors, or *microseisms*, to increase. The *properties of the rock* next to the fault may be changed by the opening of such cracks. Changes in the rock's magnetism, electrical resistivity, or seismic velocity may give some warning of an impending quake.

The opening of tiny cracks changes the rock's porosity, so *water levels in wells* often rise or fall before quakes. The cracks provide pathways for the release of radioactive radon gas from rocks (radon is a product of radioactive decay of uranium and other elements). An *increase in radon emission from wells* may be a prelude to an earthquake. The interval between eruptions of geysers may change before and after an earthquake, probably due to porosity changes within the surrounding rock.

In some areas, the *surface of Earth tilts and changes elevation* slightly before an earthquake. Scientists use highly sensitive instruments to measure this increasing strain in hopes of predicting quakes.

Chinese scientists claim successful, short-range predictions by watching *animal behavior*—horses become skittish and snakes leave their holes shortly before a quake. U.S. scientists conducted a few pilot programs along these lines, but remain skeptical because it is difficult to correlate a specific animal behavior to an impending earthquake. It is interesting that very few animals were killed by the Indian Ocean tsunami. Apparently before the tsunami hit, elephants were seen running to higher ground, flamingos left low-lying breeding areas, and dogs refused to go outdoors.

Japanese and Russian geologists were the first to predict earthquakes successfully, and Chinese geologists have made some very accurate predictions. In 1975, a 7.3-magnitude earthquake near Haicheng in northeastern China was predicted five hours before it happened. Alerted by a series of *foreshocks*, authorities evacuated about a million people from their homes; many watched outdoor movies in the open town square. Half the buildings in Haicheng were destroyed, along with many entire villages, but only a few hundred lives were lost. In grim contrast, however, the Chinese program failed to predict the 1976 Tangshan earthquake (magnitude 7.6), which struck with no warning and killed an estimated 250,000 people.

Most of these methods were once considered very promising but have since proved to be of little real help in predicting quakes. A typical quake predictor, such as tilt of the land surface, may precede one quake and then be absent for the next ten quakes. In addition, each precursor can be caused by forces unrelated to earthquakes (land tilt is also caused by mountain building, magmatic intrusion, mass wasting, and wetting and drying of the land).

Earthquake Forecasting

A fundamentally different method of determining the probability of an earthquake occurring relies on the history of earthquakes along a fault and the amount of tectonic stress building in the rock. Geologists look at the geologic record for evidence of past earthquakes using the techniques of *paleoseismology*. One technique involves digging a trench across the fault zone to examine sedimentary layers that have been offset and disrupted during past earthquakes (figure 16.27*A* and *B*). If the offset layers contain material such as volcanic ash, pollen, or organic material such as tree roots that can give a numerical age, then the average length of time between earthquakes (*recurrence interval*) can be determined. If the length of time since the last recorded earthquake far exceeds the recurrence interval, the fault is given a high probability of generating an earthquake.

Along some long-active faults are short, inactive segments called *seismic gaps* where earthquakes have not occurred for a long time. These gaps form as part of the seismic cycle and result in a zone of lowered stress, or stress shadow zone, where earthquake activity sharply decreases after a major seismic event. Such was the case after the 1906 San Francisco earthquake and after the 1857 break along the southern section of the San Andreas fault (see box 16.3).

The recurrence interval and likelihood of future earthquakes are also determined by measuring the slip rate along plate boundaries. Exciting new satellite-based techniques such as InSAR (interferometric synthetic aperture radar), in addition to GPS, have allowed seismologists to measure the vertical and horizontal movement along active faults and to determine how long it would take for sufficient stress to build up along the plate boundary to generate rupturing and slip along a fault. For example, if the slip rate along the boundary is determined to be 5 centimeters per year and the last earthquake resulted in 5 meters of slip, then you would expect the next large earthquake to occur in 100 years. Just as a rubber band will break if stretched too far, rock will also break or rupture if a critical level of stress is exceeded. In other cases, the accumulating stress is released aseismically by so-called silent earthquakes where a fault slips very slowly or creeps to gradually relieve the stress. Slip rates and recurrence intervals are used to determine the statistical probability of an earthquake occurring over a given amount of time.

By studying the seismic history of faults, geologists in the United States are sometimes able to forecast earthquakes along

FIGURE 16.27

To determine the likelihood of a large earthquake occurring again along an active fault, geologists need to know how often quakes have occurred in the past and how large the last one was. By using the techniques of paleoseismology, geologists dig a trench *(A)* across or alongside a fault and very carefully map disturbed layers of sediment and soil exposed in the upper few meters of the trench. *(B)* Trench wall exposing offset layers of sediment along the San Andreas fault. Black organic-rich layers can be age dated to determine the recurrence interval of earthquakes along the fault.
(A) Source: Jennifer Adleman, U.S. Geological Survey; (B) ©Stephen Reynolds

some segments of some faults. In 1988, the U.S. Geological Survey estimated a 50% chance of a magnitude-7 quake along the segment of the San Andreas fault near Santa Cruz. In 1989, the magnitude-7 Loma Prieta quake occurred on this very section. Since the techniques are new and in some cases only partly understood, some errors will undoubtedly be made. Many faults are not monitored or studied historically because of lack of money and personnel, so we will never have a warning of impending quakes in some regions. For large urban areas near active faults such as the San Andreas, however, earthquake risk analysis may reduce damage and loss of life.

Reducing Earthquake Damage and Risk

Another more recent approach to minimize loss of life and reduce damage in a major earthquake is to closely monitor the amount and location of strong shaking by using a dense network of broadband seismometers that digitally relay information via satellites to a central location. At this location, maps showing where the greatest amount of shaking occurred can be generated within minutes to guide emergency personnel to the areas of most damage (figure 16.28). Such a system has been developed in southern California, and there are plans for integrating other regional seismic networks into an Advanced National Seismic System (ANSS) to monitor earthquakes throughout the United States if adequate funding can be obtained.

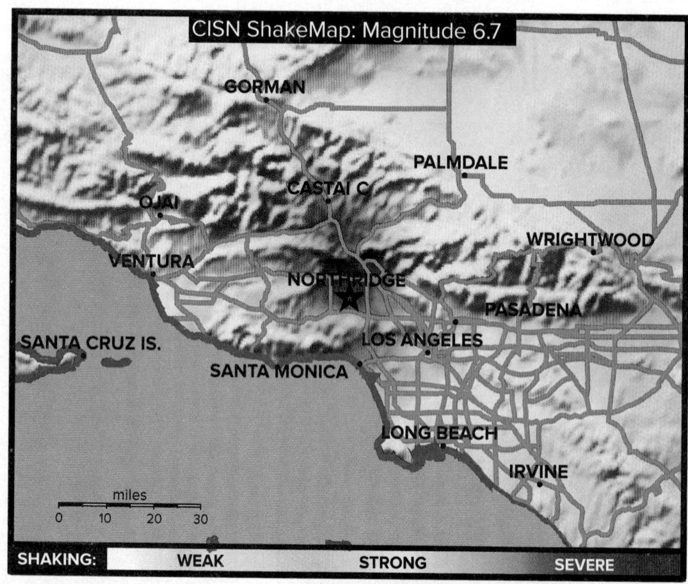

FIGURE 16.28

Map shows the amount of shaking that occurred after the 1994 Northridge earthquake. The ability to create maps within minutes after an earthquake that show the location and severity of maximum ground shaking (ShakeMap) was developed in 1995 by the U.S. Geological Survey. Had this ShakeMap been available minutes after the 1994 Northridge earthquake, emergency personnel could have been immediately directed to the most damaged areas.
Source: David Wald, U.S. Geological Survey

A major goal of the ANSS program is to locate strong-motion seismometers in buildings, bridges, canals, and pipelines to provide valuable information on how a structure moves during an earthquake to help engineers build more earthquake-resistant structures. One key to reducing damage and loss of life is to create stronger structures that resist catastrophic damage during a major earthquake.

A future goal of the program is to minimize risk by developing an early warning system. With a wide enough distribution of real-time seismometers, it is technically possible for an urban area to get an early warning of an impending earthquake if the earthquake's epicenter is far enough away from the city. For example, if an earthquake occurred 100 kilometers from downtown Los Angeles and its waves were moving at 4 kilometers per second, the system would have 25 seconds to process and analyze the data and broadcast it as an early warning. Even seconds of warning could be enough to shut off main gas pipelines, shut down subway trains, and give schoolchildren time to get under their desks. Japan has spent hundreds of millions of dollars over the past decade building an earthquake-warning system that was put to use on March 11, 2011, when the magnitude 9.0 earthquake hit offshore. The bullet trains were shut down and more than a minute of warning was issued via television and radio stations, text messages to cell phones, and e-mails to businesses before the shaking began in Tokyo. The early warning undoubtedly saved countless lives.

Summary

Earthquakes usually occur when rocks break and move along a fault to release strain that has gradually built up in the rock. Volcanic activity can also cause earthquakes. Deep quakes may be caused by mineral transformations.

Seismic waves move out from the earthquake's *focus. Body waves* (P waves and S waves) move through Earth's interior, and *surface waves* (*Love* and *Rayleigh waves*) move on Earth's surface.

Seismographs record seismic waves on *seismograms,* which can be used to determine an earthquake's strength, location, and depth of focus. Most earthquakes are shallow-focus quakes, but some occur as deep as 670 kilometers below Earth's surface.

The time interval between first arrivals of P and S waves is used to determine the distance between the seismograph and the *epicenter.* Three or more stations are needed to determine the location of earthquakes.

Earthquake *intensity* is determined by assessing damage and is measured on the *modified Mercalli scale.*

Earthquake *magnitude,* determined by the amplitude of seismic waves on a seismogram, is measured on the *Richter scale. Moment magnitudes,* determined by field work, are widely used today and often are larger than Richter magnitudes.

The most noticeable effects of earthquakes are ground motion and displacement (which destroy buildings and thereby injure or kill people), fire, landslides, and *tsunami. Aftershocks* can continue to cause damage months after the main shock.

Earthquakes are generally distributed in belts. The *circum-Pacific belt* contains most of the world's earthquakes. Earthquakes also occur on the Mediterranean-Himalayan belt, the crest of the mid-oceanic ridge, and in association with basaltic volcanoes.

Benioff zones of shallow-, intermediate-, and deep-focus earthquakes are associated with andesitic volcanoes, oceanic trenches, and the edges of continents or island arcs.

The concept of plate tectonics explains most earthquakes as being caused by interactions between two plates at their boundaries. Plate boundaries are generally defined by bands of earthquakes.

Divergent plate boundaries are marked by a narrow zone of shallow earthquakes along normal faults, usually in a rift valley. Transform boundaries are marked by shallow quakes caused by strike-slip motion along one or more faults.

Convergent boundaries where continents collide are marked by a very broad zone of shallow quakes. Convergent boundaries involving deep subduction are marked by Benioff zones of quakes caused by tension, underthrusting, and compression.

The distribution of quakes indicates subduction angles of a down-going plate. The subduction angle is probably controlled by plate density and rate of plate convergence.

To determine the probability of an earthquake occurring, scientists examine rock properties near faults, slip rate studies, and paleoseismology data to determine the recurrence interval of quakes along individual faults.

Terms to Remember

aftershock 400
Benioff zone 404
body wave 386
circum-Pacific belt 404
depth of focus 391
earthquake 386
elastic rebound theory 386
epicenter 386
focus 386
intensity 391
island arc 404
Love wave 389
magnitude 393
Mediterranean-Himalayan belt 404
modified Mercalli scale 392
moment magnitude 393
P wave 388
Rayleigh wave 389
Richter scale 393
seismic sea wave 400
seismic wave 386
seismogram 389
seismograph 389
surface wave 388
S wave 388
travel-time curve 391
tsunami (seismic sea wave) 400

Testing Your Knowledge

Use the following questions to prepare for exams based on this chapter.

1. What causes earthquakes?
2. Name and describe the various types of seismic waves.
3. Describe in detail how earthquake epicenters are located by seismograph stations.
4. Compare and contrast the concepts of intensity and magnitude of earthquakes.
5. Describe several ways that earthquakes cause damage.
6. How do earthquakes cause tsunami?
7. What are aftershocks?
8. Discuss the distribution of earthquakes with regard to location and depth of focus.
9. Show with a sketch how the concept of plate tectonics can explain the distribution of earthquakes in a Benioff zone and on the crest of the mid-oceanic ridge.
10. Describe several techniques that may help scientists forecast earthquakes.
11. Describe several precursors that have occurred before earthquakes.
12. The elastic rebound theory
    a. explains folding of rocks.
    b. explains the behavior of seismic waves.
    c. involves the sudden release of progressively stored strain in rocks, causing movement along a fault.
    d. none of the preceding.
13. The point within Earth where seismic waves originate is called the
    a. focus. b. epicenter.
    c. fault scarp. d. fold.
14. P waves are
    a. compressional. b. transverse.
    c. tensional.
15. What is the minimum number of seismic stations needed to determine the location of the epicenter of an earthquake?
    a. 1 b. 2
    c. 3 d. 5
    e. 10
16. The Richter scale measures
    a. intensity.
    b. magnitude.
    c. damage and destruction caused by the earthquake.
    d. the number of people killed by the earthquake.
17. Benioff zones are found near
    a. mid-ocean ridges. b. ancient mountain chains.
    c. interiors of continents. d. oceanic trenches.

18. Most earthquakes at divergent plate boundaries are
    a. shallow focus. b. intermediate focus.
    c. deep focus. d. all of the preceding.
19. Most earthquakes at convergent plate boundaries are
    a. shallow focus. b. intermediate focus.
    c. deep focus. d. all of the preceding.
20. A zone of shallow earthquakes along normal faults is typical of
    a. divergent boundaries. b. transform boundaries.
    c. subduction zones. d. collisional boundaries.
21. A seismic gap is
    a. the time between large earthquakes.
    b. a segment of an active fault where earthquakes have not occurred for a long time.
    c. the center of a plate where earthquakes rarely happen.
22. Which of the following is *not* true of tsunami?
    a. very long wavelength
    b. high wave height in deep water
    c. very fast moving
    d. continued flooding after wave crest hits shore

Expanding Your Knowledge

1. What are some arguments in favor of and against predicting earthquakes? What would happen in your community if a prediction were made today that within a month, a large earthquake would occur nearby?
2. Most earthquakes occur at plate boundaries where plates interact with each other. How might earthquakes be caused in the interior of a rigid plate?
3. How can you prepare for an earthquake in your own home?
4. Suppose you want to check for earthquake danger before buying a new home. How can you check the regional geology for earthquake dangers? The actual building site? The home itself?

Exploring Web Resources

http://nctr.pmel.noaa.gov

National Oceanic and Atmospheric Administration (NOAA) Tsunami website. Comprehensive site that describes NOAA's role in tsunami research, monitoring, preparedness, and warnings.

http://pubs.usgs.gov/gip/earthq3/

U.S. Geological Survey, 1990. *The San Andreas fault system.* Professional Paper 1515.

http://earthquake.usgs.gov/

U.S. Geological Survey Earthquake Hazards Program. Gives information on reducing earthquake hazards, earthquake preparedness, latest quake information, historical earthquakes, and how earthquakes are studied. Also a good starting place for links to other earthquake sites.

https://earthquake.usgs.gov/learn/faq.php

Frequently asked questions about earthquakes, maintained by the U.S. Geological Survey.

http://shakemovie.caltech.edu/

Caltech's *Shake Movie* site provides near real-time visualizations of recent seismic events in Southern California.

www.seismo.unr.edu/

University of Nevada, Reno Seismological Laboratory site contains information about recent earthquakes, earthquake preparedness, and links to other earthquake sites.

www.seismo.berkeley.edu/seismo/Homepage.html

Seismographic information page maintained by University of California–Berkeley that has many links to other earthquake sites (particularly in California), three-dimensional earthquake movie, Northridge earthquake rupture movies, and information on earthquake preparedness.

www.sciencecourseware.org/VirtualEarthquake/

California State University, Los Angeles *Virtual Earthquake.* Create and analyze an earthquake.

http://pubs.usgs.gov/gip/earthq4/severitygip.html

General information about the size of an earthquake. Discussion of Richter and Mercalli scales.

http://pubs.usgs.gov/gip/dynamic/dynamic.html

Online edition of *This Dynamic Earth* gives general information about plate tectonics.

http://pubs.usgs.gov/circ/c1187/

U.S. Geological Survey online version of *Surviving a Tsunami.*

http://walrus.wr.usgs.gov/tsunami/

U.S. Geological Survey web page gives information about tsunami and earthquake research. Includes animations from the most recent tsunami in Japan, Indonesia, and Chile.

Earth's Interior and Geophysical Properties

CHAPTER 17

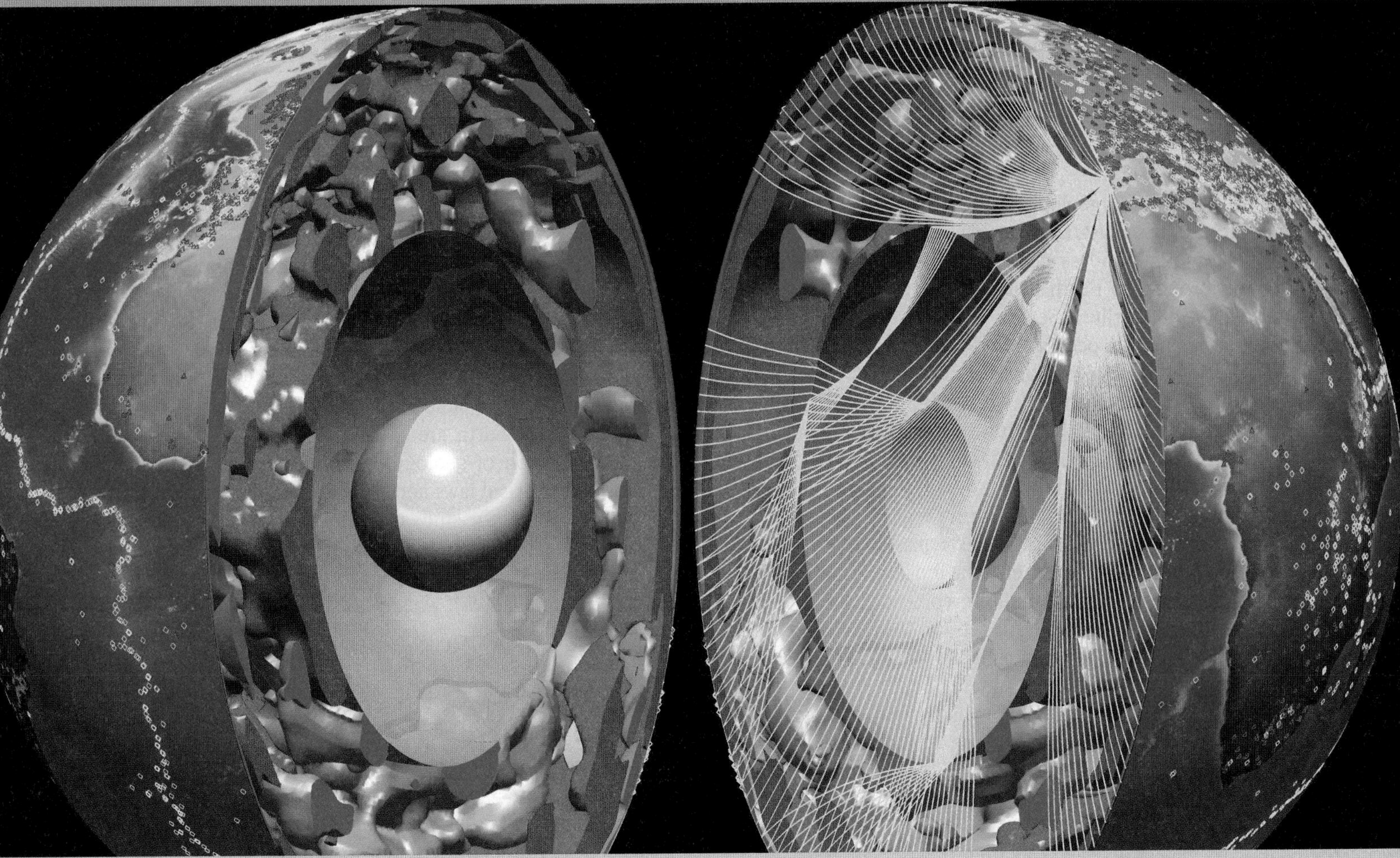

Seismic tomography image of Earth generated from seismic waves and powerful numerical models that run on supercomputers. Yellow lines represent seismic waves traveling through the interior of Earth from an earthquake epicenter in Spain and recorded by seismic stations around the world that are shown on the surface with red dots. The waves are deflected in the mantle and by the core due to differences in density and material properties. The blue and red blobs are areas where waves travel faster (blue) and slower (red) than average in the mantle and represent areas of different strength and and probable temperature. The large blue layer at the bottom of the left image is interpreted to be a large slab of a subducted tectonic plate that sank through the entire mantle and is preserved below the Indian Ocean.
Graphic by Nathan Simmons/LLNL

LEARNING OBJECTIVES

- Explain how seismic waves are used to learn about Earth's interior.
- Sketch and describe the main layers of Earth's interior. Define and label lithosphere, asthenosphere, and Mohorovičić discontinuity on your sketch.
- Explain isostatic adjustment and crustal rebound.
- Discuss how gravity and magnetic anomalies are measured and used to interpret the subsurface.
- Describe Earth's magnetic field and where it is generated. Explain the evidence for magnetic reversals.
- How does Earth's heat vary with depth and in different regions on the continents and sea floor?

The only rocks that geologists can study directly in place are those of the crust, and Earth's crust is but a thin skin of rock, making up less than 1% of Earth's total volume. Mantle rocks brought to Earth's surface in basalt flows and in diamond-bearing kimberlite pipes, as well as the tectonic attachment of lower parts of the oceanic lithosphere to the continental crust, give geologists a glimpse of what the underlying mantle might look like. Meteorites also provide clues about the possible composition of the core of Earth. But to learn more about the deep interior of Earth, geologists must study it *indirectly*, largely by using the tools of geophysics—that is, seismic waves, and measurements of gravity, heat flow, and Earth's magnetic field.

The evidence from geophysics suggests that Earth is divided into three major compositional layers—the crust, the rocky mantle beneath the crust, and the metallic core at the center of Earth. The study of plate tectonics has shown that the crust and uppermost mantle can be mechanically divided into the brittle lithosphere and the ductile asthenosphere.

You will learn in this chapter how seismic waves are used to image shallow and deep geologic structures within the Earth, and how gravity measurements can indicate where certain regions of the crust and upper mantle are being held up or held down out of their natural position of equilibrium. We will also discuss Earth's magnetic field and its history of reversals. We will show how magnetic anomalies can indicate hidden ore and geologic structures. The chapter closes with a discussion of the distribution and loss of Earth's heat.

17.1 HOW IS EARTH'S INTERIOR STUDIED?

What *do* geologists know about Earth's interior? How do they obtain information about the parts of Earth beneath the surface? Geologists, in fact, cannot sample rocks very far below Earth's surface. Some deep mines penetrate 3 kilometers into Earth, and a deep oil well may go as far as 8 kilometers beneath the surface; the deepest scientific well reached 12 kilometers in Russia. Rock samples can be brought up from a mine or a well for geologists to study.

A direct look at rocks from deeper levels can be achieved where mantle rocks have been brought up to the surface by basalt flows, by the intrusion and erosion of diamond-bearing kimberlite pipes (see chapter 5), where the lower part of the oceanic lithosphere (see chapter 18) has been tectonically attached to the continental crust at a convergent plate boundary, or where old oceanic lithosphere experiences extension and produces a "tectonic window" into the upper mantle. However, Earth has a radius of about 6,370 kilometers, so it is obvious that geologists can only scratch the surface when they try to study *directly* the rocks beneath their feet.

Deep parts of Earth are studied *indirectly*, however, largely through the branch of geology called **geophysics**, which is the application of physical laws and principles to the study of Earth. Geophysics includes the study of seismic waves and Earth's magnetic field, gravity, and heat. All of these things tell us something about the nature of the deeper parts of Earth. Together, they create a convincing picture of what makes up Earth's interior.

17.2 EVIDENCE FROM SEISMIC WAVES

Seismic waves from a large earthquake may reverberate through the entire Earth. A human-caused explosion also generates seismic waves. Geologists obtain new information about Earth's interior after every large earthquake and explosion. More recently, scientists have also been analyzing the energy waves generated by tidal friction, ocean waves, and storms to gain an even more detailed image of the crust and upper mantle. When Superstorm Sandy turned toward New York City and Long Island in October 2012, the powerful ocean waves generated microseisms (very small earthquakes) that were detected on portable seismometers set up in the eastern half of the United States as part of the EarthScope program. The energy generated by the storm waves are being used to image the crust and upper mantle under North America much like X rays are used to make CT scans of the human body (see box 17.1).

How are seismic waves used to help us visualize the subsurface? One important way of learning about Earth's interior is the study of **seismic reflection**, the return of some of the energy of seismic waves to Earth's surface after the waves bounce off a rock boundary. If two rock layers of differing densities are separated by a fairly sharp boundary, seismic waves reflect off that boundary just as light reflects off a mirror (figure 17.1). These reflected waves are recorded on a seismogram, which shows the

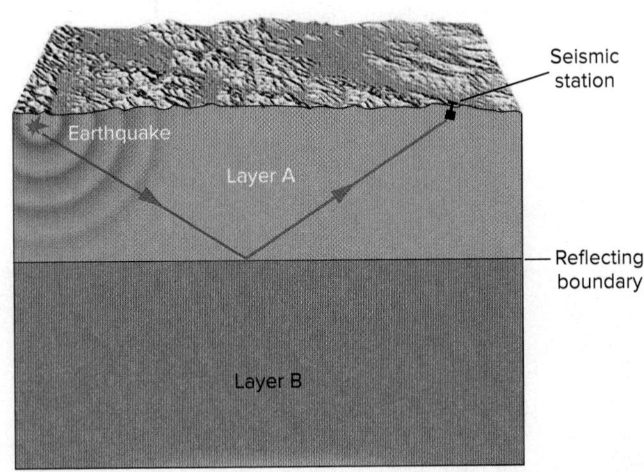

FIGURE 17.1

Seismic reflection. Seismic waves radiate outward from the focus of an earthquake and reflect off a rock boundary deep within the Earth. These reflected waves are recorded by seismic stations at the surface.

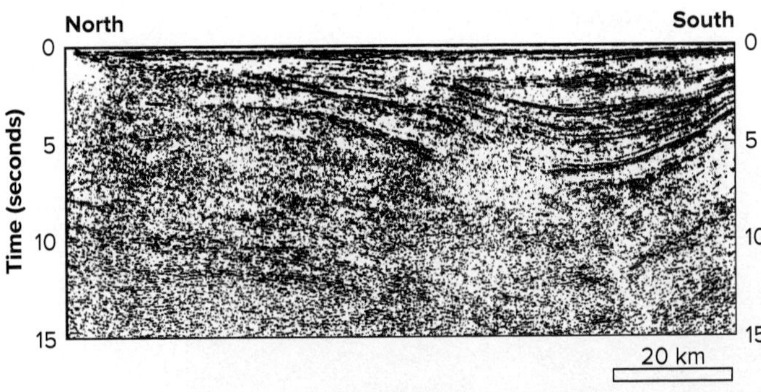

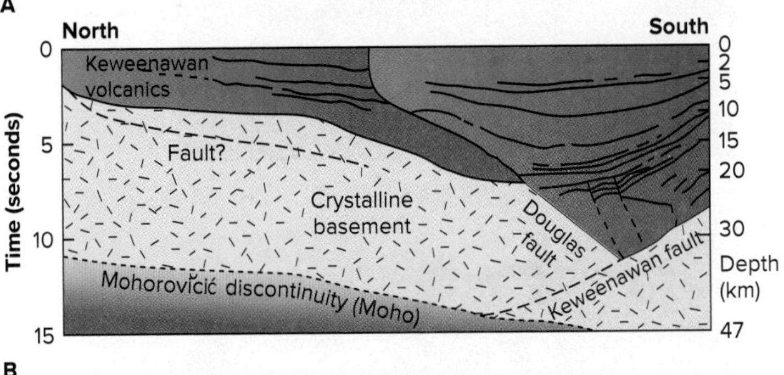

FIGURE 17.2

An example of how seismic reflection can be used to image the geologic structures in the crust and mantle. (*A*) A cleaned seismic reflection profile across the North American Midcontinent Rift System under western Lake Superior. This is a failed Precambrian rift (age ~ 1.1 billion years) that has pronounced gravity and magnetic anomalies. Strong seismic reflectors can be followed across the section, and faults can be inferred where the reflectors are interrupted. (*B*) The interpreted crustal structure from the reflection profile, which includes several major rift-forming faults that extend downward toward the boundary between the crust and mantle, also known as the Mohorovičić discontinuity, or "Moho." The rift is filled with layers of basalt and sediment.

Source: Modified from Lowrie (2007)

amount of time the waves took to travel down to the boundary, reflect off it, and return to the surface (figures 16.7 and 17.1). From the amount of time necessary for the round trip, geologists calculate the depth of the boundary.

These reflected seismic waves can be used to image geologic structures deep within the crust and mantle that are not exposed at the surface. An example is shown in figure 17.2. The uppermost image shows a typical seismic reflection profile, and the lower image shows the interpreted crustal structure. Images like these are particularly useful to petroleum geologists who are exploring for new potential oil and natural gas deposits.

Another method used to locate rock boundaries is the study of **seismic refraction**, the bending of seismic waves as they pass from one material to another, which is similar to the way that light appears to sharply bend a straw resting in a glass of water. As a seismic wave strikes a rock boundary, much of the energy of the wave passes across the boundary. As the wave crosses from one rock layer to another, it changes direction (figure 17.3). This change of direction, or refraction, occurs only if the velocity of seismic waves is different in each layer (which is generally true if the rock layers differ in density or strength).

The boundaries between such rock layers are usually distinct enough to be located by seismic refraction techniques, as shown in figure 17.4. Seismograph station 1 is receiving seismic waves that pass directly through the upper layer (*A*). Stations farther from the epicenter, such as station 2, receive seismic waves from two pathways: (1) a direct path straight through layer A and (2) a refracted path through layer A to a higher-velocity layer (*B*) and back to layer A. Station 2, therefore, receives the same wave twice.

Seismograph stations close to station 1 receive mainly the direct wave or possibly two waves, the direct (upper) wave arriving before the refracted (lower) wave. Stations near station 2 receive both the direct and the refracted waves. At some point between station 1 and station 2, there is a transformation from receiving the direct wave first to receiving the *refracted* wave first. Even though the refracted wave travels farther, it can arrive at a station first because most of its path is in the high-velocity layer (B).

The distance between this point of transformation and the epicenter of the earthquake is a function of the depth to the rock boundary between layers A and B. A series of portable seismographs can be set up in a line away from an explosion (a *seismic shot*) to find this distance, and the depth to the boundary can then be calculated. The velocities of seismic waves within the layers can also be found.

Figure 17.3 shows how waves bend as they travel downward into higher-velocity layers. But why do waves return to the surface, as shown in figure 17.4? The answer is that advancing waves give off energy in all directions. Much of this energy continues to travel horizontally within layer B (figure 17.4). This energy passes beneath station 2 and out of the figure toward the right. A small part of the energy "leaks" upward into

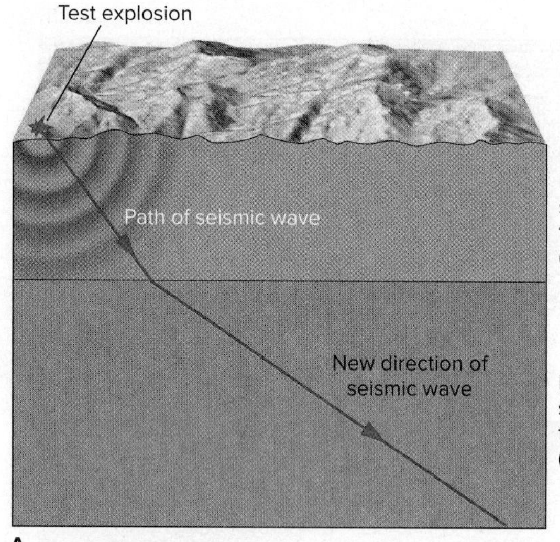

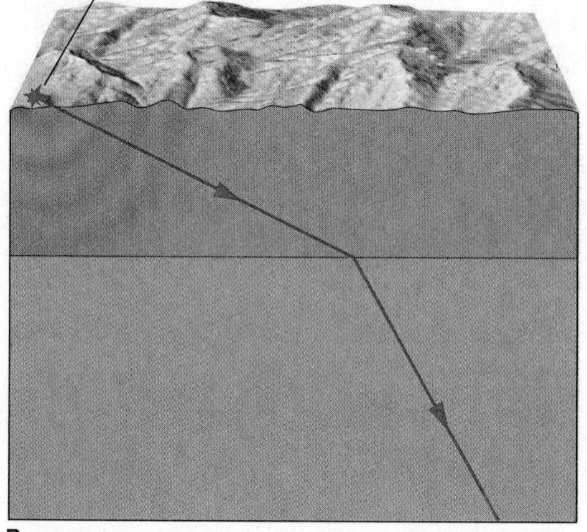

FIGURE 17.3

Seismic refraction occurs when seismic waves bend as they cross rock boundaries. At an interface, seismic (or sound or light) waves will bend toward the lower-velocity material. (*A*) Low-velocity layer above high-velocity layer. (*B*) High-velocity layer above low-velocity layer. Some of the seismic waves will also return to the surface by reflecting off the rock boundary.

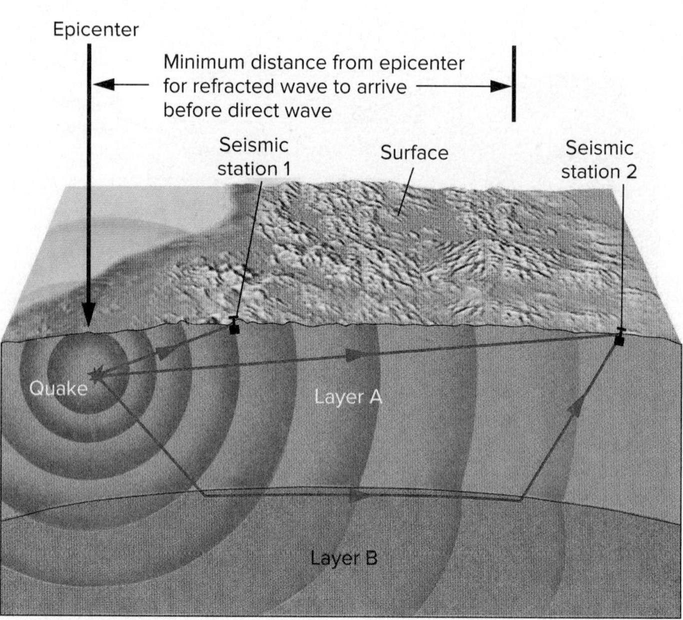

FIGURE 17.4

Seismic refraction can be used to detect boundaries between rock layers. See text for explanation.

layer A, and it is this pathway that is shown in the figure. There are many other pathways for this wave's energy that are not shown here.

Thus, in contrast to seismic reflection, a sharp rock boundary is not necessary for the refraction of seismic waves. Even in a thick layer of uniform rock, the increasing pressure with depth tends to increase the velocity of the waves. The waves follow curved paths through such a layer, as shown in figure 17.5. To understand the reason for the curving path, visualize the thick rock layer as a stack of very thin layers, each with a slightly higher velocity than the one above. The curved path results from many small changes in direction as the wave passes through the many layers.

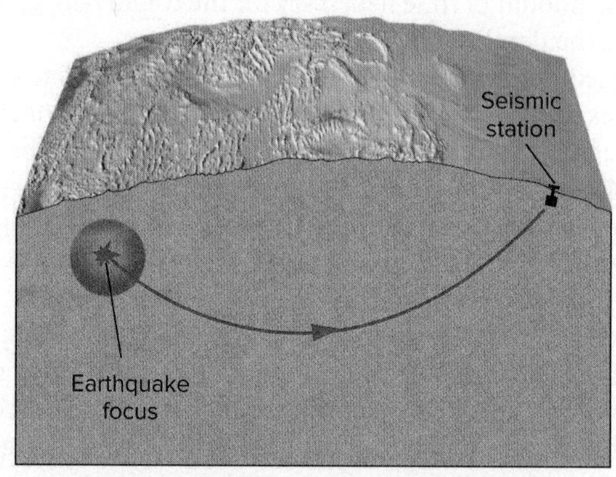

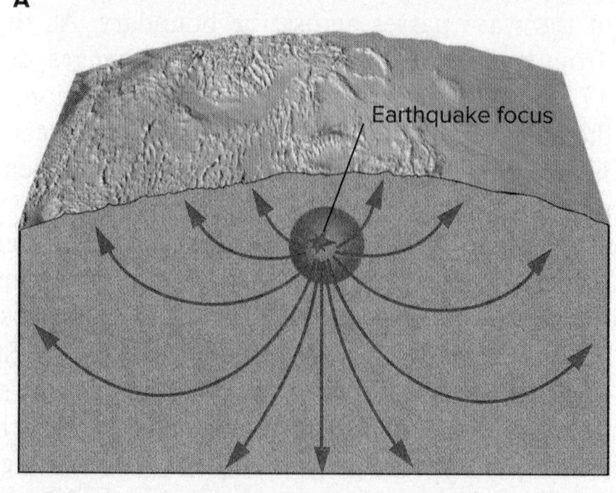

FIGURE 17.5

Curved paths of seismic waves caused by uniform rock with increasing seismic velocity with depth. (*A*) Path between earthquake and recording station. (*B*) Waves spreading out in all directions from earthquake focus.

17.3 EARTH'S INTERNAL STRUCTURE

It was the study of seismic refraction and seismic reflection that enabled scientists to plot the three main zones of Earth's interior (figure 17.6). The **crust** is the outer layer of rock, which forms a thin skin on Earth's surface. Below the crust lies the **mantle**, a thick shell of rock that separates the crust above from the core below. The **core** is the central zone of Earth. It is metallic and is the source of Earth's magnetic field.

The Crust

Studies of seismic waves have shown (1) that the crust is thinner beneath the oceans than beneath the continents (figure 17.7) and (2) that seismic waves travel faster in oceanic crust than in continental crust. Because of this velocity difference, it is assumed that the two types of crust are made up of different kinds of rock.

Seismic P waves travel through oceanic crust at about 7 kilometers per second, which is also the speed at which they travel through basalt and gabbro (the coarse-grained equivalent of basalt). Samples of rocks taken from the sea floor by oceanographic ships verify that the upper part of the oceanic crust is basalt and suggest that the lower part is gabbro. The oceanic crust averages 7 kilometers (4.3 miles) in thickness, varying from 5 to 8 kilometers (table 17.1).

Seismic P waves travel more slowly through continental crust—about 6 kilometers per second, the same speed at which they travel through granite and gneiss. Continental crust is often called "granitic," but the term should be put in quotation marks because most of the rocks exposed on land are not granite. The continental crust is highly variable and complex, consisting of a crystalline basement composed of granite, other plutonic rocks, gneisses, and schists, all capped by a layer of sedimentary rocks, like icing on a cake. Since a single rock term cannot accurately describe crust that varies so greatly in composition, some geologists use the term *felsic* (rocks high in *feldspar* and *silicon*) for continental crust and *mafic* (rocks high in magnesium and iron) for oceanic crust.

Continental crust is much thicker than oceanic crust, averaging 30 to 50 kilometers (18.6 to 31 miles) in thickness, though it varies from 20 to 70 kilometers. Seismic waves show that the crust is thickest under geologically young mountain ranges, such as the Andes and Himalayas, bulging downward as a *mountain root* into the mantle (figure 17.7). The continental crust is also less dense than oceanic crust, a fact that is important in plate tectonics (table 17.1).

The boundary that separates the crust from the mantle beneath it is called the **Mohorovičić discontinuity** (**Moho** for short). Notice on figure 17.7 that the mantle lies closer to Earth's surface

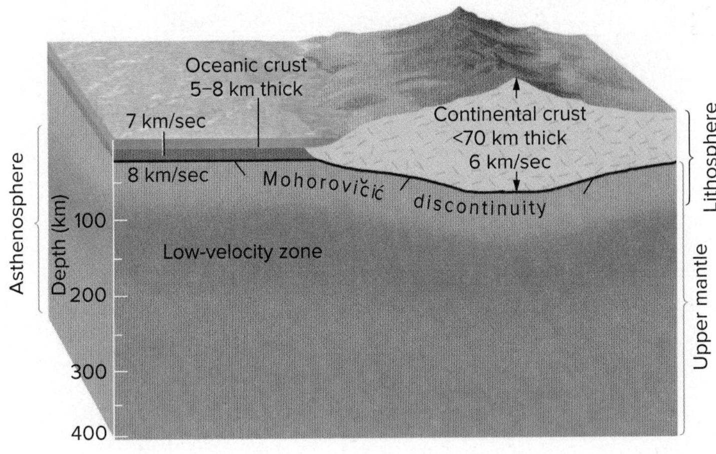

FIGURE 17.7

Thin oceanic crust has a P-wave velocity of 7 kilometers per second, whereas thick continental crust has a lower velocity. Mantle velocities are about 8 kilometers per second. The oceanic and continental crust, along with the upper rigid part of the upper mantle, form the lithosphere. The asthenosphere underlies the lithosphere and contains a low-velocity zone in its upper part that is defined by a decrease in P-wave velocities.

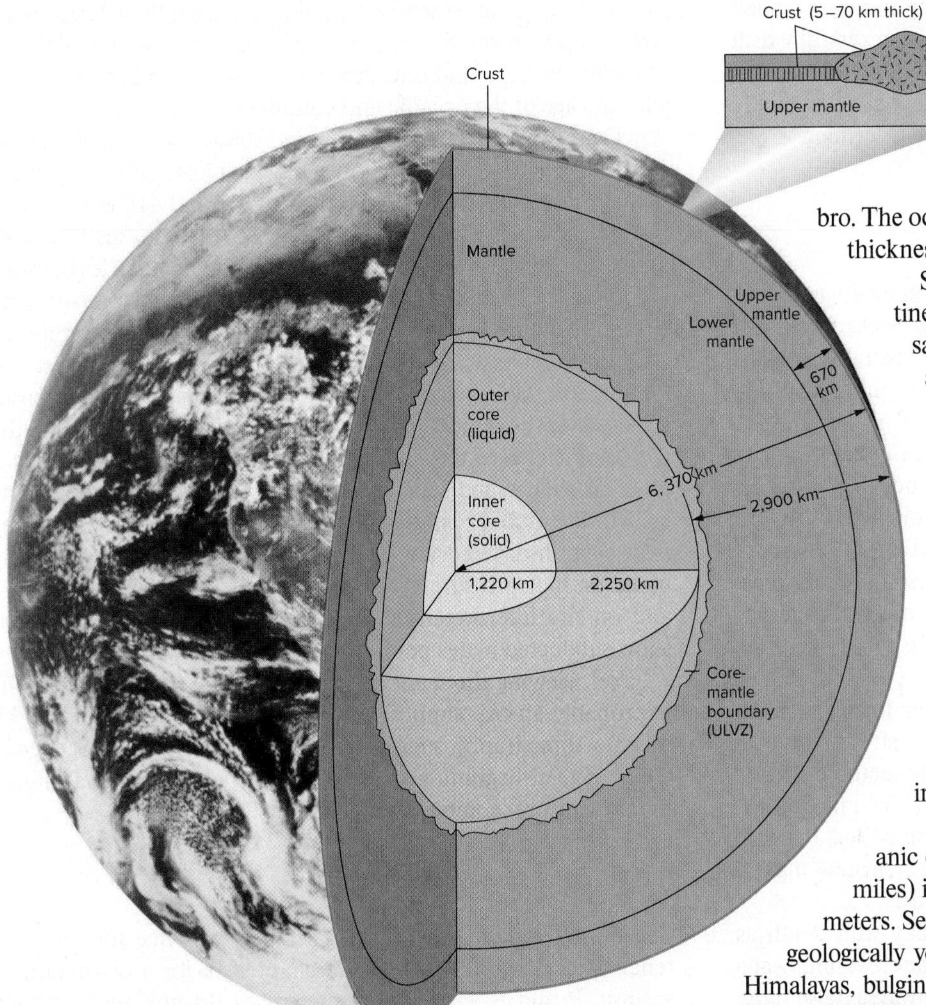

FIGURE 17.6

Earth's interior. Seismic waves show the three main divisions of Earth: the crust, the mantle, and the core.
Source: NASA

TABLE 17.1 Characteristics of Oceanic Crust and Continental Crust

	Oceanic Crust	Continental Crust
Average thickness	7 km	30 to 50 km (thickest under mountains)
Seismic P-wave velocity	7 km/second	6 km/second (higher in lower crust)
Density	3.0 g/cm^3	2.7 g/cm^3
Probable composition	Basalt underlain by gabbro	Granite, other plutonic rocks, schist, gneiss (with sedimentary rock cover)

beneath the ocean than it does beneath continents. Nearly 60 years ago, an ambitious program called *Project Mohole* was designed to use specially equipped ships to drill through the oceanic crust and obtain samples from the mantle. Although the project was abandoned because of high costs, ocean-floor drilling has become routine since then, but not to the great depth necessary to sample the mantle. A new drilling program, called the SloMo Project, began in 2015 and hopes to successfully drill to the mantle through the oceanic crust at a slow-spreading ridge in the southwestern Indian Ocean where the depth to the mantle is shallower and may be easier to reach. (Ocean drilling is discussed in more detail in chapters 18 and 19.)

The Mantle

Because both S and P waves pass through the mantle, geologists think that it, like the crust, is made of solid rock. Localized magma chambers of melted rock may occur as isolated pockets of liquid in both the crust and the upper mantle, but most of the mantle seems to be solid. Because P waves travel at about 8 kilometers per second in the upper mantle, it appears that the mantle is a different type of rock from either oceanic crust or continental crust. The best hypothesis that geologists can make about the composition of the upper mantle is that it consists of ultramafic rock such as peridotite. *Ultramafic rock* is dense igneous rock made up chiefly of ferromagnesian minerals such as olivine and pyroxene. Some ultramafic rocks contain garnet, and feldspar is extremely rare in the mantle.

The crust and uppermost mantle together form the **lithosphere**, the outer shell of Earth that is relatively strong and brittle. The lithosphere makes up the plates of plate-tectonic theory. The lithosphere averages about 70 kilometers (43.4 miles) thick beneath oceans and may be 125 to 250 kilometers thick beneath continents. Its lower boundary is marked by a curious mantle layer in which seismic waves slow down (figure 17.7).

Generally, seismic waves increase in velocity with depth as increasing pressure alters the properties of the rock. Beginning at a depth of 70 to 125 kilometers, however, seismic waves travel more slowly than they do in shallower layers, and so this zone has been called the *low-velocity zone* (figure 17.7). This zone, extending to a depth of perhaps 200 kilometers (124 miles), is also part of a larger layer called the **asthenosphere**. The rocks in this zone may be closer to their melting point than the rocks above or below the zone. (The rocks are probably not *hotter* than the rocks below—melting points are controlled by pressure as well as temperature.) Some geologists think that these rocks may actually be partially melted, forming a crystal-and-liquid slush; a very small percentage of liquid in the asthenosphere could help explain some of its physical properties.

If the rocks of the asthenosphere are close to their melting point, this zone may be important for two reasons: (1) it may represent a zone where magma is likely to be generated; and (2) the rocks here may have relatively little strength and therefore are likely to flow. If mantle rocks in the asthenosphere are weaker than they are in the overlying lithosphere, then the asthenosphere can deform easily by ductile flow. In this way, plates of brittle lithosphere gradually move over the asthenosphere, which may act as a lubricating layer below.

There is widespread agreement on the existence and depth of the asthenosphere under oceanic crust but considerable disagreement about asthenosphere under continental crust. Figure 17.7 shows asthenosphere at a depth of 125 kilometers (77.5 miles) below the continents. Some geologists think that the lithosphere is much thicker beneath older continental crust than shown in the figure and that the asthenosphere begins at a depth of 250 kilometers (or even more). Seismic evidence indicates that the thickness of the lithosphere, and thus depth to the asthenosphere, increases with the age of the oceanic and continental plates.

Data from seismic reflection and refraction indicate several concentric layers in the mantle (figure 17.8), with prominent boundaries at 400 and 670 kilometers (248 and 416 miles). (670 kilometers is also the depth of the deepest earthquakes.) Because pressure increases with depth into Earth, the boundaries between mantle layers probably represent depths at which pressure collapses the internal structure of certain minerals into denser minerals. For example, at a pressure equivalent to a depth of about 670 kilometers, the mineral *olivine* should collapse into the denser structure of the mineral *perovskite.* Some geologists think that the 670-kilometer boundary represents a chemical change as well as a physical change and separates the *upper mantle* from the chemically different *lower mantle* below. Recently, geologists think they may have discovered another, stiffer layer in the lower mantle based on high-pressure experiments on the mineral *ferropericlase* and on the interpretation of seismic tomography images that show subducted plates pooling at 1,500 kilometers (930 miles).

Yet, viewing the mantle as concentric layers (like an onion) is probably an oversimplification. Detailed images provided by seismic tomography suggest that the mantle is heterogeneous (see image at beginning of chapter), probably due to variations in temperature, composition, and density.

The Core

Seismic-wave data provide the primary evidence for the existence of the core of Earth. (See chapter 16 for a discussion of seismic P and S waves.) Seismic waves do not reach certain areas on the opposite side of Earth from a large earthquake. Figure 17.9 shows how seismic P waves spread out from a quake until, at 103° of arc (11,500 kilometers) from the epicenter, they suddenly disappear from seismograms. At more than 142°

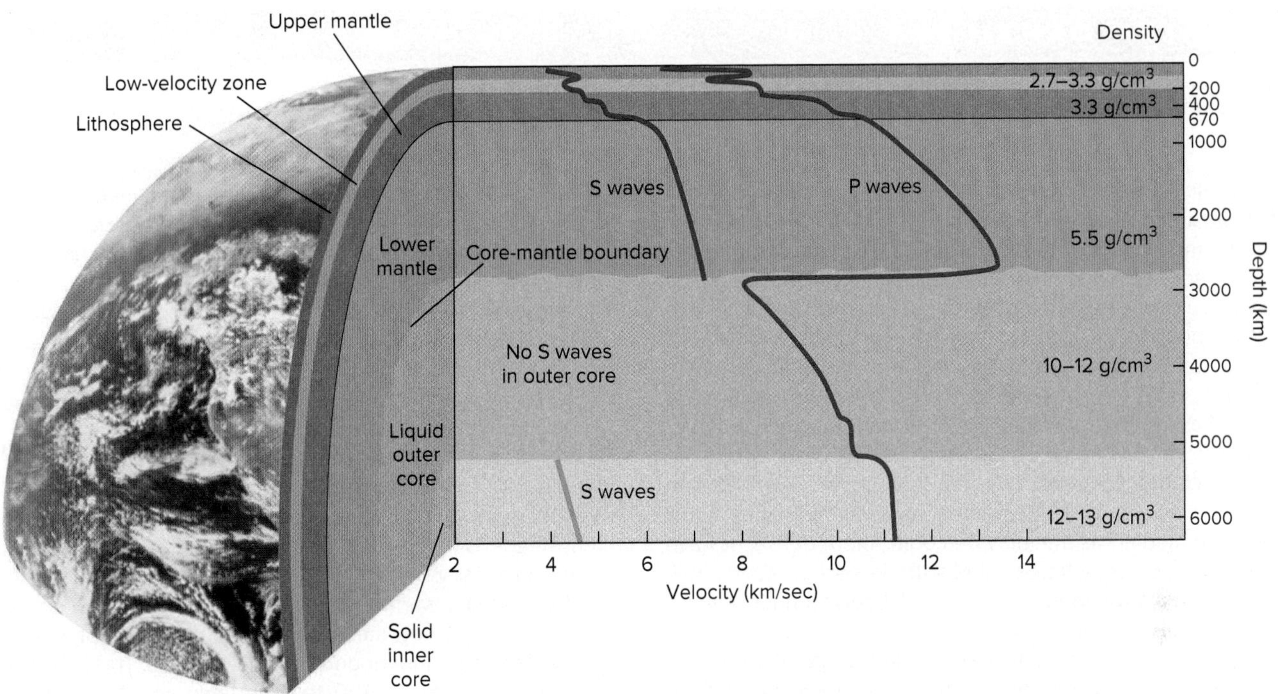

FIGURE 17.8

The concentric shell structure of Earth as defined by variation in S- and P-wave velocities and estimates of density. The velocity of seismic P and S waves generally increases with depth except in the low-velocity zone. Velocity increases at 400 and 670 kilometers may be caused by mineral collapse. S waves do not pass through the outer core but are thought to travel through the solid inner core.
Source: NASA

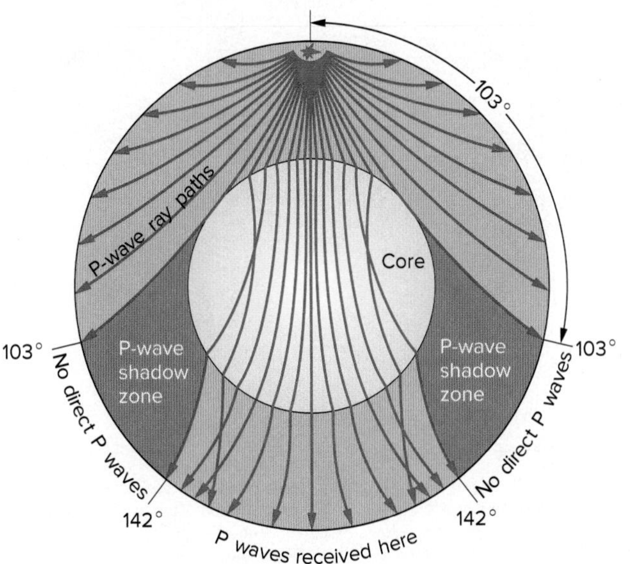

FIGURE 17.9

The P-wave shadow zone, caused by refraction of P waves within the Earth's core.

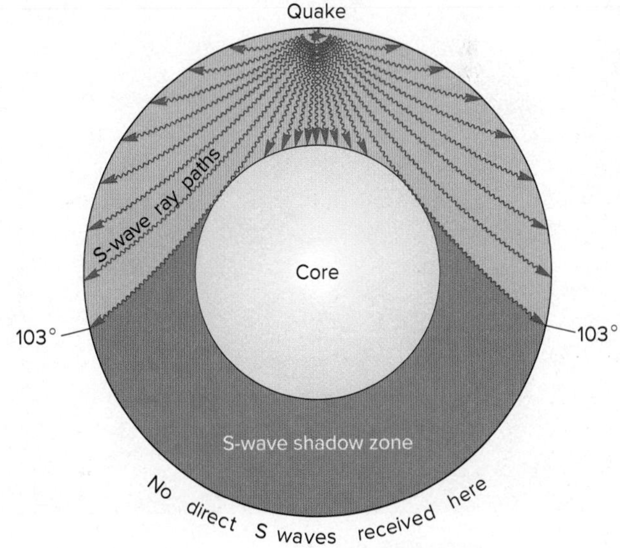

FIGURE 17.10

The S-wave shadow zone. Because no S waves pass through the core, the core is apparently a liquid (or acts like a liquid).

(15,500 kilometers) from the epicenter, P waves reappear on seismograms. The region between 103° and 142°, which lacks P waves, is called the **P-wave shadow zone**.

The P-wave shadow zone can be explained by the refraction of P waves when they encounter the core boundary deep within Earth's interior. Because the paths of P waves can be accurately calculated, the size and shape of the core can be determined also.

In figure 17.9, notice that Earth's core deflects the P waves and, in effect, "casts a shadow" where their energy does not reach the surface. In other words, P waves are missing within the shadow zone because they have been bent (refracted) by the core.

Chapter 16 on earthquakes explains that while P waves can travel through solids and fluids, S waves can travel only through solids. As figure 17.10 shows, an **S-wave shadow zone** also exists

IN GREATER DEPTH 17.1

Seismic Tomography—A Look Inside Earth

Geophysicists use a relatively new technique called **seismic tomography** to look inside Earth. Seismic tomography is similar to the medical technique of CT (or CAT—*computerized axial tomography*) scans used to build a three-dimensional picture of soft body tissue by taking a series of X-ray pictures along successive planes in the body.

Seismic tomography uses earthquake waves and powerful numerical models that run on supercomputers to study planar cross sections of Earth's interior following large earthquakes or explosions. Slight deviations in the observed arrival times are compared with the expected arrival times of seismic waves for many different travel paths through Earth. Faster-than-expected arrival times signal stronger, more rigid areas that may be cooler, more dense, or have a different composition, whereas slower-than-average waves indicate less rigid areas within Earth that are probably warmer or more deformable.

Box figure 1 shows a seismic tomography image of Earth with the outer layer of crust peeled back in the Southern Hemisphere to reveal the variation in seismic velocities in the mantle. The red areas below the southern Pacific Ocean have been interpreted as a super plume of warmer or chemically distinct mantle that has slowed seismic waves. The blue blobs indicate areas where the mantle is behaving as a more rigid, "cooler" material. Along the western coast of South America, the faster seismic waves probably reflect the subduction of the colder Pacific plate.

The remnants of the subducted Pacific plate to the north, the Farallon plate, have been imaged by seismic tomography under the United States (box figure 2). The grave of the old Farallon plate is visible under the western part of the country as a flat zone of higher seismic velocity (shown in blue and green) within the upper mantle that steepens to the east. Areas of slower seismic velocity (shown in red) represent compositional heterogeneity in the transition zone and additionally warmer, less rigid material in the uppermost mantle.

An exciting new study, called EarthScope, explores the details of the deep structure under the United States. The EarthScope project uses an array of 400 transportable seismometers that have been moved from west to east across the country. The seismometers have occupied more than 1,600 sites, and the data collected will provide an even sharper image of what lies below. Stay tuned.

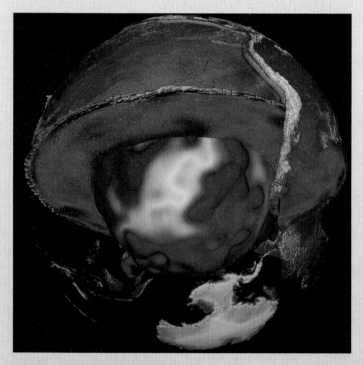

BOX 17.1 ■ FIGURE 1

Earth with the crust peeled back in Southern Hemisphere to reveal a seismic tomography image of the mantle. Areas with faster-than-average seismic waves are shown in blue and are probably cooler or differ in composition from the areas shown in red. The red area under the southern Pacific Ocean is interpreted to be an area of less rigid mantle that may represent plumes of warmer or chemically distinct mantle originating from the core–mantle boundary. The blue area just west of South America is more rigid and represents the subduction of the cooler Pacific plate.
©Thorsten W. Becker/University of Southern California

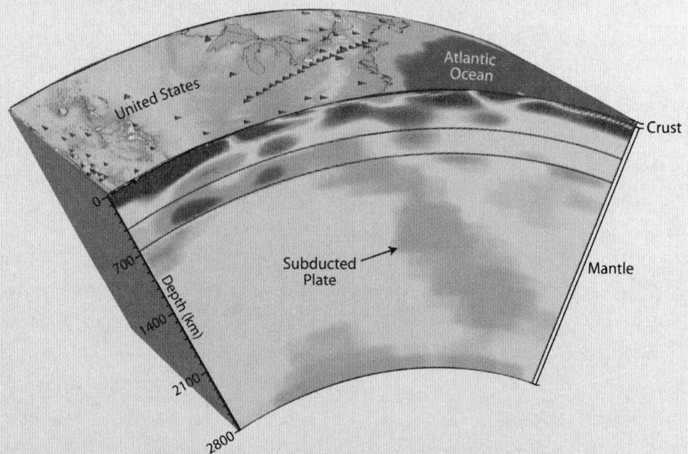

BOX 17.1 ■ FIGURE 2

Three-dimensional diagram showing a map surface of the United States (red triangles are seismic stations) and a profile view of a seismic tomograph of the lithosphere and mantle. The red areas have seismic velocities slower than average and probably represent warmer, less rigid areas, whereas the green and blue areas are more rigid and may be cooler. The green sloping anomaly is interpreted as a fossil subduction zone that contains remnants of the subducted Farallon plate.
Source: Suzan van der Lee, Northwestern University

and is larger than the P-wave shadow zone. Direct S waves are not recorded in the entire region more than 103° away from the epicenter. The S-wave shadow zone seems to indicate that S waves do not travel through the core at all. If this is true, it implies that the core of Earth is a liquid, or at least acts like a liquid.

The way in which P waves are refracted within Earth's core (as shown by careful analysis of seismograms) suggests that the core has two parts, a *liquid outer core* and a *solid inner core* (figure 17.8).

Composition of the Core

When evidence from astronomy and seismic-wave studies is combined with what we know about the properties of materials, it appears that Earth's core is made of metal—not silicate rock—and that this metal is probably a mixture of iron and nickel (along with minor amounts of oxygen, silicon, or sulfur). How did geologists arrive at this conclusion?

The overall density of Earth is 5.5 grams per cubic centimeter, based on calculations from Newton's law of gravitational attraction. The crustal rocks are relatively low density, from 2.7 grams per cubic centimeter for granite to 3.0 grams per cubic centimeter for basalt. The ultramafic rock thought to make up the mantle probably has a density of 3.3 grams per cubic centimeter in the upper mantle, although rock pressure should raise this value to about 5.5 grams per cubic centimeter at the base of the mantle (figure 17.8).

If the crust and the mantle, which have approximately 85% of Earth's volume, are at or below the average density of Earth, then the core must be very heavy to bring the average up to 5.5 grams per cubic centimeter.

Calculations show that the core has to have a density of about 10 grams per cubic centimeter at the core-mantle boundary, increasing to 12 or 13 grams per cubic centimeter at the center of Earth (figure 17.8). This great density would be enough to give Earth an average density of 5.5 grams per cubic centimeter.

Under the great pressures existing in the core, a mixture of iron and nickel would have a density slightly greater than that required in the core. The iron-nickel alloy mixed with a small amount of a lighter element, such as oxygen, potassium, sulfur, or silicon, would have the required density. Therefore, many geologists think that such a mixture makes up the core.

But a study of density by itself is hardly convincing evidence that the core is a mixture of iron and nickel, for many other heavy substances could be there instead. The choice of iron as the major component of the core comes from looking at meteorites. Meteorites are thought by scientists to be remnants of the basic material that created our own solar system. An estimated 10% of meteorites are composed of iron mixed with small amounts of nickel. Material similar to these meteorites almost certainly helped create Earth, perhaps settling to the center of Earth because of metal's high density. The composition of these meteorites, then, may provide clues to what is in the core. Nickel is denser than iron, however, so a mixture of just iron and nickel would have a density greater than that required in the core. (The other 90% of meteorites are mostly ultramafic rock and perhaps represent material that formed the mantle and crust of ancient planetesimals that formed early in the history of the solar system.)

Seismic and density data, together with assumptions based on meteorite composition, point to a core that contains iron and nickel, with at least the outer part being liquid. The existence of Earth's magnetic field, which is discussed in this chapter, also suggests a metallic core. Of course, no geologist has seen the core, but since so many lines of indirect evidence point to a liquid metal outer core, most scientists accept this theory as the best conclusion that can be made about the core's composition.

The Core-Mantle Boundary

The boundary between the core and mantle is marked by great changes in seismic velocity, density (figure 17.8), and temperature, as we will see in the section on "Heat Within the Earth" in this chapter. Here, there is a transition zone up to 200 kilometers thick, known as the *D″ layer*, at the base of the mantle where P-wave velocities decrease dramatically. The *ultralow-velocity zone* (ULVZ) (figure 17.11) that forms the undulating border at the core-mantle boundary may be due to hot core partially melting overlying mantle rock or to part of the liquid outer core reacting chemically with the adjacent mantle. The latest seismic and geodetic studies have hinted that lighter iron alloys from the liquid outer core may react with silicates in the lower mantle to form iron silicates. The less-dense iron silicate "sediment," along with liquid iron in pore spaces, rises and collects in uneven layers along the core-mantle boundary. The pressure of the accumulating

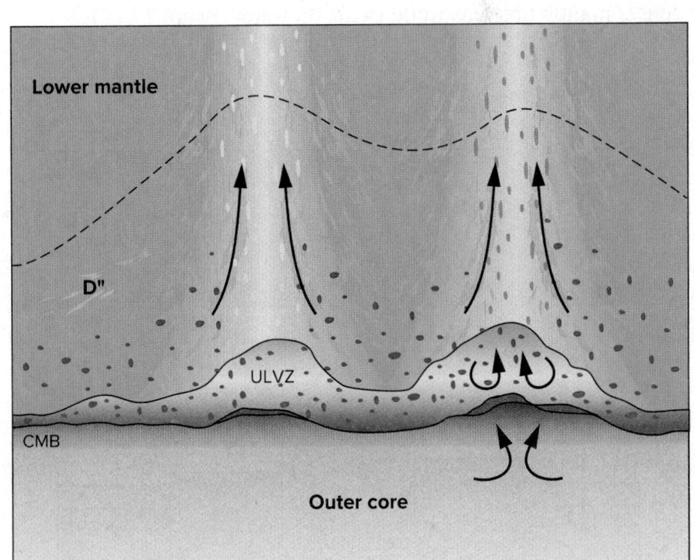

FIGURE 17.11

Seismic and geodetic studies are redefining the boundary between the lower 200 kilometers of the mantle (*D* layer) and the outer core. Iron silicate "sediments" (shown in brown) may rise from the underlying liquid core and fill pockets or inverted basins at the core–mantle boundary (CMB). Alternatively, the outer core material (shown in red) may be melting the lowermost mantle (shown in yellow) to form the ultralow-velocity zone (ULVZ).

*Source: Garnero and Jeanloz, **Science**, 2000.*

"sediment" along the boundary causes some of the liquid iron to be squeezed out of the pore spaces to form an electrically conductive layer that connects the core and mantle and explains the decrease in seismic velocities at the ULVZ. One of the grand challenges for geophysics in the near future is to explore whether the lowermost mantle is being partially melted by the core or whether the core is instead chemically reacting with the mantle.

Both the mantle and the core are undergoing **convection**, a circulation pattern in which warmer low-density material rises and cooler high-density material sinks. Based on seismic tomography studies, heavy portions of the mantle (including subducted plates) sink to its base but are unable to penetrate the denser core. Light portions of the core may rise to its top and may be incorporated into the mantle above. This is suggested by recent isotopic studies of the mantle plume that feeds the Hawaiian hot spot. The resulting Hawaiian volcanic rocks (basalts) contain a light-isotope signature that is characteristic of the core. Continent-sized blobs of liquid and liquid-crystal slush may accumulate at the core-mantle boundary, perhaps interfering with or aiding heat loss from the core to help drive mantle convection and transfer heat to the surface, and causing changes in Earth's magnetic field. This boundary is an exciting frontier for geologic study, but data, for now, are sparse and hard to obtain.

17.4 ISOSTASY

Isostasy is a balance or *equilibrium* of adjacent blocks of brittle lithosphere "floating" on the asthenosphere. Since lithosphere weighs less than mantle rocks, the crust can be thought of as floating on the denser mantle much as wood floats on water (figure 17.12).

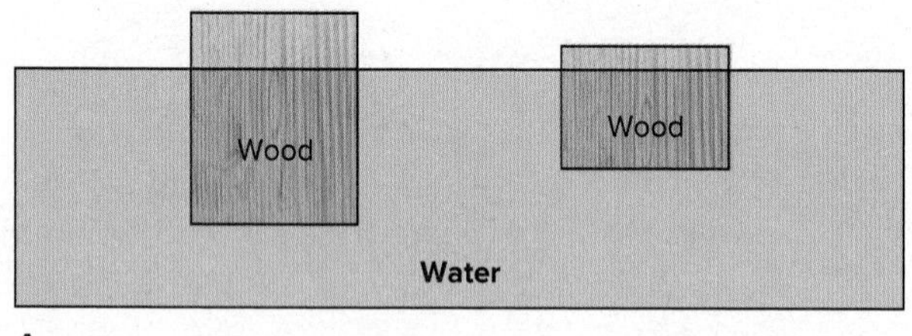

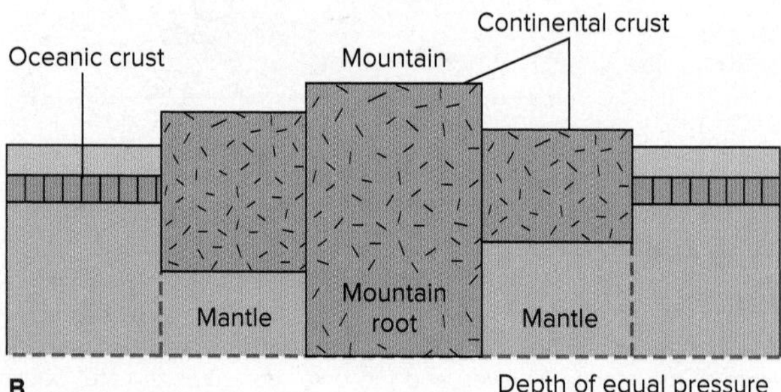

FIGURE 17.12

Isostatic balance. (*A*) Wood blocks float in water with most of their bulk submerged. (*B*) Over geologic time scales, the mantle can behave much like a fluid, where the lithosphere "floats" on the mantle, in approximately the same way as wood in water. The thicker the block, the deeper it extends into the asthenosphere.

Blocks of wood floating on water rise or sink until they displace an amount of water equal to their own weight. The weight of the displaced water buoys up the wood blocks, allowing them to float. The higher a wood block appears above the water surface, the deeper the block extends under water. Thus, a tall block has a deep "root"—much like an iceberg floating on the water.

In a greatly simplified way, the lithosphere can be thought of as tending to rise or sink gradually until it is balanced by the weight of displaced asthenosphere. This concept of vertical movement to reach equilibrium is called **isostatic adjustment**. Just as with the blocks of wood, once lithospheric "blocks" have come into isostatic balance, a tall block (a mountain range) extends deep into the mantle (a *mountain root*, as shown in figure 17.12).

Figure 17.12 shows both the blocks of wood and the blocks of lithosphere in isostatic balance. The weight of the wood is equal to the weight of the displaced water. Similarly, the weight of the lithosphere is equal to the weight of the displaced asthenosphere. As a result, the rocks (and overlying seawater) in figure 17.12 can be thought of as separated into vertical columns, each with the same pressure at its base. At some *depth of equal pressure*, each column is in balance with the other columns, for each column has the same weight. A column of thick continental lithosphere (a mountain and its root) has the same weight as a column containing thin continental lithosphere and some of the upper mantle. A column containing seawater, thin oceanic crust, and a thick section of heavy mantle weighs the same as the other two columns.

Figure 17.12 shows the lithosphere as isolated blocks free to move past each other along vertical faults, but this is not really a good picture of the structure of the lithosphere. It is more accurate to think of the lithosphere as bending in broad uplifts and downwarps without vertical faults, as shown in figures 17.13 and 17.14.

Let us look at some examples of isostatic balance (equilibrium) in continental lithosphere rocks. Suppose that two sections of crust of unequal thickness are next to each other, as in figure 17.13. Sediment from the higher part of the crust, which is subject to more rapid erosion, is deposited on the lower part. The decrease in weight from the high part of the crust causes it to rise, while the increase in weight on the low part causes it to sink. These vertical movements (isostatic adjustment) do, in fact, take place whenever large volumes of material are eroded from or deposited on parts of the crust.

Rising or sinking of the lithosphere, of course, requires ductile flow of the asthenosphere to accommodate the motion. By measuring the rate of isostatic adjustment, we can calculate the viscosity of the asthenosphere.

Another example of isostatic adjustment, caused by ductile mantle flow, is the upward movement of large areas of the crustal lithosphere since the glacial ages. The weight of the thick continental ice sheets during the Pleistocene Epoch depressed the lithosphere underneath the ice (figure 17.14). After the melting of the ice, the crust rose back upward, a process that is still going on in northeastern Canada and in the Scandinavian countries. This rise of the crustal lithosphere after the removal of the ice is known as **crustal rebound**. The process of crustal rebound can be easily demonstrated by sitting on a soft couch or bed. The indentation made from the weight of your body gradually disappears or rebounds after you stand up.

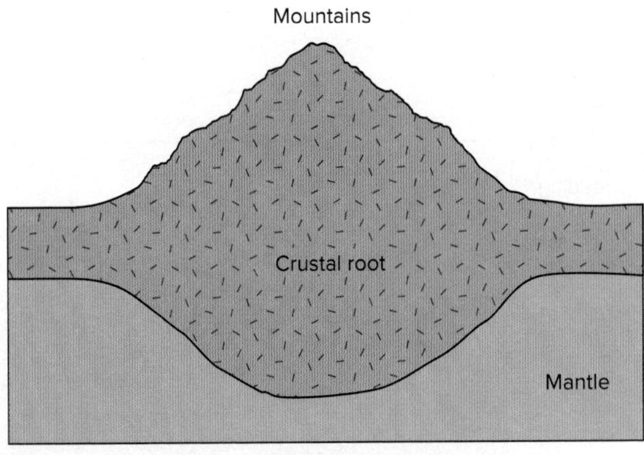

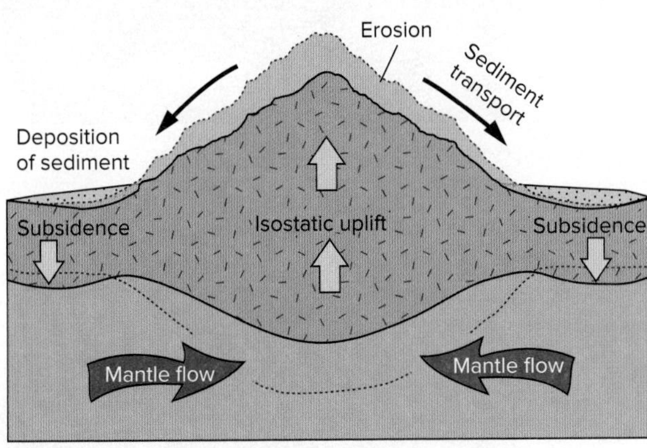

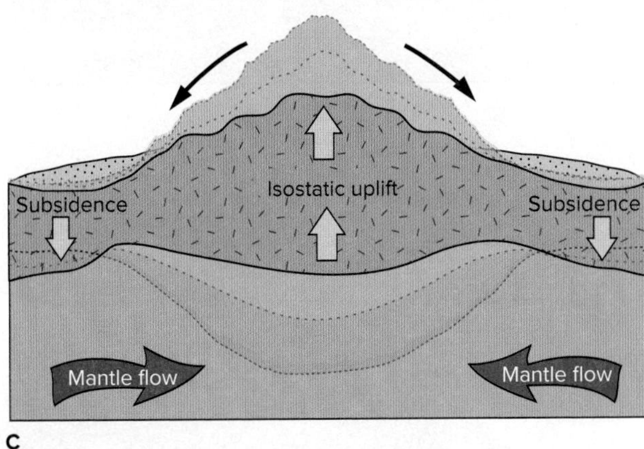

FIGURE 17.13

Isostatic adjustment due to erosion and deposition of sediment. Rock within the mantle must flow to accommodate the vertical motion of crustal blocks. Mantle flow occurs in the asthenosphere, deeper than shown in *C*.

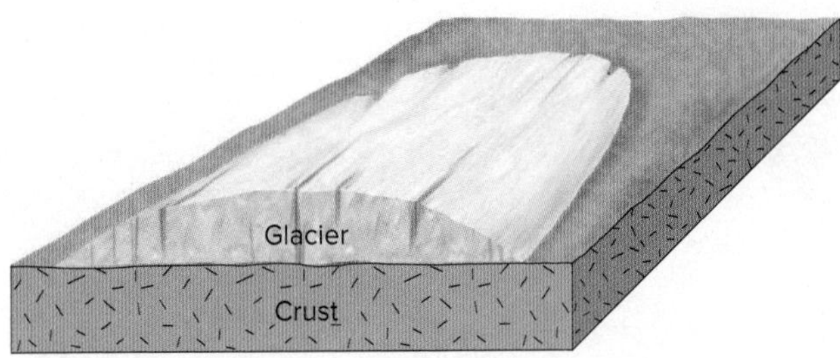

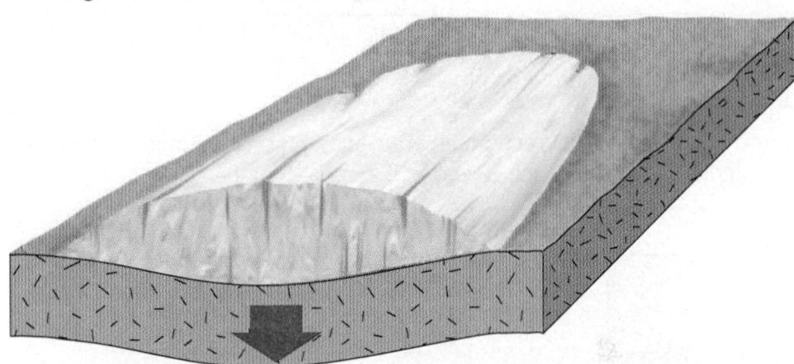

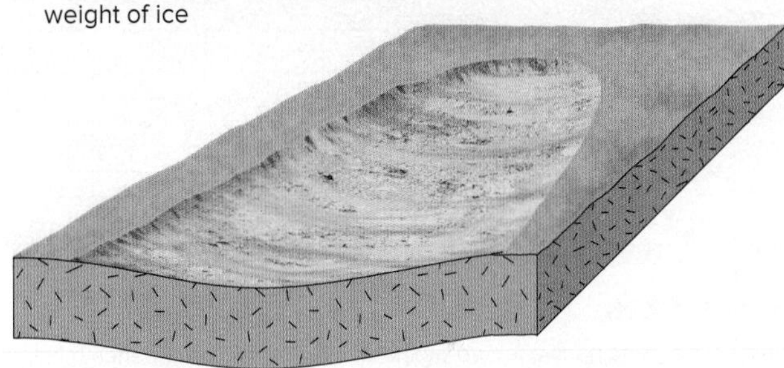

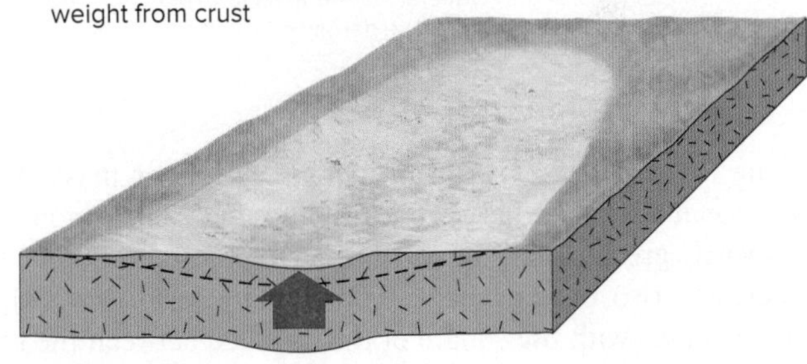

A Glacier forms, adding weight to crust

B Subsidence due to weight of ice

C Ice melts rapidly, removing weight from crust

D Crustal rebound as crust slowly rises toward original position

FIGURE 17.14

The weight of glaciers depresses the crust, and the crust rebounds when the ice melts.

Recent geophysical studies have shown that some mountains, such as the Rockies and southern Sierra Nevada, do not have thick roots and are instead buoyed by warm, less-dense mantle. It appears that the upper mantle beneath some continents is not homogeneous but has zones that are quite buoyant due to higher temperatures and less-dense mineral phases.

17.5 GRAVITY MEASUREMENTS

According to Newton's law of gravitation, the force of gravity between two objects varies with the masses of the objects and the distance between them (figure 17.15):

$$\begin{matrix}\text{Force of gravity} \\ \text{between } A \text{ and } B\end{matrix} = \text{constant}\left(\frac{\text{mass}_A \times \text{mass}_B}{\text{distance}^2}\right)$$

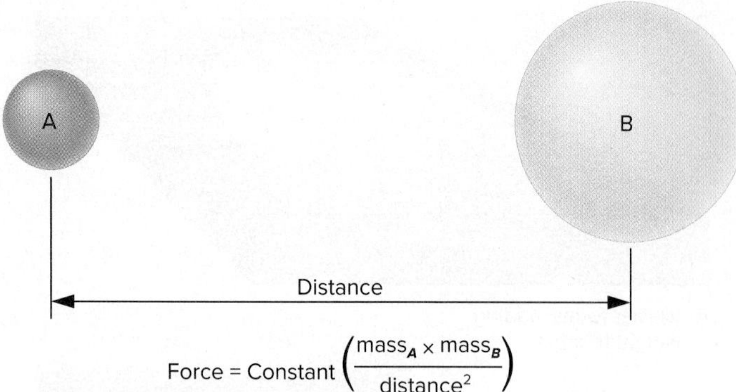

$$\text{Force} = \text{Constant}\left(\frac{\text{mass}_A \times \text{mass}_B}{\text{distance}^2}\right)$$

FIGURE 17.15

The force of gravitational attraction between two objects is a function of the masses of the objects and the distance between the centers of the objects.

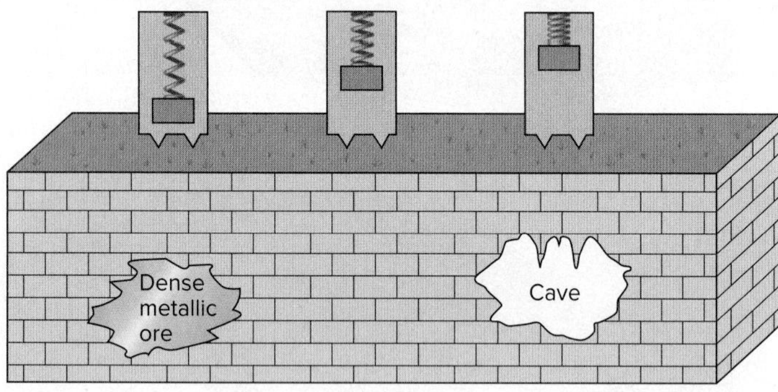

FIGURE 17.16

A gravity meter reading is affected by the density of the rocks beneath it. Dense rock pulls strongly on the mass within the meter, stretching a spring; a cavity exerts a weak pull on the mass. A gravity meter can be used to explore for hidden ore bodies, caves, and other features that have density contrasts with the surrounding rock.

The force increases with an increase in either mass. The gravitational attraction between Earth and the Moon, for example, is vastly greater than the extremely small attraction that exists between two bowling balls. The equation also shows that force decreases with the square of the distance between the two objects. In other words, as two objects are moved apart from one another, the gravitational attraction between them decreases very rapidly in a nonlinear fashion.

A useful tool for studying the crust and upper mantle is the **gravity meter**, which measures the gravitational attraction between Earth and a mass within the instrument. One use of the gravity meter is to explore for local variations in rock density (mass = density × volume). Dense rock such as metal ores and ultramafic rock pulls strongly on the mass inside the meter (figure 17.16). The strong pull stretches a spring, and the amount of stretching can be very precisely determined. So a gravity meter can be used to explore for metallic ore deposits. A cavity or a body of low-density material such as sediment causes a much weaker pull on the meter's mass (figure 17.16).

Another important use of a gravity meter is to discover whether regions are in isostatic equilibrium. If a region is in isostatic balance, as in figure 17.17*A*, each column of rock has

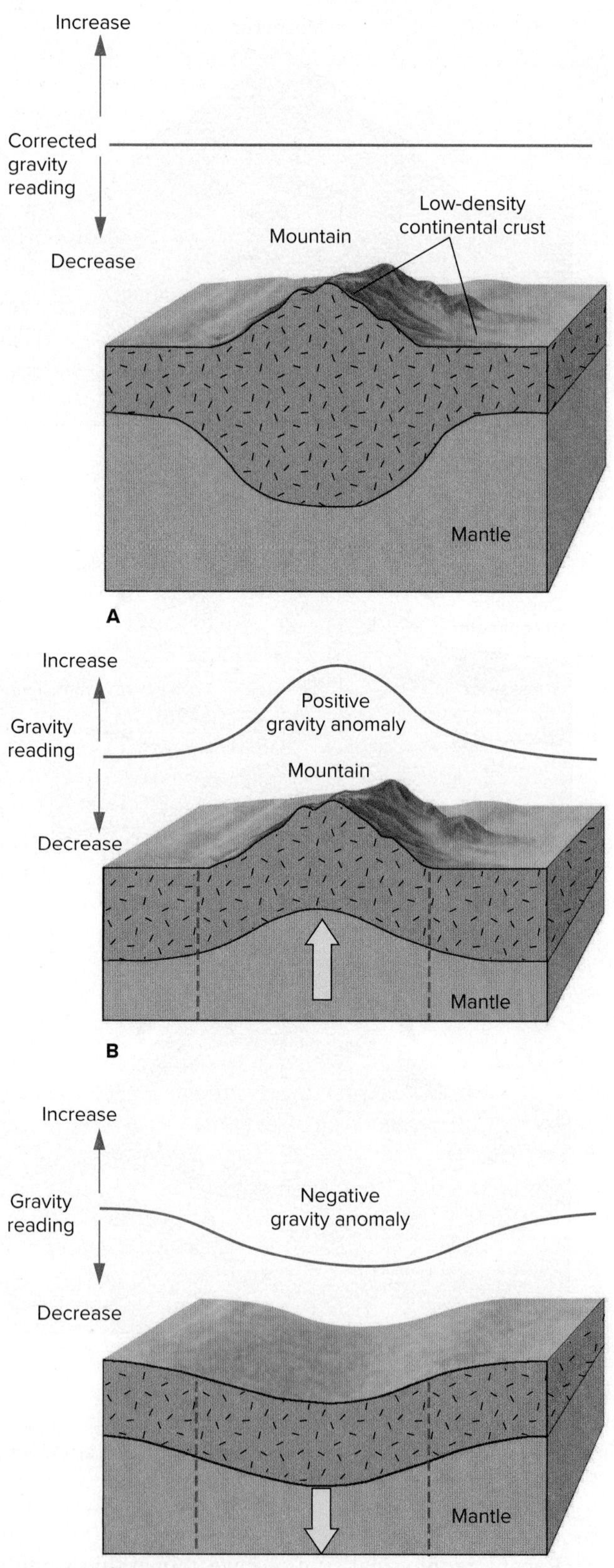

FIGURE 17.17

(*A*) A region in isostatic balance gives a uniform gravity reading (no gravity anomalies), after correcting for differences in elevation. (*B*) A region being held up out of isostatic equilibrium gives a positive gravity anomaly. (*C*) A region being held down out of isostatic equilibrium gives a negative gravity anomaly.

the same mass. If a gravity meter were carried across the rock columns, it would register the same amount of gravitational attraction for each column (after correcting for differences in elevation–gravitational attraction is less on a mountaintop than at sea level because the mountaintop is farther from the center of Earth).

Some regions, however, are held up out of isostatic equilibrium by deep tectonic forces. Figure 17.17*B* shows a region with uniformly thick crust. Tectonic forces are holding the center of the region up. This uplift creates a mountain range without a mountain root. There is a thicker section of heavy mantle rock under the mountain range than there is on either side of the mountain range. Therefore, the central "column" has more mass than the neighboring columns, and a gravity meter shows that the gravitational attraction is correspondingly greater over the central than over the side columns.

A gravity reading higher than the normal regional gravity is called a **positive gravity anomaly** (figure 17.17*B*). It can indicate that tectonic forces are holding a region up out of isostatic equilibrium, as shown in figure 17.17*B*. When the forces stop acting, the land surface sinks until it reestablishes isostatic balance. The gravity anomaly then disappears.

Positive gravity anomalies, particularly small ones, are also caused by local concentrations of dense rock such as metal ore. The gravity meter in figure 17.16 is registering a positive gravity anomaly over ore (the spring inside the meter is stretched). Since there can be more than one cause of a positive gravity anomaly, geologists usually try to combine gravity surveys with other geophysical measurements (e.g., magnetic anomalies) to refine their interpretations. Drilling into a region with a gravity anomaly ultimately discloses the true reason for the anomaly but is frequently too costly to justify.

A region can also be held down out of isostatic equilibrium, as shown in figure 17.17*C*. The mass deficiency in such a region produces a **negative gravity anomaly**–a gravity reading lower than the normal regional gravity. Negative gravity anomalies indicate either that a region is being held down (figure 17.17*C*) or that local mass deficiencies exist for other reasons (figure 17.16).

Some of the most impressive negative gravity anomalies are found in areas still experiencing isostatic rebound, such as northern Canada, which is still re-equilibrating from the melting of the glaciers of the last ice age. The gravity anomaly will disappear when isostatic rebound reaches equilibrium.

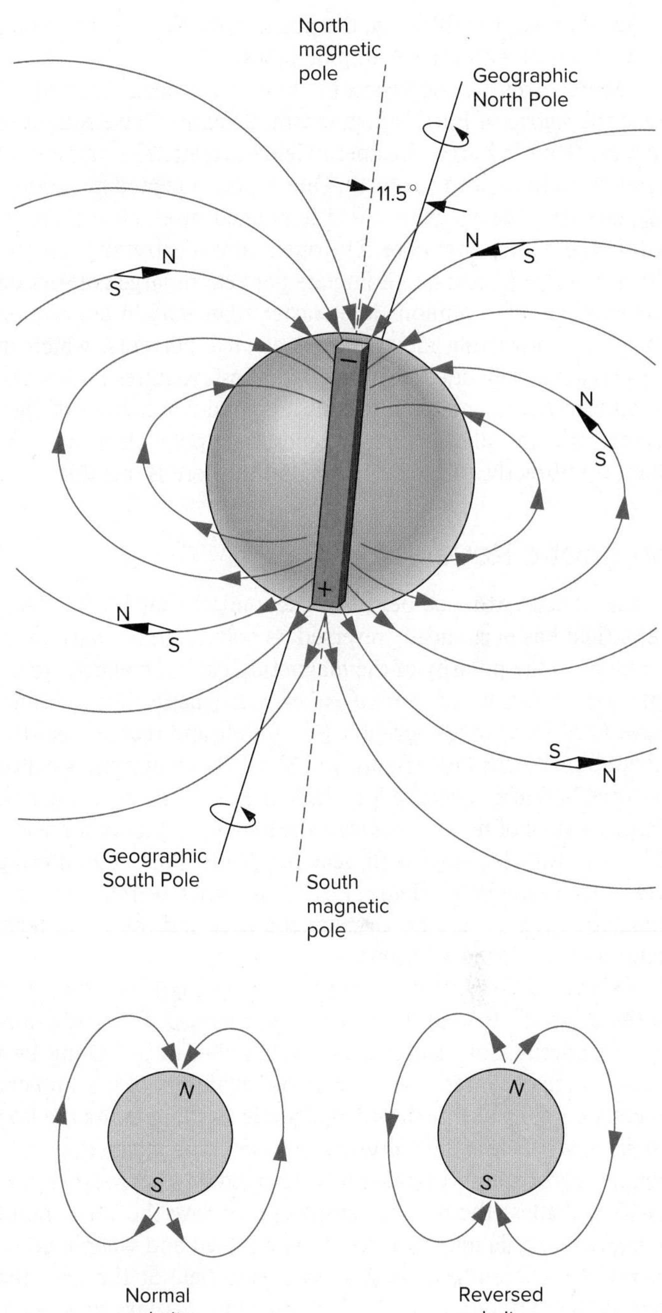

FIGURE 17.18

The Earth's magnetic field. The depiction of the internal field as a large bar magnet is a simplification of the real field, which is more complex. N and S in the two small figures indicate the *geographic* poles.

17.6 EARTH'S MAGNETIC FIELD

A region of magnetic force–a **magnetic field**–surrounds Earth. These invisible lines of magnetic force surrounding Earth deflect magnetized objects, such as compass needles, that are free to move. The field has north and south **magnetic poles**; one near the geographic North Pole and the other near the geographic South Pole. (Because it has two prominent poles, Earth's field is called *dipolar.*) The strength of the magnetic field is greatest at the magnetic poles, where magnetic lines of force appear to leave and enter Earth vertically (figure 17.18).

Because the compass is important in navigation, Earth's magnetism has been observed for centuries. Typically, there is a difference in the position of the magnetic and geographic poles (the latter of which are defined by the spin axis of the Earth). The positions of the magnetic poles are known to change on decadal time scales, and these variations have been well documented by ship captains navigating the oceans over the last 500 years. Generally, the magnetic poles appear to be moving slowly around the geographic poles. Yet, when averaged over time scales of about 10,000 years, the position of the magnetic poles coincides well with the geographic poles.

More recently, geophysical studies have been directed toward the *source* of Earth's magnetism. Scientists have sought to answer, "How is Earth's magnetic field generated?" A number of hypotheses have been put forth. One widely accepted hypothesis suggests that the magnetic field is created by electric currents within the liquid outer core. The outer core is extremely hot and flows at a rate of several kilometers per year in large convection currents, about 1 million times faster than mantle convection above it. Convecting metal creates electric currents, which in turn create a magnetic field. This hypothesis requires the core to be an electrical conductor. Metals are good conductors of electricity, whereas silicate rock is generally a poor electrical conductor. Indirectly, this is evidence that the core is metallic.

Magnetic Reversals

In the 1950s, evidence began to accumulate that Earth's magnetic field has periodically reversed its polarity in the past. Such a change in the polarity of the magnetic field is a **magnetic reversal**. During a time of *normal polarity*, magnetic lines of force leave Earth near the geographic South Pole and reenter near the geographic North Pole (figure 17.18). This orientation is called "normal" polarity because it is the same as the present polarity. During a time of *reversed polarity*, the magnetic lines of force run the other way, leaving Earth near the North Pole and entering near the South Pole (figure 17.18). In other words, during a magnetic reversal, the north magnetic pole and the south magnetic pole exchange positions.

Many rocks contain a record of the strength and direction of the magnetic field *at the time the rocks formed.* When the mineral magnetite, for example, is crystallizing in a cooling lava flow, it aligns its internal magnetization with that of the ambient magnetic field, specifically, the magnetic north pole. As the lava cools slowly below the **Curie point** (580°C for magnetite), this magnetic record is permanently trapped in the rock (figure 17.19*A*). Unless the rock is heated again above the Curie point temperature, this magnetic record is retained and when studied reveals the direction of Earth's magnetic field at the time the lava cooled. Other rock types, including sedimentary rocks such as limestones, mudstones, and red-colored sandstones stained red by iron-oxide minerals, also record former magnetic field directions. For a rock to record the Earth's magnetic field, it must contain iron-bearing minerals, such as magnetite (Fe_3O_4) and/or hematite (Fe_2O_3), which are capable of maintaining a permanent magnetization, or *remanence*, over geologic time scales (e.g., billions of years). The study of ancient magnetic fields is called **paleomagnetism**.

Most of the evidence for magnetic reversals comes from lava flows on the continents. Paleomagnetic studies of a series of stacked lava flows often show that some of the lava flows have a magnetic orientation directly opposite that of Earth's present orientation (figure 17.19*B*). That is, at the time these lava flows cooled, the magnetic poles had exchanged positions. During this time of magnetic reversal, a compass needle would have pointed

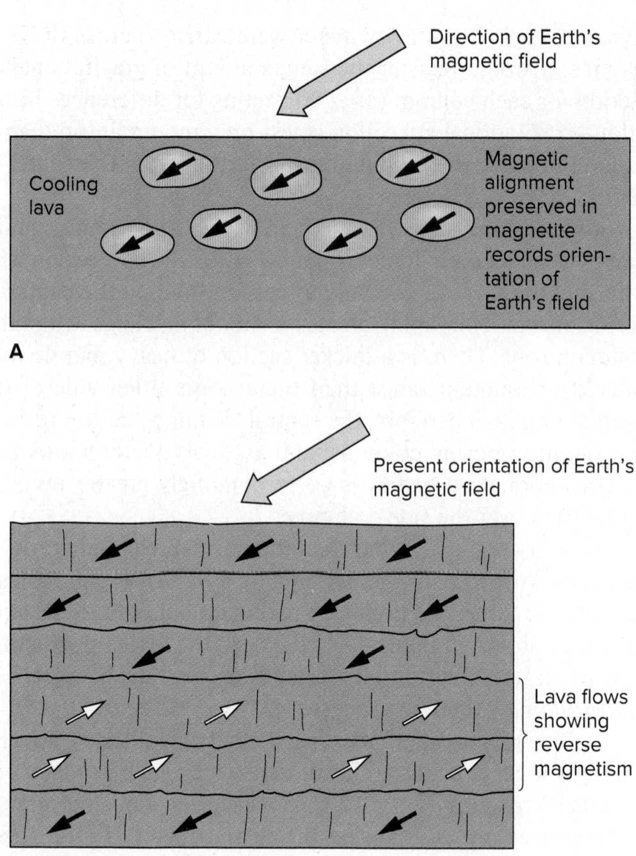

FIGURE 17.19

(*A*) Some rocks preserve a record of Earth's magnetic field. (*B*) Cross section of stacked lava flows showing evidence of magnetic reversals.

south rather than north. Many periods of normal and reverse magnetization are recorded in continental lava flows. They are worldwide events. Since lava flows can be dated isotopically, the time of these reversals in Earth's past can be determined. Although reversals appear to occur randomly (figure 17.20), records for the last ten million years suggest that Earth's field reverses on average about once every 500,000 years. The present normal orientation has lasted for the past 700,000 years. It takes time for one magnetic orientation to die out and the reverse orientation to build up. Most geologists think that it takes roughly 10,000 years for a reversal to develop, although new evidence from lavas in Nevada suggests that a reversal can occur over time scales of a few years.

A record of all the geomagnetic reversals that have occurred over the last 200 million years is well preserved within the oceanic crust (chapter 19) and forms the basis for the **geomagnetic polarity time scale** (figure 17.20). Reversals preserved within the basalts and sheeted dikes of the oceanic crust (chapter 18) were first discovered in the 1940s and 1950s by researchers who were trying to improve the way in which submarines could navigate the deep oceans. Ultimately, the normal and reversed magnetizations preserved within the oceanic crust served as the basis for the development of the modern theory of plate tectonics and the Vine-Matthews hypothesis of seafloor magnetic anomalies (chapter 19).

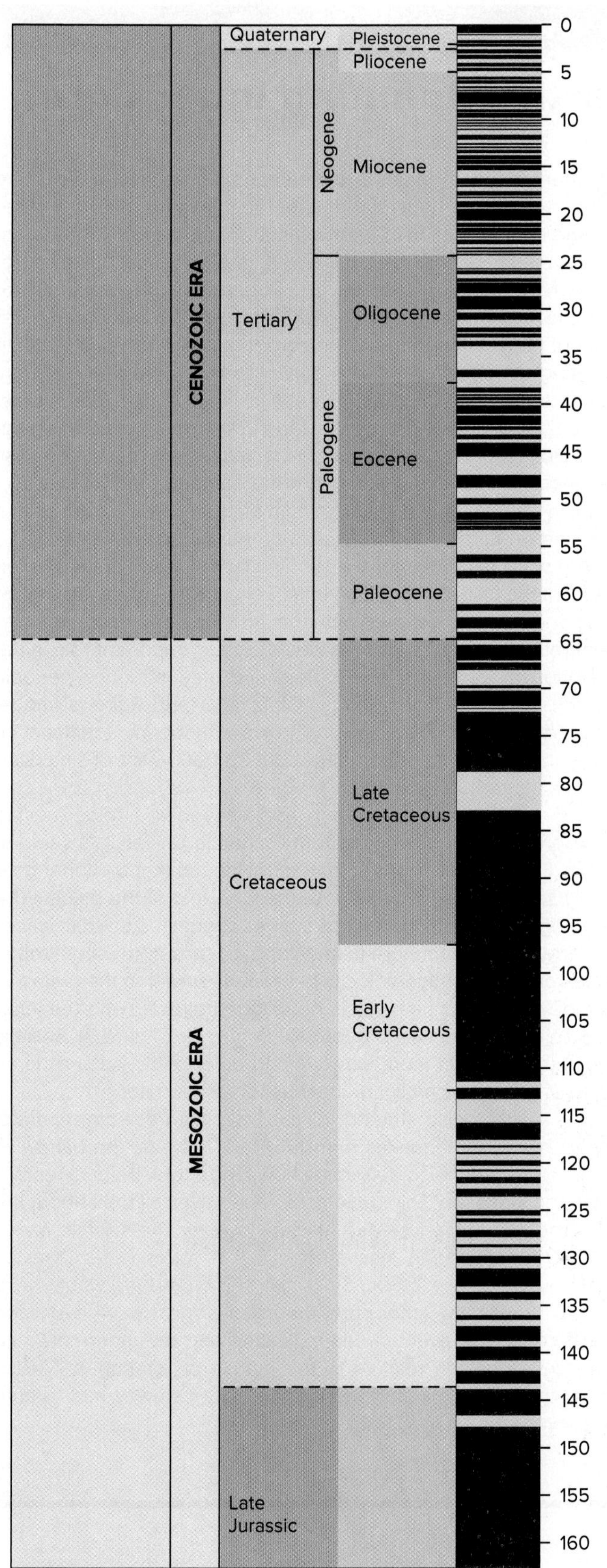

FIGURE 17.20

Worldwide magnetic polarity time scale for the Cenozoic and Mesozoic Eras. Black indicates positive anomalies (and therefore normal polarity). Tan indicates negative anomalies (reverse polarity).

What causes magnetic reversals? The question is difficult to answer because no one knows how the magnetic field is generated in the first place. Computer models developed in the past few decades and seismological research support the theory that the magnetic field is generated by convection currents in the liquid outer core (see box 17.2). If the field is caused by convection currents within the liquid outer core, perhaps a reversal is caused when the currents change direction or when a temporary current builds up and then dies out. Some geologists think that reversals may be triggered by the impact of an asteroid or comet with Earth; other geologists disagree.

A magnetic reversal can have some profound effects on Earth. In most instances, the strength of Earth's magnetic field declines to 20% of its starting value before the orientation begins to reverse; then the field strength increases to its usual value but in the opposite orientation. This collapse of the magnetic field means that harmful cosmic radiation from the Sun would be much more intense at the Earth's surface. When the magnetic field is at its usual strength, it shields Earth from these rays, but when the field collapses, this shielding is lost. Cosmic radiation negatively affects organisms, and some researchers have proposed that the increased exposure to radiation during a geomagnetic reversal would lead to increases in the rate of skin cancer among humans. Additionally, satellite communication would be severely hampered by increased levels of solar radiation, which would damage onboard instrumentation. The extinction of some species and the appearance of new species by mutation have been tentatively correlated with some magnetic reversals; however, far more reversals than mass extinctions have occurred, and it remains unclear whether there is a strong link between the Earth's magnetic field and biologic evolution.

Magnetic Anomalies

A **magnetometer** is an instrument used to measure the strength of Earth's magnetic field. A magnetometer can be carried over the land surface or flown over land or sea. At sea, magnetometers can also be towed behind ships. They are also used as metal detectors in airports.

The strength of Earth's magnetic field varies from place to place. As with gravity, a deviation from average readings is called an *anomaly.* Very broad (hundreds of km), regional magnetic anomalies may be due to *circulation patterns in the liquid outer core* or to other deep-seated causes. Smaller anomalies generally reflect *variations in rock type*, for the magnetism of near-surface rocks adds to the main magnetic field generated in the core. Rocks differ in their magnetism, depending upon their content of iron-containing minerals, particularly magnetite.

A **positive magnetic anomaly** is a reading of magnetic field strength that is higher than the regional average. Figure 17.21 shows three geologic situations that can cause positive magnetic anomalies. In figure 17.21*A*, a body of magnetite ore (a highly magnetic ore of the metal iron) has been emplaced in a bed of limestone by hot solutions rising along a fracture. The magnetism of the iron ore adds to the magnetic field of Earth, giving a stronger magnetic field measurement at the surface (a positive

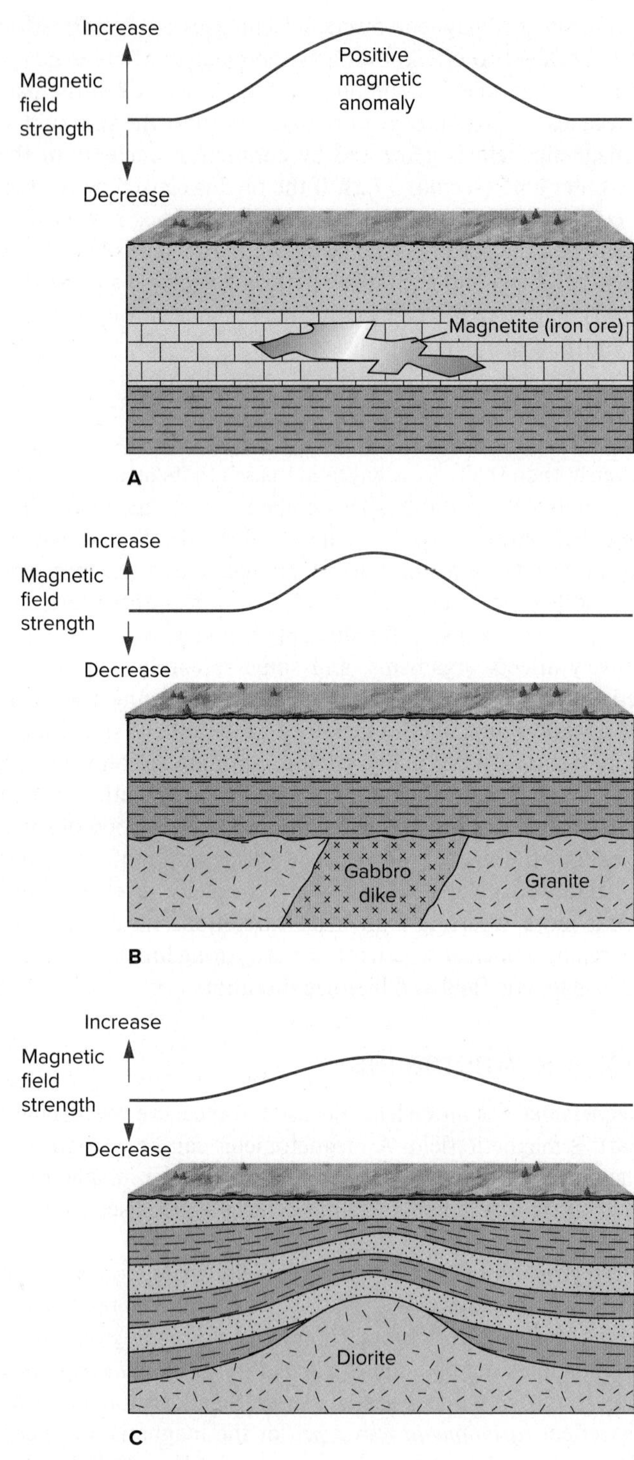

FIGURE 17.21

Positive magnetic anomalies can indicate hidden ore and geologic structures.

IN GREATER DEPTH 17.2

Earth's Spinning Inner Core

Supercomputer models and the analysis of seismic waves have led to a better understanding of the dynamics of Earth's inner core and the generation of Earth's magnetic field and periodic magnetic reversals. Gary A. Glatzmaier of Los Alamos National Laboratory in New Mexico and Paul H. Roberts of the University of California–Los Angeles developed a very sophisticated computer model of convection in the outer core that has been successful in simulating a magnetic field very similar to that measured on Earth. The model utilizes circulating metallic fluids in the outer core, caused by cooling and heat loss, as the driving force of Earth's magnetic field. The circulation of metallic fluids in the outer core has been theorized for many years, and the computer model was successful in simulating and maintaining a magnetic field similar to that measured on Earth. The model also predicted that Earth's solid inner core spins faster than the rest of the planet, gaining a full lap on the rest of the planet every 150 years. Because the magnetic lines of force penetrate and connect both the inner and outer core, a faster rate of rotation of the inner core would play an important role in the generation of Earth's magnetic field and may influence periodic magnetic reversals. Interestingly, Glatzmaier and Roberts' model produced a magnetic reversal on its own without any additional input from the experimenters after about 35,000 years of simulated time (box figure 1).

The results of this computer model inspired seismologists Xiao Dong Song and Paul Richards from Columbia University's Lamont-Doherty Earth Observatory to look for evidence that the inner core actually spins at a more rapid rate than the rest of the planet. The seismologists knew that previous studies suggested seismic waves pass through the inner core faster along a nearly north-south route. This faster route, or high-velocity pathway, is similar to the grain in a piece of wood. This pathway is not aligned directly with the inner core's spin axis but is tilted about 10° from it (box figure 2). Seismic waves tend to travel more slowly along other paths, such as in an east-west direction parallel to the inner core's equator.

The seismologists studied seismic wave records from separate, closely spaced earthquakes from 1961 to 2004 near the Sandwich Islands, off Argentina, to determine how long it took them to reach a monitoring station in College, Alaska. The waves all took about the same amount of time to reach Alaska; however, the seismic waves in the 2000s arrived in Alaska about 0.4 seconds faster than the seismic waves in the 1960s. Since the seismic waves would have traveled through the inner core, the seismologists have explained the difference in travel time as indicating that the inner core had changed its position relative to the monitoring station in Alaska. That is, the inner core and the high-velocity pathway had rotated slightly with respect to the rest of the planet.

anomaly). In figure 17.21*B*, a large dike of gabbro has intruded into granitic basement rock. Because gabbro contains more iron-based minerals than granite, gabbro is more magnetic and causes a positive magnetic anomaly. Figure 17.21*C* shows a diorite basement high (perhaps originally a hill) that has influenced later sediment deposits, causing a draping of the layers as the sediments on the hilltop compacted less than the thicker sediments to the sides. Such a structure can form an *oil trap* (see chapter 22). The diorite in the hill contains more iron in its

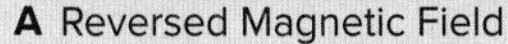

A Reversed Magnetic Field

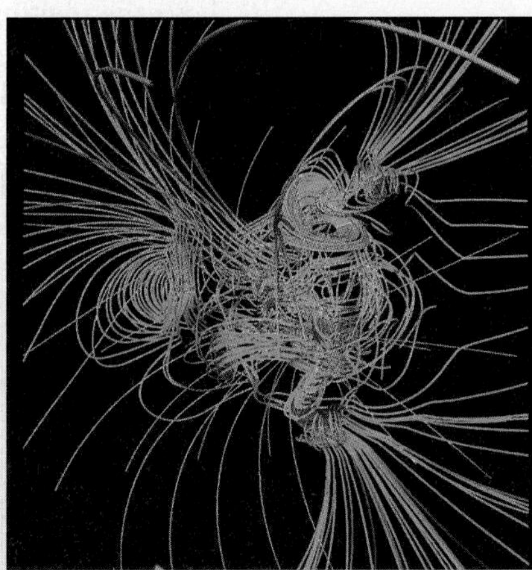

B Transitional Magnetic Field

C Normal Magnetic Field

BOX 17.2 ■ FIGURE 1

Computer simulation of Earth's magnetic field and magnetic reversal. (*A*) Reversed magnetic polarity with magnetic field lines leaving the north magnetic pole (orange) and reentering at the south pole. (*B*) Transitional magnetic field. (*C*) Normal magnetic field.
Courtesy of G. A. Glatzmaier, University of California, Santa Cruz, and P. H. Roberts, University of California, Los Angeles

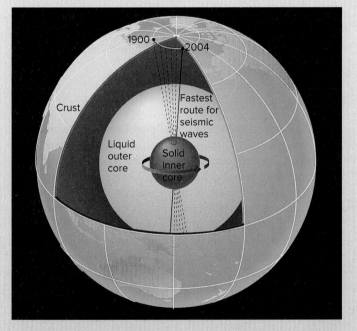

BOX 17.2 ■ FIGURE 2

Seismic waves indicate that Earth's core rotates faster than the rest of the planet by about a half degree per year. The solid line indicates the 2004 position of a point in the core relative to the surface of Earth, and the dashed lines indicate where the point has moved since 1900.
Source: Lamont-Doherty Earth Observatory

More recently, seismologists at Columbia University looked at additional earthquake records and calculated that the inner core is rotating more slowly than Glatzmaier and Roberts' model. The new records indicate that it takes 900 years for the inner core to gain a full lap on the rest of the planet. The slower rate of rotation may be caused by "clumps" in the high-velocity pathway in the inner core, much like knots in a piece of wood. This "super-rotation" of the inner core is also affected by the gravitational tug of the Moon. The balance between the super-rotation generated by the Earth's magnetic field and the damping effect of the Moon's pull on the inner core was recently used to determine the average strength of Earth's magnetic field in the liquid iron outer core (25 Gauss, or 50 times the average field strength at the Earth's surface).

This is an exciting time for Earth scientists since we may now have a better idea about the inner motion of the core and the generation of Earth's magnetic field.

Additional Resources

J. Zhang and others. 2005. Inner core differential motion confirmed by earthquake waveform doublets. *Science* 309: 1357–1360.

M. Carlowicz. 1996. Spin control. *Earth* 12(21): 62–63.

minerals than the surrounding sedimentary rocks, so a small positive magnetic anomaly occurs where the diorite is closer to the surface. Note how each example shows horizontal sedimentary rocks at the surface, with no surface hint of the subsurface geology. The magnetometer helps find hidden ores and geologic structures.

A **negative magnetic anomaly** is a reading of magnetic field strength that is lower than the regional average. Figure 17.22 shows how a negative anomaly can be produced by a down-dropped fault block (a *graben*) in basalt. The thick sedimentary fill above the graben is less magnetic than is the basalt, so a

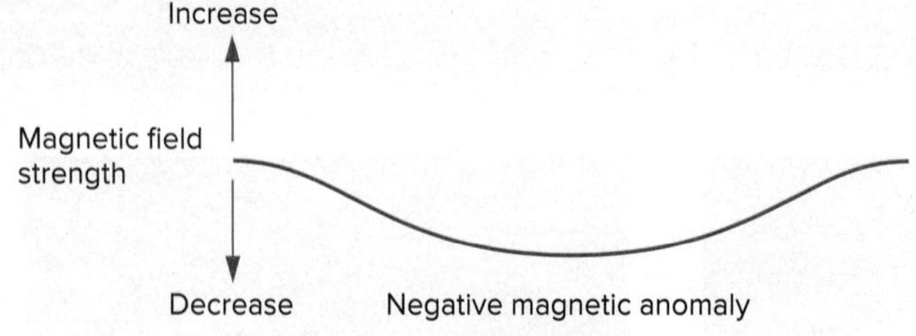

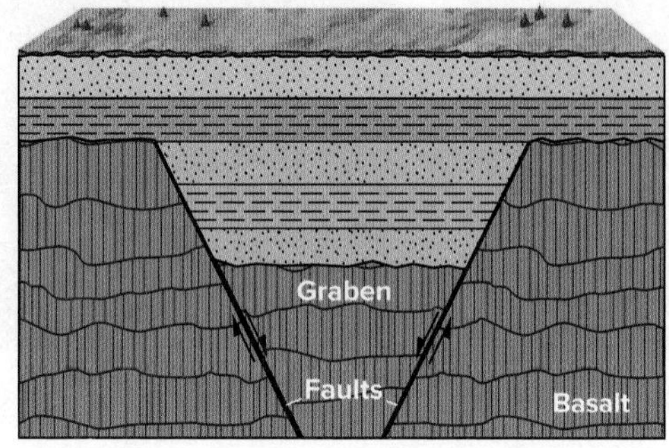

FIGURE 17.22

A graben filled with sediment can give a negative magnetic anomaly if the sediment contains fewer magnetic minerals than the rock beneath it.

weaker field (a negative magnetic anomaly) develops over the thick sediment.

Not all local magnetic anomalies are caused by variations in rock type. The linear magnetic anomalies found at sea are caused by a *variation in the direction of magnetism* recorded by the rocks, as you will see in chapter 19.

17.7 EARTH'S TEMPERATURE

Geothermal Gradient

The temperature increase with depth into Earth is called the **geothermal gradient**. The geothermal gradient can be measured on land in abandoned wells or on the sea floor by dropping special temperature-sensitive probes into the mud. The average temperature increase is 25°C per kilometer (about 75°F per mile) of depth. Some regions have a much higher gradient, indicating concentrations of heat at shallow depths. Such regions have a potential for generating *geothermal energy* (discussed in chapter 11).

The temperature increase with depth creates a problem in deep mines, such as in a 3-kilometer-deep gold mine in South Africa, where the temperature is close to the boiling point of water. Deep mines must be cooled by air-conditioning for the miners to survive. High temperatures at depth also complicate the drilling of deep oil wells. A well drilled to a depth of 7 or 8 kilometers must pass through rock with a temperature of 200°C. At such high temperatures, a tough steel drilling pipe will become soft and flexible unless it is cooled with a special mud solution pumped down the drill hole.

Geologists think that the geothermal gradient must taper off sharply a short distance into Earth. The high values of 25°C per kilometer recorded near Earth's surface could not continue very far into Earth. If they did, the temperature would be 2,500°C at the shallow depth of 100 kilometers. This temperature is above the melting point of all rocks at that depth—even though the increased pressure with depth into Earth raises the melting point of rocks. Seismic evidence seems to indicate a solid, not molten, mantle, so the geothermal gradient must drop to values as low as 1°C per kilometer within the mantle (figure 17.23*A*).

At the boundary between the inner core and the outer core, there would be some constraints on possible temperatures if the core is molten metal above the boundary and solid metal below. The weight of the thick rock layer of the mantle and the liquid metal of the outer core raises the pressure at this boundary (figure 17.23*B*) to about 3 million atmospheres. (An *atmosphere* of pressure is the force per unit area caused by the weight of the air in the atmosphere. It is about 1 kilogram per square centimeter, or 14.7 pounds per square inch.)

Using geophysical and geochemical data, in addition to computer modeling and high-pressure experiments, the internal temperature of Earth can be estimated. Recent laboratory experiments using pressure anvils and giant guns have created (for a millionth of a second) the enormous pressures found at the center of Earth. The measured temperature at this pressure was far higher than expected. New estimates of Earth's internal temperatures are 3,700°C at the core-mantle boundary, 6,300°C ± 800°C at the inner-core/outer-core boundary, and 6,400°C ± 600°C at Earth's center (hotter than the surface of the Sun!).

Heat Flow

A small but measurable amount of heat from Earth's interior is being lost gradually through the surface. This gradual loss of heat through Earth's surface is called **heat flow**. What is the origin of the heat? It could be "original" heat from the time that Earth formed as a mass of planetesimals that coalesced and compressed the inner material. Another source of heat could be the decay of radioactive isotopes inside Earth. Radioactive decay *may* actually be warming up the planet. Changes in Earth's internal temperature are extremely slow (on the order of 100 million years), and trying to work out its thermal history is a slow, often frustrating, job.

Some regions on Earth have a high heat flow. More heat is being lost through the surface in these regions than is normal. High heat flow is usually caused by the presence of a magma body or still-cooling pluton near the surface (figure 17.24). An old body of igneous rock that is rich in uranium and other radioactive isotopes can cause a high heat flow, too, because radioactive decay produces heat as it occurs. High heat flow over an extensive area may be due to the rise of warm mantle rock beneath abnormally thin crust.

The average heat flow from continents is the same as the average heat flow from the sea floor, a surprising fact if you consider the greater concentration of radioactive material in continental rock (figure 17.25). The unexpectedly high average heat flow under the ocean may be due to hot mantle rock rising

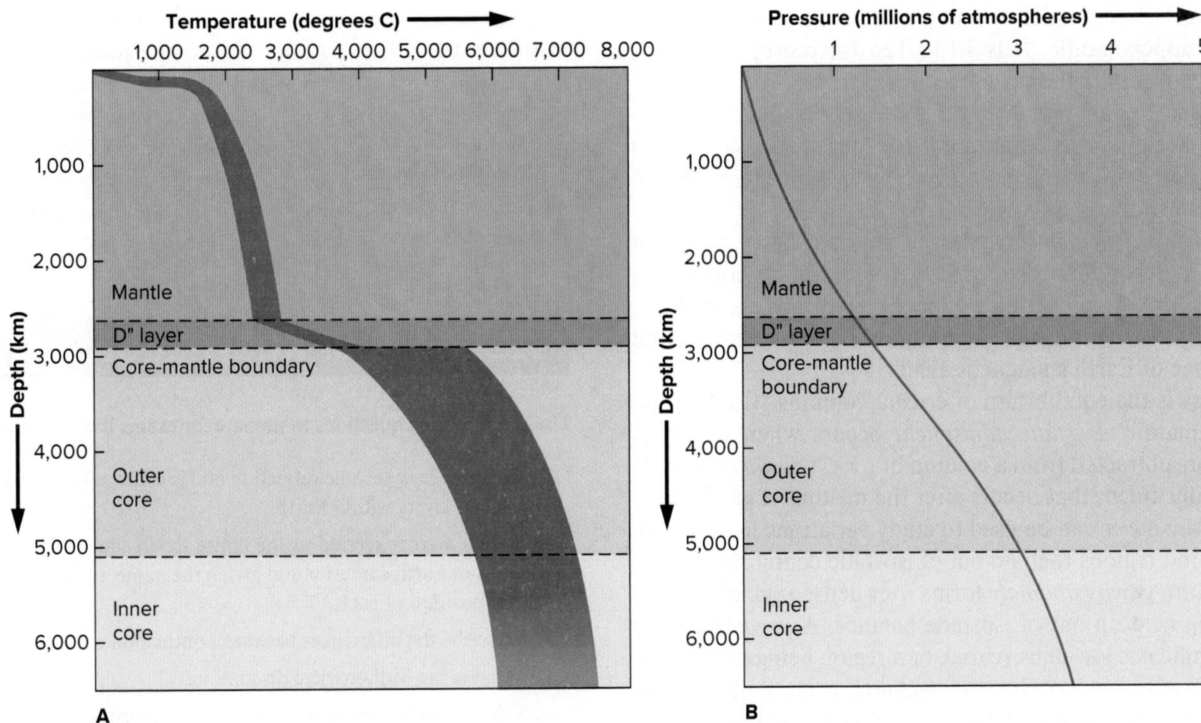

FIGURE 17.23

Estimated (*A*) temperature and (*B*) pressure with depth into Earth. The width of the red zone in graph (*A*) indicates the range of uncertainty of the estimate.

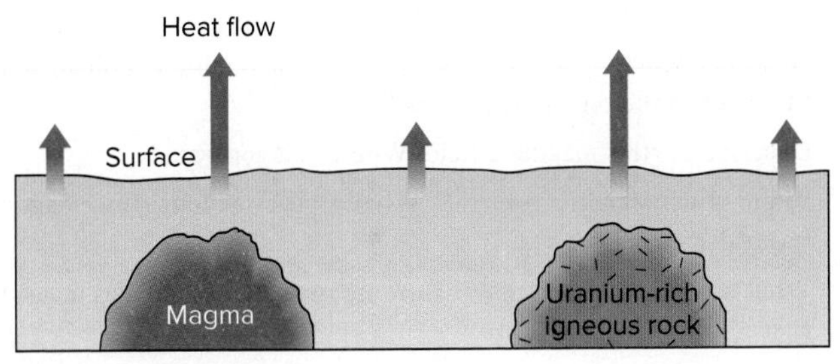

FIGURE 17.24

Some regions have higher heat flow than others; the amount of heat flow is indicated by the length of the arrow. Regions of high heat flow may be underlain by cooling magma or uranium-rich igneous rock.

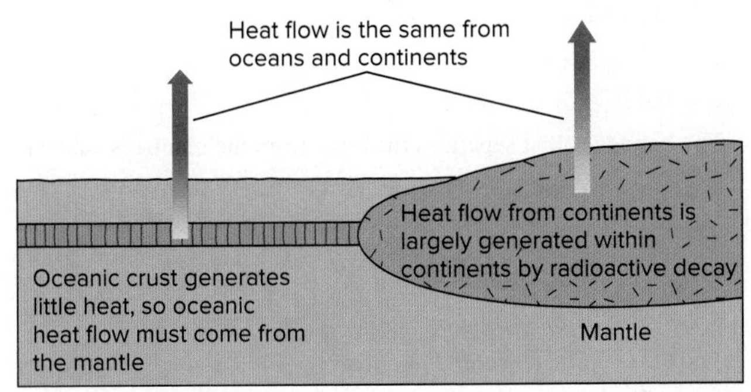

FIGURE 17.25

The average heat flow from oceans and continents is the same, but the origin of the heat differs from the ocean to continents.

slowly by convection under parts of the ocean (see chapter 1). Regional patterns of high heat flow and low heat flow on the sea floor (heat flow decreases away from the crest of the mid-oceanic ridge) may also be explained by convection of mantle rock, as we will discuss in chapter 19.

Summary

The interior of Earth is studied indirectly by *geophysics*—a study of seismic waves, gravity, Earth magnetism, and Earth heat.

Seismic reflection and *seismic refraction* can indicate the presence of boundaries between rock layers.

Earth is divided into three major zones—the *crust*, the *mantle*, and the *core*.

The crust beneath oceans is 7 kilometers thick and made of basalt on top of gabbro. Continental crust is 30 to 50 kilometers thick and consists of a crystalline basement of granite and gneiss (and other rocks) capped by sedimentary rocks.

The *Mohorovičić discontinuity* separates the crust from the mantle.

The mantle is a layer of solid rock 2,900 kilometers thick and is probably composed of an ultramafic rock such as peridotite. It makes up the majority of the volume of the Earth, and convection patterns within the mantle are likely to control plate motion at the lithosphere. Seismic waves show that the mantle has a structure of concentric shells, perhaps caused by pressure transformations of minerals.

The *lithosphere*, which forms plates, is made up of brittle crust and upper mantle. It is 70 to 125 (or more) kilometers thick and moves over the ductile asthenosphere.

The *asthenosphere* lies below the lithosphere and may represent rock close to its melting point (seismic waves slow down here). It is probably the region of most magma generation and isostatic adjustment.

Seismic-wave shadow zones show the core has a radius of 3,450 kilometers and is divided into a liquid outer core and a solid inner core. A core composition of mostly iron and nickel is suggested by Earth's density, the composition of meteorites, and the existence of Earth's magnetic field.

Isostasy is the equilibrium of crustal columns "floating" on a ductile mantle. *Isostatic adjustment* occurs when weight is added to or subtracted from a column of rock. *Crustal rebound* is isostatic adjustment that occurs after the melting of glacial ice.

A *gravity meter* can be used to study variations in rock density or to find regions that are out of isostatic equilibrium.

A *positive gravity anomaly* forms over dense rock or over regions being held up out of isostatic balance. A *negative gravity anomaly* indicates low-density rock or a region being held down.

Earth's *magnetic field* has two *magnetic poles,* probably generated by convection circulation and electric currents in the outer core.

Some rocks record Earth's magnetism at the time they form. *Paleomagnetism* is the study of ancient magnetic fields.

Magnetic reversals of polarity occurred in the past, with the north magnetic pole and south magnetic pole exchanging positions. Isotopic dating of rocks shows the ages of the reversals.

A *magnetometer* measures the strength of the magnetic field.

A *positive magnetic anomaly* develops over rock that is more magnetic than neighboring rock. A *negative magnetic anomaly* indicates rock with low magnetism.

Magnetic anomalies can also be caused by circulation patterns in Earth's core and variations in the direction of rock magnetism.

The *geothermal gradient* is about 25°C per kilometer near the surface but decreases rapidly at depth. The temperature at the center of Earth may be 6,400°C ± 600°C. *Heat flow* measurements show that heat loss per unit area from continents and oceans is about the same, perhaps because of convection of hot mantle rock beneath the oceans.

Terms to Remember

asthenosphere 420
convection 424
core 419
crust 419
crustal rebound 424
Curie point 428
geomagnetic polarity time scale 428
geophysics 416
geothermal gradient 432
gravity meter 426
heat flow 432
isostasy 424
isostatic adjustment 424
lithosphere 420
magnetic field 427
magnetic pole 427
magnetic reversal 428
magnetometer 429
mantle 419
Mohorovičić discontinuity (Moho) 419
negative gravity anomaly 427
negative magnetic anomaly 431
P-wave shadow zone 421
paleomagnetism 428
positive gravity anomaly 427
positive magnetic anomaly 429
S-wave shadow zone 421
seismic reflection 416
seismic refraction 417
seismic tomography 422

Testing Your Knowledge

Use the following questions to prepare for exams based on this chapter.

1. Describe how seismic reflection and seismic refraction show the presence of layers within Earth.
2. Sketch a cross section of the entire Earth showing the main subdivisions of Earth's interior and giving the name, thickness, and probable composition of each.
3. Describe the differences between continental crust and oceanic crust.
4. What is the Mohorovičić discontinuity?
5. What is the asthenosphere? Why is it important?
6. How does the lithosphere differ from the asthenosphere?
7. What facts make it probable that Earth's core is composed of a mixture of iron and nickel?
8. Discuss seismic-wave shadow zones and what they indicate about Earth's interior.
9. What is a gravity anomaly, and what does it generally indicate about the rocks in the region where it is found?
10. Describe Earth's magnetic field. Where is it generated?
11. What is a magnetic reversal? What is the evidence for magnetic reversals?
12. What is a magnetic anomaly? How are magnetic anomalies measured at sea?
13. What is the temperature distribution with depth into Earth?
14. Why does the heat flow from the continents equal that from the ocean?
15. Seismic refraction is caused by
    a. seismic waves bending.
    b. a change in velocity of seismic waves.
    c. sharp rock boundaries.
    d. all of the above.
    e. a and b.
16. The boundary that separates the crust from the mantle is called the
    a. lithosphere.
    b. asthenosphere.
    c. Mohorovičić discontinuity.
    d. none of the preceding.
17. The core is probably composed mainly of
    a. silicon. b. sulfur.
    c. oxygen. d. iron.

18. The principle of continents being in a buoyant equilibrium is called
    a. subsidence. b. isostasy.
    c. convection. d. rebound.
19. A positive gravity anomaly indicates that
    a. tectonic forces are holding a region up out of isostatic equilibrium.
    b. the land is sinking.
    c. local mass deficiencies exist in the crust.
    d. all of the preceding.
20. A positive magnetic anomaly could indicate
    a. a body of magnetic ore.
    b. the magnetic field strength is higher than the regional average.
    c. an intrusion of gabbro.
    d. the presence of a granitic basement high.
    e. all of the preceding.
21. Which of the following is *not* an example of the effects of isostasy?
    a. deep mountain roots
    b. magnetic reversals
    c. the postglacial rise of northeastern North America
    d. mountain ranges at subduction zones
22. The S-wave shadow zone is evidence that
    a. the core is made of iron and nickel.
    b. the inner core is solid.
    c. the outer core is fluid.
    d. the mantle behaves as ductile material.

Expanding Your Knowledge

1. What effect could glacial rebound have on global sea level rise?
2. Subsidence of Earth's surface sometimes occurs as reservoirs fill behind newly built dams. Why?
3. What geologic processes might cause the forces that can hold a region out of isostatic equilibrium?
4. Does the correlation of a species extinction with a magnetic reversal prove that the reversal caused the extinction?
5. If the upper mantle is chemically different from the lower mantle, is mantle convection possible?
6. How could you use geophysical techniques to plan the location of a subdivision in an area containing limestone caverns that could collapse if built on?

Exploring Web Resources

www.psc.edu/science/glatzmaier.html

When North Goes South website discusses the recent model for reversals of Earth's magnetic field.

http://garnero.asu.edu/research_images/index.html

Professor Ed Garnero's website contains numerous images that show how seismic waves are used to image the interior of Earth.

http://www.geotimes.org/july07/article.html?id=feature_deeper.html

Boldly going deeper in Earth. *Geotimes (Earth)* article by R. Van der Hilst

CHAPTER

18 The Sea Floor

A variety of life forms surround a coral reef in the Red Sea, Egypt.
©Richard Carey/iStock 360/Getty Images

LEARNING OBJECTIVES

- Summarize the methods used to study the sea floor.
- Draw a profile view across the sea floor and label the main features.
- Identify the major features and the activity that occurs on a mid-oceanic ridge.
- Compare and contrast a passive continental margin with an active continental margin, and include the most important features of each.
- Describe the appearance and origin of seamounts and guyots, and sketch a cross section of a fringing reef, a barrier reef, and an atoll.
- Sketch and describe the thickness and composition of the oceanic crust, and compare it with an ophiolite.
- Name and define the two types of sediment present on the sea floor.

Space travelers seeing our beautiful, blue planet would likely call it the water planet. The hydrosphere, notably the oceans, dominates the surface of Earth with over 70% of it covered by oceans. Clearly, the part of the geosphere covered by the oceans is important to our understanding of Earth systems, even though most of the sea floor is not readily accessible to direct observation.

The hydrosphere is, of course, vital to the biosphere. Primitive life began in the sea and evolved over billions of years into the rich and diverse plant and animal life that we see today. Life on land is a relative newcomer.

It has been only a few hundred million years since the first creatures ventured out of the sea and land-dwelling life evolved and flourished.

Most of what we know about the sea floor has been discovered during the second half of the twentieth century. Because of the difficulty in accessing the deep-ocean floor, our maps of its surface are not as complete as those of some of our neighboring planets. We do know that the rocks and topography of the sea floor are different from those on land. To understand the evidence for plate tectonics in chapter 19, you need to understand the nature of major seafloor features such as mid-oceanic ridges, oceanic trenches, and fracture zones, as well as the surprisingly young age of the seafloor rocks.

The material discussed in this chapter and chapter 19 is an excellent example of how the scientific method works. This chapter is concerned with the physical *description* of most seafloor features–the data-gathering part of the scientific method. Chapter 19 shows how the theory of plate tectonics explains the *origin* of many of these features. Geologists generally agree on the descriptions of features but often disagree on their interpretations. As you read, keep a clear distinction in your mind between *data* and the *hypotheses* used to explain the data.

18.1 ORIGIN OF THE OCEAN

As mentioned in chapter 17, geologists are unsure about the thermal history of Earth. The following scenario for the origin of Earth and its oceans is highly speculative. Many geologists would disagree with some of the statements; they are offered merely as an example of what could have happened in Earth's past.

According to a widely accepted theory, about 4.5 billion years ago, Earth began to form by the accretion of small, cold chunks of rock and metal that surrounded the Sun. As Earth grew, it began to heat up because of the heat of collisional impact, gravitational compaction, and radioactive decay of elements such as uranium. The temperature of Earth rose until the accretions that made up Earth melted and the iron "fell" to the center to form its core. Violent volcanic activity occurred at this time, releasing great quantities of water vapor and other gases from Earth's interior and perhaps even covering the surface with a thick, red-hot sea of lava. Earth began to cool as its growth and internal reorganization slowed down and as the amount of radioactive material was reduced by decay. Eventually, Earth's surface became solid rock, cool enough to permit the condensation of billowing clouds of volcanic water vapor to form liquid water, which fell as rain. Thus the earliest oceans were born, perhaps 4 billion years ago. Evidence for this comes from some of the oldest rocks found on Earth, including pillow basalts, which suggest eruption under water and rocks that were originally sedimentary and formed in a shallow sea. The oceans grew in size as volcanic *degassing* of Earth continued and became salty as the water picked up chlorine from other volcanic gases and sodium (and calcium and magnesium) from the chemical weathering of minerals on Earth's surface.

Some debate exists as to the role comet impacts may have played in the formation of the oceans. These extraterrestrial sources of ice bombarding Earth early in its history could have contributed a substantial amount of water.

Geologists generally agree that the oceans began to form early in Earth's history, and most would also agree that the oceans resulted mainly from degassing of Earth's interior. However, geologists were surprised to find that the present oceanic crust is geologically very young.

18.2 METHODS OF STUDYING THE SEA FLOOR

Oceans cover more than 70% of Earth's surface. Even though the rocks of the sea floor are widespread, they are difficult to study. Geologists have to rely on small samples of rock taken from the sea floor and brought to the surface, or they must study the rocks indirectly by means of instruments on board ships.

FIGURE 18.1

The JOIDES *Resolution* is a ship built for sampling both sediment and rock from the deep-ocean floor.
©ZUMA Press Inc/Alamy

FIGURE 18.2

The small research submersible *ALVIN* of Woods Hole Oceanographic Institution in Massachusetts; it is capable of taking three oceanographers to a depth of about 4,000 meters.
Source: OAR/National Undersea Research Program (NURP); Woods Hole Oceanographic Institution

Although the sea floor is difficult to study, its overall structure is relatively simple (as we discuss later in the "Features of the Sea Floor" section of this chapter), so the small number of samples is not nearly as much a problem as it would be in studying continental regions, where the structure is usually much more complex. The study of seafloor rocks, sediment, and topography provided most of the information that led to the concept of plate tectonics.

Samples of rock and sediments can be taken from the sea floor in several ways. Rocks can be broken from the sea floor by a *rock dredge*, which is an open steel container dragged over the ocean bottom at the end of a cable. Sediments can be sampled with a *corer*, a weighted steel pipe dropped vertically into the mud and sand of the ocean floor.

Both rocks and sediments can be sampled by means of *seafloor drilling*. Offshore oil platforms drill holes in the relatively shallow sea floor near shore. A ship with a drilling derrick on its deck can drill a hole in the deep-sea floor far from land (figure 18.1). The drill cuts long, rodlike rock cores from the ocean floor. Thousands of such holes have been drilled in the sea floor, and the rock and sediment cores recovered from these holes have revolutionized the field of marine geology. In the 1950s, more was known about the Moon's surface than about the floor of the sea. Seafloor drilling has been instrumental in expanding our knowledge of seafloor features and history. Small research submarines, more correctly called *submersibles*, can take geologists to many parts of the sea floor to observe, photograph, and sample rock and sediment (figure 18.2). Remotely operated submersibles and specially designed, deep-sea observation and sampling systems allow oceanographers to collect high-quality photographs and samples of the sea floor (figure 18.3).

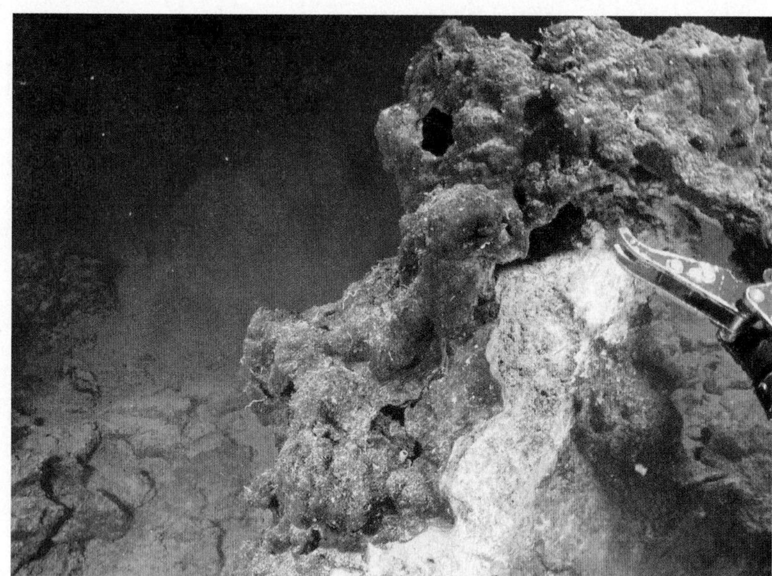

FIGURE 18.3

Remotely operated submersible *ROPOS* photographs an inactive black smoker along the Explorer Ridge in the northeast Pacific Ocean as it also prepares to take a sample. Yellow color is due to the alteration of iron-rich minerals, and the gray coating is mainly manganese.
Source: National Oceanic and Atmospheric Administration

A basic tool for indirectly studying the sea floor is the single-beam *echo sounder*, which measures water depth and draws profiles of submarine topography (figure 18.4). A sound sent downward from a ship bounces off the sea floor and returns to the ship. The water depth is determined from the time it takes the sound to make the round trip. *Multibeam sonar*, which sends out and records a variety of sound sources, is able to map the sea floor in even more detail than the single-source echo sounder. *Sidescan sonar* measures the intensity of sound (back scatter) reflected back to the tow vehicle from the sea floor to provide detailed images of the sea floor and information about sediments covering the bottom of the sea (figure 18.4).

A *seismic reflection profiler* works on essentially the same principles as echo sounders but uses a louder noise at lower

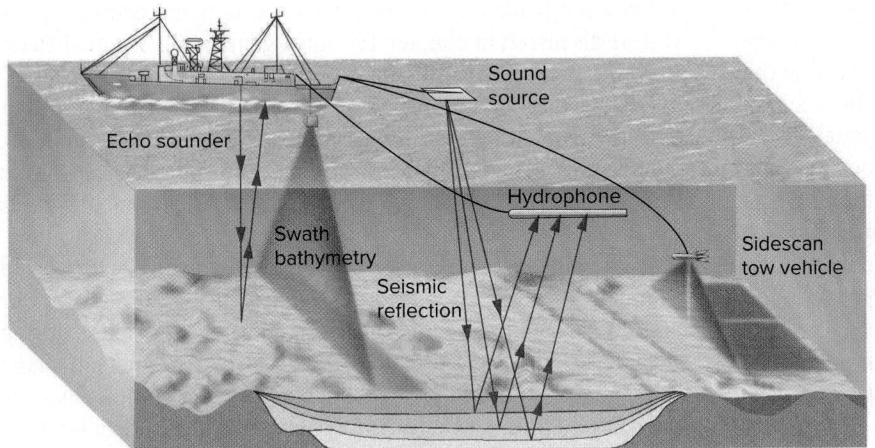

FIGURE 18.4

Diagram showing how echo sounding, seismic reflection, and sidescan sonar are used to study the sea floor.
Source: U.S. Geological Survey Fact Sheet 039-02

frequency. This sound penetrates the bottom of the sea and reflects from layers within the rock and sediment. The seismic profiler gives more information than do echo sounders. It records water depth and reveals the internal structure of the rocks and sediments of the sea floor, such as bedding planes, folds and faults, and unconformities.

Magnetic, *gravity*, and *seismic refraction* surveys (see chapter 17) can also be made at sea. Magnetometers pulled behind navy ships during World War II to look for enemy submarines provided some of the first details of the sea floor that later led to the theory of plate tectonics.

18.3 FEATURES OF THE SEA FLOOR

Figure 18.5, a simplified profile of the sea floor, shows that continents have two types of margins. A *passive continental margin*, as found on the east coast of North America, includes a continental shelf, continental slope, and continental rise. An *abyssal plain* usually forms a remarkably flat ocean floor beyond the continental rise. An *active continental margin*, found mainly around the Pacific rim, is associated with earthquakes and volcanoes and has a continental shelf and slope, but the slope

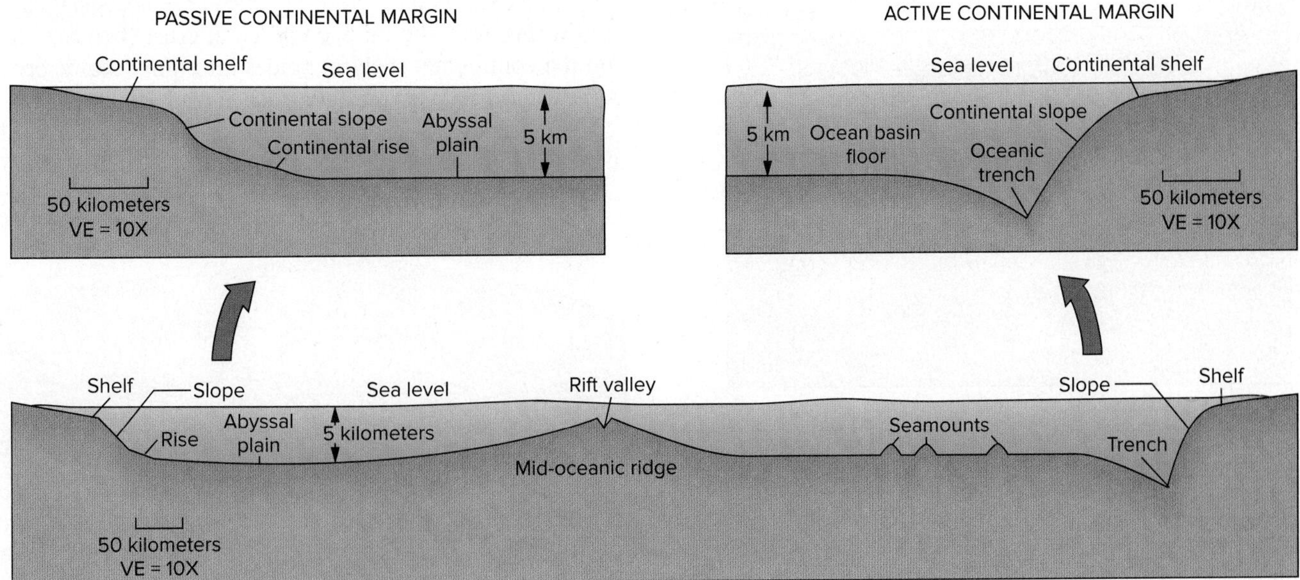

FIGURE 18.5

Profiles of seafloor topography. The vertical scales differ from the horizontal scales, causing vertical exaggeration, which makes slopes appear steeper than they really are. The bars for the horizontal scale are 50 kilometers long, while the same distance vertically represents only 5 kilometers, so the drawings have a vertical exaggeration of 10 times.

extends much deeper to form one wall of an *oceanic trench.* Abyssal plains are seldom found off active margins. The deep-ocean floor seaward of trenches is hilly and irregular, lacking the extreme flatness of abyssal plains. Encircling the globe is a *mid-oceanic ridge*, usually (but not always) near the center of an ocean. Conical *seamounts* stick up from the sea floor in some regions. Some important submarine features do not show in this profile view–in particular, fracture zones, submarine canyons, and aseismic ridges. To view these features and those detailed in the next sections of the text, you can explore the sea floor on the following website: www.virtualocean.org/

Continental Shelves and Continental Slopes

Almost all continental edges are marked by a gently sloping continental shelf near shore and a steeper continental slope that leads down to the deep ocean floor (figure 18.6). Figures such as 18.5 and 18.6 are usually drawn with great *vertical exaggeration*, which makes submarine slopes appear much steeper than they really are.

A **continental shelf**, a shallow submarine platform at the edge of a continent, inclines very gently seaward, generally at an angle of 0.1 degree. Continental shelves vary in width. On the Pacific coast of North America, the shelf is only a few kilometers wide, but off Newfoundland in the Atlantic Ocean, it is about 500 kilometers (310 miles) wide. Portions of the shelves in the Arctic Ocean off Siberia and northern Europe are even wider. Water depth over a continental shelf tends to increase regularly away from land, with the outer edge of the shelf being about 100 to 200 meters (328 to 656 feet) below sea level.

Continental shelves are *topographic features*, defined by their depth, flatness, and gentle seaward tilt. Their *geologic origin* varies from place to place and is related to plate tectonics, so it will be discussed in chapter 19. Some generalities about shelves, however, are worth noting here.

The continental shelves of the world are usually covered with relatively young sediment, in most cases derived from land. The sediment is usually sand near shore, where the bottom is shallow and influenced by wave action. Fine-grained mud is commonly deposited farther offshore in deeper, quieter water.

The outer part of a wide shelf is often covered with coarse sediment that was deposited near shore during a time of lower sea level. The advance and retreat of continental glaciers during the Pleistocene Epoch caused sea level to rise and fall many times by 100 to 200 meters. This resulted in a complex history of sedimentation for continental shelves as they were alternately covered with seawater and exposed as dry land.

Marine seismic surveys and drilling at sea have shown that the young sediments on many continental shelves are underlain by thick sequences of sandstone, shale, and less commonly limestone. These sedimentary rocks are of late Mesozoic and Cenozoic age, and they appear to have been deposited in much the same way as the modern shelf sediments. Beneath these rocks is continental crust (figure 18.7); the continental shelves are truly part of the continents, even though today the shelves are covered by seawater.

A **continental slope** is a relatively steep slope that extends from a depth of 100 to 200 meters at the edge of the continental shelf down to oceanic depths. The average angle of slope for a continental slope is 4 to 5 degrees, although locally, some parts are much steeper.

Because the continental slopes are more difficult to study than the continental shelves, less is known about them. The greater depth of water and the locally steep inclines on the continental slopes hinder rock dredging and drilling and make the results of seismic refraction and reflection harder to interpret. This is unfortunate, for the rocks that underlie the slopes are of particular interest to marine geologists, who believe that in this area, the thick continental crust (beneath the land and the continental shelves) grades into thin oceanic crust (underneath the deep-ocean floor), as shown in figure 18.7.

Although relatively little is known about continental slopes, it is clear that their character and origin vary greatly from place

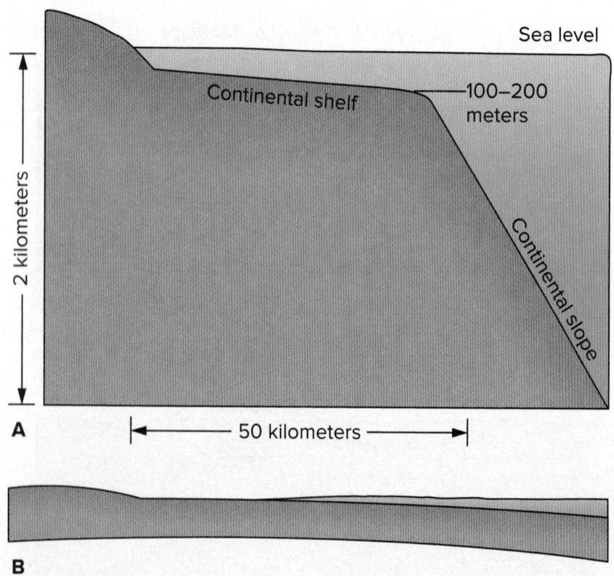

FIGURE 18.6

Continental shelf and continental slope. (*A*) Vertical exaggeration 25 times. The continental slope has an actual slope of only 4 or 5 degrees, but the great vertical exaggeration of this drawing makes it appear to be sloping about 60 degrees. (*B*) Same profile with no vertical exaggeration.

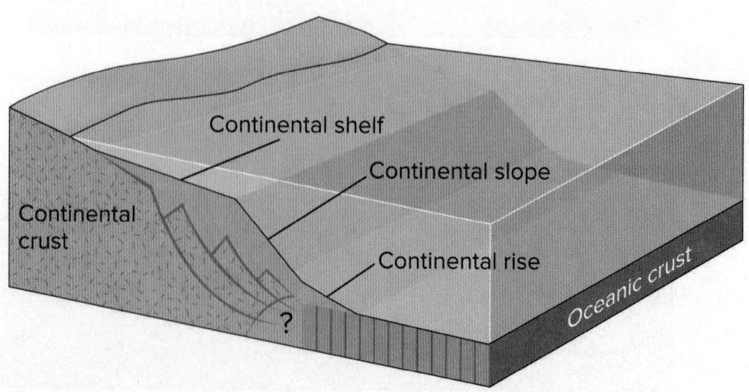

FIGURE 18.7

The continental shelf lies upon continental crust, and the continental rise lies upon oceanic crust. The complex transition from continental crust to oceanic crust lies under the continental slope.

to place. Because these variations are associated with the differences between divergent plate boundaries and convergent plate boundaries, we will discuss them in chapter 19.

Submarine Canyons

Submarine canyons are V-shaped valleys that run across continental shelves and down continental slopes (figure 18.8). On narrow continental shelves, such as those off the Pacific coast of the United States, the heads of submarine canyons may be so close to shore that they lie within the surf zone. On wide shelves, such as those off the Atlantic coast of the United States, canyon heads usually begin near the outer edge of the continental shelf tens of kilometers from shore. Great fan-shaped deposits of sediment called **abyssal fans** are found at the base of many submarine canyons (figure 18.8). Abyssal fans are made up of land-derived sediment that has moved down the submarine canyons.

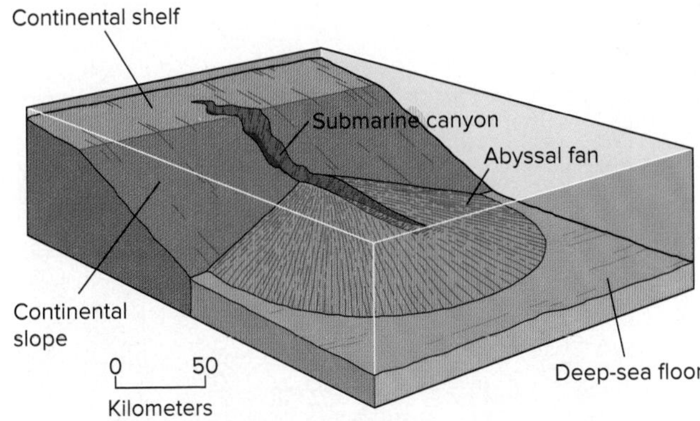

FIGURE 18.8
Submarine canyon and abyssal fan.

Along continental margins that are cut by submarine canyons, many coalescing abyssal fans may build up at the base of the continental slope (figure 18.9).

Submarine canyons are erosional features, but how rock and sediment are removed from the steep-walled canyons is controversial. Erosional agents probably vary in relative importance from canyon to canyon. Divers have filmed *down-canyon movement of sand* in slow, glacierlike flow and in more rapid sand falls (figure 18.10). This sand movement, which has been observed to cause erosion of rock, is particularly common in Pacific coast canyons, which collect great quantities of sand from longshore drift. *Bottom currents* have been measured moving up and down the canyons in a pattern of regularly alternating flow, in some cases apparently caused by ocean tides. The origin of these currents is not well understood, but they often move fast enough to erode and transport sediment. *River erosion* may have helped to cut the upper part of canyons when the drop in sea level during the Pleistocene glaciations left canyon heads above the water, such as the extension of the Hudson River into the Atlantic. Many (but not all) submarine canyons are found off land canyons or rivers, which tends to support the view that river erosion helped shape them. It is unlikely, however, that the deeper parts of submarine canyons were ever exposed as dry land.

Turbidity Currents

In addition to the canyon-cutting processes just described, turbidity currents probably play the major role in canyon erosion. **Turbidity currents** are great masses of sediment-laden water that are pulled downhill by gravity. The sediment-laden water is heavier than clear water, so the turbidity current flows down the continental slope until it comes to rest on the flat abyssal plain at the base of the slope (see figure 6.35). Turbidity currents are thought to be generated by underwater earthquakes and

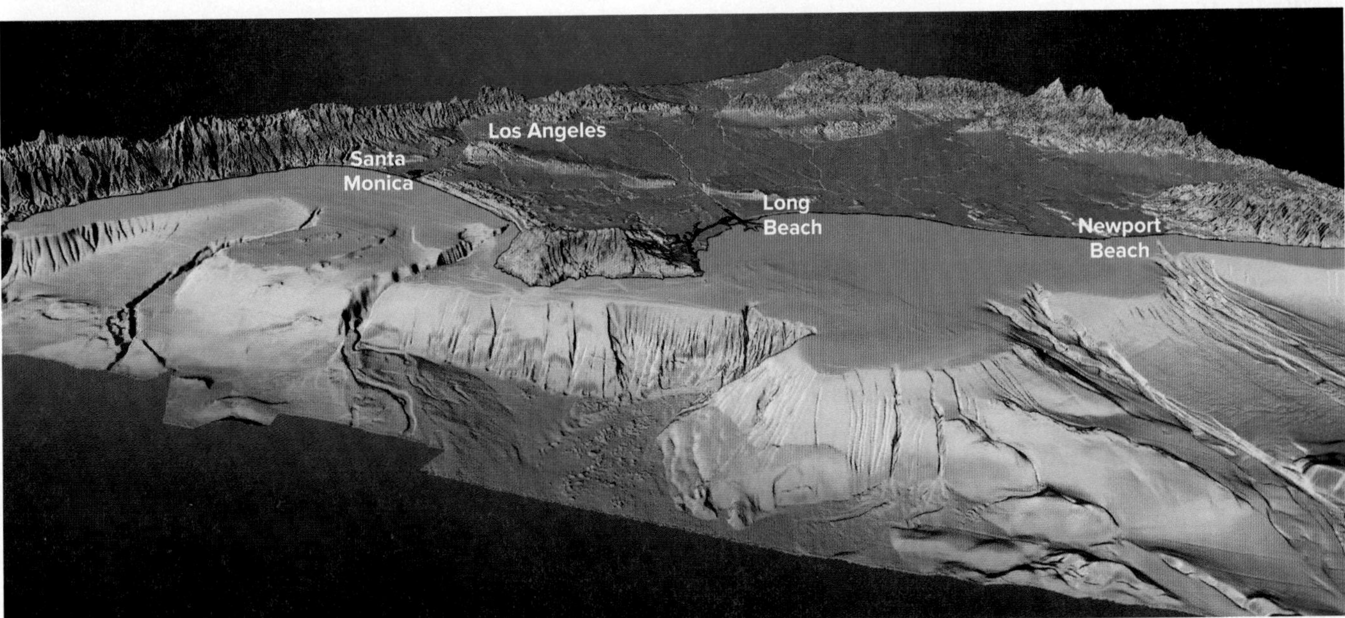

FIGURE 18.9
Continental shelf and slope off the coast of southern California are cut by submarine canyons that channel sediments to the deep ocean. Vertical exaggeration is approximately 6X. To view other high-resolution multibeam images of the coastline, visit http://walrus.wr.usgs.gov/pacmaps/la-pers1.html.
Source: Jim Gardner, U.S. Geological Survey

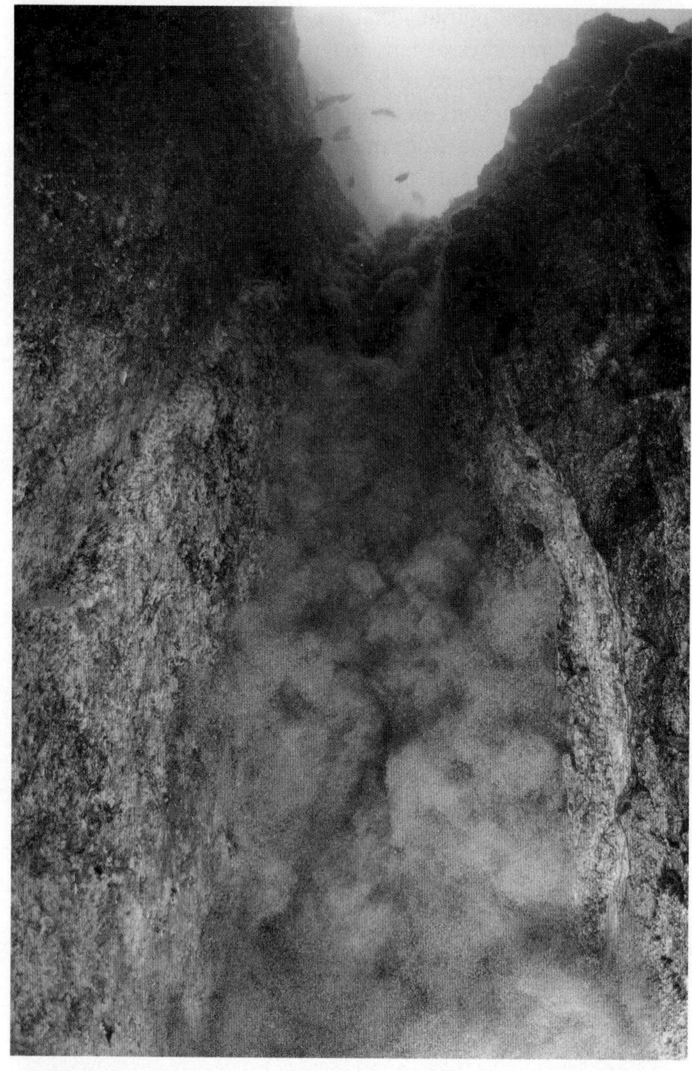

FIGURE 18.10

A sand fall in a submarine canyon near the southern tip of Baja California, Mexico. The sand is beach sand, fed into the near-shore canyon head by longshore currents. *©Luis Javier Sandoval/Oxford Scientific/Getty Images*

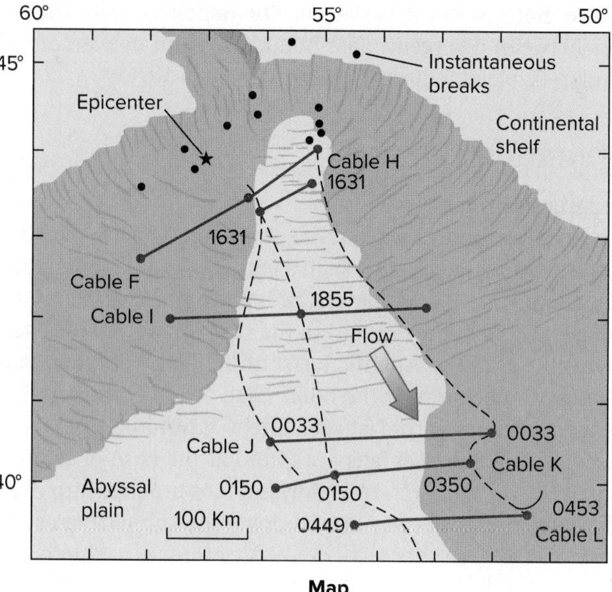

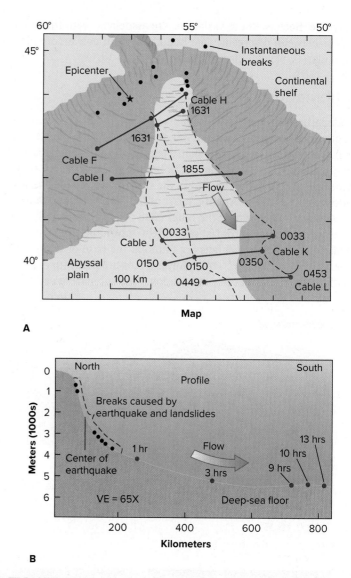

FIGURE 18.11

Submarine cable breaks following the Grand Banks earthquake of 1929. (*A*) Map view of the cable breaks. Black dots near the epicenter show locations of cable breaks that were simultaneous with the earthquake (cables not shown for these breaks). Red dots show cable breaks that followed the earthquake, with the time of each break shown (on the 24-hour clock). Segments of cable more than 100 kilometers long were broken simultaneously at both ends and then carried away. Dashes show seafloor channels that probably concentrated the flow of a turbidity current, increasing its velocity. (*B*) Profile showing the time elapsed between the quake and each cable break.
Source (A): H. W. Menard, Marine Geology of the Pacific, 1964; (B): B. C. Heezen and M. Ewing, American Journal of Science, 1952.

landslides, strong surface storms, and floods of sediment-laden rivers discharging directly into the sea on coasts with a narrow shelf. Until recently, large turbidity currents had not been directly observed in the sea. In 2012, a group of divers off the coast of Cabo San Lucas in Baja Mexico encountered and filmed a moderate-sized turbidity current https://offtheshelfedge.wordpress.com/2012/07/21/video-of-a-turbidity-current/, and in 2013, a remotely operated vehicle (ROV) captured a turbidity current near the head of a submarine canyon off the coast of Mendocino, California (https://www.youtube.com/watch?v=43Hp2ETgIXM). Small turbidity currents can also be made and studied in the laboratory (https://www.youtube.com/watch?v=8gYJJjxY8g0).

Indirect evidence also indicates that turbidity currents occur in the sea. The best evidence comes from the breaking of submarine telecommunication cables that carry digital data and voice communication messages across the ocean floor. Figure 18.11 shows a downhill sequence of cable breaks that followed the historic 1929 earthquake in the Grand Banks region of the northwest Atlantic. This sequence of cable breaks has been interpreted to be the result of an earthquake-caused turbidity current flowing rapidly down the continental slope.

Cable breaks caused by turbidity currents give good evidence of the currents' dramatic size, speed, and energy. Breaks in Grand Banks cables continued for more than thirteen hours after the 1929 earthquake, the last of the series occurring more than 700 kilometers (430 miles) from the epicenter. The velocity of the flow that caused the breaks has been calculated to be from 15 to 60 kilometers per hour. Sections of cable more than 100 kilometers (60 miles) long were broken off and carried away, both ends of a missing section being broken simultaneously. Attempts to find broken cable sections were unsuccessful, and it is assumed that they were buried by sediment.

The historic Grand Banks 1929 cable breaks are not unique. Telecommunication cables crossing submarine canyons are broken frequently, particularly after river floods and earthquakes. In 2006, an earthquake in southwestern Taiwan triggered a turbidity current that broke several submarine cables, causing a catastrophic disruption to Internet services and financial transactions in Asia. Breaks in the network of transoceanic telecommunication cables on the ocean floor is a risk to the global economy because almost all digital and voice communications worldwide are carried on these cables.

Additional indirect evidence for the existence of turbidity currents comes from the graded bedding and shallow-water fossils in the sediments that make up the continental rise and the abyssal plains, as described in the next section.

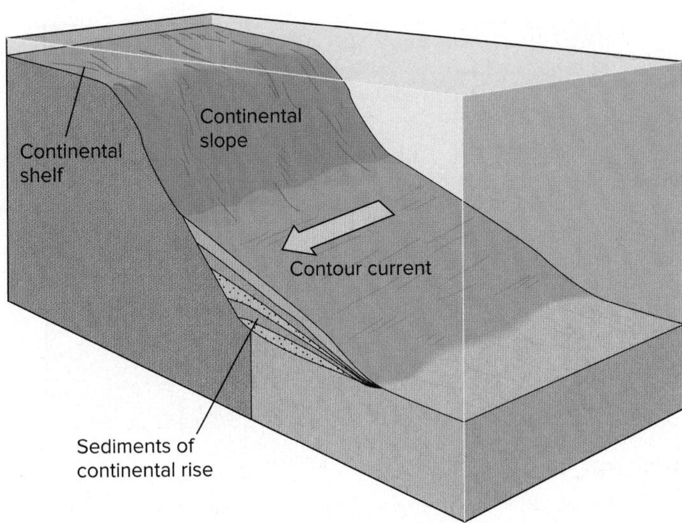

FIGURE 18.12

A contour current flowing along the continental margin shapes the continental rise by depositing fine sediment. Coarse layers within the continental rise were deposited by turbidity currents flowing down the continental slope.

Passive Continental Margins

A **passive continental margin** (figure 18.5) includes a continental shelf, continental slope, and continental rise, and generally extends down to an abyssal plain at a depth of about 5 kilometers (3.1 miles). It is called a passive margin because it usually develops on geologically quiet coasts that lack earthquakes, volcanoes, and young mountain belts.

Passive margins are found on the edges of most landmasses bordering the Atlantic Ocean. They also border most parts of the Arctic and Indian Oceans and a few parts of the Pacific Ocean.

The Continental Rise

Along the base of many parts of the continental slope lies the **continental rise**, a wedge of sediment that extends from the lower part of the continental slope to the deep-sea floor. The continental rise, which slopes at about 0.5 degree, more gently than the continental slope, typically ends in a flat abyssal plain at a depth of about 5 kilometers. The rise rests upon oceanic crust (figure 18.7).

Cores of sediment recovered from most parts of the continental rise show layers of fine sand or coarse silt interbedded with layers of fine-grained mud. The mineral grains and fossils of the coarser layers indicate that the sand and silt came from the shallow continental shelf. Some transporting agent must have carried these sediments from shallow water to deep water. The coarse layers also exhibit graded bedding, which indicates that they settled out of suspension according to size and weight; therefore, the transporting agent for these sediments was most likely *turbidity currents.* The continental rise in these locations probably formed as turbidity currents deposited abyssal fans at the base of a continental slope.

Sediments in some other parts of the continental rise, however, are uniformly fine-grained and show no graded bedding. This sediment appears to have been deposited by the regular ocean currents that flow along the sea bottom rather than by the intermittent turbidity currents that occasionally flow downslope.

A **contour current** is a bottom current that flows parallel to the slopes of the continental margin—*along* the contour rather than *down* the slope (figure 18.12). Such a current runs south along the continental margin of North America in the Atlantic Ocean. Flowing at the relatively slow speed of a few centimeters per second, this current carries a small amount of fine sediment from north to south. The current is thickest along the continental slope and gets progressively thinner seaward. The thick, landward part of the current carries and deposits the most sediment. The thinner, seaward edge of the current deposits less sediment. As a result, the deposit of sediment beneath the current is wedge shaped, becoming thinner away from land. Similar contour currents apparently shape parts of the continental rise off other continents as well.

Abyssal Plains

Abyssal plains are very flat regions usually found at the base of the continental rise. Seismic profiling has shown that abyssal plains are formed of horizontal layers of sediment. The gradual deposition of sediment buried an older, more rugged topography that can be seen on seismic profiler records as a rock basement beneath the sediment layers (figure 18.13). Samples of abyssal plain sediment show that it is derived from land. Graded bedding within sediment layers suggests deposition by turbidity currents.

Abyssal plains are the flattest features on Earth. They generally have slopes less than 1:1,000 (less than 1 meter of vertical drop for every 1,000 meters of horizontal distance) and some have slopes of only 1:10,000. Most abyssal plains are 5 kilometers (3.1 miles) deep.

Not all parts of the deep-ocean basin floor consist of abyssal plains. The deep floor is normally very rugged, broken by faults

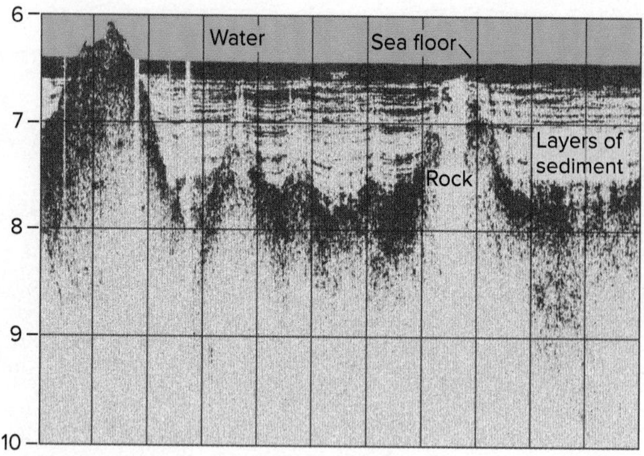

FIGURE 18.13

Seismic profiler record of an abyssal plain, showing sediment layers that have buried an irregular rock surface in the Atlantic Ocean.
From P. J. Hart (Ed.), The Earth's Crust and Upper Mantle. Washington, DC: American Geophysical Union. Copyright © 1969 by John Wiley & Sons, Inc. All rights reserved. Used with permission.

into hills and depressions and dotted with volcanic seamounts. Abyssal plains form only where turbidity currents can carry in enough sediment to bury and obscure this rugged relief. If the sediment is not available or if the bottom-hugging turbidity currents are stopped by a barrier such as an oceanic trench, then abyssal plains cannot develop.

Active Continental Margins

An **active continental margin**, typically characterized by earthquakes and a young mountain belt and volcanoes on land, consists of a continental shelf, a continental slope, and an oceanic trench (figure 18.5). An active margin usually lacks a continental rise and an abyssal plain and is associated with convergent plate boundaries.

Active margins are found on the edges of most of the landmasses bordering the Pacific Ocean and a few other places in the Atlantic and Indian Oceans (figure 18.14). A notable exception in the Pacific Ocean is most of the coast of North America. The active margin's distinctive combination of oceanic trench and land volcanoes is found in three places along the Pacific coast of North America: southern Central America; Washington, Oregon, and northernmost California (the trench here is sediment-filled); and south-central Alaska.

Oceanic Trenches

An **oceanic trench** is a narrow, deep trough parallel to the edge of a continent or an island arc (a curved line of islands like the Aleutians or Japan), as shown in figure 18.14. The continental slope on an active margin forms the landward wall of the trench, its steepness often increasing with depth. The slope is typically 4 to 5 degrees on the upper part, steepening to 10 to 15 degrees or even more near the bottom of the trench. The elongate oceanic trenches, often 8 to 10 kilometers (4.9 to 6.2 miles) deep, far exceed the average depth of abyssal plains on passive margins. The deepest spots on Earth, more than 11 kilometers (6.8 miles) below sea level, are in oceanic trenches in the southwest Pacific Ocean.

Associated with oceanic trenches are the *earthquakes of the Benioff seismic zones* (see chapter 16), which begin at a trench and dip landward under continents or island arcs (figure 18.15). *Volcanoes* are found above the upper part of the Benioff zones and typically are arranged in long belts parallel to oceanic trenches. These belts of volcanoes form island arcs or erupt within young mountain belts on the edges of continents. The

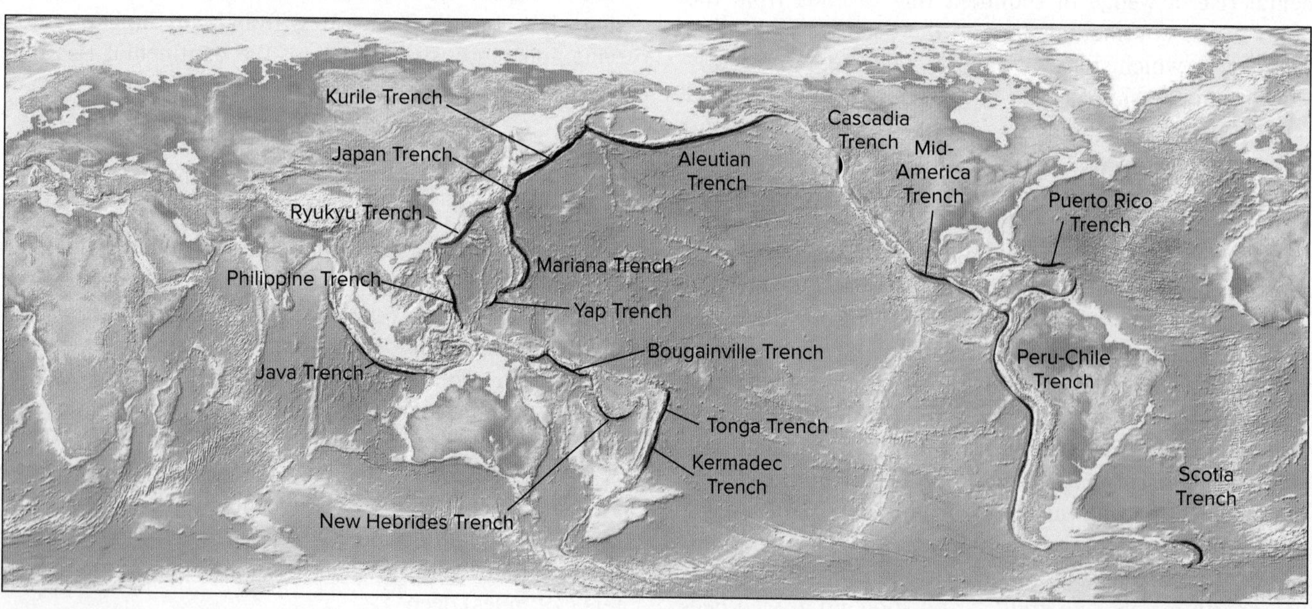

FIGURE 18.14

The distribution of oceanic trenches. Trenches next to continents mark active continental margins and convergent plate boundaries.

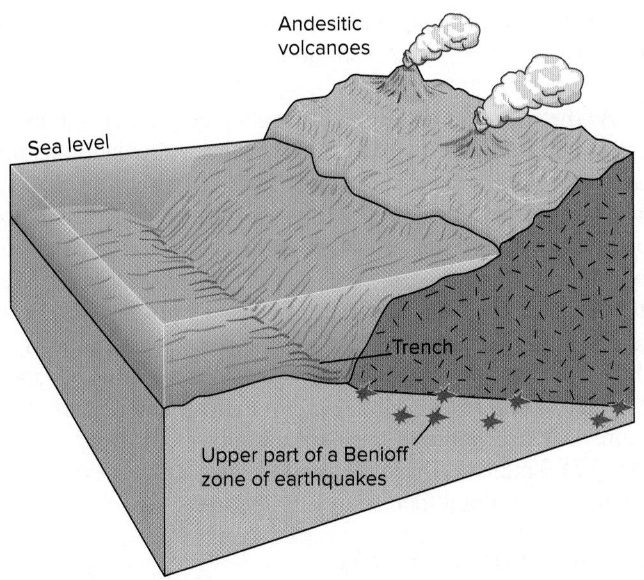

FIGURE 18.15

Active continental margin with an oceanic trench, a Benioff zone of earthquakes (only the upper part is shown), and a chain of andesitic volcanoes on land.

rock produced by these volcanoes is usually andesite, a type of extrusive rock intermediate in composition between basaltic oceanic crust and "granitic" continental crust.

Oceanic trenches are marked by abnormally *low heat flow* compared to normal ocean crust. This implies that the crust in trenches may be colder than normal crust.

As you learned in chapter 17, oceanic trenches are also characterized by very large *negative gravity anomalies*, the largest in the world. This implies that trenches are being held down, out of isostatic equilibrium. As you will learn in chapter 19, this is because the oceanic plate is being pulled down or subducted into the mantle.

Mid-Oceanic Ridges

The **mid-oceanic ridge** is a giant undersea mountain range that extends around the world like the seams on a baseball (figure 18.16). The ridge, which is made up mostly of basalt, is more than 80,000 kilometers (49,700 miles) long and 1,500 to 2,500 kilometers wide (931 to 1,552 miles). It rises 2 to 3 kilometers (1.2 to 1.8 miles) above the ocean floor.

A **rift valley**, where the crust is undergoing extension, runs down the crest of the ridge. The rift valley is 1 to 2 kilometers deep and several kilometers wide–about the dimensions of the Grand Canyon in Arizona. The rift valley on the crest of the mid-oceanic ridge is a unique feature–no mountain range on land has such a valley running along its crest. The rift valley is present in the Atlantic and Indian Oceans but is generally absent in the Pacific Ocean. (This absence is related to the faster rate of plate motion in the Pacific, as we will discuss in chapter 19.)

Geologic Activity at the Ridges

Associated with the rift valley at the crest of the mid-oceanic ridge and with the riftless crest of the ridge in the Pacific Ocean, are *shallow-focus earthquakes* from 0 to 20 kilometers (0 to 12.4 miles) below the sea floor.

Careful measurements of the heat loss from Earth's interior through the crust have shown a very *high heat flow* on the crest of the mid-oceanic ridge. The heat loss at the ridge crest is many times the normal value found elsewhere in the ocean; it decreases away from the ridge crest.

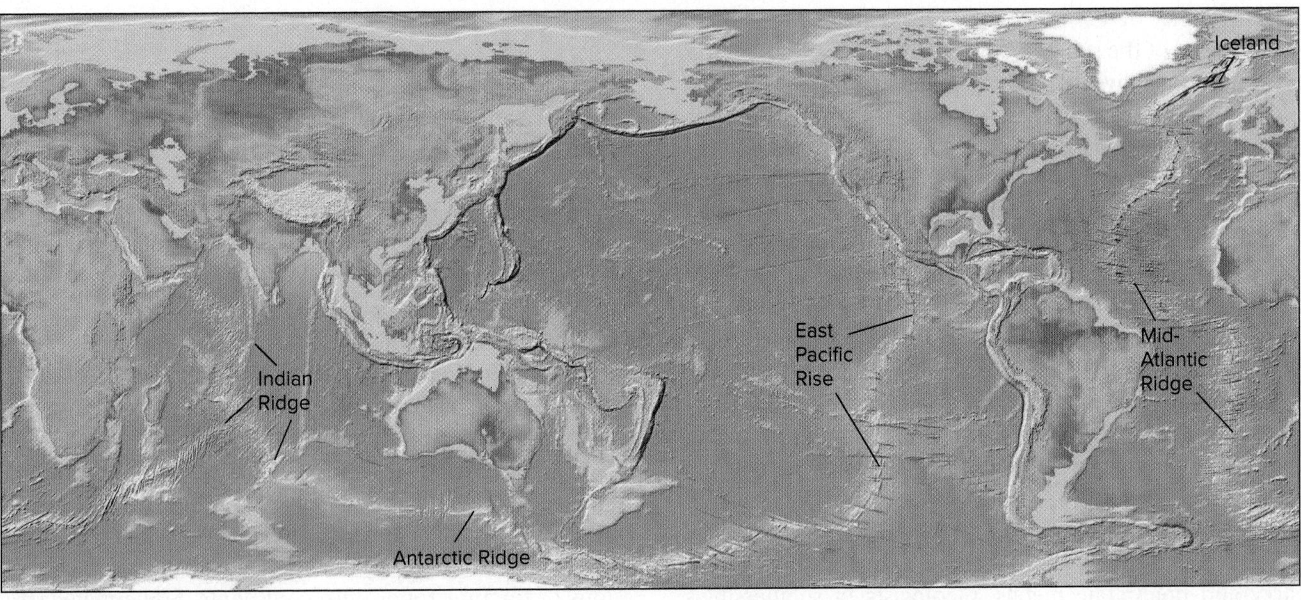

FIGURE 18.16

World map showing oceanic ridges cut by transform faults and fracture zones.

FIGURE 18.17
Underwater photograph of fresh pillow basalts on the floor of the Galapagos Rift in the Pacific Ocean.
Source: NOAA Okeanos Explorer Program, Galapagos Rift Expedition 2011

Basalt eruptions occur in and near the rift valley on the ridge crest (figure 18.17). Sometimes these eruptions build up volcanoes that protrude above sea level as oceanic islands. The large island of Iceland (figure 18.16), which is mostly basaltic, appears to be a section of the mid-oceanic ridge elevated above sea level. Many geologists have studied the active volcanoes, high heat flow, and central rift valley of Iceland to learn about the mid-oceanic ridge. Because Iceland sits on top of a hot spot, it is above sea level and may not represent a typical portion of the ridge.

In the summer of 1974, geologists were able to get a first-hand view of part of the submerged ridge and rift valley. A series of more than forty dives by submersibles, including ALVIN, carried French and American marine geologists directly into the rift valley in the North Atlantic Ocean. The project (called FAMOUS for French-American Mid-Ocean Undersea Study) allowed the ridge rock to be seen, photographed, and sampled directly, rather than indirectly from surface ships.

The geologists on the FAMOUS project saw clear evidence of extensional faults within the rift valley. These run parallel to the axis of the rift valley and range in width from hairline cracks to gaping fissures that ALVIN dived into. Fresh pillow basalts occur in a narrow band along the bottom of the rift valley, suggesting very recent volcanic activity there, although no active eruptions were observed. It appeared to the geologists that the rift was continuous and that sporadic volcanic activity occurred as a result of the rifting.

Mid-oceanic ridges are often marked by lines of hot springs that carry and precipitate metals. Geologists in submersibles have observed hot springs in several localities along the rift valleys of the mid-oceanic ridges. The hot springs, caused by the high heat flow and shallow basaltic magma beneath the rift valley, range in temperature from about 20°C up to an estimated 350°C (660°F).

As the hot water rises in the rift valley, cold water is drawn in from the sides to take its place. This creates a circulation pattern in which cold seawater is actually drawn *downward* through the cracks in the basaltic crust of the ridge flanks and then moves horizontally toward the rift valley, where it reemerges on the sea floor after being heated (figure 18.18*A*). As the seawater circulates, it dissolves metals and sulfur from the crustal rocks. When the hot, metal-rich solutions contact the cold seawater, metal sulfide particles are discharged into the cold seawater at *black smokers,* which precipitate a chimney-like mound around the hot spring (figure 18.18*B*). Some of the particles are microscopic iron sulfide (pyrite) minerals that are dispersed throughout the ocean and provide an important food source for life in the sea.

Biologic Activity at the Ridges

The occurrence of black smokers was a surprise to geologists exploring the mid-oceanic ridges, but an even bigger surprise was the presence of exotic, bottom-dwelling organisms surrounding the hot springs. The exotic organisms, including mussels, crabs, starfish, giant white clams, and giant tube worms, are able to survive toxic chemicals, high temperatures, high pressures, and total darkness at depths of more than 2 kilometers (1.2 miles) (figure 18.18*C*). The organisms live on bacteria that thrive by oxidizing hydrogen sulfide from the hot springs. It is believed that the heat-loving, or *thermophilic bacteria* normally reside beneath the sea floor but are blown out of the hot spring when it erupts. Such sulfur-digesting bacteria have also been found in acidic water in mines containing sulfide minerals. Current research in the new field of *geomicrobiology* is examining the role such bacteria may have had in the precipitation of minerals and the evolution of early life-forms on Earth and possibly other places in the solar system.

Fracture Zones

Fracture zones are major lines of weakness in Earth's crust that cross the mid-oceanic ridge at approximately right angles (figure 18.16). The rift valley of the mid-oceanic ridge is offset in many places across fracture zones (figure 18.19), and the sea floor on one side of a fracture zone is often at a different elevation than the sea floor on the other side, producing steep cliffs. Shallow-focus earthquakes occur on fracture zones but are confined to those portions of the fracture zones between segments of the rift valley. (The portion of the fracture zone that has earthquakes is known as a *transform fault*; the origin of these faults is discussed in chapter 19.)

Fracture zones extend for thousands of kilometers across the ocean floor, generally heading straight for continental margins. Although fracture zones are difficult to trace where they are buried by the sediments of the abyssal plain and the continental rise, some geologists think that they can trace the extensions of fracture zones onto continents. Some major structural trends on continents appear to lie along the hypothetical extension of fracture zones onto the continents.

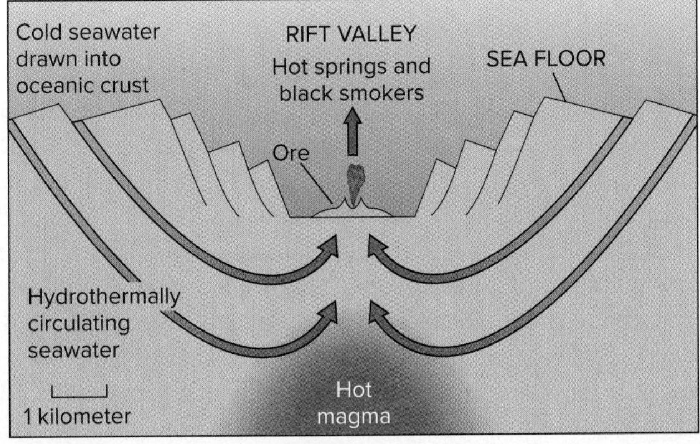

A

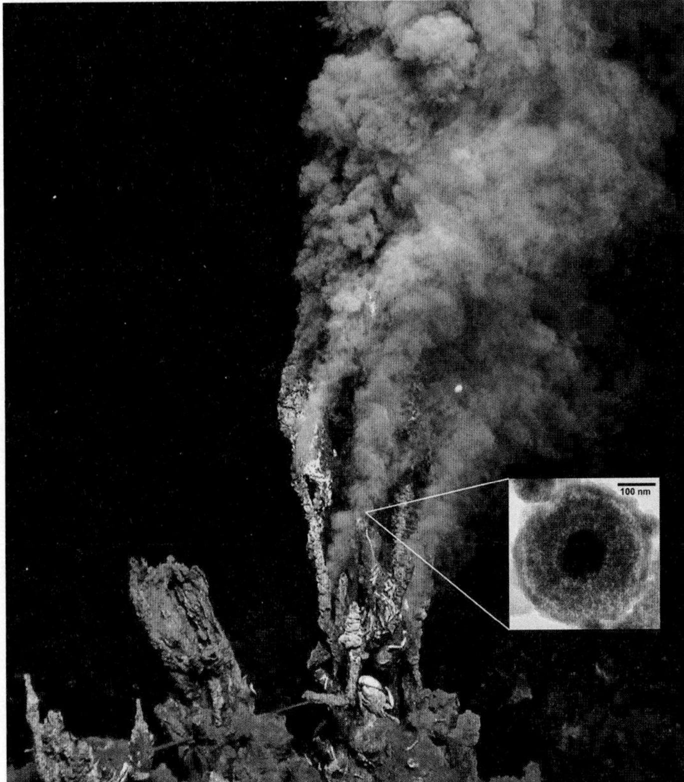

B

C

FIGURE 18.18

(*A*) Hydrothermal circulation of seawater at ridge crest creates hot springs and metallic deposits in rift valley. Cold seawater is drawn into fractured crust on ridge flanks. (Size of ore deposit is exaggerated.) (*B*) "Black smoker" or submarine hot spring from a spreading center in the Pacific Ocean. The "smoke" is a hot plume of metallic sulfide minerals being discharged into cold seawater from a chimney made of metallic deposits. Some of the sulfide minerals erupted are microscopic particles of pyrite (fool's gold) that have a diameter 1,000 times smaller than that of human hair. (*C*) Giant tube worms and crabs from the Galapagos vent.
(B) ©Dr. Mustafa Yücel/University of Delaware; (C) Source: NOAA Okeanos Explorer Program, Galapagos Rift Expedition 2011

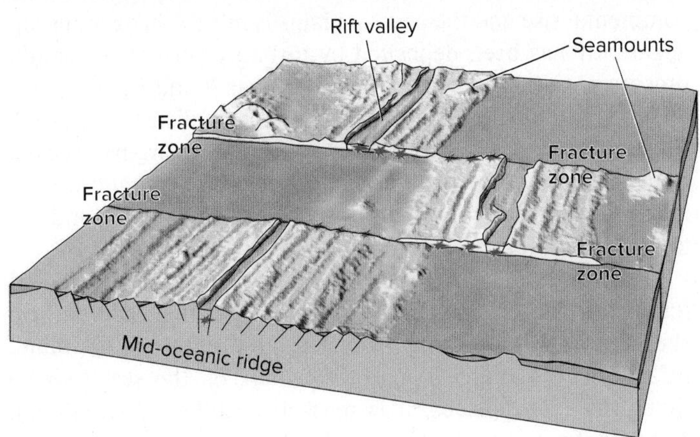

FIGURE 18.19

Fracture zones, which run perpendicular to the ridge crest and connect offset segments of the ridge, are often marked by steep cliffs up to 3 or 4 kilometers high. Shallow-focus earthquakes (shown by stars) occur below the rift valley and on the portion of the fracture zone between sides of the mid-oceanic ridge (transform faults).

Seamounts, Guyots, and Aseismic Ridges

Conical undersea mountains that rise 1,000 meters (3,280 feet) or more above the sea floor are called **seamounts** (figure 18.20). Some rise above sea level to form islands. They are scattered on the flanks of the mid-oceanic ridge and on other parts of the sea floor, including abyssal plains. One area of the sea floor with a particularly high concentration of seamounts–one estimate is 10,000–is the western Pacific. Rocks dredged from seamounts are nearly always basalt, so it is thought that most seamounts are extinct volcanoes. Of the thousands of seamounts on the sea floor, only a few are active volcanoes. Most of these are on the crests of the mid-oceanic ridges. A few others, such as the two active volcanoes on the island of Hawaii, are found at locations not associated with a ridge.

Guyots are flat-topped seamounts (figure 18.20*B*) found mostly in the western Pacific Ocean. Most geologists think that the flat summits of guyots were cut by wave action. These flat tops are now many hundreds of meters below sea level, well below the level of wave erosion. If the guyot tops were cut by waves, the guyots must have subsided after erosion took place. Evidence of such subsidence comes from the dredging of dead reef

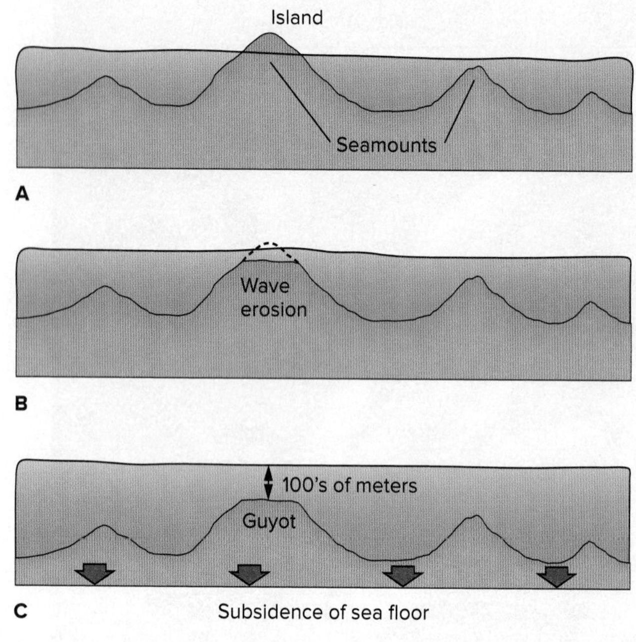

FIGURE 18.20

(*A*) Seamounts are conical mountains on the sea floor, occasionally rising above sea level to form islands. (*B*) The flat summit of a guyot was probably eroded by waves when the top of a seamount was above sea level. (*C*) The present depth of a guyot is due to subsidence.

corals from guyot tops. Since such corals grow only in shallow water, they must have been carried to their present depths as the guyots sank.

Many of the guyots and seamounts on the sea floor are aligned in chains. Such volcanic chains, together with some other ridges on the sea floor, are given the name **aseismic ridges**; that is, they are submarine ridges that are not associated with earthquakes. The name *aseismic* is used to distinguish these features from the much larger mid-oceanic ridge, where earthquakes occur along the rift valley.

Reefs

Reefs are wave-resistant ridges of coral, algae, and other calcareous organisms. They form in warm, shallow, sunlit water that is low in suspended sediment. Reefs stand above the surrounding sea floor, which is often covered with sediment derived from the reef (see figure 6.15). Three important types are *fringing reefs*, *barrier reefs*, and *atolls* (figure 18.21).

Fringing reefs are flat, tablelike reefs attached directly to shore. The seaward edge is marked by a steep slope leading down into deeper water. Many of the reefs bordering the Hawaiian Islands are of this type.

Barrier reefs parallel the shore but are separated from it by wide, deep lagoons. This type of reef is shown in figures 6.15 and 18.21*D*. The lagoon has relatively quiet water because the reef shelters it by absorbing the energy of large, breaking waves. A barrier reef lies about 8 kilometers (4.9 miles) offshore of the Florida Keys, a string of islands south of Miami. On a much grander scale is the Great Barrier Reef off northeastern Australia. It extends for about 2,000 kilometers (1,242 miles) along the coast, and its seaward edge lies up to 250 kilometers (155 miles) from shore. Another long barrier reef lies along the eastern coast of the Yucatan Peninsula in Central America, and others surround many islands in the South Pacific.

Atolls are circular reefs that rim lagoons. They are surrounded by deep water. Small islands of calcareous sand may be built by waves at places along the reef ring. The diameter of atolls varies from 1 to more than 100 kilometers. Numerous atolls dot the South Pacific. Bikini and Eniwetok atolls were used through 1958 for the testing of nuclear weapons by the United States.

Following the four-year cruise of the HMS *Beagle* in the 1830s, Charles Darwin proposed that these three types of reefs are related to one another by subsidence of a central volcanic island, as shown in figure 18.21. A fringing reef initially becomes established near the island's shore. As the volcano slowly subsides because of tectonic lowering of the sea floor, the reef becomes a barrier reef, because the corals and algae grow rapidly upward, maintaining the reef's position near sea level. Less and less of the island remains above sea level, but the reef grows upward into shallow, sunlit water, maintaining its original size and shape. Finally, the volcano disappears completely below sea level, and the reef becomes a circular atoll. Drilling through atolls in the 1950s showed that these reefs were built on deeply buried volcanic cores, thus confirming Darwin's hypothesis of 120 years before.

18.4 SEDIMENTS OF THE SEA FLOOR

The basaltic crust of the sea floor is covered in many places with layers of sediment. This sediment is either *terrigenous*, derived from land, or *pelagic*, settling slowly through seawater.

Terrigenous sediment is land-derived sediment that has found its way to the sea floor. The sediment that makes up the continental rise and the abyssal plains is mostly terrigenous and apparently has been deposited by turbidity currents or similar processes. Once terrigenous sediment has found its way down the continental slope, contour currents may distribute it along the continental rise. On active continental margins, oceanic trenches may act as traps for terrigenous sediment and prevent it from spreading out onto the deep-sea floor beyond the trenches.

Pelagic sediment is sediment that settles slowly through the ocean water. It is made up of fine-grained clay and the shells of microscopic organisms (figure 18.22). Fine-grained pelagic clay is found almost everywhere on the sea floor, although in some places, it is masked by other types of sediments that accumulate rapidly. The clay is mostly derived from land; part of it may be volcanic ash. This sediment is carried out to sea primarily by wind, although rivers and ocean currents also help to distribute it.

Microscopic shells and skeletons of plants and animals also settle slowly to the sea floor when marine organisms of the

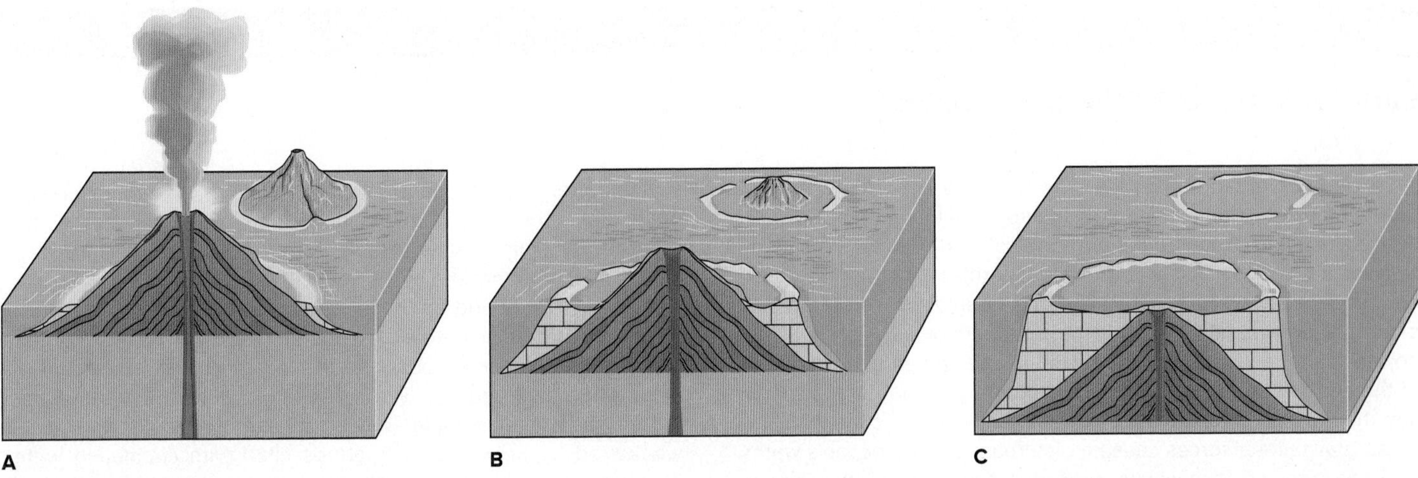

D

FIGURE 18.21

Types of coral-algal reefs. (*A*) Fringing reefs are attached directly to the island. (*B*) Barrier reefs are separated from the island by a lagoon. (*C*) Atolls are circular reefs with central lagoons. Charles Darwin proposed that the sequence of fringing, barrier, and atoll reefs forms by the progressive subsidence of a central volcano, accompanied by the rapid upward growth of corals and algae. (*D*) Reefs on the island of Mooréa in the Society Islands, south-central Pacific (Tahiti in the background behind the clouds). Living corals appear brown. Fringing reef is next to shoreline, barrier reef at breaker line; light blue lagoon between reefs is covered with carbonate sand. The island is an extinct volcano, heavily eroded by stream action.

(D) ©Sylvain Grandadam/Getty Images

EARTH SYSTEMS 18.1

Does the Earth Breathe?

Does the Earth breathe to the rhythm of the tides? Maya Tolstoy and her colleagues were the first to show that it does. Seismic data collected from a mid-oceanic ridge volcano suggest that microseismic events coincide with Earth's low tides. The link between earthquakes and Earth's tidal cycles has been discussed for nearly a century. Some studies have shown that earthquake activity along normal faults can be influenced by tides. The perpetual rise and fall of the sea are the result of the gravitational attraction of the Moon and the Sun on Earth. Because water is not firmly attached to Earth, these gravitational forces cause a disturbance of the ocean's water and generate very long waves, or tides. It is thought that the shifting of water by tides can change the overlying water load or pressure on the ocean crust, which in turn could trigger movement in the crust. For many years, this idea has been difficult to test for lack of a seismometer sensitive enough to record slight crustal movements. In addition, there is a complicated interplay of various stresses that can affect seismic activity, including water depth and crustal properties such as temperature and composition.

Maya Tolstoy and her colleagues set up an array of sensitive, ocean-bottom seismometers to monitor seismic activity continuously over a two-month period at the Axial volcano on the Juan de Fuca Ridge (box figure 1). She reasoned that the crust at mid-oceanic ridges would be vulnerable to tidal energy, because the crust is weakened by networks of fractures filled with circulating water. They recorded 402 microearthquakes (earthquakes of less than 2.5 magnitude) over the two-month period. The ocean tides at the mid-oceanic ridge were recorded as changes in water pressure; low water pressure indicates low tides and high pressure, high tides.

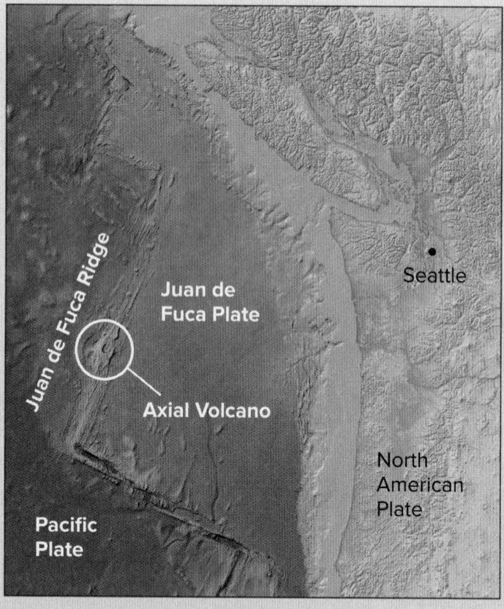

BOX 18.1 ■ FIGURE 1

(*A*) Map of Juan de Fuca Ridge showing location of Axial volcano where study was conducted. (*B*) Maya Tolstoy on research ship with ocean-bottom seismometers that will monitor the Axial volcano on the Juan de Fuca Ridge.
©Lamont-Doherty Earth Observatory

surface waters die. In some parts of the sea, such as the polar and equatorial regions, great concentrations of these shells have built unusually thick pelagic deposits.

The constant slow rain of pelagic clay and shells occurs in all parts of the sea. Although the rate of accumulation varies from place to place, pelagic sediment should be expected on all parts of the sea floor.

Surprisingly, however, pelagic sediment is almost completely absent on the crest of the mid-oceanic ridge. Pelagic sediment is found on the flanks of the mid-oceanic ridge, often thickening away from the ridge crest (figure 18.23). But its absence on the ridge crest was an unexpected discovery about seafloor sediment distribution.

18.5 OCEANIC CRUST AND OPHIOLITES

As you have seen in chapter 17, oceanic crust differs significantly from continental crust; it is both thinner and of a different composition than continental crust. Seismic reflection and seismic refraction surveys at sea have shown the oceanic crust to

Analysis of the occurrence of earthquakes and the water pressure data indicates that earthquake activity intensifies near or at low tides. Tolstoy thinks it is likely that the pressure of the overlying water causes normal faults to lock (box figure 2). Once the overlying water pressure drops in response to the ebbing (falling) tide, the reduction in pressure allows slippage along the normal faults and fractures in the Axial volcano region. Tolstoy and her colleagues also suggest that as the overlying water pressure drops during low tide, gases in the cracks will expand, increasing the pore pressure of the fractured rocks. This increased pressure may help cause slippage along the faults. The subtle changes in water pressure associated with tidal cycles appear to be strong enough to trigger microearthquakes in the susceptible midocean crust. More recently, Tolstoy and colleagues deployed ocean-bottom seismometers along the East Pacific Rise and recorded tidal-triggered microearthquakes that preceded an undersea eruption.

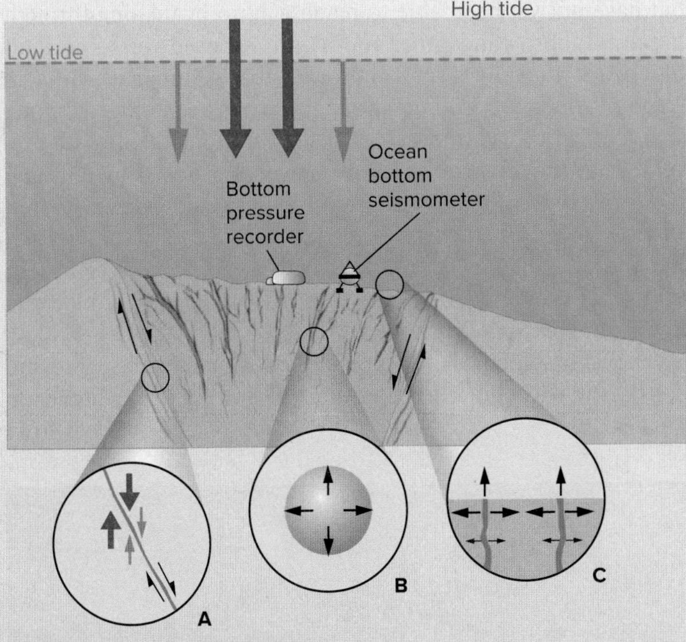

BOX 18.1 ■ FIGURE 2

Diagram of the Axial volcano area illustrating the effect of tides on earthquakes. (*A*) Because of their geometry, normal faults remain locked during high tides (red arrows) and unlocked during low tides (green arrows) when less water pressure is pushing down on the faults. (*B*) The pore pressure in cracks at hydrothermal vents increases at low tide and (*C*) allows trapped bubbles of gas to expand much like a sealed bottle with air expands on an airplane. The expansion of gas in the cracks increases the pore pressure in the fractured rocks and may help cause slippage (earthquakes) along the normal faults at low tide.

Do tidal triggered earthquakes also occur in subduction zones? Current research suggests they do. Researchers in Japan found an increase in tidal-triggered shallow earthquakes that began about ten years before the 2011 Japan and 2004 Sumatra great quakes. Tolstoy and her colleagues are currently looking at the Aleutian trench off Alaska and the Cascadia subduction zone off the Pacific Northwest of the United States to see if any segments currently show evidence of tidal triggering that might help forecast future great earthquakes.

Earthquake activity may not be the only deep-sea process that is tidally dependant. The crustal response to tidal forces may also influence the abundant life and flow of nutrients at hydrothermal systems of mid-oceanic ridges. Tolstoy and her colleagues speculate that biological cycles—life and nutrient flux—of the ridges may also be linked to the ebb and flow of the tides. Tolstoy may tell you that the Earth may, indeed, "breathe" to the rhythms of the planets.

Additional Resources

Maya Tolstoy, F. L. Vernon, J. A. Orcutt, and F. K. Wyatt. 2002. Breathing of the sea floor: Tidal correlation of seismicity at Axial volcano. *Geology* vol. 30, no. 6: 503–506.
www.ldeo.columbia.edu/~tolstoy/pdf/axial_tides.pdf

To watch Maya Tolstoy explain how tides influence earthquakes on the sea floor visit the following site and search for Maya Tolstoy:
http://tedxtalks.ted.com/video/Seafloor-Earthquakes-Maya-Tolst;TEDxCERN

Sachiko Tanaka. 2012. Tidal triggering of earthquakes prior to the 2011 Tohoku-Oki earthquake (Mw9.1). *Geophys. Res. Lett.*, 39, L0026, doi: 10.1029/2012GL051179.

T. Bhatnager, M. Tolstoy, and F. Waldhauser, 2016. Influence of fortnightly tides on earthquake triggering at the East Pacific Rise at 9o50'N. *Journal of Geophys. Res.*, v. 121, i. 3, pp. 1262-1279. doi: 10.1002/2015JBO12388

be about 7 kilometers (4.3 miles) thick and divided into three layers (figure 18.24).

The top layer (Layer 1), of variable thickness and character, is marine sediment. In an abyssal fan or on the continental rise, Layer 1 may consist of several kilometers of terrigenous sediment. On the upper flanks of the mid-oceanic ridge, there may be less than 100 meters (328 feet) of pelagic sediment. An average thickness for Layer 1 might be 0.5 kilometer (1,640 feet).

Beneath the sediment is Layer 2, which is about 1.5 kilometers (about 1 mile) thick. It has been extensively sampled. This layer consists of pillow basalt overlying dikes of basalt. The basalt pillows, rounded masses that form when hot lava erupts into cold water (figure 18.24), are highly fractured at the top of Layer 2. The dikes in the lower part of Layer 2 are closely spaced, parallel, vertical dikes ("sheeted dikes"). Widely sampled by drilling and dredging, Layer 2 has also been observed directly by geologists in submersibles diving into the rift valley of the mid-oceanic ridge during the FAMOUS expedition.

The lowest layer in the crust is Layer 3. It is about 5 kilometers (3.1 miles) thick and thought to consist of sill-like gabbro bodies. The evidence for this interpretation is suggestive but not conclusive. Geologists have drilled through 2 kilometers

FIGURE 18.22

Photograph taken with a scanning electron microscope (260× magnification) of radiolarian shells composed of silica. These single-celled marine organisms make up a portion of the pelagic sediment on the sea floor of the Pacific Ocean.
©Biophoto Associates/Science Source

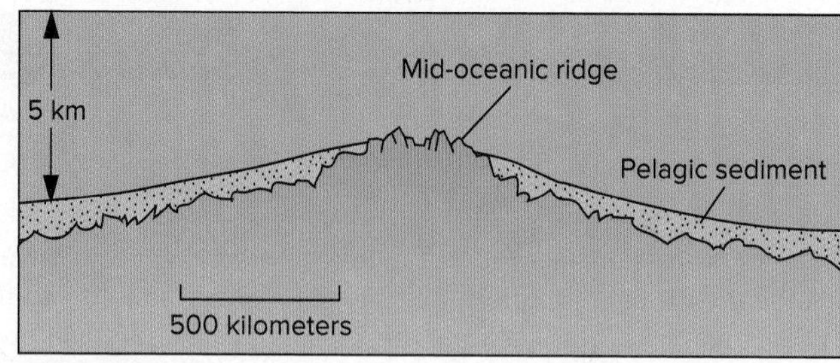

FIGURE 18.23

Pelagic sediment is thin or absent on the crest of the mid-oceanic ridge and becomes progressively thicker away from the ridge crest (distribution of sediment is highly simplified).

(1.2 miles) of pillow basalt and basaltic dikes in oceanic crust but have not yet been able to reach gabbro by drilling through basalt. Gabbro (and other rocks) are exposed on some fault blocks in rift valleys and on some steep submarine cliffs in fracture zones; 0.5 kilometer of gabbro has been drilled on one such cliff in the southern Indian Ocean. Some geologists presume that the gabbro represents Layer 3 exposed by faulting, but other geologists are not so sure. Some evidence exists that rocks exposed in fracture zones may not be representative of the entire sea floor. Even if the drilled gabbro *does* represent the upper one-tenth of Layer 3, no one has yet sampled the lower nine-tenths of the layer.

Seismic velocities of 7 kilometers (4.3 miles) per second are consistent with the choice of gabbro for Layer 3, but deep

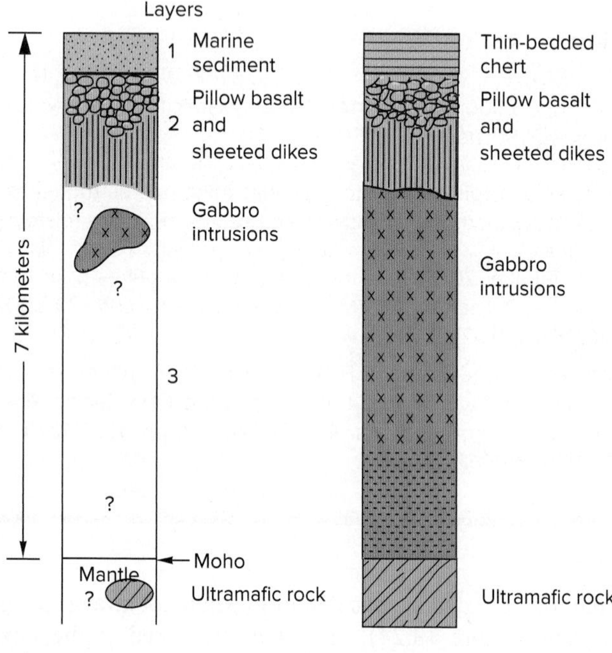

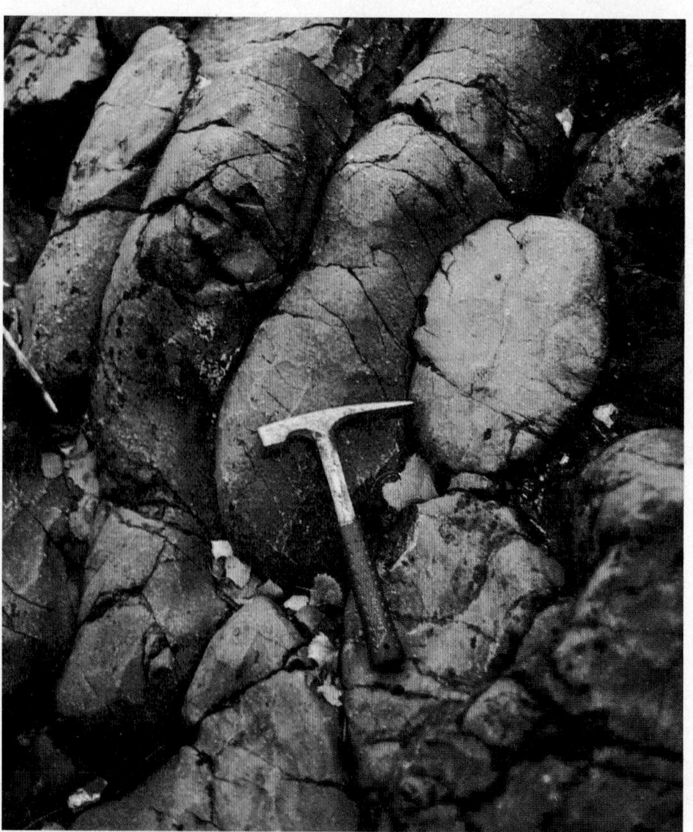

FIGURE 18.24

A comparison of oceanic crust and an ophiolite sequence. (*A*) Structure of oceanic crust, determined from seismic studies and drilling. Gabbro and ultramafic rock drilled and dredged from seafloor fault blocks may be from Layer 3 and the mantle. (*B*) Typical ophiolite sequence found in mountain ranges on land. Thickness is approximate—sequence is usually highly faulted. (*C*) Pillow basalt from the upper part of an ophiolite, northern California. These rocks formed as part of the sea floor, when hot lava cooled quickly in cold seawater.
(C) ©David McGeary

drilling on land has shown that seismic reflection and refraction records on land are routinely misinterpreted regarding rock type and depths to rock boundaries. These errors extend to oceanic crust as well. A drill hole in the eastern Pacific Ocean has been reoccupied four times in a twelve-year span and has now reached a total depth of 2,000 meters (6,560 feet) below the sea floor. Seismic evidence suggested that the Layer 2–Layer 3 boundary would be found at a depth of about 1,700 meters (5,576 feet), but the drill went well past that depth without finding the contact between the dikes of Layer 2 and the expected gabbro of Layer 3. Either the seismic interpretation or the model of Layer 3's composition must be wrong.

Geologists' ideas about the composition of oceanic crust are greatly influenced by a study of **ophiolites**, distinctive rock sequences found in many mountain chains on land (figure 18.24). The thin, top layer of an ophiolite consists of marine sedimentary rock, often including thin-bedded chert. Below the sedimentary rock lies a zone of pillow basalt, which in turn is underlain by a sheeted-dike complex that probably served as feeder dikes for the pillowed lava flows above. The similarities between the upper part of an ophiolite and Layers 1 and 2 of oceanic crust are obvious.

Below the sheeted dikes of an ophiolite is a thick layer of podlike gabbro intrusions, perhaps thick sills. Beneath the gabbro lies ultramafic rock such as peridotite, the top part of which has been converted to serpentine by metamorphism. Geologists have long thought that ophiolites represent slivers of oceanic crust somehow emplaced on land. If this is true, then the gabbro of ophiolites may represent oceanic Layer 3, and the peridotite may represent mantle rock below oceanic crust. The contact between the peridotite and the overlying gabbro would be the Mohorovičić discontinuity.

Recent work has shown that many, if not most, ophiolites are *not* typical sea floor but a special type of sea floor formed in marginal ocean basins next to continents by the process of back-arc spreading (see chapter 19). If ophiolites do not represent typical sea floor, then more extensive and deeper seafloor drilling is needed before a clear picture of oceanic crust can be formed. This is an important goal, for oceanic crust is the most common surface rock, covering 60% of Earth's surface.

18.6 THE AGE OF THE SEA FLOOR

As marine geologists began to determine the age of seafloor rocks (by isotopic dating) and sediments (by fossils), an astonishing fact was discovered. All the rocks and sediments of the sea floor proved to be younger than 200 million years. This was true only for rocks and sediments from the *deep*-sea floor, not those from the continental margins. The rocks and sediments currently found on the deep-sea floor formed during the Mesozoic and Cenozoic eras but not earlier.

By contrast, as discussed in chapter 8, Earth is estimated to be 4.5 billion years old. Every continent contains some rocks formed during the Paleozoic Era and the Precambrian. Some of the Precambrian rocks on continents are more than 3 billion years old, and a few are almost 4 billion years old. Continents, therefore, preserve rocks from most of Earth's history. In sharp contrast to the continents is the deep-sea floor, which covers more than half of Earth's surface but preserves less than one-twentieth of its history in its rocks and sediment.

18.7 THE SEA FLOOR AND PLATE TECTONICS

As we mentioned at the beginning, this chapter is concerned with the *description* of the sea floor. The *origin* of most seafloor features is related to plate tectonics. Chapter 19 shows how the theory of plate tectonics explains the existence and character of continental shelves and slopes, trenches, the mid-oceanic ridge, and fracture zones, as well as the very young age of the sea floor itself. In the workings of the scientific method, this chapter largely concerns *data.* Chapter 19 shows how *hypotheses* and *theories* account for these data.

ENVIRONMENTAL GEOLOGY 18.2

Geologic Riches in the Sea

Many resources are being extracted from the sea floor and seawater, and in some instances, there is a great potential for increased extraction. Realizing this tremendous resource, the United States established the Economic Exclusion Zone, which gives sovereign rights for resources on or below the sea floor in an area that extends 370 kilometers (200 miles) from the coastline.

Offshore oil and *gas* are the most valuable resources now being taken from the sea. More than one-sixth of U.S. oil production (and more than one-quarter of world production) comes from drilling platforms set up on the continental shelf (box figure 1). Oil and gas have been found within deeper parts of the sea floor, such as the continental slope and continental rise. Producing oil from these deeper regions is much more costly than drilling in shallow water. Oil spills from wells in deep water can be especially hard to control, as was shown by the 2010 *Deepwater Horizon*/BP oil spill in the Gulf of Mexico that took 86 days to stop and resulted in 5 million barrels of oil leaked. This was the largest marine oil spill in history, and it threatened the ecosystems and economy of the entire Gulf Coast region (see box 22.2).

Gas hydrates, which are icelike solids of natural gas and water found trapped in deep marine sediments, are another potential source of energy that is being explored.

BOX 18.2 ■ FIGURE 1

Offshore oil drilling platform Hibernia is located 320 kilometers (200 miles) off the coast of Newfoundland and began producing oil in 1997. As many as 50 different wells can be drilled from a single platform.
Courtesy of Chevron Canada Resources

Other important resources are dredged from the sea floor. *Phosphorite* can be recovered from shallow shelves and banks and used for fertilizers. *Gold*, *diamonds*, and heavy *black sands* (which are black because they contain metal-bearing minerals) are being separated from the surface sands and gravels of some continental shelves by specially designed ships.

Manganese nodules (box figure 2) cover many parts of the deep-sea floor, notably in the central Pacific. These black, potato-sized lumps contain approximately 25% manganese, 15% iron, and up to 2% nickel and 2% copper, along with smaller amounts of cobalt. Although there are international legal problems concerning who owns them, larger industrial countries such as the United States may mine them, particularly for copper, nickel, and cobalt. The United States is also interested in manganese; it imports 95% of its manganese, which is critical to producing some types of steel.

Metallic brines and *sediments*, first discovered in the Red Sea, are deposited by submarine hot springs active at the rift valley of the mid-oceanic ridge crest. The Red Sea sediments contain more than 1% copper and more than 3% zinc, together with impressive amounts of silver, gold, and lead (estimated value is $25 billion). Because of their great value, the sediments will probably be mined even though they are at great depth. Deposits similar to those of the Red Sea—although not of such great economic potential—have been found on several other parts of the ridge.

A few substances can be extracted from the salts dissolved in seawater. Approximately two-thirds of the world's production of *magnesium* is obtained from seawater, and in many regions, *sodium chloride* (table salt) is obtained by solar evaporation of seawater.

BOX 18.2 ■ FIGURE 2

Dense concentrations of manganese nodules on the floor of the abyssal plain (depth 5,350 meters) in the northeast Atlantic Ocean.
©Institute of Oceanographic Sciences/NERC/Science Source

Summary

The *continental shelf* and the steeper *continental slope* lie under water along the edges of continents. They are separated by a change in slope angle at a depth of 100 to 200 meters.

Submarine canyons are cut into the continental slope and outer continental shelf by a combination of *turbidity currents*, sand flow and fall, bottom currents, and river erosion during times of lower sea level. Graded bedding and cable breaks suggest the existence of turbidity currents in the ocean.

Abyssal fans form as sediment collects at the base of submarine canyons.

A *passive continental margin* occurs off geologically quiet coasts and is marked by a continental rise and abyssal plains at the base of the continental slope.

The *continental rise* and *abyssal plains* may form from sediment deposited by turbidity currents.

The continental rise may also form from sediment deposited by *contour currents* at the base of the continental slope.

An *active continental margin* is marked by an *oceanic trench* at the base of the continental slope; the continental rise and abyssal plains are absent.

Oceanic trenches are twice as deep as abyssal plains, which generally lie at a depth of 5 kilometers. Associated with trenches are *Benioff zones* of earthquakes and *andesitic volcanism*, forming either an island arc or a chain of volcanoes in a young mountain belt near the edge of a continent. Trenches have low heat flow and negative gravity anomalies.

The *mid-oceanic ridge* is a globe-circling mountain range of basalt, located mainly in the middle of ocean basins. The crest of the ridge is marked by a *rift valley*, shallow-focus earthquakes, high heat flow, active *basaltic volcanism*, hydrothermal activity, and exotic organisms.

Fracture zones are lines of weakness that offset the mid-oceanic ridge. Shallow-focus quakes occur on the portion of the fracture zone between the offset ridge segments.

Seamounts are conical, submarine volcanoes, now mostly extinct. *Guyots* are flat-topped seamounts, probably leveled by wave erosion before subsiding.

Chains of seamounts and guyots form *aseismic ridges.*

Corals and algae living in warm, shallow water construct *fringing reefs*, *barrier reefs*, and *atolls*.

Terrigenous sediment is composed of land-derived sediment deposited near land by turbidity currents and other processes. *Pelagic sediment* is made up of wind-blown dust and microscopic skeletons that settle slowly to the sea floor.

The crest of the mid-oceanic ridge lacks pelagic sediment.

Oceanic crust consists of marine sediments overlying basalt pillows and dikes, probably overlying gabbro.

Ophiolites in continental mountain ranges probably represent slivers of somewhat atypical oceanic crust somehow emplaced on land.

The oldest rocks on the deep-sea floor are 200 million years old. The continents, in contrast, contain some rock that is 3 to 4 billion years old.

Terms to Remember

abyssal fan 441
abyssal plain 443
active continental margin 444
aseismic ridge 448
atoll 448
barrier reef 448
continental rise 443
continental shelf 440
continental slope 440
contour current 443
fracture zone 446
fringing reef 448
guyot 447
mid-oceanic ridge 445
oceanic trench 444
ophiolite 453
passive continental margin 443
pelagic sediment 448
reef 448
rift valley 445
seamount 447
submarine canyon 441
terrigenous sediment 448
turbidity current 441

Testing Your Knowledge

Use the following questions to prepare for exams based on this chapter.

1. Describe the different methods used to study the sea floor.
2. What is a submarine canyon? How do submarine canyons form?
3. What is a turbidity current? What is the evidence that turbidity currents occur on the sea floor?
4. Describe two different origins for the continental rise.
5. Discuss the appearance, structure, and origin of abyssal plains.
6. Sketch an active continental margin and a passive continental margin, labeling all their parts. Show approximate depths.
7. In a sketch, show the association between an oceanic trench, a Benioff zone of earthquakes, and volcanoes on the edge of a continent.
8. Sketch a cross profile of the mid-oceanic ridge, showing the rift valley. Label your horizontal and vertical scales.
9. What is a fracture zone? Sketch the relation between fracture zones and the mid-oceanic ridge.
10. Describe the appearance and origin of seamounts and guyots.
11. Sketch a cross section of a fringing reef, a barrier reef, and an atoll.
12. Describe the two main types of seafloor sediment.
13. How does the age of seafloor rocks compare with the age of continental rocks? Be specific.
14. What is the thickness and composition of oceanic crust?
15. Which is true of the continental shelf?
    a. It is a shallow submarine platform at the edge of a continent.
    b. It inclines very gently seaward.
    c. It can vary in width.
    d. All of the preceding.
16. Oceanic trenches are
    a. found along a passive continental margin.
    b. associated with high heat flow.
    c. characterized by large negative gravity anomalies.
    d. none of the above.

17. Great masses of sediment-laden water that are pulled downhill by gravity are called
    a. contour currents.
    b. bottom currents.
    c. turbidity currents.
    d. traction currents.
18. Oceanic trenches
    a. are narrow, deep troughs.
    b. run parallel to the edge of a continent or an island arc.
    c. are often 8 to 10 kilometers deep.
    d. all of the preceding.
19. Which is characteristic of mid-oceanic ridges?
    a. shallow-focus earthquakes
    b. high heat flow
    c. basalt eruptions
    d. all of the preceding
20. Reefs parallel to the shore but separated from it by wide, deep lagoons are called
    a. fringing reefs.
    b. barrier reefs.
    c. atolls.
    d. lagoonal reefs.
21. Pelagic sediment could be composed of
    a. fine-grained clay.
    b. skeletons of microscopic organisms.
    c. volcanic ash.
    d. all of the preceding.
22. What part of the continental margin marks the true edge of the continent?
    a. continental shelf
    b. continental slope
    c. continental rise
    d. abyssal plain
23. Distinctive rock sequences of basalt and marine sedimentary rock that may be slices of the ocean floor are
    a. guyots.
    b. ophiolites.
    c. seamounts.
    d. fracture zones.

Expanding Your Knowledge

1. How many possible origins can you think of for the rift valley on the mid-oceanic ridge?
2. What factors could cause sea level to rise? To fall?
3. Why is the rock of the deep-sea floor (60% of Earth's surface) basalt? Where did the basalt come from?
4. How could fracture zones have formed?
5. How many hypotheses can you think of to explain the relatively young age of the sea floor?

Exploring Web Resources

http://oceancolor.gsfc.nasa.gov

Ocean Color Web contains many links to seafloor topics, such as hydrothermal vents and microbiology of deep-sea vents.

www.noaa.gov/ocean.html

National Oceanographic and Atmospheric Administration (NOAA) has many links to numerous sites relating to the oceans.

www.ngdc.noaa.gov/mgg/mggd.html

NOAA Marine Geology and Geophysics Division website contains world seafloor maps and information on ocean-drilling data and samples.

www.virtualocean.org/

Virtual Ocean website allows you to explore the details of the surface of the sea floor.

www.whoi.edu/VideoGallery/

Woods Hole Oceanographic Institution website contains video clips of black smokers, exotic organisms at mid-oceanic ridges, and oceanographic research using vessels and submersibles.

www.pmel.noaa.gov/vents/

NOAA Vents Program website provides photos, video clips, data, and research program activities about the investigation of submarine volcanoes and hydrothermal venting around the world.or more information about the likelihood of the San Andreas fault creating a large earthquake and video simulations of the shaking, visit U.S. Geological Survey websites:

- https://pubs.usgs.gov/fs/2015/3009/pdf/fs2015-3009.pdf

CHAPTER

19

Plate Tectonics—The Unifying Theory

Satellite image of the Red Sea and Arabia. Plate motion has torn the Arabian Peninsula (right center) away from Africa (left) to form the Red Sea (center).
Source: NOAA/NGDC

LEARNING OBJECTIVES

- Summarize Wegener's evidence for continental drift. Why did the scientific community not initially accept the idea of continental drift?
- Outline the evidence that revived interest in continental drift.
- Explain the concept of seafloor spreading. Discuss how seafloor spreading explains features on the sea floor.
- Describe the evidence that plates move.
- Sketch and describe the various types of plate boundaries and the different geologic features associated with each.
- Discuss the possible driving mechanisms for plate tectonics.

As you studied volcanoes; igneous, metamorphic, and sedimentary rocks; and earthquakes, you learned how these topics are related to plate tectonics. In this chapter, we take a closer look at plates and plate motion. We will pay particular attention to plate boundaries and the possible driving mechanisms for plate motion.

The history of the concept of plate tectonics is a good example of how scientists think and work and how a hypothesis can be proposed, discarded, modified, and then reborn. In the first part of this chapter, we trace the evolution of an idea–how the earlier hypotheses of moving continents (continental drift) and a moving sea floor (seafloor spreading) were combined to form the theory of plate tectonics.

Tectonics is the study of the origin and arrangement of the broad structural features of Earth's surface, including not only folds and faults but also mountain belts, continents, and earthquake belts. Tectonic models such as an expanding Earth or a contracting Earth have been used in the past to explain *some* of the surface features of Earth. Plate tectonics has come to dominate geologic thought today because it can explain so *many* features. The basic idea of **plate tectonics** is that Earth's surface is divided into a few large, thick plates that move slowly and change in size. Intense geologic activity occurs at *plate boundaries* where plates move away from one another, past one another, or toward one another. The eight large lithospheric plates shown in figure 19.1, plus a few dozen smaller plates, make up the outer shell of Earth (the crust and upper part of the mantle).

The concept of plate tectonics was born in the late 1960s by combining two preexisting ideas–continental drift and seafloor spreading. **Continental drift** is the idea that continents move freely over Earth's surface, changing their positions relative to one another. **Seafloor spreading** is a hypothesis that the sea floor forms at the crest of the mid-oceanic ridge, then moves horizontally away from the ridge crest toward an oceanic trench. The

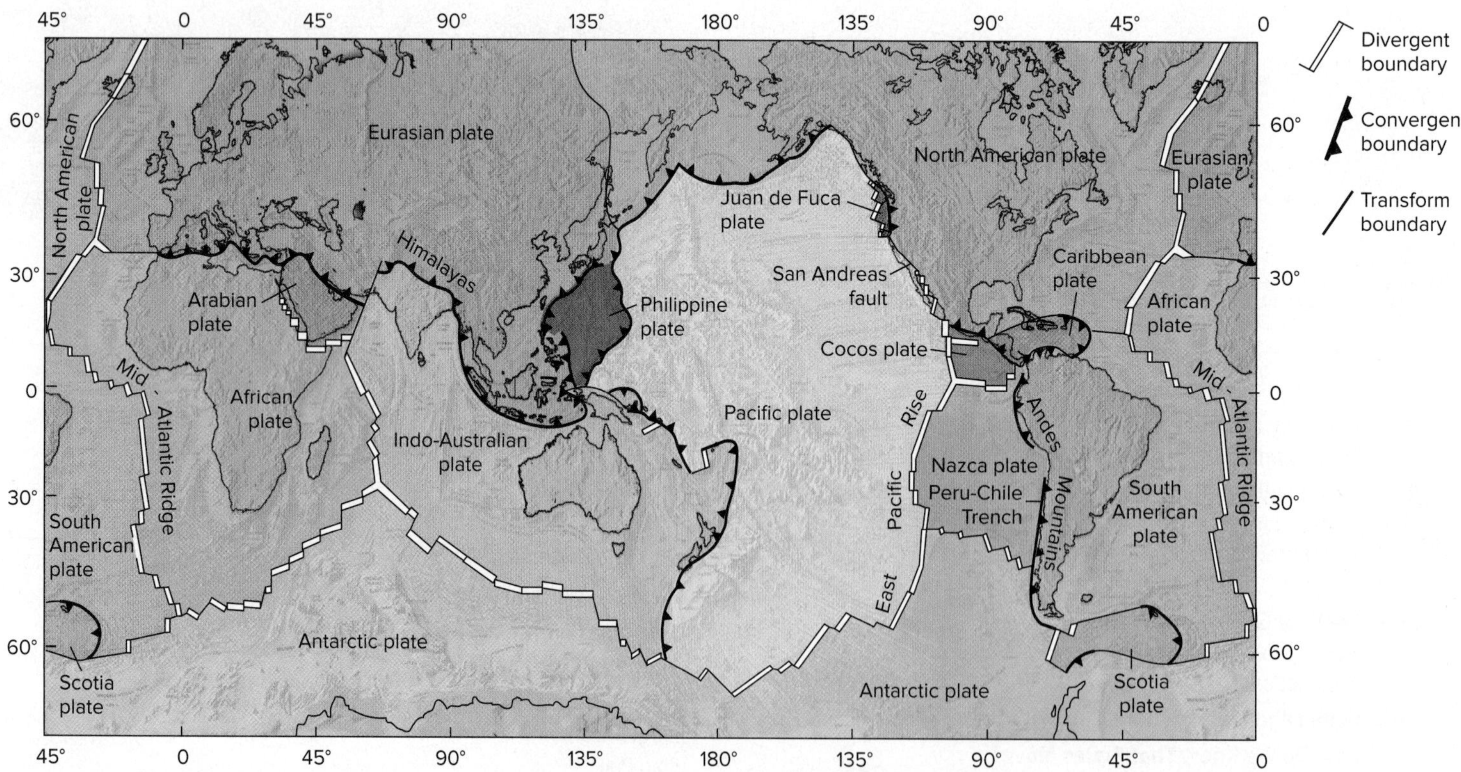

FIGURE 19.1

The major plates of the world. The western edge of the map repeats the eastern edge so that all plates can be shown unbroken. Double lines indicate spreading axes on divergent plate boundaries. Single lines show transform boundaries. Heavy lines with triangles show convergent boundaries, with triangles pointing down subduction zones. *Source: W. Hamilton, U.S. Geological Survey*

two sides of the ridge are moving in opposite directions like slow conveyor belts.

Before we take a close look at plates, we will examine the earlier ideas of moving continents and a moving sea floor because these two ideas embody the theory of plate tectonics.

19.1 THE EARLY CASE FOR CONTINENTAL DRIFT

Continents can be made to fit together like pieces of a picture puzzle. The similarity of the Atlantic coastlines of Africa and South America has long been recognized. The idea that continents were once joined together and have split and moved apart from one another has been around for more than a century (figure 19.2).

In the early 1900s, Alfred Wegener, a German meteorologist, made a strong case for continental drift. He noted that South America, Africa, India, Antarctica, and Australia had almost identical late Paleozoic rocks and fossils.

The plant *Glossopteris* is found in Pennsylvanian and Permian-age rock on all five continents, and fossil remains of *Mesosaurus,* a freshwater reptile, are found in Permian-age rocks only in Brazil and South Africa (figure 19.3). In addition, fossil remains of land-dwelling reptiles *Lystrosaurus* and *Cynognathus* are found in Triassic-age rocks on all five continents.

Wegener reassembled the continents to form a giant supercontinent, *Pangaea* (also spelled *Pangea* today). Wegener

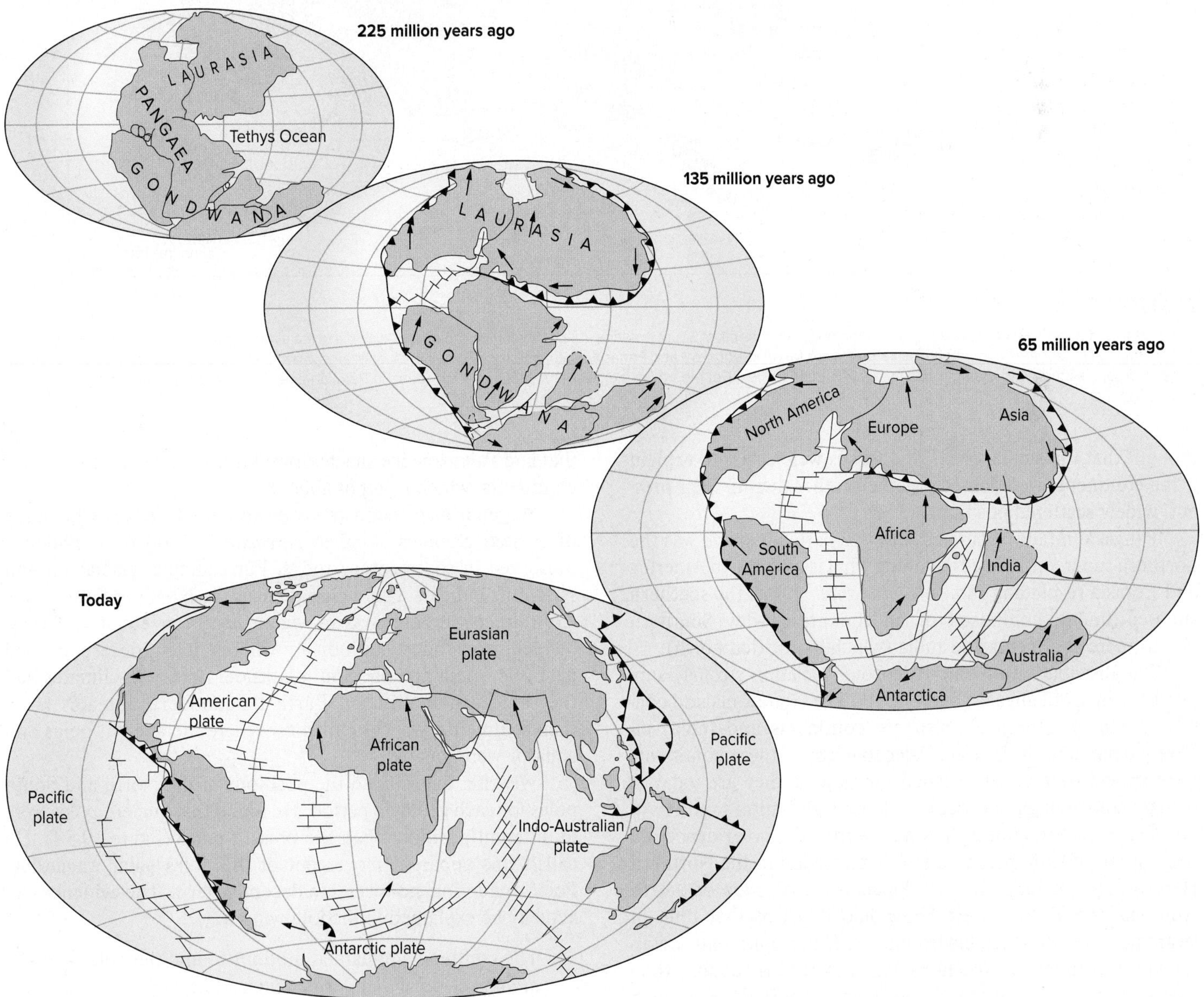

FIGURE 19.2

Pangaea breakup and continental drift.
Source: C. R. Scotese (www.scotese.com)

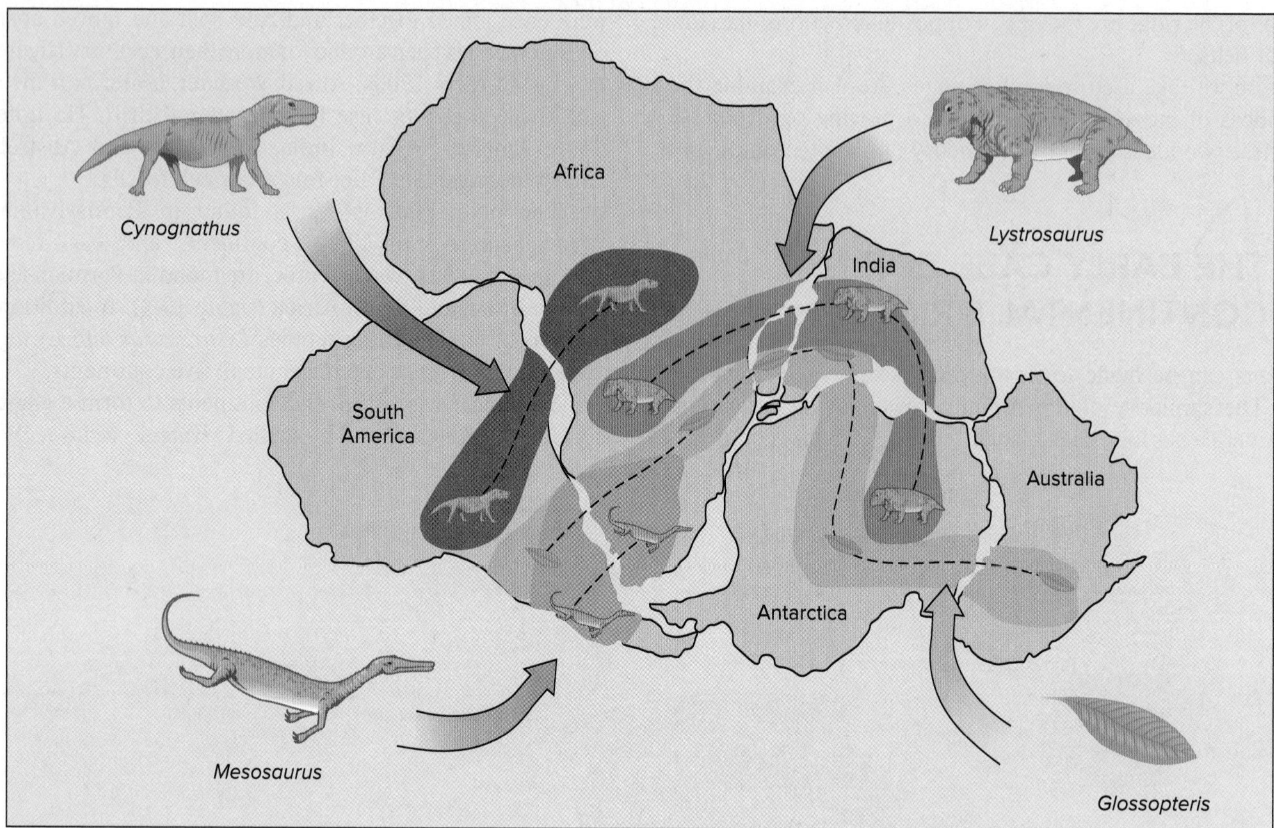

FIGURE 19.3

Distribution of plant and animal fossils that are found on the continents of South America, Africa, Antarctica, India, and Australia gives evidence for the southern supercontinent of Gondwana. *Glossopteris* and other fernlike plants are found in Permian- and Pennsylvanian-age rocks on all five continents. *Cynognathus* and *Lystrosaurus* were sheep-sized land reptiles that lived during the Early Triassic Period. Fossils of the freshwater reptile *Mesosaurus* are found in Permian-age rocks on the southern tip of Africa and South America.

thought that the similar rocks and fossils were easier to explain if the continents were joined together, rather than in their present, widely scattered positions.

Pangaea initially separated into two parts. *Laurasia* was the northern supercontinent, containing what is now North America and Eurasia (excluding India). *Gondwanaland* was the southern supercontinent, composed of all the present-day Southern Hemisphere continents and India (which has drifted north).

The distribution of Late Paleozoic glaciation strongly supports the idea of Pangaea (figure 19.4). The Gondwanaland continents (the Southern Hemisphere continents and India) all have glacial deposits of Late Paleozoic age. If these continents were spread over Earth in Paleozoic time as they are today, a climate cold enough to produce extensive glaciation would have had to prevail over almost the whole world. Yet, no evidence has been found of widespread Paleozoic glaciation in the Northern Hemisphere. In fact, the late Paleozoic coal beds of North America and Europe were being laid down at that time in swampy, probably warm environments. If the continents are arranged according to Wegener's Pangaea reconstruction, then glaciation in the Southern Hemisphere is confined to a much smaller area (figure 19.4), and the absence of widespread glaciation in the Northern Hemisphere becomes easier to explain. Also, the present arrangement of the continents would require that late Paleozoic ice sheets flowed from the oceans toward the continents, which is impossible.

Wegener also reconstructed old climate zones (the study of ancient climates is called *paleoclimatology*) from evidence preserved in sedimentary rocks. For example, glacial till and striations indicate a cold climate near the North or South Pole. Coral reefs indicate warm water near the equator. Cross-bedded sandstones can indicate where ancient deserts formed near 30° North and 30° South latitude. If ancient climates had the same distribution on Earth that modern climates have, then sedimentary rocks can show where the ancient poles and equator were located.

Wegener determined the positions of the North and South poles for each geologic period. He found that ancient poles were in different positions than the present poles (figure 19.5*A*). He called this apparent movement of the poles **polar wandering**. Polar wandering, however, is a deceptive term. The evidence can actually be explained in the following ways:

1. The continents remained motionless and the poles actually *did* move—polar wandering (figure 19.5*A*).
2. The poles stood still and the continents moved—continental drift (figure 19.5*B*).
3. Both occurred.

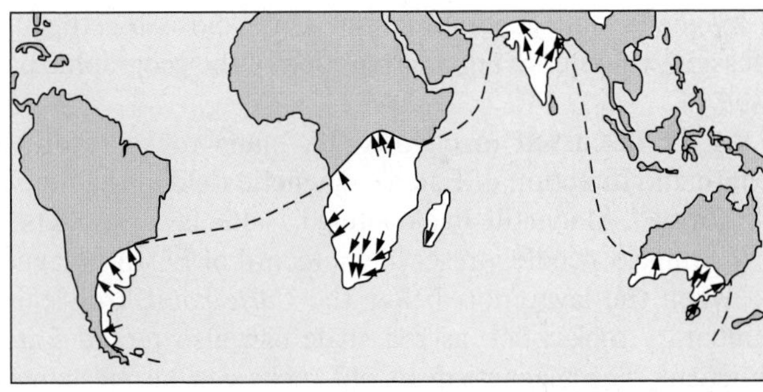

A

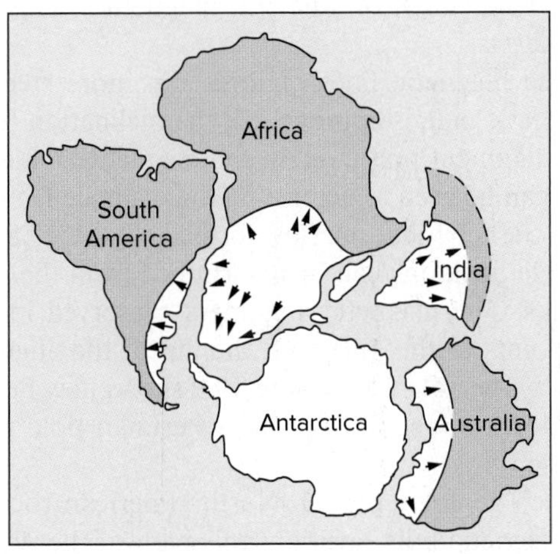

B

FIGURE 19.4

Distribution of late Paleozoic glaciation; arrows show direction of ice flow. (*A*) Continents in present positions show wide distribution of glaciation (white land areas with flow arrows). (*B*) Continents reassembled into Pangaea. Glaciated region becomes much smaller.
Source: Arthur Holmes, Principles of Physical Geology, 2nd ed., Ronald Press, 1965.

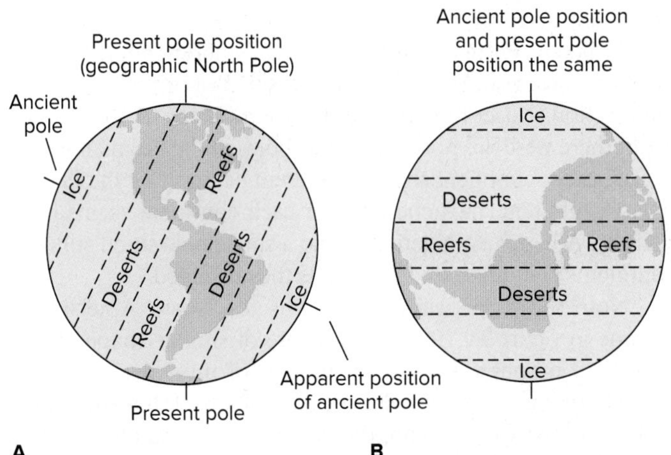

A B

FIGURE 19.5

Two ways of interpreting the distribution of ancient climate belts. (*A*) Continents are fixed, poles wander. (*B*) Poles are fixed, continents drift. For simplicity, the continents in (*B*) are shown as having moved as a unit, without changing positions relative to one another. If continents move, they should change relative positions, complicating the pattern shown.

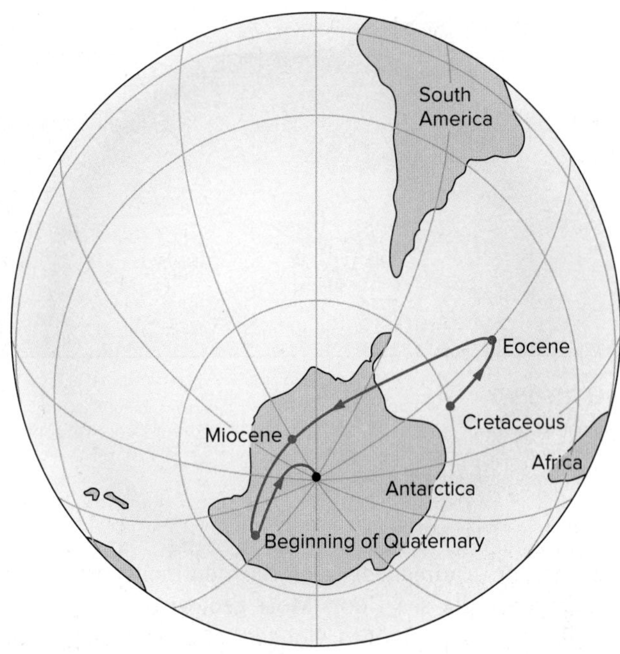

FIGURE 19.6

Apparent wandering of the South Pole since the Cretaceous Period as determined by Wegener from paleoclimate evidence. Wegener, of course, believed that *continents*, rather than the poles, moved.
Source: A. Wegener, The Origins of Continents and Oceans (1928), Dover Publications, 1968.

Wegener plotted curves of apparent polar wandering (figure 19.6). Since one interpretation of polar wandering data was that the continents moved, Wegener believed that this supported his concept of continental drift. (Notice that in only one interpretation of polar wandering do the poles actually move. You should keep in mind that when geologists use the term *polar wandering*, they are referring to an *apparent* motion of the poles, which may or may not have actually occurred.)

Skepticism about Continental Drift

Although Wegener presented the best case possible in the early 1900s for continental drift, much of his evidence was not clear-cut. The presence of land-dwelling reptiles throughout the scattered continents was explained by land bridges, which were postulated to somehow rise up from the sea floor and then subside again. The existence or nonexistence of land bridges was difficult to prove without data on the topography of the sea floor. Also, fossil plants could have been spread from one continent to another by winds or ocean currents. Their distribution over more than one continent does not *require* that the continents were all joined in the supercontinent, Pangaea. In addition, polar wandering might have been caused by moving poles rather than by moving continents. Because his evidence was not conclusive, Wegener's ideas were not widely accepted. This was particularly true in the United States, largely because of the mechanism Wegener proposed for continental drift.

Wegener proposed that continents plowed through the oceanic crust (figure 19.7), perhaps crumpling up mountain

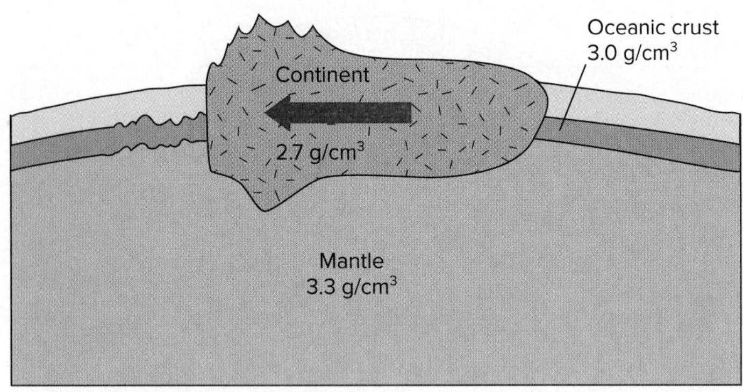

FIGURE 19.7

Wegener's concept of continental drift implied that the less-dense continents drifted *through* oceanic crust, crumpling up mountain ranges on their leading edges as they pushed against oceanic crust.

ranges on the leading edges of the continents where they pushed against the sea floor. Most geologists in the United States thought that this idea violated what was known about the strength of rocks at the time. The driving mechanism proposed by Wegener for continental drift was a combination of centrifugal force from Earth's rotation and the gravitational forces that cause tides. Careful calculations of these forces showed them to be too small to move continents. Because of these objections, Wegener's ideas received little support in the United States or much of the Northern Hemisphere (where the great majority of geologists live) in the first half of the twentieth century. The few geologists in the Southern Hemisphere, however, where Wegener's matches of fossils and rocks between continents were more evident, were more impressed with the concept of continental drift.

19.2 THE REVIVAL OF CONTINENTAL DRIFT

Much work in the 1940s and 1950s set the stage for the revival of the idea of continental drift and its later incorporation, along with seafloor spreading, into the new concept of plate tectonics. The new investigations were in two areas: (1) study of the sea floor and (2) geophysical research, especially in relation to rock magnetism.

Evidence from Paleomagnetism

Convincing new evidence about polar wandering came from the study of rock magnetism. Wegener's work dealt with the wandering of Earth's *geographic* poles of rotation. The *magnetic* poles are located close to the geographic poles, as you saw in chapter 17 on Earth's interior. Historical measurements show that the position of the magnetic poles moves from year to year but that the magnetic poles stay close to the geographic poles as they move. As we discuss magnetic evidence for polar wandering, we are referring to an apparent motion of the magnetic poles. Because the magnetic and geographic poles are close together, our discussion will refer to apparent motion of the geographic poles as well.

As we discussed in chapter 17, many rocks record the strength and direction of Earth's magnetic field at the time the rocks formed. Magnetite in a cooling basaltic lava flow acts like a tiny compass needle, preserving a record of Earth's magnetic field when the lava cools below the *Curie point.* Iron-stained sedimentary rocks such as red shale can also record Earth's magnetism. The magnetism of old rocks can be measured to determine the direction and strength of the magnetic field in the past. The study of ancient magnetic fields is called *paleomagnetism.*

Because magnetic lines of force dip more steeply as the north magnetic pole is approached, the inclination (dip) of the magnetic alignment preserved in the magnetite minerals in the lava flows can be used to determine the distance from a flow to the pole at the time that the flow formed (figure 19.8).

Old pole positions can be determined from the magnetism of old rocks. The magnetic alignment preserved in magnetite minerals points to the pole, and the dip of the alignment tells how far away the pole was. Figure 19.9 shows how Permian lava flows in North America indicate a Permian pole position in eastern Asia.

For each geologic period, North American rocks reveal a different magnetic pole position; this path of the *apparent* motion of the north magnetic pole through time is shown in figure 19.10. Paleomagnetic evidence thus verifies Wegener's idea of polar wandering (which he based on paleoclimatic evidence).

Like Wegener's paleoclimatic evidence, the paleomagnetic evidence from a *single* continent can be interpreted in two ways: either the continent stood still and the magnetic pole moved, or the pole stood still and the continent moved. At first glance, paleomagnetic evidence does not seem to be a significant advancement over paleoclimatic evidence. But when paleomagnetic evidence from *different* continents was compared, an important discovery was made.

Although Permian rocks in North America point to a pole position in eastern Asia, Permian rocks in *Europe* point to a different position (closer to Japan), as shown in figure 19.10. Does this mean there were *two* north magnetic poles in the Permian Period? In fact, every continent shows a different position for the Permian pole. A different magnetic pole for each continent seems highly unlikely. A better explanation is that a single pole stood still while continents split apart and rotated as they diverged.

Note the polar wandering paths for North America and Europe in figure 19.10. The paths are of similar shape, but the path for European poles is to the east of the North American path. If we mentally push North America back toward Europe, closing the Atlantic Ocean, the paths of polar wandering are almost identical between North America and Europe. This strongly suggests that there was one north magnetic pole and that the continents were joined together. There appear to be two north magnetic poles because the rocks of North America moved west; their magnetic minerals now point to a different polar position than they did when the minerals first formed.

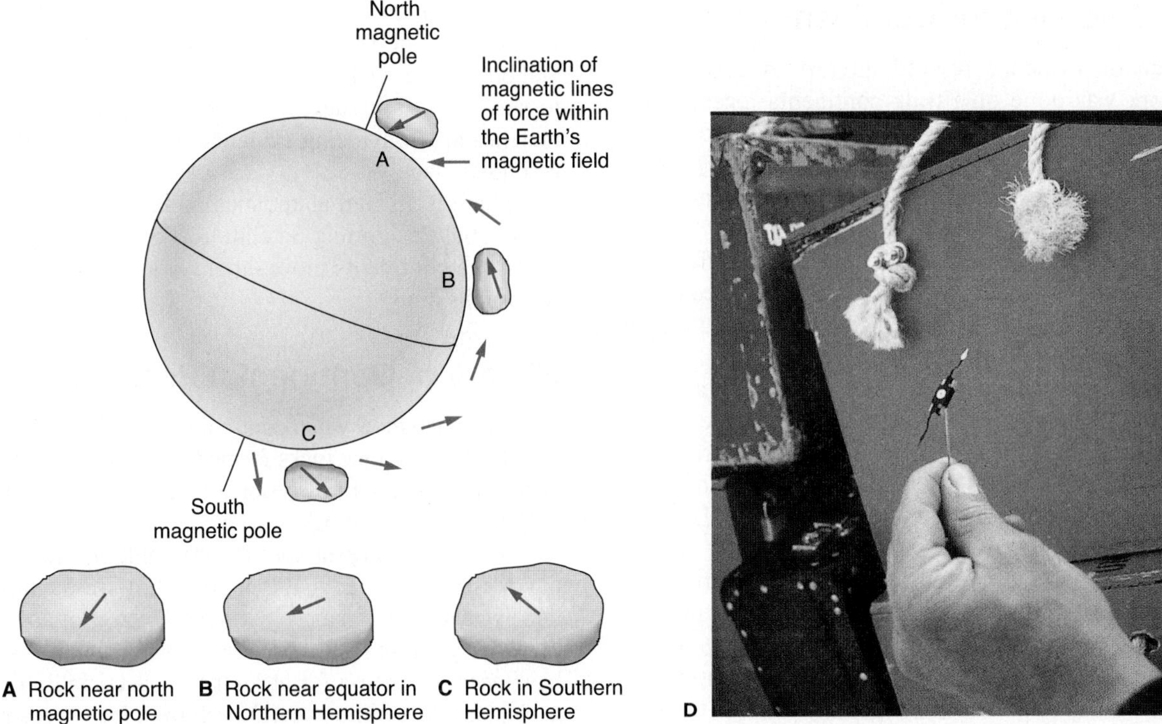

FIGURE 19.8

Magnetic dip (inclination) increases toward the north magnetic pole. Rocks in bottom part of figure are small samples viewed horizontally at locations A, B, and C on the globe. The magnetic dip can, therefore, be used to determine the distance from a rock to the north magnetic pole. (*D*) Compass needle showing steep inclination near south magnetic pole in Antarctica.

(D) ©C. C. Plummer

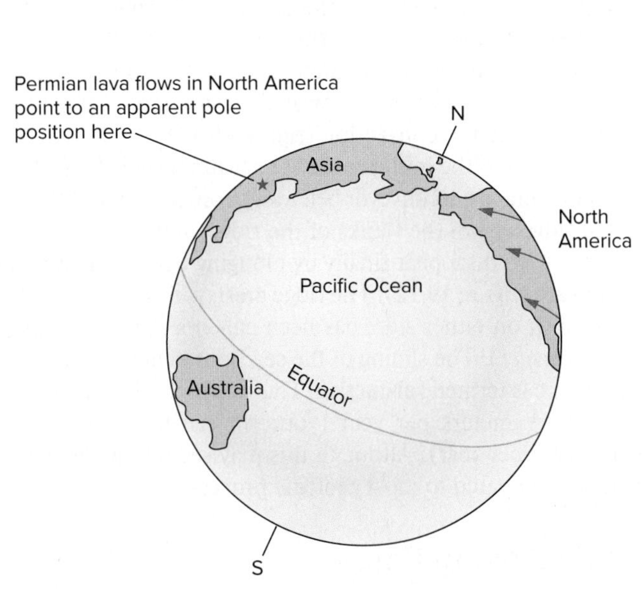

FIGURE 19.9

Paleomagnetic studies of Permian lava flows in North America indicate an apparent position for the north magnetic pole in eastern Asia.

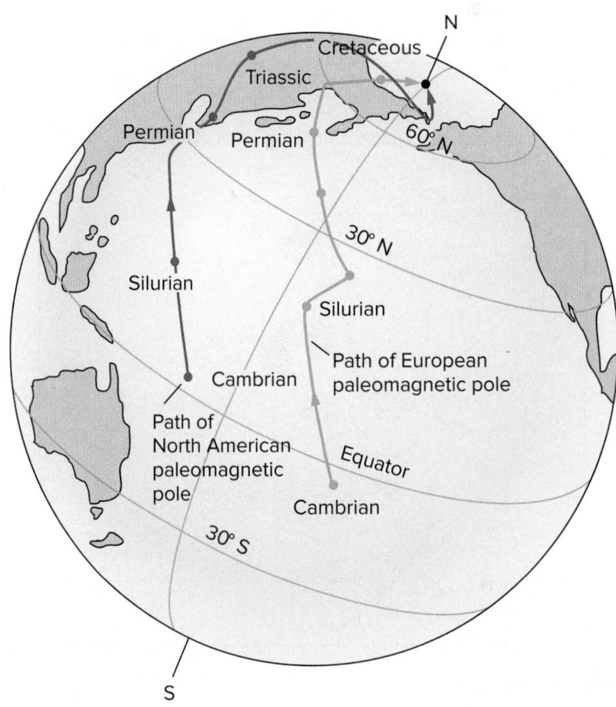

FIGURE 19.10

Apparent polar wandering of the north magnetic pole for the past 520 million years as determined from measurements of rocks from North America (red) and Europe (green).

Geologic Evidence for Continental Drift

As paleomagnetic evidence revived interest in continental drift, new work was done on fitting continents together. By defining the edge of a continent as the middle of the continental slope, rather than the present (constantly changing) shoreline, a much more precise fit has been found between continents (figure 19.11).

The most convincing evidence for continental drift came from greatly refined rock matches between now-separated continents. If continents are fitted together like pieces of a jigsaw puzzle, the "picture" should match from piece to piece.

The matches between South America and Africa are particularly striking. Some distinctive rock contacts extend out to sea along the shore of Africa. If the two continents are fitted together, the identical contacts are found in precisely the right position on the shore of South America (figure 19.11). Isotopic ages of rocks also match between these continents.

Glacial striations show that during the late Paleozoic Era, continental glaciers moved from Africa toward the present Atlantic Ocean, while similar glaciers seemingly moved *from* the Atlantic Ocean *onto* South America (figure 19.11). Continental glaciers, however, cannot move from sea onto land. If the two continents had been joined together, the ice that moved off Africa could have been the ice that moved onto South America. This hypothesis has now been confirmed; from their lithology, many of the boulders in South American tills have been traced to a source that is now in Africa.

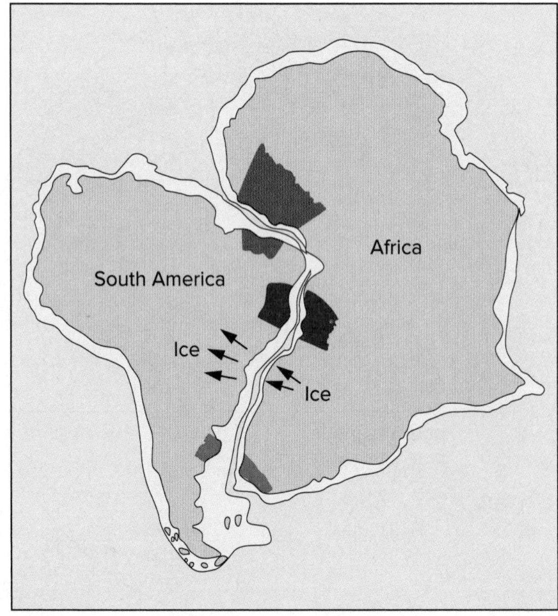

FIGURE 19.11

Jigsaw puzzle fit and matching rock types between South America and Africa. Light-blue areas around continents are continental shelves (part of continents). Colored areas within continents are broad belts of rock that correlate in type and age from one continent to another. Arrows show direction of glacier movement as determined from striations.

Some of the most detailed matches have been made between rocks in Brazil and rocks in the African country of Gabon. These rocks are similar in type, structure, sequence, fossils, ages, and degree of metamorphism. Such detailed matches are convincing evidence that continental drift did, in fact, take place.

There is also an abundance of satellite geodetic data from the Global Positioning Satellite (GPS) system, so we can now watch the continents move—about as eventful as watching your fingernails grow!

History of Continental Positions

Rock matches show when continents were together; once the continents split, the new rocks formed are dissimilar. Paleomagnetic evidence indicates the direction and rate of drift, allowing maps of old continental positions, such as figure 19.2, to be drawn.

Although Pangaea split up 200 million years ago to form our present continents, the continents were moving much earlier. Pangaea was formed by the collision of many small continents long before it split up. Recent work shows that continents have been in motion for at least the past 2 billion years (some geologists say 4 billion years), well back into Precambrian time. For more than half of Earth's history, the continents appear to have collided, welded together, then split and drifted apart, only to collide again, over and over, in an endless, slow dance.

19.3 SEAFLOOR SPREADING

At the same time that many geologists were becoming interested again in the idea of moving *continents*, Harry Hess, a geologist at Princeton University, proposed that the *sea floor* might be moving, too. This proposal contrasted sharply with the earlier ideas of Wegener, who thought that the ocean floor remained stationary as the continents plowed through it (figure 19.7). Hess's 1962 proposal was quickly named seafloor spreading, for it suggests that the sea floor moves away from the mid-oceanic ridge as a result of mantle convection (figure 19.12).

According to the initial concept of **seafloor spreading**, the sea floor is moving like a conveyor belt away from the crest of the mid-oceanic ridge, down the flanks of the ridge, and across the deep-ocean basin, to disappear finally by plunging beneath a continent or island arc (figure 19.12). The ridge crest, with sea floor moving away from it on either side, has been called a *spreading axis* (or *spreading center*). The sliding of the sea floor beneath a continent or island arc is termed **subduction**. The sea floor moves at a rate of 1 to 24 centimeters per year (your fingernail grows at about 1 centimeter per year). Although this may seem to be quite slow, it is rapid compared to most geologic processes.

Hess's Driving Force

Why does the sea floor move? Hess's original hypothesis was that seafloor spreading is driven by deep mantle convection. **Convection** is a circulation pattern driven by the rising of hot

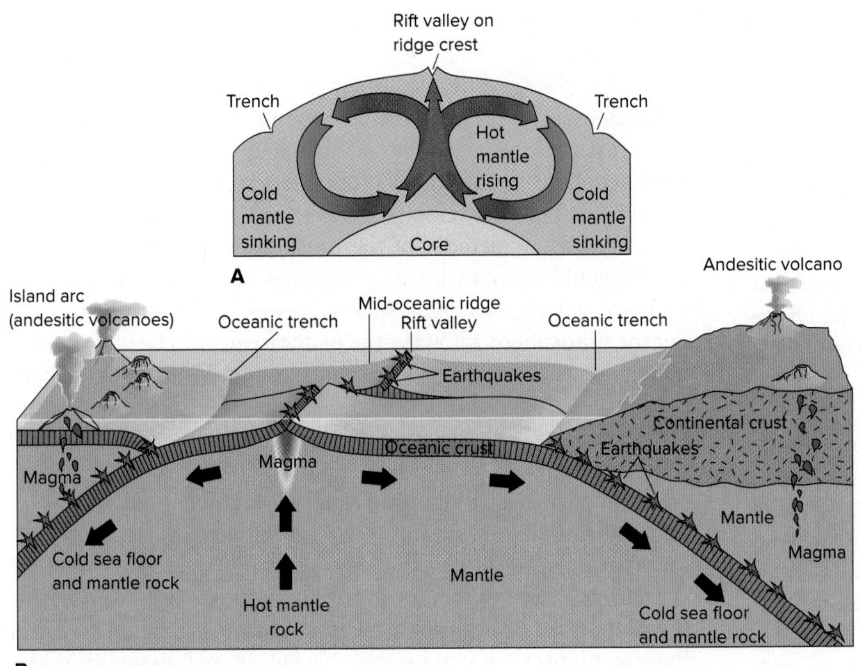

FIGURE 19.12

Seafloor spreading hypothesis of Harry Hess. (*A*) Hess proposed that convection extended throughout the mantle. (Scale of ridge and trenches is exaggerated.) (*B*) Hot mantle rock rising beneath the mid-oceanic ridge (a spreading axis) causes basaltic volcanism and high heat flow. Divergence of sea floor splits open the rift valley and causes shallow-focus earthquakes (stars on ridge). Sinking of cold rock causes subduction of older sea floor at trenches, producing Benioff zones of earthquakes and andesitic magma.

material and/or the sinking of cold material. Hot material has a lower density, so it rises; cold material has a higher density and sinks. The circulation of water heating in a pan on a stove is an example of convection. Convection in the mantle was a controversial idea in 1962; although convection can be easily demonstrated in a pan of water, it was hard to visualize the solid rock of the mantle behaving as a liquid. Over very long periods of time, however, it is possible for the hot mantle rock to flow in a ductile manner. A slow, convective circulation is set up by temperature differences in the rock, and convection can explain many seafloor features as well as the young age of the seafloor rocks. (The heat that flows outward through Earth to drive convection is both original heat from the planet's formation and heat from the decay of radioactive isotopes, as discussed in chapter 17.)

Explanations

The Mid-Oceanic Ridge

If convection drives seafloor spreading, then hot mantle rock must be rising under the mid-oceanic ridge. Hess showed how the *existence of the ridge* and its *high heat flow* are caused by the rise of this hot mantle rock. The *basalt eruptions* on the ridge crest are also related to this rising rock, for here the mantle rock is hotter than normal and begins to undergo decompression melting.

As hot rock continues to rise beneath the ridge crest, the circulation pattern splits and diverges near the surface. Mantle rock moves horizontally away from the ridge crest on each side of the ridge. This movement accompanies tension at the ridge crest, cracking open the oceanic crust to form the **rift valley** and its associated *shallow-focus earthquakes.*

Oceanic Trenches

As the mantle rock moves horizontally away from the ridge crest, it carries the sea floor (the basaltic oceanic crust) piggyback along with it. As the hot rock moves sideways, it cools and becomes denser, sinking deeper beneath the ocean surface. Hess thought it would become cold and dense enough to sink back into the mantle. This downward plunge of cold rock accounts for the *existence of the oceanic trenches* as well as their *low heat flow* values. It also explains the large *negative gravity anomalies* associated with trenches, for the sinking of the cold rock provides a force that holds trenches out of isostatic equilibrium (see chapter 17).

As the sea floor moves downward into the mantle along a subduction zone, it interacts with the rock above it. This interaction between the moving seafloor rock and the overlying crustal and mantle rock can cause the *Benioff zones of earthquakes* associated with trenches. It can also produce *andesitic volcanism,* which forms volcanoes either on the edge of a continent or in an island arc (figure 19.12).

Hess's ideas have stood up remarkably well over more than thirty years. We now think of lithospheric plates moving instead of sea floor riding piggyback on convecting mantle, and we think that several mechanisms cause plate motion, but Hess's explanation of seafloor topography, earthquakes, and age remains valid today.

Age of the Sea Floor

The *young age of seafloor rocks* (see chapter 18) is neatly explained by Hess's seafloor spreading. New, young sea floor is continually being formed by basalt eruptions at the ridge crest. This basalt is then carried sideways by convection and is subducted into the mantle at an oceanic trench. Thus, old sea floor is continually being destroyed at trenches, while new sea floor is being formed at the ridge crest. (This is also the reason for the puzzling lack of pelagic sediment at the ridge crest. Young sea floor at the ridge crest has little sediment because the basalt is newly formed. Older sea floor farther from the ridge crest has been moving under a constant rain of pelagic sediment, building up a progressively thicker layer as it goes.)

Note that seafloor spreading implies that the youngest sea floor should be at the ridge crest, with the age of the sea floor becoming progressively older toward a trench. This increase in age away from the ridge crest was not known to exist at the time of Hess's proposal but was an important prediction of his hypothesis. This prediction has been successfully tested, as you shall see in the discussion of marine magnetic anomalies in Section 19.5.

19.4 PLATES AND PLATE MOTION

By the mid-1960s, the twin ideas of moving continents and a moving sea floor were causing great excitement and emotional debate among geologists. By the late 1960s, these ideas had been combined into a single theory that revolutionized geology by providing a unifying framework for Earth science–the theory of plate tectonics.

As described earlier, a **plate** is a large, mobile slab of rock that is part of Earth's surface (figure 19.1). The surface of a plate may be made up entirely of sea floor (as is the Nazca plate), or it may be made up of both continental and oceanic rock (as is the North American plate). Some of the smaller plates are entirely continental, but all the large plates contain some sea floor.

Plate tectonics has added some new terms, based on rock behavior, to the zones of Earth's interior, as we have discussed in some previous chapters. The plates are composed of the relatively rigid outer shell of Earth called the **lithosphere**. The lithosphere includes the rocks of the crust and uppermost mantle (figure 19.13).

The lithosphere beneath oceans increases in both age and thickness with distance from the crest of the mid-oceanic ridge. Young lithosphere near the ridge crest may be only 10 kilometers thick, while very old lithosphere far from the ridge crest may be as much as 100 kilometers thick. An average thickness for oceanic lithosphere might be 70 kilometers, as shown in figure 19.13.

Continental lithosphere is thicker, varying from perhaps 125 kilometers thick to as much as 200 to 250 kilometers thick

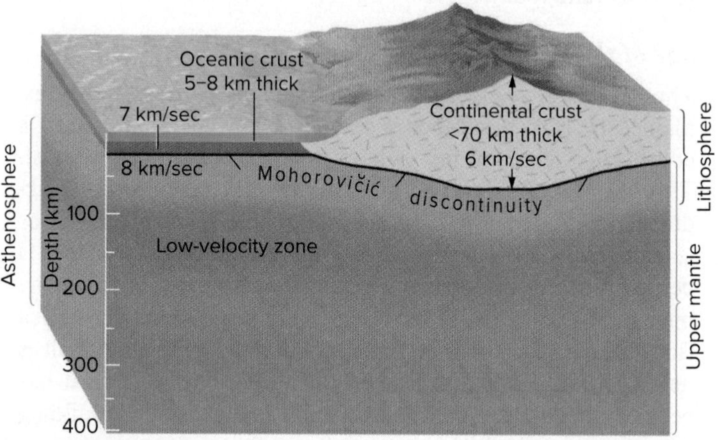

FIGURE 19.13

The rigid lithosphere includes the crust and uppermost mantle; it forms the plates. The ductile asthenosphere acts as a lubricating layer beneath the lithosphere. Oceanic lithosphere averages 70 kilometers thick; continental lithosphere varies from 125 to 250 kilometers thick. Asthenosphere may not be present under continents.

beneath the oldest, coldest, and most inactive parts of the continents.

Below the rigid lithosphere is the **asthenosphere**, a zone that behaves in a ductile manner because of increased temperature and pressure. Some geologists think that the upper part of the asthenosphere is partially molten because P and S waves slow down here. This low-velocity zone probably reflects partial melting of just a few percent of the upper asthenosphere's volume, which would account for its properties and behavior. The partially melted upper asthenosphere acts as a lubricating layer under the lithosphere, allowing the plates to move. The low-velocity zone may extend from a depth of 70 to 200 kilometers beneath oceans; its thickness, depth, and even existence under continents are vigorously debated. Below the asthenosphere is more rigid mantle rock.

The idea that plates move is widely accepted by geologists, although the reasons for this movement are debated. Plates move away from the mid-oceanic ridge crest or other spreading axes. Some plates move toward oceanic trenches. If the plate is made up mostly of sea floor (as are the Nazca and Pacific plates), the plate can be subducted down into the mantle, forming an oceanic trench and its associated features. If the leading edge of the plate is made up of continental rock (as is the South American plate), that plate will not subduct. Continental rock, being less dense (specific gravity 2.7) than oceanic rock (specific gravity 3.0), is too light to be subducted.

To a first approximation, a plate may be viewed as a rigid slab of rock that moves as a unit. As a result, the interior of a plate is relatively inactive tectonically (but see box 19.1). Plate interiors generally lack earthquakes, volcanoes, young mountain belts, and other signs of geologic activity. According to plate-tectonic theory, these features are caused by plate interactions at plate boundaries.

Plate boundaries are of three general types, based on whether the plates move away from each other, move toward each other, or move past each other. A **divergent plate boundary** is a boundary between plates that are moving apart. A **convergent plate boundary** lies between plates that are moving toward each other. A **transform plate boundary** is one at which two plates move horizontally past each other.

19.5 HOW DO WE KNOW THAT PLATES MOVE?

The proposal that Earth's surface is divided into moving plates was an exciting, revolutionary hypothesis, but it required testing to win acceptance among geologists. You have seen how the study of paleomagnetism supports the idea of moving continents. In the 1960s, two critical tests were made of the idea of a moving sea floor. These tests involved marine magnetic anomalies and the seismicity of fracture zones. These two successful tests convinced most geologists that plates do indeed move.

Marine Magnetic Anomalies

In the mid-1960s, magnetometer surveys at sea disclosed some intriguing characteristics of marine magnetic anomalies. Most magnetic anomalies at sea are arranged in bands that lie parallel to the rift valley of the mid-oceanic ridge. Alternating positive and negative anomalies (chapter 17) form a stripelike pattern parallel to the ridge crest (figure 19.14).

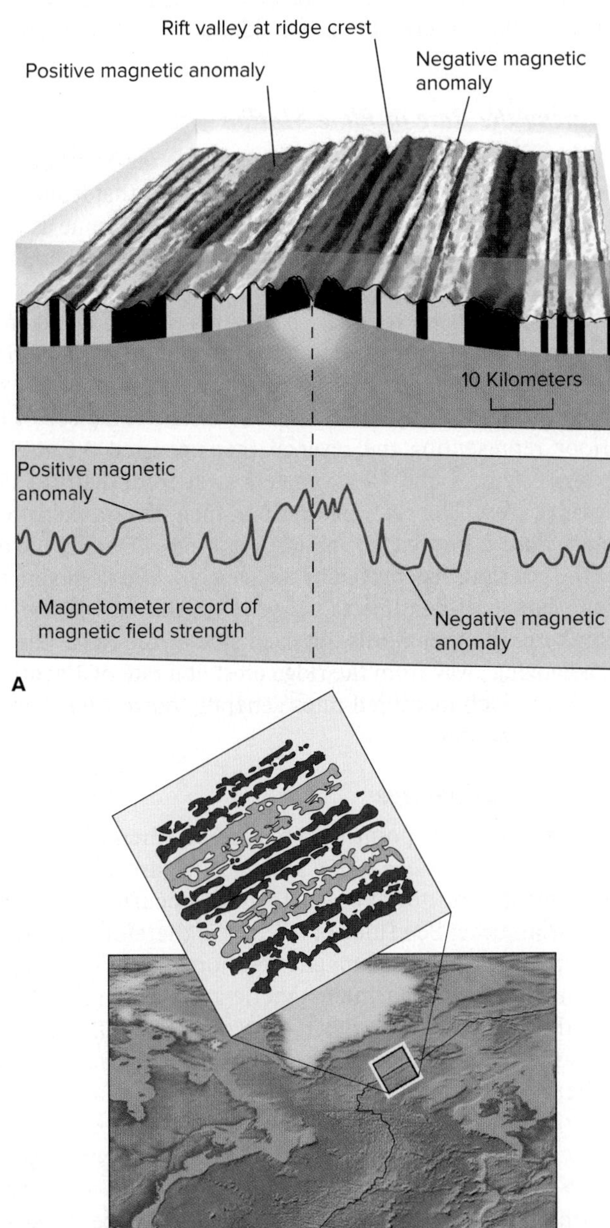

FIGURE 19.14

Marine magnetic anomalies. (*A*) The red line shows positive and negative magnetic anomalies as recorded by a magnetometer towed behind a ship. Positive anomalies are shown in black and negative anomalies are shown in tan. Notice how magnetic anomalies are parallel to the rift valley and symmetric about the ridge crest. (*B*) Symmetric magnetic anomalies ("stripes") from the mid-Atlantic ridge south of Iceland.

The Vine-Matthews Hypothesis

Two British geologists, Fred Vine and Drummond Matthews, made several important observations about these anomalies. They recognized that the pattern of magnetic anomalies was symmetrical about the ridge crest. That is, the pattern of magnetic anomalies on one side of the mid-oceanic ridge was a mirror image of the pattern on the other side (figure 19.14). Vine and Matthews also noticed that the same pattern of magnetic anomalies exists over different parts of the mid-oceanic ridge. The pattern of anomalies over the ridge in the northern Atlantic Ocean is the same as the pattern over the ridge in the southern Pacific Ocean.

The most important observation that Vine and Matthews made was that the pattern of magnetic *anomalies* at sea matches the pattern of magnetic *reversals* already known from studies of lava flows on the continents (figure 19.15 and chapter 17). This correlation can be seen by comparing the pattern of colored bands in figure 19.15 (reversals) with the pattern in figure 19.14 (anomalies).

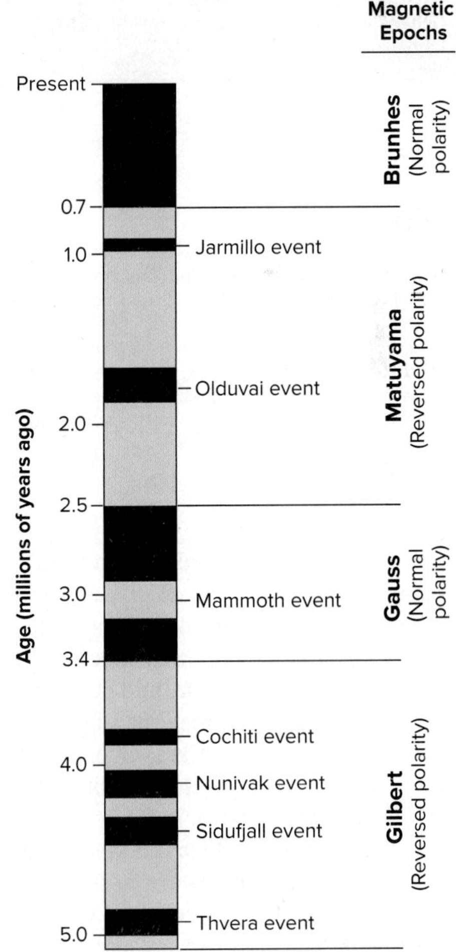

FIGURE 19.15

Magnetic reversals during the past 5 million years determined from lava flows that have been radiometrically dated. Black represents normal magnetism; tan represents reverse magnetism.

Source: Mankinen, E. A. and Dalrymple, G. B., "Revised geomagnetic polarity time scale for the interval 0–5 m.y. B.P.," Journal of Geophysical Research, v. 84, 1979, pp. 615–626.

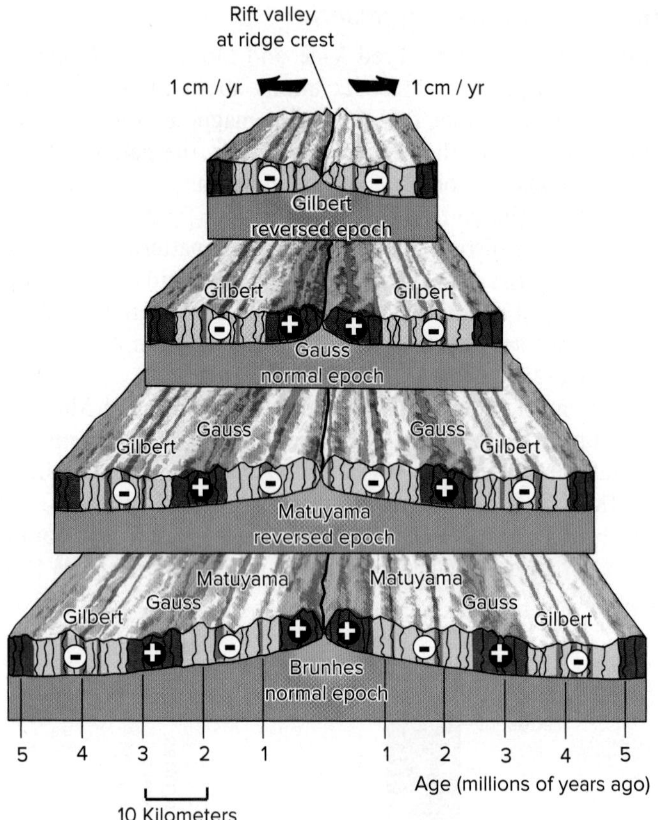

FIGURE 19.16

The origin of magnetic anomalies. During a time of reversed magnetism (Gilbert reversed epoch), a series of basaltic dikes intrudes the ridge crest, becoming reversely magnetized. The dike zone is torn in half and moved sideways, as a new group of normally magnetized dikes forms at the ridge crest. A new series of reversely magnetized dikes forms at the ridge crest. The dike pattern becomes symmetric about the ridge crest. Correlating the magnetic anomalies with magnetic reversals allows anomalies to be dated. Magnetic anomalies can therefore be used to predict the age of the sea floor and to measure the rate of seafloor spreading (plate motion).

Putting these observations together with Hess's concept of seafloor spreading, which had just been published, Vine and Matthews proposed an explanation for magnetic anomalies. They suggested that there is continual opening of tensional cracks within the rift valley on the mid-oceanic ridge crest. These cracks on the ridge crest are filled by basaltic magma from below, which cools to form dikes. Cooling magma in the dikes records Earth's magnetism at the time the magnetic minerals crystallize. The process is shown in figure 19.16.

When Earth's magnetic field has a *normal polarity* (the present orientation), cooling dikes are normally magnetized. Dikes that cool when the field is reversed (figure 19.16) are reversely magnetized. So each dike preserves a record of the polarity that prevailed during the time the magma cooled. Extension produced by the moving sea floor then cracks a dike in two, and the two halves are carried away in opposite directions down the flanks of the ridge. New magma eventually intrudes the newly opened fracture. It cools, is magnetized, and forms a new dike, which in turn is split by continued extension. In this way, a system of reversely magnetized and normally magnetized dikes forms parallel to the rift valley. These dikes, in the Vine-Matthews hypothesis, are the cause of the anomalies.

The magnetism of normally magnetized dikes adds to Earth's magnetism, and so a magnetometer carried over such dikes registers a stronger magnetism than average—a *positive* magnetic anomaly. Dikes that are reversely magnetized subtract from the present magnetic field, and so a magnetometer towed over such dikes measures a weaker magnetic field—a *negative* magnetic anomaly. Since seafloor motion separates these dikes into halves, the patterns on either side of the ridge are mirror images.

Measuring the Rate of Plate Motion

There are two important points about the Vine-Matthews hypothesis of magnetic anomaly origin. The first is that it allows us to measure the *rate of seafloor motion* (which is the same as plate motion, since continents and the sea floor move together as plates).

Because magnetic reversals have already been dated from lava flows on land (figure 19.15), the anomalies caused by these reversals are also dated and can be used to discover how fast the sea floor has moved (figure 19.16). For instance, a piece of the sea floor representing the reversal that occurred 4.5 million years ago may be found 45 kilometers away from the rift valley of the ridge crest. The piece of sea floor, then, has traveled 45 kilometers since it formed 4.5 million years ago. Dividing the distance the sea floor has moved by its age gives 10 kilometers per million years, or 1 centimeter per year, for the rate of seafloor motion here. In other words, on each side of the ridge, the sea floor is moving away from the ridge crest at a rate of 1 centimeter per year. Such measured rates generally range from 1 to 24 centimeters per year.

Predicting Seafloor Age

The other important point of the Vine-Matthews hypothesis is that it *predicts the age of the sea floor* (figure 19.16). Magnetic reversals are now known to have occurred back into Precambrian time. Sea floor of *all* ages is therefore characterized by parallel bands of magnetic anomalies. Figure 17.20 shows the pattern of marine magnetic anomalies (and the reversals that caused them) during the past 160 million years. The distinctive pattern of these anomalies through time allows them to be identified by age, a process similar to dating by tree rings.

Now, even before they sample the sea floor, marine geologists can predict the age of the igneous rock of the sea floor by measuring the magnetic anomalies at the sea surface. Most sections of the sea floor have magnetic anomalies. By matching the measured anomaly pattern with the known pattern that is shown in figure 17.20, the age of the sea floor in the region can be predicted, as shown in figure 19.17.

This is a very powerful test of the hypothesis that the sea floor moves. Suppose, for example, that the sea floor in a particular spot is predicted to be 70 million years old from a study

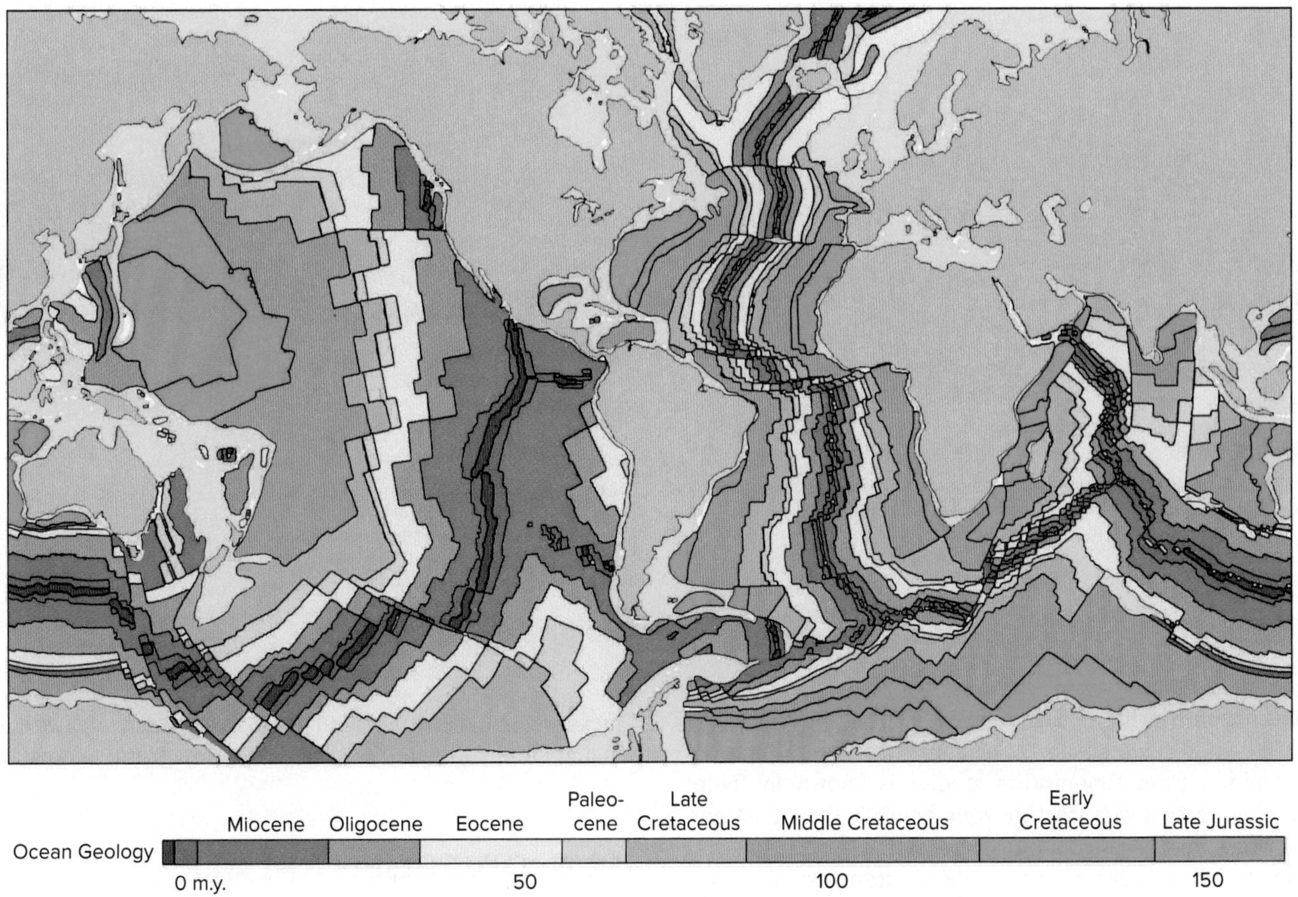

FIGURE 19.17

The age of the sea floor as determined from magnetic anomalies.
Source: R. L. Larson, W. C. Pitman, III, et al., W. H. Freeman, The Bedrock Geology of the World.

of its magnetic anomalies. If the hypothesis of seafloor motion and the Vine-Matthews hypothesis of magnetic anomaly origin are correct, a sample of igneous rock from that spot *must* be 70 million years old. If the rock proves to be 10 million years old or 200 million years old or 1.2 billion years old, or any other age except 70 million years, then both of these hypotheses are wrong. But if the rock proves to be 70 million years old, as predicted, then both hypotheses have been successfully tested.

Hundreds of rock and sediment cores recovered from holes drilled in the sea floor were used to test these hypotheses. Close correspondence has generally been found between the predicted age and the measured age of the sea floor. (The seafloor age is usually measured by fossil dating of sediment in the cores rather than by isotopic dating of igneous rock.) This evidence from deep-sea drilling has been widely accepted by geologists as verification of the hypotheses of plate motion and magnetic anomaly origin. Most geologists now think that these concepts are no longer hypotheses but can now be called theories. (A *theory*, as discussed in box 1.4 in connection with the scientific method, is a hypothesis that has been tested and found to explain observations.)

Another Test: Fracture Zones and Transform Faults

Cores from deep-sea drilling tested plate motion by allowing us to compare the actual age of the sea floor with the age predicted from magnetic anomalies. Another rigorous test of plate motion has been made by studying the seismicity of fracture zones.

The mid-oceanic ridge is offset along fracture zones (see figure 19.1). Conceivably, the mid-oceanic ridge was once continuous across a fracture zone but has been offset by strike-slip motion along the fracture zone (figure 19.18*A*). If such motion is occurring along a fracture zone, we would expect to find two things: (1) earthquakes should be distributed along the entire length of the fracture zone and (2) the motion of the rocks on either side of the fracture zone should be in the direction shown by the arrows in figure 19.18*A*.

In fact, these things are not true about fracture zones. Earthquakes do occur along fracture zones, but only in those segments between offset sections of ridge crest. In addition, first-motion studies of earthquakes (see chapter 16) along fracture zones show that the motion of the rocks on either side of

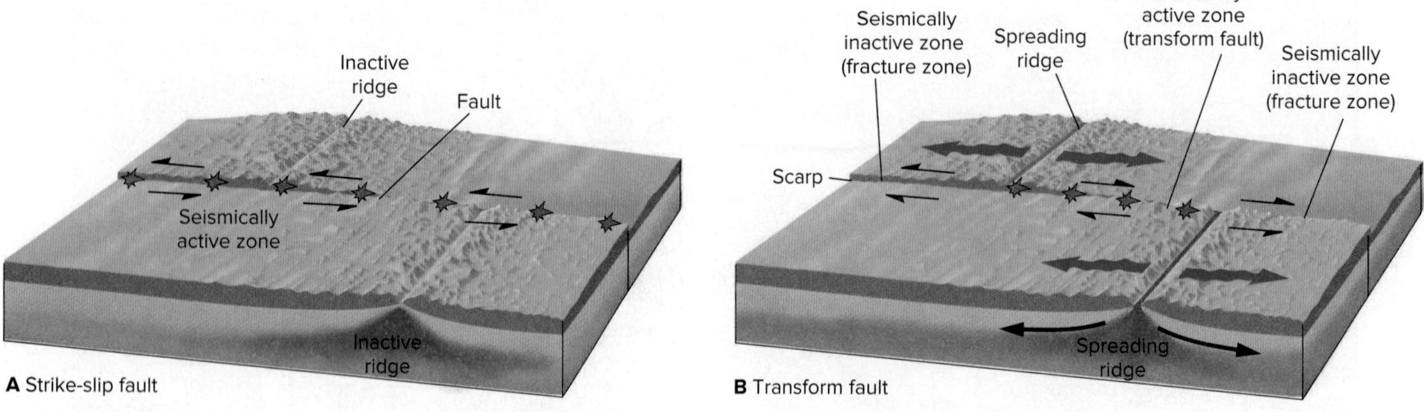

FIGURE 19.18

Two possible explanations for the relationship between fracture zones and the mid-oceanic ridge. (*A*) The expected rock motions and earthquake distribution, assuming that the ridge was once continuous across the fracture zone. (*B*) The expected rock motions and earthquake distribution, assuming that the two ridge segments were never joined together and that the sea floor moves away from the rift valley segments. Only explanation (*B*) fits the data. The portion of the fracture zone between the ridge segments is a transform fault.

the fracture zone during an earthquake is exactly opposite to the motion shown in figure 19.18*A*. The actual motion of the rocks as determined from first-motion studies is shown in figure 19.18*B*. The portion of a fracture zone between two offset portions of ridge crest is called a **transform fault**.

The motion of rocks on either side of a transform fault was predicted by the hypothesis of a moving sea floor. Note that sea floor moves away from the two segments of ridge crest (figure 19.18*B*). Looking along the length of the fracture zone, you can see that blocks of rock move in opposite directions only on that section of the fracture zone between the two segments of ridge crest. Earthquakes, therefore, occur only on this section of the fracture zone, the transform fault. The direction of motion of rock on either side of the transform fault is exactly predicted by the assumption that rock is moving away from the ridge crests. Verification by first-motion studies of this predicted motion along fracture zones was another successful test of plate motion.

Measuring Plate Motion Directly

In recent years, the motion of plates has been directly measured using satellites, radar, lasers, and the Global Positioning System (GPS). These techniques can measure the distance between two widely separated points to within 1 centimeter. GPS is now routinely used to measure the relative motion between plates because of its accuracy and because the receivers are relatively inexpensive and fairly portable (figure 19.19*A*). Plate motions are now recorded on a yearly basis throughout the world (figure 19.19*B*).

If two plates move toward each other at individual rates of 2 centimeters per year and 6 centimeters per year, the combined rate of convergence is 8 centimeters per year. The measurement techniques are sensitive enough to easily measure such a rate if measurements are repeated each year. Such measured rates match closely the predicted rates from magnetic anomalies.

19.6 DIVERGENT PLATE BOUNDARIES

Divergent plate boundaries, where plates move away from each other, can occur in the middle of the ocean or in the middle of a continent. The result of divergent plate boundaries is to create, or open, new ocean basins. This dynamic process has occurred throughout the geologic past.

When a supercontinent such as Pangaea breaks up, a divergent boundary can be found in the middle of a continent. The divergent boundary is marked by rifting, basaltic volcanism, and uplift. During rifting, the continental crust is stretched and thinned. This extension produces shallow-focus earthquakes on normal faults, and a *rift valley* forms as a central *graben* (a downdropped fault block). The faults act as pathways for basaltic magma that forms by decompression melting and rises from the mantle to erupt on the surface as cinder cones and basalt flows. Uplift at a divergent boundary is usually caused by the upwelling of hot mantle beneath the crust; the surface is elevated by the thermal expansion of the hot, rising rock and of the surface rock as it is warmed from below.

Figure 19.20 shows how a continent might rift to form an ocean. The figure shows rifting before uplift, because recent work indicates that this was the sequence for the opening of the Red Sea. The crust is initially stretched and thinned. Numerous normal faults break the crust, and the surface subsides into a central graben (figure 19.20*A*). Shallow earthquakes and basalt eruptions occur in this rift valley, which also has high heat flow. An example of a boundary at this stage is the African Rift Valleys in eastern Africa (figure 19.21). The valleys are grabens that may mark the

site of the future breakup of Africa. A dramatic example of rifting occurred in September 2005 when a 60-kilometer-long fissure or crack opened in just three weeks after a series of earthquakes shook the Afar region. The opening cracks swallowed goats and camels, and nomads in the area reported black smoke that smelled of sulfur venting out of the fissures. They also saw what looked like "large black birds" flying out of the linear vents. What they were witnessing was the largest single rip in the crust since the advent of satellite monitoring, and the associated injection of mafic magma (enough to fill a football stadium 2,000 times) along a vertical crack as new crust was being formed (figure 19.21*C*). The stretching apart of this area may eventually tear northeastern Africa away from the rest of the continent.

As divergence continues, the continental crust on the upper part of the plate clearly separates, and seawater floods into the linear basin between the two divergent continents (figure 19.20*B*). A series of fault blocks have rotated along curved fault planes at the edges of the continents, thinning the continental crust. The rise of hot mantle rock beneath the thinned crust causes continued basalt eruptions that create true oceanic crust between the two continents. The center of the narrow ocean is marked by a rift valley with its typical high heat flow and shallow earthquakes. The Red Sea is an example of a divergent margin at this stage (figure 19.21).

After modest widening of the new ocean, uplift of the continental edges may occur. As continental crust thins by stretching and faulting, the surface initially subsides. At the same time, hot mantle rock wells up beneath the stretched crust (figure 19.20*B*).

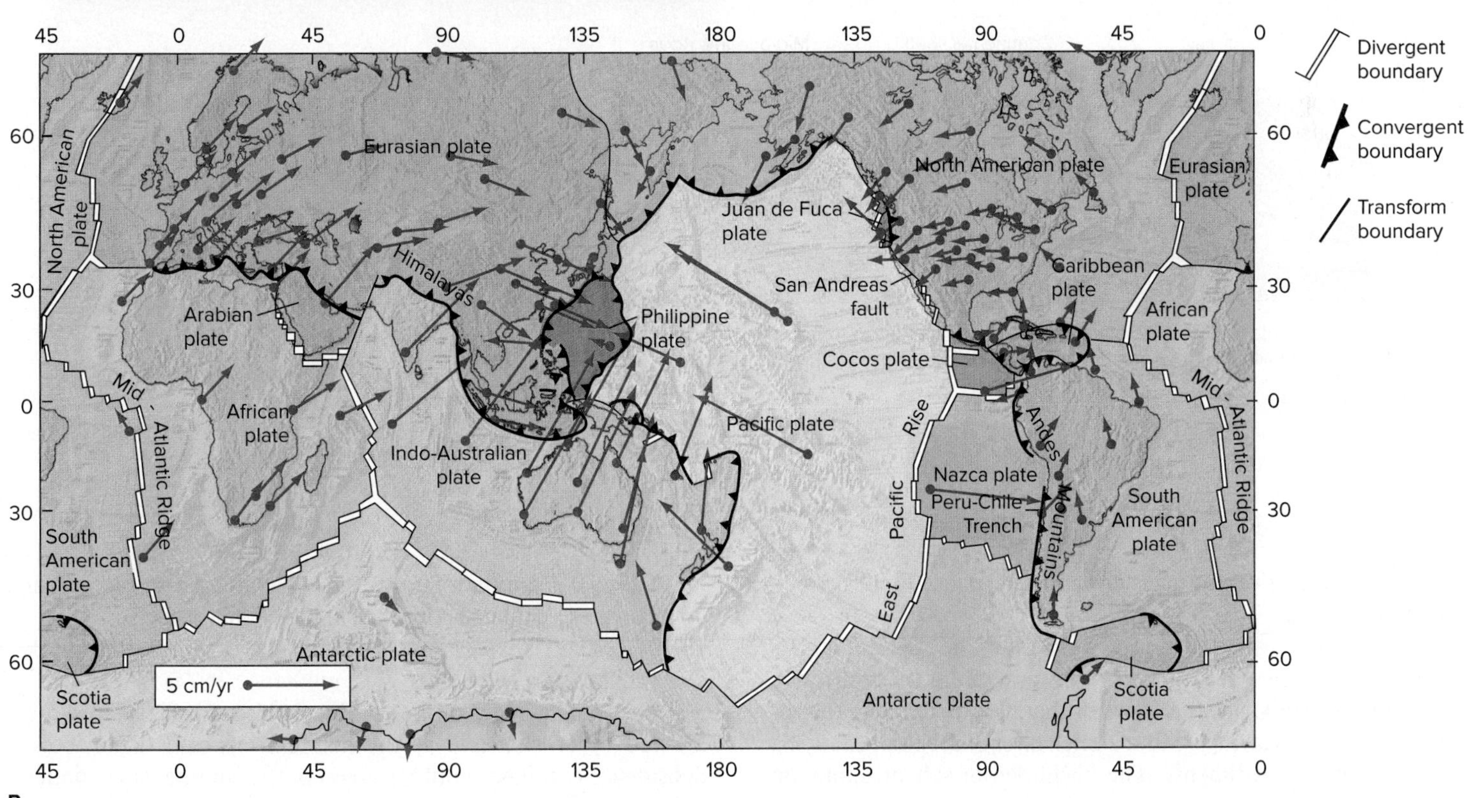

FIGURE 19.19

(*A*) Global Positioning System (GPS) station being installed in Iceland that will collect signals from orbiting GPS satellites to determine plate motions. (*B*) Yearly plate motions from stations around the world as measured by GPS.

Photo (A) ©Icelandic Met Office; (B) Source: NASA. http://sideshow.jpl.nasa.gov/post/series.html.

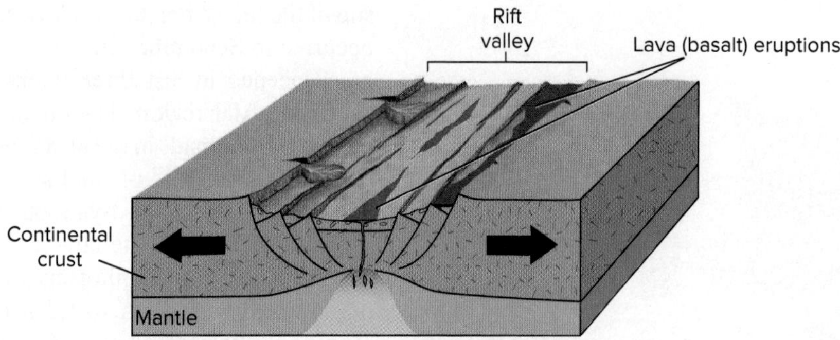

A Continent undergoes extension. The crust is thinned and a rift valley forms.

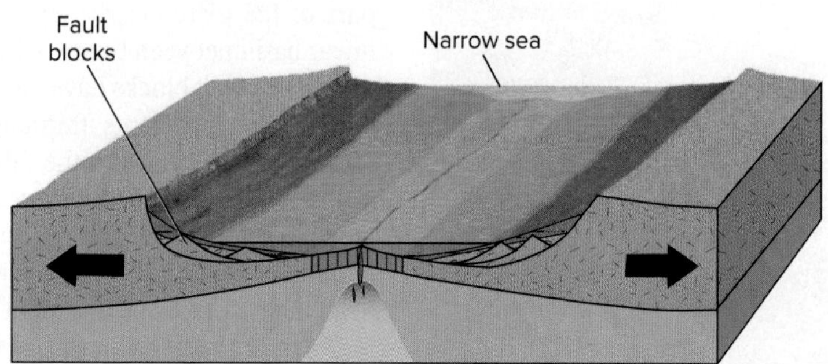

B Continent tears in two. Continent edges are faulted and uplifted. Basalt eruptions form oceanic crust.

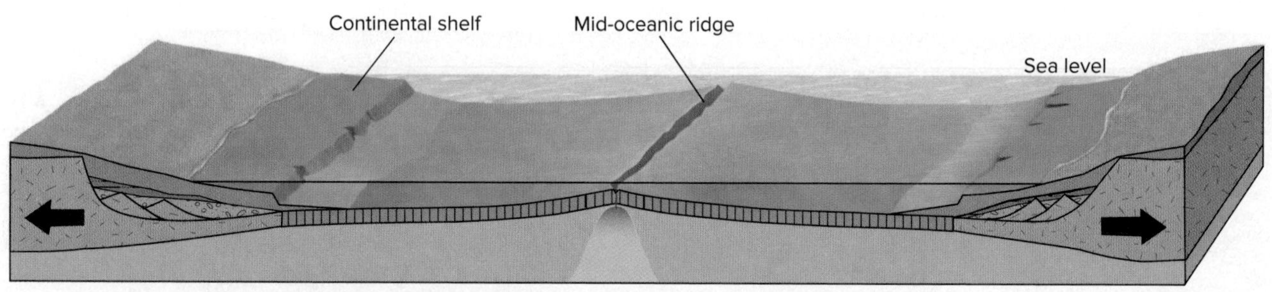

C Continental sediments blanket the subsiding margins to form continental shelves. The ocean widens and a mid-oceanic ridge develops, as in the Atlantic Ocean.

FIGURE 19.20

A divergent plate boundary forming in the middle of a continent will eventually create a new ocean.

The rising diapir of hot mantle rock would cause uplift by thermal expansion.

The new ocean is narrow, and the tilt of the adjacent land is away from the new sea, so rivers flow away from the sea (figure 19.20*B*). At this stage, the seawater that has flooded into the rift may evaporate, leaving behind a thick layer of rock salt overlying the continental sediments. The likelihood of salt precipitation increases if the continent is in one of the desert belts or if one or both ends of the new ocean should become temporarily blocked, perhaps by volcanism. Not all divergent boundaries contain rock salt, however.

The plates continue to diverge, widening the sea. Thermal uplift creates a mid-oceanic ridge in the center of the sea (figure 19.20*C*). The flanks of the ridge subside as the seafloor rock cools as it moves.

The trailing edges of the continents also subside as they are lowered by erosion and as the hot rock beneath them cools. Subsidence continues until the edges of the continents are under water. A thick sequence of marine sediment blankets the thinned continental rock, forming a *passive continental margin* (figures 19.20*C* and 19.22; see also chapter 18). The sediment forms a shallow continental shelf, which may contain a deeply buried salt

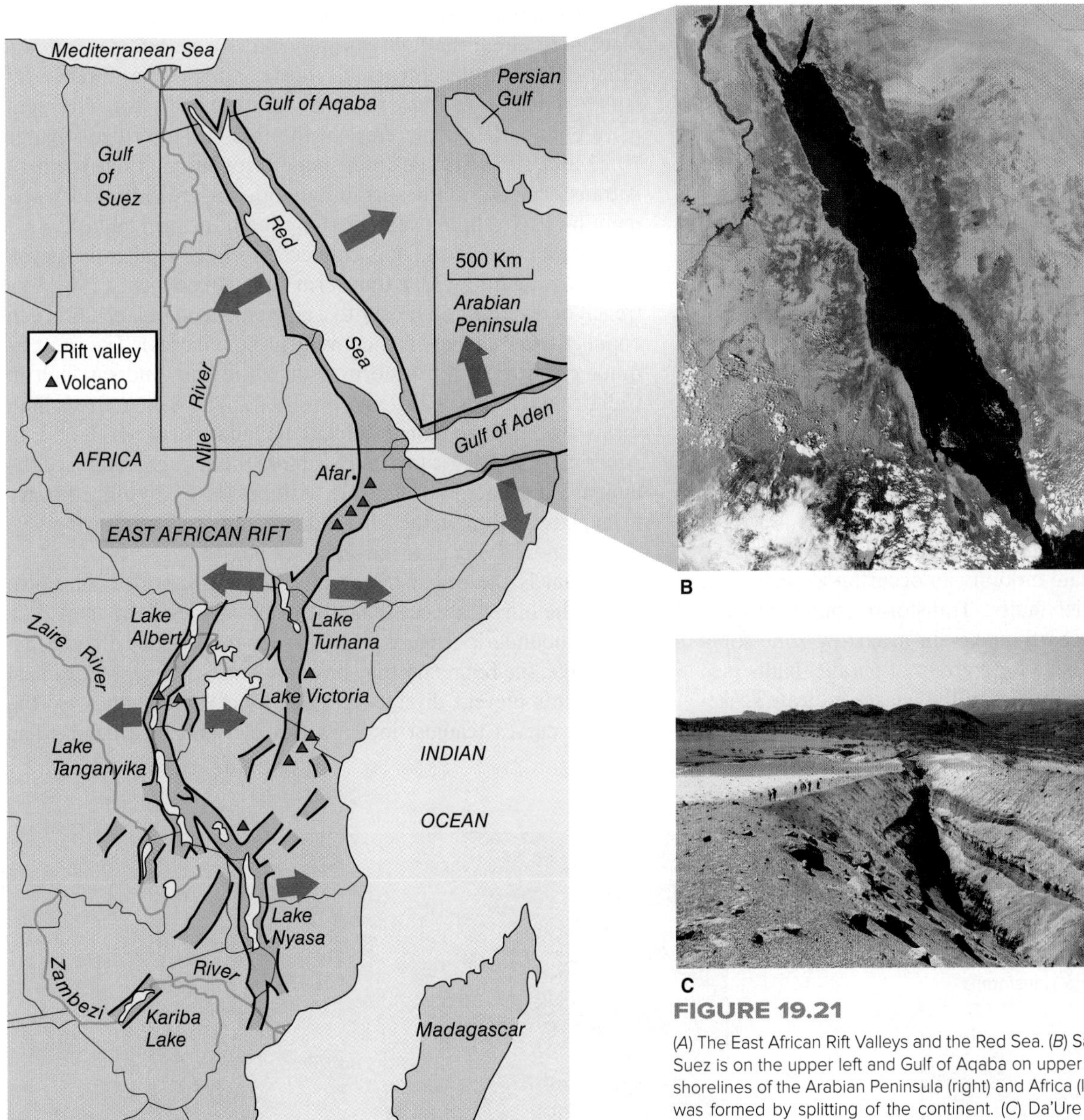

FIGURE 19.21

(*A*) The East African Rift Valleys and the Red Sea. (*B*) Satellite photo of Red Sea. Gulf of Suez is on the upper left and Gulf of Aqaba on upper right. Note the similarities in the shorelines of the Arabian Peninsula (right) and Africa (left) suggesting that the Red Sea was formed by splitting of the continent. (*C*) Da'Ure volcanic vent and fracture that opened during the September 200 5 rifting event in Afar, Ethiopia. This rifting event was the largest ever observed on land, and it will eventually lead to eastern Ethiopia being torn away from the rest of Africa and the birth of a new sea. Note people for scale.

(B) Source: Jeff Schmaltz, MODIS Rapid Response Team, NASA/GSFC; (C) Photograph by Anthony R. Philpotts

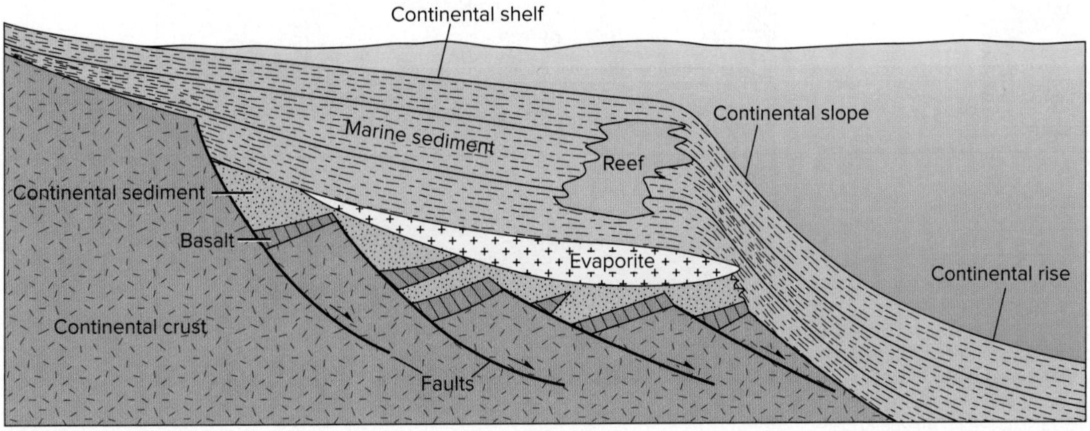

FIGURE 19.22

A passive continental margin formed by continental breakup and divergence. Downfaulted continental crust forms basins, which fill with basalt and sediment. A layer of rock salt may form if a narrow ocean evaporates. A thick sequence of marine sediments covers these rocks and forms the continental shelf, slope, and rise. A reef may form at the shelf edge if the water is warm; buried reefs occur on many parts of the Atlantic shelf of North America.

layer. The deep continental rise is formed as sediment is carried down the continental slope by turbidity currents and other mechanisms. The Atlantic Ocean is currently at this stage of divergence.

A divergent boundary on the sea floor is located on the crest of the mid-oceanic ridge. If the spreading rate is slow, as it is in the Atlantic Ocean (1 centimeter per year), the crest has a rift valley. Fast spreading, as along the East Pacific Rise (18 centimeters per year) and along other ridges in the Pacific Ocean, prevents a rift from forming. A divergent boundary at sea is marked by the same features as a divergent boundary on land—tensional cracks, normal faults, shallow earthquakes, high heat flow, and basaltic eruptions. The basalt forms dikes within the cracks and pillow lavas on the sea floor, creating new oceanic crust on the trailing edges of plates.

19.7 TRANSFORM BOUNDARIES

At transform boundaries, where one plate slides horizontally past another plate, the plate motion can occur on a single fault or on a group of parallel faults. Transform boundaries are marked by shallow-focus earthquakes in a narrow zone for a single fault or in a broad zone for a group of parallel faults (see figure 16.26). First-motion studies of the quakes indicate strike-slip movement parallel to the faults.

The name *transform fault* comes from the fact that the displacement along the fault abruptly ends or transforms into another kind of displacement. The most common type of transform fault occurs along fracture zones and connects two divergent plate boundaries at the crest of the mid-oceanic ridge (figures 19.23 and 19.18*B*). The spreading motion at one ridge segment is transformed into the spreading motion at the other ridge segment by strike-slip movement along the transform fault.

Not all transform faults connect two ridge segments. As you can see in figure 19.23, a transform fault can connect a ridge to a trench (a divergent boundary to a convergent boundary), or it can connect two trenches (two convergent boundaries). The most famous example of a transform fault is the San Andreas fault in California (figure 19.23*D* and box 19.2). The San Andreas fault forms a ridge-ridge transform plate boundary between the North American and Pacific plates. To explore the surface features of the fault as it cuts across California, visit box 15.2. Continued transform plate motion on the San Andreas fault creates an earthquake risk for those living near the fault (see box 16.3).

What is the origin of the offset in a ridge-ridge transform fault? The offsets appear to be the result of irregularly shaped divergent boundaries (figure 19.24). When two oceanic plates begin to diverge, the boundary may be curved on a sphere. Mechanical constraints prevent divergence along a curved boundary, so the original curves readjust into a series of right-angle bends. The

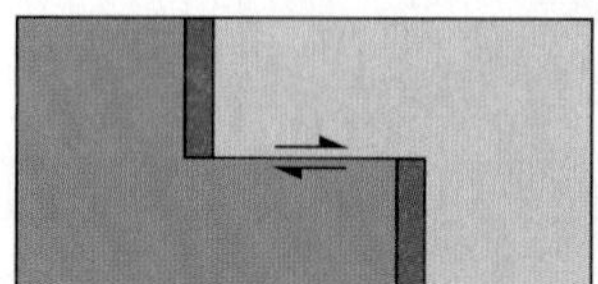

A Ridge-Ridge Transform

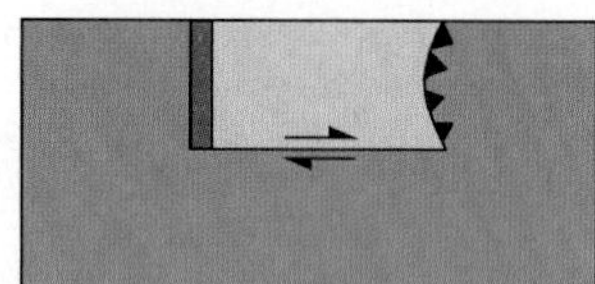

B Ridge-Trench Transform

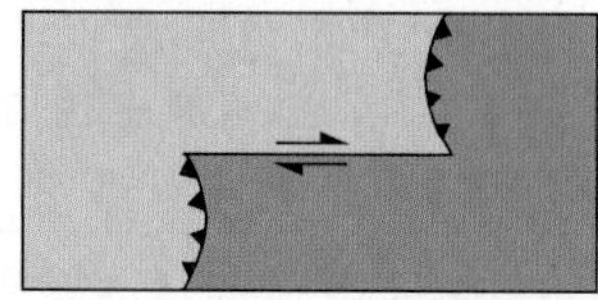

C Trench-Trench Transform

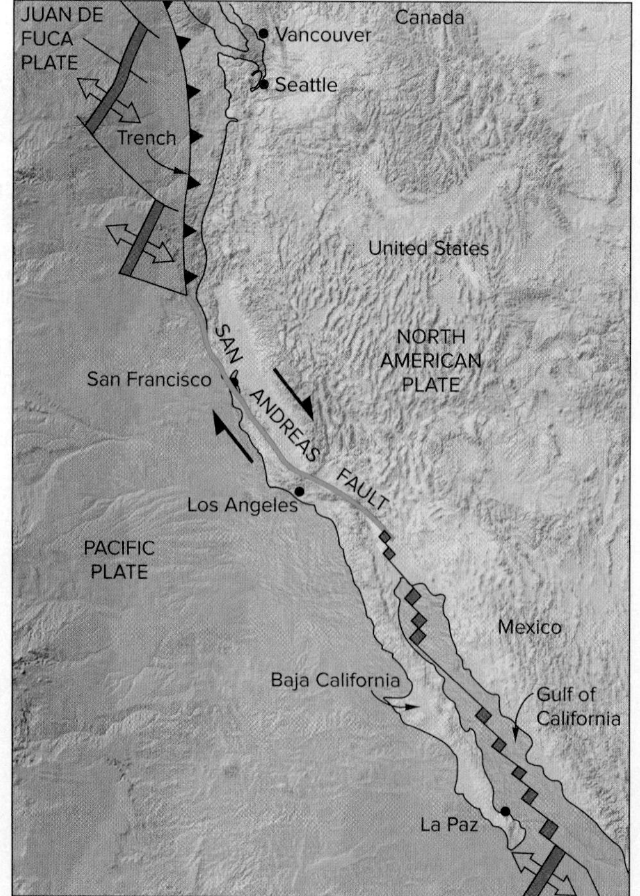

D

FIGURE 19.23

Transform boundaries (*A*) between two ridges; (*B*) between a ridge and a trench; and (*C*) between two trenches. Triangles on trenches point down subduction zones. Trench-trench transform boundaries are common in the southeast Pacific. Color tones show two plates in each case. (*D*) The San Andreas fault is a ridge-ridge transform plate boundary between the North American plate and the Pacific plate. The south end of the San Andreas fault is a ridge segment (shown in red) near the United States-Mexico border. The north end of the fault is a "triple junction" where three plates meet at a point. The relative motion along the San Andreas fault is shown by the large black arrows, as the Pacific plate slides horizontally past the North American plate. *(D) Source: U.S. Geological Survey*

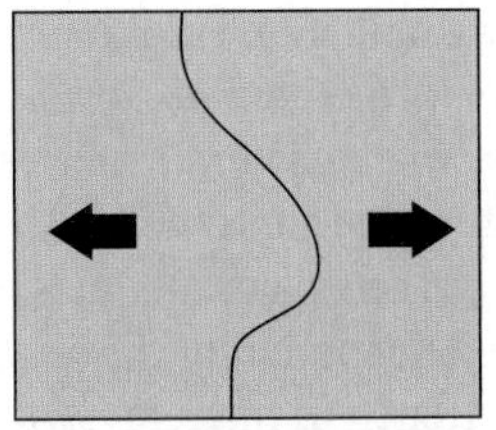

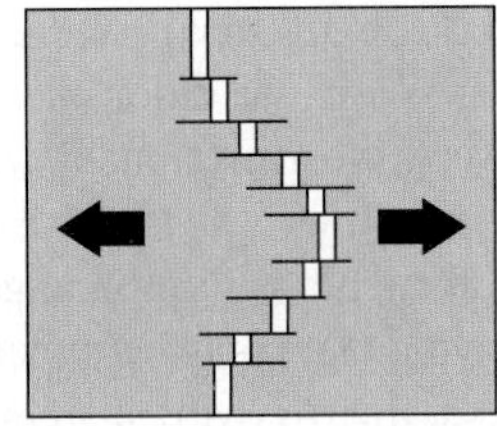

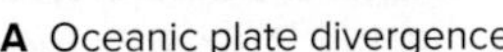

A Oceanic plate divergence

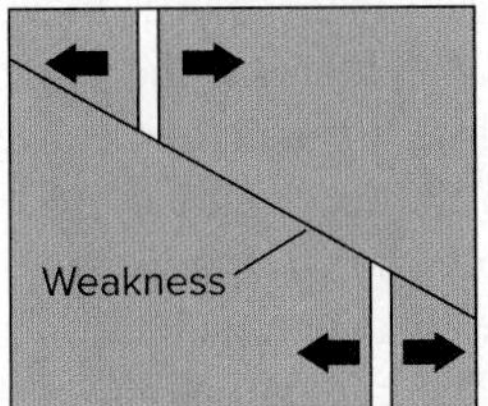

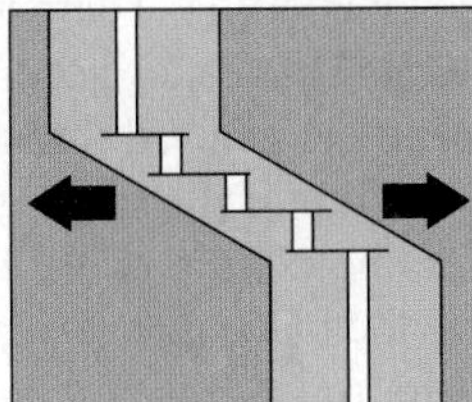

B Continental plate divergence

FIGURE 19.24

Divergent boundaries form ridge crests perpendicular to the spreading direction and transform faults parallel to the spreading direction. (*A*) Oceanic plates. (*B*) Continental plates.

ridge crests align perpendicular to the spreading direction, and the transform faults align parallel to the spreading direction. An old line of weakness in a continent may cause the initial divergent boundary to be oblique to the spreading direction when the continent splits. The boundary will then readjust into a series of transform faults parallel to the spreading direction.

19.8 CONVERGENT PLATE BOUNDARIES

At convergent plate boundaries, two plates move toward each other (often obliquely). The character of the boundary depends partly on the types of plates that converge. A plate capped by oceanic crust can move toward another plate capped by oceanic crust, in which case one plate dives (subducts) under the other. If an oceanic plate converges with a plate capped by a continent, the dense oceanic plate subducts under the continental plate. If the two approaching plates are both carrying continents, the continents collide and crumple, but neither is subducted.

Ocean-Ocean Convergence

Where two plates capped by sea floor converge, one plate subducts under the other (the Pacific plate sliding under the western Aleutian Islands is an example). The subducting plate bends downward, forming the outer wall of an oceanic trench, which usually forms a broad curve convex to the subducting plate (figures 19.25 and 19.26).

As one plate subducts under another, a Benioff zone of shallow-, intermediate-, and deep-focus earthquakes is created within the upper portion of the down-going lithosphere (see figure 16.23). The reasons for these quakes are discussed in chapter 16. The existence of deep-focus earthquakes to a depth of 670 kilometers tells us that brittle plates continue to (at least) that depth. The pattern of quakes shows that the angle of subduction changes with depth, usually becoming steeper (figure 19.25). Some plates crumple or break into segments as they descend.

As the descending plate reaches depths of at least 100 kilometers, magma is generated in the overlying asthenosphere (figure 19.25). The magma probably forms by partial melting of the asthenosphere, perhaps triggered by dewatering of the down-going oceanic crust as it is subducted, as described in chapter 3. Differentiation and assimilation may also play an important role in the generation of the magma, which is typically andesitic to basaltic in composition.

The magma works its way upward to erupt as an **island arc**, a curved line of volcanoes that form a string of islands parallel to the oceanic trench (figure 19.25). Beneath the volcanoes are large plutons in the thickened arc crust.

The distance between the island arc and the trench can vary, depending upon where the subducting plate reaches the 100-kilometer depth. If the subduction angle is steep, the plate reaches this magma-generating depth at a location close to the trench, so the horizontal distance between the arc and trench is short. If the subduction angle is gentle, the arc-trench distance is greater. A thick, buoyant plate (such as a subducting aseismic ridge) may subduct at such a gentle angle that it merely slides horizontally along under another plate. Because the top of the subducting plate never reaches the 100-kilometer depth, such very shallow subduction zones lack volcanism.

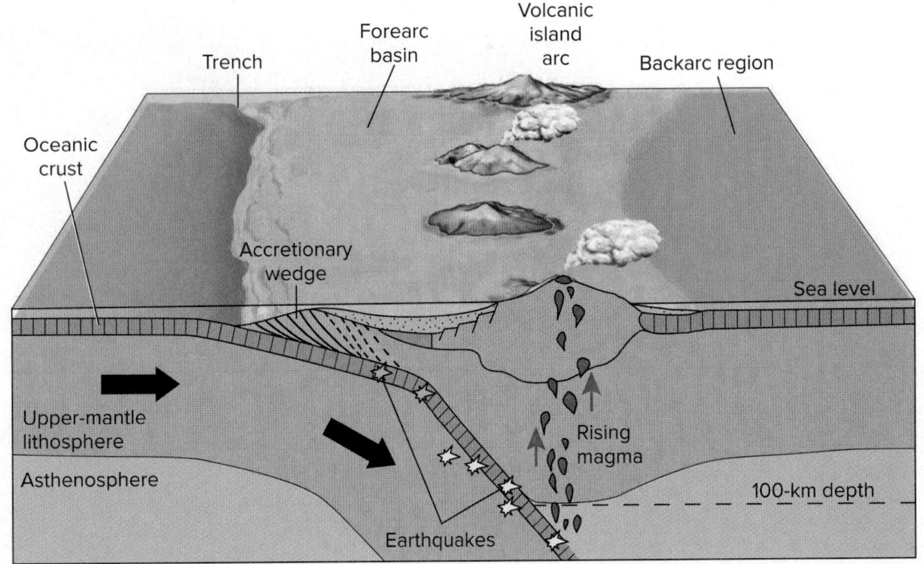

FIGURE 19.25

Ocean-ocean convergence forms a trench, a volcanic island arc, and a Benioff zone of earthquakes.

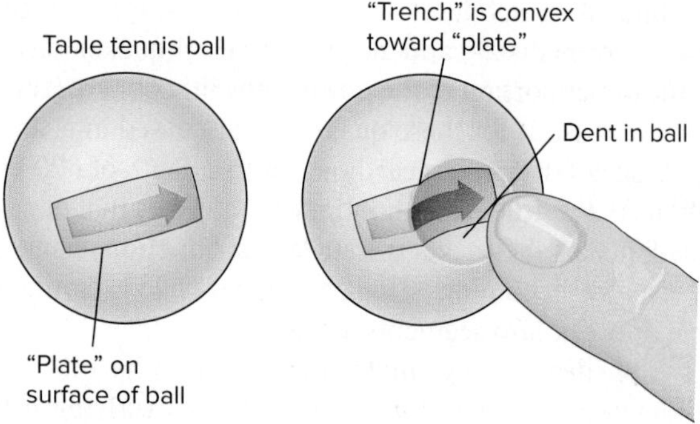

FIGURE 19.26

A dented table tennis ball can show why trenches are curved on a sphere.

When a plate subducts far from a mid-oceanic ridge, the plate is cold, with a low heat flow. Oceanic plates form at ridge crests, then cool and sink as they spread toward trenches. Eventually, they become cold and dense enough to sink back into the mantle. Oceanic trenches are marked by strong negative gravity anomalies. These show that trenches are not currently in isostatic equilibrium but are being actively pulled down. Hess thought that this pulling was caused by a down-turning convection current in the mantle. Today, most geologists think that the pulling is caused by the sinking of cold, dense lithosphere.

The inner wall of a trench (toward the arc) consists of an *accretionary wedge* (or *subduction complex*) of thrust-faulted and folded marine sediment and pieces of oceanic crust (figure 19.25). The sediment is "snowplowed" off the subducting plate by the overlying plate. New slices of sediment are continually added to the bottom of the accretionary wedge, pushing it upward to form a ridge on the sea floor. A relatively undeformed *forearc basin* lies between the accretionary wedge and the volcanic arc. (The trench side of an arc is the forearc; the other side of the arc is the backarc.)

Trench positions change with time. As one plate subducts, the overlying plate may be moving toward it. The motion of the leading edge of the overlying plate will force the trench to migrate horizontally over the subducting plate. The Peru-Chile Trench is moving over the Nazca plate in this manner as South America moves westward (figure 19.1). There is another reason that trenches move. A subducting plate may not sink in a direction parallel to the length of the plate but may fall through the mantle at an angle that is *steeper* than the dip of the down-going plate. This steep sinking pulls the subducting plate progressively away from the overlying plate and causes the hinge line of bending and the oceanic trench to migrate seaward onto the subducting plate. The migration may cause stretching or extension in the backarc region of the overlying plate, a process called *backarc spreading*. The location at which the subducting plate contacts the 100-kilometer depth where magmas are generated in the asthenosphere also migrates seaward toward the subducting plate and may cause the position of the island arc to migrate toward the subducting plate as well.

Ocean-Continent Convergence

When a plate capped by oceanic crust is subducted under the *continental* lithosphere, an accretionary wedge and forearc basin form an *active continental margin* between the trench and the continent (figure 19.27). A Benioff zone of earthquakes dips under

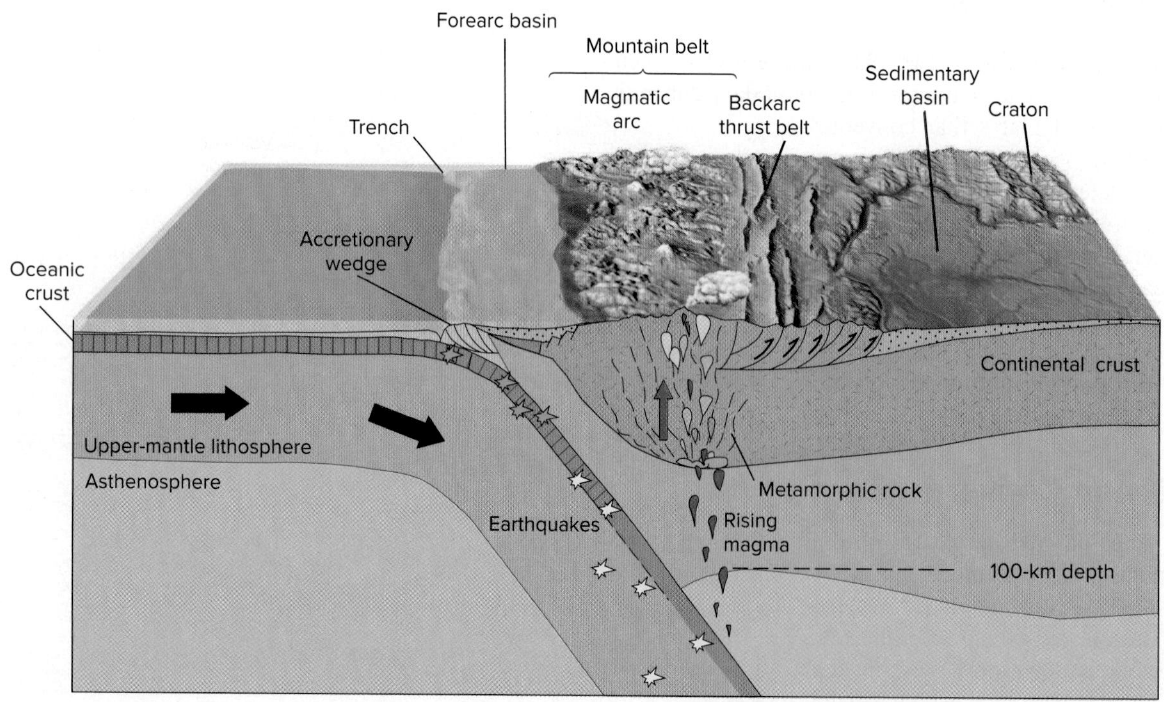

FIGURE 19.27

Ocean-continent convergence forms an active continental margin with a trench, a Benioff zone, a magmatic arc, and a young mountain belt on the edge of the continent.

the edge of the continent, which is marked by andesitic volcanism and a young mountain belt. Examples of this type of boundary are the subduction of the Nazca plate under western South America and the Juan de Fuca plate under North America.

The magma that is created by ocean-continent convergence forms a **magmatic arc**, a broad term used both for island arcs at sea and for belts of igneous activity on the edges of continents. The surface expression of a magmatic arc is either a line of andesitic islands (such as the Aleutian Islands) or a line of andesitic continental volcanoes (such as the Cascade volcanoes of the Pacific Northwest). Beneath the volcanoes are large plutons in thickened crust. We see these plutons as batholiths on land when they are exposed by deep erosion. The igneous processes that form the granitic and intermediate magmas of batholiths are described in chapter 3.

The hot magma rising from the subduction zone thickens the continental crust and makes it weaker and more mobile than cold crust. Regional metamorphism takes place within this hot, mobile zone. Crustal thickening causes uplift, so a young mountain belt forms here as the thickened crust rises isostatically.

Another reason for the growth of the mountain belt is the stacking up of thrust sheets on the continental (backarc) side of the magmatic arc (figure 19.27). The thrust faults, associated with folds, move slivers of mountain-belt rocks landward over the continental interior (the *craton*). Underthrusting of the rigid craton beneath the hot, mobile core of the mountain belt may help form the fold-thrust belt.

Inland of the backarc fold-thrust belt, the craton subsides to form a sedimentary basin (sometimes called a *foreland basin*). The weight of the stacked thrust sheets depresses the craton isostatically. The basin receives sediment, some of which may be marine if the craton is forced below sea level. This basin extends the effect of subduction far inland. Subduction of the sea floor off California during the Mesozoic Era produced basin sedimentation as far east as the central Great Plains.

Continent-Continent Convergence

Two continents may approach each other and collide. They must be separated by an ocean floor that is being subducted under one continent and that lacks a spreading axis to create new oceanic crust (figure 19.28). The edge of one continent will initially have a magmatic arc and all the other features of ocean-continent convergence.

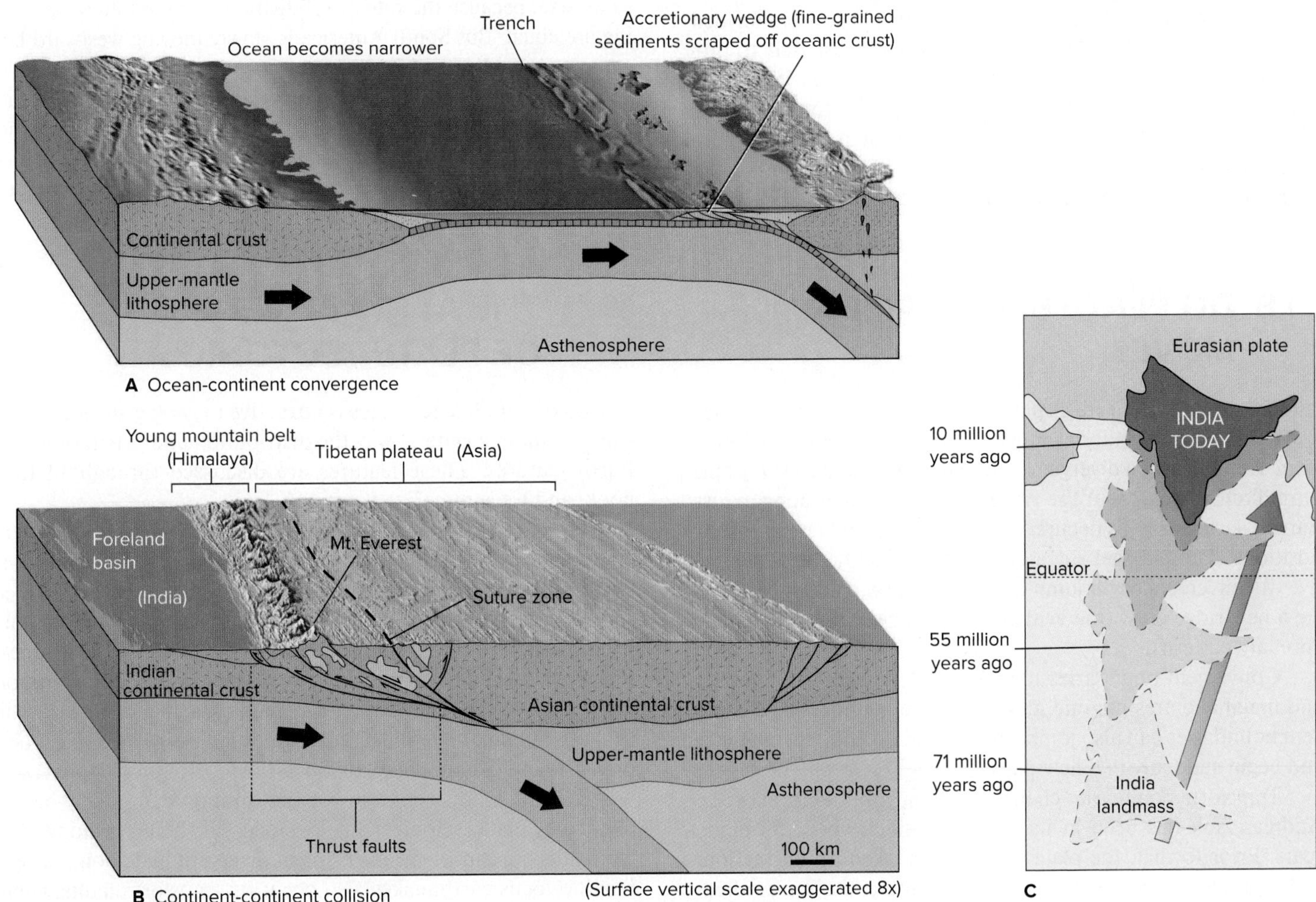

FIGURE 19.28

The collision of two continents forms a young mountain belt in the interior of a new, larger continent. The most famous example of continent-continent collision is the collision of India with Asia. (*A*) India is moving toward Asia due to ocean-continent convergence. (*B*) India collides with Asia to form the Himalayas, the highest mountain range on Earth. (*C*) Map view of the northward movement of India through time.

As the sea floor is subducted, the ocean becomes narrower and narrower until the continents eventually collide and destroy or close the ocean basin. Oceanic lithosphere is heavy and can sink into the mantle, but continental lithosphere is less dense and does not sink easily. One continent may slide a short distance under another and, under most instances, will not go down a subduction zone. After collision, the heavy oceanic lithosphere breaks off the continental lithosphere and continues to sink, leaving most of the continent behind.

The two continents are welded together along a dipping *suture zone* that marks the old site of subduction (figure 19.28*B*). Thrust belts and subsiding basins occur on both sides of the original magmatic arc, which is now inactive. The presence of the original arc thickens the crust in the region of impact. The crust is thickened further by the shallow underthrusting of one continent beneath the other and by the stacking of thrust sheets in the two thrust belts. The result is a mountain belt in the interior of a continent (a new, large continent formed by the collision of the two, smaller continents). The entire region of impact is marked by a broad belt of shallow-focus earthquakes along the numerous faults, as shown in figure 16.26. A few deeper quakes may occur within the sinking oceanic lithosphere beneath the mountain range.

The Himalayas in central Asia are thought to have formed in this way, as India collided with and underthrust Asia to produce exceptionally thick crust and high elevations. Paleomagnetic studies show that India was once in the Southern Hemisphere and moved north to its present position (figure 19.28*C*). The collision with Asia occurred after an intervening ocean was destroyed by subduction (figure 19.2).

19.9 DO PLATE BOUNDARIES MOVE?

Almost nothing is fixed in plate tectonics. Not only do plates move, but plate boundaries move as well. Plates may move away from each other at a divergent boundary on a ridge crest for tens of millions of years, but the ridge crest can be migrating across Earth's surface as this occurs. Ridge crests can also jump to new positions. The original ridge crest may suddenly become inactive; the divergence will jump quickly to a new position and create a new ridge crest (the evidence lies in the seafloor magnetic anomaly pattern).

Convergent boundaries migrate, also. As they do, trenches and magmatic arcs migrate along with the boundaries. Convergent boundaries can also jump; subduction can stop in one place and begin suddenly in a new place.

Transform boundaries change position, also. California's San Andreas fault has been in its present position about 5 million years. Prior to that, the plate motion was taken up on seafloor faults parallel to the San Andreas. In the future, the San Andreas may shift eastward again. The 1992 Landers earthquake, on a new fault in the Mojave Desert, and its pattern of aftershocks extending an astonishing 500 miles northward, suggest that the San Andreas is trying to jump inland again. Geodetic studies have shown that more than 25% of the plate motion between the Pacific and North American plates is accommodated along faults in eastern California and western Nevada (see box 19.1, figure 2). If more motion is taken up along this zone, most of California will be newly attached to the Pacific plate instead of the North American plate, and California will slide northwestward relative to the rest of North America.

19.10 CAN PLATES CHANGE IN SIZE?

Plates can change in size. For example, new sea floor is being added onto the trailing edge of the North American plate at the spreading axis in the central Atlantic Ocean. Most of the North American plate is not being subducted along its leading edge because this edge is made up of lightweight continental rock. Thus, the North American plate is growing in size as it moves slowly westward.

The Nazca plate is getting smaller. The spreading axis is adding new rock along the trailing edge of the Nazca plate, but the leading edge is being subducted down the Peru-Chile Trench. If South America were stationary, the Nazca plate might remain the same size, because the rate of subduction and the rate of spreading are equal. But South America is slowly moving westward because of spreading on the Atlantic Ridge, pushing the Peru-Chile Trench in front of it. This means that the site of subduction of the Nazca plate is gradually coming closer to its spreading axis to the west, and so the Nazca plate is getting smaller. The same thing is probably happening to the Pacific plate as the Eurasian plate moves eastward into the Pacific Ocean.

19.11 THE ATTRACTIVENESS OF PLATE TECTONICS

The theory of plate tectonics is attractive to geologists because it can explain in a general way the distribution and origin of many Earth features. These features are discussed throughout this book, and we summarize them here.

The distribution and composition of the world's *volcanoes* can be explained by plate tectonics. *Basaltic* volcanoes and lava flows form at divergent plate boundaries when hot mantle rock rises at a spreading axis. *Andesitic* volcanoes, particularly those in the circum-Pacific belt, result from subduction of an oceanic plate beneath either a continental plate or another oceanic plate. Although most of the world's volcanoes occur at plate margins, some do not (Hawaii being an example). We will discuss some of these isolated volcanoes in the "Mantle Plumes and Hot Spots" section of this chapter.

Earthquake distribution and first motion can largely be explained by plate tectonics. Shallow-focus earthquakes along normal faults are caused by extension at divergent plate boundaries. Shallow-focus earthquakes also occur on transform faults when plates slide past one another. Broad zones of shallow-focus earthquakes are located where two continents collide. Dipping Benioff zones of shallow-, intermediate-, and deep-focus quakes are found along the giant thrust faults formed when an oceanic plate is subducted beneath another plate. Most of the world's earthquakes (like most volcanoes) occur along plate boundaries, although a few take place within plates and are difficult to explain in terms of plate tectonics.

Young mountain belts–with their associated igneous intrusions, metamorphism, and fold-thrust belts–form at convergent boundaries. "Subduction mountains" form at the edges of continents where sea floor is sliding under continents. Examples include the Andes and Cascade Mountains. "Continental-collision" mountains such as the Himalayas form in continental interiors when two continents collide to form a larger continent. Old mountain belts such as the Urals in Russia mark the position of old, now inactive, plate boundaries.

The major features of the sea floor can also be explained by plate tectonics. The *mid-oceanic ridge* with its rift valley forms at divergent boundaries. *Oceanic trenches* are found where oceanic plates are subducted at convergent boundaries. *Fracture zones* are created at transform boundaries.

19.12 WHAT CAUSES PLATE MOTIONS?

A great deal of speculation currently exists about why plates move. There may be several reasons for plate motion. Any mechanism for plate motion has to explain why:

1. Mid-oceanic ridge crests are hot and elevated, while trenches are cold and deep;
2. Ridge crests have tensional cracks; and
3. The leading edges of some plates are subducting sea floor, while the leading edges of other plates are continents (which cannot subduct).

Possible driving mechanisms for plate tectonics include: mantle convection, ridge push, slab pull, trench suction, and mantle plumes.

Mantle Convection

There is no doubt that convection in the mantle is linked in some crucial way to plate motions (see figure 19.12). Mantle convection–the slow overturning of Earth's hot, ductile interior as heated rock wells up from below, cools near the surface, and sinks back down again–could take place as a series of giant cells, individually extending all the way from the heat source at the core-mantle boundary to the base of the lithosphere itself. However, studies using seismic tomography and computer modeling indicate that this idea of "whole mantle convection" is too simplistic (figure 19.29). Change in density with depth in the Earth, the property of large continents to trap mantle heat, and the "stirring" of the mantle from the sinking of subducted oceanic lithosphere all contribute to a more complex pattern of convective heat loss. Cold lithospheric plates may subduct down to the core-mantle boundary, whereas other, less-dense (younger) plates may only reach the 670-kilometer boundary. One of the models suggests that the lowermost part of the mantle does not mix with the upper and middle mantle but acts like a lava lamp turned on low, fueled by internal heating and heat flow across the core-mantle boundary. Variation in the thickness of this dense layer may control where mantle plumes rise and subducted plates ultimately rest.

Some geologists think that mantle convection is a *result* of plate motion rather than a cause of it. The sinking of a cold, subducting plate can create mantle convection (convection can be driven by either hot, rising material or by cold, sinking

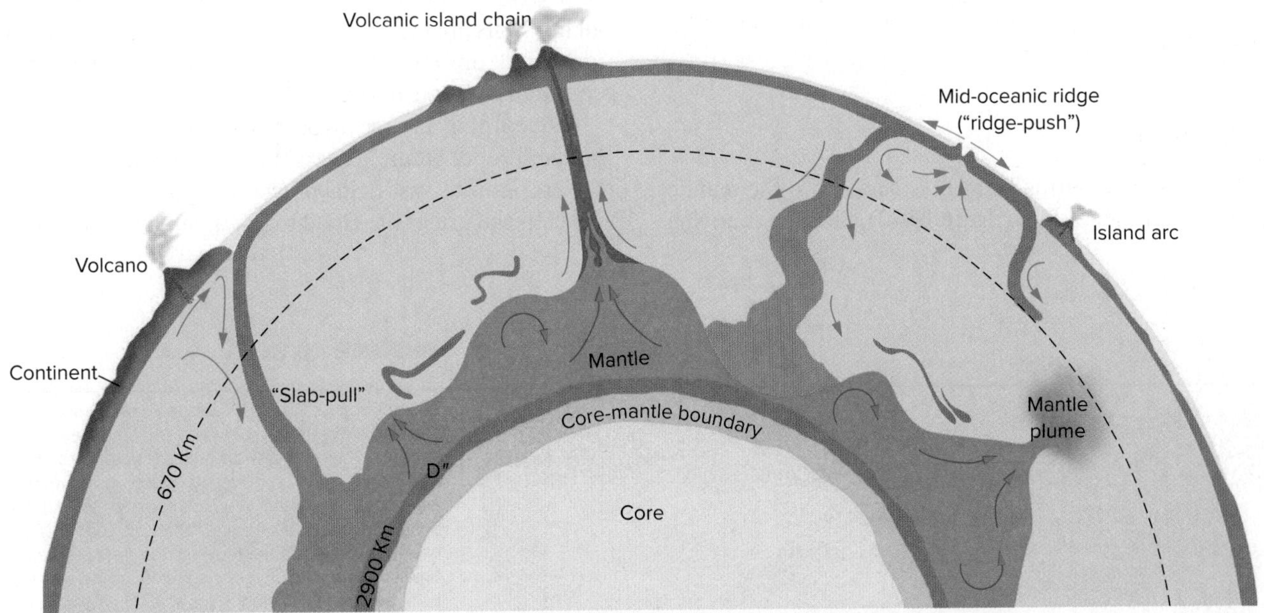

FIGURE 19.29

A possible model of mantle convection.

*Source: L. H. Kellogg, B. H. Hager, and R. D. van der Hilst, **Science**, 1999, 283, pp. 1881–84.*

material). Hot mantle rock rises at divergent boundaries to take the place of the diverging plates; however, such plate-caused convection would be shallow rather than mantle-deep.

The basic question in plate motion is, why do plates diverge and sink? Several mechanisms may be at work here.

Ridge Push

One proposal is called *ridge push.* As a plate moves away from a divergent boundary, it cools and thickens. Cooling sea floor subsides as it moves, and this subsidence forms the broad side slopes of the mid-oceanic ridge. An even more important slope forms on the base of the lithosphere mantle. The mantle thickens as cooling converts asthenospheric mantle to lithospheric mantle. Therefore, the boundary between them is a slope down which the lithosphere slides (figure 19.30). The oceanic plate is thought to slide down this slope at the base of the lithosphere, which may have a relief of 80 to 100 kilometers.

Slab Pull

Another mechanism is called *slab pull* (figure 19.30). Cold lithosphere sinking at a steep angle through hot mantle should pull the surface part of the plate away from the ridge crest and then down into mantle as it cools. A subducting plate sinks because it is denser than the surrounding mantle. This density contrast is partly due to the fact that the sinking lithosphere is cold. The subducting plate may also increase its density while it sinks, as low-density materials such as water are lost and as plate minerals collapse into denser forms during subduction. Slab pull is thought to be at least twice as important as ridge push in moving an oceanic plate away from a ridge crest. Slab pull causes rapid plate motion. The bigger the plate and the longer the subduction zone, the faster the plates move. This can be observed in figure 19.19, where the largest plates, the Pacific, Nazca, and Indo-Australian plates, have the fastest motion and subduct into very long zones of convergence.

Trench Suction

If a subducting plate (slab) falls, or *rolls back*, into the mantle at an angle steeper than its dip (figure 19.30), then the overlying plate is pulled horizontally toward the trench. This mechanism has been termed *trench suction* and is thought to be caused by small-scale convection in the wedge of mantle above the subducting plate. It is probably a minor force, but it may result in stretching or extension in the backarc region of the overlying volcanic arc.

All three of these mechanisms (ridge push, slab pull, and trench suction), particularly in combination, are compatible with high, hot ridges; cold, deep trenches; and tensional cracks at the ridge crest. They can account for the motion of both oceanic and continental plates. In this scheme, plate motions are controlled by variations in lithosphere density and thickness, which, in turn, are controlled largely by cooling. In other words, the reasons for plate motions are the properties of the plates themselves and the pull of gravity. This idea is in sharp contrast to most convection models, which assume that plates are dragged along by the movement of mantle rock beneath the plates.

Mantle Plumes and Hot Spots

A modification of the convection process was suggested early on by W. Jason Morgan of Princeton University. Morgan proposed that convection occurs in the form of **mantle plumes**, which he defined as narrow columns of hot mantle rock that rise through the mantle, much like smoke rising from a chimney (figure 19.31). Mantle plumes are thought to have large spherical or mushroom-shaped heads above a narrow, rising tail. They are essentially stationary with respect to moving plates and to each other. New seismic tomography images suggest that plumes may be much wider and rooted in the base of the mantle.

Plumes may form **hot spots** of active volcanism at Earth's surface. Note in figure 19.32 that many hot spots are located in volcanic regions such as Iceland, Yellowstone, and Hawaii. Recent seismic tomography images of the mantle suggest that not all hot spots are fed by mantle plumes. Of the forty-five hot spots identified on Earth, only twelve show evidence of a deep, continuous plume in the underlying mantle.

According to one hypothesis, when the head of a large plume ("super plume") nears the surface, it may cause uplift and the eruption of vast fields of flood basalts. As the head widens beneath the crust, the flood-basalt area widens and the crust is

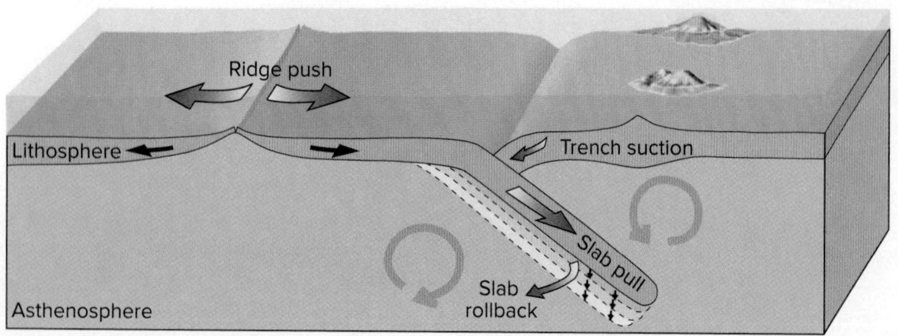

FIGURE 19.30

Other possible mechanisms for plate motion. Plates are pushed apart at the ridge (*ridge push*) by sliding downhill on the sloping boundary between the lithosphere and asthenosphere. Plates may also be pulled (*slab pull*) as the dense leading edge of a subducting plate sinks down into the asthenosphere. If the subducting plate falls into the asthenosphere at an angle steeper than its dip (*slab rollback*) then the overlying plate is pulled toward the trench by *trench suction*. Small-scale convection in the mantle associated with slab rollback and trench suction is shown with brown arrows.

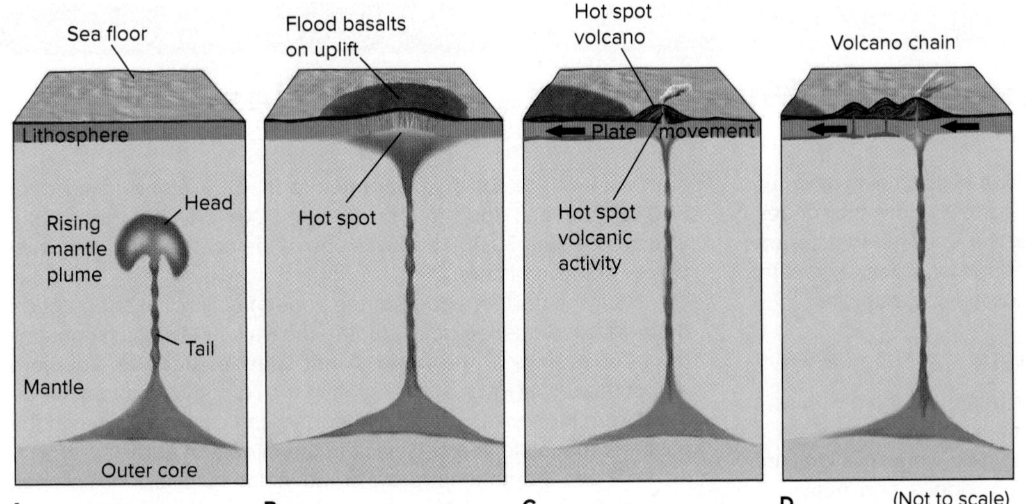

FIGURE 19.31

Model of mantle plume rising upward through the mantle to form a hot spot and associated flood basalts and volcanic chain. (*A*) Rising mantle plume contains a hot, mushroom-shaped plume head and a narrow tail. (*B*) Plume head forms a broad hot spot when it reaches the top of the mantle and causes uplift and stretching of the crust and eruption of flood basalts. (*C*) When the tail rises to the surface, a narrower hot spot forms a volcano. (*D*) Continued plate motion over the hot spot creates a trail or chain of volcanoes.

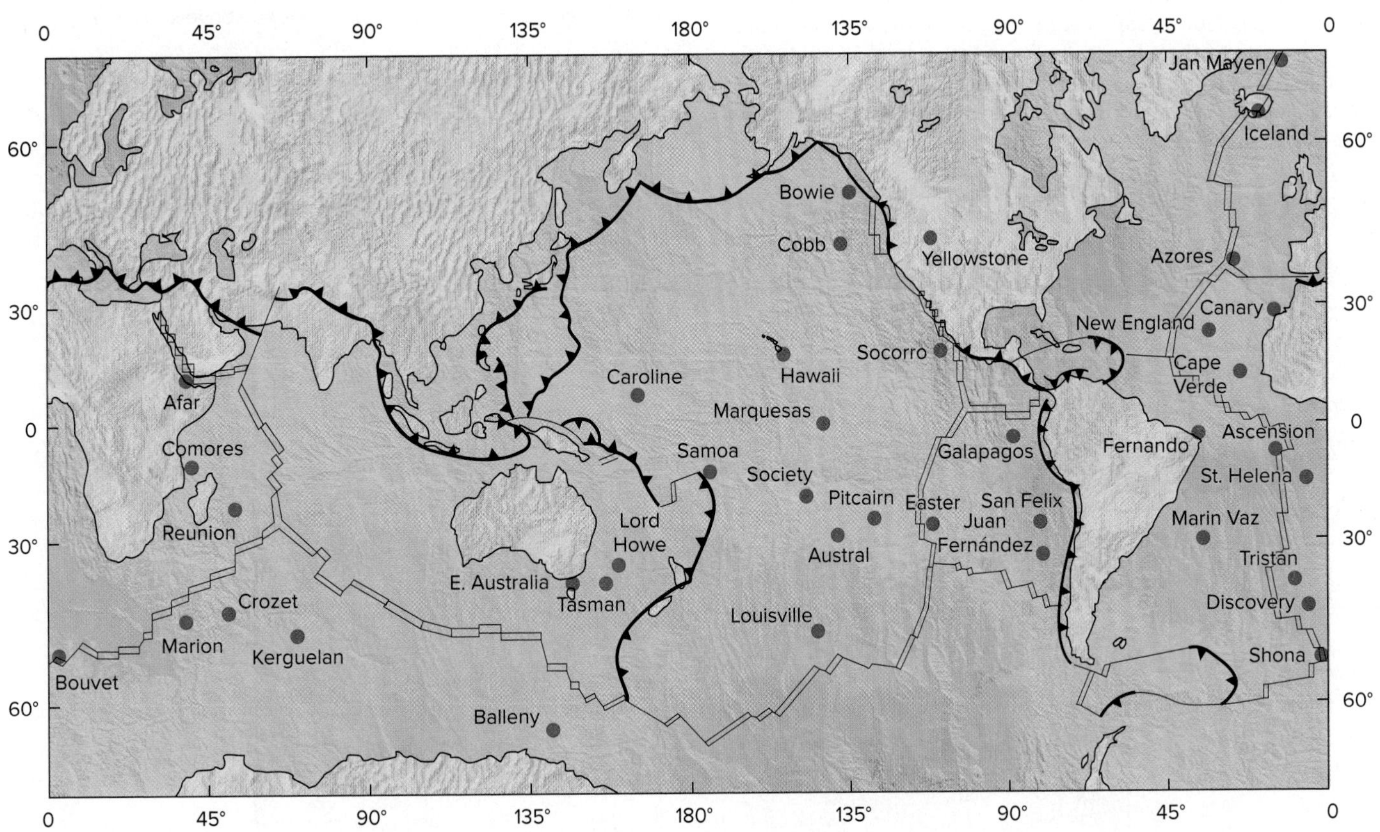

FIGURE 19.32

Distribution of hot spots, identified by volcanic activity and structural uplift within the past few million years. The hot spots near the poles are not shown.

stretched. The tail that follows the head produces a narrow spot of volcanic activity, much smaller than the head.

The outward, radial flow of the expanding head may be strong enough to break the lithosphere and start plates moving. In Morgan's view, a few plumes, such as those underlying some of the hot spots on the mid-oceanic ridge in the Atlantic Ocean in figure 19.32, are strong enough to drive plates apart (in this case, to push the American plates westward). Lithospheric tension set up by trench suction or slab pull could combine with mantle plume action to break a large plate (such as the former Pangaea) into smaller, diverging fragments. New studies suggest that plumes can also push a plate and speed up its movement. The Indian plate may have gotten an initial push northward from the plume under the Reunion hot spot (see figures 19.28 and 19.32).

IN GREATER DEPTH 19.1

Indentation Tectonics and "Mushy" Plate Boundaries

While it is easiest to conceive of Earth's skin as being made up of rigid tectonic plates that interact narrowly along their edges, the reality is more complicated—and interesting. The forces that cause plates to be geologically active along their boundaries may extend far into their interiors as well. Consider the following two examples:

The Collision of India with Asia

Approximately 40 million years ago, India began colliding with Asia to form the Himalaya Mountains, the biggest mountain system in the world. The sea that once separated India from Asia drained away as the former ocean floor rose into ridges and peaks as much as 5 miles high. The stresses of the continent-continent convergence extend far to the north of the Himalaya plate boundary, however—perhaps as far as 5,000 kilometers into Central Asia. Huge strike-slip fault systems with roughly east-west orientation break China apart (box figure 1). These have formed as Central Asia shifts out of the way of India, with the lithosphere moving primarily eastward to override the Pacific and Philippine plates in a series of very active subduction zones. India, in other words, has greatly "indented" Asia by colliding with it. The world's most destructive earthquakes have occurred in China, far from any plate boundaries. The Shaanxi earthquake in 1556 alone killed over 800,000 people, and the more recent 2008 Sichuan earthquake killed more than 70,000 and was one of the costliest natural disasters in Chinese history.

Near the northern edge of the zone of collisional stress, huge grabens have opened up, including the Baikal Rift, which contains the deepest lake in the world. A hot spot lies near the southern end of the Baikal Rift system, creating the Hangay Mountain Range in western Mongolia, which has been volcanically active within the past few thousand years—smack in the middle of Eurasia.

In Tibet, the northern prow of the Indian landmass has slipped beneath Central Asia along a series of Himalayan thrust faults, causing a doubling up of the lithosphere and uplift of the largest high-elevation plateau in the world. Lhasa, the capital of Tibet, lies in a fertile valley at an elevation of 3,700 meters (12,000 feet) above sea level.

The San Andreas Transform Boundary

The San Andreas fault is a 1,100-kilometer-long rupture marking the border between the Pacific and North American plates in California. But only about *one-third* of the approximately 2,000 kilometers of total slip between the two places, the biggest

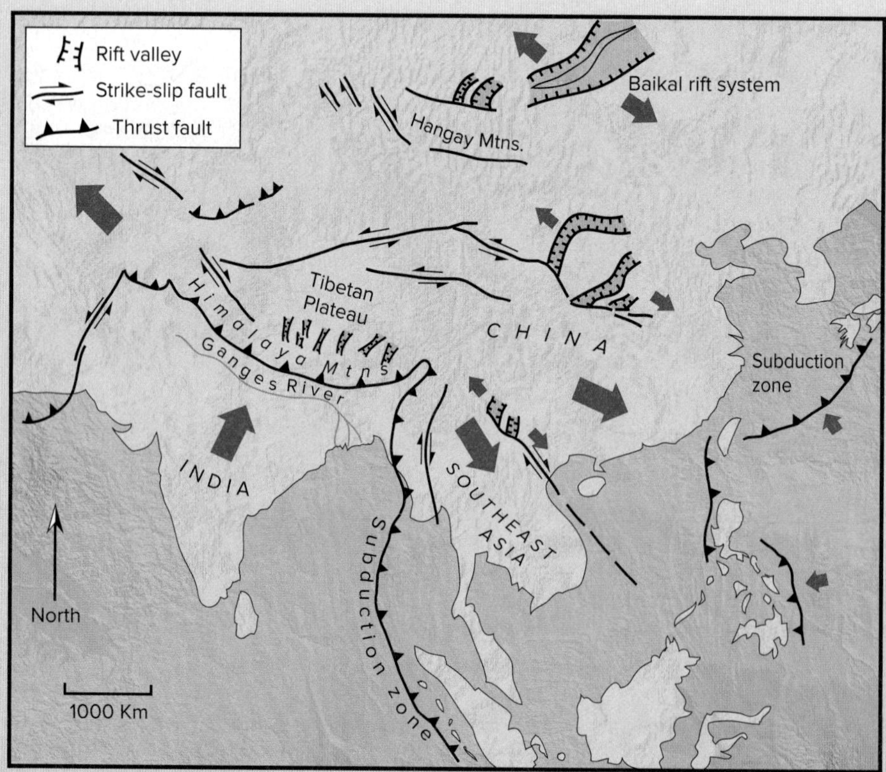

BOX 19.1 ■ FIGURE 1

Central Asia has adjusted to the broadside collision of India through uplift of the Tibetan plateau, and by stretching and slipping along major "intraplate" faults.

plates on Earth, has taken place along the fault during its 25- to 30-million-year history. How then do the plates actually move past one another in this region? The answer is that the San Andreas belongs to a much larger system of related parallel faults. Other ruptures, such as the Death Valley fault in eastern California and the Brothers fault zone in Oregon, also take up components of plate motion, so that the plate boundary is actually a *zone* of slippage about 600 kilometers wide rather than the single line you see on a map in an introductory geology textbook (box figure 2). The western side of North America is sliced up like a giant stack of dominoes—in other words, each domino slides past another in a right-lateral sense.

The San Andreas itself bends in places, so that it is not always parallel to the vectors of plate movement. In southern California, local plate convergence along the fault has shoved up the mountains bordering Los Angeles. Here, the San Andreas is a dynamic, evolving structure that will almost certainly wane as new faults inland more efficiently ease the plates past one another in the not-so-distant geological future.

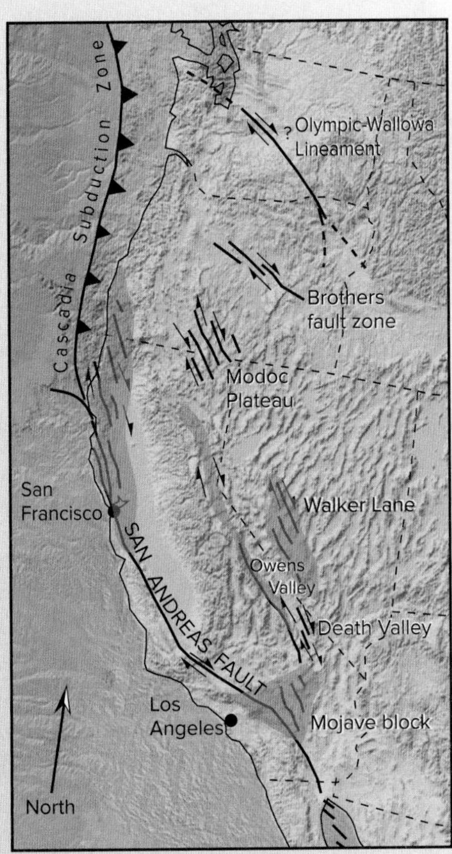

BOX 19.1 ■ FIGURE 2

Many faults participate in easing North America past the Pacific plate.

A mantle plume rising beneath a continent should heat the land and bulge it upward to form a dome marked by volcanic eruptions. As the dome forms, the stretched crust typically fractures in a three-pronged pattern (figure 19.33). Continued radial flow outward from the rising plume eventually separates the crust along two of the three fractures but leaves the third fracture inactive. In this model of continental breakup, the two active fractures become continental edges as new sea floor forms between the divergent continents. The third fracture is a *failed rift* (or *aulacogen*), an inactive rift that becomes filled with sediment.

An example of this type of fracturing can be seen in the vicinity of the Red Sea (figure 19.34). The Red Sea and the Gulf of Aden are active diverging boundaries along which the Arabian

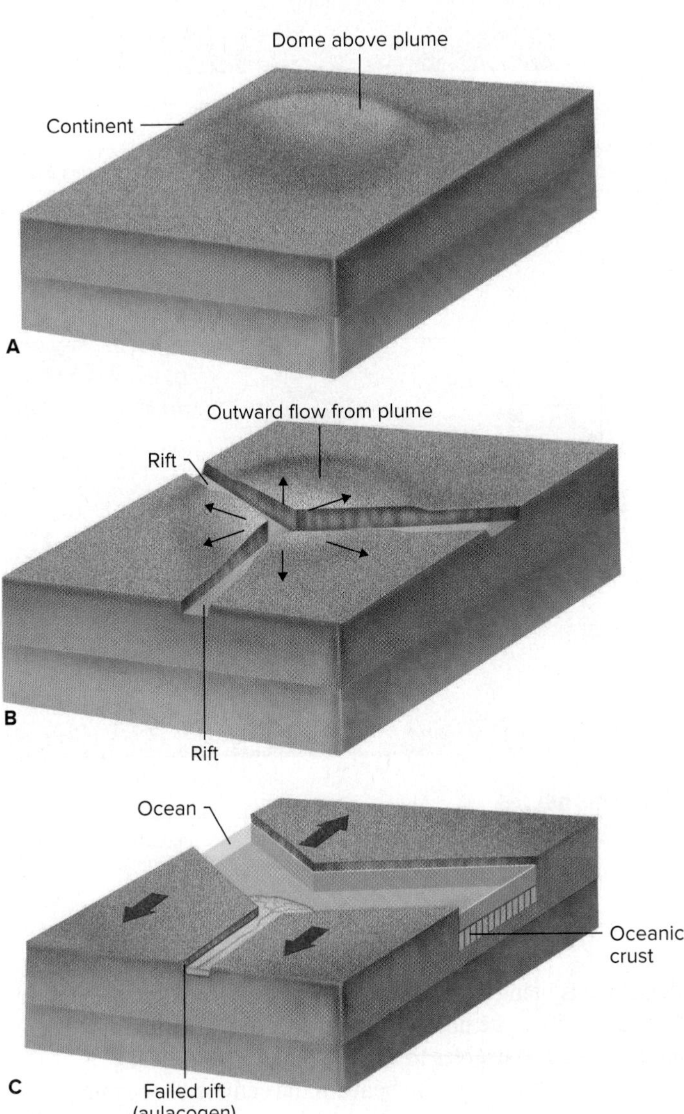

FIGURE 19.33

Continental breakup caused by a mantle plume. (*A*) A dome forms over a mantle plume rising beneath a continent. (*B*) Three radial rifts develop due to outward radial flow from the top of the mantle plume. (*C*) Continent separates into two pieces along two of the three rifts, with new ocean floor forming between the diverging continents. The third rift becomes an inactive "failed rift" (or aulacogen) filled with continental sediment.

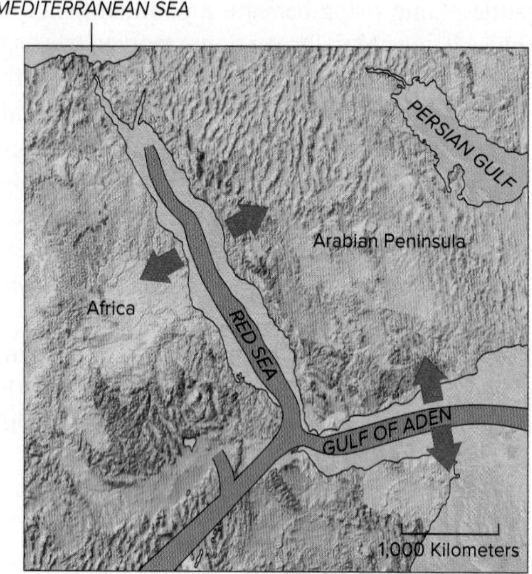

FIGURE 19.34

An example of radial rifts. The Red Sea and the Gulf of Aden are the active rifts, as the Arabian peninsula moves away from Africa. The Gulf of Aden contains a mid-oceanic ridge and central rift valley. The less active, failed rift (aulacogen) is the rift valley shown in Africa.

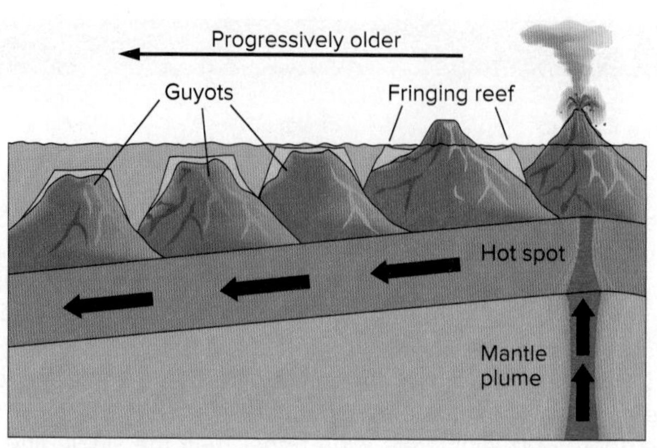

FIGURE 19.35

Fringing reef forms around volcanic island as it moves off hot spot. Waves erode and flatten top of volcanic islands to form guyots that progressively sink and become submerged away from hot spot.

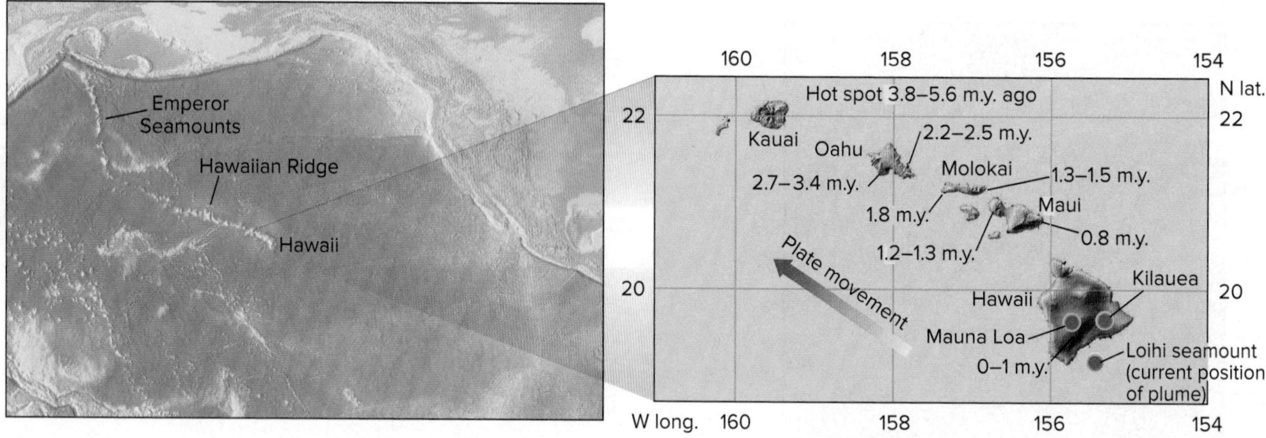

FIGURE 19.36

Ages of volcanic rock of the Hawaiian island group. Ages increase to northwest. Two active volcanoes on Hawaii are shown by red dots. The plume is currently offshore under the Loihi seamount (red dot), where recent underwater eruptions have been documented.

Peninsula is being separated from northeastern Africa. The third, less-active rift is the northernmost African Rift Valley, lying at an angle about 120 degrees to each of the narrow seaways.

Some plumes may rise beneath the centers of oceanic plates. A suspected plume under Hawaii rises in the center of the Pacific plate. As the plate moves over the plume, a line of volcanoes forms, creating an aseismic ridge (figure 19.35). The volcanoes are gradually carried away from the eruptive center, isostatically sinking as they go because of cooling. The result is a line of extinct volcanoes (seamounts and guyots) increasing in age away from an active volcano directly above the plume.

In the Hawaiian island group, the only two active volcanoes are in the extreme southeastern corner (figure 19.36). The isotopic ages of the Hawaiian basalts increase regularly to the northwest, and a long line of submerged volcanoes forms an aseismic ridge to the northwest of Kauai. Most aseismic ridges on the sea floor appear to have active volcanoes at one end, with ages increasing away from the eruptive centers. Deep-sea drilling has shown, however, that not all aseismic ridges increase in age along their lengths. This evidence has led to alternate hypotheses for the origin of aseismic ridges. It may pose difficulties for the plume hypothesis itself.

EARTH SYSTEMS 19.2

The Relationship between Plate Tectonics and Ore Deposits

The plate-tectonic theory provides an overall model for the origin of metallic ore deposits that has been used to explain the occurrence of known deposits and to explore for new deposits. Because many ore deposits are associated with igneous activity, a close relationship exists between plate boundaries and metallic ore deposits.

As discussed in chapter 18, *divergent plate boundaries* are often marked by lines of active hot springs in rift valleys that carry and precipitate metallic minerals in mounds around the hot springs. The metals in rift-valley hot springs are predominantly iron, copper, and zinc, with smaller amounts of manganese, gold, and silver. Although the mounds are nearly solid metal sulfide, they are small and widely scattered on the sea floor, so commercial mining of them may not be practical. Occasionally, the ore minerals may be concentrated in richer deposits. On the floor of the Red Sea, metallic sediments have precipitated in basins filled with hot-spring solutions. Although the solutions are hot (up to 60°C or 140°F), they are very dense because of their high salt content (they are seven times saltier than seawater), so they collect in seafloor depressions instead of mixing with the overlying seawater. Although not currently mined, the metallic sediments have an estimated value of $25 billion.

Hot metallic solutions are also found along some divergent continental boundaries. Near the Salton Sea in southern California, which lies along the extension of the mid-oceanic ridge inland, hot water very similar to the Red Sea brines has been discovered underground. The hot water is currently being used to run a geothermal power plant. The high salt and metal content is corrosive to equipment, but metals such as copper and silver may one day be recovered as valuable by-products.

Seafloor spreading carries the metallic ores away from the ridge crest (box figure 1), perhaps to be subducted beneath island arcs or continents at *convergent plate boundaries.* Slivers of *ophiolite* on land may contain these rich ore minerals in relatively intact form. A notable example of such ores occurs on the island of Cyprus in the Mediterranean Sea (box figure 2). Banded chromite ores may also be contained in the serpentinized ultramafic rock at the bottom of ophiolites.

Volcanism at *island arcs* can also produce hot-spring deposits on the flanks of the andesitic volcanoes. Pods of very rich ore collect above local bodies of magma, and the ore is sometimes distributed as sedimentary layers in shallow basins (box figure 3).

Many important ore deposits are also found at *convergent plate boundaries*, where metals from the subducting plate or overlying mantle are released and are concentrated in magmas or hydrothermal fluids. More than half of the world's supply of copper is mined from deposits associated with plate convergence.

It is tempting to think that *mantle plumes* might cause ore deposition, for plumes provide a source of both magma and hydrothermal solutions. The locations of supposed plumes, however, such as Yellowstone and Hawaii, are notable for their *lack* of ore deposits.

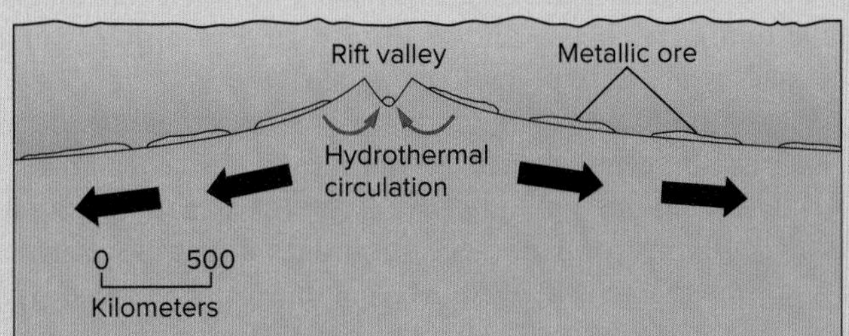

BOX 19.2 ■ FIGURE 1

Divergent oceanic plates carry metallic ores away from rift valley. (Size of ore deposits is exaggerated.)

BOX 19.2 ■ FIGURE 2

Second-largest copper mine in Cyprus, Greece, mines copper from the Troodos ophiolite. Copper deposits were initially formed on the sea floor where active hot springs precipitated metallic minerals in rift valleys where the oceanic lithosphere was being pulled apart.
©*Jonathan Blair/Corbis/Getty Images*

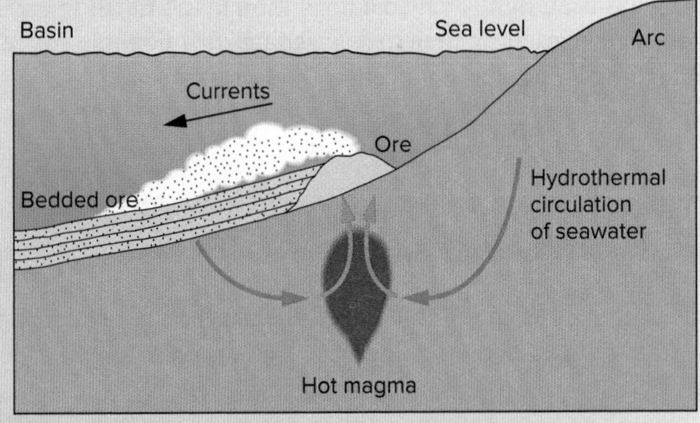

BOX 19.2 ■ FIGURE 3

On island arcs, metallic ores can form over hot springs and be redistributed into layers by currents in shallow basins.

19.13 A FINAL NOTE

Objections were raised to plate tectonics after it was proposed in the late 1960s. Some seafloor features did not seem compatible with a moving sea floor. The geology of many continental regions did not seem to fit into the theory of plate tectonics—in some cases, not even slightly. But a revolutionary idea in science is always controversial. As it progresses from an "outrageous hypothesis" to a more widely accepted theory, after much discussion and testing, a new idea evolves and changes. The newness of the idea wears off, and successful tests and predictions convert skeptics to supporters (sometimes grudgingly). Perhaps equally important, dissenters die off.

As refinements were made to plate tectonics and as more was learned about the puzzling seafloor features and continental regions, they began to seem more compatible with plate tectonics. Objections died out, and plate tectonics became widely accepted.

It is wise to remember that at the time of Wegener, most geologists vehemently disagreed with continental drift. Because Wegener proposed that continents plow through seafloor rock and because his proposed forces for moving continents proved inadequate, most geologists thought that continental drift was wrong. Although these geologists had sound reasons for their dissent, we now know, due to overwhelming evidence, that lithospheric plates move and the early geologists were wrong.

The evidence for plate tectonics is very convincing. The theory has been rightly called a revolution in the Earth sciences. It has been an exciting time to be a geologist. Our whole concept of Earth dynamics has changed in the last fifty years.

Summary

Plate tectonics is the idea that Earth's surface is divided into several large plates that change position and size. Intense geologic activity occurs at plate boundaries.

Plate tectonics combines the concepts of *seafloor spreading* and *continental drift.*

Alfred Wegener proposed continental drift in the early 1900s. His evidence included coastline fit, similar fossils and rocks in now-separated continents, and paleoclimatic evidence for *apparent polar wandering.* Wegener proposed that all continents were once joined together in the supercontinent *Pangaea.*

Wegener's ideas were not widely accepted until the 1950s, when work in paleomagnetism revived interest in polar wandering.

Evidence for continental drift includes careful fits of continental edges and detailed rock matches between now-separated continents. The positions of continents during the past 200 million years have been mapped.

Hess's hypothesis of *seafloor spreading* suggests that the sea floor moves away from the ridge crest and toward trenches as a result of mantle convection.

According to the concept of seafloor spreading, the high heat flow and volcanism of the ridge crest are caused by hot mantle rock rising beneath the ridge. Divergent *convection* currents in the mantle cause the rift valley and earthquakes on the ridge crest, which is a *spreading axis* (or *center*). New sea floor near the rift valley has not yet accumulated pelagic sediment.

Seafloor spreading explains trenches as sites of seafloor *subduction,* which causes low heat flow and negative gravity anomalies. Benioff zones and andesitic volcanism are caused by interaction between the subducting sea floor and the rocks above.

Seafloor spreading also explains the young age of the rock of the sea floor as caused by the loss of old sea floor through subduction into the mantle.

Plates are composed of blocks of *lithosphere* riding on a ductile *asthenosphere.* Plates move away from spreading axes, which add new sea floor to the trailing edges of the plates.

An apparent confirmation of plate motion came in the 1960s with the correlation of marine *magnetic anomalies* to *magnetic reversals* by Vine and Matthews. The origin of magnetic anomalies at sea apparently is due to the recording of normal and reverse magnetization by dikes that intrude the crest of the mid-oceanic ridge, then split and move sideways to give anomaly patterns a mirror symmetry.

The Vine-Matthews hypothesis gives the rate of plate motion and can predict the age of the sea floor before it is sampled.

Deep-sea drilling has apparently verified plate motions and the age predictions made from magnetic anomalies.

Earthquake distribution and first-motion studies on *transform faults* on fracture zones also verify plate motions.

Divergent plate boundaries are marked by rift valleys, shallow-focus earthquakes, high heat flow, and basaltic volcanism.

Transform boundaries between plates sliding past one another are marked by strike-slip (transform) faults and shallow-focus earthquakes.

Convergent plate boundaries can cause *subduction* or *continental collision.* Subducting plate boundaries are marked by trenches, low heat flow, Benioff zones, andesitic volcanism, and young mountain belts or island arcs. Continental-collision boundaries have shallow-focus earthquakes and form young mountain belts in continental interiors.

The distribution and origin of most volcanoes, earthquakes, young mountain belts, and major seafloor features can be explained by plate tectonics.

Plate motion was once thought to be caused by *mantle convection* but is now also attributed to the cold, dense leading edge of a subducting plate pulling the rest of the plate along with it (*slab pull*). Plates near mid-oceanic ridges also slide down the sloping lithosphere-asthenosphere boundary at the ridge (*ridge push*). *Trench suction* may help continents diverge.

Mantle plumes are narrow columns of hot, rising mantle rock. They cause flood basalts and may split continents, causing plate divergence.

An aseismic ridge may form as an oceanic plate moves over a mantle plume acting as an eruptive center (hot spot).

Terms to Remember

asthenosphere 466
continental drift 458
convection 464
convergent plate boundary 466
divergent plate boundary 466
hot spots 480
island arc 475
lithosphere 466
magmatic arc 477
mantle plume 480
plate 466
plate tectonics 458
polar wandering 460
rift valley 465
seafloor spreading 458
subduction 464
transform fault 470
transform plate boundary 466

Testing Your Knowledge

Use the following questions to prepare for exams based on this chapter.

1. Summarize Wegener's evidence for continental drift.
2. Define polar wandering. What is the paleoclimatic evidence for polar wandering? What is the magnetic evidence for polar wandering? Does polar wandering require the poles to move?
3. Describe the evidence that South America and Africa were once joined.
4. In a series of sketches, show how the South Atlantic Ocean might have formed by the movement of South America and Africa.
5. What is Pangaea?
6. In a single cross-sectional sketch, show the concept of seafloor spreading and how it relates to the mid-oceanic ridge and oceanic trenches.
7. How does seafloor spreading account for the age of the sea floor?
8. What is a plate in the concept of plate tectonics?
9. Define *lithosphere* and *asthenosphere.*
10. Discuss the origin of marine magnetic anomalies according to Vine and Matthews.
11. Why does the pattern of magnetic anomalies at sea match the pattern of magnetic reversals (recorded in lava flows on land)?
12. How has deep-sea drilling tested the concept of plate motion?
13. How has the study of fracture zones tested the concept of plate motion?
14. Explain how plate tectonics can account for the existence of the mid-oceanic ridge and its associated rift valley, earthquakes, high heat flow, and basaltic volcanism.
15. What is a transform fault? What type of transform fault is the San Andreas fault?
16. Explain how plate tectonics can account for the existence of oceanic trenches as well as their low heat flow, their negative gravity anomalies, the associated Benioff zones of earthquakes, and andesitic volcanism.
17. Describe the various types of plate boundaries and the geologic features associated with them.
18. Discuss possible driving mechanisms for plate tectonics.
19. What is a mantle plume? What is the geologic significance of mantle plumes?
20. The ancient southern supercontinent is called
    a. Gondwanaland. b. Pangaea.
    c. Laurasia. d. Glossopteris.
21. The sliding of oceanic lithosphere beneath a continent or island arc is called
    a. rotation. b. tension.
    c. subduction. d. polar wandering.
22. In cross section, the plates are part of a rigid outer shell of the Earth called the
    a. lithosphere. b. asthenosphere.
    c. crust. d. mantle.
23. The Vine-Matthews hypothesis explains the origin of
    a. polar wandering.
    b. seafloor magnetic anomalies.
    c. continental drift.
    d. mid-oceanic ridges.
24. The San Andreas fault in California is a
    a. normal fault. b. reverse fault.
    c. transform fault. d. thrust fault.
25. What would you most expect to find at ocean-ocean convergence?
    a. suture zone b. island arc
    c. midocean ridge d. none of the above
26. What would you most expect to find at ocean-continent convergence?
    a. magmatic arc b. suture zone
    c. island arc d. midocean ridge
27. What would you most expect to find at continent-continent convergence?
    a. magmatic arc b. suture zone
    c. island arc d. midocean ridge
28. Passive continental margins are created at
    a. divergent plate boundaries.
    b. transform faults.
    c. convergent plate boundaries.
29. The Hawaiian Islands are thought to be the result of
    a. subduction.
    b. midocean ridge volcanics.
    c. mantle plumes.
    d. ocean-ocean convergence.

Expanding Your Knowledge

1. Plate tectonics helps cool Earth as hot mantle rock rises near the surface at ridge crests and mantle plumes. What can we assume about the internal temperature of other planets that do not seem to have plate tectonics? What would happen to Earth's internal temperature if the plates stopped moving?
2. Are ridge offsets along fracture zones easier to explain with mantle-deep convection *causing* plate motion or with shallow convection occurring as a *result* of plate motion?

3. Why are mantle plumes narrow? What conditions at the core-mantle boundary could cause the formation and rise of a mushroom-shaped plume?
4. The slab pull and ridge push mechanisms of plate motion may operate only after a plate starts to move. What starts plate motion?
5. If subducting plates can penetrate the 670-kilometer mantle boundary and sink all the way to the base of the mantle, why are there no earthquakes deeper than 670 kilometers?

Exploring Web Resources

http://pubs.usgs.gov/gip/dynamic/dynamic.html

This Dynamic Earth: The Story of Plate Tectonics. U.S. Geological Survey online book by W. J. Kious and R. Tilling provides general information about plate tectonics.

http://www.earthscope.org

EarthScope website includes information about the Plate Boundary Observatory, which uses seismic and geodetic studies to study the details of the Pacific and North American plate boundary in the western United States.

http://vulcan.wr.usgs.gov/Glossary/PlateTectonics/framework.html

USGS website with links to maps, publications, and other items of interest regarding plate tectonics, seafloor spreading, and subduction zones.

http://pubs.usgs.gov/gip/dynamic/historical.html

USGS publication describing the historical perspective of plate tectonics, beginning with Alfred Wegener's theory of "continental drift."

www.ucmp.berkeley.edu/geology/tectonics.html

Find information on the history and mechanisms of plate tectonics, as well as animations and related links.

www.scotese.com

View images and animations by C. R. Scotese of the assembly and breakup of Pangaea through time.

CHAPTER 20

Mountain Belts and the Continental Crust

The setting sun catches the highest peaks of the Himalaya range in Nepal. The three peaks still lit are the first (Mt. Everest, on the left), fourth (Lhotse, just to the right of Everest), and fifth (Makalu, center distance) tallest mountains in the world. In the foreground you can see the surface of the Ngozumba Glacier.
©Alisha Wenzel

LEARNING OBJECTIVES

- Distinguish between mountain, mountain range, and mountain belt.
- Compare the major features of mountain belts to cratons.
- Describe the major factors that control the growth and development of mountain ranges.
- Define an orogeny, and describe the tectonic settings associated with orogenies.
- Explain the Wilson Cycle, and relate it to the formation of mountain belts and cratons.
- Distinguish between the processes occurring during an orogeny and those believed to be responsible for post-orogenic uplift.

Mountain belts are among the most spectacular of Earth's landforms, inspiring works of art and inviting the intrepid to ascend their lofty peaks for thousands of years. But mountains are not permanent features. Under the enormous forces of plate tectonics, mountain belts can evolve from marine-deposited rocks to towering peaks during periods of tens of millions of years. Ultimately, through the influence of weathering and erosion, towering peaks are worn down to plains and become part of the stable interior of a continent. To appreciate the long and complex process of mountain building, you need to know much of the material covered in previous chapters. For instance, you must understand structural geology to appreciate what a particular pattern of folds and faults can tell us about the history of mountain building in a particular region. To understand how the rocks formed during the various stages of a mountain belt's history, you must know about volcanism, plutonism, sedimentation, and metamorphism. Your earlier study of weathering and erosion will help you understand how mountains are worn away. Plate-tectonic theory has been strikingly effective in helping geologists make sense of often complex aspects of mountain belts and the continental crust. For this reason, you may need to go back to some of the material in chapter 19, in particular, the section on convergent boundaries, to appreciate how continents evolve.

In this chapter, we first point out what geologists have observed of mountain belts. Next, we describe how these observations are interpreted, particularly in light of plate-tectonic theory. Finally, we discuss current perceptions of how continents change and grow.

20.1 MOUNTAINS AND MOUNTAIN BUILDING

A mountain, as you know, is a large terrain feature that rises more or less abruptly from surrounding levels. Volcanoes are mountains; so are erosional remnants of plateaus (mesas). In this chapter, we will not focus on individual mountains; rather, we are concerned here with Earth's **major mountain belts**, chains thousands of kilometers long composed of numerous mountain ranges. A **mountain range** is a group of closely spaced mountains or parallel ridges (figure 20.1). A mountain range is likely to be composed of tectonically deformed sedimentary, volcanic, or metamorphic rocks. It may also show a history of intrusive igneous activity and volcanism.

The map in figure 20.2 shows that most of the world's mountains are in long mountain belts that extend for thousands of kilometers. The Himalaya, the Andes, the Alps, and the Appalachians are examples of major mountain belts, each comprising numerous mountain ranges.

Geologists find working in mountain ranges to be physically challenging and intellectually stimulating. High mountains have steep faces and broad exposures of bedrock. This is good because they allow a geologist to decipher complex interrelationships between rock units. But the geologist may have to become a proficient mountain climber to access the good exposures. (Conversely, mountain climbers who develop an interest in the rocks they climb sometimes become geologists.) On the other hand, exposures of bedrock critical to interpreting the local geology may be buried beneath glaciers or talus from rockfall. Furthermore, even in the highest and best exposed mountains, we never see bedrock representative of all of a mountain range. Significant amounts of rock (usually thousands of meters) once overlying the rocks we now see have been eroded away. Moreover, the present exposures are like the proverbial tips of icebergs—there is much more rock below the exposed mountain range that we cannot observe. For instance, the Himalaya, Earth's highest mountain range, rise to 8,000 meters above sea level; yet their roots (Earth's crust beneath the mountains) extend downward 65,000 meters. In other words, at best, less than one-eighth of the thickness of a mountain range is exposed to us.

Our models of how major mountain belts evolve use data from over a century of studying the geologic structures and rocks exposed in the world's many mountain ranges. Often, a particular study aims to piece together the geologic history of a single mountain range or part of a range. In other field studies, a geologist focuses on a particular type of rock exposed in a mountain range. For instance, a geologist may study the variations in metamorphic rocks in a mountainous area to determine the temperature, depth of burial, and nature of deformation during metamorphism. Geologists working on the "big picture," developing hypotheses of how major mountain belts evolve, might employ the published results of hundreds or thousands of local studies, using them as pieces of a puzzle. (Science works largely because scientists build on the work of others.) Models that currently are widely accepted regarding the evolution of mountain belts are cast within the broader framework of plate-tectonic theory and will be described later in this chapter.

Mountain belts differ from one another because each has undergone a unique combination of events that contributed to

FIGURE 20.1

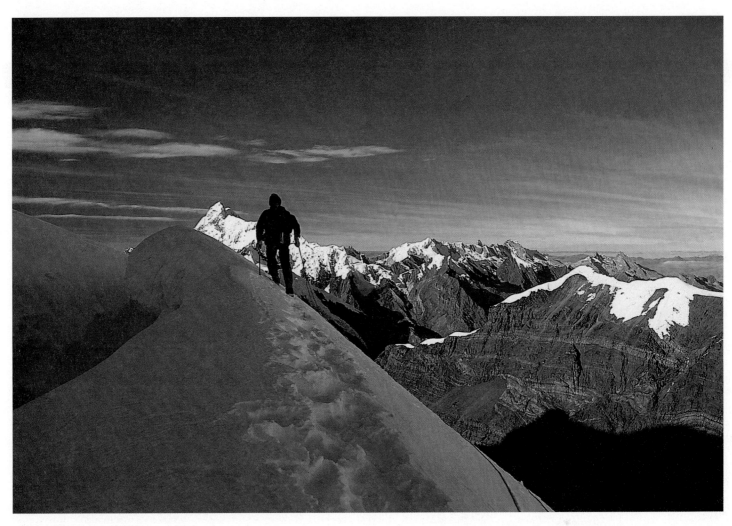

View of glaciated peaks in one of the mountain ranges in the Andes mountain belt. A parallel but much lower range is visible at the extreme right skyline of the picture. *©C. C. Plummer*

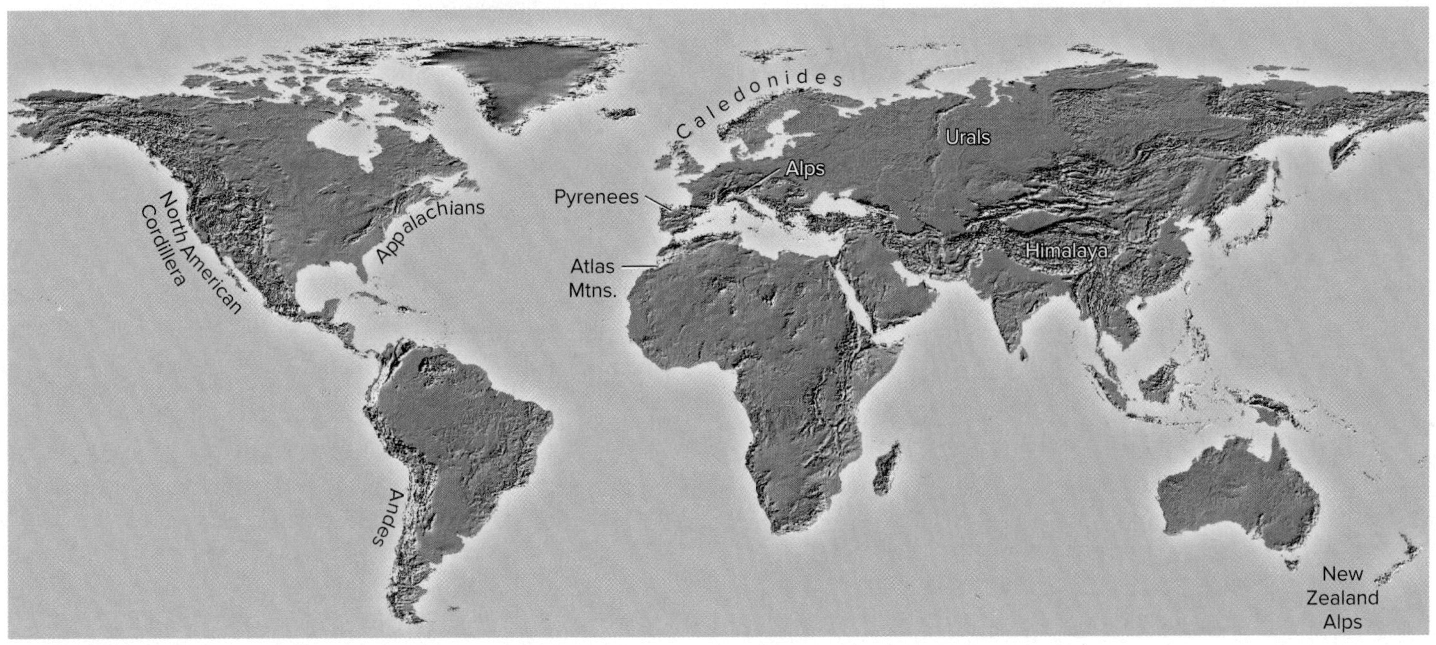

FIGURE 20.2

Map of the world showing major mountain belts.

EARTH SYSTEMS 20.1

An Earth Systems Approach to Understanding Mountains

During the last few decades, geologists have used an Earth systems approach to gain insight into the growth and wearing away of mountains. This approach regards mountains as products of three closely interdependent components: (1) tectonics (plate tectonics and isostasy), (2) climate, and (3) erosion. In other words, the atmosphere, hydrosphere, and geosphere all play a role in mountain building.

The tendency in the past has been to concentrate on tectonics to explain the growth of a mountain belt and to relegate climate and erosion to relatively minor roles. Through mountain system analyses, we gain insight into the extent to which each of the three components interacts with and changes the other two components. Climate influences erosion in obvious ways. For instance, if there is a wet climate, there will be erosion due to abundant running water at lower elevation and heavy glaciation at higher elevation. If the climate is arid, erosion will be much slower. Tectonics affects climate because if a region is uplifted to a high elevation, the climate there will be cold and glaciers can develop. With less uplift and lower mountains, erosion will be mainly due to running water. A moist climate can also result in heavy vegetation at lower elevations, which would tend to retard erosion.

Erosion and climate can influence tectonics as well. For example, the extent and type of erosion can help determine whether a highland grows higher or lower with time. If a high plateau, dissected by only a few valleys, undergoes erosion, the plateau is eroded downward uniformly (box figure 1). Following erosion, isostasy results in the plateau floating upward but not up to its original level. Its average surface, which essentially is its actual surface, is at a lower elevation than before erosion took place. If erosion carves many deep valleys and leaves relatively few mountains between the valleys, the entire regional block will have less mass and will float isostatically upward. As in the case of the plateau, its average surface would rise to a level lower than before erosion; however, its average surface is somewhere between the peaks and bottoms of valleys. Although the average height of the block rises to a level below its previous average height, the mountains rise to heights greater than before.

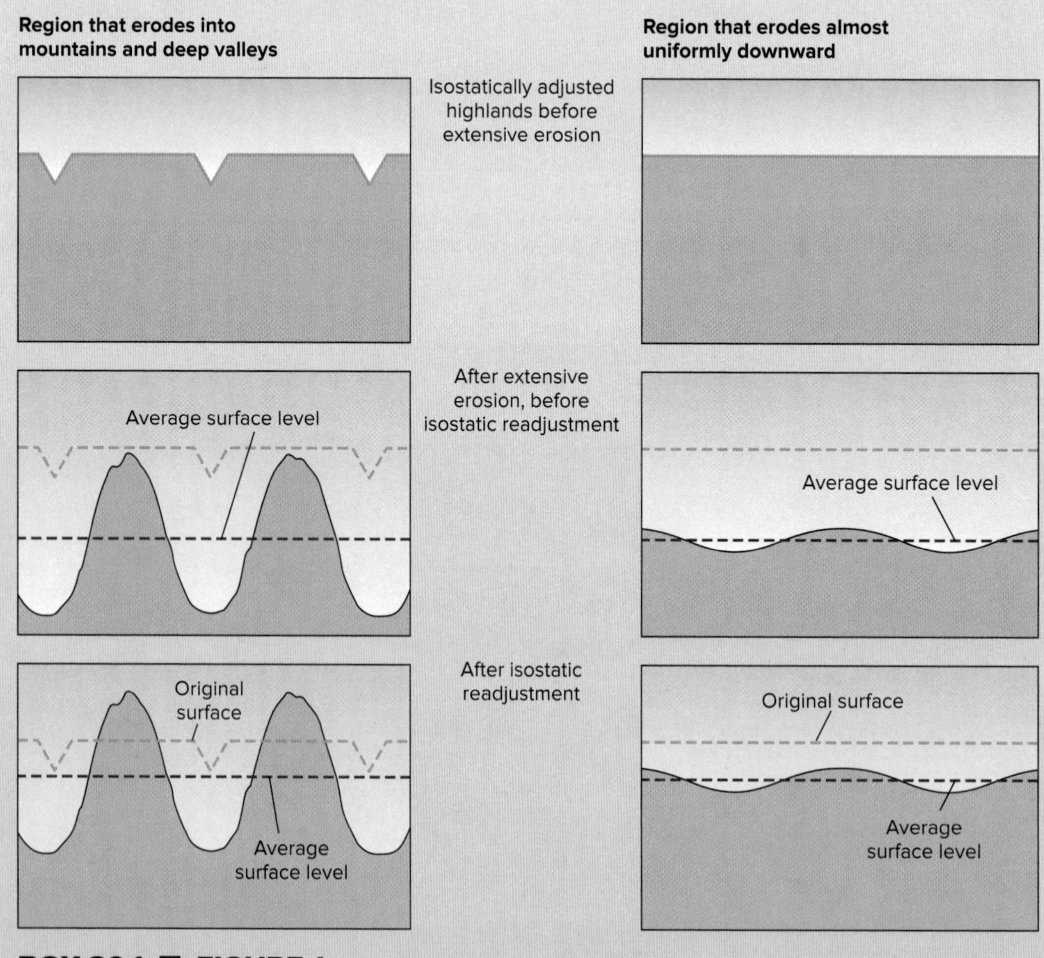

BOX 20.1 ■ FIGURE 1

Comparison between two regions before extensive erosion, after extensive erosion, and after isostatic readjustment. Region on the left erodes into mountains and deep valleys. Region on the right remains a plateau after approximately uniform erosion. Steepness of mountains is exaggerated.

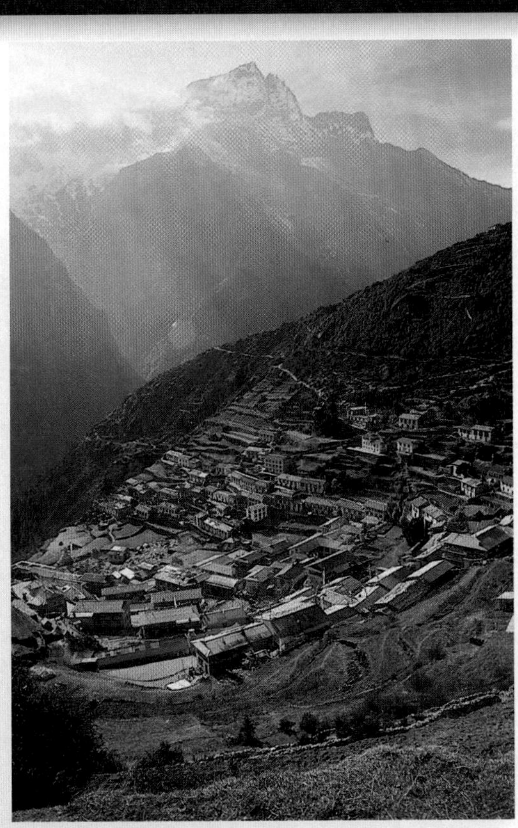

BOX 20.1 ■ FIGURE 2

Namche Bazaar in the Nepalese Himalaya. Mountain peaks rise thousands of meters above the town. Streams have carved deep valleys below the town.
©C. C. Plummer

Climate enters the picture because, interacting with tectonics, it helps control the type and extent of weathering that takes place. For instance, heavy precipitation takes place in the Himalaya because of the flow of very humid air from the south during the summer monsoon. At higher elevations, the precipitation in the form of heavy snowfall contributes to extensive and very active glaciation. As described in chapter 12 glaciers are extremely effective agents of erosion. Sharp peaks are separated by glacially carved valleys. At a lower level, streams fed by meltwater (and rainfall) deepen stream-carved valleys (box figure 2). The mountains will grow higher during isostatic adjustment at the same time the region as a whole is lowered by erosion. The Tibetan Plateau is north of the Himalaya. It is the highest, largest plateau in the world, with an average elevation of around 5 kilometers—higher than any mountain in the United States except in the state of Alaska. The plateau has not been carved into mountains because the climate is quite different from that of the Himalaya. The moist air from the south is blocked by the Himalaya, and the Tibetan Plateau is in its rain shadow (see chapter 13). Without water, there are no glaciers or large rivers to carve the plateau into mountains and deep valleys. So this region is slowly being eroded downward, getting progressively lower as erosion and isostasy balance each other out.

Additional Resource

N. Pinter and M. T. Brandon. How erosion builds mountains. *Scientific American* (April 1997): 74–79.

its present characteristics. The major controlling factors that interact with one another during a mountain belt's long history are:

- *Intense deformation.* This is mainly compression and results in intense folding and faulting of sedimentary and volcanic rocks. At depth, deformation results in foliation accompanying metamorphism. Such an episode (usually lasting millions of years) of intense deformation is known as an **orogeny**. We now attribute orogenies mainly to plate convergence.
- *Isostasy.* Vertical movement of mountain belts, both during and after an orogeny, is accounted for by isostasy (described in chapters 1 and 17). Isostatic adjustment means that thicker continental crust tends to "float" higher on the mantle than does thinner crust.
- *Weathering and erosion.* The rate and nature of weathering and erosion are affected by many factors, such as the climate, type of rock, and height of a landmass above sea level. See box 20.1 for a discussion of how the interaction of the geosphere, atmosphere, and hydrosphere plays a role in forming mountains.

20.2 CHARACTERISTICS OF MAJOR MOUNTAIN BELTS

Size and Alignment

Major mountain belts are very long compared to their width. Figure 20.3 shows the two major mountain belts of North America, the *Appalachian Mountains* along the East Coast and the *North American Cordillera* in the West. Some of the better known ranges in the North American Cordillera, such as the Sierra Nevada and the Rocky Mountains, are labeled. Note that the mountain belts in North America tend to be parallel to the coastlines. However, some mountain belts elsewhere in the world (most notably the Himalaya) are not parallel to a coastline.

Ages of Mountain Belts and Continents

Although individual ranges within a mountain belt may vary considerably in height, major mountain belts with higher mountain ranges tend to be geologically younger than those where mountains are lower. The two major mountain belts of the North American continent, the Appalachian Mountains and the North American Cordillera (figure 20.3), provide a good example. The Appalachians are topographically much less prominent than the ranges of the North American Cordillera, which have many peaks of over 4,000 meters. Fossils and isotopic ages of rocks indicate that the Appalachian mountains began to evolve much earlier than the North American Cordillera. Other than isostatic adjustment, mountain building in the Appalachians ceased around 250 million years ago, while uplift continues today in some parts of the North American Cordillera.

Mountain regions commonly show evidence that they were once high above sea level during an orogeny, were eroded to hills

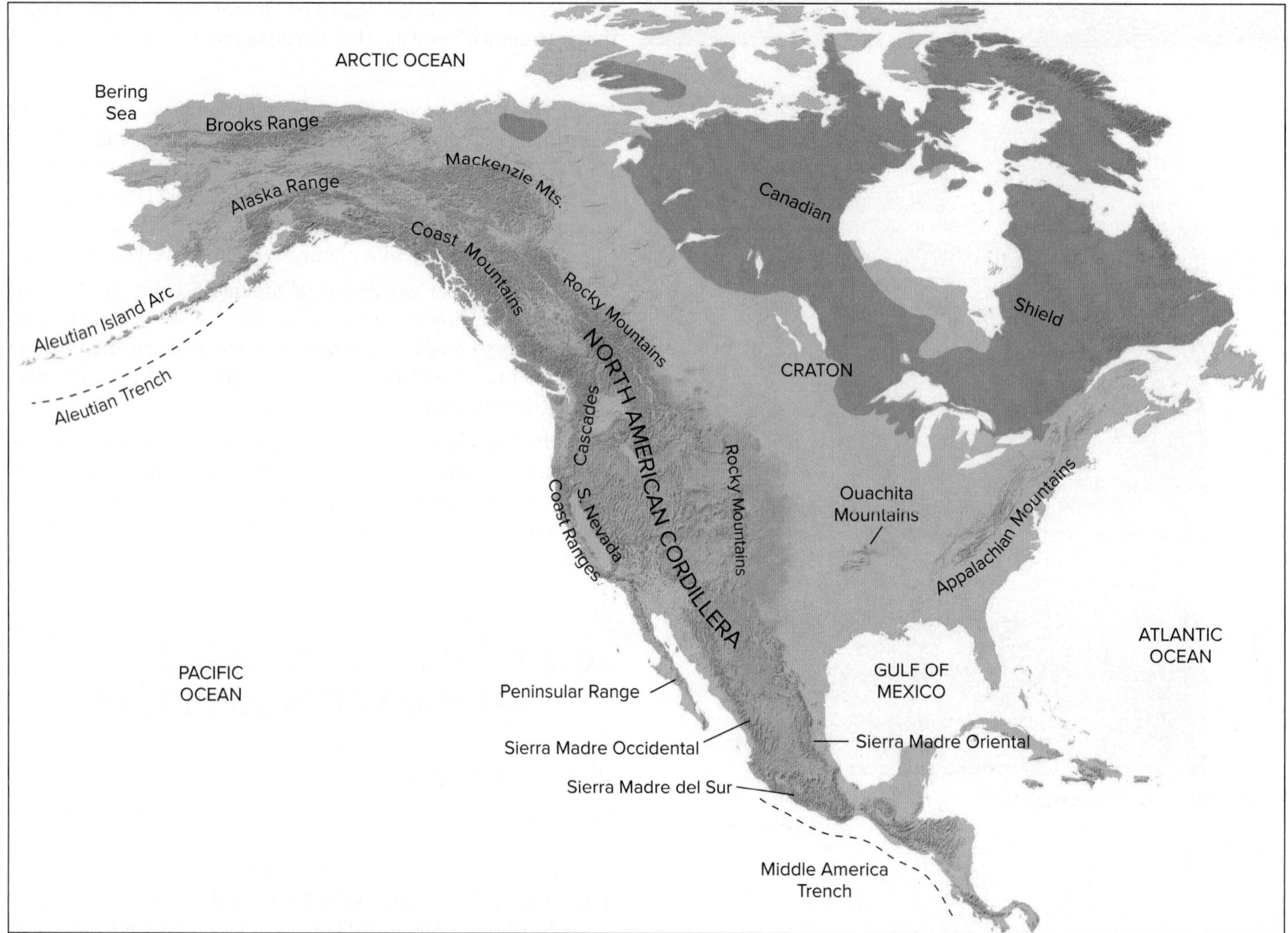

FIGURE 20.3

The mountain belts and craton (including Canadian Shield) of North America. Major ranges in the Cordillera are labeled.

or low plains, and then rose again in a later episode of isostatic uplift. Such episodes of uplift and erosion may occur a number of times during the long history of a mountain range. Ultimately, mountain ranges stabilize and are eroded to plains.

The interior plains between the Appalachians and the Cordillera are considered to have evolved from mountain belts in the very distant geologic past (during the Precambrian). The once deep-seated roots of the former Precambrian mountain belts are the *basement* rock for the now stable, central part of the continent. Layers of Paleozoic and younger sedimentary rock cover most of that basement. The great age of the orogenic episodes that preceded the Paleozoic sedimentation is confirmed by isotopically determined dates of over 1 billion years obtained from plutonic and metamorphic rocks in the few scattered locations where the basement is exposed. (The most noteworthy are the Grand Canyon in Arizona, the Ozark dome in Missouri, the Black Hills of South Dakota, some ranges in the Rocky Mountains, and the Adirondacks in New York.) The region of a continent that has been structurally stable for a prolonged period of time is called a **craton** (figures 20.3, 20.4, and 20.5). The central part of the United States and Canada is all part of a craton. Other continents similarly have a craton at their core.

Most of the craton in the central United States has a very thin blanket–only 1,000 to 2,000 meters–of sedimentary rock layers overlying its basement. Sediment was mostly deposited in shallow inland seas during Paleozoic time; however, for the craton in much of eastern and northern Canada, no sedimentary rocks cover the eroded remnants of old mountain ranges. This region is a **Precambrian shield**–that is, a complex of Precambrian metamorphic and plutonic rocks exposed over a large area. Such shields and basement complexes of cratons represent the roots of mountain ranges that completed the deformation process more than a billion years ago.

Thickness and Characteristics of Rock Layers

In sharp contrast to the relatively thin cover of sedimentary rock overlying the basement in a craton is the thick sedimentary sequence typical of mountain belts (figure 20.4). In mountain

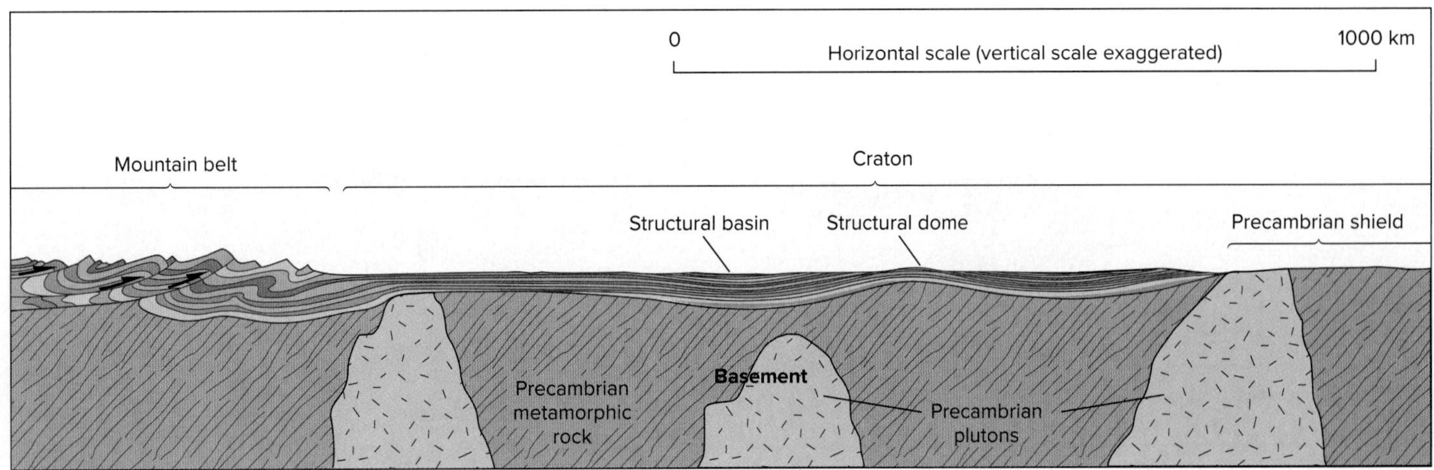

FIGURE 20.4

Schematic cross section through part of a mountain belt (left) and part of a continental interior (craton). Vertical scale is exaggerated.

FIGURE 20.5

Satellite image of part of a craton in Western Australia. Metamorphic rock (dark gray) that is 3 billion to 3.5 billion years old surrounds oval-shaped domes of granite and gneiss (white) that are 2.8 billion to 3.3 billion years old. Gently dipping sedimentary and volcanic rocks (tan and reddish) unconformably overlie the granite-metamorphic basement complex. The area is 400 kilometers across. *Landsat mosaic produced by Satellite Remote Sensing Services, Landgate, Western Australia*

belts, layered sedimentary rock commonly is more than 10 kilometers thick. We now know that these thick sequences of mostly marine-deposited sedimentary rock were originally deposited on continental margins (continental shelf and continental slope–see chapter 18). If the sedimentary rock is mainly shale, sandstone, and limestone, we can infer that marine deposition was at a passive continental margin. If the sediment has a significant component of volcanic material, the depositional environment was an active continental margin.

The sedimentary rock in cratons may show no deformation, or it may have been gently warped into basins and domes above the basement (figure 20.4). By contrast, mountain belts are characterized by a variety of folds and faults that indicate moderate to very intense orogenic deformation.

Patterns of Folding and Faulting

Reconstructing the original position and determining the original thickness of layers of sedimentary and volcanic rock in mountain belts are complicated because, in most instances, the layered rocks have been folded and faulted after they were deposited. As you learned in chapter 15, geologic structures can be used to interpret the direction and magnitude of the tectonic forces that formed them. (Refer to figure 20.6 as you read through the following paragraphs.) Folds will be open in those parts of a mountain belt where deformation was not very intense. Tighter folds (figure 20.7) indicate greater deformation. Large overturned and recumbent folds (figure 20.8) may be exposed in more intensely deformed portions of mountain belts. Reverse faults are common, particularly in the intensely folded regions. Especially noteworthy are the **fold and thrust belts** found in many mountainous regions. These are characterized by large thrust faults (reverse faults at a low angle to horizontal), stacked one upon another; the intervening rock usually was folded while it was being transported during faulting.

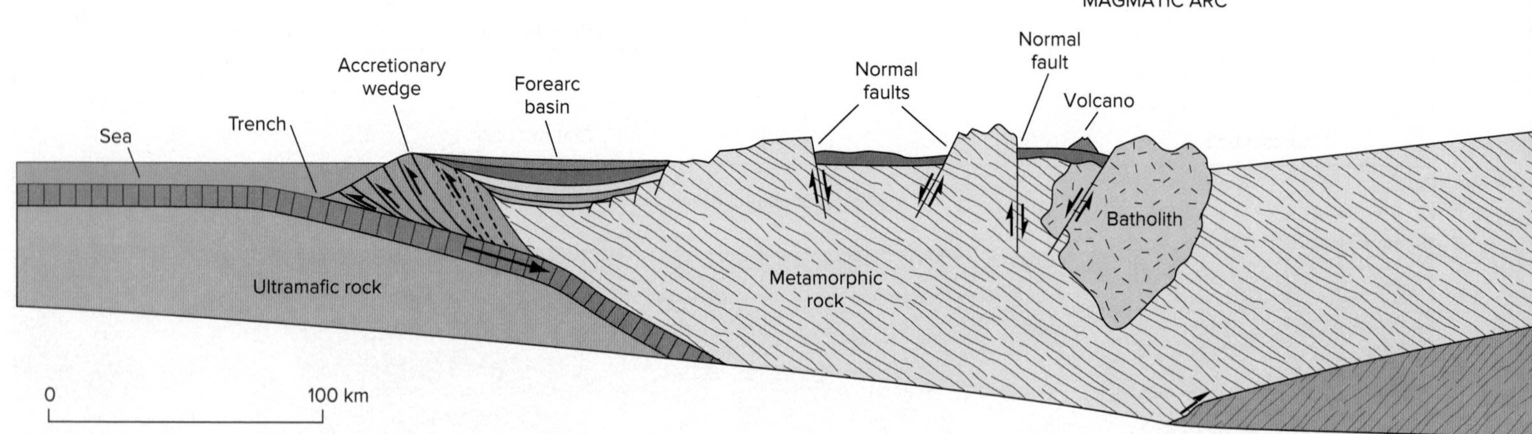

FIGURE 20.6

Cross section of an "Andean type" mountain belt; that is, one whose orogeny is due to oceanic-continental convergence. For simplicity, only a few of the many layers of sedimentary rock are shown. The size of some features is exaggerated for illustrative purposes.

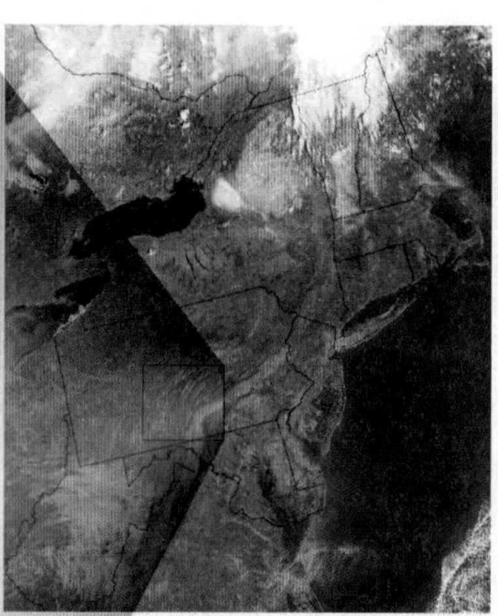

FIGURE 20.7

False-color satellite image of part of the Valley and Ridge province of the Appalachian mountain belt, near Harrisburg, Pennsylvania. The ridges are sedimentary beds resistant to erosion. The pattern indicates tight (ridges close together) and open (ridges farther apart) folding occurred prior to erosion.

Source: Jacques Descloitres, MODIS Land Rapid Response Team, NASA/GSFC

Overall, the folds and thrust faults in a mountain belt suggest tremendous squeezing or *crustal shortening* and *crustal thickening.* The sedimentary rocks of the Alps, for instance, are estimated to have covered an area of ocean floor about 500 kilometers wide when they were deposited. During the Alpine orogeny, they were compressed into the present width of the Alps, which is less than 200 kilometers (see figure 20.13).

Metamorphism and Plutonism

A complex of regional metamorphic and plutonic rock is generally found in the mountain ranges of the most intensely deformed portions of major mountain belts. Most of the metamorphic rocks were originally sedimentary and volcanic rocks that had been deeply buried and subjected to intense stress and high temperature during an orogeny. *Migmatites* (intermixed granitic and metamorphic rock, such as that shown in figure 7.19) may represent those parts of the mountain belts that were once at even deeper levels in the crust, where higher temperatures caused partial melting of the rocks (as described in chapters 3 and 7). These must have been transported into much higher levels of the crust during and after an orogeny. Batholiths, largely granitic, also have their origin in the lower crust (or uppermost mantle). Rather than remaining behind and forming migmatites, the magma generated from partial melting collects in large blobs (diapirs) that work their way upward into an upper level of Earth's crust.

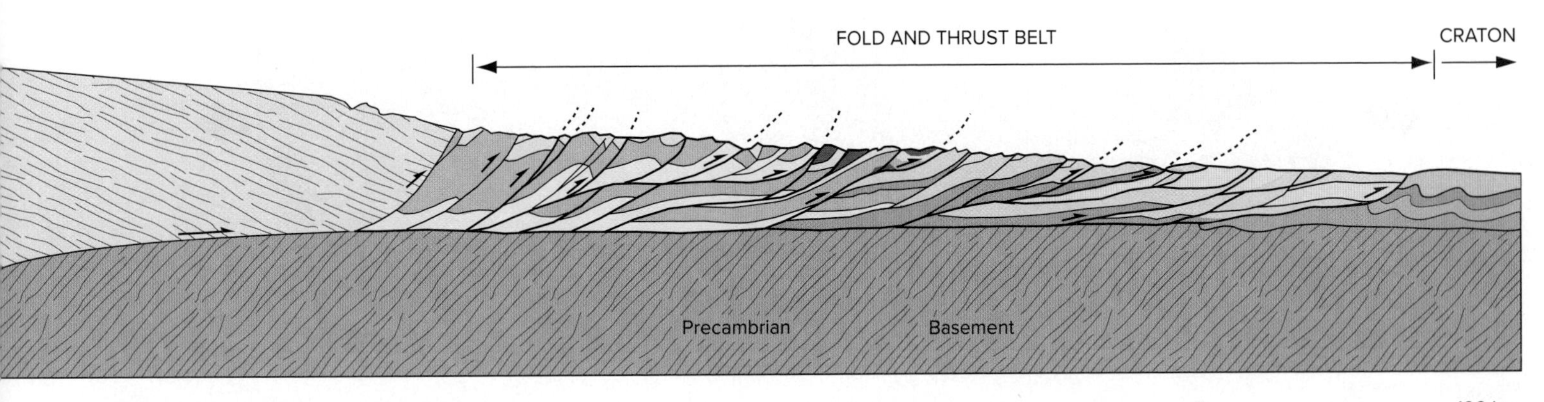

Recumbent folds
Geologist's View

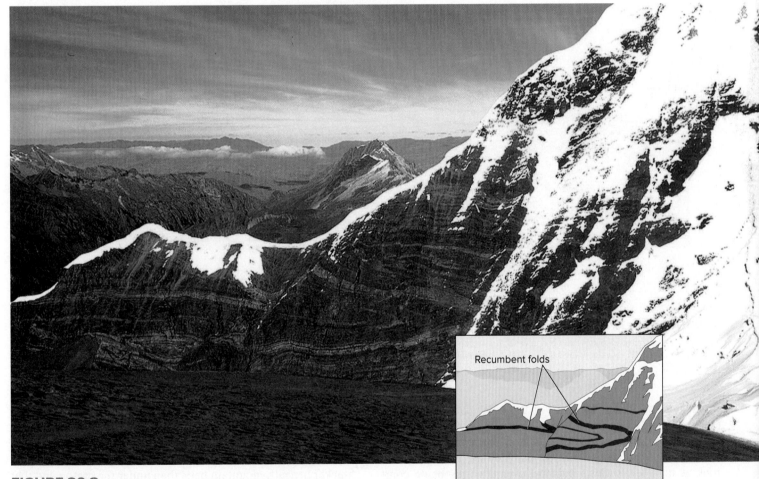

FIGURE 20.8
Recumbent folds exposed on a mountainside in the Andes.
©C. C. Plummer

Normal Faulting

Older portions of some major mountain belts have undergone normal faulting (figure 20.9). Cross-cutting relationships show that the normal faulting occurs after the orogeny that resulted in tight folding, thrust faulting, and metamorphism, and after most batholiths had formed. This late stage of normal faulting (described in chapter 15 on geologic structures) is a result of *vertical uplift* or *horizontal extension.* Either of these contrasts with the overall shortening that prevailed during the orogeny.

Normal faulting may also take place in the high, central part of a mountain belt during an orogeny, while folding and

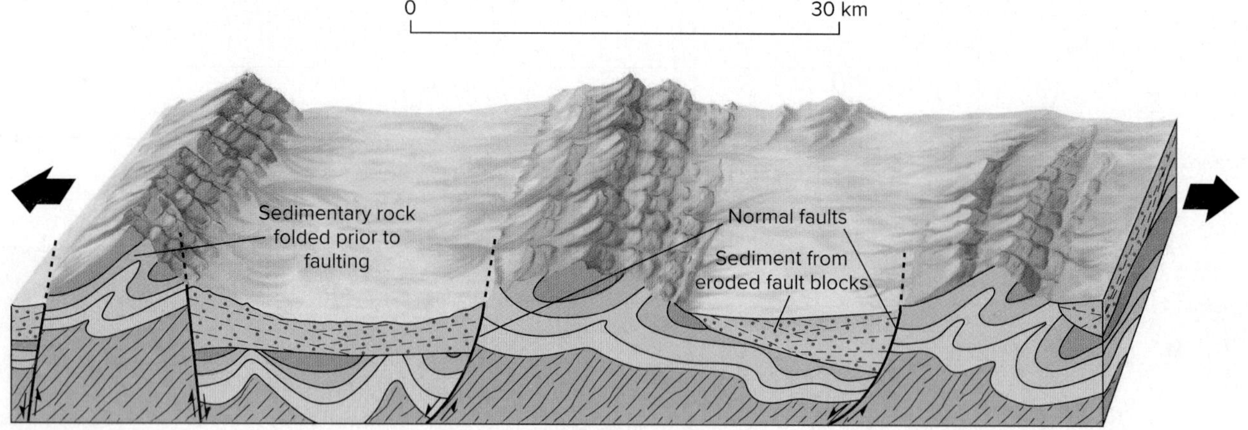

FIGURE 20.9

Fault-block mountains with movement along normal faults.

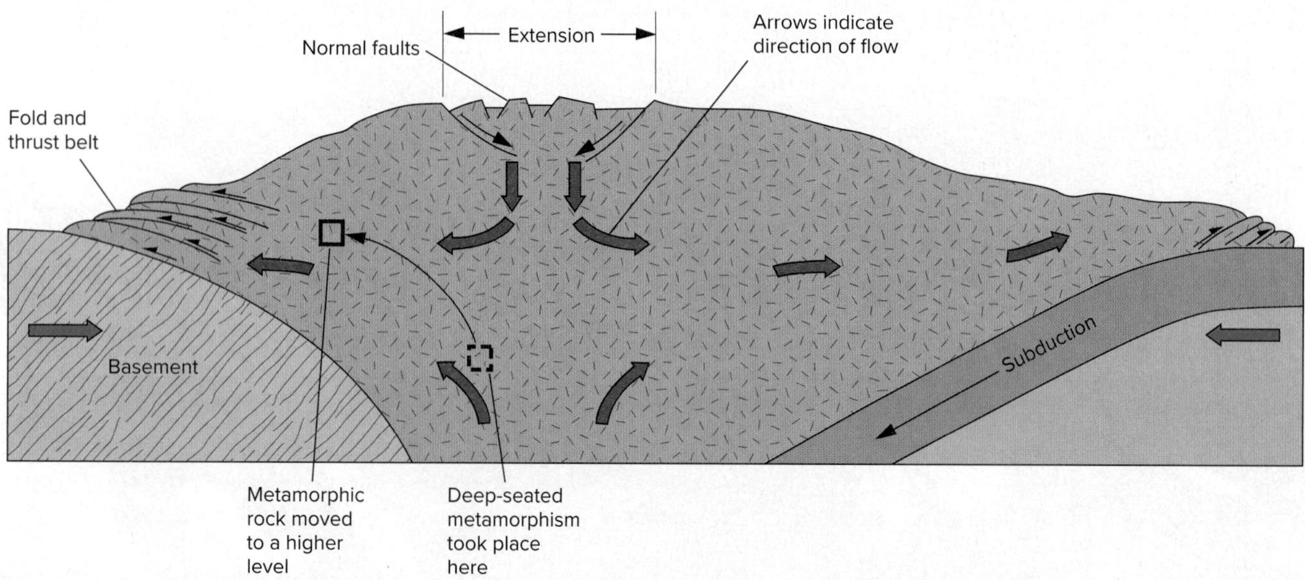

FIGURE 20.10

Schematic cross section of a mountain belt in which gravitational collapse and spreading are taking place during plate convergence. Red arrows indicate flowage of rock. Faulting occurs in brittle rock near the surface. Rock that was metamorphosed at depth flows to a higher level in the mountain belt. Not drawn to scale.

thrust faulting are taking place at the outer parts of the belt (see figure 20.10). As the mountain belt is being compressed and shortening takes place, the central portion is pushed upward. Extension, along with normal faulting, takes place as rock at high levels flows outward over the rock being compressed at the lower level.

Thickness and Density of Rocks

Geophysical investigations yield additional information about mountains and the continental crust. As discussed in chapter 17 about Earth's interior, gravity measurements indicate that the rocks of the continental crust (including mountain belts) are lighter (less dense) than those of the oceanic crust. Seismic velocities indicate a composition approximating that of granite for continental crust. Furthermore, evidence from seismic studies supports the view that this lighter crust is much thicker beneath mountain belts than under the craton and that the crust is thicker under younger mountain belts than under older ones.

Features of Active Mountain Ranges

Frequent earthquakes are characteristic of portions of mountain belts that are geologically young and considered still active. Also, deep-ocean trenches are found parallel to many young

IN GREATER DEPTH 20.2

Ultramafic Rocks in Mountain Belts—From the Mantle to Talcum Powder

Ultramafic rocks (described in chapter 3 on intrusive rocks) occur commonly in the portions of mountain belts occupied by metamorphic and plutonic rocks. Ultramafic rocks tend to crop out in long, narrow zones that parallel the trend of a mountain belt. Geologists regard most bodies of ultramafic rocks as being former mantle material that was faulted into the crust during the mountain-building process. Some of the ultramafic bodies are found associated with marine-deposited volcanic and sedimentary rocks in an *ophiolite sequence* (described in chapter 18 about the sea floor). An ophiolite is regarded as a segment of a former oceanic crust together with its underlying mantle.

Ultramafic rocks in mountain belts commonly show the effect of the metamorphism that has altered adjacent rock units. Two of the foliated metamorphic products of ultramafic rocks are of special interest. One is *serpentinite*, a rock composed of the mineral serpentine. Another is a rock composed mostly of the mineral talc, commonly known as *soapstone*. Serpentine and talc are both hydrated magnesium silicate minerals. They are products of metamorphic recrystallization of ultramafic rock when water is present. Metamorphism takes place in the crust under cooler and lower-pressure conditions than those under which the original ultramafic rock formed in the mantle.

Serpentinite is a shiny, mottled, dark green and black rock that looks rather like a snake's skin. It splits apart easily along irregular, slippery, foliation surfaces. Hillsides or slopes with serpentinite as bedrock are sparsely vegetated because constant sliding prevents soil and vegetation from building up. Houses built on serpentinite hillsides (by people without a knowledge of geology) also slide downslope. Serpentinite is the official state rock of California—a state in which a large number of homes have been destroyed because they were built on sliding hillsides. (Serpentinite, however, is seldom to blame.)

BOX 20.2 ■ FIGURE 1

Soapstone (talc) sculpture, seal hunter, by Inuit artist Paulossie Weetaluktuk. *©Danita Delimont/Alamy*

Soapstone, which is less common than serpentinite, is valuable mainly because of talc's softness (number 1 on Mohs scale). Many sculptures (most notably, Inuit carvings; box figure 1) are made from soapstone because of the ease with which it can be cut. The best-known product of talc, however, is talcum powder.

mountain belts (the Andes, for example). Trenches lie off the coasts of island arcs, which can be regarded as very young mountain ranges. Isolated active volcanoes perched on top of older rock in a mountain range suggest that melting is still taking place at depth.

20.3 EVOLUTION OF MOUNTAIN BELTS

Orogenies and Plate Convergence

As described earlier, an orogeny is an episode of intense deformation of the rocks in a region; the deformation is usually accompanied by metamorphism and igneous activity. Layered rocks are compressed into folds. Reverse faulting (especially thrust faulting) is widespread during an orogeny. Normal faulting may also occur but is not as widespread.

The more deeply buried rocks are subjected to regional metamorphism and are converted to schists and gneisses (see chapter 7). Magma generated in the deep crust or the upper mantle works its way upward to erupt in volcanoes or form batholiths (see chapter 3).

One important aspect of an orogeny is that the continental crust becomes thicker. This is achieved by the intense compression that results in tight folds and reverse faults. The addition of batholiths in the crust also helps thicken the crust and make it more buoyant. The thicker crust will isostatically "float" higher on the underlying mantle, resulting in higher mountains.

Each mountain belt has its own characteristics and history. However, by understanding which one of three kinds of plate convergence took place (described in chapter 19), we can better understand the mountain-building processes that a mountain belt underwent during an orogeny. The three types of convergence are discussed next.

Orogenies and Ocean-Continent Convergence

Figure 20.6 shows a hypothetical mountain belt that has ongoing ocean-continent convergence. The Andes, in which the South American plate is overriding the Nazca plate, is an example, and this type of mountain belt is often referred to as *Andean-type.*

Figures 3.28 and 7.21 show igneous and metamorphic processes during oceanic-continental convergence. Plate convergence also accounts for the folded and reverse-faulted layered rocks found in mountain belts. An *accretionary wedge* develops where newly formed layers of marine sediment are folded and faulted as they are scraped off the subducting oceanic plate (see figure 19.27 and explanation in chapter 19).

Rock caught in and pulled down the subduction zone is subjected to intense shearing. If rock is carried farther down the subduction zone, it becomes metamorphosed (as described in chapter 7).

Fold and thrust belts may develop on the craton (backarc) side of the mountain belt (figures 20.6 and 20.10). Thrusting is away from the magmatic arc toward the craton. The magmatic arc is at a high elevation because the crust is thicker and composed largely of hot igneous and metamorphic rocks. The large thrust sheets move toward and sometimes over the craton. (In the Rocky Mountains, thrust faulting of the craton itself has taken place.) The thrusting probably is largely due to the crustal shortening caused by convergence. There is, however, some controversy among geologists over additional processes that may take place. Some geologists regard gravity flow (from the high and mobile magmatic arc outward over the low and rigid craton) as contributing significantly to the process. Others think that the expanding magmatic arc pushes the sedimentary (and sometimes metamorphic and igneous) rocks outward to become the fold and thrust belt. (The magmatic arc is likened to a bulldozer pushing a wedge of loose material outward.)

In the late 1980s and early 1990s, geologists developed a model that explains (1) fold and thrust belts, (2) simultaneous normal faulting, and (3) how once deep-seated metamorphic rocks rise to an upper level in a mountain belt. What is believed to occur is that the thick and high part of the mountain belt becomes too high and gravitationally unstable, resulting in **gravitational collapse and spreading**. The mobile portion becomes increasingly elevated during plate convergence. This is due to compression of sedimentary and metamorphic rocks, as well as to volcanic eruptions and emplacement of plutons. After some time, the welt in the mountain belt becomes too high to be supported by the underlying rocks, and collapse begins. (Geologist John Dewey, then at Oxford University, suggested that collapse begins when the welt exceeds 3 kilometers above sea level.) As shown in figure 20.10, the gravitational collapse forces rock outward as well as downward. At deeper levels in the mountain belt, the rock is *ductile* (or *plastic*) and flows; nearer the surface, rock fractures, so movement is through faulting. The rock is pushed outward and helps create, along with crustal shortening, the fold and thrust belt.

In the high part, the outward flowing rock results in extension (figure 20.10). Therefore, the brittle, near-surface rocks fracture, and normal faulting takes place.

The flowage pattern (as shown in figure 20.10) can also explain how once deep-seated metamorphic rocks (migmatites, for example) are found in upper levels of a mountain belt. Lower crustal rocks are squeezed, forcing them to flow upward and outward, bringing them closer to the surface.

Arc-Continent Convergence

Sometimes an island arc collides with a continent (figure 20.11). As the arc and continent converge, the intervening ocean floor is destroyed by subduction. When collision occurs, the arc, like a continent, is too buoyant to be subducted. Continued convergence of the two plates may cause the remaining sea floor to break away from the arc and create a new site of subduction and

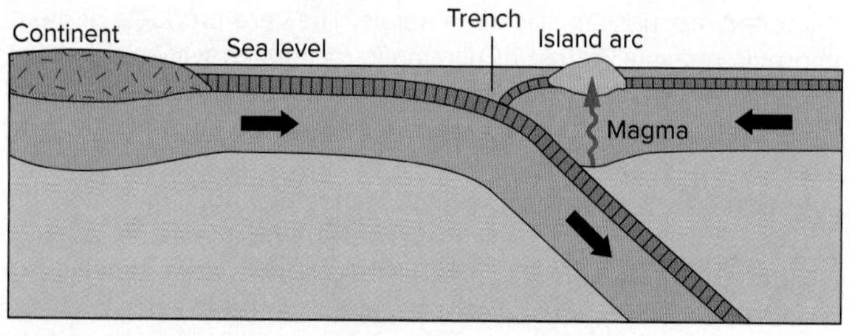

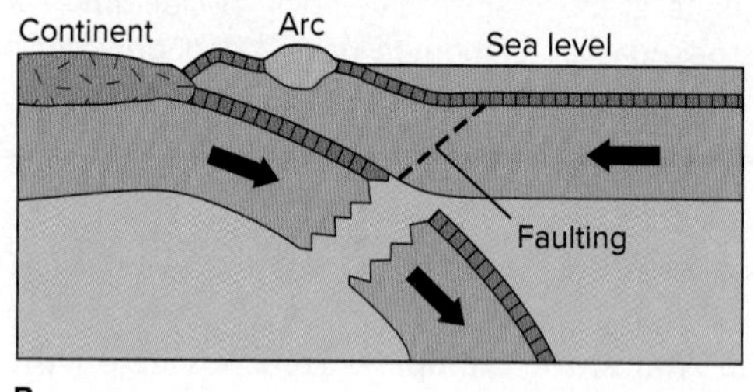

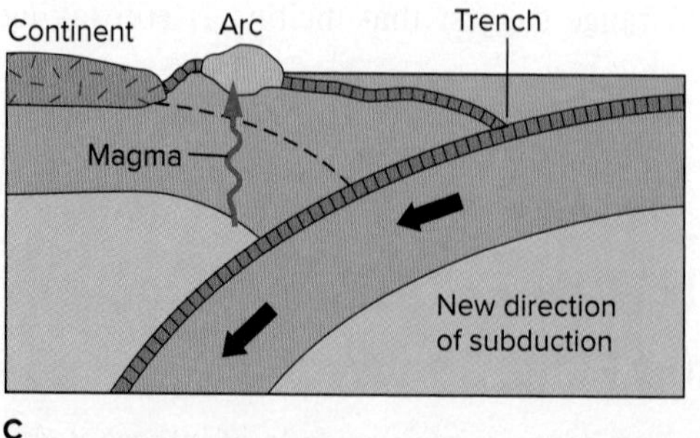

FIGURE 20.11

Arc-continent convergence can weld an island arc onto a continent. The direction of subduction can change after impact.

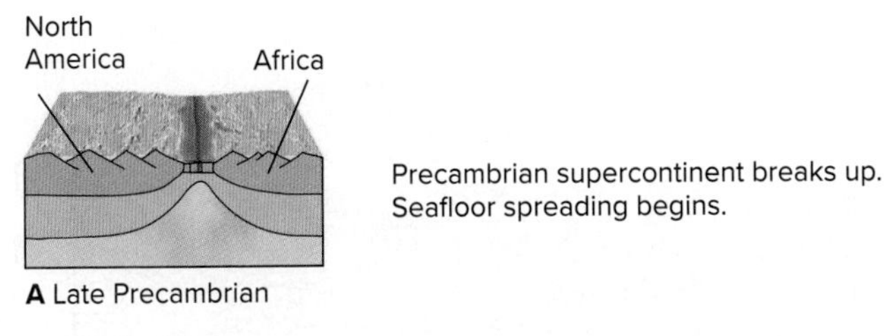

A Late Precambrian

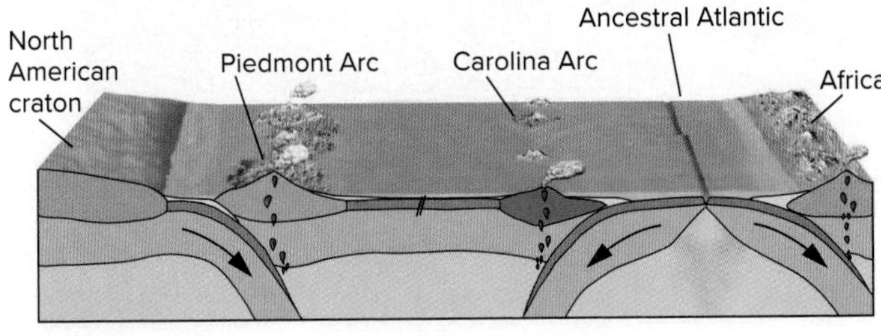

B Cambrian

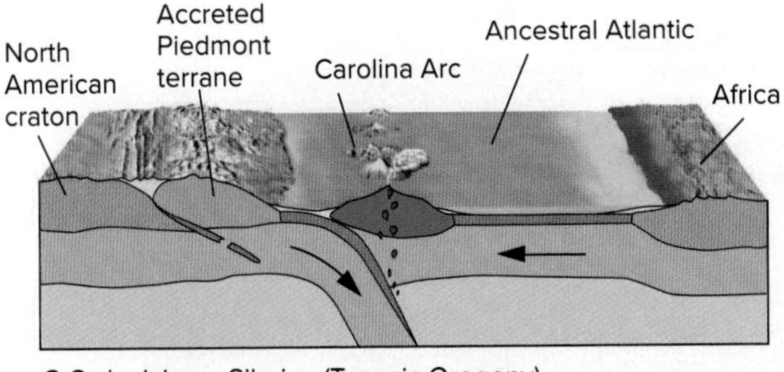

C Ordovician – Silurian (Taconic Orogeny)

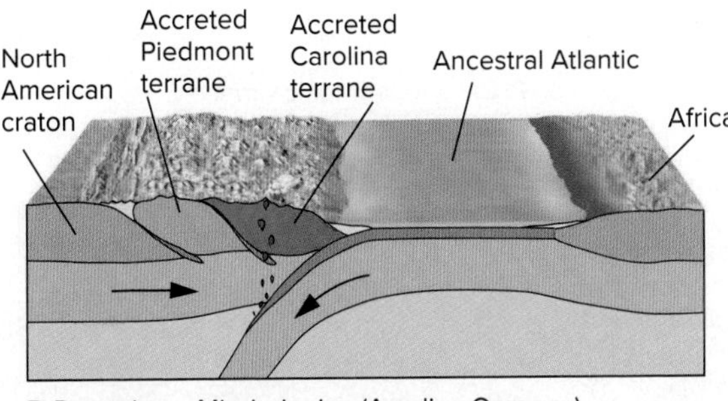

D Devonian – Mississippian (Acadian Orogeny)

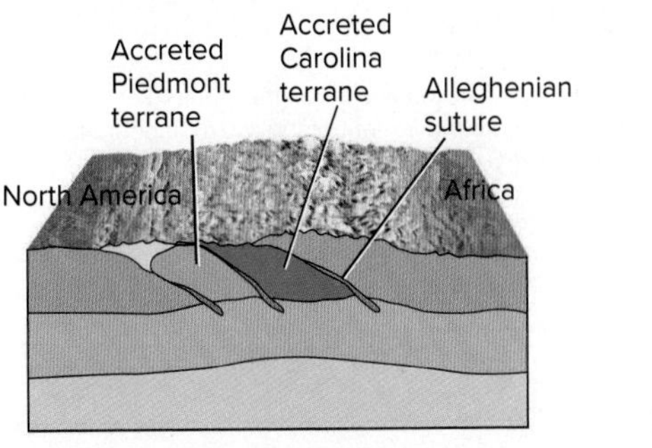

E Pennsylvanian – Permian (Alleghenian Orogeny)
North America and Africa joined

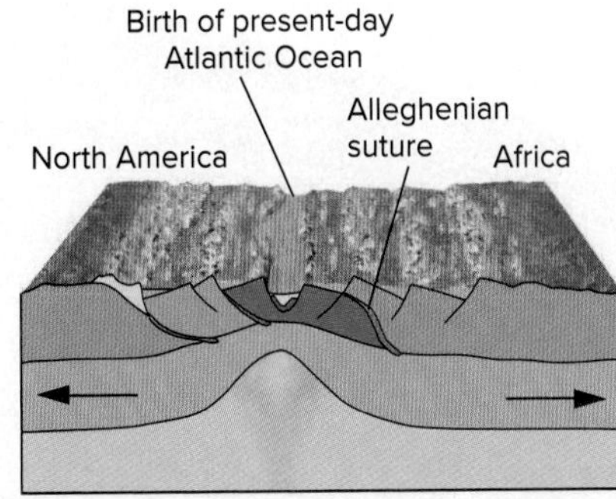

F Triassic – Rifting begins. Breakup of Pangaea starts

FIGURE 20.12

The geologic evolution of the southern Appalachians. *R. D. Hatcher, Jr. "Tectonic synthesis of the U.S. Appalachians," Geology of North America*: An Overview, vol. F-2, 1989, Figure 9 (modified). *See also U.S. Geological Survey pamphlet* Birth of the Mountains. The Geologic Story of the Southern Appalachian Mountains. pdf version available through http://pubs.usgs.gov/gip/birth/birth.pdf

a new trench seaward of the arc (figure 20.11*C*). Note that the direction of the new subduction is opposite the direction of the original subduction (this is sometimes called a *flipping subduction zone*), but it still may supply the arc with magma. The arc has now become welded to the continent, increasing the size of the continent.

This type of collision apparently occurred during recent geologic time in northern New Guinea (north of Australia). A similar collision may have added an island arc to the Sierra Nevada complex in California during Mesozoic time, when a subduction zone may have existed in what now is central California. Many geologists think that much of westernmost North America has formed from a series of arcs colliding with North America (discussed later in this chapter in the "Displaced Terranes" section). During the Paleozoic, arc-continent collision played a significant role in the building of the Appalachians (figure 20.12).

Orogenies and Continent-Continent Convergence

As described in chapter 19 (see figure 19.28), some mountain belts form when an ocean basin closes and continents collide along a suture zone. Mountain belts that we find within continents (with cratons on either side) are believed to be products of continent-continent convergence. The Ural Mountains resulted from the collision of Asia and Europe. Convergence of the African and European plates created the Alps (figure 20.13). Our highest mountains are in the Himalayan belt. The Himalayan orogeny started around 45 million years ago as India began colliding with Asia. (India was originally in the Southern Hemisphere.) The thick sequences of sedimentary rocks that had built up on both continental margins were intensely faulted and folded. Fold and thrust belts developed and were carved by erosion into the mountain ranges that make up the Himalaya. The mountains are still rising, and frequent earthquakes indicate continuing tectonic activity. The magnitude 7.8 Nepal earthquake that killed almost 9,000 people in April 2015 occurred on the Main Frontal thrust fault (see figure 20.14) and is an example of this ongoing seismic activity. North of the Himalaya, Tibet rose to become what is now the highest plateau in the world. Normal faults in the Tibetan Plateau tell us that gravitational collapse is taking place.

The rate and type of erosion that takes place during an orogeny influences the height as well as the shape of mountains. The influence of climate during *isostatic uplift* of the Himalaya and Tibetan Plateau is described in box 20.1. Intense erosion also

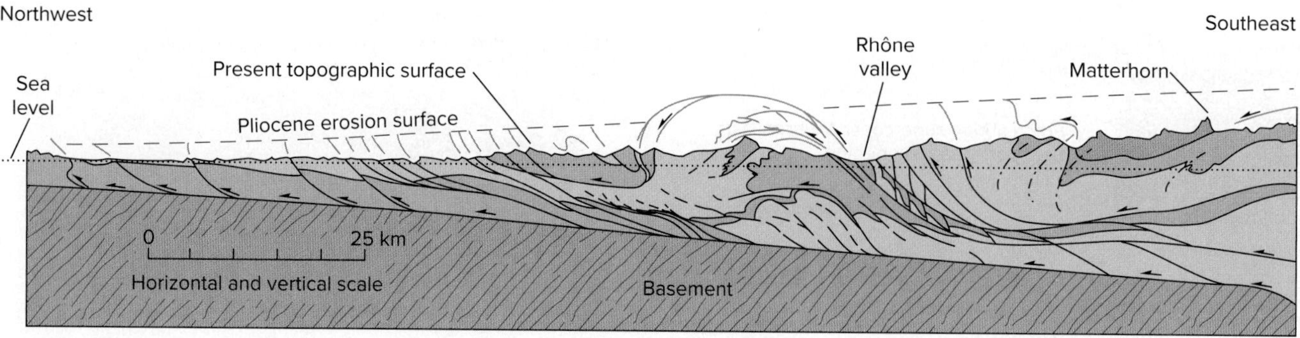

FIGURE 20.13

Cross section through part of the Alps. Thicker lines are thrust faults. Lesser folds are not shown. Movement is from the right to the left of the diagram (southeast to northwest). Only a few arrows are shown to indicate movement of the overriding thrust block.
Source: S. E. Boyer and D. Elliot, AAPG Bulletin, 1982.

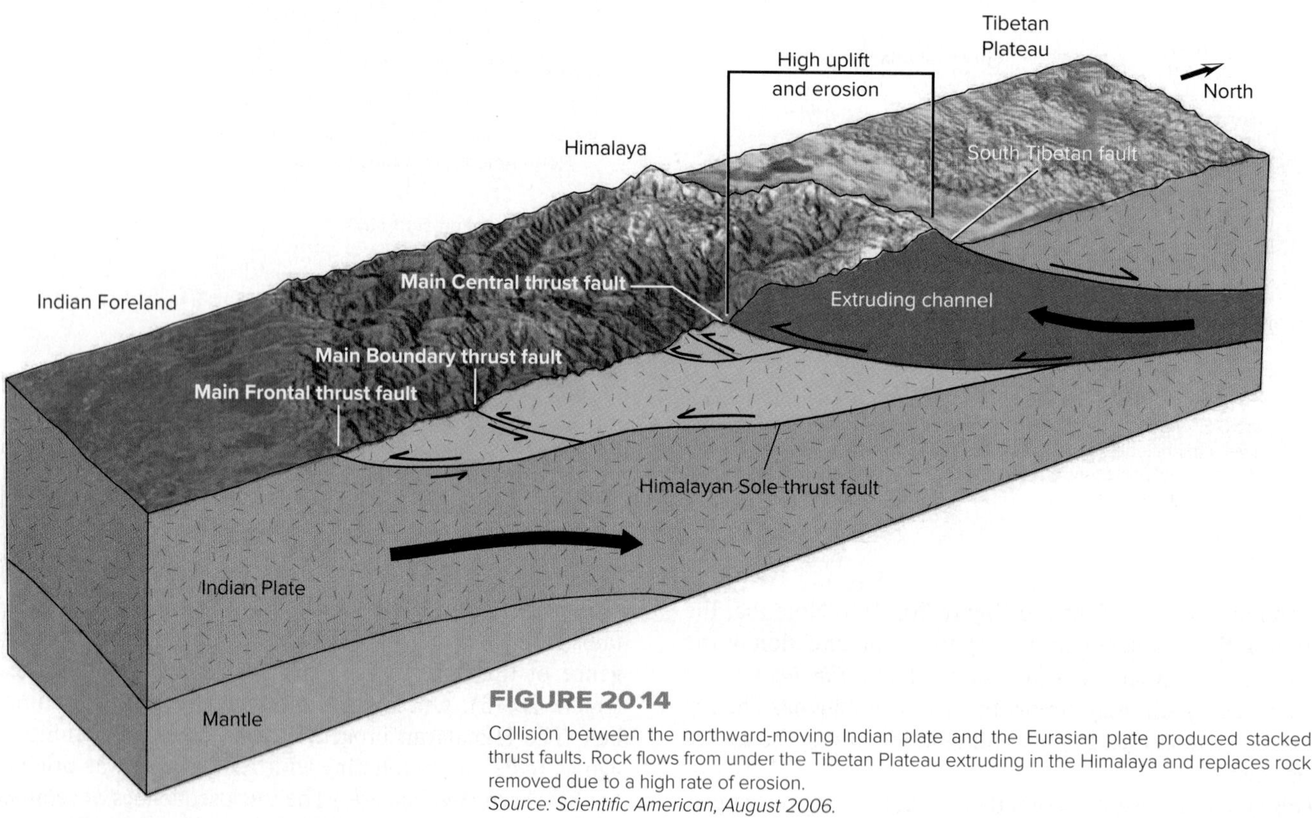

FIGURE 20.14

Collision between the northward-moving Indian plate and the Eurasian plate produced stacked thrust faults. Rock flows from under the Tibetan Plateau extruding in the Himalaya and replaces rock removed due to a high rate of erosion.
Source: Scientific American, August 2006.

interacts with *tectonic forces* to expedite the growth of the high Himalaya, according to a study published in 2006. Figure 20.14 shows thrust faults stacked one upon another. These were produced by the collision of India with Asia. The overall motion of the Himalaya at present is northward, as shown in figure 20.15. However, in the high Himalaya, rock is flowing underground and replacing rock eroded away at the surface. This is the block between the South Tibetan fault and the Main Central thrust fault shown in figure 20.14. According to the hypothesis, the rock here is being extruded from deep under the Tibetan Plateau due to the ongoing collapse of the Tibetan Plateau. Rock that is ductile (or plastic) at depth is squeezed and extruded toward a zone of least resistance. The zone of least resistance is where erosion is taking place at an exceptionally high rate because of the high precipitation of rain and snow during monsoons on the steep slopes of the high Himalaya.[1]

The collision of India and Asia has affected parts of Asia well beyond the Himalaya and Tibet. Figure 20.15 shows present-day crustal motion in and around the Himalaya and Tibet. As India continues to push northward, some of the motion is deflected from Tibet eastward into China. Box 19.1, figure 1 in the plate tectonics chapter shows major fault systems extending through China and neighboring countries and their relationships to the ongoing plate collision. China's disastrous 2008 earthquake was

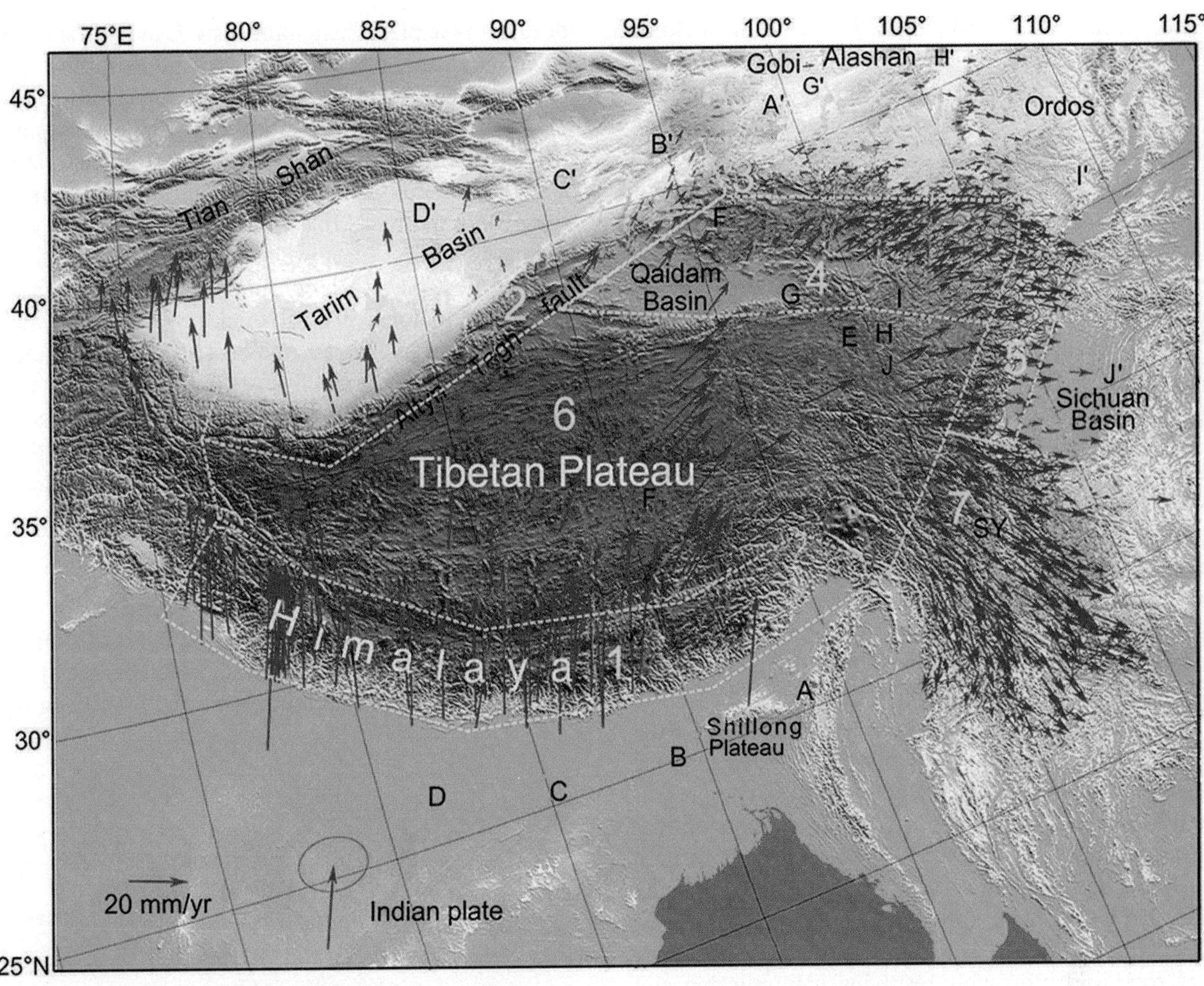

FIGURE 20.15

Motion in and around Tibetan Plateau as determined through Global Positioning System (GPS) measurements. Blue arrows point in the direction of motion, and their lengths indicate velocities (millimeters/year). The scale in the lower left corner indicates the length of an arrow for 20 millimeters/year. Dashed yellow polygons show regions—for instance, 1 is the Himalaya, 6 the Tibetan Plateau.

From Zhang, et al. Continuous deformation of the Tibetan Plateau from global positioning system data. Geology 2004 Vol. 32, No. 9, pp. 809–812. Fig 1 on pg. 810; Pei-Zhen Zhang, et al. "Continuous deformation of the Tibetan Plateau from global positioning system data," Geology, vol. 32, no. 9, p. 810, Figure 1.

caused by fault motion along the Sichuan Basin (right side of figure 20.15) and was indirectly caused by the collision of India and Asia.

The Appalachian Mountains are an example of continent-continent convergence but with a more complicated history. Arc-continent convergence and oceanic-continental convergence were also involved, and the mountain belt was later split apart by plate divergence.

A condensed version of orogeny in the Appalachians is as follows (figure 20.12): During late Precambrian and earliest Paleozoic time, the ancestral Atlantic Ocean developed as seafloor spreading forced the passive margins of North America, Europe, and Africa away from one another. During the Paleozoic, plate motion shifted, subduction began, and the ocean basin began closing. Island arcs developed between the continents. These became plastered onto the North American craton as the ancestral Atlantic basin closed. A couple hundred million years after subduction began, the ocean basin closed completely, first with Europe and later with Africa crashing into North America. By the end of the Paleozoic, the three continents were sutured together. The Appalachians and what is now the Caledonide mountain belt of Great Britain and Norway and the Atlas Mountains of North Africa were part of a single mountain belt within the supercontinent *Pangaea.* The mountain belt was comparable to the present-day Himalaya.

Early in the Mesozoic Era, the supercontinent split, roughly parallel to the old suture zone. The present continents moved (and continue to move) farther and farther away from their present divergent boundary, the mid-Atlantic ridge.

What happened to the Appalachians seems implausible. Yet, if one accepts the principles of plate-tectonic theory and examines the rocks and structures in the Appalachians (and their counterparts in Europe and Africa), the argument for this sequence of events is not only plausible but convincing.

The cycle of splitting of a supercontinent, opening of an ocean basin, followed by closing of the basin and collision of continents is known as the *Wilson Cycle.* Canadian geologist

J. Tuzo Wilson proposed the cycle in the 1960s for the tectonic history of the Appalachians.

The Wilson Cycle apparently has occurred before. A question raised is why would a continent split apart more or less along a suture zone where one would expect the crust to be thickest? One recently proposed hypothesis is that this zone is weakened and thinned somewhat by outward flow of rock during gravitational collapse and spreading. (Another hypothesis involves *delamination*, detachment and sinking downward of the underlying lithospheric mantle, as described later in this section.)

Post-Orogenic Uplift and Block-Faulting

After an orogeny ceases to affect part or all of a mountain belt and the prevailing compressive force is relaxed, there is a long period of uplift accompanied by erosion. Isostatic adjustment takes place during orogeny, but it is typically overshadowed by the compressive horizontal forces of plate convergence. When horizontal forces become insignificant at the end of an orogeny, isostatic adjustment takes over as the dominant process. For many millions of years, large regions in the mountain belt move vertically upward. Erosion may keep pace with uplift, and the area remains low. Alternatively, uplift may temporarily outpace erosion, resulting in plateaus or mountain ranges. The present Appalachian Mountains are the result of uplift and erosion that have taken place long after the last orogeny ended more than 250 million years ago. It is likely that the Appalachian Mountains eroded down to a plain after the last Paleozoic orogeny. The coastal plain east of the Appalachians is made of young sedimentary rock unconformably overlying metamorphic and igneous rocks that were part of the original mountain belt. This region has remained a plain. The present Appalachians represent rejuvenation following relatively recent uplift in late Tertiary time. The uplift may have been due to reactivation of ancient thrust faults caused by compressive stress within the westward-moving North American plate. The coastal plains have not moved upward, probably due to a lack of thrust faults. So the topography of the Appalachians is geologically quite young, while the original structures due to orogenic deformation are quite old. The Adirondack Mountains of northern New York also participated in the uplift and rejuvenation, but the orogeny that they went through is Precambrian–much older than the Appalachian orogenies. Eventually, the entire Appalachian mountain belt will be eroded to a plain and become part of the North American craton.

Isostasy

According to the concept of isostasy, lighter, less-dense continental crust "floats" higher on the mantle than the denser oceanic crust. A craton has achieved an equilibrium and is floating at the proper level for its thickness. Mountains, being thicker continental crust, "float" higher than the stable craton. As material is removed from mountains by erosion, the range floats upward to regain its isostatic balance (figure 20.16). This process can be thought of as "the pull of erosion." Isostatic adjustment does not take place instantaneously. Usually, there will be a considerable time lag between erosion and isostatic adjustment. As the mountains wear down to a low plain, erosion becomes virtually ineffective and the now thin crust achieves isostatic balance; the former mountain belt becomes part of the craton. The reason a craton consists of plutonic and metamorphic rock is that these were the rocks that formed the deep roots of the former mountain belt.

At most places on continents, the altitude above sea level is related to local crustal thickness. Beneath the 5-kilometer-high Tibetan Plateau, the crust is 75 kilometers thick. Under Kansas, the crust is 44 kilometers thick, and beneath Denver, the "mile-high city," the crust is 50 kilometers thick. (If the United States ever joins the rest of the world and goes metric, Denver will be known as the "1.6 kilometer-high city.") Just west of Denver, the altitude of the Rocky Mountains jumps to 2 kilometers higher than that at Denver. Scientists expected to find a corresponding thickening of the crust beneath these mountains. They were surprised by 1995 seismic studies that indicated that the crust is no thicker beneath the Rockies than at Denver. (Similar discrepancies between crustal thickness and mountain elevations have been reported for

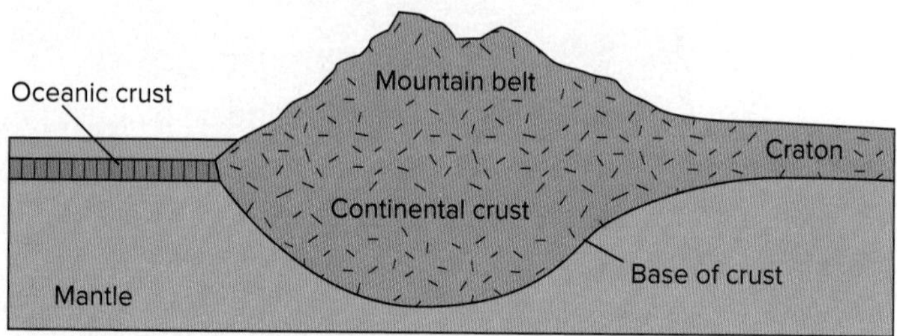

A At end of orogenic stage

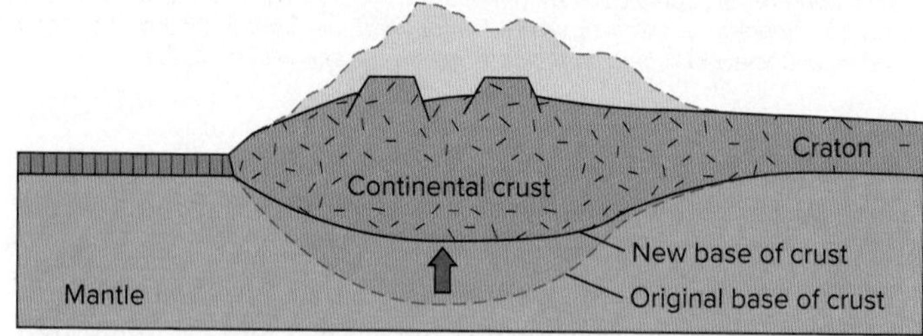

B Mountain belt moves upward

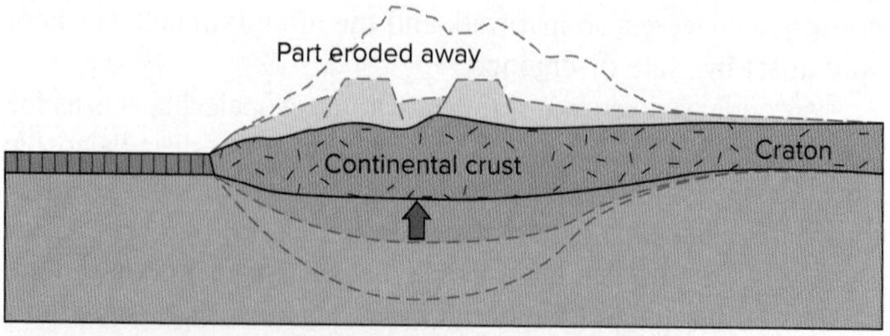

C Erosion and renewed uplift continue until crust beneath mountain belt is the same thickness as that of the craton

FIGURE 20.16

Isostasy in a mountain belt. The thickness of the continental crust is exaggerated.

the southern Sierra Nevada.) Geologists explain the higher elevations by regarding the mantle as hotter and therefore less dense beneath that part of the Rockies. The crust plus less-dense mantle are floating on deeper, denser mantle. Seismic wave studies verify that the mantle here is hot and appears to be asthenosphere that is at a shallower level in Earth than usual.

Normal Faulting

Normal faults develop after orogenies in several settings. One is when a continent is split and a divergent boundary forms (see chapter 19). An example is the breakup of Pangaea in the early Mesozoic, after the final orogeny in the Appalachians (figure 20.12*F*) and counterpart mountains in Africa and Europe. A normal fault will also develop if part of the crust moves upward isostatically more than does adjoining crust.

In parts of some mountain belts, the crust breaks into fault-bounded blocks. If an upthrown block is large enough, it becomes a **fault-block mountain range**. The normal faulting implies a *horizontal extension strain*, the regional pulling apart of the crust. Isostatic vertical adjustment of a fault block probably occurs at the same time.

Fault-block mountain ranges are bounded by normal faults on either side of the range, or, more commonly, are tilted fault blocks in which the uplift has been great along one side of the range, while the other side of the range has pivoted as if hinged (figure 20.17). The Sierra Nevada (California) and Teton (Wyoming) Range are tilted fault-block mountains (figure 20.18).

Isolated volcanic activity may be associated with some faults. Eruptions occur along faults extending deep into the crust or the upper mantle.

Uplift is neither rapid nor continuous. Part of a mountain range may suddenly move upward a few centimeters (or, more rarely, a few meters) and then not move again for hundreds of years. Erosion works relentlessly on newly uplifted mountains, carving the block into peaks during the long, spasmodic rise. Over the long time period, the later episodes of renewed faulting and uplift involve successively less and less vertical movement.

Block-faulting is taking place in much of the western United States–the Basin and Range province (also called the Great Basin) of Nevada and parts of Utah, Arizona, New Mexico, Idaho, and California (figure 20.19). Hundreds of small, block-faulted mountain ranges are in evidence. They are separated by valleys that are filling with debris eroded from the mountains.

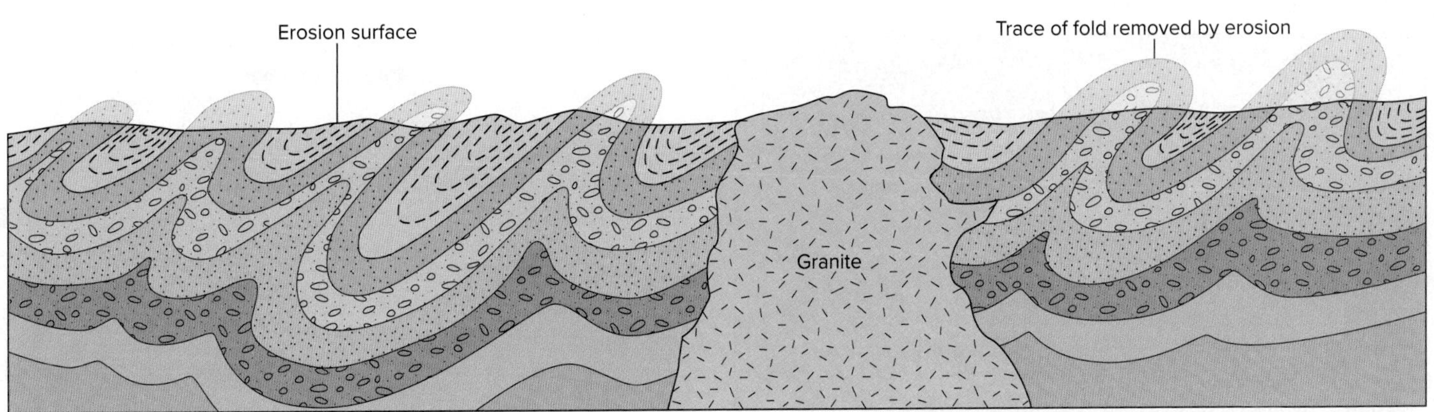

A Before block-faulting. Folding and intrusion of a pluton during an orogeny has been followed by a period of erosion.

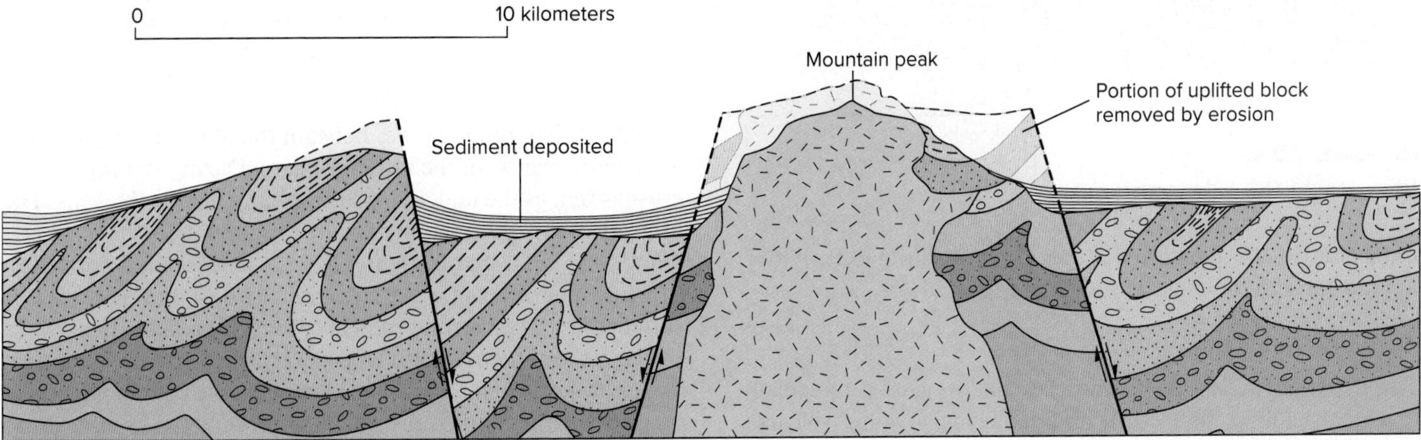

B The same area after block-faulting. Tilted fault-block mountain range on left. Range to right is bounded by normal faults.

FIGURE 20.17

Development of fault-block mountain ranges.

FIGURE 20.18

The Teton Range, Wyoming, a tilted fault-block range. The rocks exposed are Precambrian metamorphic and igneous rocks that were faulted upward. Extensive past glaciation is largely responsible for their rugged nature. For more information, go to https://www.nps.gov/grte/learn/nature/geology.htm

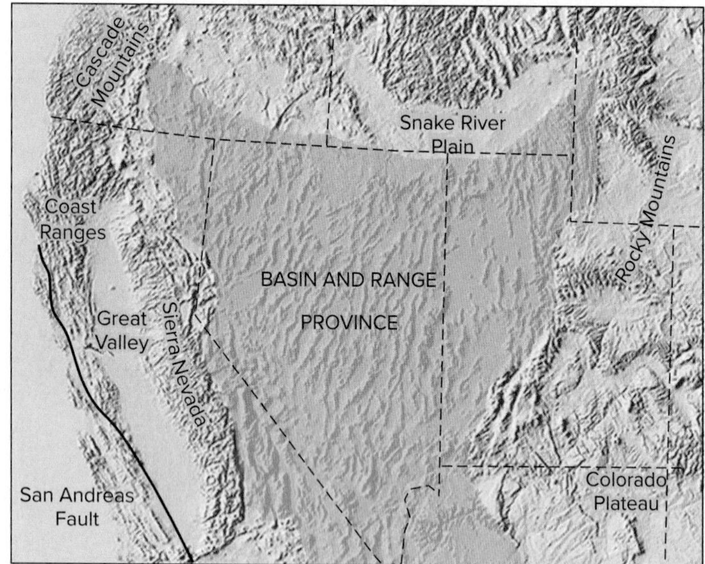

FIGURE 20.19

The Basin and Range and adjoining geological provinces.

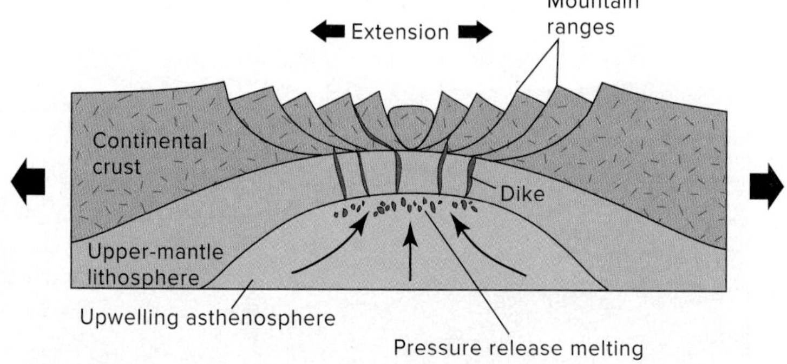

FIGURE 20.20

Upwelling, hot, buoyant mantle (asthenosphere) causes extension, thinning, and block-faulting of the overlying crust. Go to https://www.youtube.com/watch?v=Owjt_WJKrBo *to see an animated description of the formation of the Basin and Range.*

Extension in the Basin and Range is probably due to hotter mantle beneath the crust as shown in figure 20.20.

Delamination

The hypothesis of lithospheric delamination is used to explain the block-faulting, thin crust, and geologically young volcanic activity of the Basin and Range. **Lithospheric delamination** (or simply **delamination**) is the detachment of part of the mantle portion of the lithosphere beneath a mountain belt (figure 20.21). As you know, the lithosphere consists of the crust and the underlying, rigid mantle. Beneath the lithosphere is the hotter, ductile mantle of the asthenosphere. During an orogeny, the crust as well as the underlying lithospheric mantle thickens. The lithospheric mantle is cooler and denser than the asthenospheric mantle. As indicated in figure 20.21, the thickened portion of the lithospheric mantle is gravitationally unstable, so after it is softened from convecting asthenosphere, it breaks off and sinks through the asthenosphere to a lower level in the mantle. Hot asthenospheric mantle flows in to replace the foundered, colder mantle. Heating of the crust follows, allowing the lower crust to flow. The once-thick crust becomes thinner than that of adjoining regions of the mountain belt. Extension results in block-faulting in the upper part of the crust (as in figures 20.20 and 20.21*C*).

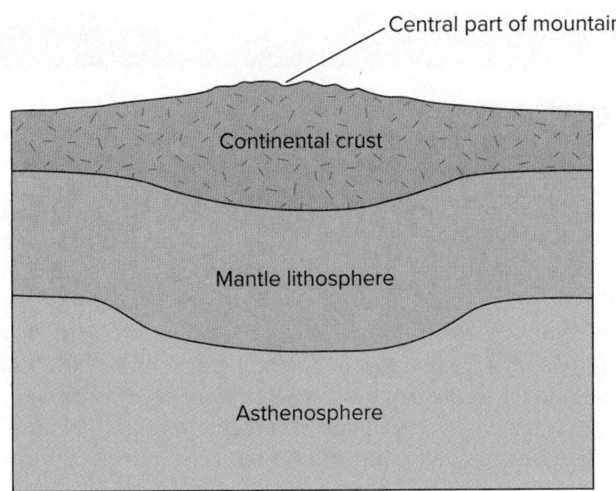

A Thick continental crust of a mountain belt produced during orogeny.

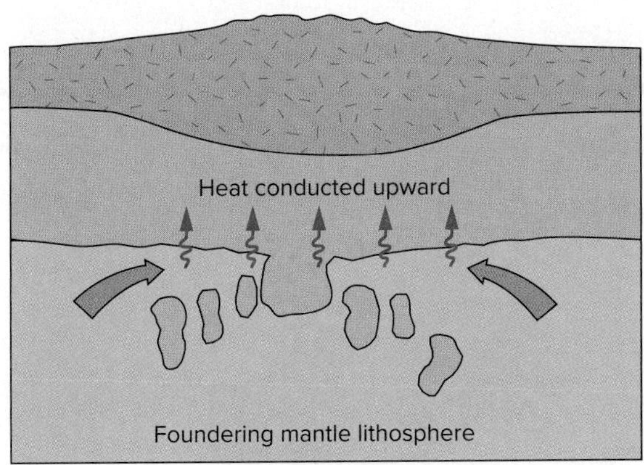

B Delamination of gravitationally unstable mantle lithosphere. Hot asthenosphere flows and replaces foundering lithosphere, heating overlying lithosphere.

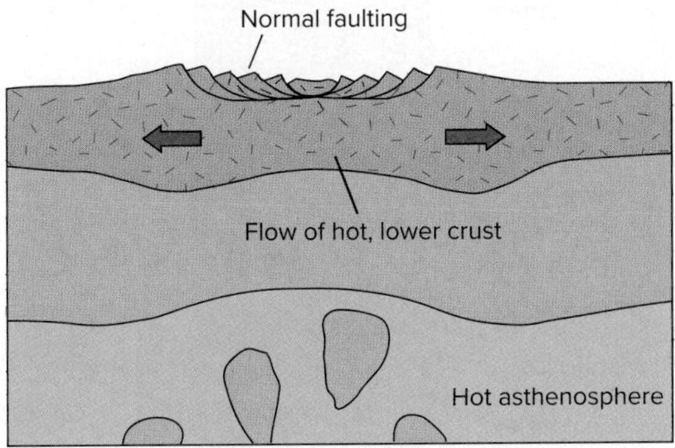

C Extension with hot, lower crust flows outward.

FIGURE 20.21

Delamination and thinning of continental crust following orogeny. (Not drawn to scale.)

Delamination beneath the Basin and Range helps explain the extensive rhyolitic and basaltic eruptions that occurred tens of millions of years after the end of the last orogeny. Heating in the lower part of the crust to 700°C would have generated silicic magma that erupted as rhyolite. Basaltic magma would have formed from partial melting of the asthenosphere when it moved upward (replacing the foundered lithospheric mantle) and pressure was reduced (as explained in chapter 3). That the crust was once thicker in the Basin and Range is supported by recent studies of fossil plants indicating that the Basin and Range was 3 kilometers higher than at present.

Delamination is also being invoked to help explain why, when Pangaea broke up, North America split from Europe and Africa more or less along the old suture zone. *Gravitational collapse* could have contributed to the weakening and thinning of this once-thick part of the mountain belt during its last orogeny. The breakup of the supercontinent began around 30 million years after the late Paleozoic orogeny ended. Delamination of the underlying lithospheric mantle would have resulted in heating and thinning of the overlying, remaining lithosphere. Rifting of the supercontinent began with normal faulting (see chapter 19 and figure 19.20) and was accompanied by basaltic eruptions and intrusions. The Appalachians split from the European Caledonides. Europe, Africa, and North America went their separate ways as the Atlantic opened and widened.

Recently, delamination has been used to explain part of what takes place during an orogeny (see box 20.3).

Delamination (like gravity collapse) is an example of a hypothesis that builds on plate-tectonic theory. It was proposed because it explains data better than other concepts do. It still needs further testing to become widely accepted as a theory.

20.4 THE GROWTH OF CONTINENTS

Continents grow bigger as mountain belts evolve along their margins. Accumulation of sediment and igneous activity add new continental crust beyond former coastlines. In the Paleozoic Era, the Appalachians were added to eastern North America, and during the Mesozoic and Cenozoic Eras, the continent grew westward because of accumulation and orogenic processes in many parts of what is now the Cordillera. Therefore, if we isotopically dated rocks that had been through an orogeny, starting in the Canadian Shield and working toward the east and west coasts, we should find the rocks to be progressively younger. In a very general way, this seems to be the case; however, there are some rather glaring exceptions.

Displaced Terranes

In some regions of mountain belts, the age and characteristics of the bedrock appear unrelated to that of adjacent regions. To better understand the geology of mountain belts, geologists have in recent decades begun dividing major mountain belts into

IN GREATER DEPTH 20.3

Rise of the Andes during Plate Convergence

The Andes is Earth's second-highest mountain belt. It is the product of oceanic-continental convergence between the western-moving South American plate and subduction of the Nazca oceanic plate. The resulting orogeny has been ongoing for the last 50 million years. A group of scientists, using a multidisciplinary approach, determined the rate of increase in elevation for the Central Andes through time. In 2008, Carmata Garzione and coauthors described their research and conclusions:

During the first 40 million years of the orogeny, crustal shortening and thickening of the lithosphere took place, resulting in a slow but steady rise in surface elevation. But, until approximately 10 million years ago (Ma), the surface elevations were anomalously low. The rate of vertical growth increased beginning around 10 Ma. The high rate of uplift continued until 6 Ma when the Andes approached their present heights.

The authors attribute the slow rising prior to 10 Ma to dense rock in the lower crust and lithospheric mantle. These denser rocks kept the thickening upper crust from reaching the isostatic level expected from less-dense crust. Between 10 and 6 Ma, the denser rock was removed from the overlying, less-dense crust, likely through lithospheric delamination. The less-dense crust, now overlying hotter and less-dense mantle, rose isostatically at a more rapid rate. Mountain topography became considerably more elevated as it approached heights of the present Andes.

Source: C.N. Garzione, G.D. Hoke, J.C. Libarkin, S. Withers, B. MacFadden, J. Eiler, P. Ghosh, A. Mulch. Rise of the Andes. 2008. *Science*, vol. 320, pp. 1304–1307. *http://science.sciencemag.org/content/320/5881/1304.full*

tectonostratigraphic terranes (or, more simply, **terranes**), regions within which there is geologic continuity. The geology in one terrane is markedly different from that of a neighboring terrane. Terrane boundaries are usually faults. Typically, a terrane covers thousands of square kilometers, but some terranes are considerably smaller. Alaska and western Canada have been subdivided by some geologists into over fifty terranes (figure 20.22). Terranes are named after major geographic features; for instance, Wrangellia, parts of which are now in Alaska and Canada (with fragments in Washington and Idaho, according to some geologists) was named after the Wrangell Mountains of Alaska.

Many terranes appear to have formed essentially in place as a result of accumulation and orogeny along the continent's margin. Other terranes have rock types and ages that do not seem related to the rest of the geology of the mountain belt and have been called **suspect terranes**, that is, terranes that may not have formed at their present site. If evidence indicates that a terrane did not form at its present site on a continent, it is regarded as an **accreted terrane**. Accreted terranes that can be shown to have traveled great distances are known as *exotic terranes.*

A suspect terrane will have rock types and ages different from adjoining terranes, but to prove that it came from elsewhere in the world (and therefore is an accreted terrane), geologists may compare fossil assemblages or determine the paleomagnetic poles (see chapter 19) of the terrane's rocks. If the terrane is exotic, its fossil assemblage should indicate a very different climatic or environmental setting compared to that of the adjoining terrane. (For example, the Cache Creek terrane in western Canada contains fossils from the Permian period that indicate a marine equatorial environment.) For an exotic terrane, the paleomagnetic poles for the rocks in the terrane will plot at some part of the world very distant from poles of adjoining

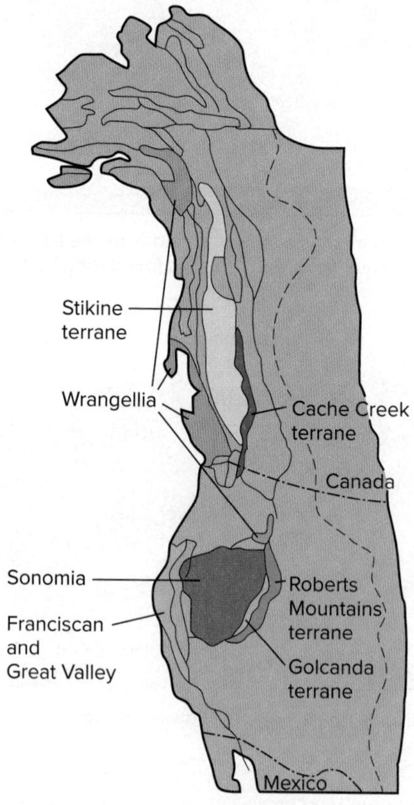

FIGURE 20.22

Some terranes in western North America. Note: Wrangellia is in Alaska, British Columbia, and Idaho.
Source: U.S. Geological Survey Open File Map 83–716

terranes that formed in place. This indicates that a particular terrane formed in a different part of Earth and drifted into the continent of which it is now a part. Some accreted terranes were island arcs, and some might have been *microcontinents* (such as present-day New Zealand) that moved considerable distances before crashing into other landmasses. Others may have been fragments of distant continents that split off and moved a long distance because of transform faulting. Imagine what might happen if the San Andreas fault remains active for another 100 million years or so. Not only would Los Angeles continue northward toward San Francisco and bypass it in about 25 million years (see box 15.2), but the block of coastal California west of the fault would continue moving out to sea, becoming a large island with continental crust that drifts northward across the Pacific. Ultimately, it would crash into and suture onto Alaska.

The Appalachians as well as mountain belts in other continents have also been divided into terranes. Even the Canadian Shield has been subdivided into terranes. Some geologists think they can determine, despite the great age and complexity of the shield's rocks, the extent to which some terranes traveled before crashing together.

We should caution the reader that geologists do not always agree on the nature and boundaries of terranes. While most would probably agree that some terranes are exotic, many geologists think the subdividing of Alaska and western Canada into fifty terranes is overdoing it and not supported by sufficient evidence. Only time and more painstaking gathering of evidence will allow geologists to determine the history of each alleged terrane.

Concluding Comment

Only a few decades ago, many geologists thought that through the application of plate-tectonic theory, we could easily determine the processes at work in each mountain belt and work back in time to understand the history of each of the continents. Some suggested that there would hardly be any major problems for Earth scientists to solve in the future. Plate tectonics was a breakthrough, and a great many problems were solved; but with this great leap forward in the science, we have identified new problems. New generations of geologists will have no shortage of challenges and no less excitement from solving newly discovered problems than did their predecessors who saw the dawn of the plate-tectonics breakthrough. Science present builds on science past.

Summary

Major mountain belts are made up of a number of *mountain ranges.* Mountain belts are generally several thousand kilometers long but only a few hundred kilometers wide.

The major factors that control the growth and development of mountain ranges are *intense deformation* (during an *orogeny*), *isostasy*, and *weathering and erosion.* An orogeny involves folding and faulting of sedimentary and volcanic rock, regional metamorphism, and igneous activity. Orogenies are associated with plate convergence. After an orogeny, there is a long period of uplift, often with block-faulting, and erosion. Eventually, the mountain belt is eroded down to a plain and incorporated into the *craton*, or stable interior of the continent.

According to plate-tectonic theory, mountains on the edges of continents are formed by continent-oceanic convergence, and mountains in the interior of continents are formed by continent-continent collisions.

The uplift of a region following termination of an orogeny is generally attributed to isostatic adjustment of continental crust.

Continents grow larger when new mountain belts evolve along continental margins. They may also grow by the addition of terranes that may have traveled great distances before colliding with a continent.

Terms to Remember

accreted terrane 508
craton 494
fault-block mountain range 505
fold and thrust belts 495
gravitational collapse and spreading 500
lithospheric delamination (or delamination) 506
major mountain belt 490
mountain range 490
orogeny 493
Precambrian shield 494
suspect terrane 508
terrane (tectonostratigraphic terrane) 508

Testing Your Knowledge

Use the following questions to prepare for exams based on this chapter.

1. What does a fold and thrust belt tell us about what occurred during an orogeny?
2. What differences would you expect to find between a young, active mountain belt and an older mountain belt that is no longer undergoing uplift?
3. Explain how erosion and isostasy eventually produce stable, relatively thin continental crust.
4. In what ways do the rocks and topography of cratons differ from those of mountain belts?
5. How is it possible to form mountain belts in a region undergoing extensional forces? What is the name for this type of mountain belt? Give an example of a mountain belt formed under extension.
6. Using the Appalachian Mountains as an example, describe the Wilson Cycle.
7. The mountain belt that forms the eastern part of North America is called the

   a. Appalachians. b. North American Cordillera.
   c. Himalaya. d. Andes.
   e. Rockies.
8. The portion of a continent that has been structurally stable for a prolonged period of time is called a(n)

   a. orogeny. b. basin.
   c. mountain belt. d. craton.

9. Folds and reverse faults in a mountain belt suggest
   a. crustal shortening.
   b. tensional stress.
   c. deep-water deposition of the sediment.
   d. all of the preceding.
10. What type of plate boundary is most commonly associated with mountain building?
   a. divergent
   b. convergent
   c. transform
11. The likely reason that the Himalaya is much higher than the adjoining Tibetan Plateau is:
   a. Heavy precipitation results in rapid erosion in the Himalaya.
   b. The crust of the Himalaya is less dense than that of the Tibetan Plateau.
   c. The Tibetan Plateau is being subducted.
   d. The mantle beneath the Himalaya is hotter than that beneath the Tibetan Plateau.
12. To explain fold and thrust belts, simultaneous normal faulting, and how once deep-seated metamorphic rocks rise to an upper level in a mountain belt, geologists use a model called
   a. tectonism.
   b. gravitational collapse and spreading.
   c. orogeny.
   d. faulting.
13. Which of the following is *not* characteristic of a mountain belt that formed through ocean-continent convergence?
   a. fold and thrust belts
   b. thick accumulations of marine sediment
   c. prevalence of normal faults over reverse faults
   d. metamorphism
14. Regional metamorphic rocks and plutonic rocks are most commonly associated with
   a. the outer edge of the fold and thrust belt.
   b. the most intensely deformed portions of major mountain belts.
   c. fault-block mountain ranges.
   d. the forearc basin of a convergent plate boundary.
15. Which of the following is *not* an example of a mountain belt formed by continent-continent convergence?
   a. Urals
   b. Himalaya
   c. Rockies
   d. Alps
16. The concept of isostasy suggests that higher-elevation mountain ranges are underlain by
   a. a thick root of continental crust.
   b. dense oceanic crust.
   c. a shallow root of continental crust.
   d. a complex of normal faults.
17. The Wilson Cycle describes
   a. the cycle of uplift and erosion of mountains.
   b. the movement of asthenosphere.
   c. the block-faulting that occurs at mountains.
   d. the cycle of splitting of a supercontinent, opening of an ocean basin, followed by closing of the basin and collision of continents.
18. The detachment of part of the mantle portion of the lithosphere beneath a mountain belt is called
   a. gravitational collapse.
   b. rifting.
   c. lithospheric delamination.
   d. none of the preceding.
19. Which is *not* a type of terrane?
   a. accumulated
   b. exotic
   c. suspect
   d. accreted
20. Which is a source for terranes?
   a. microcontinents
   b. fragments of distant continents
   c. island arcs
   d. all of the preceding

Expanding Your Knowledge

1. How are unconformities used to determine when orogenies occurred?
2. How has seismic tomography contributed to our understanding of mountain belts?
3. How do basalt and ultramafic rocks from the oceanic lithosphere become part of mountain belts?
4. Why is a craton locally warped into basins and domes?
5. How could fossils in a terrane's rocks be used to indicate that it is an exotic terrane?

Exploring Web Resources

http://pubs.usgs.gov/gip/birth/birth.pdf

Birth of the Mountains. The Geologic Story of the Southern Appalachian Mountains. This is the pdf version of a publication by the U.S. Geologic Survey.

https://www.youtube.com/watch?v=mR8n1vkF9lU

Continents Collide: The Appalachians and the Himalayas. Video produced by the McClung Museum of Natural History and Culture.

http://www.see.leeds.ac.uk/structure/virtualfield/npfront.htm

Nanga Parbat–Mountain Uplift and Tectonics. A virtual field trip to the Nanga Parbat area in the western part of the Himalaya.

CHAPTER 21

Global Climate Change

by Christopher Cappa and Delphine Farmer

This spectacular satellite image shows clouds over the Earth's surface and the thin film of atmosphere on the horizon. Using a collection of satellite-based observations, scientists and visualizers stitched together months of observations of the land surface, oceans, sea ice, and clouds into a seamless, true-color mosaic of every square kilometer (.386 square mile) of our planet.
Source: Image Science and Analysis Laboratory, NASA-Johnson Space Center

LEARNING OBJECTIVES

- Distinguish between weather and climate.
- Describe the composition and structure of the atmosphere.
- Explain the processes that can drive major variations in Earth's climate over time scales of tens of thousands of years.
- Understand the general mechanism by which greenhouse gases in planetary atmospheres (including that of Earth) cause planetary warming.
- Describe the nature of feedbacks within the Earth's climate system and identify at least two key feedbacks.
- Describe the evidence for (1) recent global warming, (2) its link to greenhouse gases, and (3) its link to human activities.
- Identify at least three major impacts on the Earth system of modern climate change and/or the emission of carbon dioxide into the atmosphere.

Earth's climate is determined by interactions between the atmosphere, biosphere, hydrosphere, and lithosphere, encompassing all aspects of geology that you have learned up to this point. When we view Earth from space, as in the satellite image shown here, this becomes apparent. Life on Earth is possible because we have a climate that is hospitable to our existence and that has allowed humans to thrive over thousands of years. Although Earth's climate has changed throughout geologic history, in recent decades it has become apparent that there is an unprecedented and rapid change occurring that is primarily the result of human activities. In this chapter, we discuss the various factors that determine Earth's climate, investigate the evidence for modern climate change, and look at the consequences of modern climate change for life on the planet.

21.1 WEATHER, CLIMATE, AND CLIMATE CHANGE

Weather describes what the atmosphere is doing over short time scales; it is what you see when you go outside or look out your window. Weather is extremely variable, changing from hot to cold, rainy to sunny, windy to calm from hour to hour, day to day, or season to season. **Climate** is the *average* weather pattern in a region over long periods of time, typically 30 years or more. For example, California has a Mediterranean climate; that is, Californians can expect long, hot, dry summers and cooler, wet winters. The weather changes with the seasons and may vary from year to year (for example, an unusually wet winter), but the general climate pattern is consistent. One way to think about this is that weather is what is happening now, while climate is what you might expect to happen based on many past experiences.

Climate, like weather, is variable, only on much longer time scales. Although much of the recent debate about climate change has revolved around global warming, it is important to remember that climate refers not only to temperature, but to precipitation, cloudiness, winds, and relative humidity, among other variables. As we will see, climate has fluctuated in the past and will continue to do so, due to a variety of natural factors. However, current concerns over climate change relate to how human activities are causing Earth's climate to change, often referred to as **anthropogenic climate change** or anthropogenic global warming. Anthropogenic refers to the influence that humans can have on the natural world. As you will learn later in this chapter, average global temperatures have been increasing since before the Industrial Revolution in the nineteenth century, and they are projected to increase significantly in the future. The most important reason for this temperature increase is the increase in concentrations of greenhouse gases in the atmosphere due to human activities. If concentrations of greenhouse gases continue to increase, we expect this warming trend to continue. However, the term "climate change" is considered more accurate than "global warming," as the climate is not expected to warm uniformly around the globe: while most regions will get warmer, a few may actually get cooler. Just as important, the dry areas of the world are expected to get drier and experience more droughts, while the wetter parts of the world are projected to receive more rainfall.

Concerns Over Modern Climate Change

Given that climate has always changed and that life on Earth has somehow survived, why should we be concerned about anthropogenic climate change? There are a few factors that are of great concern to scientists. For one, the rate of climate change today is unprecedented when compared to climate variations that have occurred over the last million or so years. As a consequence, organisms that might normally have been able to adapt to changes are less able to do so today. Second, the human population has exploded from fewer than 10 million people 12,000 years ago to 250 million people 2,000 years ago to more than 7 billion people today. Political and economic realities mean that people have limited abilities to move away from drought, famine, rising sea levels, and other consequences of climate change.

21.2 UNDERSTANDING THE ATMOSPHERE

In order to understand climate, the evidence for current and past climate change, the factors that drive climate change, and the potential consequences of future climate change, we must first learn about our atmosphere and its important role in determining the energy budget of the Earth.

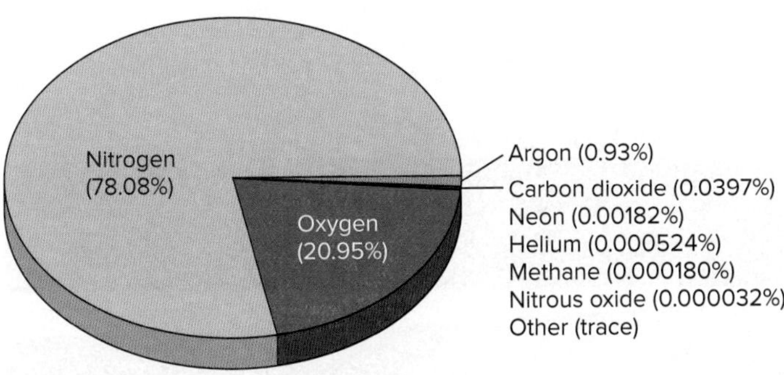

FIGURE 21.1

Composition of the Earth's dry atmosphere, by volume (as of 2009). Water is a highly variable component, and it depends on the altitude and region. Note that the three major components account for 99.96% of all the molecules in the atmosphere, while the remainder of the atmosphere is composed of thousands of trace gases, including organic and inorganic compounds.

Composition

The Earth's atmosphere is composed of many different gases (figure 21.1). Some of these derive from natural processes, some from anthropogenic processes, and some from both. *Nitrogen* (N_2) and *oxygen* (O_2) make up over 99% by volume of the entire atmosphere, not counting water vapor. The rest is made up of less abundant gases, such as *argon* (Ar), *carbon dioxide* (CO_2), *neon* (Ne), and *helium* (He). Most of the compounds relevant to air pollution and climate change are present in very small quantities, in the parts-per-million by volume (1 part in 10^6) to the parts-per-quadrillion (1 part in 10^{12}) range.

Water (H_2O) has a unique role in the atmosphere because the amount of water can vary dramatically depending on location, in terms of both latitude and longitude. Water can also condense to form clouds and precipitation. Water vapor concentrations vary from about 0.0001% to 4%. As you learned in chapter 13, the amount of water vapor that can be held in the atmosphere is dependent on temperature. Warm air can hold more water vapor than cold air. When the air is saturated with water vapor, the water will condense, forming clouds and precipitation.

Structure

The atmosphere is divided into four layers, defined by how temperature varies with altitude (figure 21.2). In the lowest layer, the **troposphere**, temperature decreases with altitude. You may have experienced this when climbing a mountain. At the base of the mountain it may be quite warm, but when you reach the top, you notice that it is distinctly cooler. The weather we experience, such as rain and wind, occurs within the troposphere. Above the troposphere, the **stratosphere** is defined by an increase in temperature with altitude. The stratosphere is where the *ozone layer* occurs. The ozone layer has high concentrations of ozone (O_3), and it plays an important role in protecting the surface from the Sun's harmful ultraviolet radiation. The increase in temperature in the stratosphere is due to the absorption of ultraviolet sunlight by the ozone layer. Above the stratosphere is the

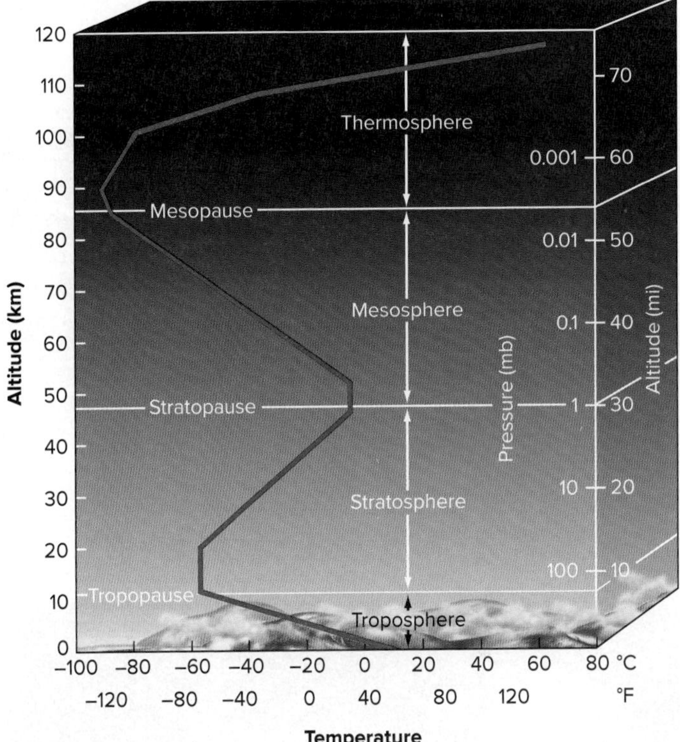

FIGURE 21.2

The vertical structure of the Earth's temperature.
Source: http://3.bp.blogspot.com/-hcCmIJUqJYY/TcIGwBHjgOI/AAAAAAAAAAQ/LVbdyr9LUXU/s1600/structure.jpg

mesosphere, and above that the *thermosphere* (also called the *ionosphere*). These last two regions play little role in determining climate. The thickness of any of these layers can vary significantly with latitude and season, especially the troposphere.

Energy from the Sun

The Sun is the Earth's primary source of energy. Every second, the Earth absorbs five orders of magnitude (100,000 times) more energy from the Sun than is produced globally from fossil fuel combustion over an entire year. A key number to remember is that the average amount of solar energy incident on the Earth is 342 W/m^2, which describes the rate of energy input per square meter of the Earth's surface. The Sun transmits its energy to the Earth in the form of *electromagnetic radiation* (figure 21.3). Most of the Sun's radiation and energy are in the visible region of the electromagnetic spectrum, although radiation in both the ultraviolet and infrared regions of the spectrum is also important. Electromagnetic radiation is characterized by its wavelength; ultraviolet (UV) radiation has a shorter wavelength than the visible spectrum and infrared (IR) radiation has a longer wavelength.

Figure 21.4 shows what happens to the Sun's radiation when it reaches the Earth. About 50% makes its way through the atmosphere and is absorbed by land and sea at the Earth's surface, and 20% is absorbed by the Earth's atmosphere. The

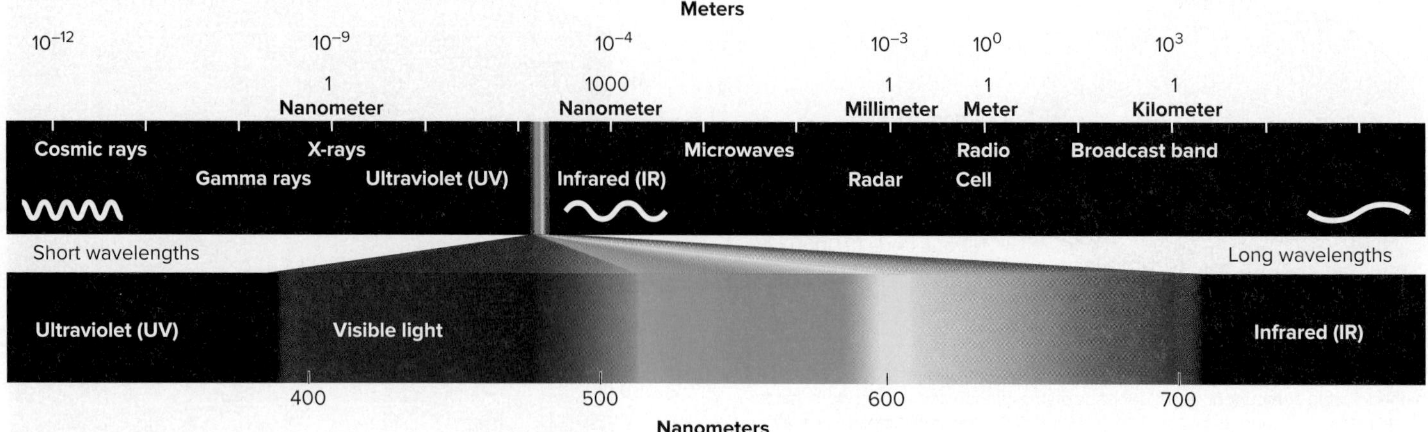

FIGURE 21.3

The electromagnetic spectrum. Earth's climate depends primarily on radiation in the UV, visible, and IR regions.
Source: http://kirraweehighphysics.files.wordpress.com/2011/04/spectrum.jpg

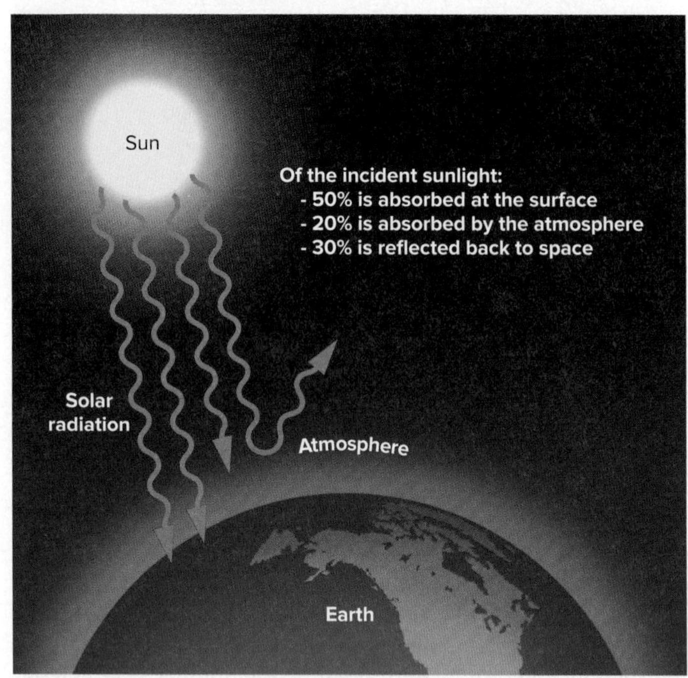

FIGURE 21.4

Shortwave solar radiation, illustrated by the orange arrows, is either reflected back to space by clouds, aerosols, gases, or the Earth's surface, absorbed by the atmosphere, or absorbed by the Earth's surface. The absorbed solar radiation is the primary energy input to the Earth's climate system.
Source: Modified composite of figures taken from NOAA: www.esrl.noaa.gov/gmd/outreach/carbon_toolkit/basics.html

remaining 30% is reflected back into space by the atmosphere, clouds, and areas of the Earth's surface that are reflective, such as regions covered in snow and ice. The solar radiation that is absorbed at the Earth's surface warms the land and oceans. The percentage of radiation reflected from a surface is referred to as its **albedo**. Lighter-colored objects have a higher albedo than darker ones (figure 21.5). The polar regions of the Earth have a much higher albedo than heavily forested areas near the equator. Clouds also have a high albedo; the cloudier the day, the more of the Sun's radiation is reflected back into space. The 30% of solar radiation reflected back into space represents the global average albedo.

Blackbody Radiation

To a reasonable approximation, both the Sun and the Earth can be considered **blackbody radiators**. That is, the intensity and wavelength of the electromagnetic radiation they emit is dependent on their temperature. In fact, all objects give off radiation in a manner that is proportional to their temperature. Hotter objects radiate more energy, and at shorter wavelengths, than cooler objects. The radiation emitted by the Sun can be approximated as if the Sun were a blackbody radiator with a temperature of 5,500°C (~10,000°F). At this temperature, the peak in the wavelength distribution is in the visible spectrum, around 0.53 micrometers (figure 21.6).

The average surface temperature of the Earth is much cooler, a very comfortable 15°C (59°F). The radiation emitted by the Earth therefore has a longer wavelength, with most being in the infrared region of the electromagnetic spectrum, around 10 micrometers. This infrared radiation is commonly referred to as **terrestrial radiation** or long-wave radiation.

Importantly, the radiation emitted by the Earth is not simply radiated directly back into space. Because terrestrial radiation has a longer wavelength than the incoming solar radiation, gases in the atmosphere that do not absorb solar radiation can absorb the infrared radiation. This is the basis of the greenhouse effect, discussed in the next section. However, ultimately some of this infrared radiation is emitted to space, allowing the Earth system to get rid of energy. Without this emission of infrared radiation to space, the Earth would rapidly warm up due to the energy input from the absorption of solar radiation.

The Greenhouse Effect

If all of the energy emitted by the Earth was radiated directly into space, the planet's average surface temperature would be −19°C (−2.2°F), well below freezing. Why is the average

FIGURE 21.5

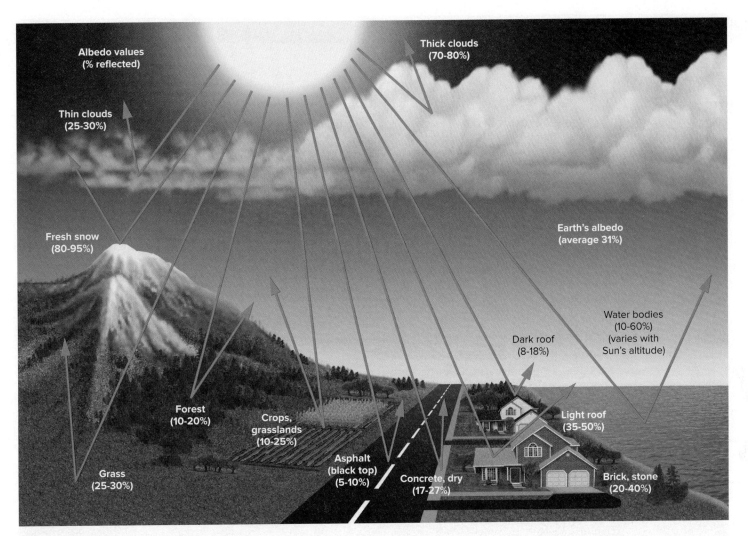

The albedo of different types of surfaces on the Earth.
Source: www.sci.uidaho.edu/scripter/geog100/lect/03-atmos-energy-global-temps/03-06-albedo-values.htm

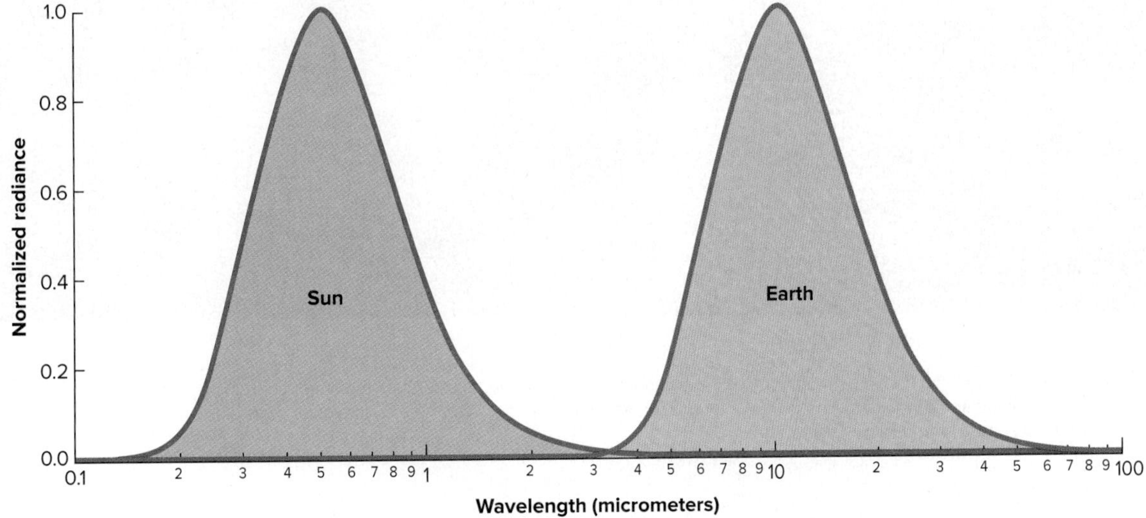

FIGURE 21.6

Idealized blackbody curves for the Sun and for the Earth. Note that the curves have been normalized to 1. In reality, the amount of blackbody radiation given off by the Sun is much larger than that given off by the Earth due to the Sun's much higher temperature.
Source: Modified composite of figures taken from NOAA: www.esrl.noaa.gov/gmd/outreach/carbon_toolkit/basics.html

PLANETARY GEOLOGY 21.1

Venus versus Earth versus Mars

The greenhouse effect on Earth is responsible for raising the global average temperature by about 34°C. How does this compare with some of our neighboring planets (box figure 1)? Although it is closer to the Sun, and thus receives more intense sunlight, Venus has an albedo of 90%, meaning that only 10% of the incident sunlight is absorbed. Therefore, Venus has an equilibrium blackbody temperature of only −89°C. However, the true temperature of Venus is around 477°C! The reason why the surface of Venus is so hot is not because it is close to the Sun, but because Venus has a very large greenhouse effect. As you learned in chapter 5, the atmosphere of Venus is 96.5% CO_2 compared to just 0.039% in the Earth's atmosphere. Also, the atmospheric pressure on Venus is ninety-two times greater than that on the Earth, which means the amount of CO_2 is extremely high.

Mars has an albedo of 0.25, similar to the Earth. However, because Mars is farther away from the Sun it receives less sunlight, and its blackbody temperature is only −63°C. Similar to Venus, the Martian atmosphere is 95.3% CO_2. However, the Martian atmosphere has a pressure that is only 1/1000th that of the Earth. Because of the low atmospheric pressure, the greenhouse effect on Mars is very weak, and the mean surface temperature is very similar to the blackbody temperature. Thus, by examining our neighboring planets, we can see that the greenhouse effect is very sensitive to concentrations of greenhouse gases in the atmosphere.

BOX 21.1 ■ FIGURE 1

The equilibrium blackbody temperatures of Venus, Earth, and Mars compared to the actual surface temperatures.
Source: USRA/Lunar and Planetary Institute

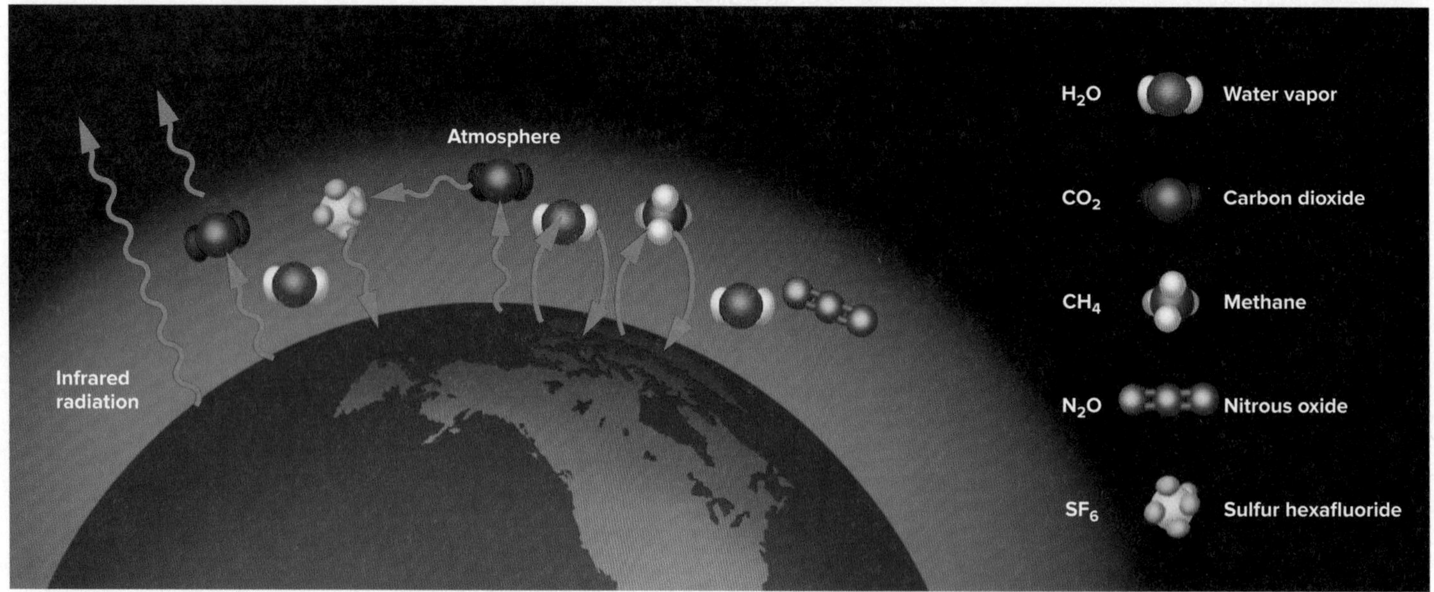

FIGURE 21.7

Terrestrial infrared radiation emitted from the surface is readily absorbed by greenhouse gases, such as CO_2, in the atmosphere. The absorbed energy is then emitted by the atmosphere as radiation both upward and downward. This downward radiation essentially acts to "trap" heat in the lower parts of the atmosphere, leading to warming at the surface. The flow of infrared radiation is indicated by the orange arrows.
Source: Modified composite of figures taken from NOAA: www.esrl.noaa.gov/gmd/outreach/carbon_toolkit/basics.html

temperature of the surface so much warmer? Much of the infrared energy radiated from the Earth's surface is absorbed by gases in the atmosphere, such as water vapor, carbon dioxide, methane, nitrous oxide, and other trace gases that are collectively known as greenhouse gases (figure 21.7 and table 21.1). **Greenhouse gases** are those that efficiently absorb long-wave

TABLE 21.1 Greenhouse Gas Atmospheric Lifetimes, Global Warming Potentials (GWP), and Major Anthropogenic Sources

GHG	Lifetime (Years)	GWP (100 years)	Concentration in 1850	Concentration today	Activity and/or Source
Carbon dioxide (CO_2)	100s–1000s	1	280 ppb	400 ppm	Fossil fuel and biofuel combustion; transportation; energy production; biomass burning; deforestation; cement production
Methane (CH_4)	12	25	700 ppb	1,874 ppb	Livestock enteric fermentation; manure management; rice cultivation; landfills
Nitrous oxide (N_2O)	114	298	270 ppb	324 ppb	Fertilizer use related to agriculture
Halocarbons	Variable	Variable			Refrigeration/air-conditioning systems; fire suppression systems; electrical equipment; various manufacturing processes
(SF_6)	3,200	22,800	0	7.5 ppt	
(CFC-12)	100	10,900	0	530 ppt	
Tropospheric ozone (O_3)	$<<1$	n/a	Variable	Variable	Combustion; solvent manufacturing and use; ethanol production

radiation. Not all gases in the atmosphere are classified as greenhouse gases. In fact, nitrogen and oxygen, the most abundant atmospheric gases, are not considered greenhouse gases.

The infrared radiation absorbed by greenhouse gases heats the atmosphere. The atmosphere itself then acts as a blackbody, emitting infrared radiation. Half of this emitted energy is radiated back toward the surface of the Earth, where it can be absorbed, thus warming the surface further. Greenhouse gases in the atmosphere therefore effectively act as a blanket, trapping heat and keeping it close to the surface. Without greenhouse gases in the atmosphere, the Earth would be a gigantic ball of ice and completely uninhabitable!

You feel a similar effect when getting into a car that has been parked in the sun. The car's windows allow the Sun's short-wavelength energy into the car where it heats up objects inside, such as the seats and the dashboard. The heat is converted to longer-wavelength infrared radiation, but the windows do not allow IR radiation to efficiently pass through, and some heat is trapped inside the car.

21.3 CAUSES OF CLIMATE CHANGE

The Earth's average climate is determined by its **radiative balance**, the balance between the amount of energy entering and exiting the Earth system. Changes in the radiative balance result from changes to (1) the amount of energy entering the Earth system, (2) the amount of solar energy absorbed by the Earth system, and (3) the amount of long-wave radiation emitted to space. Factors that change the radiative balance, and thus climate, are called **forcings**, and can be either natural or anthropogenic. A positive forcing leads to global warming, and a negative forcing leads to global cooling. These forcings act on a variety of different time scales, from decades to millions of years. Some of the most important factors influencing the Earth's radiative balance, and thus climate, are discussed in the following sections.

Solar Variability and the Orbital Theory of Climate

The amount of solar energy reaching the Earth varies on many different time scales. For example, the tilt of the Earth with respect to the Sun determines our annual seasons. In 1921, a Serbian engineer and mathematician, Milutin Milankovitch, theorized that periodic variations in the Earth's orbit and tilt with respect to the Sun determine climate on time scales of thousands of years. These variations are driven by gravitational interactions with the Sun, with other planets, notably Jupiter and Saturn, and with the Moon.

The orbit of the Earth around the Sun is somewhat elliptical. The roundness of the Earth's orbit, termed *eccentricity*, varies on approximately 100,000- and 400,000-year cycles (see figure 21.8*A*). The month at which the Earth makes its closest approach to the Sun, termed *perihelion*, varies on a 19,000-year cycle. If perihelion occurs during January, then Northern Hemisphere winters will be milder than if perihelion occurs during July.

Variations in the Earth's *axial tilt* or *obliquity* occur on a 41,000-year time scale (see figure 21.8*B*). The tilt ranges from 22.1° to 24.5°. More tilt means greater contrast between the seasons: warmer summers and colder winters. Less tilt means the opposite: colder summers and warmer winters.

Finally, the Earth's axis of rotation "wobbles" over time in response to the pull of the Sun and Moon and to the equatorial bulge of the Earth. This axial *precession* occurs on a period of about 23,000 years (see figure 21.8*C*).

The combination of the 100,000-year eccentricity, 41,000-year obliquity, and 23,000-year precession cycles determines the overall input and distribution of solar energy to the Earth, and thus is a major driver of Earth's climate on long time scales. This *orbital theory of climate* is known as **Milankovitch theory**.

The amount of energy radiated by the Sun also varies over time, mostly due to sunspot cycles. **Sunspots** are dark areas on the Sun's surface caused by extreme magnetic activity. Individ-

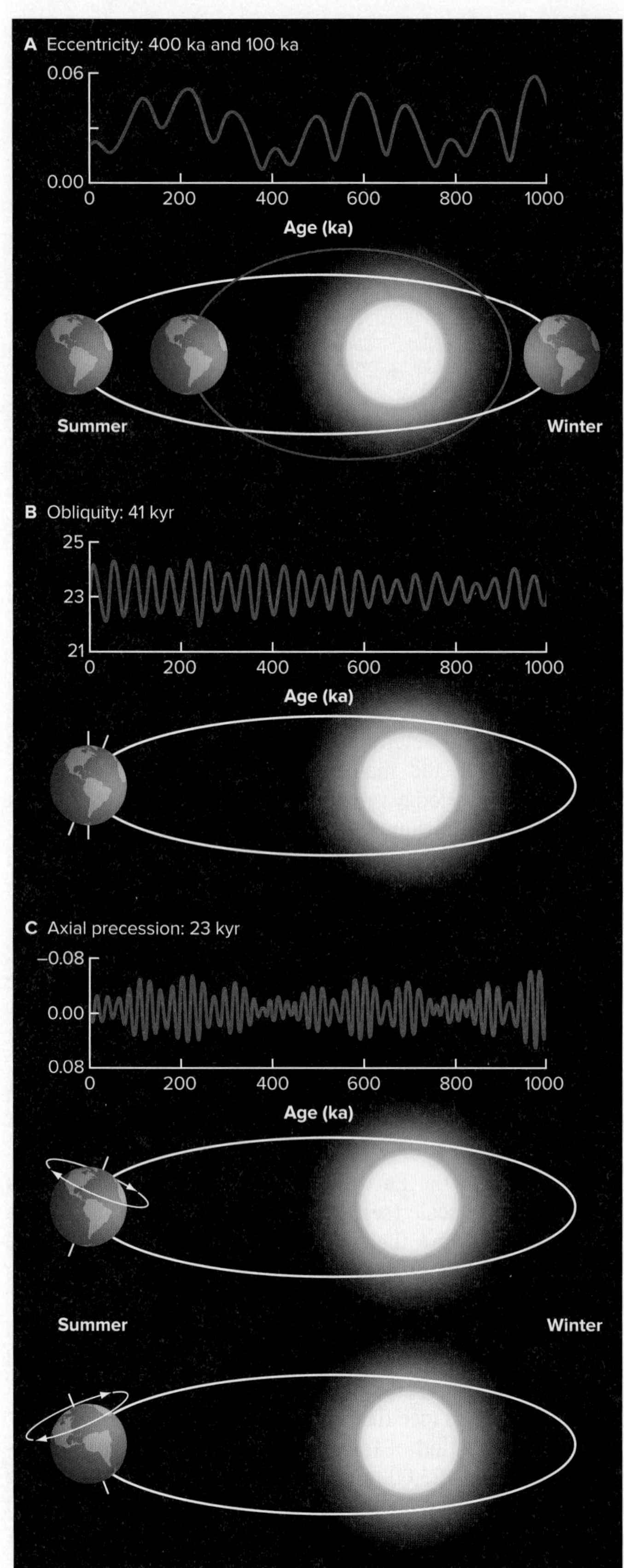

FIGURE 21.8

The Earth's orbital motions around the Sun. The graphs show the periodicity of the different orbital variations.

Source: "Trends, Rhythms, and Aberrations in Global Climate 65 Ma to Present," Science, April 27, 2001: Vol. 292, No. 5517, pp. 686–693. DOI: 10.1126/science.1059412.

ual sunspots last from days to weeks, but the overall frequency of sunspot activity typically varies on an eleven-year cycle. Periods of time with fewer sunspots tend to correspond to colder time periods on Earth; this is thought to be the case for the "Maunder Minimum" during the "Little Ice Age" that occurred in the late seventeenth century (figure 21.9). During the Maunder Minimum, sunspot activity became exceedingly rare, and this corresponded with lower-than-average global temperatures. With respect to current climate, the connection between sunspots and global temperatures is less certain, but it is nonetheless apparent that solar variability alone cannot explain the observed warming over the last sixty years.

Variations in the Earth's Albedo

As you have already learned, not all of the sunlight that reaches the Earth is absorbed by the Earth system. Instead, approximately 30% of the energy is reflected back to space. Surfaces reflect sunlight back to space with different efficiencies, determined by their albedo. Very bright surfaces, such as snow, ice, or clouds, have albedo values greater than 30%. Darker surfaces, such as oceans and forests, have lower albedo values. Changes that increase the Earth's overall albedo can lead to less energy reaching the Earth's surface and a global cooling. Changes that decrease the Earth's albedo can lead to more energy reaching the surface and global warming.

Snow and ice have high albedos, making the albedo at high latitudes (closer to the poles), on average, higher than at low latitudes (closer to the Equator). Melting of snow and refreezing of ice decreases the overall albedo, as does deposition of soot or dust on the ice surface. The conversion of high-albedo snow and ice to lower-albedo surfaces such as water or rock by melting causes an ice-albedo climate **feedback**. Feedbacks are important concepts in climate science. They are factors that amplify (*positive feedback*) or decrease (*negative feedback*) the rate of a process. One of the most dramatic examples of a positive feedback occurs when melting sea ice exposes the lower-albedo ocean surface, which absorbs more solar radiation, leading to local warming and further melting

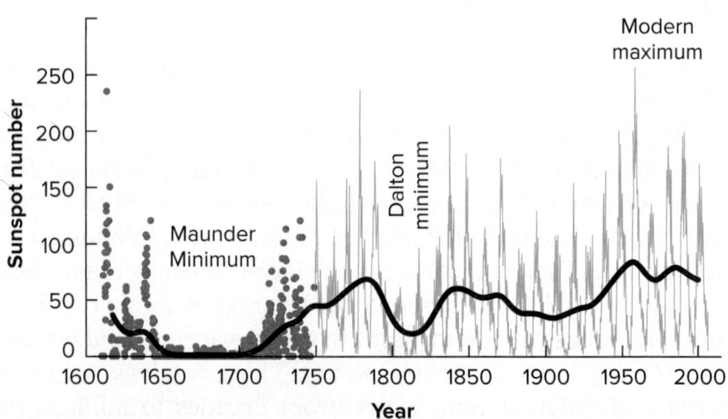

FIGURE 21.9

Sunspot cycles over the last four hundred years. The black line shows the ten-year running mean of the red and blue data. Note that there is no trend in the sunspot number since 1950.

Source: Global Warming Art Project

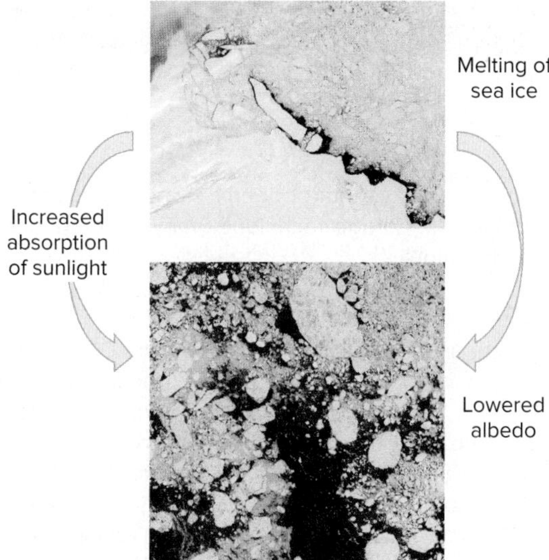

FIGURE 21.10

Schematic of the ice-albedo positive feedback. Note that melting of the sea ice both decreases the intrinsic albedo of the ice and exposes the darker seawater. *(top) Source: Jacques Descloitres, MODIS Rapid Response Team, NASA/GSFC; (bottom) Source: Robert Simmon, based on Landsat-7 data from the Global Land Cover Facility*

of the ice (figure 21.10). Melting of glaciers similarly exposes lower-albedo surfaces, creating a local feedback.

Changes to the Earth's landscape can affect the Earth's overall albedo. For example, replacing forest and grassland with cities typically increases the Earth's albedo, and if we were to paint the roofs of all the buildings in the cities white we might further increase the overall reflectivity. Conversion of forests to meadows or deserts (or vice versa), due either to direct anthropogenic activities or as a result of climate change, will also change the albedo of an area. For example, desertification, which is the continuous degradation of dry-land ecosystems, either due to human activities (e.g., overgrazing) or due to climate variation, tends to increase the Earth's albedo. While desertification increases albedo, a cooling effect, it also removes plants that take up CO_2, causing a net warming. The net effect of land-use change is to increase the Earth's albedo and to cool the atmosphere, although melting of snow and ice, and the resulting positive feedback, is a warming effect and is projected to increase in the future.

At any given time, much of the Earth is covered by clouds. Clouds therefore play a very important role in determining the global average albedo. When clouds exist over the oceans they have a large effect on the global average albedo by effectively making the dark oceans look bright. Conversely, if clouds occur over ice, their impact is smaller because they are covering an already bright surface.

There are multiple potential feedbacks associated with clouds, both positive and negative. One possible negative feedback mechanism is related to the behavior of low-level clouds. As the planet warms, increased evaporation from the oceans could lead to the formation of more low clouds in the atmosphere, which would increase the Earth's albedo, which would counter the initial warming.

Greenhouse Gases

Greenhouse gases play an important role in warming the Earth's surface. Changes in the amount of greenhouse gases in the atmosphere will change the amount of terrestrial radiation that is absorbed by the atmosphere and re-radiated back to the Earth's surface. Table 21.1 shows the most common greenhouse gases. Carbon dioxide (CO_2) is the most abundant and generally most important greenhouse gas, although nitrous oxide (N_2O) and methane (CH_4) are also important due to their strong absorption of terrestrial radiation and changing concentrations. Another important greenhouse gas is tropospheric ozone (O_3). Note that while the ozone layer in the stratosphere shields the planet from harmful UV radiation, ozone in the troposphere is considered a greenhouse gas. An important class of greenhouse gases is the halocarbons, which include the chlorine, bromine, and fluorine-containing molecules that are also responsible for damaging the ozone layer. Unlike CO_2, N_2O, CH_4, and O_3, there are essentially no natural sources of these compounds.

Each of the greenhouse gases has a different ability to affect the Earth's climate. This is for two main reasons: (1) each greenhouse gas absorbs radiation with varying efficiencies and (2) each greenhouse gas can survive in the atmosphere for different amounts of time, which is known as the **atmospheric lifetime** (table 21.1). For example, CO_2 has an atmospheric lifetime of hundreds to thousands of years, whereas CH_4 has an atmospheric lifetime of only twelve years. However, CH_4 currently absorbs infrared radiation much more efficiently than CO_2 in our atmosphere. The **global warming potential (GWP)** is a parameter that accounts for these differences between greenhouse gases to provide a relative measure of how much heat a particular gas will trap in the atmosphere (table 21.1). All GWPs are defined relative to CO_2, so the GWP of CO_2 is always equal to 1. The definition of GWP also accounts for differences in atmospheric lifetime, and thus the GWP must be specified over different time horizons. For our example, we find that the overall hundred-year GWP for methane is about 25, meaning that for equal emissions of methane and CO_2 by mass today, the heat-trapping ability of methane is 25 times as large as for CO_2 over the next one hundred years.

To understand the impacts of different greenhouse gases, we need to know how their concentrations in the atmosphere have changed with time. In the last century, human activities have caused dramatic increases in concentrations of many greenhouse gases (figure 21.11). For example, increases in CO_2 concentrations correspond extremely well with increases in fossil fuel usage (figure 21.12), and nitrous oxide increases occur alongside the rise in the use of synthetic nitrogen fertilizers. This increase in greenhouse gas concentrations corresponds to a *positive forcing*, meaning that it causes an increase in the average global temperature. While fluctuations in greenhouse gas concentrations in the atmosphere have occurred throughout geologic history, the rise observed throughout the twentieth century occurred rapidly, and the relative changes were large. Some of the major anthropogenic activities that

IN GREATER DEPTH 21.2

The Global Carbon Cycle

The carbon cycle (box figure 1) describes the way that carbon is exchanged between the hydrosphere, lithosphere, biosphere, and atmosphere. Each of these components of the Earth system act as carbon *pools*, which are reservoirs for carbon. Carbon moves between these pools through both natural and anthropogenic processes. The rate at which carbon leaves a pool determines its lifetime in that pool. The longer carbon atoms stay in the atmosphere as CO_2, the greater the effect on climate. In the atmosphere, CO_2 has a lifetime on the order of centuries, although recent studies suggest that a substantial fraction of CO_2 emitted into the atmosphere may remain for tens of thousands of years. The amount of CO_2 in the atmosphere is defined by its sources and sinks, while the lifetime depends only on the sinks. Natural sources of CO_2 include the biosphere (respiration by soils, plants, and animals; decay of organic matter) and lithosphere (volcanic eruptions, metamorphism). The oceans are currently a net sink for CO_2, although the capacity of the ocean to take up CO_2 decreases as temperature increases and, in certain regions, ocean waters are a CO_2 source. Anthropogenic sources of CO_2 include combustion of fossil and biofuels and cement production. Only a small fraction of the total available fossil fuel carbon has so far been put into the atmosphere as CO_2. Forest fires also release CO_2 from the biosphere to the atmosphere, and they can be either natural or anthropogenic. A variety of natural processes remove CO_2 from the atmosphere on different time scales. These include photosynthesis by terrestrial plants and oceanic microorganisms, dissolution in the ocean, and weathering of silicate minerals in rocks. Anthropogenic carbon sinks are minor, but carbon capture and storage techniques may become increasingly important in the future.

The Oceans

The atmosphere and the oceans readily exchange CO_2. In the oceans, dissolved CO_2 at the surface converts to bicarbonate (HCO_3^-) and carbonate (CO_3^{2-}) ions. These ions are used by organisms to produce shells made of the mineral calcite ($CaCO_3$). When these organisms die, they sink slowly to the deep ocean floor. As the calcite shells accumulate on the ocean floor, they form the sedimentary rock limestone (see chapter 5). Limestone is the largest carbon pool on the planet. Ocean circulation brings deep ocean waters to the surface and surface water down to greater depths. This is important because the deep waters have not been in contact with the atmosphere for hundreds of years, and therefore they have a lower CO_2 content and can absorb more CO_2 from the atmosphere. This replenishment is key to keeping the ocean as a net carbon sink; if oceanic mixing were to slow down, then the surface waters could become saturated in CO_2, dramatically slowing uptake from the atmosphere. If that were to happen, the CO_2 levels in the atmosphere would start to increase at an even faster rate than they are today, increasing the rate of global climate change.

The Biosphere

In the biosphere, carbon plays an important role as a building block for life: it accounts for about half the dry weight of most living

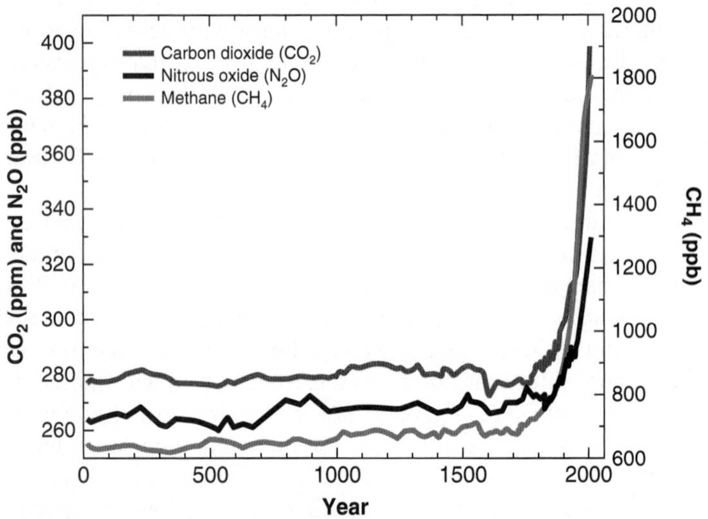

FIGURE 21.11

Atmospheric concentrations of the important long-lived greenhouse gases CO_2, methane (CH_4), and nitrous oxide (N_2O) over the last 2,000 years in the Southern Hemisphere.
Source: NOAA National Climatic Data Center and NOAA Earth System Research Laboratory Global Monitoring Division

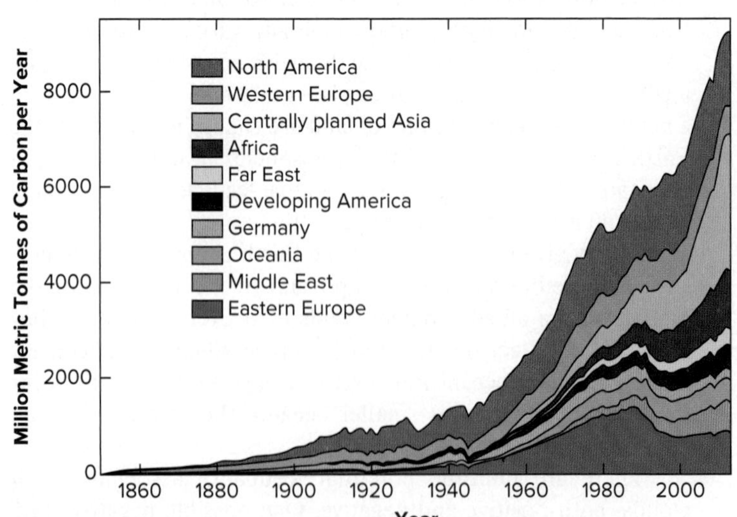

FIGURE 21.12

Emissions of carbon from fossil fuel usage, in millions of metric tonnes per year, colored by global region. Data are from the Carbon Dioxide Information Analysis Center.
Source: Data from http://cdiac.ornl.gov/trends/emis/meth_reg.html

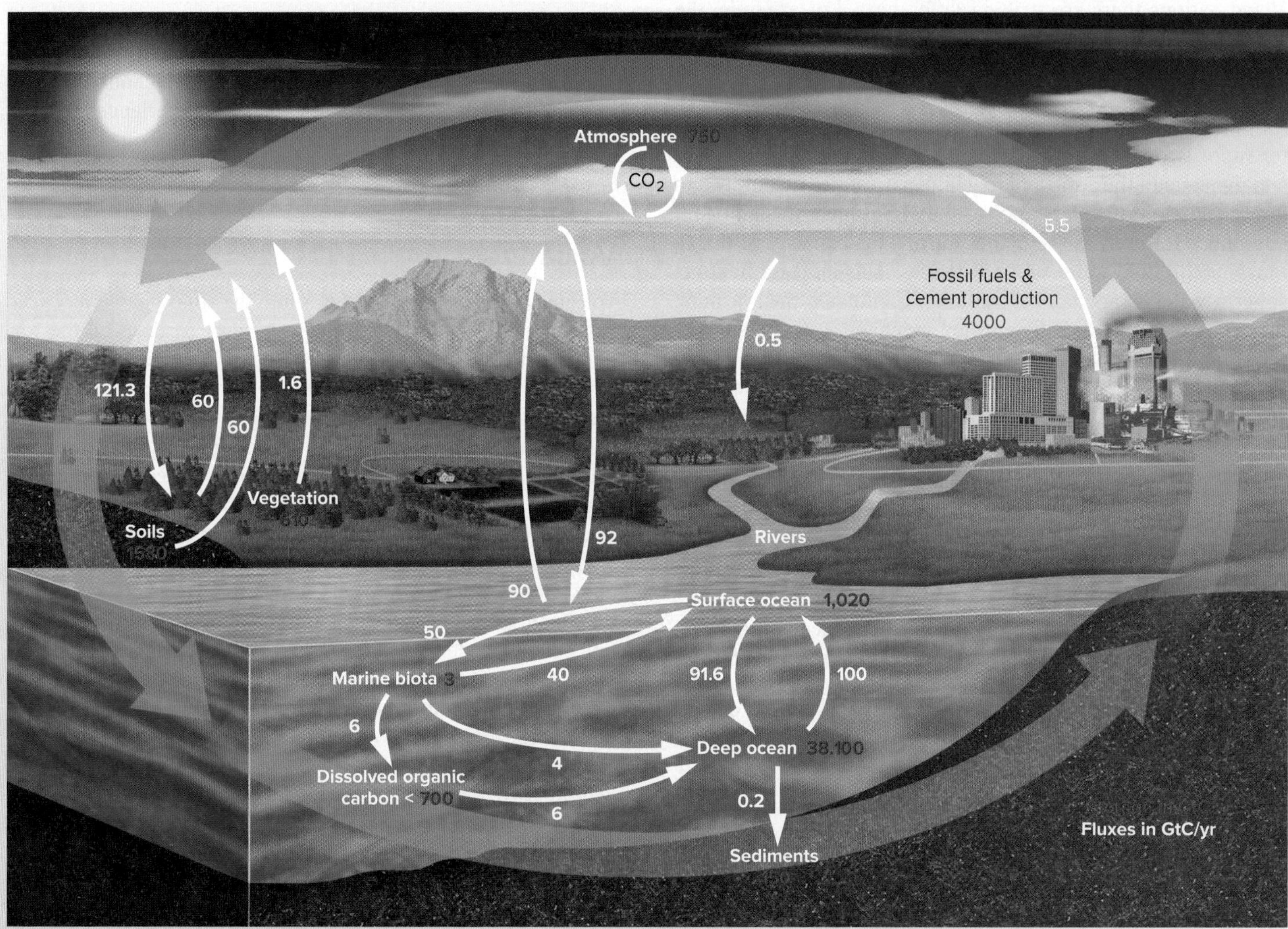

BOX 21.2 ■ FIGURE 1

The carbon cycle. The amount of carbon stored in the different pools is shown as orange numbers in gigatons. The amount of carbon exchanged between pools by different processes is shown as white numbers in gigatons per year. The flux of carbon from anthropogenic sources is 5.5 gigatons per year. While this is smaller than some other fluxes, note that it is not an exchange; that is, no carbon is absorbed from the atmosphere by burning fossil fuels.
Source: http://en.wikipedia.org/wiki/File:Carbon_cycle-cute_diagram.svg

organisms. *Photosynthesis* is the biochemical process by which plants convert CO_2 into sugar molecules using solar radiation:

$$6CO_2 + 6H_2O + \text{sunlight} \rightarrow C_6H_{12}O_6 + 6O_2$$

CO_2 + water+ sunlight *goes to* sugar + oxygen

Plants use these sugars to grow, and CO_2 is removed from the atmosphere and stored as biomass. As plants are eaten and used for energy, CO_2 is released in a process that reverses the photosynthesis reaction, *respiration*. Eventually, as the organic matter from plants or animals decays, it becomes part of the soil. Soil carbon eventually decays to emit CO_2 over a time scale of decades to centuries, making it a more permanent sink for carbon than aboveground biomass, such as crops and forests. However, soil carbon lifetime depends on the physical and chemical properties of the soil, and is subject to land use. For example, tilling soil for crop cultivation releases carbon and reduces its lifetime in the soil. Also, as temperatures increase, the rate of soil matter decay may increase, leading to a more rapid return of carbon to the atmosphere.

lead to emissions of greenhouse gases are given in table 21.1. Emissions of greenhouse gases are not distributed equally around the planet. In general, people in more industrialized countries tend to emit more greenhouse gases per person than those in developing countries. In fact, even among the industrialized nations there are striking differences. For example, the average person in the United States is responsible for the emission of twice as much carbon per year as the average person in Western Europe.

Water vapor is an extremely important greenhouse gas. Water vapor strongly absorbs IR radiation and is highly abundant and dynamic in the atmosphere. Human activities do not lead to direct emission of water vapor to the atmosphere in significant amounts. However, as the planet warms, the capacity of the atmosphere to hold water vapor increases. This increase in atmospheric water vapor concentrations accentuates the warming influence of other, directly emitted greenhouse gases (such as CO_2), and thus serves as a positive feedback to the climate system. Further, increasing the amount of water vapor in the atmosphere will alter the hydrologic cycle, likely increasing the amount of precipitation and the amount of energy in the atmosphere.

Clouds and Particles

Cloud droplets and atmospheric **aerosols**, which are small particles suspended in the atmosphere, together reflect around 23% of the incident solar radiation back to space, which is approximately two-thirds of the total reflected solar radiation.

As you learned in chapter 13, clouds and cloud droplets are formed in the atmosphere when the atmosphere cools sufficiently to become super-saturated with water. This water will condense on preexisting aerosols to grow into cloud droplets. Cloud formation is intimately linked with large-scale atmospheric motions.

There are both natural and anthropogenic sources of aerosols. Natural sources of aerosols include sea spray, biological material (e.g., fungal spores, pollen, bacteria), volcanic activity, and forest fires caused by lightning. Anthropogenic sources include combustion and industrial activity. Aerosols are among the most short-lived radiative forcing agents; they have an atmospheric lifetime of about a week. Therefore, the forcing by aerosol particles requires a continuous input, without which their concentrations would rapidly drop toward zero.

The net effect of particles in the atmosphere is to cool the planet, although particles affect the Earth's radiative balance in multiple ways:

1. Aerosol particles can directly reflect incoming sunlight. This is termed the *direct effect* and is a net negative forcing (cooling).
2. Aerosols can influence cloud formation. All cloud droplets form on preexisting aerosols, and thus cloud properties depend on the number of aerosol particles in the atmosphere. Increasing aerosol concentrations generally cause clouds to be made of more and smaller droplets that reflect more solar radiation overall. This is termed the cloud albedo effect, and it is a net negative forcing.
3. Aerosols can absorb solar radiation as part of the direct effect. Certain aerosol types, in particular soot (aka black carbon), strongly absorb solar radiation. Absorption by soot aerosols is a positive forcing and causes warming.
4. Cloud droplets and large dust particles will absorb long-wave IR radiation, essentially acting like greenhouse gases and warming the planet. You may have experienced this last effect; it is the reason why it is usually colder on clear nights than on cloudy nights.

Volcanoes

Volcanic activity releases particles into the atmosphere every year. Particularly large volcanic events can inject particles all the way into the stratosphere, where the particles have lifetimes of many months or even years. By scattering solar radiation back to space, these particles increase the albedo of the Earth. Large volcanic events can lead to observable cooling of the entire Earth system. Mount Tambora is an active composite volcano located in Sumbawa, Indonesia. In April 1815, Tambora erupted explosively, ejecting 160 cubic kilometers' worth of material and sending around 60 tons of particulate-forming sulfur into the atmosphere up to altitudes of 43 kilometers. It was the largest observed eruption in recorded history. Much of the ejected material in the eruption column reached the stratosphere. The largest particles settled out due to gravity over the next few weeks, but smaller particles remained in the stratosphere for several years, eventually spreading out over the entire globe. As a result of the stratospheric injection, brilliant sunsets and twilights, which result from scattering of sunlight by particles in the atmosphere, were observed as far away as London. More importantly, the scattering of solar radiation by these particles also had an influence on global climate by decreasing the amount of sunlight coming into the Earth system. The year following the Tambora eruption is often referred to as the "Year Without a Summer" due to the anomalously low temperatures experienced in Europe and North America. It even snowed in New York and Maine in June! It has been estimated that temperatures in 1816 were 0.5°C lower than normal due to the eruption. The lower temperatures, coupled with the decreased availability of sunlight for photosynthesis by plants, led to large-scale crop failures and livestock death and resulted in the worst famine of the nineteenth century. The pall of darkness cast across Europe perhaps not surprisingly led to the writing of *Frankenstein* by Mary Shelley, *Darkness* by Lord Byron, and *The Vampyre* by J. W. Polidori.

Tambora is not the only volcanic eruption that has had global consequences. Many of the more recent large volcanic eruptions, including Agung (1964), El Chichón (1982), and Pinatubo (1991), led to global temperature decreases in the year that followed the eruption.

Active volcanoes are also continuously releasing CO_2 into the atmosphere. Today, the contribution of volcanoes to overall CO_2 levels is extremely small; globally, volcanoes emit around the same amount of CO_2 annually as does a state the size of

Michigan, and there is no indication that the emissions have changed over the recent past. Even exceptionally large volcanic events do not by themselves lead to noticeable changes in atmospheric CO_2 levels. While this effect is minor today, CO_2 emissions from volcanoes may have played an important role in pulling the planet out of Snowball Earth conditions 650 million years ago (discussed further in chapter 12). Because the Earth was almost completely covered by snow and ice, weathering of rocks, which tends to trap CO_2, was inhibited, and thus the continuous volcanic emissions of CO_2 over millions of years could lead to a buildup in the atmosphere of very high concentrations of CO_2. It is thought that CO_2 levels roughly 350 times higher than they are today would have been necessary to thaw the Earth. Such high CO_2 levels would have led to some melting near the equator and the triggering of a positive feedback that pulled the Earth out of the Snowball Earth period.

Plate Tectonics

On million-year time scales, plate tectonics alters the surface albedo of the Earth and affects ocean circulation, which together control the distribution and transfer of heat around the planet. Movements of the plates may affect the cycling of carbon between the atmosphere and the biosphere and oceans, thus changing the abundance and lifetime of CO_2 and methane in the atmosphere. During periods when large supercontinents existed, there were likely to be extreme temperature swings between seasons in the continental interior and the occurrence of "megamonsoons." Tectonic uplift and the formation of mountain ranges also strongly affect climate by changing atmospheric circulation patterns and the potential for glacier formation. Although extremely important to understanding climates of the past, the movement of plates is too slow to effect climate changes on scales that humans are typically concerned with.

21.4 A BRIEF HISTORY OF EARTH'S CLIMATE

Over long time scales, the Earth's climate is determined by a few primary factors: the solar input, the concentration of greenhouse gases, and the average global albedo. All of these have varied dramatically over geologic time. In order to understand how climate depends on these factors, and thus be able to predict how climate is likely to change in the near and long-term future, we need to look back in time at ancient climate, a field of study known as **paleoclimatology**.

Paleoclimate studies require that scientists find paleoclimate *proxies*–materials that have retained a unique signature of the ancient climate conditions. The signatures that can be captured are not limited to temperature. Proxies have been found that provide information on ice volume, sea level, wind speed, precipitation, and greenhouse gas levels. These proxies can take many forms. A few examples among many are: the number of pollen spores per volume of sediment; the width of tree rings; the concentrations of noble gases in groundwaters; the stable isotopic composition of many different items (e.g., ice cores, speleothems, carbonate minerals, fossils–see box 2.1); and the composition of air bubbles trapped in ice cores. Each of these proxies provides a record at a given location going back a certain amount of time. As you might guess, the resolution, precision, and accuracy of these proxy methods tend to get worse the further back in time we go. Nonetheless, by combining multiple paleoclimate proxies, scientists have been able to build up a relatively comprehensive picture of what the Earth's climate looked like in the past.

Climate Millions of Years Ago

On really long time scales, we need to account for changes in the solar input and greenhouse gas concentrations and for changes to the continental geography. Global temperatures during pre-Quaternary periods, prior to 2.6 million years ago, were typically higher than they are today (figure 21.13) and were generally associated with atmospheric CO_2 levels that were significantly higher than they are today. For example, during the Early Eocene (52 to 48 million years ago), atmospheric CO_2 concentrations exceeded about 1,000 ppm, more than three times as large as in recent preindustrial times, and temperatures were correspondingly higher. And during the Paleocene–Eocene Thermal Maximum (PETM), a brief period at the boundary between the Paleocene and Eocene lasting only a few tens of thousands of years, there was a massive release of carbon into the atmosphere that precipitated a large increase in global temperatures; the origins of this massive carbon release are still debated by scientists. This indicates that there is a connection between higher greenhouse gas levels and warmer climates. On such long time scales the carbon cycle, and thus CO_2 levels, are thought to be controlled primarily by tectonic processes.

Concentrations of atmospheric CO_2 millions of years ago can be reconstructed from the isotopic content of carbon in fossils and from the density of stomata in fossilized leaves (stomata in plants control how CO_2 and H_2O move in and out of plants).

Climate Over the Last Million Years

The Milankovitch theory of climate was not put to the test until 1976, when Hays, Imbrie, and Shackleton used measurements of oxygen isotopes in planktonic foraminifera in deep-sea sediment cores to extract a signature of global climate over the past 450,000 years. The relative abundance of different oxygen isotopes in the foraminifera is determined in large part by the temperature of the water where they formed (see box 2.1). Variations in the oxygen isotopes therefore tell us about how past temperatures varied. The sediment cores revealed clear signatures of periodicity in past climate corresponding to 23,000, 41,000, and 100,000 years, in accordance with Milankovitch's prediction. Since then there have been many more investigations, over even longer time scales, confirming that variations in orbital parameters play a key role in determining climate on long time scales (figure 21.14).

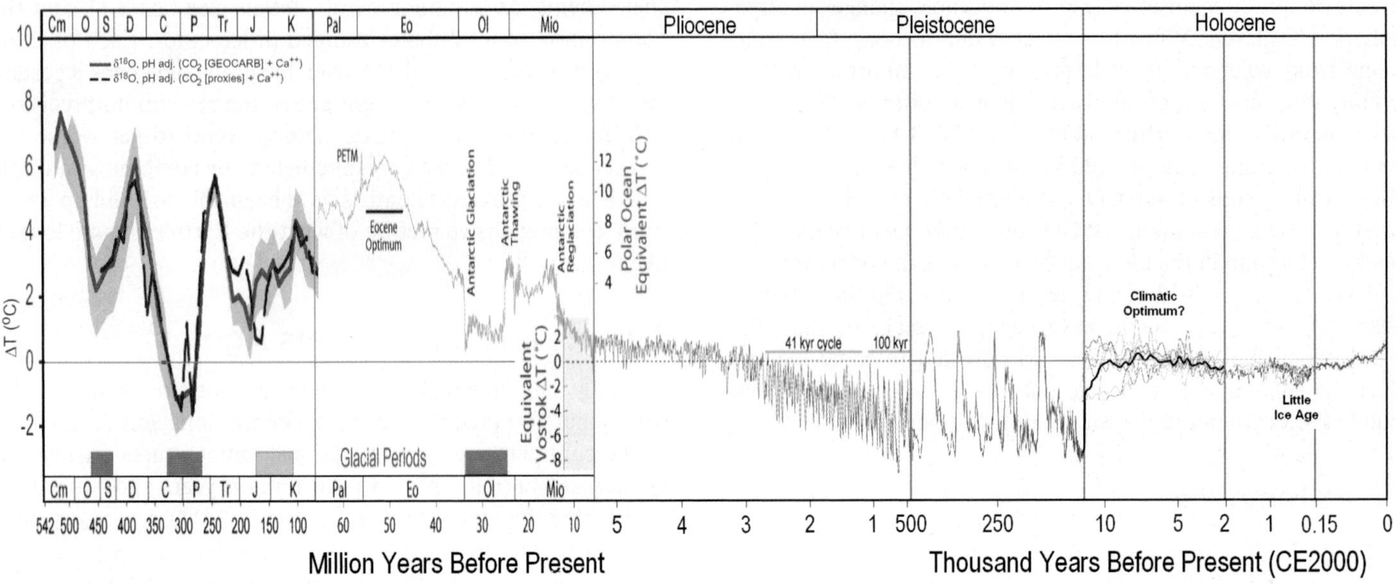

FIGURE 21.13

Variations in Earth's temperature over the last 550 million years, relative to modern times. At various points in the geologic record, the temperature was substantially higher than it is today. These periods of high temperature corresponded to periods when the atmospheric concentrations of greenhouse gases were particularly large.

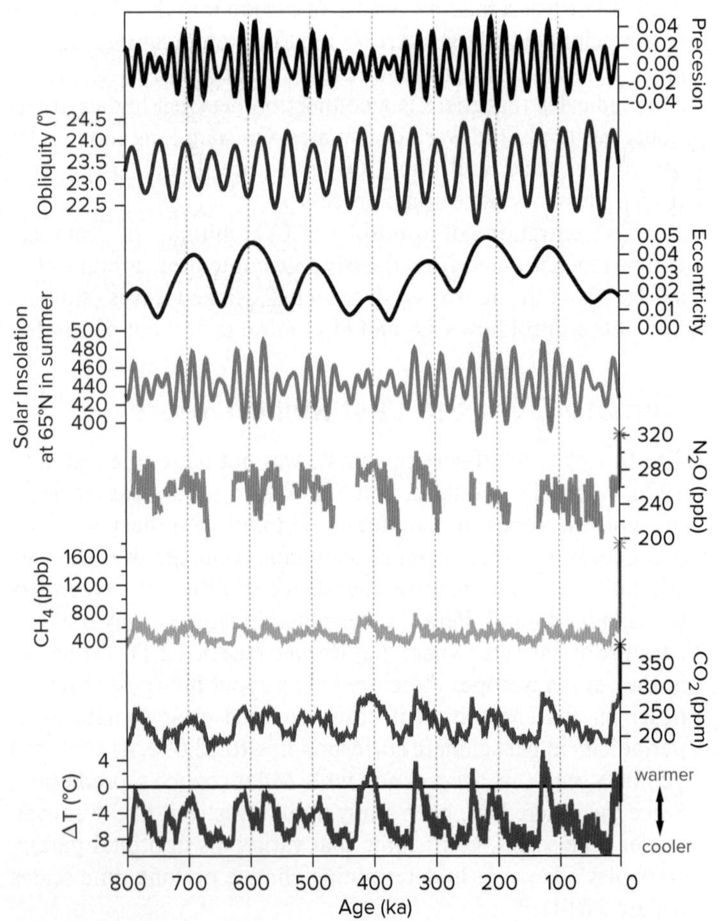

FIGURE 21.14

Global temperatures depend on the orbital forcing on millennial time scales and co-vary with atmospheric concentrations of the greenhouse gases carbon dioxide, methane, and nitrous oxide. The reconstructed temperature is shown as a difference from the average of the period 1950 to 1980. The temperature and greenhouse gas records were obtained from an ice core collected at EPICA Dome C in Antarctica. *Source:* *www.ncdc.noaa.gov/paleo/icecore/antarctica/domec/domec_epica_data.html* and *ftp://ftp.ncdc.noaa.gov/pub/data/paleo/insolation/*

However, variations in the orbital parameters are not the only important considerations. Greenhouse gas concentrations have also varied quite dramatically over the geologic record. How do we know this? One important way is that scientists have been able to extract ice cores thousands of meters long from central Greenland and Antarctica (figure 21.15 and box 12.3). Like the sediment cores discussed previously, these ice cores contain signatures of climate over hundreds of thousands of years. One unique aspect of the ice cores is that ancient air may be trapped as bubbles within the ice. From these bubbles, it has been possible to measure how concentrations of greenhouse gases varied in the past. Additionally, the oxygen and hydrogen isotope ratios of the ice itself provide information on how temperature has varied.

Figure 21.14 also shows the variations in CO_2, methane, nitrous oxide, and temperature over the last 800,000 years. The first thing to notice is that temperature has gone up and down, with major changes occurring on approximately 100,000-year cycles. The cold times correspond to periods of extensive glaciation, while the warm times correspond to interglacial periods. This figure shows us that we are currently in a warm interglacial period. Second, we see that there is a strong correspondence

FIGURE 21.15

Ice cores obtained from ice sheets contain a record of hundreds of thousands of years of past climate. Some of these ice cores are over 3,500 meters (11,000 feet) deep and contain information on the last million years of Earth history. These photographs are from the Greenland Ice Cap in 2005. For more details on ice cores and drilling, see http://www.icecores.org/.
©Reto Stöckli

between changes in temperature and changes in CO_2 and methane concentrations. Although the Milankovitch theory can help us to understand the timing of past glacial and interglacial periods, the magnitude of the temperature changes can only be understood if we allow for feedbacks within the Earth system that influence atmospheric greenhouse gas concentrations. It is well established that, over at least the last five glacial cycles, the deglaciations began in response to changes in solar forcing. This initial warming likely affected global biogeochemical cycles and ocean circulation patterns, causing atmospheric concentrations of CO_2 and methane to increase, thus amplifying the effect of the orbital forcing–a positive feedback.

Climate Over the Last Few Thousand Years to Present

The *last glacial maximum* ended approximately 15,000 years ago, driven by orbital forcing but amplified by increases in greenhouse gas concentrations. As we get closer and closer to the present, a greater number of proxies can be used to assess the ancient climate, although ice cores and fossils trapped in sediments are still important.

On somewhat shorter time scales (about 50,000 years before present), *borehole thermometry* can be used to determine past temperatures on any land surface, not just where ice sheets exist. Borehole thermometry works by measuring temperatures deep in the ground and using our understanding of heat transfer in materials to infer what the surface temperature must have been in the past. On short time scales (a few thousand years) borehole thermometry can be highly accurate.

An additional important proxy as we move closer to the present is *tree rings.* Local climate (temperature, precipitation, etc.) influences the growth of trees. By counting the number of rings we can move through time, and the width, density, and isotopic composition of the rings tell us about climate at the time they were formed.

As we move even closer to the present day, temperature records from direct measurement become available, known as the *instrumental temperature record.* We have a reliable record of global temperature starting from around 1850, and the spatial coverage has continued to improve.

When we look at the modern temperature record compared to the past, we find something important happens around the time of the Industrial Revolution (figure 21.16). At this point, global mean temperatures begin to increase substantially and quite rapidly. Although there have been some ups and downs in the global temperature record, the trend over the last century is clear–the Earth is warming. The *rate* of this warming is unprecedented when we compare it with what we know from the geologic record. Over the last fifty years, the rate of increase has been about 0.16°C per decade (0.26°F per decade).

21.5 CLIMATE CHANGE IN THE MODERN AGE

We have now seen that many factors work together to influence and determine the Earth's climate, and key among these are orbital variations and atmospheric concentrations of greenhouse gases. Since 1750, the net radiative forcing for the Earth has been positive, with the sum of anthropogenic activities up to 2005 being 1.6 (±1) W/m^2, and this has led to a clear and rapid increase in the Earth's mean temperature (figure 21.16). The units W/m^2 indicate the amount of energy per unit time per unit area. This net forcing value is the sum of the positive forcing by greenhouse gases and the net negative forcing by clouds and aerosols (figure 21.17). The net positive forcing is leading to changes in Earth's climate. But how do we know that modern climate is changing? Or what is driving current changes in climate? And what is expected to occur in the future? This is explored in the following sections.

Temperature

Global temperature trends are typically taken as the average over the entire planet's surface. Numerous analyses of observational data sets indicate that near-surface air temperatures have risen globally, by about 1.2°C between 1901 and 2016, which corresponds to a rate of 0.11°C per decade (figure 21.16). Importantly, the rate of warming has been increasing, to an average of 0.15 per decade from 1951 to 2016. While the rate of increase slowed somewhat from around 2001 to 2012, temperatures in recent years have increased rapidly, and 2016 was the hottest on record. Some variability in the rate of warming is expected on

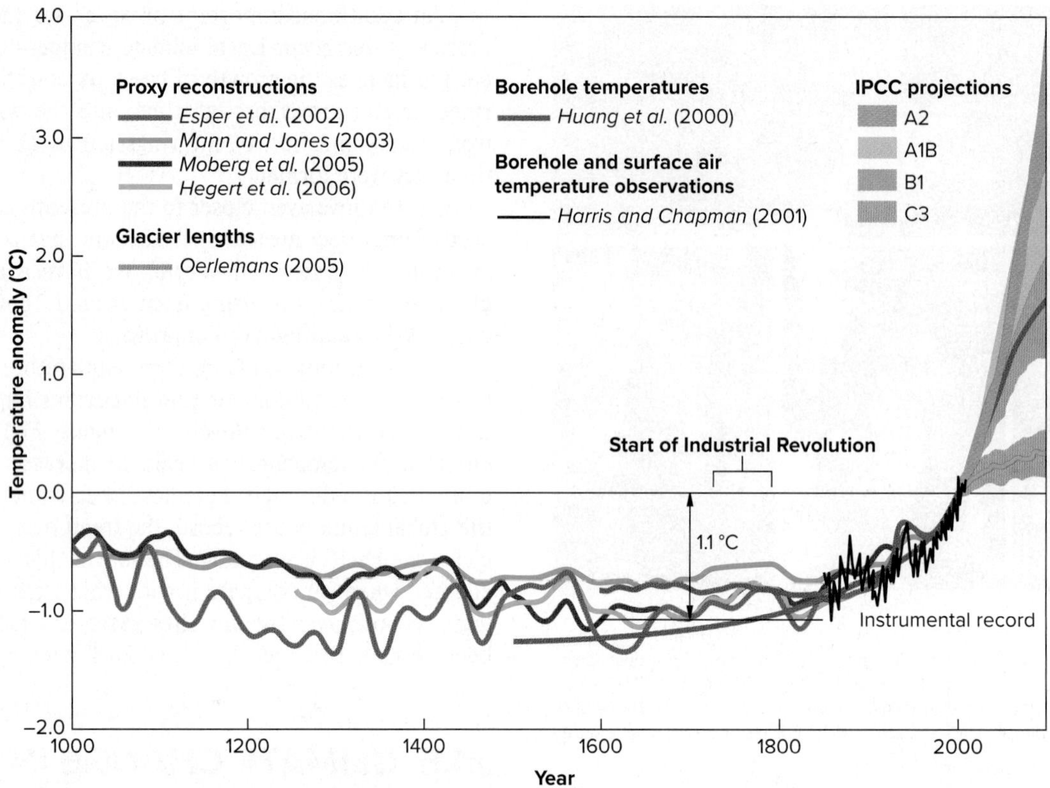

FIGURE 21.16

Temperatures over the last 1,000 years as determined using numerous proxies, the instrumental record, and projected into the future. The black lines show the instrumental record. IPCC stands for Intergovernmental Panel on Climate Change.

Modified from: D. S. Chapman and M. G. Davis, "Climate Change: Past, Present, and Future," Eos, Vol. 91, No. 37, September 14, 2010, p. 325.

decadal time scales, given the inherent variability of the climate system. Regardless, an overall upward trajectory in global temperatures is evident in the modern record.

If the concentration of CO_2 in the atmosphere doubles, it is expected that the Earth will warm by around 2.5°C to 4°C (4.5°F–7.2°F), although much higher values are possible. Scientists give ranges for warming rather than an exact figure because the complexity of the climate system and uncertainties associated with the strength of various feedbacks, such as the water vapor or carbon cycle feedbacks, make prediction very difficult. It is expected that by 2100 the Earth's temperature will increase by around 1.5°C to 4°C, although this depends on how rapidly greenhouse gases are emitted in the future (figure 21.16).

The changes in surface temperature are not evenly spaced across the globe. For example, air temperatures above land surfaces are warming faster than over oceans due to differences in their heat capacities; it takes more energy to heat up an equivalent volume of water than it does land. Also, temperatures at high latitudes in the Northern Hemisphere are increasing at twice the rate of the rest of the globe, which has been contributing to melting of sea ice and glaciers. It is also possible to find certain locations where the temperature has actually decreased over the last century. However, these regions are few and far between, and the planet is overall becoming warmer. Temperatures are changing not only at the surface, but in the atmosphere above. Since the mid-twentieth century, the troposphere as a whole has been warming, whereas the lower stratosphere has been cooling. This distinction is important because if the recent warming was due to increased solar radiation, we would expect to see the stratosphere warming, not cooling. Instead, this pattern—increasing surface and tropospheric temperatures, but decreasing stratospheric temperatures—is consistent with greenhouse gases being the primary drivers of current climate change. In fact, models of climate are only able to reproduce the warming over the last century if greenhouse gases are included.

Precipitation

Global climate change involves more than just changing temperatures: precipitation levels are also changing globally. Precipitation is an important part of the hydrologic cycle, returning water from the atmosphere to the surface. Water is provided to the atmosphere through evaporation from surface reservoirs, such as the oceans, lakes, rivers, and the biosphere, and it eventually condenses to form clouds and precipitation. Because temperature affects evaporation, convection, and movement of air

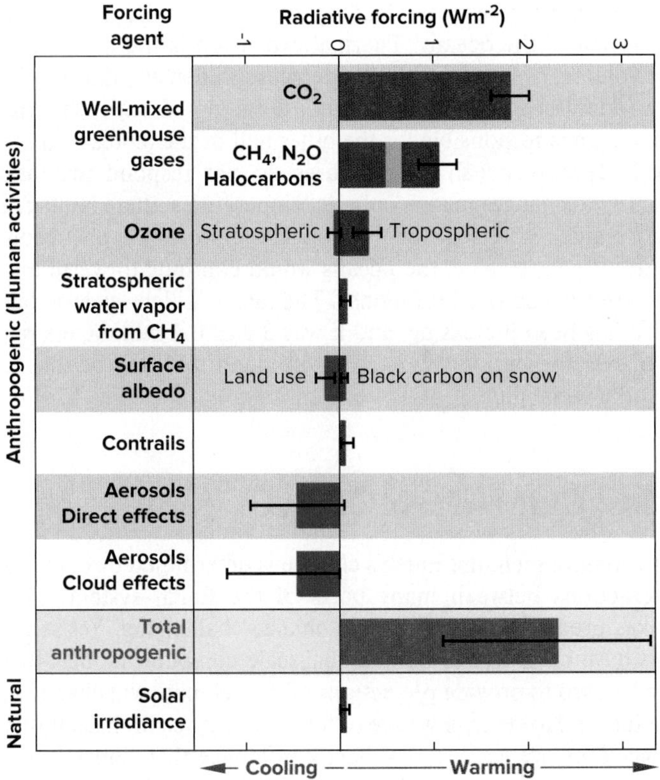

FIGURE 21.17

The radiative forcing of climate between 1750 and 2011 for different forcing agents, such as greenhouse gases and aerosols. Red bars indicate positive forcings (warming) while blue bars indicate negative forcings (cooling). The net total anthropogenic forcing, shown in orange, is positive.

Data adapted from Fig. 8.15 in Climate Change 2013: The Physical Science Basis. Contribution of Working Group I to the Fifth Assessment Report of the Intergovernmental Panel on Climate Change

and because warmer air can hold more moisture, precipitation changes are expected on a warming planet. Over the twentieth century, observations indicate that some areas have become wetter, including eastern North America, southern South America, and much of Europe, while other areas have become drier, including the Sahel desert, southern Europe, and southern Africa. It is expected that these regional differences will only grow larger into the future.

Sea Ice and Glacier Melt

Snow and ice cover is currently decreasing on the Earth's surface. Most glaciers are shrinking, and sea ice levels are decreasing, particularly in the Northern Hemisphere. Since 1979, when satellite records began, the summer-minimum sea ice coverage in the Northern Hemisphere has receded by an average of 3.4% per decade (figure 21.18). Most projections suggest that the sea ice will be completely gone by the end of the century, possibly as soon as 2030. Until 2009, the famed sea route through the Arctic, the Northwest Passage, was essentially unnavigable, but the retreat and thinning of sea ice over the last decades has now made it possible for ships to more reliably travel through this region.

Ice sheets are also showing evidence of changes. Over the last century, the ice sheets covering Greenland and Antarctica have lost mass because the amount of new snow and ice accumulated in the winter has not kept up with loss to melting in the summers, and many glaciers around the world have been exhibiting melting (see box 21.3). Melting of glaciers in Norway, for example, has led to some recent archaeological finds of items that were buried in the snow thousands of years ago. This melting and loss of ice has been responsible for around 50% of the sea-level rise over the last century. If the entire Antarctic ice sheet were to melt, global sea level would rise by around 70 meters (230 ft). Thankfully, such a scenario is unlikely, but it is possible that significantly greater ice sheet loss will occur in the future. Such loss was dramatically seen in 2002 when the Larson B ice shelf in Antarctica broke apart over the course of a month (figure 21.19).

Permafrost regions are also beginning to thaw. Many activities in the northernmost latitudes rely on the permafrost to allow for travel and infrastructure development. Melting of the permafrost has long-term implications for the people who live in these areas, as melting permafrost affects infrastructure such as roads, houses, and pipelines. Further, melting permafrost may lead to the release of methane and CO_2 that are trapped within the permafrost.

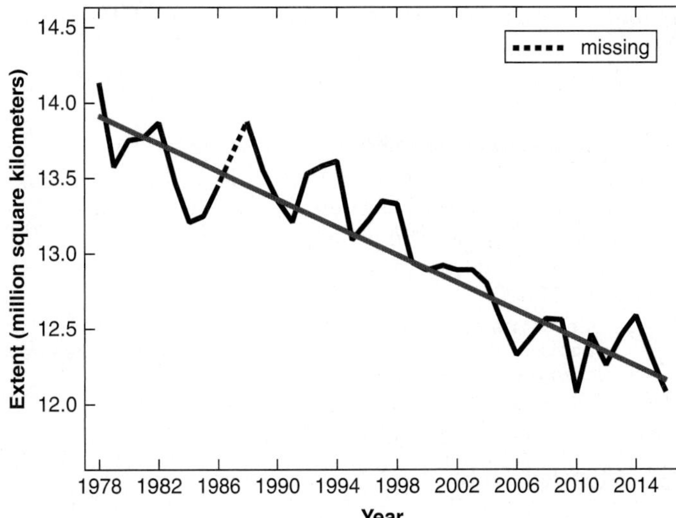

FIGURE 21.18

Arctic sea ice has been decreasing over the last 30 years. Sea ice extent at the end of the melt season in September is determined by conditions in the atmosphere, the ocean, and of the ice cover itself. Data are from the National Snow and Ice Data Center. For more details on sea ice, see http://nsidc.org/data/seaice_index/.
Source: (NSIDC)

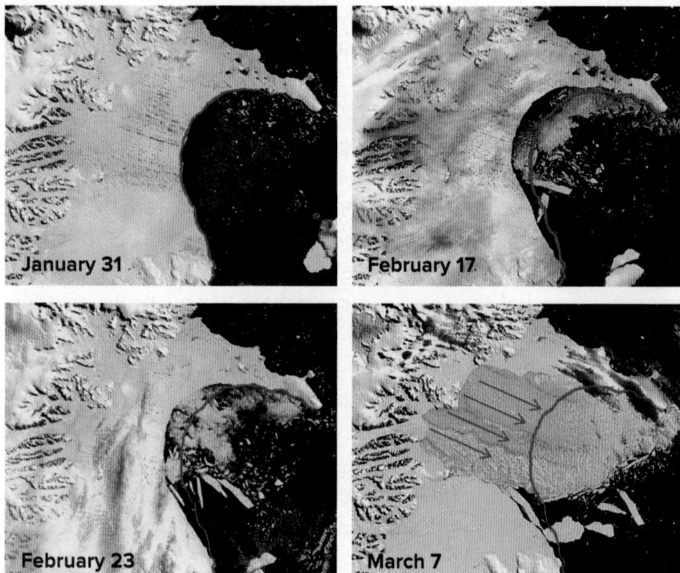

FIGURE 21.19

Satellite images showing the breakup of a 3,275 km^2 section of the Larson B ice shelf over the period January 31–March 7, 2002. The light blue region on March 7 is an area with lots of finely divided small icebergs. For more details, see http://earthobservatory.nasa.gov/Features/WorldOfChange/larsenb.php.
Source: NASA/Goddard Space Flight Center, Scientific Visualization Studio

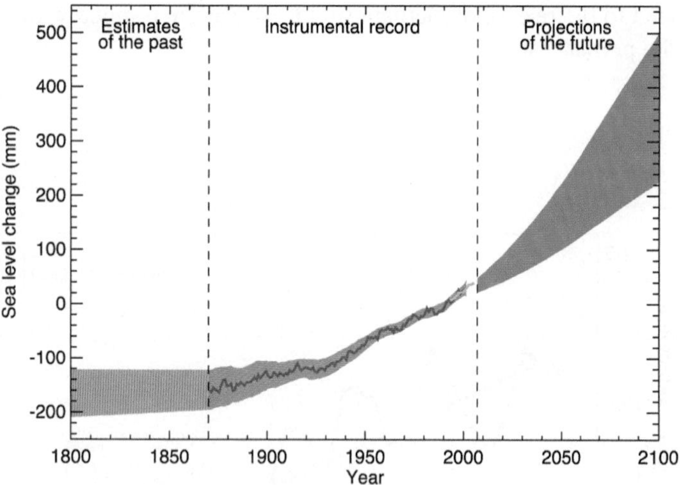

FIGURE 21.20

Sea-level rise over the last two hundred years and projected into the future. Note that the projections do not include any potential catastrophic collapse of ice sheets that could rapidly raise sea level by many meters.
Source: www.ipcc.ch/graphics/ar4-wg1/jpg/faq-5-1-fig-1.jpg

Sea-Level Rise

Between 1901 and 2010, global mean sea level rose by 1.7 ± 0.2 mm/yr^{-1}, corresponding to an increase of 0.19 meters over this period (figure 21.20). This is due to two main factors. First, the melting of glaciers, ice caps, and ice sheets has caused approximately half this increase. Note that the melting of sea ice does not contribute significantly to sea-level rise because the sea ice is already floating on the oceans, much like ice cubes melting in a glass of lemonade will not cause the glass to overflow. The second physical process leading to sea-level rise is **thermal expansion** of the oceans. Thermal expansion is the increase in the volume, or equivalently the decrease in density, that results from increased ocean temperature. On a global scale, thermal expansion is responsible for the other half of the increase in sea level. Because ocean temperatures do not respond instantaneously to changes in atmospheric temperatures, there is a lag in the oceanic response. Even if future temperatures were held constant at today's value, the oceans would continue to warm and sea levels would rise in response. The rate at which sea level has risen has been increasing, and it was 3.2 ± 0.4 millimeters per year over the period 1993-2010, although this may be due to decadal variability.

21.6 CLIMATE MODELS

As you have seen, the Earth's climate is determined by complex interactions between many parts of the Earth system. This makes predicting future climate change challenging. Yet scientists have been able to build global-scale computer models that can be used to provide projections of how climate will change in the future. However, if we are to have confidence in these future projections, we need to first have confidence that climate models can replicate the previous century's climate. These models are, by necessity, simplified representations of the climate system, and this can lead to differences between individual models. By combining results from numerous models, scientists have found that the average global temperature over the last century can be well represented. An important finding is that this is only true if the increases in greenhouse gas concentrations are included in the models. If only variations in natural phenomena, such as volcanic eruptions or sunspots, are included, the models significantly underpredict past temperatures (figure 21.21). This provides very strong evidence that increases in greenhouse gas concentrations are responsible for the vast majority of warming over the last century.

Despite the successes of climate models, important uncertainties remain. Much scientific effort goes into constraining these uncertainties and, by understanding which components of the climate system are better or more poorly understood, scientists can decide which components to focus on. For example, while greenhouse gases have the largest warming effects, they are also among the best understood components.

Even if we had a perfect model of the climate system, a limitation in any projection of future climate is that we cannot know with certainty how emissions will change into the future; scientists can only make educated guesses. These guesses are used to create *emissions scenarios*, which are representations of future greenhouse gas emissions. These scenarios range from "business as usual" (the A1B and A2 scenarios in figure 21.16), in which few changes are made to current technologies and the global population continues to grow, to scenarios in which population growth slows or even declines and where energy use moves strongly toward alternative energy, non-fossil sources (the B1 scenario in figure 21.16). By choosing a range of emissions

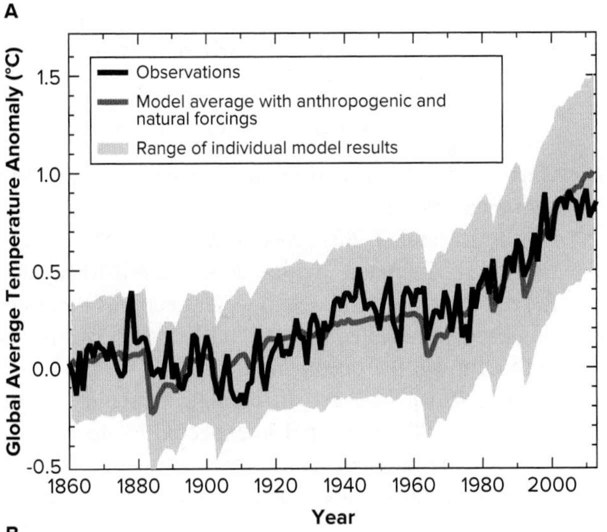

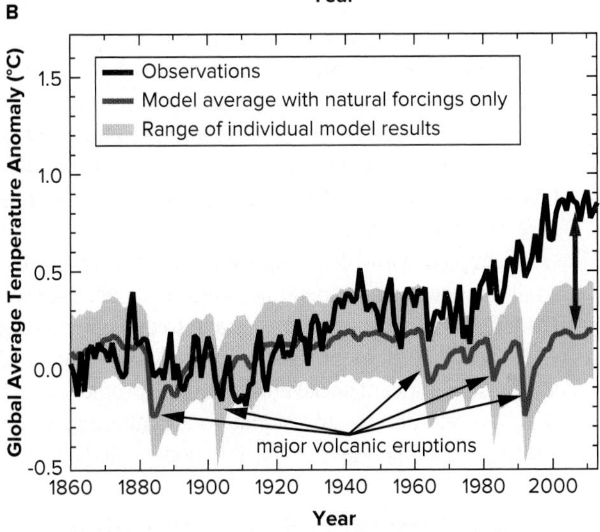

FIGURE 21.21

The observational temperature record since 1860 shows a clear increase with time. Computer model results that include the increasing concentrations of greenhouse gases (shown in panel A) do a good job of reproducing the observations, while model results when greenhouse gases are not allowed to increase (shown in panel B) are significantly lower than the observations after 1960. The average results from many different models are shown as thick colored lines, and the range of results from individual models are shown as colored bands (http://cmip-pcmdi.llnl.gov/cmip5/index.html). The observations are from the HadCRUT4 surface temperature data set (http://www.cru.uea.ac.uk/cru/data/temperature/).
Adapted from http://www.climatechange2013.org/images/figures/WGI_AR5_Fig10-1.jpg

scenarios, scientists and policy makers can consider a range of possible futures.

The IPCC

The **Intergovernmental Panel on Climate Change (IPCC)** is an international panel of scientists that assesses the evidence for and effects of climate change, with the aim of reaching a global consensus on the state of the science. In its 2013 report, the IPCC concluded first and foremost that "Warming of the climate system is unequivocal" and that "Human influence on the climate system is clear . . . It is extremely likely that human influence has been the dominant cause of the observed warming since the mid-20th century." Further, the IPCC warns that "Continued emissions of greenhouse gases will cause further warming and changes in all components of the climate system. Limiting climate change will require substantial sustained reductions of greenhouse gas emissions." For its work, the IPCC was awarded the 2008 Nobel Peace Prize jointly with former Vice President Al Gore. For more information on the IPCC, visit its website at http://www.ipcc.ch/.

21.7 IMPACTS AND CONSEQUENCES OF GLOBAL CLIMATE CHANGE

Climate change will affect human populations in many ways. For example, there is a potential for famine due to crop failure, particularly in areas with reduced precipitation. Because many of the areas projected to experience drought include already unstable regions of the world, climate change raises social issues of equality. Many of the low-lying countries particularly susceptible to rising sea level are also among the least economically developed, including southeast Asia and islands of the South Pacific. Risks associated with climate change have been evaluated in terms of their overall impacts on broad categories, such as the risks to threatened species, extreme weather, or large-scale discontinuities such as rapid sea-level rise due to the melting of the ice sheets (figure 21.22). In this section, specific impacts and consequences of climate change are discussed.

Biodiversity and Agriculture

Changing climate affects the biosphere in multiple ways. Most plants and animals are adapted to specific temperature and precipitation regimes and, as these change, will attempt to migrate to maintain those regimes. One example is the slow migration of bird populations up the sides of mountains to maintain their temperature zone. Natural topography, however, will limit how long these migrations are sustainable. One specific concern is that population shifts of insects will lead to the spread of vector-borne diseases, such as malaria carried by mosquitos.

Plants respond in varying ways to changes in CO_2, temperature, and precipitation. Increased CO_2 and temperature may initially stimulate plant growth–and CO_2 uptake–but eventually nutrient availability becomes a limiting factor. In terms of agriculture, the potential effects of climate change include greater susceptibility of crops to extreme weather events; altered need for irrigation, particularly in areas undergoing decreased precipitation; and, in parts of the world that are already near maximal temperatures for crop growth, increased likelihood of heat stress caused by increasing temperatures. For example, in 2002, record high temperatures and drought caused grain shortages in India and the United States; the result was a rise in the cost of

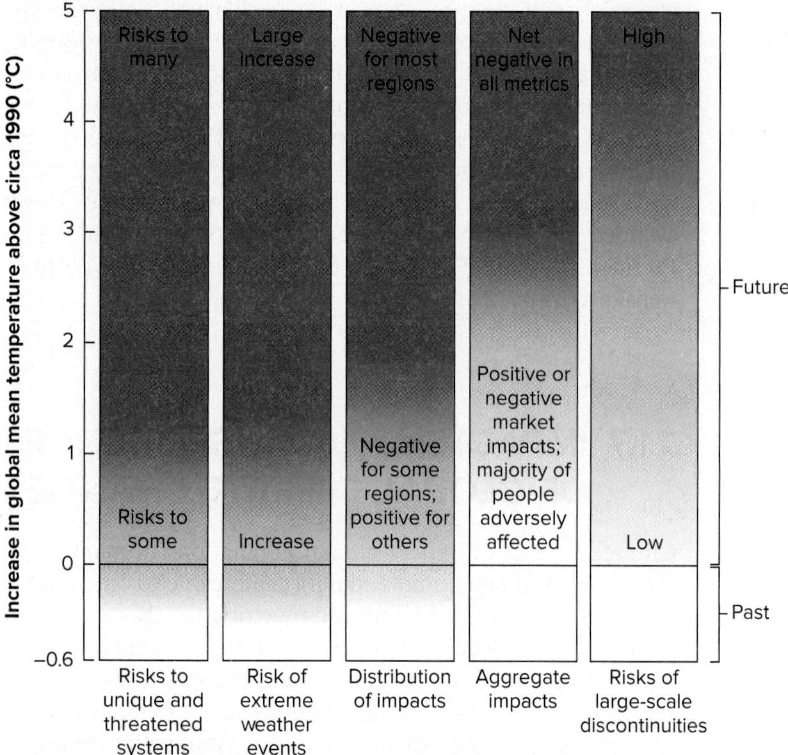

FIGURE 21.22

The "burning embers" plot shows the likelihood of particular risks as a function of a change in temperature above the 1990 value, where red colors indicate large impacts across broad regions of the world. Recall that a doubling of atmospheric CO_2 concentrations is likely to lead to a 2.5–4°C temperature change. "Distribution of impacts" refers to climate change harming some regions while benefiting others. "Aggregate impacts" refers to the overall economic impacts of climate change. "Large-scale discontinuities" refers to the probability of calamitous events, such as the complete melting of the Greenland ice sheet.

Source: Joel B. Smith et al., "Assessing dangerous climate change through an update of the Intergovernmental Panel on Climate Change (IPCC) 'reasons for concern,'" Proceedings of the National Academy of Sciences, *December 9, 2008, figure 1 (modified)*

food and shortages in many of the poorer regions of the world. In regions of the world that get fresh water from mountain runoff, such as in Central and Southeast Asia and South America, there is growing concern that not enough food will be produced. In the American West, snowmelt is occurring earlier and more rapidly, leaving agricultural areas susceptible to late summer droughts. However, while agricultural productivity will likely decrease in some regions, it will increase in other areas of the world, particularly those where cold temperatures currently limit the growing season. For countries such as Russia and Canada, potential food-producing areas may increase.

Ocean Acidification

As you learned in chapter 5, CO_2 is soluble in water. As atmospheric CO_2 levels increase, the oceans respond by taking up more CO_2. Currently, the oceans take up around half of the CO_2 emitted each year. When CO_2 dissolves in the ocean, it reacts to form a balance of different dissolved ions and molecules:

$$CO_{2(aq)} + H_2O \Leftrightarrow H_2CO_{3(aq)} \Leftrightarrow H^+ + HCO_3^- \Leftrightarrow 2H^+ + CO_3^{2-}$$

Carbon dioxide + water + carbonic acid *is in equilibrium with* hydrogen + bicarbonate *is in equilibrium with* hydrogen + carbonate

$$CO_2 + CO_3^{2-} + H_2O \Leftrightarrow 2HCO_3^-$$

Carbon dioxide + carbonate + water *is in equilibrium with* bicarbonate

As more CO_2 dissolves in the ocean, the concentration of H^+ increases, which is equivalent to saying that the pH decreases and ocean water becomes more acidic; this process is called **ocean acidification**. Over the twentieth century, the ocean pH decreased by about 0.1 pH units, corresponding to a change in acidity of nearly 30%. By 2100, ocean pH is expected to decrease by at least 0.2 pH units, or nearly 60%.

Numerous ocean creatures, including shellfish and corals, have a structure based on the mineral calcite ($CaCO_3$). These creatures can survive only in a very narrow range of pH conditions, as $CaCO_3$ is soluble in weakly acidic water, forming the calcium ion (Ca^+) and carbonate ion (CO_3^2) upon dissolution. The addition of CO_2 to the oceans can therefore lead to dissolution of $CaCO_3$, including $CaCO_3$ that forms the shells or structure of many oceanic organisms.

Coupled with bleaching events that result from even small increases in temperature, coral reefs are expected to diminish as CO_2 concentrations increase and to become more sensitive to other local environmental factors. The loss of coral reefs is expected to cause significant social and economic problems: millions of people rely on corals for subsistence food gathering, while the coral reef tourism and fisheries are worth billions of dollars annually.

Coastal Regions and Sea-Level Rise

While rising sea levels on the order of tens of centimeters may not seem significant, these numbers are large and worrisome to the more than 600 million people living along coastlines that are less than 10 meters above sea level. **Sea-level rise** will affect wetlands, putting numerous plant and animal species at risk. Countries such as Bangladesh and Vietnam have large populations living in deltas that will be subject to flooding, and these populations will be displaced (figure 21.23). Island nations such as the Philippines and Tuvalu have the additional problem of being too small to allow the displaced people to move to higher ground. Some island nations are further concerned because rising sea levels may flood aquifers, significantly reducing freshwater supplies. Furthermore, there are some concerns that catastrophic melting of ice sheets could lead to rapid sea-level rise of a few meters, which would inundate most coastal regions and displace millions.

21.8 GEOENGINEERING

Geoengineering refers to large-scale intentional efforts to modify Earth's climate, with the goal of counteracting the warming influence of greenhouse gas-driven climate change. Geoengineering

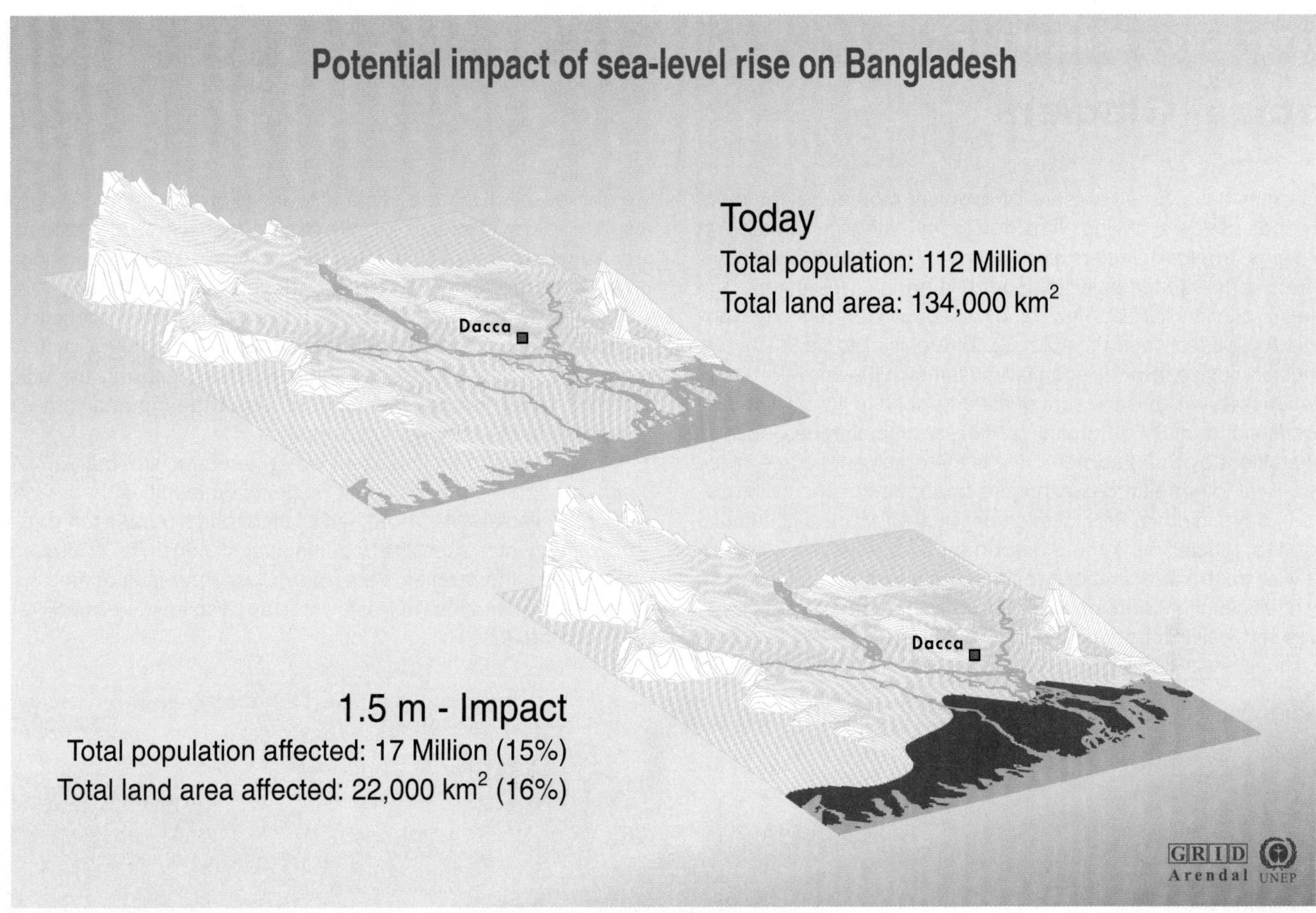

FIGURE 21.23

Impacts of a 1.5 m rise in sea level on the coastal regions of Bangladesh.
Source: http://maps.grida.no/library/files/storage/321_large.jpg

remains a contentious topic. Some think that simply by considering geoengineering we are providing ourselves an excuse to take no action to reduce greenhouse gas emissions. Others think that attempts to engineer our climate will inevitably come with unintended, most likely negative, consequences. Still others think that it is important to give serious consideration to geoengineering ideas now, either thinking of it as a last resort or an economically viable alternative to greenhouse gas reductions. And finally, some think that it is dangerous to believe that we could fine-tune such a complex system as the Earth's atmosphere. Regardless, there are significant ethical issues associated with geoengineering that need to be considered well before any action is to take place. We will consider two general schemes for climate geoengineering: solar radiation management and carbon management.

Solar Radiation Management

It has been stated that solar radiation management has three essential characteristics: it is cheap, fast, and economical. **Solar radiation management** refers to intentional efforts to control the amount of solar radiation being absorbed by the Earth system by reflecting it back to space. With a decrease in the amount of sunlight being absorbed, the Earth should settle toward a lower mean temperature.

It is important to understand that solar radiation management does not fundamentally address the root cause of climate change, which is the increasing concentrations of greenhouse gases in the atmosphere. All solar radiation management schemes should be considered as, at best, temporary fixes to a long-term problem. One of the major concerns over any solar radiation management scheme is that modification of the solar radiation budget, in particular the geographic distribution of sunlight, could have dramatic consequences for the global hydrologic cycle. Precipitation patterns could be changed, potentially leading to drought in some regions and flooding in others. A number of different schemes have been proposed for solar radiation management. These include injection of particles into the stratosphere, ocean cloud modification, and placement of a "space sunshade" between the Sun and the Earth.

Carbon Management

Carbon dioxide added to the atmosphere from fossil fuel combustion will be removed through natural processes over

EARTH SYSTEMS 21.3

Tropical Glaciers

Glaciers have been present on tropical mountains for thousands of years, having withstood many swings in climate. Yet today many tropical glaciers are receding, including Quelccaya, Hascaran, and Chacaltaya in South America, Rwenzori and Kilimanjaro glaciers in East Africa, and Dasuopu in the Himalaya (box figure 1). As you learned in chapter 12, for a glacier to be sustained, accumulation of ice through snowfall must match the snow loss rate, or ablation rate, which is the sum of melting (solid to liquid) and sublimation (solid to gas). For tropical glaciers at high altitudes, absorption of sunlight by the glacier surface is the dominant source of ice loss, while at lower altitudes exposure to above-freezing temperatures can drive melting. Precipitation not only affects the accumulation of the glacier but, more importantly, affects the glacier's reflectivity: fresh snow is more reflective than aged snow and so absorbs less sunlight and melts at a slower rate. Thus a glacier that receives snowfall throughout the year can melt at a slower rate than a glacier receiving the same amount of precipitation all in one season. Healthy glaciers typically have a higher-altitude accumulation zone where ice accumulates, and a lower-altitude ablation zone, with the boundary between these the *equilibrium line* (see figure 12.6). The altitude at which the equilibrium line occurs determines the size of the glacier: under a warming climate, the equilibrium line will rise to higher altitudes, causing the glacier to shrink. When the line reaches the top of the mountain, the glacier will eventually completely disappear.

Glaciers are an important water resource (see box 12.1). As glaciers disappear, the melting will affect communities that rely on runoff from glaciers. As about half of the world's population gets water from meltwater from the mountains, including the Andes and the Himalayas, the problems are enormous. The result of melting is not only a long-term loss of fresh water, but also an increased likelihood of flooding.

BOX 21.3 ■ FIGURE 1

Qori Kalis glacier in the Peruvian Andes. Left: July 1978. Right: July 2011.

(A) Source: Dr. Lonnie G. Thompson, Distinguished University Professor, Byrd Polar and Climate Research Center, The Ohio State University/National Snow and Ice Data Center; (B) Source: Dr. Lonnie G. Thompson, Distinguished University Professor, Byrd Polar and Climate Research Center, The Ohio State University

thousands of years. **Carbon management** refers to methods for accelerating the reduction in CO_2 concentrations to time scales that matter to humans, that is, years to decades. Carbon management can include processes that either directly remove greenhouse gases or that enhance natural removal processes. Unlike solar radiation management, carbon management *does* address the root cause of climate change. Carbon management is sometimes referred to more generally as greenhouse gas remediation. Some of the various carbon management ideas that have been proposed include carbon dioxide air capture, ocean fertilization, and carbon capture and sequestration.

Carbon capture and sequestration involves the capture of CO_2 when it is produced and its long-term storage in the Earth's surface. Current major challenges include identifying appropriate geologic locations for storage, separating the CO_2 from the rest of the combustion gases, and transporting the CO_2 to an appropriate location for storage. Suitable storage locations are generally either geologic formations (depleted oil or gas reservoirs, deep aquifers, unminable coal seams) or the deep ocean. With all of these, there are concerns about their suitability for long-term storage that are still being addressed.

21.9 DECREASING EMISSIONS OF GREENHOUSE GASES

Ultimately, any solar radiation management or carbon management geoengineering scheme will need to be coupled with an immense effort to decrease new emissions of CO_2 and other greenhouse gases. This is the only true "solution" to climate change. It has been estimated that in order to avoid dramatic climate change, the world must avoid emitting about 200 billion tons of carbon over the next fifty years, which corresponds to limiting greenhouse gas emissions to be at or even below today's values starting immediately. The longer emissions reductions are delayed, the greater the reductions will need to be.

Most new emissions of CO_2 result from energy production and usage. The most straightforward way to decrease emissions of CO_2 is to move away from using fossil fuels to produce energy and toward using less carbon-intensive or alternative energy sources, such as solar, wind, geothermal, or hydro. Efficiency improvements to existing technologies, such as increasing the fuel economy of vehicles or widespread usage of low-energy appliances, will also be necessary. Such technological innovations must occur alongside cultural shifts wherein people's demand for energy and natural resources is decreased.

Fate of Atmospheric Carbon Dioxide

Carbon dioxide emitted today will persist in the atmosphere for many years. Different processes act on various time scales to remove CO_2 from the atmosphere. Uptake of CO_2 by the land biosphere, through the process of photosynthesis, occurs on time scales of years to decades. Dissolution of CO_2 into the oceans and the eventual transport of dissolved carbon from surface waters to the deep ocean, a process called ocean invasion, removes carbon from the atmosphere on time scales ranging from decades to centuries or more. On millennial time scales, CO_2 is removed from the atmosphere through reactions with calcium carbonate or through the process of silicate weathering, that is, the conversion of $CaSiO_3$ to $CaCO_3$.

For a pulse of CO_2 emitted today, approximately a third to a half is taken up by the biosphere or ocean within a few decades, and the rest remains in the atmosphere. After about 1,000 years, and depending on how much carbon is initially released, approximately 15% to 40% of the emitted CO_2 still remains in the atmosphere, and after 10,000 years about 10% to 25% still remains. The long time scales associated with atmospheric CO_2 removal indicate that decisions made today regarding the continued use of fossil fuels–and the use of our atmosphere as a dump for the primary waste product of combustion, CO_2–will have consequences that will persist for a very long time.

Scientific Consensus

Although Earth's climate system is complex and there remain details to be fully understood, the fundamental scientific link between greenhouse gases and planetary temperatures is well established. Based on the preponderance of evidence, there is strong consensus among climate science experts that the modern climate is warming, that the rate of warming is unprecedented in geologic history, and that the warming is primarily due to human activities. There is also broad agreement among scientists that negative consequences are already being felt the world over that will only become stronger in the future if global temperatures continue to rise. Although the future cannot be predicted with certainty, scientific understanding of the physical world can allow us to state with confidence that if near-term actions are not taken to reduce emissions of greenhouse gases, then the climate of the next century will be quite different than it is today.

Summary

Climate is a long-term average of weather and is the result of interactions between the atmosphere, geosphere, hydrosphere, biosphere, and solar radiation. While Earth's climate has varied naturally in the past and will continue to do so in the future, greenhouse gas emissions due to human activities over the last century have caused an unprecedented increase in global temperatures that will continue well into the future. Humans have had an undeniable influence on the composition of the atmosphere. These changes are well established as the root cause of the current planetary warming, and they are also leading to shifts in precipitation patterns, sea-level rise, ocean acidification, melting of glaciers, and widespread effects throughout the biosphere.

CO_2 is the most important greenhouse gas, although methane, nitrous oxide, and the halocarbons are of growing importance. All greenhouse gases absorb infrared radiation in the atmosphere, affecting the radiative balance of the Earth. This absorbed infrared radiation warms the lower layers of the atmosphere, warming the surface of the Earth.

Various geoengineering schemes have been proposed, and they vary from injection of particles into the stratosphere to carbon capture and sequestration techniques. However, many of these techniques are expensive and, unless carbon is removed from the atmosphere, will not solve the problem of ocean acidification. Because a large fraction of CO_2 emitted today will persist in the atmosphere for hundreds to thousands of years, the impacts of climate change will be felt long into the future.

Terms to Remember

Testing Your Knowledge

1. What is the difference between weather and climate?
2. What is albedo? Give an example of an object with (*a*) high albedo and (*b*) low albedo.
3. In what way do greenhouse gases warm the planet?
4. How do volcanoes affect the Earth's temperature?
5. Describe three pieces of evidence that climate is changing.
6. What are the two main causes of sea-level rise?
7. How do scientists use ice cores to determine past temperatures?
8. Why doesn't increasing CO_2 lead to increased plant growth in the long term?
9. What are the two approaches to geoengineering?
10. What is a climate feedback? Give an example of a negative feedback.
11. Which of the following is the correct order of the layers of the atmosphere from the ground up?
    a. stratosphere, mesosphere, troposphere, thermosphere
    b. thermosphere, troposphere, stratosphere, mesosphere
    c. troposphere, stratosphere, mesosphere, thermosphere
    d. stratosphere, troposphere, mesosphere, thermosphere
12. What would the temperature of the Earth be if there were no atmosphere?
    a. 15°C b. 545°C
    c. 0°C d. −19°C
13. Which of the following is a greenhouse gas?
    a. nitrogen b. oxygen
    c. methane d. argon
14. Ocean acidification causes
    a. sea-level rise.
    b. loss of calcium carbonate in seashells.
    c. coral bleaching.
15. Which of the following is an example of a positive climate feedback?
    a. Warming causes melting of glaciers, affecting the albedo of the Earth.
    b. Increased CO_2 causes increased uptake by the oceans, causing acidification.
    c. Rising sea levels decrease coastal regions, affecting freshwater supplies.
16. The study of ancient climate is called
    a. geochronology.
    b. paleoclimatology.
    c. meteorology.
    d. proxy study.
17. Which of the following is *not* an example of a Milankovitch cycle?
    a. perihelion
    b. precession
    c. obliquity
    d. eccentricity
18. The Maunder Minimum or "Little Ice Age" of the late seventeenth century is believed to have been caused by:
    a. a large volcanic eruption.
    b. a period of very low sunspot activity.
    c. a decrease in the concentration of greenhouse gases in the atmosphere.
    d. greater global albedo due to heavy cloud cover in the Northern Hemisphere.
19. The "Year Without a Summer" in the 1800s was caused by:
    a. a large volcanic eruption.
    b. a period of very low sunspot activity.
    c. a decrease in the concentration of greenhouse gases in the atmosphere.
    d. greater global albedo due to heavy cloud cover in the Northern Hemisphere.
20. Which of the following energy sources does *not* directly produce carbon dioxide?
    a. coal b. nuclear power
    c. natural gas d. petroleum

Expanding Your Knowledge

1. The amount of water vapor in the atmosphere ranges from 0 to 4%. You experience this as relative humidity, which can range from 0 to 100%. What is the difference between the absolute amount of water in the atmosphere and relative humidity?
2. Why is ozone at the surface considered harmful and ozone in the stratosphere considered beneficial?
3. Think about your daily schedule. In what ways could you reduce your greenhouse gas emissions? What about your city?
4. What is the climate like in your state? How could global climate change affect your region?

Exploring Web Resources

www.ipcc.ch/

The home website of the *Intergovernmental Panel on Climate Change* contains full copies of its reports.

www.grist.org/article/series/skeptics

The Grist.org *How to Talk to a Climate Skeptic* web page contains straightforward answers to many questions about climate change.

www.realclimate.org/

RealClimate is a commentary site on climate science by working climate scientists for the interested public and journalists. They aim to provide a quick response to developing stories and provide the context sometimes missing in mainstream commentary.

http://cmi.princeton.edu/wedges/

The Carbon Mitigation Initiative website explores what steps can be taken using current-day or near-term technologies to reduce carbon emissions, one billion metric tons at a time.

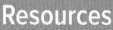

CHAPTER

22 Resources

The Mt. Whaleback iron ore mine, located in western Australia, is over 5 km long and 1.5 km wide.
©Peter Hendrie/Photographer's Choice/Getty Images

LEARNING OBJECTIVES

- Explain the difference between a reserve and a resource and the factors that determine whether a resource is regarded as a reserve.
- Compare the geologic processes that lead to the formation of coal, oil, and natural gas deposits.
- Understand how a nuclear reactor works, and discuss the pros and cons of nuclear energy.
- Describe each of the renewable sources of energy, and compare their importance to fossil-based energy sources.
- Name the important metallic and nonmetallic resources, their primary uses, and the geologic processes through which they form.
- Describe the mining techniques used to extract metallic and nonmetallic resources.
- Consider the environmental impact of resource extraction and consumption.

Geologic resources sustain life, and the most fundamental of these resources are soil and water. Industrial civilization, however, draws a much greater variety of resources from the Earth. In many ways, our modern lives have come to depend upon dozens of different kinds of Earth materials, from the coal and oil that power our electricity to the neodymium used to make the tiny powerful batteries that allow us to have small portable electronics such as cell phones and laptop computers. In early times, people knew what resources they needed to support their lives, and where to find them. Today, few people fully appreciate the multitude of resources and the complex web of supply and demand that enables each of us to live. Consider your typical morning routine. Your digital alarm clock is made of plastic produced from petroleum, and its LED display is made of materials such as silica and aluminum. When you brush your teeth, you are using a plastic toothbrush, and you are using toothpaste containing silica as an abrasive and fluoride from the mineral fluorite. Your bathroom mirror is made of glass (which is made from beach sand) and coated with silver or aluminum to make it reflective. It hangs on a wall that is most likely made of gypsum wallboard. You travel to work in a car or on a bus or bicycle—all made of a mixture of metal, plastic, and rubber components. These resources, and the many others that you use within the first few hours of your day, have to be mined and processed by machinery made of iron in factories powered by electricity.

Excluding soil and water, which we discuss elsewhere in the book, we group geologic resources into three general categories: (1) energy resources, (2) metallic resources, and (3) nonmetallic resources. *Energy resources*, like petroleum and uranium, provide the power that drives the modern world. *Metallic resources*, such as iron ore, enable us to create the metals which provide strength for modern construction and help many technologies operate—for example, by conducting electricity or sparking motors. *Nonmetallic resources*, including building stone and road gravels, have a long history in the development of civilization and are still vital to the modern world. Consider a highway overpass. It consists of a nonmetallic resource exterior (the concrete) with a metallic core (the rebar and girders). The overpass supports a road of asphalt—fossil organic matter mixed with nonmetallic aggregates—that provides access to cars made of metals (body and engine) and biological matter (rubber), and powered by petroleum.

Each year, the average American uses approximately 3,501 liters (925 gallons) of oil; 2,544 m^3 (89,847 ft^3) of natural gas; 156 kg (344 lb) of iron ore, aluminum ore, and copper; and 7,536 kg (16,615 lb) of crushed stone, sand, and gravel. Try to imagine how much mining, blasting, drilling, and pumping all of this requires, and it is easy to believe the fact that human activity now moves more earth—perhaps as much as 3 trillion tons annually—than all of the rivers of the world combined (a *mere* 24 billion tons per year of transported sediment).

Some geologic resources are *renewable*, that is, they are replenished by natural processes fast enough that people can use them continuously. Water is the best example. Under sustainable conditions, the supply of water is never ending, provided that we extract water no faster than it is replenished naturally by precipitation, runoff, and infiltration. Most geologic resources, however, are **nonrenewable resources**. They form very slowly, often over millions of years under unusual conditions in restricted geographic settings. Humans extract nonrenewable resources much faster than nature replaces them. The annual rate of extraction of crude oil, for example, is on the order of a million times faster than natural rates of replenishment.

In this chapter, we will first discuss the difference between a reserve and a resource. We will then explore energy resources, both nonrenewable and renewable, in terms of their origin and how they are exploited. We will move on to metallic and nonmetallic resources and a discussion of mining practices. Finally, the balance between our need for resources and our responsibilities toward protecting the environment and conserving resources for future generations will be discussed.

22.1 RESERVES AND RESOURCES

Resource is the term used to describe the total amount of any given geologic material of *potential economic interest*, whether discovered or not. A resource can be measured directly through mining or drilling ("demonstrated," "measured," or "indicated" resources), or simply inferred based upon reasonable geologic guesswork and statistical modeling ("inferred," "hypothetical," or "speculative" resources). The size of a nonrenewable resource does not change in time; it is a value that is fixed and theoretically determinable. **Reserve** is the term used to describe the portion of a resource that has been discovered or inferred with some

IN GREATER DEPTH 22.1

Copper and Reserve Growth

As the prices of metals and the energy used to extract them fluctuate, so do the potential profits from minerals. For example, in 1900, copper could be mined at a profit only if its concentration in ore exceeded 5%. By the early 1980s, this profit level dropped to 0.5%, and the world's recoverable copper reserves rose to a half billion tons. Since then, the world has consumed about 400 million tons of copper, but the introduction of recycling (which now provides the United States with almost as much copper as direct mining), the introduction of substitutes for copper (such as fiber-optic cable), and the discovery of new reserves actually increased world copper reserves to 720 million tons by 2016. Mineral markets and reserves are volatile and erratic, with reserves generally shrinking as market value declines and swelling as prices rise, most often due to investor speculation and temporary supply shortages and gluts. Nevertheless, there are some persistent trends: The world almost always appears to have large reserves of iron and aluminum, moderate reserves of copper, lead, and zinc, and scanty reserves of gold and silver. These levels reflect the relative abundance of these resources in nature. There is no sign that we are about to "run out" of any of these metals.

Other challenges, however, loom on the horizon because the sizes of mineral reserves are tied critically to the price of energy. It takes very large amounts of energy to mine, refine, process, and transport minerals for use. Mineral extraction, in fact, is the most energy-intensive industry in the world. Over most of the past 120 years, overall unit energy costs (adjusted for inflation) have not grown appreciably and have held generally steady, providing a reliable platform for industrial growth. If the long-term cost of energy increases, however, we might expect the sizes of mineral reserves to drop in response, simply because it will become so much more expensive to mine at a profit.

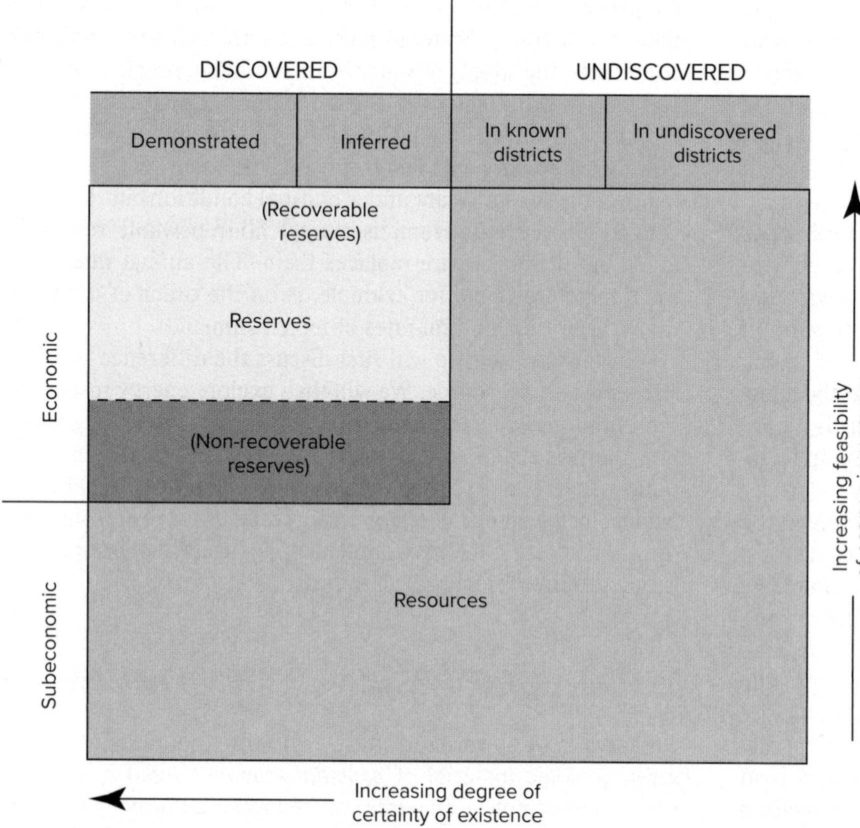

FIGURE 22.1
Important factors in the classification of reserves and resources. (Reserves are subsets of resources.)

degree of certainty and can be extracted for a profit. Figure 22.1 shows the relationship between resources and reserves. Unlike a resource, the size of a reserve can change over time, depending upon a variety of factors (box 22.1). For example, mining or drilling of a substance will cause the reserve to shrink, especially if no new discoveries are made. Wage increases and a drop in market price could make a deposit of a material too expensive to continue mining, which would also reduce the reserve size. On the other hand, new discoveries and new technologies making it easier to locate and mine a resource make a reserve larger. From 2000 to 2011, the price of gold rose precipitously from a little over $250 per ounce to almost $2,000 per ounce. This rise in the price of gold led to a revival of gold mining in California, with a number of old mines preparing to reopen. You can check the daily price of gold at http://www.kitco.com/charts/livegold.html. Since 2011, the price of gold has dropped, although at this writing it remains above $1,000 per ounce. What do you think will happen to plans for mining in California if the price of gold continues to fall? Changes in laws can also affect reserve sizes. Large areas of government-owned land are off-limits to mining and drilling, so any geologic materials under these areas are not legally extractable and cannot be included in reserves. Opening more land to extraction would therefore increase reserves.

Not all of a reserve may be recoverable. For example, in the United States, the total coal *resource* is on the order of 3.54 trillion tons. Only about 11% of this (434 billion metric tons) makes up the U.S. coal *reserve*, however. But not all of this reserve can actually be extracted. Some has to be left behind during mining for safety reasons–to support mine pillars, to prevent landslides, to avoid water pollution problems, and so on. In fact, only about 60% of the coal in any bed that is mined below ground can be removed. The value for strip and open-pit mines is somewhat greater–80% to 90%. The *recoverable reserve* for coal in the United States, hence, is only about 6% (232 billion metric tons) of the total known coal resource in the country. We will never be able to exhaust all the coal in the world, but this is only because we will never be able to mine it all safely *at a profit.*

22.2 ENERGY RESOURCES

Energy is simply "the ability to do work." Without energy, nothing could exist; in fact, it is basic to everything. There are many different forms of energy, and these are divided into two basic categories. **Kinetic energy** involves movement, while **potential energy** is stored energy. Kinetic energy forms include electrical energy (the movement of electrical charges), radiant energy (the movement of electromagnetic rays such as visible light or X rays), thermal energy (heat), motion energy (the movement of objects from one place to another), and sound. Potential energy forms include chemical energy (energy stored in atomic bonds), stored mechanical energy (such as the energy stored in a compressed spring), nuclear energy, and gravitational energy. We fuel our bodies with chemical energy stored in the food we eat. That energy ultimately comes from the radiant energy provided by the Sun. **Energy resources** are the materials we use to produce heat and electricity or as fuel for transport. Modern society is dependent upon energy resources, not only for heat and fuel but to produce all of the items we use daily in our homes and offices. Energy consumption in the United States makes up about 18% of world energy consumption, although the U.S. population is only about 5% of world population. A single person in the United States uses approximately twice as much energy as a person in Europe, three times as much energy as a person in China, and just over eleven times as much energy as a person in India.

Although interest in alternative, renewable energy resources is increasing, the majority of our energy needs are met by nonrenewable *fossil fuel resources* such as coal and oil. The state of our energy supplies is very much on the minds of many people today. How much oil do we have left? What will power the airplane of the future? Is a new "hydrogen economy" on the way? To answer some of these questions, we must examine how and where our energy resources are formed.

Nonrenewable Energy Resources

Nonrenewable energy resources are those that cannot be replenished naturally in a short period of time. Coal, petroleum, natural gas, and propane are all considered **fossil fuels** because they formed from the buried remains of plants and animals that lived millions of years ago. Uranium ore is an important energy resource that is *not* a fossil fuel. Geothermal power, although discussed in the "Renewable Energy Resources" section, can be regarded as nonrenewable because it can be exhausted readily, even though water and heat are both replenished rapidly over time.

Coal

Coal, as described in chapter 6, is a sedimentary rock that forms from the compaction of plant material that has not completely decayed. The most abundant fossil fuel in the Earth's crust, coal became a major substitute for wood as a source of energy in Europe beginning in the fifteenth century. Efforts to mine coal from greater depths led Englishman Thomas Newcomen to invent a pump in 1712 to drain deep coal mines that were below the water table. This pump was the ancestor of the steam engine that started the Industrial Revolution four decades later. Coal was the main fuel of industrial civilization until people discovered that large amounts of petroleum could be pumped from the Earth, and that petroleum provided a less dirty, more transportable fuel with all sorts of new and exciting uses. The Coal Age gave way to the Petroleum Interval almost a century ago. In 1900, more than 90% of American energy needs were satisfied by burning coal. Today, the United States uses coal to provide less than 20% of its energy demand and, in recent years, as coal-fired power plants were converted to natural gas, the reliance on coal for electricity has declined. About 26% of the world's total coal reserves lie in the United States (figure 22.2). Russia has the second-most-abundant reserve base (18%), with China and India making up another 20%. At current rates of consumption, U.S. recoverable reserves will be exhausted in about 250 years. A similar level of depletion is occurring worldwide. Of course, the many factors just mentioned may alter the sizes of the world's

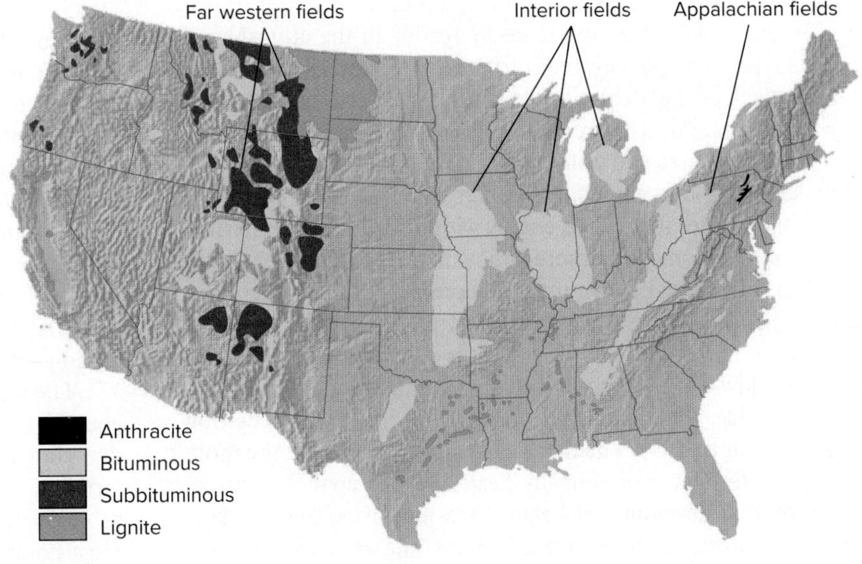

FIGURE 22.2

Coal fields of the United States. Alaska also has coal. *Source: U.S. Geological Survey*

TABLE 22.1 **Varieties (Ranks) of Coal**

	Color	Water Content (%)	Other Volatiles (%)	Fixed Carbon[2] (%)	Approximate Heat Value (BTUs of heat per pound of dry coal)
Peat[1]	Brown	75	10	15	Varies
Lignite	Brown to brownish-black	45	25	30	7,000
Subbituminous coal	Black	25	35	40	10,000
Bituminous coal (soft coal)	Black	5 to 15	20 to 30	45 to 86	10,500 to 15,000
Anthracite (hard coal)	Black	5 to 10	5	86 to 98	14,000 to 15,000

1. Peat is not truly a coal, but may be thought of as "pre-coal."
2. "Fixed carbon" means solid combustible material left after water, volatiles, and ash (noncombustible solids) are removed.

coal reserves, and continued rapid economic growth will significantly impact the longevity of any resource supply. The 250-year estimate for the United States is based on current consumption rates. However, a report released in 2011 predicted that U.S. coal consumption would increase by about 1% per year during the next twenty-five years. If that growth rate were to continue, the U.S. reserves would be exhausted in about 120 years.

How does coal form? Imagine a swampy, coastal environment in a tropical setting. Sunlight filters through the still trees onto dark, stagnant water below, providing the energy for photosynthesis to take place even in the shadows. Photosynthesis converts the Sun's radiant energy into chemical energy stored in molecular bonds, holding together hydrocarbon (hydrogen + carbon) molecules, the building blocks of cellulose and other plant tissues. This is the photosynthetic reaction:

$$CO_2 + H_2O \rightarrow CH_2O + O_2$$

(Atmospheric carbon dioxide + water from the soil combine in sunlight to make cellulose and other plant tissues, releasing oxygen by plant respiration.)

When a plant dies, it will decay readily in the atmosphere, and its stored energy will return to the atmosphere. The reaction taking place during decay is essentially the reverse of the photosynthetic reaction, with oxygen reacting with dead tissues to release pungent gases and water. But if the dead plant matter settles into stagnant, oxygen-depleted water and becomes buried by sediment, it takes that stored energy with it. In time, the inherited energy may become even more concentrated as the molecules in the dead plant break down into less-complex forms. Under pressure and heat, the fossil plant remains transform into coal.

There is a succession of stages in coal development, from relatively low-energy forms with a small amount of concentrated carbon inside, to higher-energy forms with high relative carbon contents (table 22.1). The more carbon that is present, the more combustible—and economically desirable—the coal. The initial stage of coal development begins as a mat of densely packed, spongy, moist, unconsolidated plant material called *peat* (figure 22.3). When dried out, peat can be burned as a fuel, as in Britain and ancient Rome. With compaction, peat transforms into solid *lignite (brown coal)*, which may still contain visible pieces of wood. Lignite is soft and often crumbles as it dries in air. It is subject to spontaneous combustion as it oxidizes in air, and this somewhat limits its use as a fuel. *Subbituminous coal* and *bituminous coal (soft coal)* are black and often banded with layers of different plant material. They are dusty to handle, ignite easily, and burn with a smoky flame. *Anthracite (hard coal)*, the highest "grade" or "rank" of coal, has the most concentrated stored solar energy and is hard to ignite, but it is dust-free and smokeless. If the coal is squeezed and heated any further, its hydrocarbon molecules break down altogether under essentially metamorphic conditions, and all that remains is pure carbon—graphite, the stuff we put in pencil leads.

FIGURE 22.3
A layer of peat being cut and dried for fuel on the island of Mull, Scotland. Coal often forms from peat.
©David McGeary

The scientific unit of energy is the *joule.* One joule is roughly equivalent to the amount of energy needed to heat one gram of dry, cool air by 1°C. In the United States, energy is measured in terms of BTUs, or *British Thermal Units.* One BTU is equivalent to 1,055 joules. A kilogram of ordinary bituminous coal, the most common type of coal in the United States, typically contains 45–86% carbon and releases 25–35 million joules, or megajoules of heat energy (10,500–15,000 BTU per pound). This is sufficient to produce electricity and is equivalent to two to three times the food energy consumed by the average person every

FIGURE 22.4

Coal embedded with sandstone.
©Don Hadden/ardea.com/Panthe/Pantheon/Superstock

FIGURE 22.5

Mountaintop strip mined for coal in West Virginia.
©Robb Kendrick/National Geographic Magazines/Getty Images

day. Anthracite, the highest-grade coal (86–98% carbon) will produce 26–35 megajoules per kilogram (14,000–15,000 BTU/lb). In comparison, average-grade gasoline produces 47 megajoules per kilogram (20,200 BTU/lb).

Coal beds are typically interlayered with ordinary sedimentary rocks, including sandstones and shales (figure 22.4). Beds typically range in thickness from a few centimeters to 30 meters or more. Miners dig up beds that lie close to the surface—within a few tens of meters—by **strip mining**, the complete removal of overlying rock and vegetation (figure 22.5). Strip mining is an environmentally harmful activity that destroys topsoil and leaves behind open pits that must be filled back in and replanted to curb further erosion and water pollution. But strip mining is the only way much of the world's coal supply can be safely mined. Shaft and tunnel mining provide access to deeper coal deposits (figure 22.6). This form of "deep-rock" mining is especially dangerous because of the weakness of coal beds and high concentrations of flammable gas and coal dust. In the decade before Congress established the U.S. Bureau of Mines in 1910, 2,000 persons died each year in coal-mining accidents in the United States alone. From 2000 to 2015, the average number of U.S. coal-mining fatalities was 25 per year. In comparison, during the same time period the average number of coal-mining fatalities in China was approximately 2,000 per year. The most recent coal-mining accident in the United States that generated a lot of attention was the Upper Big Branch accident in 2010, which resulted in 29 deaths.

Once the coal is mined, it is shipped as raw rubble by train, barge, or freighter to power plants, foundries, smelters, and other distributors; little additional processing is needed to make it usable. When lightly burned, the most volatile ingredients in coal—particularly noxious sulfur fumes—escape to leave a new form of coal called *coke.* Coke releases more intense heat in a furnace than does ordinary coal, and it is hardly smoky at all. Because of these fortunate properties, it has become one of the most important substances in our industrial civilization, serving as the main fuel for producing steel in foundries. Without coke, our metals would be too brittle to use in building skyscrapers, bridges, and other infrastructure.

FIGURE 22.6

Coal miner working at a coal seam in a deep mine.
©Monty Rakusen/Getty Images RF

Coal has also been converted into liquid fuels. The South Africans and Chinese, in fact, are looking at their immense coal reserves as a potential source of future automobile fuels. Less-refined, liquefied coal—*slurry*—can be flushed through pipelines stretching up to several hundred miles from mine to factory or power plant.

Petroleum and Natural Gas

In recent years, an argument has raged between economists, who believe that abundant supplies of new petroleum can yet be discovered, and many resource geologists, who caution that most of the regions that contain "new" oil have already been explored and, in fact, are being rapidly depleted. Who is right? We won't know for sure until we complete global exploration and experience a peak in global petroleum production, but odds are that the geologists know something that many more optimistic business people don't: The geologic factors responsible for creating a rich petroleum reservoir are special indeed, and they greatly limit the chances for petroleum to form under natural conditions.

Petroleum and **natural gas**, like coal, are formed from the partially decayed remains of organic matter. The origin of petroleum and natural gas, however, differs significantly from that of coal. Instead of a coastal swamp, imagine well-lit coastal seawater, or a sparkling, tropical lagoon, light-green, with suspended microscopic life-forms including plankton, foraminifera, diatoms, and other organisms. These life-forms thrive continuously in waters well supplied with nutrients from upwelling marine currents and rivers entering the sea nearby. This type of marine environment is typically rich in oxygen, and dead organic matter is readily decayed before it can settle on the sea floor. Oil forms when rapid accumulation of mud and sand bury dead organic matter and separate it from the oxidized seawater. In this anoxic (oxygen-deprived) environment, the organic remains break down slowly. With continued accumulation of sediment, the organic remains are buried more and more deeply (figure 22.7).

The buried hydrocarbons break down or "crack" into simpler molecules with increasing pressure and temperature as the organic-rich sediments are buried more and more deeply by continued sediment accumulation. Initially, chemical reactions with the clay minerals in the sediment produce a gooey, hydrocarbon-rich sediment known as *sapropel.* As it is buried deeper, the sapropel heats up, at a rate of 25°C for every thousand meters (44°F for every 1,000 feet). In order to form oil, the sapropel must be buried deeply enough for temperatures to reach 50°C to 100°C—approximately 2,000–4,500 meters (6,500–13,000 feet). This is known as the oil window (figure 22.7). At temperatures between 100°C and 200°C (200–400°F), the liquid petroleum will break down to natural gas. Beyond 200°C, the hydrocarbons will break down completely.

The result of this process is a petroleum-bearing **source rock**, such as oil shale. Once petroleum has formed, it must next accumulate in concentrations that can be drilled and pumped. Under deep-burial conditions, pressure easily squeezes the oil and gas up into overlying, permeable **reservoir rocks**. The upwelling petroleum may continue to migrate all the way to the surface to issue from the Earth as tar and oil seeps (*breas*) (figure 22.8).

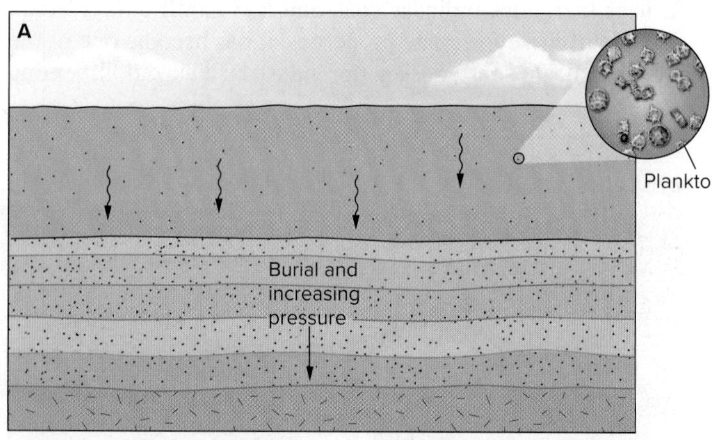

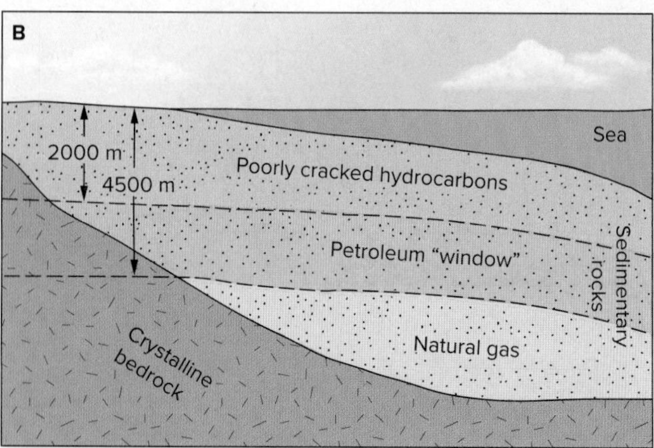

FIGURE 22.7

Formation of oil and typical depths of hydrocarbon cracking. (*A*) Remains of organisms collect on the sea floor and are buried by sediment. (*B*) The "oil window" lies between 2,000 and 4,500 meters (6,500–13,000 feet). Depth will vary somewhat, depending on the geothermal gradient.

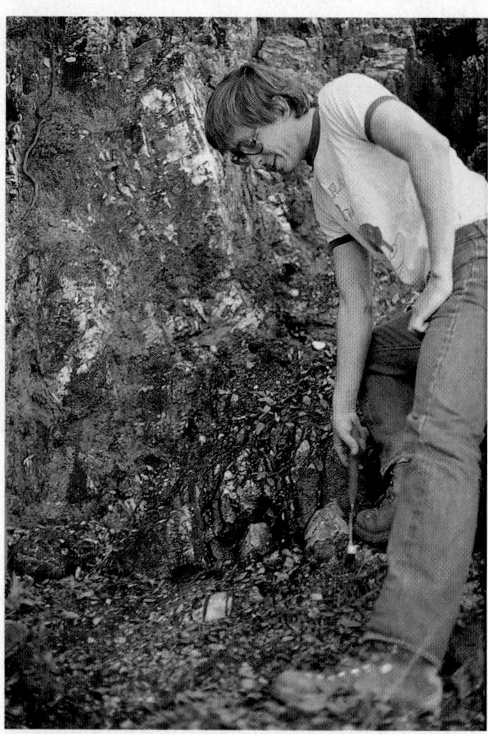

FIGURE 22.8

A brea, or natural oil seep, in a hill slope near Santa Paula, California. ©*Richard Hazlett*

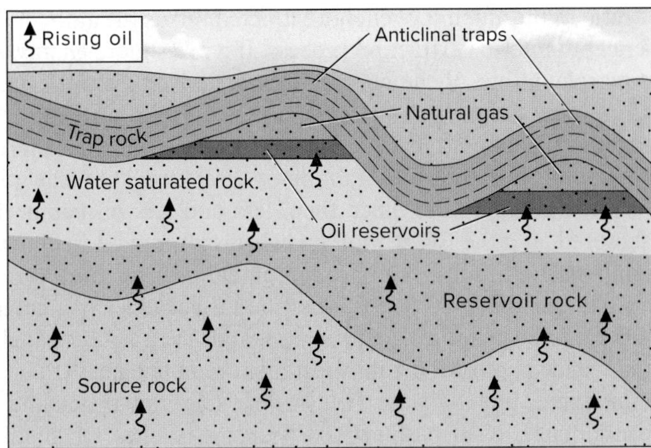

Accumulation of petroleum into reservoirs

FIGURE 22.9

Features related to petroleum reservoirs.

The first uses of oil, in providing mortar for mud bricks in ancient Sumeria, exploited such sites.

Natural oil seeps do not concentrate enough oil, however, to be of interest in the modern economy. Instead, petroleum geologists look for places where upward-infiltrating oil has encountered a structural **oil trap**, a place where impermeable rock (called "trap rock") prevents any further upward percolation of petroleum (figure 22.9). Natural gas requires the same conditions as oil for accumulation, and drillers can never be quite certain how much natural gas they may encounter when they first begin exploiting a potential petroleum deposit. In fact, as you might suppose, some prospects yield nothing but natural gas.

Figure 22.10 depicts several types of structural traps for oil and gas. Some types of traps are more abundant in particular regions than in others. For example, anticlines and domes (described in chapter 15) create the most common oil traps in the Persian Gulf; anticlines and faults are important trap-formers in southern California's oil field; and salt domes account for most of the petroleum reserve in the Gulf of Mexico. Where oil and water occur together in folded sandstone beds, the oil droplets, being less dense than water, rise within the permeable sandstones toward the top of the fold. There, the oil may be trapped by impermeable shale overlying the sandstone reservoir rocks. Because natural gas is less dense than oil, the gas collects in a pocket under fairly high pressure, on top of the oil. It is important to bear in mind that, as with aquifers, this layered pool of fluids does not fill a hollow underground chamber, like a flooded cave, but is merely *filling all of the pore spaces* in a highly permeable sedimentary rock (figure 22.10).

As you can see, petroleum requires a set of very special circumstances in order for it to be found in large enough accumulations to be economically useful. The accumulation of organic matter requires a warm, tropical sea that can support large numbers of organisms. Rapid sediment accumulation is required in order for the organic matter to be buried and protected from decay in the highly oxygenated ocean waters. The hydrocarbon-bearing sediment must then be buried to just the right depth to turn the organic matter into oil or natural gas contained in a source rock. The upwelling petroleum must then encounter a porous and permeable layer in which it can become concentrated—the reservoir rock. An impermeable oil trap must exist in order

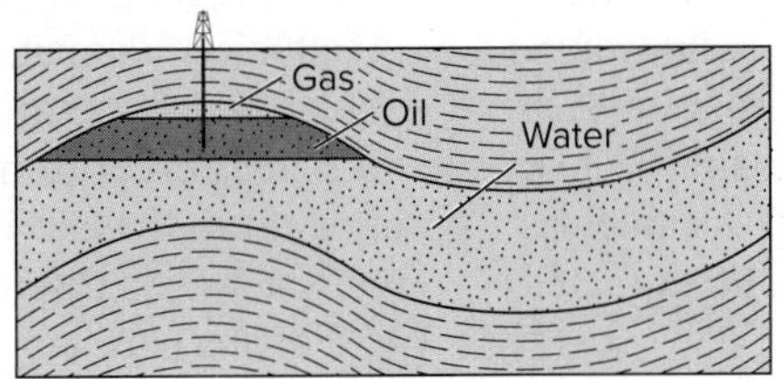

A Anticline

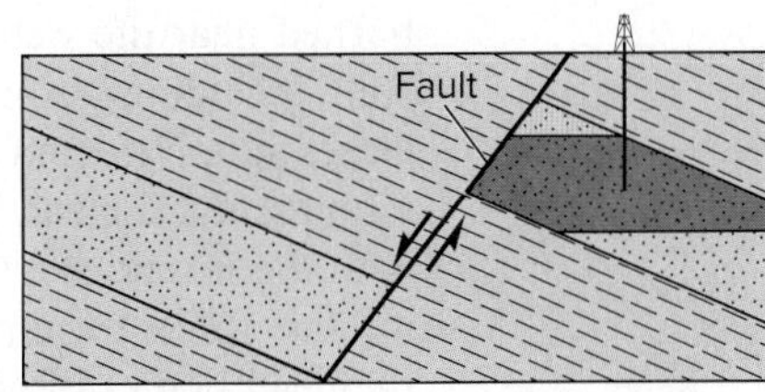

B Normal fault

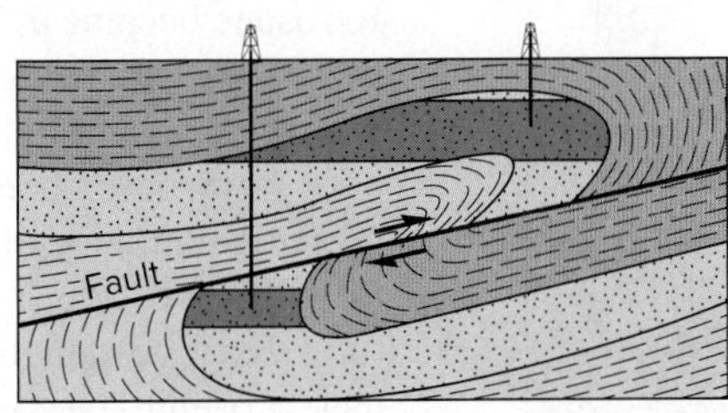

C Thrust fault

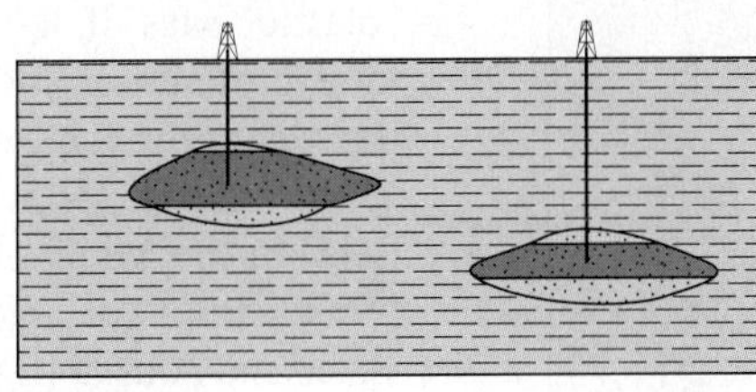

D Sandstone lenses

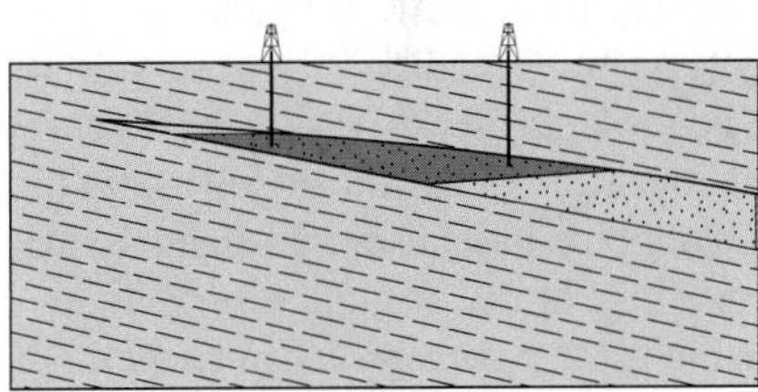

E Sandstone pinchout

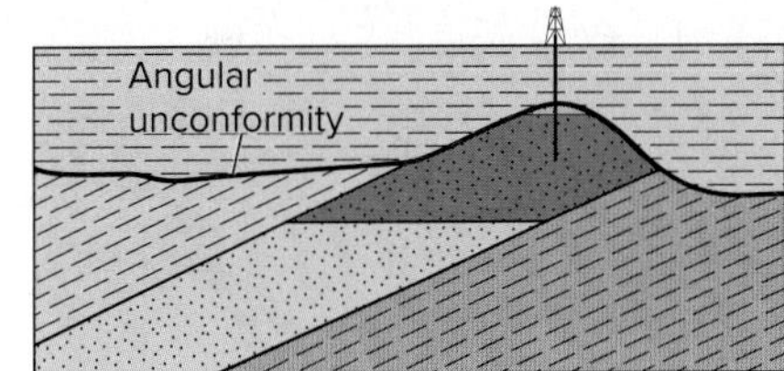

F Unconformity

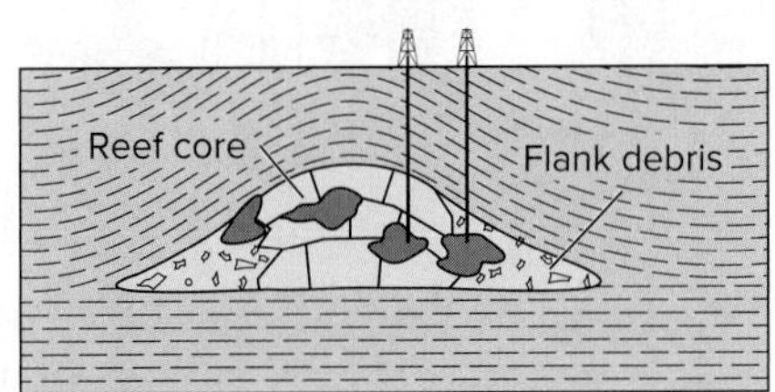

G Reef (a small "patch" reef)

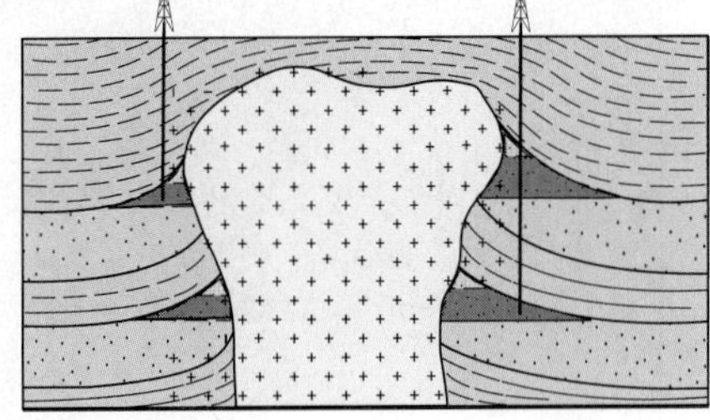

H Salt dome

FIGURE 22.10

Major types of petroleum traps. In all cases, impermeable rock encloses or caps the petroleum.

to stop the petroleum from reaching the surface. This may involve tectonic activity to fold or fault the rocks and produce trapping structures. The entire process requires millions of years. When you consider all of this, it is easy to understand why the world's petroleum is a finite resource and why geologists are unsure whether we will be able to continue to make new oil discoveries to meet the world's demand for petroleum products.

Oil is exploited by drilling down into an oil reservoir. **Oil fields** are regions underlain by one or more oil reservoirs. When a drill hole first penetrates an oil reservoir, pressurized natural gas within may drive the petroleum all the way to the surface so that no pumping has to be done. This "fluid-pressure effect" saves oil companies a tremendous amount of money and, in fact, may make all the difference between a successful drilling operation and one that needs to be abandoned. The celebrated gushers of many oil photos showing the discovery of new oil reservoirs illustrate the very high fluid pressures that gas may help generate in some deposits. With continued extraction, fluid pressure in the reservoir will diminish, and an oil field becomes less economical to operate. Remaining oil may be flushed out of the ground by "flooding" the reservoir with injected groundwater. The groundwater drives the petroleum ahead of it from the area of injection wells toward oil wells for removal (figure 22.11). Developers have also used steam to drive out the oil. As much as one-third of the original reserve in an oil field may be extracted using these *secondary recovery* methods.

One important consideration in operating an oil field is a factor called *energy return on energy invested (EROEI)*. This is

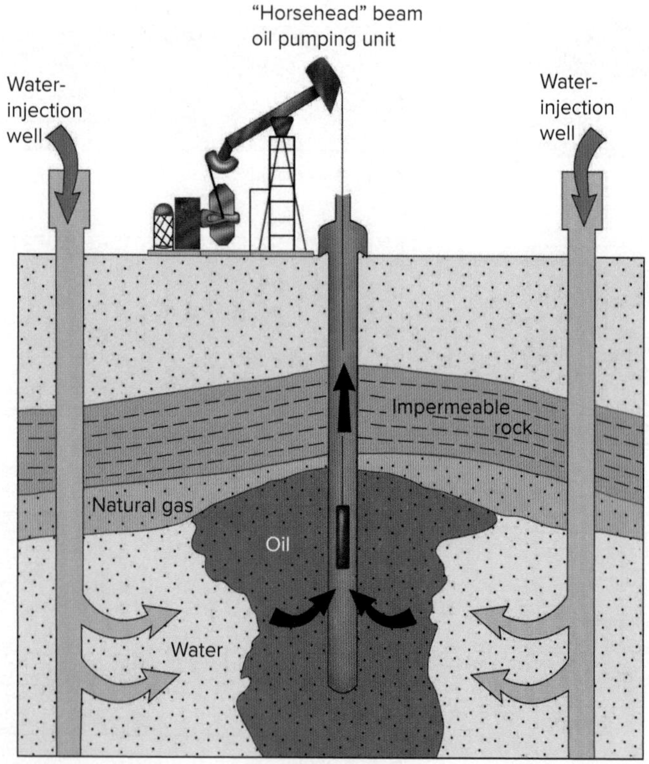

FIGURE 22.11
Water is often injected to drive additional, hard-to-get oil out of the ground.

the ratio of the amount of energy extracted versus the amount of energy put into the extraction process. If it takes more energy to get petroleum out of the ground than is derived from the sale and consumption of that petroleum, then it is no longer worth operating the oil field. During the heyday of petroleum exploration in the 1940s, the EROEI for newly discovered oil fields was typically around 100:1. Since then, less and less oil has been discovered, and the EROEI for a newly discovered oil field now is only 8:1 on average. The peak in global new discoveries occurred in the mid-1960s. Because of the surface tension effect in petroleum-bearing pores, and because oil drilling is increasingly taking place at greater depths or in more remote places (thanks to the exhaustion of older fields and rising global demand), mean global EROEI has declined. When it reaches 1:1, the industrial world will have to find a new energy source. Long before this happens, one would hope we will have turned to new energy technologies and a much different kind of civilization. One of the implications of EROEI is that we will never really run out of oil. However, it will become too expensive for us to continue exploiting oil in large quantities, perhaps within your lifetime.

Society demands a wide range of oil and gas products. Table 22.2 lists the various types of hydrocarbons that we artificially crack from oil delivered to refineries. In the early days of oil, from the late 1850s until around 1900, the sole product of interest was kerosene for lighting lamps. No more than about 40% of any barrel of oil pumped would go to market as this product. Since there was no use for the heavier compounds of oil, such as asphalt and diesel fuel, this material was often simply burned near the well, creating awful palls of smog. The lighter stuff, including gasoline, was often dumped in rivers and streams. Gases simply vented to the air, worsening the already severe environmental impact. Subsequent demand for oil products arose with the invention of the automobile and the conversion of the military forces to petroleum-based transport. Asphalt—essentially the dead bodies of countless, tiny marine organisms—ended up paving roads to minimize dust and facilitate high-speed driving. Kerosene became aviation fuel. Of the gasolines, octane (C_8H_{18}) proved best for performance (speed, power) in car engines and produced the least exhaust upon combustion. But because refineries have never been able to produce enough pure octane to meet demand, we have introduced substitutes ("reformulated fuels") to provide the same fuel services. Some of these substitutes (e.g., leaded fuels and MTBE—methyl tributyl alcohol) have proven to be costly environmental hazards. Highly complex hydrocarbons, such as polyethylene, end up in supermarket "plastic" bags. Indeed, from nylon and computer components to food production and pharmaceuticals, it is hard to see where petroleum products *aren't* used in the modern world.

Over 30% of the world's oil comes from exploitation of oil fields in the Middle East, much less than in the recent past. Although the two largest oil fields in the United States, in East Texas and Alaska (figure 22.12) are in decline, at this writing the United States is experiencing an oil boom, which has decreased its dependency on foreign oil sources. This boom is the result of increased production of oil from oil-bearing shales (also referred to as tight oil) in North Dakota and Montana through the use of

TABLE 22.2 Types of Cracked Petroleum-Related Hydrocarbons and Their Uses (Listed in Order of Increasing Complexity)

Name	Chemical Formula	Type of Hydrocarbon	Use
Methane	CH_4	Natural gas	Fertilizer manufacture; source of hydrogen for fuel cells
Ethane	C_2H_6	Natural gas	Fertilizer manufacture; source of hydrogen for fuel cells
Propane	C_3H_8	Gas condensates*	Cooking stoves, home heating, lanterns
Butane	C_4H_{10}	Gas condensates*	Cooking stoves, lanterns, lighters, home heating, soldering irons
Hexane	C_6H_{14}	Gasoline	
Heptane	C_7H_{16}	Gasoline	
Octane	C_8H_{18}	Gasoline	Isooctane, a form of octane, is the best kind of gasoline for internal combustion engines
Nonane	C_9H_{20}		
Decane	$C_{10}H_{22}$		

Successively Heavier Hydrocarbon Molecules:
1 Kerosenes and heating oils—aviation fuel, home heating
2 Diesel fuels—transportation fuel for trucks, trains, ships
3 Heavy crude oils ($C_{17}H_{36}$-$C_{22}H_{46}$)—lubricating and engine oils
4 Asphalts, waxes, greases, paraffins—paving, machinery lubricating
5 Plastics, polyethylene—computer frames, shopping bags, toys, CDs, etc.
*Also called "natural gas liquids," "drip gases," or "white gold"

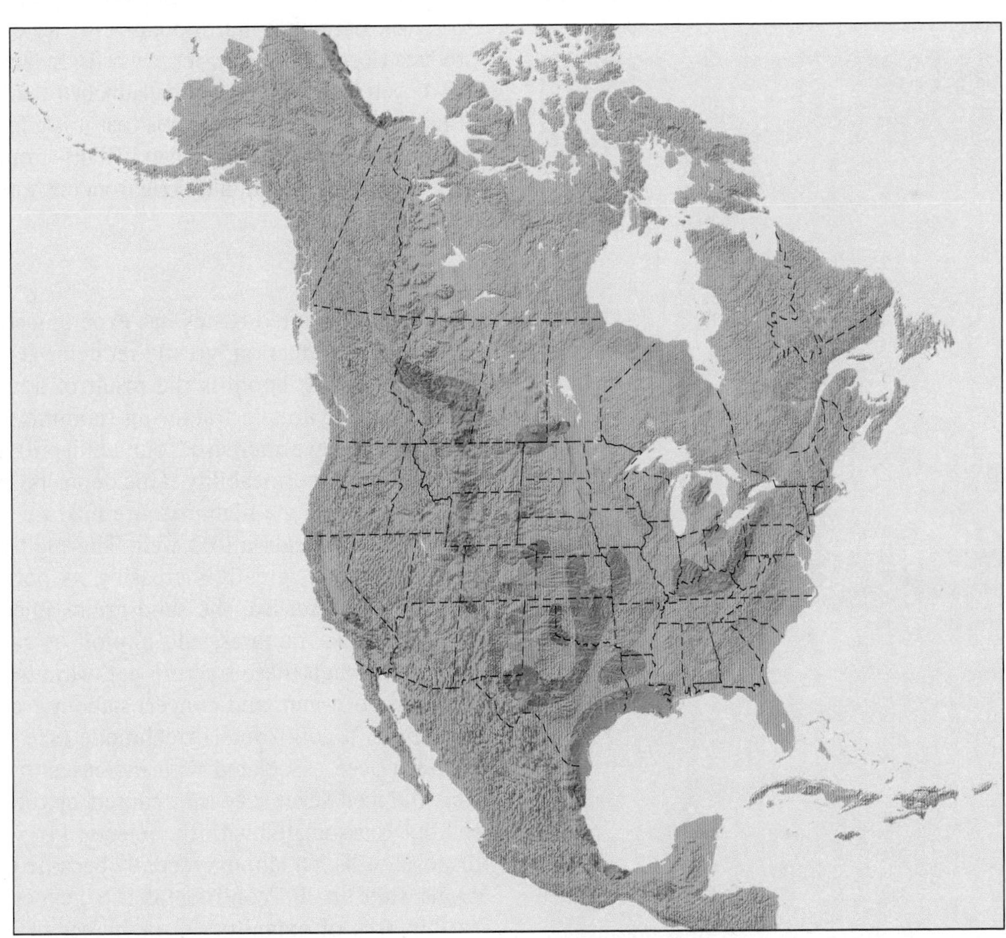

FIGURE 22.12

Major oil fields in North America. The amount of oil in a field is not necessarily related to its areal extent on a map. It is also governed by the vertical "thickness" of the oil pools in the field. The fields with the most oil are in Alaska and east Texas. *Source: U.S. Geological Survey and other sources*

horizontal drilling and hydraulic fracturing techniques. In addition, the exploitation of unconventional oil sources such as oil sands and oil shale has increased. However, unless demand for oil decreases, the vision of a world scrambling to meet demand for a shrinking resource is an ominous portent of conflict. The peace and organization of human society are basically dictated by the choice of the resources we exploit.

While coal resources and reserves may be estimated or ascertained with reasonable confidence, geologists are less confident about the amount of petroleum in the world because it is more widely dispersed through the crust and difficult to locate. Current estimates of the world's recoverable oil reserves average approximately 2,090 billion barrels (a barrel contains 159 liters, or 42 gallons, of oil). At present, the world consumes approximately 95 million barrels of oil every day, giving us approximately 60 years before the current reserve disappears. These numbers, however, do not take into account discoveries of new reserves or the increased size of reserves due to new technology. In addition, as you will read in the following sections, there are alternative sources of petroleum that, although less economical to exploit, are becoming more important as the reserves of light, easily recoverable oil are depleted.

The total resource, of course, is much larger than the reserve. Much of the resource, however, is in difficult-to-reach localities under the sea floor or in subarctic settings (figure 22.13). Exploration in these areas can lead to the degradation of sensitive ecosystems such as the subarctic or major environmental disasters such as the 2010 oil spill in the Gulf of Mexico (see box 22.2).

FIGURE 22.13

Drilling rig on Alaska's North Slope. *©BP America, Inc.*

Natural gas *resources* are even more difficult to estimate. The Energy Information Agency estimates that there are approximately 64 trillion cubic meters (2,276 trillion cubic feet or tcf) of technically recoverable natural gas in the United States. Current U.S. levels of consumption amount to around 0.75 trillion cubic meters (27 tcf) per year, around 20% of the world's total–giving approximately 84 years of natural gas from domestic sources alone. The largest producers of natural gas in the United State are Texas, Pennsylvania, and Oklahoma. Current U.S. *reserves* are estimated to be about 338 tcf, giving only 12.5 years of natural gas from domestic sources alone. Reserve estimates for natural gas in the western United States would certainly increase–though it's not known by how much–by opening certain public lands to gas extraction, including some national forests and monuments. This has made the further development of gas reserves a controversial issue.

It is much harder to transport natural gas than petroleum, requiring pipelines and LNG (liquid-natural gas) tankers for widespread distribution. Nevertheless, natural gas has become vital in the modern world. It is used to heat homes, cook food, make synthetic NPK (nitrogen-phosphorus-potassium) fertilizers for agriculture, produce electricity, and power fuel cells. In fact, the much-discussed "Hydrogen Revolution" may be launched thanks to cheap supplies of natural gas. Another bonus is that it is a less-polluting fuel than coal or petroleum, and has a high EROEI–from 10.2 to 6.3, depending on whether extraction is done from onshore wells or offshore.

Fracking

Recently, the United States has experienced a boom in oil and natural gas production, greatly reducing its reliance on foreign oil sources. This boom is the result of advances in horizontal drilling and hydraulic fracturing (commonly referred as *fracking*), which have increased our ability to exploit oil and gas trapped in low-permeability shale deposits. Hydraulic fracturing works by injecting a high-pressure mixture of water mixed with sand and chemicals into a well. The high pressure causes the rock to fracture, greatly increasing its permeability. Once the pressure is decreased, the sand grains injected with the water hold open the fractures, and any oil or gas present is able to migrate through these fracture pathways to the well. Due to its heavy use of water, and concerns about groundwater pollution, fracking is a controversial technique (see box 11.3). Fracking has also been associated with increases in earthquake activity. This induced seismicity isn't caused by the fracking process itself but is associated with the injection of wastewater into deep disposal wells. Oklahoma recently became the most earthquake prone state in the continental U.S., experiencing almost 700 earthquakes of magnitude 3 or higher in 2015. Compare this with only 20 earthquakes of magnitude 3 or higher in 2009. On September 3, 2016, a magnitude 5.8 earthquake–the strongest Oklahoma has ever recorded–shook the area around Pawnee. Residents of the town later filed a class-action lawsuit against 27

ENVIRONMENTAL GEOLOGY 22.2

The Gulf of Mexico Spill—The Cost of Oil Exploration in Ever More Difficult-to-Reach Areas

On April 20, 2010, at approximately 9:45 P.M., a massive explosion caused by methane gas rising up from great depths engulfed the Deepwater Horizon drilling rig, killing eleven workers (box figure 1). The platform eventually sank after 36 hours. Almost a mile of drill pipe collapsed onto the sea floor. Oil rose from depths of almost 3.5 miles below the sea floor and began gushing from the broken pipe. Over the next three months, as engineers struggled to find a way to stop the leak, an estimated 4.9 billion barrels of crude oil gushed from the broken drill pipes on the sea floor—the largest marine oil spill in history (box figure 2). The spill impacted the ecology of the Gulf of Mexico, affecting marine creatures, birds, and the delicate ecosystems of the wetlands along the Gulf Coast. The economic impact on the fishing industry was enormous.

Although it has now been attributed to shortcuts taken by the drilling company and human error, this disaster highlights the enormous risks involved as we search for new sources of petroleum. One of the frontiers in oil exploration is deep-sea drilling—exploiting oil reservoirs that are not only buried deep in the Earth, but are also under great depths of ocean water. In the Gulf of Mexico, it is estimated that there are 60 billion barrels of oil beneath the sea floor, enough to supply U.S. demand for a decade. The Deepwater Horizon platform was drilling an exploratory well into the Macondo Prospect about 41 miles off the coast of Louisiana under 1,500 meters (5,000 feet—almost a mile) of water and about 5,500 meters (18,000 feet—almost 3.5 miles) of sediment. These conditions made containment of the spill very difficult for engineers who had to use deep-water submersibles to work due to the high pressures at such great depths. After a number of failed attempts, the spill was eventually halted on July 15, 2010, using a large containment cap lowered over the well head. It took until September 2010 to drill a relief well that intersected the original well and then to plug it by pumping in cement.

Six years after the Deepwater Horizon disaster, the environmental impacts of the spill are still being evaluated. Studies of dolphins off the coast of Louisiana show continuing high rates of sickness and disease. Immediately following the disaster, the U.S. Department of the Interior ordered a six-month moratorium on offshore drilling, and the safety practices and emergency plans of oil companies were questioned. The number of new permits being issued drastically declined. However, offshore drilling accounts for over 15% of U.S. oil production, and with concerns about dependence on foreign oil and large reserves in the Gulf of Mexico, deep water drilling will most likely remain the next big frontier in oil production.

What do you think? Is exploration in ever deeper water worth the risk to the environment? How much risk is acceptable? Can deep-sea drilling ever truly be safe?

Additional Resources

- http://ocean.nationalgeographic.com/ocean/critical-issues/gulf-oil-spill/

This *National Geographic* site includes a number of interesting features on the oil spill and its effects.

- www.nature.com/news/2010/100901/full/467022a.html

This *Nature* article examines the ecological effects of the spill.

- www.bp.com

BP, the owners of the drill rig, have a series of items regarding the accident, their efforts to contain the spill, and the restoration effort. Click on the Gulf of Mexico Restoration link.

BOX 22.2 ■ FIGURE 1
Fire boat response crews battle the blazing remnants of the offshore oil rig Deepwater Horizon on April 21, 2010.
Source: U.S. Coast Guard

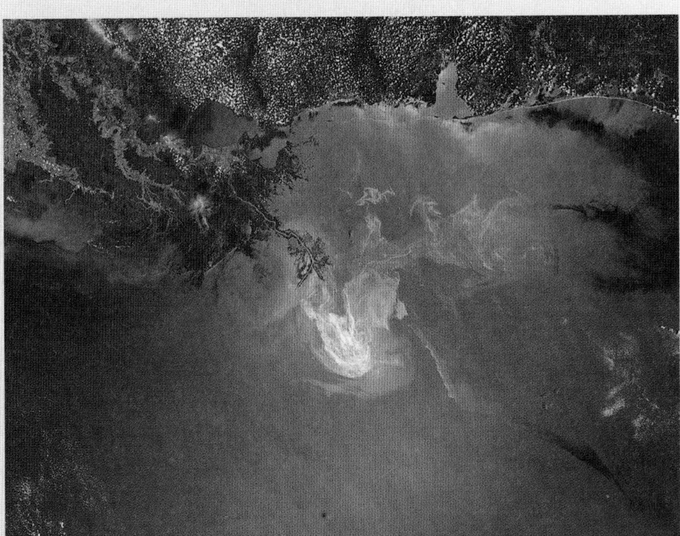

BOX 22.2 ■ FIGURE 2
NASA satellite image captures sunlight illuminating the oil slick off the Mississippi Delta on May 24, 2010.
Source: Jeff Schmaltz, MODIS Rapid Response Team/NASA

fracking companies, demanding they pay for damages caused by the earthquake.

Coal Bed Methane

Coal beds themselves may prove to be a major source of natural gas in the future. When coal forms, water and natural gas in the form of methane are trapped in the fine pores, pockets, and fractures that speckle and lace the interior of the coal. Pumping the water out lowers pressure and releases gas in huge quantities. Coal can store six to seven times more gas than an equivalent amount of rock in an ordinary natural gas field. A problem arises with respect to the water removed during pumping. Coal-water gets saltier the deeper the deposit, and disposal of salty water into surface watersheds seriously degrades water quality. Groundwater supplies may also be contaminated during gas extraction. In any event, there is a considerable amount of **coal bed methane** in the United States and, according to the Energy Information Administration, it accounted for approximately 4% of U.S. natural gas production in 2015. The overall resource may exceed 20 trillion cubic meters (700 tcf), but only about 2.8 trillion cubic meters (100 tcf) is likely to be economically recoverable. The U.S. Geological Survey estimates that the total coal bed methane resource worldwide might be as high as 200 trillion cubic meters (7,500 tcf).

Heavy Crude and Oil Sands

Heavy crude is dense, viscous petroleum. It may flow into a well, but its rate of flow is too slow to be economical. As a result, heavy crude is left out of reserve and resource estimates of less viscous "light oil" or regular oil. Heavy crude can be made to flow faster by injecting steam or solvents down wells, and if it can be recovered, it can be refined into gasoline and many other products just as light oil is. Most California oil is heavy crude.

Oil sands (or **tar sands**) are bitumen-cemented sand or sandstone deposits. The bitumen is solid, so oil sands are often mined rather than drilled into, although the techniques for reducing the viscosity of heavy crude often work on oil sands as well.

The origin of heavy crude and oil sands is uncertain. They may form from regular oil if the lighter components are lost by evaporation or other processes. Oil sands and asphalt seeps at Earth's surface (such as the Rancho La Brea Tar Pits in Los Angeles) probably formed from evaporating oil. But some heavy crude and oil sands are found as far as 4,000 meters underground. Most have much higher concentrations of sulfur and metals, such as nickel and vanadium, than does regular oil. Some geologists believe that oil sands represent oil that arose from source rocks but never became trapped and concentrated by structural traps.

The best-known oil sand deposit in the world is the Athabasca Tar Sand in northern Alberta, Canada. Approximately 50% of Canada's present oil production comes from these oil sands. Counting this unconventional resource, this gives Canada the third-largest total oil reserves in the world, after Saudi Arabia and Venezuela. Oil-hungry countries such as the United States view these deposits with keen interest. Unfortunately, the EROEI of extracting petroleum from oil sands is substantially lower than that of conventional oil–close to 3:1 by some estimates. This is mostly due to the need to dilute the viscous heavy oil ("bitumen") to get it out of the sand. Natural gas, gas condensates, hot water, steam, and naptha (an aromatic solvent) are all used—and each of these flushing compounds is, in and of itself, an energy material or requires energy to produce.

The time-tested method of extracting the bitumen requires mining the ground directly. For every 2 tons of earth processed, only one barrel of oil can be made. Disposing of this material, not to mention dealing with environmental concerns such as natural habitat destruction and water pollution, has raised serious questions about the future of the industry. New technology more recently has allowed miners to extract the oil from deep underground without disturbing the surface. This process involves mixing the oil sands in place with hot water and then pumping the slurry through a pipeline to the processing plant. Questions of pollution and low EROEI remain, but this is a definite improvement, and it appears that the oil sand industry has established a firm future for itself in the world economy.

Oil Shale

Oil shale is a black or brown sedimentary rock with a high content of solid organic matter or *kerogen.* Oil shale, not to be confused with oil-bearing shale deposits, is both a source of oil and a fuel as it can be burned with little or no processing. Oil shale is formed in much the same way as conventional oil deposits, but the sapropel is not buried as deeply and doesn't reach the temperatures required for cracking the hydrocarbons into lighter fractions. The kerogen in oil shale is solid and, as a result, will not separate from the sedimentary rock. Thus, oil shales must be processed by distillation to convert the solid oil to an extractable substance.

Oil shale formations are found in many locations around the world. The largest are found in the United States. The best-known oil shale in the United States is the Green River Formation, which covers more than 40,000 square kilometers in Colorado, Wyoming, and Utah, with deposits up to 650 meters (2,100 feet) thick (figure 22.14). The oil shale, which includes numerous fossils of fish skeletons, formed from mud deposited on the bottom of large, shallow Eocene lakes. The organic matter came from algae and other organisms that lived in the lake. The Green River Formation is estimated to contain up to 3 trillion barrels of

FIGURE 22.14

Cliffs of oil shale that have been mined near Rifle, Colorado.
©William W. Atkinson, Jr.

oil, up to half of which may be recoverable, although this estimate has been disputed. Relatively low-grade oil shales in Montana contain another 180 billion barrels of recoverable oil in shale that should be economical to mine because of its high content of vanadium, nickel, and zinc. Therefore, oil shale can supply potentially vast amounts of oil in the future as our liquid petroleum runs out.

A few distillation plants extract shale oil, but the current price for oil makes shale oil uneconomical. Interest in oil shales increases whenever oil prices rise, and large-scale production of shale oil may be feasible in the future.

The mining of oil shale can create environmental problems. Mined oil shale is crushed prior to processing, which increases the volume of the rock. During distillation, the shale expands. This increase in volume creates a space problem–the solid by-products will take up more space than the hole from which they were mined. Spent shale could be piled in valleys and compacted, but land reclamation would be troublesome. A great amount of water is required, for both distillation and reclamation, and water supply is always a problem in the arid western United States. New processing techniques that extract the oil in place without bringing the shale to the surface may eventually help solve some of the problems and lower the water requirements. One method involves using heaters placed in deep holes to heat the oil in place. After one to two years, the surrounding rock has reached high enough temperatures for the oil to be fluid enough for extraction. All of the methods for underground extraction are currently viewed as experimental.

Uranium

The metal *uranium*, which powers nuclear reactors, occurs as uraninite (more commonly referred to as *pitchblende*), a black uranium oxide mineral found in hydrothermal veins, or, much more commonly in the United States, as yellow *carnotite*, a complex, hydrated oxide mineral found as incrustations in sedimentary rocks. Oxidized uranium is relatively soluble and easily transported by water. It precipitates in association with organic matter where bacterial activity reduced the oxygen.

Most of the easily recoverable uranium in the United States is found in sandstone in New Mexico, Utah, Colorado, and Wyoming, some of it in and near petrified wood. During the 1950s uranium boom, western prospectors looked for petrified logs and checked them with Geiger counters. Some individual logs contained tens of thousands of dollars worth of uranium. Some petrified logs have so much uranium that it would be dangerous to keep them as souvenirs. Most of the uranium, however, is in sandstone channels that contain plant fragments.

Organic phosphorite deposits of marine origin in Idaho and Florida also contain uranium. The uranium is not very concentrated, but the deposits are so large that, overall, they contain a substantial amount of uranium. The black Devonian shales of the eastern United States also contain uranium. These shales are really low-grade oil shales (figure 22.14). Uranium may be recovered from phosphorites or shales as a by-product of another mining operation.

The principal use of uranium at present is to provide power for electricity-generating nuclear reactors, although uranium was also used to make tens of thousands of nuclear warheads during the Cold War. Some naval craft use uranium to power ship-borne nuclear engines, but concerns about accidents and radioactive pollution greatly limit expansion of this form of transportation. At present, ninety-nine nuclear reactors produce almost 20% of the energy needs of the United States. France is the industrialized nation most dependent upon nuclear power, which generates over 75% of its electricity. However, the problems at Fukushima in Japan following the 2011 earthquake (see box 22.3) have led many countries to reduce their nuclear power production, and France is now planning on reducing its nuclear power production by a third over the next twenty years.

Electricity is generated in nuclear power plants in much the same way as in coal-fired plants. The fuel (in this case, uranium) is used to boil water which generates steam which, in turn, powers turbines which generate electricity. Unlike coal, the heat generation in a nuclear power plant does not come from burning the fuel but from a process called **fission**. Fission is the splitting apart of the nucleus of an atom into lighter components such as the nuclei of lighter atoms and neutrons. When a heavy element is split apart (or fissioned), this process also generates heat.

In chapter 8, we discussed isotopes and radioactivity. Uranium has three natural isotopes, all of which are unstable. Uranium-238 (^{238}U) is the most stable and the most abundant, making up 99.8% of all uranium isotopes. Uranium-235 (^{235}U) and uranium 234 (^{234}U) are less abundant (0.7% and 0.005%, respectively). ^{235}U is the most important isotope of uranium as it is the only one known to be fissile–that is, it can be split apart when bombarded by neutrons. To be useful in a nuclear reactor, the uranium minerals must be processed to concentrate the unstable isotope uranium-235 (^{235}U), the main reactor fuel. Only 0.7% of the mass of natural uranium consists of ^{235}U. The refining of uranium ore to concentrate ^{235}U requires considerable amounts of energy, metals, and other resources.

Once processed, the ^{235}U goes into *fuel rods*, which are inserted into the cores of nuclear reactors (figure 22.15). A reactor core consists of a *containment vessel*, or reservoir, lined with

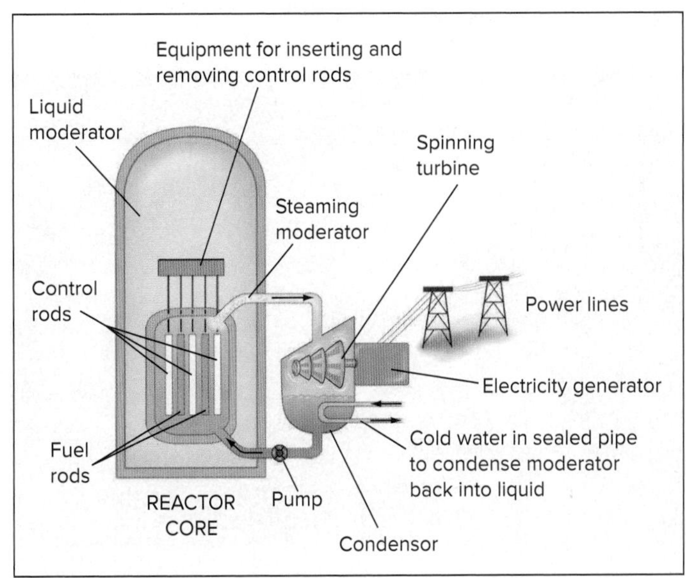

FIGURE 22.15

Basic design of a nuclear reactor. This shows operation of a BWR- (boiling water) type reactor, the simplest design.

ENVIRONMENTAL GEOLOGY 22.3

The Nuclear Crisis in Japan—The Future of Nuclear Power Put to the Test

The magnitude 9.0 earthquake and tsunami that hit eastern Japan on March 11, 2011, was devastating. Entire coastal towns and villages were destroyed, leaving hundreds of thousands of people homeless. As of 2012, the official death toll was 15,854 with another 3,203 missing. The disaster also caused a nuclear crisis that has global implications for the future of nuclear power.

Nuclear power has been viewed as a way to reduce reliance on foreign sources of fossil fuels, but there are many risks associated with nuclear reactors that have been highlighted by the events in Japan. Prior to the earthquake and tsunami, Japan relied on nuclear power for 30% of its electricity and had plans to expand its nuclear capacity to 50% by 2030. Located in a region known for powerful earthquakes and tsunami, the Fukushima Daiichi plant was designed with many safety features. When the earthquake hit, the plant's electricity was cut off, but backup generators quickly kicked in and began an emergency shutdown of the active reactors. Cooling systems pumping water around the fuel rods kept the reactors from overheating. One hour later, a 14-meter (46-foot) tsunami reached the plant, which was protected by a seawall. Unfortunately, while planners had assumed a 10-meter seawall would provide adequate protection against tsunami, they did not anticipate an event of this magnitude or the 1-meter drop in elevation of the coast caused by the massive earthquake. The seawall was easily overtopped by the massive wave. Water flooded the facility and disabled the backup generators, which were all located at or below ground level. All power was lost, and without cooling water circulating through them, the reactors began to overheat and partially melt down. In the following weeks, efforts to get the reactors under control were hampered by explosions and fires as well as flooding with radioactive water (box figure 1). Spent fuel rods stored in pools in each reactor building also began to overheat as water levels in the pools dropped. It took months to fully stabilize the plant, and it could take decades to remove the melted core material and complete a cleanup of the area.

BOX 22.3 ■ FIGURE 1

Smoke from the damaged Fukushima Daiichi nuclear plant in Okuma, Japan. *©DigitalGlobe/Getty Images*

Radiation leaked from the plant as a result of explosions, planned steam releases to relieve pressure in the reactors, and discharge of coolant water into the sea. The Japanese government evacuated people living within 30 km (18 miles) of the plant (box figure 2). Radiation was detected in milk and food produced in the area as well as in water in Tokyo. Trace amounts of radiation were observed around the world, although the levels were extremely low and were not considered to pose any threat. Cleanup efforts continue as of this writing at Fukushima and will likely take decades to complete. Of grave concern is the continued leak of radioactive waters into the ocean and the groundwater.

The crisis at Fukushima caused many countries to reconsider the benefits of nuclear power. Japan is moving to reduce its dependence on nuclear energy. Germany made a decision to close the last of its seventeen nuclear reactors. The Swiss government also recommended phasing out its nuclear power plants. Italy put an indefinite hold on plans to build new power plants. France, which currently relies on nuclear power for 75% of its electricity, recently announced plans to cut nuclear output by a third in twenty years. In the United States, plans to expand nuclear power capacity may be slowed or even halted.

What will be the implications of a worldwide decrease in nuclear power production? Without a decrease in demand for energy, it will most likely lead to increased reliance on fossil fuels like coal and oil. What will be the impact on the climate of increased use of carbon-based fuel? How do we weigh the risk of a nuclear disaster against long-term climate change?

Additional Resources

- www.iaea.org/newscenter/focus/fukushima/

The International Atomic Energy Agency has set up a special site focused on the Fukushima nuclear accident.

- http://news.nationalgeographic.com/news/energy/2011/03/1103165-japan-nuclear-chernobyl-three-mile-island/

This article by *National Geographic* compares the disaster in Japan to the accidents at Chernobyl and Three Mile Island.

- http://news.nationalgeographic.com/news/energy/2011/03/110314-japan-nuclear-power-plant-disaster/

Another article by National Geographic.

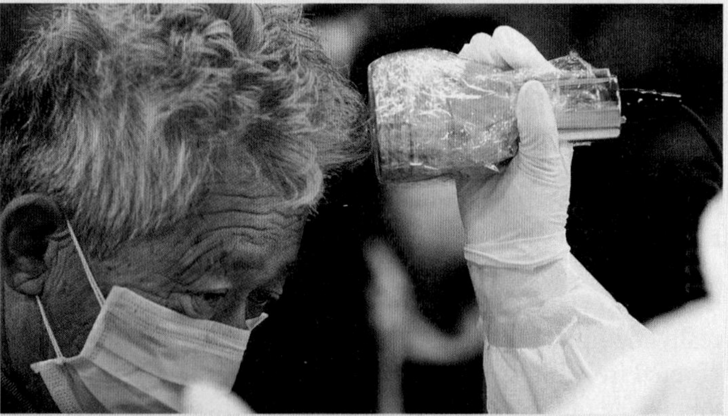

BOX 22.3 ■ FIGURE 2

A man being screened for radiation exposure. *©Gregory Bull/AP Images*

steel and concrete to protect outsiders from harmful radiation. The vessel is filled with a fluid called a *moderator*, usually water. Each fuel rod is about the size of an automobile, but because of the high atomic weight of its constituent materials, weighs around 30,000 kilograms (65,000 pounds). It contains only about 2–3% ^{235}U (most of the uranium is harmless ^{238}U).

As the nucleus of a ^{235}U atom disintegrates, heat and neutrons escape and the uranium atom transforms into daughter nuclei. These, in turn, produce more neutrons through radioactive decay. The escaping neutrons may be slow-moving or fast. The slower neutrons penetrate the nuclei of neighboring ^{235}U atoms, triggering their decay as well and, in short order, a full, heat-generating chain reaction is underway. The moderator fluid helps to slow the neutrons, thus stimulating the chain reaction even further. The fluid also gets very hot, reaching the boiling point and generating steam to drive turbines for electricity. (In some nuclear reactors, the moderator is passed through a heat exchanger to boil a secondary supply of water for the turbines. This keeps radioactivity out of the turbines.)

To control the nuclear reaction, *control rods* made of carbon, or cadmium (which is quite poisonous) must be inserted into the moderator. These soak up neutrons and can stop a chain reaction altogether. Careful manipulation of control and fuel rods brings a reactor to just the right level of heat production for energy-generation purposes. Nuclear accidents may take place if this balance in control is lost (box 22.3).

There are tremendous environmental problems associated with the disposal of fuel rods once their ^{235}U has depleted. A typical fuel rod has a service lifetime of only about three years. Even though most of its ^{235}U may be gone, neutrons released during a chain reaction will transform ^{238}U into deadly plutonium-239 (^{239}Pu). Spent fuel rods, hence, are quite dangerous. Temporary storage of used fuel rods takes place in a pool near the reactor core at a plant site. Later, the spent rods may be transported and reprocessed to extract the ^{239}Pu for nuclear weapons development. This is a special concern for persons who monitor nuclear proliferation—"rogue states" that have built nuclear reactors, ostensibly for "peaceful purposes," can easily create the fuel to build nuclear weapons. The ^{239}Pu may also be used as a fuel in a different kind of reactor, a *breeder reactor*, which greatly enhances the ability of uranium to produce energy. Full conversion of nuclear reactors to breeder designs could extend the effective uranium reserve lifetime sixty-fold at current consumption rates, providing the world with electricity for several millennia to come. Breeder reactors, unfortunately, have notably higher potential for disastrous accidents. The moderator in conventional breeder reactors, liquid sodium, is very sensitive and explosive, and reactor core blasts into the kiloton range are possible.

Spent fuel rods and other nuclear waste, particularly ^{239}Pu, must be stored someplace out of contact from people for a long period of time—as long as 250,000 years. Many proposals have arisen to do this, including "science-fiction" scenarios that require shooting this deadly waste into the Sun or placing it in subduction zones. In the United States, the proposed "permanent" waste repository was Yucca Mountain, located in southern Nevada. The nation's reactor waste was to be housed in welded pyroclastic rocks nearly 300 meters (1,000 feet) underground. The rocks contain zeolite minerals, which are natural sponges that absorb escaping radioactivity. The proposed Yucca Mountain site was controversial however, and in 2010 the plan was scrapped. In Scandinavia, a similar repository stores nuclear wastes in a granite bedrock vault underneath the Baltic Sea. The advantage of this site is that any escaping waste will remain confined in saline groundwater beneath the ocean floor, rather than infiltrate water supplies tapped by people on surrounding lands. Many other waste facilities have been established, but it is beyond the scope of this book to consider them all. In all cases, they involve the shallow, underground storage of nuclear materials.

Due to uncertainties in calculating the costs associated with nuclear power (refining of uranium ore, building and shutting down power plants), estimates of the EROEI for nuclear vary widely, from 5 to 75. In all, about 65,000 tons of uranium ore are produced every year to satisfy the needs of the world's 450 nuclear reactors, which supply about 10% of the world's total energy needs. Present world reserves of uranium (at about $130 per kilogram of uranium) are around 6 million tons. Canada is currently the largest producer of uranium in the world. Some of the world's largest reserves are located in Australia and Kazakhstan. Reserves at current consumption rates would last about eighty years. The size of the total uranium resource is unknown given incomplete exploration and the wide dispersion of uranium-bearing minerals, even in ordinary rocks such as granite. In any case, it is likely that the largest, most accessible uranium deposits have already been identified and are undergoing exploitation.

Renewable Energy Sources

Some energy resources are unquestionably renewable and easily tapped. These include solar, wind, wave, tidal, and hydroelectric power. Geothermal energy is also regarded as a renewable energy source. Together they provide the world with about 10% of its electricity. At present, the growth in renewable electricity supply is about 5% per year.

Unlike fossil fuels, which require huge industries to mine, transport, and distribute to users, solar and wind power can be generated locally—even in a backyard or atop the roof of one's own home. The development of solar and wind power is stimulating a transformation from highly centralized power production to much more distributed, smaller-scale, "neighborhood-scale" sources of power.

Geothermal Energy

Geothermal energy is heat energy from beneath the Earth's surface. The word *geothermal* comes from the Greek words *geo* (earth) and *therme* (heat). There is an enormous amount of heat escaping from the Earth's interior every day. In the upper 10 kilometers of Earth's surface within the United States alone, there is the heat equivalent of 1,000 trillion tons of coal—enough energy to satisfy the country's needs for 100,000 years! But in most locations it would take an area about the size of a football field to provide the escaping heat energy needed to power a single 60-watt, high-efficiency (20%) light bulb.

FIGURE 22.16
A geothermal station about 50 kilometers to the east of Reykjavik, Iceland, extracts hot groundwater heated by shallow magma intrusions and pumps it to Reykjavik for commercial and residential use.
©Richard Hazlett

In some parts of the world, however, there are areas of unusually high heat flow around young, cooling plutons and volcanic areas that can be exploited for energy development. The temperature of the crust in these areas rises to as high as 350°C or more at depths of 1 to 3 kilometers—enough to turn groundwater into steam if it is pumped all the way to the surface. (Water under great pressures will not boil.) The escaping steam, in turn, can be channeled into turbines to generate electricity. If the hot groundwater occurs in a confined aquifer, so much the better; the groundwater will require little, if any, effort to extract as high fluid pressure drives it surfaceward through boreholes.

Geothermal energy is used to generate electricity in over twenty nations worldwide. In Iceland (figure 22.16), geothermal energy accounts for 25% of all electricity needs (the remaining 75% is produced using hydroelectric power) and 90% of all heating needs. The world's largest geothermal power plant is at the Geysers in the Coast Ranges of northern California. This 750-megawatt facility provides the energy needs for 750,000 people, though its level of production has been declining steadily since 1980. The most important reason for this is that groundwater supplies are withdrawn faster than nature can replenish—and reheat—them. Water is an excellent natural carrier of heat energy. Dry rock acts as an insulator, or a slow conductor of heat. In areas where the groundwater supply has been depleted, or where there is not much groundwater due to arid climate conditions, water can be injected into hot, dry rocks, and artificial fracturing of these rocks can create reservoir space for a considerable volume of water. At the Geysers, 34 million liters (9 million gallons) of treated wastewater, carried through a 66-kilometer (41-mile) underground pipeline from the town of Santa Rosa, is injected into the ground every day. This water maintains the pressure in the aquifer, stabilizing the energy production level from the field.

Even where water is recycled by pumping it directly back into the Earth after passing through turbines, costs of operating a geothermal power plant can be quite high. Hot groundwater often carries with it dissolved minerals and acids that corrode metal pipes and turbines or clog them with precipitated minerals. Maintenance costs to keep a plant in operation may be prohibitive for this reason alone. Furthermore, there is a cost involved in scrubbing and purifying vapors that may contain natural contaminants. Despite all of these limitations, energy production from geothermal power is on the whole quite competitive with that of fossil fuels. EROEI levels as high as 13 exist, and for lightly populated regions such as Iceland and New Zealand, there is great incentive to develop geothermal energy rather than import oil or build nuclear reactors. The world production of geothermal power provides over 11,000 megawatts of electricity—equivalent to that of around twenty-four conventional nuclear reactors. In some nations, including parts of the United States, geothermal waters provide heat as well as electricity. The hot water passes through pipes and walls to warm home interiors, and can even be used directly in showers and taps. In Iceland, instead of turning on the hot water tap and waiting for it to run hot, you can find yourself turning on the cold water tap and waiting for it to run cold!

Because groundwater and geothermal heat are considered to be renewable resources, geothermal energy is technically a renewable energy resource. However, it is important to point out that if heated groundwater is drawn from the ground at a faster rate than it can be replenished and heated, then this energy resource can be regarded as nonrenewable.

Solar Energy

The Sun's energy drives many of Earth's systems, including the hydrologic cycle, the winds, and the ocean currents. The amount of energy provided by the Sun is vast, amounting to as much energy in an hour as humans use in a year. However, the Sun's energy is diffuse, and harnessing it efficiently is a major challenge to widespread use of solar power.

Solar energy must be concentrated to produce heat or electricity. Three strategies for using solar energy are *passive*, *active*, and the use of *photovoltaic cells* to generate electricity. **Passive solar heating** can be achieved by including large Sun-facing windows and efficient insulation in a building's design. **Active solar heating** uses solar panels to heat water which can be used to provide hot water and space heating. The solar panels consist of a black surface beneath a glass panel. The Sun's energy is absorbed by the black surface, which heats up. Water passed through the panel will be heated and can then be stored in a central hot water tank. **Photovoltaic cells** (figure 22.17) convert sunlight directly into electricity. The cells are made of thin wafers of silicon that has been doped with other elements to form an electric field. One side is treated with an element that will

FIGURE 22.17

Photovoltaic cells convert sunlight directly into electricity.
©Mick Roessler/Corbis/Getty Images RF

produce electrons, the other is treated with an element that will capture electrons. When sunlight hits the wafer, electrons are able to travel from one side to the other, producing an electrical current. Photovoltaic cells are still relatively inefficient, converting only 10–15% of the Sun's energy into electricity, and the cells are expensive to produce. New technologies may increase this efficiency in the future, with estimated EROEIs around 30.

FIGURE 22.18

Altamont Pass Wind Farm, California.
©Doug Sherman/Geofile RF

Wind Power

Winds are generated by the Sun. The energy of the wind has been harnessed by humankind for many years in the form of windmills. Today, modern wind turbines are used to generate electricity. The wind turns the blades on a turbine that is connected to a generator that produces electricity. Wind turbines must be placed in an area with constant strong winds–and, to generate large amounts of electricity, hundreds of turbines are needed. These fields of turbines, such as the one at Altamont Pass in California, are called wind farms (figure 22.18).

The Danish government, which has invested heavily in developing wind power for its national energy supply, estimates that EROEI may climb as high as 50 in harvesting the strong sea breezes of that country. Currently, wind power produces only about 3% of the world's energy use; in Denmark, wind power accounts for 42% of energy production.

Hydroelectric Power

Hydroelectric power facilities transform gravitational energy in the form of falling water into electrical energy. Most hydroelectric power stations lie at the foot of dams, where water spilling from reservoirs into rivers downstream spins turbines. These stations produce electricity somewhat more cheaply than fossil fuels (EROEI is around 100), especially in regions where oil, coal, or gas has to be imported from afar. Downstream bank erosion, disruption of fish migrations, the flooding of land, and displacement of populations by filling reservoirs are major environmental concerns.

Hydroelectric power is by far the largest of the developed renewable energy resources at present, accounting for 84% of all renewable electricity production. Hydropower provides 6% of the electricity consumed in the United States today. It can be generated locally; a small station placed on a creek or stream can provide the power needs of a home, farm, or ranch. Hydropower does not directly produce any air pollution, and can even be tapped during times of low energy demand to recycle water back upriver by pumping into reservoirs.

Tidal Power

Tidal power is a variation of hydropower. A barrier, called a *barrage*, must be constructed across the mouth of an estuary or bay (figure 22.19). Gates in the barrage allow water to pass through as the tide rides, spinning turbines to produce electricity. The gates close when the tide is in, capturing the water inshore from the barrage. The gates reopen after the tide falls on the seaward side, and the water pouring out spins the turbines again in reverse.

The world's largest tidal-generating station, at Rance in France, generates 320 megawatts, enough to supply several hundred thousand users. High costs; concerns about impacts on fish, bird life, and ecosystem health; and irregular power supply have greatly impeded development of tidal mills elsewhere in the world. In fact, only six countries at present–France, China, Russia, South Korea, Northern Ireland, and Canada–have tidal power facilities.

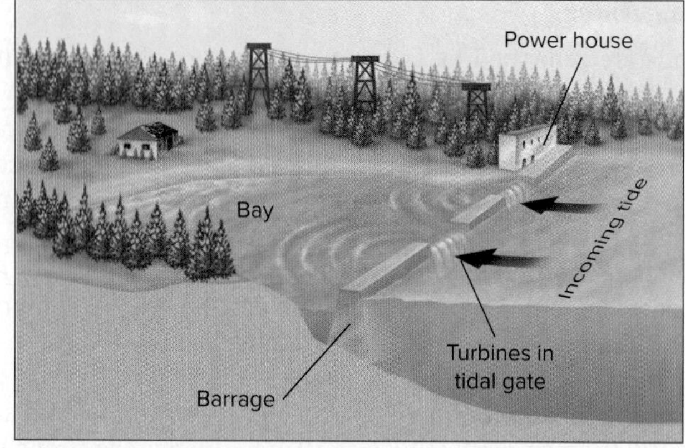

A

B

FIGURE 22.19

Generation of tidal power.

Wave Power

Wave power is often confused with tidal power. Instead of capturing the energy of the tides, wave power captures the energy held in waves. This energy is a combination of kinetic energy and potential energy. The energy is transferred from the wind to the surface of the ocean. The technology to harness wave energy is still mostly experimental, and there are a number of different devices that work in different ways. One device consists of a segmented tube floating on the surface. As the wave passes along the length of the tube, the different segments move up and down relative to each other. The resistance between them is used to pump oil though hydraulic motors, which in turn drive electrical generators. Electricity is then fed to a cable that is connected to land.

Wave power does not contribute much to the world's energy needs and, to date, there are only a couple of wave farms in production or in development. The world's first commercial wave farm opened in 2008 off the coast of Portugal. It consists of three 140-meter-long segmented tubes that can generate 2.25 megawatts of energy, enough to supply the annual needs of 1,500 homes. Another wave farm is currently being developed off the coast of southwest England. In the United States, small wave-energy projects, using different technologies, are being tested off New Jersey and Hawaii, with another project being planned for Oregon.

Biofuels

Biofuel is defined as fuel derived from biologic matter. It differs from fossil fuels in that the biologic matter is recently dead. In recent years, interest has increased in the use of plant matter to produce fuel. The two most commonly used types of biofuel are ethanol and vegetable oil. Ethanol is generated by yeast fermentation from plants that contain a lot of sugar (sugar cane) or starch (corn). Vegetable oil comes from plants such as soybeans that produce a lot of oil. The oil can be used to fuel a diesel engine, or it can be processed to form biodiesel. Although the production of biofuels does produce the greenhouse gas carbon dioxide, this is offset by the uptake of carbon dioxide by the plants grown to produce the fuel.

One of the problems associated with the increased use of biofuels is the competing demands for food and fuel. A World Bank report released in 2008 suggested that biofuels may have contributed to world food price increases. As demand for alternative fuel sources continues to rise, the "fuel versus food" debate is likely to intensify.

22.3 METALLIC RESOURCES

Modern industrial society stands upon two feet–one of fossil fuels and the other of metal. While civilizations have always had access to basic construction materials–and always will–the production of high-quality, high-strength metals and the exploitation of fossil fuels make our times stand out in all of human history. Table 22.3 shows common metallic resources and some of their uses.

The successful search for metals depends on finding **ores**, which are naturally occurring materials that can be profitably mined (table 22.3). It is important to recognize that the local concentration of a metal must be greater (usually much greater) than its average crustal abundance to be a potential ore body. Metals must be concentrated in a particular place in a large enough amount to be viable ore bodies. Take gold in seawater. You could become fabulously wealthy if you could extract a fraction of the gold in seawater. There are over 10^{11} troy ounces of gold–around \$52 trillion worth in the world's oceans. But the concentration is 4 grams per 1 million tons of water. It would cost you far more to remove that gold than you could sell it for.

Whether or not a mineral (or rock) is considered a metal ore depends on its chemical composition, the percentage of extractable metal, and the market value of the metal. The mineral hematite (Fe_2O_3), for example, is usually a good *iron ore* because it contains 70% iron by weight; this high percentage is profitable to extract at current prices for iron. Limonite ($FeO(OH) \cdot nH_2O$) contains less iron than hematite and, hence, is not as extensively mined. Even a mineral containing a high percentage of metal is not described as an *ore* if the metal is too difficult to extract or the site is too far from a market; profit is part of what defines an ore.

Many different kinds of geologic processes can accumulate ores, from weathering and sedimentation to the settling of crystals deep within magma chambers (table 22.4). We'll survey the

TABLE 22.3 Common Metallic Resources

		Ore Minerals and Their Uses	
Metal	**Ore Mineral**	**Composition**	**Uses**
Aluminum	Bauxite	$AlO(OH)$ and $Al(OH)_3$	Manufacture of beer and soft drink cans, airplanes, electrical cables
Chromium	Chromite	$FeCr_2O_4$	Essential ingredient in stainless steel, coating on automobile parts
Copper	Native copper	Cu	Electrical wire and equipment, production of brass
	Chalcocite	Cu_2S	
	Chalcopyrite	$CuFeS_2$	
Gold	Native gold	Au	Coins, jewelry, dentistry, electronics
Iron	Hematite	Fe_2O_3	Essential ingredient of steel
	Magnetite	Fe_3O_4	
Lead	Galena	PbS	Batteries
Manganese	Pyrolusite	MnO_2	Alloy in steel
Mercury	Cinnabar	HgS	Thermometers, silent electrical switches, batteries
Nickel	Pentlandite	$(Fe,Ni)S$	Important alloy in steel
Silver	Native silver	Ag	Coins, tableware, jewelry, photographic film
	Argentite	Ag_2S	
Tin	Cassiterite	SnO_2	Solder, anti-corrosion plating, bronze
Zinc	Sphalerite	ZnS	Galvanized steel, brass

TABLE 22.4 Some Ways Ore Deposits Form

Type of Ore Deposit	Some Metals Found in This Type of Ore Deposit
Crystal settling within cooling magma	Chromium, iron
Hydrothermal deposits (contact metamorphism, hydrothermal veins, disseminated deposits, hot-spring deposits)	Copper, lead, zinc, gold, silver, iron, molybdenum, tungsten, tin, mercury, cobalt
Pegmatites	Lithium, rare metals
Chemical precipitation as sediment	Iron, manganese, copper
Placer deposits	Gold, tin, platinum, titanium
Concentration by weathering and groundwater	Aluminum, nickel, copper, silver, uranium, iron, manganese, lead, tin, mercury

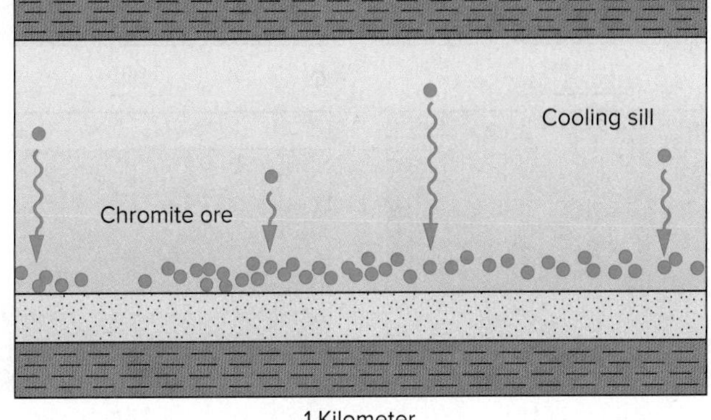

FIGURE 22.20

Early-forming minerals such as chromite may settle through magma to collect in layers near the bottom of a cooling sill.

possibilities, moving from the Earth's interior to the surface, in the pages that follow. In all cases, note that people mined ore minerals long before we understood how they form. The field of *economic geology* developed, in large part, to study the origin of ore deposits and to expand our ability to locate and develop new reserves more easily.

Ores Formed by Igneous Processes

Crystal Settling

Crystal settling occurs as early-forming minerals crystallize and settle to the bottom of a cooling body of magma (figure 22.20). This process was described under differentiation in chapter 3. The metal chromium comes from chromite bodies near the base of sills and other intrusions. Most of the world's chromium comes from a single intrusion, the huge Bushveldt Complex in South Africa. In Montana, another huge Precambrian sill called the Stillwater Complex contains similar, but lower-grade deposits.

Hydrothermal Fluids

Hydrothermal fluids, also discussed in chapter 7, are the most important source of metallic ore deposits other than for iron and aluminum. The hot water and other fluids are part of the magma itself, injected into the surrounding country rock during the last stages of magma crystallization (figure 22.21). Atoms of metals such as copper and gold, which do not fit into the growing crystals of feldspar and other minerals in the cooling pluton, are concentrated residually in the remaining water-rich magma. Eventually, a hot solution, rich in metals and silica (quartz is the lowest-temperature mineral on Bowen's reaction series), moves

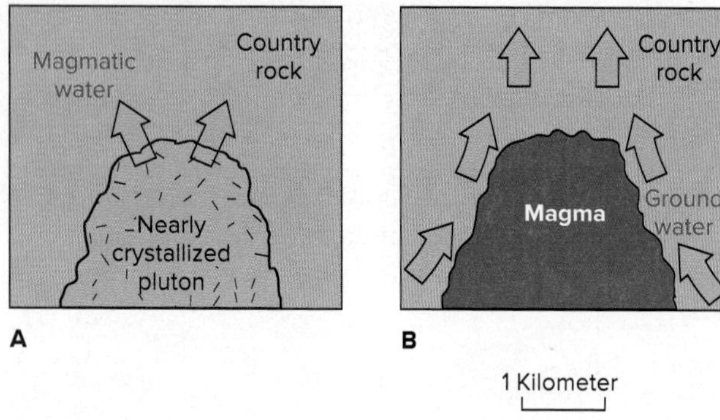

FIGURE 22.21

Two possible origins of hydrothermal fluids. (*A*) Residually concentrated magmatic water moves into country rock when magma is nearly all crystallized. (*B*) Groundwater becomes heated by magma (or by a cooling solid pluton), and a convective circulation is set up.

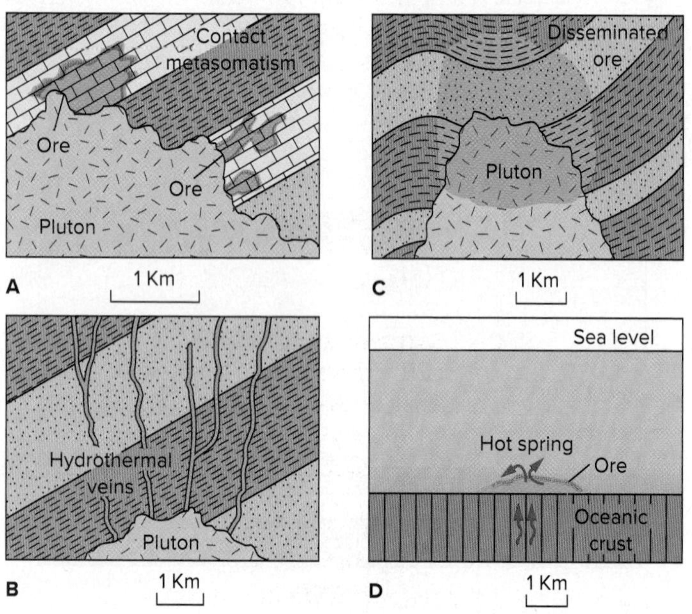

FIGURE 22.22

Hydrothermal ore deposits. (*A*) Contact metamorphism in which ore replaces limestone. (*B*) Ore emplaced in hydrothermal veins. (*C*) Disseminated ore within and above a pluton (porphyry copper deposits, for example). (*D*) Ore precipitated around a submarine hot spring (size of ore deposit is exaggerated).

into the country rock to create ore deposits. Most hydrothermal ores are metallic sulfides, often mixed with quartz. The origin of the sulfur is widely debated.

A magma body or hot rock may heat groundwater and cause convection circulation. This water may mix with water given off from solidifying magma, or it may leach metals from solid rock and deposit metallic minerals elsewhere as the water cools. However the hydrothermal solutions form, they tend to create four general types of hydrothermal ore deposits: (1) contact metamorphic deposits, (2) hydrothermal veins, (3) disseminated deposits, and (4) hot-springs deposits.

Contact metamorphism can create ores of iron, tungsten, copper, lead, zinc, silver, and other metals in country rock. The country rock may be completely or partially removed and replaced by ore (figure 22.22*A*). This is particularly true of limestone beds, which react readily with hydrothermal solutions. (The metasomatic addition of ions to country rock is described in chapter 7.) The ore bodies can be quite large and very rich.

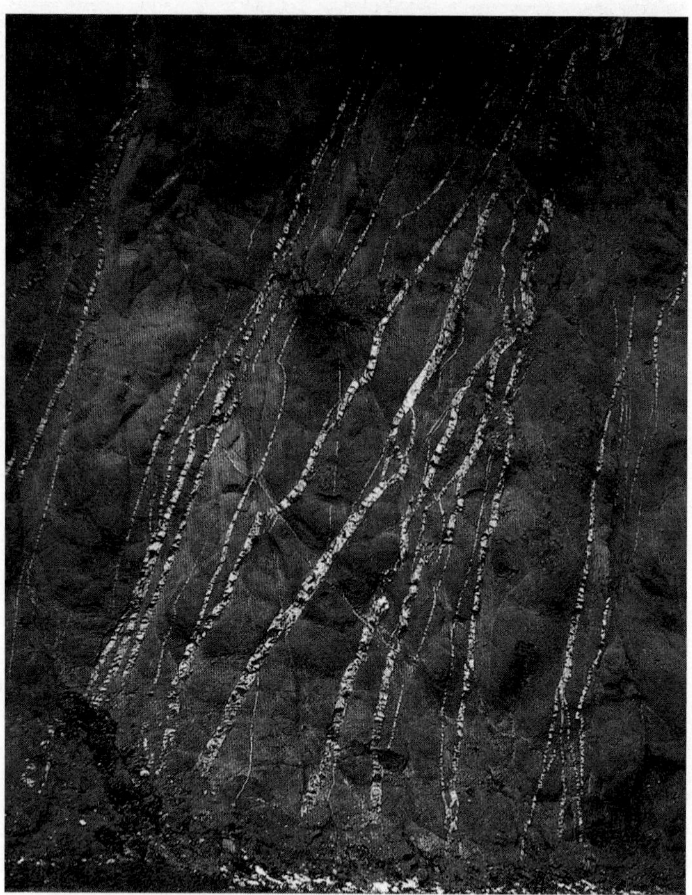

FIGURE 22.23

Hydrothermal quartz veins in granite.
©David McGeary

Hydrothermal veins are narrow ore bodies formed along joints and faults (figure 22.22*B*). They can extend great distances from their apparent plutonic sources. Some extend so far that it is questionable whether they are even associated with plutons. The fluids can precipitate ore (and quartz) within cavities along the fractures and may replace the wall rock of the fractures with ore. Hydrothermal veins (figure 22.23) form most of the world's great deposits of lead, zinc, silver, gold, tungsten, tin, mercury, and, to some extent, copper.

Hot solutions can also form *disseminated deposits* in which metallic sulfide ore minerals are distributed in very low concentration through large volumes of rock, both above and within a pluton (figure 22.22*C* and box 22.4). Most of the world's copper comes from disseminated deposits (also called *porphyry copper deposits* because the associated pluton is usually porphyritic). Along with the copper are deposited many other metals, such as lead, zinc, molybdenum, silver, and gold (and iron, though not in commercial quantities).

Where hot solutions rise to Earth's surface, *hot springs* form (see chapters 11 and 18). Hot springs on land may contain large amounts of dissolved metals. More impressive are hot springs on the sea floor (figure 22.22*D*), which can precipitate large mounds of metallic sulfides, sometimes in commercial quantities (chapters 7 and 19).

Pegmatites (see box 3.1), very coarse-grained plutonic rocks, are another type of ore deposit associated with igneous activity. They may contain important concentrations of minerals containing lithium, beryllium, and other rare metals, as well as gemstones such as emeralds and sapphires.

Ores Formed by Surface Processes

Chemical precipitation in layers is the most common origin for ores of iron and manganese. A few copper ores form in this way, too. Banded iron ores, usually composed of alternating layers of iron minerals and chert, formed as sedimentary rocks in many parts of the world during the Precambrian, apparently in shallow, water-filled basins (figure 22.24). Later folding, faulting, metamorphism, and solution have destroyed many of the original features of the ore, so its origin is difficult to interpret. The water may have been fresh or marine, and the iron may have come from volcanic activity or deep weathering of the surrounding continents. The alternating bands may have been created by some rhythmic variation in volcanic activity, river runoff, basic water circulation, growth of organisms, or some other factor. Since banded iron formations are all Precambrian, their origin might be connected to an ancient atmosphere or ocean chemically different from today's.

Placer deposits in which streams have concentrated heavy sediment grains in a river bar are described in chapter 10. Wave action can also form placers at beaches. Placers include gold nuggets and dust, native platinum, diamonds, and other gemstones, and worn pebbles or sand grains composed of the heavy oxides of titanium and tin.

Ore deposits due to *concentration by weathering* were described in chapter 5. Aluminum (in bauxite) forms through weathering in tropical climates.

Another type of concentration by weathering is the *supergene enrichment* of disseminated ore deposits. Through supergene enrichment, low-grade ores of 0.3% copper in rock can be enriched to a minable 1% copper. The major ore mineral in a disseminated copper deposit is chalcopyrite, a copper-iron sulfide containing about 35% copper. Near Earth's surface, downward-moving groundwater can leach copper and sulfur from the ore, leaving the iron behind (figure 22.25). At or below the water table, the dissolved copper can react with chalcopyrite in the lower part of the disseminated deposit, forming a richer ore mineral such as chalcocite, which is about 80% copper:

$$\underset{\text{Copper dissolved in groundwater}}{3Cu^{++}} + \underset{\text{Chalcopyrite}}{CuFeS_2} \rightarrow \underset{\text{Chalcocite}}{2Cu_2S} + \underset{\text{Iron in solution}}{Fe^{++}}$$

In this way, copper is removed from the top of the deposit and added to the lower part (figure 22.25). The ore below the water table may be several times richer than the ore in the rest of the deposit, with silver concentrating as readily as copper. The iron remains behind, staining the surface as it oxidizes to form a gossan (defined in the next section).

FIGURE 22.24

A 2.1 billion-year-old banded iron formation from Michigan. *©Doug Sherman/Geofile RF*

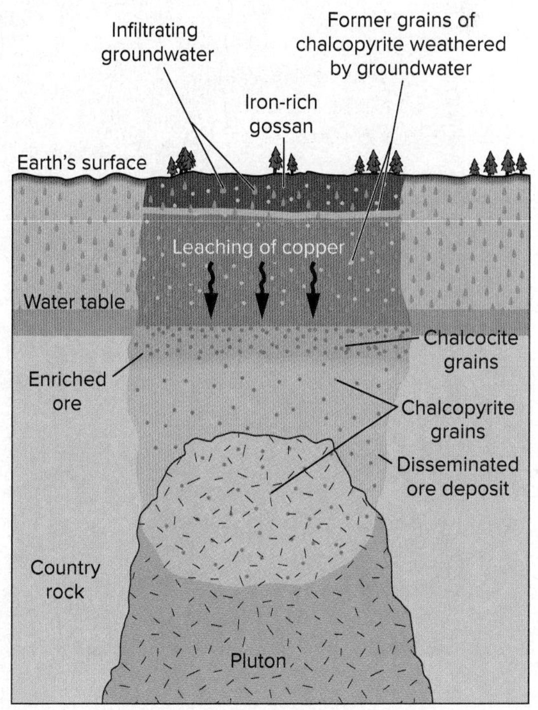

FIGURE 22.25

Supergene enrichment. Groundwater leaches copper from upper part of disseminated deposit and precipitates it at or below the water table, forming rich ore.

ENVIRONMENTAL GEOLOGY 22.4

The World's Largest Human-Made Hole—The Bingham Canyon Copper Mine

The Bingham Canyon mine near Salt Lake City, Utah, is thought to be the biggest single human-made hole in the world (box figure 1). (The Morenci mine in Arizona is volumetrically larger, but is not a single pit.) The 800-meter (½-mile) deep open-pit mine is 4 kilometers (2½ miles) wide at the top and continues to be enlarged. The reason for this hole is copper.

About 40,000 kilograms of explosives are used per day to blast apart over 60,000 tons of ore (copper-bearing rock) and an equal amount of waste rock. An 8-kilometer-long conveyor belt system moves up to 10,000 tons of crushed rock per hour through a tunnel out of the pit for processing.

Mining began here as a typical underground operation in 1863. The shafts and tunnels of the mine followed a series of veins. Originally, ores of silver and lead were mined. Later, it was discovered that fine-grained, copper-bearing minerals (chalcopyrite and other copper sulfide minerals) were disseminated in tiny veinlets throughout a granite stock. Although the percentage of copper in the rock was small, the total volume of copper was recognized as huge. With efficient earth-moving techniques, large volumes of ore-bearing rock can be moved and processed. Today, mining is still going on, and the company can make a profit even though only 0.6% of the rock being mined is copper. Since 1904, over 12 million tons of copper have been mined, processed, and sold. The mine has also produced impressive amounts of gold, silver, and molybdenum.

Such an operation is not without environmental problems. Some people regard the huge hole in the mountains as an eyesore (but it is a popular tourist attraction). Disposing of the waste—over 99% of the rock material mined—creates problems. Wind stirs up dust storms from the piles of finely crushed waste rock unless it is kept wet. The nearby smelter that extracts the pure copper from the sulfide minerals has created a toxic smoke containing sulfuric acid fumes. During most of the twentieth century, the toxic smoke was released into the atmosphere; occasionally, wind blew polluted air to Salt Lake City. Now, over 99% of the sulfur fumes are removed at the smelter.

Additional Resource

Bingham Canyon Mine Site

- www.infomine.com/minesite/minesite.asp?site=bingham

BOX 22.4 ■ FIGURE 1

Bingham Canyon copper mine in Utah.
©Royce Bair/Hemera 360/Getty Images

22.4 MINING

As in the case of coal, miners use both surface and underground techniques to extract ore minerals (figure 22.26). Strip mining–the wholesale removal of large areas of soil and shallow rock cover–has already been mentioned in connection with coal beds. Aluminum ore (bauxite), which forms in weathered soil beds under tropical conditions, is often most easily extracted this way. *Open-pit mining* is related to strip mining, but concentrates on the removal of valuable deposits from a specific, relatively small area (figure 22.27). Open-pit mines often dig much deeper than strip mines.

Placer mines are localized to ancient or modern river bar or beach deposits. In some parts of the world, gold is found concentrated in placer deposits (California's Gold Rush of 1849 was triggered by discoveries of placer gold). Gold nuggets, flakes, and dust can be separated from the other sediments by (1) panning; (2) *sluice boxes* (figure 22.28), which catch the heavy gold on the bottom of a box as gravel is washed through it; (3) *hydraulic mining* (figure 22.29), which washes gold-bearing gravel from a hillside into a sluice box; or (4) floating *dredges* (figure 22.30), which separate gold from gravel aboard a large barge, piling the spent gravel behind.

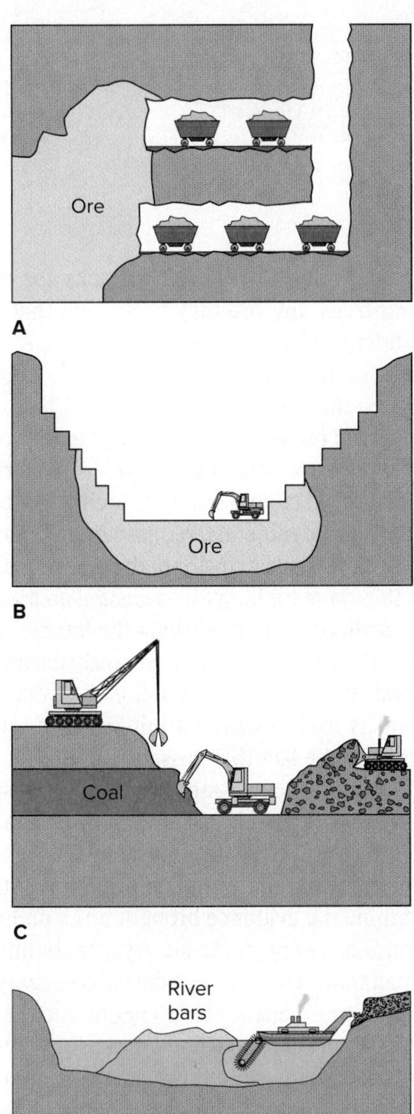

FIGURE 22.26

Types of mines: *(A)* Underground. *(B)* Open pit. *(C)* Strip. *(D)* Placer (being mined by a floating dredge).

FIGURE 22.27

Open-pit copper mine in Morenci, Arizona. Heavy equipment in the bottom of the pit gives a sense of scale.
©David McGeary

FIGURE 22.28

Sluice box used to separate gold from gravel in Alaska.
Source: D. J. Miller/U.S. Geological Survey

Underground, or *bedrock mining*, must be done to excavate many valuable mineral deposits. Bedrock mining of ores typically extends to much greater depths than ordinary coal mines, and this presents its own set of technical challenges. The world's deepest mines, in South Africa, extend to depths of 1,500–2,500 meters. The walls grow hot to the touch so deep underground, and pumping of fresh, cool air and water must be done to make working conditions tolerable. Mines have notoriously poor air circulation, and the use of dynamite to blast openings releases toxic gases that must be removed. Ammonium nitrate (NH_4NO_3) mixed with fuel oil (CH_2) is a typical blasting agent. The explosive

FIGURE 22.29

Hydraulic mining for gold in Alaska.
Source: T. L. Pewe/U.S. Geological Survey

FIGURE 22.30

A gold dredge separates gold from gravel.
©David McGeary

reaction generates poisonous carbon monoxide whenever there is a slight excess of oil in the mixture. Carbon monoxide (CO) is a heavier-than-air gas that sinks deep into the mine. This is one of the reasons why abandoned mines should never be explored.

Where the mines extend beneath the water table, the water that seeps in must be pumped out to avoid flooding. An active mine consumes large amounts of energy as well as material resources.

The design of a mine takes into consideration three vital factors: (1) the geometry of the underground ore body; (2) the need for safety; and (3) the need to maximize profit. It is typically easiest for miners to construct a set of vertical and horizontal passages to access and remove the ore (figure 22.31). The vertical openings, called *shafts* (or *winzes*, if they do not open all the way to the surface), allow elevators to take miners underground and bring ore up to the surface. Shafts are also conduits for electrical cables, water hoses, and air lines. Miners can blast and dig open a shaft at a rate of 30 feet every day. The horizontal tunnels, termed *adits* (or *drifts*, if they do not open to the surface), are the pathways through which ore is directly excavated. In larger mines, multiple drifts radiate off of shafts. *Ramps* are

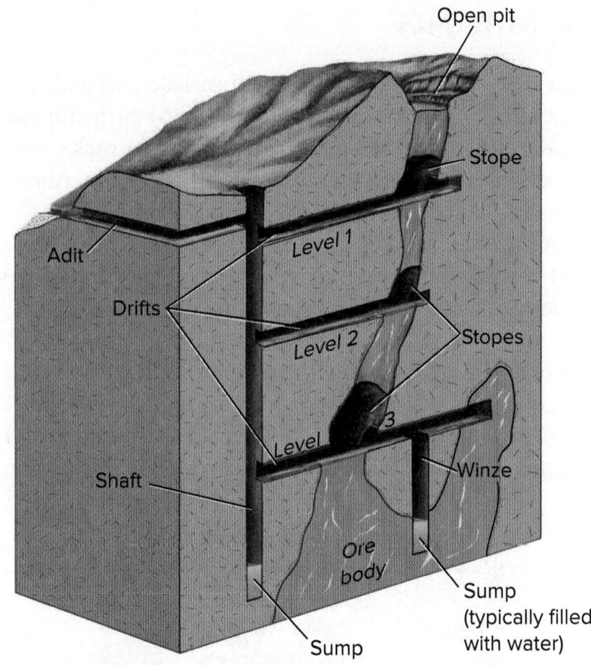

FIGURE 22.31

Design of a typical bedrock mine.

slanted tunnels, many of which have tracks for winching ore carts. In some places, the ore may be so rich that miners excavate a giant underground chamber called a *stope*. To avoid collapse, the walls of the stope may have to be shored up with timbers or other construction materials.

In earlier times, prospectors located potential ore bodies by looking at the eroded rock fragments ("float") in streams and on hillsides, hoping to find telltale minerals such as white quartz with gold or oxidized sulfide minerals, red jasper, or bright blue chrysocholla. Following this debris upslope, the treasure seekers might discover what they were looking for—a *gossan*, an area of yellow or ruddy orange, oxidized ground marking the intersection of an ore body or vein with the surface. The Spanish termed such areas "colorados," and the name of the state, in fact, derives from this origin. "Gossan" is itself a Cornish mining term, meaning "iron hat," in reference to the fact that gossans cap deeper ore bodies.

Today, more sophisticated—and expensive—prospecting techniques are applied to locating and determining the shape of an underground ore body. Exploration geologists must study the structural geology (stratigraphy and deformation of the surrounding rocks), examine the evidence brought up in preliminary boreholes, and conduct geophysical surveys, including the use of gravimeters, magnetometers, and electrical-resistivity equipment. Some ore bodies are excellent electrical conductors and may be highly magnetic. Exploration geologists have the ultimate say on whether or not a company should proceed with mining.

Given the dangers and economic factors involved, whole technical schools have been established to train mining engineers (e.g., the Colorado School of Mines). Today, these schools must also consider environmental factors because, in the past, mines have been terrible sources of watershed pollution, among other problems. Groundwater running or being pumped out of a mine causes *acid mine drainage*.

Sulfide ore minerals (table 22.3) and pyrite (FeS_2) are most often the source of the trouble. Groundwater transports oxygen to the sulfides, which are then oxidized to iron oxide and sulfuric acid. In some mines, expensive programs of holding and neutralizing drainage water in ponds or artificial wetlands prevent pollution of surface streams and harm to forests and wildlife. The worst problem is with long-abandoned mines that are still draining acid waters. Many of these may never be neutralized.

22.5 NONMETALLIC RESOURCES

Nonmetallic resources are Earth resources that are not mined to extract a metal or as a source of energy. Most rocks and minerals contain metals, but when nonmetallic resources are mined, it is usually to use the rock (or mineral) as is (for example, gravel and sand for construction projects), whereas metallic ores are processed to extract metal. With the exception of the gemstones such as diamonds and rubies, nonmetallic resources do not have the glamour of many metals or energy resources. Nonmetallic resources are generally inexpensive and are needed in large quantities (again, except for gemstones); however, their value exceeds that of all mined metals. The large demand and low unit prices mean that these resources are best taken from local sources. Transportation over long distances would add significantly to the cost.

Construction Materials

Sand and *gravel* are both needed for the manufacture of concrete for building and highway construction. Sand is also used in mortar, which holds bricks and cement blocks together. The demand for sand and gravel in the United States has more than doubled in the last twenty-five years. Sand dunes, river channel and bar deposits, glacial outwash, and beach deposits are common sources for sand and gravel. Cinder cones are mined for "gravel" in some areas. Sand and gravel are ordinarily mined in open pits (figure 22.32).

FIGURE 22.32

Sand and gravel pit in a glacial esker near Saranac Lake, New York.
©Randy Schaetzl/Michigan State University

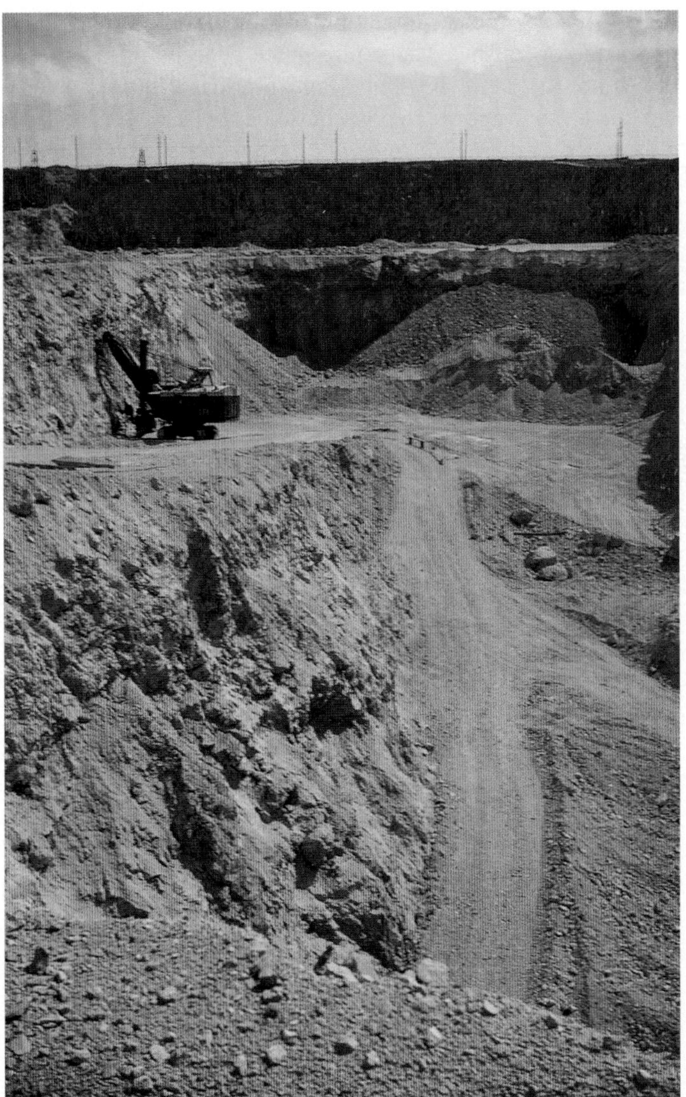

FIGURE 22.33

A limestone quarry in northern Illinois. The horizon marks the original land surface before the rock was removed.
©David McGeary

Stone refers to rock used in blocks to construct buildings or crushed to form roadbed. Most stone in buildings is limestone or granite, and most crushed stone is limestone. Huge quantities of stone are used each year in the United States. Stone is removed from open pits called *quarries* (figure 22.33).

Limestone has many uses other than building stone or crushed roadbed. Cement, used in concrete and mortar, is made from limestone and is vital to an industrial economy. Pulverized limestone is in demand as a soil conditioner and is the principal ingredient of many chemical products.

Fertilizers and Evaporites

Fertilizers (phosphate, nitrate, and potassium compounds) are extremely important to agriculture today, so much so that they are one of the few nonmetallic resources transported across the sea. *Phosphate* is produced from phosphorite, a sedimentary rock formed by the accumulation and alteration of the remains of marine organisms. Major phosphate deposits in the United States are in Idaho, Wyoming, and Florida. *Nitrate* can form directly as an evaporite deposit but today is usually made from atmospheric nitrogen. *Potassium compounds* are often found as evaporites.

Rock salt is coarsely crystalline halite formed as an evaporite. Salt beds are mined underground in Ohio and Michigan; underground salt domes are mined in Texas and Louisiana. (Some salt is also extracted from seawater by evaporation.) Rock salt is used in many ways–deicing roads in winter, preserving food, as table salt, and in manufacturing hydrochloric acid and sodium compounds for baking soda, soap, and other products. Rock salt is heavily used by industry.

Gypsum forms as an evaporite. Beds of gypsum are mined in many states, notably California, Michigan, Iowa, and Texas. Gypsum, the essential ingredient of plaster and wallboard (Sheetrock®), is used mainly by the construction industry, although there are other uses.

Sulfur occurs as bright yellow deposits of elemental sulfur. Most of its commercial production comes from the cap rock of salt domes. Sulfur is widely used in agriculture as a fungicide and fertilizer and by industry to manufacture sulfuric acid, matches, and many other products.

Other Nonmetallics

Gemstones (called *gems* when cut and polished) include precious stones such as diamonds, rubies, emeralds, and sapphires and semiprecious stones such as beryl, garnet, jade, spinel, topaz, turquoise, and zircon. Gems (see box 2.6) are used for jewelry, bearings, and abrasives (most are above 7 on Mohs' scale of hardness). Diamond drills and diamond saws are used to drill and cut rock. Old watches and other instruments often have hard gems at bearing points of friction ("17-jewel watches"). Gemstones are often found in pegmatites or in close association with other igneous intrusives. Some are recovered from placer deposits.

Asbestos is a fibrous variety of serpentine or chain silicate minerals. The fibers can be separated and woven into fireproof fabric used for firefighters' clothes and theater curtains. Asbestos is also used in manufacturing ceiling and sound insulation, shingles, and brake linings, although the use of asbestos has been rapidly curtailed because of concern about its connection with lung cancer (see box 2.4). The United States and Canada no longer produce asbestos. *Talc*, used in talcum powder and other products, is often found associated with asbestos (see box 20.2).

Other nonmetallic resources are important. *Mica* is used in electrical insulators. *Barite* ($BaSO_4$), because of its high specific gravity, is used to make heavy drilling mud to prevent oil gushers. *Borates* are boron-containing evaporites used in fiberglass, cleaning compounds, and ceramics. *Fluorite* (CaF_2) is used in toothpaste, Teflon finishes, and steel smelting. *Clays* are used in ceramics, manufacturing paper, and as filters and absorbents. *Diatomite* is used in swimming pool filters and to filter out yeast in beer and wine. *Glass sand*, which is over 95% quartz, is the main component of glass. *Graphite* is used in foundries, lubricants, steelmaking, batteries, and pencil "lead."

22.6 RESOURCES, THE ENVIRONMENT, AND SUSTAINABILITY

There is a saying, "If it can't be grown, it must be mined." As we described at the beginning of this chapter, every day you interact with materials that are either directly mined or produced from materials that are mined. The demand for fossil fuels, metallic ores, and nonmetallic materials is only likely to increase as world population grows and developing nations continue to industrialize and become more wealthy. The extraction and transportation of resources can have an enormous impact on the environment, leaving enormous holes in the ground (figure 22.27), removing entire mountaintops (figure 22.5), or resulting in massive oil spills (box 22.2). Release of greenhouse gases through the burning of fossil fuels and during extraction, processing, and transportation of resources may be causing global climate change (chapter 21). Additionally, many of these resources take many millions of years to form, and unchecked exploitation of them now may deprive future generations of the resources that they need.

It is a complicated issue. It is not feasible to ban all mining and drilling because we need those materials to sustain our lifestyles. Nor is it acceptable to exploit with no regard for the effect on the environment or for the needs of future generations. Almost everyone agrees that we need to find a middle ground that includes both mining practices that minimize environmental impact and reducing consumption and increasing recycling to ensure a supply of resources for future generations. The challenge is in finding the right balance and one that all nations–developed and developing–can agree upon. Your understanding of geology is an important step in your being able to help resolve moral dilemmas that we face to which there is no ideal solution.

Summary

Most people interact with the Earth primarily through their interaction with *geologic resources*: soil, water, metals, nonmetals, and fuels, most of which are *nonrenewable* and form under particular and transient natural conditions.

Coal, *petroleum*, and *natural gas* are fossil fuels that are the main sources of energy in the modern world. They are also important in nonenergy applications, such as making fertilizer, steel, and many other products. These resources are essentially ancient *solar energy*, unlocked by combustion in power plants after mining or drilling. *Coal beds* occur in areas of ancient swamps and marshes, and derive from the accumulation of dead land-based plant matter. *Oil* (petroleum) and natural gas derive from certain nearshore marine settings where dead, microscopic, floating organisms accumulate. Coal and oil both require heat and burial to develop. Oil and natural gas also need to be sealed into subsurface reservoirs as they percolate toward the surface. Anticlines, faults, and other structural traps provide this lid.

Reserves are known as deposits that can be legally and economically recovered now—the short-term supply. Resources include reserves as well as other known and undiscovered deposits that might be economically extractable in the future. There is evidence that the world's reserves of petroleum are nearing critical depletion, given the high level of demand. This will force the world into finding energy alternatives, including the burning of more coal and increased development of *nuclear power.*

Uranium-235 is the primary fuel of *nuclear reactors.* Through breeder designs, nuclear energy could provide us with electricity long into the future. But risk of accidents and waste-disposal issues raise questions about the long-term viability of this energy source.

Geothermal power benefits only a few localized areas around the world. This resource can be easily exhausted, and it is never likely to become a principal source of world energy, despite the enormous amount of heat contained inside the Earth.

Renewable energy strategies, such as *solar, wind,* and *wave power,* are an appealing, environmentally clean substitute for "conventional" energy sources. *Hydroelectric power* is the most successful and extensively used type of renewable energy, though it is localized to areas with abundant flowing water. Biofuels provide a carbon-neutral source of fuel for our vehicles, but the cost of producing the biofuel is more expensive food.

Metallic ores, which can be profitably mined, are often associated with igneous rocks, particularly their *hydrothermal fluids,* which can form in contact metamorphic deposits, hydrothermal veins, disseminated deposits, and submarine hot springs at divergent plate boundaries, on the flanks of island arcs, and in belts on the edges of continents above subduction zones.

Ores are mined at the surface in *strip* and *open-pit mining,* and in costly, potentially dangerous, and carefully executed underground mining. *Placer* mining takes advantage of the sedimentary reworking and concentration of ore minerals.

Nonmetallic resources, such as sand and gravel and limestone for crushed rock and cement, are used in huge quantities. Fertilizers, rock salt, gypsum, sulfur, and clays are also widely used.

Terms to Remember

active solar heating 552
biofuel 554
coal 539
coal bed methane 548
energy resources 539
fission 549
fossil fuels 539
heavy crude 548
hydrothermal veins 556
kinetic energy 539
natural gas 542
nonrenewable resources 537
oil field 544
oil (tar) sands 548
oil shale 548
ore 554
passive solar heating 552
petroleum 542
photovoltaic cells 552
placer deposits 557
potential energy 539
reserve 537
reservoir rock 542
resource 537
source rock 542
strip mining 541
structural (or oil) trap 543

Testing Your Knowledge

1. What is the difference between a reserve and a resource? Under what circumstances would a resource come to be regarded as a reserve?
2. Coal is often referred to as "clean coal." What is that in reference to? Is coal a truly clean energy source?
3. Is it likely that oil resources will be completely depleted one day? Are we likely to stop exploiting oil? What factors would contribute to us ceasing oil production, and what factors may lead to us continuing to exploit oil?
4. Contrast the geologic conditions responsible for the formation of coal, oil, and natural gas.
5. Describe how a nuclear reactor works.
6. Discuss the pros and cons of exploiting geothermal and hydroelectric power.
7. Discuss the ways in which solar, wind, wave, and tidal energy are harnessed.
8. Describe the ways in which igneous processes can form metallic ore deposits.
9. How can surface processes create ore deposits?
10. Describe two ways in which resources are mined, and discuss the pros and cons of each.
11. Discuss common uses for metallic resources and nonmetallic resources.
12. Discuss the tension between our need to exploit resources for energy and materials and the environmental impact of that exploitation.
13. Which is *not* a type of coal?
    a. lignite b. bituminous
    c. sulfite d. anthracite
14. Which metal would most likely be found in an ore deposit formed by crystal settling?
    a. copper b. gold
    c. silver d. chromium
15. Which metal would *not* be found in hydrothermal veins?
    a. lead b. aluminum
    c. silver d. gold
16. Coal forms
    a. by crystal settling.
    b. through hydrothermal processes.
    c. by compaction of plant material.
    d. on the ocean floor.
17. The world's largest oil reserves are found in
    a. Venezuela. b. the Middle East.
    c. the United States. d. Canada.

18. What factors can increase reserves of Earth resources (choose all that apply)?
    a. decrease in cost of extraction
    b. increased demand
    c. price decreases
    d. new mining technology
19. The largest use of sand and gravel is
    a. glassmaking. b. extraction of quartz.
    c. construction. d. ceramics.
20. Oil accumulates when the following conditions are met (choose all that apply):
    a. source rock where oil forms
    b. permeable reservoir rock
    c. impermeable oil trap
    d. shallow burial

Expanding Your Knowledge

1. Many underdeveloped countries would like to have the standard of living enjoyed by the United States, which has 6% of the world population and uses 15% to 40% of the world's production of many resources. As these countries become industrialized, what happens to the world demand for geologic resources? Where will these needed resources come from?
2. If driven 12,000 miles per year, how many more gallons of gasoline per year does a sport utility vehicle or pickup truck rated at 12 miles per gallon use than a minicompact car rated at 52 mpg? Over five years, how much more does it cost to buy gasoline at $4 per gallon for the low-mileage car? At $8 per gallon (the price in many European countries)?

Exploring Web Resources

www.NRCan.gc.ca/

Natural Resources Canada. Use this site to get information on Canada's mineral and energy resources.

http://minerals.usgs.gov/

U.S. Geological Survey Mineral Resources Program. Provides current information on occurrence, quality, quantity, and availability of mineral resources.

www.eia.doe.gov/

U.S. Energy Information Administration. Provides data, analysis, and forecasts of energy and issues related to energy.

www.api.org

American Petroleum Institute. Information on all aspects of petroleum from the industry's perspective.

CHAPTER 23

The Earth's Companions

by Tom Arny and Steve Kadel

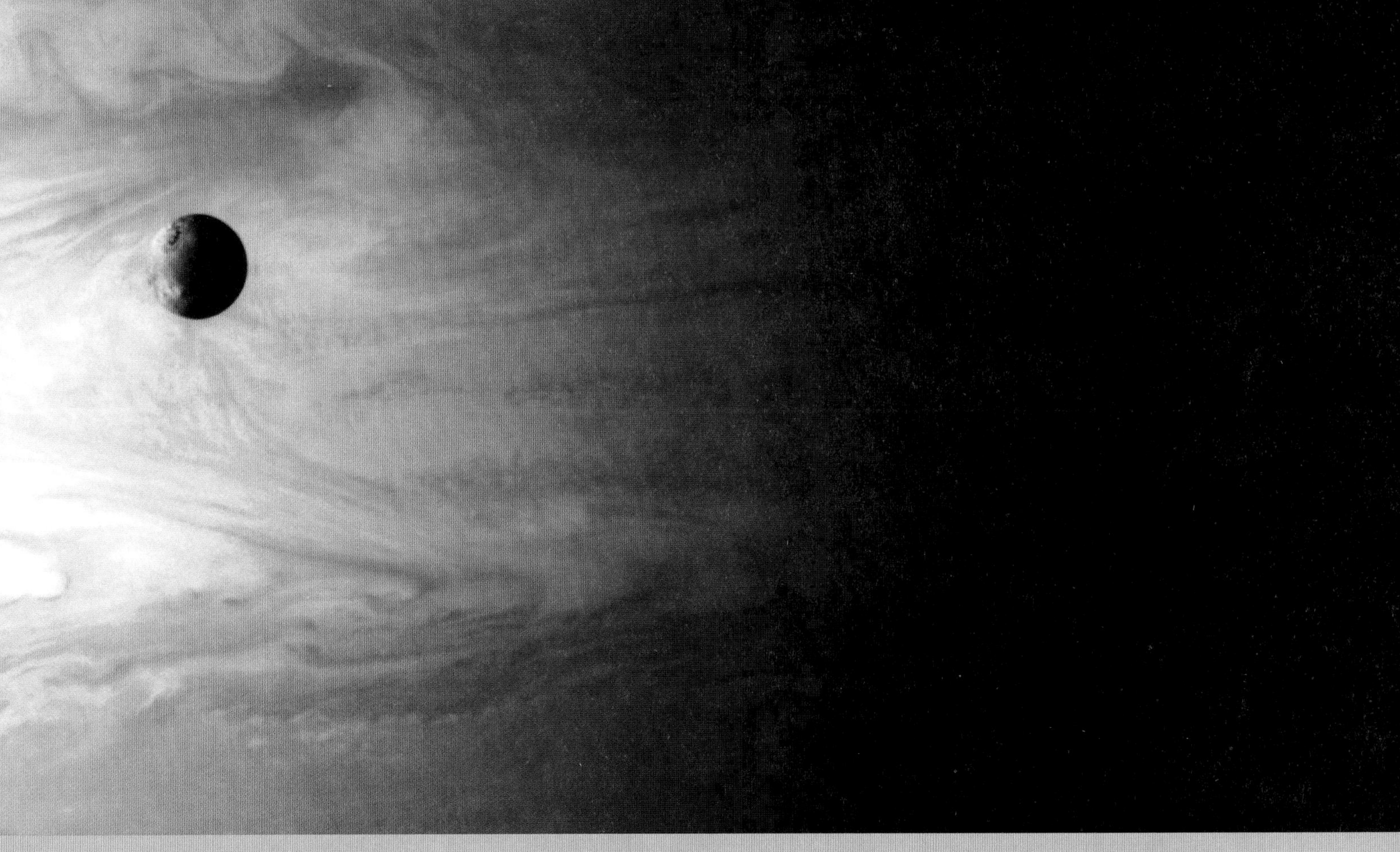

Jupiter's moon Io seen against the cloud tops of the giant planet by the *Cassini* spacecraft. It is about the same size as our moon—about one-quarter the size of Earth.
Source: JPL/NASA/University of Arizona

LEARNING OBJECTIVES

- Summarize important surface and compositional features of the terrestrial planets, and describe how they are similar to and different from one another.
- Summarize important features of the Jovian planets, and describe how they are similar to and different from one another.
- Explain how the important features are consequences of the natural development of these two different groups of planets according to the Nebular Hypothesis.
- Summarize important features that distinguish the minor bodies of the solar system—satellites, asteroids, comets, and Kuiper Belt objects—from planets and each other.
- Analyze what compositional differences among the smaller solar system objects imply about the nature of their formation.
- Justify the exploration of the objects in our solar system as a worthwhile endeavor for humankind.

23.1 THE EARTH IN SPACE

Earth is unique, but not alone, in space. It is but one of eight **planets** and innumerable smaller bodies that orbit the Sun. Figure 23.1 shows pictures of these eight distinctive objects and illustrates their relative size and appearance. Some are far smaller and others vastly larger than Earth, but all are dwarfed by the Sun. Humans have set foot on only one other body–Earth's

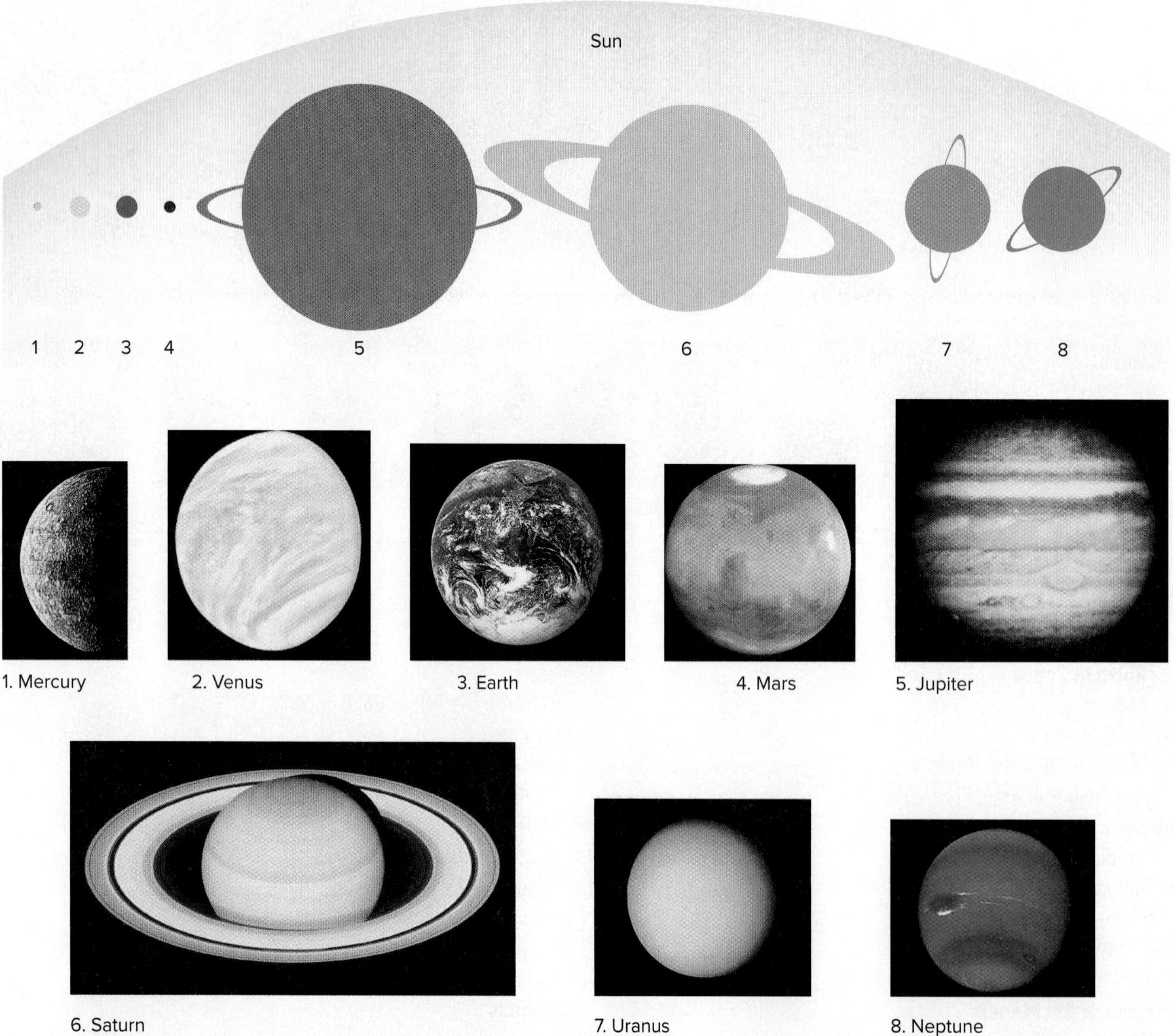

FIGURE 23.1

Images of the eight planets and a sketch at top showing them and Sun to approximately correct relative size.
(Mercury) Source: JPL/NASA; (Venus) Source: NASA; (Earth) Source: NASA; (Mars) Philip James (University of Toledo), Steven Lee (University of Colorado), NASA; (Jupiter) Source: Courtesy of NASA/JPL; (Saturn) ©StockTrek/Digital Vision/Getty Images RF; (Uranus) Source: NASA; (Neptune) Source: Courtesy of NASA/JPL

Moon—but we have explored many others with ever more powerful telescopes and robots. These instruments provide us with information that geologists use to understand how the similarities and differences among these bodies came to be.

The Sun is at the center of our solar system. Moving outward from the Sun, we find the Earth-like or terrestrial planets, Mercury, Venus, Earth, and Mars. A belt of **asteroids** separates the terrestrial planets from the gas giants Jupiter, Saturn, Uranus, and Neptune.[1] Outside of Neptune's orbit is another belt of objects known as trans-Neptunian or Kuiper Belt objects (KBOs). All of these are enclosed within a shell of icy bodies making up the Oort cloud, beyond which is the vastness of relatively empty space known as the interstellar medium.

The Sun

The Sun is a **star**, a huge ball of gas over 100 times the diameter of the Earth (figure 23.2) and over 300,000 times more massive. If the Sun were a volleyball, the Earth would be about the size of a pinhead, and Jupiter roughly the size of a nickel. The Sun's great mass pins the planets in their orbits, but belies the fact that for the most part a star is a low-density gas, made of about 75% hydrogen and about 24% helium.

A star differs from a planet in that a star generates energy that is given off as visible light. This radiation is released when hydrogen fuses in its core, forming helium atoms in a process that converts mass into radiant energy. Fusion occurs in the core because the star's enormous mass creates incredibly high temperatures and pressures there. The Earth is about 150 million kilometers (an important distance called an *astronomical unit*, or AU) from the Sun, so Earth receives only a tiny fraction of that energy, yet that's enough to drive Earth's weather and the hydrologic cycle!

The Solar System

Our **solar system** is made of objects of all shapes and sizes, including dust, meteoroids, asteroids, comets, satellites, planets, and the Sun. Despite the diversity of its members, the solar system shows much regularity. For example, all the planets move around the Sun in the same direction. They all follow approximately circular orbits centered on the Sun (figure 23.3), and these orbits all lie approximately in the same plane. Only Mercury's orbit is tilted strongly, and even that is inclined by only 7 degrees to Earth's orbit. Thus, the planets lie in a disk (figure 23.4) that is flatter, or thinner for a given diameter, than that of a U.S. twenty-five-cent piece.

The materials found in the solar system can be divided into three categories, based on their physical states under the conditions found in space. Gases (mainly hydrogen and helium) are the most common materials. Rocky materials (mainly silicate minerals and iron) are generally in solid form. Ices of water (H_2O), carbon

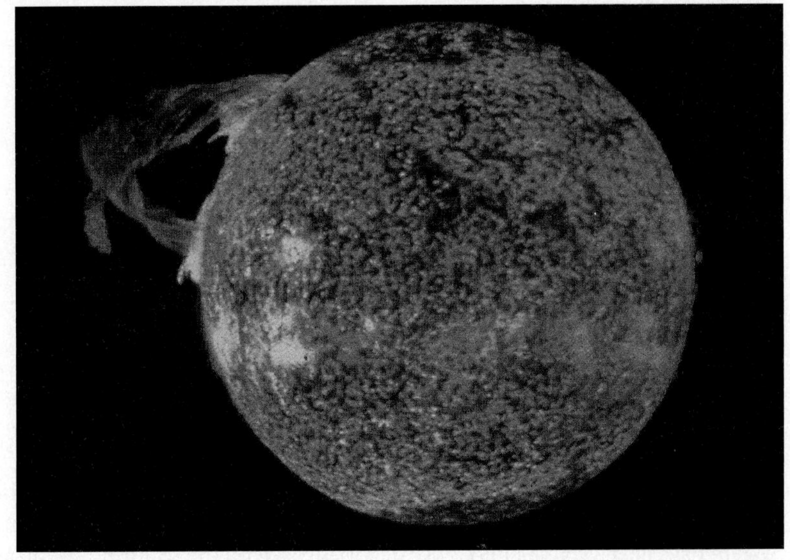

FIGURE 23.2

The Sun as viewed through a special filter that allows its outer gases to be seen. The Earth and Jupiter are shown to scale beside it.
Source: Naval Research Laboratory/NASA

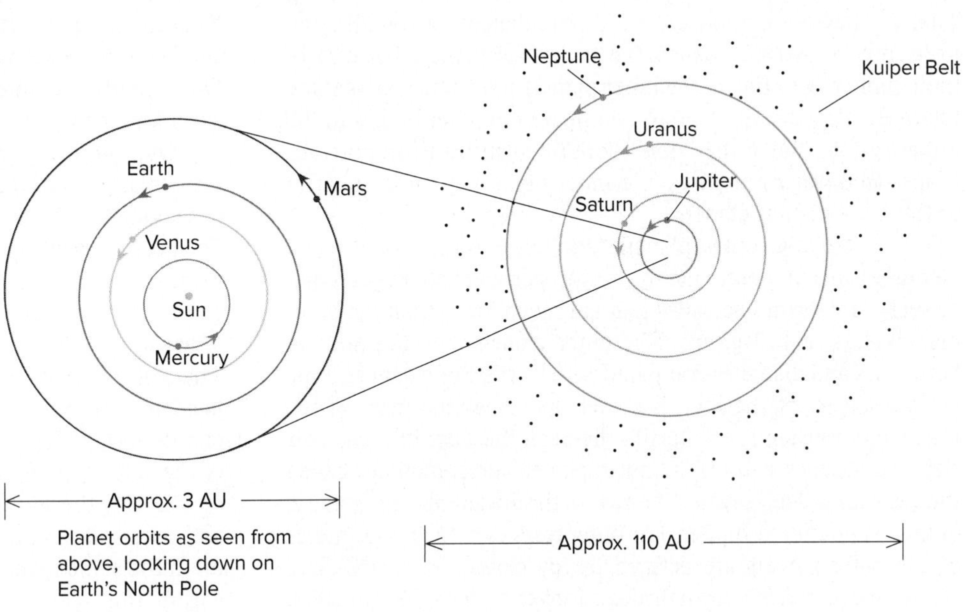

FIGURE 23.3

An artist's view of the solar system from above. The orbits are shown at the correct relative scale in the two drawings. *Note: 1 AU is the average Earth-Sun distance.*

[1]Pluto was recently reclassified as an "ice dwarf," along with the other large icy bodies in the outer solar system.

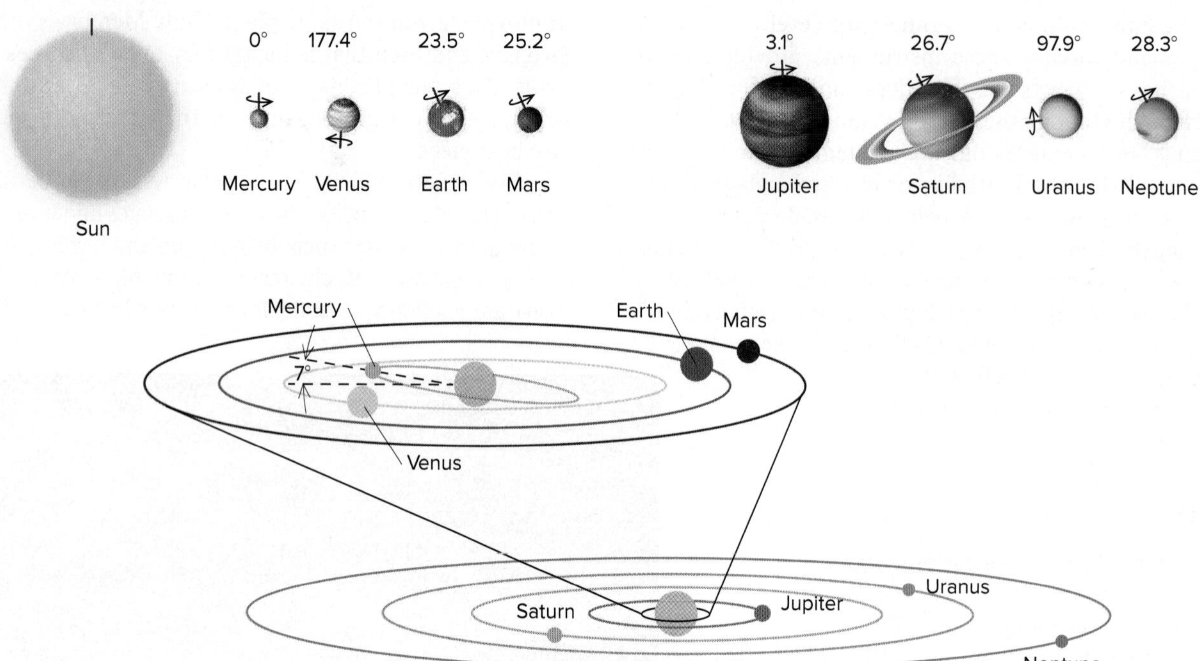

FIGURE 23.4

Planets and their orbits from the side. Sketches also show the orientation of the rotation axes of the planets and Sun. Orbits and bodies are not to the same scale. (Technically, Venus's tilt angle is about 177°, but it can also be described as 3° [180°–177°], with a "backwards," or retrograde spin.)

dioxide (CO_2), methane (CH_4), and ammonia (NH_3) are solid when cold but gaseous when warm. The eight planets of the Solar System have been grouped into two categories based on the relative proportions of these materials: the **inner planets** (because they lie close to the Sun) and the **outer planets**. Figure 23.5 shows cutaway views of the eight planets and illustrates their structural differences.

The inner planets are the four planets nearest the Sun (Mercury, Venus, Earth, and Mars) and are basically balls of rock. That is, they are composed mainly of silicates surrounding an iron core. Sometimes called the terrestrial planets because of their similarity to Earth, the inner planets have densities that are relatively high, around 5 g/cm^3, compared to other bodies in the Solar System. Since they don't contain significant proportions of ices (too warm) or gases, the inner planets are small relative to the more distant planets.

The four big, outer planets (Jupiter, Saturn, Uranus, and Neptune) are composed mainly of hydrogen and helium gases and as such have overall densities that are much lower than the inner planets—less than 2 g/cm^3. The outer planets are enormous in both mass and diameter compared with Earth. For example, Jupiter, the largest of the four, has over 300 times the mass of the Earth and over ten times Earth's diameter. Because of their large size and gaseous nature, the outer planets are sometimes called the gas giants. They are also known as the Jovian planets because of their similarity to Jupiter. The outer reaches of the atmospheres of the outer planets are defined by icy clouds. When NASA's *Galileo* spacecraft fell down through Jupiter's atmosphere in 2003, it found that the ices sublimate to become gases as the pressure and temperature build (sublimation is the transition of a substance directly from the solid to the gas phase, without passing through an intermediate liquid phase). Eventually the pressures would be so great that those gases would behave as a liquid; this means that the gas giants have no surface to land on. Their moons, on the other hand, have many extremely curious surface features, some of which are as yet not well understood.

The inner planets and outer planets are separated by the asteroid belt, a region occupied by thousands of rocky bodies. Beyond the outer planets lies a region referred to as the trans-Neptunian region. It contains the Kuiper Belt, a ring of debris similar to the asteroid belt but composed mainly of ice, and the Oort cloud, a remote frigid zone named in honor of the Dutch astronomer Jan Oort, who hypothesized its existence.

Our knowledge of the solar system would be more complete if the planets were not so remote. For example, Neptune's orbit lies about 4.5 billion kilometers (2.8 billion miles) from the Sun. To measure such a vast distance in miles or kilometers is as meaningless as to measure the distance between New York and Tokyo in inches or centimeters. Thus, astronomers use a much larger unit based on the Earth's average distance from the Sun. That distance, about 150 million kilometers, we call the **astronomical unit (AU)**. The solar system out to Neptune has a diameter of about 60 AU. If we include the Oort cloud at the fringes of the solar system, it is approximately 100,000 AU across.

So far, the groups of objects we have explored—the inner and outer planets, the asteroid belt, and the Kuiper Belt—all orbit in the same plane called the **ecliptic**. Many **comets** also move within this plane, and those are thought to come from the Kuiper Belt at a distance of about 30 to 50 AU (figure 23.6). Comets have about the same composition as KBOs: a mixture of mostly ices and some rocky (or "dusty") components.

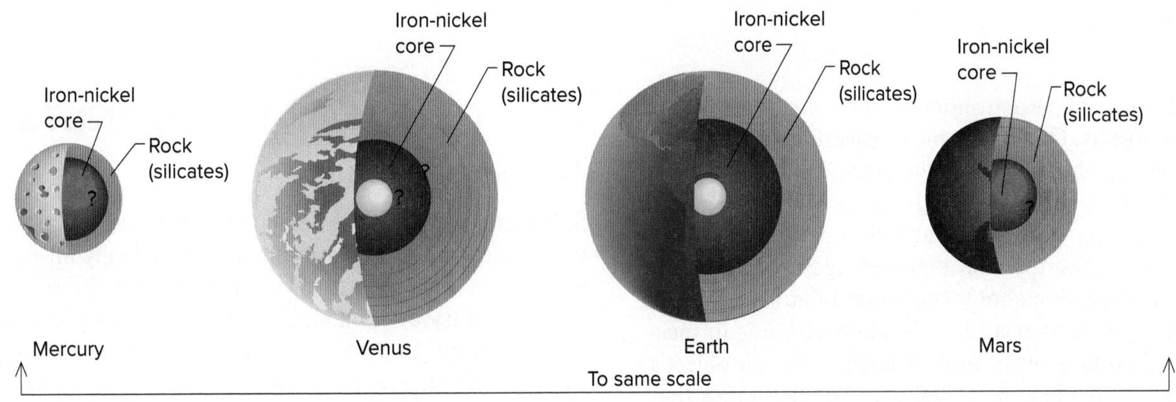

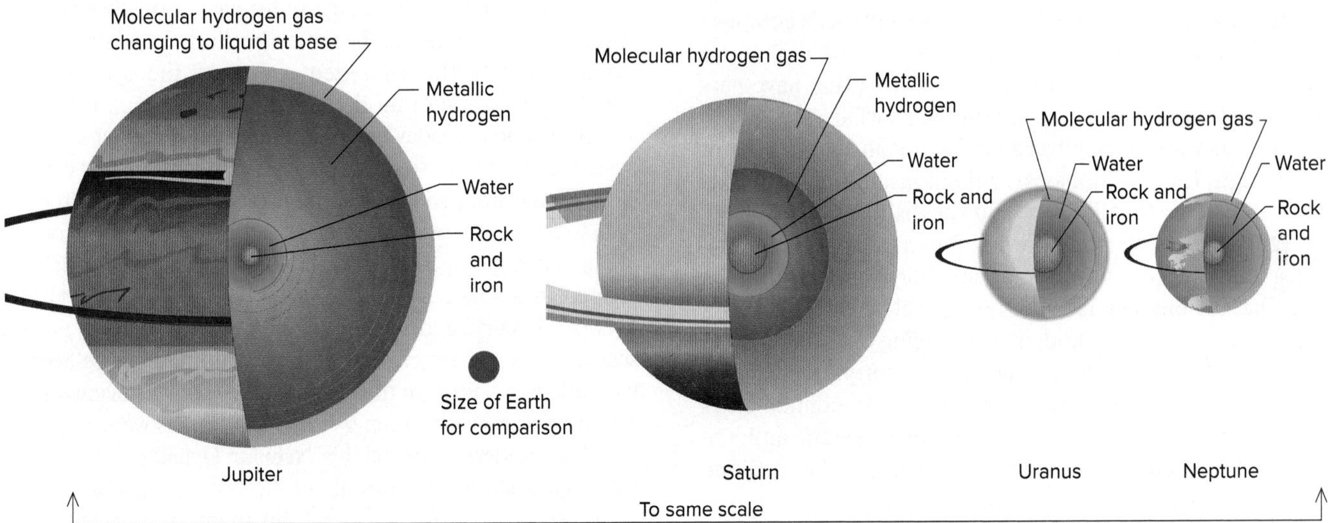

FIGURE 23.5

Sketches of the interiors of the planets. Details of sizes and composition of inner regions are uncertain for many of the planets.

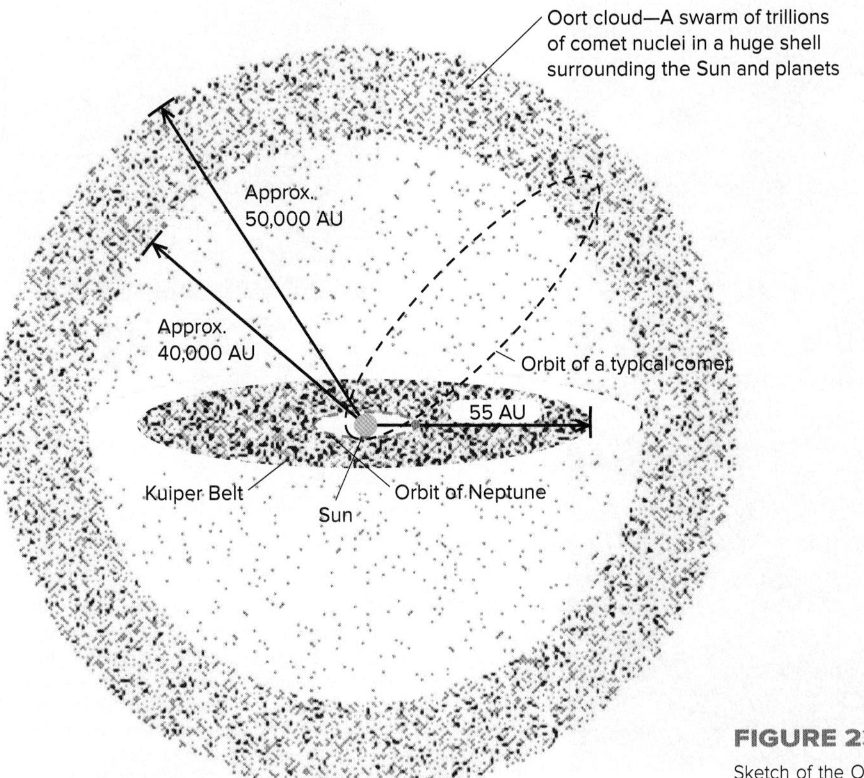

FIGURE 23.6

Sketch of the Oort cloud and the Kuiper Belt. The dimensions shown are known only approximately. Orbits and bodies are not to scale.

You might wonder how astronomers know the composition of these distant objects. One of the most powerful ways to learn that information is from analysis of the sunlight reflected from their surface or atmosphere. When light passes through a gas or reflects from a surface, atoms there absorb some of the colors. The missing colors may then be matched against laboratory samples to give information about what kinds of atoms are present. Yet another way to infer a planet's composition is to measure its density. You may recall from chapter 2 that an object's density is its mass divided by its volume. We saw there that density is an important clue to the identity of minerals. Similarly, astronomers can use density to deduce a planet's composition and structure. In the case of Venus, Mars, and Jupiter, we have additional confirmation from space probes that have penetrated their atmospheres or landed on their surface.

Because they are so small and far away, comets are difficult to observe until their motions carry them inside the snow line. Then the ices begin to sublimate and dust is liberated, making them much more visible and allowing for analysis of their compositions. Some comets, however, do not follow paths within the ecliptic. These orbits led Jan Oort to speculate that there is a nebulous shell of cometary bodies surrounding our solar system at about 50,000 AU—the Oort cloud. Comets originating in the Oort cloud are of special interest since they may contain clues about the earliest composition of the solar system and the boundary between our solar system and the interstellar medium.

The Milky Way and the Universe

Just as the Earth is but one of many planets orbiting the Sun, so too is the Sun but one of a vast swarm of stars orbiting within our galaxy, the **Milky Way**. Astronomers estimate that the Milky Way contains some 100 billion stars spread throughout a roughly disk-shaped volume. That volume is so huge it is difficult to imagine, even by analogy. For example, if the solar system were the size of a cookie, the Milky Way would be the size of the Earth. In fact, even the astronomical unit is too small to be a sensible unit for measuring the size of galaxies. Accordingly, astronomers use a far larger unit of distance, the light-year, to discuss such sizes.

One light-year is the distance light travels in a year—about 10 trillion kilometers. In these immense units, the Milky Way is roughly 100,000 light-years in diameter. But the Milky Way is just one among myriads of other galaxies. Together, these galaxies with all their stars and planets form the visible **universe**. The universe is believed to have formed 13.75 billion years ago starting with the Big Bang.

23.2 ORIGIN OF THE PLANETS

The Solar Nebula

Speculation about how the planets formed dates back to prehistoric times and the creation myths of virtually all peoples. But among the first scientific hypotheses were those of the eighteenth century by Immanuel Kant, a German philosopher, and Pierre Simon Laplace, a French mathematician. Kant and Laplace independently proposed that the solar system originated from a rotating, flattened disk of gas and dust called the **solar nebula**. In this model, the outer part of the disk became the planets, and the center became the Sun. This simple picture explains nicely the flattened shape of the system and the common direction of motion of the planets around the Sun.

Astronomers today have additional evidence that supports this basic picture. First, we now know that the inner planets are rich in silicates and iron, and the outer planets are rich in hydrogen. Second, to the best of our knowledge, the Earth, Sun, and Moon are all about the same age—about 4.6 billion years old—suggesting they formed in a single event. Third, the surfaces of objects like our Moon, Mercury, and the moons of the outer planets are heavily scarred with ancient craters, suggesting that they were bombarded with infalling objects in the distant past. With this evidence, we can now discuss how astronomers think the planets were born.

The modern form of this **Nebular Hypothesis** proposes that our solar system was born about 4.6 billion years ago, roughly 9 billion years after the Big Bang, from an interstellar cloud, an enormous aggregate of gas and dust like the one shown in figure 23.7*A*.

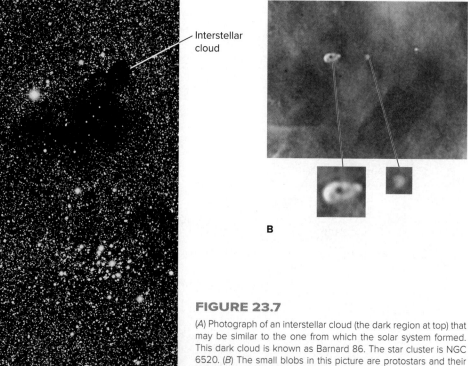

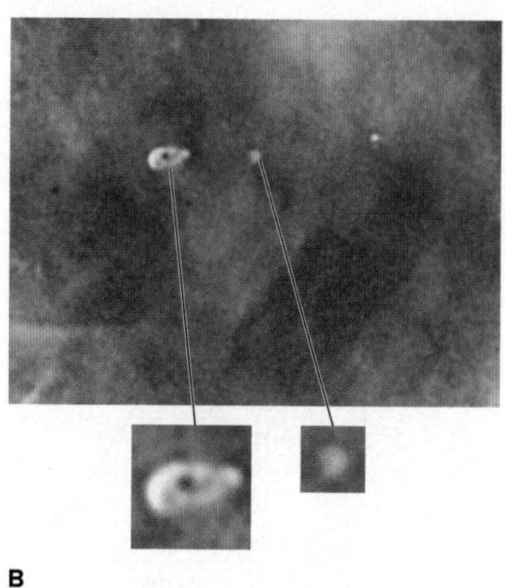

FIGURE 23.7

(*A*) Photograph of an interstellar cloud (the dark region at top) that may be similar to the one from which the solar system formed. This dark cloud is known as Barnard 86. The star cluster is NGC 6520. (*B*) The small blobs in this picture are protostars and their surrounding disk of dust and gas. They are in the Orion Nebula, a huge cloud of gas and dust about 1,500 light-years from Earth.
(A) ©Australian Astronomical Observatory/David Malin Images; (B) Source: C.R. O'Dell/Rice University/NASA

Such clouds are common between the stars in our galaxy even to-day, and astronomers now think most, if not all, stars and their planets have formed from them. Although our main concern in this chapter is with the birth of our own solar system, we should bear in mind that our hypothesis applies more broadly and implies that most stars could have planets or at least surrounding disks of dust and gas from which planets might form. In fact, astronomers observe just such disks around many young stars (figure 23.7*B*), and new data suggest that there are perhaps 40 billion Earth-like planets in our galaxy.

Although interstellar clouds are found in many shapes and sizes, the one that became the Sun and its planets was probably a few light-years in diameter and was made mostly of hydrogen (71%) and helium (27%) gas, with tiny traces of other chemical elements. In addition to gas, the cloud also contained microscopic dust particles. From analysis of the radiation absorbed and emitted by such dust particles, astronomers know they are made of a mixture of silicates, iron compounds, carbon compounds, and ices, especially of water, carbon dioxide, methane, and ammonia.

The cloud that became our solar system began its transformation into the Sun and planets when the gravitational attraction between the particles in the cloud caused it to collapse inward, as shown in figure 23.8*A*. The presence of radioactive daughter products in meteorites suggests that the collapse may have been triggered by a nearby supernova. But, because the cloud was rotating, it flattened, as shown in figure 23.8*B*. Flattening occurred because rotation retarded the collapse perpendicular to the cloud's rotation axis.

It probably took a few million years for the cloud to collapse and form a rotating disk with a warm bulge in the center that

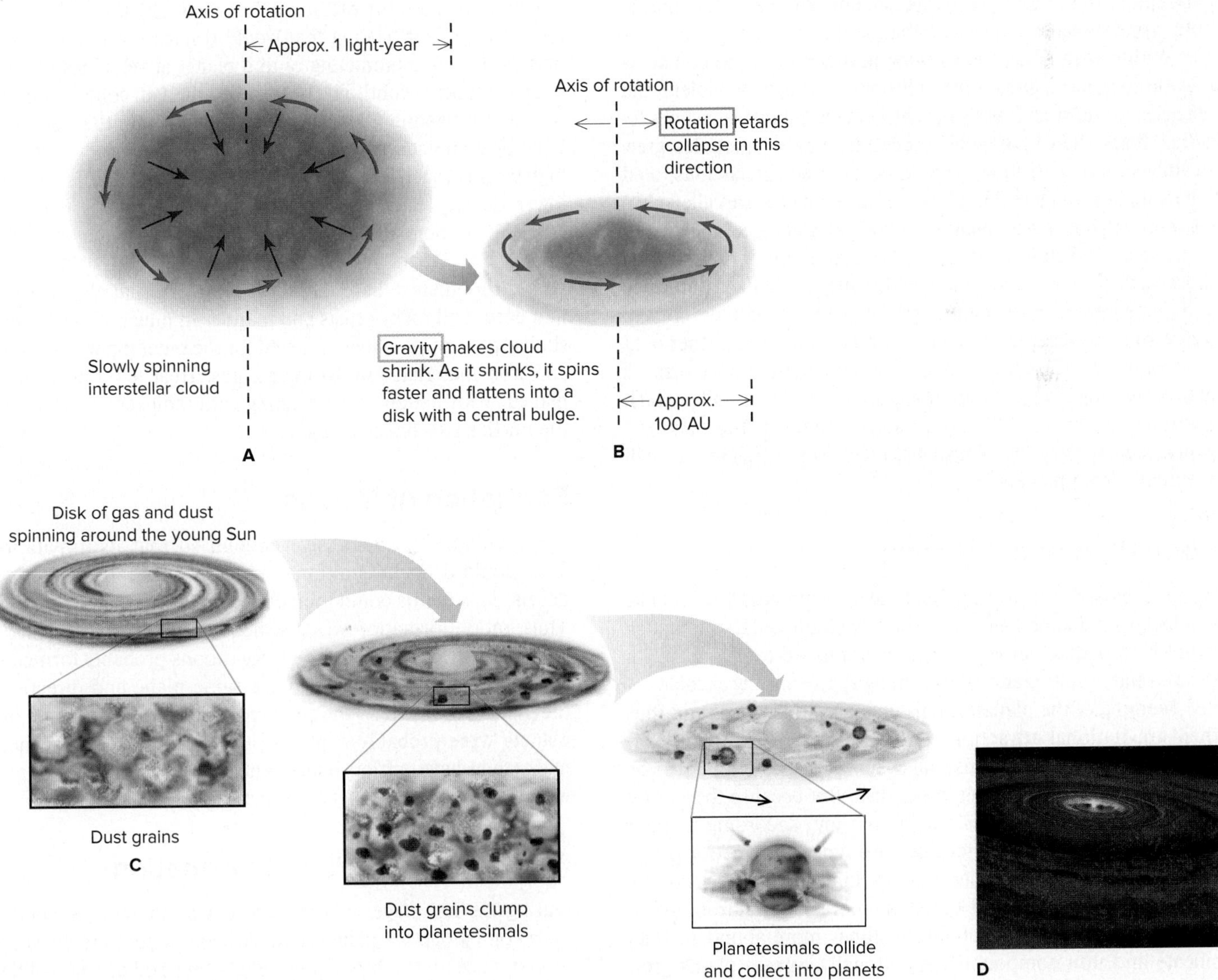

FIGURE 23.8

Sketch illustrating the (*A*) collapse of an interstellar cloud and (*B*) its flattening. (*C*) Artist's depiction of the condensation of dust grains in the solar nebula and the formation of planetesimals. (*D*) Artist's conception of infrared data indicating planet-building, as detected by the Spitzer Space Telescope.
(D) Source: T. Pyle (SSC)/JPL-Caltech/NASA

became the Sun. Over a few more million years, dust particles in the disk began to stick together, perhaps helped by static electric forces such as those that make lint cling to clothes in a dryer. As the particles slowly grew in size, they were more likely to collide with other particles, hastening their growth. But the composition of these particles depended strongly on where in the disk they formed. In the inner part of the disk (where the particles were near the Sun), it was so warm that the ices were sublimated. Solid particles there were thus composed almost entirely of silicate and iron-rich material. Only at about Jupiter's distance from the Sun was the disk cold enough for the ices to remain as solid particles. Thus, particles in those outer regions consisted of silicate and iron-rich material *plus* frozen water. Although water molecules were relatively abundant throughout the cloud, they could only accumulate with other particles in the outer disk. As a result, those particles grew much larger than the particles in the inner disk. Thus, the nebula became divided into two regions: an inner zone of silicate and iron particles, and an outer zone that also contained abundant ices.

Within each zone, the growing particles began to collide as they orbited in the disk. If the collisions were not too violent, the particles would stick (much as gently squeezing two snowballs together fuses them). By such processes, smaller particles grew steadily in size until they were kilometers across, as illustrated schematically in figure 23.8*C*. These larger objects are called **planetesimals** (that is, small, planet-like bodies). Because the planetesimals near the Sun formed from silicate and iron particles, while those farther out were cold enough that they could incorporate ice and frozen gases, two main types of planetesimals developed: rocky-iron ones near the Sun and icy-rocky-iron ones farther out.

Figure 23.8D shows an artist's conception of a disk of planet-building dust surrounding a brown dwarf (or failed star). Astronomers believe this "mini solar system in the making," which was spotted by NASA's Spitzer Space Telescope, will eventually form planets.

Formation of the Planets

As time passed, the planetesimals themselves began to collide. Computer simulations show that some collisions led to the shattering of both bodies, but less-violent collisions led to merging, with the planetary orbits gradually becoming approximately circular.

Merging of the planetesimals increased their mass and thus their gravitational attraction. That, in turn, helped them grow even more massive. At this stage, objects large enough to be called planets were orbiting in the disk. But because there were two types of planetesimals (rocky and icy) according to their location in the inner or outer disk, two types of planets formed.

Planet growth was especially rapid in the outer parts of the solar nebula. Planetesimals there had more material from which to grow because ice was about ten times more abundant than silicate and iron compounds. Additionally, once a planet grew somewhat larger than the diameter and mass of the Earth, it was able to attract and retain gas by its own gravity. Because hydrogen was overwhelmingly the most abundant material in the solar nebula, planets large enough to tap that reservoir could grow vastly larger than those that formed only from solid material. Thus, Jupiter, Saturn, Uranus, and Neptune may have begun as slightly larger than Earth-sized bodies of ice and rock, but their gravitational attraction resulted in their becoming surrounded by the huge hydrogen-rich atmospheres that we see today. In the inner solar system, the smaller and warmer bodies could not capture hydrogen directly, and therefore they remained small.

As planetesimals struck the growing planets, their impact released gravitational energy that heated both the planetesimal and the planet. Gravitational energy is liberated whenever something falls. For example, if you drop a bowling ball into a box of tennis balls, the impact scatters the tennis balls in all directions, giving them kinetic energy—energy of motion. In much the same manner, planetesimals falling onto a planet's surface give energy to the atoms in the crustal layers, energy that appears as heating. You can easily demonstrate that motion can generate heat by hitting an iron nail a dozen or so times with a hammer and then carefully touching the nail: the metal will feel distinctly hot. Imagine now the vastly greater heating that would result from mountain-sized masses of rock plummeting onto a planet at velocities of tens of kilometers per second. The heat so liberated, in combination with radioactive heating (as occurs even today within the Earth), partially or completely melted the planets and allowed matter with high density (such as iron) to sink to their cores, while matter with lower density (silicate minerals such as plagioclase feldspar) "floated" to their surfaces, a process known as planetary **differentiation**. The Earth's iron core probably formed by this process, and astronomers believe that the other terrestrial planets formed iron cores and rocky crusts and mantles in much the same way. A similar process probably occurred for the outer planets when rock and iron material sank to their cores. Heat left over from this planetary formation process, along with radioactive decay, drives the Earth's internal heat engine.

Formation of Moons

Once a planet grew massive enough so that its gravitational force could draw in additional material, it became ringed with debris. This debris could then collect into lumps to form moons. Thus, moon formation was a scaled-down version of planet formation. Although most of the large moons probably formed this way (they revolve and spin in the same plane and direction as the planets they orbit), the smaller moons of Mars and the outer planets were probably captured asteroids and small planetesimals (they have orbital planes and directions that are random with respect to the planets they orbit).

Final Stages of Planet Formation

During the last stage of planetary formation, leftover planetesimals bombarded the planets and blasted out huge craters such as those we see on the Moon and on all other bodies with solid surfaces in the solar system. Occasionally, an impacting body was so large that it did more than simply leave a crater. For example, Earth's Moon may have been created when an object a few times the size of Mars struck the Earth, as we will discuss further in the section titled "Portraits of the Planets." Likewise, Mercury may have suffered a massive impact that blasted away much of its crust. The extreme tilt of the rotational axis of Venus and Uranus may also have arisen from large planetesimal collisions. In short, planets and satellites were brutally battered by the remaining planetesimals early in the history of our solar system.

Formation of Atmospheres

Atmosphere formation was perhaps the last part of the planet-forming process. The inner and outer planets are thought to have formed atmospheres differently, a concept that explains their very different atmospheric compositions. The outer planets probably captured most of their atmospheres directly from the solar nebula, as mentioned in the "Formation of the Planets" section. Because the nebula was rich in hydrogen and helium, so are the atmospheres of the outer planets.

The inner planets were not massive enough to capture gas directly from the solar nebula and are therefore deficient in hydrogen and helium. The atmospheres of Venus, Earth, and Mars probably formed by a combination of processes: volcanic eruptions releasing gases from their interiors; vaporization of comets and icy planetesimals that have struck them; and gases freed from within the planetesimals out of which the planets themselves formed. Objects such as Mercury and the Moon, which have only traces of atmosphere today, show few signs of volcanic activity within the last few billion years. These smaller bodies have such weak gravity that any atmospheric gases they originally possessed escaped easily from them.

Other Planetary Systems

This theory for the origin of the solar system explains many of its features, but astronomers still have many questions about the Sun and its family of planets and moons. Answers to some of these questions are beginning to emerge from the discovery of planets around thousands of other stars. These planets are typically far too dim to be seen directly with present telescopes, although a number have been imaged directly in recent years (figure 23.9). Some planets are detected by the slight gravitational tug they exert on their parent stars. That tug can be detected by analyzing the light from such stars. Others are detected by the periodic dimming of the stars they orbit. Thousands of these *extrasolar* planets have been detected so far especially by the Kepler Space Telescope, and recently, Kepler data have been interpreted to indicate that there may be 40 *billion* Earth-like planets orbiting other stars in our galaxy. Such observations have delighted astronomers by confirming that planets have indeed formed around other stars. However, they have also led to surprises. For example, many of these systems have huge, Jupiter-like planets very close to their parent stars. One hypothesis for this puzzling observation is that the planets formed farther out and "migrated" inward toward the star. Such shifts in orbits may also have happened in our solar system. For example, many astronomers think that Neptune has migrated outward from an original orbit that was nearer the Sun. Thus, although we know that our solar system is not unique, we have yet to fully understand its properties and origin.

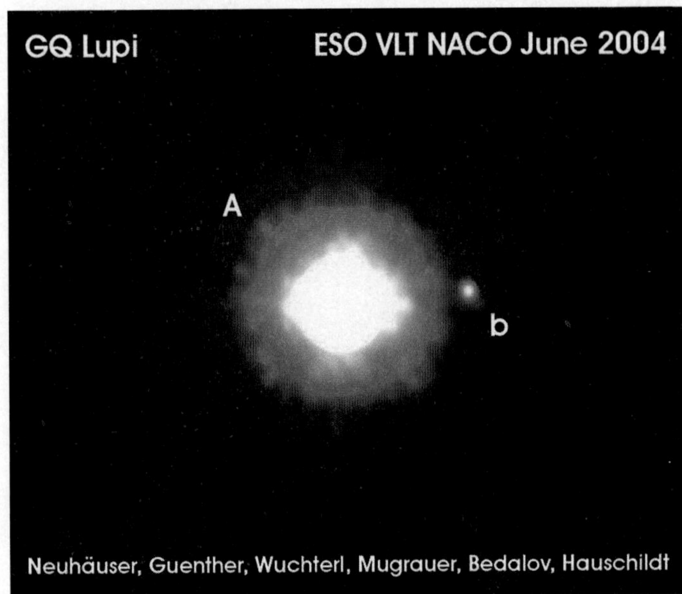

FIGURE 23.9
First confirmed image of a planet (labeled *b* in the image) orbiting a star (GQ Lupi) other than the Sun.
Courtesy of ESO

23.3 PORTRAITS OF THE PLANETS

What is the nature of these planets whose birth we have discussed? In this section, we will work our way through our solar system, describing each of the terrestrial planets in turn. Table 23.1 summarizes some important physical properties, such as size and mass, of the eight planets and our Moon. Table 23.2 summarizes their orbital properties. The four large gas giant planets (Jupiter, Saturn, Uranus, and Neptune) have no solid surfaces to describe geologically. However, the surfaces of their rocky and icy moons will be discussed in turn. Before we consider planets far from Earth, however, it will help us to look at the only object beyond Earth that has been visited by humans and from which rock samples have been returned and studied—our Moon.

Our Moon

The Moon is our nearest neighbor in space, a natural satellite orbiting the Earth. It is the frontier of direct human exploration and of great interest to astronomers. Unlike Earth, it is essentially a dead world, with neither plate tectonic nor current volcanic activity. Although extensive volcanism occurred early in the Moon's geologic history, it was largely confined to lower elevations of the lunar *nearside* (the side that permanently faces toward the Earth). This localized volcanism, coupled with the Moon's lack of atmosphere, means that many of its surface features are virtually unaltered since its youth. Its surface thus bears a record of events in the early solar system that gives clues not only to the Moon's birth but to that of the solar system as well.

TABLE 23.1 Physical Properties of the Planets and Our Moon

Name	Radius (compared to Earth)	Radius (Equator) (km)	Mass (compared to Earth)	Mass (kg)	Average Density (g/cm^3)
Mercury	0.382	2,439	0.055	3.30×10^{23}	5.43
Venus	0.949	6,051	0.815	4.87×10^{24}	5.25
Earth	1.00	6,378	1.00	5.97×10^{24}	5.52
Our Moon	0.27	1,738	0.012	7.35×10^{22}	3.3
Mars	0.533	3,397	0.107	6.42×10^{23}	3.93
Jupiter	11.19	71,492	317.9	1.90×10^{27}	1.33
Saturn	9.46	60,268	95.18	5.68×10^{26}	0.69
Uranus	3.98	25,559	14.54	8.68×10^{25}	1.32
Neptune	3.81	24,764	17.13	1.02×10^{26}	1.64

TABLE 23.2 Orbital Properties of the Planets

Name	Distance from Sun*		Period		Orbital Inclination†	Orbital Eccentricity
	(AU)	(10^6 km)	Years	(Earth Days)	Degrees	
Mercury	0.387	57.9	0.2409	(87.97)	7.00	0.206
Venus	0.723	108.2	0.6152	(224.7)	3.39	0.007
Earth	1.00	149.6	1.0	(365.26)	0.00	0.017
Mars	1.524	227.9	1.8809	(686.98)	1.85	0.093
Jupiter	5.203	778.3	11.8622	(4,332.59)	1.31	0.048
Saturn	9.539	1,427.0	29.4577	(10,759.22)	2.49	0.056
Uranus	19.19	2,869.6	84.014	(30,685.4)	0.77	0.046
Neptune	30.06	4,496.6	164.793	(60,189)	1.77	0.010

*These values are half the long diameter of the orbit.
†With respect to the Earth's orbit.

Origin and History

Lunar rocks brought back to Earth by the *Apollo* astronauts have led astronomers to radically revise their ideas of how the Moon formed. Before the *Apollo* program, lunar scientists had three hypotheses of the Moon's origin. In one, the Moon was originally a small planet orbiting the Sun that approached the Earth and was captured by its gravity (capture hypothesis). In another, the Moon and Earth were "twins," forming side by side from a common cloud of dust and gas (twin formation hypothesis). In the third, the Earth initially spun enormously faster than now and formed a bulge that ripped away from it to become the Moon (fission hypothesis).

The failure of evidence based on lunar surface samples to confirm any of the three hypotheses led astronomers to consider alternatives, and a completely different picture of the Moon's origin has emerged. According to the new hypothesis, the Moon formed from debris blasted out of the Earth by the impact of an object about the size of Mars, as shown in figure 23.10*A*. The great age of lunar rocks (returned samples have ages of up to 4.45 billion years) and the absence of any enormous impact feature on the Earth indicate that this event must have occurred simultaneously with the Earth's own formation, at least 4.5 billion years ago. The colliding body melted and vaporized millions of cubic kilometers of the Earth's crust and mantle and hurled it into space in an incandescent plume. Although most of this debris would have rained back down on Earth, what remained in orbit was drawn together by gravity into what we now see as the Moon.

After the Moon's birth, stray fragments of the ejected rock pelted its surface, creating the craters that blanket the highlands. A few craters were so large that they excavated enough crust to allow the underlying mantle to melt. The resulting basaltic magma escaped to the surface, flooding the floors of those craters and forming the lunar maria. Since about 3.5 billion years ago, the Moon has experienced no major changes. It has been a virtually dead world for all but the earliest times in its history, experiencing only a much decreased rate of ongoing meteor impact events.

Description

General Features

The Moon is about one-fourth the diameter of the Earth and is a barren ball of rock possessing no air, water,[2] or life. In the

[2]Astronomers have found some evidence for extremely small amounts of water frozen in perpetually shadowed craters near the Moon's poles.

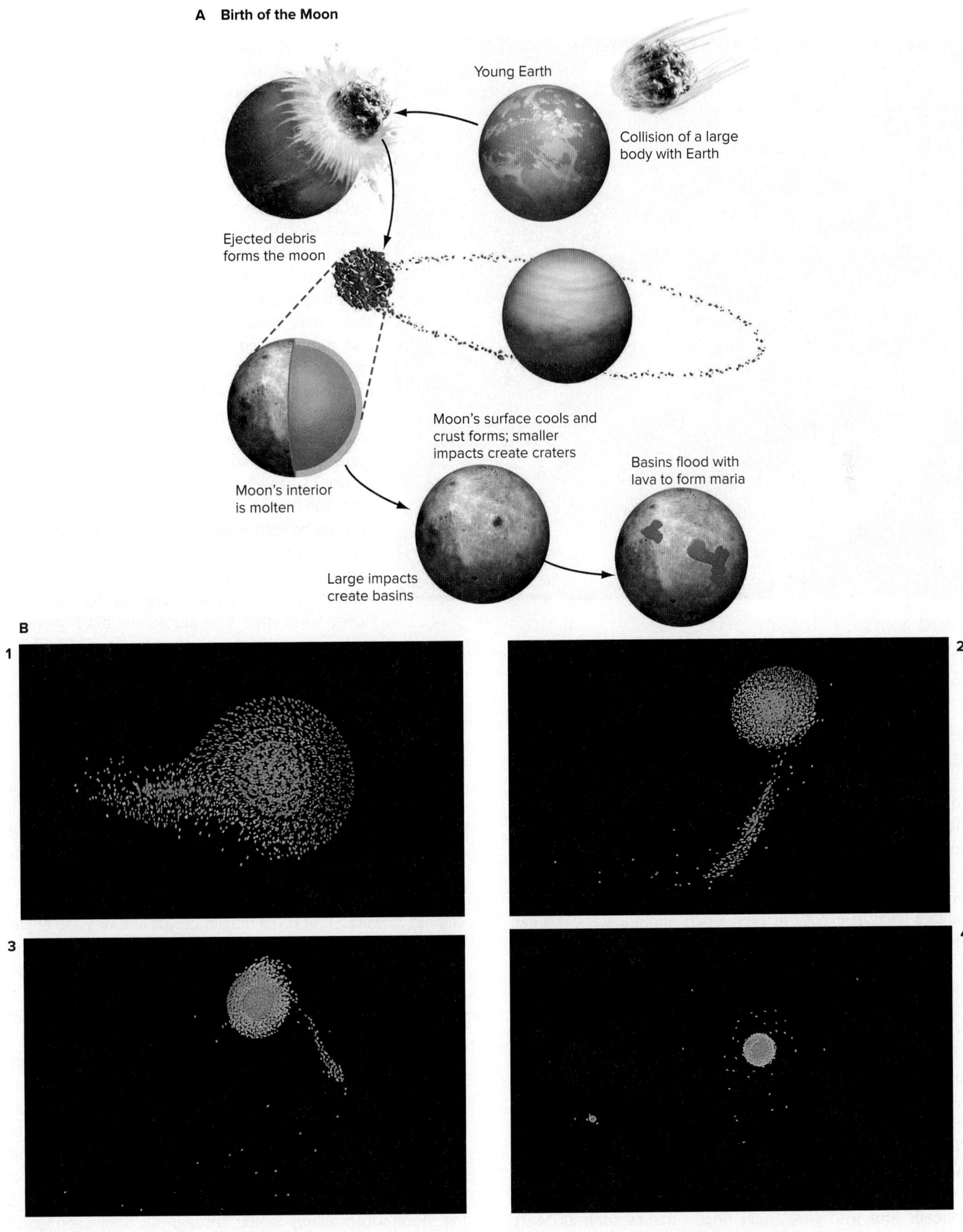

FIGURE 23.10

(A) A sketch of how the Moon may have formed. *(B)* A computer simulation showing how the collision of a roughly Mars-sized body with the young Earth can splash out debris, which then is drawn together by its own gravity to make Earth's Moon. The scale changes between the frames: (1) is a close-up of the impact; (4) shows the newly assembled Moon (lower left) orbiting Earth, which is still surrounded by debris.

(B) Courtesy of A. G. W. Cameron, Lunar and Planetary Laboratory, University of Arizona

FIGURE 23.11

Photograph showing the different appearance of the lunar highlands and maria. The highlands are bright, rough, and heavily cratered. The maria are dark, smooth, and have fewer craters.

words of lunar astronaut Buzz Aldrin, the Moon is a place of "magnificent desolation." But you don't have to walk on the Moon to see its desolation. To the naked eye, the Moon is a world of grays; yet even gray has its variety, as you can see where the dark, roughly circular areas stand out from the lighter background, as shown in figure 23.11.

Surface Features

Through a small telescope or even a pair of binoculars, you can see that the dark areas of the Moon are smooth while the bright areas are covered with numerous large, circular pits called *craters,* as illustrated in figure 23.12*A* and *B*. Craters usually have a raised rim and range in size from microscopic pits to gaping scars in the Moon's crust as much as 240 kilometers across. The larger craters have mountain peaks or peak rings at their center (figure 23.12*C*). Still larger lunar *impact basins* range up to 2,500 kilometers in diameter, with multiple concentric rings of mountain peaks and fault scarps (figure 23.12*D*).

The large, smooth, dark areas are called *maria* (pronounced *MAH-ree-a*), from the Latin word for "seas." This usage comes from early observers who believed the maria looked like oceans and who gave them poetic names such as Mare (pronounced *MAH-ray*) Serenitatis (Sea of Serenity) and Mare Tranquillitatis (Sea of Tranquility), the site where astronauts first landed on the Moon. These regions were formed by extensive volcanic activity. The maria are essentially low areas (primarily impact craters and basins) flooded by dark-colored basaltic lava flows.

The bright areas that surround the maria are called *highlands.* The highlands and maria differ in brightness because they are composed of different rock types. The maria are composed of basalt, a dark, congealed lava rich in iron, magnesium, and titanium silicates. The highlands, on the other hand, are mainly anorthosite, a rock type rich in calcium and aluminum-rich plagioclase feldspar. This difference has been verified from rock samples obtained by astronauts. Moreover, the samples also show that the highland material is considerably older than the mare material. The highlands are also more rugged than the maria, being pitted with more and larger craters. In fact, highland craters are so abundant that they often overlap (figure 23.12A).

Craters are not the only type of lunar surface feature. From many craters, long, light streaks of pulverized rock called rays radiate outward, as shown in figure 23.13*A*. A particularly bright set spreads out from the crater Tycho near the Moon's South Pole and can be seen easily with a pair of binoculars when the Moon is full. A small telescope reveals still other surface features. Numerous lunar canyons known as sinuous rilles, channels created by ancient lava flows, wind their way across the mare plains away from some craters, as shown in figure 23.13*B*. Elsewhere, straight rilles break the surface, probably representing faults created by local tectonic activity.

Origin of Lunar Surface Features

Most of the surface features we see on the Moon—craters, rays, and basins (which bound the larger maria)—were made by the impact of solid objects on its surface. When such an object hits a solid surface at high speed (up to 70 kilometers per second!), most of the object disintegrates in a cloud of vaporized rock and fragments. The resulting explosion blasts a hole whose diameter depends on the mass and velocity of the impacting object. The shape of the hole is circular, however, unless the impact occurs at a very low angle.

As the force of compression migrates away from the point of impact, it forces surrounding rock outward, piling and overturning it into a raised rim. Pulverized rock is ejected in all directions, forming rays in the blanket of ejected material that is deposited around the crater. Bigger impacts compress the crust so much that it rebounds upward, creating a central peak (figure 23.12*C*) or even a ring inside the crater (figure 23.12*D*).

The maria are the largest impact features, but to understand their formation, we must briefly describe the early history of the Moon. The great age of the highland rocks (in some cases, as old as 4.45 billion years) leads astronomers to deduce that these rugged uplands formed shortly after the Moon's birth. At that time, the Moon was probably molten throughout, allowing dense, iron-rich material to sink to its interior while less-dense material floated to the lunar surface. On reaching the surface, the less-dense rock cooled and crystallized, forming the Moon's crust. A similar process probably formed the ancient crust of the Earth. The

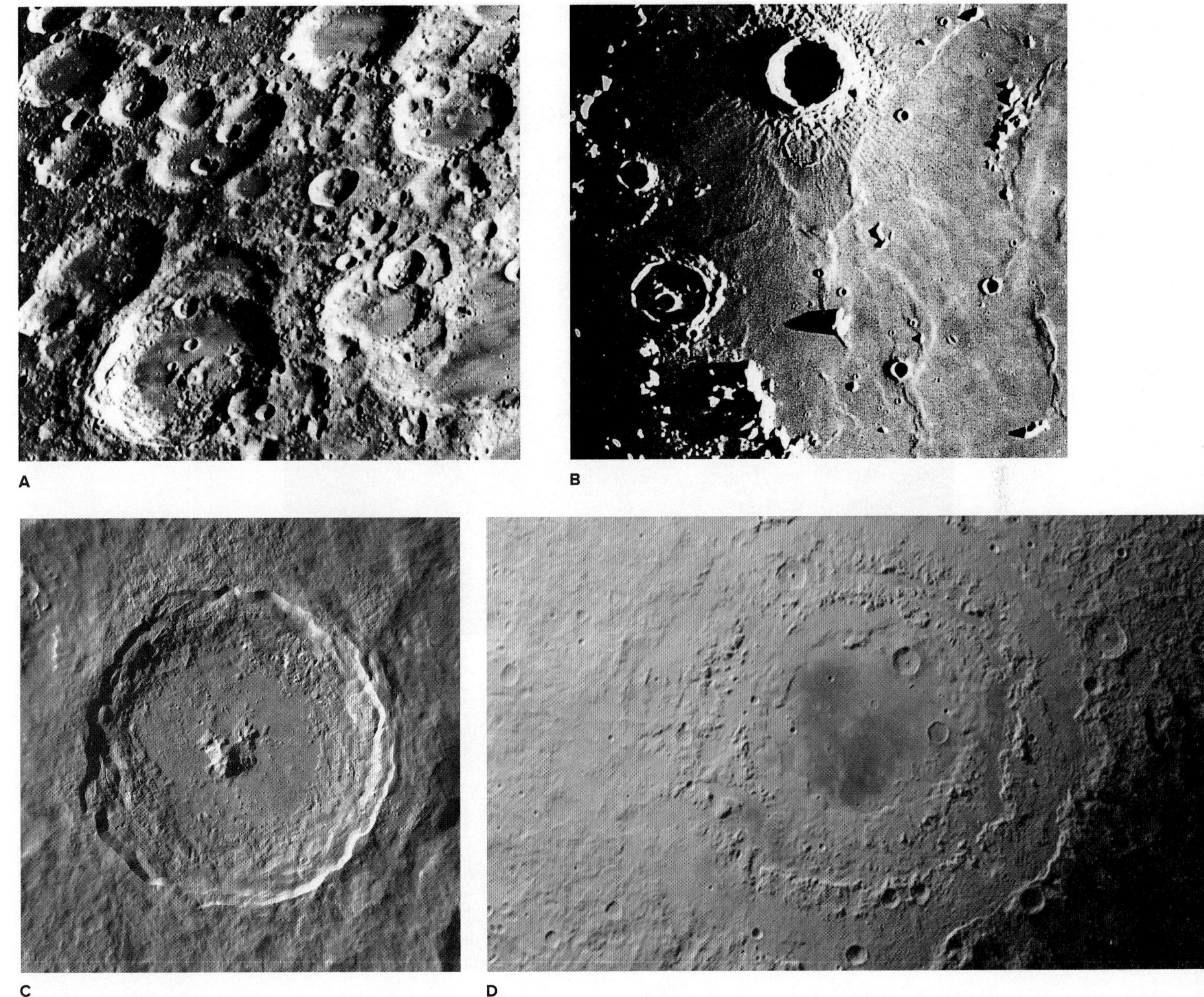

FIGURE 23.12

(*A*) Overlapping craters in the Moon's highlands. (*B*) Isolated craters in the smooth mare. (*C*) Tycho—an 85-kilometer crater with a central peak, located on the south-central nearside of the Moon. (*D*) The 930-kilometer impact basin Orientale, located on the western limb of the lunar nearside.
(A and B) ©UC Regents/Lick Observatory. Unauthorized use prohibited.; (C) Source: Lunar Reconnaissance Orbiter/NASA; (D) Source: NASA's Goddard Space Flight Center Scientific Visualization Studio

Moon's highlands were then heavily bombarded by solid bodies from space, creating the numerous craters we see there.

As the Moon continued to cool, its crust thickened. But before the crust grew too thick and the underlying mantle became too cool, a small number of exceptionally large objects–about 100 kilometers across–struck the surface, blasting huge holes to create the largest lunar basins. Since the underlying mantle rock was still very warm, the removal of overlying material allowed the mantle to melt, forming basaltic magma. Over the next billion years or so, this molten material from deep within the Moon flooded these vast craters and crystallized to form the smooth, dark lava plains that we see now. Because the lava flowed more recently, the mare material is younger than the highlands. Moreover, by the time the maria formed, most of the small, numerous impacting bodies were gone–swept up and captured by the Earth and Moon in earlier collisions. Thus, few bodies remained to crater the maria, which therefore remain mostly smooth (that is, far less heavily cratered than the highlands) to this day.

Our home planet Earth furnishes additional evidence that most lunar craters formed early in the Moon's history. Like the Moon, Earth too was presumably battered by many impacts in its youth. Although the vast multitude of these craters have been obliterated by weathering, erosion, volcanism, and plate tectonics, a few remain in ancient rock whose measured age is typically

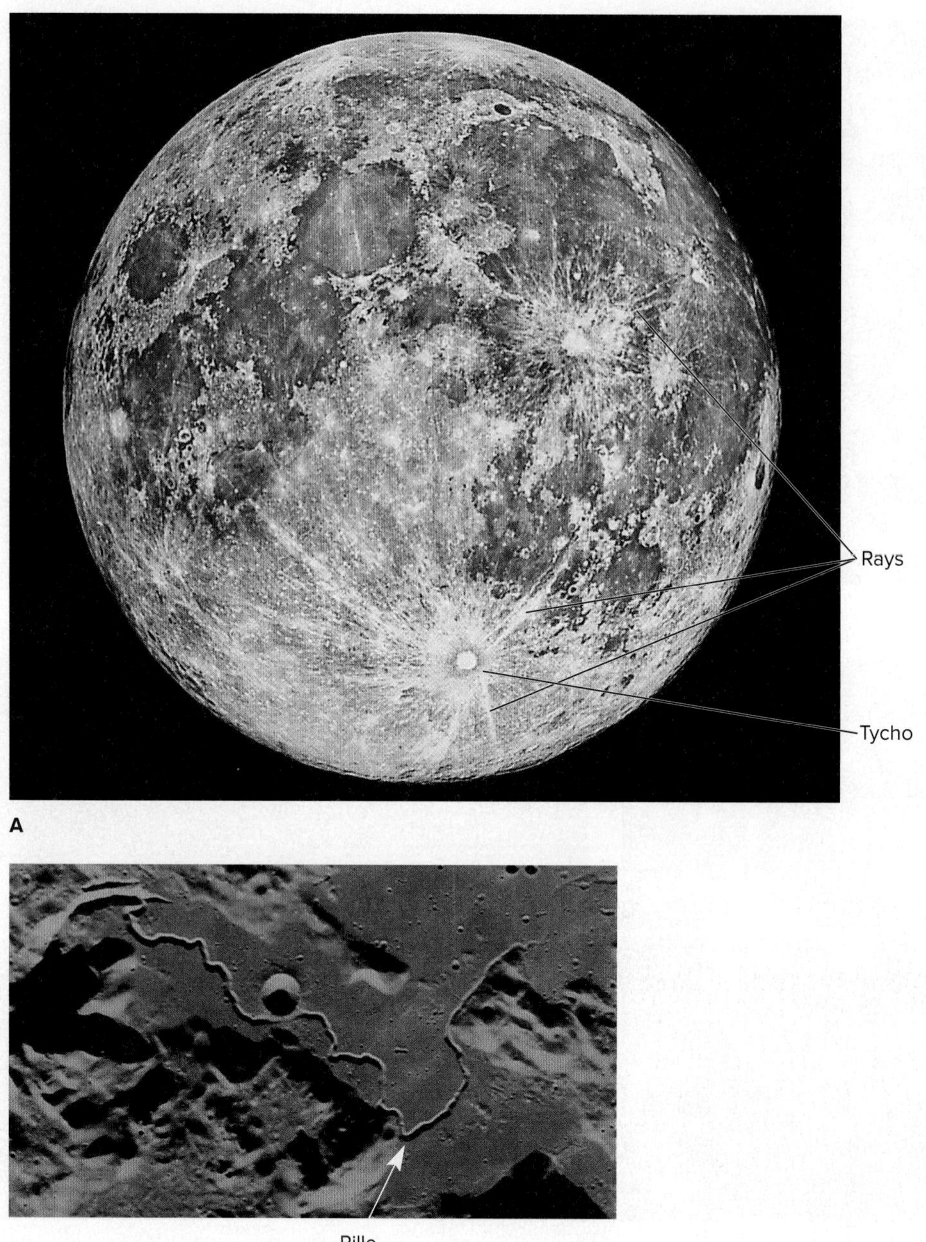

FIGURE 23.13

(*A*) The long, narrow, white streaks radiating away from the craters at the top are lunar rays. (*B*) Photograph of Hadley Rille, a large lava channel at the *Apollo 15* landing site (just to the right of the arrow).
(A) ©UC Regents/Lick Observatory. Unauthorized use prohibited.; (B) Source: The Lunar & Planetary Institute/NASA

hundreds of millions or billions of years. From the scarcity of such craters on Earth, astronomers can deduce that the main bombardment must have ended almost 4 billion years ago.

Craters and maria so dominate the lunar landscape that we might not notice the absence of folded mountain ranges and the great rarity of volcanic peaks, landforms that are common on Earth. Why have such features not formed on the Moon?

Structure

The Moon's small size relative to the Earth explains the differences between the two. The Moon, like Earth, was born hot and, as we mentioned before, was perhaps completely molten at some point in its youth. However, because the Moon is so small relative to the Earth, heat escapes from it more rapidly. Thus, the Moon has cooled far more than the Earth. (Think of how a small baked potato taken from a hot oven cools much faster than a big one.) Being cooler, the Moon lacks the internal convection currents that drive plate-tectonic activity on the Earth. Confirmation of this comes from studies of the Moon's interior using data obtained from instruments deployed on the lunar surface by *Apollo* astronauts between 1969 and 1972.

Crust and Interior

The Moon's interior can be studied by seismic waves just as Earth's can. One of the first instruments set up on the Moon by the *Apollo* astronauts was a seismic detector. Measurements from that and other seismic detectors placed by later *Apollo* crews show that the Moon's interior is essentially inactive and has a much simpler structure than Earth's.

The Moon's surface layer is shattered rock from countless impacts that forms a regolith (meaning "blanket of rock") about 10 meters deep. Below this rocky rubble is the Moon's solid crust, which is about 60 to 70 kilometers thick, on the average. The crust of the lunar highlands is composed of silicate rocks, relatively rich in aluminum and poor in iron. Beneath the crust is a thick mantle of solid rock, extending down about 1,000 kilometers. Unlike the Earth's mantle, however, the lunar mantle appears too cold and rigid to be stirred by the Moon's feeble remaining heat.

The Moon's lower average density (3.3 grams per cubic centimeter) tells us its interior contains proportionately less iron than Earth's interior. Moreover, because its small size allowed heat to escape relatively rapidly, its core is cool compared to Earth's. These deductions are also consistent with the Moon's lack of a magnetic field. Thus, a compass would be of no use to an astronaut lost on the Moon.

The Absence of a Lunar Atmosphere

The Moon's surface is never hidden by lunar clouds or haze. In fact, astronomers detect only the barest traces of an atmosphere, less than a hundred-trillionth (10^{-14}) the density of Earth's atmosphere. While early studies seemed to indicate that the lunar mantle was devoid of water, recent studies of samples brought back by *Apollo* astronauts have shown that the Moon's mantle may contain as much water as the Earth's. Tiny inclusions of molten rock contained in volcanic glass returned from the Moon show that lunar magma contains 100 times more water than previous studies have suggested. Volcanic activity on the moon would therefore have released significant amounts of water, so why is the Moon's atmosphere so thin? First, the Moon's interior is too cool to cause ongoing volcanic activity and thus release water. Second, and more important, even if volcanoes had created an atmosphere in its youth, the Moon's small mass creates too weak a gravitational force for it to retain the erupted gas.

Mercury

Mercury is the smallest planet and resembles our Moon in both size and appearance. It is very difficult to study using Earth-based telescopes, and it has been visited by only two missions. *Mariner 10* flew by three times in 1974–75, and *Messenger,* launched by NASA in 2004, orbited Mercury between 2011 and 2015, when it was allowed to impact the surface. To find out more about the *Messenger* mission, visit http://messenger.jhuapl.edu/index.html. Mercury is covered with circular craters caused by impacts like those on the Moon (figure 23.14*A*), but its surface is not totally moonlike. About 27% of Mercury's surface is covered with smooth plains of ancient basaltic lava flows, but these flows do not appear to be associated with large craters, as

A

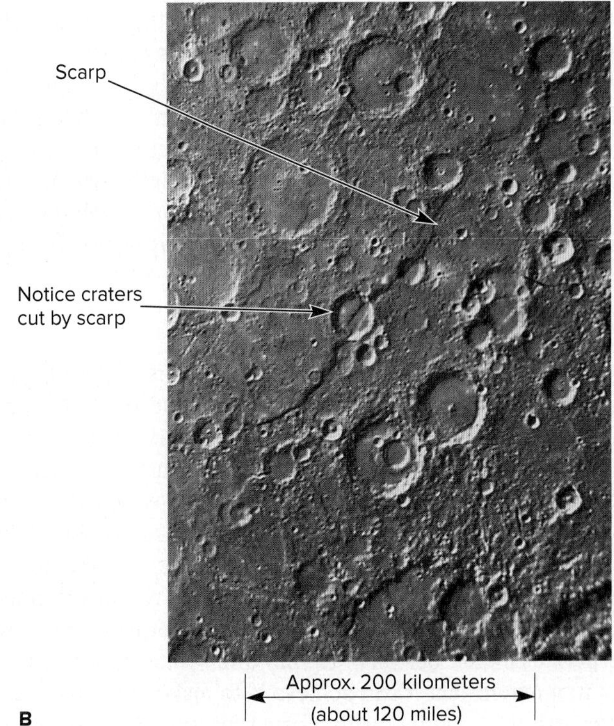

B

FIGURE 23.14

(*A*) Mosaic of pictures of Mercury's cratered surface (taken by the *Mariner 10* spacecraft). (*B*) Photograph of a scarp, (a cliff), running across Mercury's surface.
(A) Source: NASA/Johns Hopkins University Applied Physics Laboratory/Carnegie Institution of Washington; (B) Source: NASA

was the case for the lunar maria. Enormous scarps–formed where the crust has shifted along faults–run for hundreds of kilometers across Mercury's surface, as seen in figure 23.14*B*. These scarps may have formed as the planet cooled and shrank, wrinkling like a dried apple.

Mercury's Temperature and Atmosphere

Mercury's surface is one of the hottest places in the solar system. At its equator, noon temperatures reach approximately 430°C (about 800°F). On the other hand, nighttime temperatures are among the coldest in the solar system, dropping to approximately −173°C (about −280°F). These extremes result from Mercury's closeness to the Sun, its very slow rotation, and its almost complete lack of atmosphere: Only traces of gas have been detected. Without any effective atmosphere, nothing moderates the inflow of sunlight during the day or retains heat during the night.

Mercury lacks an atmosphere for the same reason the Moon does: its small mass makes its gravity too weak to retain much gas around it. Moreover, its proximity to the Sun makes keeping an atmosphere especially difficult because the resulting high temperature causes gas molecules to move so fast that they readily escape into space. The intense **solar wind** (rapid outflow of ionized gas from the Sun) at Mercury's proximity to the Sun would also tend to strip away any existing atmosphere quite rapidly.

Mercury's Interior

Mercury probably has an unusually large (~75% of its diameter) iron core beneath its silicate crust and proportionally thinner mantle. While no spacecraft has landed there to deploy seismic detectors, the *Messenger* spacecraft has confirmed the *Mariner* mission's determination of Mercury's high density, which indicates an iron-rich interior with only a thin rock (silicate) mantle. In addition, *Messenger* has returned data indicating a stronger-than-expected magnetic field, which has been interpreted to mean that a small portion of Mercury's core is still molten. Why Mercury is so relatively rich in iron is unclear.

Mercury's Rotation

Mercury spins very slowly. Its rotation period is 58.646 Earth days, exactly two-thirds its orbital period around the Sun. This means that it spins exactly three times for each two trips it makes around the Sun. This proportion is not just coincidence. The Sun's gravity exerts a force on Mercury, which tends to twist the planet and make it rotate with the same period in which it orbits the Sun. Mercury's orbit, however, is very elliptical, so its orbital speed changes as it moves around the Sun. Because of that changing speed, the Sun cannot lock Mercury into a perfect match between its spin and orbital motion: The closest to a match that it can achieve is to make three spins for each two orbits. A model for the formation and early evolution of Mercury that explains the planet's compositional and orbital characteristics involves collision with an enormous planetesimal. The collision might have blasted off some of a previously thick crust and might have knocked Mercury into its inclined and elliptical orbit.

Venus

Of all the planets, Venus is most like the Earth in diameter and mass. We might therefore expect it to be like the Earth in other ways. However, Venus and the Earth have radically different surfaces and atmospheres.

The Venusian Atmosphere

The atmosphere of Venus is mainly (96%) carbon dioxide, as deduced from analysis of reflected sunlight and from measurements with space probes. These methods also show that, in addition to carbon dioxide, Venus's atmosphere contains about 3.5% nitrogen and very small amounts of water vapor and other gases. (For comparison, Earth's atmosphere is about 78% nitrogen and 21% oxygen, with CO_2 being a mere 0.04%.)

Dense clouds perpetually cloak Venus and prevent us from ever seeing its surface directly. These clouds are composed of sulfuric acid droplets that formed when sulfur compounds–perhaps ejected from volcanoes–combined with the traces of water in the atmosphere. Robotic spacecraft that have landed on Venus show that below the clouds, the Venusian atmosphere is relatively clear, and some sunlight (about 10%) penetrates to the surface. The light is tinged orange, however, because the blue colors are absorbed in the deep cloud layer.

Venus's atmosphere is extremely dense. It exerts a pressure roughly ninety times that of the Earth. Its lower atmosphere is also extremely hot: the surface temperature is more than 480°C (about 900°F)–hotter than the daylit surface of Mercury! What makes the atmosphere of a planet so similar to the Earth in size and only slightly nearer the Sun so very hot? The answer is the greenhouse effect.

Venus's Surface Temperature and the Greenhouse Effect

A planet's surface temperature is set by a balance between the energy it receives from the Sun and the energy it re-radiates back into space. Certain gases, such as carbon dioxide, freely transmit visible light but strongly absorb infrared (heat) radiation. Thus, if a planet's atmosphere contains carbon dioxide (or other gases such as water or methane), sunlight can reach the surface to warm the planet, but the radiation of that heat back into space is impeded, which makes the planet warmer than it would be if it had no atmosphere. This phenomenon, whereby an atmosphere warms a planet's surface, is called the **greenhouse effect**. Here on Earth, water vapor and, to a lesser degree, carbon dioxide create a weak greenhouse effect. Water vapor content in Earth's atmosphere is highly variable, but it reaches a maximum of only about 4%, while the carbon dioxide content is only about 0.04% (400 parts per million). Venus's atmosphere, however, has about 300,000 times more carbon dioxide than Earth, and so its greenhouse effect is much stronger.

The Surface of Venus

Because the surface of Venus is hidden by its clouds (see figure 23.1), scientists cannot see it directly. However, just as aircraft

Congealed lava domes at the eastern edge of the Alpha Regio Highlands

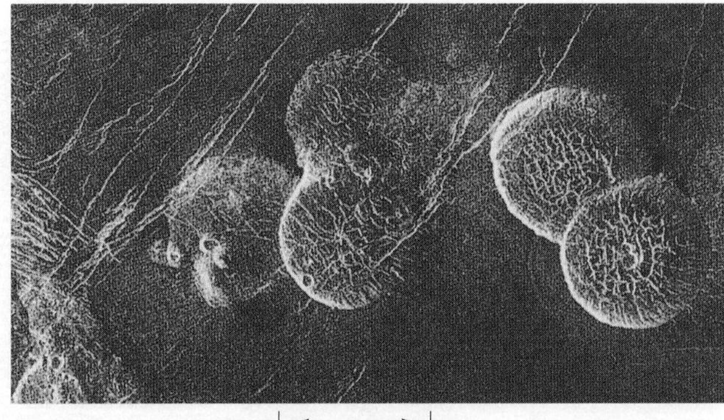

20 kilometers (about 12 miles)

Maat Mons volcano (vertical scale exaggerated)

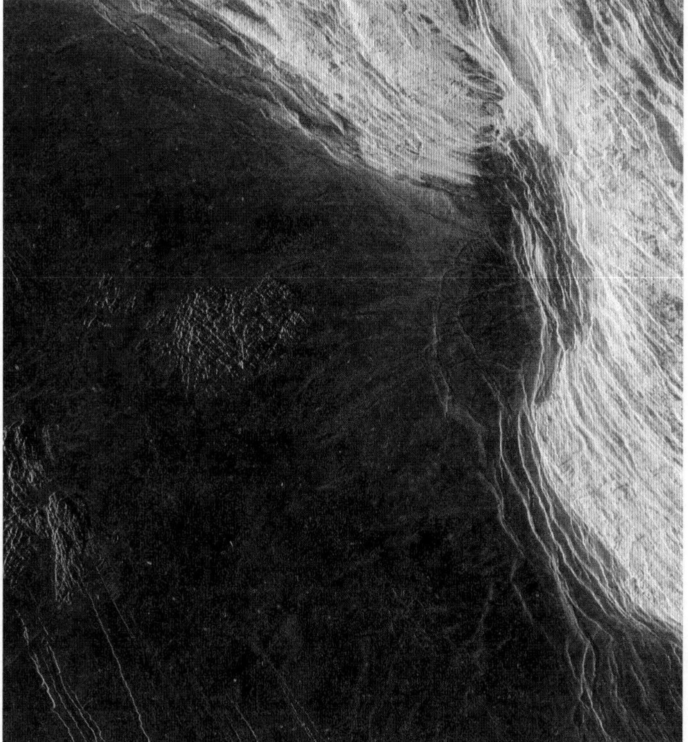

Craters in the Lavinia region

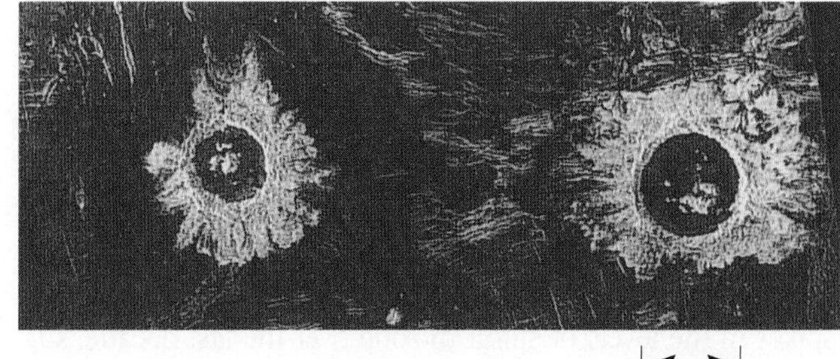

Approx. 50 kilometers (about 31 miles)

Fractured plains in the Lakshmi region

Approx. 37 kilometers (about 23 miles)

FIGURE 23.15

Venus surface features imaged by *Magellan* radar. The orange color has been added to match the color of the landscape observed by the Russian *Venera* landers. *Source: JPL/NASA*

pilots use radar to penetrate terrestrial clouds and fog to show a runway, astronomers use radar to penetrate the Venusian clouds and map the planet's surface. Such maps show that Venus is generally less mountainous and rugged than Earth. Most of its surface is low, gently rolling plains, but two major highland regions, Ishtar and Aphrodite, rise above the lowlands to form continent-size landmasses. Ishtar is largely composed of mountain belts, the highest part of which, Maxwell Montes, rises more than 11 kilometers above the average level of the planet. That's a little less than the height that Mount Everest is above the ocean floor here on Earth. The parallel ridges, troughs, and fractures in these mountain belts indicate they were formed by tectonic processes resulting from both compressional and tensional forces. However, structures indicative of rigid plate tectonics, as found on Earth, are missing.

Venus is so similar in diameter and mass to the Earth that planetary geologists expected to see landforms there similar to those on the Earth. Yet Venus has a surface almost totally unlike our planet's. Although Venus has some folded mountains, volcanic landforms dominate. Surface features include peaks with immense lava flows, "blisters" of uplifted rock, grids of long, narrow faults, and continent-scale uplands bounded by mountain belts (figure 23.15). All of these features indicate a young and active surface. From the small number of impact craters on Venus, planetary geologists deduce that virtually all of Venus's original old surface has been covered over by volcanic activity. (Small numbers of craters indicate a young surface just as few potholes in a road indicate a recent paving job.) The surface we see is probably at most half a billion years old, significantly younger than much of Earth's continental regions. Whether Venus's volcanoes are still active is unknown, but ongoing analyses of the atmospheric composition by the European Space Agency's Venus Express show spikes in the levels of sulfur dioxide over the last decade, suggesting that volcanoes *could* be still erupting.

The numerous volcanic features on Venus suggest that heat flows less uniformly within Venus than within the Earth. Although a couple of dozen locations on Earth (Yellowstone National Park, for example) are heated by "plumes" of rising hot rock, such plumes seem far more common on Venus. We may be viewing on our sister planet what Earth looked like as its lithosphere began to form and before smooth heat flows were established. Why do Venus and Earth have such different surface features and subsurface activity? One set of explanations is based on differences in the amount of water in the rocks of these two worlds, and surface temperature may also play a role.

Water trapped in rock lowers the rock's melting temperature and makes a magma less viscous than magma with little or no water. These effects in turn lead to differences in the thickness of a planet's lithosphere. In particular, on a planet whose rocks are rich with water, the molten rock flows more easily, leading to a thinner lithosphere than if the rocks contained little water. Other things being equal, we therefore expect wet Earth to have a thinner lithosphere than dry Venus. This thinness of Earth's outer layers allows a steady loss of heat, so our planet's interior cools steadily. Moreover, a thin lithosphere breaks easily into small plates. This breaking allows crustal motion and activity to occur more or less continuously and at many places on the surface, cycling water back into Earth's mantle and helping convection to continue. On Venus, lavas may not have cooled quickly against a hot atmosphere, and the crust would not become brittle enough to break. Compression due to downward motion of cooler mantle may have pushed such a hot lithosphere into folded mountain ranges, rather than subducting any water-bearing rocks back into the mantle. The thicker lithosphere would hold heat in, keeping the interior hot and limiting the crust's breakup.

Ultimately, however, the trapped heat must escape, and it is here that the hypotheses differ strongly. According to one hypothesis, at points where the hot, rising material reaches the crust, the surface bulges upward and weakens, and volcanoes may form. Where cooler material sinks, the thick crust crumples into a continent-like region. This creates a planet with few and small continents and whose surface is tectonically active in a manner that is different from Earth's, and volcanically active in only isolated spots at any given point in time.

According to another hypothesis, the trapped heat gradually melts the bottom of the thick crust, thinning it and allowing it to break up. This may happen over widespread areas, flooding almost the entire planetary surface with lava in a brief time. The heat then rapidly escapes to space, the interior cools, and the crust again thickens. This process could occur cyclically.

Although most of what planetary scientists know of the Venusian surface comes from radar maps, several Russian *Venera* spacecraft have landed there and transmitted pictures to Earth (figure 23.16). These pictures show a barren surface covered with flat, broken rocks and lit by the pale, orange glow of sunlight diffusing through the deep clouds. The robotic spacecraft have also sampled the rocks, showing them to be volcanic (primarily basaltic) in nature.

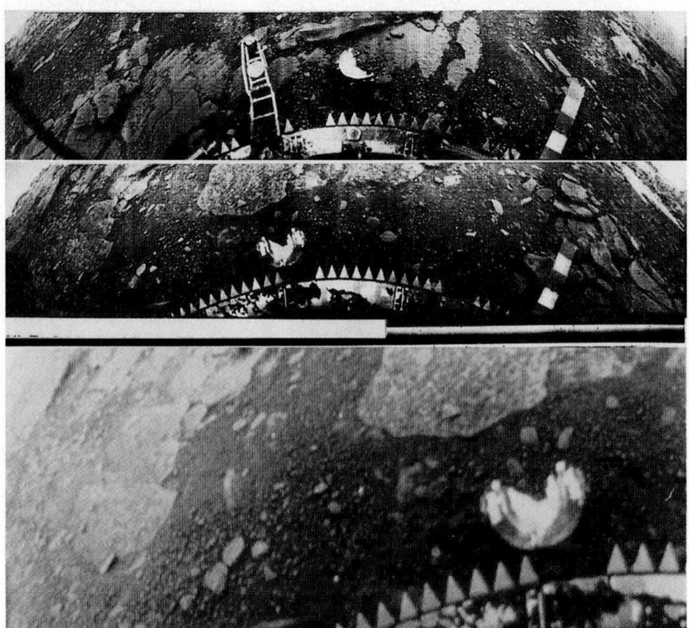

FIGURE 23.16

Images of the surface of Venus returned by the Soviet *Venera* 13 lander in 1982. The upper two images show panoramic views of flat surface rocks, soil, and the sampling arm and ejected camera cover fragments (semicircular objects in each frame). The bottom image is an approximately true color image of an area within the second panorama.
Source: NSSDC/NASA/the Soviet Venera 13 Mission

The Interior of Venus

The deep interior of Venus is probably like the Earth's, an iron core and silicate rock mantle. However, astronomers have no seismic information to confirm this conjecture and must rely on deductions from its gravity and density, which are similar to the Earth's. However, Venus's lack of a global magnetic field suggests significant differences between the interior dynamics of these two sister planets, perhaps largely as a result of Venus's knuckle-ball-like slow rotation compared with Earth's.

The next planet out from the Sun after Venus is our own Earth, the subject of the earlier chapters of this book. Let us, therefore, continue our journey to the next rocky planet, Mars.

Mars

Mars is named for the Roman god of war, presumably because of its distinctly reddish-orange color.[3] Although Martian winds sweep dust and patchy clouds of ice crystals through its sky, its atmosphere is generally clear enough for astronomers to view its surface directly. Such views from Earth and from spacecraft orbiting Mars show a world of mostly familiar features.

[3]Mars' reddish-orange color comes from the abundance of oxidized iron on its surface, as occurs in deserts on Earth.

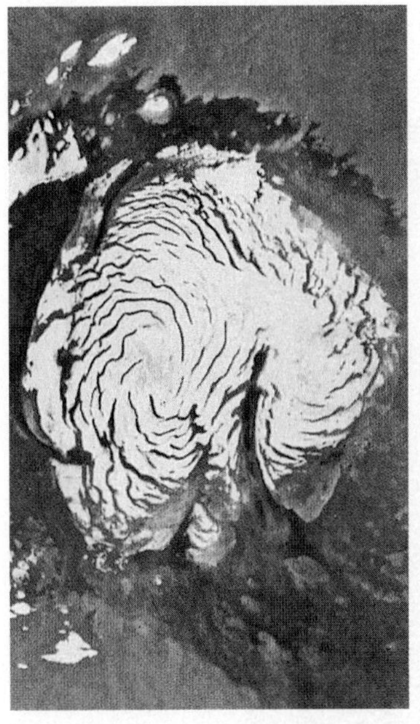
A

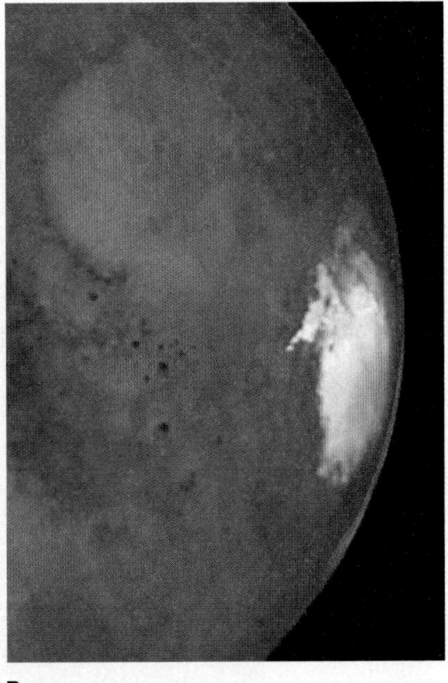
B

C

FIGURE 23.17

(*A*) The north Martian polar cap and (*B*) the south Martian polar cap. (*C*) Layers of water-ice in the north Martian polar cap.
(A) Source: NSASA/JPL/USGS; (B) ©NASA, J. Bell (Cornell U.) and M. Wolff (SSI); (C) Source: JPL/NASA-Caltech/University of Arizona

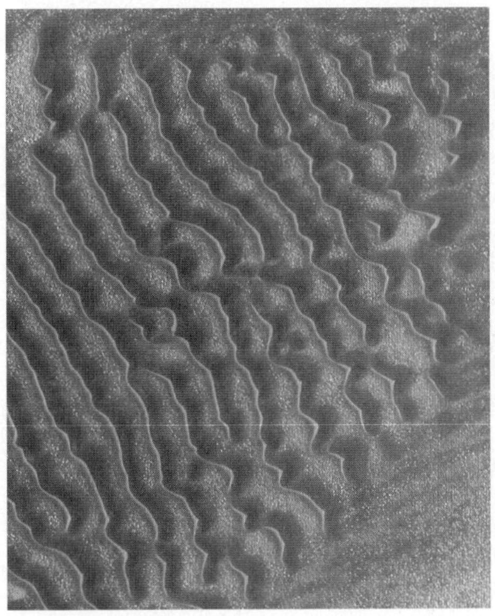
A

B

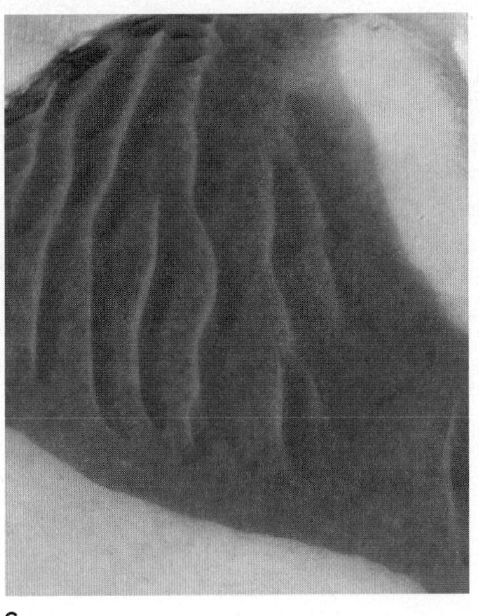
C

FIGURE 23.18

(*A*) Transverse and barchanoid dunes in a crater in the southern hemisphere of Mars. (*B*) Barchan dunes near the edge of Mars' north polar cap. (*C*) Small field of transverse dunes in the southern hemisphere of Mars. Note that, unlike on Earth where most sand dunes are light colored, dunes on Mars are typically dark. This is due to different minerals present in the dunes—primarily light quartz on Earth versus darker ferromagnesian minerals on Mars.
©NASA/JPL/Malin Space Science Systems

Perhaps the most obvious of Mars' surface features are its polar caps (figure 23.17*A*,*B*). These frigid regions are not just frozen water, as on Earth. The south polar cap is frozen carbon dioxide–dry ice–and the north cap has a surface layer of frozen CO_2 over deep layers of ordinary water-ice. These numerous separate layers (easily seen in figure 23.17*C*) indicate that the Martian climate changes cyclically.

Many locations on Mars, particularly around the northern polar area and in craters in the southern hemisphere, exhibit dunes blown by the Martian winds into parallel ridges and crescentic barchans, as illustrated in figure 23.18. At mid-latitudes, a huge upland called the Tharsis bulge is dotted with volcanic peaks. One of these peaks, Olympus Mons (figure 23.19), rises about 25 kilometers above its surroundings, two-and-a-half times

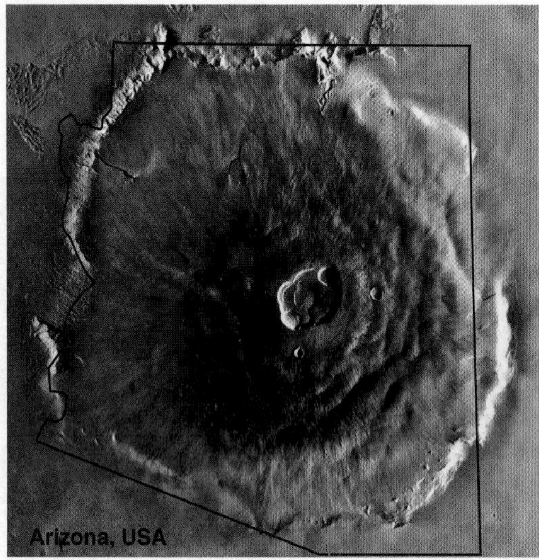

FIGURE 23.19

Olympus Mons, the largest known volcano (probably inactive) in the solar system, is approximately the size of Arizona.
Source: JPL/NASA

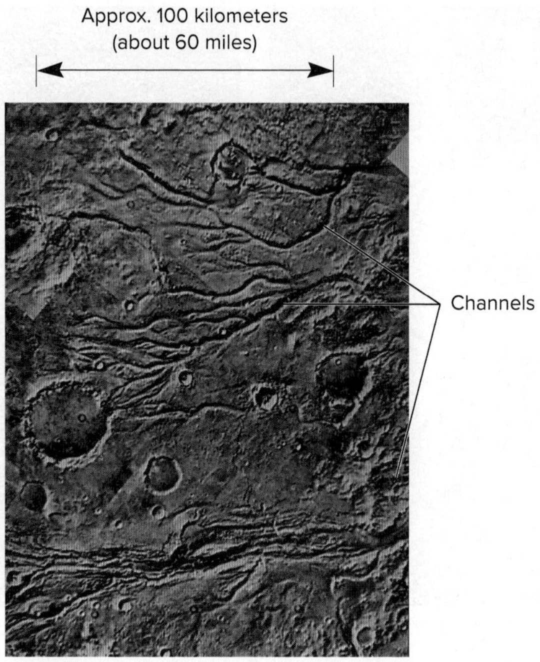

FIGURE 23.21

Picture of channels probably carved by running water on Mars.
Source: NASA

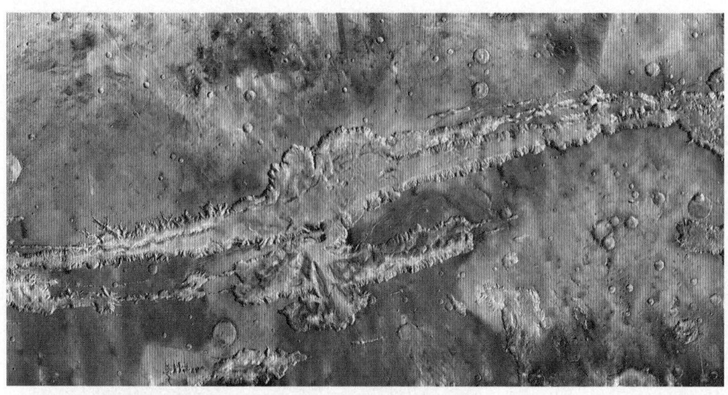

FIGURE 23.20

Colored mosaic of infrared images of Valles Marineris, the Grand Canyon of Mars. This enormous gash in Mars' crust may be a rift that began to split apart the Martian crust but failed to open farther. The canyon is more than 4,000 kilometers long. Were it on Earth, it would stretch from Los Angeles to New York City. The canyon was named for the *Mariner* spacecraft, which sent the first pictures of it back to Earth.
Source: JPL/NASA/Arizona State University

the height of Earth's highest peaks, and is the highest known mountain in the solar system. The base of this immense peak spans a distance (about 350 km) slightly greater than the width of Arizona.

Some planetary geologists think that the same event that created the Tharsis bulge may also have created the gigantic Valles Marineris (figure 23.20), which lies to the southeast, near the Martian equator. This enormous canyon is 4,000 kilometers long, 100 kilometers wide, and 10 kilometers deep. It dwarfs Earth's Grand Canyon and would span the continental United States. Some planetary geologists think it formed as the Tharsis region swelled, stretching and cracking the crust. Others believe that this vast chasm is evidence for plate-tectonic activity, and that the Martian crust began to split, but the motion ceased as the planet aged and cooled. In either case, unlike most canyons on Earth, this immense canyon system is primarily the result of tectonic activity rather than erosional forces.

Perhaps the most surprising features revealed by spacecraft are the huge channels and what look like dry riverbeds, such as those seen in figure 23.21. Planetary geologists infer from these features that liquid water once flowed on Mars, even though no surface liquid is present now. In fact, many planetary scientists now believe that huge lakes or small oceans once existed on Mars. The strongest evidence for ancient bodies of water is layered rocks, some of which have been recently confirmed by surface rovers to have formed in the presence of standing and running water (figure 23.22), within many craters and basins. Moreover, some of these features are cut by narrow canyons (figure 23.23) that breach their rims and appear to "drain" into lowland areas. But the atmospheric pressure on Mars now is so low that liquid water is not stable at the surface. However, recently acquired images suggest it may even today flow across the surface briefly (perhaps beneath blankets of water frost) before boiling off into the thin atmosphere (figure 23.24). We thus are faced with at least two puzzles: where has the water gone, and what has led to the change of conditions that makes liquid water no longer stable there? A first step toward solving these puzzles lies in examining the Martian atmosphere.

The Martian Atmosphere

Clouds and wind-blown dust (figure 23.25) are visible evidence that Mars has an atmosphere. Analysis of sunlight reflected from

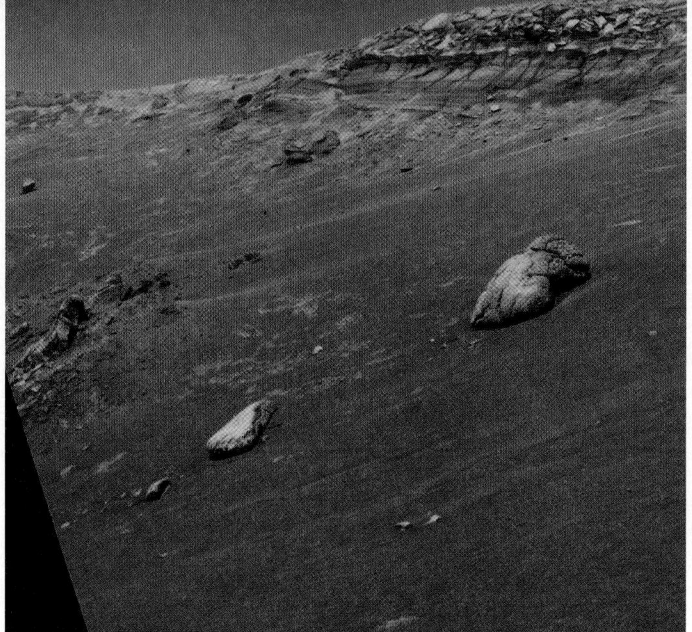

A

B

FIGURE 23.22

(*A*) This approximate true-color image taken by NASA's Mars *Exploration Rover Opportunity* (October 2004) shows an unusual, lumpy rock informally named "Wopmay" on the lower slopes of Endurance Crater. The finely layered rocks in the 5-meter cliff in the background of this image appear to represent sediments either laid down in, or later saturated with, liquid water. (*B*) This image, taken by NASA's Mars rover *Curiosity* in December 2012, shows a layered outcrop called "Shaler" in Mars' Gale Crater. The cross-bedding visible in the outcrop is indicative of sediment transport by a stream.
(A) Source: JPL/NASA/Cornell; (B) Source: JPL/NASA-Caltech/Malin Space Science Systems

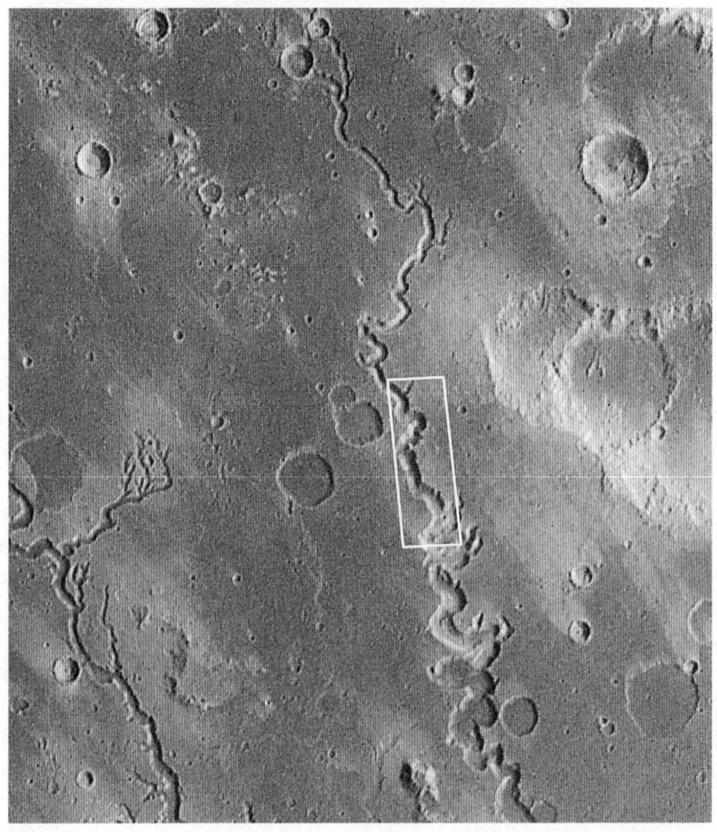

A

134 kilometers
(83 miles)

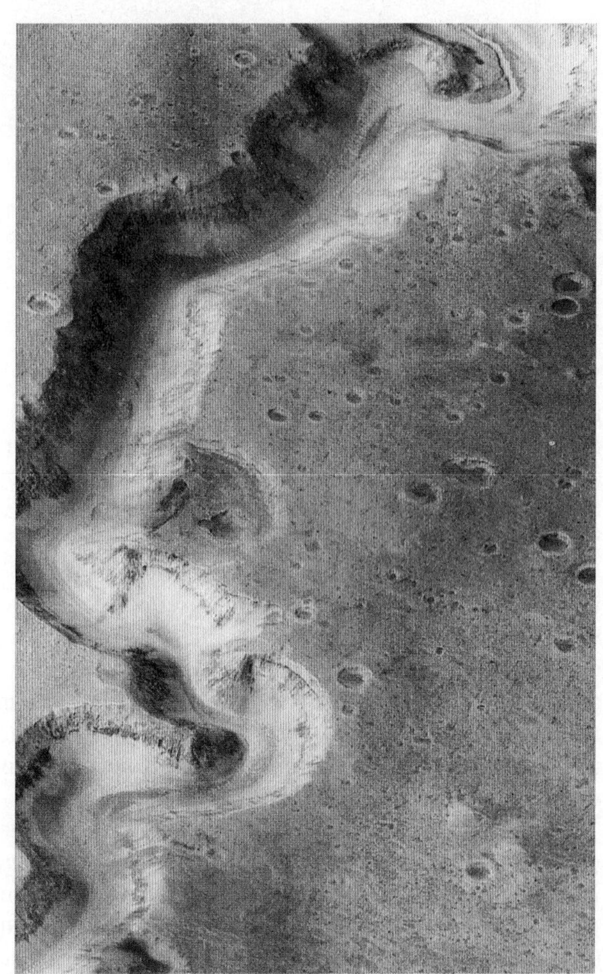

B

9.8 kilometers
(6.1 miles)

FIGURE 23.23

Martian valley systems cutting through the cratered plains of Mars. (*A*) *Viking* orbiter image showing valleys that drained water toward the northern Martian lowlands. (*B*) *Mars Global Surveyor* image showing close-up view of part of one of these valleys, outlined by the white box in (*A*).
©NASA/JPL/Malin Space Science Systems

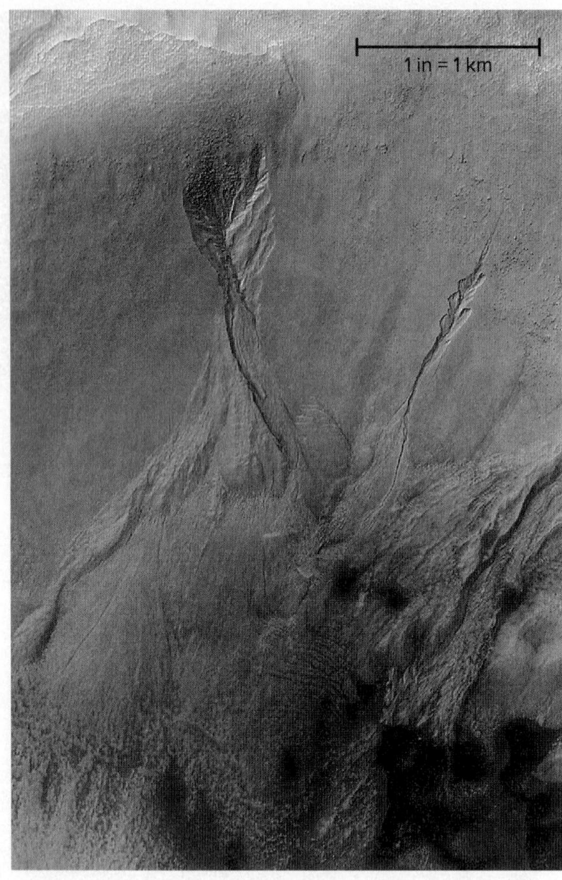

FIGURE 23.24

Mars Global Surveyor image of recently formed gullies in a crater wall on Mars. These gullies appear to have been formed by the transient flow of liquid water on the surface. Their downslope deposits overlap dark (dust-free) sand dune deposits at bottom right, indicating a young age for the gullies.
©NASA/JPL/Malin Space Science Systems

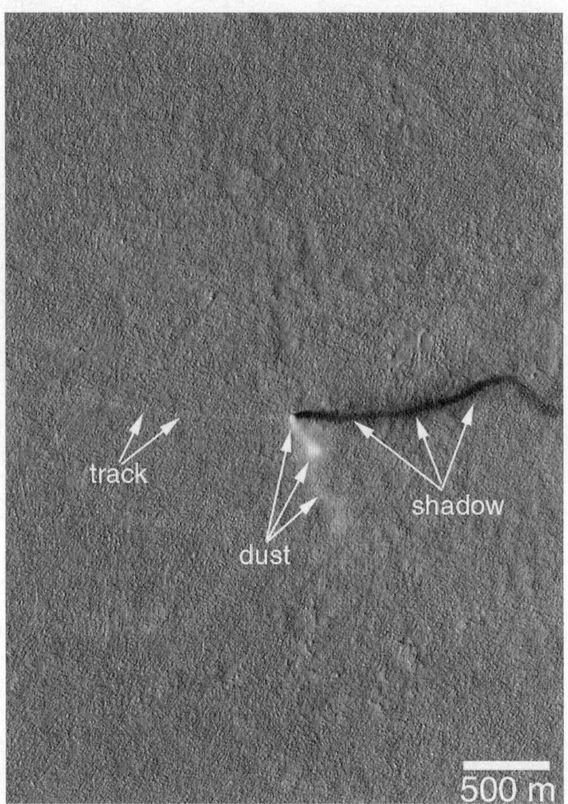

FIGURE 23.25

Mars Global Surveyor image of a dust devil on the surface of Mars.
©NASA/JPL/Malin Space Science Systems

the planet shows that this atmosphere is mostly (95%) carbon dioxide with small amounts (3%) of nitrogen and traces of oxygen and water. The atmosphere is very tenuous–only about 1% the density of Earth's–and its composition and low density have been verified by spacecraft that have landed on Mars.

The Martian atmosphere is so tenuous that the carbon dioxide creates only a very weak greenhouse effect, which, combined with Mars' greater distance from the Sun, makes the planet very cold. Temperatures at noon at the equator may reach a little above the freezing point of water, but at night they plummet to far below zero. Thus, although water exists on Mars, most of it is frozen, locked up either below the surface in the form of permafrost or in the polar caps as solid water-ice.

Clouds of dry ice (frozen CO_2) and water-ice crystals (H_2O) drift through the atmosphere carried by the Martian winds. The Martian winds are generally gentle, but seasonally and near the poles, they become gales, which sometimes pick up large amounts of dust from the surface. The resulting vast dust storms occasionally envelop the planet completely and turn its sky pink.

No rain falls from the Martian sky, despite its clouds, because the atmosphere is too cold and contains too little water. Mars has not always been so dry, as the numerous channels in its highlands show. The existence of such channels carved by liquid water is therefore strong evidence that in its past, Mars was warmer and had a thicker atmosphere. But that milder climate must have ended billions of years ago. From the large number of craters on Mars, planetary scientists deduce that its surface has not been significantly eroded by rain or flowing water for about 3 billion years. Where, then, has its water gone?

Some water is buried below the Martian surface as ice. If the Martian climate was once warmer and then cooled drastically, water would condense from its atmosphere and freeze, forming sheets of surface ice. Wind might then bury this ice under protective layers of dust, as happens in polar and high mountain regions of Earth. In fact, recent observations suggest widespread deposits of such dust-covered ice, and figure 23.26 shows indirect evidence of it. In this image, you can see an area where planetary geologists think that subsurface ice has melted, perhaps from the heating associated with an impact, and drained away in a catastrophic flood. The removal of the ice has then caused the ground to collapse and leave the lumpy, fractured terrain visible at the broad end of the channel. The released meltwater then flowed downstream, carving the 20-kilometer-wide canyon. But how did Mars' water come to be buried and frozen, and why did the planet cool so?

If Mars had a denser atmosphere in the past, then the greenhouse effect might have made the planet significantly warmer

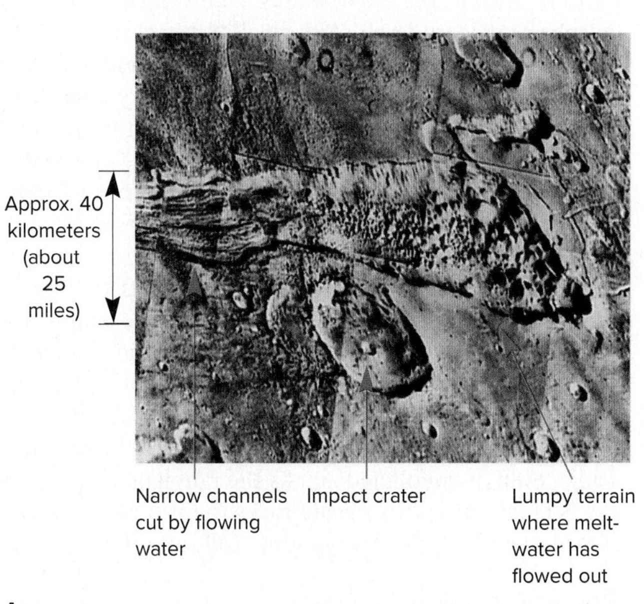

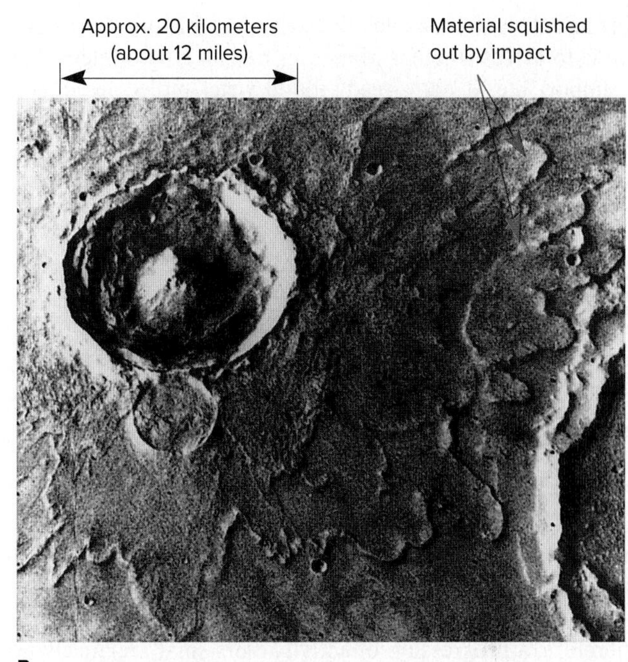

FIGURE 23.26

Evidence for buried ice on Mars. (*A*) A channel cut by water released as subsurface ice melted. Note the chaotic terrain at the broad end of the channel where the surface collapsed as the water drained away. (*B*) Crater with surrounding flow patterns. Heat released by the impact that formed the crater has melted subsurface ice. The thawed material has oozed out.
Source: NASA

than it is now. The loss of such an atmosphere would then weaken the greenhouse effect and permanently cool the planet. One way such a loss could happen is that Mars may have been struck by a huge asteroid whose impact blasted its atmosphere off into space. Such impacts, although rare, do occur, and our own planet may have been struck about 65 million years ago with results nearly as dire, as we shall learn in the section on "Giant Impacts." Alternatively, Mars may have gradually lost its atmospheric gases by solar wind erosion. Mars' current lack of a magnetic field allows solar wind particles to directly impact its atmosphere. Over billions of years, much of Mars' atmosphere could have been lost to space in this way.

The Martian Interior

Planetary scientists believe that the interior of Mars, like that of Earth, consists of a crust, mantle, and iron core. Because Mars is smaller than Earth, however, it would be expected to lose heat more rapidly, and thus, its interior is now probably cooler than Earth's. Unfortunately, scientists have no direct seismic confirmation of Mars' interior structure, although recent analysis of the orbital motion of the *Mars Global Surveyor* satellite suggests that Mars' core may be partly molten.

Mars has no folded mountains, so planetary geologists think that the interior of Mars has cooled and its lithosphere thickened to perhaps twice the thickness of the Earth's lithosphere. As a result, its now-weak interior heat sources can no longer drive crustal motions. Mars' current low level of crustal activity is also demonstrated by its many craters, whose numbers imply that large parts of Mars have been relatively geologically quiet for billions of years.

The Martian Moons

Mars has two tiny moons, Phobos and Deimos. They are named for the minor gods of fear and panic, respectively (in Greek mythology), and the horses of Mars' chariot (in Roman mythology). First discovered in 1877, these irregular and cratered bodies are only about 20 kilometers across and are probably captured asteroids.

Life on Mars?

Scientists have long wondered whether life developed on Mars. To search for life there, the United States landed two *Viking* spacecraft on Mars in 1976. These craft carried instruments to look for signs of carbon chemistry in the soil and metabolic activity in soil samples that were put in a nutrient broth carried on the lander. All tests were either negative or ambiguous.

Then, in 1996, a group of American and English scientists reported possible signs of life in meteorites blasted out of Mars. These objects had traveled through interplanetary space and happened to land on Earth. Over the last few years, however, scientists have shown that ordinary chemical weathering can form structures similar to the supposed Martian "fossils." As a result, most scientists today are unconvinced that any meteorite yet studied shows persuasive evidence of Martian life.

Since August 2013, the *Curiosity* rover has been exploring the Gale Crater, a 3.5 to 3.8 billion-year-old impact crater located just south of the Martian equator. The Gale Crater was selected as the landing site for *Curiosity* because it contains a mountain named Mount Sharp, which is made up of many sedimentary layers, allowing scientists to look back through Mars' history. *Curiosity's* mission is to assess whether Mars ever had an environment that could support life. It has already found water molecules in soil samples and completed chemical analyses of rocks. You can follow *Curiosity's* progress by visiting NASA's Mars Science Laboratory website at http://mars.nasa.gov/msl/.

Why Are the Terrestrial Planets So Different?

We have seen that the four terrestrial planets have little in common apart from being rocky spheres. Astronomers think their great differences arise from their different masses, radii, and distance from the Sun.

Role of Mass and Radius

As discussed earlier, a planet's mass and radius affect its interior temperature and thus its level of tectonic activity, with low-mass, small-radius planets being cooler inside than larger bodies. We see, therefore, a progression of activity from small, relatively inert Mercury, to slightly larger and once-active Mars, to the larger and far more active interiors and surfaces of Venus and Earth. Mercury's surface still bears the craters from its earliest history. In contrast, Earth and Venus retain essentially none of their original crust: Their surfaces have been enormously modified by activity in their interiors over the lifetimes of these planets. Such internal activity in turn affects a planet's atmosphere. Recall that volcanoes probably supplied much of our atmosphere. Thus, small, volcanically inactive[4] Mercury probably never had much atmosphere, and Mars, active once but now quiescent, likewise could create and retain only a thin atmosphere. Moreover, the low mass of these planets produces a weak surface gravity relative to Venus and Earth, which makes retaining an atmosphere difficult. On the other hand, larger and highly active Venus and Earth have extensive atmospheres.

Role of Distance from the Sun

A planet's distance from the Sun, other things being equal, would determine its surface temperature. Thus, we expect that Venus will be warmer than Earth and that Earth will be warmer than Mars. These expected temperature differences are increased, however, by the atmospheres of these bodies, which trap heat to differing extents. Even relatively small differences in temperature can lead to large differences in physical behavior and chemical reactions within an atmosphere. For example, Venus is just enough nearer the Sun than the Earth that, even without a strong greenhouse effect, most of its atmosphere would be so warm that water would have difficulty condensing and turning to rain. Moreover, with the atmosphere being warmer throughout, water vapor can rise to great heights in the Venusian atmosphere, whereas on cooler Earth, water vapor condenses to ice at about 10 kilometers, making our upper atmosphere almost totally devoid of water. The great difference in the water content of the upper atmospheres of Earth and Venus has led to a drastic difference between their atmospheres at lower levels. At high altitudes, ultraviolet light from the Sun breaks apart any water molecules present into their component oxygen and hydrogen atoms. Being very light, the hydrogen atoms so liberated escape into space, while the heavier oxygen atoms remain. Because water can rise to great heights in the Venusian atmosphere, over billions of years it has been almost completely lost from our sister planet. In our atmosphere, water has survived.

Water makes possible chemical reactions that profoundly alter the composition of our atmosphere. For example, CO_2 dissolves in liquid water and creates carbonic acid, which in dilute form we drink as "soda water." In fact, the bubbles in soda water are just carbon dioxide that is coming out of solution. As rain falls through our atmosphere, it picks up CO_2, making it slightly acidic, even in unpolluted air. As the rain falls on the ground, it reacts chemically with silicate rocks to form carbonates (as discussed in chapter 5), locking some CO_2 into the rock.

Role of Biological Processes

Biological processes also remove some CO_2 from the atmosphere. For example, plants use it to make cellulose, the large organic molecule that is their basic structural material. This CO_2 is usually stored for only short periods of time, before decay or burning releases it back into the atmosphere. Sometimes the plant material is buried quickly enough that it is not released, and instead may turn to coal. More permanent removal occurs when rain carrying dissolved CO_2 runs off into the oceans, where sea creatures use it to make shells of calcium carbonate. As these creatures die, their shells sink to the bottom and form sediment that eventually is changed to rock (see chapter 6). Thus, carbon dioxide is swept from our atmosphere and locked up both chemically and biologically in the crust of our planet. With most of the carbon dioxide removed from our atmosphere, mostly nitrogen is left. In fact, our atmosphere contains roughly the same total amount of nitrogen as the atmosphere of Venus.

Our atmosphere is also rich in free oxygen, a gas found in such relative abundance nowhere else in the solar system. Our planet's oxygen is almost entirely the product of green plants breaking down the H_2O molecule during photosynthesis. What evidence supports the idea that liquid water and living things removed most of the early Earth's carbon dioxide? It is the carbonate rock we mentioned earlier. If the carbon dioxide locked chemically in that rock were released back into our atmosphere and the oxygen were removed, our planet's atmosphere would closely resemble that of Venus.

If water is so effective at removing carbon dioxide from our atmosphere, why does *any* CO_2 remain? Although human activities add small amounts of CO_2 by burning wood and fossil fuels, the major contribution is from natural processes. In particular, geologic activity gradually releases CO_2 from rock back into our atmosphere. At plate boundaries, all rock, including sedimentary material, is carried downward into the mantle, where it is melted. Heating breaks down the carbonate rock and releases

[4]Some smooth plains on Mercury may have been produced by volcanism very early in the planet's history

carbon dioxide, which then rises with the magmas to the surface and reenters the atmosphere. A similar process may once have occurred on Mars to remove carbon dioxide from its atmosphere, locking it up in rock there. Mars' present absence of tectonic activity, however, prevents its CO_2 from being recycled. The small mass of Mars is not as capable of holding atmospheric gases, and much of its original atmosphere has escaped into space. Thus, with so little of its original carbon dioxide left, Mars has grown progressively colder. Our Earth, because it is an active recycler, has retained enough CO_2 in its atmosphere to maintain a moderate greenhouse effect, making our planet habitable. Thus, poised between one planet that is too hot and another that is too cold, Earth is fortunate, indeed, to have a relatively stable atmosphere, one factor in the complex and fragile web of an environment on which our existence depends.

Jupiter

To the ancient Romans, Jupiter was the king of the gods. Although they did not know how immense this planet is, they nevertheless chose its name appropriately. Jupiter is the largest planet in both radius and mass. In fact, its mass is larger than that of all other planets in the solar system combined. It is about eleven times the Earth's diameter and more than 300 times its mass. Dense, richly colored parallel bands of clouds cloak the planet, as shown in figure 23.27. Analysis of the sunlight reflected from these clouds shows that Jupiter's atmosphere consists mostly of hydrogen, helium, and hydrogen-rich gases such as methane (CH_4), ammonia (NH_3), and water (H_2O); various forms of sulfur also play a role. These gases were also directly detected in December 1995, when the *Galileo* spacecraft parachuted a probe into Jupiter's atmosphere. The clouds themselves are harder to analyze, but their bright colors may come from complex organic molecules whose composition is still uncertain.

Jupiter's Moons

When the seventeenth-century Italian scientist Galileo first viewed Jupiter with his telescope, he saw four moons orbiting the planet. Today, more than sixty are known, although most of these are too small to be readily seen from Earth and are probably captured asteroids. Jupiter also has rings of much smaller, rocky particles in orbit around its equator.

The four large satellites discovered by Galileo—Io, Europa, Ganymede, and Callisto—are very large: All but Europa are larger than our Moon, and Ganymede (with a diameter bigger than Mercury) is the largest moon in the solar system. The large moons all orbit approximately in Jupiter's equatorial plane, forming a flattened disk, rather like a miniature solar system, suggesting they condensed in orbit around Jupiter. Many of the smaller moons move along highly inclined orbits, suggesting they were captured later on.

Io (pronounced *EYE-oh*) is the nearest to Jupiter of the Galilean moons and therefore is subject to a strong distorting force created by Jupiter's gravity, which heats Io by internal friction. (You can see the effect of such distorting on a smaller scale if you repeatedly bend a metal paperclip until it breaks and then touch the freshly broken end: It will feel hot.) As molten matter oozes to the surface, it erupts, creating volcanic plumes and lava flows, as shown in figure 23.28*A*. In fact, Io is the most volcanically active object in our solar system at this time. Io's volcanic gases are very rich in sulfur, which colors its surface with reds, yellows, and whites. The black areas on the surface appear to be basaltic lava flows.

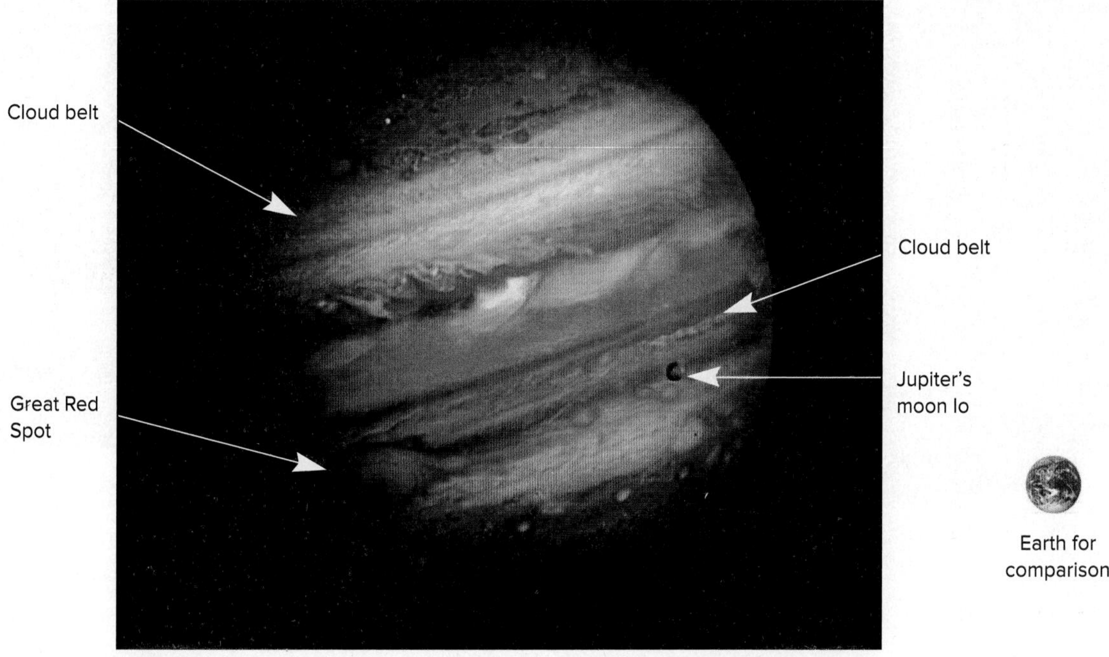

FIGURE 23.27

Jupiter's Red Spot and cloud belts.
Source: NASA

Europa, the smallest of the Galilean moons, looks rather like a cracked egg. Long, thin lines score its surface, as shown in figure 23.28*B*. The bluish-white material is a crust of water ice, while the red material is probably mineral-rich salts that have oozed to the surface (in water) through the cracks. Europa, too, may be heated by Jupiter's gravitational forces, although not as strongly as is Io. That heat, in combination with a small amount from radioactive decay of rocky material in its core, may be sufficient to keep a layer of water melted beneath Europa's icy crust, forming an ocean. Some scientists have suggested that this ocean might harbor primitive life-forms, in a manner similar to the hydrothermal vent communities at the bottom of the oceans of Earth, but no evidence exists so far to support this idea.

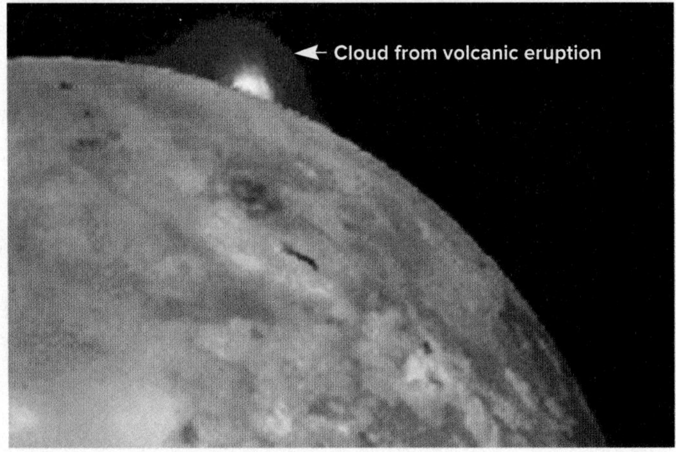

Ganymede and Callisto lie still farther from Jupiter and show proportionately less evidence of internal heating and geologic activity (figure 23.29). Ganymede has numerous faults and fractures slicing across its heavily cratered surface and a magnetic field stronger than an icy moon should have, implying the presence of a salty brine under its icy crust similar to Europa's. The surface of Callisto shows almost nothing but impact scars.

Saturn

Saturn is the second-largest planet and lies about 10 AU from the Sun. Surrounded by its lovely rings, Saturn bears the name of an ancient Roman harvest god. In later mythology, Saturn came to be identified with Cronus (also spelled Kronos), whom the ancient Greeks considered the father of the gods.

Saturn's diameter is about nine times larger than Earth's. Saturn's mass is about ninety-five times that of the Earth, and its average density is very low—only 0.7 grams per cubic centimeter. Such a low density suggests that Saturn, like Jupiter, is composed mostly of hydrogen and hydrogen-rich compounds.

Saturn's Rings

Saturn's rings were first seen by Galileo. Through his small, primitive telescope, however, they looked like "handles" on each side of the planet, and it was not until 1659 that Christiaan Huygens, a Dutch scientist, observed that the rings were detached from Saturn and encircled it.

The rings are very wide but astonishingly thin. The main band is 70,000 kilometers across and extends to a little more

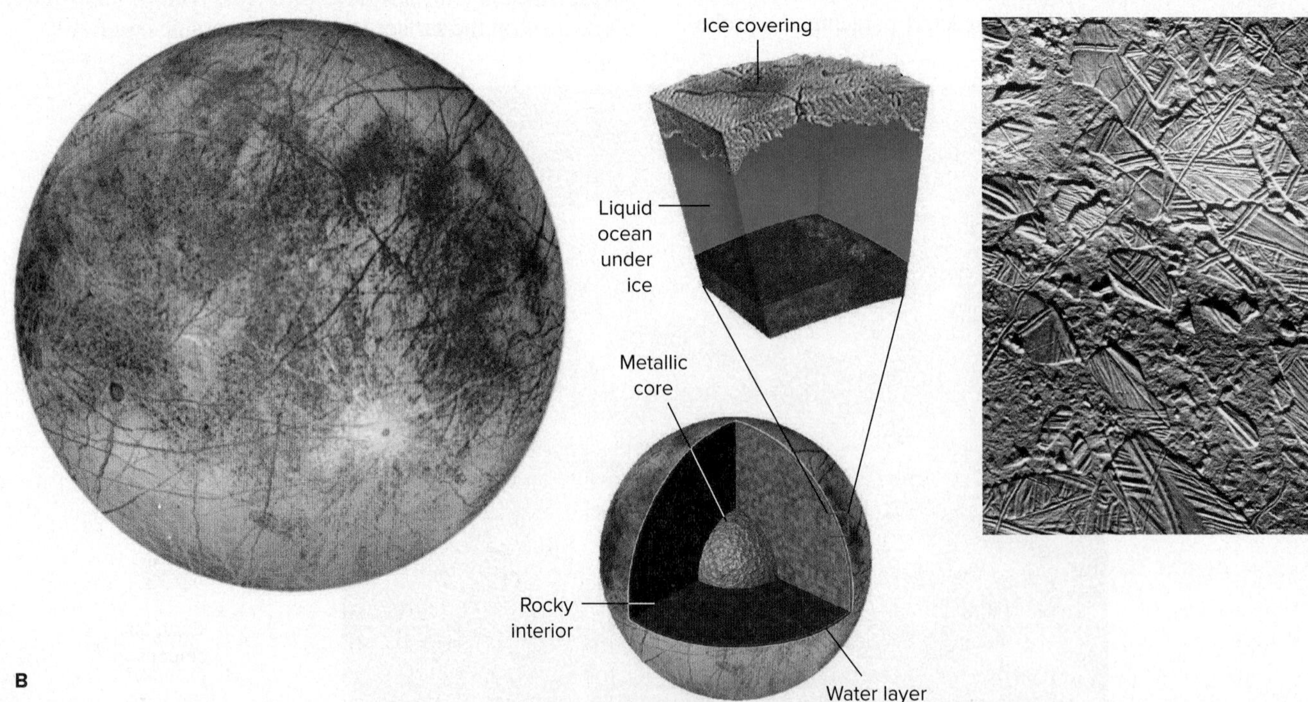

FIGURE 23.28

(*A*) Jupiter's moon Io during an eruption of one of its volcanoes. (*B*) Jupiter's moon Europa as imaged by the *Voyager* and *Galileo* spacecrafts; a model of Europa's interior. (Note possible liquid water ocean below icy crust and close-up of ice chunks on Europa's surface.) All photos are color enhanced to bring out details.
(A) Source: JPL/NASA/USGS; (B-1) Source: JPL/NASA/DLR; (B-3) Source: JPL/NASA/University of Arizona

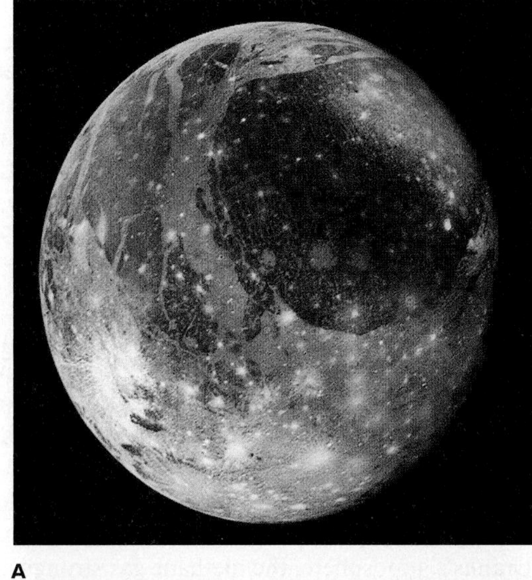

A

B

FIGURE 23.29

(*A*) Global color view of Ganymede, Jupiter's (and the solar system's) largest moon. Note the many linear features, which are sets of tectonic fractures, amongst the bright, circular impact craters. (*B*) Global color view of Callisto, Jupiter's second-largest moon. Note how its ancient surface is dominated by almost innumerable impact craters.
©Calvin J. Hamilton

than twice the planet's radius. Yet despite the rings' immense breadth, they are less than 100 meters thick–thin enough to allow stars to be seen through them.

The rings are not solid structures but a swarm of small, solid chunks of mostly water-ice and bits of rock in individual orbits. These pieces are relatively small, averaging only a few centimeters to a few meters across, as deduced from radar signals bounced off them. What formed the rings? Present thinking is that some are debris from collisions between tiny moons orbiting Saturn or between a moonlet and a comet that strayed too close to the giant planet, but much of the ice is now thought to be derived from a kind of volcanic activity occurring on the larger moons.

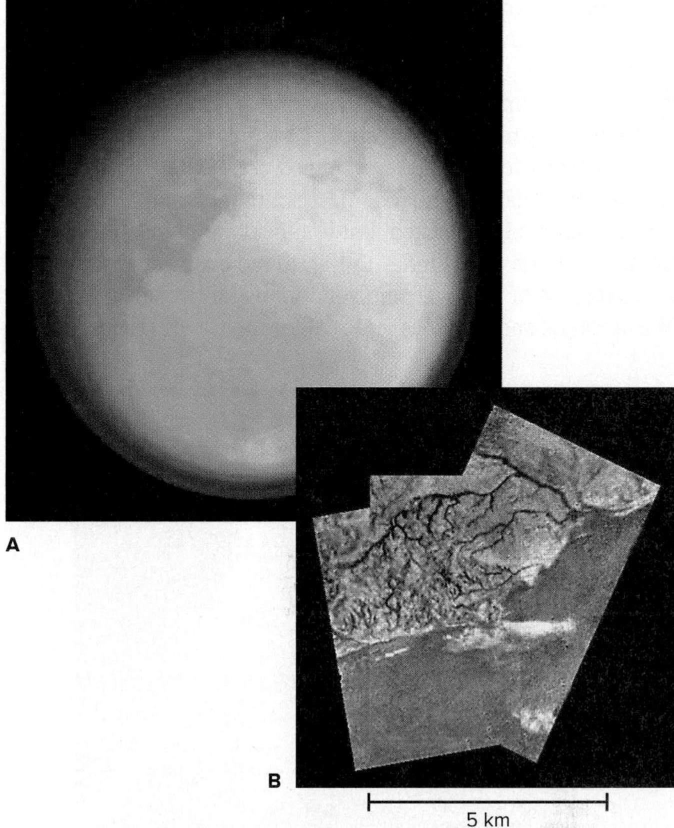

FIGURE 23.30

(*A*) The icy surface of Titan, as seen by the *Cassini* spacecraft in October 2004. The bright patch toward the bottom of the image is clouds over the southern polar region. Note the sharp boundaries between light and dark (carbon-rich?) areas on the surface, and the lack of obvious impact craters. (*B*) Close-up of dendritic channel system on Titan, probably carved by liquid methane (mosaic of three images taken by Huygens Probe as it descended to Titan's frozen surface).
(A) Source: JPL/NASA/Space Science Institute; (B) Source: JPL/NASA/ESA/University of Arizona

Saturn's Moons

Saturn has several large moons and about fifty known smaller ones. Its largest moon is Titan, whose diameter is about 5,000 kilometers, making it slightly larger than Mercury. Because Titan is so far from the Sun and is thus very cold, gas molecules that leak from its interior move relatively slowly and are unable to escape Titan's gravitational attraction. Titan thus possesses its own atmosphere, which analysis shows to be mostly nitrogen. Clouds in Titan's atmosphere perpetually hide its surface, which is very cold (−184°C or −300°F). On the basis of chemical calculations, some astronomers argued that Titan's surface may have lakes or oceans of liquid methane or the hydrocarbon ethane (C_2H_6), and "continents" of water-ice, and that ethane "rain" may fall from its clouds. Images of Titan's surface, recently returned by the *Cassini* spacecraft, show sharp boundaries between light and dark terrain, very few impact craters, evidence for erosion by flowing liquid, and, in fact, lakes and some evidence for a kind of icy volcanism, suggesting a relatively young (and apparently geologically active) surface (figure 23.30).

Another of Saturn's many interesting moons is Enceladus. Orbiting near the outermost of Saturn's rings (the E-ring), Enceladus is the most reflective object in the solar system, since its surface is dominated by fresh, clean water-ice. While it had long been speculated that there was a relationship between Enceladus and the E-ring, the orbiting *Cassini* spacecraft captured pictures and compositional data showing a strange kind of "water-volcanism" (cryovolcanism) as plumes of water-ice, vapor, and trace organic compounds were ejected from Enceladus's south polar region. The heat source for this highly active region is still being debated, but it appears to be related to some form of tidal flexing, similar to that found on Jupiter's moon Io. The emission of water-ice and vapor constantly coats the surface of Enceladus and provides material for the E-ring (figure 23.31).

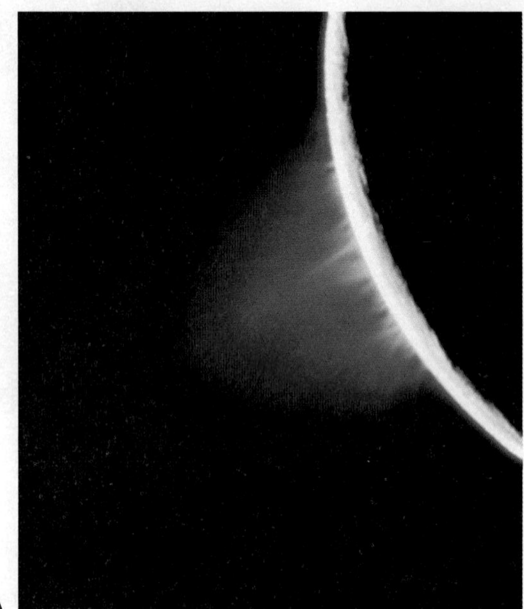
A

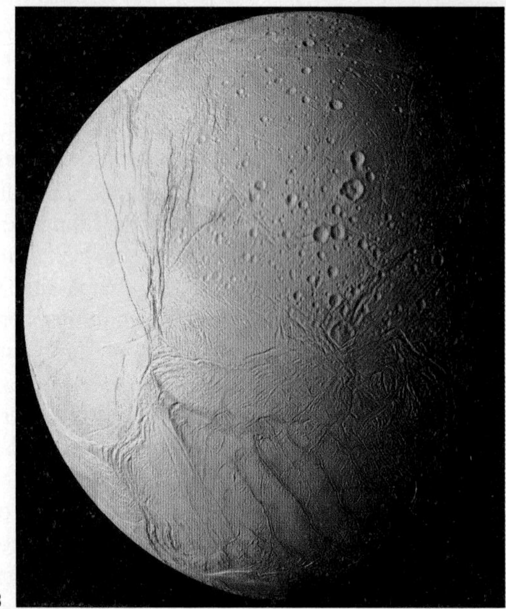
B

FIGURE 23.31

(*A*) Plumes of water vapor and ice particles emanating from the south polar region of Enceladus, as seen by the *Cassini* spacecraft in 2005. The image was taken when the spacecraft was on the side of Enceladus opposite the Sun. (*B*) A full-disk composite image of Enceladus, showing its north pole at the top and the "tiger stripes" of the south polar region at the bottom.
Source: JPL/NASA/Space Science Institute

Uranus

Uranus, although small compared with Jupiter and Saturn, is nonetheless much larger than the Earth. Its diameter is about four times that of the Earth, and its mass is about fifteen Earth masses. Lying more than twice Saturn's distance from the Sun, Uranus is difficult to study from Earth. Even pictures of it taken by the *Voyager 2* spacecraft, which left Earth in 1977 and is the only spacecraft to fly by Uranus and Neptune, show few details.

Uranus was unknown to the ancients, even though it is just visible to the naked eye. It was discovered in 1781 by Sir William Herschel. The planet's name comes from ancient Greek mythology, where Uranus was god of the heavens and the father of Cronus.

Uranus's atmosphere is rich in hydrogen and methane. In fact, methane gives the planet its deep blue color. When sunlight falls on Uranus's atmosphere, the methane gas strongly absorbs the red light. The remaining light, now blue, scatters from cloud particles in the Uranian atmosphere and is reflected into space. Scientists have conducted experiments showing that at high pressure, methane can convert to diamond. It is believed that diamonds may be "raining" down through the atmosphere of Uranus toward its core.

Astronomers rely on theoretical models to understand the interior of Uranus. Such models suggest that the planet has a deep, liquid-water ocean below its methane-rich clouds. During its formation, Uranus must have collected heavy elements such as silicon and iron. These denser materials probably sank to its center (as they did in the other giant planets), where they now form a core of rock and iron a little larger than Earth.

An interesting property is that Uranus's rotation axis is tipped so that its equator is nearly perpendicular to its orbit. Thus, it spins nearly "on its side." Moreover, the orbits of its moons are also similarly tilted, and its magnetic field is far removed from its rotational axis. To explain the tilt and offset magnetic field, some astronomers hypothesize that during its formation, Uranus was struck by an enormous planetesimal whose impact tilted the planet and splashed out material to create its family of moons. It may be in the midst of a magnetic polar reversal, but more data are needed to confirm which of these hypotheses is correct. In addition to its moons (figure 23.32), Uranus is encircled by a set of narrow rings that, like those of Saturn, are composed of myriad small particles.

Neptune

Neptune is the outermost of the large planets, and is a closer twin to Uranus than Earth is to Venus. Neptune is very similar to Uranus in size, with a diameter about 3.9 times that of the Earth and a mass about seventeen times the Earth's. Like Uranus, it is a lovely blue color, and for the same reason: strong absorption of red light by the methane in its atmosphere. Pictures show that cloud bands encircle it, and it even has briefly shown a Great "Dark" Spot, a huge, dark blue atmospheric vortex similar to Jupiter's Great Red Spot.

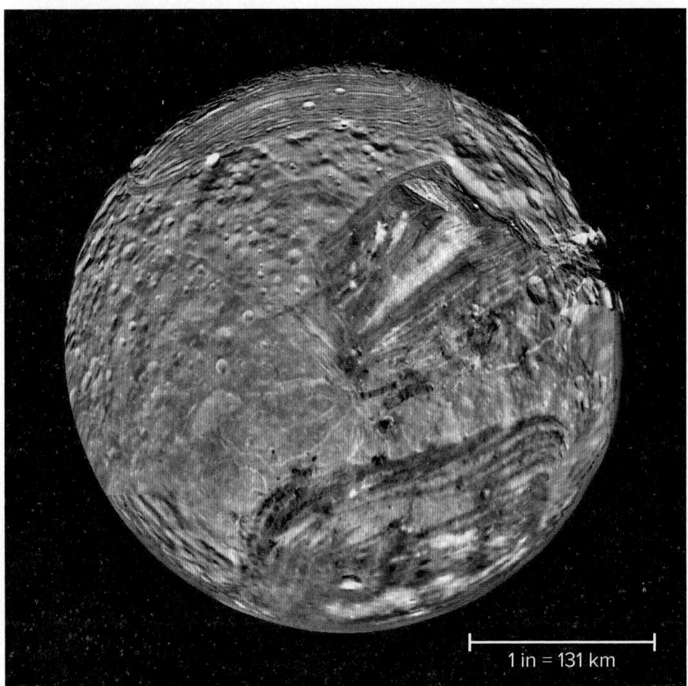

FIGURE 23.32
View of the south polar region of Uranus's moon Miranda. This moon has some of the most bizarre and least understood surface features in our solar system. The patchwork of chevron-shaped terrains includes fault scarps over 10 kilometers high!
Source: JPL/NASA

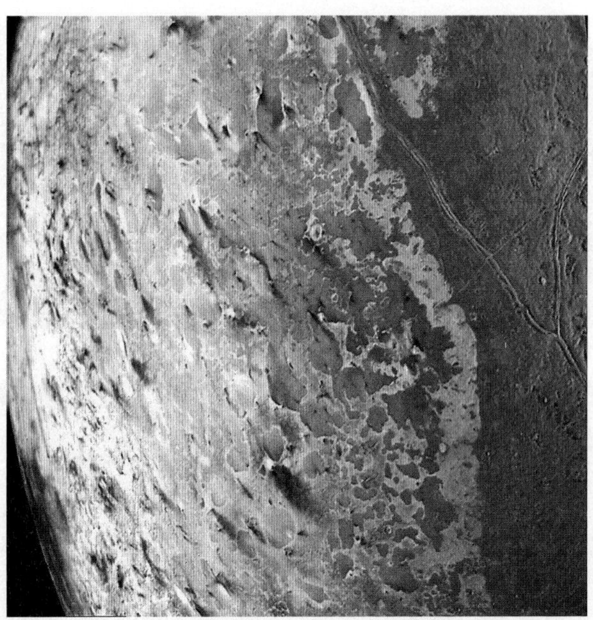

FIGURE 23.33
The strange icy surface of Neptune's largest moon, Triton, shows evidence of icy geysers whose deposits have been blown downwind (dark streaks) by winds in its thin atmosphere.
Source: JPL/NASA

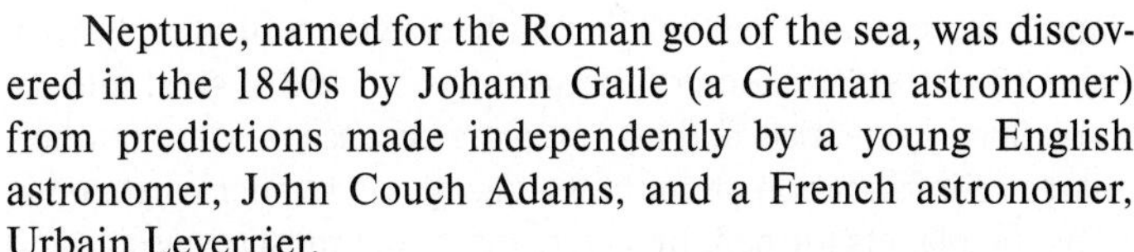

Neptune, named for the Roman god of the sea, was discovered in the 1840s by Johann Galle (a German astronomer) from predictions made independently by a young English astronomer, John Couch Adams, and a French astronomer, Urbain Leverrier.

Neptune's structure is probably similar to Uranus's. That is, the planet is composed mostly of ordinary water (with a small rock and iron core) surrounded by a thin atmosphere rich in hydrogen and hydrogen compounds, such as methane and ammonia.

Neptune, like the other giant planets, has rings, but they are very narrow, more like those of Uranus than those of Saturn. It also has a number of small moons and one immense moon, Triton. Triton orbits "backwards" (counter to Neptune's rotation) compared with the motion of most other satellites, and its orbit is highly tilted with respect to Neptune's equator (most large moons follow orbits approximately parallel to their planet's equator). These orbital peculiarities suggest that Triton is a surviving icy planetesimal or Kuiper Belt object captured from what was originally a more distant orbit around the Sun.

Triton intrigues astronomers for more than its orbital oddity. It is massive enough that its gravity, in combination with its low temperature, allows it to retain a thin but noticeable envelope of gases around it. Triton, along with Saturn's moon Titan, is thus one of two moons in the solar system with an atmosphere. Moreover, it shows signs of eruptions of gases (mostly nitrogen) from its interior (figure 23.33).

23.4 PLUTO AND THE ICE DWARVES

Pluto, named for the Greek and Roman god of the underworld, was the last of the nine planets to be discovered, but it was demoted in 2006 to **dwarf planet** status (a class that includes the largest asteroid, Ceres, and the larger KBOs) by the International Astronomical Union (IAU) due to its similarity in size, orbital characteristics, and composition to the numerous KBOs swarming about in orbits between 30 and 55 AUs from the Sun. It was found in 1930 by the American astronomer, Clyde Tombaugh. Pluto's great distance from the Sun combined with its small size makes it very dim. Even in the largest telescopes on the ground, Pluto looks like a dim star.

Pluto has a large moon, Charon (pronounced *KAIR-en*), named for the mythological boatman who ferries dead souls across the river Styx to the underworld, and two tiny moons (Nix and Hydra) (figure 23.34*A*,*B*). Charon's diameter is approximately one-half Pluto's. From Charon's motion around Pluto, astronomers can calculate Pluto's mass using the law of gravity. They find that Pluto has only about 0.002 times the Earth's mass, making it even smaller than the largest KBO, Eris.

From Pluto's mass and radius, astronomers can deduce its density to be about 2.1 grams per cubic centimeter, a value suggesting that the dwarf planet must be a mix of water, ice, and rock. They know little, however, about its surface. Images of Pluto made with the *Hubble* Space Telescope (figure 23.34*C*) show only that Pluto's polar regions are much brighter than its equator, implying the presence of polar caps there, probably composed of frozen methane, carbon monoxide, or nitrogen.

A

B

C

FIGURE 23.34

Best images from *Hubble* space telescope of (*A*) Pluto and its smaller moon, Charon. (*B*) Recent *Hubble* space telescope image showing two additional tiny moons in the Pluto system—Nix and Hydra, and (*C*) surface brightness patterns on Pluto. *(A) Source: Dr. R. Albrecht, ESA/ESO Space Telescope European Coordinating Facility/NASA; (B) Source: H. Weaver (JHUAPL)/A. Stern (SwRI)/HST Pluto Companion Search Team/ESA/NASA; (C) Source: NASA Jet Propulsion Laboratory (NASA-JPL)*

By analyzing the spectrum of sunlight reflected from Pluto, astronomers have detected that it has a very tenuous atmosphere composed mostly of nitrogen and carbon monoxide, with traces of methane.

Pluto's orbit is peculiar. It crosses Neptune's, but because it is also more highly tilted than the orbits of the other planets, Pluto passes well above or below Neptune when the crossings occur. Its odd orbit in combination with its small mass once led some astronomers to hypothesize that Pluto was originally a satellite of Neptune that escaped and now orbits the Sun independently. Today, however, astronomers think almost the reverse—that Neptune has "captured" Pluto. The reason for this hypothesis is that Pluto's orbital period of 247.7 years is almost exactly one-and-one-half times Neptune's. Thus, Pluto makes two orbits around the Sun for every three made by Neptune. This match of orbital periods gives Neptune's gravitational attraction on Pluto a cumulative effect that has probably "tugged" it into its current orbit. In fact, several dozen other icy KBOs, objects a few hundred kilometers to slightly larger than Pluto in diameter, have been found orbiting at nearly the same distance from the Sun as Pluto.

23.5 MINOR OBJECTS OF THE SOLAR SYSTEM

Orbiting the Sun and scattered throughout the solar system are numerous bodies much smaller than the planets—asteroids and comets. Asteroids are generally rocky objects in the inner solar system. Comets are icy bodies that spend most of their time in the outer solar system. These small members of the Sun's family, remnants from the formation of the solar system, are of great interest to planetary scientists and astronomers because they are our best source of information about how long ago and under what conditions the planets formed. In fact, many (if not most) asteroids and comets may be planetesimals—the solid bodies from which the planets were assembled—that have survived nearly unchanged from the birth of the solar system. Apart from their scientific value, asteroids and comets are fascinating objects both for their beauty and for their potential to endanger life on Earth.

Meteors and Meteorites

If you have spent even an hour looking at the night sky in a dark location away from city lights, you have probably seen a "shooting star," a streak of light that appears in a fraction of a second and as quickly fades. Astronomers call this brief but lovely phenomenon a **meteor**.

A meteor is the glowing trail of hot gas and vaporized debris left by a solid object that is heated by friction as it moves through the Earth's atmosphere. The solid body itself, while in space and before it reaches the atmosphere, is called a *meteoroid.*

Heating of Meteors

Meteors heat up on entering the atmosphere for the same reason a reentering spacecraft does. When an object plunges from space into the upper layers of our atmosphere, it collides with atmospheric molecules and atoms at an initial velocity of 10 to 70 kilometers per second. These collisions convert some of the body's energy of motion (kinetic energy) into heat and vaporize matter from its surface. Most of the heating occurs between about 100 and 50 kilometers altitude, in the outer fringes of the atmosphere. The trail of hot evaporated matter and atmospheric gas emits light, making the glow that we see.

Most meteors that we see last only a few seconds and are made by meteoroids the size of a pea or smaller. These tiny objects are heated so strongly that they completely vaporize. Larger pieces, though heated and partially vaporized, are so drastically slowed by air resistance that they may survive the ordeal and reach the ground. We call these fragments found on the Earth **meteorites**.

Meteorites

Meteorites are divided into three broad categories: stony (composed mainly of silicate minerals), iron (composed mainly of metallic iron-nickel), and stony iron (a mixture of both metallic and silicate material). Ninety-four percent of all meteorites that fall on Earth are stony and of these, 86% are classed as chondrites, named for the small rounded particles known as chondrules that they contain (figure 23.35). Chondrules appear to have originated as molten droplets formed by rapid melting and cooling of rocky material. Where this may have occurred is not known.

Chondrules contain traces of radioactive material, which can be used to measure their age. They are extremely old, 4.56 billion years, and are believed to be among the first solids to have formed in the solar nebula.

Some chondritic meteorites contain small amounts of organic compounds such as amino acids and kerogen and are called carbonaceous chondrites. Amino acids are the same complex molecules used by living things for the construction of their proteins and genetic material. Thus, the presence of amino acids in meteoritic matter indicates that the raw material of life can form in space and that it is likely to have been available right from the start within the solar system. It is believed that most meteors are fragments of asteroids, although very few have been found to be fragments from the Moon or even Mars, probably blasted off the surface by a significant impact event. Iron meteorites may represent the core material of a body that was large enough to undergo core formation. Their iron-nickel composition gives us a model for the composition of Earth's core.

FIGURE 23.35

A chondrite meteorite. Note the many circular chondrules of which it is composed. Such meteorites are made primarily of dust to sand-sized grains of silicate minerals. *Source: JPL/NASA*

Asteroids

Asteroids are small, generally rocky bodies that orbit the Sun. Most lie in the asteroid belt, a region between the orbits of Mars and Jupiter. Asteroids range tremendously in diameter, from Ceres–about 930 kilometers (about the size of Texas) across–down to bodies only a few meters in diameter. Most are irregularly shaped (figure 23.36). Only Ceres and a few dozen other large asteroids are approximately spherical. Because of their large size, their gravitational force is strong enough to compact their material into a sphere. Small bodies with weak gravities remain irregular and are made more so by collisions adding or blasting away pieces. Collisions leave the parent body pitted and irregularly shaped, and the fragments become smaller asteroids in their own right.

Origin of Asteroids

The properties of asteroids that we have discussed (composition, size, and their location between Mars and Jupiter) give us clues to their origin and support the solar nebula hypothesis for

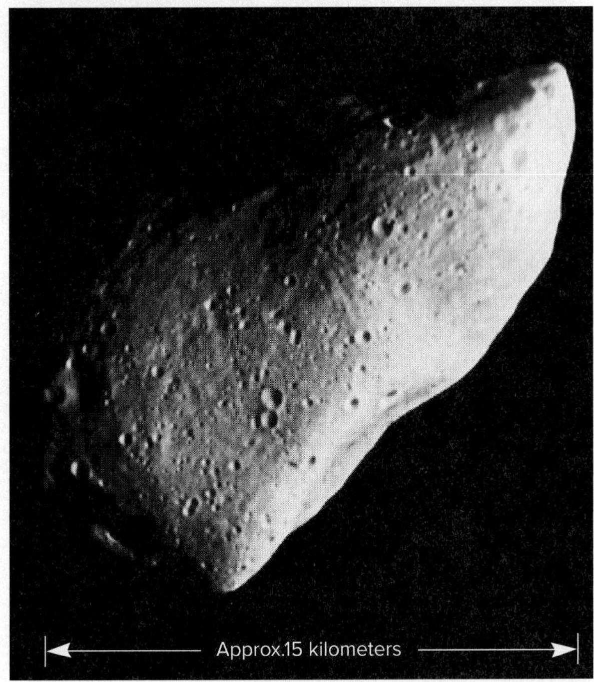

FIGURE 23.36

The asteroid Gaspra, as imaged by the *Galileo* spacecraft. *Source: JPL/NASA* the origin of the solar system. In fact, the asteroids are probably fragments of planetesimals, the bodies from which the planets were built. Although science-fiction movies and TV shows like to show the asteroids as a dense swarm, they are actually very widely separated—typically, thousands of kilometers apart. The ones we see today failed to be incorporated into a planet, probably as the result of their nearness to Jupiter, whose immense gravity disturbed their paths sufficiently to keep them from aggregating into a planet.

Farther from the main belt are the Apollo asteroids, whose orbits carry them into the inner solar system across the Earth's orbit. Although there are about 1,000 such Earth-crossing asteroids larger than 1 kilometer in diameter, the chance of collision in the near future is slim. Nevertheless, on the average, one such body hits the Earth about every 10,000 years.

Comets

A bright comet is a stunning sight, as you can see in figure 23.37*A*. But such sights are now sadly rare for most people because light pollution from our cities drowns the view. Comets have long been held in fear and reverence, and their sudden appearance and equally sudden disappearance after a few days or so have added to their mystery.

Structure of Comets

Comets consist of two main parts, as illustrated in figure 23.37*B*. The largest part is the long tail, a narrow column of dust and gas that may stretch across the inner solar system for as much as 100 million kilometers (nearly an AU!).

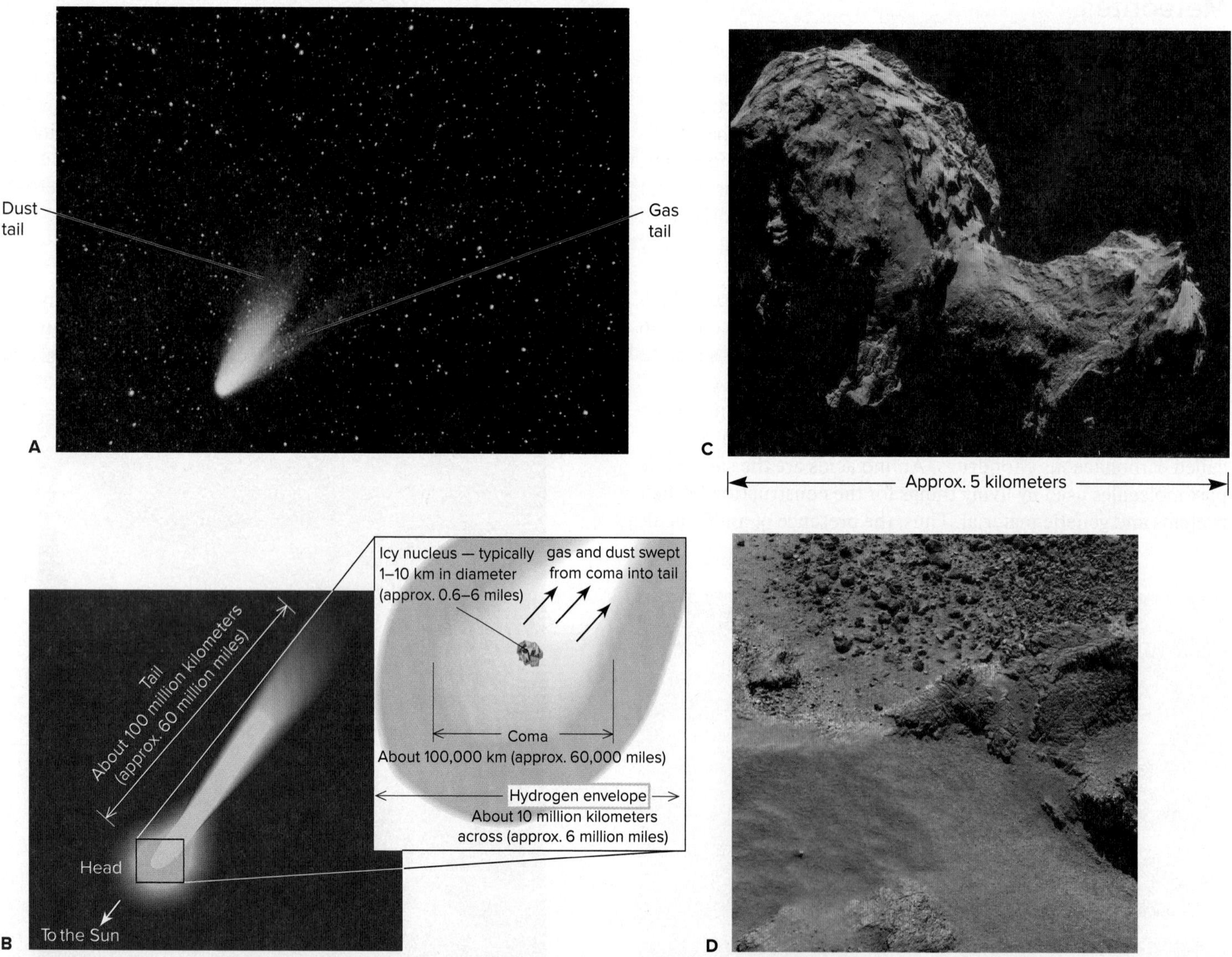

FIGURE 23.37

(*A*) Photograph of Comet Hale-Bopp in the evening sky. Note the blue tail (gas) and the whitish tail (sunlight reflected from tiny dust particles). (*B*) Artist's depiction of the structure of a comet showing the tiny nucleus, surrounding coma, and the long tail. (*C*) Image of the nucleus of comet 67P/Churyumov-Gerasimenko taken by the *Rosetta space probe. (D) The surface of comet 67P/Churyumov-Gerasimenko taken by Rosetta from a distance of 8 km.*

(A) ©Mike Skrutskie/University of Virginia; (C) Source: ESA/Rosetta/NAVCAM; (D) Source: ESA/Rosetta/MPS for OSIRIS Team MPS/UPD/LAM/IAA/SSO/INTA/UPM/DASP/IDA

The tail emerges from a cloud of gas called the *coma,* which may be some 100,000 kilometers in diameter (more than ten times or so the size of the Earth). Despite the great volume of the coma and the tail, these parts of the comet contain very little mass. The gas and dust are extremely tenuous, and so a cubic centimeter of the gas contains only a few thousand atoms and molecules. By terrestrial standards, this would be considered a superb vacuum. This extremely rarified gas is matter that the Sun's heat has sublimated off the icy heart of the comet, its nucleus. Therefore the coma and tail only appear when the comet has come close enough to the Sun for sublimation to occur; once outside the orbit of Jupiter, the tail ceases to exist, and the nucleus is no longer shrouded by the coma.

The comet nucleus is a block of ices and dust that have frozen in the extreme cold of the outer solar system into an irregular mass averaging perhaps a bit less than 10 kilometers in diameter. The nucleus of a comet (figure 23.37*C* and figure 23.37*D*) has been described as a giant "iceberg" or "dirty snowball," and it contains most of the comet's mass. In 2014, the *Rosetta* space probe, launched in 2004 by the European Space Agency, entered into orbit with comet 67P/Churyumov-Gerasimenko. It was the first spacecraft to orbit a comet. Three months later, its lander module *Philae* performed the first successful landing on a comet, although unfortunately, due to where it landed, it was unable to collect solar power and lost contact with *Rosetta* after two days. In September 2016, *Rosetta* was allowed to strike the comet's surface, collecting data as it did so. The mission provided a lot of important data about the comet, much of which confirmed what we already knew about comets. The surface of the comet is covered with a layer of dust as much as 20 centimeters thick. Below that is hard ice mixed with dust. Several organic compounds were detected by instruments on the *Philae* lander, although no link to organisms was found. One important find was that the composition of water vapor from the comet is substantially different from water found on Earth, suggesting that Earth's water did not come from comets like this one.

Origin of Comets

Astronomers think that most comets come from the Oort cloud, the swarm of trillions of icy bodies believed to lie far beyond the Kuiper Belt orbit of Pluto, as we discussed in the section titled "The Solar System." Astronomers think the Oort cloud formed from planetesimals that originally orbited near the giant planets and were tossed into the outer parts of the solar system by the gravitational force of those planets. Some comets also seem to come from the Kuiper Belt, which begins near the orbit of Neptune and appears to extend to about 55 AU from the Sun. Planetary astronomers are greatly interested in this region because it appears to be the birthplace of Pluto, Charon, and Neptune's moon Triton.

Each comet nucleus moves along its own path, and those in the Oort cloud take millions of years to complete an orbit. With orbits so far from the Sun, these icy bodies receive essentially no heat from the Sun, so their gases and ices remain deeply frozen.

These distant objects are invisible to us on Earth, but sometimes a comet's orbit changes, carrying it closer to us and the Sun. Such orbital changes may arise from the chance passage of a star far beyond the outskirts of the solar system during its orbit around the center of the Milky Way galaxy.

As the comet falls inward toward the inner solar system, the Sun's radiation heats it and begins to sublimate the ices. At a distance of about 5 AU from the Sun (Jupiter's orbit), the heat is enough to vaporize the ices, forming gas that escapes to make a coma around the comet nucleus. The escaping gas carries tiny dust grains that were frozen into the nucleus with it. The comet then appears through a telescope as a dim, fuzzy ball. As the comet falls ever nearer the Sun, its gas escapes even faster, but now the Sun begins to exert additional forces on the cometary gas and dust.

Formation of the Comet's Tail

Sunlight striking dust grains imparts a tiny force to them, a process known as radiation pressure. We don't feel radiation pressure when sunlight falls on us because the force is tiny and the human body is far too massive to be shoved around by sunlight. However, the microscopic dust grains in the comet's coma do respond to radiation pressure and are pushed away from the Sun.

The tail pushed out by radiation pressure is made of dust particles, but figure 23.37*A* shows that comets often have a second tail. That tail is created by the interaction of the solar wind with the comet's released gases.

The solar wind blows away from the Sun at about 400 kilometers per second. It is very tenuous, containing only a few atoms per cubic centimeter. But the material in the comet's coma is tenuous, too, and the solar wind is dense enough by comparison to blow it into a long plume. Thus, two forces, radiation pressure and the solar wind, act on the comet to drive out a tail. Because those forces are directed away from the Sun, the comet's tail always points more or less away from the Sun.

Although most comets that we see from Earth swing by the Sun on orbits that will bring them back to the inner solar system only after millions of years, a small number (including Halley's) reappear at time intervals of less than 200 years. Astronomers think that these objects come from the Kuiper Belt. They may also be responsible for meteor showers because as they repeatedly orbit the Sun, its heat gradually whittles them away: All the ices and gases eventually evaporate, and only the small amount of solid matter, dust and grit, remains. This fate is like that of a snowball made from snow scooped up alongside the road, where small amounts of gravel have been packed into it. If such a snowball is brought indoors, it melts and evaporates, leaving behind only the grit accidentally incorporated in it. So, too, the evaporated comet leaves behind in its orbit grit that continues to circle the Sun. The material left by the comet produces a delightful benefit: It is a source of meteor showers. Meteor showers tend to recur at the same time each year, as the Earth passes through cometary dust trails crossing its orbit.

23.6 GIANT IMPACTS

Every few thousand years, the Earth is hit by a huge meteoroid, a body tens of meters or more in size. A large meteoroid will produce not only a spectacular glare as it passes through the atmosphere but also an enormous blast on impact. The violence of such events arises because a meteoroid has a very large kinetic energy, that is, energy of motion, which is released when it hits the ground or when it breaks up in the atmosphere.

The energy so released can be huge. A body 20 meters in diameter, about the size of a small house, would have on impact the explosive power of a thermonuclear bomb and make a crater about one-half kilometer in diameter. Were such a body to hit a heavily populated area, it would obviously have catastrophic results. In March 2013, a small example was provided when a meteor exploded over the Russian city of Chelyabinsk. Find out more on the web at http://news.nationalgeographic.com/news/2013/11/131106-russian-meteor-chelyabinsk-airburst-500-kilotons/.

Giant Meteor Impacts

One of the most famous meteor impacts was the event that formed the giant crater in northern Arizona. About 50,000 years ago, a meteoroid estimated to have been some 50 meters in diameter hit the Earth about 40 miles east of what is now Flagstaff. Its impact vaporized tons of rock, which expanded and peeled back the ground, creating a crater about 1.2 kilometers across and 200 meters deep (see box 7.1 figure 1 in chapter 7).

More recently, in 1908, a small comet or asteroid broke up in our atmosphere over a largely uninhabited part of north-central Siberia. This so-called *Tunguska event*, named for the region where it hit, leveled trees radially outward from the blast point to a distance of some 30 kilometers. The blast was preceded by a brilliant fireball in the sky and was followed by clouds of dust that rose to the upper atmosphere. Casualties were few because the area was so remote.

Not all impacts have so few casualties, however. In the distant past, about 65 million years ago, at the end of the Cretaceous period, an asteroid hit the Earth. Its impact and the subsequent disruption of the atmosphere and climate are blamed for exterminating the dinosaurs and many less conspicuous but widespread creatures and plants (see box 8.2).

Other mass extinctions have occurred earlier and later than the Cretaceous event. These may have resulted from similar impact events, but many scientists believe that massive volcanic eruptions or drastic changes in sea level may have played a role as well. Thus, like so many of the most interesting issues in science, this story has no definitive explanation at this time.

Summary

The Earth is just one of eight planets orbiting our Sun. The planets closest to the Sun (the inner planets–Mercury, Venus, Earth, and Mars) are rocky and are similar in size. They differ greatly from the outer planets (Jupiter, Saturn, Uranus, and Neptune). Pluto is small and icy and is now considered a *dwarf planet,* along with Ceres and the largest Kuiper Belt objects. Smaller objects–asteroids and comets–also orbit the Sun, and all these objects formed in our solar system.

The motion of the planets around the Sun within a flat, disk-shaped region and their compositional differences give clues to their origin. Astronomers think that our solar system formed from the collapse of a slowly spinning interstellar cloud of dust and gas. Rotation flattened the cloud into a disk–the *solar nebula.* Dust particles within the disk stuck together and gradually grew in size, eventually becoming solid objects a few kilometers across—*planetesimals.* The planetesimals, in accordance with the gravitational attraction of their masses, drew together and formed planets. Near the Sun, it was too warm for the planetesimals to capture or incorporate much water or other gases. The objects near the Sun, therefore, are composed mainly of silicates and iron particles. Farther from the Sun, it was cold enough for water-ice to be captured, which allowed larger planetesimals to form. These eventually grew big enough to capture gas directly from the solar nebula, producing the four gas giant planets.

In the late stages of their growth, the planets were bombarded with surviving planetesimals still orbiting the Sun. The impact of these objects created craters on their surfaces. Some planetesimals (or their fragments) exist even today as *asteroids* and *comets.*

Size strongly affects a planet's history. All planets were probably born hot, but small ones (such as Mercury) cooled quickly and are inactive today, lacking volcanic and tectonic activity. Larger planets (such as Earth and Venus, and a few of the larger moons) remain hot enough to have geologically active surfaces. Their surfaces are altered now mainly by processes driven by their internal heat or gravitational forces, but impacts of asteroidal or cometary objects still occur and on rare occasions have devastating results.

Terms to Remember

asteroid 567
astronomical unit 568
comet 568
differentiation 572
dwarf planet 593
ecliptic 568
greenhouse effect 580
inner planet 568
meteor 594
meteorite 595
Milky Way 570
Nebular Hypothesis 570
outer planet 568
planet 566
planetesimal 572
solar nebula 570
solar system 567
solar wind 580
star 567
universe 570

Testing Your Knowledge

1. What is the approximate shape of the solar system?
2. What evidence supports the hypothesis that the solar system formed from a rotating disk of dust and gas?

3. Why is the surface of Venus hotter than the surface of Mercury, despite Mercury being closer to the Sun?
4. Why is the Earth much less cratered than the Moon?
5. Why do scientists believe there was once liquid water on the surface of Mars?
6. Compare and contrast the major features of the terrestrial and Jovian planets.
7. What has led to the formation of two very different types of planets in the solar system?
8. Where do comets come from? What are they composed of?
9. What created the rings of Saturn and the other large planets?
10. Why was Pluto reclassified as a dwarf planet, reducing the number of planets in the Solar System from nine to eight?
11. A planetesimal is
    a. the rocky core of a comet.
    b. a fragment of a planet destroyed by a major impact.
    c. a star surrounded by a disk of cosmic dust that can form planets.
    d. a small, planet-like body formed by collisions in a solar nebula.
12. Which of the following is evidence that Jupiter's composition is rich in hydrogen?
    a. Its large mass b. Its large density
    c. Its low density d. Its low temperature
13. Why do Mercury and the Moon lack an atmosphere?
    a. They formed after all the gas had been used up.
    b. They are so cold that all their gases have frozen into deposits below their surface.
    c. They formed before the solar nebula had captured any gas.
    d. They are so small that their gravity is too weak to retain an atmosphere.
14. Why would it be difficult to land a spacecraft on Jupiter?
    a. Jupiter has no solid surface.
    b. Jupiter's immense gravity would squash it.
    c. Jupiter's intense magnetic field would destroy it.
    d. The clouds are so thick it would be hard to navigate to a safe spot.
15. Mercury's inclined and elliptical orbit is believed to be the result of
    a. its proximity to the Sun.
    b. collision with an enormous planetesimal.
    c. the gravitational pull of Jupiter.
    d. its very slow rotation.
16. Uranus's blue color is due to
    a. liquid water on its surface.
    b. its methane-rich atmosphere.
    c. its distance from the Sun.
    d. dust particles trapped in its atmosphere.
17. The core of a comet is composed of
    a. molten iron. b. frozen gases and dust.
    c. liquid hydrogen. d. uranium.
18. An astronomical unit (AU) is equivalent to approximately
    a. 50 million miles.
    b. 150 million kilometers.
    c. 100 billion kilometers.
    d. 250 million miles.
19. Why are most asteroids not spherical?
    a. Their gravity is too weak to pull them into a sphere, and they have been fragmented by impacts.
    b. The Sun's gravity distorts them.
    c. Strong magnetic fields in their molten core make them lumpy.
    d. The statement is false. Nearly all asteroids are spherical.
20. What causes meteor showers such as those that occur around August 12?
    a. The breakup of an asteroid in our upper atmosphere
    b. Bursts of particles ejected from the Sun
    c. A comet being captured into orbit around the Earth
    d. The Earth passing through the trail of debris left by a comet

Expanding Your Knowledge

1. What obvious evidence suggests that the lunar highlands are older than the maria?
2. What has led to the inner planets having iron cores?
3. What is the likely origin of the main asteroid belt and the Kuiper Belt?
4. Why is it not surprising that the Moon lacks folded mountain ranges like we have on Earth?
5. Did Venus ever have Earth-like plate tectonics? What evidence is there?
6. How can volcanism occur on small bodies like Io or Enceladus?

Exploring Web Resources

photojournal.jpl.nasa.gov
NASA Planetary Photojournal

http://nssdc.gsfc.nasa.gov/photo_gallery/
National Space Science Data Center, NASA's archive for space science mission data

www.seds.org
Students for the Exploration and Development of Space

http://www.solarviews.com/eng/homepage.htm
Views of the Solar System, by Calvin Hamilton

http://www.esa.int/Our_Activities/Space_Science/Rosetta/
European Space Agency's Rosetta Page

APPENDIX A
Identification of Minerals

Each mineral is identified by a unique set of physical or chemical properties. Determining some of these properties requires specialized equipment and techniques. Most common minerals, however, can be distinguished from one another by tests involving simple observations. Cleavage is an especially useful property. If cleavage is present, you should determine the number of cleavage directions, estimate the angles between cleavage directions, and note the quality of each direction of cleavage. Other easily performed tests and observations check for hardness (abbreviated H), luster, and color, and determining crystal form (if present). A simple chemical test can be made using dilute hydrochloric acid to see if the mineral effervesces.

The identification tables included here can be used to identify the most common minerals (the rock-forming minerals) and some of the most common ore minerals. For identifying less common minerals, refer to one of the websites on mineralogy listed at the end of chapter 2. Mineral identification takes practice, and you will probably want to verify your mineral identifications with a geology instructor.

Because the common rock-forming minerals are the ones you are most likely to encounter, we have included a simple key for identifying them. The key is based on first determining whether or not the mineral is harder than glass and then checking other properties that should lead to identification of the mineral. You should verify your identification by seeing whether other properties of your sample correspond to those listed for the mineral in table A.1.

Ore minerals are usually distinctive enough that a key is unnecessary. To identify an ore mineral, read through table A.2 and determine which set of properties best fits the unknown mineral.

Key for Identifying Common Rock-Forming Minerals

Determine whether a fresh surface of the mineral is harder or softer than glass. If you can scratch the mineral with a knife blade, the mineral is softer than glass.

I. Harder than glass—knife will not scratch mineral. (If softer than glass, go to II of this key.)
   A. Determine if cleavage is present or absent (this may require careful examination). If cleavage is present, proceed to B. If cleavage is absent, determine the luster and proceed to 1, 2, or 3 below.
      1. Vitreous luster
         a. Olive green or brown—*olivine*
         b. Reddish brown or in equidimensional crystals with twelve or more faces—*garnet*
         c. Usually light-colored or clear—*quartz*
      2. Metallic luster
         a. Bright yellow—*pyrite*
      3. Greasy or waxy luster
         a. Mottled green and black—*serpentine*
   B. Cleavage present. Determine the number of directions of cleavage in an individual grain or crystal.
      1. Two directions, good, at or near 90°—*feldspar*
         a. If striations are visible on cleavage surfaces—*plagioclase*
         b. If pink or salmon-colored—*potassium feldspar*
         c. If white or light gray without striations, it could be either type of feldspar
      2. Two directions, fair, at 90°
         a. Dark green to black—*pyroxene*(usually augite)
      3. Two directions, excellent, not near 90°
         a. Dark green to black—*amphibole* (usually hornblende)

II. Softer than glass—knife scratches mineral. Determine whether or not the mineral has cleavage.
   A. No cleavage detectable
      1. Earthy luster, in masses too fine to distinguish individual grains—*clay group* (for instance, *kaolinite*)
   B. Cleavage present
      1. One direction
         a. Perfect cleavage in flexible sheets—*mica*:
            Clear or white—*muscovite mica*
            Black or dark brown—*biotite mica*
      2. Three directions
         a. All three perfect and at 90° to each other (cubic cleavage)—*halite*
         b. All three perfect and not near 90° to each other:
            If effervesces in dilute acid—*calcite*
            If effervesces in dilute acid only after being pulverized—*dolomite*

Table A.1 Diagnostic Properties of the Common Rock-Forming Minerals

Name (mineral groups shown in capitals)	Chemical Composition	Chemical Group	Diagnostic Properties	Other Properties
AMPHIBOLE (A mineral group in which *hornblende* is the most common member.)	$XSi_8O_{22}(OH)_2$ (*X* is a combination of Ca, Na, Fe, Mg, Al)	Chain silicate	2 good cleavage directions at 60° (120°) to each other.	H = 5–6 (barely scratches glass). Hornblende is dark green to black; tends to form in needlelike or elongate crystals; vitreous luster.
Augite (*see* Pyroxene)				
Biotite (*see* Mica)				
Calcite	$CaCO_3$	Carbonate	3 excellent cleavage directions, *not* at right angles (they define a rhombohedron). H = 3. Effervesces vigorously in weak acid.	Usually white, gray, or colorless; vitreous luster. Clear crystals show double refraction.
CLAY MINERALS (*Kaolinite* is a common example of this large mineral group.)	Compositions include $XSi_4O_{10}(OH)_8$ (*X* is Al, Mg, Fe, Ca, Na, K)	Sheet silicate	Generally microscopic crystals. Masses of clay minerals are softer than fingernail. Earthy luster. Claylike smell when damp.	Seen as a chemical weathering product of feldspars and most other silicate minerals. A constituent of most soils.
Dolomite	$CaMg(CO_3)_2$	Carbonate	Identical to calcite (rhombohedral cleavage, H = 3) except effervesces in weak acid only when pulverized.	Usually white, gray, or colorless. Vitreous luster.
FELDSPAR (Most common group of minerals.)	Framework	Framework silicates	H = 6 (scratch glass). 2 good cleavage directions at about 90° to each other.	Vitreous luster but surface may be weathered to clay, giving an earthy luster. Perfect crystal, shaped like an elongated box.
The group includes:				
Potassium feldspar	$KAlSi_3O_8$		White, pink, or salmon-colored.	Never has striations on cleavage surfaces.
Plagioclase (sodium and calcium feldspar)	Mixture of: $CaAl_2Si_2O_8$ and $NaAlSi_3O_8$		White, light to dark gray, rarely other colors. *May* have striations on cleavage surfaces.	Calcium-rich varieties generally a darker gray and may show an iridescent play of colors.
GARNET	$XAl_2Si_3O_{12}$ (*X* is a combination of Ca, Mg, Fe, Al, Mn)	Isolated silicate	No cleavage. Usually reddish brown. Tends to occur in perfect equidimensional crystals, usually 12 sided. H = 7.	Rarely yellow, green, or black. Usually found in metamorphic rocks. Vitreous luster.
Gypsum	$CaSO_4 \cdot 2H_2O$	Sulfate	H = 2. 1 good and 2 perfect cleavage directions. Vitreous or silky luster.	Clear, white, or pastel colors. Flexible cleavage fragments.

Table A.1 Diagnostic Properties of the Common Rock-Forming Minerals *(continued)*

Name (mineral groups shown in capitals)	Chemical Composition	Chemical Group	Diagnostic Properties	Other Properties
Halite	NaCl	Halide	3 excellent cleavage directions at 90° to each other (cubic). $H = 2\frac{1}{2}$. Salty taste. Soluble in water.	Usually clear or white.
Hematite (*see* Table A.2)				
Hornblende (*see* Amphibole)				
Kaolinite (*see* Clay)				
MICA	$K(X)(AlSi_3O_{10})(OH)_2$	Sheet silicate	1 perfect cleavage direction (splits easily into flexible sheets).	H = 2–3. Vitreous luster.
The group includes:				
Biotite	(*X* is Mg, Fe, and Al)		Black or dark brown.	
Muscovite	(*X* is Al)		White or transparent.	
Olivine	X_2SiO_4 (*X* is Fe, Mg)	Isolated silicate	No cleavage. Generally olive green or brown. H = 6–7 (scratches glass). Vitreous luster.	Usually as small grains in mafic or ultramafic igneous rocks.
Orthoclase (*see* Feldspar)				
Plagioclase (*see* Feldspar)				
Pyrite ("fool's gold")	FeS_2	Sulfide	H = 6 (scratches glass). Bright, yellow, metallic luster. Black streak.	Commonly occurs as perfect crystals: cubes or crystals with five-sided faces. Weathers to brown.
PYROXENE (A mineral group; *Augite* is most common member.)	$XSiO_3$ (*X* is Fe, Mg, Al, Ca)	Chain silicate	2 fair cleavage directions at 90° to each other.	H = 6. Augite is dark green to black. Vitreous luster; usually stubby crystals.
Quartz	SiO_2	Framework silicate	H = 7. No cleavage. Vitreous luster. Does not weather to clay.	Almost any color but commonly white or clear. Good crystals have six-sided "column" with complex "pyramid" on top.
Serpentine	$Mg_6Si_4O_{10}(OH)_8$	Sheet silicate	Hardness variable but softer than glass. Mottled green and black. Greasy luster. Fractures along smooth curved surfaces.	Sometimes fibrous (asbestos).

Table A.2 Diagnostic Properties of the Most Common Ore Minerals

Name	Chemical Composition	Diagnostic Properties	Other Properties
Azurite	$Cu_3(CO_3)_2(OH)_2$	Azure blue; effervesces in weak acid.	H = 3–4.
Bauxite	$Al_2O_3 \cdot nH_2O$	Earthy luster. A variety of clay. Generally pea-sized spheres included in a fine-grained mass.	
Bornite	Cu_3FeS_4	Metallic luster, tarnishes to iridescent purple color.	Gray streak; H = 3 (softer than glass).
Chalcopyrite	$CuFeS_2$	Metallic luster, brass-yellow. Softer than glass.	Black streak.
Cinnabar	HgS	Scarlet red, bright red streak.	Softer than glass. Generally an earthy luster.
Galena	PbS	Metallic luster, gray; 3 directions of cleavage at 90° (cubic). High specific gravity.	Softer than glass; gray streak.
Gold	Au	Metallic luster, yellow. H = 3 (softer than glass, can be pounded into thin sheets, easily deformed).	Yellow streak; high specific gravity.
Halite	NaCl	Salty taste; 3 cleavage directions at 90° (cubic).	Clear or white; easily soluble in water.
Hematite	Fe_2O_3	Red-brown streak.	Either in earthy reddish masses or in metallic, silver-colored flakes or crystals.
Limonite	$Fe_2O_3 \cdot nH_2O$	Earthy luster; yellow-brown streak.	Yellow to brown color; softer than glass.
Magnetite	Fe_3O_4	Metallic luster, black; magnetic.	Harder than glass; black streak.
Malachite	$Cu_2(CO_3)(OH)_2$	Bright-green color and streak.	Softer than glass; effervesces in weak acid.
Sphalerite	ZnS	Brown to yellow color; 6 directions of cleavage.	Lusterlike resin; yellow or cream-colored streak; softer than glass.
Talc	$Mg_3Si_4O_{10}(OH)_2$	White, gray, or green; softer than fingernail (H = 1).	Greasy feel.

APPENDIX B
Identification of Rocks

IGNEOUS ROCKS

Igneous rocks are classified on the basis of texture and composition. For some rocks, texture alone suffices for naming the rock. For most igneous rocks, composition as well as texture must be taken into account. Ideally, the mineral content of the rock should be used to determine composition; but for fine-grained igneous rocks, accurate identification of minerals may require a polarizing microscope or other special equipment. In the absence of such equipment, we rely on the color of fine-grained rocks and assume the color is indicative of the minerals present.

To identify a common igneous rock, use either table 3.1 or follow the key given below.

Key for Identifying Common Igneous Rocks

I. What is the texture of the rock?
   - A. Is it glassy (a very vitreous luster)? If so, it is *obsidian*, regardless of its chemical composition. Obsidian exhibits a pronounced conchoidal fracture.
   - B. Does it have a frothy appearance? If so, it is *pumice.* Pumice is light in weight and feels abrasive (it probably will float on water).
   - C. Does it have angular fragments of rock embedded in a volcanic-derived matrix? If so, it is a *volcanic breccia.* If the precise nature of the rock fragments and matrix can be identified, modifiers may be used; for instance, the rock may be an *andesite* breccia or a *rhyolite* breccia.
   - D. Is the rock composed of interlocking, very coarse-grained minerals? (The minerals should be more than 5 centimeters across.) If so, the rock is a *pegmatite.* Most pegmatites are mineralogically equivalent to granite, with feldspars and quartz being the predominant minerals.
   - E. Is the rock entirely phaneritic? (That is, does it have a coarse-grained, interlocking crystalline texture in which nearly all grains are easily distinguished by the naked eye?) If so, go to part II of this key.
   - F. Is the rock *entirely* aphanitic? (Are grains too fine to distinguish with the naked eye?) If so, go to part III of this key.
   - G. Is the matrix fine-grained with some coarse-grained minerals visible in the rock? If so, go to part III and add the adjective *porphyritic* to the name of the rock.

II. Igneous rocks composed of interlocking coarse-grained minerals.
   - A. Is quartz present? If so, the rock is a *granite.* Confirmation: Granite should be composed predominantly of feldspar–generally white, light gray, or pink (indicating high amounts of potassium or sodium in the feldspar). Rarely are there more than 20% ferromagnesian minerals in a granite.
   - B. Are quartz and feldspar absent? If so, the rock should be composed entirely of ferromagnesian minerals and is ultramafic, likely a *peridotite.* Confirmation: Identify the minerals as being olivine or pyroxene (or less commonly, amphibole or biotite).
   - C. Does the rock have less than 50% feldspar and no quartz? If so, the rock should be a *gabbro.* Confirmation: Most of the rock should be ferromagnesian minerals. Plagioclase can be medium or dark gray. There would be no pink feldspars.
   - D. Is the rock composed of 30% to 60% feldspar (and no quartz)? If so, the rock is a *diorite.* Confirmation: Feldspar (plagioclase) is usually white to medium gray but never pink.

III. Igneous rocks that are fine-grained.
   - A. Can small grains of quartz be identified in the rock? If so, the rock is a *rhyolite.*
   - B. If the rock is too fine-grained for you to determine whether quartz is present but is white, light gray, pink, or pale green, the rock is most likely a *rhyolite.*
   - C. Is the rock composed predominantly of ferromagnesian minerals? If so, the rock is *basalt.*
   - D. If the rock is too fine-grained to identify ferromagnesian minerals but is black or dark gray, the rock is probably a *basalt.*
     1. Does the rock have rounded holes in it? If so, it is a *vesicular basalt* (some vesicles) or *scoria* (mostly vesicles).
   - E. Is the rock composed of roughly equal amounts of white or gray feldspar and ferromagnesian minerals (but no quartz)? If so, the rock is an *andesite.* Confirmation: Most andesite is porphyritic, with numerous identifiable crystals of white or light gray feldspar and lesser amounts of hornblende crystals within the darker, fine-grained matrix. Andesite is usually medium to dark gray or green.

SEDIMENTARY ROCKS

The following key shows how sedimentary rocks are classified on the basis of texture and composition. The descriptions of the rocks in the main body of the text provide additional information, such as common rock colors. *Equipment* needed for identification of sedimentary rocks includes a bottle of dilute hydrochloric acid, a hand lens or magnifying glass, a millimeter scale, a glass plate for hardness tests, and a pocketknife or rock hammer.

Begin by testing the rock for carbonate minerals by applying a small amount of dilute hydrochloric acid to the surface of the rock.

1. The rock does not effervesce (fizz) in acid, or effervesces weakly, but when powdered by a knife or hammer, the powder effervesces strongly. If so, the rock is *dolomite.*
2. The rock does not effervesce at all, even when powdered, or effervesces only in some places, such as the cement between grains. Go to part I of this key.
3. The rock effervesces strongly. The rock is *limestone.* Go to part II of this key to determine limestone type.

I. With a hand lens or magnifying glass, determine if the rock has a clastic texture (grains cemented together) or a crystalline texture (visible, interlocking crystals).

 A. If clastic:
  1. Most grains are more than 2 millimeters in diameter.
   a. Angular grains–*sedimentary breccia.*
   b. Rounded grains–*conglomerate.*
  2. Most grains are between 1/16 and 2 millimeters in diameter. Rock feels gritty to the fingers. *Sandstone.*
   a. More than 90% of the grains are quartz–*quartz sandstone.*
   b. More than 25% of the grains are feldspar–*arkose.*
   c. More than 25% of the grains are fine-grained rock fragments, such as shale, slate, and basalt–*lithic sandstone.*
   d. More than 15% of the rock is fine-grained matrix–*graywacke.*
  3. Rock is fine-grained (grains less than 1/16 millimeter in diameter). Feels smooth to fingers.
   a. Grains visible with a hand lens–*siltstone.*
   b. Grains too small to see, even with a hand lens.
    1. Rock is laminated, fissile–*shale.*
    2. Rock is unlayered, blocky–*mudstone.*

 B. If crystalline:
  1. Crystals fine to coarse, hardness of 2–*rock gypsum.*
  2. Coarse crystals that dissolve in water–*rock salt.*

 C. Hard to determine if clastic or crystalline:
  1. Very fine-grained, smooth to touch, conchoidal fracture, hardness of 6 (scratches glass), nonporous–*chert* (*flint* if dark).
  2. Very fine-grained, smooth to touch, breaks into flat chips–*shale.*
  3. Black or dark brown, readily broken, soils fingers–*coal.*

II. *Limestone* may be clastic or crystalline, fine- or coarse-grained, and may or may not contain visible fossils. Usually gray, tan, buff, or white. Some distinctive varieties are:

 A. *Bioclastic limestone*–clastic texture, grains are whole or broken fossils. Two relatively rare varieties are:
  1. *Coquina*–very coarse, recognizable shells, much open pore space.
  2. *Chalk*–very fine-grained, white or tan, soft and powdery.

 B. *Oolitic limestone*–grains are small spheres (less than 2 millimeters in diameter), all about the same size.

 C. *Travertine*–coarsely crystalline, no pore space, often contains different-colored layers (bands).

METAMORPHIC ROCKS

The characteristics of a metamorphic rock are largely governed by (1) the composition of the parent rock and (2) the particular combination of temperature, confining pressure, and directed pressure. These factors cause different textures in rocks formed under different sets of conditions. For this reason, texture is usually the main basis for naming a metamorphic rock. Determining the composition (e.g., mineral content) is necessary for naming some rocks (e.g., *quartzite*), but for others, the minerals present are used as adjectives to describe the rock completely (e.g., *biotite* schist).

Metamorphic rocks are identified by determining first whether the rock has a *foliated* or *nonfoliated* texture.

Key for Identifying Metamorphic Rocks

I. If the rock is *nonfoliated,* then it is identified on the basis of its mineral content.

 A. Does the rock consist of mostly quartz? If so, the rock is a *quartzite.* A quartzite has a mosaic texture of interlocking grains of quartz and will easily scratch glass.

 B. Is the rock composed of interlocking coarse grains of calcite? If so, it is *marble.* (The individual grains should exhibit rhombohedral cleavage; the rock is softer than glass.)

 C. Is the rock a dense, dark mass of grains mostly too fine to identify with the naked eye? If so, it probably is a *hornfels.* A hornfels may have a few larger crystals of uncommon minerals enclosed in the fine-grained mass.

II. If the rock is *foliated,* determine the type of foliation and then, if possible, identify the minerals present.

 A. Is the rock very fine-grained, and does it split into sheetlike slabs? If so, it is *slate.* Most slate is composed of extremely fine-grained sheet silicate minerals, and the rock has an earthy luster.

 B. Does the rock have a silky sheen but otherwise appear similar to slate? If so, it is a *phyllite.*

 C. Is the rock composed mostly of visible grains of platy or needlelike minerals that are approximately parallel to one another? If so, the rock is a *schist.* If the rock is composed mainly of mica, it is a *mica schist.* If it also contains garnet, it is called a *garnet mica schist.* If hornblende is the predominant mineral, the rock is a *hornblende schist.* If talc prevails, it is a *talc schist* (sometimes called soapstone).

 D. Are dark and light minerals found in separate lenses or layers? If so, the rock is a *gneiss.* The light layers are composed of feldspars and perhaps quartz, whereas the darker layers commonly are formed of biotite, amphibole, or pyroxene. A gneiss may appear similar to granite or diorite but can be distinguished from the igneous rocks by the foliation.

APPENDIX C
The Elements Most Significant to Geology

Table C.1

Atomic Number	Name	Symbol	Atomic Weight	Usual Atomic Charge of Ions
1	Hydrogen	H	1.0	+1
2	Helium	He	4.0	0 inert
3	Lithium	Li	6.9	+1
4	Beryllium	Be	9.0	+2
5	Boron	B	10.8	+3
6	Carbon	C	12.0	+4
7	Nitrogen	N	14.0	+5
8	Oxygen	O	16.0	–2
9	Fluorine	F	19.0	–1
10	Neon	Ne	20.2	0 inert
11	Sodium	Na	23.0	+1
12	Magnesium	Mg	24.3	+2
13	Aluminum	Al	27.0	+3
14	Silicon	Si	28.1	+4
15	Phosphorus	P	31.0	+5
16	Sulfur	S	32.1	–2
17	Chlorine	C	35.5	–1
18	Argon	Ar	40.0	0 inert
19	Potassium	K	39.1	+1
20	Calcium	Ca	40.1	+2
22	Titanium	Ti	47.9	+4
23	Vanadium	V	50.9	–
24	Chromium	Cr	52.0	–
25	Manganese	Mn	54.9	+4, +3
26	Iron	Fe	55.8	+2, +3
27	Cobalt	Co	58.9	–
28	Nickel	Ni	58.7	+2
29	Copper	Cu	63.6	+2
30	Zinc	Zn	65.4	+2
33	Arsenic	As	74.9	+3
35	Bromine	Br	79.9	–
37	Rubidium	Rb	85.5	+1
38	Strontium	Sr	87.6	+2
40	Zirconium	Zr	91.2	–
42	Molybdenum	Mo	96.0	+4
47	Silver	Ag	107.9	+1
48	Cadmium	Cd	112.4	–
50	Tin	Sn	118.7	+4
51	Antimony	Sb	121.8	+3
52	Tellurium	Te	127.6	–
55	Cesium	Cs	132.9	–
56	Barium	Ba	137.3	+2
60	Neodymium	Nd	144.2	+3
62	Samarium	Sm	150.4	+3
74	Tungsten	W	183.8	–
78	Platinum	Pt	195.1	–
79	Gold	Au	197.0	–
80	Mercury	Hg	200.6	+2
82	Lead	Pb	207.2	+2
83	Bismuth	Bi	209.0	–
86	Radon	Rn	222.0	0 inert
88	Radium	Ra	226.0	–
90	Thorium	Th	232.0	–
92	Uranium	U	238.0	–
94	Plutonium	Pu	244.0	–

APPENDIX D
Periodic Table of Elements

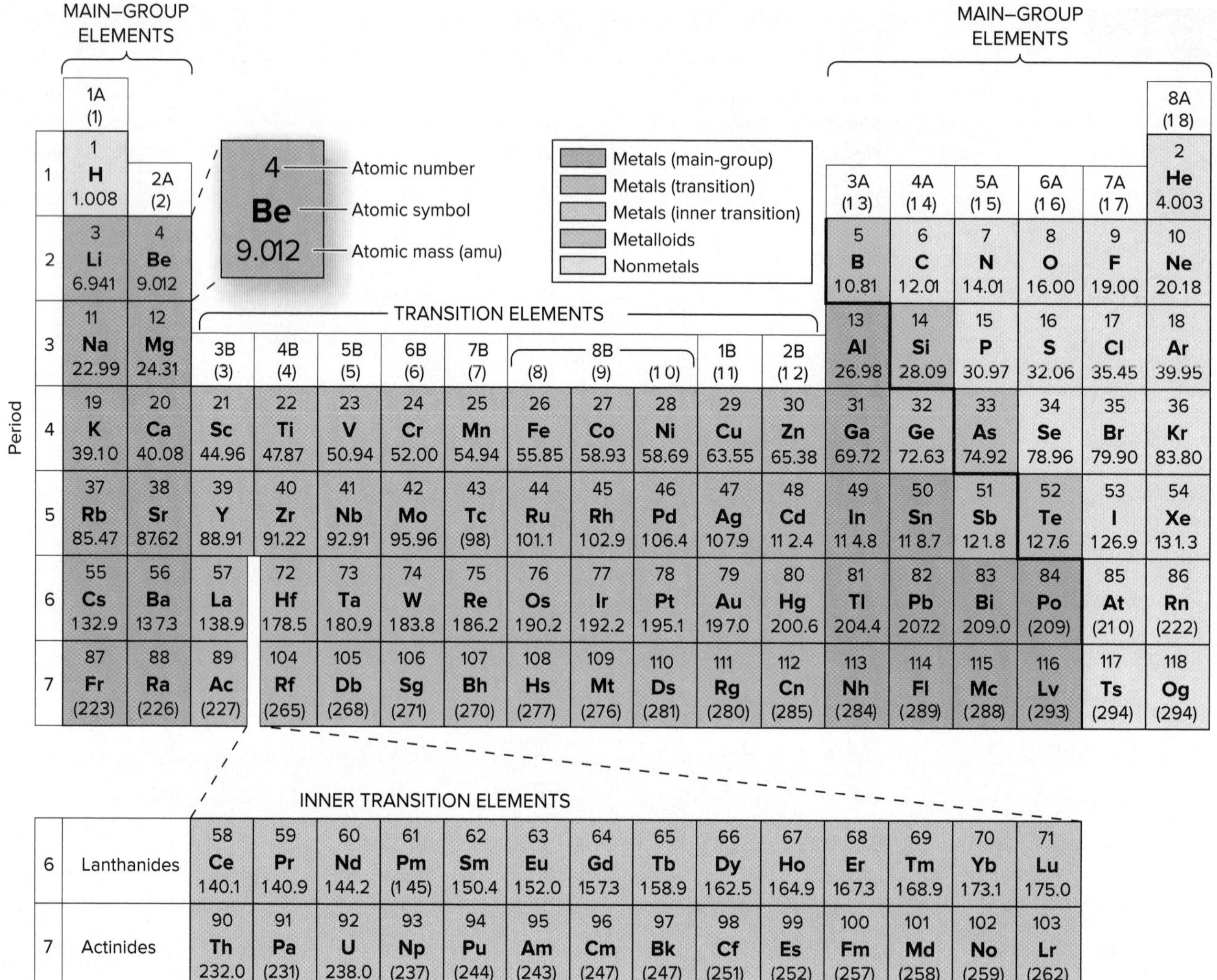

Period	1A (1)	2A (2)	3B (3)	4B (4)	5B (5)	6B (6)	7B (7)	8B (8)	8B (9)	8B (10)	1B (11)	2B (12)	3A (13)	4A (14)	5A (15)	6A (16)	7A (17)	8A (18)
1	1 H 1.008																	2 He 4.003
2	3 Li 6.941	4 Be 9.012											5 B 10.81	6 C 12.01	7 N 14.01	8 O 16.00	9 F 19.00	10 Ne 20.18
3	11 Na 22.99	12 Mg 24.31											13 Al 26.98	14 Si 28.09	15 P 30.97	16 S 32.06	17 Cl 35.45	18 Ar 39.95
4	19 K 39.10	20 Ca 40.08	21 Sc 44.96	22 Ti 47.87	23 V 50.94	24 Cr 52.00	25 Mn 54.94	26 Fe 55.85	27 Co 58.93	28 Ni 58.69	29 Cu 63.55	30 Zn 65.38	31 Ga 69.72	32 Ge 72.63	33 As 74.92	34 Se 78.96	35 Br 79.90	36 Kr 83.80
5	37 Rb 85.47	38 Sr 87.62	39 Y 88.91	40 Zr 91.22	41 Nb 92.91	42 Mo 95.96	43 Tc (98)	44 Ru 101.1	45 Rh 102.9	46 Pd 106.4	47 Ag 107.9	48 Cd 112.4	49 In 114.8	50 Sn 118.7	51 Sb 121.8	52 Te 127.6	53 I 126.9	54 Xe 131.3
6	55 Cs 132.9	56 Ba 137.3	57 La 138.9	72 Hf 178.5	73 Ta 180.9	74 W 183.8	75 Re 186.2	76 Os 190.2	77 Ir 192.2	78 Pt 195.1	79 Au 197.0	80 Hg 200.6	81 Tl 204.4	82 Pb 207.2	83 Bi 209.0	84 Po (209)	85 At (210)	86 Rn (222)
7	87 Fr (223)	88 Ra (226)	89 Ac (227)	104 Rf (265)	105 Db (268)	106 Sg (271)	107 Bh (270)	108 Hs (277)	109 Mt (276)	110 Ds (281)	111 Rg (280)	112 Cn (285)	113 Nh (284)	114 Fl (289)	115 Mc (288)	116 Lv (293)	117 Ts (294)	118 Og (294)

6	Lanthanides	58 Ce 140.1	59 Pr 140.9	60 Nd 144.2	61 Pm (145)	62 Sm 150.4	63 Eu 152.0	64 Gd 157.3	65 Tb 158.9	66 Dy 162.5	67 Ho 164.9	68 Er 167.3	69 Tm 168.9	70 Yb 173.1	71 Lu 175.0
7	Actinides	90 Th 232.0	91 Pa (231)	92 U 238.0	93 Np (237)	94 Pu (244)	95 Am (243)	96 Cm (247)	97 Bk (247)	98 Cf (251)	99 Es (252)	100 Fm (257)	101 Md (258)	102 No (259)	103 Lr (262)

Go to www.webelements.com for an excellent periodic table on which you can click any element and get a detailed description of its properties.

Elements with an atomic number greater than 92 are not naturally occurring.

APPENDIX E
Selected Conversion Factors

Table E.1

	English Unit	Conversion Factor	Metric Unit	Conversion Factor	English Unit
Length and Distance	inch (in)	2.54	centimeters (cm)	0.4	inch (in)
	foot (ft)	0.3048	meter (m)	3.28	feet (ft)
	inch (in)	0.026	meter (m)	39.4	inches (in)
	mile, statute (mi)	1.61	kilometers (km)	0.62	mile (mi)
Area	square inch (in^2)	6.45	square centimeters (cm^2)	0.16	square inch (in^2)
	square foot (ft^2)	0.093	square meter (m^2)	10.8	square feet (ft^2)
	square mile (mi^2)	2.59	square kilometers (km^2)	0.39	square mile (mi^2)
	acre	0.4	hectare	2.47	acres
Volume	cubic inch (in^3)	16.4	cubic centimeters (cm^3)	0.06	cubic inch (in^3)
	cubic yard (yd^3)	0.76	cubic meter (m^3)	1.3	cubic yards (yd^3)
	cubic foot (ft^3)	0.0283	cubic meter (m^3)	35.5	cubic feet (ft^3)
	quart (qt)	0.95	liter	1.06	quarts (qt)
Weight	ounce (oz)	28.3	grams (g)	0.04	ounce (oz)
	pound (lb)	0.45	kilogram (kg)	2.2	pounds (lb)
	ton, short (2,000 lb)	907	kilograms (kg)	0.001	ton, short
	ton, short	0.91	ton, metric	1.1	ton, short
Temp.	degrees Fahrenheit (°F) $- 32° \times 5/9$		degrees Celsius (°C)	$\times 1.8 + 32°$	degrees Fahrenheit (°F)

APPENDIX F
Rock Symbols

Shown below are the rock symbols used in the text. In general, these symbols are used by all geologists, although they sometimes are modified slightly.

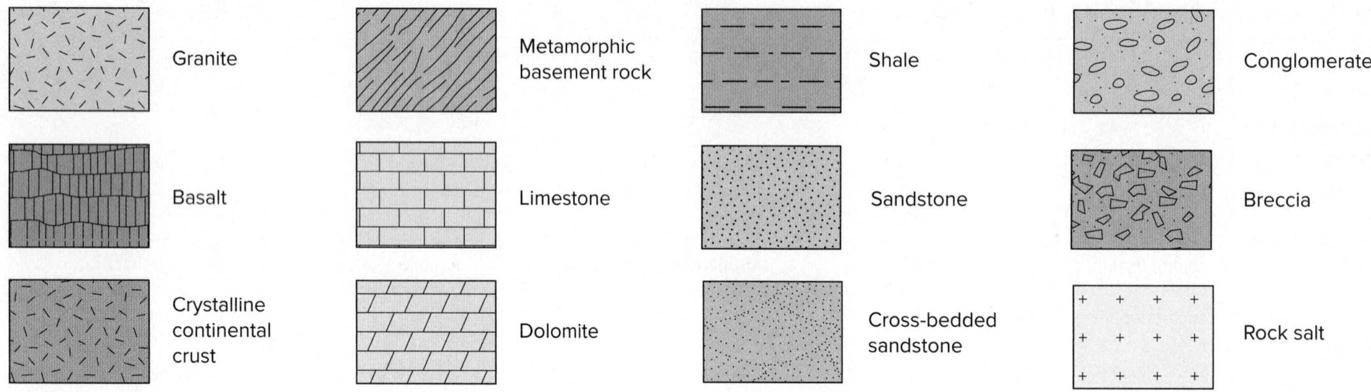

APPENDIX G
Commonly Used Prefixes, Suffixes, and Roots

abyss deep (Greek)
alluvium deposited by flowing water (Latin)
anti- opposite (Greek)
archea (archaeo)- ancient (Greek)
astheno- weak, lack of strength (Greek)
ceno recent (Greek)
circum- about, around, round about (Latin)
clast broken (Greek)
-cline tilted, gradient (Greek)
de- lower, reduce, take away (Latin)
dis- separation, opposite of (Latin)
ex- out of, away from (Greek)
feld field (Swedish, German)
folium leaf (Latin)
geo- Earth (Greek)
glomero- cluster (Latin)
hydro- water (Greek)
iso- equal (Greek)
-lith stone or rock (Greek)
meso- middle (Greek)
meta- change (Greek)
-morph form, shape (Greek)
paleo- ancient (Greek)
ped- foot (Latin)
pelagic pertaining to the ocean (Greek)
petro- stone or rock (Greek)
phanero- visible, evident (Greek)
pheno- large, conspicuous ("to show" in Greek)
pluto- deep-seated (from Roman god of the underworld or infernal regions)
pre- before, in front (Latin)
proto- first, primary, primitive (Greek)
pyro- fire (Greek)
spar crystalline material (German)
-sphere ball (Greek)
strat- layer or layered (Latin)
stria small groove, streak, band (Latin)
sub- under, less than (Latin)
super- above, more than, in addition to (Latin)
syn- together, at the same time (Latin, Greek)
tecto- means building or constructing in Greek and Latin; in geology, it means movement of structures caused by internal forces.
terra, terre pertaining to Earth (Latin)
thermal pertaining to heat (Greek)
trans- over, beyond, through, across (Latin)
xeno- strange, foreign (Greek)
zoo, zoic- animal (Greek)

APPENDIX H

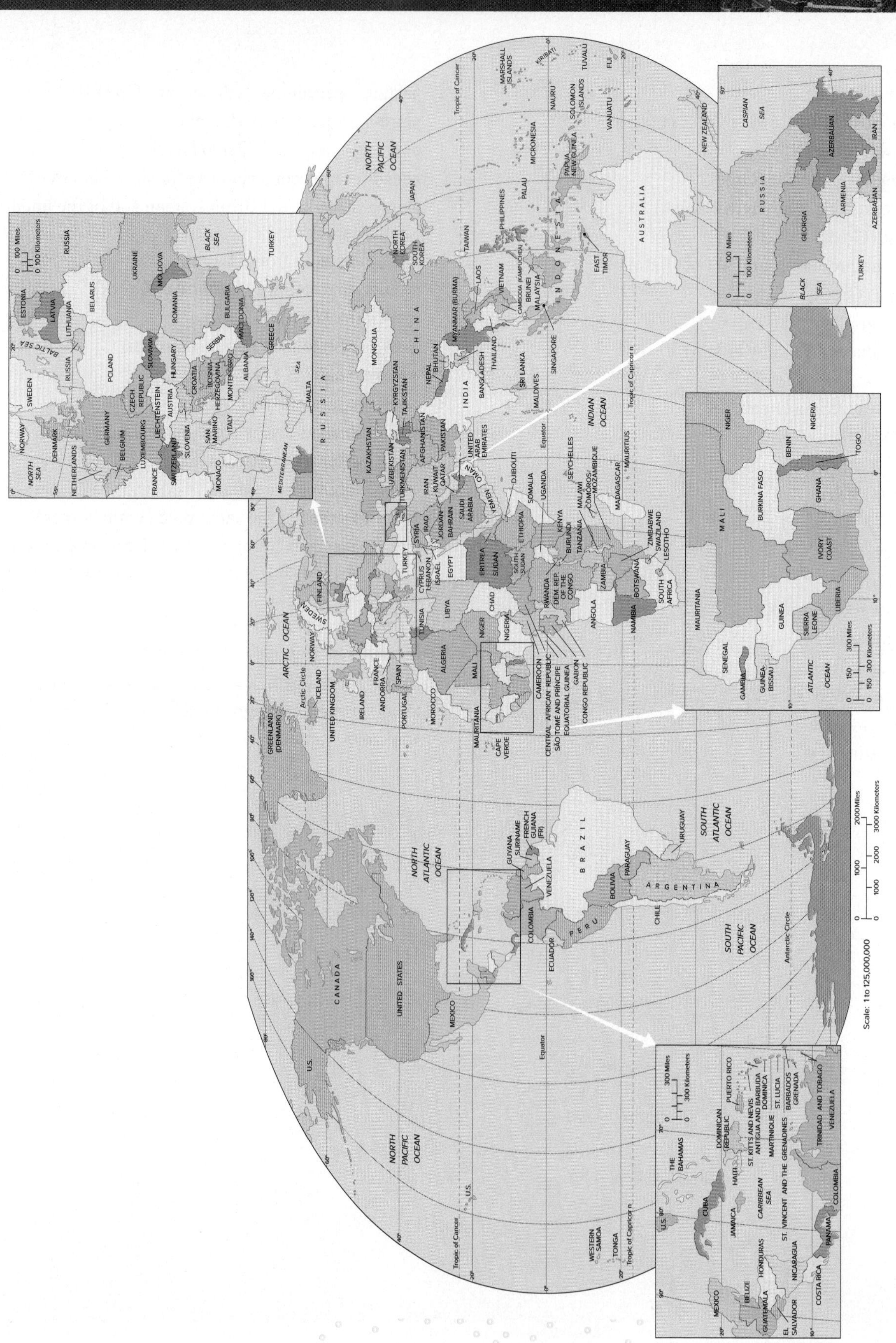

ARCTIC OCEAN
NORTH PACIFIC OCEAN
NORTH ATLANTIC OCEAN
SOUTH ATLANTIC OCEAN
SOUTH PACIFIC OCEAN
INDIAN OCEAN
CANADA
UNITED STATES
U.S.
MEXICO
GREENLAND (DENMARK)
ICELAND
UNITED KINGDOM
IRELAND
FRANCE
ANDORRA
SPAIN
PORTUGAL
NORWAY
SWEDEN
FINLAND
RUSSIA
MOROCCO
ALGERIA
TUNISIA
LIBYA
EGYPT
MALI
NIGER
CHAD
NIGERIA
MAURITANIA
CAPE VERDE
CAMEROON
CENTRAL AFRICAN REPUBLIC
SÃO TOMÉ AND PRÍNCIPE
EQUATORIAL GUINEA
GABON
CONGO REPUBLIC
SUDAN
SOUTH SUDAN
ERITREA
ETHIOPIA
DJIBOUTI
SOMALIA
UGANDA
KENYA
RWANDA
BURUNDI
DEM. REP. OF THE CONGO
TANZANIA
ANGOLA
ZAMBIA
MALAWI
COMOROS
MOZAMBIQUE
MADAGASCAR
ZIMBABWE
SWAZILAND
LESOTHO
BOTSWANA
NAMIBIA
SOUTH AFRICA
SEYCHELLES
MAURITIUS
TURKEY
CYPRUS
LEBANON
ISRAEL
SYRIA
IRAQ
JORDAN
BAHRAIN
SAUDI ARABIA
KUWAIT
QATAR
IRAN
UNITED ARAB EMIRATES
OMAN
YEMEN
KAZAKHSTAN
UZBEKISTAN
TURKMENISTAN
KYRGYZSTAN
TAJIKISTAN
AFGHANISTAN
PAKISTAN
INDIA
NEPAL
BHUTAN
BANGLADESH
SRI LANKA
MALDIVES
MONGOLIA
CHINA
NORTH KOREA
SOUTH KOREA
JAPAN
TAIWAN
MYANMAR (BURMA)
LAOS
THAILAND
VIETNAM
CAMBODIA (KAMPUCHEA)
BRUNEI
MALAYSIA
SINGAPORE
PHILIPPINES
PALAU
INDONESIA
EAST TIMOR
PAPUA NEW GUINEA
AUSTRALIA
NEW ZEALAND
MICRONESIA
MARSHALL ISLANDS
KIRIBATI
NAURU
SOLOMON ISLANDS
TUVALU
VANUATU
FIJI
WESTERN SAMOA
TONGA
GUYANA
SURINAME
FRENCH GUIANA (FR)
VENEZUELA
COLOMBIA
ECUADOR
PERU
BRAZIL
BOLIVIA
PARAGUAY
CHILE
ARGENTINA
URUGUAY
Arctic Circle
Tropic of Cancer
Equator
Tropic of Capricorn
Antarctic Circle
Scale: 1 to 125,000,000
2000 Miles
3000 Kilometers
NORTH SEA
BALTIC SEA
BLACK SEA
MEDITERRANEAN SEA
NETHERLANDS
DENMARK
BELGIUM
LUXEMBOURG
GERMANY
SWITZERLAND
LIECHTENSTEIN
AUSTRIA
CZECH REPUBLIC
POLAND
SLOVAKIA
HUNGARY
SLOVENIA
CROATIA
BOSNIA HERZEGOVINA
SERBIA
MONTENEGRO
ALBANIA
MACEDONIA
GREECE
BULGARIA
ROMANIA
MOLDOVA
UKRAINE
BELARUS
LITHUANIA
LATVIA
ESTONIA
ITALY
SAN MARINO
MONACO
MALTA
100 Miles
100 Kilometers
CASPIAN SEA
GEORGIA
ARMENIA
AZERBAIJAN
SENEGAL
GAMBIA
GUINEA-BISSAU
GUINEA
SIERRA LEONE
LIBERIA
IVORY COAST
BURKINA FASO
GHANA
TOGO
BENIN
ATLANTIC OCEAN
300 Miles
300 Kilometers
THE BAHAMAS
CUBA
JAMAICA
HAITI
DOMINICAN REPUBLIC
PUERTO RICO
ST. KITTS AND NEVIS
ANTIGUA AND BARBUDA
DOMINICA
MARTINIQUE
ST. LUCIA
BARBADOS
GRENADA
ST. VINCENT AND THE GRENADINES
TRINIDAD AND TOBAGO
CARIBBEAN SEA
BELIZE
HONDURAS
GUATEMALA
EL SALVADOR
NICARAGUA
COSTA RICA
PANAMA

USGS
science for a changing world
U.S. DEPARTMENT OF THE INTERIOR
U.S. GEOLOGICAL SURVEY
The North America Tapestry of Time and Terrain
Cobertura de Tiempo y Terrenos de Norte América
L'Amérique du Nord : un collage de terrains d'âges différents
Tapestry Compiled by
Kate E. Barton, David G. Howell, and José F. Vigil
Geology Compiled by
John C. Reed, Jr. and John O. Wheeler
U.S. Geological Survey, Geological Survey of Canada
2003
Sedimentary Rocks
Rocas sedimentarias
Roches sédimentaires
Volcanic Rocks
Rocas volcánicas
Roches volcaniques
Plutonic Rocks
Rocas plutónicas
Roches plutoniques
Metamorphic Rocks
Rocas metamórficas
Roches métamorphiques
Hawaiian Islands
Geologic Time Scale

Generalized Geologic and Tectonic Map of North America

SEDIMENTARY UNITS

Thick deposits in structurally negative areas

Paleozoic and Mesozoic active margin deposits

Synorogenic and postorogenic deposits

Paleozoic and Mesozoic passive margin deposits

Late Precambrian deposits
Of Middle and Upper Proterozoic ages

VOLCANIC AND PLUTONIC UNITS

Postorogenic volcanic cover

Ultramafic rocks

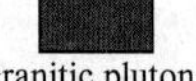

Granitic plutons
Ages are generally within the span of the tectonic cycle of the foldbelt in which they lie

SPECIAL UNITS

Former subduction complex rocks of the Pacific border

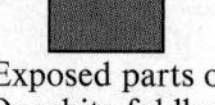

Exposed parts of Ouachita foldbelt

Probable western extension of Innuitian foldbelt
In cores of northern Alaska ranges

PLATFORM AREAS

Ice cap of Quaternary age
On Precambrian and Paleozoic basement

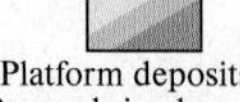

Platform deposits on Precambrian basement
In central craton

Platform deposits on Paleozoic basement
In Atlantic and Gulf coastal plains

Platform deposits within the Precambrian
Mainly in the Canadian Shield

PRECAMBRIAN

Basement igneous and metamorphic complexes mainly of Precambrian age

Grenville foldbelt
Deformed 880–1,000 m.y. ago

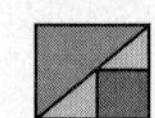

Hudsonian foldbelts
Deformed 1,640–1,820 m.y. ago

Kenoran foldbelts
Deformed 2,390–2,600 m.y. ago

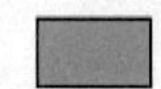

Anorthosite bodies
Plutons composed almost entirely of plagioclase

STRUCTURAL SYMBOLS

Normal fault
Hachures on downthrown side

Subsea fault

Salt domes and salt diapirs
In Gulf coastal plain and Gulf of Mexico

Strike-slip fault
Arrows show relative lateral movement

Volcano

Thrust fault
Barbs on upthrown side

World's oldest rock

Contours on basement surfaces beneath platform areas
All contours are below sea level except where marked with plus symbols. Interval is 1,000 meters

Axes of seafloor spreading

Modified from the Generalized Tectonic Map of North America by P.B. King and Gertrude J. Edmonston, U.S. Geological Survey Map I-688

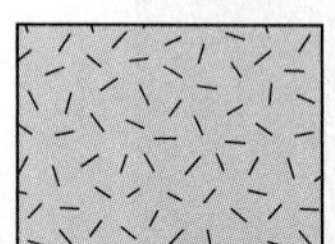

Granite

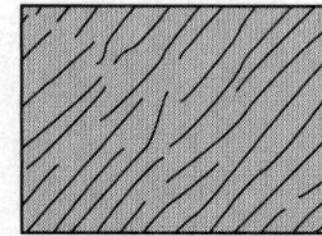

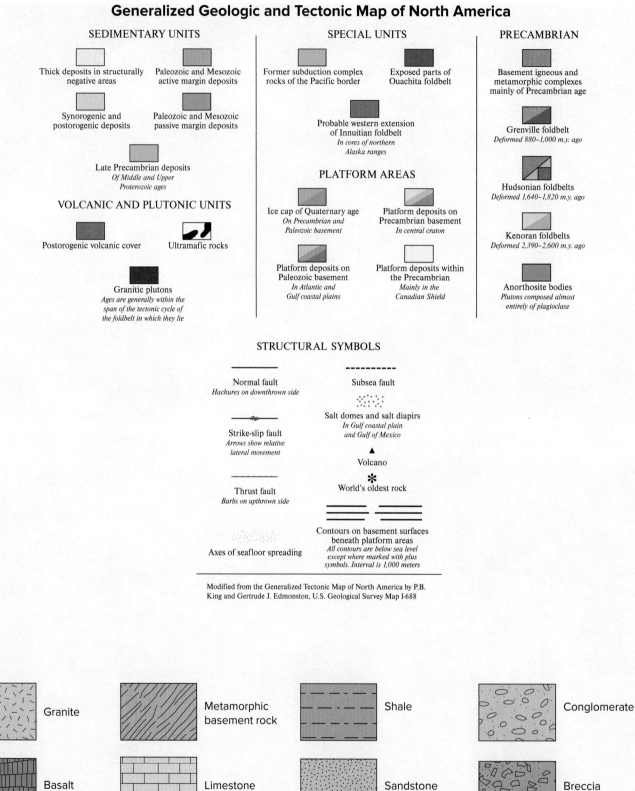

Metamorphic basement rock

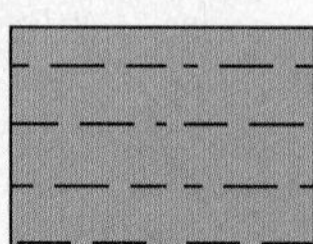

Shale

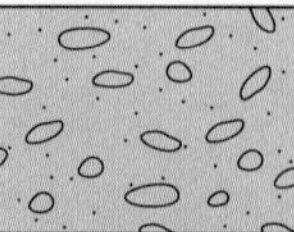

Conglomerate

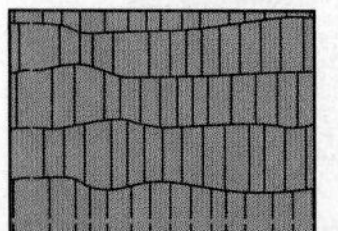

Basalt

Limestone

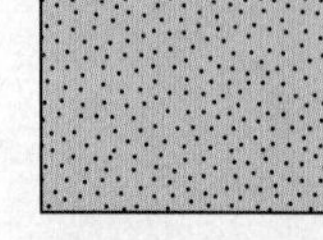

Sandstone

Breccia

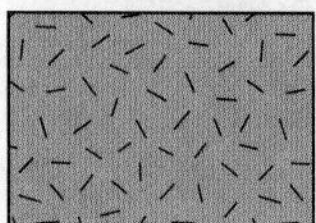

Crystalline continental crust

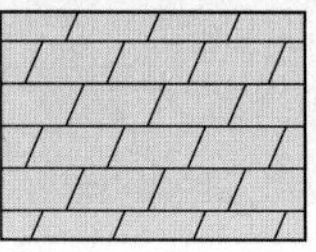

Dolomite

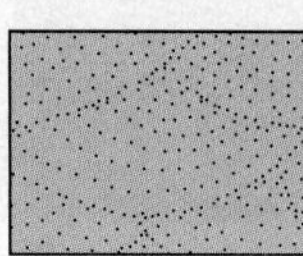

Cross-bedded sandstone

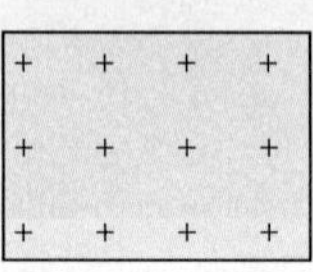

Rock salt

World's oldest
rock found here

GLOSSARY

A

A horizon The top layer of soil, characterized by the downward movement of water. (Also called *zone of leaching.*)

a'a A lava flow that solidifies with a spiny, rubbly surface.

ablation The loss of the glacial ice or snow by melting, evaporation, or breaking off into icebergs. (Also called *wastage.*)

abrasion The grinding away of rock by friction and impact during transportation.

absolute age Age given in years or some other unit of time. (Also known as *numerical age.*)

abyssal fan Great fan-shaped deposit of sediment on the deep-sea floor at the base of many submarine canyons.

abyssal plain Very flat, sediment-covered region of the deep-sea floor, usually at the base of the continental rise.

accreted terrane Terrane that did not form at its present site on a continent.

accretionary wedge (subduction complex) A wedge of thrust-faulted and folded sediment scraped off a subducting plate by the overlying plate.

active continental margin A margin consisting of a continental shelf, a continental slope, and an oceanic trench.

active solar heating The use of solar panels to heat water for hot water and heating.

active volcano A volcano that is currently erupting or has erupted recently.

actualism The principle that the same processes and natural laws that operated in the past are those we can actually observe or infer from observations as operating at present. Under present usage, *uniformitarianism* has the same meaning as actualism for most geologists.

advancing glacier Glacier with a positive budget, so that accumulation results in the lower edges being pushed outward and downward.

aerosols Small particles suspended in the atmosphere.

aftershock Small earthquake that follows a main shock.

albedo The percentage of radiation reflected from a surface.

alkali soil Soil containing such a great quantity of sodium salts precipitated by evaporating ground water that it is generally unfit for plant growth.

alluvial fan Large, fan-shaped pile of sediment that usually forms where a stream's velocity decreases as it emerges from a narrow canyon onto a flat plain at the foot of a mountain range.

alpine glaciation Glaciation of a mountainous area.

amphibole group Mineral group in which all members are double-chain silicates.

amphibolite Amphibole (hornblende), plagioclase schist.

andesite Fine-grained igneous rock of intermediate composition. Up to half of the rock is plagioclase feldspar with the rest being ferromagnesian minerals.

angle of dip A vertical angle measured downward from the horizontal plane to an inclined plane.

angular Sharp-edged; lacking rounded edges or corners.

angular unconformity An unconformity in which younger strata overlie an erosion surface on tilted or folded layered rock.

anion A negatively charged ion.

anorthosite A crystalline rock composed almost entirely of calcium-rich plagioclase feldspar.

antecedent stream A stream that maintains its original course despite later deformation of the land.

anthracite Coal that has undergone low-grade metamorphism. Burns dust-free and smokeless.

anthropogenic climate change (anthropogenic global warming) Changes to the Earth's climate caused by human activities.

anticline A fold shaped like an arch in which the rock layers usually dip away from the axis of the fold and the oldest rocks are in the center of the fold.

aphanitic Texture in which the crystals in an igneous rock are too small to distinguish with the naked eye.

aquifer A body of saturated rock or sediment through which water can move readily.

arête A sharp ridge that separates adjacent glacially carved valleys.

arch (sea arch) Bridge of rock left above an opening eroded in a headland by waves.

Archean Eon The oldest eon of Earth's geologic history, marked by formation of the crust.

arid region An area with less than 25 centimeters of rain per year.

arkose A sandstone in which more than 25% of the grains are feldspar.

artesian aquifer *See* confined aquifer.

artesian well A well in which water rises above the aquifer.

artificial recharge Groundwater recharge increased by engineering techniques.

aseismic ridge Submarine ridge with which no earthquakes are associated.

ash (volcanic) Fine pyroclasts (less than 4 millimeters).

assimilation The process in which very hot magma melts country rock and assimilates the newly molten material.

asteroid A small, generally rocky, solid body orbiting the Sun and ranging in diameter from a few meters to hundreds of kilometers.

asteroid belt Separates terrestrial from giant planets. Most asteroids are confined to the asteroid belt, which exists in an orbit between Mars and Jupiter.

asthenosphere A region of Earth's outer shell beneath the lithosphere. The asthenosphere is of indeterminate thickness and behaves plastically.

astronomical unit (AU) A distance unit based on the average distance of the Earth from the Sun.

atmosphere Gases that envelop Earth.

atmospheric lifetime The amount of time a gas or particle persists in the atmosphere.

atoll A circular reef surrounding a deeper lagoon.

atom Smallest possible particle of an element that retains the properties of that element.

atomic mass The sum of the weight of the subatomic particles in an average atom of an element, given in atomic mass units.

atomic mass number The total number of neutrons and protons in an atom.

atomic number The total number of protons in an atom.

atomic weight The sum of the weight of the subatomic particles in an average atom of an element, given in atomic mass units.

augite Mineral of the pyroxene group found in mafic igneous rocks.

aulacogen *See* failed rift.

aureole Zone of contact metamorphism adjacent to a pluton.

axial plane A plane containing all of the hinge lines of a fold.

axis *See* hinge line.

B

B horizon A soil layer characterized by the accumulation of material leached downward from the A horizon above; also called *zone of accumulation.*

backarc spreading A type of seafloor spreading that moves an island arc away from a continent, or tears an island arc in two, or splits the edge of a continent, in each case forming new sea floor.

backshore Upper part of the beach, landward of the high-water line.

bajada A broad, gently sloping, depositional surface formed at the base of a mountain range in a dry region by the coalescing of individual alluvial fans.

bar A ridge of sediment, usually sand or gravel, that has been deposited in the middle or along the banks of a stream by a decrease in stream velocity.

barchan A crescent-shaped dune with the horns of the crescent pointing downwind.

barrier island Ridge of sand paralleling the shoreline and extending above sea level.

barrier reef A reef separated from the shoreline by the deeper water of a lagoon.

basal sliding Movement in which the entire glacier slides along as a single body on its base over the underlying rock.

basalt A fine-grained, mafic, igneous rock composed predominantly of ferromagnesian minerals and with lesser amounts of calcium-rich plagioclase feldspar.

base level A theoretical downward limit for stream erosion of Earth's surface.

batholith A large, discordant pluton with an areal extent at Earth's surface greater than 100 square kilometers.

bauxite The principal ore of aluminum; $Al_2O_3 \cdot nH_2O$.

baymouth bar A ridge of sediment that cuts a bay off from the ocean.

beach Strip of sediment, usually sand but sometimes pebbles, boulders, or mud, that extends from the low-water line inland to a cliff or zone of permanent vegetation.

beach face The section of the beach exposed to wave action.

bed load Heavy or large sediment particles in a stream that travel near or on the streambed.

bedding An arrangement of layers or beds of rock.

bedding plane A nearly flat surface separating two beds of sedimentary rock.

bedrock Solid rock that underlies soil.

Benioff zone Distinct earthquake zone that begins at an oceanic trench and slopes landward and downward into Earth at an angle of about 30° to 60°.

bergschrund The crevasse that develops where a glacier is pulling away from a cirque wall.

berm Platform of wave-deposited sediment that is flat or slopes slightly landward.

biochemical Precipitated by the action of organisms.

bioclastic limestone A limestone consisting of fragments of shells, corals, and algae.

biofuel Fuel derived from biologic matter.

biosphere All of the living or once-living material on Earth.

biotite Iron/magnesium-bearing mica.

blackbody radiator Theoretical body that absorbs all energy that hits it. The intensity and wavelength of radiation emitted by a blackbody radiator is dependent upon its temperature.

block Large angular pyroclast.

blowout A depression on the land surface caused by wind erosion.

body wave Seismic wave that travels through Earth's interior.

bomb Large spindle- or lens-shaped pyroclast.

bonding Attachment of an atom to one or more adjacent atoms.

bottomset bed A delta deposit formed from the finest silt and clay, which are carried far out to sea by river flow or by sediments sliding downhill on the sea floor.

boulder A sediment particle with a diameter greater than 256 millimeters.

Bowen's reaction series The sequence in which minerals crystallize from a cooling basaltic magma.

braided stream A stream that flows in a network of many interconnected rivulets around numerous bars.

breaker A wave that has become so steep that the crest of the wave topples forward, moving faster than the main body of the wave.

breakwater An offshore structure built to absorb the force of large breaking waves and provide quiet water near shore.

brittle strain Cracking or rupturing of a body under stress.

butte A narrow pinnacle of resistant rock with a flat top and very steep sides.

C

C horizon A soil layer composed of incompletely weathered parent material.

calcareous Containing calcium carbonate.

calcite Mineral with the formula $CaCO_3$.

caldera A volcanic depression much larger than the original crater.

capacity (of stream) The total load that a stream can carry.

capillary action The drawing of water upward into small openings as a result of surface tension.

capillary fringe A thin zone near the water table in which capillary action causes water to rise above the zone of saturation.

caption Text that describes an image. It appears outside the image, usually below it.

carbon capture and sequestration Efforts to capture anthropogenic carbon dioxide at its source and store it within the Earth's crust or oceans.

carbon management Efforts to reduce carbon dioxide concentrations in the atmosphere.

carbonaceous chondrite Stony meteorite containing chondrules and composed mostly of serpentine and large quantities of organic materials.

carbonic acid H_2CO_3, a weak acid common in rain and surface waters.

cation A positively charged ion.

cave (cavern) Naturally formed underground chamber.

cement The solid material that precipitates in the pore space of sediments, binding the grains together to form solid rock.

cementation The chemical precipitation of material in the spaces between sediment grains, binding the grains together into a hard rock.

Cenozoic Era The most recent of the eras; followed the Mesozoic Era.

chain silicate structure Silicate structure in which two of each tetrahedron's oxygen ions are shared with adjacent tetrahedrons, resulting in a chain of tetrahedrons.

chalk A very fine-grained bioclastic limestone.

channel *(Mars)* Feature on the surface of the planet Mars that very closely resembles certain types of stream channels on Earth.

chaotic terrain *(Mars)* Patch of jumbled and broken angular slabs and blocks on the surface of Mars.

chemical sedimentary rock A rock composed of material precipitated directly from solution.

chemical weathering The decomposition of rock resulting from exposure to water and atmospheric gases.

chert A hard, compact, fine-grained sedimentary rock formed almost entirely of silica.

chill zone In an intrusion, the finer-grained rock adjacent to a contact with country rock.

chondrule Round silicate grain within some stony meteorites.

cinder (volcanic) Pyroclast approximately the size of a sand grain. Sometimes defined as between 4 and 32 millimeters in diameter.

cinder cone A volcano constructed of loose rock fragments ejected from a central vent.

circum-Pacific belt Major belt around the edge of the Pacific Ocean on which most composite volcanoes are located and where many earthquakes occur.

cirque A steep-sided, amphitheaterlike hollow carved into a mountain at the head of a glacial valley.

citation The information that documents the source from which an author took content. Citations that are in a print book's footnotes and endnotes are stored in Remarks.

clastic texture An arrangement of rock fragments bound into a rigid network by cement.

clay Sediment composed of particles with diameter less than 1/256 millimeter.

clay mineral A hydrous aluminum silicate that occurs as a platy grain of microscopic size with a sheet-silicate structure.

clay mineral group Collective term for clay minerals.

cleavage The ability of a mineral to break along preferred planes.

climate Average weather pattern in a region over a long period of time.

coal A sedimentary rock formed from the consolidation of plant material. It is rich in carbon, usually black, and burns readily.

coal bed methane Gas trapped in coal.

coarse-grained rock Rock in which most of the grains are larger than 1 millimeter (igneous) or 2 millimeters (sedimentary).

coast The land near the sea, including the beach and a strip of land inland from the beach.

coastal straightening The gradual straightening of an irregular shoreline by wave erosion of headlands and wave deposition in bays.

cobble A sediment particle with a diameter of 64 to 256 millimeters.

column A dripstone feature formed when a stalactite growing downward and a stalagmite growing upward meet and join.

columnar jointing Volcanic rock in parallel, usually vertical columns, mostly six-sided; also called *columnar structure.*

comet Small object in space, no more than a few kilometers in diameter, composed of frozen methane, frozen ammonia, and water-ice, with small solid particles and dust imbedded in the ices.

compaction A loss in overall volume and pore space of a rock as the particles are packed closer together by the weight of overlying material.

competence The largest particle that a stream can carry.

composite volcano (stratovolcano) A volcano constructed of alternating layers of pyroclastics and rock solidified from lava flows.

compressive stress A stress due to a force pushing together on a body.

conchoidal fracture Curved fracture surfaces.

concordant Parallel to layering or earlier developed planar structures.

concretion Hard, rounded mass that develops when a considerable amount of cementing material precipitates locally in a rock, often around an organic nucleus.

cone of depression A depression of the water table formed around a well when water is pumped out; it is shaped like an inverted cone.

confined aquifer (artesian aquifer) An aquifer completely filled with pressurized water and separated from the land surface by a relatively impermeable confining bed, such as shale.

confining pressure Pressure applied equally on all surfaces of a body; also called *lithostatic pressure.*

conglomerate A coarse-grained sedimentary rock (grains coarser than 2 millimeters) formed by the cementation of rounded gravel.

consolidation Any process that forms firm, coherent rock from sediment or from liquid.

contact Boundary surface between two different rock types or ages of rocks.

contact (thermal) metamorphism Metamorphism under conditions in which high temperature is the dominant factor.

continent-continent convergence A convergent plate boundary where two continental plates are colliding

continental crust The thick, granitic crust under continents.

continental drift A concept suggesting that continents move over Earth's surface.

continental glaciation The covering of a large region of a continent by a sheet of glacial ice.

continental rise A wedge of sediment that extends from the lower part of the continental slope to the deep-sea floor.

continental shelf A submarine platform at the edge of a continent, inclined very gently seaward generally at an angle of less than 1°.

continental slope A relatively steep slope extending from a depth of 100 to 200 meters at the edge of the continental shelf down to oceanic depths.

contour current A bottom current that flows parallel to the slopes of the continental margin (along the contour rather than down the slope).

contour line A line on a topographic map connecting points of equal elevation.

convection (convection current) The circulation of a substance driven by differences in temperature and density within that substance.

convergent plate boundary A boundary between two plates that are moving toward each other.

coquina A limestone consisting of coarse shells.

core The central zone of Earth.

correlation In geology, correlation usually means determining time equivalency of rock units. Rock units may be correlated within a region, a continent, and even between continents.

country rock Any rock that was older than and intruded by an igneous body.

covalent bonding Bonding due to the sharing of electrons by adjacent atoms.

crater (of a volcano) A basinlike depression over a vent at the summit of a volcanic cone.

craton Portion of a continent that has been structurally stable for a prolonged period of time.

creep (soil creep) Very slow, continuous downslope movement of soil or debris.

crest (of wave) The high point of a wave.

crevasse Open fissure in a glacier.

cross section *See* geologic cross section.

cross-bedding An arrangement of relatively thin layers of rock inclined at an angle to the more nearly horizontal bedding planes of the larger rock unit.

cross-cutting relationship A principle or law stating that a disrupted pattern is older than the cause of disruption.

crude oil A liquid mixture of naturally occurring hydrocarbons.

crust The outer layer of rock, forming a thin skin over Earth's surface.

crustal rebound The rise of Earth's crust after the removal of glacial ice.

crystal A homogeneous solid with an orderly internal atomic arrangement.

crystal form Arrangement of various faces on a crystal in a definite geometric relationship to one another.

crystal settling The process whereby the minerals that crystallize at a high temperature in a cooling magma move downward in the magma chamber because they are denser than the magma.

crystalline Describing a substance in which the atoms are arranged in a regular, repeating, orderly pattern.

crystalline texture An arrangement of interlocking crystals.

crystallization Crystal development and growth.

cuesta A ridge with a steep slope on one side and a gentle slope on the other side.

Curie point The temperature below which a material becomes magnetized.

D

data What scientists regard as facts.

daughter product The isotope produced by radioactive decay.

debris Unconsolidated material (soil) in which coarse-grained fragments predominate.

debris avalanche Very rapid and turbulent mass wasting of debris, air, and water.

debris flow Mass wasting involving the flow of soil (unconsolidated material) in which coarse material (gravel, boulders) is predominant.

decompression melting Partial melting of hot mantle rock when it moves upward and the pressure is reduced to the extent that the melting point drops to the temperature of the body.

deflation The removal of clay, silt, and sand particles from the land surface by wind.

delamination *See* lithospheric delamination.

delta A body of sediment deposited at the mouth of a river when the river velocity decreases as it flows into a standing body of water.

dendritic pattern Drainage pattern of a river and its tributaries that resembles the branches of a tree or veins in a leaf.

density Weight per given volume of a substance.

deposition The settling or coming to rest of transported material.

depth of focus Distance between the focus and the epicenter of an earthquake.

desert A region with low precipitation (usually defined as less than 25 centimeters per year).

desert pavement A thin layer of closely packed gravel that protects the underlying sediment from deflation; also called *pebble armor.*

desertification The expansion of barren deserts into once-populated regions.

detachment fault Major fault in a mountain belt above which rocks have been intensely folded and faulted.

detrital sedimentary rock A sedimentary rock composed of fragments of preexisting rock.

diapir Bodies of rock (e.g., rock salt) or magma that ascend within Earth's interior because they are less dense than the surrounding rock.

differential stress When pressures on a body are not of equal strength in all directions.

differential weathering Varying rates of weathering resulting from some rocks in an area being more resistant to weathering than others.

differentiation Separation of different ingredients from an originally homogeneous mixture.

dike A tabular, discordant intrusive structure.

diorite Coarse-grained igneous rock of intermediate composition. Up to half of the rock is plagioclase feldspar and the rest is ferromagnesian minerals.

dip *See* angle of dip, direction of dip.

dip-slip fault A fault in which movement is parallel to the dip of the fault surface.

directed pressure *See* differential stress.

direction of dip The compass direction in which the angle of dip is measured.

discharge In a stream, the volume of water that flows past a given point in a unit of time.

disconformity A surface that represents missing rock strata but beds above and below that surface are parallel to one another.

discordant Not parallel to any layering or parallel planes.

dissolved load The portion of the total sediment load in a stream that is carried in solution.

distributary Small shifting river channel that carries water away from the main river channel and distributes it over a delta's surface.

divergent plate boundary Boundary separating two plates moving away from each other.

divide Line dividing one drainage basin from another.

dolomite A sedimentary rock composed mostly of the mineral dolomite.

dolomitic marble Marble in which dolomite, rather than calcite, is the prevalent mineral.

dome *See* structural dome.

dormant volcano A volcano that has not erupted in many thousands of years but is expected to at some point in the future.

double refraction The splitting of light into two components when it passes through certain crystalline substances.

downcutting A valley-deepening process caused by erosion of a streambed.

drainage basin Total area drained by a stream and its tributaries.

drainage pattern The arrangement in map view of a river and its tributaries.

drawdown The lowering of the water table near a pumped well.

dripstone Deposits of calcite (and, rarely, other minerals) built up by dripping water in caves.

drumlin A long, streamlined hill made of till.

ductile Capable of being molded and bent under stress.

ductile (plastic) strain Strain in which a body is molded or bent under stress and does not return to its original shape after the stress is removed.

dust (volcanic) Finest-sized pyroclasts.

dwarf planet An object in our solar system that is in orbit around the Sun, has sufficient mass for its self-gravity to overcome rigid body forces so that it assumes a hydrostatic equilibrium (nearly round) shape, has not cleared the neighborhood around its orbit, and is not a satellite.

E

E horizon Soil horizon that is the zone of leaching, characterized by the downward movement of water and removal of fine-grained soil components.

earth In mass wasting, soil in which fine-grained particles are predominant.

Earth systems Study of Earth by analyzing how its components, or subsystems, interrelate.

earthflow Slow-to-rapid mass wasting in which fine-grained soil moves downslope as a very viscous fluid.

earthquake A trembling or shaking of the ground caused by the sudden release of energy stored in the rocks beneath the surface.

earthy luster A luster giving a substance the appearance of unglazed pottery.

echo sounder An instrument used to measure and record the depth to the sea floor.

ecliptic The plane containing most of the objects that orbit the Sun.

effusive eruption A volcanic eruption characterized by the outpouring of lava.

elastic Deformed material goes back to its original shape after stress is removed, like a rubber band that returns to its original shape after being stretched.

elastic limit The maximum amount of stress that can be applied to a body before it deforms in a permanent way by bending or breaking.

elastic rebound theory The sudden release of progressively stored strain in rocks results in movement along a fault.

elastic strain Strain in which a deformed body recovers its original shape after the stress is released.

electron A single, negative electric charge that contributes virtually no mass to an atom.

element A substance that cannot be broken down to other substances by ordinary chemical methods. Each atom of an element possesses the same number of protons.

emergent coast A coast in which land formerly under water has recently been placed above sea level, either by uplift of the land or by a drop in sea level.

end moraine A ridge of till piled up along the front edge of a glacier.

energy resources The materials we use to produce heat and electricity or as fuel for transport.

enhancement Interactive elements within an Inkling title. The majority enhancements are created using widgets, but in its broadest use, the term can also refer to poptips, inline audio, and other improvements made possible by the Inkling platform.

environment of deposition The location in which deposition occurs, usually marked by characteristic physical, chemical, or biological conditions.

eon The largest unit of geological time.

epicenter The point on Earth's surface directly above the focus of an earthquake.

epoch Each period of the standard geologic time scale is divided into epochs (e.g., Pleistocene Epoch of the Quaternary Period).

equilibrium Material is in equilibrium if it is adjusted to the physical and chemical conditions of its environment so that it does not change with time.

equilibrium line An irregular line marking the highest level to which the winter snow cover on a glacier is lost during a melt season; also called *snow line.*

era Major subdivision of the standard geologic time scale (e.g., Mesozoic Era).

erosion The physical removal of rock by an agent such as running water, glacial ice, or wind.

erratic An ice-transported boulder that does not derive from bedrock near its present site.

esker A long, sinuous ridge of sediment deposited by glacial meltwater.

estuary Drowned river mouth.

etch-pitted terrain *(Mars)* A terrain on the surface of Mars characterized by small pits.

evaporite Rock that forms from crystals precipitating during evaporation of water.

exfoliation The stripping of concentric rock slabs from the outer surface of a rock mass.

exfoliation dome A large, rounded landform developed in a massive rock, such as granite, by the process of exfoliation.

exotic terrane Terrane that did not form at its present site on a continent and traveled a great distance to get to its present site.

expansive clay Clay that increases in volume when water is added to it.

explosive eruption A volcanic eruption characterized by the violent ejection of pyroclastic material.

extension Strain involving an increase in length. Extension can cause crustal thinning and faulting.

extinct volcano A volcano that has not erupted for a very long time and shows no signs of erupting ever again.

extrusive rock Any igneous rock that forms at Earth's surface, whether it solidifies directly from a lava flow or is pyroclastic.

F

faceted A rock fragment with one or more flat surfaces caused by erosive action.

failed rift (aulacogen) An inactive, sediment-filled rift that forms above a mantle plume. The rift becomes inactive as two other rifts widen to form an ocean.

fall The situation in mass wasting that occurs when material free-falls or bounces down a cliff.

fault A fracture in bedrock along which movement has taken place.

fault-block mountain range A range created by uplift along normal or vertical faults.

faunal succession A principle or law stating that fossil species succeed one another in a definite and recognizable order; in general, fossils in progressively older rock show increasingly greater differences from species living at present.

feedback Factor that amplifies (positive feedback) or decreases (negative feedback) the rate of a process.

feldspar Group of most common minerals of Earth's crust. All feldspars contain silicon, aluminum, and oxygen and may contain potassium, calcium, and sodium.

felsic Silica-rich igneous rock or magma with a relatively high content of potassium and sodium.

ferromagnesian mineral Iron/magnesium-bearing mineral, such as augite, hornblende, olivine, or biotite.

fine-grained rock A rock in which most of the mineral grains are less than 1 millimeter across (igneous) or less than 1/16 millimeter (sedimentary).

fiord A coastal inlet that is a glacially carved valley, the base of which is submerged.

firn A compacted mass of granular snow, transitional between snow and glacier ice.

firn limit *See* equilibrium line.

fissility The ability of a rock to split into thin layers.

fission The splitting apart of the nucleus of an atom into lighter components.

flank eruption An eruption in which lava erupts out of a vent on the side of a volcano.

flash flood Flood of very high discharge and short duration; sudden and local in extent.

flood basalts (plateau basalts) Layers of basalt flows that have built up to great thicknesses.

flood plain A broad strip of land built up by sedimentation on either side of a stream channel.

flow A type of movement that implies that a descending mass is moving downslope as a viscous fluid.

flowstone Calcite precipitated by flowing water on cave walls and floors.

flux melting Melting process where the addition of a compound to a material can lower the melting temperature of the material.

focus The point within Earth from which seismic waves originate in an earthquake.

fold Bend in layered bedrock.

fold and thrust belt A portion of a major mountain belt characterized by large thrust faults, stacked one upon another. Layered rock between the faults was folded when faulting was taking place.

fold axis *See* hinge line.

foliation Parallel alignment of textural and structural features of a rock.

footwall The underlying surface of an inclined fault plane.

forcings Factors that affect the radiative balance of the Earth leading to global warming (positive forcings) or global cooling (negative forcings).

foreland basin A basin that forms parallel to a back-arc fold and thrust belt and is associated with bending of the lithosphere during ocean-continent convergence.

foreset bed A sediment layer in the main part of a delta, deposited at an angle to the horizontal.

foreshock Small earthquake that precedes a main shock.

foreshore The zone that is regularly covered and uncovered by the rise and fall of tides.

formation A body of rock of considerable thickness that has a recognizable unity or similarity making it distinguishable from adjacent rock units. Usually composed of one bed or several beds of sedimentary rock, although the term is also applied to units of metamorphic and igneous rock. A convenient unit for mapping, describing, or interpreting the geology of a region.

fossil Traces of plants or animals preserved in rock.

fossil fuel Energy resources formed from the buried remains of plants and animals.

fracture The way a substance breaks where not controlled by cleavage.

fracture zone Major line of weakness in Earth's crust that crosses the mid-oceanic ridge at approximately right angles.

fracturing Cracking or rupturing of a body under stress.

fragmental rock A term used to describe an igneous rock made up of volcanic fragments such as ash and lapilli.

framework silicate structure Crystal structure in which all four oxygen ions of a silica tetrahedron are shared by adjacent ions.

fretted terrain *(Mars)* Flat lowland with some scattered high plateaus on the surface of Mars.

fringing reef A reef attached directly to shore. *See* barrier reef.

frost action Mechanical weathering of rock by freezing water.

frost heaving The lifting of rock or soil by the expansion of freezing water.

frost wedging A type of frost action in which the expansion of freezing water pries a rock apart.

G

gabbro A mafic, coarse-grained igneous rock composed predominantly of ferromagnesian minerals and with lesser amounts of calcium-rich plagioclase feldspar.

gaining stream A stream that receives water from the zone of saturation.

geode Partly hollow, globelike body found in limestone or other cavernous rock.

geoengineering Large-scale efforts to modify Earth systems such as the climate.

geologic cross section A representation of a portion of Earth in a vertical plane.

geologic map A map representing the geology of a given area.

geologic resources Valuable materials of geologic origin that can be extracted from Earth.

geology The scientific study of the planet Earth.

geomagnetic polarity time scale Time scale that shows the onset and duration of reversal of Earth's magnetic field.

geophysics The application of physical laws and principles to a study of Earth.

geosphere (solid Earth system) The rock and other inorganic material that make up the bulk of the planet.

geothermal energy Energy produced by harnessing naturally occurring steam and hot water.

geothermal gradient Rate of temperature increase associated with increasing depth beneath the surface of Earth (normally about 25°C per kilometer).

geyser A type of hot spring that periodically erupts hot water and steam.

geyserite A deposit of silica that forms around many geysers and hot springs.

glacier A large, long-lasting mass of ice, formed on land by the compaction and recrystallization of snow, which moves because of its own weight.

glacier ice The mosaic of interlocking ice crystals that form a glacier.

glassy (or vitreous) luster A luster that gives a substance a glazed, porcelainlike appearance.

glassy rocks Igneous rocks primarily composed of glass and containing few, if any, crystals.

global warming potential (GWP) Relative measure of how much heat a particular gas will trap in the atmosphere over a period of time. By definition, carbon dioxide has a GWP of 1.

gneiss A metamorphic rock composed of light and dark layers or lenses.

gneissic The texture of a metamorphic rock in which minerals are separated into light and dark layers or lenses.

goethite The commonest mineral in the limonite group; $Fe_2O_3 \cdot nH_2O$.

Gondwanaland The southern part of *Pangaea* (*see definition*) that formed South America, Africa, India, Australia, and Antarctica.

graben A downdropped block bounded by normal fault.

graded bed A single bed with coarse grains at the bottom of the bed and progressively finer grains toward the top of the bed.

graded stream A stream that exhibits a delicate balance between its transporting capacity and the sediment load available to it.

granite A felsic, coarse-grained, intrusive igneous rock containing quartz and composed mostly of potassium- and sodium-rich feldspars.

gravel Rounded particles coarser than 2 millimeters in diameter.

gravitational collapse and spreading When part of a mountain belt becomes too high, it moves vertically downward, forcing rock at depth to spread out laterally.

gravity The force of attraction that two bodies exert on each other that is proportional to the product of their masses and inversely proportional to the square of the distance from the centers of the two bodies.

gravity anomaly A deviation from the average gravitational attraction between Earth and an object. *See* negative gravity anomaly, positive gravity anomaly.

gravity meter An instrument that measures the gravitational attraction between Earth and a mass within the instrument.

graywacke A sandstone with more than 15% fine-grained matrix between the sand grains.

greenhouse effect The trapping of heat by a planet's atmosphere, making the planet warmer than would otherwise be expected. Generally, the greenhouse effect operates if visible sunlight passes freely through a planet's atmosphere, but the infrared radiation produced by the warm surface cannot escape readily into space.

greenhouse gases Gases in the atmosphere that efficiently absorb long-wave (terrestrial) radiation.

groin Short wall built perpendicular to shore to trap moving sand and widen a beach.

ground moraine A blanket of till deposited by a glacier or released as glacier ice melted.

groundwater The water that lies beneath the ground surface, filling the cracks, crevices, and pore space of rocks.

guyot Flat-topped seamount.

H

Hadean Eon The oldest eon, characterized by Earth's initial formation.

half-life The time it takes for a given amount of a radioactive isotope to be reduced by one-half.

hanging valley A smaller valley that terminates abruptly high above a main valley.

hanging wall The overlying surface of an inclined fault plane.

hardness The relative ease or difficulty with which a smooth surface of a mineral can be scratched; commonly measured by Mohs scale.

headland Point of land along a coast.

headward erosion The lengthening of a valley in an uphill direction above its original source by gullying, mass wasting, and sheet erosion.

heat engine A device that converts heat energy into mechanical energy.

heat flow Gradual loss of heat (per unit of surface area) from Earth's interior out into space.

heavy crude Dense, viscous petroleum that flows slowly or not at all.

hematite A type of iron oxide that has a brick-red color when powdered; Fe_2O_3.

highland *(Moon)* A rugged region of the lunar surface representing an early period in lunar history when intense meteorite bombardment formed craters.

hinge line Line about which a fold appears to be hinged. Line of maximum curvature of a folded surface.

hinge plane *See* axial plane.

hogback A sharp-topped ridge formed by the erosion of steeply dipping beds.

Holocene (Recent) Epoch The youngest epoch, which began around 10,000 years ago and is continuing presently.

horn A sharp peak formed where cirques cut back into a mountain on several sides.

hornblende Common amphibole frequently found in igneous and metamorphic rocks.

hornfels A fine-grained, unfoliated metamorphic rock.

horst An upraised block bounded by normal faults.

hot spot An area of volcanic eruptions and high heat flow above a rising mantle plume.

hot spring Spring with a water temperature warmer than human body temperature.

hydraulic action The ability of water to pick up and move rock and sediment.

hydrologic cycle The movement of water and water vapor from the sea to the atmosphere, to the land, and back to the sea and atmosphere again.

hydrology The study of water's properties, circulation, and distribution.

hydrosphere The water on or near Earth's surface.

hydrothermal metamorphism Metamorphism that involves the interaction of rock with high-temperature fluids.

hydrothermal vein Quartz or other minerals that have been deposited in a crack by hot fluids.

hypocenter Synonym for the focus of an earthquake.

hypothesis A tentative theory.

I

ice cap A glacier covering a relatively small area of land but not restricted to a valley.

ice sheet A glacier covering a large area (more than 50,000 square kilometers) of land.

iceberg Block of glacier-derived ice floating in water.

icefall A chaotic jumble of crevasses that split glacier ice into pinnacles and blocks.

igneous rock A rock formed or apparently formed from solidification of magma.

incised meander A meander that retains its sinuous curves as it cuts vertically downward below the level at which it originally formed.

inclusion A fragment of rock that is distinct from the body of igneous rock in which it is enclosed.

inclusion, principle of Fragments included in a host rock are older than the host rock.

index fossil A fossil from a very short-lived species known to have existed during a specific period of geologic time.

inline image An image that appears on a line of text. Special symbols or equations in the text that cannot be rendered in HTML are rendered as inline images.

inner planet A planet orbiting in the inner part of the Solar System. Sometimes taken to mean Mercury, Venus, Earth, and Mars.

intensity A measure of an earthquake's size by its effect on people and buildings.

Intergovernmental Panel on Climate Change (IPCC) An international panel of scientists that assesses the evidence for and effects of climate change.

intermediate Igneous rock or magma with a chemical content between felsic and mafic compositions.

intrusion (intrusive structure) A body of intrusive rock classified on the basis of size, shape, and relationship to surrounding rocks.

intrusive rock Rock that appears to have crystallized from magma emplaced in surrounding rock.

ion An electrically charged atom or group of atoms.

ionic bonding Bonding due to the attraction between positively charged ions and negatively charged ions.

iron meteorite A meteorite composed principally of iron-nickel alloy.

island arc A curved line of islands.

isoclinal fold A fold in which the limbs are parallel to one another.

isolated silicate structure Silicate minerals that are structured so that none of the oxygen atoms are shared by silica tetrahedrons.

isostasy The balance or equilibrium between adjacent blocks of crust resting on a plastic mantle.

isostatic adjustment Concept of vertical movement of sections of Earth's crust to achieve balance or equilibrium.

isotherm A line along which the temperature of rock (or other material) is the same.

isotopes Atoms (of the same element) that have different numbers of neutrons but the same number of protons.

isotopic dating Determining the age of a rock or mineral through its radioactive elements and decay products (previously and somewhat inaccurately called *radiometric* or *radioactive dating*).

J

jetty Rock wall protruding above sea level, designed to protect the entrance of a harbor from sediment deposition and storm waves; usually built in pairs.

joint A fracture or crack in bedrock along which essentially no displacement has occurred.

joint set Joints oriented in one direction approximately parallel to one another.

K

kame Low mound or irregular ridge formed of outwash deposits on a stagnating glacier.

kame and kettle topography Irregular, bumpy landscape of hills and depressions associated with many moraines.

karst topography An area with many sinkholes and a cave system beneath the land surface and usually lacking a surface stream.

kettle A depression caused by the melting of a stagnant block of ice that was surrounded by sediment.

kimberlite An ultramafic rock that contains olivine along with mica, garnet, or both. Diamonds are found in some kimberlite bodies.

kinetic energy The energy that an object possesses due to its motion.

komatiite An ultramafic extrusive igneous rock.

L

laccolith A concordant intrusive structure, similar to a sill, with the central portion thicker and domed upward. Laccoliths are not common.

lahar A volcanic mudflow.

laminar flow Slow, smooth flow, with each drop of water traveling a smooth path parallel to its neighboring drops.

laminated terrain *(Mars)* Area where series of alternating light and dark layers can be seen on the surface of Mars.

lamination A thin layer in sedimentary rock (less than 1 centimeter thick).

landform A characteristically shaped feature of Earth's surface, such as a hill or a valley.

landslide The general term for a slowly to very rapidly descending mass of rock or debris.

lapilli (plural) Pyroclasts in the 2–64 millimeter size range (singular, *lapillus*).

lateral continuity Principle that states that an original sedimentary layer extends laterally until it tapers or thins at its edges.

lateral erosion Erosion and undercutting of stream banks caused by a stream swinging from side to side across its valley floor.

lateral moraine A low, ridgelike pile of till along the side of a glacier.

lava Magma on Earth's surface.

lava dome Steep-sided dome- or spine-shaped mass of volcanic rock.

lava flow Flow of lava from a crater or fissure.

lava tube Tunnel-like cave within a lava flow. It forms during the late stages of solidification of a mafic lava flow.

left-lateral fault A strike-slip fault in which the block seen across the fault appears displaced to the left.

limb Portion of a fold shared by an anticline and a syncline.

limestone A sedimentary rock composed mostly of calcite.

limonite A type of iron oxide that is yellowish-brown when powdered; $Fe_2O_3{\cdot}nH_2O$.

liquefaction A type of ground failure in which water-saturated sediment turns from a solid to a liquid as a result of shaking, often caused by an earthquake.

lithification The consolidation of sediment into sedimentary rock.

lithosphere The rigid outer shell of Earth, 70 to 125 or more kilometers thick.

lithospheric delamination The detachment of part of the mantle portion of the lithosphere beneath a mountain belt.

lithostatic pressure Confining pressure due to the weight of overlying rock.

loam Soil containing approximately equal amounts of sand, silt, and clay.

loess A fine-grained deposit of wind-blown dust.

longitudinal dune (seif) Large, symmetrical ridge of sand parallel to the wind direction.

longitudinal profile A line showing a stream's slope, drawn along the length of the stream as if it were viewed from the side.

longshore current A moving mass of water that develops parallel to a shoreline.

longshore drift Movement of sediment parallel to shore when waves strike a shoreline at an angle.

losing stream Stream that loses water to the zone of saturation.

Love waves A type of surface seismic wave that causes the ground to move side to side in a horizontal plane perpendicular to the direction the wave is traveling.

low-velocity zone Mantle zone at a depth of about 100 kilometers where seismic waves travel more slowly than in shallower layers of rock.

luster The quality and intensity of light reflected from the surface of a mineral.

M

mafic Silica-deficient igneous rock or magma with a relatively high content of magnesium, iron, and calcium.

magma Molten rock, usually mostly silica. The liquid may contain dissolved gases as well as some solid minerals.

magmatic arc A line of batholiths or volcanoes. Generally the line, as seen from above, is curved.

magmatic underplating *See* underplating.

magnetic anomaly A deviation from the average strength of Earth's magnetic field. *See* negative magnetic anomaly, positive magnetic anomaly.

magnetic field Region of magnetic force that surrounds Earth.

magnetic pole An area where the strength of the magnetic field is greatest and where the magnetic lines of force appear to leave or enter Earth.

magnetic reversal A change in Earth's magnetic field between normal polarity and reversed polarity. In normal polarity, the north magnetic pole, where magnetic lines of force enter Earth, lies near the geographic North Pole. In reversed polarity, the south magnetic pole, where lines of force leave Earth, lies near the geographic North Pole (the magnetic poles have exchanged positions).

magnetite An iron oxide that is attracted to a magnet.

magnetometer An instrument that measures the strength of Earth's magnetic field.

magnitude A measure of the energy released during an earthquake.

major mountain belt A long chain (thousands of kilometers) of mountain ranges.

mantle A thick shell of rock that separates Earth's crust above from the core below.

mantle diapir A body of mantle rock, hotter than its surroundings, that ascends because it is less dense than the surrounding rock.

mantle plume Narrow column of hot mantle rock that rises and spreads radially outward.

marble A coarse-grained rock composed of interlocking calcite (or dolomite) crystals.

maria *(Moon)* Lava plains on Moon's surface (singular, *mare*).

marine terrace A broad, gently sloping platform that may be exposed at low tide.

mass wasting (or mass movement) Movement, caused by gravity, in which bedrock, rock debris, or soil moves downslope in bulk.

matrix Fine-grained material found in the pore space between larger sediment grains.

meander A pronounced sinuous curve along a stream's course.

meander cutoff A new, shorter channel across the narrow neck of a meander.

meander scar An abandoned meander filled with sediment and vegetation.

mechanical weathering The physical disintegration of rock into smaller pieces.

medial moraine A single long ridge of till on a glacier, formed by adjacent lateral moraines joining and being carried downglacier.

Mediterranean-Himalayan belt (Mediterranean belt) A major concentration of earthquakes and composite volcanoes that runs through the Mediterranean Sea, crosses the Mideast and the Himalaya, and passes through the East Indies.

melt Liquid rock resulting from melting in a laboratory.

Mercalli scale *See* modified Mercalli scale.

mesa A broad, flat-topped hill bounded by cliffs and capped with a resistant rock layer.

Mesozoic Era The era that followed the Paleozoic Era and preceded the Cenozoic Era.

metallic bonding Bonding, as in metals, whereby atoms are closely packed together and electrons move freely among atoms.

metallic luster Luster giving a substance the appearance of being made of metal.

metamorphic facies Metamorphic rocks that contain the same set of pressure- or temperature-sensitive minerals are regarded as belonging to the same facies, implying that they formed under broadly similar pressure and temperature conditions.

metamorphic facies an assemblage of metamorphic minerals that indicate the pressure and temperature conditions under which metamorphism took place

metamorphic rock A rock produced by metamorphism.

metamorphism The transformation of preexisting rock into texturally or mineralogically distinct new rock as a result of high temperature, high pressure, or both but without the rock melting in the process.

metasomatism Metamorphism coupled with the introduction of ions from an external source.

meteor Fragment that passes through Earth's atmosphere, heated to incandescence by friction; sometimes incorrectly called "shooting" or "falling" stars.

meteorite Meteor that strikes Earth's surface.

meteoroid Small, solid particles of stone and/or metal orbiting the Sun.

mica group Group of minerals with a sheet-silicate structure.

microcline (potassium) feldspar A feldspar with the formula $KAlSi_3O_8$.

mid-oceanic ridge A giant mountain range that lies under the ocean and extends around the world.

migmatite A mixed igneous and metamorphic rock, formed when a metamorphic rock partially melts

migmatite Mixed igneous and metamorphic rock.

Milankovich cycles Periodic changes in the shape of Earth's orbit around the sun as well as variations in position of Earth's spin axis. The changes affect the amount of solar energy reaching different parts of Earth. They are thought to largely control the timing of glacial and interglacial ages.

Milankovitch theory Theory of how variations in the Earth's orbit and tilt with respect to the Sun determine climate on time scales of thousands of years.

Milky Way galaxy The galaxy to which the Sun belongs. Seen from Earth, the galaxy is a pale, milky band in the night sky.

mineral An inorganic crystalline substance that is naturally occurring and is chemically and physically distinctive.

mineraloid A substance that is not crystalline but otherwise would be considered a mineral.

model In science, a model is an image–graphic, mathematical, or verbal–that is consistent with the known data.

modified Mercalli scale Scale expressing intensities of earthquakes (judged on amount of damage done) in Roman numerals ranging from I to XII.

Mohorovičić discontinuity The boundary separating the crust from the mantle beneath it (also called Moho).

Mohs' hardness scale Scale on which ten minerals are designated as standards of hardness.

molecule The smallest possible unit of a substance that has the properties of that substance.

moment magnitude An earthquake magnitude calculated from the strength of the rock, surface area of the fault rupture, and the amount of rock displacement along the fault.

monocline A local steeping in a gentle regional dip; a steplike fold in rock.

moraine A body of till either being carried on a glacier or left behind after a glacier has receded.

mountain range A group of closely spaced mountains or parallel ridges.

mud Term loosely used for silt and clay, usually wet.

mud crack Polygonal crack formed in very fine-grained sediment as it dries.

mudflow A flowing mixture of debris and water, usually moving down a channel.

mudstone A fine-grained sedimentary rock that lacks shale's laminations and fissility.

muscovite Transparent or white mica that lacks iron and magnesium.

N

natural gas A gaseous mixture of naturally occurring hydrocarbons.

natural levee Low ridges of flood-deposited sediment formed on either side of a stream channel, which thin away from the channel.

nebula A large volume of interstellar gas and dust.

Nebular Hypothesis The hypothesis that the Solar System formed from a rotating cloud of gas and dust, the solar nebula.

negative gravity anomaly Less than normal gravitational attraction.

negative magnetic anomaly Less than average strength of Earth's magnetic field.

neutron A subatomic particle that contributes mass to an atom and is electrically neutral.

nonconformity An unconformity in which an erosion surface on plutonic or metamorphic rock has been covered by younger sedimentary or volcanic rock.

nonmetallic luster Luster that gives a substance the appearance of being made of something other than metal (e.g., glassy).

nonrenewable resource A resource that forms at extremely slow rates compared to its rate of consumption.

normal fault A fault in which the hanging-wall block moved down relative to the footwall block.

nucleus Protons and neutrons form the nucleus of an atom. Although the nucleus occupies an extremely tiny fraction of the volume of the entire atom, practically all the mass of the atom is concentrated in the nucleus.

numerical age Age given in years or some other unit of time.

O

O horizon Dark-colored soil layer that is rich in organic material and forms just below surface vegetation.

oblique-slip fault A fault with both strike-slip and dip-slip components.

obsidian Volcanic glass.

ocean acidification Increase in acidity of ocean water as a result of increasing amounts of carbon dioxide absorbed by the oceans.

ocean-continent convergence A plate boundary where oceanic crust meets continental crust.

ocean-ocean convergence A convergent plate boundary where oceanic crust meets oceanic crust.

oceanic crust The thin, basaltic crust under oceans.

oceanic trench A narrow, deep trough parallel to the edge of a continent or an island arc.

oil *See* crude oil.

oil (tar) sand Asphalt-cemented sand deposit.

oil field An area underlain by one or more oil pools.

oil pool Underground accumulation of oil.

oil shale Shale with a high content of organic matter from which oil may be extracted by distillation.

oil trap A set of conditions that hold petroleum in a reservoir rock and prevent its escape by migration.

olivine A ferromagnesian mineral with the formula $(Fe, Mg)_2SiO_4$.

oolite (ooid) A small sphere of calcite precipitated from seawater.

oolitic limestone A limestone formed from oolites.

opal A mineraloid composed of silica and water.

open fold A fold with gently dipping limbs and the angle between the limbs is large.

open-pit mine Mine in which ore is exposed at the surface in a large excavation.

ophiolite A distinctive rock sequence found in many mountain ranges on continents.

ore Naturally occurring material that can be profitably mined.

ore mineral A mineral of commercial value.

organic sedimentary rock Rock composed mostly of the remains of plants and animals.

original horizontality The deposition of most water-laid sediment in horizontal or near-horizontal layers that are essentially parallel to Earth's surface.

orogeny An episode of intense deformation of the rocks in a region, generally accompanied by metamorphism and plutonic activity.

orthoclase (potassium) feldspar A feldspar with the formula $KAlSi_3O_8$.

outcrop A surface exposure of bare rock, not covered by soil or vegetation.

outer planet A planet whose orbit lies in the outer part of the solar system. Jupiter, Saturn, Uranus, and Neptune are outer planets.

outwash Material deposited by debris-laden meltwater from a glacier.

overburden The upper part of a sedimentary deposit. Its weight causes compaction of the lower part.

overturned fold A fold in which both limbs dip in the same direction.

oxbow lake A crescent-shaped lake occupying the abandoned channel of a stream meander that is isolated from the present channel by a meander cutoff and sedimentation.

P

P wave A compressional wave (seismic wave) in which rock vibrates parallel to the direction of wave propagation.

P-wave shadow zone The region on Earth's surface, 103° to 142° away from an earthquake epicenter, in which P waves from the earthquake are absent.

pahoehoe A lava flow characterized by a ropy or billowy surface.

paired terraces *Stream terraces* (*see definition*) that occur at the same elevation on each side of a river.

paleoclimatology The study of ancient climate.

paleomagnetism A study of ancient magnetic fields.

Paleozoic Era The era that followed the Precambrian and began with the appearance of complex life, as indicated by fossils.

Pangaea A supercontinent that broke apart 200 million years ago to form the present continents.

parabolic dune A deeply curved dune in a region of abundant sand. The horns point upwind and are often anchored by vegetation.

parent rock (protolith) Original rock before being metamorphosed.

partial melting Melting of the components of a rock with the lowest melting temperatures.

passive continental margin A margin that includes a continental shelf, continental slope, and continental rise that generally extends down to an abyssal plain at a depth of about 5 kilometers.

passive solar heating Heating achieved through efficient building design.

paternoster lakes A series of rock-basin lakes carved by glacial erosion.

pattern Inkling term for a snippet of HTML code that includes a reference to CSS. Readymade patterns are available in Habitat's Pattern Picker and Inkling's Pattern Library.

peat A brown, lightweight, unconsolidated or semi-consolidated deposit of plant remains.

pebble A sediment particle with a diameter of 2 to 64 millimeters.

pediment A gently sloping erosional surface cut into the solid rock of a mountain range in a dry region; usually covered with a thin veneer of gravel.

pegmatite Extremely coarse-grained igneous rock.

pelagic sediment Sediment made up of fine-grained clay and the skeletons of microscopic organisms that settle slowly down through the ocean water.

perched water table A water table separated from the main water table beneath it by a zone that is not saturated.

peridotite Ultramafic rock composed of pyroxene and olivine.

period Each era of the standard geologic time scale is subdivided into periods (e.g., the Cretaceous Period).

permafrost Ground that remains permanently frozen for many years.

permeability The capacity of a rock to transmit a fluid such as water or petroleum.

petrified wood A material that forms as the organic matter of buried wood is either filled in or replaced by inorganic silica carried in by groundwater.

petroleum Crude oil and natural gas. (Some geologists use petroleum as a synonym for oil.)

phaneritic Texture in which the crystals making up an igneous rock are distinguishable without using a microscope.

Phanerozoic Eon Eon of geologic time. Includes all time following the Precambrian.

phenocryst Any of the large crystals in porphyritic igneous rock.

photovoltaic cells Thin wafers of silicon used to convert sunlight directly into electricity.

phyllite A metamorphic rock in which clay minerals have recrystallized into microscopic micas, giving the rock a silky sheen.

physical continuity Being able to physically follow a rock unit between two places.

physical geology A large division of geology concerned with Earth materials, changes of the surface and interior of Earth, and the forces that cause those changes.

pillow basalts (pillow structure) Rocks, generally basalt, formed in pillow-shaped masses fitting closely together; caused by underwater lava flows.

placer deposit Valuable mineral grains concentrated in sedimentary deposit by streamflow or wave action.

plagioclase feldspar A feldspar containing sodium, calcium, or both, in addition to aluminum, silicon, and oxygen.

planet A body in orbit around a star.

planetesimal Small, planet-like body.

plastic Capable of being molded and bent under stress.

plastic flow Movement within a glacier in which the ice is not fractured.

plate A large, mobile slab of rock making up part of Earth's surface.

plate tectonics A theory that Earth's surface is divided into a few large, thick plates that are slowly moving and changing in size. Intense geologic activity occurs at the plate boundaries.

plateau Broad, flat-topped area elevated above the surrounding land and bounded, at least in part, by cliffs.

playa A very flat surface underlain by hard, mud-cracked clay.

playa lake A shallow temporary lake (following a rainstorm) on a flat valley floor in a dry region.

Pleistocene Epoch An epoch of the Quaternary Period characterized by several glacial ages.

plunging fold A fold in which the hinge line (or axis) is not horizontal.

pluton An igneous body that crystallized deep underground.

plutonic rock Igneous rock formed at great depth.

pluvial lake A lake formed during an earlier time of abundant rainfall.

point bar A stream *bar* (*see definition*) deposited on the inside of a curve in the stream, where the water velocity is low.

polar wandering An apparent movement of the Earth's poles.

polarity *See* magnetic reversal.

polymorphs Substances having the same chemical composition but different crystal structures (e.g., diamond and graphite).

pore space The total amount of space taken up by openings between sediment grains.

porosity The percentage of a rock's volume that is taken up by openings.

porphyritic rock An igneous rock in which large crystals are enclosed in a matrix (or groundmass) of much finer-grained minerals or obsidian.

positive gravity anomaly Greater than normal gravitational attraction.

positive magnetic anomaly Greater than average strength of the Earth's magnetic field.

potassium feldspar A feldspar with the formula $KAlSi_3O_8$.

potential energy The stored energy of position.

pothole Depression eroded into the hard rock of a streambed by the abrasive action of the stream's sediment load.

Precambrian The vast amount of time that preceded the Paleozoic Era.

Precambrian shield A complex of old Precambrian metamorphic and plutonic rocks exposed over a large area.

pressure release A significant type of mechanical weathering that causes rocks to crack when overburden is removed.

prograde metamorphism Metamorphism in which progressively greater pressure and temperature act on a rock type with increasing depth in Earth's crust.

Proterozoic Eon Eon of Precambrian time.

proton A subatomic particle that contributes mass and a single positive electrical charge to an atom.

pumice A frothy volcanic glass.

pyroclast Fragment of rock formed by volcanic explosion.

pyroclastic debris Rock fragments produced by volcanic explosion.

pyroclastic flow Turbulent mixture of pyroclastics and gases flowing down the flank of a volcano.

pyroxene group Mineral group, all members of which are single-chain silicates.

Q

quartz Mineral with the formula SiO_2.

quartz sandstone A sandstone in which more than 90% of the grains are quartz.

quartzite A rock composed of sand-sized grains of quartz that have been welded together during metamorphism.

Quaternary Period The youngest geologic period; includes the present time.

R

radial pattern A drainage pattern in which streams diverge outward like spokes of a wheel.

radiative balance The balance between the amount of energy entering and exiting the Earth system.

radiative forcing A change in the energy balance of the Earth system.

radioactive decay The spontaneous nuclear disintegration of certain isotopes.

radioactivity The spontaneous nuclear disintegration of atoms of certain isotopes.

radon A radioactive gas produced by the radioactive decay of uranium.

rain shadow A region on the downwind side of mountains that has little or no rain because of the loss of moisture on the upwind side of the mountains.

rampart crater *(Mars)* Meteorite crater that is surrounded by material that appears to have flowed from the point of impact.

rayed crater *(Moon)* Crater with bright streaks radiating from it on the Moon's surface.

Rayleigh waves A type of surface seismic wave that behaves like a rolling ocean wave and causes the ground to move in an elliptical path.

receding glacier A glacier with a negative budget, which causes the glacier to grow smaller as its edges melt back.

Recent (Holocene) Epoch The present epoch of the Quaternary Period.

recessional moraine An end moraine built during the retreat of a glacier.

recharge The addition of new water to an aquifer or to the zone of saturation.

reclamation Restoration of the land to usable condition after mining has ceased.

recrystallization The development of new crystals in a rock, often of the same composition as the original grains.

rectangular pattern A drainage pattern in which tributaries of a river change direction and join one another at right angles.

recumbent fold A fold overturned to such an extent that the limbs are essentially horizontal.

reef A resistant ridge of calcium carbonate formed on the sea floor by corals and coralline algae.

regional metamorphism Metamorphism that takes place at considerable depth underground.

regolith Loose, unconsolidated rock material resting on bedrock.

relative time The sequence in which events took place (not measured in time units).

relief The vertical distance between points on Earth's surface.

reserves The discovered deposits of a geologic material that are economically and legally feasible to recover under present circumstances.

reservoir rock A rock that is sufficiently porous and permeable to store and transmit petroleum.

residual clay Fine-grained particles left behind as insoluble residue when a limestone containing clay dissolves.

residual soil Soil that develops directly from weathering of the rock below.

resources The total amount of a geologic material in all its deposits, discovered and undiscovered. *See also* reserves.

reverse fault A fault in which the hanging-wall block moved up relative to the footwall block.

revision Every time a project team member saves a change to content in Habitat, that revision of the file is archived with a unique number. A revision automatically generates analyses of the content, and the results are presented in the Reports area of Habitat. Because every revision is given a unique number and is archived, you can view the state of a project at any moment in its history.

rhyolite A fine-grained, felsic, igneous rock made up mostly of feldspar and quartz.

Richter scale A numerical scale of earthquake magnitudes.

ridge push The concept that oceanic plates diverge as a result of sliding down the sloping lithosphere-asthenosphere boundary.

rift valley A tensional valley bounded by normal faults. Rift valleys are found at diverging plate boundaries on continents and along the crest of the mid-oceanic ridge.

right-lateral fault A strike-slip fault in which the block seen across the fault appears displaced to the right.

rigid zone Upper part of a glacier in which there is no plastic flow.

rille *(Moon)* Elongate trenched or cracklike valley on the lunar surface.

rip current Narrow currents that flow straight out to sea in the surf zone, returning water seaward that has been pushed ashore by breaking waves.

ripple mark Any of the small ridges formed on sediment surfaces exposed to moving wind or water. The ridges form perpendicularly to the motion.

rock Naturally formed, consolidated material composed of grains of one or more minerals. (There are a few exceptions to this definition.)

rock avalanche A very rapidly moving, turbulent mass of broken-up bedrock.

rock cycle A theoretical concept relating tectonism, erosion, and various rock-forming processes to the common rock types.

rock flour A powder of fine fragments of rock produced by glacial abrasion.

rock gypsum An evaporite composed of gypsum.

rock salt An evaporite composed of halite.

rock-basin lake A lake occupying a depression caused by glacial erosion of bedrock.

rockfall Rock falling freely or bouncing down a cliff.

rockslide Rapid sliding of a mass of bedrock along an inclined surface of weakness.

rotational slide In mass wasting, movement along a curved surface in which the upper part moves vertically downward while the lower part moves outward. Also called a slump.

rounded knobs (glacial) Bedrock that is more resistant to glacial erosion stands out as rounded knobs, usually elongated parallel to the direction of glacier flow. These are also known as *roche moutonnées* (French for "rock sheep").

rounding The grinding away of sharp edges and corners of rock fragments during transportation.

rubble Angular sedimentary particles coarser than 2 millimeters in diameter.

S

S wave A seismic wave propagated by a shearing motion, which causes rock to vibrate perpendicular to the direction of wave propagation.

S-wave shadow zone The region on Earth's surface (at any distance more than 103° from an earthquake epicenter) in which S waves from the earthquake are absent.

saltation A mode of transport that carries sediment downcurrent in a series of short leaps or bounces.

sand Sediment composed of particles with a diameter between 1/16 and 2 millimeters.

sand dune A mound of loose sand grains heaped up by the wind.

sandbox chapter A chapter that is part of a Habitat project but is not meant to be part of your final Inkling title. It can contain examples of content, reusable HTML patterns, or exploratory material. When you create a such chapter in Table of Contents, identify it as a sandbox chapter in the Inspector.

sandstone A medium-grained sedimentary rock (grains between 1/16 and 2 millimeters) formed by the cementation of sand grains.

saturated zone A subsurface zone in which all rock openings are filled with water.

scale The relationship between distance on a map and the distance on the terrain being represented by that map.

schist A metamorphic rock characterized by coarse-grained minerals oriented approximately parallel.

schistose The texture of a rock in which visible platy or needle-shaped minerals have grown essentially parallel to each other under the influence of directed pressure.

scientific method A means of gaining knowledge through objective procedures.

scoria A basalt that is highly vesicular.

sea cave A cavity eroded by wave action at the base of a sea cliff.

sea cliff Steep slope that retreats inland by mass wasting as wave erosion undercuts it.

sea-level rise Increase in the volume of water in the oceans, resulting in inundation of coastal areas. Occurs from both melting of glaciers and from thermal expansion.

seafloor metamorphism Metamorphism of rock along a mid-oceanic ridge caused by circulating hot water.

seafloor spreading The concept that the ocean floor is moving away from the mid-oceanic ridge and across the deep-ocean basin, to disappear beneath continents and island arcs.

seamount Conical mountain rising 1,000 meters or more above the sea floor.

seawall A wall constructed along the base of retreating cliffs to prevent wave erosion.

sediment Loose, solid particles that can originate by (1) weathering and erosion of preexisting rocks, (2) chemical precipitation from solution, usually in water, and (3) secretion by organisms.

sedimentary breccia A coarse-grained sedimentary rock (grains coarser than 2 millimeters) formed by the cementation of angular rubble.

sedimentary facies Significantly different rock types occupying laterally distinct parts of the same layered rock unit.

sedimentary rock Rock that has formed from (1) lithification of any type of sediment, (2) precipitation from solution, or (3) consolidation of the remains of plants or animals.

sedimentary structure A feature found within sedimentary rocks, usually formed during or shortly after deposition of the sediment and before lithification.

seismic gap A segment of a fault that has not experienced earthquakes for a long time; such gaps may be the site of large future quakes.

seismic profiler An instrument that measures and records the subbottom structure of the sea floor.

seismic reflection The return of part of the energy of seismic waves to Earth's surface after the waves bounce off a rock boundary.

seismic refraction The bending of seismic waves as they pass from one material to another.

seismic sea wave *See* tsunami.

seismic tomography A technique similar to CT scans to image the interior of Earth using seismic waves from earthquakes and numerical models run on supercomputers.

seismic wave A wave of energy produced by an earthquake.

seismogram Paper record of Earth vibration.

seismograph A seismometer with a recording device that produces a permanent record of Earth motion.

seismometer An instrument designed to detect seismic waves or Earth motion.

serpentine A magnesium silicate mineral. Most asbestos is a variety of serpentine.

shale A fine-grained sedimentary rock (grains finer than 1/16 millimeter in diameter) formed by the cementation of silt and clay (mud). Shale has thin layers (laminations) and an ability to split (fissility) into small chips.

shear force In mass wasting, the component of gravitational force that is parallel to an inclined surface.

shear strength In mass wasting, the resistance to movement or deformation of material.

shear stress Stress due to forces that tend to cause movement or strain parallel to the direction of the forces.

shearing Movement in which parts of a body slide relative to one another and parallel to the forces being exerted.

sheet erosion The removal of a thin layer of surface material, usually topsoil, by a flowing sheet of water.

sheet joints Cracks that develop parallel to the outer surface of a large mass of expanding rock, as pressure is released during unloading.

sheet silicate structure Crystal structure in which each silica tetrahedron shares three oxygen ions.

sheetwash Water flowing down a slope in a layer.

shield volcano Broad, gently sloping cone constructed of solidified lava flows.

silica A term used for oxygen plus silicon.

silica tetrahedron *See* silicon-oxygen tetrahedron.

silicate A substance that contains silica as part of its chemical formula.

silicon-oxygen tetrahedron Four-sided, pyramidal object that visually represents the four oxygen atoms surrounding a silicon atom; the basic building block of silicate minerals. Also called a silica tetrahedron or a silicon tetrahedron.

sill A tabular intrusive structure concordant with the country rock.

silt Sediment composed of particles with a diameter of 1/256 to 1/16 millimeter.

siltstone A sedimentary rock consisting mostly of silt grains.

sinkhole A closed depression found on land surfaces underlain by limestone.

sinter A deposit of silica that forms around some hot springs and geysers.

slab pull The concept that subducting plates are pulled along by their dense leading edges.

slate A fine-grained rock that splits easily along flat, parallel planes.

slaty Describing a rock that splits easily along nearly flat and parallel planes.

slaty cleavage The ability of a rock to break along closely spaced parallel planes.

slide In mass wasting, movement of a relatively coherent descending mass along one or more well-defined surfaces.

slip face The steep, downwind slope of a dune; formed from loose, cascading sand that generally keeps the slope at the angle of repose (about 34°).

slump In mass wasting, movement along a curved surface in which the upper part moves vertically downward while the lower part moves outward. Also called a *rotational slide.*

snow line *See* equilibrium line.

soil A layer of weathered, unconsolidated material on top of bedrock; often also defined as containing organic matter and being capable of supporting plant growth. In mass wasting, *soil* means unconsolidated material, regardless of particle size or composition (also called *engineering soil*).

soil horizon Any of the layers of soil that are distinguishable by characteristic physical or chemical properties.

solar nebula The rotating disk of gas and dust from which the Sun and planets formed.

solar radiation management Efforts to control the amount of solar energy absorbed by the Earth by reflecting it back to space.

solar system The Sun, planets, their moons, and other bodies that orbit the Sun.

solar wind The outflow of low-density, hot gas from the Sun's upper atmosphere. It is partially this wind that creates the tail of a comet by blowing dust and gas away from the comet's immediate surroundings.

solid solution The substitution of atoms of one element for those of another element in a particular mineral.

solifluction Flow of water-saturated soil over impermeable material.

solution Usually slow but effective process of weathering and erosion in which rocks are dissolved by water.

sorting Process of selection and separation of sediment grains according to their grain size (or grain shape or specific gravity).

source area The locality that eroded to provide sediment to form a sedimentary rock.

source rock A rock containing organic matter that is converted to petroleum by burial and other postdepositional changes.

spatter cone A small, steep-sided cone built from lava spattering out of a vent.

specific gravity The ratio of the mass of a substance to the mass of an equal volume of water, determined at a specified temperature.

speleothem Dripstone deposit of calcite that precipitates from dripping water in caves.

spheroidal weathering Weathering effects that produce rounded rock from an initial blocky shape.

spit A fingerlike ridge of sediment attached to land but extending out into open water.

spreading axis (or spreading center) The crest of the mid-oceanic ridge, where sea floor is moving away in opposite directions on either side.

spring A place where water flows naturally out of rock onto the land surface.

stable Describing a mineral that will not react with or convert to a new mineral or substance, given enough time.

stack A small rock island that is an erosional remnant of a headland left behind as a wave-eroded coast retreats inland.

stalactite Iciclelike pendant of dripstone formed on cave ceilings.

stalagmite Cone-shaped mass of dripstone formed on cave floors, generally directly below a stalactite.

standard geologic time scale A worldwide relative scale of geologic time divisions.

star A massive, gaseous body held together by gravity and generally emitting light. Normal stars generate energy by nuclear reactions in their interiors.

static pressure *See* confining pressure.

stock A small, discordant pluton with an areal extent at Earth's surface of less than 100 square kilometers.

stony meteorite A meteorite made up mostly of plagioclase and iron-magnesium silicates.

stony-iron meteorite A meteorite composed of silicate minerals and iron-nickel alloy in approximately equal amounts.

stoping Upward movement of a body of magma by fracturing of overlying country rock. Magma engulfs the blocks of fractured country rock as it moves upward.

storm surge High sea level caused by the low pressure and high winds of hurricanes.

strain Change in size (volume) or shape of a body (or rock unit) in response to stress.

stratigraphy The field of geology concerning layered rocks and their interrelationships.

stratosphere The layer of the atmosphere directly above the troposphere. Characterized by an increase in temperature with altitude.

stratovolcano *See* composite volcano.

streak Color of a pulverized substance; a useful property for mineral identification.

stream A moving body of water, confined in a channel and running downhill under the influence of gravity.

stream capture *See* stream piracy.

stream channel A long, narrow depression, shaped and more or less filled by a stream.

stream discharge Volume of water that flows past a given point in a unit of time.

stream gradient Downhill slope of a stream's bed or the water surface, if the stream is very large.

stream headwaters The upper part of a stream near the source.

stream mouth The place where the stream enters the sea, a large lake, or a larger stream.

stream piracy The natural diversion of the headwaters of one stream into the channel of another.

stream terrace Steplike landform found above a stream and its flood plain.

stream velocity The speed at which water in a stream travels.

stream-dominated delta A delta with fingerlike distributaries formed by the dominance of stream sedimentation; also called a birdfoot delta.

stress A force acting on a body, or rock unit, that tends to change the size or shape of that body, or rock unit. Force per unit area within a body.

striations (1) On minerals, extremely straight, parallel lines; (2) glacial straight scratches in rock caused by abrasion by a moving glacier.

strike The compass direction of a line formed by the intersection of an inclined plane (such as a bedding plane) with a horizontal plane.

strike-slip fault A fault in which movement is parallel to the strike of the fault surface.

strip mine A mine in which the valuable material is exposed at the surface by removing a strip of overburden.

structural [or oil] trap *See* oil trap.

structural basin A structure in which the beds dip toward a central point and the youngest rock layers are in the center or core of the structure.

structural dome A structure in which beds dip away from a central point.

structural geology The branch of geology concerned with the internal structure of bedrock and the shapes, arrangement, and interrelationships of rock units.

subduction The sliding of the sea floor beneath a continent or island arc.

subduction complex *See* accretionary wedge.

subduction zone Elongate region in which subduction takes place.

submarine canyon V-shaped valley that runs across the continental shelf and down the continental slope.

submergent coast A coast in which formerly dry land has been recently drowned, either by land subsidence or a rise in sea level.

subsidence Sinking or downwarping of a part of the Earth's surface.

sunspots Dark spots on the Sun's surface caused by extreme magnetic activity.

superposed stream A river let down onto a buried geologic structure by erosion of overlying layers.

superposition A principle or law stating that within a sequence of undisturbed sedimentary rocks, the oldest layers are on the bottom, the youngest on the top.

surf Breaking waves.

surface wave A seismic wave that travels on Earth's surface.

suspect terrane A terrane that may not have formed at its present site.

suspended load Sediment in a stream that is light enough in weight to remain lifted indefinitely above the bottom by water turbulence.

swelling clay *See* expansive clay.

syncline A fold shaped like a trough with the youngest rock in the center of the fold. Usually the rock layers dip toward the hinge line of the fold.

T

talus An accumulation of broken rock at the base of a cliff.

tarn *See* rock-basin lake.

tectite Small, rounded bits of glass formed from rock melting and being thrown into the air due to a meteorite impact.

tectonic forces Forces generated from within Earth that result in uplift, movement, or deformation of part of Earth's crust.

tectonostratigraphic terrane *See* terrane.

tensional stress A stress due to a force pulling away on a body.

tephra *See* pyroclastic debris.

terminal moraine An end moraine marking the farthest advance of a glacier.

terminus The lower edge of a glacier.

terrane (tectonostratigraphic terrane) A region in which the geology is markedly different from that in adjoining regions.

terrestrial radiation Energy emitted by the Earth at infrared (long-wave) wavelengths of the electromagnetic spectrum. This radiation is selectively absorbed by greenhouse gases.

terrigenous sediment Land-derived sediment that has found its way to the sea floor.

texture A rock's appearance with respect to the size, shape, and arrangement of its grains or other constituents.

theory An explanation for observed phenomena that has a high possibility of being true.

theory of glacial ages At times in the past, colder climates prevailed during which significantly more of the land surface of Earth was glaciated than at present.

thermal expansion The increase in volume of a material as a result of increased temperature.

thermal metamorphism *See* contact (thermal) metamorphism.

thrust fault A reverse fault in which the dip of the fault plane is at a low angle to horizontal.

ticket Tickets are comments or instructions about content. Create tickets in Habitat Proof mode. Use them to identify errors in content, to suggest improvements, and so on. Assign a ticket to yourself or another team member who can respond to your comment or fix an error.

tidal delta A submerged body of sediment formed by tidal currents passing through gaps in barrier islands.

tidal wave An incorrect name for a tsunami.

tide-dominated delta A delta formed by the reworking of sand by strong tides.

tight fold A fold that has a small angle between the limbs.

till Unsorted and unlayered rock debris carried by a glacier.

tillite Lithified till.

time-transgressive rock unit An apparently continuous rock layer in which different portions formed at different times.

tombolo A bar of marine sediment connecting a former island or stack to the mainland.

topographic map A map on which elevations are shown by means of contour lines.

topset bed In a delta, a nearly horizontal sediment bed of varying grain size formed by distributaries shifting across the delta surface.

trace fossil Trail, track, or burrow resulting from animal movement preserved in sedimentary rock.

traction Movement by rolling, sliding, or dragging of sediment fragments along a stream bottom.

transform fault The portion of a fracture zone between two offset segments of a mid-oceanic ridge crest.

transform plate boundary Boundary between two plates that are sliding past each other.

translational slide In mass wasting, movement of a descending mass along a plane approximately parallel to the slope of the surface.

transportation The movement of eroded particles by agents such as rivers, waves, glaciers, or wind.

transported soil Soil not formed from the local rock but from parent material brought in from some other region and deposited, usually by running water, wind, or glacial ice.

transverse dune A relatively straight, elongate dune oriented perpendicular to the wind.

travel-time curve A plot of seismic-wave arrival times against distance.

travertine A porous deposit of calcite that often forms around hot springs.

trellis pattern A drainage pattern consisting of parallel main streams with short tributaries meeting them at right angles.

trench *See* oceanic trench.

trench suction The concept that overlying plates move horizontally toward oceanic trenches as subducting plates sink at an angle steeper than their dip.

tributary Small stream flowing into a large stream, adding water to the large stream.

trigger (Associated with mass wasting) is the immediate cause of failure of already unstable ground. Examples of landslide triggers are earthquakes and periods of heavy rainfall.

troposphere The lowermost layer of the atmosphere, defined by a decrease in temperature with altitude.

trough (of wave) The low point of a wave.

truncated spur Triangular facet where the lower end of a ridge has been eroded by glacial ice.

tsunami Huge ocean wave produced by displacement of the sea floor; also called seismic sea wave.

tufa A deposition of calcite that forms around a spring, lake, or percolating groundwater.

tuff A rock formed from fine-grained pyroclastic particles (ash and dust).

turbidity current A flowing mass of sediment-laden water that is heavier than clear water and therefore flows downslope along the bottom of the sea or a lake.

turbulent flow Eddying, swirling flow in which water drops travel along erratically curved paths that cross the paths of neighboring drops.

U

U-shaped valley Characteristic cross-profile of a valley carved by glacial erosion.

ultramafic Magma or igneous rock that contains less than 45% silica and is enriched in calcium, iron, and magnesium.

unconfined aquifer A partially filled aquifer exposed to the land surface and marked by a rising and falling water table.

unconformity A surface that represents a break in the geologic record, with the rock unit immediately above it being considerably younger than the rock beneath.

unconsolidated In referring to sediment grains, loose, separate, or unattached to one another.

underplating The pooling of magmas at the base of the continental crust.

uniformitarianism Principle that geologic processes operating at present are the same processes that operated in the past. The principle is stated more succinctly as, "The present is the key to the past." *See also* actualism.

universe The largest astronomical structure we know of. The universe contains all matter and radiation and encompasses all space.

unloading The removal of a great weight of rock.

unpaired terraces *Stream terraces* (*see definition*) that do not have the same elevation on opposite sides of a river.

unsaturated zone A subsurface zone in which rock openings are generally unsaturated and filled partly with air and partly with water; above the saturated zone.

upright fold A fold with a vertical axial plane.

V

valley glacier A glacier confined to a valley. The ice flows from a higher to a lower elevation.

varve Two thin layers of sediment, one dark and the other light in color, representing one year's deposition in a lake.

vein *See* hydrothermal vein.

vent The opening in Earth's surface through which a volcanic eruption takes place.

ventifact Boulder, cobble, or pebble with flat surfaces caused by the abrasion of wind-blown sand.

vertical exaggeration An artificial steepening of slope angles on a topographic profile caused by using a vertical scale that differs from the horizontal scale.

vesicle A cavity in volcanic rock caused by gas in a lava.

vesicular texture An igneous rock containing many vesicles.

viscosity Resistance to flow.

vitreous luster *See* glassy luster.

volcanic breccia Rock formed from large pieces of volcanic rock (cinders, blocks, bombs).

volcanic dome A steep-sided, dome- or spine-shaped mass of volcanic rock formed from viscous lava that solidifies in or immediately above a volcanic vent.

Volcanic Explosivity Index An index used to measure, on a scale of 1–10, the power of a volcanic eruption.

volcanic neck An intrusive structure that apparently represents magma that solidified within the throat of a volcano.

volcanism Volcanic activity, including the eruption of lava and rock fragments and gas explosions.

volcano A hill or mountain constructed by the extrusion of lava or rock fragments from a vent.

W

wastage *See* ablation.

water table The upper surface of the zone of saturation.

wave crest *See* crest (of wave).

wave height The vertical distance between the crest (the high point of a wave) and the trough (the low point).

wave refraction Change in direction of waves due to slowing as they enter shallow water.

wave trough *See* trough (of wave).

wave-cut platform A horizontal bench of rock formed beneath the surf zone as a coast retreats because of wave erosion.

wave-dominated delta A delta formed by the reworking of sand by wave action.

wavelength The horizontal distance between two wave crests (or two troughs).

weather Short-term variation of atmospheric conditions, such as cloudiness, temperature, and wind.

weathering The group of processes that change rock at or near Earth's surface.

welded tuff A rock composed of pyroclasts welded together.

well A hole, generally cylindrical and usually walled or lined with pipe, that is dug or drilled into the ground to penetrate an aquifer below the zone of saturation.

widget Widgets enable you to put javascript within <object> tags to provide interactive content in Inkling pages.

Wilson cycle The cycle of splitting of a continent, opening of an ocean basin, followed by closing of the basin and collision of the continents.

wind ripple Small, low ridge of sand produced by the saltation of wind-blown sand.

wrinkle ridge *(Moon)* Wrinkle on lunar maria surface.

X

xenolith Fragment of rock distinct from the igneous rock in which it is enclosed.

XHTML XHTML is a stricter version of HTML. For example, in XHTML all tags must be closed, so use <br /> rather than <br> and <img /> rather than <img>. Habitat's well-formed HTML is a subset of the XHTML standards; Habitat requires code that validates to XHTML standards to ensure that content displays correctly on all supported devices. See Inkling standards for HTML and alsoXHTML Reference.

Z

zone of ablation That portion of a glacier in which ice is lost.

zone of accumulation (1) That portion of a glacier with a perennial snow cover; (2) *see* B horizon (a soil layer).

zone of leaching *See* A horizon (a soil layer).

zoning Orderly variation in the chemical composition within a single crystal.

zone of saturation *See* saturated zone.

INDEX

Note: Figures and tables are indicated by *f* and *t* respectively. Figures are cited only when they appear outside related text discussions.

A

D

E

F

G

H

N

O

P

Q

R

S

T

U

V

W

X

Y

Z